Introductory Algebra

Sixth Edition

Elayn Martin-Gay
University of New Orleans

Director, Portfolio Management: *Michael Hirsch*
Courseware Portfolio Manager: *Rachel Ross*
Courseware Portfolio Management Assistant: *Shannon Slocum*
Managing Producer: *Karen Wernholm*
Content Producer: *Lauren Morse*
Media Producer: *Audra Walsh*
Manager, Courseware QA: *Mary Durnwald*
Manager Content Development, Math: *Eric Gregg*
Product Marketing Manager: *Alicia Wilson*
Field Marketing Manager: *Lauren Schur*
Product Marketing Assistant: *Brooke Imbornone*
Senior Author Support/Technology Specialist: *Joe Vetere*
Manager, Rights and Permissions: *Gina Cheselka*
Manufacturing Buyer: *Carol Melville, LSC Communications*
Text Design: *Tamara Newnam*
Composition and Production Coordination: *Integra*
Illustrations: *Scientific Illustrators*
Senior Designer: *Barbara T. Atkinson*
Cover Design: *Tamara Newman*
Cover Image: *Mesiats/Shutterstock*

Library of Congress Cataloging-in-Publication Data

Names: Martin-Gay, K. Elayn, author.
Title: Introductory algebra / Elayn Martin-Gay.
Description: Sixth edition. | Hoboken, NJ : Pearson Education, Inc., [2020] | Includes index.
Identifiers: LCCN 2019003450 (print) | LCCN 2019005405 (ebook) | ISBN 9780135176306 (epub) | ISBN 9780135169377 (se : alk. paper) | ISBN 9780135173299 (aie : alk. paper)
Subjects: LCSH: Algebra–Textbooks.
Classification: LCC QA152.3 (ebook) | LCC QA152.3 .M37 2020 (print) | DDC 512.9—dc23
LC record available at https://lccn.loc.gov/2019003450

14 2024

ISBN-13: 978-0-13-516937-7 (Student Edition)
ISBN-10: 0-13-516937-2

This book is dedicated to students everywhere—
and we should all be students. After all, is there anyone among
us who truly knows too much? Take that hint and continue
to learn something new every day of your life.

Best wishes from a fellow student:
Elayn Martin-Gay

Contents

Preface xi

Applications Index xxi

R Prealgebra Review R-1

R.1 Factors and the Least Common Multiple R-2
R.2 Fractions and Mixed Numbers R-8
R.3 Decimals and Percents R-19
R.4 Reading Pictographs and Bar, Line, and Circle Graphs R-30
Vocabulary Check R-48
Chapter Highlights R-48
Chapter Review R-52
Getting Ready for the Test R-57
Chapter Test R-59

1 Real Numbers and Introduction to Algebra 1

1.1 Study Skill Tips for Success in Mathematics 2
1.2 Symbols and Sets of Numbers 8
1.3 Exponents, Order of Operations, and Variable Expressions 19
1.4 Adding Real Numbers 30
1.5 Subtracting Real Numbers 39
Integrated Review—Operations on Real Numbers 49
1.6 Multiplying and Dividing Real Numbers 51
1.7 Properties of Real Numbers 64
1.8 Simplifying Expressions 72
Vocabulary Check 81
Chapter Highlights 81
Chapter Review 85
Getting Ready for the Test 89
Chapter Test 90

2 Equations, Inequalities, and Problem Solving 92

2.1 The Addition Property of Equality 93
2.2 The Multiplication Property of Equality 102
2.3 Further Solving Linear Equations 111
Integrated Review—Solving Linear Equations 120
2.4 An Introduction to Problem Solving 122
2.5 Formulas and Problem Solving 136
2.6 Percent and Mixture Problem Solving 149

2.7 Linear Inequalities and Problem Solving 162
Vocabulary Check 173
Chapter Highlights 174
Chapter Review 177
Getting Ready for the Test 182
Chapter Test 183
Cumulative Review 185

3 Exponents and Polynomials 188

3.1 Exponents 189
3.2 Negative Exponents and Scientific Notation 201
3.3 Introduction to Polynomials 211
3.4 Adding and Subtracting Polynomials 221
3.5 Multiplying Polynomials 228
3.6 Special Products 235
Integrated Review—Exponents and Operations on Polynomials 243
3.7 Dividing Polynomials 245
Vocabulary Check 252
Chapter Highlights 253
Chapter Review 256
Getting Ready for the Test 261
Chapter Test 262
Cumulative Review 264

4 Factoring Polynomials 267

4.1 The Greatest Common Factor and Factoring by Grouping 268
4.2 Factoring Trinomials of the Form $x^2 + bx + c$ 278
4.3 Factoring Trinomials of the Form $ax^2 + bx + c$ 285
4.4 Factoring Trinomials of the Form $ax^2 + bx + c$ by Grouping 292
4.5 Factoring Perfect Square Trinomials and the Difference of Two Squares 297
Integrated Review—Choosing a Factoring Strategy 306
4.6 Solving Quadratic Equations by Factoring 308
4.7 Quadratic Equations and Problem Solving 316
Vocabulary Check 326
Chapter Highlights 326
Chapter Review 329
Getting Ready for the Test 333
Chapter Test 334
Cumulative Review 336

5 Rational Expressions 338

5.1 Simplifying Rational Expressions 339
5.2 Multiplying and Dividing Rational Expressions 350

5.3 Adding and Subtracting Rational Expressions with the Same Denominator and Least Common Denominator **360**
5.4 Adding and Subtracting Rational Expressions with Different Denominators **368**
5.5 Solving Equations Containing Rational Expressions **375**
Integrated Review—Summary on Rational Expressions 384
5.6 Proportions and Problem Solving with Rational Equations **386**
5.7 Simplifying Complex Fractions **400**
Vocabulary Check **408**
Chapter Highlights **409**
Chapter Review **413**
Getting Ready for the Test **417**
Chapter Test **418**
Cumulative Review **420**

6 Graphing Equations and Inequalities **423**

6.1 The Rectangular Coordinate System **424**
6.2 Graphing Linear Equations **437**
6.3 Intercepts **447**
6.4 Slope and Rate of Change **457**
6.5 Equations of Lines **474**
Integrated Review—Summary on Linear Equations 486
6.6 Introduction to Functions **488**
6.7 Graphing Linear Inequalities in Two Variables **500**
Vocabulary Check **510**
Chapter Highlights **510**
Chapter Review **514**
Getting Ready for the Test **521**
Chapter Test **522**
Cumulative Review **525**

7 Systems of Equations **528**

7.1 Solving Systems of Linear Equations by Graphing **529**
7.2 Solving Systems of Linear Equations by Substitution **540**
7.3 Solving Systems of Linear Equations by Addition **548**
Integrated Review—Summary on Solving Systems of Equations 556
7.4 Systems of Linear Equations and Problem Solving **557**
Vocabulary Check **570**
Chapter Highlights **570**
Chapter Review **573**
Getting Ready for the Test **576**
Chapter Test **577**
Cumulative Review **579**

8 Roots and Radicals 582

8.1 Introduction to Radicals 583
8.2 Simplifying Radicals 591
8.3 Adding and Subtracting Radicals 599
8.4 Multiplying and Dividing Radicals 603
Integrated Review—Simplifying Radicals 612
8.5 Solving Equations Containing Radicals 614
8.6 Radical Equations and Problem Solving 620
8.7 Direct and Inverse Variation Including Radical Applications 626
Vocabulary Check 637
Chapter Highlights 637
Chapter Review 640
Getting Ready for the Test 644
Chapter Test 645
Cumulative Review 647

9 Quadratic Equations 650

9.1 Solving Quadratic Equations by the Square Root Property 651
9.2 Solving Quadratic Equations by Completing the Square 658
9.3 Solving Quadratic Equations by the Quadratic Formula 663
Integrated Review—Summary on Solving Quadratic Equations 672
9.4 Graphing Quadratic Equations in Two Variables 675
Vocabulary Check 685
Chapter Highlights 685
Chapter Review 687
Getting Ready for the Test 691
Chapter Test 692
Cumulative Review 694

Appendices

Appendix A Tables 697
A.1 Table of Percents, Decimals, and Fraction Equivalents 697
A.2 Table on Finding Common Percents of a Number 698
A.3 Table of Squares and Square Roots 699
A.4 Geometric Formulas 700
Appendix B Factoring Sums and Differences of Cubes 701
Appendix C Mean, Median, Mode, Range, and Introduction to Statistics 703
Appendix D Sets 713

Appendix E Review of Angles, Lines, and Special Triangles **717**
Appendix F Interval Notation and Finding Domains and Ranges from Graphs **725**
Student Resources **731**
Study Skills Builders **731**
Bigger Picture—Study Guide Outline **740**
Practice Final Exam **744**

Answers to Selected Exercises **A1**

Subject Index **SI-1**
Photo Credits **P-1**

Preface

***Introductory Algebra*, Sixth Edition** was written to provide a solid foundation in algebra and is intended for basic for students who might not have previous experience in algebra. Specific care was taken to make sure students have the most up-to-date relevant text preparation for their next mathematics course or for non-mathematical courses that require an understanding of algebraic fundamentals. I have tried to achieve this by writing a user-friendly text that is keyed to objectives and contains many worked-out examples. As suggested by AMATYC and the NCTM Standards (plus Addenda), real-life and real-data applications, data interpretation, conceptual understanding, problem solving, writing, cooperative learning, appropriate use of technology, number sense, estimation, critical thinking, and geometric concepts are emphasized and integrated throughout the book.

The many factors that contributed to the success of the previous edition have been retained. In preparing the Sixth Edition, I considered comments and suggestions of colleagues, students, and many users of the prior edition throughout the country.

What's New in the Sixth Edition?

- **The Martin-Gay Program** has been revised and enhanced with a new design in the text and MyLab Math to actively encourage students to use the text, video program, and Video Organizer as an integrated learning system.
- **New Section Chapter R** includes a new section (R.4 Reading Pictographs and Bar, Line, and Circle Graphs), and the end of Chapter R contains updated Highlights, Review Questions, and Test Questions.
- **Updated Section Chapter 8** has been updated to include a section previously covered in Chapter 6 (8.7 Direct Inverse Variation Including Radical Applications).
- **Appendices** The newly numbered appendices now contain two new tables (A.2 Table on Finding Common Percents of a Number and A.4 Geometric Formulas), as well as a new appendix (Appendix F Interval Notation and Finding Domains and Ranges from Graphs).

    In addition, Appendix C covers an expanded section entitled "Mean, Median, Mode, Range and Intro to Statistics." This is a new robust section that not only reviews mean, median, and mode, but includes an introduction of Range, which is a measure of dispersion. This section also includes frequency distribution tables and graphs and a formula for finding the position of the median.

    Appendices B–F have been updated so that all are now organized by objectives and all contain practice problems that accompany the examples.
- **New Getting Ready for the Test** can be found before each Chapter Test. These exercises can increase student success by helping students prepare for their Chapter Test. The purpose of these exercises is to check students' conceptual understanding of the topics in the chapter as well as common student errors. It is suggested that students complete and check these exercises before taking a practice Chapter Test. All Getting Ready for the Test exercises are either Multiple Choice or Matching, and all answers can be found in the answer section of this text.

    Video Solutions of all exercises can be found in MyLab Math. These video solutions contain brief explanations and reminders of material in the chapter. Where applicable, incorrect choices contain explanations.

    Getting Ready for the Test exercise numbers marked in blue indicate that the exercise is available in **Learning Catalytics.** LC

- **New Learning Catalytics** is an interactive student response tool that uses students' smartphones, tablets, or laptops to engage them in more sophisticated tasks and thinking. Generate class discussion, guide your lecture, and promote peer-to-peer learning with real-time analytics. Accessible through MyLab Math, instructors can use Learning Catalytics to:
  - Pose a variety of open-ended questions that help your students develop critical thinking skills.
  - Monitor responses to find out where students are struggling.
  - Use real-time data to adjust your instructional strategy and try other ways of engaging students during class.
  - Manage student interactions by automatically grouping students for discussion, teamwork, and peer-to-peer learning.
  - Pearson-created questions for developmental math topics are available to allow you to take advantage of this exciting technology. Additionally, "Getting Ready for the Test" exercises (marked in blue) are available in Learning Catalytics. Search the question library for "MGIntro" and the chapter number, for example, MGIntro7 would be the questions from Chapter 7.
- **New Key Concept Activity Lab Workbook** includes Extension Exercises, Exploration Activities, Conceptual Exercises, and Group Activities. These activities are a great way to engage students in conceptual projects and exploration as well as group work. This workbook is available in MyLab Math, or can be packaged with a text or MyLab code.
- **Exercise Sets** have been carefully examined and revised. Special focus was placed on making sure that even- and odd-numbered exercises are carefully paired and that real-life applications are updated.
- **The Martin-Gay MyLab Math** course has been updated and revised to provide more exercise coverage, including assignable Video Check questions and an expanded video program. There are Lecture Videos for every section, which students can also access at the specific objective level; Student Success Tips videos; and an increased number of video clips at the exercise level to help students while doing homework in MyLab Math. Suggested homework assignments have been premade for assignment at the instructor's discretion.

Key Pedagogical Features

The following key features have been retained and/or updated for the Sixth Edition of the text:

- **Problem-Solving Process** This is formally introduced in Chapter 2 with a four-step process that is integrated throughout the text. The four steps are **Understand, Translate, Solve,** and **Interpret.** The repeated use of these steps in a variety of examples shows their wide applicability. Reinforcing the steps can increase students' comfort level and confidence in tackling problems.
- **Exercise Sets Revised and Updated** The exercise sets have been carefully examined and extensively revised. Special focus was placed on making sure that even- and odd-numbered exercises are paired and that real-life applications were updated.
- **Examples** Detailed, step-by-step examples were added, deleted, replaced, or updated as needed. Many examples reflect real life. Additional instructional support is provided in the annotated examples.
- **Practice Exercises** Throughout the text, each worked-out example has a parallel Practice exercise. These invite students to be actively involved in the learning process. Students should try each Practice Exercise after finishing the corresponding example. Learning by doing will help students grasp ideas before moving on to other concepts. Answers to the Practice Exercises are provided at the bottom of each page.
- **Helpful Hints** Helpful Hints contain practical advice on applying mathematical concepts. Strategically placed where students are most likely to need immediate

reinforcement, Helpful Hints help students avoid common trouble areas and mistakes.

- **Concept Checks** This feature allows students to gauge their grasp of an idea as it is being presented in the text. Concept Checks stress conceptual understanding at the point-of-use and help suppress misconceived notions before they start. Answers appear at the bottom of the page. Exercises related to Concept Checks are included in the exercise sets.
- **Mixed Practice Exercises** In the section exercise sets, these exercises require students to determine the problem type and strategy needed to solve it just as they would need to do on a test.
- **Integrated Reviews** This unique mid-chapter exercise set helps students assimilate new skills and concepts that they have learned separately over several sections. These reviews provide yet another opportunity for students to work with "mixed" exercises as they master the topics.
- **Vocabulary Check** This feature provides an opportunity for students to become more familiar with the use of mathematical terms as they strengthen their verbal skills. These appear at the end of each chapter before the Chapter Highlights.
- **Vocabulary, Readiness & Video Check Questions** are assignable for each section of the text and in MyLab Math. **Vocabulary** exercises check student understanding of new terms. The **Readiness** exercises center on a student's understanding of a concept that is necessary in order to continue to the exercise set. The **Video Check questions** correlate to the videos in MyLab Math, and are a great way to assess whether students have viewed and understood the key concepts presented in the videos. Answers to all Video Check questions are available in an answer section at the back of the text.
- **Chapter Highlights** Found at the end of every chapter, these contain key definitions and concepts with examples to help students understand and retain what they have learned and help them organize their notes and study for tests.
- **Chapter Review** The end of every chapter contains a comprehensive review of topics introduced in the chapter. The Chapter Review offers exercises keyed to every section in the chapter, as well as Mixed Review exercises that are not keyed to sections.
- **Chapter Test and Chapter Test Prep Videos** The Chapter Test is structured to include those problems that involve common student errors. The **Chapter Test Prep Videos** give students instant access to a step-by-step video solution of each exercise in the Chapter Test.
- **Cumulative Review** This review follows every chapter in the text (except Chapter 1). Each odd-numbered exercise contained in the Cumulative Review is an earlier worked example in the text that is referenced in the back of the book along with the answer.
- **Writing Exercises** These exercises occur in almost every exercise set and require students to provide a written response to explain concepts or justify their thinking.
- **Applications** Real-world and real-data applications have been thoroughly updated, and many new applications are included. These exercises occur in almost every exercise set and show the relevance of mathematics and help students gradually and continuously develop their problem-solving skills.
- **Review Exercises** These exercises occur in each exercise set (except in Chapter 1) and are keyed to earlier sections. They review concepts learned earlier in the text that will be needed in the next section or chapter.
- **Exercise Set Resource Icons** Located at the opening of each exercise set, these icons remind students of the resources available for extra practice and support:

See Student Resources descriptions on page xv for details on the individual resources available.

- **Exercise Icons** These icons facilitate the assignment of specialized exercises and let students know what resources can support them.

    Video icon: exercise worked in the Interactive Lecture Series.

    △ Triangle icon: identifies exercises involving geometric concepts.

    Pencil icon: indicates a written response is needed.

    Calculator icon: optional exercises intended to be solved using a scientific or graphing calculator.

- **Group Activities** Found at the end of each chapter, these activities are for individual or group completion, and are usually hands-on or data-based activities that extend the concepts found in the chapter, allowing students to make decisions and interpretations and to think and write about algebra.

- **Optional: Calculator Exploration Boxes and Calculator Exercises** The optional Calculator Explorations provide keystrokes and exercises at appropriate points to give students an opportunity to become familiar with these tools. Section exercises that are best completed by using a calculator are identified by for ease of assignment.

- **The Video Organizer** workbook is designed to help students take notes and work practice exercises while watching the Interactive Lecture Series videos in MyLab Math, making it easy for students to create a course notebook and build good study habits.
    - Covers all of the video examples in order.
    - Provides ample space for students to write down key definitions and properties.
    - Includes "Play" and "Pause" button icons to prompt students to follow along with the author for some exercises while they try others on their own.

The Video Organizer is available in a loose-leaf, notebook-ready format, or can be downloaded from the MyLab Math course.

- **Interactive Lecture Series,** featuring Elayn Martin-Gay, provides students with learning at their own pace. The videos offer the following resources and more:
    - **A complete lecture for each section of the text** highlights key examples and exercises from the text. "Pop-ups" reinforce key terms, definitions, and concepts.
    - **An interface with menu navigation features** allows students to quickly find and focus on the examples and exercises they need to review.
    - **Interactive Concept Check** exercises measure students' understanding of key concepts and common trouble spots.
    - **Student Success Tip Videos** are in short segments designed to be daily reminders to be organized and to study.
    - **New! Getting Ready for the Test video solutions** cover every Getting Ready for the Test exercise. These appear at the end of each chapter and give students an opportunity to assess whether they understand the big picture concepts of the chapter, and help them focus on avoiding common errors.
    - **The Chapter Test Prep Videos** help students during their most teachable moment—when they are preparing for a test. This innovation provides step-by-step solutions for the exercises found in each Chapter Test.
    - **The Practice Final Exam Videos** help students prepare for an end-of-course final. Students can watch full video solutions to each exercise in the Practice Final Exam at the end of this text.

Resources for Success

Get the Most Out of MyLab Math for *Introductory Algebra*, Sixth Edition by Elayn Martin-Gay

Elayn Martin-Gay believes that every student can succeed, and every MyLab course that accompanies her texts is infused with her student-centric approach. The seamless integration of Elayn's signature support with the #1 choice in digital learning for developmental math gives students a completely consistent experience from print to MyLab.

A Comprehensive and Dynamic Video Program

The **Martin-Gay video program** is 100% presented by Elayn Martin-Gay in her clear, approachable style. The video program includes full section lectures and smaller objective level videos. Within many section lecture videos, **Interactive Concept Checks** measure students' understanding of concepts and common trouble spots—students are asked to try a question within the video in order, after which correct and incorrect answers are explained.

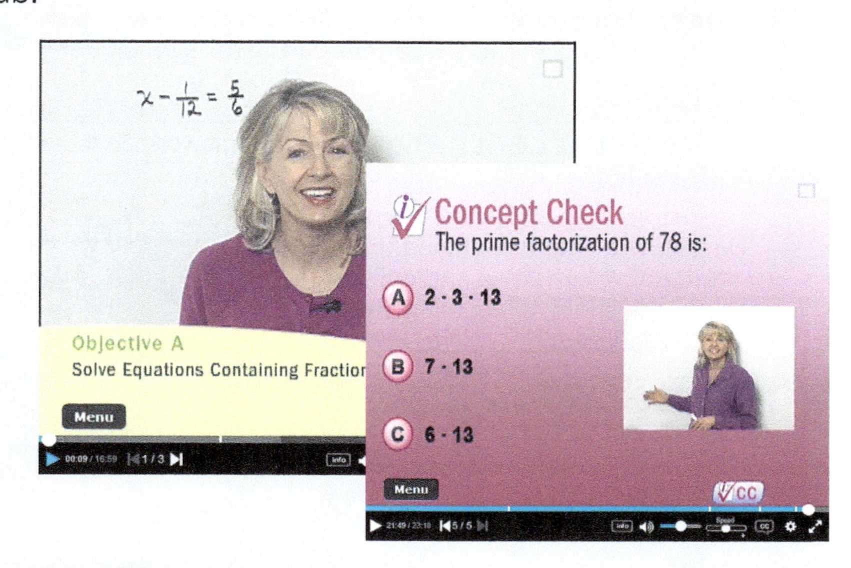

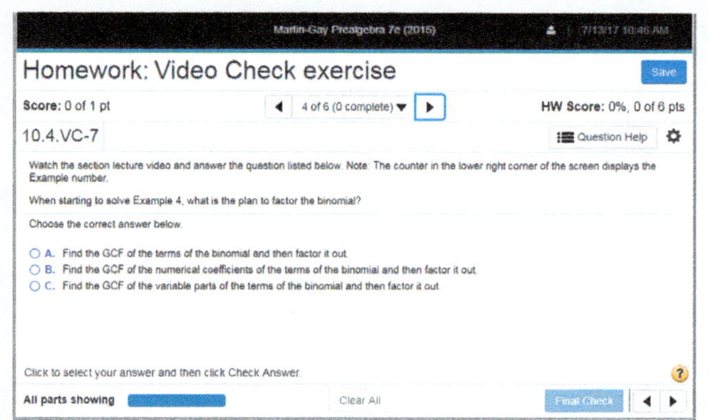

Assignable **Video Check questions** ensure that students have viewed and understood the key concepts from the section lecture videos.

Supporting College Success

Other hallmark Martin-Gay videos include **Student Success Tip videos**, which are short segments designed to be daily reminders to stay organized and to study. Additionally in keeping with Elayn's belief that every student can succeed, a new **Mindset module** is available in the course, with mindset-focused videos and exercises that encourage students to maintain a positive attitude about learning, value their own ability to grow, and view mistakes as a learning opportunity.

pearson.com/mylab/math

Pearson MyLab

Resources for Success

Resources for Review

New! Getting Ready for the Test video solutions cover every Getting Ready for the Test exercise. These appear at the end of each chapter and give students an opportunity to assess whether they understand the big picture concepts of the chapter, and help them focus on avoiding common errors. Students also have **Chapter Test Prep videos**, a Martin-Gay innovation, to help during their most teachable moment —when preparing for a test.

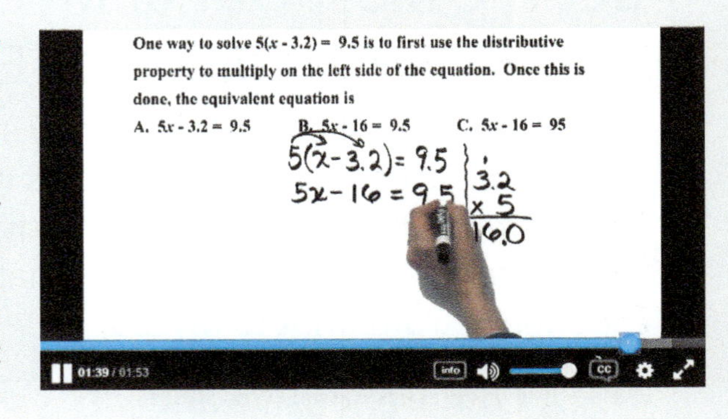

Personalize Learning

New! Skill Builder exercises offer just-in-time additional adaptive practice. The adaptive engine tracks student performance and delivers questions to each individual that adapt to his or her level of understanding.This new feature allows instructors to assign fewer questions for homework, allowing students to complete as many or as few questions as they need.

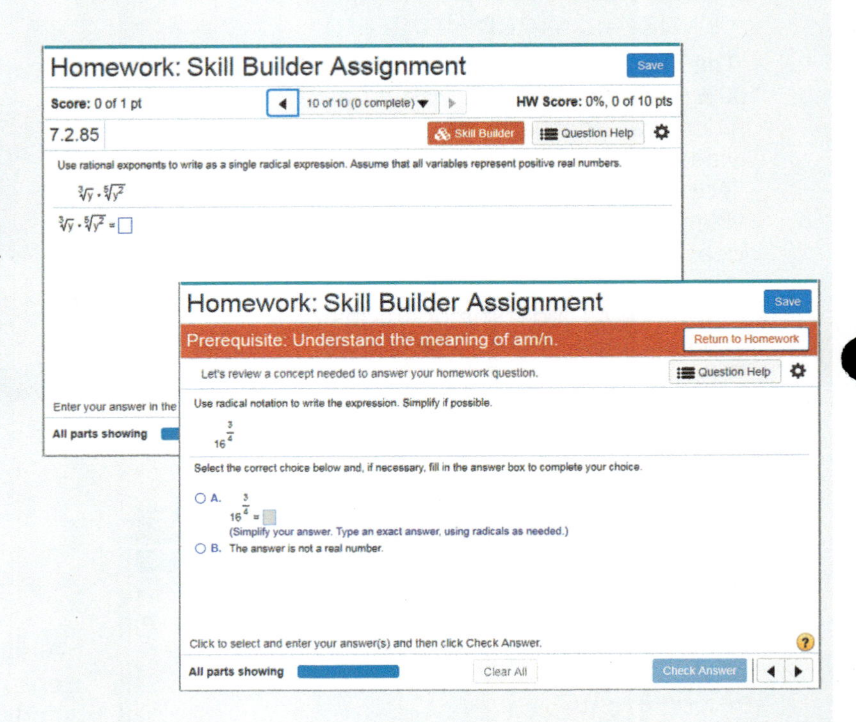

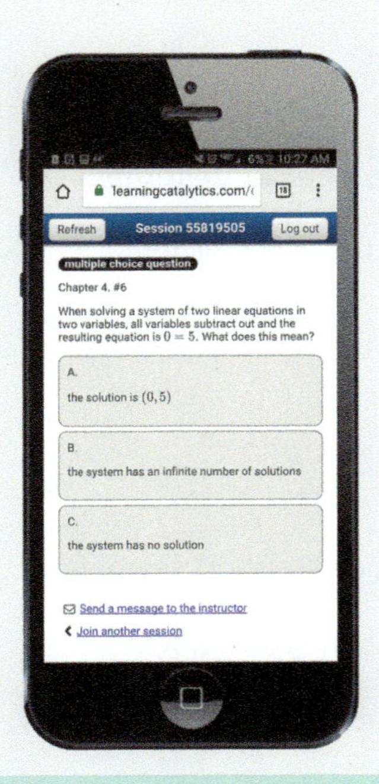

Get Students Engaged

New! Learning Catalytics Martin-Gay-specific questions are pre-built and available through MyLab Math. Learning Catalytics is an interactive student response tool that uses students' smartphones, tablets, or laptops to engage them in more sophisticated tasks and thinking. **Getting Ready for the Test** exercises marked in blue in the text are pre-built in Learning Catalytics to use in class. These questions can be found in Learning Catalytics by searching for "MGIntro#" where # is the chapter number.

pearson.com/mylab/math

Resources for Success

Instructor Resources

Annotated Instructor's Edition

Contains all the content found in the student edition, plus answers to even and odd exercises on the same text page, and Teaching Tips throughout the text placed at key points.

The resources below are available through Pearson's Instructor Resource Center, or from MyLab Math.

Instructor's Resource Manual with Tests and Mini-Lectures

Includes mini-lectures for each text section, additional practice worksheets for each section, several forms of tests per chapter—free response and multiple choice, and answers to all items.

Instructor's Solutions Manual

Contains detailed, worked-out solutions to evennumbered exercises in the text.

TestGen®

Enables instructors to build, edit, print, and administer tests using a computerized bank of questions developed to cover all the objectives of the text. TestGen is algorithmically based, allowing instructors to create multiple but equivalent versions of the same question or test with the click of a button. Instructors can also modify test bank questions or add new questions.

Instructor-to-Instructor Videos

Provide instructors with suggestions for presenting course material as well as time-saving teaching tips.

PowerPoint Lecture Slides

Available for download only, these slides present key concepts and definitions from the text.

Student Resources

Video Organizer

Designed to help students take notes and work practice exercises while watching the Interactive Lecture Series videos.

- Covers all of the video examples in order.
- Provides prompts with ample space for students to write down key definitions and rules.
- Includes "Play" and "Pause" button icons to prompt students to follow along with the author for some exercises while they try others on their own.
- Includes Student Success Tips Outline and Questions.

Available printed in a loose-leaf, notebook-ready format and to download in MyLab Math. All answers are available in Instructor Resources in MyLab Math.

New! Key Concept Activity Lab Workbook

Includes Extension Exercises, Exploration Activities, Conceptual Exercises, and Group Activities. This workbook is available in MyLab Math, or can be packaged in printed form with a text or MyLab Math code. All answers available in Instructor Resources in MyLab Math.

Student Solutions Manual

Provides completely worked-out solutions to the odd-numbered section exercises; all exercises in the Integrated Reviews, Chapter Reviews, Chapter Tests, and Cumulative Reviews.

Acknowledgments

There are many people who helped me develop this text, and I will attempt to thank some of them here. Courtney Slade and Cindy Trimble were *invaluable* for contributing to the overall accuracy of the text. Gina Linko and Patty Bergin provided guidance throughout the production process and Suellen Robinson provided many suggestions for updating applications during the writing of this Sixth Edition.

A very special thank you goes to my editor, Rachel Ross. And, my thanks to the staff at Pearson for all their support: Barbara Atkinson, Alicia Wilson, Michael Hirsch, Chris Hoag, Paul Corey, Michelle Renda, and Lauren Schur among many others.

I would like to thank the following reviewers for their input and suggestions that have affected this and previous editions:

Lisa Angelo, *Bucks Community College*
Victoria Baker, *Nicholls State College*
Teri Barnes, *McLennan Community College*
Laurel Berry, *Bryant & Stratton*
Thomas Blackburn, *Northeastern Illinois University*
Gail Burkett, *Palm Beach State College*
Anita Collins, *Mesa Community College*
Lois Colpo, *Harrisburg Area Community College*
Fay Dang, *Joliet Junior College*
Robert Diaz, *Fullerton College*
Tamie Dickson, *Reading Area Community College*
Latonya Ellis, *Gulf Coast Community College*
Sonia Ford, *Midland College*
Cheryl Gibby, *Cypress College*
Kathryn Gunderson, *Three Rivers Community College*
Elizabeth Hamman, *Cypress College*
Craig Hardesty, *Hillsborough Community College*
Lloyd Harris, *Gulf Coast Community College*
Teresa Hasenauer, *Indian River State College*
Julia Hassett, *Oakton Community College*
Jeff Koleno, *Lorain County Community College*
Judy Langer, *Westchester Community College*
Sandy Lofstock, *St. Petersburg College*
Stan Mattoon, *Merced College*
Dr. Kris Mudunuri, *Long Beach City College*
Carol Murphy, *San Diego Miramar College*
Greg Nguyen, *Fullerton College*
Jean Olsen, *Pikes Peak Community College*
Darlene Ornelas, *Fullerton College*
Warren Powell, *Tyler Junior College*
Jeanette Shea, *Central Texas College*
Katerina Vishnyakova, *Collin College*
Corey Wadlington, *West Kentucky Community and Technical College*
Edward Wagner, *Central Texas College*
Jenny Wilson, *Tyler Junior College*

I would also like to thank the following dedicated group of instructors who participated in our focus groups, Martin-Gay Summits, and our design review for the series. Their feedback and insights have helped to strengthen this edition of the text. These instructors include:

Billie Anderson, *Tyler Junior College*
Cedric Atkins, *Mott Community College*
Lois Beardon, *Schoolcraft College*
Laurel Berry, *Bryant & Stratton*
John Beyers, *University of Maryland*
Bob Brown, *Community College of Baltimore County–Essex*
Lisa Brown, *Community College of Baltimore County–Essex*
NeKeith Brown, *Richland College*
Gail Burkett, *Palm Beach Community College*
Cheryl Cantwell, *Seminole Community College*
Jackie Cohen, *Augusta State College*
Julie Dewan, *Mohawk Valley Community College*
Janice Ervin, *Central Piedmont Community College*
Richard Fielding, *Southwestern College*
Cindy Gaddis, *Tyler Junior College*
Nita Graham, *St. Louis Community College*
Pauline Hall, *Iowa State College*

Pat Hussey, *Triton College*
Dorothy Johnson, *Lorain County Community College*
Sonya Johnson, *Central Piedmont Community College*
Irene Jones, *Fullerton College*
Paul Jones, *University of Cincinnati*
Kathy Kopelousous, *Lewis and Clark Community College*
Nancy Lange, *Inver Hills Community College*
Judy Langer, *Westchester Community College*
Lisa Lindloff, *McLinnan Community College*
Sandy Lofstock, *St. Petersburg College*
Kathy Lovelle, *Westchester Community College*
Jean McArthur, *Joliet Junior College*
Kevin McCandless, *Evergreen Valley College*
Daniel Miller, *Niagra County Community College*
Marica Molle, *Metropolitan Community College*
Carol Murphy, *San Diego Miramar College*
Greg Nguyen, *Fullerton College*
Eric Oilila, *Jackson Community College*
Linda Padilla, *Joliet Junior College*
Davidson Pierre, *State College of Florida*
Marilyn Platt, *Gaston College*
Ena Salter, *Manatee Community College*
Carole Shapero, *Oakton Community College*
Janet Sibol, *Hillsborough Community College*
Anne Smallen, *Mohawk Valley Community College*
Barbara Stoner, *Reading Area Community College*
Jennifer Strehler, *Oakton Community College*
Ellen Stutes, *Louisiana State University Elinice*
Tanomo Taguchi, *Fullerton College*
MaryAnn Tuerk, *Elsin Community College*
Walter Wang, *Baruch College*
Leigh Ann Wheeler, *Greenville Technical Community College*
Valerie Wright, *Central Piedmont Community College*

A special thank you to those students who participated in our design review: Katherine Browne, Mike Bulfin, Nancy Canipe, Ashley Carpenter, Jeff Chojnachi, Roxanne Davis, Mike Dieter, Amy Dombrowski, Kay Herring, Todd Jaycox, Kaleena Levan, Matt Montgomery, Tony Plese, Abigail Polkinghorn, Harley Price, Eli Robinson, Avery Rosen, Robyn Schott, Cynthia Thomas, and Sherry Ward.

Elayn Martin-Gay

Personal Acknowledgements

I would like to personally thank my extended family. Although this list has grown throughout the years, it still warrants mentioning in my texts as each of these family members has contributed to my work in one way or another – from suggesting application exercises with data and updating/upgrading my computer to understanding that I usually work on "Vacations." I am deeply grateful to them all:

Clayton, Bryan (in heaven), Eric, Celeste, and Tové Gay; Leo and Barbara Miller; Mark and Madison Martin and Carrie Howard; Stuart and Earline Martin; Karen Martin Callac Pasch (in heaven); Michael, Christopher, Matthew, Nicole, and Jessica Callac; Dan Kirk; Keith, Mandy, Erin, and Clayton McQueen, Bailey Martin, Ethan, Avery, and Mia Barnes; Melissa and Belle Landrum.

About the Author

Elayn Martin-Gay has taught mathematics at the University of New Orleans for more than 25 years. Her numerous teaching awards include the local University Alumni Association's Award for Excellence in Teaching, and Outstanding Developmental Educator at University of New Orleans, presented by the Louisiana Association of Developmental Educators.

Prior to writing textbooks, Elayn Martin-Gay developed an acclaimed series of lecture videos to support developmental mathematics students in their quest for success. These highly successful videos originally served as the foundation material for her texts. Today, the videos are specific to each book in the Martin-Gay series. The author has also created Chapter Test Prep Videos to help students during their most "teachable moment"—as they prepare for a test—along with Instructor-to-Instructor videos that provide teaching tips, hints, and suggestions for each developmental mathematics course, including basic mathematics, prealgebra, beginning algebra, and intermediate algebra.

Elayn is the author of 13 published textbooks, and a new Interactive Assignment MyLab Math course, all specializing in developmental mathematics courses. She has also published series in Algebra 1, Algebra 2, and Geometry. She has participated as an author across the broadest range of educational materials: textbooks, videos, tutorial software, and courseware. This provides an opportunity of various combinations for an integrated teaching and learning package offering great consistency for the student.

Applications Index

Advertising/marketing
billboard dimensions, 179
Coca-Cola sign, 143
sign measurements, 143
spending by media type, 555

Agriculture
apple production, 130
corn production, 18
cranberry production, 158
cropland price in U.S., 473
farm size in U.S., 159
farms in Texas and Missouri, 578
lettuce cost, 159
milk cow operations in U.S., 158
number of farms in U.S., 434
nut production, 159
orange production, 277, R-58

Animals
cheetah running speed, 355
cricket chirps, 135, 145, 148
falcon diving speed, 358
fencing needed for pasture, 568
number of fish in tank, 145
sides of pens, 147
speed of flying fish, 148
time for one animal to overtake another, 400

Astronomy/space
alignment of planets, 367
day of week for certain dates, 407
diameter of Milky Way, 257
distance light travels, 209
elevation of optical telescope, 208
energy converted by Sun, 208
Julian day number, 407
magnitude of stars, 19
meteorite weight, 100, 133
rope length needed to encircle Earth, 146
satellites of planets, 133
surface temperature of planets, 63
time for space plane to circle Earth, 144
volume of Jupiter, 257
wavelengths observed by optical telescope, 210
weight of elephant on Pluto, 394
weight of person above Earth, 635
weight of satellite on Mars, 394

Automotive/motor vehicles
car rental fees and mileage charges, 565
cost of owning small SUV, 471
fuel economy, 472
hybrid vehicle sales, 484
manufacturing losses, 63
median automobile age in U.S., 471
miles driven on a budget, 130, 132
motor vehicle production in U.S., R-28
motorcycle exports by Japan, R-28
solar-powered-car land speed, 358
used car prices, 159
vehicles sold in each category, 134

Aviation
runway length, 147
vertical change in flying altitudes, 46

Business
Barnes & Noble stores in operation, 457
break-even point, 569
consulting fees, 415
cost to manufacture headphones, 635
cost to manufacture medicine, 642
cost to manufacture notebooks, 635
cost to operate delivery service, 598
Costco membership, 436
daily sales predictions, 485
defective products, 419
depreciation, 429
Dick's Sporting Goods net sales, 426
discount, 151, 156
employee layoffs, 160
gross profit margin, 349
hourly minimum wage, 497
hourly wage, 434
hours worked, 125–126, R-55
IKEA annual revenue, 445
labor estimates, 396
manufacturing costs, 418
manufacturing time, 415, 457
markup, 159, 180
net income, 38
new price after decrease, 160
number of units manufactured, 324
original price, 152–153, 157
pay for designated hours, 635
percent increase, 152–153, 159, 180, 181
price after discount, 151, 156
price decrease, 160
price of each item, 564, 574
production costs, 434
profit, 484
quantity pricing, 432
quarterly income, 38
quarterly losses, 63
restaurant food and drink sales, 671
restaurant sales, 446
revenue from artificial hearts, 663
salary before pay raise, 157
sale price, 160
sales needed to attain monthly salary, 181
Starbucks locations in U.S., 172
Target's revenue, 671
time for one person to complete job, 398
time for two conveyor belts to move items, 395
time for two working together to finish job, 391–392, 395, 398, 399, 416, 422
tipping calculations, 156

Chemistry/physics
angstrom value, 257
Avogadro's number, 208
nuclear-generated electricity in France, R-28
period of pendulum, 590
resistors in parallel, 407
solutions/mixtures, 154, 157, 159, 176, 180, 397, 561–562, 566, 567, 568, 578

Demographics/populations
Americans by age groups, R-62
average number of children born per woman, 160
counties in Montana and California, 133
fastest-growing occupations, R-29
households with television in U.S., 470
Internet-crime complaints, 157
job predictions, 408, 555
money spent promoting tourism, 134
motion picture/TV businesses with fewer than 10 employees, 180
national debt of selected countries, 208
number of restaurants in U.S., 484
overnight stays in national parks, 156
pet-related expenditures, 432
population per square mile in U.S., 484
projected demand for audiologists, 555
purpose of trips made by Americans, 647
registered nurses in U.S., 441
restaurant employment in U.S., 465
tourism expenditures, 470
trips made by travelers, 150–151
visitors to National Park sites, 218, 277, 323, 324, 688
world population by continent, R-28

Education
book prices, 159
college tuition and fees, 152–153, 445
female and male students enrolled, 134
final exam scores, 172
floor space needed for each student, 396
high school graduates, 276
majors of college freshmen, R-18
maximum number of students in room, 396
number of colleges in U.S., 85
number of freshman students, 17
number of graduate students, 17
persons over age 25 with four or more years of college, R-54

Education (*continued*)
students pursuing higher education, 276
students taking SAT exam, 445
time spent studying, 432

Electronics/computers
cell phone dimensions, 133
cell phone towers in U.S., 160
cell phone use while driving, 180
consumer spending on DVDs or Blu-rays, 547
cost of movie rental, 388
cost to produce DVDs, 349
cost to produce fax machines, 349
daily text message users by age groups, 291
Google annual searches, 209
households with cell phones, R-28
online shopping, 158
virtual reality devices, 159
wireless data usage, 657
wireless households in U.S., 160, 209

Finance
annual inflation rate in U.S., R-61
car rental budget, 509
credit card balances, 46
estate calculations, 132
gold prices, 663
money exchange, 356
monthly expenses within budget, R-56
savings account balances, 17
stock market losses/gains, 62, 85, 87, 91
stock prices, 564
types of stock shares owned, 564
weekly revenue from fundraiser, R-60

Food/nutrition
beverage consumption, 547
blended coffee beans, 566
bottled water consumption in U.S., 276, 547
calories from fat, 161
calories from protein, 161
calories in foods, 395, 398
candy mixture, 568
coffee blend, 158
hot dogs and buns amounts to purchase, 367
length and width of chocolate bar, 670
nut mixture, 158, 566
nut types in can, 398
nutrition labels, 161
pepper hotness measure, 160
pizza piece, 146
pounds of dairy consumed per person in U.S., R-28
salmon imports/exports, 539
sodium from food, 161
trail mix ingredients, 160
yogurt production in U.S., 487

Geography/geology
alloys with copper, 159
desert areas, 100, 133
distance glacier travels, 148
diving depths, 17, 37, 62
elevation differences, 42–43, 46, 47, 525
elevations below sea level, 37
flow rate of water discharged into stream, 356
freshwater distribution on Earth, R-60
geothermal power in U.S., 227
mountain heights, 17
National Park Service areas, R-19
National Park units in U.S., 179
solar-generated power in U.S., 688
sunrise times monthly, 491–492
sunset times monthly, 496
surface area of oceans, 209
time for glacier to reach lake, 137, 187
tornadoes reported in U.S., 184, 426
volcano heights, 175
volcano lava flow time, 146
water flowing over Niagara Falls, 209
wind-generated power in U.S., 227

Geometry
angle measures, 101, 126–127, 130–131, 134, 135, 566
area of circle, 27, 198, 631
area of geometric figures, 242, 277, 304
area of parallelogram, 198
area of rectangle, 198, 219, 233, 234, 241, 259, 260, 356, 374, 436, 610, R-19, R-52, R-60
area of square, 27, 29, 198, 219, 233, 241, 259, 356
area of triangle, 208, 233, R-19, R-52, R-60
base and height of geometric figures, 321
base and height of parallelogram, 580
base and height of triangle, 692
base length of triangular sail, 331
base of triangle, 323, 335, 396, 398, 626
circumference of circle, 141
complementary angles, 46, 47, 100, 132, 178, 375, 383, 566
diagonals of polygon, 322
dimensions of base of square, 656
dimensions of rectangle, 136, 322, 323, 332, 335
golden triangle, 135, 136
height of parallelogram, 251
height of triangle, 146, 323
height of triangular sail, 317–318
length and width of geometric figures, 321
length and width of rectangle, 670
length of rectangle, 171, 277, 619, 641, 643
number of sides of polygon, 693
percent increase in areas, 157, 159
perimeter of geometric figures, 219, 226, 292, 602
perimeter of parallelogram, 523
perimeter of pentagon, 119, 296
perimeter of rectangle, 29, 80, 284, 292, 329, 436, 447, 602
perimeter of square, 367, 499
perimeter of trapezoid, 367, 447
perimeter of triangle, 80, 118, 119, 296, 329, 499
Pythagorean theorem, 319–320
radius of circle, 321, 646
radius of cylinder, 619
radius of sphere, 642
side length of right triangle, 620, 623, 643, 646
side length of square-based pyramid, 619, 656
side lengths of cube, 590, 597, 598
side lengths of geometric figures, 110, 146, 319–320, 330
side lengths of original geometric figures, 323
side lengths of similar triangles, 395, 398, 399, 415, 416, 419
side lengths of squares, 251, 324, 331, 589, 656
side lengths of triangles, 146, 171, 260, 323, 331, 332, 335, 389, 580, 639, 641
supplementary angles, 46, 47, 100, 132, 178, 375, 383, 566
surface area of box, 258
surface area of cube, 200, 631–632
surface area of sphere, 642
unknown angles, 86
volume of cube, 29, 198, 200, 208, 233
volume of cylinder, 198
volume of square-based pyramid, 624

Home improvement
area of room, 144
area of rug, 241
area of tabletop, 263
beam/molding lengths, 226
board lengths, 100, 118, 124, 130, 133, 179, 374, 420
carpet cleaning charges, 130
deck length, 138
distance between marks to drill holes, 407
fencing needed, 144, 266, 568
fertilizer needed for lawn, 146
flower bed dimensions, 145
garden dimensions, 135, 138
length of ladder, 639
perimeter of room, 144
pitch of roof, 465, 469, 472
plumbing estimates, 130
roof area, 588
room area, 610
sewer pipe slope, 470
time needed to paint house, 396
wall area and perimeter, 144
water seal needed for deck, 183

Medicine/health
blood types, R-29
body surface area of human, 598
body-mass index, 349
bone components, R-29

calories used in exercise, 171
cephalic index, 349
cost to manufacture medicine, 642
crutch adjustments, R-18
diameter of DNA strand, 210
diameter of red blood cell, 210
eye blinking rate, 135
heart transplants in U.S., 473
heights of females given femur lengths, 499
heights of students, R-62
heights of two persons, 177
medicine dosage for children, 349, 374
medicine dosage for dogs, 499
nurse practitioners in U.S., 159
organ transplants in U.S., 517
population expected to have flu shot, 388
registered nurses in U.S., 441
revenue from artificial hearts, 663
smoking and pulse rate, 424

Miscellaneous
anniversary party budgeting, 171
area codes, 127–128, 131, 134, 184
area of canvas, 241
beam lengths, 130
billboard dimensions, 146
blueprint measurements, 395
brace length needed for frame, 624
Burj Khalifa Tower curtain wall area, 358
coin types, 564, 575
coin values, 80
combination lock code, 131
connecting pipe length for pipeline, 623
consecutive integers, 110, 127–128, 131, 133, 178, 179, 318–319, 322, 331, 420
distance apart of two people, 641
door numbers, 110, 131
flag dimensions, 331
floor spaces in building, 132
fundraiser attendance, 566
guy wire length, 322, 623
items sold at original price and reduced price, 566
ladder leaning against wall, 322
land measurements, 118
magazines sold in U.S., 483
magic squares, 80
movie-screens in U.S., 465–466
newspaper circulation, 517
page numbers in book, 131, 322
Pentagon office and storage space area, 358
perimeter of picture frame, 602
picture frame measurements, 144
pool dimensions, 325
postage for large envelopes, 497
print newspaper employment, 457
public libraries in selected states, 184
retirement party budgeting, 171
room numbers, 322
rope lengths, 129
same day in New Orleans for two workers, R-8
same night off for two workers, R-8
size of gambling area in casino, 355
stamp types, 564, 575
steel lengths, 129
string lengths, 100, 132, 181
swimming pool dimensions, 179
swimming pool length, 251
tabletop dimensions, 670
tanning lotion mixture, 160
telephone connections handled by switchboard, 324
top tourism destinations, 435
total area of water trough, 603
total lengths of braces for bin sides, 603
unknown numbers, 123, 129, 132, 133, 179, 180, 181, 183, 187, 264, 265, 317, 325, 390, 395, 396, 397, 399, 415, 416, 419, 557–558, 564, 574, 575, 578, 579, 694
volume of a building, 355
volume of Hoberman Sphere, 147
volume of microwave, 610
volume of shipping crate, 147
Washington Monument height, 179
wedding budgeting, 167, 509
wheelchair ramp slope, 470
width of walk around garden, 325
wire lengths, 124, 132

Politics
ballots cast in election, 257
Democratic and Republican governors in U.S., 133
Democrats and Republicans in U.S. House, 124–125, 187
electoral votes California vs. Texas, 124

Real estate
home ownership rate in U.S., R-53
new home construction by region, 158, R-54
perimeter of plot of land, 118, 602
value of a building, 485

Recreation/entertainment
area of ping pong table top, 251
base length of triangular sail, 331
box office sales for U.S. and Canada, 524
card game scores, 46, 58
cinema ticket prices, 180, 434
digital 3-D movie screens, 160
digital and analog cinema screens in U.S., 538
domestic box office for U.S. movie industry, 431
fair attendance, 558
growth in 3-D theater screens, 153
hang glider flight rate, 144
height of triangular sail, 146
indoor cinemas in U.S., 484
kite height, 624
pool toy price, 478, 485
preferences for music genre, R-62
side length of Rubik's Cube, 589, 598
streaming music revenue, 524
theater seating, 574
ticket prices for adults and children, 558–559
vinyl album sales, 218

Sports
baseball game attendance, 578
baseball runs batted in, 564
baseball slugging percentage, 348
basketball player heights, 171
basketball points scored, 564
bowling average, 171
dimensions of roof of Water Cube, 597
dimensions of Water Cube, 597
distance between home plate and second base, 588
football scores, 133
football yards lost/gained, 62, 91
frisbee sales, 478–479
golf scores, 38, 58, 181
ice hockey penalty killing percentage, 374
jogging speeds, 396
medals won in Olympics, 130, 134, 355, R-55
race times, 180
radius of baseball, 619
radius of tennis ball, 619
shooting averages, 173
side length of Water Cube, 656
women's tennis prize money, 406

Temperature/weather
average low temperatures, 38, 46, 47–48
conversions of temperature, 138–139, 144, 147, 148, 180
monthly precipitation, R-61
rainfall calculations, R-18
record highs and lows, 34, 37, 46, R-55
temperature differences, 47–48, 91
temperature drops, 62

Time/distance
conversions of distance, 80
cycling speeds in two parts of trip, 392
distance across lake/pond, 621
distance apart of two vehicles over time, 626
distance from starting point, 398
distance glacier travels, 148
distance light travels, 209
distance parachuter falls, 635
distance seen from top of building, 625
distance spring stretches, 635, 642
distance swimming across river with current, 597
driving speeds in flatland and mountains, 392
driving time of car before receiving speeding ticket, 397
driving two speeds on trip, 399
height of falling object over time, 213–214, 258, 263, 305

Time/distance (*continued*)
height of object fired/thrown upward over time, 218, 284, 315, 321
hiking times of two groups/persons, 567, 578
how far object falls in 5 seconds, 632
jogging speeds, 396
maximum height of fireball from Roman candle, 683
miles driven by taxi for specific cost, 130
miles driven on a budget, 130, 132
rowing speed in still water, 565
rowing time, 396
safe speed on curved road, 625
skidding distance and car speed, 624
solar-powered-car land speed, 358
speed of boat in still water, 396, 397, 415, 419, 574
speed of current, 565, 574
speed of dropped/falling object, 484
speed of one of two vehicles, 399
speed of plane in still air, 397, 399, 565
speeds of two boats traveling apart, 324
speeds of two hikers walking toward each other, 397
speeds of two streetcars, 649
speeds of two trains, 132, 145, 397
speeds of two trains traveling toward each other, 565
speeds of two vehicles, 392–393, 397, 398, 400, 415, 567
time for car to stop when brakes applied, 635
time for cliff diver to reach ocean, 316, 337
time for each pump to fill tank, 398
time for glacier to reach lake, 137, 187
time for object launched/thrown upward/dropped to hit ground, 323, 335, 624, 671
time for rocket to reach certain height, 331
time for rocket to reach ground, 331
time for third pipe to fill pool, 398
time for two pumps to fill tank, 399, 419
time for vehicle to overtake another, 398
time of free-falling dive/fall, 654, 656, 687, 693
time of travel in each vehicle, 397
time spent driving, 137
time spent jogging and walking, 574
time spent on bicycle during trip, 565
time to clean beach, 382
time to fill tank, 382
time to make return trip, 635
time when fireball from Roman candle at maximum height, 683
time when fireball from Roman candle hits ground, 683
travel time, 145, 147
velocity of object after fallen selected distances, 622
walking speeds of each person, 560–561, 649
wind speed, 397, 399, 565

Transportation
grade of road, 465, 470
height of bulge in train tracks, 626
highway bridges in U.S., 657
highway lengths, 100
length of side of parking lot, 589
parking lot length, 144
radius of curvature of road, 625
road sign dimensions, 139–140, 567
safe speed on curved road, 625
time for car to stop when brakes applied, 635
train fares for adults and children, 564
train track slope, 470
travel time, 145, 147
yield sign dimensions, 145

World records
coldest temperatures, 46
fastest trains, 132
highest dive from diving board, 693
highest dive into lake, 654
highest elevations, 42
largest building, 355
largest meteorite, 100
longest interstate highway, 100
lowest elevations, 37
steepest street, 470
tallest building, 305, 358
temperature differences, 46

Prealgebra Review

This optional review chapter covers basic topics and skills from prealgebra, such as fractions, decimals, percents, and reading graphs. Knowledge of these topics is needed for success in algebra.

Sections

R.1 Factors and the Least Common Multiple

R.2 Fractions and Mixed Numbers

R.3 Decimals and Percents

R.4 Reading Pictographs and Bar, Line, and Circle Graphs

Check Your Progress

Vocabulary Check

Chapter Highlights

Chapter Review

Getting Ready for the Test

Chapter Test

How Many Times per Month Do You Usually Eat Pizza?

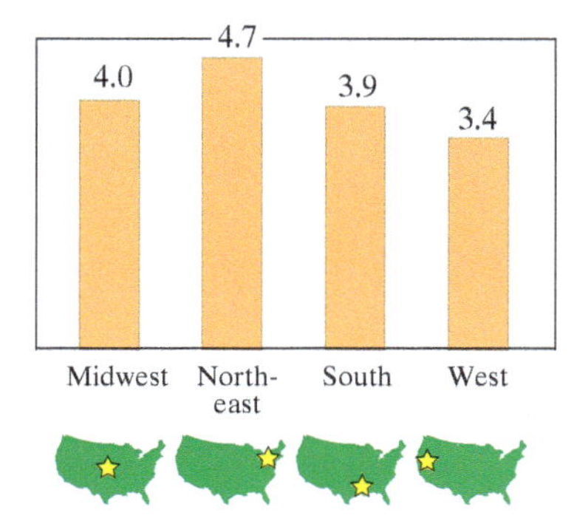

The **average** is 4 times a month.

How Many Slices Do You Usually Eat When Eating Pizza?

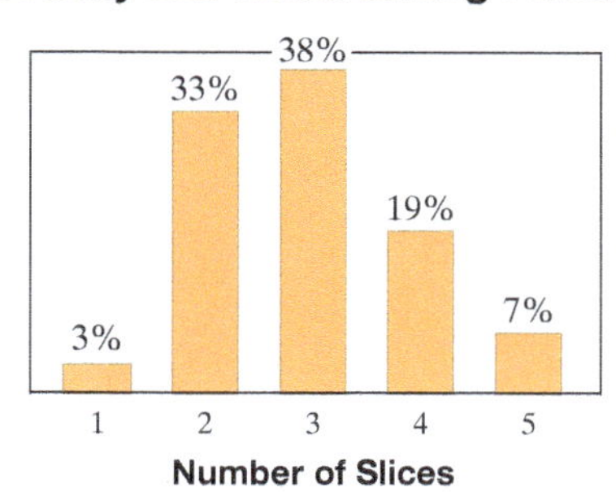

The **average** is about 3 slices.

All About Pizza!

The word *pizza* was first documented in AD 997 in Italy, and it literally means "pie." Foods similar to pizza can be traced back to ancient Greece, Persia, and other countries, although modern-day pizza is believed to have been invented in Naples, Italy. Pizza was mainly eaten in Italy until immigrants brought the idea of pizza to the United States.

Surveys show that pizza is ranked number 1 among comfort foods. Below are a few more interesting facts and graphs about pizza. For example:

- the average price of a slice of pizza is $3.26, and
- the average cost of a pie (or whole pizza) is $16.73.
- Depending on the survey, New York City is usually ranked first as the city with the best pizza, followed by Chicago.

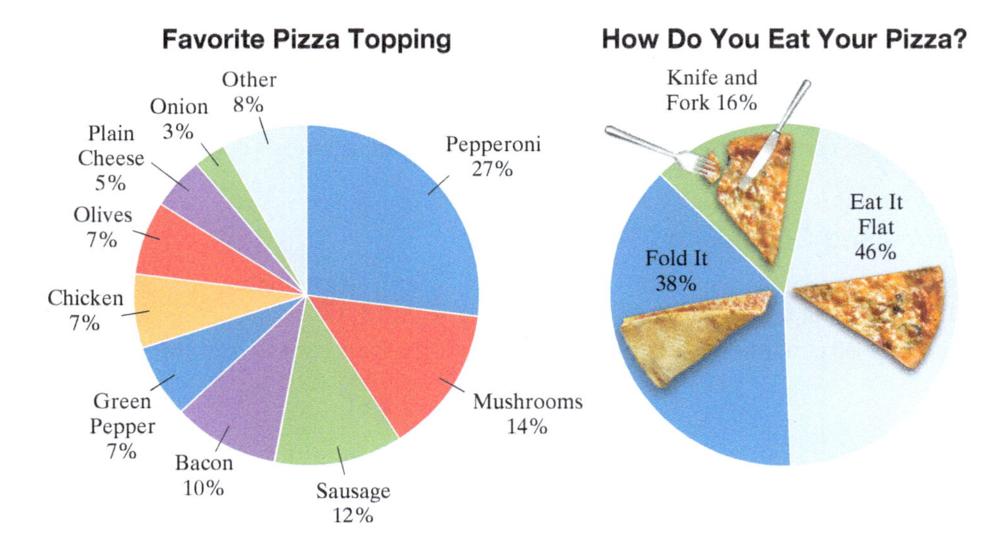

Sources: Zagat, Statista, Harris Poll

Throughout this chapter, we review basic topics from prealgebra including reading all types of graphs.

R.1 Factors and the Least Common Multiple

Objectives

A Write the Factors of a Number.

B Write the Prime Factorization of a Number.

C Find the LCM of a List of Numbers.

Objective A Factoring Numbers

In arithmetic we factor numbers, and in algebra we factor expressions containing variables.

To **factor** means to write as a product.

Throughout this text, you will encounter the word *factor* often.

Always remember that **factoring means writing as a product.**

Since $2 \cdot 3 = 6$, we say that 2 and 3 are **factors** of 6. Also, $2 \cdot 3$ is a **factorization** of 6.

Practice 1

List the factors of 10.

Example 1 List the factors of 6.

Solution: First we write the different factorizations of 6.

$$6 = 1 \cdot 6, \quad 6 = 2 \cdot 3$$

The factors of 6 are 1, 2, 3, and 6.

Work Practice 1

Practice 2

List the factors of 18.

Example 2 List the factors of 20.

Solution: $20 = 1 \cdot 20, \quad 20 = 2 \cdot 10, \quad 20 = 4 \cdot 5$

The factors of 20 are 1, 2, 4, 5, 10, and 20.

Work Practice 2

In this section, we will concentrate on **natural numbers** only. The natural numbers (also called counting numbers) are

Natural Numbers: 1, 2, 3, 4, 5, 6, 7, and so on

Every natural number except 1 is either a prime number or a composite number.

Prime and Composite Numbers

A **prime number** is a natural number greater than 1 whose only factors are 1 and itself. The first few prime numbers are 2, 3, 5, 7, 11, 13, 17, ...

A **composite number** is a natural number greater than 1 that is not prime.

Practice 3

Identify each number as prime or composite: 5, 16, 23, 42.

Example 3 Identify each number as prime or composite: 3, 28, 19, 35

Solution:

3 is a prime number. Its factors are 1 and 3 only.
28 is a composite number. Its factors are 1, 2, 4, 7, 14, and 28.
19 is a prime number. Its factors are 1 and 19 only.
35 is a composite number. Its factors are 1, 5, 7, and 35.

Work Practice 3

Answers

1. 1, 2, 5, 10 **2.** 1, 2, 3, 6, 9, 18
3. 5, 23 prime; 16, 42 composite

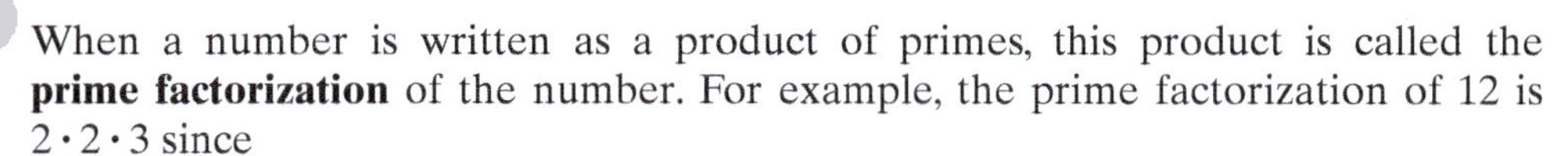

Objective B Writing Prime Factorizations

When a number is written as a product of primes, this product is called the **prime factorization** of the number. For example, the prime factorization of 12 is $2 \cdot 2 \cdot 3$ since

$$12 = 2 \cdot 2 \cdot 3$$

and all the factors are prime numbers.

Example 4

Write the prime factorization of 45.

Solution: We can begin by writing 45 as the product of two numbers, say 9 and 5.

$$45 = 9 \cdot 5$$

The number 5 is prime, but 9 is not. So we write 9 as $3 \cdot 3$.

$$45 = 9 \cdot 5$$
$$= 3 \cdot 3 \cdot 5$$

Each factor is now a prime number, so the prime factorization of 45 is $3 \cdot 3 \cdot 5$.

Work Practice 4

Practice 4

Write the prime factorization of 44.

Helpful Hint

Recall that order is not important when multiplying numbers. For example,

$$3 \cdot 3 \cdot 5 = 3 \cdot 5 \cdot 3 = 5 \cdot 3 \cdot 3 = 45$$

For this reason, any of the products shown can be called *the* prime factorization of 45, and we say that the prime factorization of a number is unique.

Example 5

Write the prime factorization of 80.

Solution: We first write 80 as a product of two numbers. We continue this process until all factors are prime.

$$80 = 8 \cdot 10$$
$$= 4 \cdot 2 \cdot 2 \cdot 5$$
$$= 2 \cdot 2 \cdot 2 \cdot 2 \cdot 5$$

All factors are now prime, so the prime factorization of 80 is

$$2 \cdot 2 \cdot 2 \cdot 2 \cdot 5.$$

Work Practice 5

Practice 5

Write the prime factorization of 60.

✓**Concept Check** Suppose that you choose $80 = 4 \cdot 20$ as your first step in Example 5 and another student chooses $80 = 5 \cdot 16$. Will you both end up with the same prime factorization as in Example 5? Explain.

Answers

4. $2 \cdot 2 \cdot 11$ **5.** $2 \cdot 2 \cdot 3 \cdot 5$

✓**Concept Check Answer**

yes; answers may vary

Helpful Hint

There are a few quick **divisibility tests** to determine if a number is divisible by the primes 2, 3, or 5.

A whole number is divisible by

- **2** if the ones digit is 0, 2, 4, 6, or 8.

  ↓

  132 is divisible by 2
- **3** if the sum of the digits is divisible by 3.

  144 is divisible by 3 since $1 + 4 + 4 = 9$ is divisible by 3
- **5** if the ones digit is 0 or 5.

  ↓

  1115 is divisible by 5

When finding the prime factorization of larger numbers, you may want to use the procedure shown in Example 6.

Practice 6

Write the prime factorization of 297.

Example 6 Write the prime factorization of 252.

Solution: Since the ones digit of 252 is 2, we know that 252 is divisible by 2.

```
   126
2)252
```

126 is divisible by 2 also.

```
    63
2)126
2)252
```

63 is not divisible by 2 but is divisible by 3. We divide 63 by 3 and continue in this same manner until the quotient is a prime number.

```
     7
3)  21
3)  63
2)126
2)252
```

The prime factorization of 252 is $2 \cdot 2 \cdot 3 \cdot 3 \cdot 7$.

Work Practice 6

Objective C Finding the Least Common Multiple

A **multiple** of a number is the product of that number and any natural number. For example, the multiples of 3 are

$3 \cdot 1$	$3 \cdot 2$	$3 \cdot 3$	$3 \cdot 4$	$3 \cdot 5$	$3 \cdot 6$	$3 \cdot 7$	
3,	6,	9,	12,	15,	18,	21,	and so on.

The multiples of 2 are

$2 \cdot 1$	$2 \cdot 2$	$2 \cdot 3$	$2 \cdot 4$	$2 \cdot 5$	$2 \cdot 6$	$2 \cdot 7$	
2,	4,	6,	8,	10,	12,	14,	and so on.

Answer

6. $3 \cdot 3 \cdot 3 \cdot 11$

Notice that 2 and 3 have multiples that are common to both.

Multiples of 2: 2, 4, 6, 8, 10, 12, 14, 16, 18, and so on

Multiples of 3: 3, 6, 9, 12, 15, 18, 21, and so on

Common multiples of 2 and 3: 6, 12, 18, . . .

The least or smallest common multiple of 2 and 3 is 6. The number 6 is called the **least common multiple** or **LCM** of 2 and 3. It is the smallest number that is a multiple of both 2 and 3.

The **least common multiple (LCM)** of a list of numbers is the smallest number that is a multiple of all the numbers in the list.

Finding the LCM by the method above can sometimes be time-consuming. Let's look at another method that uses prime factorization.

To find the LCM of 4 and 10, for example, we write the prime factorization of each.

$$4 = 2 \cdot 2$$
$$10 = 2 \cdot 5$$

If the LCM is to be a multiple of 4, it must contain the factors $2 \cdot 2$. If the LCM is to be a multiple of 10, it must contain the factors $2 \cdot 5$. Since we decide whether the LCM is a multiple of 4 and 10 separately, the LCM does not need to contain three factors of 2. The LCM only needs to contain a factor the greatest number of times that the factor appears in any **one** prime factorization.

$$\text{LCM} = \overbrace{2 \cdot 2}^{\text{The LCM is a multiple of 4.}} \cdot 5 = 20$$

The LCM is a multiple of 4. ($2 \cdot 2$)
The LCM is a multiple of 10. ($2 \cdot 5$)

The number 2 is a factor twice since that is the greatest number of times that 2 is a factor in either of the prime factorizations.

To Find the LCM of a List of Numbers

Step 1: Write the prime factorization of each number.

Step 2: Write the product containing each different prime factor (from Step 1) the greatest number of times that it appears in any one factorization. This product is the LCM.

Example 7 Find the LCM of 18 and 24.

Solution: First we write the prime factorization of each number.

$$18 = 2 \cdot 3 \cdot 3$$
$$24 = 2 \cdot 2 \cdot 2 \cdot 3$$

Now we write each factor the greatest number of times that it appears in any **one** prime factorization.

The greatest number of times that 2 appears is **3** times.

The greatest number of times that 3 appears is **2** times.

$$\text{LCM} = \underbrace{2 \cdot 2 \cdot 2}_{\text{2 is a factor 3 times.}} \cdot \underbrace{3 \cdot 3}_{\text{3 is a factor 2 times.}} = 72$$

Work Practice 7

Practice 7

Find the LCM of 14 and 35.

Answer

7. 70

Practice 8

Find the LCM of 5 and 9.

Example 8 Find the LCM of 11 and 10.

Solution: 11 is a prime number, so we simply rewrite it. Then we write the prime factorization of 10.

$$11 = 11$$
$$10 = 2 \cdot 5$$
$$\text{LCM} = 2 \cdot 5 \cdot 11 = 110$$

Work Practice 8

Practice 9

Find the LCM of 4, 15, and 10.

Example 9 Find the LCM of 5, 6, and 12.

Solution:

$$5 = 5$$
$$6 = 2 \cdot 3$$
$$12 = 2 \cdot 2 \cdot 3$$
$$\text{LCM} = 2 \cdot 2 \cdot 3 \cdot 5 = 60.$$

Work Practice 9

Answers

8. 45 **9.** 60

Vocabulary, Readiness & Video Check

Use the choices below to fill in each blank.

least common multiple	composite	multiple
prime factorization	prime	factor

1. The number 40 equals $2 \cdot 2 \cdot 2 \cdot 5$. Since each factor is prime, we call $2 \cdot 2 \cdot 2 \cdot 5$ the ______________ of 40.
2. A natural number, other than 1, that is not prime is called a(n) ______________ number.
3. A natural number that has exactly two different factors, 1 and itself, is called a(n) ______________ number.
4. The ______________ of a list of numbers is the smallest number that is a multiple of all the numbers in the list.
5. To ______________ means to write as a product.
6. A(n) ______________ of a number is the product of that number and any natural number.

Martin-Gay Interactive Videos

See Video R.1

Watch the section lecture video and answer the following questions.

Objective A **7.** From the lecture before Example 2, are all natural numbers either prime or composite?

Objective B **8.** Complete this statement based on Example 4: We may write factors in different ______________, but every natural number has only ______________ prime factorization.

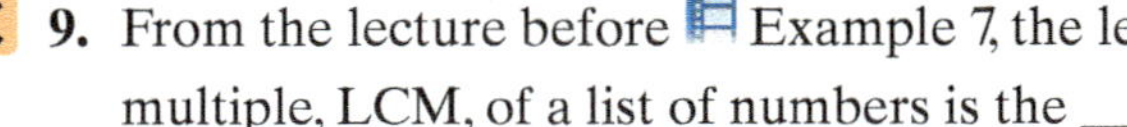

Objective C **9.** From the lecture before Example 7, the least common multiple, LCM, of a list of numbers is the ______________ number that is a multiple of each number in the list.

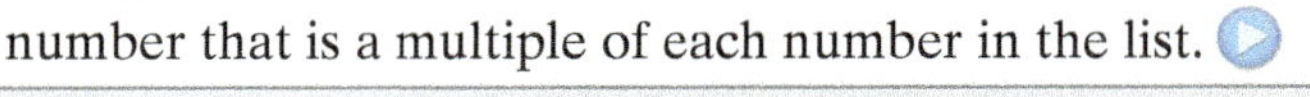

R.1 Exercise Set MyLab Math 

Objective A *List the factors of each number. See Examples 1 and 2.*

1. 9 **2.** 8 **3.** 24 **4.** 36 **5.** 42

6. 63 **7.** 80 **8.** 50 **9.** 19 **10.** 31

Identify each number as prime or composite. See Example 3.

11. 13 **12.** 21 **13.** 39 **14.** 53 **15.** 41

16. 51 **17.** 201 **18.** 307 **19.** 2065 **20.** 1798

Objective B *Write each prime factorization. See Examples 4 through 6.*

21. 18 **22.** 28 **23.** 20 **24.** 30

25. 56 **26.** 48 **27.** 81 **28.** 64

29. 300 **30.** 500 **31.** 588 **32.** 315

Multiple choice. Select the best choice to complete each statement.

33. The factors of 48 are
- **a.** $2 \cdot 2 \cdot 2 \cdot 6$
- **b.** $2 \cdot 2 \cdot 2 \cdot 3$
- **c.** $2 \cdot 2 \cdot 2 \cdot 2 \cdot 3$
- **d.** 1, 2, 3, 4, 6, 8, 12, 16, 24, 48

34. The prime factorization of 63 is
- **a.** 1, 3, 7, 9, 63
- **b.** 1, 3, 7, 9, 21, 63
- **c.** $3 \cdot 3 \cdot 7$
- **d.** 1, 3, 21, 63

Objective C *Find the LCM of each list of numbers. See Examples 7 through 9.*

35. 3, 4 **36.** 4, 5 **37.** 6, 14 **38.** 9, 15

39. 20, 30 **40.** 30, 40 **41.** 5, 7 **42.** 2, 11

43. 9, 12 **44.** 4, 18 **45.** 16, 20 **46.** 18, 30

47. 40, 90 **48.** 50, 70 **49.** 24, 36 **50.** 21, 28

51. 2, 8, 15 **52.** 3, 9, 20 **53.** 2, 3, 7 **54.** 3, 5, 7

55. 8, 24, 48 **56.** 9, 36, 72 **57.** 8, 18, 30 **58.** 4, 14, 35

Concept Extensions

59. Solve. See the Concept Check in the section.

a. Write the prime factorization of 40 using 2 and 20 as the first pair of factors.

b. Write the prime factorization of 40 using 4 and 10 as the first pair of factors.

c. Explain any similarities or differences found in parts a and b.

60. The LCM of 6 and 7 is 42. In general, describe when the LCM of two numbers is equal to their product.

61. Craig Campanella and Edie Hall both have night jobs. Craig has every fifth night off, and Edie has every seventh night off. How often will they have the same night off?

62. Elizabeth Kaster and Lori Sypher are both publishing company representatives in Louisiana. Elizabeth spends a day in New Orleans every 35 days, and Lori spends a day in New Orleans every 20 days. How often are they in New Orleans on the same day?

Find the LCM of each pair of numbers.

63. 315, 504

64. 1000, 1125

R.2 Fractions and Mixed Numbers

Objectives

A Discover Fraction Properties Having to Do with 0 and 1.

B Write Equivalent Fractions.

C Write Fractions in Simplest Form.

D Multiply and Divide Fractions.

E Add and Subtract Fractions.

F Perform Operations on Mixed Numbers.

A quotient of two numbers such as $\frac{2}{9}$ is called a **fraction**. The parts of a fraction are:

Fraction bar $\rightarrow$ $\frac{2}{9}$ $\begin{matrix}\leftarrow \text{Numerator}\\ \leftarrow \text{Denominator}\end{matrix}$

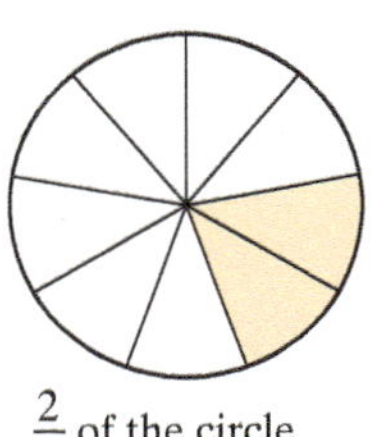

$\frac{2}{9}$ of the circle is shaded.

A fraction may be used to refer to part of a whole. For example, $\frac{2}{9}$ of the circle is shaded. The denominator 9 tells us how many equal parts the whole circle is divided into, and the numerator 2 tells us how many equal parts are shaded.

In this section, we will use numerators that are **whole numbers** and denominators that are nonzero whole numbers. The whole numbers consist of 0 and the natural numbers.

Whole Numbers: 0, 1, 2, 3, 4, 5, and so on

Objective A Discovering Fraction Properties with 0 and 1

Before we continue further, don't forget that the fraction bar indicates division. For example,

$$\frac{8}{4} = 8 \div 4 = 2 \quad \text{since} \quad 2 \cdot 4 = 8$$

Thus, we may simplify some fractions by recalling that the fraction bar means division.

$$\frac{6}{6} = 6 \div 6 = 1 \quad \text{and} \quad \frac{3}{1} = 3 \div 1 = 3$$

Examples Simplify by dividing the numerator by the denominator.

1. $\frac{3}{3} = 1$ Since $3 \div 3 = 1$.

2. $\frac{4}{2} = 2$ Since $4 \div 2 = 2$.

3. $\frac{7}{7} = 1$ Since $7 \div 7 = 1$.

4. $\frac{8}{1} = 8$ Since $8 \div 1 = 8$.

5. $\frac{0}{6} = 0$ Since $0 \cdot 6 = 0$.

6. $\frac{6}{0}$ is undefined because there is no number that when multiplied by 0 gives 6.

Work Practice 1–6

Practice 1–6

Simplify by dividing the numerator by the denominator.

1. $\frac{4}{4}$ 2. $\frac{9}{3}$ 3. $\frac{10}{10}$

4. $\frac{5}{1}$ 5. $\frac{0}{11}$ 6. $\frac{11}{0}$

From Examples 1 through 6, we can say the following:

Let a be any number other than 0.

$$\frac{a}{a} = 1, \quad \frac{0}{a} = 0,$$

$$\frac{a}{1} = a, \quad \frac{a}{0} \text{ is undefined}$$

Objective B Writing Equivalent Fractions

More than one fraction can be used to name the same part of a whole. Such fractions are called **equivalent fractions**.

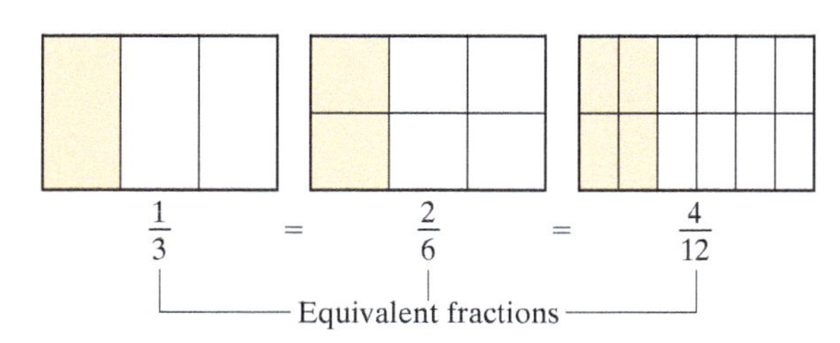

Equivalent Fractions

Fractions that represent the same portion of a whole are called **equivalent fractions**.

Answers
1. 1 **2.** 3 **3.** 1 **4.** 5 **5.** 0 **6.** undefined

For example, let's write $\frac{1}{3}$ as an equivalent fraction with a denominator of 12. To do so, notice the denominator of 3, multiplied by 4, gives a denominator of 12. Thus let's multiply by 1 in the form of $\frac{4}{4}$.

$$\frac{1}{3} = \frac{1}{3}\cdot 1 = \frac{1}{3}\cdot\frac{4}{4} = \frac{1\cdot 4}{3\cdot 4} = \frac{4}{12}$$

$\frac{4}{4} = 1$

So $\frac{1}{3} = \frac{4}{12}$.

To Write an Equivalent Fraction

$$\frac{a}{b} = \frac{a}{b}\cdot\frac{c}{c} = \frac{a\cdot c}{b\cdot c}$$

Since $\frac{a}{b} = \frac{a}{b}\cdot 1$

where a, b, and c are nonzero numbers.

Practice 7

Write $\frac{1}{4}$ as an equivalent fraction with a denominator of 20.

Example 7 Write $\frac{2}{5}$ as an equivalent fraction with a denominator of 15.

Solution: In the denominator, since $5\cdot 3 = 15$, we multiply the fraction $\frac{2}{5}$ by 1 in the form of $\frac{3}{3}$.

$$\frac{2}{5} = \frac{2}{5}\cdot\frac{3}{3} = \frac{2\cdot 3}{5\cdot 3} = \frac{6}{15}$$

Then $\frac{2}{5}$ is equivalent to $\frac{6}{15}$. They both represent the same part of a whole.

Work Practice 7

Objective C Simplifying Fractions

A special equivalent fraction is one that is simplified or in lowest terms. A fraction is said to be **simplified** or in **lowest terms** when the numerator and the denominator have no factors in common other than 1. For example, the fraction $\frac{5}{11}$ is in lowest terms since 5 and 11 have no common factors other than 1.

To simplify a fraction, we write an equivalent fraction, but one with no common factors in the numerator and denominator. Since we are writing an equivalent fraction, we use the same method as before, except we are "removing" factors of 1 instead of "inserting" factors of 1. We call this the Fundamental Principle of Fractions.

Helpful Hint We call this principle the Fundamental Principle of Fractions.

To Write a Simplified, Equivalent Fraction

$$\frac{a\cdot c}{b\cdot c} = \frac{a}{b}\cdot\frac{c}{c} = \frac{a}{b}$$

Since $\frac{a}{b}\cdot 1 = \frac{a}{b}$

Answer

7. $\frac{5}{20}$

Example 8 Simplify: $\frac{42}{49}$

Solution: To help us see common factors in the numerator and denominator, or factors of 1, we write the numerator and the denominator as products of primes.

$$\frac{42}{49} = \frac{2 \cdot 3 \cdot 7}{7 \cdot 7} = \frac{2 \cdot 3}{7} \cdot \frac{7}{7} = \frac{2 \cdot 3}{7} = \frac{6}{7}$$

Work Practice 8

Practice 8

Simplify: $\frac{20}{35}$

✓**Concept Check** Explain the error in the following steps.

a. $\frac{15}{55} = \frac{15}{55} = \frac{1}{5}$ b. $\frac{6}{7} = \frac{5+1}{5+2} = \frac{1}{2}$

Examples Simplify each fraction.

9. $\frac{11}{27} = \frac{11}{3 \cdot 3 \cdot 3}$ There are no common factors in the numerator and denominator other than 1, so $\frac{11}{27}$ is already simplified.

10. $\frac{88}{20} = \frac{2 \cdot 2 \cdot 2 \cdot 11}{2 \cdot 2 \cdot 5} = \frac{2}{2} \cdot \frac{2}{2} \cdot \frac{2 \cdot 11}{5} = \frac{22}{5}$

Work Practice 9–10

Practice 9–10

Simplify each fraction.

9. $\frac{7}{20}$ 10. $\frac{12}{40}$

Below are two important notes about simplifying fractions.

Note 1: When simplifying, we can use a shortcut notation if desired. From Example 8,

$$\frac{42}{49} = \frac{2 \cdot 3 \cdot \overset{1}{\cancel{7}}}{7 \cdot \underset{1}{\cancel{7}}} = \frac{2 \cdot 3}{7} = \frac{6}{7}$$

Note 2: Also, feel free to save time if you immediately notice common factors. In Example 10, notice that the numerator and denominator of $\frac{88}{20}$ have a common factor of 4.

$$\frac{88}{20} = \frac{\overset{1}{\cancel{4}} \cdot 22}{\underset{1}{\cancel{4}} \cdot 5} = \frac{22}{5}$$

A **proper fraction** is a fraction whose numerator is less than its denominator. The fraction $\frac{22}{5}$ from Example 10 is called an improper fraction. An **improper fraction** is a fraction whose numerator is greater than or equal to its denominator.

The improper fraction $\frac{22}{5}$ may be written as the mixed number $4\frac{2}{5}$. Notice that a **mixed number** has a whole number part and a fraction part. We review operations on mixed numbers in objective F in this section. First, let's review operations on fractions.

Objective D Multiplying and Dividing Fractions

To multiply two fractions, we multiply numerator times numerator to obtain the numerator of the product. Then we multiply denominator times denominator to obtain the denominator of the product.

Answers

8. $\frac{4}{7}$ 9. $\frac{7}{20}$ 10. $\frac{3}{10}$

✓**Concept Check Answers**

a. $\frac{15}{55} = \frac{3 \cdot 5}{11 \cdot 5} = \frac{3}{11}$

b. $\frac{6}{7}$ can't be simplified

Helpful Hint The symbol "≠" to the right means "is not equal to."

Multiplying Fractions

$$\frac{a}{b}\cdot\frac{c}{d}=\frac{a\cdot c}{b\cdot d}\quad \text{if } b\neq 0 \text{ and } d\neq 0$$

Practice 11

Multiply: $\frac{3}{4}\cdot\frac{8}{9}$. Simplify the product if possible.

Example 11 Multiply: $\frac{2}{15}\cdot\frac{5}{13}$. Simplify the product if possible.

Solution: $\frac{2}{15}\cdot\frac{5}{13}=\frac{2\cdot 5}{15\cdot 13}$ Multiply numerators. Multiply denominators.

To simplify the product, we divide the numerator and the denominator by any common factors.

$$\frac{2}{15}\cdot\frac{5}{13}=\frac{2\cdot\overset{1}{\cancel{5}}}{3\cdot\underset{1}{\cancel{5}}\cdot 13}=\frac{2}{39}$$

Work Practice 11

Before we divide fractions, we first define **reciprocals**. Two numbers are reciprocals of each other if their product is 1.

The reciprocal of $\frac{2}{3}$ is $\frac{3}{2}$ because $\frac{2}{3}\cdot\frac{3}{2}=\frac{6}{6}=1$.

The reciprocal of 5 is $\frac{1}{5}$ because $5\cdot\frac{1}{5}=\frac{5}{1}\cdot\frac{1}{5}=\frac{5}{5}=1$.

To divide fractions, we multiply the first fraction by the reciprocal of the second fraction. For example,

$$\frac{1}{2}\div\frac{5}{7}=\frac{1}{2}\cdot\frac{7}{5}=\frac{1\cdot 7}{2\cdot 5}=\frac{7}{10}$$

To divide, multiply by the reciprocal.

Dividing Fractions

$$\frac{a}{b}\div\frac{c}{d}=\frac{a}{b}\cdot\frac{d}{c},\quad \text{if } b\neq 0,\ d\neq 0, \text{ and } c\neq 0$$

Practice 12–14

Divide and simplify.

12. $\frac{2}{9}\div\frac{3}{4}$

13. $\frac{8}{11}\div 24$

14. $\frac{5}{4}\div\frac{15}{8}$

Examples Divide and simplify.

12. $\frac{4}{5}\div\frac{5}{16}=\frac{4}{5}\cdot\frac{16}{5}=\frac{4\cdot 16}{5\cdot 5}=\frac{64}{25}$ The numerator and denominator have no common factors.

13. $\frac{7}{10}\div 14=\frac{7}{10}\div\frac{14}{1}=\frac{7}{10}\cdot\frac{1}{14}=\frac{\overset{1}{\cancel{7}}\cdot 1}{2\cdot 5\cdot 2\cdot\underset{1}{\cancel{7}}}=\frac{1}{20}$

14. $\frac{3}{8}\div\frac{3}{10}=\frac{3}{8}\cdot\frac{10}{3}=\frac{\overset{1}{\cancel{3}}\cdot\overset{1}{\cancel{2}}\cdot 5}{\underset{1}{\cancel{2}}\cdot 2\cdot 2\cdot\underset{1}{\cancel{3}}}=\frac{5}{4}$

Work Practice 12–14

Answers

11. $\frac{2}{3}$ **12.** $\frac{8}{27}$ **13.** $\frac{1}{33}$ **14.** $\frac{2}{3}$

Objective E Adding and Subtracting Fractions 

To add or subtract fractions with the same denominator, we combine numerators and place the sum or difference over the common denominator.

Adding and Subtracting Fractions with the Same Denominator

$$\frac{a}{b} + \frac{c}{b} = \frac{a + c}{b}, \quad \text{if } b \neq 0$$

$$\frac{a}{b} - \frac{c}{b} = \frac{a - c}{b}, \quad \text{if } b \neq 0$$

Examples Add or subtract as indicated. Then simplify if possible.

15. $\frac{2}{7} + \frac{4}{7} = \frac{2 + 4}{7} = \frac{6}{7}$ ← Add numerators. ← Keep the common denominator.

16. $\frac{3}{10} + \frac{2}{10} = \frac{3 + 2}{10} = \frac{5}{10} = \frac{\overset{1}{\cancel{5}}}{2 \cdot \underset{1}{\cancel{5}}} = \frac{1}{2}$

17. $\frac{5}{3} - \frac{1}{3} = \frac{5 - 1}{3} = \frac{4}{3}$ ← Subtract numerators. ← Keep the common denominator.

18. $\frac{9}{7} - \frac{2}{7} = \frac{9 - 2}{7} = \frac{7}{7} = 1$

Work Practice 15–18

Practice 15–18

Add or subtract as indicated. Then simplify if possible.

15. $\frac{2}{11} + \frac{5}{11}$ **16.** $\frac{1}{8} + \frac{3}{8}$

17. $\frac{7}{6} - \frac{2}{6}$ **18.** $\frac{13}{10} - \frac{3}{10}$

To add or subtract with different denominators, we first write the fractions as **equivalent fractions** with the same denominator. We use the smallest or **least common denominator**, or **LCD**. The LCD is the same as the least common multiple of the denominators (see Section R.1).

Example 19 Add: $\frac{2}{5} + \frac{1}{4}$

Solution: We first must find the least common denominator before the fractions can be added. The least common multiple of the denominators 5 and 4 is 20. This is the LCD we will use.

We write both fractions as equivalent fractions with denominators of 20. Since

$$\frac{2}{5} = \frac{2}{5} \cdot 1 = \frac{2}{5} \cdot \frac{4}{4} = \frac{2 \cdot 4}{5 \cdot 4} = \frac{8}{20} \quad \text{and} \quad \frac{1}{4} = \frac{1}{4} \cdot 1 = \frac{1}{4} \cdot \frac{5}{5} = \frac{1 \cdot 5}{4 \cdot 5} = \frac{5}{20}$$

then

$$\frac{2}{5} + \frac{1}{4} = \frac{8}{20} + \frac{5}{20} = \frac{13}{20}$$

Work Practice 19

Pracsice 19

Add: $\frac{3}{8} + \frac{1}{20}$

Answers

15. $\frac{7}{11}$ **16.** $\frac{1}{2}$ **17.** $\frac{5}{6}$ **18.** 1 **19.** $\frac{17}{40}$

Practice 20

Subtract and simplify: $\frac{8}{15} - \frac{1}{3}$

Example 20 Subtract and simplify: $\frac{19}{6} - \frac{23}{12}$

Solution: The LCD is 12. We write both fractions as equivalent fractions with denominators of 12.

$$\frac{19}{6} - \frac{23}{12} = \frac{19}{6}\cdot\frac{2}{2} - \frac{23}{12} = \frac{19\cdot 2}{6\cdot 2} - \frac{23}{12} = \frac{38}{12} - \frac{23}{12} = \frac{15}{12} = \frac{\overset{1}{\cancel{3}}\cdot 5}{2\cdot 2\cdot \underset{1}{\cancel{3}}} = \frac{5}{4}$$

Work Practice 20

Objective F Performing Operations on Mixed Numbers

To perform operations on mixed numbers, first write each mixed number as an improper fraction. To recall how this is done, let's write $3\frac{1}{5}$ as an improper fraction.

$$3\frac{1}{5} = 3 + \frac{1}{5} = \frac{15}{5} + \frac{1}{5} = \frac{16}{5}$$

Because of the steps above, notice we can use a shortcut process for writing a mixed number as an improper fraction.

$$3\frac{1}{5} = \frac{5\cdot 3 + 1}{5} = \frac{16}{5}$$

Practice 21

Multiply: $5\frac{1}{6}\cdot 4\frac{2}{5}$

Example 21 Divide: $2\frac{1}{8} \div 1\frac{2}{3}$

Solution: First write each mixed number as an improper fraction.

$$2\frac{1}{8} = \frac{8\cdot 2 + 1}{8} = \frac{17}{8}; \quad 1\frac{2}{3} = \frac{3\cdot 1 + 2}{3} = \frac{5}{3}$$

Now divide as usual.

$$2\frac{1}{8} \div 1\frac{2}{3} = \frac{17}{8} \div \frac{5}{3} = \frac{17}{8}\cdot\frac{3}{5} = \frac{51}{40}$$

The fraction $\frac{51}{40}$ is improper. To write it as an equivalent mixed number, remember that the fraction bar means division, and divide.

$$\begin{array}{r} 1\frac{11}{40} \\ 40\overline{)51} \\ -40 \\ \hline 11 \end{array}$$

Thus, the quotient is $\frac{51}{40}$ or $1\frac{11}{40}$.

Work Practice 21

Answers

20. $\frac{1}{5}$ **21.** $\frac{341}{15}$ or $22\frac{11}{15}$

As a general rule, if the original exercise contains mixed numbers, write the result as a mixed number, if possible.

Example 22 Add: $2\frac{1}{8} + 1\frac{2}{3}$

Solution: $2\frac{1}{8} + 1\frac{2}{3} = \frac{17}{8} + \frac{5}{3} = \frac{17 \cdot 3}{8 \cdot 3} + \frac{5 \cdot 8}{3 \cdot 8} = \frac{51}{24} + \frac{40}{24} = \frac{91}{24}$ or $3\frac{19}{24}$

Work Practice 22

Practice 22

Add: $7\frac{3}{8} + 6\frac{3}{4}$

When adding or subtracting larger mixed numbers, you might want to use the following method.

Example 23 Subtract: $50\frac{1}{6} - 38\frac{1}{3}$

Solution:

$$\begin{array}{rcrcr} 50\frac{1}{6} & = & 50\frac{1}{6} & = & 49\frac{7}{6} \\ -38\frac{1}{3} & = & -38\frac{2}{6} & = & -38\frac{2}{6} \\ \hline & & & & 11\frac{5}{6} \end{array}$$

$50\frac{1}{6} = 49 + 1 + \frac{1}{6} = 49\frac{7}{6}$

Work Practice 23

Practice 23

Subtract: $76\frac{1}{12} - 35\frac{1}{4}$

Answers

22. $14\frac{1}{8}$ **23.** $40\frac{5}{6}$

Vocabulary, Readiness & Video Check

Use the choices below to fill in each blank.

improper	fraction	proper	reciprocals	mixed number	least common denominator (LCD)	$\frac{a \cdot d}{b \cdot c}$	$\frac{a \cdot c}{b \cdot d}$
equivalent	denominator	24	simplest form	numerator	$\frac{a - c}{b}$	$\frac{a + c}{b}$	

1. The number $\frac{17}{31}$ is called a(n) ______________. The number 31 is called its ______________ and 17 is called its ______________.

2. The fraction $\frac{8}{3}$ is called a(n) ______________ fraction, the fraction $\frac{3}{8}$ is called a(n) ______________ fraction, and $10\frac{3}{8}$ is called a(n) ______________.

3. In $\frac{11}{48}$, since 11 and 48 have no common factors other than 1, $\frac{11}{48}$ is in ______________.

4. Fractions that represent the same portion of a whole are called ______________ fractions.

5. To multiply two fractions, we write $\frac{a}{b} \cdot \frac{c}{d} =$ ______________.

6. Two numbers are ______________ of each other if their product is 1.

7. To divide two fractions, we write $\frac{a}{b} \div \frac{c}{d} =$ ______________.

8. $\frac{a}{b} + \frac{c}{b} =$ ______________ and $\frac{a}{b} - \frac{c}{b} =$ ______________.

9. The smallest positive number divisible by all the denominators of a list of fractions is called the ______________.

10. The LCD for $\frac{1}{6}$ and $\frac{5}{8}$ is ______________.

Martin-Gay Interactive Videos

See Video R.2

Watch the section lecture video and answer the following questions.

Objective A **11.** From the lecture before Example 1, what can we conclude about any fraction where the numerator and denominator are the same nonzero number?

Objective B **12.** Complete this statement based on Example 4. Equivalent fractions represent the same ________.

Objective C **13.** Describe the first step taken to simplify the fraction in Example 5.

Objective D **14.** What is the reciprocal of the second fraction in Example 7?

Objective E **15.** In Example 8, what is the main difference between adding or subtracting fractions and multiplying or dividing fractions?

Objective F **16.** In Example 10, why isn't our original sum in proper form?

R.2 Exercise Set MyLab Math

Objective A *Simplify by dividing the numerator by the denominator. See Examples 1 through 6.*

1. $\frac{14}{14}$ **2.** $\frac{19}{19}$ **3.** $\frac{20}{2}$ **4.** $\frac{30}{5}$ **5.** $\frac{13}{1}$

6. $\frac{21}{1}$ **7.** $\frac{0}{9}$ **8.** $\frac{0}{15}$ **9.** $\frac{9}{0}$ **10.** $\frac{15}{0}$

Objective B *Write each fraction as an equivalent fraction with the given denominator. See Example 7.*

11. $\frac{7}{10}$ with a denominator of 30

12. $\frac{2}{3}$ with a denominator of 9

13. $\frac{2}{9}$ with a denominator of 18

14. $\frac{8}{7}$ with a denominator of 56

15. $\frac{4}{5}$ with a denominator of 20

16. $\frac{4}{5}$ with a denominator of 25

Objective C *Simplify each fraction. See Examples 8 through 10.*

17. $\frac{2}{4}$ **18.** $\frac{3}{6}$ **19.** $\frac{10}{15}$ **20.** $\frac{15}{20}$

21. $\frac{3}{7}$ **22.** $\frac{5}{9}$ **23.** $\frac{18}{30}$ **24.** $\frac{42}{45}$

25. $\frac{16}{20}$ **26.** $\frac{8}{40}$ **27.** $\frac{66}{48}$ **28.** $\frac{64}{24}$

29. $\frac{120}{244}$ **30.** $\frac{360}{700}$ **31.** $\frac{192}{264}$ **32.** $\frac{455}{525}$

Objectives D F Mixed Practice *Multiply or divide as indicated. See Examples 11 through 14 and 21.*

33. $\frac{1}{2} \cdot \frac{3}{4}$ **34.** $\frac{7}{11} \cdot \frac{3}{5}$ **35.** $\frac{2}{3} \cdot \frac{3}{4}$ **36.** $\frac{7}{8} \cdot \frac{3}{21}$

37. $\frac{1}{2} \div \frac{7}{12}$ **38.** $\frac{7}{12} \div \frac{1}{2}$ **39.** $\frac{3}{4} \div \frac{1}{20}$ **40.** $\frac{3}{5} \div \frac{9}{10}$

41. $5\frac{1}{9} \cdot 3\frac{2}{3}$ **42.** $2\frac{3}{4} \cdot 1\frac{7}{8}$ **43.** $8\frac{3}{5} \div 2\frac{9}{10}$ **44.** $1\frac{7}{8} \div 3\frac{8}{9}$

Objectives E F Mixed Practice *Add or subtract as indicated. See Examples 15 through 20, 22, and 23.*

45. $\frac{4}{5} + \frac{1}{5}$ **46.** $\frac{6}{7} + \frac{1}{7}$ **47.** $\frac{4}{15} - \frac{1}{12}$ **48.** $\frac{11}{12} - \frac{1}{16}$

49. $\frac{2}{3} + \frac{3}{7}$ **50.** $\frac{3}{4} + \frac{1}{6}$ **51.** $\frac{10}{3} - \frac{5}{21}$ **52.** $\frac{11}{7} - \frac{3}{35}$

53. $8\frac{1}{8} - 6\frac{3}{8}$ **54.** $5\frac{2}{5} - 3\frac{4}{5}$ **55.** $1\frac{1}{2} + 3\frac{2}{3}$ **56.** $7\frac{3}{20} + 2\frac{13}{15}$

Objectives D E F Mixed Practice *Perform the indicated operations. See Examples 11 through 23.*

57. $\frac{23}{105} + \frac{4}{105}$ **58.** $\frac{13}{132} + \frac{35}{132}$ **59.** $\frac{17}{21} - \frac{10}{21}$ **60.** $\frac{18}{35} - \frac{11}{35}$

61. $\frac{7}{10} \cdot \frac{5}{21}$ **62.** $\frac{3}{35} \cdot \frac{10}{63}$ **63.** $\frac{9}{20} \div 12$ **64.** $\frac{25}{36} \div 10$

65. $\frac{5}{22} - \frac{5}{33}$ **66.** $\frac{7}{15} - \frac{7}{25}$ **67.** $17\frac{2}{5} + 30\frac{2}{3}$ **68.** $26\frac{11}{20} + 40\frac{7}{10}$

69. $7\frac{2}{5} \div \frac{1}{5}$ **70.** $9\frac{5}{6} \div \frac{1}{6}$ **71.** $4\frac{2}{11} \cdot 2\frac{1}{2}$ **72.** $6\frac{6}{7} \cdot 3\frac{1}{2}$

73. $\frac{12}{5} - 1$ **74.** $2 - \frac{3}{8}$ **75.** $8\frac{11}{12} - 1\frac{5}{6}$ **76.** $4\frac{7}{8} - 2\frac{3}{16}$

Concept Extensions

Perform the indicated operations.

77. $\frac{2}{3} - \frac{5}{9} + \frac{5}{6}$ **78.** $\frac{8}{11} - \frac{1}{4} + \frac{1}{2}$

For Exercises 79 through 82, determine whether the work is correct or incorrect. If incorrect, find the error and correct. See the Concept Check in this section.

79. $\frac{12}{24} = \frac{2 + 4 + 6}{2 + 4 + 6 + 12} = \frac{1}{12}$

80. $\frac{30}{60} = \frac{2 \cdot 3 \cdot 5}{2 \cdot 2 \cdot 3 \cdot 5} = \frac{1}{2}$

81. $\frac{2}{7} + \frac{9}{7} = \frac{11}{14}$

82. $\frac{16}{28} = \frac{2 \cdot 5 + 6 \cdot 1}{2 \cdot 5 + 6 \cdot 3} = \frac{1}{3}$

83. In your own words, describe how to divide fractions.

84. In your own words, describe how to add or subtract fractions.

Each circle below represents a whole, or 1. Determine the unknown part of the circle.

85.

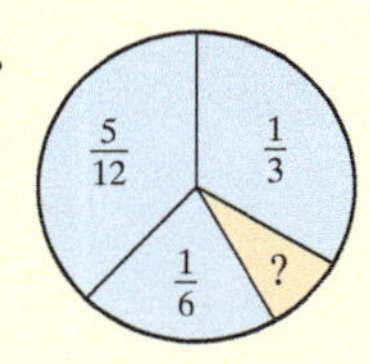

86.

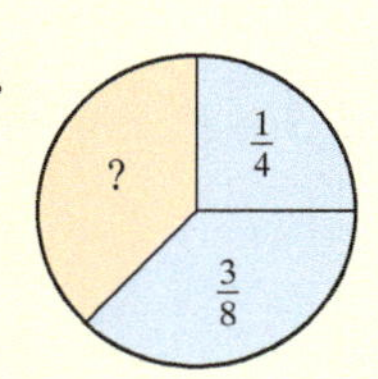

87.

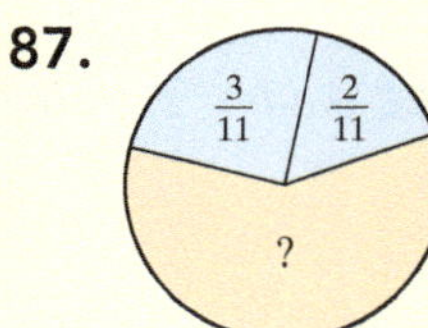

88.

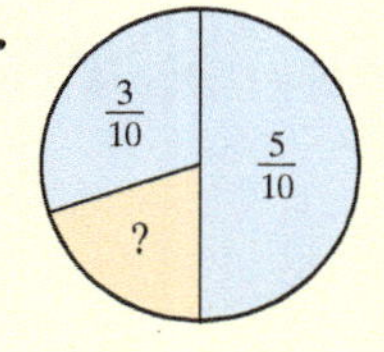

89. If Tucson's average rainfall is $11\frac{1}{4}$ inches and Yuma's is $3\frac{3}{5}$ inches, how much more rain, on average, does Tucson get than Yuma?

90. A pair of crutches needs adjustment. One crutch is 43 inches and the other is $41\frac{5}{8}$ inches. Find how much the short crutch should be lengthened to make both crutches the same length.

The following graph is called a circle graph or pie chart. Each sector (shaped like a piece of pie) shows the fraction of entering college freshmen who choose to major in each discipline shown. The whole circle represents the entire class of college freshmen. Use this graph to answer Exercises 91 through 94. Write each fraction answer in simplest form.

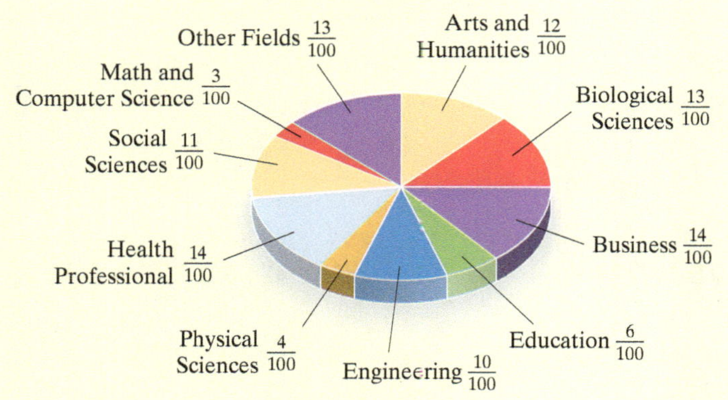

Source: The Higher Education Research Institute

91. What fraction of entering college freshmen plan to major in education?

92. What fraction of entering college freshmen plan to major in engineering?

93. Why is the business sector the same size as the health professional sector?

94. Why is the physical sciences sector smaller than the business sector?

Use this circle graph to answer Exercises 95 through 98. Write each fraction answer in simplest form.

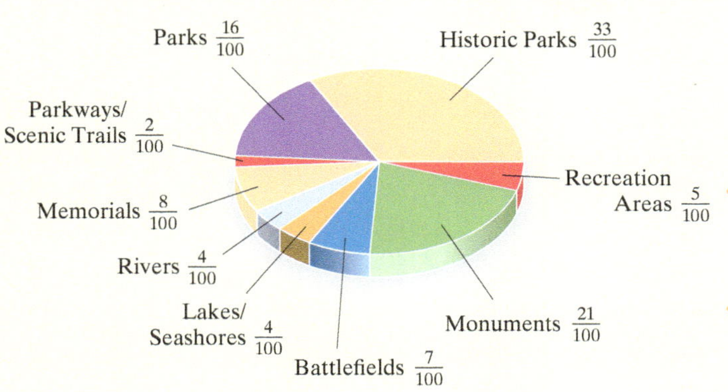

Source: National Park Service

95. What fraction of National Park Service areas are Memorials?

96. What fraction of National Park Service areas are Parks?

97. Why is the Battlefields sector smaller than the Monuments sector?

98. Why is the Lakes/Seashores sector the same size as the Rivers sector?

The area of a plane figure is a measure of the amount of surface of the figure. Find the area of each figure. (The area of a rectangle is the product of its length and width. The area of a triangle is $\frac{1}{2}$ the product of its base and height. Recall that area is measured in square units.)

 **99.**

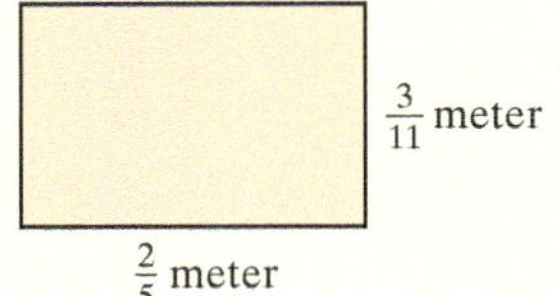

 **100.**

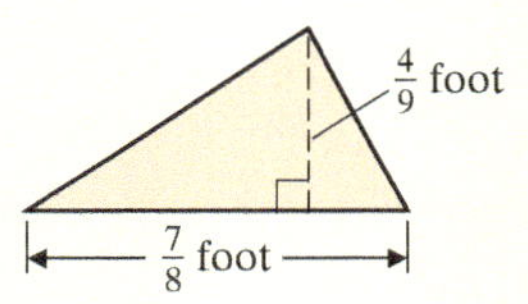

R.3 Decimals and Percents

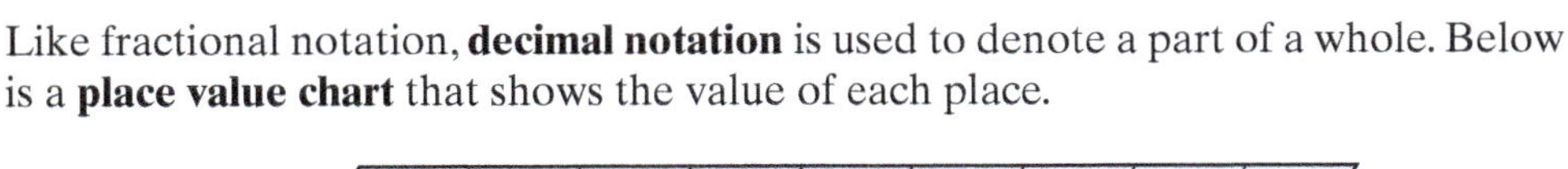

Objective A Writing Decimals as Fractions

Like fractional notation, **decimal notation** is used to denote a part of a whole. Below is a **place value chart** that shows the value of each place.

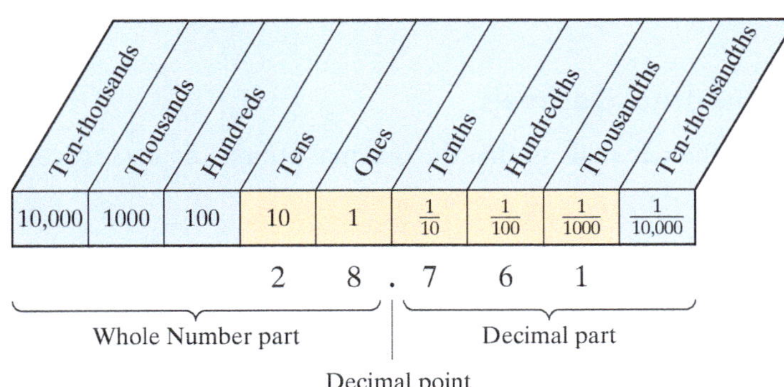

✓**Concept Check** Fill in the blank: In the number 52.634, the 3 is in the _________ place.

a. Tens **b.** Ones **c.** Tenths
d. Hundredths **e.** Thousandths

Objectives

A Write Decimals as Fractions.

B Add, Subtract, Multiply, and Divide Decimals.

C Round Decimals to a Given Decimal Place.

D Write Fractions as Decimals.

E Write Percents as Decimals and Decimals as Percents.

✓**Concept Check Answer**
d

The next chart shows decimals written as fractions.

Decimal Form	Fractional Form
0.1 (tenths)	$\frac{1}{10}$
0.07 (hundredths)	$\frac{7}{100}$
2.31 (hundredths)	$\frac{231}{100}$
0.9862 (ten-thousandths)	$\frac{9862}{10,000}$

To write a decimal as a fraction, use place values.

Practice 1–3

Write each decimal as a fraction. Do not simplify.

1. 0.27 **2.** 5.1 **3.** 7.685

Examples Write each decimal as a fraction. Do not simplify.

1. $0.37 = \frac{37}{100}$

2 decimal places, 2 zeros

2. $1.3 = \frac{13}{10}$

1 decimal place, 1 zero

3. $2.649 = \frac{2649}{1000}$

3 decimal places, 3 zeros

Work Practice 1–3

Objective B Adding, Subtracting, Multiplying, and Dividing Decimals

To **add** or **subtract** decimals, follow the steps below.

To Add or Subtract Decimals

Step 1: Write the decimals so that the decimal points line up vertically.

Step 2: Add or subtract as for whole numbers.

Step 3: Place the decimal point in the sum or difference so that it lines up vertically with the decimal points in the problem.

Notice that these steps simply ensure that we add or subtract digits with the same place value.

Answers

1. $\frac{27}{100}$ **2.** $\frac{51}{10}$ **3.** $\frac{7685}{1000}$

Example 4 Add.

a. $5.87 + 23.279 + 0.003$ **b.** $7 + 0.23 + 0.6$

Solution:

a.
$$\begin{array}{r} 5.87 \\ 23.279 \\ +\ 0.003 \\ \hline 29.152 \end{array}$$

b.
$$\begin{array}{r} 7. \\ 0.23 \\ +0.6 \\ \hline 7.83 \end{array}$$

Work Practice 4

Practice 4

Add.

a. $7.19 + 19.782 + 1.006$

b. $12 + 0.79 + 0.03$

Example 5 Subtract.

a. $32.15 - 11.237$ **b.** $70 - 0.48$

Solution:

a.
$$\begin{array}{r} \overset{1}{3}\overset{11}{\not 2}.\overset{4}{\not 1}\overset{10}{\not 5}\not 0 \\ -\ 1\ 1.\ 2\ 3\ 7 \\ \hline 2\ 0.\ 9\ 1\ 3 \end{array}$$

b.
$$\begin{array}{r} \overset{6}{\not 7}\overset{9}{\not 0}.\overset{9}{\not 0}\overset{10}{\not 0} \\ -\ 0.\ 4\ 8 \\ \hline 6\ 9.\ 5\ 2 \end{array}$$

Work Practice 5

Practice 5

Subtract.

a. $84.23 - 26.982$

b. $90 - 0.19$

Now let's study the following product of decimals. Notice the pattern in the decimal points.

$$0.03 \times 0.6 = \frac{3}{100} \times \frac{6}{10} = \frac{18}{1000} \text{ or } 0.018$$

0.03: 2 decimal places; 0.6: 1 decimal place; 0.018: 3 decimal places

In general, to **multiply** decimals, follow the steps below.

To Multiply Decimals

Step 1: Multiply the decimals as though they are whole numbers.

Step 2: The decimal point in the product is placed so that the number of decimal places in the product is equal to the **sum** of the number of decimal places in the factors.

Example 6 Multiply.

a. 0.072×3.5 **b.** 0.17×0.02

Solution:

a.
$$\begin{array}{rl} 0.072 & \text{3 decimal places} \\ \times\ 3.5 & \text{1 decimal place} \\ \hline 360 & \\ 216\ & \\ \hline 0.2520 & \text{4 decimal places} \end{array}$$

b.
$$\begin{array}{rl} 0.17 & \text{2 decimal places} \\ \times\ 0.02 & \text{2 decimal places} \\ \hline 0.0034 & \text{4 decimal places} \end{array}$$

Work Practice 6

Practice 6

Multiply.

a. 0.31×4.6

b. 1.26×0.03

Answers

4. a. 27.978 **b.** 12.82 **5. a.** 57.248 **b.** 89.81 **6. a.** 1.426 **b.** 0.0378

To divide a decimal by a whole number using long division, we place the decimal point in the quotient directly above the decimal point in the dividend. For example,

```
   2.47
3)7.41
 -6
  1 4
 -1 2
    21
   -21
     0
```

To check, see that $2.47 \times 3 = 7.41$

Helpful Hint Don't forget the names of the numbers in a division problem.

```
       quotient
divisor)dividend
```

In general, to **divide** decimals, use the steps below.

To Divide Decimals

Step 1: Move the decimal point in the divisor to the right until the divisor is a whole number.

Step 2: Move the decimal point in the dividend to the right the **same number of places** as the decimal point was moved in Step 1.

Step 3: Divide. The decimal point in the quotient is directly over the moved decimal point in the dividend.

Practice 7

Divide.

a. $21.75 \div 0.5$
b. $15.6 \div 0.006$

Example 7 Divide.

a. $9.46 \div 0.04$ **b.** $31.5 \div 0.007$

Solution:

a.

```
      236.5
004.)946.0
     -8
      14
     -12
       26
      -24
        2 0
       -2 0
          0
```

A zero is inserted to continue dividing.

b.

```
      4500.
0007.)31500.
     -28
       35
      -35
        0
```

Zeros are inserted in order to move the decimal point three places to the right.

Work Practice 7

Objective C Rounding Decimals

We **round** the decimal part of a decimal number in nearly the same way as we round whole numbers. The only difference is that we drop digits to the right of the rounding place, instead of replacing these digits by 0s. For example,

24.954 rounded to the nearest hundredth is 24.95

↑ hundredths place

Answers

7. a. 43.5 **b.** 2600

To Round Decimals to a Place Value to the Right of the Decimal Point

Step 1: Locate the digit to the right of the given place value.

Step 2:
- If this digit is 5 or greater, add 1 to the digit in the given place value and drop all digits to its right.
- If this digit is less than 5, drop all digits to the right of the given place.

Example 8 Round 7.8265 to the nearest hundredth.

Solution: 7.8265 (hundredths place: 2)

Step 1. Locate the digit to the right of the hundredths place.
Step 2. This digit is 5 or greater, so we add 1 to the hundredths place digit and drop all digits to its right.

Thus, 7.8265 rounded to the nearest hundredth is 7.83.

Work Practice 8

Practice 8

Round 12.9187 to the nearest hundredth.

Example 9 Round 19.329 to the nearest tenth.

Solution: 19.329 (tenths place: 3)

Step 1. Locate the digit to the right of the tenths place.
Step 2. This digit is less than 5, so we drop this digit and all digits to its right.

Thus, 19.329 rounded to the nearest tenth is 19.3.

Work Practice 9

Practice 9

Round 245.348 to the nearest tenth.

Objective D Writing Fractions as Decimals

To write fractions as decimals, interpret the fraction bar as division and find the quotient.

To Write a Fraction as a Decimal

Divide the numerator by the denominator.

Example 10 Write $\frac{1}{4}$ as a decimal.

Solution:

$$\begin{array}{r} 0.25 \\ 4\overline{)1.00} \\ \underline{-8} \\ 20 \\ \underline{-20} \\ 0 \end{array} \qquad \frac{1}{4} = 0.25$$

Work Practice 10

Practice 10

Write $\frac{2}{5}$ as a decimal.

Answers

8. 12.92 **9.** 245.3 **10.** 0.4

Practice 11

Write $\frac{5}{6}$ as a decimal.

Example 11 Write $\frac{2}{3}$ as a decimal.

Solution:

$$\begin{array}{r} 0.666 \\ 3\overline{)2.000} \\ \underline{-1\,8} \\ 20 \\ \underline{-18} \\ 20 \\ \underline{-18} \\ 2 \end{array}$$

This division pattern will continue so that $\frac{2}{3} = 0.6666\ldots$.

A bar can be placed over the digit 6 to indicate that it repeats. We call this a **repeating decimal**.

$$\frac{2}{3} = 0.666\ldots = 0.\overline{6}$$

Work Practice 11

We can also write a decimal approximation for $\frac{2}{3}$. For example, $\frac{2}{3}$ rounded to the nearest hundredth is 0.67. This can be written as $\frac{2}{3} \approx 0.67$. The $\approx$ sign means "is approximately equal to."

✓ **Concept Check** The notation $0.5\overline{2}$ is the same as

a. $\frac{52}{100}$ **b.** $\frac{52\ldots}{100}$ **c.** $0.52222222\ldots$

Practice 12

Write $\frac{1}{9}$ as a decimal. Round to the nearest thousandth.

Example 12 Write $\frac{22}{7}$ as a decimal. Round to the nearest hundredth.

Solution:

$$\begin{array}{r} 3.142 \\ 7\overline{)22.000} \\ \underline{-21} \\ 1\,0 \\ \underline{-7} \\ 30 \\ \underline{-28} \\ 20 \\ \underline{-14} \\ 6 \end{array} \approx 3.14$$

If rounding to the nearest hundredth, carry the division process out to one more decimal place, the thousandths place.

The fraction $\frac{22}{7}$ in decimal form is approximately 3.14. (The fraction $\frac{22}{7}$ is an approximation for π.)

Work Practice 12

Objective E Writing Percents as Decimals and Decimals as Percents

The word **percent** comes from the Latin phrase *per centum,* which means **"per 100."** The % symbol is used to denote percent. Thus, 53% means 53 per 100, or

$$53\% = \frac{53}{100}$$

Answers

11. $0.8\overline{3}$ **12.** 0.111

✓ **Concept Check Answer**

c

When solving problems containing percents, it is often necessary to write a percent as a decimal. To see how this is done, study the chart below.

Percent	Fraction	Decimal
7%	$\frac{7}{100}$	0.07
63%	$\frac{63}{100}$	0.63
109%	$\frac{109}{100}$	1.09

To convert directly from a percent to a decimal, notice that

7% = 0.07

To Write a Percent as a Decimal

Drop the percent symbol, %, and move the decimal point two places to the left.

Example 13

Write each percent as a decimal.

a. 25% **b.** 2.6% **c.** 195%

Solution: We drop the % and move the decimal point two places to the left. Recall that the decimal point of a whole number is to the right of the ones place digit.

a. 25% = 25.% = 0.25

b. 2.6% = 02.6% = 0.026

c. 195% = 195.% = 1.95

Work Practice 13

Practice 13

Write each percent as a decimal.

a. 20%

b. 1.4%

c. 465%

To write a decimal as a percent, we simply reverse the preceding steps. That is, we move the decimal point two places to the right and attach the percent symbol, %.

To Write a Decimal as a Percent

Move the decimal point two places to the right and attach the percent symbol, %.

Example 14

Write each decimal as a percent.

a. 0.85 **b.** 1.25 **c.** 0.012 **d.** 0.6

Solution: We move the decimal point two places to the right and attach the percent symbol, %.

a. 0.85 = 0.85 = 85%

b. 1.25 = 1.25 = 125%

c. 0.012 = 0.012 = 1.2%

d. 0.6 = 0.60 = 60%

Work Practice 14

Practice 14

Write each decimal as a percent.

a. 0.42

b. 0.003

c. 2.36

d. 0.7

Answers

13. a. 0.20 **b.** 0.014 **c.** 4.65
14. a. 42% **b.** 0.3% **c.** 236% **d.** 70%

Vocabulary, Readiness & Video Check

Fill in each blank with one of the choices listed below. Some choices may be used more than once and some not at all.

vertically	decimal	right	100%	percent
left	0.01	sum	denominator	numerator

1. Like fractional notation, ____________ notation is used to denote a part of a whole.

2. To write fractions as decimals, divide the ____________ by the ____________.

3. To add or subtract decimals, write the decimals so that the decimal points line up ____________.

4. When multiplying decimals, the decimal point in the product is placed so that the number of decimal places in the product is equal to the ____________ of the number of decimal places in the factors.

5. ____________ means "per hundred."

6. ____________ = 1.

7. The % symbol is read as ____________.

8. To write a percent as a *decimal,* drop the % symbol and move the decimal point two places to the ____________.

9. To write a decimal as a *percent*, move the decimal point two places to the ____________ and attach the % symbol.

Martin-Gay Interactive Videos

See Video R.3

Watch the section lecture video and answer the following questions.

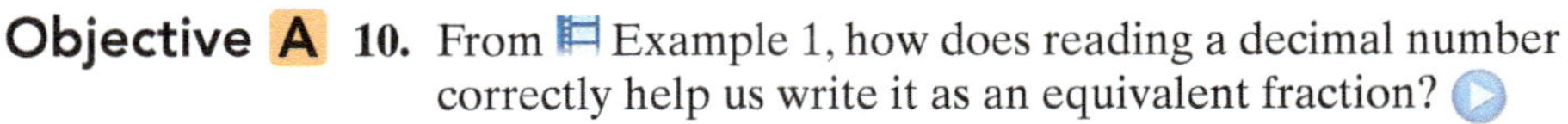

Objective A **10.** From Example 1, how does reading a decimal number correctly help us write it as an equivalent fraction?

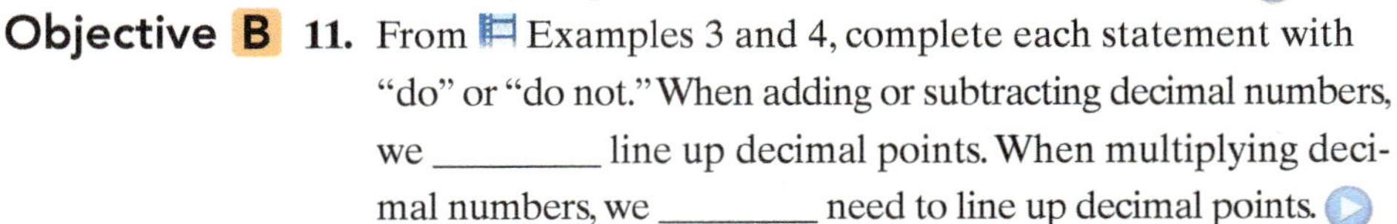

Objective B **11.** From Examples 3 and 4, complete each statement with "do" or "do not." When adding or subtracting decimal numbers, we ________ line up decimal points. When multiplying decimal numbers, we ________ need to line up decimal points.

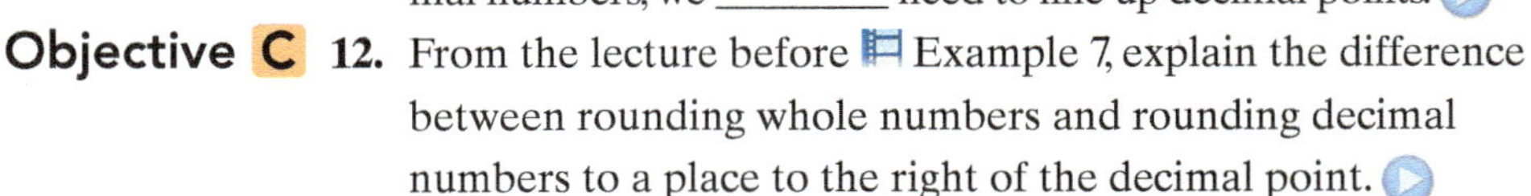

Objective C **12.** From the lecture before Example 7, explain the difference between rounding whole numbers and rounding decimal numbers to a place to the right of the decimal point.

Objective D **13.** Complete this statement based on the lecture before Example 8. To write a fraction as a decimal, we divide the ________ by the ________.

Objective E **14.** In Example 13, 100% equals what natural number?

R.3 Exercise Set MyLab Math

Objective A *Write each decimal as a fraction. Do not simplify. See Examples 1 through 3.*

1. 0.6 **2.** 0.9 **3.** 1.86 **4.** 7.23

5. 0.114 **6.** 0.239 **7.** 123.1 **8.** 892.7

Objective B *Add or subtract as indicated. See Examples 4 and 5.*

9. 5.7 + 1.13 **10.** 2.31 + 6.4 **11.** 24.6 + 2.39 + 0.0678 **12.** 32.4 + 1.58 + 0.0934

13. 8.8 − 2.3 **14.** 7.6 − 2.1 **15.** 18 − 2.78 **16.** 28 − 3.31

Multiply or divide as indicated. See Examples 6 and 7.

17. $\begin{array}{r} 0.2 \\ \times\ 0.6 \\ \hline \end{array}$

18. $\begin{array}{r} 0.7 \\ \times\ 0.9 \\ \hline \end{array}$

19. $\begin{array}{r} 0.063 \\ \times\ \ \ 4.2 \\ \hline \end{array}$

20. $\begin{array}{r} 0.079 \\ \times\ \ \ 3.6 \\ \hline \end{array}$

21. $5\overline{)8.4}$

22. $2\overline{)11.7}$

23. $0.82\overline{)4.756}$

24. $0.92\overline{)3.312}$

Mixed Practice

Perform the indicated operation. See Examples 4 through 7.

25. $\begin{array}{r} 45.02 \\ 3.006 \\ +\ 8.405 \\ \hline \end{array}$

26. $\begin{array}{r} 65.0028 \\ 5.0903 \\ +\ 6.9 \\ \hline \end{array}$

27. $\begin{array}{r} 6.75 \\ \times\ \ \ 10 \\ \hline \end{array}$

28. $\begin{array}{r} 8.91 \\ \times\ 100 \\ \hline \end{array}$

29. $0.6\overline{)42}$

30. $0.9\overline{)36}$

31. $\begin{array}{r} 654.9 \\ -\ 56.67 \\ \hline \end{array}$

32. $\begin{array}{r} 863.2 \\ -\ 39.45 \\ \hline \end{array}$

33. $\begin{array}{r} 5.62 \\ \times\ 7.7 \\ \hline \end{array}$

34. $\begin{array}{r} 8.03 \\ \times\ 5.5 \\ \hline \end{array}$

35. $0.063\overline{)52.92}$

36. $0.054\overline{)51.84}$

37. $\begin{array}{r} 16.003 \\ \times\ 5.31 \\ \hline \end{array}$

38. $\begin{array}{r} 31.006 \\ \times\ \ \ 3.71 \\ \hline \end{array}$

Objective C

Round each decimal to the given place value. See Examples 8 and 9.

39. 0.57, nearest tenth

40. 0.75, nearest tenth

41. 0.234, nearest hundredth

42. 0.452, nearest hundredth

43. 0.5945, nearest thousandth

44. 63.4529, nearest thousandth

45. 98,207.23, nearest tenth

46. 68,936.543, nearest tenth

47. 12.347, nearest hundredth

48. 42.9878, nearest thousandth

Objective D

Write each fraction as a decimal. If the decimal is a repeating decimal, write using the bar notation and then round to the nearest hundredth. See Examples 10 through 12.

49. $\frac{3}{4}$

50. $\frac{9}{25}$

51. $\frac{1}{3}$

52. $\frac{7}{9}$

53. $\frac{7}{16}$

54. $\frac{5}{8}$

55. $\frac{6}{11}$

56. $\frac{1}{6}$

57. $\frac{29}{6}$

58. $\frac{34}{9}$

Objective E Write each percent as a decimal. See Example 13.

59. 28% **60.** 36% **61.** 3.1% **62.** 2.2%

63. 135% **64.** 417% **65.** 200% **66.** 700%

67. 96.55% **68.** 81.49% **69.** 0.1% **70.** 0.6%

71. In the United States recently, 15.8% of households had no landlines, just cell phones. (*Source:* CTIA—The Wireless Association)

72. Japan exports 73.2% of all motorcycles manufactured there. (*Source:* Japan Automobile Manufacturers Association)

Write each decimal as a percent. See Example 14.

73. 0.68 **74.** 0.32 **75.** 0.876 **76.** 0.521

77. 1 **78.** 3 **79.** 0.5 **80.** 0.1

81. 1.92 **82.** 2.15 **83.** 0.004 **84.** 0.005

85. In a recent year, 0.781 of all electricity produced in France was nuclear generated.

86. The United States' share of the total world motor vehicle production is 0.142. (*Source: World Almanac*)

Concept Extensions

Write the percent from the circle graph as a decimal and a fraction.

World Population by Continent

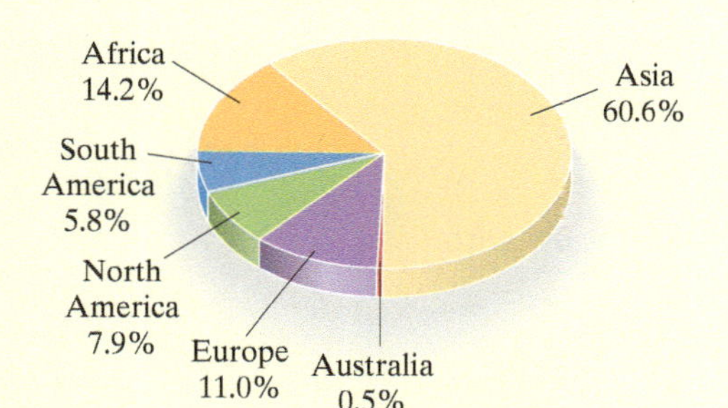

87. Australia: 0.5% **88.** Europe: 11%

89. Africa: 14.2% **90.** Asia: 60.6%

Solve. See the Concept Checks in this section.

91. In the number 3.659, identify the place value of the
a. 6 **b.** 9 **c.** 3

92. The notation $0.\overline{67}$ is the same as
a. 0.6777 . . . **b.** 0.67666 . . . **c.** 0.6767 . . .

93. In your own words, describe how to multiply decimal numbers.

94. In your own words, describe how to add or subtract decimal numbers.

*The chart shows the average number of pounds of various dairy products consumed by each U.S. citizen. Use this chart for Exercises **95** and **96**. (Source: Dairy Information Center)*

Dairy Product	Pounds
Fluid Milk	149
Cheese	39.1
Butter	5.7

95. How much more fluid milk products than cheese products does the average U.S. citizen consume?

96. What is the total amount of these milk products consumed by the average U.S. citizen annually?

97. Given the percent 52.8647%, round as indicated.
 a. Round to the nearest tenth percent.
 b. Round to the nearest hundredth percent.

98. Given the percent 0.5269%, round as indicated.
 a. Round to the nearest tenth percent.
 b. Round to the nearest hundredth percent.

99. Which of the following are correct?
 a. 6.5% = 0.65
 b. 7.8% = 0.078
 c. 120% = 0.12
 d. 0.35% = 0.0035

100. Which of the following are correct?
 a. 0.231 = 23.1%
 b. 5.12 = 0.0512%
 c. 3.2 = 320%
 d. 0.0175 = 0.175%

*Recall that 1 = 100%. This means that 1 whole is 100%. Use this for Exercises **101** and **102**. (Source:* Some Body, *by Dr. Pete Rowen)*

101. The four blood types are A, B, O, and AB. (Each blood type can also be further classified as Rh-positive or Rh-negative depending upon whether your blood contains protein or not.) Given the percent blood types for people in the United States below, calculate the percent of the U.S. population with AB blood type.

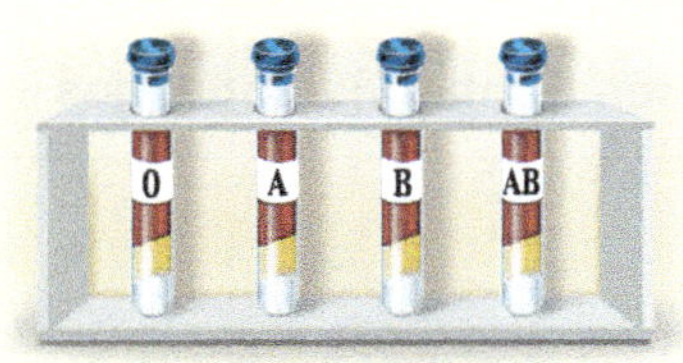

45% 40% 11% ?%

102. The top four components of bone are below. Find the missing percent.
 1. Minerals—45%
 2. Living tissue—30%
 3. Water—20%
 4. Other—?

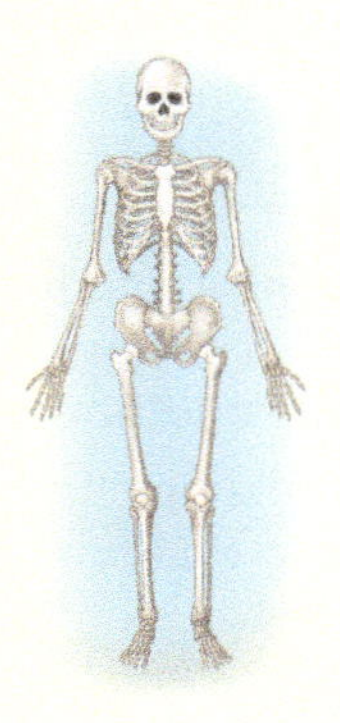

The bar graph shows the predicted fastest-growing occupations by percent that require an associate degree or more education. Use this graph for Exercises 103 through 106.

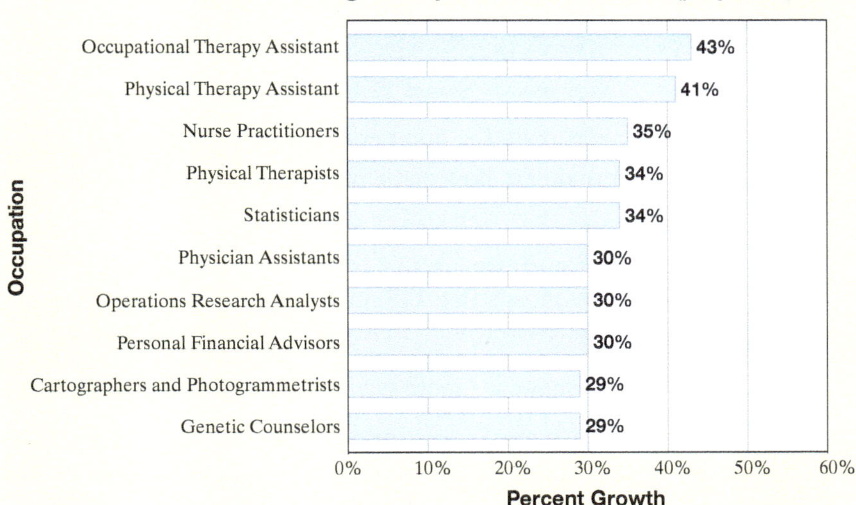

Source: Bureau of Labor Statistics

103. What occupation is predicted to be the fastest growing?

104. What occupation is predicted to be the second fastest growing?

105. Write the percent change for physician assistants as a decimal.

106. Write the percent change for statisticians as a decimal.

107. In your own words, explain how to write a percent as a decimal.

108. In your own words, explain how to write a decimal as a percent.

R.4 Reading Pictographs and Bar, Line, and Circle Graphs

Often data are presented visually in a graph. In this section, we practice reading several kinds of graphs including pictographs, bar graphs, and line graphs.

Objectives

A Read Pictographs.

B Read and Construct Bar Graphs.

C Read and Construct Histograms (or Frequency Distribution Graphs).

D Read Line Graphs.

E Read Circle Graphs.

Objective A Reading Pictographs

A **pictograph** such as the one below is a graph in which pictures or symbols are used. This type of graph contains a key that explains the meaning of the symbol used. An advantage of using a pictograph to display information is that comparisons can easily be made. A disadvantage of using a pictograph is that it is often hard to tell what fractional part of a symbol is shown. For example, in the pictograph below, Portuguese shows a part of a symbol, but it's hard to read with any accuracy what fractional part of a symbol is shown.

Example 1 Calculating Languages Spoken

The following pictograph shows the top eight most-spoken (primary) languages. Use this pictograph to answer the questions.

Top 8 Most-Spoken (Primary) Languages

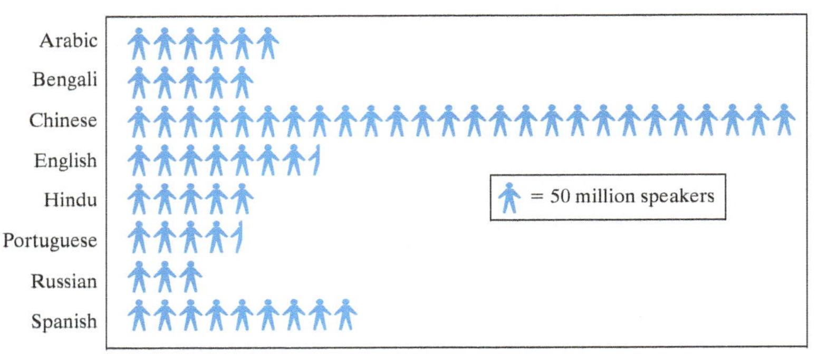

Source: www.ethnologue.com

a. Approximate the number of people who primarily speak Russian.

b. Approximate how many more people primarily speak English than Russian.

Solution:

a. Russian corresponds to 3 symbols, and each symbol represents 50 million speakers. This means that the number of people who primarily speak Russian is approximately $3 \cdot (50 \text{ million})$ or 150 million people.

b. English shows $4\frac{1}{2}$ more symbols than Russian. This means that $4\frac{1}{2} \cdot (50 \text{ million})$ or 225 million more people primarily speak English than Russian.

Work Practice 1

Practice 1

Use the pictograph shown in Example 1 to answer the following questions:

a. Approximate the number of people who primarily speak Spanish.

b. Approximate how many more people primarily speak Spanish than Arabic.

Objective B Reading and Constructing Bar Graphs

Another way to visually present data is with a **bar graph**. Bar graphs can appear with vertical bars or horizontal bars and we now practice reading the height or length of the bars contained in a bar graph. An advantage to using bar graphs is that a scale is usually included for greater accuracy. Care must be taken when reading bar graphs as well as other types of graphs—they may be misleading, as shown later in this section.

Answers

1. a. 450 million people **b.** 150 million people

Example 2 Finding the Number of Endangered Species

The following bar graph shows the number of endangered species in the United States in 2017. Use this graph to answer the questions.

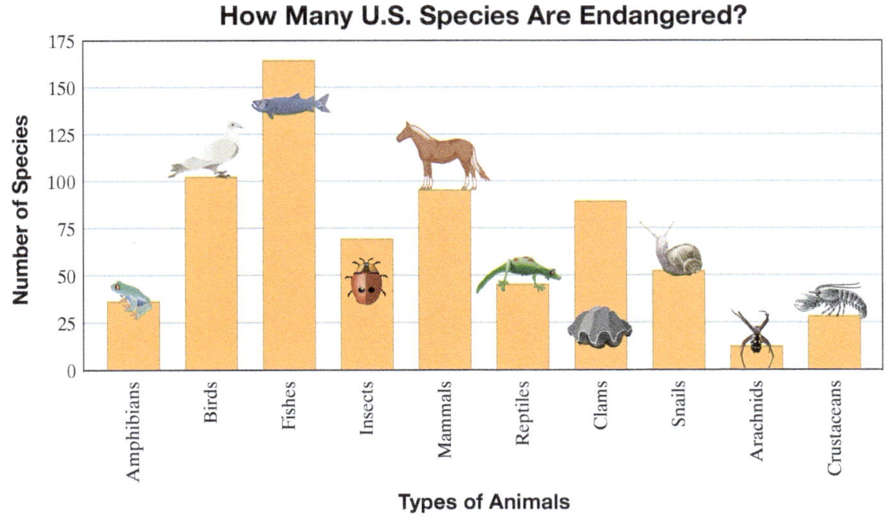

Source: U.S. Fish and Wildlife Service

a. Approximate the number of endangered species that are reptiles.

b. Which category has the most endangered species?

Solution:

a. To approximate the number of endangered species that are reptiles, we go to the top of the bar that represents reptiles. From the top of this bar, we move horizontally to the left until the scale is reached. We read the height of the bar on the scale as approximately 45. There are approximately 45 reptile species that are endangered, as shown. (See the graph below.)

b. The most endangered species is represented by the tallest (longest) bar. The tallest bar corresponds to fishes.

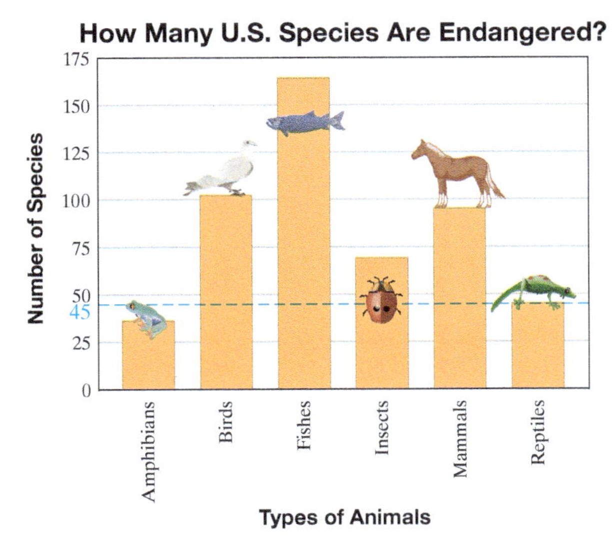

Source: U.S. Fish and Wildlife Service

Work Practice 2

Practice 2

Use the bar graph in Example 2 to answer the following questions:

a. Approximate the number of endangered species that are fishes.

b. Which category shows the fewest endangered species?

Answers

2. a. 164 (exact number) **b.** arachnids

Next, we practice constructing a bar graph.

Practice 3

Draw a vertical bar graph using the information in the table about electoral votes for selected states.

Total Electoral Votes for Selected States

State	Electoral Votes
Texas	38
California	55
Florida	29
Nebraska	5
Illinois	20
Georgia	15
(Source: World Almanac)	

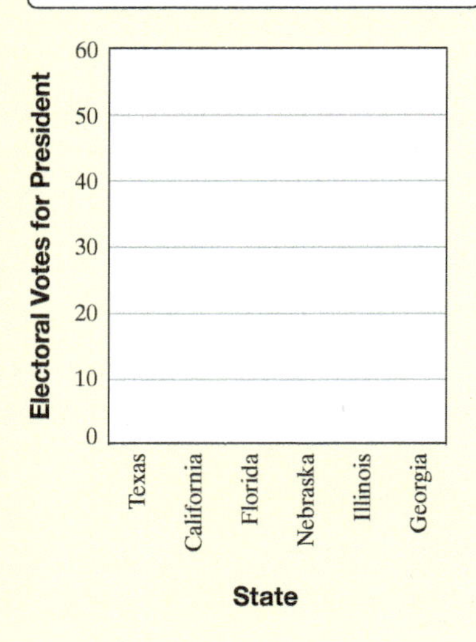

Example 3

Draw a vertical bar graph using the information in the table below, which gives the caffeine content of selected foods.

Average Caffeine Content of Selected Foods

Food	Milligrams	Food	Milligrams
Brewed Coffee (8 ounces)	145	Brewed decaffeinated coffee (8 ounces)	6
Brewed black tea (8 ounces)	42	Pepsi Cola (12 ounces)	38
Red Bull (8.46 ounces)	80	Monster energy drink (16 ounces)	160
Dark Chocolate (semisweet, 1 ounce)	22	Jolt gum (2 pieces)	145
(*Source:* Caffeineinformer.com)			

Solution: We draw and label a vertical line and a horizontal line as shown below on the left. These lines are also called axes. We place the different food categories along the horizontal axis. Along the vertical axis, we place a scale.

There are many choices of scales that would be appropriate. Notice that the milligrams range from a low of 6 to a high of 160. From this information, we use a scale that starts at 0 and then shows multiples of 20 so that the scale is not too cluttered. The scale stops at 180, the smallest multiple of 20 that is greater that the high of 160 milligrams. It may also be helpful to draw horizontal lines along the scale markings to help draw the vertical bars at the correct height. The finished bar graph is shown below on the right.

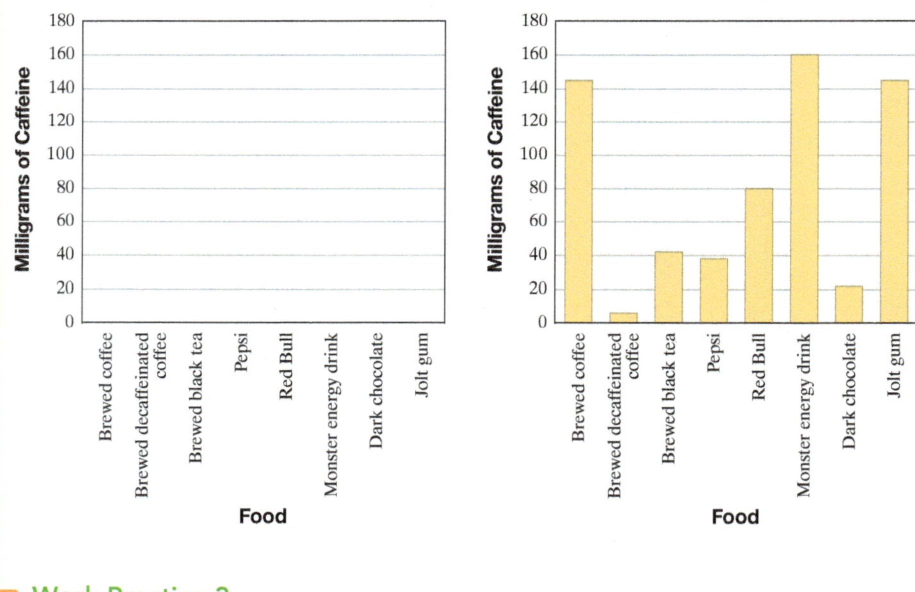

Work Practice 3

As mentioned previously, graphs can be misleading. Both graphs on the next page show the same information but with different scales. Special care should be taken when forming conclusions from the appearance of a graph.

Answer

3.

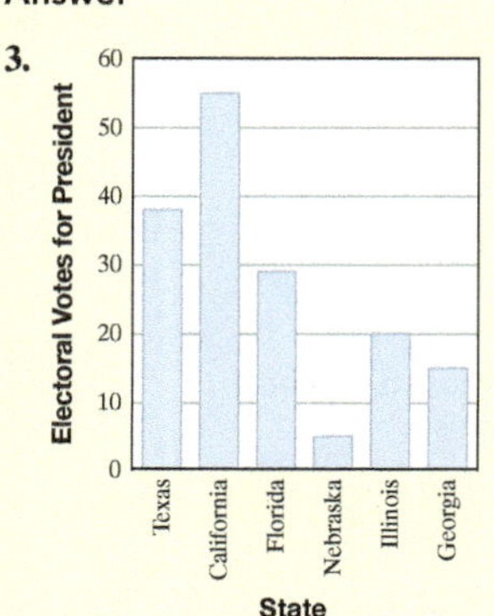

Notice the ⸗ symbol on each vertical scale on the graphs below. This symbol alerts us that numbers are missing from that scale

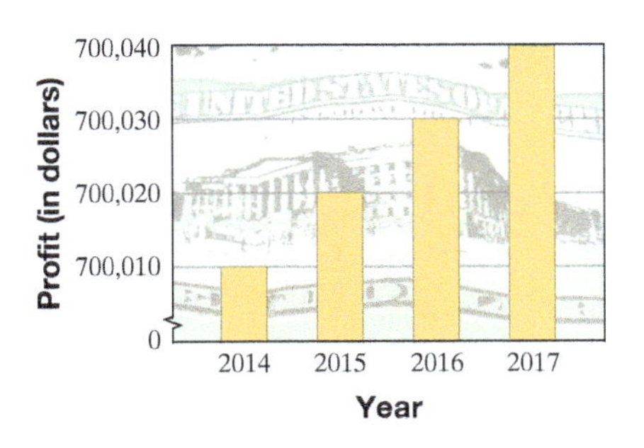

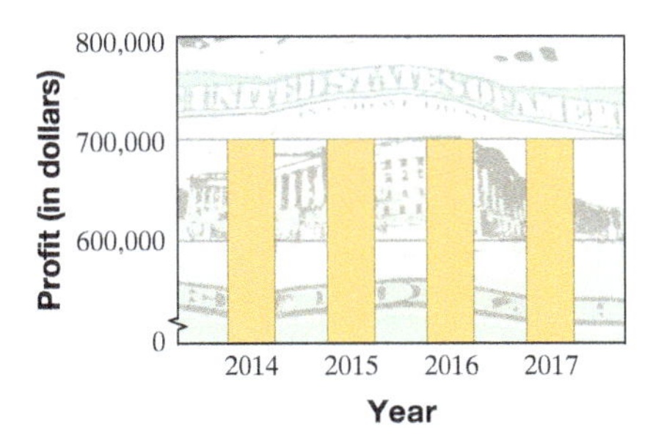

Are profits shown in the graphs above greatly increasing, or are they remaining about the same?

Objective C Reading and Constructing Histograms 

Suppose that the test scores of 36 students are summarized in the table below. We call this table a **frequency distribution table** since one column gives the frequency or number of times the event in the other column occurred.

Student Scores	Frequency (Number of Students)
40–49	1
50–59	3
60–69	2
70–79	10
80–89	12
90–99	8

The results in this frequency distribution table can be displayed in a histogram. A **histogram** is a special bar graph. The width of each bar represents a range of numbers called a **class interval**. The height of each bar corresponds to how many times a number in the class interval occurs and is called the **class frequency**. The bars in a histogram lie side by side with no space between them. Note: Another name for this histogram is a **frequency distribution graph**.

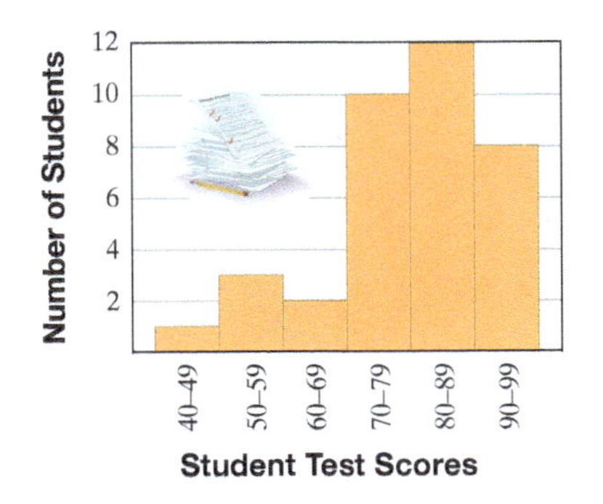

Example 4 Reading a Histogram on Student Test Scores

Use the preceding histogram to determine how many students scored 50–59 on the test.

Solution: We find the bar representing 50–59. The height of this bar is 3, which means 3 students scored 50–59 on the test.

Work Practice 4

Practice 4

Use the histogram above Example 5 to determine how many students scored 80–89 on the test.

Answer

4. 12

Practice 5

Use the histogram above Example 4 to determine how many students scored less than 80 on the test.

Practice 6

Complete the frequency distribution table for the data below. Each number represents a credit card owner's unpaid balance for one month.

0	53	89	125
265	161	37	76
62	201	136	42

Class Intervals (Credit Card Balances)	Tally	Class Frequency (Number of Months)
\$0–\$49	____	____
\$50–\$99	____	____
\$100–\$149	____	____
\$150–\$199	____	____
\$200–\$249	____	____
\$250–\$299	____	____

Practice 7

Construct a histogram from the frequency distribution table for Practice 6.

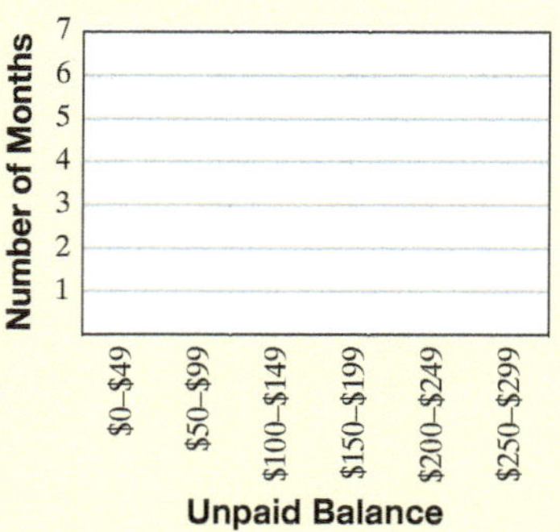

Answers

5. 16

6.

Tally	Class Frequency (Number Months)	Tally	Class Frequency (Number Months)
|||	3	|	1
||||	4	|	1
||	2	|	1

7.

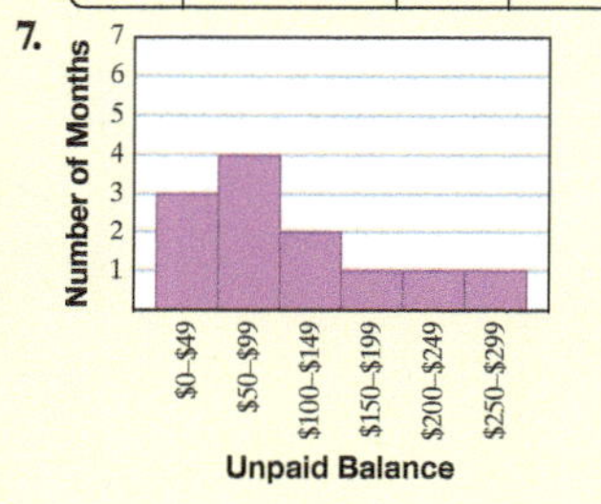

Example 5 Reading a Histogram on Student Test Scores

Use the histogram above Example 4 to determine how many students scored 80 or above on the test.

Solution: We see that two different bars fit this description. There are 12 students who scored 80–89 and 8 students who scored 90–99. The sum of these two categories is $12 + 8$ or 20 students. Thus, 20 students scored 80 or above on the test.

Work Practice 5

Now we will look at a way to construct histograms.

The daily high temperatures for 1 month in New Orleans, Louisiana, are recorded in the following list:

85°	90°	95°	89°	88°	94°
87°	90°	95°	92°	95°	94°
82°	92°	96°	91°	94°	92°
89°	89°	90°	93°	95°	91°
88°	90°	88°	86°	93°	89°

The data in this list have not been organized and can be hard to interpret. One way to organize the data is to place them in a **frequency distribution table**. We will do this in Example 6.

Example 6 Completing a Frequency Distribution on Temperature

Complete the frequency distribution table for the preceding temperature data.

Solution: Go through the data and place a tally mark in the second column of the table next to the class interval. Then count the tally marks and write each total in the third column of the table.

Class Intervals (Temperatures)	Tally	Class Frequency (Number of Days)
82°–84°	|	1
85°–87°	|||	3
88°–90°	~~||||~~ ~~||||~~ |	11
91°–93°	~~||||~~ ||	7
94°–96°	~~||||~~ |||	8

Work Practice 6

Example 7 Constructing a Histogram

Construct a histogram from the frequency distribution table in Example 6.

Solution:

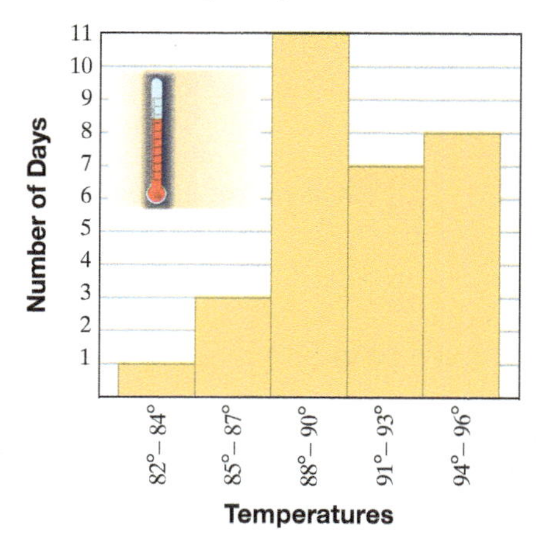

Work Practice 7

✓**Concept Check** Which of the following sets of data is better suited to representation by a histogram? Explain.

Set 1		Set 2	
Grade on Final	# of Students	Section Number	Avg. Grade on Final
51–60	12	150	78
61–70	18	151	83
71–80	29	152	87
81–90	23	153	73
91–100	25		

Objective D Reading Line Graphs 

Another common way to display information with a graph is by using a **line graph.** An advantage of a line graph is that it can be used to visualize relationships between two quantities. A line graph can also be very useful in showing changes over time.

Example 8 Reading Temperatures from a Line Graph

The following line graph shows the average daily temperature for each month in Omaha, Nebraska. Use this graph to answer the questions below.

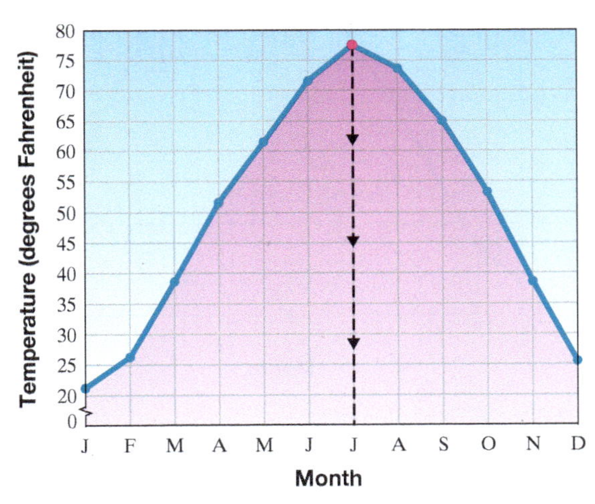

Source: National Climatic Data Center

a. During what month is the average daily temperature the highest?

b. During what month, from July through December, is the average daily temperature 65°F?

c. During what months is the average daily temperature less than 30°F?

Solution:

a. The month with the highest temperature corresponds to the highest point. This is the red point shown on the graph above. We follow this highest point downward to the horizontal month scale and see that this point corresponds to July.

(*Continued on next page*)

Practice 8

Use the temperature graph in Example 8 to answer the following questions:

a. During what month is the average daily temperature the lowest?

b. During what month is the average daily temperature 25°F?

c. During what months is the average daily temperature greater than 70°F?

Answers

8. a. January **b.** December **c.** June, July, and August

✓**Concept Check Answer**

Set 1; the grades are arranged in ranges of scores.

b. The months July through December correspond to the right side of the graph. We find the 65°F mark on the vertical temperature scale and move to the right until a point on the right side of the graph is reached. From that point, we move downward to the horizontal month scale and read the corresponding month. During the month of September, the average daily temperature is 65°F.

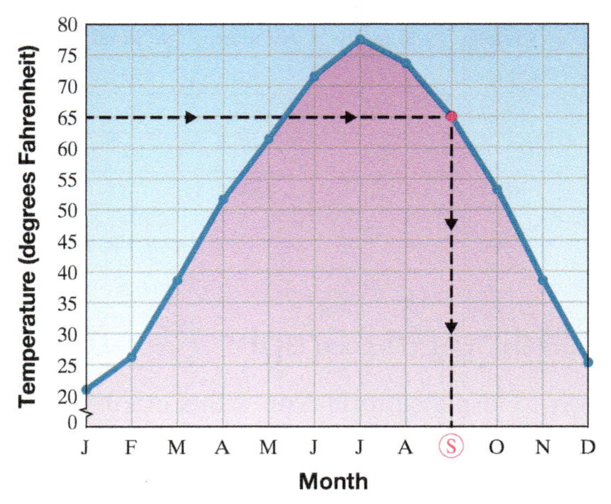

Source: National Climatic Data Center

c. To see what months the temperature is less than 30°F, we find what months correspond to points that fall below the 30°F mark on the vertical scale. These months are January, February, and December.

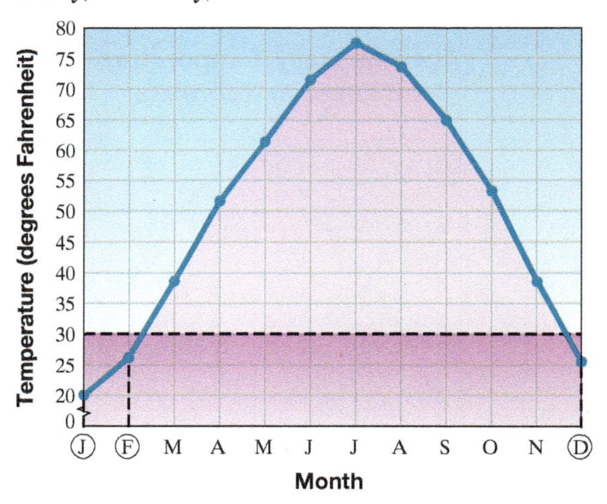

Source: National Climatic Data Center

Work Practice 8

Objective E Reading Circle Graphs 

Thus far in Chapter R we have seen various **circle graphs**.

This particular graph below shows the favorite sport for 100 adults.

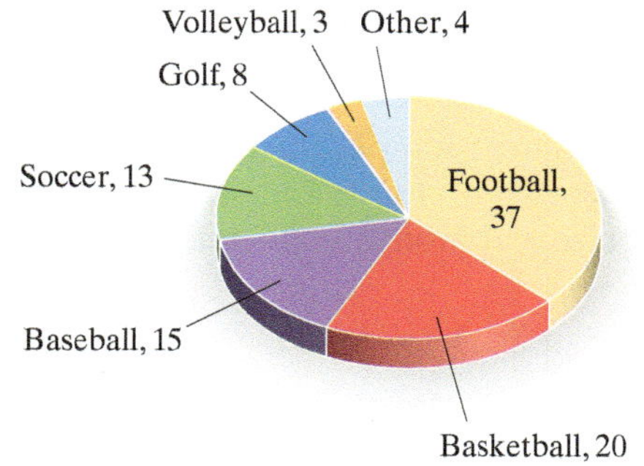

Each sector of the graph (shaped like a piece of pie) shows a category and the relative size of the category. In other words, the most popular sport is football, and it is represented by the largest sector.

Example 9 Find the ratio of adults preferring basketball to total adults. Write the ratio as a fraction in simplest form.

Solution: The ratio is

$$\frac{\text{people preferring basketball}}{\text{total adults}} = \frac{20}{100} = \frac{1}{5}$$

Work Practice 9

Practice 9

Find the ratio of adults preferring golf to total adults. Write the ratio as a fraction in simplest form.

A circle graph is often used to show percents in different categories, with the whole circle representing 100%.

Example 10 Using a Circle Graph

The following graph shows the percent of visitors to the United States in a recent year by various regions. Using the circle graph shown, determine the percent of visitors who came to the United States from Mexico or Canada.

Visitors to U.S. by Region

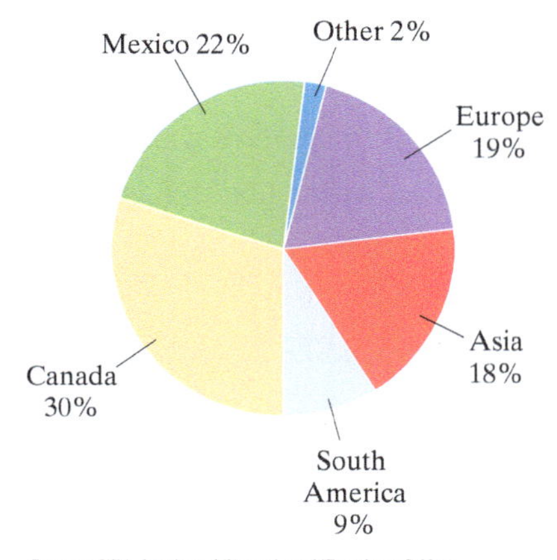

Source: ITA National Travel and Tourism Office

Solution: To find this percent, we add the percents corresponding to Mexico and Canada. The percent of visitors to the United States that came from Mexico or Canada is

$$22\% + 30\% = 52\%$$

Work Practice 10

Practice 10

Using the circle graph shown in Example 10, determine the percent of visitors to the United States that came from Europe, Asia, or South America.

Helpful Hint

Since a circle graph represents a whole, the percents should add to 100% or 1. Notice this is true for Example 10.

Example 11 Finding Percent of Population

The U.S. Department of Commerce forecasts 89 million international visitors to the United States in 2022. Use the circle graph from Example 10 and predict the number of tourists that might be from Europe.

Solution: We use the percent equation.

$$\begin{aligned} \text{amount} &= \text{percent} \cdot \text{base} \\ \text{amount} &= 0.19 \cdot 89{,}000{,}000 \\ &= 0.19(89{,}000{,}000) \\ &= 16{,}910{,}000 \end{aligned}$$

Thus, 16,910,000 tourists might come from Europe in 2022.

Work Practice 11

Practice 11

Use the information in Example 11 and the circle graph from Example 10 to predict the number of tourists from Mexico in 2022.

Answers

9. $\frac{2}{25}$ **10.** 46%
11. 19,580,000 tourists from Mexico

✓**Concept Check** Can the following data be represented by a circle graph? Why or why not?

Responses to the Question "In Which Activities Are You Involved?"	
Intramural sports	60%
On-campus job	42%
Fraternity/sorority	27%
Academic clubs	21%
Music programs	14%

✓**Concept Check Answer**

no; the percents add up to more than 100%

Vocabulary, Readiness & Video Check

Fill in each blank with one of the choices below.

pictograph	bar	class frequency	sector	1
histogram	line	class interval	circle	100

1. A ____________ graph presents data using vertical or horizontal bars.

2. A ____________ is a graph in which pictures or symbols are used to visually present data.

3. A ____________ graph displays information with a line that connects data points.

4. A ____________ is a special bar graph in which the width of each bar represents a ____________ and the height of each bar represents the ____________.

5. In a ____________ graph, each section (shaped like a piece of pie) shows a category and the relative size of the category.

6. A circle graph contains pie-shaped sections, each called a ____________.

7. If a circle graph has percent labels, the percents should add to ____________%.

8. If a circle graph has fraction labels, the fractions should add to ____________.

Martin-Gay Interactive Videos

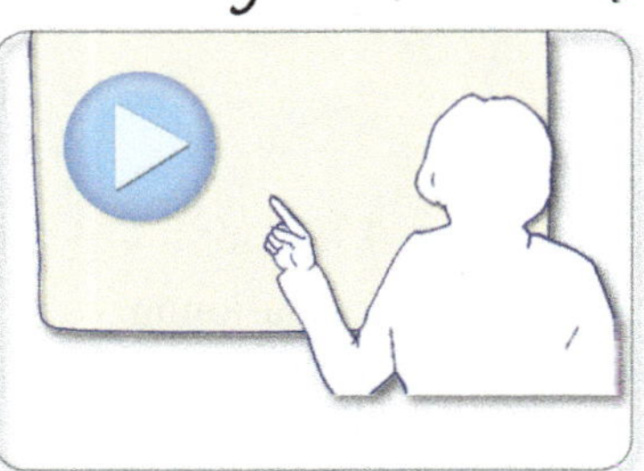

See Video R.4

Watch the section lecture video and answer the following questions.

Objective A **9.** From the pictograph in Example 1, how would you approximate the number of wildfires for any given year?

Objective B **10.** What is one advantage of displaying data in a bar graph?

Objective C **11.** Complete this statement based on the lecture before Example 6: A histogram is a special kind of ______.

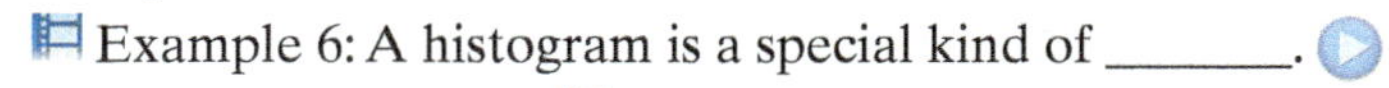

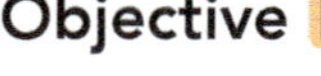

Objective D **12.** From the line graph in Examples 10–13 , what year averaged the greatest number of goals per game average and what was this average?

Objective E **13.** From Example 16, when a circle graph shows different parts or percents of some whole category, what is the sum of the percents in the whole circle graph?

R.4 Exercise Set MyLab Math 

Objective A *The following pictograph shows the number of acres devoted to wheat production in selected states in 2017. Use this graph to answer Exercises 1 through 8. See Example 1. (Source: U.S. Department of Agriculture)*

1. Which state plants the greatest acreage in wheat?

2. Which of the states shown plant the least amount of wheat acreage?

3. Approximate the number of acres of wheat planted in Oklahoma.

4. Approximate the number of acres of wheat planted in Kansas.

5. Which state(s) plant less than 3,000,000 acres of wheat?

6. Which state(s) plant more than 7,000,000 acres of wheat?

7. Which state plants more wheat: North Dakota or Oklahoma?

Annual Wheat Acreage in Selected Top States

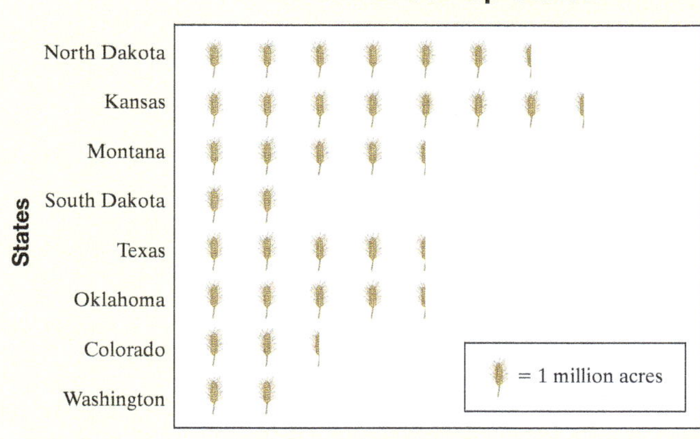

8. Which state plants more wheat: Colorado or Washington?

The following pictograph shows the average number of wildfires in the United States between 2011 and 2017. Use this graph to answer Exercises 9 through 16. See Example 1. (Source: National Interagency Fire Center)

9. Approximate the number of wildfires in 2013.

10. Approximately how many wildfires were there in 2012?

11. Which year, of the years shown, had the most wildfires?

12. In what years were the number of wildfires greater than 72,000?

Wildfires in the United States

13. What was the amount of decrease in wildfires from 2012 to 2013?

14. What was the amount of increase in wildfires from 2015 to 2016?

15. What was the average annual number of wildfires from 2013 to 2015? (*Hint:* How do you calculate the average?)

16. Give a possible explanation for the sharp increase in the number of wildfires in 2016.

Objective B *The National Weather Service has exacting definitions for hurricanes; they are tropical storms with winds in excess of 74 mph. The following bar graph shows the number of hurricanes, by month, that have made landfall on the mainland United States between 1851 and 2017. Use this graph to answer Exercises 17 through 22. See Example 2. (Source: National Weather Service: National Hurricane Center)*

17. In which month did the most hurricanes make landfall in the United States?

18. In which month did the fewest hurricanes make landfall in the United States?

19. Approximate the number of hurricanes that made landfall in the United States during the month of August.

20. Approximate the number of hurricanes that made landfall in the United States in September.

21. In 2005 alone, three hurricanes made landfall during the month of October. What fraction of all the 54 hurricanes that made landfall during October is this?

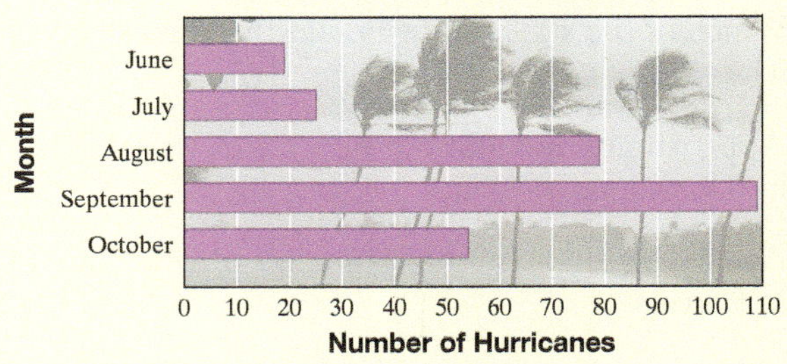

22. In 2007, only one hurricane made landfall in the United States during the entire season, in the month of September. If there have been 109 hurricanes to make landfall in the month of September since 1851, approximately what percent of these arrived in 2007?

The following horizontal bar graph shows the recent population of the world's largest cities (including their suburbs). Use this graph to answer Exercises 23 through 28. See Example 2. (Source: World Atlas)

23. Name the city with the largest population, and estimate its population.

24. Name the city whose population is between 26 million and 27 million.

25. Name the city in the United States with the largest population, and estimate its population.

26. Name the two cities that have approximately a population of 24 million.

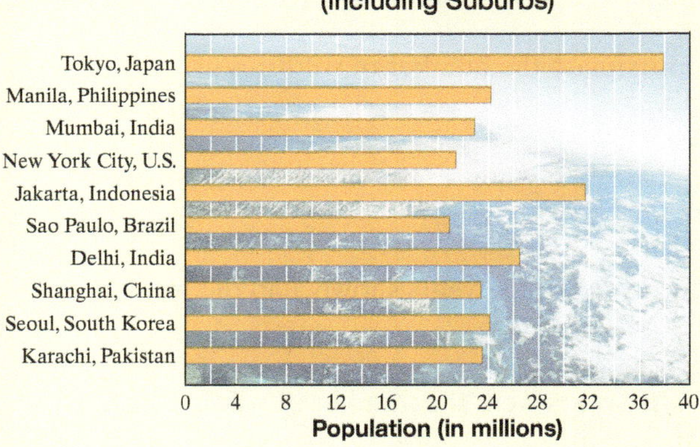

27. How much larger (in terms of population) is Manila, Philippines, than Sao Paulo, Brazil?

28. How much larger (in terms of population) is Jakarta, Indonesia, than Delhi, India?

Use the information given to draw a vertical bar graph. Clearly label the bars. See Example 3.

29.

Fiber Content of Selected Foods

Food	Grams of Total Fiber
Kidney beans $\left(\frac{1}{2}\text{ c}\right)$	4.5
Oatmeal, cooked $\left(\frac{3}{4}\text{ c}\right)$	3.0
Peanut butter, chunky (2 tbsp)	1.5
Popcorn (1 c)	1.0
Potato, baked with skin (1 med)	4.0
Whole wheat bread (1 slice)	2.5

(*Sources:* American Dietetic Association and National Center for Nutrition and Dietetics)

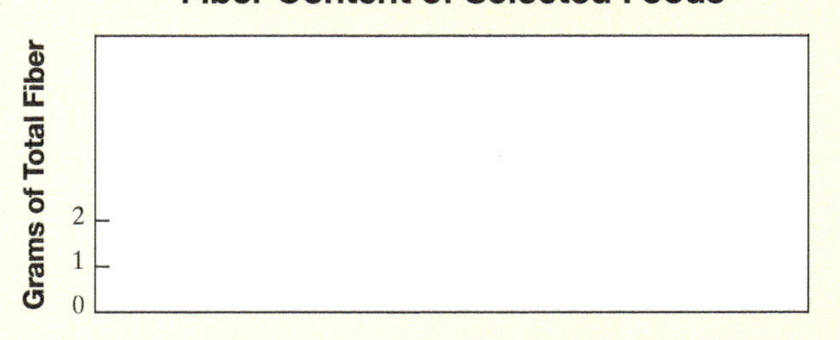

30.

U.S. Annual Food Sales	
Year	**Sales in Billions of Dollars**
2012	1345
2013	1410
2014	1462
2015	1512
2016	1584
2017	2123

(*Source:* U.S. Department of Agriculture)

31.

Best-Selling Albums of All Time (U.S. Sales)	
Album	**Estimated Sales (in millions)**
Pink Floyd: *The Wall* (1979)	23
Michael Jackson: *Thriller* (1982)	29
Billy Joel: *Greatest Hits Volumes I&II* (1985)	23
Eagles: *Their Greatest Hits* (1976)	29
Led Zeppelin: *Led Zeppelin IV* (1971)	23

(*Source*: Recording Industry Association of America)

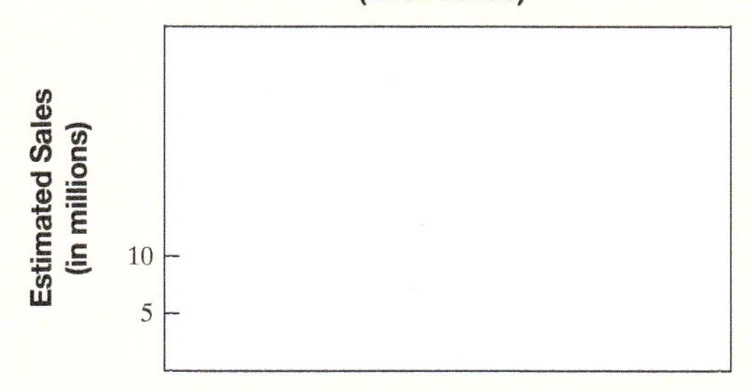

32.

Selected Worldwide Commercial Space Launches	
Location or Name	**Total Commercial Space Launches 1990–2017**
United States	231
Russia	200
Europe	187
China	60
Sea Launch*	48

*Sea Launch is an international venture involving four countries that uses its own launch facility outside national borders.

(*Source:* Space Launch Report and NASA)

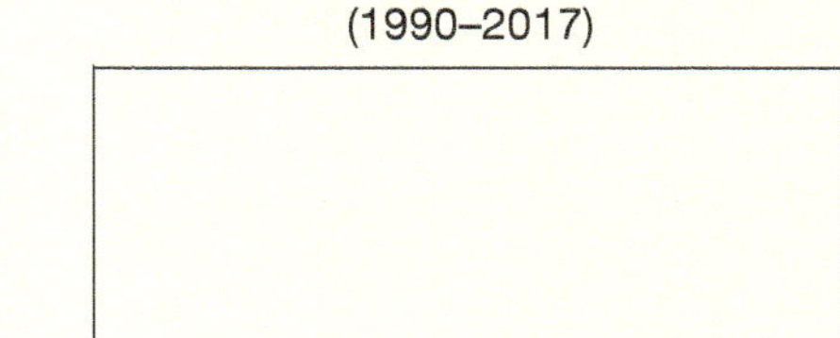

Objective C *The following histogram shows the number of miles that each adult, from a survey of 100 adults, drives per week. Use this histogram to answer Exercises 33 through 42. See Examples 4 and 5.*

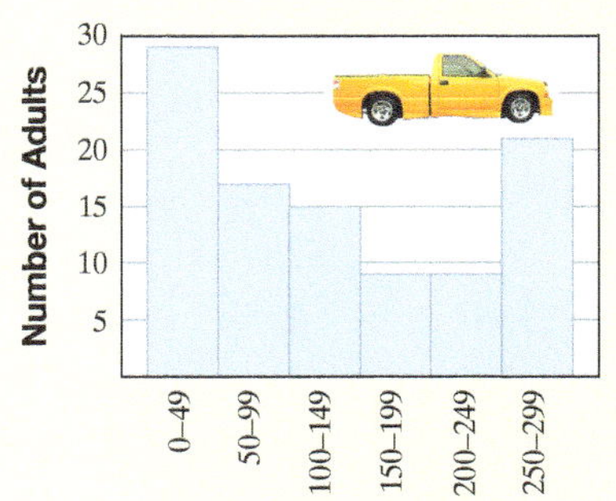

33. How many adults drive 100–149 miles per week?

34. How many adults drive 200–249 miles per week?

35. How many adults drive fewer than 150 miles per week?

36. How many adults drive 200 miles or more per week?

37. How many adults drive 100–199 miles per week?

38. How many adults drive 150–249 miles per week?

39. How many more adults drive 250–299 miles per week than 200–249 miles per week?

40. How many more adults drive 0–49 miles per week than 50–99 miles per week?

41. What is the ratio of adults who drive 150–199 miles per week to the total number of adults surveyed?

42. What is the ratio of adults who drive 50–99 miles per week to the total number of adults surveyed?

The following histogram shows the projected population (in millions), by age groups, for the United States for the year 2020. Use this histogram to answer Exercises 43 through 50. For Exercises 45 through 48, estimate to the nearest whole million. See Examples 4 and 5.

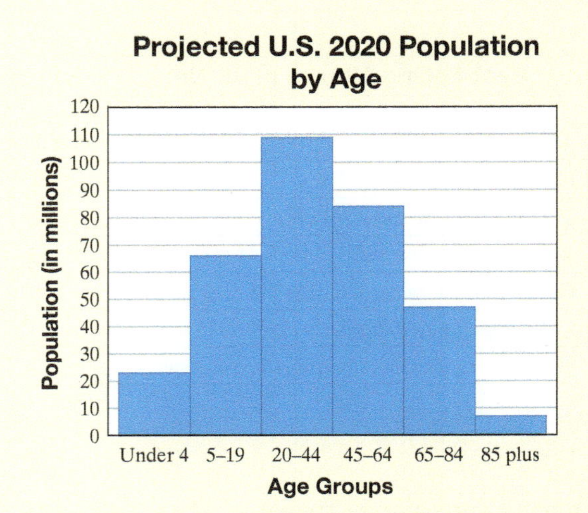

43. What age range will be the largest population group in 2020?

44. What age range will be the smallest population group in 2020?

45. How large is the population of 20- to 44-year-olds expected to be in 2020?

46. How large is the population of 45- to 64-year-olds expected to be in 2020?

47. How large is the population of those less than 4 years old expected to be in 2020?

48. How large is the population of 5- to 19-year-olds expected to be in 2020?

49. Which bar represents the age range you expect to be in during 2020?

50. How many more 20- to 44-year-olds are there expected to be than 45- to 64-year-olds in 2020?

The following list shows the golf scores for an amateur golfer. Use this list to complete the frequency distribution table to the right. See Example 6.

78	84	91	93	97
97	95	85	95	96
101	89	92	89	100

	Class Intervals (Scores)	Tally	Class Frequency (Number of Games)
51.	70–79		
52.	80–89		
53.	90–99		
54.	100–109		

Twenty-five people in a survey were asked to give their current checking account balances. Use the balances shown in the following list to complete the frequency distribution table to the right. See Example 6.

$53	$105	$162	$443	$109
$468	$47	$259	$316	$228
$207	$357	$15	$301	$75
$86	$77	$512	$219	$100
$192	$288	$352	$166	$292

	Class Intervals (Account Balances)	Tally	Class Frequency (Number of People)
55.	$0–$99		
56.	$100–$199		
57.	$200–$299		
58.	$300–$399		
59.	$400–$499		
60.	$500–$599		

61. Use the frequency distribution table from Exercises **51** through **54** to construct a histogram. See Example 7.

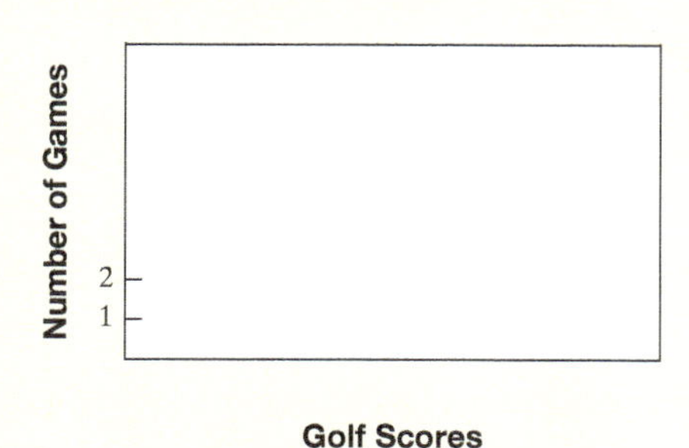

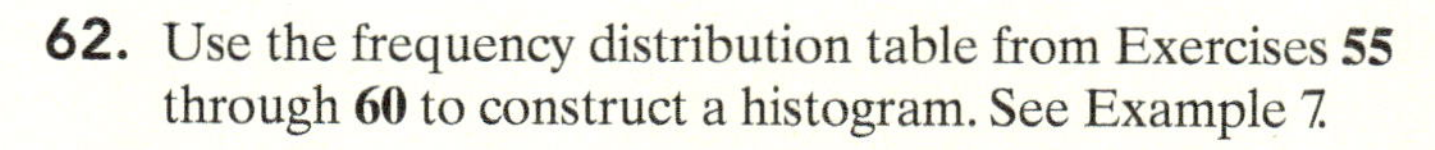

62. Use the frequency distribution table from Exercises **55** through **60** to construct a histogram. See Example 7.

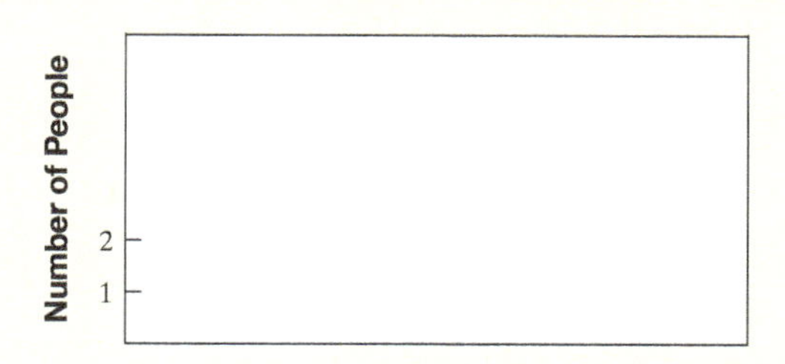

Objective D *Beach Soccer World Cup is now held every two years. The following line graph shows the World Cup goals per game average for beach soccer during the years shown. Use this graph to answer Exercises 63 through 70. See Example 8.*

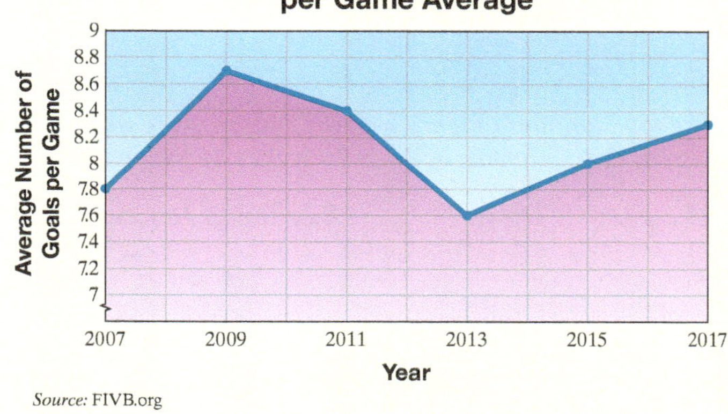

Source: FIVB.org

63. Find the average number of goals per game in 2017.

64. Find the average number of goals per game in 2013.

65. During what year shown was the average number of goals per game the highest?

66. During what year shown was the average number of goals per game the lowest?

67. From 2013 to 2015, did the average number of goals per game increase or decrease?

68. From 2011 to 2013, did the average number of goals per game increase or decrease?

69. During what year(s) shown were the average goals per game less than 8?

70. During what year(s) shown were the average goals per game greater than 8?

Objective E *The following circle graph is a result of surveying 700 college students. They were asked where they live while attending college. Use this graph to answer Exercises 71 through 76. Write all ratios as fractions in simplest form. See Example 9.*

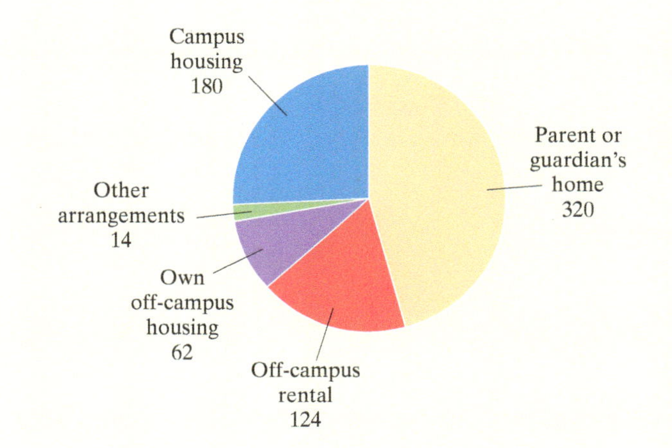

71. Where do most of these college students live?

72. Besides the category "other arrangements," where do the fewest of these college students live?

73. Find the ratio of students living in campus housing to total students.

74. Find the ratio of students living in off-campus rentals to total students.

75. Find the ratio of students living in campus housing to students living in a parent or guardian's home.

76. Find the ratio of students living in off-campus rentals to students living in a parent or guardian's home.

The following circle graph shows the percent of the land area of the continents of Earth. Use this graph for Exercises 77 through 84. See Example 10.

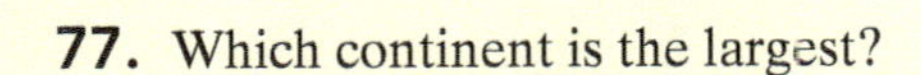

77. Which continent is the largest?

78. Which continent is the smallest?

79. What percent of the land on Earth is accounted for by Asia and Europe together?

80. What percent of the land on Earth is accounted for by North and South America?

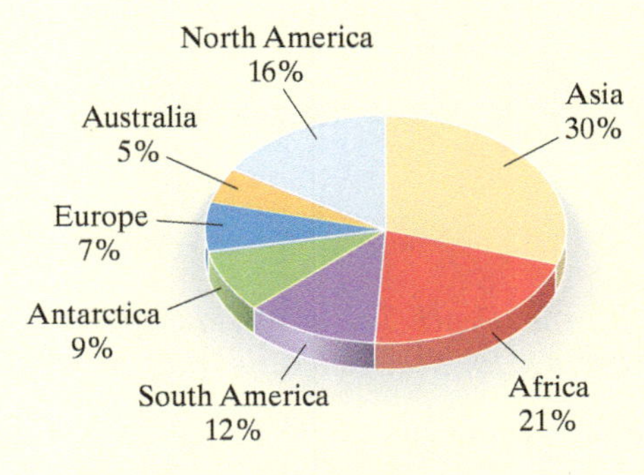

Source: National Geographic Society

The total amount of land from the continents is approximately 57,000,000 square miles. Use the graph above to find the area of the continents given in Exercises 81 through 84. See Example 11.

81. Asia

82. South America

83. Australia

84. Europe

The following circle graph shows the percent of the types of books available at Midway Memorial Library. Use this graph for Exercises 85 through 94. See Example 10.

85. What percent of books are classified as some type of fiction?

86. What percent of books are nonfiction or reference?

87. What is the second-largest category of books?

88. What is the third-largest category of books?

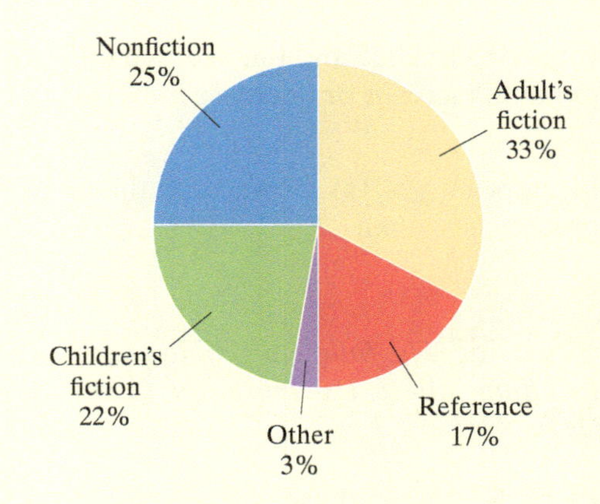

If this library has 125,600 books, find how many books are in each category given in Exercises 89 through 94. See Example 11.

89. Nonfiction

90. Reference

91. Children's fiction

92. Adult's fiction

93. Reference or other

94. Nonfiction or other

Concept Extensions

The following double line graph shows temperature highs and lows for a week. Use this graph to answer Exercises 95 through 100.

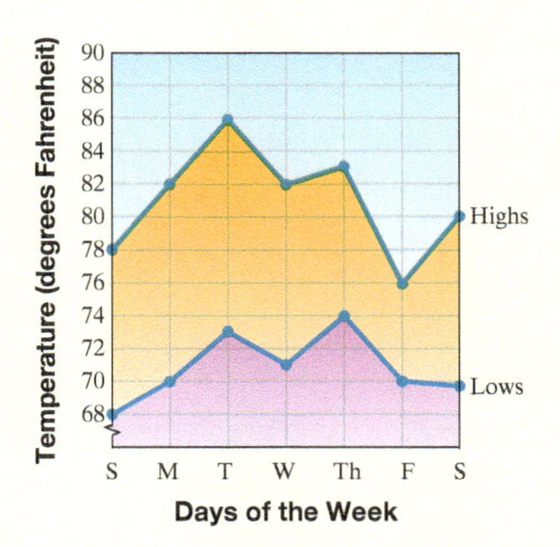

95. What was the high temperature reading on Thursday?

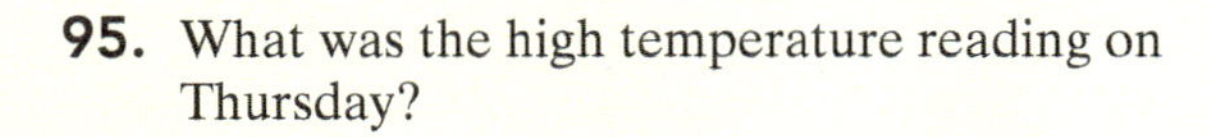

96. What was the low temperature reading on Thursday?

97. What day was the temperature the lowest? What was this low temperature?

98. What day of the week was the temperature the highest? What was this high temperature?

99. On what day of the week was the difference between the high temperature and the low temperature the greatest? What was this difference in temperature?

100. On what day of the week was the difference between the high temperature and the low temperature the least? What was this difference in temperature?

101. True or false? With a bar graph, the width of the bar is just as important as the height of the bar. Explain your answer.

102. Kansas plants about 24% of the wheat acreage in the United States. About how many acres of wheat are planted in the United States, according to the pictograph for Exercises **1** through **8**? Round to the nearest million acre.

The following circle graph shows the relative sizes of the great oceans.

103. Without calculating, determine which ocean is the largest. How can you answer this question by looking at the circle graph?

104. Without calculating, determine which ocean is the smallest. How can you answer this question by looking at the circle graph?

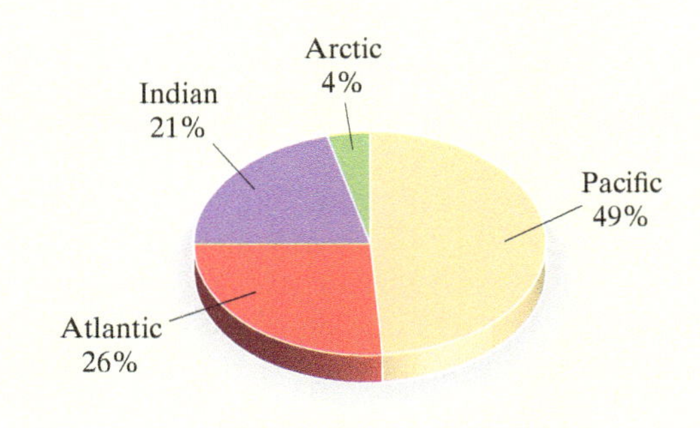

Source: Philip's World Atlas

These oceans together make up 264,489,800 square kilometers of Earth's surface. Find the square kilometers for each ocean.

105. Pacific Ocean

106. Atlantic Ocean

107. Indian Ocean

108. Arctic Ocean

The following circle graph summarizes the results of a survey of online spending in America. Let's use these results to make predictions about the online spending behavior of a community of 2800 Internet users age 18 and over. Use this graph for Exercises 109 through 114. Round to the nearest whole. (Note: Because of rounding, these percents do not have a sum of 100%.)

109. How many would you predict to spend \$0 online each month?

110. How many would you predict to spend \$1–\$100 online each month?

111. How many would you predict to spend \$0 to \$100 online each month?

112. How many would you predict to spend \$1 to \$1000 online each month?

Online Spending per Month

\$101–\$1000 14%
> \$1000 0.2%
\$0 24%
\$1–\$100 62%

Source: The Digital Future Report

113. Find the ratio of predicted *number* who spend \$0 online to predicted *number* who spend \$1–\$100 online. Write the ratio as a fraction. Simplify the fraction if possible.

114. Find the ratio of *percent* of respondents who spend \$101–\$1000 online to *percent* of those who spend \$1–\$100. Write the ratio as a fraction with integers in the numerator and denominator. Simplify the fraction if possible.

See the Concept Checks in this section.

115. Can the data below be represented by a circle graph? Why or why not?

Responses to the Question "What Classes Are You Taking?"	
Math	80%
English	72%
History	37%
Biology	21%
Chemistry	14%

116. True or false? The smaller a sector in a circle graph, the smaller the percent of the total it represents. Explain why.

Study the Chapter Opener circle graphs below and conduct surveys with at least 30 people.

117. Using the "Favorite Pizza Topping" circle graph as a guide, ask each person in your survey to choose his or her favorite pizza topping. Tally the results, draw a circle graph and compare your circle graph to the one shown.

118. Using the "How Do You Eat Your Pizza?" circle graph as a guide, ask each person in your survey to choose how he or she eats pizza. Then follow the directions in Exercise **117**.

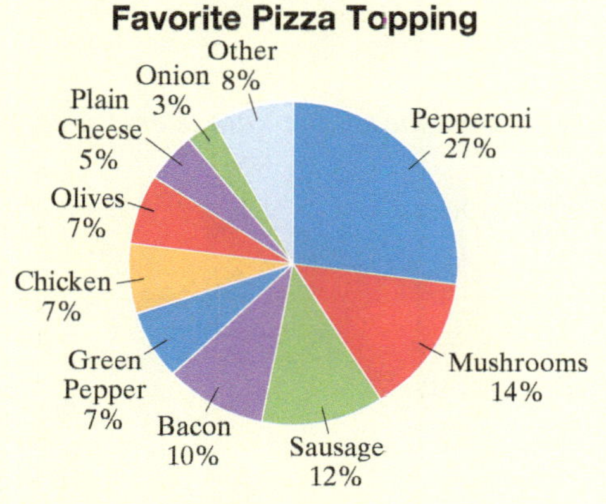

Sources: Zagot, Statista, Harris Poll

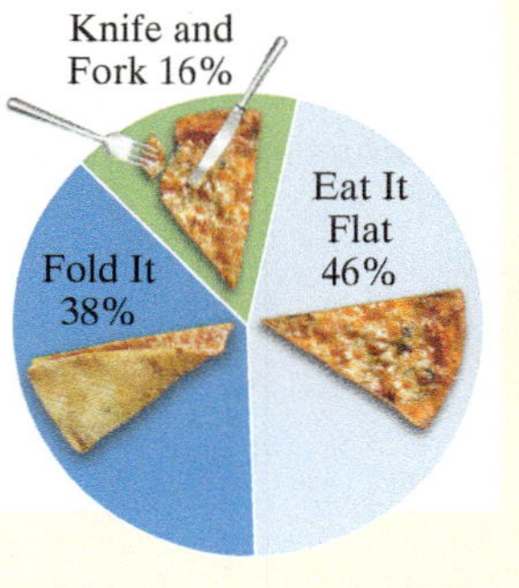

Sources: Zagot, Statista, Harris Poll

Chapter R Group Activity

Interpreting Survey Results

This activity may be completed by working in groups or individually.

Conduct the following survey with 12 students in one of your classes and record the results.

a. What is your age?
Under 20 20s 30s 40s 50s 60 and older

b. What is your gender?
Female Male

c. How did you arrive on campus today?
Walked Drove Bicycled
Took public transportation Other

1. For each survey question, tally the results for each category.

Age	
Category	**Tally**
Under 20	
20s	
30s	
40s	
50s	
60+	
Total	

Gender	
Category	**Tally**
Female	
Male	
Total	

Mode of Transportation	
Category	**Tally**
Walk	
Drive	
Bicycle	
Public Transit	
Other	
Total	

2. For each survey question, find the fraction of the total number of responses that fall in each answer category. Use the tallies from Question 1 to complete the Fraction columns of the tables at the right.

3. For each survey question, convert the fraction of the total number of responses that fall in each answer category to a decimal number. Use the fractions from Question 2 to complete the Decimal columns of the tables below.

4. For each survey question, find the percent of the total number of responses that falls in each answer category. Complete the Percent columns of the tables below.

5. Study the tables. What may you conclude from them? What do they tell you about your survey respondents? Write a paragraph summarizing your findings.

Age			
Category	**Fraction**	**Decimal**	**Percent**
Under 20			
20s			
30s			
40s			
50s			
60+			

Gender			
Category	**Fraction**	**Decimal**	**Percent**
Female			
Male			

Mode of Transportation			
Category	**Fraction**	**Decimal**	**Percent**
Walk			
Drive			
Bicycle			
Public Transit			
Other			

Chapter R Vocabulary Check

Fill in each blank with one of the words or phrases listed below.

mixed number	factor	improper fraction	class frequency	percent	circle	bar
multiple	composite number	proper fraction	class interval	simplified	line	pictograph
prime number	equivalent	histogram				

1. To ______________ means to write as a product.
2. A(n) ______________ of a number is the product of that number and any natural number.
3. A(n) ______________ is a natural number greater than 1 that is not prime.
4. The word ______________ means per 100.
5. Fractions that represent the same portion of a whole are called ______________ fractions.
6. A(n) ______________ is a fraction whose numerator is greater than or equal to its denominator.
7. A(n) ______________ is a natural number greater than 1 whose only factors are 1 and itself.
8. A fraction is ______________ when the numerator and the denominator have no factors in common other than 1.
9. A(n) ______________ is one whose numerator is less than its denominator.
10. A(n) ______________ contains a whole number part and a fraction part.
11. A(n) ______________ graph presents data using vertical or horizontal bars.
12. A(n) ______________ is a graph in which pictures or symbols are used to visually present data.
13. A(n) ______________ graph displays information with a line that connects data points.
14. In a(n) ______________ graph, each section (shaped like a piece of pie) shows a category and the relative size of the category.
15. A(n) ______________ is a special bar graph in which the width of each bar represents a(n) ______________ and the height of each bar represents the ______________.

Helpful Hint

Are you preparing for your test? To help, don't forget to take these:

- Chapter R Getting Ready for the Test on page R-57
- Chapter R Test on page R-59

Then check all of your answers at the back of this text. For further review, the step-by-step video solutions to any of these exercises are located in MyLab Math.

R Chapter Highlights

Definitions and Concepts	Examples
Section R.1 Factors and the Least Common Multiple	
To **factor** means to write as a product. Since $2 \cdot 6 = 12$, 2 and 6 are factors of 12.	The factors of 12 are 1, 2, 3, 4, 6, 12
When a number is written as a product of primes, this product is called the **prime factorization** of a number.	Write the prime factorization of 60. $60 = 6 \cdot 10$ $= 2 \cdot 3 \cdot 2 \cdot 5$ The prime factorization of 60 is $2 \cdot 2 \cdot 3 \cdot 5$.
The **least common multiple (LCM)** of a list of numbers is the smallest number that is a multiple of all the numbers in the list.	

Definitions and Concepts	Examples
Section R.1 Factors and the Least Common Multiple (Continued)	
To Find the LCM of a List of Numbers **Step 1:** Write the prime factorization of each number. **Step 2:** Write the product containing each different prime factor (from Step 1) the greatest number of times that it appears in any one factorization. This product is the LCM.	Find the LCM of 12 and 40. $12 = 2 \cdot 2 \cdot 3$ $40 = 2 \cdot 2 \cdot 2 \cdot 5$ $\text{LCM} = 2 \cdot 2 \cdot 2 \cdot 3 \cdot 5 = 120$
Section R.2 Fractions and Mixed Numbers	
Fractions that represent the same portion of a whole are called **equivalent fractions**.	$\frac{1}{5} = \frac{1 \cdot 4}{5 \cdot 4} = \frac{4}{20}$ $\frac{1}{5}$ and $\frac{4}{20}$ are equivalent fractions.
To write an equivalent fraction, $\frac{a}{b} = \frac{a}{b} \cdot \frac{c}{c} = \frac{a \cdot c}{b \cdot c}$	Write $\frac{8}{21}$ as an equivalent fraction with a denominator of 63. $\frac{8}{21} = \frac{8}{21} \cdot \frac{3}{3} = \frac{8 \cdot 3}{21 \cdot 3} = \frac{24}{63}$
A fraction is **simplified** when the numerator and the denominator have no factors in common other than 1.	$\frac{13}{17}$ is simplified.
To simplify a fraction, $\frac{a \cdot c}{b \cdot c} = \frac{a}{b} \cdot \frac{c}{c} = \frac{a}{b}$	Simplify. $\frac{6}{14} = \frac{2 \cdot 3}{2 \cdot 7} = \frac{2}{2} \cdot \frac{3}{7} = \frac{3}{7}$
Two fractions are **reciprocals** if their product is 1. The reciprocal of $\frac{a}{b}$ is $\frac{b}{a}$, as long as a and b are not 0.	The reciprocal of $\frac{6}{25}$ is $\frac{25}{6}$.
To multiply fractions, multiply numerator times numerator to find the numerator of the product and denominator times denominator to find the denominator of the product.	$\frac{2}{5} \cdot \frac{3}{7} = \frac{6}{35}$
To divide fractions, multiply the first fraction by the reciprocal of the second fraction.	$\frac{5}{9} \div \frac{2}{7} = \frac{5}{9} \cdot \frac{7}{2} = \frac{35}{18}$
To add fractions with the same denominator, add the numerators and place the sum over the common denominator.	$\frac{5}{11} + \frac{3}{11} = \frac{8}{11}$
To subtract fractions with the same denominator, subtract the numerators and place the difference over the common denominator.	$\frac{13}{15} - \frac{3}{15} = \frac{10}{15} = \frac{2}{3}$
To add or subtract fractions with different denominators, first write each fraction as an equivalent fraction with the LCD as denominator.	$\frac{2}{9} + \frac{3}{6} = \frac{2 \cdot 2}{9 \cdot 2} + \frac{3 \cdot 3}{6 \cdot 3} = \frac{4 + 9}{18} = \frac{13}{18}$

Definitions and Concepts	Examples
Section R.3 Decimals and Percents	
To write decimals as fractions, use place values.	$0.11 = \frac{11}{100}$
To Add or Subtract Decimals	
Step 1: Write the decimals so that the decimal points line up vertically.	Subtract: $2.8 - 1.04$ Add: $25 + 0.02$
Step 2: Add or subtract as for whole numbers. **Step 3:** Place the decimal point in the sum or difference so that it lines up vertically with the decimal points in the problem.	$\begin{array}{r} \overset{7\,10}{2.\cancel{8}\cancel{0}} \\ -1.04 \\ \hline 1.76 \end{array}$ $\begin{array}{r} 25. \\ +\ 0.02 \\ \hline 25.02 \end{array}$
To Multiply Decimals	
Step 1: Multiply the decimals as though they are whole numbers. **Step 2:** The decimal point in the product is placed so that the number of decimal places in the product is equal to the **sum** of the number of decimal places in the factors.	Multiply: 1.48×5.9 $\begin{array}{r} 1.48 \\ \times\ 5.9 \\ \hline 1332 \\ 740 \\ \hline 8.732 \end{array}$ $\begin{array}{l} \leftarrow 2 \text{ decimal places} \\ \leftarrow 1 \text{ decimal place} \\ \\ \\ \leftarrow 3 \text{ decimal places} \end{array}$
To Divide Decimals	
Step 1: Move the decimal point in the divisor to the right until the divisor is a whole number. **Step 2:** Move the decimal point in the dividend to the right the **same number of places** as the decimal point was moved in Step 1. **Step 3:** Divide. The decimal point in the quotient is directly over the moved decimal point in the dividend.	Divide: $1.118 \div 2.6$ $\begin{array}{r} 0.43 \\ 26.\overline{)11.18} \\ -10\,4 \\ \hline 78 \\ -78 \\ \hline 0 \end{array}$
To write fractions as decimals, divide the numerator by the denominator.	Write $\frac{3}{8}$ as a decimal. $\begin{array}{r} 0.375 \\ 8\overline{)3.000} \\ -2\,4 \\ \hline 60 \\ -56 \\ \hline 40 \\ -40 \\ \hline 0 \end{array}$
To write a percent as a decimal, drop the percent symbol, %, and move the decimal point two places to the left.	$25\% = .25.\% = 0.25$
To write a decimal as a percent, move the decimal point two places to the right and attach the percent symbol, %.	$0.7 = 0.70. = 70\%$

Definitions and Concepts	Examples

Section R.4 Reading Pictographs and Bar, Line, and Circle Graphs

A **pictograph** is a graph in which pictures or symbols are used to visually present data.

A **bar graph** presents data using vertical or horizontal bars.

The bar graph on the right shows the number of acres of corn harvested in a recent year for leading states.

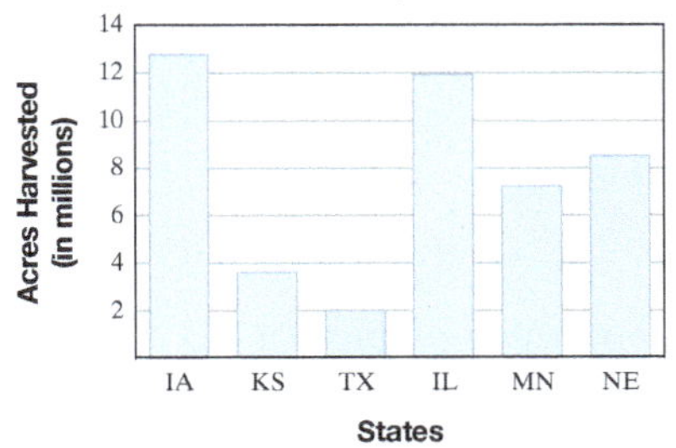

Source: U.S. Department of Agriculture

1. Approximately how many acres of corn were harvested in Iowa?

   12,800,000 acres

2. About how many more acres of corn were harvested in Illinois than Nebraska?

$$\begin{array}{r} 12 \text{ million} \\ -\ 8.5 \text{ million} \\ \hline 3.5 \text{ million} \end{array} \quad \text{or } 3{,}500{,}000 \text{ acres}$$

A **histogram** is a special bar graph in which the width of each bar represents a **class interval** and the height of each bar represents the **class frequency**.

The histogram on the right shows student quiz scores.

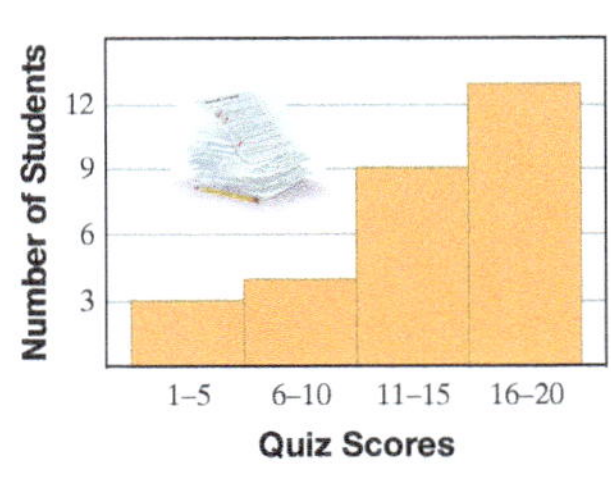

1. How many students received a score of 6–10?

   4 students

2. How many students received a score of 11–20?

   $9 + 13 = 22$ students

A **line graph** displays information with a line that connects data points.

In a **circle graph**, each section (shaped like a piece of pie) shows a category and the relative size of the category.

The circle graph on the right classifies tornadoes by wind speed.

Tornado Wind Speeds

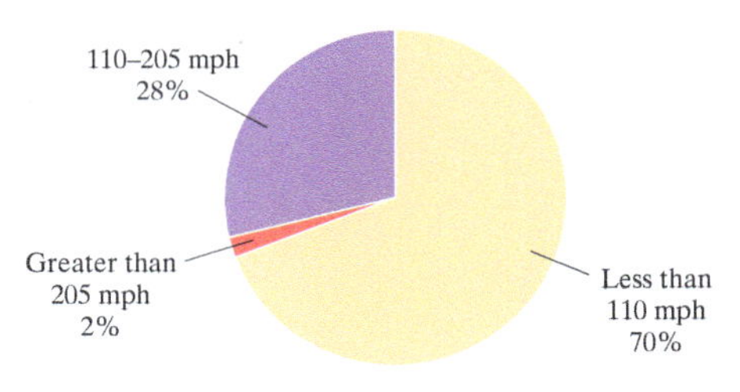

Source: National Oceanic and Atmospheric Administration

1. What percent of tornadoes have wind speeds of 110 mph or greater?

   $28\% + 2\% = 30\%$

2. If there were 939 tornadoes in the United States in 2012, how many of these might we expect to have had wind speeds less than 110 mph? Find 70% of 939.

   $70\%(939) = 0.70(939) = 657.3 \approx 657$

   Around 657 tornadoes would be expected to have had wind speeds of less than 110 mph.

Chapter R Review

(R.1) *Write the prime factorization of each number.*

1. 42

2. 800

Find the least common multiple (LCM) of each list of numbers.

3. 12, 30

4. 7, 42

5. 4, 6, 10

6. 2, 5, 7

(R.2) *Write each fraction as an equivalent fraction with the given denominator.*

7. $\frac{5}{8}$ with a denominator of 24

8. $\frac{2}{3}$ with a denominator of 60

Simplify each fraction.

9. $\frac{8}{20}$

10. $\frac{15}{100}$

11. $\frac{12}{6}$

12. $\frac{8}{8}$

Perform each indicated operation and simplify.

13. $\frac{1}{7} \cdot \frac{8}{11}$

14. $\frac{5}{12} + \frac{2}{15}$

15. $\frac{3}{10} \div 6$

16. $\frac{7}{9} - \frac{1}{6}$

17. $3\frac{3}{8} \cdot 4\frac{1}{4}$

18. $2\frac{1}{3} - 1\frac{5}{6}$

19. $16\frac{9}{10} + 3\frac{2}{3}$

20. $6\frac{2}{7} \div 2\frac{1}{5}$

The area of a plane figure is a measure of the amount of surface of the figure. Find the area of each figure below. (The area of a rectangle is the product of its length and width. The area of a triangle is $\frac{1}{2}$ the product of its base and height.)

21.

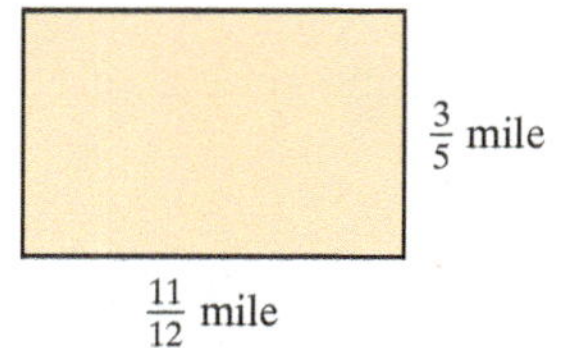

22.

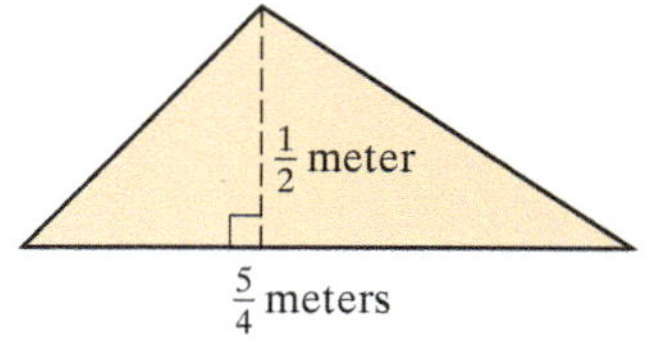

(R.3) *Write each decimal as a fraction. Do not simplify.*

23. 1.81

24. 0.035

Perform each indicated operation.

25. $\begin{array}{r} 76.358 \\ +18.76 \\ \hline \end{array}$

26. $35 + 0.02 + 1.765$

27. $18 - 4.62$

28. $\begin{array}{r} 804.062 \\ -112.489 \\ \hline \end{array}$

29. $\begin{array}{r} 7.6 \\ \times\ 12 \\ \hline \end{array}$

30. $\begin{array}{r} 14.63 \\ \times\ \ 3.2 \\ \hline \end{array}$

31. $27\overline{)772.2}$

32. $0.06\overline{)13.8}$

Round each decimal to the given place value.

33. 0.7652, nearest hundredth

34. 25.6293, nearest tenth

Write each fraction as a decimal. If the decimal is a repeating decimal, write it using the bar notation and then round to the nearest thousandth.

35. $\frac{1}{2}$

36. $\frac{3}{8}$

37. $\frac{4}{11}$

38. $\frac{5}{6}$

Write each percent as a decimal.

39. 29%

40. 1.4%

Write each decimal as a percent.

41. 0.39

42. 1.2

43. In 2018, the home ownership rate in the United States was 64.3%. Write this percent as a decimal.

44. Choose the true statement.

a. 2.3% = 0.23
b. 5 = 500%
c. 40% = 4

(R.4) *The following pictograph shows the number of new homes whose construction was completed in 2017 by region. Use this graph to answer Exercises 45 through 50.*

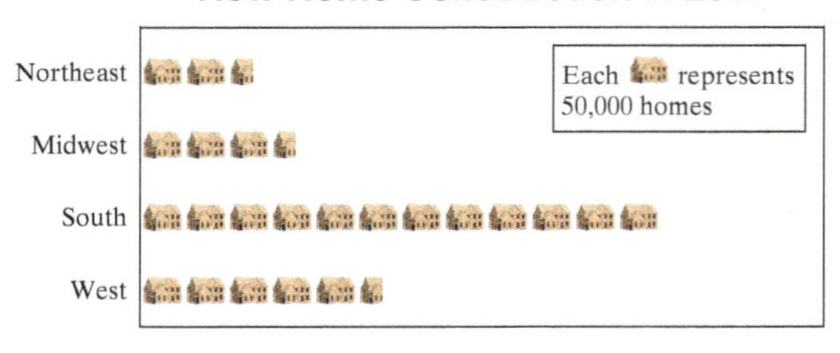

Source: U.S. Census Bureau

45. How many new homes were constructed in the Midwest during 2017?

46. How many new homes were constructed in the South during 2017?

47. Which region had the most new homes constructed?

48. Which region had the fewest new homes constructed?

49. Which region(s) had 250,000 or more new homes constructed?

50. Which region(s) had fewer than 200,000 new homes constructed?

The following bar graph shows the percent of persons age 25 or over who completed four or more years of college. Use this graph to answer Exercises 51 through 54.

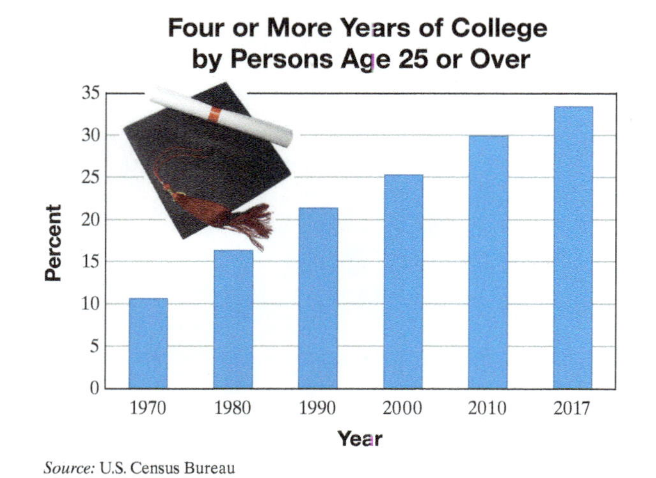

Source: U.S. Census Bureau

51. Approximate the percent of persons who had completed four or more years of college by 2010.

52. What year shown had the greatest percent of persons completing four or more years of college?

53. What years shown had 20% or more of persons completing four or more years of college?

54. Describe any patterns you notice in this graph.

The following line graph shows the total number of Olympic medals awarded during the Summer Olympics since 1996. Use this graph to answer Exercises 55 through 60.

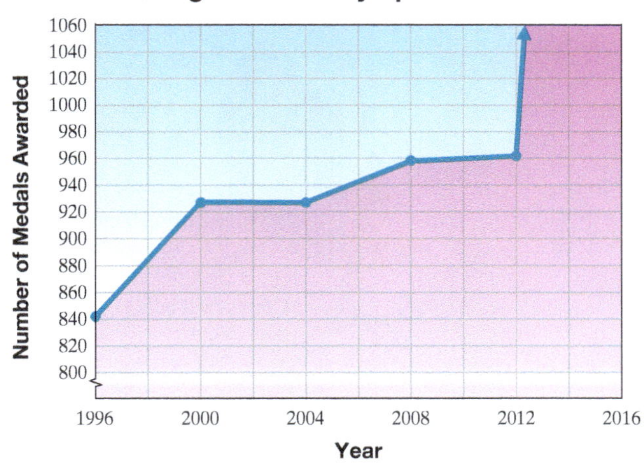

Source: International Olympic Committee

55. Approximate the number of medals awarded during the Summer Olympics of 2012.

56. Approximate the number of medals awarded during the Summer Olympics of 2000.

57. Approximate the number of medals awarded during the Summer Olympics of 2008.

58. Approximate the number of medals awarded during the Summer Olympics of 1996.

59. How many more medals were awarded at the Summer Olympics of 2008 than at the Summer Olympics of 2004?

60. The number of medals awarded at the Summer Olympics of 2016 was 2102. This was more than twice the number of medals awarded at previous Summer Olympics. Why do you think this is so? Give your explanation in complete sentences.

The following histogram shows the hours worked per week by the employees of Southern Star Furniture. Use this histogram to answer Exercises 61 through 64.

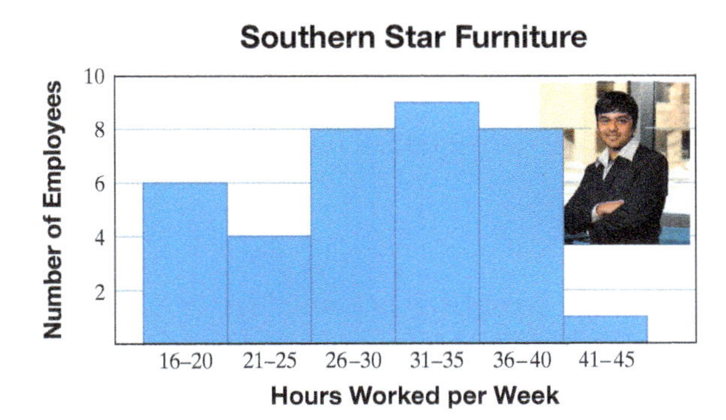

61. How many employees work 41–45 hours per week?

62. How many employees work 21–25 hours per week?

63. How many employees work 30 hours or less per week?

64. How many employees work 36 hours or more per week?

Following is a list of monthly record high temperatures for New Orleans, Louisiana. Use this list to complete the frequency distribution table below.

83	96	101	92
85	100	92	102
89	101	87	84

	Class Intervals (Temperatures)	Tally	Class Frequency (Number of Months)
65.	80°–89°		
66.	90°–99°		
67.	100°–109°		

68. Use the table from Exercises **65** through **67** to draw a histogram.

Temperatures

The following circle graph shows a family's $4000 monthly budget. Use this graph to answer Exercises 69 through 74. Write all ratios as fractions in simplest form.

69. What is the largest budget item?

70. What is the smallest budget item?

71. How much money is budgeted for the mortgage payment and utilities?

72. How much money is budgeted for savings and contributions?

73. Find the ratio of the mortage payment to the total monthly budget.

74. Find the ratio of food to the total monthly budget.

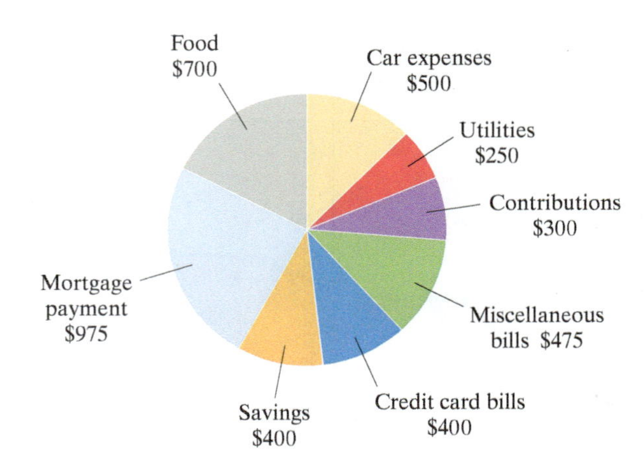

In 2017, there were 133 buildings 200 meters or taller completed in the world. The following circle graph shows the percent of these buildings by region. Use this graph to answer Exercises 75 through 78. Round each answer to the nearest whole.

Percent of Tall Buildings Completed in 2017 200 Meters or Tall by Region

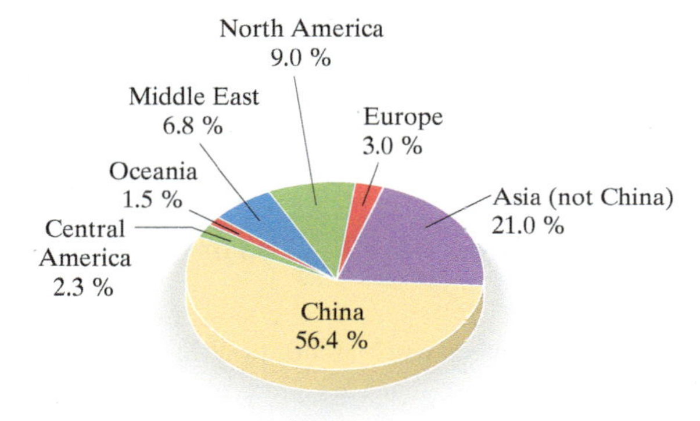

Source: Council on Tall Building and Urban Habitats

75. How many completed tall buildings were located in China?

76. How many completed tall buildings were located in the rest of Asia?

77. How many completed tall buildings were located in Oceania?

78. How many completed tall buildings were located in the Middle East?

Chapter R Getting Ready for the Test LC

MULTIPLE CHOICE Exercises 1 through 10 are **Multiple Choice**. Choose the correct letter.

For Exercises 1 through 4, choose whether the expression simplifies to

A. 1 **B.** -1 **C.** 0 **D.** undefined

1. $\frac{-2}{-2}$ **2.** $\frac{-2}{2}$ **3.** $\frac{2}{0}$ **4.** $\frac{0}{-2}$

5. The mixed number $4\frac{3}{5}$ written as a fraction is:

A. $\frac{12}{5}$ **B.** $\frac{5}{23}$ **C.** $\frac{23}{5}$ **D.** $\frac{23}{3}$

6. The improper fraction $\frac{23}{8}$ written as a mixed number is:

A. 2.3 **B.** $2\frac{7}{8}$ **C.** $8\frac{2}{3}$ **D.** $2\frac{8}{7}$

For Exercises 7 through 10, the exercises statement and correct answer are given. Choose whether the correct operation in the box should be:

A. $+$ (addition) **B.** $-$ (subtraction) **C.** $\cdot$ (multiplication) **D.** $\div$ (division)

7. $\frac{8}{11}\square\frac{2}{11}$; Answer: $\frac{16}{121}$

8. $\frac{8}{11}\square\frac{2}{11}$; Answer: $\frac{8}{2}$ or 4

9. $\frac{8}{11}\square\frac{2}{11}$; Answer: $\frac{6}{11}$

10. $\frac{8}{11}\square\frac{2}{11}$; Answer: $\frac{10}{11}$

MATCHING *For Exercises 11 through 14, **match** each operation of fractions in the first column with the correct answer in the second or third column.*

11. $\frac{5}{7}+\frac{1}{7}$

12. $\frac{5}{7}\cdot\frac{1}{7}$

13. $\frac{5}{7}\div\frac{1}{7}$

14. $\frac{5}{7}-\frac{1}{7}$

A. $\frac{5}{7}$

B. $\frac{5}{1}$ or 5

C. $\frac{6}{14}$ or $\frac{3}{7}$

D. $\frac{4}{7}$

E. undefined

F. $\frac{6}{7}$

G. $\frac{6}{49}$

H. $\frac{5}{49}$

MATCHING *For Exercises 15 through 18, the number 8603.2855 is rounded to different place values. **Match** the rounded number in the left column to the correct place it is rounded to in the columns to the right.*

15. 8603.3

16. 8600

17. 8603.286

18. 8603.29

A. 8603.2855 rounded to ones

B. 8603.2855 rounded to tens

C. 8603.2855 rounded to tenths

D. 8603.2855 rounded to hundredths

E. 8603.2855 rounded to thousandths

MULTIPLE CHOICE *Exercises 19 through 30 are **Multiple Choice**. Choose the correct answer.*

19. Find $10 - 0.08$.

A. 2 **B.** 9.2 **C.** 9.02 **D.** 9.92

20. Find $10 + 0.08$.

A. 10.08 **B.** 10.8 **C.** 18 **D.** 10.008

21. Find 37 + 2.1

A. 58 **B.** 39.1 **C.** 37.21 **D.** 3.91

22. A product of decimal numbers below is completed except for placement of the decimal point in the product. Choose the correct product.

```
   2.326
 ×   1.5
   11630
   2326
   34890
```

A. 348.90 **B.** 34.890 **C.** 3.4890 **D.** 3489.0

23. A quotient of decimal numbers below is completed except for placement of the decimal point in the quotient. Choose the correct quotient.

$$0.38\overline{)7.068}$$ with quotient 186

A. 0.186 **B.** 1.86 **C.** 18.6 **D.** 186

24. Since "percent" means "per hundred," choose the number that does *not* equal 12%.

A. $\frac{12}{100}$ **B.** 0.12 **C.** $\frac{3}{25}$ **D.** 1.2

25. Choose the number that does *not* equal 100%.

A. 10 **B.** 1 **C.** 1.00 **D.** $\frac{100}{100}$

For Exercises 26 through 29, use the graph below.

Bushels of Oranges Picked

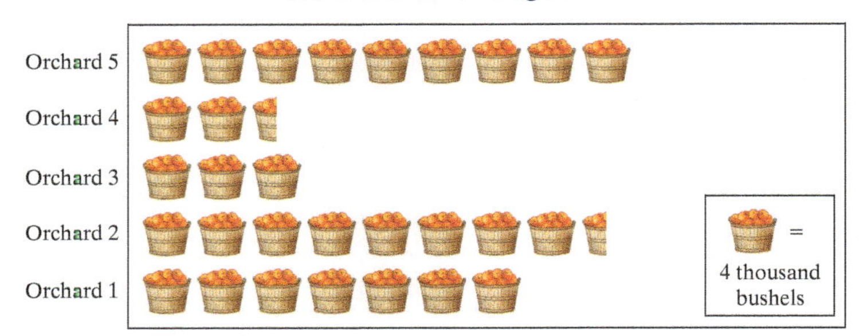

26. How many bushels of oranges did Orchard 1 produce?

A. 7 bushels **B.** 28 bushels **C.** 28,000 bushels

27. Which orchard above produced the most bushels?

A. Orchard 1 **B.** Orchard 2 **C.** Orchard 5 **D.** Orchards 2 and 5 produced the same.

28. How many bushels of oranges did Orchard 4 produce?

A. 3 bushels **B.** $2\frac{1}{2}$ bushels **C.** 10 bushels **D.** 10 thousand bushels

29. How many more bushels did Orchard 2 produce than Orchard 4?

A. 24 bushels **B.** $6\frac{1}{2}$ bushels **C.** 24,000 bushels **D.** 6000 bushels

For Exercise 30, choose the correct letter.

30. Choose the degrees in the unknown sector of this circle graph.

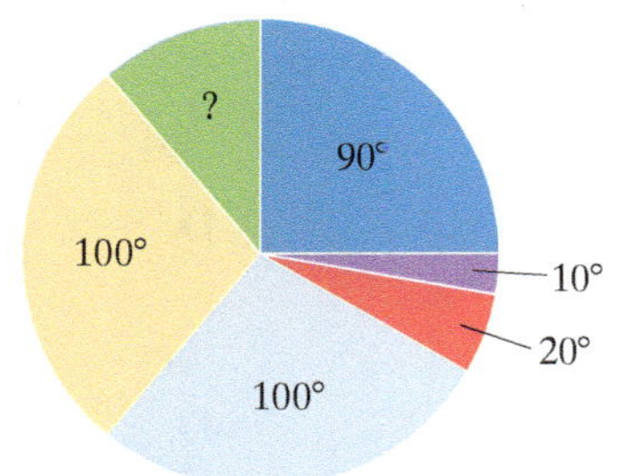

A. 60° **B.** 50° **C.** 40° **D.** 80°

Answers

1. Write the prime factorization of 72.

2. Find the LCM of 5, 18, 20.

3. Write $\frac{5}{12}$ as an equivalent fraction with a denominator of 60.

Simplify each fraction.

4. $\frac{15}{20}$

5. $\frac{48}{100}$

6. Write 1.3 as a fraction.

Perform each indicated operation and simplify.

7. $\frac{5}{8} + \frac{7}{10}$

8. $\frac{2}{3} \cdot \frac{27}{49}$

9. $\frac{9}{10} \div 18$

10. $\frac{8}{9} - \frac{1}{12}$

11. $1\frac{2}{9} + 3\frac{2}{3}$

12. $5\frac{6}{11} - 3\frac{7}{22}$

13. $6\frac{7}{8} \div \frac{1}{8}$

14. $2\frac{1}{10} \cdot 6\frac{1}{2}$

Perform each indicated operation.

15. $43 + 0.21 + 1.9$

16. $123.6 - 57.72$

17. 7.93×1.6

18. $0.25\overline{)80}$

19. Round 23.7272 to the nearest hundredth.

20. Write $\frac{7}{8}$ as a decimal.

21. Write $\frac{1}{6}$ as a repeating decimal. Then approximate the result to the nearest thousandth.

22. Write 63.2% as a decimal.

1. ________
2. ________
3. ________
4. ________
5. ________
6. ________
7. ________
8. ________
9. ________
10. ________
11. ________
12. ________
13. ________
14. ________
15. ________
16. ________
17. ________
18. ________
19. ________
20. ________
21. ________
22. ________

23. ______

24. ______

25. ______

26. ______

27. ______

28. ______

29. ______

30. ______

31. ______

32. ______

33. ______

23. Write 0.09 as a percent.

24. Write $\frac{3}{4}$ as a percent. (*Hint:* Write $\frac{3}{4}$ as a decimal, and then write the decimal as a percent.)

Most of the water on Earth is in the form of oceans. Only a small part is fresh water. This circle graph shows the distribution of fresh water on Earth. Use this graph to answer Exercises 25 through 28. (Source: Philip's World Atlas)

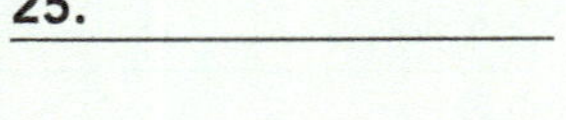

Fresh Water Distribution

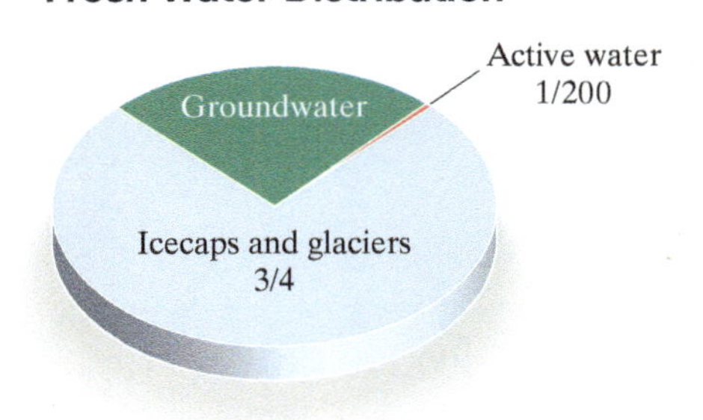

25. What fractional part of fresh water is icecaps and glaciers?

26. What fractional part of fresh water is active water?

27. What fractional part of fresh water is groundwater?

28. What fractional part of fresh water is groundwater or icecaps and glaciers?

Find the area of each figure. (The area of a rectangle is the product of its length and width. The area of a triangle is $\frac{1}{2}$ the product of its base and height.)

29.

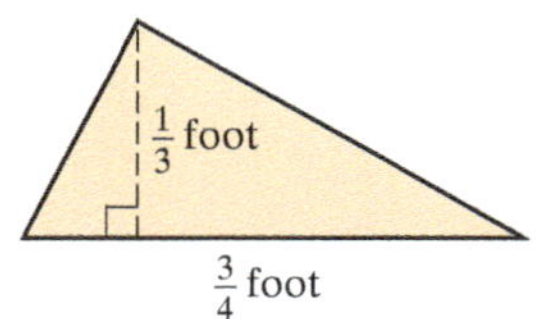

30.

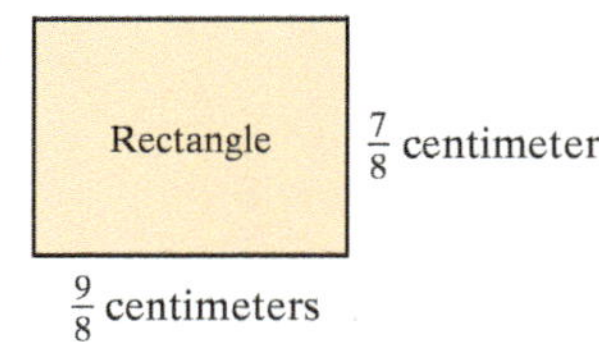

The following pictograph shows the money collected each week from a wrapping paper fundraiser. Use this graph to answer Exercises 31 through 33.

Weekly Wrapping Paper Sales

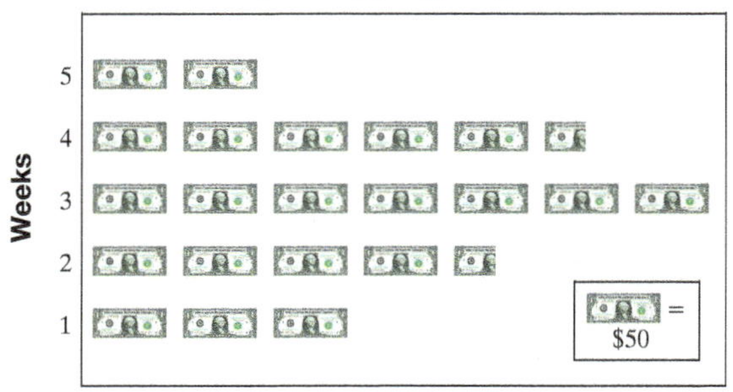

31. How much money was collected during the second week?

32. During which week was the most money collected? How much money was collected during that week?

33. What was the total money collected for the fundraiser?

The bar graph shows the normal monthly precipitation in centimeters for Chicago, Illinois. Use this graph to answer Exercises 34 through 36.

34. ______

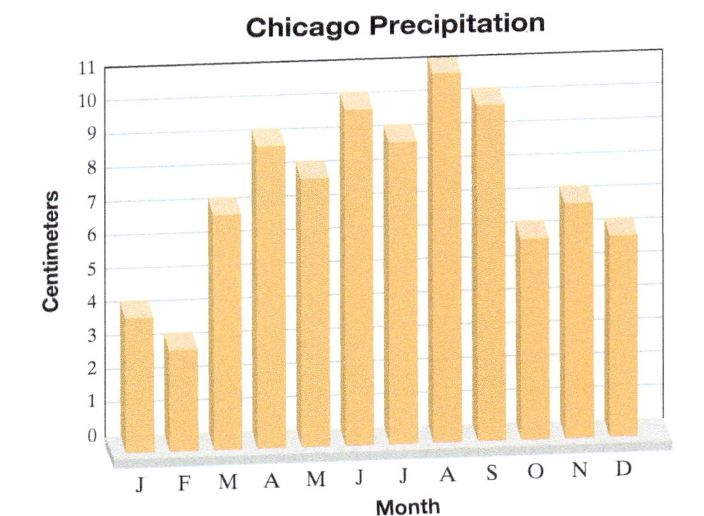

Source: U.S. National Oceanic and Atmospheric Administration, *Climatography of the United States*, No. 81

35. ______

36. ______

34. During which month(s) does Chicago normally have more than 9 centimeters of precipitation?

35. During which month does Chicago normally have the least amount of precipitation? How much precipitation occurs during that month?

36. During which month(s) does 7 centimeters of precipitation normally occur?

37. ______

The following line graph shows the annual inflation rate in the United States for the years 2008–2017. Use this graph to answer Exercises 37 through 39.

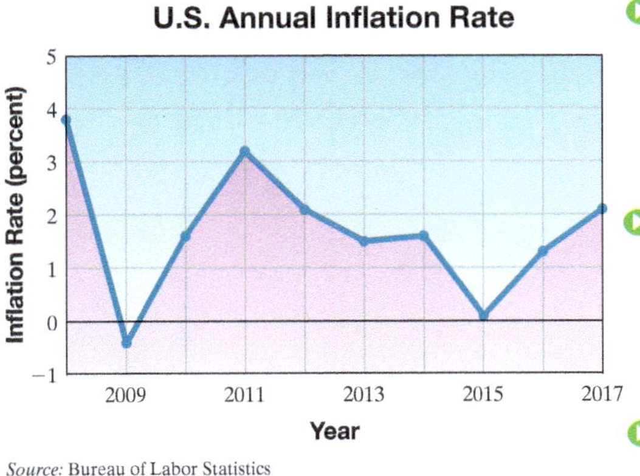

Source: Bureau of Labor Statistics

37. Approximate the annual inflation rate in 2014.

38. ______

38. During which of the years shown was the inflation rate greater than 3%?

39. During which sets of years was the inflation rate decreasing?

39. ______

The result of a survey of 200 people is shown in the following circle graph. Each person was asked to tell his or her favorite type of music. Use this graph to answer Exercises 40 and 41.

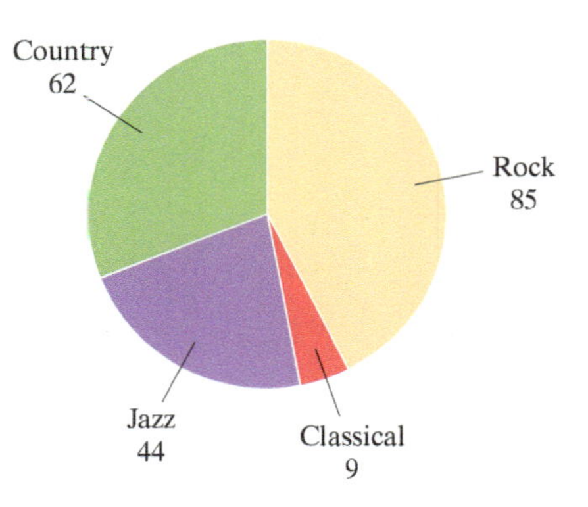

40. Find the ratio of those who prefer rock music to the total number surveyed.

41. Find the ratio of those who prefer country music to those who prefer jazz.

40. ______

41. ______

The following circle graph shows the projected age distribution of the population of the United States in 2020. There are projected to be 335 million people in the United States in 2020. Use the graph to find how many people are expected to be in the age groups given.

U.S. Population in 2020 by Age Groups

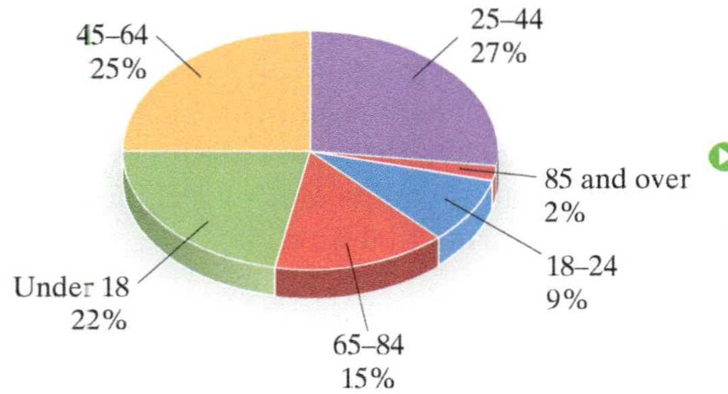

Source: U.S. Census Bureau

42. Under 18 (Round to nearest whole million.)

43. 25–44 (Round to nearest whole million.)

42. ______

43. ______

A professor measures the heights of the students in her class. The results are shown in the following histogram. Use this histogram to answer Exercises 44 and 45.

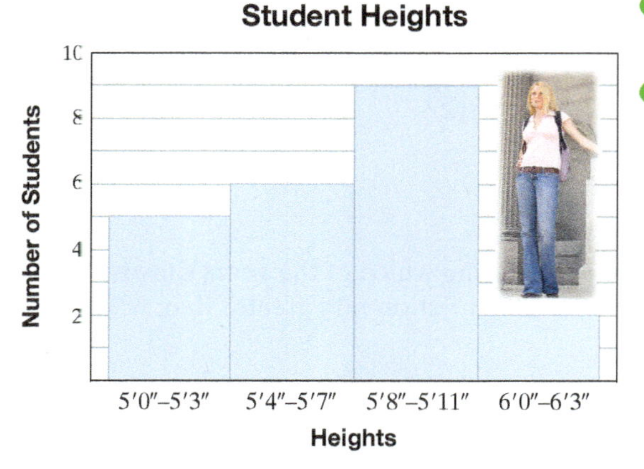

44. How many students are 5′8″–5′11″ tall?

45. How many students are 5′7″ or shorter?

44. ______

45. ______

Real Numbers and Introduction to Algebra

1

A Selection of Resources for Success in This Mathematics Course

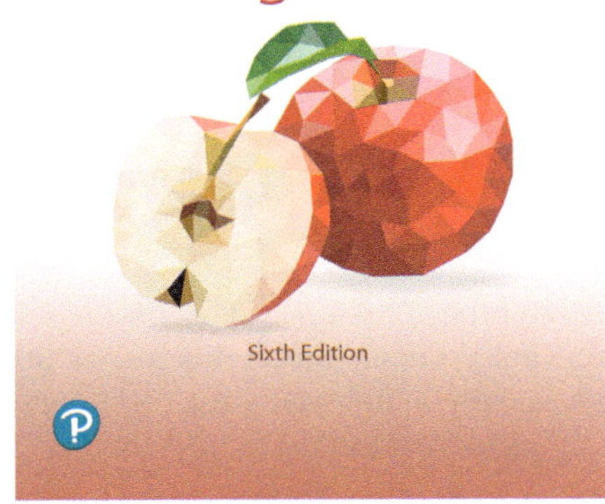

Textbook

Instructor

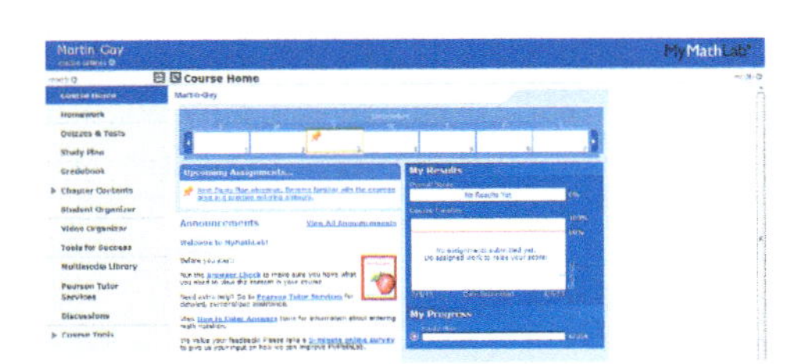

MyLab Math and MathXL

Video Organizer

Interactive Lecture Series

For more information about the resources illustrated above, read Section 1.1.

In this chapter, we begin with a review of the basic symbols—the language—of mathematics. We then introduce algebra by using a variable in place of a number. From there, we translate phrases to algebraic expressions and sentences to equations. This is the beginning of problem solving, which we formally study in Chapter 2.

Sections

1.1 Study Skill Tips for Success in Mathematics

1.2 Symbols and Sets of Numbers

1.3 Exponents, Order of Operations, and Variable Expressions

1.4 Adding Real Numbers

1.5 Subtracting Real Numbers

Integrated Review—Operations on Real Numbers

1.6 Multiplying and Dividing Real Numbers

1.7 Properties of Real Numbers

1.8 Simplifying Expressions

Check Your Progress

Vocabulary Check

Chapter Highlights

Chapter Review

Getting Ready for the Test

Chapter Test

1.1 Study Skill Tips for Success in Mathematics

Objectives

A Get Ready for This Course.

B Understand Some General Tips for Success.

C Know How to Use This Text.

D Know How to Use Text Resources.

E Get Help as Soon as You Need It.

F Learn How to Prepare for and Take an Exam.

G Develop Good Time Management.

Before reading this section, ask yourself a few questions.

1. Were you satisfied—really satisfied—with your performance in your last math course? In other words, do you feel that your outcome represented your best effort?

2. When you took your last math course, were your notes and materials from that course organized and easy to find, or were they disorganized and hard to find—if you saved them at all?

If the answer is "no" to these questions, then it is time to make a change. To begin, continue reading this section.

Objective A Let's Get Ready for This Course 

1. *Start with a Positive Attitude.*

Now that you have decided to take this course, remember that a *positive attitude* will make all the difference in the world. Your belief that you can succeed is just as important as your commitment to this course. Make sure you are ready for this course by having the time and positive attitude that it takes to succeed.

2. *Understand How Your Course Material Is Presented—Lecture by Instructor, Online with Computer, or Both?*

Make sure that you are familiar with the way that this course is being taught. Is it a traditional course, in which you have a printed textbook and meet with an instructor? Is it taught totally online, and your textbook is electronic and you e-mail your instructor? Or is your course structured somewhere in between these two methods? (Not all of the tips that follow will apply to all forms of instruction.)

3. *Schedule Your Class So That It Does Not Interfere with Other Commitments.*

Make sure that you have scheduled your math course for a time that will give you the best chance for success. For example, if you are also working, you may want to check with your employer to make sure that your work hours will not conflict with your course schedule.

Objective B Here Are a Few General Tips for Success 

Below are some general tips that will increase your chance for success in a mathematics class. Many of these tips will also help you in other courses you may be taking.

1. *Most Important! Organize Your Class Materials. Unless Told Otherwise, Use a 3-Ring Binder Solely for Your Mathematics Class.*

MyLab Math and MathXL When assignments are turned in online, keep a hard copy of your complete written work. You will need to refer to your written work to be able to ask questions and to study for tests later.

In the next couple pages, many ideas will be presented to help you organize your class materials—notes, any handouts, completed homework, previous tests, etc. In general, you MUST have these materials organized. All of them will be valuable references throughout your course and when studying for upcoming tests and the final exam. One way to make sure you can locate these materials when you need them is to use a three-ring binder. This binder should be used solely for your mathematics class and should be brought to each and every class or lab. This way, any material can be immediately inserted in a section of this binder and will be there when you need it.

2. *Choose to Attend All Class Periods.*

If possible, sit near the front of the classroom. This way, you will see and hear the presentation better. It may also be easier for you to participate in classroom activities.

3. *Complete Your Homework. This Means: Attempt All of It, Check All of It, Correct Any Mistakes, and Ask for Help If Needed.*

You've probably heard the phrase "practice makes perfect" in relation to music and sports. It also applies to mathematics. You will find that the more time you spend solving mathematics exercises, the easier the process becomes. Be sure to schedule enough time to complete your assignments before the due date assigned by your instructor.

Review the steps you took while working a problem. Learn to check your answers in the original exercises. You may also compare your answers with the "Answers to Selected Exercises" section in the back of the book. If you have made a mistake, try to figure out what went wrong. Then correct your mistake. If you can't find what went wrong, **don't** erase your work or throw it away. Show your work to your instructor, a tutor in a math lab, or a classmate. It is easier for someone to find where you had trouble if he or she looks at your original work.

> **Helpful Hint** **MyLab Math and MathXL** If you are doing your homework online, you can work and re-work those exercises that you struggle with until you master them. Try working through all the assigned exercises twice before the due date.

It's all right to ask for help. In fact, it's a good idea to ask for help whenever there is something that you don't understand. Make sure you know when your instructor has office hours and how to find his or her office. Find out whether math tutoring services are available on your campus. Check on the hours, location, and requirements of the tutoring service.

> **Helpful Hint** **MyLab Math and MathXL** If you are completing your homework online, it's important to work each exercise on paper before submitting the answer. That way, you can check your work and follow your steps to find and correct any mistakes.

4. *Learn from Your Mistakes and Be Patient with Yourself.*

Everyone, even your instructor, makes mistakes. (That definitely includes me—Elayn Martin-Gay.) Use your errors to learn and to become a better math student. The key is finding and understanding your errors.

Was your mistake a careless one, or did you make it because you can't read your own math writing? If so, try to work more slowly or write more neatly and make a conscious effort to carefully check your work.

Did you make a mistake because you don't understand a concept? Take the time to review the concept or ask questions to better understand it.

Did you skip too many steps? Skipping steps or trying to do too many steps mentally may lead to preventable mistakes.

5. *Turn In Assignments on Time.*

This way, you can be sure that you will not lose points for being late. Show every step of a problem and be neat and organized. Also be sure that you understand which problems are assigned for homework. If allowed, you can always double-check the assignment with another student in your class.

> **Helpful Hint** **MyLab Math and MathXL** Be aware of assignments and due dates set by your instructor. Don't wait until the last minute to submit work online.

Objective C Knowing and Using Your Text or e-Text

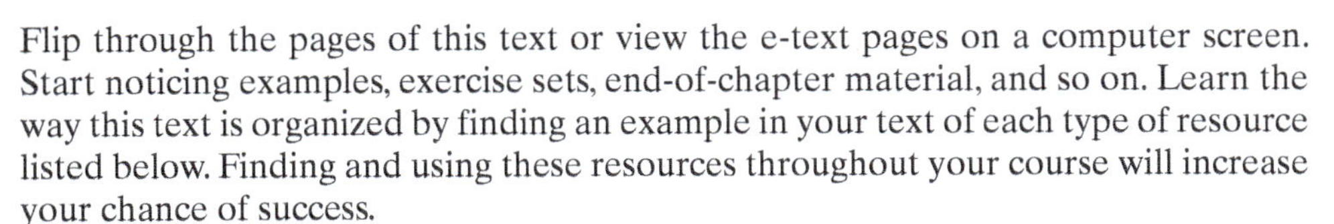

Flip through the pages of this text or view the e-text pages on a computer screen. Start noticing examples, exercise sets, end-of-chapter material, and so on. Learn the way this text is organized by finding an example in your text of each type of resource listed below. Finding and using these resources throughout your course will increase your chance of success.

- ***Practice Exercises.*** Each example in every section has a parallel Practice exercise. Work each Practice exercise after you've finished the corresponding example. Answers are at the bottom of the page. This "learn-by-doing" approach will help you grasp ideas before you move on to other concepts.
- ***Objectives.*** Every section of this text is divided into objectives, such as **A** or **B**. They are listed at the beginning of the section and noted in that section. The main section of exercises in each exercise set is also referenced by an objective, such as **A** or **B**, and also an example(s). There is also often a section of exercises titled "Mixed Practice," which is referenced by two or more objectives or sections. These are mixed exercises written to prepare you for your next exam. Use all of this referencing if you have trouble completing an assignment from the exercise set.

- ***Icons (Symbols).*** Make sure that you understand the meaning of the icons that are beside many exercises. ▶ tells you that the corresponding exercise may be viewed on the video Lecture Series that corresponds to that section. ✎ tells you that this exercise is a writing exercise in which you should answer in complete sentences. △ tells you that the exercise involves geometry.
- ***Integrated Reviews.*** Found in the middle of each chapter, these reviews offer you a chance to practice—in one place—the many concepts that you have learned separately over several sections.
- ***End-of-Chapter Opportunities.*** There are many opportunities at the end of each chapter to help you understand the concepts of the chapter.

  Vocabulary Checks contain key vocabulary terms introduced in the chapter.

  Chapter Highlights contain chapter summaries and examples.

  Chapter Reviews contain review problems. The first part is organized section by section and the second part contains a set of mixed exercises.

  Getting Ready for the Tests are multiple choice or matching exercises designed to check your knowledge of chapter concepts, before you attempt the chapter test. Video solutions are available for all these exercises.

  Chapter Tests are sample tests to help you prepare for an exam. The Chapter Test Prep Videos found in MyLab Math provide the video solution to each question on each Chapter Test.

  Cumulative Reviews start at Chapter 2 and are reviews consisting of material from the beginning of the book to the end of that particular chapter.

- ***Student Resources in Your Textbook.*** You will find a **Student Resources** section at the back of this textbook. It contains the following to help you study and prepare for tests:

  Study Skill Builders contain study skills advice. To increase your chance for success in the course, read these study tips, and answer the questions.

  Bigger Picture—Study Guide Outline provides you with a study guide outline of the course, with examples.

  Practice Final provides you with a Practice Final Exam to help you prepare for a final.

- ***Resources to Check Your Work.*** The **Answers to Selected Exercises** section provides answers to all odd-numbered section exercises and to all integrated review, chapter review, getting ready for the test, chapter test, and cumulative review exercises.

Objective D Knowing and Using Video and Notebook Organizer Resources ▶

Helpful Hint **MyLab Math** In MyLab Math, you have access to the following video resources:

- Lecture Videos for each section
- Getting Ready for the Test Videos
- Chapter Test Prep Videos
- Final Exam Videos

Use these videos provided by the author to prepare for class, review, and study for tests.

Video Resources

Below is a list of video resources that are all made by me—the author of your text, Elayn Martin-Gay. By making these videos, I can be sure that the methods presented are consistent with those in the text. All video resources may be found in MyLab Math.

- ***Interactive Video Lecture Series.*** Exercises marked with a ▶ are fully worked out by the author. The lecture series provides approximately 20 minutes of instruction per section and is organized by Objective.
- ***Getting Ready for the Test Videos.*** These videos provide solutions to all of the Getting Ready for the Test exercises.
- ***Chapter Test Prep Videos.*** These videos provide solutions to all of the Chapter Test exercises worked out by the author. They can be found in MyLab Math. This supplement is very helpful before a test or exam.

- ***Tips for Success in Mathematics.*** These video segments are about 3 minutes long and are daily reminders to help you continue practicing and maintaining good organizational and study habits.
- ***Final Exam Videos.*** These video segments provide solutions to each question.

Video Organizer

This organizer is in three-ring notebook ready form. It is to be inserted in a three-ring binder and completed. This organizer is numbered according to the sections in your text to which it refers.

It is closely tied to the Interactive (Video) Lecture Series. Each section should be completed while watching the lecture video on the same section. Once completed, you will have a set of notes to accompany the (Video) Lecture Series section by section.

Objective E Getting Help 

If you have trouble completing assignments or understanding the mathematics, get help as soon as you need it! This tip is presented as an objective on its own because it is so important. In mathematics, usually the material presented in one section builds on your understanding of the previous section. This means that if you don't understand the concepts covered during a class period, there is a good chance that you will not understand the concepts covered during the next class period. If this happens to you, get help as soon as you can.

Where can you get help? Try your instructor, a tutoring center, or a math lab, or you may want to form a study group with fellow classmates. If you do decide to see your instructor or go to a tutoring center, make sure that you have a neat notebook and are ready with your questions.

> **Helpful Hint** **MyLab Math and MathXL**
>
> - Use the **Help Me Solve This** button to get step-by-step help for the exercise you are working. You will need to work an additional exercise of the same type before you can get credit for having worked it correctly.
> - Use the **Video** button to view a video clip of the author working a similar exercise.

Objective F Preparing for and Taking an Exam 

Make sure that you allow yourself plenty of time to prepare for a test. If you think that you are a little "math anxious," it may be that you are not preparing for a test in a way that will ensure success. The way that you prepare for a test in mathematics is important. To prepare for a test:

1. Review your previous homework assignments.
2. Review any notes from class and section-level quizzes you have taken. (If this is a final exam, also review chapter tests you have taken.)
3. Review concepts and definitions by reading the Chapter Highlights at the end of each chapter.
4. Practice working out exercises by completing the Chapter Review found at the end of each chapter. (If this is a final exam, go through a Cumulative Review. There is one found at the end of each chapter except Chapter 1. Choose the review found at the end of the latest chapter that you have covered in your course.) *Don't stop here!*
5. Take the Chapter Getting Ready for the Test. All answers to these exercises are available to you as well as video solutions.
6. Take a sample test with no notes, etc., available for help. It is important that you place yourself in conditions similar to test conditions to find out how you will perform. There is a Chapter Test available at the end of each chapter, or you can work selected problems from the Chapter Review. Your instructor may also provide you with a review sheet. Then check your sample test. If your sample test is the Chapter Test in the text, don't forget that the video solutions are in MyLab Math.
7. On the day of the test, allow yourself plenty of time to arrive at where you will be taking your exam.

> **Helpful Hint** **MyLab Math and MathXL** Review your written work for previous assignments. Then, go back and rework previous assignments. Open a previous assignment, and click **Similar Exercise** to generate new exercises. Rework the exercises until you fully understand them and can work them without help features.

When taking your test:

1. Read the directions on the test carefully.
2. Read each problem carefully as you take the test. Make sure that you answer the question asked.
3. Watch your time and pace yourself so that you can attempt each problem on your test.
4. If you have time, check your work and answers.
5. Do not turn your test in early. If you have extra time, spend it double-checking your work.

Objective G Managing Your Time

As a college student, you know the demands that classes, homework, work, and family place on your time. Some days you probably wonder how you'll ever get everything done. One key to managing your time is developing a schedule. Here are some hints for making a schedule:

1. Make a list of all of your weekly commitments for the term. Include classes, work, regular meetings, extracurricular activities, etc. You may also find it helpful to list such things as laundry, regular workouts, grocery shopping, etc.
2. Next, estimate the time needed for each item on the list. Also make a note of how often you will need to do each item. Don't forget to include time estimates for the reading, studying, and homework you do outside of your classes. You may want to ask your instructor for help estimating the time needed.
3. In the exercise set that follows, you are asked to block out a typical week on the schedule grid given. Start with items with fixed time slots like classes and work.
4. Next, include the items on your list with flexible time slots. Think carefully about how best to schedule items such as study time.
5. Don't fill up every time slot on the schedule. Remember that you need to allow time for eating, sleeping, and relaxing! You should also allow a little extra time in case some items take longer than planned.
6. If you find that your weekly schedule is too full for you to handle, you may need to make some changes in your workload, classload, or other areas of your life. You may want to talk to your advisor, manager or supervisor at work, or someone in your college's academic counseling center for help with such decisions.

1.1 Exercise Set MyLab Math 

1. What is your instructor's name?

2. What are your instructor's office location and office hours?

3. What is the best way to contact your instructor?

4. Do you have the name and contact information of at least one other student in class?

5. Will your instructor allow you to use a calculator in this class?

6. Why is it important that you write step-by-step solutions to homework exercises and keep a hard copy of all work submitted?

7. Is there a tutoring service available on campus? If so, what are its hours? What services are available?

8. Have you attempted this course before? If so, write down ways that you might improve your chances of success during this attempt.

9. List some steps that you can take if you begin having trouble understanding the material or completing an assignment. If you are completing your homework in MyLab Math and MathXL, list the resources you can use for help.

10. How many hours of studying does your instructor advise for each hour of instruction?

11. What does the ✎ icon in this text mean?

12. What does the △ icon in this text mean?

13. What does the ▶ icon in this text mean?

14. Search the minor columns in your text. What are Practice exercises?

15. When might be the best time to work a Practice exercise?

16. Where are the answers to Practice exercises?

17. What answers are contained in this text and where are they?

18. What are Tips for Success in Mathematics and where are they located?

19. What and where are Integrated Reviews?

20. How many times is it suggested that you work through the homework exercises in MyLab Math or MathXL before the submission deadline?

21. How far in advance of the assigned due date is it suggested that homework be submitted online? Why?

22. Chapter Highlights are found at the end of each chapter. Find the Chapter 1 Highlights and explain how you might use it and how it might be helpful.

23. Chapter Reviews are found at the end of each chapter. Find the Chapter 1 Review and explain how you might use it and how it might be helpful.

24. Chapter Tests are found at the end of each chapter. Find the Chapter 1 Test and explain how you might use it and how it might be helpful when preparing for an exam on Chapter 1. Include how the Chapter Test Prep Videos may help. If you are working in MyLab Math and MathXL, how can you use previous homework assignments to study?

25. What is the Video Organizer? Explain the contents and how it might be used.

26. Read or reread objective **G** and fill out the schedule grid on the next page.

	Monday	Tuesday	Wednesday	Thursday	Friday	Saturday	Sunday
4:00 a.m.							
5:00 a.m.							
6:00 a.m.							
7:00 a.m.							
8:00 a.m.							
9:00 a.m.							
10:00 a.m.							
11:00 a.m.							
12:00 p.m.							
1:00 p.m.							
2:00 p.m.							
3:00 p.m.							
4:00 p.m.							
5:00 p.m.							
6:00 p.m.							
7:00 p.m.							
8:00 p.m.							
9:00 p.m.							
10:00 p.m.							
11:00 p.m.							
Midnight							
1:00 a.m.							
2:00 a.m.							
3:00 a.m.							

1.2 Symbols and Sets of Numbers

Objectives

A Define the Meaning of the Symbols $=, \neq, <, >, \leq,$ and $\geq$.

B Translate Sentences into Mathematical Statements.

C Identify Integers, Rational Numbers, Irrational Numbers, and Real Numbers.

D Find the Absolute Value of a Real Number.

We begin with a review of the set of natural numbers and the set of whole numbers and how we use symbols to compare these numbers. A **set** is a collection of objects, each of which is called a **member** or **element** of the set. A pair of brace symbols { } encloses the list of elements and is translated as "the set of" or "the set containing."

Natural Numbers

$\{1, 2, 3, 4, 5, 6, \ldots\}$

Whole Numbers

$\{0, 1, 2, 3, 4, 5, 6, \ldots\}$

Helpful Hint

The three dots (an ellipsis) at the end of the list of elements of a set means that the list continues in the same manner indefinitely.

Objective A Equality and Inequality Symbols 

Picturing natural numbers and whole numbers on a number line helps us to see the order of the numbers. Symbols can be used to describe in writing the order of two quantities. We will use equality symbols and inequality symbols to compare quantities.

Below is a review of these symbols. The letters a and b are used to represent quantities. Letters such as a and b that are used to represent numbers or quantities are called **variables**.

Equality and Inequality Symbols

		Meaning
Equality symbol:	$a = b$	a is equal to b.
Inequality symbols:	$a \neq b$	a is not equal to b.
	$a < b$	a is less than b.
	$a > b$	a is greater than b.
	$a \leq b$	a is less than or equal to b.
	$a \geq b$	a is greater than or equal to b.

These symbols may be used to form **mathematical statements** such as

$$2 = 2 \quad \text{and} \quad 2 \neq 6$$

Recall that on a number line, we see that a number **to the right of** another number is **larger**. Similarly, a number **to the left of** another number is **smaller**. For example, 3 is to the left of 5 on the number line, which means that 3 is less than 5, or $3 < 5$. Similarly, 2 is to the right of 0 on the number line, which means that 2 is greater than 0, or $2 > 0$. Since 0 is to the left of 2, we can also say that 0 is less than 2, or $0 < 2$.

Helpful Hint

Recall that $2 > 0$ has exactly the same meaning as $0 < 2$. Switching the order of the numbers and reversing the direction of the inequality symbol does not change the meaning of the statement.

$6 > 4$ has the same meaning as $4 < 6$.

Also notice that when the statement is true, the inequality arrow points to the smaller number.

Our discussion above can be generalized in the order property below.

Order Property for Real Numbers

For any two real numbers a and b, a is less than b if a is to the left of b on a number line.

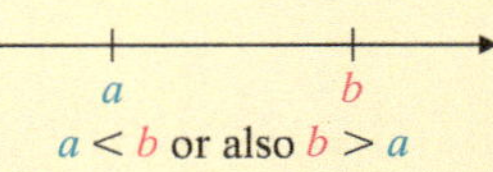

$a < b$ or also $b > a$

Practice 1–6

Determine whether each statement is true or false.

1. $8 < 6$ **2.** $100 > 10$
3. $21 \leq 21$ **4.** $21 \geq 21$
5. $0 \geq 5$ **6.** $25 \geq 22$

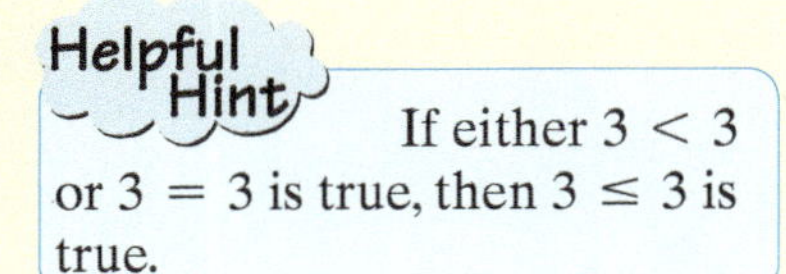

If either $3 < 3$ or $3 = 3$ is true, then $3 \leq 3$ is true.

Examples

Determine whether each statement is true or false.

1. $2 < 3$ True. Since 2 is to the left of 3 on a number line
2. $72 < 27$ False. 72 is to the right of 27 on a number line, so $72 > 27$.
3. $8 \geq 8$ True. Since $8 = 8$ is true
4. $8 \leq 8$ True. Since $8 = 8$ is true
5. $23 \leq 0$ False. Since neither $23 < 0$ nor $23 = 0$ is true
6. $0 \leq 23$ True. Since $0 < 23$ is true

Work Practice 1–6

Objective B Translating Sentences into Mathematical Statements

Now, let's use the symbols discussed above to translate sentences into mathematical statements.

Practice 7

Translate each sentence into a mathematical statement.

a. Fourteen is greater than or equal to fourteen.
b. Zero is less than five.
c. Nine is not equal to ten.

Example 7

Translate each sentence into a mathematical statement.

a. Nine is less than or equal to eleven.
b. Eight is greater than one.
c. Three is not equal to four.

Solution:

a.

nine	is less than or equal to	eleven
↓	↓	↓
9	$\leq$	11

b.

eight	is greater than	one
↓	↓	↓
8	$>$	1

c.

three	is not equal to	four
↓	↓	↓
3	$\neq$	4

Work Practice 7

Answers

1. false **2.** true **3.** true **4.** true **5.** false **6.** true **7. a.** $14 \geq 14$ **b.** $0 < 5$ **c.** $9 \neq 10$

Objective C Identifying Common Sets of Numbers

Whole numbers are not sufficient to describe many situations in the real world. For example, quantities smaller than zero must sometimes be represented, such as temperatures less than 0 degrees.

Recall that we can place numbers less than zero on a number line as follows: Numbers less than 0 are to the left of 0 and are labeled −1, −2, −3, and so on. The numbers we have labeled on the number line below are called the set of **integers**.

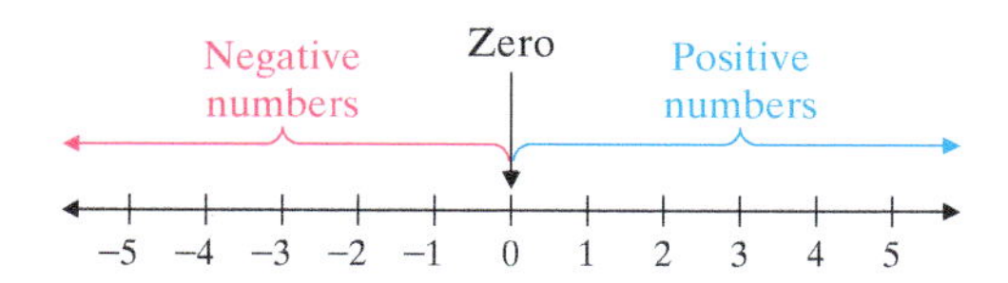

Integers to the left of 0 are called **negative integers**; integers to the right of 0 are called **positive integers**. The integer 0 is neither positive nor negative.

Integers

$\{\ldots, -3, -2, -1, 0, 1, 2, 3, \ldots\}$

Helpful Hint

A − sign, such as the one in −2, tells us that the number is to the left of 0 on a number line.

−2 is read "negative two."

A + sign or no sign tells us that the number lies to the right of 0 on a number line. For example, 3 and +3 both mean positive three.

Example 8 Use an integer to express the number in the following. "The lowest temperature ever recorded at South Pole Station, Antarctica, occurred during the month of June. The record-low temperature was 117 degrees Fahrenheit below zero." (*Source*: National Oceanic and Atmospheric Administration)

Solution: The integer −117 represents 117 degrees Fahrenheit below zero.

Work Practice 8

Practice 8

Use an integer to express the number in the following. "The elevation of Laguna Salada in Mexico is 10 meters below sea level." (*Source: The World Almanac*)

Answer

8. −10

A problem with integers in real-life settings arises when quantities are smaller than some integer but greater than the next smallest integer. On a number line, these quantities may be visualized by points between integers. Some of these quantities between integers can be represented as a quotient of integers. For example,

The point on the number line halfway between 0 and 1 can be represented by $\frac{1}{2}$, a quotient of integers.

The point on the number line halfway between 0 and -1 can be represented by $-\frac{1}{2}$. Other quotients of integers and their graphs are shown below.

$-\frac{7}{4}$ $-\frac{1}{2}$ $\frac{1}{2}$ $\frac{4}{3}$ $\frac{5}{2}$

−5 −4 −3 −2 −1 0 1 2 3 4 5

These numbers, each of which can be represented as a quotient of integers, are examples of **rational numbers**. It's not possible to list the set of rational numbers using the notation that we have been using. For this reason, we will use a different notation.

Rational Numbers

$$\left\{ \frac{a}{b} \middle| a \text{ and } b \text{ are integers and } b \neq 0 \right\}$$

We read this set as "the set of numbers $\frac{a}{b}$ such that a and b are integers and ***b* is not equal to 0**."

Helpful Hint

We commonly refer to rational numbers as fractions.

Notice that every integer is also a rational number since each integer can be written as a quotient of integers. For example, the integer 5 is also a rational number since $5 = \frac{5}{1}$. For the rational number $\frac{5}{1}$, recall that the top number, 5, is called the numerator and the bottom number, 1, is called the denominator.

Let's practice **graphing** numbers on a number line.

Practice 9

Graph the numbers on the number line.

$-2\frac{1}{2}, \ -\frac{2}{3}, \ \frac{1}{5}, \ \frac{5}{4}, \ 2.25$

−5 −4 −3 −2 −1 0 1 2 3 4 5

Example 9 Graph the numbers on a number line.

$-\frac{4}{3}, \ \frac{1}{4}, \ \frac{3}{2}, \ -2\frac{1}{8}, \ 3.5$

Solution: To help graph the improper fractions in the list, we first write them as mixed numbers.

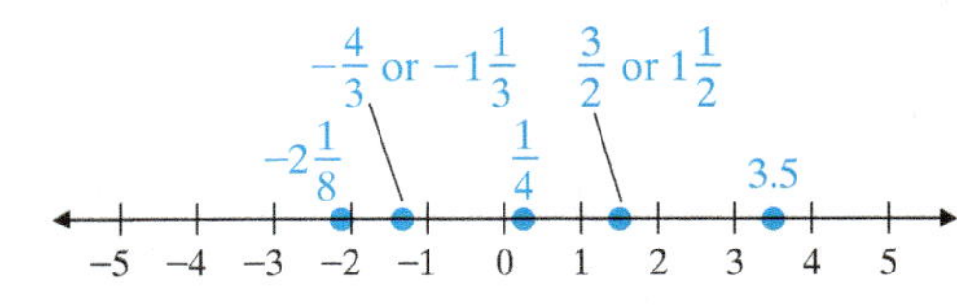

Work Practice 9

Every rational number has a point on the number line that corresponds to it. But not every point on the number line corresponds to a rational number. Those points that do not correspond to rational numbers correspond instead to **irrational numbers**.

Answer

9.

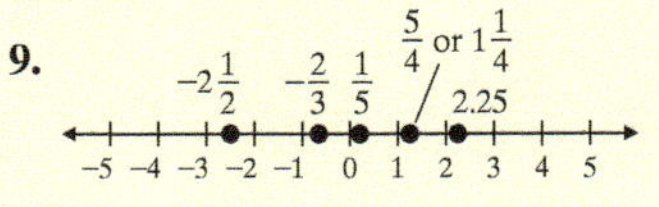

Irrational Numbers

{Nonrational numbers that correspond to points on a number line}

An irrational number that you have probably seen is π. Also, $\sqrt{2}$, the length of the diagonal of the square shown below, is an irrational number.

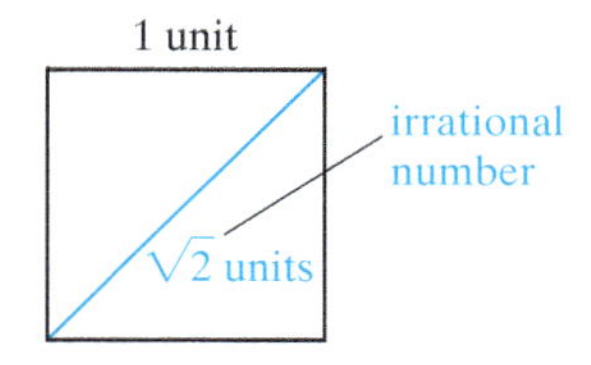

Both rational and irrational numbers can be written as decimal numbers. The decimal equivalent of a rational number will either terminate or repeat in a pattern. For example, upon dividing we find that

$$\frac{3}{4} = 0.75 \qquad \text{(Decimal number terminates, or ends.)}$$

$$\frac{2}{3} = 0.66666\ldots \qquad \text{(Decimal number repeats in a pattern.)}$$

The decimal representation of an irrational number will neither terminate nor repeat. (For further review of decimals, see Section R.3.)

The set of numbers, each of which corresponds to a point on a number line, is called the set of **real numbers**. One and only one point on a number line corresponds to each real number.

Real Numbers

{All numbers that correspond to points on a number line}

Several different sets of numbers have been discussed in this section. The following diagram shows the relationships among these sets of real numbers. Notice that, together, the rational numbers and the irrational numbers make up the real numbers.

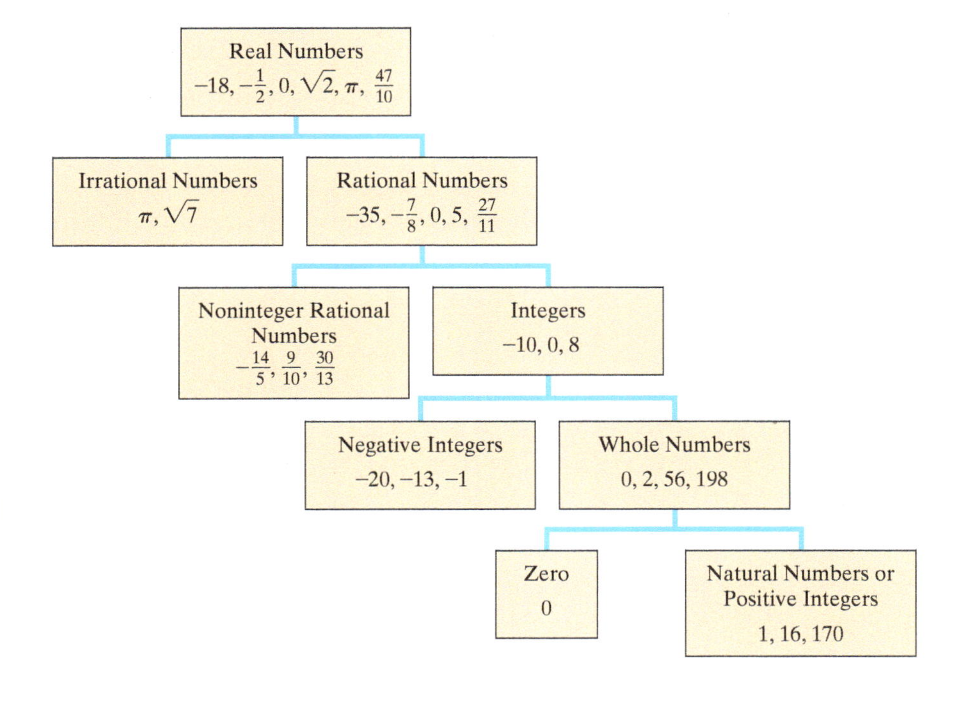

Now that other sets of numbers have been reviewed, let's continue our practice of comparing numbers.

Practice 10

Insert $<$, $>$, or $=$ between the pairs of numbers to form true statements.

a. $-11 \quad -9$ **b.** $4.511 \quad 4.151$

c. $\frac{7}{8} \quad \frac{2}{3}$

Example 10

Insert $<$, $>$, or $=$ between the pairs of numbers to form true statements.

a. $-5 \quad -6$ **b.** $3.195 \quad 3.2$ **c.** $\frac{1}{4} \quad \frac{1}{3}$

Solution:

a. $-5 > -6$ since -5 lies to the right of -6 on a number line.

b. By comparing digits in the same place values, we find that $3.195 < 3.2$, since $0.1 < 0.2$.

c. By dividing, we find that $\frac{1}{4} = 0.25$ and $\frac{1}{3} = 0.33\ldots$. Since $0.25 < 0.33\ldots$, $\frac{1}{4} < \frac{1}{3}$.

Work Practice 10

Practice 11

Given the set $\left\{-\frac{2}{5}, 0, \pi, -100, 6, 913\right\}$, list the numbers in this set that belong to the set of:

a. Natural numbers
b. Whole numbers
c. Integers
d. Rational numbers
e. Irrational numbers
f. Real numbers

Example 11

Given the set $\left\{-2, 0, \frac{1}{4}, 112, -3, 11, \sqrt{2}\right\}$, list the numbers in this set that belong to the set of:

a. Natural numbers **b.** Whole numbers **c.** Integers
d. Rational numbers **e.** Irrational numbers **f.** Real numbers

Solution:

a. The natural numbers are 11 and 112.
b. The whole numbers are 0, 11, and 112.
c. The integers are $-3, -2, 0, 11$, and 112.
d. Recall that integers are rational numbers also. The rational numbers are $-3, -2, 0, \frac{1}{4}, 11$, and 112.
e. The only irrational number is $\sqrt{2}$.
f. All numbers in the given set are real numbers.

Work Practice 11

Objective D Finding the Absolute Value of a Number

A number line not only gives us a picture of the real numbers, it also helps us visualize the distance between numbers. The distance between a real number a and 0 is given a special name called the **absolute value** of a. "The absolute value of a" is written in symbols as $|a|$.

Helpful Hint

Since $|a|$ is a distance, $|a|$ is always either positive or 0. It is never negative. That is, **for any real number** a, $|a| \geq 0$.

Absolute Value

The **absolute value** of a real number a, denoted by $|a|$, is the distance between a and 0 on a number line.

For example, $|3| = 3$ and $|-3| = 3$ since both 3 and -3 are a distance of 3 units from 0 on a number line.

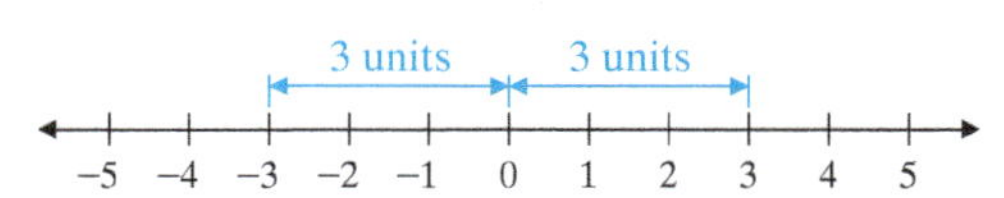

Answers

10. a. $<$ **b.** $>$ **c.** $>$ **11. a.** 6, 913 **b.** 0, 6, 913 **c.** $-100, 0, 6, 913$ **d.** $-100, -\frac{2}{5}, 0, 6, 913$ **e.** π **f.** all numbers in the given set

Example 12 Find the absolute value of each number.

a. $|4|$ **b.** $|-5|$ **c.** $|0|$

d. $\left|-\frac{2}{9}\right|$ **e.** $|4.93|$

Solution:

a. $|4| = 4$ since 4 is 4 units from 0 on a number line.

b. $|-5| = 5$ since -5 is 5 units from 0 on a number line.

c. $|0| = 0$ since 0 is 0 units from 0 on a number line.

d. $\left|-\frac{2}{9}\right| = \frac{2}{9}$

e. $|4.93| = 4.93$

Work Practice 12

Practice 12

Find the absolute value of each number.

a. $|7|$ **b.** $|-8|$ **c.** $\left|\frac{2}{3}\right|$

d. $|0|$ **e.** $|-3.06|$

Example 13 Insert $<$, $>$, or $=$ in the appropriate space to make each statement true.

a. $|0| \quad 2$ **b.** $|-5| \quad 5$ **c.** $|-3| \quad |-2|$

d. $|-9| \quad |-9.7|$ **e.** $\left|-7\frac{1}{6}\right| \quad |7|$

Solution:

a. $|0| < 2$ since $|0| = 0$ and $0 < 2$.

b. $|-5| = 5$.

c. $|-3| > |-2|$ since $3 > 2$.

d. $|-9| < |-9.7|$ since $9 < 9.7$.

e. $\left|-7\frac{1}{6}\right| > |7|$ since $7\frac{1}{6} > 7$.

Work Practice 13

Practice 13

Insert $<$, $>$, or $=$ in the appropriate space to make each statement true.

a. $|-4| \quad 4$

b. $-3 \quad |0|$

c. $|-2.7| \quad |-2|$

d. $|-6| \quad |-16|$

e. $|10| \quad \left|-10\frac{1}{3}\right|$

Answers

12. a. 7 **b.** 8 **c.** $\frac{2}{3}$ **d.** 0 **e.** 3.06

13. a. $=$ **b.** $<$ **c.** $>$ **d.** $<$ **e.** $<$

Vocabulary, Readiness & Video Check

Use the choices below to fill in each blank. Not all choices will be used.

real	natural	absolute value	$\frac{1}{2}$	$\frac{1}{4}$	$\|a\|$	whole
rational	inequality	integers	0	1	$\|-1\|$	

1. The ________ numbers are $\{0, 1, 2, 3, 4, \ldots\}$.

2. The ________ numbers are $\{1, 2, 3, 4, 5, \ldots\}$.

3. The symbols $\neq$, $\leq$, and $>$ are called ________ symbols.

4. The ________ are $\{\ldots, -3, -2, -1, 0, 1, 2, 3, \ldots\}$.

5. The ________ numbers are $\{$all numbers that correspond to points on a number line$\}$.

6. The ________ numbers are $\left\{\frac{a}{b} \,\middle|\, a \text{ and } b \text{ are integers}, b \neq 0\right\}$.

7. The integer ________ is neither positive nor negative.

8. The point on a number line halfway between 0 and $\frac{1}{2}$ can be represented by ____________.

9. The distance between a real number a and 0 is called the ____________ of a.

10. The absolute value of a is written in symbols as ____________.

Martin-Gay Interactive Videos

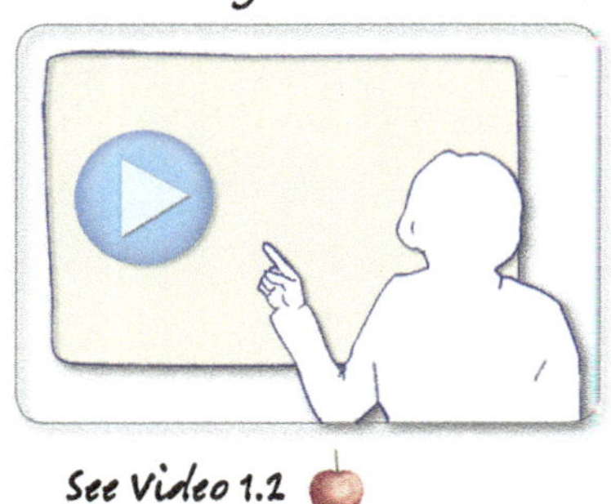

See Video 1.2

Watch the section lecture video and answer the following questions.

Objective A **11.** In Example 2, why is the symbol $<$ inserted between the two numbers?

Objective B **12.** Write the sentence given in Example 4 and translate it to a mathematical statement, using symbols.

Objective C **13.** Which sets of numbers does the number in Example 6 belong to? Why is this number not an irrational number?

Objective D **14.** Complete this statement based on the lecture given before Example 8. The ____________ of a real number a, denoted by $|a|$, is the distance between a and 0 on a number line.

1.2 Exercise Set MyLab Math 

Objectives A C Mixed Practice *Insert $<$, $>$, or $=$ in the space between the paired numbers to make each statement true. See Examples 1 through 6 and 10.*

1. 4   10

2. 8   5

3. 7   3

4. 9   15

5. 6.26   6.26

6. 1.13   1.13

7. 0   7

8. 20   0

9. The freezing point of water is 32° Fahrenheit. The boiling point of water is 212° Fahrenheit. Write an inequality statement using $<$ or $>$ comparing the numbers 32 and 212.

10. The freezing point of water is 0° Celsius. The boiling point of water is 100° Celsius. Write an inequality statement using $<$ or $>$ comparing the numbers 0 and 100.

△ **11.** An angle measuring 30° and an angle measuring 45° are shown. Write an inequality statement using $\le$ or $\ge$ comparing the numbers 30 and 45.

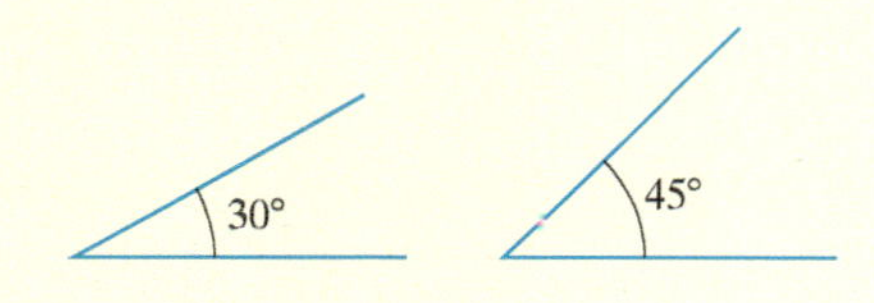

△ **12.** The sum of the measures of the angles of a parallelogram is 360°. The sum of the measures of the angles of a triangle is 180°. Write an inequality statement using $\le$ or $\ge$ comparing the numbers 360 and 180.

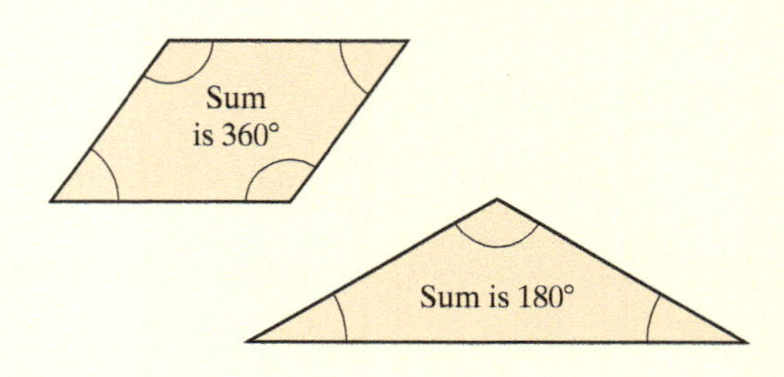

Determine whether each statement is true or false. See Examples 1 through 6 and 10.

13. $11 \le 11$

14. $8 \ge 9$

15. $-11 > -10$

16. $-16 > -17$

17. $5.092 < 5.902$

18. $1.02 > 1.021$

19. $\frac{9}{10} \le \frac{8}{9}$

20. $\frac{4}{5} \le \frac{9}{11}$

Rewrite each inequality so that the inequality symbol points in the opposite direction and the resulting statement has the same meaning as the given one. See Examples 1 through 6 and 10.

21. $25 \ge 20$

22. $-13 \le 13$

23. $0 < 6$

24. $5 > 3$

25. $-10 > -12$

26. $-4 < -2$

Objectives B C Mixed Practice—Translating *Write each sentence as a mathematical statement. See Example 7.*

27. Seven is less than eleven.

28. Twenty is greater than two.

29. Five is greater than or equal to four.

30. Negative ten is less than or equal to thirty-seven.

31. Fifteen is not equal to negative two.

32. Negative seven is not equal to seven.

Use integers to represent the value(s) in each statement. See Example 8.

33. The highest elevation in California is Mt. Whitney, with an altitude of 14,494 feet. The lowest elevation in California is Death Valley, with an altitude of 282 feet below sea level. (*Source:* U.S. Geological Survey)

34. Driskill Mountain, in Louisiana, has an altitude of 535 feet. New Orleans, Louisiana, lies 8 feet below sea level. (*Source:* U.S. Geological Survey)

35. The number of graduate students at the University of Florida at Gainesville was 17,813 fewer than the number of undergraduate students. (*Source:* University of Florida at Gainesville, 2017)

36. The number of students admitted to the class of 2020 at UCLA was 79,647 fewer students than the number that had applied. (*Source:* UCLA)

37. A community college student deposited $475 in her savings account. She later withdrew $195.

38. A deep-sea diver ascended 17 feet and later descended 15 feet.

Graph each set of numbers on the number line. See Example 9.

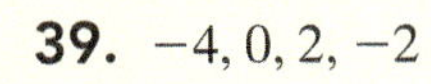

39. $-4, 0, 2, -2$

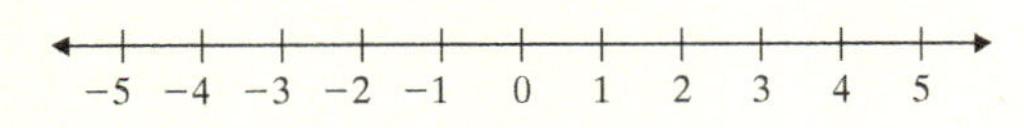

40. $-3, 0, 1, -5$

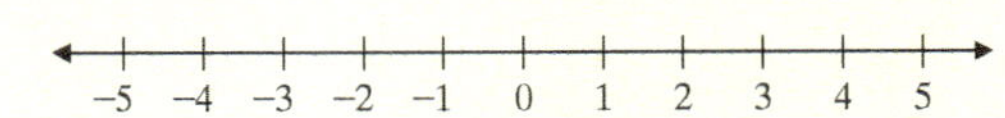

41. $-2, 4, \frac{1}{3}, -\frac{1}{4}$

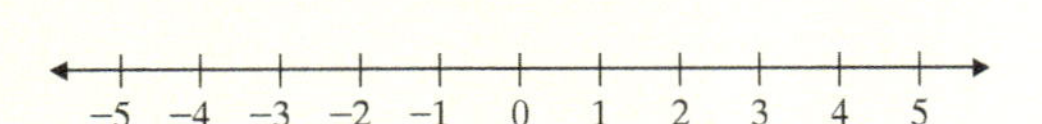

42. $-5, 3, -\frac{1}{3}, \frac{7}{8}$

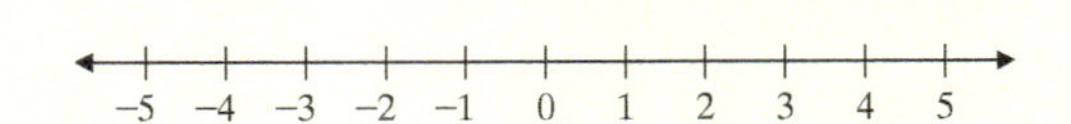

43. $-4.5, \frac{7}{4}, 3.25, -\frac{3}{2}$

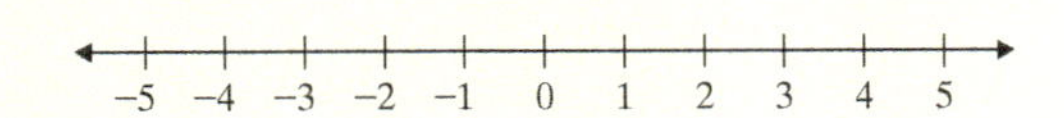

44. $4.5, -\frac{9}{4}, 1.75, -\frac{7}{2}$

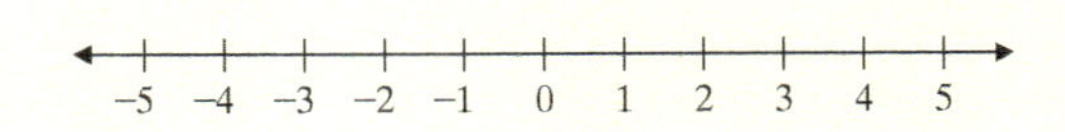

Tell which set or sets each number belongs to: natural numbers, whole numbers, integers, rational numbers, irrational numbers, or real numbers. See Example 11.

45. 0 **46.** $\frac{1}{4}$ **47.** -7 **48.** $-\frac{1}{7}$

49. 265 **50.** 7941 **51.** $\frac{2}{3}$ **52.** $\sqrt{3}$

Determine whether each statement is true or false.

53. Every rational number is also an integer.

54. Every natural number is positive.

55. 0 is a real number.

56. $\frac{1}{2}$ is an integer.

57. Every negative number is also a rational number.

58. Every rational number is also a real number.

59. Every real number is also a rational number.

60. Every whole number is an integer.

Objective D *Find each absolute value. See Example 12.*

61. $|8.9|$ **62.** $|11.2|$ **63.** $|-20|$ **64.** $|-17|$

65. $\left|\frac{9}{2}\right|$ **66.** $\left|\frac{10}{7}\right|$ **67.** $\left|-\frac{12}{13}\right|$ **68.** $\left|-\frac{1}{15}\right|$

Insert $<$, $>$, or $=$ in the appropriate space to make each statement true. See Examples 12 and 13.

69. $|-5| \quad -4$ **70.** $|-12| \quad |0|$ **71.** $\left|-\frac{5}{8}\right| \quad \left|\frac{5}{8}\right|$ **72.** $\left|\frac{2}{5}\right| \quad \left|-\frac{2}{5}\right|$

73. $|-2| \quad |-2.7|$ **74.** $|-5.01| \quad |-5|$ **75.** $|0| \quad |-8|$ **76.** $|-12| \quad \frac{-24}{2}$

Concept Extensions

The bar graph shows corn production from the top six corn-producing states. Use this graph to answer Exercises 77 through 80.

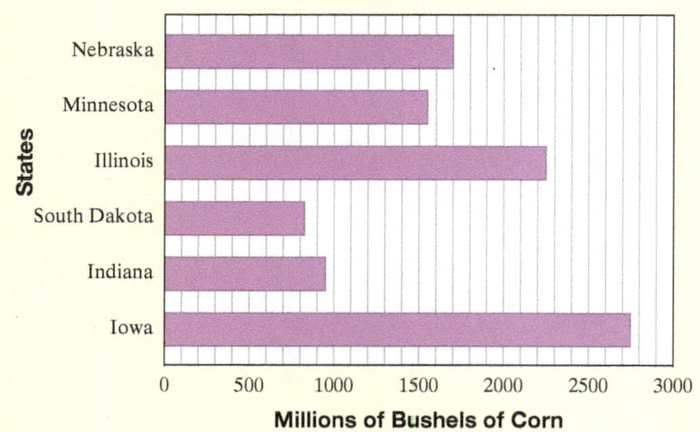

Source: U. S. Dept. of Agriculture

77. Write an inequality comparing the corn production in Illinois with the corn production in Iowa.

78. Write an inequality comparing the corn production in Minnesota with the corn production in South Dakota.

79. Determine the difference between the corn production in Nebraska and the corn production in Illinois.

80. Determine the difference between the corn production in Indiana and the corn production in Minnesota.

The apparent magnitude of a star is the measure of its brightness as seen by someone on Earth. The smaller the apparent magnitude, the brighter the star. Below, the apparent magnitudes of some stars are listed. Use this table to answer Exercises 81 through 86.

Star	Apparent Magnitude	Star	Apparent Magnitude
Arcturus	−0.04	Spica	0.98
Sirius	−1.46	Rigel	0.12
Vega	0.03	Regulus	1.35
Antares	0.96	Canopus	−0.72
Sun (Sol)	−26.7	Hadar	0.61

(*Source: Norton's 2000.0: Star Atlas and Reference Handbooks*, 18th ed., Longman Group, UK, 1989)

81. The apparent magnitude of the sun is −26.7. The apparent magnitude of the star Arcturus is −0.04. Write an inequality statement comparing the numbers −0.04 and −26.7.

82. The apparent magnitude of Antares is 0.96. The apparent magnitude of Spica is 0.98. Write an inequality statement comparing the numbers 0.96 and 0.98.

83. Which is brighter, the sun or Arcturus?

84. Which is dimmer, Antares or Spica?

85. Which star listed is the brightest?

86. Which star listed is the dimmest?

87. In your own words, explain how to find the absolute value of a number.

88. Give an example of a real-life situation that can be described with integers but not with whole numbers.

1.3 Exponents, Order of Operations, and Variable Expressions

Objective A Exponents and the Order of Operations

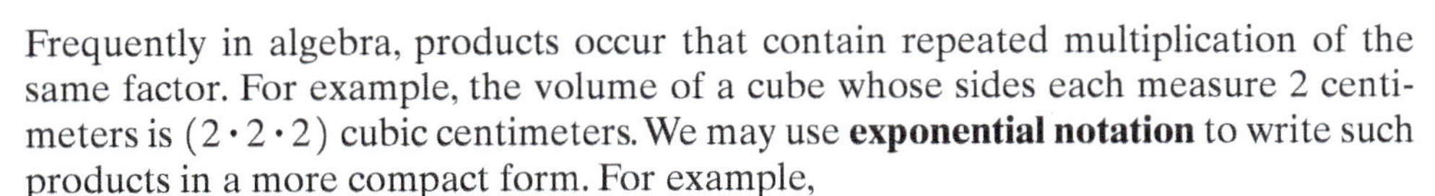

Frequently in algebra, products occur that contain repeated multiplication of the same factor. For example, the volume of a cube whose sides each measure 2 centimeters is $(2 \cdot 2 \cdot 2)$ cubic centimeters. We may use **exponential notation** to write such products in a more compact form. For example,

$2 \cdot 2 \cdot 2$ may be written as 2^3.

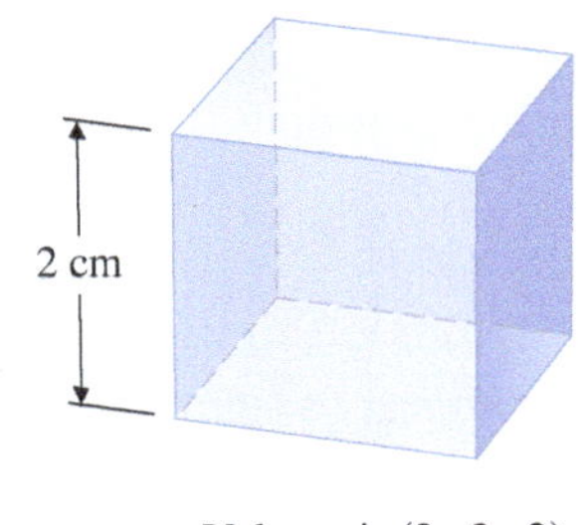

Volume is $(2 \cdot 2 \cdot 2)$ cubic centimeters.

The 2 in 2^3 is called the **base**; it is the repeated factor. The 3 in 2^3 is called the **exponent** and is the number of times the base is used as a factor. The expression 2^3 is called an **exponential expression**.

exponent

$$2^3 = 2 \cdot 2 \cdot 2 = 8$$

base — 2 is a factor 3 times.

Objectives

A Define and Use Exponents and the Order of Operations.

B Evaluate Algebraic Expressions Given Replacement Values for Variables.

C Determine Whether a Number Is a Solution of a Given Equation.

D Translate Phrases into Expressions and Sentences into Equations.

Practice 1

Evaluate each expression.

a. 4^2
b. 2^2
c. 3^4
d. 9^1
e. $\left(\frac{2}{5}\right)^3$
f. $(0.8)^2$

Helpful Hint

$2^3 \neq 2 \cdot 3$ since 2^3 indicates **repeated multiplication of the same factor.**

$$2^3 = 2 \cdot 2 \cdot 2 = 8,$$

whereas $2 \cdot 3 = 6$

Example 1 Evaluate (find the value of) each expression.

a. 3^2 [read as "3 squared" or as "3 to the second power"]
b. 5^3 [read as "5 cubed" or as "5 to the third power"]
c. 2^4 [read as "2 to the fourth power"]
d. 7^1
e. $\left(\frac{3}{7}\right)^2$
f. $(0.6)^2$

Solution:

a. $3^2 = 3 \cdot 3 = 9$
b. $5^3 = 5 \cdot 5 \cdot 5 = 125$
c. $2^4 = 2 \cdot 2 \cdot 2 \cdot 2 = 16$
d. $7^1 = 7$
e. $\left(\frac{3}{7}\right)^2 = \left(\frac{3}{7}\right)\left(\frac{3}{7}\right) = \frac{3 \cdot 3}{7 \cdot 7} = \frac{9}{49}$
f. $(0.6)^2 = (0.6)(0.6) = 0.36$

Work Practice 1

Using symbols for mathematical operations is a great convenience. The more operation symbols presented in an expression, the more careful we must be when performing the indicated operation. For example, in the expression $2 + 3 \cdot 7$, do we add first or multiply first? To eliminate confusion, **grouping symbols** are used. Examples of grouping symbols are parentheses ( ), brackets [ ], braces { }, absolute value bars | |, and the fraction bar. If we wish $2 + 3 \cdot 7$ to be simplified by adding first, we enclose $2 + 3$ in parentheses.

$$(2 + 3) \cdot 7 = 5 \cdot 7 = 35$$

If we wish to multiply first, $3 \cdot 7$ may be enclosed in parentheses.

$$2 + (3 \cdot 7) = 2 + 21 = 23$$

To eliminate confusion when no grouping symbols are present, we use the following agreed-upon order of operations.

Order of Operations

1. Perform all operations within grouping symbols first, starting with the innermost set.
2. Evaluate exponential expressions.
3. Multiply or divide in order from left to right.
4. Add or subtract in order from left to right.

Using this order of operations, we now simplify $2 + 3 \cdot 7$. There are no grouping symbols and no exponents, so we multiply and then add.

$$2 + 3 \cdot 7 = 2 + 21 \quad \text{Multiply.}$$
$$= 23 \quad \text{Add.}$$

Answers

1. a. 16 **b.** 4 **c.** 81 **d.** 9 **e.** $\frac{8}{125}$ **f.** 0.64

Examples Simplify each expression.

2. $6 \div 3 + 5^2 = 6 \div 3 + 25$ Evaluate 5^2

$= 2 + 25$ Divide.

$= 27$ Add.

3. $20 \div 5 \cdot 4 = 4 \cdot 4$

$= 16$

Helpful Hint Remember to multiply or divide in order from left to right.

4. $\frac{3}{2} \cdot \frac{1}{2} - \frac{1}{2} = \frac{3}{4} - \frac{1}{2}$ Multiply.

$= \frac{3}{4} - \frac{2}{4}$ The least common denominator is 4.

$= \frac{1}{4}$ Subtract.

5. $1 + 2[5(2 \cdot 3 + 1) - 10] = 1 + 2[5(7) - 10]$ Simplify the expression in the innermost set of parentheses. $2 \cdot 3 + 1 = 6 + 1 = 7$.

$= 1 + 2[35 - 10]$ Multiply 5 and 7.

$= 1 + 2[25]$ Subtract inside the brackets.

$= 1 + 50$ Multiply 2 and 25.

$= 51$ Add.

Work Practice 2–5

Practice 2–5

Simplify each expression.

2. $3 \cdot 2 + 4^2$

3. $28 \div 7 \cdot 2$

4. $\frac{9}{5} \cdot \frac{1}{3} - \frac{1}{3}$

5. $5 + 3[2(3 \cdot 4 + 1) - 20]$

In the next example, the fraction bar serves as a grouping symbol and separates the numerator and denominator. Simplify each separately.

Example 6 Simplify: $\frac{3 + |4 - 3| + 2^2}{6 - 3}$

Solution:

$\frac{3 + |4 - 3| + 2^2}{6 - 3} = \frac{3 + |1| + 2^2}{6 - 3}$ Simplify the expression inside the absolute value bars.

$= \frac{3 + 1 + 2^2}{3}$ Find the absolute value and simplify the denominator.

$= \frac{3 + 1 + 4}{3}$ Evaluate the exponential expression.

$= \frac{8}{3}$ Simplify the numerator.

Work Practice 6

Practice 6

Simplify: $\frac{1 + |7 - 4| + 3^2}{8 - 5}$

Helpful Hint

Be careful when evaluating an exponential expression.

$3 \cdot 4^2 = 3 \cdot 16 = 48$ — Base is 4.

$(3 \cdot 4)^2 = (12)^2 = 144$ — Base is $3 \cdot 4$.

Answers

2. 22 **3.** 8 **4.** $\frac{4}{15}$ **5.** 23 **6.** $\frac{13}{3}$

Objective B Evaluating Algebraic Expressions 

Recall that letters used to represent quantities are called **variables**. An **algebraic expression** is a collection of numbers, variables, operation symbols, and grouping symbols. For example,

$$2x, \quad -3, \quad 2x - 10, \quad 5(p^2 + 1), \quad xy, \quad \text{and} \quad \frac{3y^2 - 6y + 1}{5}$$

are algebraic expressions.

Expressions	Meaning
$2x$	$2 \cdot x$
$5(p^2 + 1)$	$5 \cdot (p^2 + 1)$
$3y^2$	$3 \cdot y^2$
xy	$x \cdot y$

If we give a specific value to a variable, we can **evaluate an algebraic expression**. To evaluate an algebraic expression means to find its numerical value once we know the values of the variables.

Algebraic expressions are often used in problem solving. For example, the expression

$$16t^2$$

gives the distance in feet (neglecting air resistance) that an object will fall in t seconds.

Practice 7

Evaluate each expression when $x = 1$ and $y = 4$.

a. $3y^2$

b. $2y - x$

c. $\frac{11x}{3y}$

d. $\frac{x}{y} + \frac{6}{y}$

e. $y^2 - x^2$

Example 7

Evaluate each expression when $x = 3$ and $y = 2$.

a. $5x^2$ **b.** $2x - y$ **c.** $\frac{3x}{2y}$ **d.** $\frac{x}{y} + \frac{y}{2}$ **e.** $x^2 - y^2$

Solution:

a. Replace x with 3. Then simplify.

$$5x^2 = 5 \cdot (3)^2 = 5 \cdot 9 = 45$$

b. Replace x with 3 and y with 2. Then simplify.

$$\begin{aligned} 2x - y &= 2(3) - 2 && \text{Let } x = 3 \text{ and } y = 2. \\ &= 6 - 2 && \text{Multiply.} \\ &= 4 && \text{Subtract.} \end{aligned}$$

c. Replace x with 3 and y with 2. Then simplify.

$$\frac{3x}{2y} = \frac{3 \cdot 3}{2 \cdot 2} = \frac{9}{4} \quad \text{Let } x = 3 \text{ and } y = 2.$$

d. Replace x with 3 and y with 2. Then simplify.

$$\frac{x}{y} + \frac{y}{2} = \frac{3}{2} + \frac{2}{2} = \frac{5}{2}$$

e. Replace x with 3 and y with 2. Then simplify.

$$x^2 - y^2 = 3^2 - 2^2 = 9 - 4 = 5$$

Work Practice 7

Answers

7. **a.** 48 **b.** 7 **c.** $\frac{11}{12}$ **d.** $\frac{7}{4}$ **e.** 15

Objective C Solutions of Equations 

Many times a problem-solving situation is modeled by an equation. An **equation** is a mathematical statement that two expressions have equal value. The equal symbol "=" is used to equate the two expressions. For example,

$$3 + 2 = 5,\ 7x = 35,\ \frac{2(x - 1)}{3} = 0, \text{ and } I = PRT \text{ are all equations.}$$

Helpful Hint

An equation contains the equal symbol "=". An algebraic expression does not.

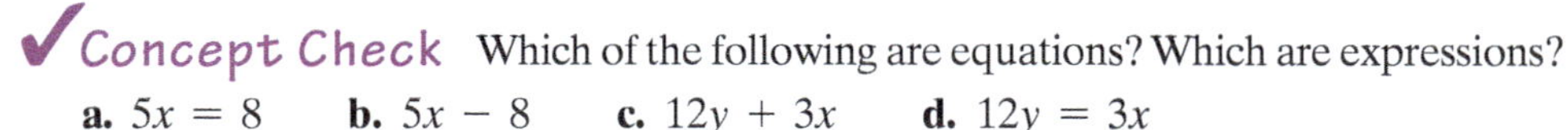

✓Concept Check Which of the following are equations? Which are expressions?

a. $5x = 8$ **b.** $5x - 8$ **c.** $12y + 3x$ **d.** $12y = 3x$

When an equation contains a variable, deciding which value(s) of the variable make the equation a true statement is called **solving** the equation for the variable. A **solution** of an equation is a value for the variable that makes the equation a true statement. For example, 3 is a solution of the equation $x + 4 = 7$ because if x is replaced with 3 the statement is true.

$$x + 4 = 7$$
$$\downarrow$$
$$3 + 4 \stackrel{?}{=} 7 \quad \text{Replace } x \text{ with 3.}$$
$$7 = 7 \quad \text{True}$$

Similarly, 1 is not a solution of the equation $x + 4 = 7$ because $1 + 4 = 7$ is **not** a true statement.

Example 8 Decide whether 2 is a solution of $3x + 10 = 8x$.

Solution: Replace x with 2 and see if a true statement results.

$$3x + 10 = 8x \quad \text{Original equation}$$
$$3(2) + 10 \stackrel{?}{=} 8(2) \quad \text{Replace } x \text{ with 2.}$$
$$6 + 10 \stackrel{?}{=} 16 \quad \text{Simplify each side.}$$
$$16 = 16 \quad \text{True}$$

Since we arrived at a true statement after replacing x with 2 and simplifying both sides of the equation, 2 is a solution of the equation.

Work Practice 8

Practice 8

Decide whether 3 is a solution of $5x - 10 = x + 2$.

Objective D Translating Words to Symbols 

Now that we know how to represent an unknown number by a variable, let's practice translating phrases into algebraic expressions (no "=" symbol) and sentences into equations (with "=" symbol). Oftentimes solving problems involves the ability to translate word phrases and sentences into symbols. A list of key words and phrases to help us translate is on the next page.

Answer

8. It is a solution.

✓Concept Check Answer

equations: **a, d;** expressions: **b, c**

Helpful Hint

Order matters when subtracting and also dividing, so be especially careful with these translations.

Addition (+)	Subtraction (−)	Multiplication (·)	Division (÷)	Equality (=)
Sum	Difference of	Product	Quotient	Equals
Plus	Minus	Times	Divide	Gives
Added to	Subtracted from	Multiply	Into	Is/was/should be
More than	Less than	Twice	Ratio	Yields
Increased by	Decreased by	Of	Divided by	Amounts to
Total	Less			Represents
				Is the same as

Practice 9

Write an algebraic expression that represents each phrase. Let the variable x represent the unknown number.

a. The product of 5 and a number

b. A number added to 7

c. A number divided by 11.2

d. A number subtracted from 8

e. Twice a number, plus 1

Example 9

Write an algebraic expression that represents each phrase. Let the variable x represent the unknown number.

a. The sum of a number and 3

b. The product of 3 and a number

c. The quotient of 7.3 and a number

d. 10 decreased by a number

e. 5 times a number, increased by 7

Solution:

a. $x + 3$ since "sum" means to add

b. $3 \cdot x$ and $3x$ are both ways to denote the product of 3 and x

c. $7.3 \div x$ or $\frac{7.3}{x}$

d. $10 - x$ because "decreased by" means to subtract

e. $\underbrace{5x}_{\text{5 times a number}} + 7$

Work Practice 9

Helpful Hint

Make sure you understand the difference when translating phrases containing "decreased by," "subtracted from," and "less than."

Phrase	Translation	
A number decreased by 10	$x - 10$	Notice the order.
A number subtracted from 10	$10 - x$	
10 less than a number	$x - 10$	
A number less 10	$x - 10$	

Now let's practice translating sentences into equations.

Answers

9. a. $5 \cdot x$ or $5x$ **b.** $7 + x$ **c.** $x \div 11.2$ or $\frac{x}{11.2}$ **d.** $8 - x$ **e.** $2x + 1$

Example 10 Write each sentence as an equation. Let x represent the unknown number.

a. The quotient of 15 and a number is 4.
b. Three subtracted from 12 is a number.
c. 17 added to four times a number is 21.

Solution:

a. In words: the quotient of 15 and a number | is | 4
↓ ↓ ↓
Translate: $\frac{15}{x} = 4$

b. In words: three subtracted **from** 12 | is | a number
↓ ↓ ↓
Translate: $12 - 3 = x$

Care must be taken when the operation is subtraction. The expression $3 - 12$ would be incorrect. Notice that $3 - 12 \neq 12 - 3$.

c. In words: 17 | added to | four times a number | is | 21
↓ ↓ ↓ ↓ ↓
Translate: $17 + 4x = 21$

Work Practice 10

Practice 10

Write each sentence as an equation. Let x represent the unknown number.

a. The ratio of a number and 6 is 24.
b. The difference of 10 and a number is 18.
c. One less than twice a number is 99.

Answers

10. a. $\frac{x}{6} = 24$ **b.** $10 - x = 18$ **c.** $2x - 1 = 99$

Calculator Explorations Exponents

To evaluate exponential expressions on a calculator, find the key marked $\boxed{y^x}$ or $\boxed{\wedge}$. To evaluate, for example, 6^5, press the following keys: $\boxed{6}\ \boxed{y^x}\ \boxed{5}\ \boxed{=}$ or $\boxed{6}\ \boxed{\wedge}\ \boxed{5}\ \boxed{=}$.

↕ or

$\boxed{\text{ENTER}}$

The display should read $\boxed{7776}$

Order of Operations

Some calculators follow the order of operations, and others do not. To see whether or not your calculator has the order of operations built in, use your calculator to find $2 + 3 \cdot 4$. To do this, press the following sequence of keys:

$\boxed{2}\ \boxed{+}\ \boxed{3}\ \boxed{\times}\ \boxed{4}\ \boxed{=}$

↕ or

$\boxed{\text{ENTER}}$

The correct answer is 14 because the order of operations is to multiply before we add. If the calculator displays $\boxed{14}$, then it has the order of operations built in.

Even if the order of operations is built in, parentheses must sometimes be inserted. For example, to simplify $\frac{5}{12 - 7}$, press the keys

$\boxed{5}\ \boxed{\div}\ \boxed{(}\ \boxed{1}\ \boxed{2}\ \boxed{-}\ \boxed{7}\ \boxed{)}\ \boxed{=}$.

↕ or

$\boxed{\text{ENTER}}$

The display should read $\boxed{1}$.

Use a calculator to evaluate each expression.

1. 5^3
2. 7^4
3. 9^5
4. 8^6
5. $2(20 - 5)$
6. $3(14 - 7) + 21$
7. $24(862 - 455) + 89$
8. $99 + (401 + 962)$
9. $\frac{4623 + 129}{36 - 34}$
10. $\frac{956 - 452}{89 - 86}$

Vocabulary, Readiness & Video Check

Use the choices below to fill in each blank. Some choices may be used more than once.

addition	multiplication	exponent	expression	solution	evaluating the expression	
subtraction	division	base	equation	variable(s)	true	false

1. In 2^5, the 2 is called the ______________ and the 5 is called the ______________.
2. True or false: 2^5 means $2 \cdot 5$. ______________.
3. To simplify $8 + 2 \cdot 6$, which operation should be performed first? ______________
4. To simplify $(8 + 2) \cdot 6$, which operation should be performed first? ______________
5. To simplify $9(3 - 2) \div 3 + 6$, which operation should be performed first? ______________
6. To simplify $8 \div 2 \cdot 6$, which cperation should be performed first? ______________
7. A combination of operations on letters (variables) and numbers is a(n) ______________.
8. A letter that represents a number is a(n) ______________.
9. $3x - 2y$ is called a(n) ______________ and the letters x and y are ______________.
10. Replacing a variable in an expression by a number and then finding the value of the expression is called ______________.
11. A statement of the form "expression = expression" is called a(n) ______________.
12. A value for the variable that makes the equation a true statement is called a(n) ______________.

Martin-Gay Interactive Videos

See Video 1.3

Watch the section lecture video and answer the following questions.

Objective A 13. In Example 3 and the lecture before, what is the main point made about the order of operations?

Objective B 14. What happens with the replacement value for z in Example 6 and why?

Objective C 15. Is the value 0 a solution of the equation given in Example 9? How is this determined?

Objective D 16. Earlier in this video it was noted that equations have =, while expressions do not. In the lecture before Example 10, translating from English to math is discussed and another difference between expressions and equations is explained. What is it?

1.3 Exercise Set MyLab Math 

Objective A *Evaluate. See Example 1.*

1. 3^5
2. 5^4
3. 3^3
4. 4^4
5. 1^5
6. 1^8
7. 5^1
8. 8^1
9. 7^2
10. 9^2
11. $\left(\frac{2}{3}\right)^4$
12. $\left(\frac{6}{11}\right)^2$
13. $\left(\frac{1}{5}\right)^3$
14. $\left(\frac{1}{2}\right)^5$
15. $(1.2)^2$
16. $(1.5)^2$
17. $(0.7)^3$
18. $(0.4)^3$

△ **19.** The area of a square whose sides each measure 5 meters is $(5 \cdot 5)$ square meters. Write this area using exponential notation.

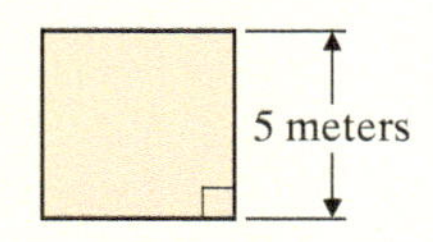

△ **20.** The area of a circle whose radius is 9 meters is $(9 \cdot 9 \cdot \pi)$ square meters. Write this area using exponential notation.

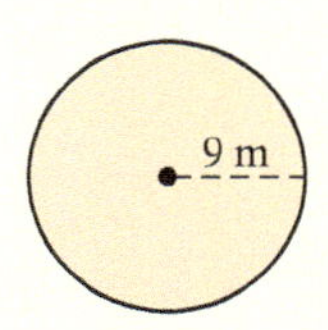

Simplify each expression. See Examples 2 through 6.

21. $5 + 6 \cdot 2$

22. $8 + 5 \cdot 3$

23. $4 \cdot 8 - 6 \cdot 2$

24. $12 \cdot 5 - 3 \cdot 6$

25. $18 \div 3 \cdot 2$

26. $48 \div 6 \cdot 2$

27. $2 + (5 - 2) + 4^2$

28. $6 - 2 \cdot 2 + 2^5$

29. $5 \cdot 3^2$

30. $2 \cdot 5^2$

31. $\frac{1}{4} \cdot \frac{2}{3} - \frac{1}{6}$

32. $\frac{3}{4} \cdot \frac{1}{2} + \frac{2}{3}$

33. $\frac{6 - 4}{9 - 2}$

34. $\frac{8 - 5}{24 - 20}$

35. $2[5 + 2(8 - 3)]$

36. $3[4 + 3(6 - 4)]$

37. $\frac{19 - 3 \cdot 5}{6 - 4}$

38. $\frac{14 - 2 \cdot 3}{12 - 8}$

39. $\frac{|6 - 2| + 3}{8 + 2 \cdot 5}$

40. $\frac{15 - |3 - 1|}{12 - 3 \cdot 2}$

41. $\frac{4 + 3(5 + 3)}{3^2 + 1}$

42. $\frac{3 + 6(8 - 5)}{4^2 + 2}$

43. $\frac{6 + |8 - 2| + 3^2}{18 - 3}$

44. $\frac{16 + |13 - 5| + 4^2}{17 - 5}$

45. $2 + 3[10(4 \cdot 5 - 16) - 30]$

46. $3 + 4[9(5 \cdot 5 - 20) - 41]$

47. $\left(\frac{2}{3}\right)^3 + \frac{1}{9} + \frac{1}{3} \cdot \frac{4}{3}$

48. $\left(\frac{3}{8}\right)^2 + \frac{1}{4} + \frac{1}{8} \cdot \frac{3}{2}$

Objective B *Evaluate each expression when $x = 1$, $y = 3$, and $z = 5$. See Example 7.*

49. $3y$

50. $4x$

51. $\frac{z}{5x}$

52. $\frac{y}{2z}$

53. $3x - 2$

54. $6y - 8$

55. $|2x + 3y|$

56. $|5z - 2y|$

57. $xy + z$

58. $yz - x$

59. $5y^2$

60. $2z^2$

Evaluate each expression when $x = 12$, $y = 8$, and $z = 4$. See Example 7.

61. $\frac{x}{z} + 3y$

62. $\frac{y}{z} + 8x$

63. $x^2 - 3y + x$

64. $y^2 - 3x + y$

65. $\frac{x^2 + z}{y^2 + 2z}$

66. $\frac{y^2 + x}{x^2 + 3y}$

Objective C *Decide whether the given number is a solution of the given equation. See Example 8.*

67. $3x - 6 = 9; 5$

68. $2x + 7 = 3x; 6$

69. $2x + 6 = 5x - 1; 0$

70. $4x + 2 = x + 8; 2$

71. $2x - 5 = 5; 8$

72. $3x - 10 = 8; 6$

73. $x + 6 = x + 6; 2$

74. $x + 6 = x + 6; 10$

75. $x = 5x + 15; 0$

76. $4 = 1 - x; 1$

77. $\frac{1}{3}x = 9; 27$

78. $\frac{2}{7}x = \frac{3}{14}; 6$

Objective D *Write each phrase as an algebraic expression. Let x represent the unknown number. See Example 9.*

79. Fifteen more than a number

80. A number increased by 9

81. Five subtracted from a number

82. Five decreased by a number

83. The ratio of a number and 4

84. The quotient of a number and 9

85. Three times a number, increased by 22

86. Twice a number, decreased by 72

Write each sentence as an equation or inequality. Use x to represent any unknown number. See Example 10.

87. One increased by two equals the quotient of nine and three.

88. Four subtracted from eight is equal to two squared.

89. Three is not equal to four divided by two.

90. The difference of sixteen and four is greater than ten.

91. The sum of 5 and a number is 20.

92. Seven subtracted from a number is 0.

93. The product of 7.6 and a number is 17.

94. 9.1 times a number equals 4

95. Thirteen minus three times a number is 13.

96. Eight added to twice a number is 42.

Concept Extensions

97. Are parentheses necessary in the expression $2 + (3 \cdot 5)$? Explain your answer.

98. Are parentheses necessary in the expression $(2 + 3) \cdot 5$? Explain your answer.

For Exercises 99 and 100, match each expression in the first column with its value in the second column.

99. **a.** $(6 + 2) \cdot (5 + 3)$ 19
b. $(6 + 2) \cdot 5 + 3$ 22
c. $6 + 2 \cdot 5 + 3$ 64
d. $6 + 2 \cdot (5 + 3)$ 43

100. **a.** $(1 + 4) \cdot 6 - 3$ 15
b. $1 + 4 \cdot (6 - 3)$ 13
c. $1 + 4 \cdot 6 - 3$ 27
d. $(1 + 4) \cdot (6 - 3)$ 22

△ *Recall that perimeter measures the distance around a plane figure and area measures the amount of surface of a plane figure. The expression $2l + 2w$ gives the perimeter of the rectangle below (measured in units), and the expression lw gives its area (measured in square units). Complete the chart below for the given lengths and widths. Be sure to include units.*

	Length: l	**Width: w**	**Perimeter of Rectangle: $2l + 2w$**	**Area of Rectangle: lw**
101.	4 in.	3 in.		
102.	6 in.	1 in.		
103.	5.3 in.	1.7 in.		
104.	4.6 in.	2.4 in.		

105. Study the perimeters and areas found in the chart to the left. Do you notice any trends?

106. In your own words, explain the difference between an expression and an equation.

107. Insert one set of parentheses so that the following expression simplifies to 32.

$$20 - 4 \cdot 4 \div 2$$

108. Insert one set of parentheses so that the following expression simplifies to 28.

$$2 \cdot 5 + 3^2$$

Determine whether each is an expression or an equation. See the Concept Check in this section.

109. **a.** $5x + 6$
b. $2a = 7$
c. $3a + 2 = 9$
d. $4x + 3y - 8z$
e. $5^2 - 2(6 - 2)$

110. **a.** $3x^2 - 26$
b. $3x^2 - 26 = 1$
c. $2x - 5 = 7x - 5$
d. $9y + x - 8$
e. $3^2 - 4(5 - 3)$

△ **111.** Why is 4^3 usually read as "four cubed"? (*Hint:* What is the volume of the **cube** below?)

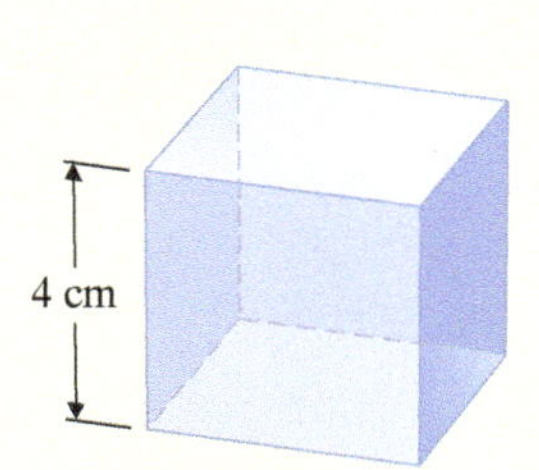

△ **112.** Why is 8^2 usually read as "eight squared"? (*Hint:* What is the area of the **square** below?)

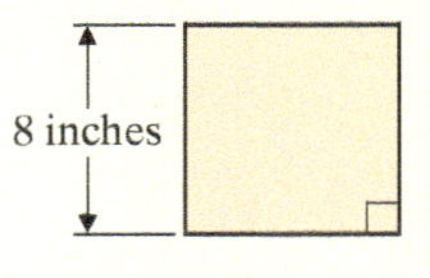

113. Write any expression, using 3 or more numbers, that simplifies to -11.

114. Write any expression, using 4 or more numbers, that simplifies to 7.

1.4 Adding Real Numbers

Objectives

A Add Real Numbers.

B Find the Opposite of a Number.

C Evaluate Algebraic Expressions Using Real Numbers.

D Solve Applications That Involve Addition of Real Numbers.

Real numbers can be added, subtracted, multiplied, divided, and raised to powers, just as whole numbers can.

Objective A Adding Real Numbers

Adding real numbers can be visualized by using a number line. A positive number can be represented on a number line by an arrow of appropriate length pointing to the right, and a negative number by an arrow of appropriate length pointing to the left.

Both arrows represent 2 or +2.

They both point to the right, and they are both 2 units long.

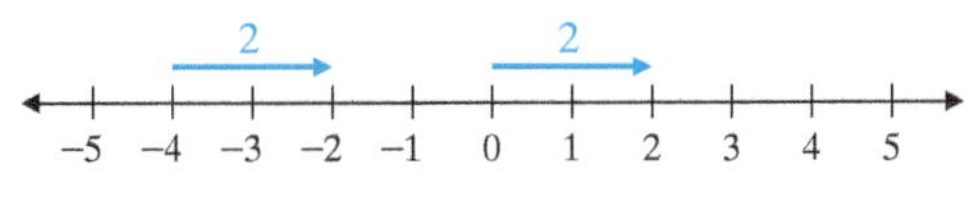

Both arrows represent −3.

They both point to the left, and they are both 3 units long.

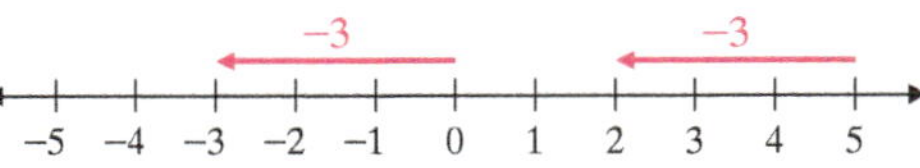

To add signed numbers such as $5 + (-2)$ on a number line, we start at 0 on a number line and draw an arrow representing 5. From the tip of this arrow, we draw another arrow representing −2. The tip of the second arrow ends at their sum, 3.

$$5 + (-2) = 3$$

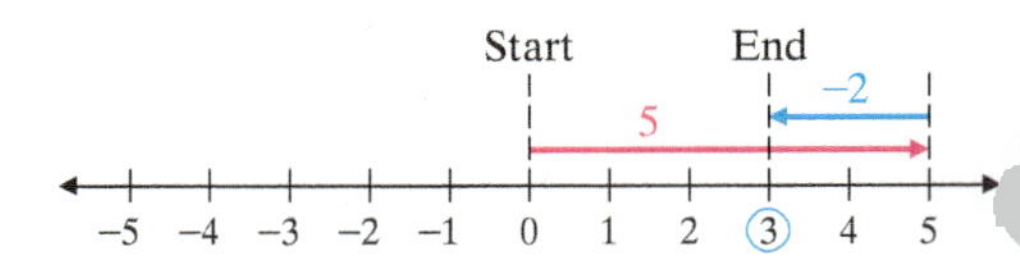

To add $-1 + (-4)$ on a number line, we start at 0 and draw an arrow representing −1. From the tip of this arrow, we draw another arrow representing −4. The tip of the second arrow ends at their sum, −5.

$$-1 + (-4) = -5$$

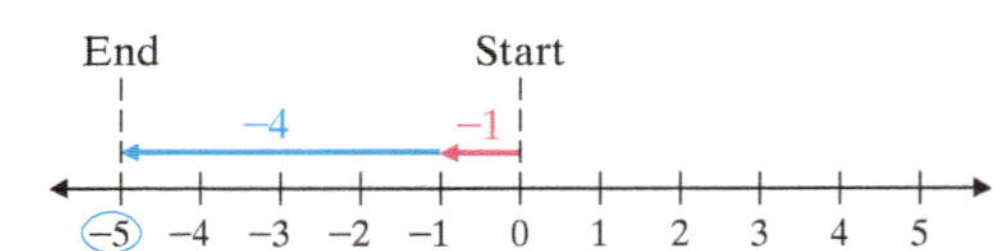

Practice 1

Add using a number line:

$-2 + (-4)$

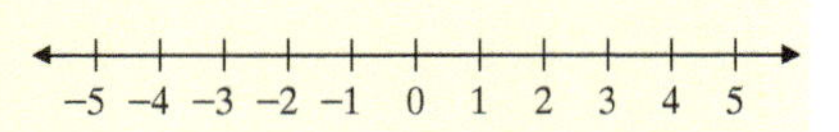

Example 1 Add: $-1 + (-2)$

Solution:

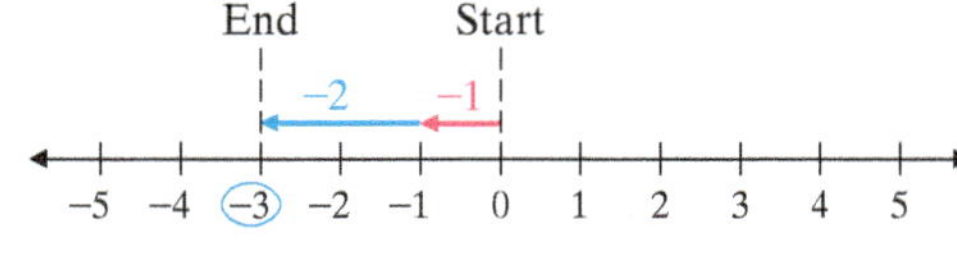

$$-1 + (-2) = -3$$

Work Practice 1

Thinking of integers as money earned or lost might help make addition more meaningful. Earnings can be thought of as positive numbers. If \$1 is earned and later another \$3 is earned, the total amount earned is \$4. In other words, $1 + 3 = 4$.

On the other hand, losses can be thought of as negative numbers. If \$1 is lost and later another \$3 is lost, a total of \$4 is lost. In other words, $(-1) + (-3) = -4$.

In Example 1, we added numbers with the same sign. Adding numbers whose signs are not the same can be pictured on a number line also.

Answer

1. −6

Example 2 Add: $-4 + 6$

Solution:

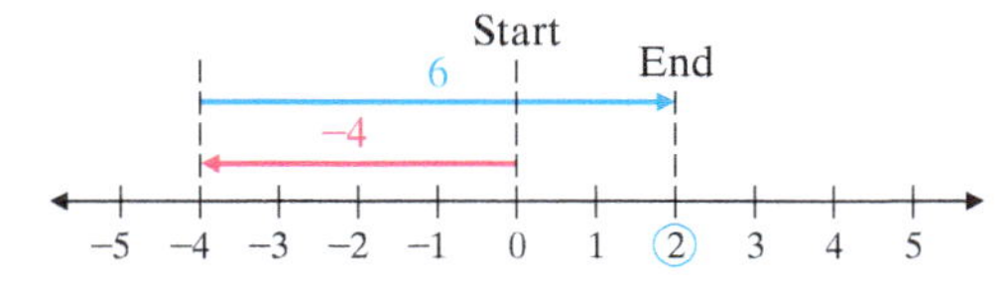

$$-4 + 6 = 2$$

■ Work Practice 2

Practice 2

Add using a number line:
$-5 + 8$

−5 −4 −3 −2 −1 0 1 2 3 4 5

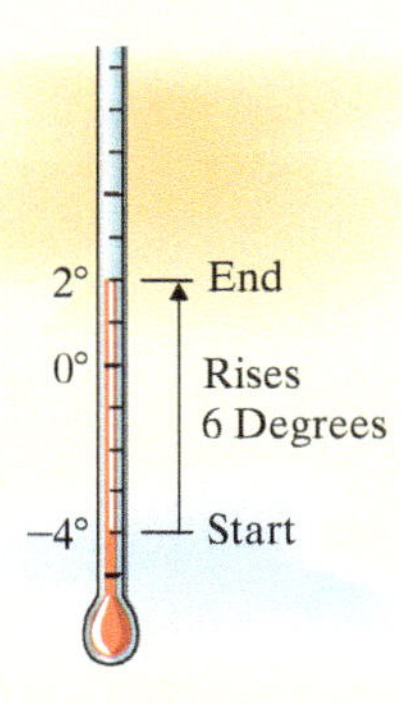

Let's use temperature as an example. If the thermometer registers 4 degrees below 0 degrees and then rises 6 degrees, the new temperature is 2 degrees above 0 degrees. Thus, it is reasonable that $-4 + 6 = 2$. (See the diagram in the margin.)

Example 3 Add: $4 + (-6)$

Solution:

$$4 + (-6) = -2$$

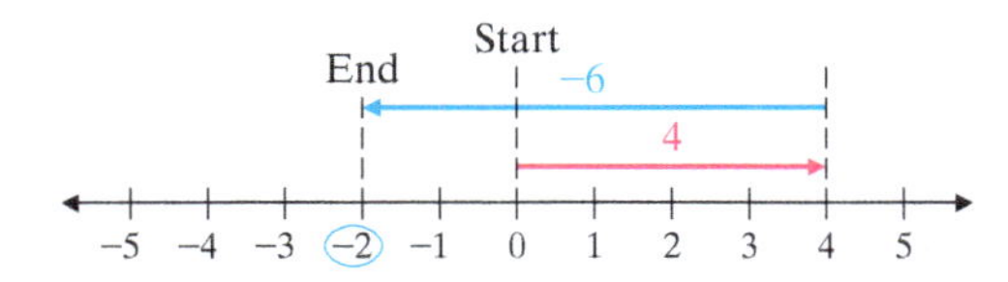

■ Work Practice 3

Practice 3

Add using a number line:
$5 + (-4)$

−5 −4 −3 −2 −1 0 1 2 3 4 5

Using a number line each time we add two numbers can be time consuming. Instead, we can notice patterns in the previous examples and write rules for adding real numbers.

Adding Real Numbers

To add two real numbers

1. with the *same sign,* add their absolute values. Use their common sign as the sign of the answer.
2. with *different signs,* subtract their absolute values. Give the answer the same sign as the number with the larger absolute value.

Example 4 Add without using a number line: $(-7) + (-6)$

Solution: Here, we are adding two numbers with the same sign.

$$(-7) + (-6) = -13$$

↑ same sign ↖ sum of absolute values ($|-7| = 7, |-6| = 6, 7 + 6 = 13$)

■ Work Practice 4

Practice 4

Add without using a number line: $(-8) + (-5)$

Example 5 Add without using a number line: $(-10) + 4$

Solution: Here, we are adding two numbers with different signs.

$$(-10) + 4 = -6$$

↑ sign of number with larger absolute value, -10 ↖ difference of absolute values ($|-10| = 10, |4| = 4, 10 - 4 = 6$)

■ Work Practice 5

Practice 5

Add without using a number line: $(-14) + 6$

Answers

2. 3 **3.** 1 **4.** −13 **5.** −8

Practice 6–11

Add without using a number line.

6. $(-17) + (-10)$
7. $(-4) + 12$
8. $1.5 + (-3.2)$
9. $-\frac{5}{12} + \left(-\frac{1}{12}\right)$
10. $12.1 + (-3.6)$
11. $-\frac{4}{5} + \frac{2}{3}$

Examples Add without using a number line.

6. $(-8) + (-11) = -19$
7. $(-2) + 10 = 8$
8. $0.2 + (-0.5) = -0.3$
9. $-\frac{7}{10} + \left(-\frac{1}{10}\right) = -\frac{8}{10} = -\frac{\overset{1}{\cancel{2}} \cdot 4}{\underset{1}{\cancel{2}} \cdot 5} = -\frac{4}{5}$
10. $11.4 + (-4.7) = 6.7$
11. $-\frac{3}{8} + \frac{2}{5} = -\frac{15}{40} + \frac{16}{40} = \frac{1}{40}$

Work Practice 6–11

In Example 12a, we add three numbers. Remember that by the associative and commutative properties for addition, we may add numbers in any order that we wish. For Example 12a, let's add the numbers from left to right.

Practice 12

Find each sum.

a. $16 + (-9) + (-9)$

b. $[3 + (-13)] + [-4 + (-7)]$

Example 12 Find each sum.

a. $3 + (-7) + (-8)$

b. $[7 + (-10)] + [-2 + (-4)]$

Solution:

a. Perform the additions from left to right.

$3 + (-7) + (-8) = -4 + (-8)$ Adding numbers with different signs

$= -12$ Adding numbers with like signs

b. Simplify inside the brackets first.

$[7 + (-10)] + [-2 + (-4)] = [-3] + [-6]$

$= -9$ Add.

Helpful Hint Don't forget that brackets are grouping symbols. We simplify within them first.

Work Practice 12

Objective B Finding Opposites

To help us subtract real numbers in the next section, we first review what we mean by opposites. The graphs of 4 and -4 are shown on the number line below.

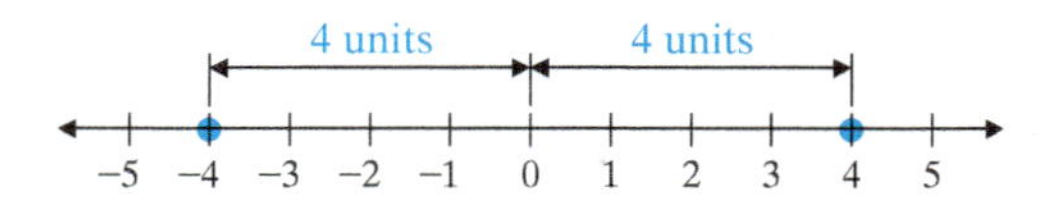

Notice that the graphs of 4 and -4 lie on opposite sides of 0, and each is 4 units away from 0. Such numbers are known as **opposites** or **additive inverses** of each other.

Opposite or Additive Inverse

Two numbers that are the same distance from 0 but lie on opposite sides of 0 are called **opposites** or **additive inverses** of each other.

Answers

6. -27 7. 8 8. -1.7 9. $-\frac{1}{2}$ 10. 8.5 11. $-\frac{2}{15}$ 12. a. -2 b. -21

Examples Find the opposite of each number.

13. 10 The opposite of 10 is -10.

14. -3 The opposite of -3 is 3.

15. $\frac{1}{2}$ The opposite of $\frac{1}{2}$ is $-\frac{1}{2}$.

16. -4.5 The opposite of -4.5 is 4.5.

Work Practice 13–16

We use the symbol "$-$" to represent the phrase "the opposite of" or "the additive inverse of." In general, if a is a number, we write the opposite or additive inverse of a as $-a$. We know that the opposite of -3 is 3. Notice that this translates as

the opposite of	-3	is	3
↓	↓	↓	↓
$-$	(-3)	$=$	**3**

This is true in general.

If a is a number, then $-(-a) = a$.

Example 17 Simplify each expression.

a. $-(-10)$ **b.** $-\left(-\frac{1}{2}\right)$ **c.** $-(-2x)$

d. $-|-6|$ **e.** $-|4.1|$

Solution:

a. $-(-10) = 10$

b. $-\left(-\frac{1}{2}\right) = \frac{1}{2}$

c. $-(-2x) = 2x$

d. $-|-6| = -6$ Since $|-6| = 6$.

e. $-|4.1| = -4.1$ Since $|4.1| = 4.1$

Work Practice 17

Let's discover another characteristic about opposites. Notice that the sum of a number and its opposite is always 0.

$10 + (-10) = 0$ (opposites) $-3 + 3 = 0$ (opposites)

$\frac{1}{2} + \left(-\frac{1}{2}\right) = 0$ (opposites)

In general, we can write the following:

The sum of a number a and its opposite $-a$ is 0.

$a + (-a) = 0$ Also, $-a + a = 0$.

Notice that this means that the opposite of 0 is then 0 since $0 + 0 = 0$.

Practice 13–16

Find the opposite of each number.

13. -35 **14.** 12

15. $-\frac{3}{11}$ **16.** 1.9

Practice 17

Simplify each expression.

a. $-(-22)$

b. $-\left(-\frac{2}{7}\right)$

c. $-(-x)$

d. $-|-14|$

e. $-|2.3|$

Answers

13. 35 **14.** -12 **15.** $\frac{3}{11}$ **16.** -1.9

17. a. 22 **b.** $\frac{2}{7}$ **c.** x **d.** -14 **e.** -2.3

Practice 18–19

Add.

18. $30 + (-30)$

19. $-81 + 81$

Examples Add.

18. $-56 + 56 = 0$

19. $17 + (-17) = 0$

Work Practice 18–19

✓**Concept Check** What is wrong with the following calculation?

$5 + (-22) = 17$

Objective C Evaluating Algebraic Expressions 

We can continue our work with algebraic expressions by evaluating expressions given real-number replacement values.

Practice 20

Evaluate $x + 3y$ for $x = -6$ and $y = 2$.

Example 20 Evaluate $2x + y$ for $x = 3$ and $y = -5$.

Solution: Replace x with 3 and y with -5 in $2x + y$.

$$\begin{aligned} 2x + y &= 2 \cdot 3 + (-5) \\ &= 6 + (-5) \\ &= 1 \end{aligned}$$

Work Practice 20

Practice 21

Evaluate $x + y$ for $x = -13$ and $y = -9$.

Example 21 Evaluate $x + y$ for $x = -2$ and $y = -10$.

Solution: $x + y = (-2) + (-10)$ Replace x with -2 and y with -10.

$= -12$

Work Practice 21

Objective D Solving Applications That Involve Addition 

Positive and negative numbers are used in everyday life. Stock market returns show gains and losses as positive and negative numbers. Temperatures in cold climates often dip into the negative range, commonly referred to as "below zero" temperatures. Bank statements report deposits and withdrawals as positive and negative numbers.

Practice 22

If the temperature was $-7°$ Fahrenheit at 6 a.m., and it rose 11 degrees by 8 a.m., what was the temperature at 8 a.m.?

Example 22 Calculating Temperature

In Philadelphia, Pennsylvania, the record extreme high temperature is 104°F. Decrease this temperature by 111 degrees, and the result is the record extreme low temperature. Find this temperature. (*Source:* National Climatic Data Center)

Solution:

In words:	extreme low temperature	=	extreme high temperature	+	decrease of 111°
	↓		↓		↓
Translate:	extreme low temperature	=	104	+	(−111)
		=	−7		

The record extreme low temperature in Philadelphia, Pennsylvania, is -7°F.

Work Practice 22

Answers

18. 0 **19.** 0 **20.** 0 **21.** −22 **22.** 4°F

✓**Concept Check Answer**

$5 + (-22) = -17$

Vocabulary, Readiness & Video Check

Use the choices below to fill in each blank. Not all choices will be used.

$-a$ a 0 commutative associative

1. If n is a number, then $-n + n =$ ________.
2. Since $x + n = n + x$, we say that addition is ________.
3. If a is a number, then $-(-a) =$ ________.
4. Since $n + (x + a) = (n + x) + a$, we say that addition is ________.

Martin-Gay Interactive Videos

See Video 1.4 

Watch the section lecture video and answer the following questions.

Objective A 5. Complete this statement based on the lecture given before Example 1. To add two numbers with the same sign, add their ________ and use their common sign as the sign of the sum.

6. What is the sign of the sum in Example 6 and why?

Objective B 7. Example 11 illustrates the idea that if a is a real number, the opposite of $-a$ is a. Example 12 looks similar to Example 11, but it's actually quite different. Explain the difference.

Objective C 8. Explain the difference between the algebraic expression for Example 13 and the algebraic expression for Example 14.

Objective D 9. What is the real-life application of negative numbers used in Example 15? The answer to Example 15 is -231. What does this number mean in the context of the problem?

1.4 Exercise Set MyLab Math

Objectives A B Mixed Practice *Add. See Examples 1 through 12, 18, and 19.*

1. $6 + (-3)$
2. $9 + (-12)$
3. $-6 + (-8)$
4. $-6 + (-14)$
5. $8 + (-7)$
6. $16 + (-4)$
7. $-14 + 2$
8. $-10 + 5$
9. $-2 + (-3)$
10. $-7 + (-4)$
11. $-9 + (-3)$
12. $-11 + (-5)$
13. $-7 + 3$
14. $-5 + 9$
15. $10 + (-3)$
16. $8 + (-6)$
17. $5 + (-7)$
18. $3 + (-6)$
19. $-16 + 16$
20. $23 + (-23)$
21. $27 + (-46)$
22. $53 + (-37)$
23. $-18 + 49$
24. $-26 + 14$

25. $-33 + (-14)$

26. $-18 + (-26)$

27. $6.3 + (-8.4)$

28. $9.2 + (-11.4)$

29. $117 + (-79)$

30. $144 + (-88)$

31. $|-8| + (-16)$

32. $|-6| + (-61)$

33. $-\frac{3}{8} + \frac{5}{8}$

34. $-\frac{5}{12} + \frac{7}{12}$

35. $-\frac{7}{16} + \frac{1}{4}$

36. $-\frac{5}{9} + \frac{1}{3}$

37. $-\frac{7}{10} + \left(-\frac{3}{5}\right)$

38. $-\frac{5}{6} + \left(-\frac{2}{3}\right)$

39. $-9\frac{3}{5} + \left(-3\frac{1}{2}\right)$

40. $-6\frac{7}{10} + \left(-7\frac{3}{5}\right)$

41. $-15 + 9 + (-2)$

42. $-9 + 15 + (-5)$

43. $-21 + (-16) + (-22)$

44. $-18 + (-6) + (-40)$

45. $-23 + 16 + (-2)$

46. $-14 + (-3) + 11$

47. $|5 + (-10)|$

48. $|7 + (-17)|$

49. $6 + (-4) + 9$

50. $8 + (-2) + 7$

51. $[-17 + (-4)] + [-12 + 15]$

52. $[-2 + (-7)] + [-11 + 22]$

53. $|9 + (-12)| + |-16|$

54. $|43 + (-73)| + |-20|$

55. $-13 + [5 + (-3) + 4]$

56. $-30 + [1 + (-6) + 8]$

57. Find the sum of -38 and 12.

58. Find the sum of -44 and 16.

Objective B *Find each additive inverse or opposite. See Examples 13 through 17.*

59. 6

60. 4

61. -2

62. -8

63. 0

64. $-\frac{1}{4}$

65. $|-6|$

66. $|-11|$

Simplify each of the following. See Example 17.

67. $-|-2|$

68. $-|-5|$

69. $-(-7)$

70. $-(-14)$

71. $-(-7.9)$

72. $-(-8.4)$

73. $-(-5z)$

74. $-(-7m)$

75. $\left|-\frac{2}{3}\right|$

76. $-\left|-\frac{2}{3}\right|$

Objective C *Evaluate $x + y$ for the given replacement values. See Examples 20 and 21.*

77. $x = -20$ and $y = -50$

78. $x = -1$ and $y = -29$

Evaluate $3x + y$ for the given replacement values. See Examples 20 and 21.

79. $x = 2$ and $y = -3$

80. $x = 7$ and $y = -11$

Objective D Translating *Translate each phrase; then simplify. See Example 22.*

81. Find the sum of -6 and 25.

82. Find the sum of -30 and 15.

83. Find the sum of -31, -9, and 30.

84. Find the sum of -49, -2, and 40.

Solve. See Example 22.

85. Suppose a deep-sea diver dives from the surface to 215 feet below the surface. He then dives down 16 more feet. Use positive and negative numbers to represent this situation. Then find the diver's present depth.

86. Suppose a diver dives from the surface to 248 meters below the surface and then swims up 8 meters, down 16 meters, down another 28 meters, and then up 32 meters. Use positive and negative numbers to represent this situation. Then find the diver's depth after these movements.

87. The lowest temperature ever recorded in Massachusetts was $-35°$F. The highest recorded temperature in Massachusetts was 142° higher than the record low temperature. Find Massachusetts' highest recorded temperature. (*Source:* National Climatic Data Center)

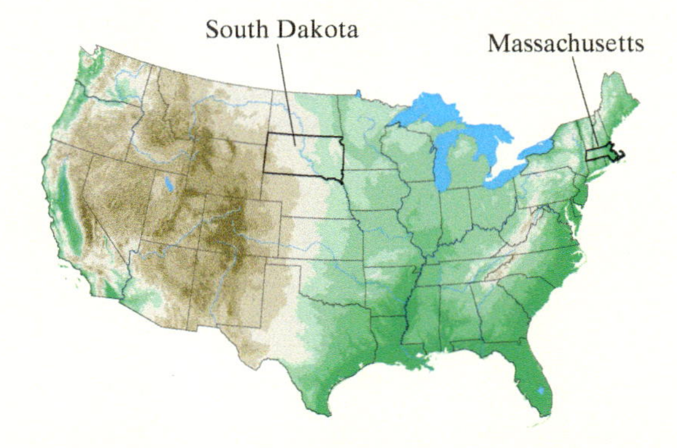

88. On January 2, 1943, the temperature was $-4°$ at 7:30 a.m. in Spearfish, South Dakota. Incredibly, it got 49° warmer in the next 2 minutes. To what temperature did it rise by 7:32?

89. The lowest elevation on Earth is -411 meters (that is, 411 meters below sea level) at the Dead Sea. If you are standing 316 meters above the Dead Sea, what is your elevation? (*Source:* National Geographic Society)

90. The lowest elevation in Australia is -52 feet at Lake Eyre. If you are standing at a point 439 feet above Lake Eyre, what is your elevation? (*Source:* National Geographic Society)

91. During the 2018 PGA Masters Tournament, the winner, Patrick Reed, had scores of -3, -6, -5, and -1 over four rounds of golf. What was his total score for the tournament? (*Source:* Professional Golfer's Association)

92. Brooke Henderson won the 2018 LPGA Lotte Championship Tournament with scores of -4, $+1$, -6, and -3 over four rounds of golf. What was her total score for the tournament? (*Source:* LPGA of America)

93. A negative net income results when a company spends more money than it brings in. Mattel Inc. had the following quarterly net incomes during its 2017 fiscal year. (*Source:* Mattel, Inc.)

Quarter of Fiscal 2017	Net Income (in millions)
First	−113.2
Second	−56.1
Third	−603.2
Fourth	−281.3

What was the total net income for fiscal year 2017?

94. Barnes & Noble Inc. had the following quarterly net incomes during their fiscal 2018. (*Source:* Market Watch, Inc.)

Quarter of Fiscal 2018	Net Income (in millions)
ended July 29, 2017	−10.8
ended October 28, 2017	−30.1
ended January 27, 2018	−63.5
ended April 28, 2018	−21.1

What was the total net income shown in the table?

Concept Extensions

The following bar graph shows each month's average daily low temperature in degrees Fahrenheit for Barrow, Alaska. Use this graph to answer Exercises 95 through 100.

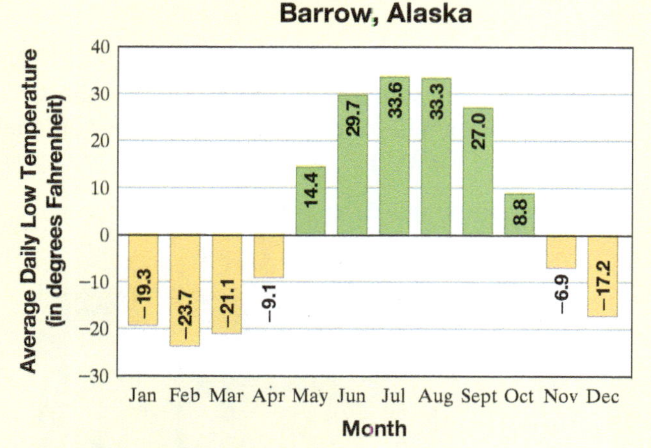

Source: National Climatic Data Center

95. For what month is the graphed temperature the highest?

96. For what month is the graphed temperature the lowest?

97. For what month is the graphed temperature positive *and* closest to 0°?

98. For what month is the graphed temperature negative *and* closest to 0°?

99. Find the average of the temperatures shown for the months of April, May, and October. (To find the average of three temperatures, find their sum and divide by 3.)

100. Find the average of the temperatures shown for the months of January, September, and October. (To find the average of three temperatures, find their sum and divide by 3.)

101. Name 2 numbers whose sum is -17.

102. Name 2 numbers whose sum is -30.

Each calculation below is incorrect. Find the error and correct it. See the Concept Check in this section.

103. $7 + (-10) \stackrel{?}{=} 17$

104. $-4 + 14 \stackrel{?}{=} -18$

105. $-10 + (-12) \stackrel{?}{=} -120$

106. $-15 + (-17) \stackrel{?}{=} 32$

For Exercises 107 through 110, determine whether each statement is true or false.

107. The sum of two negative numbers is always a negative number.

108. The sum of two positive numbers is always a positive number.

109. The sum of a positive number and a negative number is always a negative number.

110. The sum of zero and a negative number is always a negative number.

111. In your own words, explain how to add two negative numbers.

112. In your own words, explain how to add a positive number and a negative number.

1.5 Subtracting Real Numbers

Objective A Subtracting Real Numbers

Now that addition of real numbers has been discussed, we can explore subtraction. We know that $9 - 7 = 2$. Notice that $9 + (-7) = 2$, also. This means that

$$9 - 7 = 9 + (-7)$$

Notice that the *difference* of 9 and 7 is the same as the *sum* of 9 and the opposite of 7. This is how we can subtract real numbers.

Subtracting Real Numbers

If a and b are real numbers, then $a - b = a + (-b)$.

In other words, to find the difference of two numbers, we add the opposite of the number being subtracted.

Objectives

A Subtract Real Numbers.

B Evaluate Algebraic Expressions Using Real Numbers.

C Determine Whether a Number Is a Solution of a Given Equation.

D Solve Applications That Involve Subtraction of Real Numbers.

E Find Complementary and Supplementary Angles.

Practice 1

Subtract.

a. $-20 - 6$
b. $3 - (-5)$
c. $7 - 17$
d. $-4 - (-9)$

Example 1

Subtract.

a. $-13 - 4$ **b.** $5 - (-6)$ **c.** $3 - 6$ **d.** $-1 - (-7)$

Solution:

add
a. $-13 - 4 = -13 + (-4)$ Add −13 to the opposite of 4, which is −4.
opposite

$= -17$

add
b. $5 - (-6) = 5 + (6)$ Add 5 to the opposite of −6, which is 6.
opposite

$= 11$

c. $3 - 6 = 3 + (-6)$ Add 3 to the opposite of 6, which is −6.

$= -3$

d. $-1 - (-7) = -1 + (7) = 6$

Work Practice 1

Helpful Hint

Study the patterns indicated.

No change / Change to addition. / Change to opposite.

$$5 - 11 = 5 + (-11) = -6$$
$$-3 - 4 = -3 + (-4) = -7$$
$$7 - (-1) = 7 + (1) = 8$$

Practice 2–4

Subtract.

2. $9.6 - (-5.7)$

3. $-\frac{4}{9} - \frac{2}{9}$

4. $-\frac{1}{4} - \left(-\frac{2}{5}\right)$

Examples

Subtract.

2. $5.3 - (-4.6) = 5.3 + (4.6) = 9.9$

3. $-\frac{3}{10} - \frac{5}{10} = -\frac{3}{10} + \left(-\frac{5}{10}\right) = -\frac{8}{10} = -\frac{4}{5}$

4. $-\frac{2}{3} - \left(-\frac{4}{5}\right) = -\frac{2}{3} + \left(\frac{4}{5}\right) = -\frac{10}{15} + \frac{12}{15} = \frac{2}{15}$

Work Practice 2–4

Practice 5

Write each phrase as an expression and simplify.

a. Subtract 7 from −11.
b. Decrease 35 by −25.

Example 5

Write each phrase as an expression and simplify.

a. Subtract 8 from −4. **b.** Decrease 10 by −20.

Solution: Be careful when interpreting these. The order of numbers in subtraction is important.

a. 8 is to be subtracted **from** −4.

$$-4 - 8 = -4 + (-8) = -12$$

b. To decrease 10 by −20, we find 10 **minus** −20.

$$10 - (-20) = 10 + 20 = 30$$

Work Practice 5

If an expression contains additions and subtractions, just write the subtractions as equivalent additions. Then simplify from left to right.

Answers

1. a. −26 **b.** 8 **c.** −10 **d.** 5 **2.** 15.3 **3.** $-\frac{2}{3}$ **4.** $\frac{3}{20}$ **5. a.** −18 **b.** 60

Example 6 Simplify each expression.

a. $-14 - 8 + 10 - (-6)$ **b.** $1.6 - (-10.3) + (-5.6)$

Solution:

a. $-14 - 8 + 10 - (-6) = -14 + (-8) + 10 + 6 = -6$

b. $1.6 - (-10.3) + (-5.6) = 1.6 + 10.3 + (-5.6) = 6.3$

Work Practice 6

Practice 6

Simplify each expression.

a. $-20 - 5 + 12 - (-3)$

b. $5.2 - (-4.4) + (-8.8)$

When an expression contains parentheses and brackets, remember the order of operations. Start with the innermost set of parentheses or brackets and work your way outward.

Example 7 Simplify each expression.

a. $-3 + [(-2 - 5) - 2]$ **b.** $2^3 - 10 + [-6 - (-5)]$

Solution:

a. Start with the innermost set of parentheses. Rewrite $-2 - 5$ as an addition.

$$\begin{aligned} -3 + [(-2 - 5) - 2] &= -3 + [(-2 + (-5)) - 2] \\ &= -3 + [(-7) - 2] && \text{Add: } -2 + (-5). \\ &= -3 + [-7 + (-2)] && \text{Write } -7 - 2 \text{ as an addition.} \\ &= -3 + [-9] && \text{Add.} \\ &= -12 && \text{Add.} \end{aligned}$$

b. Start simplifying the expression inside the brackets by writing $-6 - (-5)$ as an addition.

$$\begin{aligned} 2^3 - 10 + [-6 - (-5)] &= 2^3 - 10 + [-6 + 5] \\ &= 2^3 - 10 + [-1] && \text{Add.} \\ &= 8 - 10 + (-1) && \text{Evaluate } 2^3. \\ &= 8 + (-10) + (-1) && \text{Write } 8 - 10 \text{ as an addition.} \\ &= -2 + (-1) && \text{Add.} \\ &= -3 && \text{Add.} \end{aligned}$$

Work Practice 7

Practice 7

Simplify each expression.

a. $-9 + [(-4 - 1) - 10]$

b. $5^2 - 20 + [-11 - (-3)]$

Objective B Evaluating Algebraic Expressions

It is important to be able to evaluate expressions for given replacement values. This helps, for example, when checking solutions of equations.

Example 8 Find the value of each expression when $x = 2$ and $y = -5$.

a. $\dfrac{x - y}{12 + x}$ **b.** $x^2 - y$

Solution:

a. Replace x with 2 and y with -5. Be sure to put parentheses around -5 to separate signs. Then simplify the resulting expression.

$$\frac{x - y}{12 + x} = \frac{2 - (-5)}{12 + 2} = \frac{2 + 5}{14} = \frac{7}{14} = \frac{1}{2}$$

b. Replace x with 2 and y with -5 and simplify.

$$x^2 - y = 2^2 - (-5) = 4 - (-5) = 4 + 5 = 9$$

Work Practice 8

Practice 8

Find the value of each expression when $x = 1$ and $y = -4$.

a. $\dfrac{x - y}{14 + x}$ **b.** $x^2 - y$

Answers

6. a. -10 **b.** 0.8 **7. a.** -24 **b.** -3

8. a. $\frac{1}{3}$ **b.** 5

Helpful Hint

For additional help when replacing variables with replacement values, first place parentheses about any variables.

For Example 8b on the previous page, we have

$$x^2 - y = \underbrace{(x)^2 - (y)}_{\text{Place parentheses about variables}} = \underbrace{(2)^2 - (-5)}_{\text{Replace variables with values}} = 4 - (-5) = 4 + 5 = 9$$

Objective C Solutions of Equations

Recall from Section 1.3 that a solution of an equation is a value for the variable that makes the equation true.

Practice 9

Determine whether -2 is a solution of $-1 + x = 1$.

Example 9 Determine whether -4 is a solution of $x - 5 = -9$.

Solution: Replace x with -4 and see if a true statement results.

$x - 5 = -9$ Original equation

$-4 - 5 \stackrel{?}{=} -9$ Replace x with -4.

$-4 + (-5) \stackrel{?}{=} -9$

$-9 = -9$ True

Thus -4 is a solution of $x - 5 = -9$.

Work Practice 9

Objective D Solving Applications That Involve Subtraction

Another use of real numbers is in recording altitudes above and below sea level, as shown in the next example.

Practice 10

The highest point in Asia is the top of Mount Everest, at a height of 29,028 feet above sea level. The lowest point is the Dead Sea, which is 1312 feet below sea level. How much higher is Mount Everest than the Dead Sea? (*Source:* National Geographic Society)

Example 10 Finding a Change in Elevation

The highest point in the United States is the top of Denali, at a height of 20,320 feet above sea level. The lowest point is Death Valley, California, which is 282 feet below sea level. How much higher is Denali than Death Valley? (*Source:* U.S. Geological Survey)

Solution: To find "how much higher," we subtract. Don't forget that since Death Valley is 282 feet *below* sea level, we represent its height by -282. Draw a diagram to help visualize the problem.

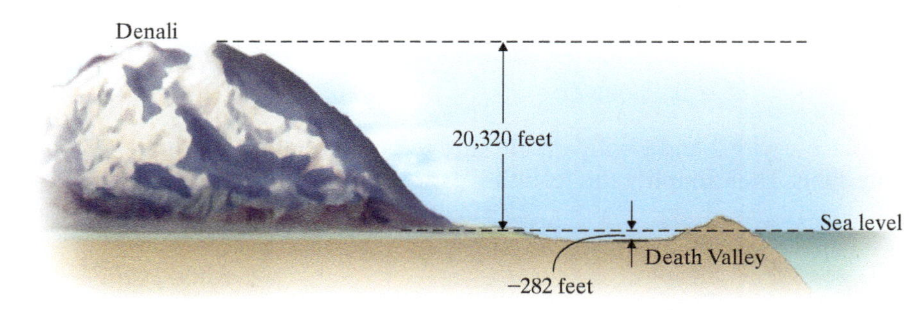

Answers

9. -2 is not a solution. **10.** 30,340 ft

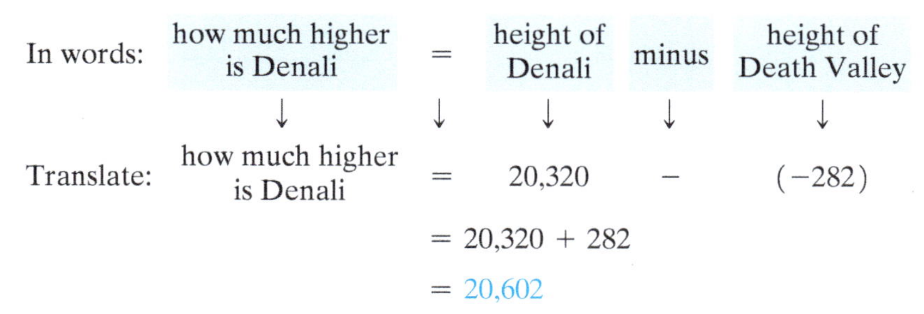

In words:	how much higher is Denali	=	height of Denali	minus	height of Death Valley
	↓	↓	↓	↓	↓
Translate:	how much higher is Denali	=	20,320	−	(−282)

$$= 20{,}320 + 282$$
$$= 20{,}602$$

Thus, Denali is 20,602 feet higher than Death Valley.

Work Practice 10

Objective E Finding Complementary and Supplementary Angles

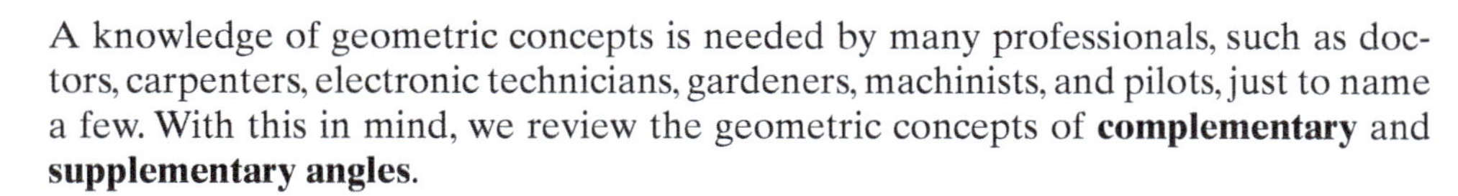

A knowledge of geometric concepts is needed by many professionals, such as doctors, carpenters, electronic technicians, gardeners, machinists, and pilots, just to name a few. With this in mind, we review the geometric concepts of **complementary** and **supplementary angles**.

Complementary and Supplementary Angles

Two angles are **complementary** if the sum of their measures is 90°.

$m\angle x + m\angle y = 90°$

Two angles are **supplementary** if the sum of their measures is 180°.

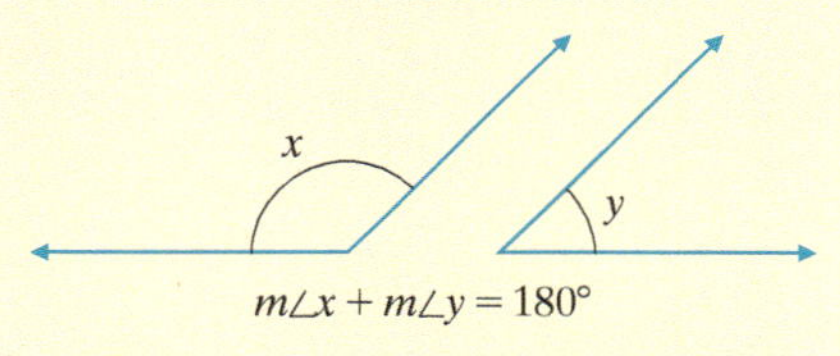

Example 11 Find the measure of each unknown complementary or supplementary angle.

a.

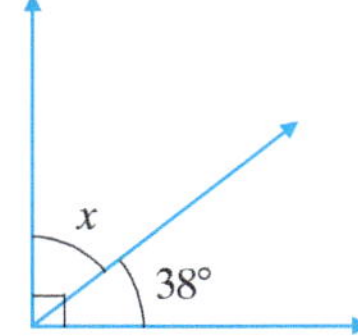

b.

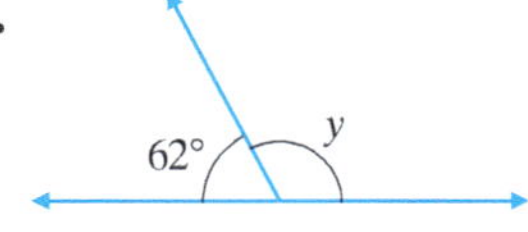

Solution:

a. These angles are complementary, so their sum is 90°. This means that the measure of angle x, $m\angle x$, is $90° - 38°$.

$$m\angle x = 90° - 38° = 52°$$

b. These angles are supplementary, so their sum is 180°. This means that $m\angle y$ is $180° - 62°$.

$$m\angle y = 180° - 62° = 118°$$

Work Practice 11

Practice 11

Find the measure of each unknown complementary or supplementary angle.

a.

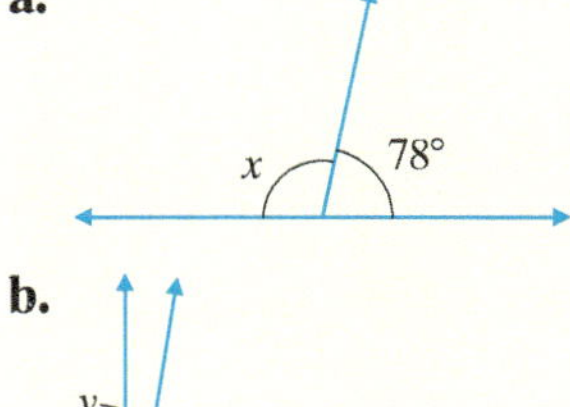

b.

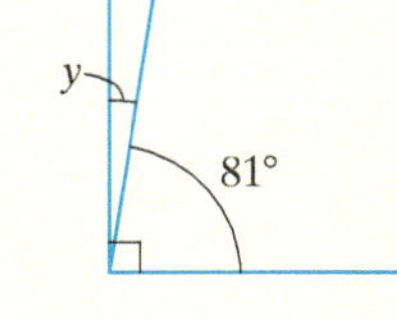

Answers

11. a. 102° **b.** 9°

Vocabulary, Readiness & Video Check

Multiple choice: Select the correct lettered response following each exercise.

1. It is true that $a - b =$ ________.

a. $b - a$ **b.** $a + (-b)$ **c.** $a + b$

2. The opposite of n is ________.

a. $-n$ **b.** $-(-n)$ **c.** n

3. To evaluate $x - y$ for $x = -10$ and $y = -14$, we replace x with -10 and y with -14 and evaluate ________.

a. $10 - 14$ **b.** $-10 - 14$ **c.** $-14 - 10$ **d.** $-10 - (-14)$

4. The expression $-5 - 10$ equals ________.

a. $5 - 10$ **b.** $5 + 10$ **c.** $-5 + (-10)$ **d.** $10 - 5$

Martin-Gay Interactive Videos

See Video 1.5

Watch the section lecture video and answer the following questions.

Objective A **5.** Complete this statement based on the lecture given before Example 1. To subtract two real numbers, change the operation to ________ and take the ________ of the second number.

6. When simplifying Example 5, what is the result of the first step and why is the expression rewritten in this way?

Objective B **7.** In Example 7, why are you told to be especially careful when working with the replacement value in the numerator?

Objective C **8.** In Example 8, we learned that what number is NOT a solution of what equation?

Objective D **9.** For Example 9, why is the overall vertical change represented as a negative number?

Objective E **10.** The definition of supplementary angles is given just before Example 10. Explain how this definition is used to solve Example 10.

1.5 Exercise Set MyLab Math

Objective A *Subtract. See Examples 1 through 4.*

1. $-6 - 4$ **2.** $-12 - 8$ **3.** $4 - 9$ **4.** $8 - 11$ **5.** $16 - (-3)$

6. $12 - (-5)$ **7.** $7 - (-4)$ **8.** $3 - (-6)$ **9.** $-26 - (-18)$ **10.** $-60 - (-48)$

11. $-6 - 5$ **12.** $-8 - 4$ **13.** $16 - (-21)$ **14.** $15 - (-33)$ **15.** $-6 - (-11)$

16. $-4 - (-16)$ **17.** $-44 - 27$ **18.** $-36 - 51$ **19.** $-21 - (-21)$ **20.** $-17 - (-17)$

21. $-\frac{3}{11} - \left(-\frac{5}{11}\right)$ **22.** $-\frac{4}{7} - \left(-\frac{1}{7}\right)$ **23.** $9.7 - 16.1$ **24.** $8.3 - 11.2$ **25.** $-2.6 - (-6.7)$

26. $-6.1 - (-5.3)$ **27.** $\frac{1}{2} - \frac{2}{3}$ **28.** $\frac{3}{4} - \frac{7}{8}$ **29.** $-\frac{1}{6} - \frac{3}{4}$ **30.** $-\frac{1}{10} - \frac{7}{8}$

31. $8.3 - (-0.62)$ **32.** $4.3 - (-0.87)$ **33.** $0 - 8.92$ **34.** $0 - (-4.21)$

Translating *Translate each phrase to an expression and simplify. See Example 5.*

35. Subtract −5 from 8.

36. Subtract −2 from 3.

37. Find the difference between −6 and −1.

38. Find the difference between −17 and −1.

39. Subtract 8 from 7.

40. Subtract 9 from −4.

41. Decrease −8 by 15.

42. Decrease 11 by −14.

Mixed Practice (*Sections 1.3, 1.4, and 1.5*) *Simplify each expression. (Remember the order of operations.) See Examples 6 and 7.*

43. $-10 - (-8) + (-4) - 20$

44. $-16 - (-3) + (-11) - 14$

45. $5 - 9 + (-4) - 8 - 8$

46. $7 - 12 + (-5) - 2 + (-2)$

47. $-6 - (2 - 11)$

48. $-9 - (3 - 8)$

49. $3^3 - 8 \cdot 9$

50. $2^3 - 6 \cdot 3$

51. $2 - 3(8 - 6)$

52. $4 - 6(7 - 3)$

53. $(3 - 6) + 4^2$

54. $(2 - 3) + 5^2$

55. $-2 + [(8 - 11) - (-2 - 9)]$

56. $-5 + [(4 - 15) - (-6) - 8]$

57. $|-3| + 2^2 + [-4 - (-6)]$

58. $|-2| + 6^2 + (-3 - 8)$

Objective B *Evaluate each expression when $x = -5$, $y = 4$, and $t = 10$. See Example 8.*

59. $x - y$

60. $y - x$

61. $\frac{9 - x}{y + 6}$

62. $\frac{15 - x}{y + 2}$

63. $|x| + 2t - 8y$

64. $|y| + 3x - 2t$

65. $y^2 - x$

66. $t^2 - x$

67. $\frac{|x - (-10)|}{2t}$

68. $\frac{|5y - x|}{6t}$

Objective C *Decide whether the given number is a solution of the given equation. See Example 9.*

69. $x - 9 = 5;\ -4$

70. $x - 10 = -7;\ 3$

71. $-x + 6 = -x - 1;\ -2$

72. $-x - 6 = -x - 1;\ -10$

73. $-x - 13 = -15;\ 2$

74. $4 = 1 - x;\ 5$

Objectives D E Mixed Practice *Solve. See Examples 10 and 11.*

75. The coldest temperature ever recorded on Earth was −129°F in Antarctica. The warmest temperature ever recorded was 134°F in Death Valley, California. How many degrees warmer is 134°F than −129°F? (*Source: The World Almanac*, 2013)

76. The coldest temperature ever recorded in the United States was −80°F in Alaska. The warmest temperature ever recorded was 134°F in California. How many degrees warmer is 134°F than −80°F? (*Source: The World Almanac*, 2013)

77. A commercial jetliner hits an air pocket and drops 250 feet. After climbing 120 feet, it drops another 178 feet. What is its overall vertical change?

78. A woman received a statement of her charge account at Old Navy. She spent $93 on purchases last month. She returned an $18 top because she didn't like the color. She also returned a $26 nightshirt because it was damaged. What does she actually owe on her account?

79. Find x if the angles below are complementary angles.

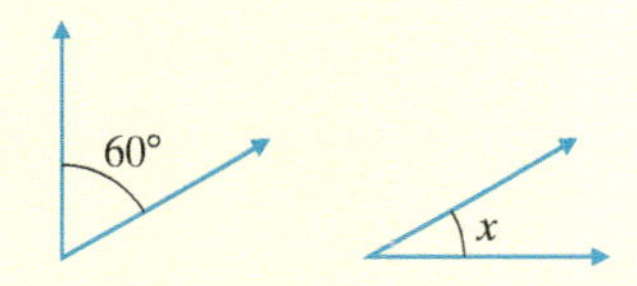

80. Find y if the angles below are supplementary angles.

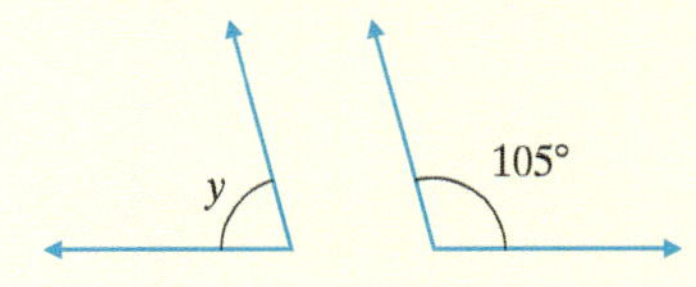

81. Mauna Kea in Hawaii has an elevation of 13,796 feet above sea level. The Mid-America Trench in the Pacific Ocean has an elevation of 21,857 feet below sea level. Find the difference in elevation between those two points. (*Source:* National Geographic Society and Defense Mapping Agency)

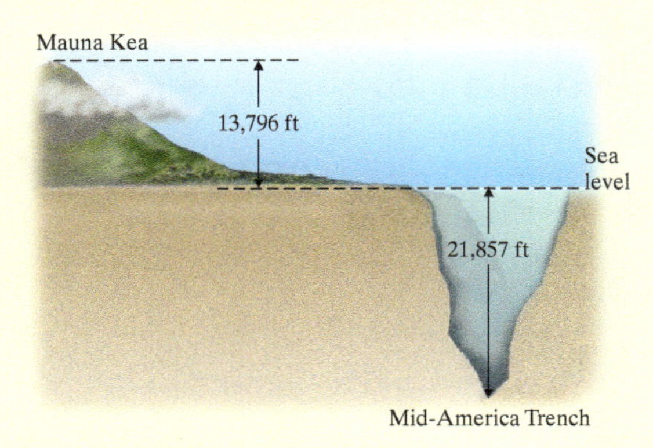

82. In some card games, it is possible to have a negative score. Lavonne Schultz currently has a score of 15 points. She then loses 24 points. What is her new score?

83. The highest point in Africa is Mt. Kilimanjaro, Tanzania, at an elevation of 19,340 feet. The lowest point is Lake Assal, Djibouti, at 512 feet below sea level. How much higher is Mt. Kilimanjaro than Lake Assal? (*Source:* National Geographic Society)

84. The airport in Bishop, California, is at an elevation of 4101 feet above sea level. The nearby Furnace Creek Airport in Death Valley, California, is at an elevation of 226 feet below sea level. How much higher in elevation is the Bishop Airport than the Furnace Creek Airport? (*Source:* National Climatic Data Center)

Find each unknown complementary or supplementary angle.

85.

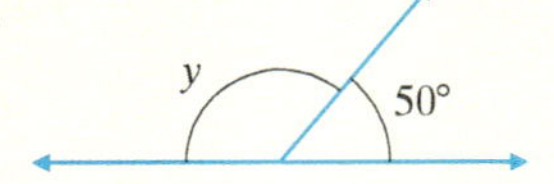

△ **86.**

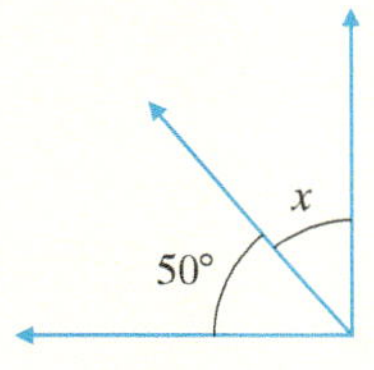

Mixed Practice—Translating (*Sections 1.4 and 1.5*)

Translate each phrase to an algebraic expression. Use "x" to represent "a number."

87. The sum of −5 and a number.

88. The difference of −3 and a number.

89. Subtract a number from −20.

90. Add a number and −36.

Concept Extensions

Recall the bar graph from Section 1.4. It shows each month's average daily low temperature in degrees Fahrenheit for Barrow, Alaska. Use this graph to answer Exercises 91 through 94 on the next page.

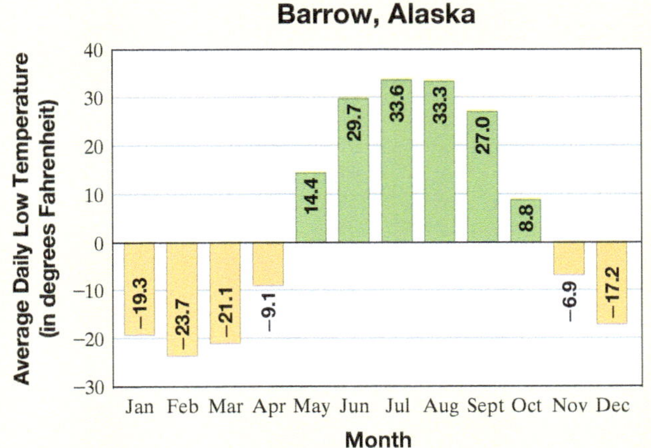

Source: National Climatic Data Center

91. Record the monthly increases and decreases in the low temperature from the previous month.

Month	Monthly Increase or Decrease (from the previous month)
February	
March	
April	
May	
June	

92. Record the monthly increases and decreases in the low temperature from the previous month.

Month	Monthly Increase or Decrease (from the previous month)
July	
August	
September	
October	
November	
December	

93. Use the tables in Exercises **91** and **92** to determine which month had the greatest increase in temperature?

94. Use the tables in Exercises **91** and **92** to determine which month had the greatest decrease in temperature?

Solve.

95. Find two numbers whose difference is -5.

96. Find two numbers whose difference is -9.

Each calculation below is ***incorrect.*** *Find the error and correct it.*

97. $9 - (-7) \stackrel{?}{=} 2$

98. $-4 - 8 \stackrel{?}{=} 4$

99. $10 - 30 \stackrel{?}{=} 20$

100. $-3 - (-10) \stackrel{?}{=} -13$

If p is a positive number and n is a negative number, determine whether each statement is true or false. Explain your answer.

101. $p - n$ is always a positive number.

102. $n - p$ is always a negative number.

103. $|n| - |p|$ is always a positive number.

104. $|n - p|$ is always a positive number.

Without calculating, determine whether each answer is positive or negative. Then use a calculator to find the exact difference.

105. 56,875 − 87,262

106. 4.362 − 7.0086

Operations on Real Numbers

Answers

1. ______
2. ______
3. ______
4. ______
5. ______
6. ______
7. ______
8. ______
9. ______
10. ______
11. ______
12. ______
13. ______
14. ______
15. ______
16. ______
17. ______
18. ______
19. ______
20. ______
21. ______
22. ______
23. ______
24. ______

Answer the following with positive, negative, or 0.

1. The opposite of a positive number is a ______ number.

2. The sum of two negative numbers is a ______ number.

3. The absolute value of a negative number is a ______ number.

4. The absolute value of zero is ______.

5. The sum of two positive numbers is a ______ number.

6. The sum of a number and its opposite is ______.

7. The absolute value of a positive number is a ______ number.

8. The opposite of a negative number is a ______ number.

Fill in the chart:

	Number	Opposite	Absolute Value
9.	$\frac{1}{7}$		
10.	$-\frac{12}{5}$		
11.		-3	
12.		$\frac{9}{11}$	

Perform each indicated operation and simplify. Don't forget to use order of operations if needed.

13. $-19 + (-23)$ **14.** $7 - (-3)$ **15.** $-15 + 17$ **16.** $-8 - 10$

17. $18 + (-25)$ **18.** $-2 + (-37)$ **19.** $-14 - (-12)$ **20.** $5 - 14$

21. $4.5 - 7.9$ **22.** $-8.6 - 1.2$ **23.** $-\frac{3}{4} - \frac{1}{7}$ **24.** $\frac{2}{3} - \frac{7}{8}$

25. ______
26. ______
27. ______
28. ______
29. ______
30. ______
31. ______
32. ______
33. ______
34. ______
35. ______
36. ______
37. ______
38. ______
39. ______
40. ______
41. ______
42. ______
43. ______
44. ______
45. ______
46. ______

25. $-9 - (-7) + 4 - 6$

26. $11 - 20 + (-3) - 12$

27. $24 - 6(14 - 11)$

28. $30 - 5(10 - 8)$

29. $(7 - 17) + 4^2$

30. $9^2 + (10 - 30)$

31. $|-9| + 3^2 + (-4 - 20)$

32. $|-4 - 5| + 5^2 + (-50)$

33. $-7 + [(1 - 2) + (-2 - 9)]$

34. $-6 + [(-3 + 7) + (4 - 15)]$

35. Subtract 5 from 1.

36. Subtract -2 from -3.

37. Subtract $-\frac{2}{5}$ from $\frac{1}{4}$.

38. Subtract $\frac{1}{10}$ from $-\frac{5}{8}$.

39. $2(19 - 17)^3 - 3(-7 + 9)^2$

40. $3(10 - 9)^2 + 6(20 - 19)^3$

Evaluate each expression when $x = -2$, $y = -1$, and $z = 9$.

41. $x - y$

42. $x + y$

43. $y + z$

44. $z - y$

45. $\dfrac{|5z - x|}{y - x}$

46. $\dfrac{|-x - y + z|}{2z}$

1.6 Multiplying and Dividing Real Numbers 

Objective A Multiplying Real Numbers

Objectives

A Multiply Real Numbers.

B Find the Reciprocal of a Real Number.

C Divide Real Numbers.

D Evaluate Expressions Using Real Numbers.

E Determine Whether a Number Is a Solution of a Given Equation.

F Solve Applications That Involve Multiplication or Division of Real Numbers.

Multiplication of real numbers is similar to multiplication of whole numbers. We just need to determine when the answer is positive, when it is negative, and when it is zero. To discover sign patterns for multiplication, recall that multiplication is repeated addition. For example, 3(2) means that 2 is added to itself three times, or

$$3(2) = 2 + 2 + 2 = 6$$

Also,

$$3(-2) = (-2) + (-2) + (-2) = -6$$

Since $3(-2) = -6$, this suggests that the product of a positive number and a negative number is a negative number.

What about the product of two negative numbers? To find out, consider the following pattern.

Factor decreases by 1 each time.

$-3 \cdot 2 = -6$

$-3 \cdot 1 = -3$ Product increases by 3 each time.

$-3 \cdot 0 = 0$

$-3 \cdot -1 = 3$

$-3 \cdot -2 = 6$

This suggests that the product of two negative numbers is a positive number. Our results are given below.

Multiplying Real Numbers

1. The product of two numbers with the *same* sign is a positive number.
2. The product of two numbers with *different* signs is a negative number.

Examples Multiply.

1. $-7(6) = -42$ Different signs, so the product is negative.
2. $2(-10) = -20$
3. $-2(-14) = 28$ Same sign, so the product is positive.
4. $-\frac{2}{3} \cdot \frac{4}{7} = -\frac{2 \cdot 4}{3 \cdot 7} = -\frac{8}{21}$
5. $5(-1.7) = -8.5$
6. $-18(-3) = 54$

Work Practice 1–6

Practice 1–6

Multiply.

1. $-8(3)$ 2. $5(-30)$ 3. $-4(-12)$
4. $-\frac{5}{6} \cdot \frac{1}{4}$ 5. $6(-2.3)$ 6. $-15(-2)$

Answers

1. -24 2. -150 3. 48 4. $-\frac{5}{24}$ 5. -13.8 6. 30

We already know that the product of 0 and any whole number is 0. This is true of all real numbers.

Products Involving Zero

If b is a real number, then $b \cdot 0 = 0$. Also $0 \cdot b = 0$.

Practice 7

Multiply.

a. $5(0)(-3)$
b. $(-1)(-6)(-7)$
c. $(-2)(4)(-8)(-1)$

Example 7

Multiply.

a. $7(0)(-6)$ b. $(-2)(-3)(-4)$ c. $(-1)(-5)(-9)(-2)$

Solution:

a. By the order of operations, we multiply from left to right. Notice that because one of the factors is 0, the product is 0.

$$7(0)(-6) = 0(-6) = 0$$

b. Multiply two factors at a time, from left to right.

$$(-2)(-3)(-4) = (6)(-4) \quad \text{Multiply } (-2)(-3).$$
$$= -24$$

c. Multiply from left to right.

$$(-1)(-5)(-9)(-2) = (5)(-9)(-2) \quad \text{Multiply } (-1)(-5).$$
$$= -45(-2) \quad \text{Multiply } 5(-9).$$
$$= 90$$

Work Practice 7

✓**Concept Check** What is the sign of the product of five negative numbers? Explain.

Helpful Hint

Have you noticed a pattern when multiplying signed numbers?

If we let $(-)$ represent a negative number and $(+)$ represent a positive number, then

$(-)(-) = (+)$
$(-)(-)(-) = (-)$
$(-)(-)(-)(-) = (+)$
$(-)(-)(-)(-)(-) = (-)$

The product of an even number of negative numbers is a positive result.

The product of an odd number of negative numbers is a negative result.

Now that we know how to multiply positive and negative numbers, let's see how we find the values of $(-5)^2$ and -5^2, for example. Although these two expressions look similar, the difference between the two is the parentheses. In $(-5)^2$, the parentheses tell us that the base, or repeated factor, is -5. In -5^2, only 5 is the base. Thus,

$$(-5)^2 = (-5)(-5) = 25 \quad \text{The base is } -5.$$
$$-5^2 = -(5 \cdot 5) = -25 \quad \text{The base is } 5.$$

Practice 8

Evaluate.

a. $(-2)^4$ b. -2^4
c. $(-1)^5$ d. -1^5
e. $\left(-\frac{7}{9}\right)^2$

Example 8

Evaluate.

a. $(-2)^3$ b. -2^3 c. $(-3)^2$ d. -3^2 e. $\left(-\frac{2}{3}\right)^2$

Solution:

a. $(-2)^3 = (-2)(-2)(-2) = -8$ The base is -2.
b. $-2^3 = -(2 \cdot 2 \cdot 2) = -8$ The base is 2.
c. $(-3)^2 = (-3)(-3) = 9$ The base is -3.
d. $-3^2 = -(3 \cdot 3) = -9$ The base is 3.
e. $\left(-\frac{2}{3}\right)^2 = \left(-\frac{2}{3}\right)\left(-\frac{2}{3}\right) = \frac{4}{9}$ The base is $-\frac{2}{3}$.

Work Practice 8

Answers

7. a. 0 b. -42 c. -64 8. a. 16 b. -16 c. -1 d. -1 e. $\frac{49}{81}$

✓**Concept Check Answer**

negative

Helpful Hint

Be careful when identifying the base of an exponential expression.

$(-3)^2$	-3^2
Base is -3	Base is 3
$(-3)^2 = (-3)(-3) = 9$	$-3^2 = -(3 \cdot 3) = -9$

Objective B Finding Reciprocals

Addition and subtraction are related. Every difference of two numbers $a - b$ can be written as the sum $a + (-b)$. Multiplication and division are related also. For example, the quotient $6 \div 3$ can be written as the product $6 \cdot \frac{1}{3}$. Recall that the pair of numbers 3 and $\frac{1}{3}$ has a special relationship. Their product is 1 and they are called **reciprocals** or **multiplicative inverses** of each other.

Reciprocal or Multiplicative Inverse

Two numbers whose product is 1 are called **reciprocals** or **multiplicative inverses** of each other.

Example 9 Find the reciprocal of each number.

a. 22 Reciprocal is $\frac{1}{22}$ since $22 \cdot \frac{1}{22} = 1$.

b. $\frac{3}{16}$ Reciprocal is $\frac{16}{3}$ since $\frac{3}{16} \cdot \frac{16}{3} = 1$.

c. -10 Reciprocal is $-\frac{1}{10}$ since $-10 \cdot -\frac{1}{10} = 1$.

d. $-\frac{9}{13}$ Reciprocal is $-\frac{13}{9}$ since $-\frac{9}{13} \cdot -\frac{13}{9} = 1$.

e. 1.7 Reciprocal is $\frac{1}{1.7}$ since $1.7 \cdot \frac{1}{1.7} = 1$.

Work Practice 9

Practice 9

Find the reciprocal of each number.

a. 13 **b.** $\frac{7}{15}$ **c.** -5

d. $-\frac{8}{11}$ **e.** 7.9

Helpful Hint

The fraction $\frac{1}{1.7}$ is not simplified since the denominator is a decimal number. For the purpose of finding a reciprocal, we will leave the fraction as is.

Does the number 0 have a reciprocal? If it does, it is a number n such that $0 \cdot n = 1$. Notice that this can never be true since $0 \cdot n = 0$. This means that 0 has no reciprocal.

Helpful Hint

Remember that 0 has no reciprocal.

Quotients Involving Zero

The number 0 does not have a reciprocal.

Answers

9. a. $\frac{1}{13}$ **b.** $\frac{15}{7}$ **c.** $-\frac{1}{5}$ **d.** $-\frac{11}{8}$ **e.** $\frac{1}{7.9}$

Objective C Dividing Real Numbers 

We may now write a quotient as an equivalent product.

Quotient of Two Real Numbers

If a and b are real numbers and b is not 0, then

$$a \div b = \frac{a}{b} = a \cdot \frac{1}{b}$$

In other words, the quotient of two real numbers is the product of the first number and the multiplicative inverse or reciprocal of the second number.

Practice 10

Use the definition of the quotient of two numbers to find each quotient.

a. $-12 \div 4$ **b.** $\frac{-20}{-10}$ **c.** $\frac{36}{-4}$

Example 10 Use the definition of the quotient of two numbers to find each quotient. $\left(a \div b = a \cdot \frac{1}{b}\right)$

a. $-18 \div 3$ **b.** $\frac{-14}{-2}$ **c.** $\frac{20}{-4}$

Solution:

a. $-18 \div 3 = -18 \cdot \frac{1}{3} = -6$

b. $\frac{-14}{-2} = -14 \cdot -\frac{1}{2} = 7$

c. $\frac{20}{-4} = 20 \cdot -\frac{1}{4} = -5$

Work Practice 10

Since the quotient $a \div b$ can be written as the product $a \cdot \frac{1}{b}$, it follows that sign patterns for dividing two real numbers are the same as sign patterns for multiplying two real numbers.

Dividing Real Numbers

1. The quotient of two numbers with the *same* sign is a positive number.
2. The quotient of two numbers with *different* signs is a negative number.

Practice 11

Divide.

a. $\frac{-25}{5}$ **b.** $\frac{-48}{-6}$ **c.** $\frac{50}{-2}$ **d.** $\frac{-72}{0.2}$

Example 11 Divide.

a. $\frac{-30}{-10} = 3$ Same sign, so the quotient is positive.

b. $\frac{-100}{5} = -20$

c. $\frac{20}{-2} = -10$

d. $\frac{42}{-0.6} = -70$

Different signs, so the quotient is negative. $0.6\overline{)42.0}$ = 70.

Work Practice 11

Answers

10. a. −3 **b.** 2 **c.** −9 **11. a.** −5 **b.** 8 **c.** −25 **d.** −360

✓Concept Check What is wrong with the following calculation?

$$\frac{-36}{-9} = -4$$

In the examples on the previous page, we divided mentally or by long division. When we divide by a fraction, it is usually easier to multiply by its reciprocal.

Examples Divide.

12. $\frac{2}{3} \div \left(-\frac{5}{4}\right) = \frac{2}{3} \cdot \left(-\frac{4}{5}\right) = -\frac{8}{15}$

13. $-\frac{1}{6} \div \left(-\frac{2}{3}\right) = -\frac{1}{6} \cdot \left(-\frac{3}{2}\right) = \frac{3}{12} = \frac{\overset{1}{\cancel{3}}}{\underset{1}{\cancel{3}} \cdot 4} = \frac{1}{4}$

Work Practice 12–13

Practice 12–13

Divide.

12. $-\frac{5}{9} \div \frac{2}{3}$ **13.** $-\frac{2}{7} \div \left(-\frac{1}{5}\right)$

Our definition of the quotient of two real numbers does not allow for division by 0 because 0 does not have a reciprocal. How then do we interpret $\frac{3}{0}$? We say that an expression such as this one is **undefined**. Can we divide 0 by a number other than 0? Yes; for example,

$$\frac{0}{3} = 0 \cdot \frac{1}{3} = 0$$

Division Involving Zero

If a is a nonzero number, then $\frac{0}{a} = 0$ and $\frac{a}{0}$ is undefined.

Example 14 Divide, if possible.

a. $\frac{1}{0}$ is undefined. **b.** $\frac{0}{-3} = 0$

Work Practice 14

Practice 14

Divide if possible.

a. $\frac{-7}{0}$ **b.** $\frac{0}{-2}$

Notice that $\frac{12}{-2} = -6$, $-\frac{12}{2} = -6$, and $\frac{-12}{2} = -6$. This means that

$$\frac{12}{-2} = -\frac{12}{2} = \frac{-12}{2}$$

In other words, a single negative sign in a fraction can be written in the denominator, in the numerator, or in front of the fraction without changing the value of the fraction.

If a and b are real numbers, and $b \neq 0$, then $\frac{a}{-b} = \frac{-a}{b} = -\frac{a}{b}$.

Objective D Evaluating Expressions 

Examples combining basic arithmetic operations along with the principles of the order of operations help us to review these concepts of multiplying and dividing real numbers.

Answers

12. $-\frac{5}{6}$ **13.** $\frac{10}{7}$ **14. a.** undefined **b.** 0

✓Concept Check Answer

$\frac{-36}{-9} = 4$

Practice 15

Use order of operations to evaluate each expression.

a. $\frac{0(-5)}{3}$

b. $-3(-9) - 4(-4)$

c. $(-3)^2 + 2[(5 - 15) - |-4 - 1|]$

d. $\frac{-7(-4) + 2}{-10 - (-5)}$

e. $\frac{5(-2)^3 + 52}{-4 + 1}$

Example 15

Use order of operations to evaluate each expression.

a. $\frac{0(-8)}{2}$

b. $-4(-11) - 5(-2)$

c. $(-2)^2 + 3[(-3 - 2) - |4 - 6|]$

d. $\frac{(-12)(-3) + 4}{-7 - (-2)}$

e. $\frac{2(-3)^2 - 20}{|-5| + 4}$

Solution:

a. $\frac{0(-8)}{2} = \frac{0}{2} = 0$

b. $(-4)(-11) - 5(-2) = 44 - (-10)$ Find the products.

$= 44 + 10$ Add 44 to the opposite of -10.

$= 54$ Add.

c. $(-2)^2 + 3[(-3 - 2) - |4 - 6|] = (-2)^2 + 3[(-5) - |-2|]$ Simplify within innermost sets of grouping symbols.

$= (-2)^2 + 3[-5 - 2]$ Write $|-2|$ as 2.

$= (-2)^2 + 3(-7)$ Combine.

$= 4 + (-21)$ Evaluate $(-2)^2$ and multiply $3(-7)$.

$= -17$ Add.

For parts **d** and **e**, first simplify the numerator and denominator separately; then divide.

d. $\frac{(-12)(-3) + 4}{-7 - (-2)} = \frac{36 + 4}{-7 + 2}$

$= \frac{40}{-5}$

$= -8$ Divide.

e. $\frac{2(-3)^2 - 20}{|-5| + 4} = \frac{2 \cdot 9 - 20}{5 + 4} = \frac{18 - 20}{9} = \frac{-2}{9} = -\frac{2}{9}$

Work Practice 15

Using what we have learned about multiplying and dividing real numbers, we continue to practice evaluating algebraic expressions.

Practice 16

Evaluate each expression when $x = -1$ and $y = -5$.

a. $\frac{3y}{45x}$

b. $x^2 - y^3$

c. $\frac{x + y}{3x}$

Example 16

Evaluate each expression when $x = -2$ and $y = -4$.

a. $\frac{3x}{2y}$ b. $x^3 - y^2$ c. $\frac{x - y}{-x}$

Solution: Replace x with -2 and y with -4 and simplify.

a. $\frac{3x}{2y} = \frac{3(-2)}{2(-4)} = \frac{-6}{-8} = \frac{6}{8} = \frac{\overset{1}{\cancel{2}} \cdot 3}{\underset{1}{\cancel{2}} \cdot 4} = \frac{3}{4}$

Answers

15. a. 0 **b.** 43 **c.** -21 **d.** -6 **e.** -4 **16. a.** $\frac{1}{3}$ **b.** 126 **c.** 2

b. $x^3 - y^2 = (-2)^3 - (-4)^2$ Substitute the given values for the variables.
$= -8 - (16)$ Evaluate $(-2)^3$ and $(-4)^2$.
$= -8 + (-16)$ Write as a sum.
$= -24$ Add.

c. $\dfrac{x - y}{-x} = \dfrac{-2 - (-4)}{-(-2)} = \dfrac{-2 + 4}{2} = \dfrac{2}{2} = 1$

Work Practice 16

Helpful Hint

Remember: For additional help when replacing variables with replacement values, first place parentheses about any variables.

Evaluate $3x - y^2$ when $x = 5$ and $y = -4$.

$3x - y^2 = 3(x) - (y)^2$ Place parentheses about variables only.
$= 3(5) - (-4)^2$ Replace variables with values.
$= 15 - 16$ Simplify.
$= -1$

Objective E Solutions of Equations

We use our skills in multiplying and dividing real numbers to check possible solutions of an equation.

Example 17 Determine whether -10 is a solution of $\dfrac{-20}{x} + 15 = 2x$.

Solution:

$\dfrac{-20}{x} + 15 = 2x$ Original equation

$\dfrac{-20}{-10} + 15 \stackrel{?}{=} 2(-10)$ Replace x with -10.

$2 + 15 \stackrel{?}{=} -20$ Divide and multiply.

$17 = -20$ False

Since we have a false statement, -10 is *not* a solution of the equation.

Work Practice 17

Practice 17

Determine whether -8 is a solution of $\dfrac{x}{4} - 3 = x + 3$.

Objective F Solving Applications That Involve Multiplying or Dividing Numbers

Many real-life problems involve multiplication and division of numbers.

Answer

17. -8 is a solution

Practice 18

A card player had a score of -13 for each of four games. Find the total score.

Answer
18. -52

Example 18 Calculating a Total Golf Score

A professional golfer finished seven strokes under par (-7) for each of three days of a tournament. What was her total score for the tournament?

Solution: Although the key word is "total," since this is repeated addition of the same number, we multiply.

In words:	golfer's total score	=	number of days	·	score each day
	↓	↓	↓	↓	↓
Translate:	golfer's total	=	3	·	(-7)
		=	-21		

Thus, the golfer's total score was -21, or 21 strokes under par.

Work Practice 18

Calculator Explorations

Entering Negative Numbers on a Scientific Calculator

To enter a negative number on a scientific calculator, find a key marked [+/−]. (On some calculators, this key is marked [CHS] for "change sign.") To enter -8, for example, press the keys [8] [+/−]. The display will read [−8].

Entering Negative Numbers on a Graphing Calculator

To enter a negative number on a graphing calculator, find a key marked [(−)]. Do not confuse this key with the key [−], which is used for subtraction. To enter -8, for example, press the keys [(−)] [8]. The display will read [−8].

Operations with Real Numbers

To evaluate $-2(7 - 9) - 20$ on a calculator, press the keys

[2] [+/−] [×] [(] [7] [−] [9] [)] [−] [2] [0] [=],

or

[(−)] [2] [(] [7] [−] [9] [)] [−] [2] [0] [ENTER].

The display will read [−16] or [$-2(7-9)-20$ / -16]

Use a calculator to simplify each expression.

1. $-38(26 - 27)$
2. $-59(-8) + 1726$
3. $134 + 25(68 - 91)$
4. $45(32) - 8(218)$
5. $\dfrac{-50(294)}{175 - 205}$
6. $\dfrac{-444 - 444.8}{-181 - (-181)}$
7. $9^5 - 4550$
8. $5^8 - 6259$
9. $(-125)^2$ (Be careful.)
10. -125^2 (Be careful.)

Vocabulary, Readiness & Video Check

Use the choices below to fill in each blank. Each choice may be used more than once.

negative	0
positive	undefined

1. The product of a negative number and a positive number is a(n) __________ number.
2. The product of two negative numbers is a(n) __________ number.
3. The quotient of two negative numbers is a(n) __________ number.
4. The quotient of a negative number and a positive number is a(n) __________ number.
5. The product of a negative number and zero is __________.
6. The reciprocal of a negative number is a __________ number.
7. The quotient of 0 and a negative number is __________.
8. The quotient of a negative number and 0 is __________.

Martin-Gay Interactive Videos

See Video 1.6

Watch the section lecture video and answer the following questions.

Objective A 9. Explain the significance of the use of parentheses when comparing Examples 6 and 7.

Objective B 10. In Example 9, why is the reciprocal equal to $\frac{3}{2}$ and not $-\frac{3}{2}$?

Objective C 11. Before Example 11, the sign rules for division of real numbers are discussed. Are the sign rules for division the same as for multiplication? Why or why not?

Objective D 12. In Example 17, the importance of placing the replacement values in parentheses when evaluating is emphasized. Why?

Objective E 13. In Example 18, is 5 a solution of $-3x - 5 = -20$? Why or why not?

Objective F 14. In Example 19, explain why each loss of 4 yards is represented by -4 and not 4.

1.6 Exercise Set MyLab Math

Objective A *Multiply. See Examples 1 through 7.*

1. $-6(4)$
2. $-8(5)$
3. $2(-1)$
4. $7(-4)$
5. $-5(-10)$
6. $-6(-11)$
7. $-3 \cdot 15$
8. $-2 \cdot 37$
9. $-\frac{1}{2}\left(-\frac{3}{5}\right)$
10. $-\frac{1}{8}\left(-\frac{1}{3}\right)$
11. $5(-1.4)$
12. $6(-2.5)$
13. $(-1)(-3)(-5)$
14. $(-2)(-4)(-6)$
15. $(2)(-1)(-3)(0)$
16. $(3)(-5)(-2)(0)$

Evaluate. See Example 8.

17. $(-4)^2$ **18.** $(-3)^3$ **19.** -4^2 **20.** -6^2

21. $\left(-\frac{3}{4}\right)^2$ **22.** $\left(-\frac{2}{7}\right)^2$ **23.** -0.7^2 **24.** -0.8^2

Objective B *Find each reciprocal. See Example 9.*

25. $\frac{2}{3}$ **26.** $\frac{1}{7}$ **27.** -14 **28.** -8

29. $-\frac{3}{11}$ **30.** $-\frac{6}{13}$ **31.** 0.2 **32.** 1.5

Objective C *Divide. See Examples 10 through 14.*

33. $\frac{18}{-2}$ **34.** $\frac{36}{-9}$ **35.** $-48 \div 12$ **36.** $-60 \div 5$

37. $\frac{0}{-4}$ **38.** $\frac{0}{-9}$ **39.** $\frac{5}{0}$ **40.** $\frac{8}{0}$

41. $\frac{6}{7} \div \left(-\frac{1}{3}\right)$ **42.** $\frac{4}{5} \div \left(-\frac{1}{2}\right)$ **43.** $-3.2 \div -0.02$ **44.** $-4.9 \div -0.07$

Objectives A C Mixed Practice *Perform the indicated operation. See Examples 1 through 14.*

45. $(-8)(-8)$ **46.** $(-7)(-7)$ **47.** $\frac{2}{3}\left(-\frac{4}{9}\right)$ **48.** $\frac{2}{7}\left(-\frac{2}{11}\right)$

49. $\frac{-12}{-4}$ **50.** $\frac{-45}{-9}$ **51.** $\frac{30}{-2}$ **52.** $\frac{14}{-2}$

53. $(-5)^3$ **54.** $(-2)^5$ **55.** $(-0.2)^3$ **56.** $(-0.3)^3$

57. $-\frac{3}{4}\left(-\frac{8}{9}\right)$ **58.** $-\frac{5}{6}\left(-\frac{3}{10}\right)$ **59.** $-\frac{5}{9} \div \left(-\frac{3}{4}\right)$ **60.** $-\frac{1}{10} \div \left(-\frac{8}{11}\right)$

61. $-2.1(-0.4)$ **62.** $-1.3(-0.6)$ **63.** $\frac{-48}{1.2}$ **64.** $\frac{-86}{2.5}$

65. $(-3)^4$ **66.** -3^4 **67.** -1^7 **68.** $(-1)^7$

69. Multiply -11 by 11. **70.** Multiply -12 by 12.

71. Find the quotient of $-\frac{4}{9}$ and $\frac{4}{9}$. **72.** Find the quotient of $-\frac{5}{12}$ and $\frac{5}{12}$.

Mixed Practice (*Sections 1.4, 1.5, and 1.6*) *Perform the indicated operation.*

73. $-9 - 10$ **74.** $-8 - 11$ **75.** $-9(-10)$ **76.** $-8(-11)$

77. $7(-12)$ **78.** $6(-15)$ **79.** $7 + (-12)$ **80.** $6 + (-15)$

Objective D *Evaluate each expression. See Example 15.*

81. $\frac{-9(-3)}{-6}$ **82.** $\frac{-6(-3)}{-4}$ **83.** $-3(2 - 8)$

84. $-4(3 - 9)$ **85.** $-7(-2) - 3(-1)$ **86.** $-8(-3) - 4(-1)$

87. $2^2 - 3[(2 - 8) - (-6 - 8)]$ **88.** $3^2 - 2[(3 - 5) - (2 - 9)]$ **89.** $\frac{-6^2 + 4}{-2}$

90. $\frac{-3^2 + 4}{5}$ **91.** $\frac{-3 - 5^2}{2(-7)}$ **92.** $\frac{-2 - 4^2}{3(-6)}$ **93.** $\frac{22 + (3)(-2)^2}{-5 - 2}$

94. $\frac{-20 + (-4)^2(3)}{1 - 5}$ **95.** $\frac{(-4)^2 - 16}{4 - 12}$ **96.** $\frac{(-2)^2 - 4}{4 - 9}$ **97.** $\frac{6 - 2(-3)}{4 - 3(-2)}$

98. $\frac{8 - 3(-2)}{2 - 5(-4)}$ **99.** $\frac{|5 - 9| + |10 - 15|}{|2(-3)|}$ **100.** $\frac{|-3 + 6| + |-2 + 7|}{|-2 \cdot 2|}$

101. $\frac{-7(-1) + (-3)4}{(-2)(5) + (-6)(-8)}$ **102.** $\frac{8(-7) + (-2)(-6)}{(-9)(3) + (-10)(-11)}$

Evaluate each expression when $x = -5$ and $y = -3$. See Example 16.

103. $\frac{2x - 5}{y - 2}$ **104.** $\frac{2y - 12}{x - 4}$ **105.** $\frac{6 - y}{x - 4}$ **106.** $\frac{10 - y}{x - 8}$

107. $\frac{4 - 2x}{y + 3}$ **108.** $\frac{2y + 3}{-5 - x}$ **109.** $\frac{x^2 + y}{3y}$ **110.** $\frac{y^2 - x}{2x}$

Objective E *Decide whether the given number is a solution of the given equation. See Example 17.*

111. $-3x - 5 = -20$; 5 **112.** $17 - 4x = x + 27$; -2 **113.** $\frac{x}{5} + 2 = -1$; 15

114. $\frac{x}{6} - 3 = 5$; 48 **115.** $\frac{x - 3}{7} = -2$; -11 **116.** $\frac{x + 4}{5} = -6$; -30

Objective F Translating *Translate each phrase to an expression. Use x to represent "a number." See Example 18.*

117. The product of -71 and a number

118. The quotient of -8 and a number

119. Subtract a number from -16.

120. The sum of a number and -12

121. -29 increased by a number

122. The difference of a number and -10

123. Divide a number by -33.

124. Multiply a number by -17.

Solve. See Example 18.

125. A football team lost four yards on each of three consecutive plays. Represent the total loss as a product of signed numbers and find the total loss.

126. A stockbroker lost $400 on each of seven consecutive days in the stock market. Represent his total loss as a product of signed numbers and find his total loss.

127. A deep-sea diver must move up or down in the water in short steps in order to keep from getting a physical condition called the "bends." Suppose a diver moves down from the surface in five steps of 20 feet each. Represent his total movement as a product of signed numbers and find the depth.

128. A weather forecaster predicts that the temperature will drop five degrees each hour for the next six hours. Represent this drop as a product of signed numbers and find the total drop in temperature.

Concept Extensions

State whether each statement is true or false. See the first Concept Check in this section.

129. The product of seven negative integers is negative.

130. The product of seven positive integers is positive.

131. The product of eight negative integers is negative.

132. The product of eight positive integers is positive.

Study the bar graph below showing the average surface temperatures of planets. Use Exercises 133 and 134 to complete the planet temperatures on the graph. (Pluto is now classified as a dwarf planet.)

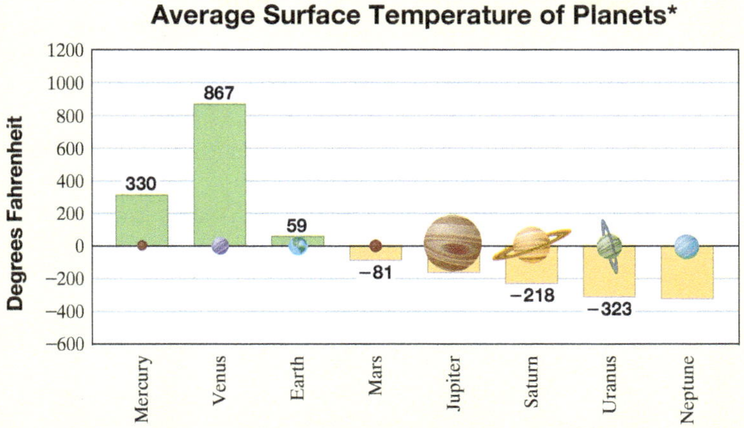

Average Surface Temperature of Planets*

*For some planets, the temperature given is the temperature where the atmospheric pressure equals 1 Earth atmosphere. *Source: The World Almanac*

133. The surface temperature of Jupiter is twice the temperature of Mars. Find this temperature.

134. The surface temperature of Neptune is equal to the temperature of Mercury divided by -1. Find this temperature.

135. For the second quarter of 2018, Wal-Mart Stores, Inc. posted a loss of \$51 million on investment. If this trend was consistent for each month of the quarter, how much would you expect this loss to have been for each month? (*Source:* Wal-Mart Stores, Inc.)

136. For the first quarter of 2018, Ford Motor Company, posted a loss of \$54 million in the Middle East and Africa. If this trend was consistent for each month of the quarter, how much would you expect this loss to have been for each month? (*Source:* Ford Motor Company)

137. Explain why the product of an even number of negative numbers is a positive number.

138. If a and b are any real numbers, is the statement $a \cdot b = b \cdot a$ always true? Why or why not?

139. Find two real numbers that are their own reciprocal. Explain why there are only two.

140. Explain why 0 has no reciprocal.

Mixed Practice (*1.4, 1.5, and 1.6*)

Write each as an algebraic expression. Then simplify the expression.

141. 7 subtracted from the quotient of 0 and 5

142. Twice the sum of -3 and -4

143. -1 added to the product of -8 and -5

144. The difference of -9 and the product of -4 and -6

1.7 Properties of Real Numbers

Objectives

A Use the Commutative and Associative Properties.

B Use the Distributive Property.

C Use the Identity and Inverse Properties.

Objective A Using the Commutative and Associative Properties

In this section we review properties of real numbers with which we are already familiar. Throughout this section, the variables a, b, and c represent real numbers.

We know that order does not matter when adding numbers. For example, we know that $7 + 5$ is the same as $5 + 7$. This property is given a special name—the **commutative property of addition**. We also know that order does not matter when multiplying numbers. For example, we know that $-5(6) = 6(-5)$. This property means that multiplication is commutative also and is called the **commutative property of multiplication**.

Commutative Properties

Addition: $a + b = b + a$

Multiplication: $a \cdot b = b \cdot a$

These properties state that the *order* in which any two real numbers are added or multiplied does not change their sum or product. For example, if we let $a = 3$ and $b = 5$, then the commutative properties guarantee that

$$3 + 5 = 5 + 3 \quad \text{and} \quad 3 \cdot 5 = 5 \cdot 3$$

Helpful Hint

Is subtraction also commutative? Try an example. Is $3 - 2 = 2 - 3$? **No!** The left side of this statement equals 1; the right side equals -1. There is no commutative property of subtraction. Similarly, there is no commutative property of division. For example, $10 \div 2$ does not equal $2 \div 10$.

Practice 1

Use a commutative property to complete each statement.

a. $7 \cdot y =$ _____

b. $4 + x =$ _____

Example 1 Use a commutative property to complete each statement.

a. $x + 5 =$ _____ **b.** $3 \cdot x =$ _____

Solution:

a. $x + 5 = 5 + x$ By the commutative property of addition

b. $3 \cdot x = x \cdot 3$ By the commutative property of multiplication

Work Practice 1

Concept Check Which of the following pairs of actions are commutative?

a. "raking the leaves" and "bagging the leaves"

b. "putting on your left glove" and "putting on your right glove"

c. "putting on your coat" and "putting on your shirt"

d. "reading a novel" and "reading a newspaper"

Answers

1. a. $y \cdot 7$ **b.** $x + 4$

Concept Check Answer

b, d

Let's now discuss grouping numbers. When we add three numbers, the way in which they are grouped or associated does not change their sum. For example, we know that $2 + (3 + 4) = 2 + 7 = 9$. This result is the same if we group the numbers differently. In other words, $(2 + 3) + 4 = 5 + 4 = 9$, also. Thus, $2 + (3 + 4) = (2 + 3) + 4$. This property is called the **associative property of addition**.

In the same way, changing the grouping of numbers when multiplying does not change their product. For example, $2 \cdot (3 \cdot 4) = (2 \cdot 3) \cdot 4$ (check it). This is the **associative property of multiplication**.

Associative Properties

Addition: $(a + b) + c = a + (b + c)$

Multiplication: $(a \cdot b) \cdot c = a \cdot (b \cdot c)$

These properties state that the way in which three numbers are *grouped* does not change their sum or their product.

Example 2 Use an associative property to complete each statement.

a. $5 + (4 + 6) =$ _____ **b.** $(-1 \cdot 2) \cdot 5 =$ _____

c. $(m + n) + 9 =$ _____ **d.** $(xy) \cdot 12 =$ _____

Solution:

a. $5 + (4 + 6) = (5 + 4) + 6$ By the associative property of addition

b. $(-1 \cdot 2) \cdot 5 = -1 \cdot (2 \cdot 5)$ By the associative property of multiplication

c. $(m + n) + 9 = m + (n + 9)$ By the associative property of addition

d. $(xy) \cdot 12 = x \cdot (y \cdot 12)$ Recall that xy means $x \cdot y$.

Work Practice 2

Practice 2

Use an associative property to complete each statement.

a. $5 \cdot (-3 \cdot 6) =$ _____

b. $(-2 + 7) + 3 =$ _____

c. $(q + r) + 17 =$ _____

d. $(ab) \cdot 21 =$ _____

Helpful Hint

Remember the difference between the commutative properties and the associative properties. The commutative properties have to do with the *order* of numbers and the associative properties have to do with the *grouping* of numbers.

Examples Determine whether each statement is true by an associative property or a commutative property.

3. $(7 + 10) + 4 = (10 + 7) + 4$ Since the order of two numbers was changed and their grouping was not, this is true by the commutative property of addition.

4. $2 \cdot (3 \cdot 1) = (2 \cdot 3) \cdot 1$ Since the grouping of the numbers was changed and their order was not, this is true by the associative property of multiplication.

Work Practice 3–4

Practice 3–4

Determine whether each statement is true by an associative property or a commutative property.

3. $5 \cdot (4 \cdot 7) = 5 \cdot (7 \cdot 4)$

4. $-2 + (4 + 9) = (-2 + 4) + 9$

Let's now illustrate how these properties can help us simplify expressions.

Answers

2. a. $(5 \cdot -3) \cdot 6$ **b.** $-2 + (7 + 3)$ **c.** $q + (r + 17)$ **d.** $a \cdot (b \cdot 21)$ **3.** commutative **4.** associative

Practice 5–6

Simplify each expression.

5. $(-3 + x) + 17$

6. $4(5x)$

Examples Simplify each expression.

5. $10 + (x + 12) = 10 + (12 + x)$ By the commutative property of addition
$= (10 + 12) + x$ By the associative property of addition
$= 22 + x$ Add.

6. $-3(7x) = (-3 \cdot 7)x$ By the associative property of multiplication
$= -21x$ Multiply.

Work Practice 5–6

Objective B Using the Distributive Property 

The **distributive property of multiplication over addition** is used repeatedly throughout algebra. It is useful because it allows us to write a product as a sum or a sum as a product.

We know that $7(2 + 4) = 7(6) = 42$. Compare that with

$$7(2) + 7(4) = 14 + 28 = 42$$

Since both original expressions equal 42, they must equal each other, or

$$7(2 + 4) = 7(2) + 7(4)$$

This is an example of the distributive property. The product on the left side of the equal sign is equal to the sum on the right side. We can think of the 7 as being distributed to each number inside the parentheses.

Distributive Property of Multiplication over Addition

$$a(b + c) = ab + ac$$

Since multiplication is commutative, this property can also be written as

$$(b + c)a = ba + ca$$

The distributive property can also be extended to more than two numbers inside the parentheses. For example,

$$3(x + y + z) = 3(x) + 3(y) + 3(z)$$
$$= 3x + 3y + 3z$$

Since we define subtraction in terms of addition, the distributive property is also true for subtraction. For example,

$$2(x - y) = 2(x) - 2(y)$$
$$= 2x - 2y$$

Practice 7–12

Use the distributive property to write each expression without parentheses. Then simplify the result.

7. $5(x + y)$

8. $-3(2 + 7x)$

9. $4(x + 6y - 2z)$

10. $-1(3 - a)$

11. $-(8 + a - b)$

12. $\frac{1}{2}(2x + 4) + 9$

Examples Use the distributive property to write each expression without parentheses. Then simplify the result.

7. $2(x + y) = 2(x) + 2(y)$
$= 2x + 2y$

8. $-5(-3 + 2z) = -5(-3) + (-5)(2z)$
$= 15 - 10z$

9. $5(x + 3y - z) = 5(x) + 5(3y) - 5(z)$
$= 5x + 15y - 5z$

Answers

5. $14 + x$ 6. $20x$ 7. $5x + 5y$ 8. $-6 - 21x$ 9. $4x + 24y - 8z$ 10. $-3 + a$ 11. $-8 - a + b$ 12. $x + 11$

10. $-1(2 - y) = (-1)(2) - (-1)(y)$
$= -2 + y$

11. $-(3 + x - w) = -1(3 + x - w)$
$= (-1)(3) + (-1)(x) - (-1)(w)$
$= -3 - x + w$

Helpful Hint Notice in Example 11 that $-(3 + x - w)$ can be rewritten as $-1(3 + x - w)$.

12. $\frac{1}{2}(6x + 14) + 10 = \frac{1}{2}(6x) + \frac{1}{2}(14) + 10$ Apply the distributive property.
$= 3x + 7 + 10$ Multiply.
$= 3x + 17$ Add.

Work Practice 7–12

The distributive property can also be used to write a sum as a product.

Examples Use the distributive property to write each sum as a product.

13. $8 \cdot 2 + 8 \cdot x = 8(2 + x)$

14. $7s + 7t = 7(s + t)$

Work Practice 13–14

Practice 13–14

Use the distributive property to write each sum as a product.

13. $9 \cdot 3 + 9 \cdot y$

14. $4x + 4y$

Objective C Using the Identity and Inverse Properties

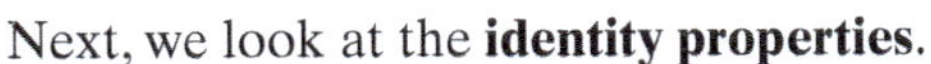

Next, we look at the **identity properties**.

The number 0 is called the identity for addition because when 0 is added to any real number, the result is the same real number. In other words, the *identity* of the real number is not changed.

The number 1 is called the identity for multiplication because when a real number is multiplied by 1, the result is the same real number. In other words, the *identity* of the real number is not changed.

Identities for Addition and Multiplication

0 is the identity element for addition.

$a + 0 = a$ and $0 + a = a$

1 is the identity element for multiplication.

$a \cdot 1 = a$ and $1 \cdot a = a$

Notice that 0 is the *only* number that can be added to any real number with the result that the sum is the same real number. Also, 1 is the *only* number that can be multiplied by any real number with the result that the product is the same real number.

Additive inverses or **opposites** were introduced in Section 1.4. Two numbers are called additive inverses or opposites if their sum is 0. The additive inverse or opposite of 6 is -6 because $6 + (-6) = 0$. The additive inverse or opposite of -5 is 5 because $-5 + 5 = 0$.

Reciprocals or **multiplicative inverses** were introduced in Section 1.6. Two nonzero numbers are called reciprocals or multiplicative inverses if their product is 1. The reciprocal or multiplicative inverse of $\frac{2}{3}$ is $\frac{3}{2}$ because $\frac{2}{3} \cdot \frac{3}{2} = 1$. Likewise, the reciprocal of -5 is $-\frac{1}{5}$ because $-5\left(-\frac{1}{5}\right) = 1$.

Answers
13. $9(3 + y)$ **14.** $4(x + y)$

Additive or Multiplicative Inverses

The numbers a and $-a$ are additive inverses or opposites of each other because their sum is 0; that is,

$$a + (-a) = 0$$

The numbers b and $\frac{1}{b}$ (for $b \neq 0$) are reciprocals or multiplicative inverses of each other because their product is 1; that is,

$$b \cdot \frac{1}{b} = 1$$

✓**Concept Check** Which of the following is

a. the opposite of $-\frac{3}{10}$, and

b. the reciprocal of $-\frac{3}{10}$?

$1, -\frac{10}{3}, \frac{3}{10}, 0, \frac{10}{3}, -\frac{3}{10}$

Examples Name the property illustrated by each true statement.

15. $3(x + y) = 3 \cdot x + 3 \cdot y$ Distributive property

16. $(x + 7) + 9 = x + (7 + 9)$ Associative property of addition (grouping changed)

17. $(b + 0) + 3 = b + 3$ Identity element for addition

18. $2 \cdot (z \cdot 5) = 2 \cdot (5 \cdot z)$ Commutative property of multiplication (order changed)

19. $-2 \cdot \left(-\frac{1}{2}\right) = 1$ Multiplicative inverse property

20. $-2 + 2 = 0$ Additive inverse property

21. $-6 \cdot (y \cdot 2) = (-6 \cdot 2) \cdot y$ Commutative and associative properties of multiplication (order and grouping changed)

Work Practice 15–21

Practice 15–21

Name the property illustrated by each true statement.

15. $7(a + b) = 7 \cdot a + 7 \cdot b$

16. $12 + y = y + 12$

17. $-4 \cdot (6 \cdot x) = (-4 \cdot 6) \cdot x$

18. $6 + (z + 2) = 6 + (2 + z)$

19. $3\left(\frac{1}{3}\right) = 1$

20. $(x + 0) + 23 = x + 23$

21. $(7 \cdot y) \cdot 10 = y \cdot (7 \cdot 10)$

Answers

15. distributive property
16. commutative property of addition
17. associative property of multiplication
18. commutative property of addition
19. multiplicative inverse property
20. identity element for addition
21. commutative and associative properties of multiplication

✓**Concept Check Answers**

a. $\frac{3}{10}$ **b.** $-\frac{10}{3}$

Vocabulary, Readiness & Video Check

Use the choices below to fill in each blank.

distributive property	associative property of multiplication	commutative property of addition
opposites or additive inverses	associative property of addition	
reciprocals or multiplicative inverses	commutative property of multiplication	

1. $x + 5 = 5 + x$ is a true statement by the ______________.

2. $x \cdot 5 = 5 \cdot x$ is a true statement by the ______________.

3. $3(y + 6) = 3 \cdot y + 3 \cdot 6$ is true by the ______________.

4. $2 \cdot (x \cdot y) = (2 \cdot x) \cdot y$ is a true statement by the ______________.

5. $x + (7 + y) = (x + 7) + y$ is a true statement by the ______________.

6. The numbers $-\frac{2}{3}$ and $-\frac{3}{2}$ are called ______________________.

7. The numbers $-\frac{2}{3}$ and $\frac{2}{3}$ are called ______________________.

Martin-Gay Interactive Videos *Watch the section lecture video and answer the following questions.*

See Video 1.7 

Objective A 8. The commutative properties are discussed in Examples 1 and 2 and the associative properties are discussed in Examples 3–7. What's the one word used again and again to describe the commutative property? The associative property?

Objective B 9. In Example 10, what point is made about the term 2?

Objective C 10. Complete these statements based on the lecture given before Example 12. 

- The identity element for addition is ______ because if we add ______ to any real number, the result is that real number.
- The identity element for multiplication is ______ because any real number times ______ gives a result of that original real number.

1.7 Exercise Set MyLab Math

Objective A *Use a commutative property to complete each statement. See Examples 1 and 3.*

1. $x + 16 =$ ______
2. $8 + y =$ ______
3. $-4 \cdot y =$ ______
4. $-2 \cdot x =$ ______
5. $xy =$ ______
6. $ab =$ ______
7. $2x + 13 =$ ______
8. $19 + 3y =$ ______

Use an associative property to complete each statement. See Examples 2 and 4.

9. $(xy) \cdot z =$ ______
10. $3 \cdot (x \cdot y) =$ ______
11. $2 + (a + b) =$ ______
12. $(y + 4) + z =$ ______
13. $4 \cdot (ab) =$ ______
14. $(-3y) \cdot z =$ ______
15. $(a + b) + c =$ ______
16. $6 + (r + s) =$ ______

Use the commutative and associative properties to simplify each expression. See Examples 5 and 6.

17. $8 + (9 + b)$
18. $(r + 3) + 11$
19. $4(6y)$
20. $2(42x)$
21. $\frac{1}{5}(5y)$
22. $\frac{1}{8}(8z)$
23. $(13 + a) + 13$
24. $7 + (x + 4)$
25. $-9(8x)$
26. $-3(12y)$
27. $\frac{3}{4}\left(\frac{4}{3}s\right)$
28. $\frac{2}{7}\left(\frac{7}{2}r\right)$
29. $-\frac{1}{2}(5x)$
30. $-\frac{1}{3}(7x)$

Objective B *Use the distributive property to write each expression without parentheses. Then simplify the result, if possible. See Examples 7 through 12.*

31. $4(x + y)$
32. $7(a + b)$
33. $9(x - 6)$
34. $11(y - 4)$
35. $2(3x + 5)$
36. $5(7 + 8y)$
37. $7(4x - 3)$
38. $3(8x - 1)$
39. $3(6 + x)$
40. $2(x + 5)$
41. $-2(y - z)$
42. $-3(z - y)$
43. $-\frac{1}{3}(3y + 5)$
44. $-\frac{1}{2}(2r + 11)$
45. $5(x + 4m + 2)$
46. $8(3y + z - 6)$
47. $-4(1 - 2m + n) + 4$
48. $-4(4 + 2p + 5r) + 16$
49. $-(5x + 2)$
50. $-(9r + 5)$
51. $-(r - 3 - 7p)$
52. $-(q - 2 + 6r)$
53. $\frac{1}{2}(6x + 7) + \frac{1}{2}$
54. $\frac{1}{4}(4x - 2) - \frac{7}{2}$
55. $-\frac{1}{3}(3x - 9y)$
56. $-\frac{1}{5}(10a - 25b)$
57. $3(2r + 5) - 7$
58. $10(4s + 6) - 40$
59. $-9(4x + 8) + 2$
60. $-11(5x + 3) + 10$
61. $-0.4(4x + 5) - 0.5$
62. $-0.6(2x + 1) - 0.1$

Use the distributive property to write each sum as a product. See Examples 13 and 14.

63. $4 \cdot 1 + 4 \cdot y$
64. $14 \cdot z + 14 \cdot 5$
65. $11x + 11y$
66. $9a + 9b$
67. $(-1) \cdot 5 + (-1) \cdot x$
68. $(-3)a + (-3)y$
69. $30a + 30b$
70. $25x + 25y$

Objectives A C Mixed Practice *Name the property illustrated by each true statement. See Examples 15 through 21.*

71. $3 \cdot 5 = 5 \cdot 3$
72. $4(3 + 8) = 4 \cdot 3 + 4 \cdot 8$
73. $2 + (x + 5) = (2 + x) + 5$
74. $9 \cdot (x \cdot 7) = (9 \cdot x) \cdot 7$
75. $(x + 9) + 3 = (9 + x) + 3$
76. $1 \cdot 9 = 9$
77. $(4 \cdot y) \cdot 9 = 4 \cdot (y \cdot 9)$
78. $-4 \cdot (8 \cdot 3) = (8 \cdot 3) \cdot (-4)$
79. $0 + 6 = 6$
80. $(a + 9) + 6 = a + (9 + 6)$
81. $-4(y + 7) = -4 \cdot y + (-4) \cdot 7$
82. $(11 + r) + 8 = (r + 11) + 8$

83. $6 \cdot \frac{1}{6} = 1$

84. $r + 0 = r$

85. $-6 \cdot 1 = -6$

86. $-\frac{3}{4}\left(-\frac{4}{3}\right) = 1$

Concept Extensions

Fill in the table with the opposite (additive inverse), the reciprocal (multiplicative inverse), or the expression. Assume that the value of each expression is not 0.

	87.	88.	89.	90.	91.	92.
Expression	8	$-\frac{2}{3}$	x	$4y$		
Opposite						$7x$
Reciprocal					$\frac{1}{2x}$	

Decide whether each statement is true or false. See the second Concept Check in this section.

93. The opposite of $-\frac{a}{2}$ is $-\frac{2}{a}$.

94. The reciprocal of $-\frac{a}{2}$ is $\frac{a}{2}$.

Determine which pairs of actions are commutative. See the first Concept Check in this section.

95. "taking a test" and "studying for the test"

96. "putting on your shoes" and "putting on your socks"

97. "putting on your left shoe" and "putting on your right shoe"

98. "reading the sports section" and "reading the comics section"

99. "mowing the lawn" and "trimming the hedges"

100. "baking a cake" and "eating the cake"

101. "feeding the dog" and "feeding the cat"

102. "dialing a number" and "turning on the cell phone"

Name the property illustrated by each step.

103. **a.** $\triangle + (\square + \bigcirc) = (\square + \bigcirc) + \triangle$

b. $= (\bigcirc + \square) + \triangle$

c. $= \bigcirc + (\square + \triangle)$

104. **a.** $(x + y) + z = x + (y + z)$

b. $= (y + z) + x$

c. $= (z + y) + x$

105. Explain why 0 is called the identity element for addition.

106. Explain why 1 is called the identity element for multiplication.

107. Write an example that shows that division is not commutative.

108. Write an example that shows that subtraction is not commutative.

1.8 Simplifying Expressions 

Objectives

A Identify Terms, Like Terms, and Unlike Terms.

B Combine Like Terms.

C Simplify Expressions Containing Parentheses.

D Write Word Phrases as Algebraic Expressions.

As we explore in this section, we will see that an expression such as $3x + 2x$ is not written as simply as possible. This is because—even without replacing x by a value—we can perform the indicated addition.

Objective A Identifying Terms, Like Terms, and Unlike Terms

Before we practice simplifying expressions, we must learn some new language. A **term** is a number or the product of a number and variables raised to powers.

Terms

$$-y, \quad 2x^3, \quad -5, \quad 3xz^2, \quad \frac{2}{y}, \quad 0.8z$$

The **numerical coefficient** of a term is the numerical factor. The numerical coefficient of $3x$ is 3. Recall that $3x$ means $3 \cdot x$.

Term	Numerical Coefficient
$3x$	3
$\frac{y^3}{5}$	$\frac{1}{5}$ since $\frac{y^3}{5}$ means $\frac{1}{5} \cdot y^3$
$-0.7ab^3c^5$	-0.7
z	1
$-y$	-1
-5	-5

Helpful Hint

The term z means $1z$ and thus has a numerical coefficient of 1.
The term $-y$ means $-1y$ and thus has a numerical coefficient of -1.

Practice 1

Identify the numerical coefficient of each term.

a. $-4x$ **b.** $15y^3$ **c.** x

d. $-y$ **e.** $\frac{z}{4}$

Example 1 Identify the numerical coefficient of each term.

a. $-3y$ **b.** $22z^4$ **c.** y **d.** $-x$ **e.** $\frac{x}{7}$

Solution:

a. The numerical coefficient of $-3y$ is -3.

b. The numerical coefficient of $22z^4$ is 22.

c. The numerical coefficient of y is 1, since y is $1y$.

d. The numerical coefficient of $-x$ is -1, since $-x$ is $-1x$.

e. The numerical coefficient of $\frac{x}{7}$ is $\frac{1}{7}$, since $\frac{x}{7}$ is $\frac{1}{7} \cdot x$.

Work Practice 1

Answers

1. a. -4 **b.** 15 **c.** 1 **d.** -1 **e.** $\frac{1}{4}$

Terms with the same variables raised to exactly the same powers are called **like terms**. Terms that aren't like terms are called **unlike terms**.

Like Terms	Unlike Terms	Reason Why
$3x, 2x$	$5x, 5x^2$	Why? Same variable x, but different powers of x and x^2
$-6x^2y, 2x^2y, 4x^2y$	$7y, 3z, 8x^2$	Why? Different variables
$2ab^2c^3, ac^3b^2$	$6abc^3, 6ab^2$	Why? Different variables and different powers

Helpful Hint

In like terms, each variable and its exponent must match exactly, but these factors don't need to be in the same order.

$2x^2y$ and $3yx^2$ are like terms.

Example 2 Determine whether the terms are like or unlike.

a. $2x, 3x^2$ **b.** $4x^2y, x^2y, -2x^2y$ **c.** $-2yz, -3zy$
d. $-x^4, x^4$ **e.** $-8a^5, 8a^5$

Solution:

a. Unlike terms, since the exponents on x are not the same.
b. Like terms, since each variable and its exponent match.
c. Like terms, since $zy = yz$ by the commutative property of multiplication.
d. Like terms. The variable and its exponent match.
e. Like terms. The variable and its exponent match.

Work Practice 2

Practice 2

Determine whether the terms are like or unlike.

a. $7x^2, -6x^3$
b. $3x^2y^2, -x^2y^2, 4x^2y^2$
c. $-5ab, 3ba$
d. $2x^3, 4y^3$
e. $-7m^4, 7m^4$

Objective B Combining Like Terms 

An algebraic expression containing the sum or difference of like terms can be simplified by applying the distributive property. For example, by the distributive property, we rewrite the sum of the like terms $6x + 2x$ as

$$6x + 2x = (6 + 2)x = 8x$$

Also,

$$-y^2 + 5y^2 = (-1 + 5)y^2 = 4y^2$$

Simplifying the sum or difference of like terms is called **combining like terms**.

Example 3 Simplify each expression by combining like terms.

a. $7x - 3x$ **b.** $10y^2 + y^2$
c. $8x^2 + 2x - 3x$ **d.** $9n^2 - 5n^2 + n^2$

Solution:

a. $7x - 3x = (7 - 3)x = 4x$
b. $10y^2 + y^2 = (10 + 1)y^2 = 11y^2$
c. $8x^2 + 2x - 3x = 8x^2 + (2 - 3)x = 8x^2 - 1x$ or $8x^2 - x$
d. $9n^2 - 5n^2 + n^2 = (9 - 5 + 1)n^2 = 5n^2$

Work Practice 3

Practice 3

Simplify each expression by combining like terms.

a. $9y - 4y$
b. $11x^2 + x^2$
c. $5y - 3x + 4x$
d. $14m^2 - m^2 + 3m^2$

Answers

2. a. unlike **b.** like **c.** like **d.** unlike **e.** like **3. a.** $5y$ **b.** $12x^2$ **c.** $5y + x$ **d.** $16m^2$

The preceding examples suggest the following.

Combining Like Terms

To **combine like terms**, combine the numerical coefficients and multiply the result by the common variable factors.

Practice 4–7

Simplify each expression by combining like terms.

4. $7y + 2y + 6 + 10$
5. $-2x + 4 + x - 11$
6. $3z - 3z^2$
7. $8.9y + 4.2y - 3$

Examples Simplify each expression by combining like terms.

4. $2x + 3x + 5 + 2 = (2 + 3)x + (5 + 2)$
$= 5x + 7$

5. $-5a - 3 + a + 2 = -5a + 1a + (-3 + 2)$
$= (-5 + 1)a + (-3 + 2)$
$= -4a - 1$

6. $4y - 3y^2$
These two terms cannot be combined because they are unlike terms.

7. $2.3x + 5x - 6 = (2.3 + 5)x - 6$
$= 7.3x - 6$

Work Practice 4–7

Objective C Simplifying Expressions Containing Parentheses

In simplifying expressions we make frequent use of the distributive property to remove parentheses.

It may be helpful to study the examples below.

$+(3a + 2) = +1(3a + 2) = +1(3a) + (+1)(2) = 3a + 2$

(+ means +1)

$-(3a + 2) = -1(3a + 2) = -1(3a) + (-1)(2) = -3a - 2$

(− means −1)

Practice 8–10

Find each product by using the distributive property to remove parentheses.

8. $3(11y + 6)$
9. $-4(x + 0.2y - 3)$
10. $-(3x + 2y + z - 1)$

Examples Find each product by using the distributive property to remove parentheses.

8. $5(3x + 2) = 5(3x) + 5(2)$ Apply the distributive property.
$= 15x + 10$ Multiply.

9. $-2(y + 0.3z - 1) = -2(y) + (-2)(0.3z) - (-2)(1)$ Apply the distributive property.
$= -2y - 0.6z + 2$ Multiply.

10. $-(9x + y - 2z + 6) = -1(9x + y - 2z + 6)$ Distribute −1 over each term.
$= -1(9x) + (-1)(y) - (-1)(2z) + (-1)(6)$
$= -9x - y + 2z - 6$

Work Practice 8–10

Answers

4. $9y + 16$ 5. $-x - 7$ 6. $3z - 3z^2$ 7. $13.1y - 3$ 8. $33y + 18$ 9. $-4x - 0.8y + 12$ 10. $-3x - 2y - z + 1$

Helpful Hint

If a "−" sign precedes parentheses, the sign of each term inside the parentheses is changed when the distributive property is applied to remove the parentheses.

Examples:

$$-(2x + 1) = -2x - 1$$
$$-(x - 2y) = -x + 2y$$
$$-(-5x + y - z) = 5x - y + z$$
$$-(-3x - 4y - 1) = 3x + 4y + 1$$

When simplifying an expression containing parentheses, we often use the distributive property first to remove parentheses and then again to combine any like terms.

Examples

Simplify each expression.

11. $3(2x - 5) + 1 = 6x - 15 + 1$ Apply the distributive property.
$= 6x - 14$ Combine like terms.

12. $8 - (7x + 2) + 3x = 8 - 7x - 2 + 3x$ Apply the distributive property.
$= -7x + 3x + 8 - 2$
$= -4x + 6$ Combine like terms.

13. $-2(4x + 7) - (3x - 1) = -8x - 14 - 3x + 1$ Apply the distributive property.
$= -11x - 13$ Combine like terms.

14. $9 + 3(4x - 10) = 9 + 12x - 30$ Apply the distributive property.
$= -21 + 12x$ Combine like terms.
or $12x - 21$

Helpful Hint

Don't forget to use the distributive property and multiply before adding or subtracting like terms.

Work Practice 11–14

Practice 11–14

Simplify each expression.

11. $4(4x - 6) + 20$
12. $5 - (3x + 9) + 6x$
13. $-3(7x + 1) - (4x - 2)$
14. $8 + 11(2y - 9)$

Example 15

Subtract $4x - 2$ from $2x - 3$.

Solution: We first note that "subtract $4x - 2$ **from** $2x - 3$" translates to $(2x - 3) - (4x - 2)$. Notice that parentheses were placed around each given expression. This is to ensure that the entire expression after the subtraction sign is subtracted. Next, we simplify the algebraic expression.

$(2x - 3) - (4x - 2) = 2x - 3 - 4x + 2$ Apply the distributive property.
$= -2x - 1$ Combine like terms.

Work Practice 15

Practice 15

Subtract $9x - 10$ from $4x - 3$.

Answers

11. $16x - 4$ **12.** $3x - 4$ **13.** $-25x - 1$ **14.** $-91 + 22y$ **15.** $-5x + 7$

Objective D Writing Algebraic Expressions 

To prepare for problem solving, we next practice writing word phrases as algebraic expressions.

Examples Write each phrase as an algebraic expression and simplify if possible. Let x represent the unknown number.

16. Twice a number, plus 6

$2x + 6$

This expression cannot be simplified.

17. The difference of a number and 4, divided by 7

$(x - 4) \div 7$ or $\frac{x-4}{7}$

This expression cannot be simplified.

18. Five plus the sum of a number and 1

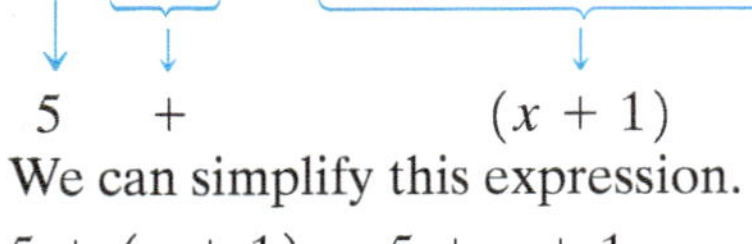

$5 + (x + 1)$

We can simplify this expression.

$$5 + (x + 1) = 5 + x + 1$$
$$= 6 + x$$

19. Four times the sum of a number and 3

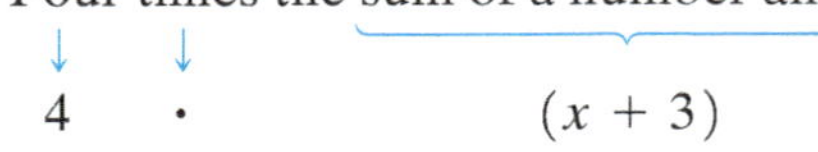

$4 \cdot (x + 3)$

Use the distributive property to simplify the expression.

$$4 \cdot (x + 3) = 4(x + 3)$$
$$= 4 \cdot x + 4 \cdot 3$$
$$= 4x + 12$$

Work Practice 16–19

Practice 16–19

Write each phrase as an algebraic expression and simplify if possible. Let x represent the unknown number.

16. Three times a number, subtracted from 10

17. The sum of a number and 2, divided by 5

18. Three times the sum of a number and 6

19. Seven times the difference of a number and 4.

Answers

16. $10 - 3x$

17. $(x + 2) \div 5$ or $\frac{x+2}{5}$

18. $3x + 18$ **19.** $7x - 28$

Vocabulary, Readiness & Video Check

Use the choices below to fill in each blank. Some choices may be used more than once.

numerical coefficient	expression	unlike	distributive
combine like terms	like	term	

1. $14y^2 + 2x - 23$ is called a(n) ____________ while $14y^2$, $2x$, and -23 are each called a(n) ____________.

2. To multiply $3(-7x + 1)$, we use the ____________ property.

3. To simplify an expression like $y + 7y$, we ____________.

4. The term z has an understood ____________ of 1.

5. The terms $-x$ and $5x$ are ____________ terms and the terms $5x$ and $5y$ are ____________ terms.

6. For the term $-3x^2y$, -3 is called the ____________.

Martin-Gay Interactive Videos

See Video 1.8

Watch the section lecture video and answer the following questions.

Objective A 7. Example 7 shows two terms with exactly the same variables. Why are these terms not considered like terms?

Objective B 8. Example 8 shows us that when combining like terms, we are actually applying what property?

Objective C 9. The expression in Example 11 shows a minus sign before parentheses. When using the distributive property to multiply and remove parentheses, what number are we actually distributing to each term within the parentheses?

Objective D 10. Write the phrase given in Example 14, translate it to an algebraic expression, then simplify it. Why are we able to simplify it?

1.8 Exercise Set MyLab Math

Objective A *Identify the numerical coefficient of each term. See Example 1.*

1. $-7y$
2. $3x$
3. x
4. $-y$
5. $17x^2y$
6. $1.2xyz$

Indicate whether the terms in each list are like or unlike. See Example 2.

7. $5y, -y$
8. $-2x^2y, 6xy$
9. $2z, 3z^2$
10. $ab^2, -7ab^2$
11. $8wz, \frac{1}{7}zw$
12. $7.4p^3q^2, 6.2p^3q^2r$

Objective B *Simplify each expression by combining any like terms. See Examples 3 through 7.*

13. $7y + 8y$
14. $3x + 2x$
15. $8w - w + 6w$
16. $c - 7c + 2c$
17. $3b - 5 - 10b - 4$
18. $6g + 5 - 3g - 7$
19. $m - 4m + 2m - 6$
20. $a + 3a - 2 - 7a$
21. $5g - 3 - 5 - 5g$
22. $8p + 4 - 8p - 15$
23. $6.2x - 4 + x - 1.2$
24. $7.9y - 0.7 - y + 0.2$
25. $2k - k - 6$
26. $7c - 8 - c$
27. $-9x + 4x + 18 - 10x$
28. $5y - 14 + 7y - 20y$
29. $6x - 5x + x - 3 + 2x$
30. $8h + 13h - 6 + 7h - h$
31. $7x^2 + 8x^2 - 10x^2$
32. $8x^3 + x^3 - 11x^3$
33. $3.4m - 4 - 3.4m - 7$
34. $2.8w - 0.9 - 0.5 - 2.8w$
35. $6x + 0.5 - 4.3x - 0.4x + 3$
36. $0.4y - 6.7 + y - 0.3 - 2.6y$

Objective C *Simplify each expression. Use the distributive property to remove any parentheses. See Examples 8 through 10.*

37. $5(y + 4)$

38. $7(r + 3)$

39. $-2(x + 2)$

40. $-4(y + 6)$

41. $-5(2x - 3y + 6)$

42. $-2(4x - 3z - 1)$

43. $-(3x - 2y + 1)$

44. $-(y + 5z - 7)$

Objectives B C Mixed Practice *Remove parentheses and simplify each expression. See Examples 8 through 14.*

45. $7(d - 3) + 10$

46. $9(z + 7) - 15$

47. $-4(3y - 4) + 12y$

48. $-3(2x + 5) + 6x$

49. $3(2x - 5) - 5(x - 4)$

50. $2(6x - 1) - (x - 7)$

51. $-2(3x - 4) + 7x - 6$

52. $8y - 2 - 3(y + 4)$

53. $5k - (3k - 10)$

54. $-11c - (4 - 2c)$

55. $(3x + 4) - (6x - 1)$

56. $(8 - 5y) - (4 + 3y)$

57. $5(x + 2) - (3x - 4)$

58. $4(2x - 3) - (x + 1)$

59. $\frac{1}{3}(7y - 1) + \frac{1}{6}(4y + 7)$

60. $\frac{1}{5}(9y + 2) + \frac{1}{10}(2y - 1)$

61. $2 + 4(6x - 6)$

62. $8 + 4(3x - 4)$

63. $0.5(m + 2) + 0.4m$

64. $0.2(k + 8) - 0.1k$

65. $10 - 3(2x + 3y)$

66. $14 - 11(5m + 3n)$

67. $6(3x - 6) - 2(x + 1) - 17x$

68. $7(2x + 5) - 4(x + 2) - 20x$

69. $\frac{1}{2}(12x - 4) - (x + 5)$

70. $\frac{1}{3}(9x - 6) - (x - 2)$

Perform each indicated operation. Don't forget to simplify if possible. See Example 15.

71. Add $6x + 7$ to $4x - 10$.

72. Add $3y - 5$ to $y + 16$.

73. Subtract $7x + 1$ from $3x - 8$.

74. Subtract $4x - 7$ from $12 + x$.

75. Subtract $5m - 6$ from $m - 9$.

76. Subtract $m - 3$ from $2m - 6$.

Objective D *Write each phrase as an algebraic expression and simplify if possible. Let x represent the unknown number. See Examples 16 through 19.*

77. Twice a number, decreased by four

78. The difference of a number and two, divided by five

79. Three-fourths of a number, increased by twelve

80. Eight more than triple a number

81. The sum of 5 times a number and -2, added to 7 times the number

82. The sum of 3 times a number and 10, *subtracted from* 9 times the number

83. Eight times the sum of a number and six

84. Six times the difference of a number and five

85. Double a number minus the sum of the number and ten

86. Half a number minus the product of the number and eight

Concept Extensions

Given the following information, determine whether each scale is balanced.

1 cone balances 1 cube

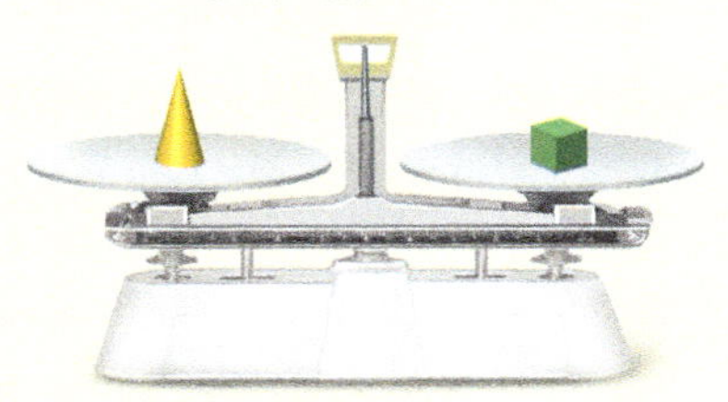

1 cylinder balances 2 cubes

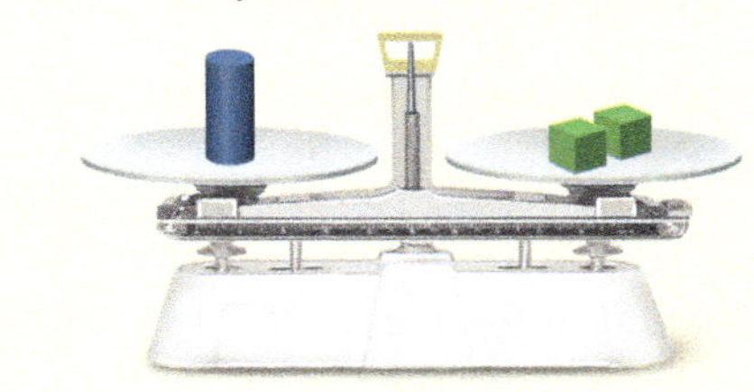

87.

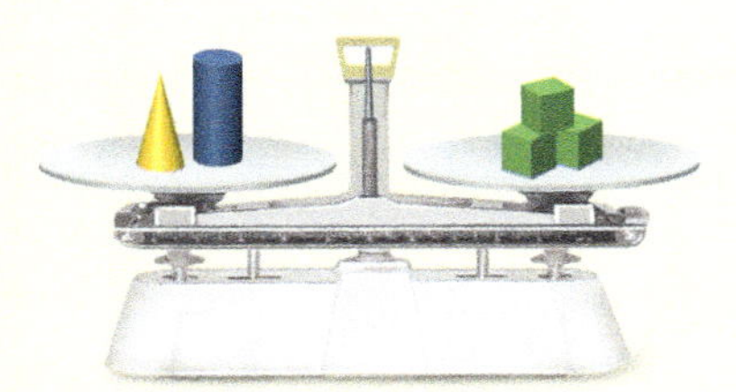

88.

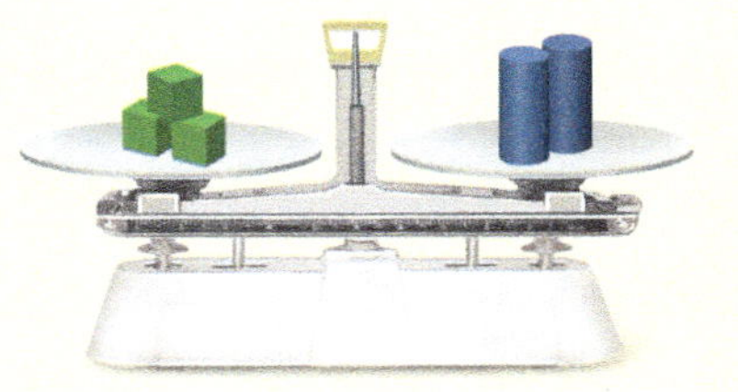

89.

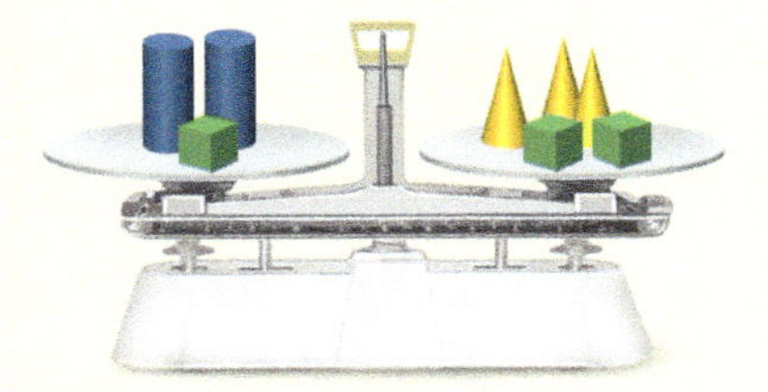

90.

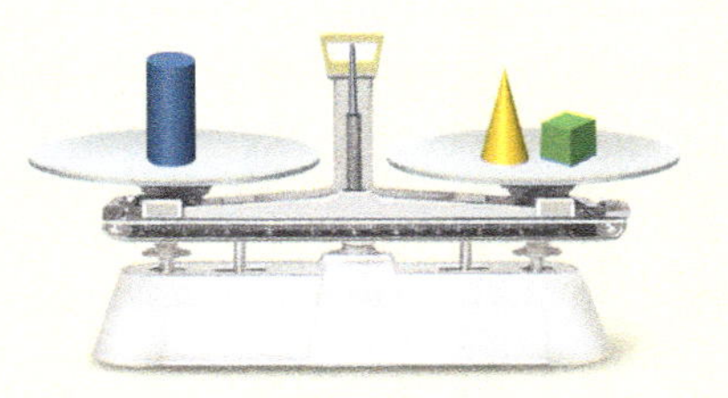

Write each algebraic expression described.

91. Write an expression with 4 terms that simplifies to $3x - 4$.

92. Write an expression of the form _____(_____ + _____) whose product is $6x + 24$.

△ **93.** Recall that the perimeter of a figure is the total distance around the figure. Given the following rectangle, express the perimeter as an algebraic expression containing the variable x.

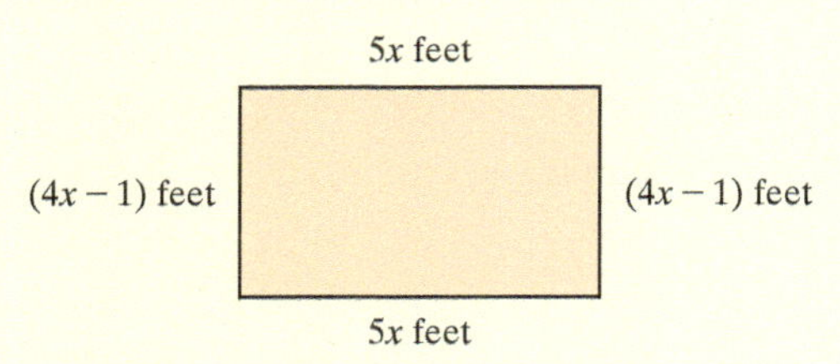

△ **94.** Recall that the perimeter of a figure is the total distance around the figure. Given the following triangle, express its perimeter as an algebraic expression containing the variable x.

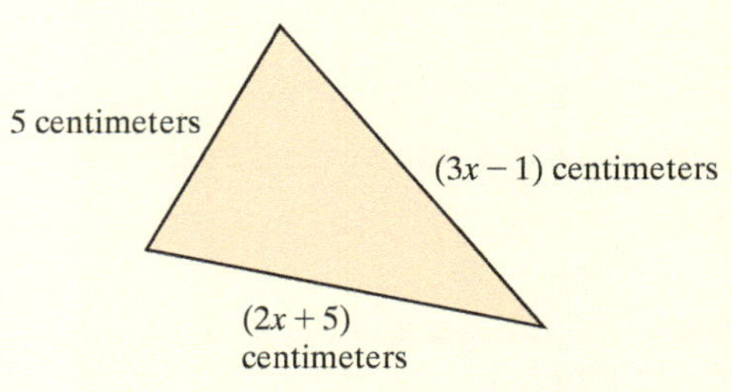

△ **95.** To convert from feet to inches, we multiply by 12. For example, the number of inches in 2 feet is $12 \cdot 2$ inches. If one board has a length of $(x + 2)$ *feet* and a second board has a length of $(3x - 1)$ *inches,* express their total length in inches as an algebraic expression.

96. The value of 7 nickels is $5 \cdot 7$ cents. Likewise, the value of x nickels is $5x$ cents. If the money box in a drink machine contains x *nickels,* $3x$ *dimes,* and $(30x - 1)$ *quarters,* express their total value in cents as an algebraic expression.

97. In your own words, explain how to combine like terms.

98. Do like terms always contain the same numerical coefficients? Explain your answer.

Chapter 1 Group Activity

Magic Squares

Sections 1.3, 1.4, 1.5

A magic square is a set of numbers arranged in a square table so that the sum of the numbers in each column, row, and diagonal is the same. For instance, in the magic square below, the sum of each column, row, and diagonal is 15. Notice that no number is used more than once in the magic square.

2	9	4
7	5	3
6	1	8

The properties of magic squares have been known for a very long time and once were thought to be good luck charms. The ancient Egyptians and Greeks understood their patterns. A magic square even made it into a famous work of art. The engraving titled *Melencolia I,* created by German artist Albrecht Dürer in 1514, features the following four-by-four magic square on the building behind the central figure.

16	3	2	13
5	10	11	8
9	6	7	12
4	15	14	1

Group Exercises

1. Verify that what is shown in the Dürer engraving is, in fact, a magic square. What is the common sum of the columns, rows, and diagonals?
2. Negative numbers can also be used in magic squares. Complete the following magic square:

		−2
	−1	
0		−4

3. Use the numbers −12, −9, −6, −3, 0, 3, 6, 9, and 12 to form a magic square.

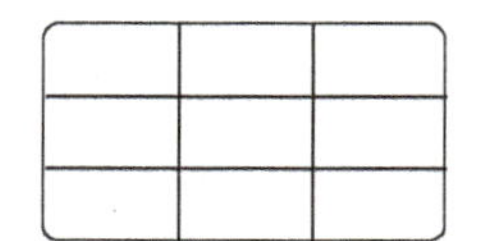

Chapter 1 Vocabulary Check

Fill in each blank with one of the words or phrases listed below.

inequality symbols	exponent	term	numerical coefficient
grouping symbols	solution	like terms	unlike terms
equation	absolute value	numerator	denominator
opposites	base	reciprocals	variable

1. The symbols $\neq$, $<$, and $>$ are called ________________.

2. A mathematical statement that two expressions are equal is called a(n) ________________.

3. The ________________ of a number is the distance between that number and 0 on a number line.

4. A symbol used to represent a number is called a(n) ________________.

5. Two numbers that are the same distance from 0 but lie on opposite sides of 0 are called ________________.

6. The number in a fraction above the fraction bar is called the ________________.

7. A(n) ________________ of an equation is a value for the variable that makes the equation a true statement.

8. Two numbers whose product is 1 are called ________________.

9. In 2^3, the 2 is called the ________________ and the 3 is called the ________________.

10. The ________________ of a term is its numerical factor.

11. The number in a fraction below the fraction bar is called the ________________.

12. Parentheses and brackets are examples of ________________.

13. A(n) ________________ is a number or the product of a number and variables raised to powers.

14. Terms with the same variables raised to the same powers are called ________________.

15. If terms are not like terms, then they are ________________.

Helpful Hint

Are you preparing for your test?
To help, don't forget to take these:

- Chapter 1 Getting Ready for the Test on page 89
- Chapter 1 Test on page 90

Then check all of your answers at the back of this text. For further review, the step-by-step video solutions to any of these exercises are located in MyLab Math.

1 Chapter Highlights

Definitions and Concepts	Examples
Section 1.2 Symbols and Sets of Numbers	
A **set** is a collection of objects, called **elements,** enclosed in braces.	$\{a, c, e\}$ Given the set $\left\{-3.4, \sqrt{3}, 0, \frac{2}{3}, 5, -4\right\}$ list the numbers that belong to the set of
Natural numbers: $\{1, 2, 3, 4, \ldots\}$	Natural numbers: 5
Whole numbers: $\{0, 1, 2, 3, 4, \ldots\}$	Whole numbers: 0, 5
Integers: $\{\ldots, -3, -2, -1, 0, 1, 2, 3, \ldots\}$	Integers: 0, 5, -4
Rational numbers: {real numbers that can be expressed as quotients of integers}	Rational numbers: $-3.4, 0, \frac{2}{3}, 5, -4$

(continued)

Definitions and Concepts	Examples
Section 1.2 Symbols and Sets of Numbers *(continued)*	
Irrational numbers: {real numbers that cannot be expressed as quotients of integers} A line used to picture numbers is called a **number line**.	Irrational numbers: $\sqrt{3}$ 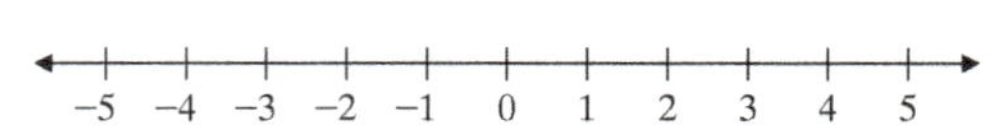
Real numbers: {all numbers that correspond to points on the number line} The **absolute value** of a real number a denoted by $\|a\|$ is the distance between a and 0 on a number line.	Real numbers: $-3.4, \sqrt{3}, 0, \frac{2}{3}, 5, -4$ $\|5\| = 5 \quad \|0\| = 0 \quad \|-2\| = 2$
Symbols: $=$ is equal to $\neq$ is not equal to $>$ is greater than $<$ is less than $\leq$ is less than or equal to $\geq$ is greater than or equal to	$-7 = -7$ $3 \neq -3$ $4 > 1$ $1 < 4$ $6 \leq 6$ $18 \geq -\frac{1}{3}$
Order Property for Real Numbers For any two real numbers a and b, a is less than b if a is to the left of b on a number line.	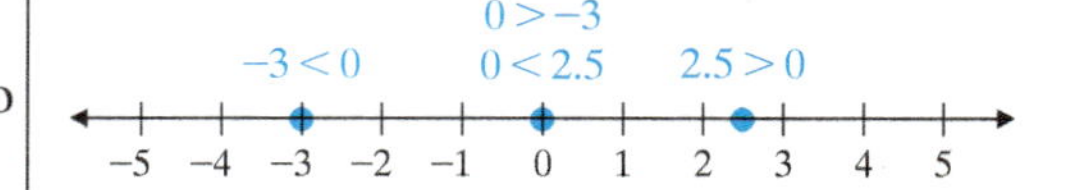
Section 1.3 Exponents, Order of Operations, and Variable Expressions	
The expression a^n is an **exponential expression.** The number a is called the **base;** it is the repeated factor. The number n is called the **exponent;** it is the number of times that the base is a factor.	$4^3 = 4 \cdot 4 \cdot 4 = 64$ $7^2 = 7 \cdot 7 = 49$
Order of Operations 1. Perform all operations within grouping symbols first, starting with the innermost set. 2. Evaluate exponential expressions. 3. Multiply or divide in order from left to right. 4. Add or subtract in order from left to right.	$\frac{8^2 + 5(7-3)}{3 \cdot 7} = \frac{8^2 + 5(4)}{21}$ $= \frac{64 + 5(4)}{21}$ $= \frac{64 + 20}{21}$ $= \frac{84}{21}$ $= 4$
A symbol used to represent a number is called a **variable.**	Examples of variables are q, x, z
An **algebraic expression** is a collection of numbers, variables, operation symbols, and grouping symbols.	Examples of algebraic expressions are $5x, \quad 2(y-6), \quad \frac{q^2 - 3q + 1}{6}$
To **evaluate an algebraic expression** containing a variable, substitute a given number for the variable and simplify.	Evaluate $x^2 - y^2$ when $x = 5$ and $y = 3$. $x^2 - y^2 = (5)^2 - 3^2$ $= 25 - 9$ $= 16$
A mathematical statement that two expressions are equal is called an **equation.**	Equations: $3x - 9 = 20$ $A = \pi r^2$

Definitions and Concepts	Examples
Section 1.3 Exponents, Order of Operations, and Variable Expressions *(continued)*	
A **solution** of an equation is a value for the variable that makes the equation a true statement.	Determine whether 4 is a solution of $5x + 7 = 27$. $5x + 7 = 27$ $5(4) + 7 \stackrel{?}{=} 27$ $20 + 7 \stackrel{?}{=} 27$ $27 = 27$ True 4 is a solution.
Section 1.4 Adding Real Numbers	
To Add Two Numbers with the Same Sign **1.** Add their absolute values. **2.** Use their common sign as the sign of the sum.	Add. $10 + 7 = 17$ $-3 + (-8) = -11$
To Add Two Numbers with Different Signs **1.** Subtract their absolute values. **2.** Use the sign of the number whose absolute value is larger as the sign of the sum.	$-25 + 5 = -20$ $14 + (-9) = 5$
Two numbers that are the same distance from 0 but lie on opposite sides of 0 are called **opposites** or **additive inverses**. The opposite of a number a is denoted by $-a$.	The opposite of -7 is 7. The opposite of 123 is -123.
Section 1.5 Subtracting Real Numbers	
To subtract two numbers a and b, add the first number a to the opposite of the second number, b. $a - b = a + (-b)$	Subtract. $3 - (-44) = 3 + 44 = 47$ $-5 - 22 = -5 + (-22) = -27$ $-30 - (-30) = -30 + 30 = 0$
Section 1.6 Multiplying and Dividing Real Numbers	
Multiplying Real Numbers The product of two numbers with the same sign is a positive number. The product of two numbers with different signs is a negative number.	Multiply. $7 \cdot 8 = 56$ $\quad -7 \cdot (-8) = 56$ $-2 \cdot 4 = -8$ $\quad 2 \cdot (-4) = -8$
Products Involving Zero The product of 0 and any number is 0. $b \cdot 0 = 0$ and $0 \cdot b = 0$	$-4 \cdot 0 = 0$ $\quad 0 \cdot \left(-\frac{3}{4}\right) = 0$
Quotient of Two Real Numbers $\frac{a}{b} = a \cdot \frac{1}{b}, b \neq 0$	Divide. $\frac{42}{2} = 42 \cdot \frac{1}{2} = 21$
Dividing Real Numbers The quotient of two numbers with the same sign is a positive number. The quotient of two numbers with different signs is a negative number.	$\frac{90}{10} = 9$ $\quad \frac{-90}{-10} = 9$ $\frac{42}{-6} = -7$ $\quad \frac{-42}{6} = -7$
Quotients Involving Zero Let a be a nonzero number. $\frac{0}{a} = 0$ and $\frac{a}{0}$ is undefined.	$\frac{0}{18} = 0$ $\quad \frac{0}{-47} = 0$ $\quad \frac{-85}{0}$ is undefined.

Definitions and Concepts	Examples
Section 1.7 Properties of Real Numbers	
Commutative Properties	
Addition: $a + b = b + a$	$3 + (-7) = -7 + 3$
Multiplication: $a \cdot b = b \cdot a$	$-8 \cdot 5 = 5 \cdot (-8)$
Associative Properties	
Addition: $(a + b) + c = a + (b + c)$	$(5 + 10) + 20 = 5 + (10 + 20)$
Multiplication: $(a \cdot b) \cdot c = a \cdot (b \cdot c)$	$(-3 \cdot 2) \cdot 11 = -3 \cdot (2 \cdot 11)$
Two numbers whose product is 1 are called **multiplicative inverses** or **reciprocals**. The reciprocal of a nonzero number a is $\frac{1}{a}$ because $a \cdot \frac{1}{a} = 1$.	The reciprocal of 3 is $\frac{1}{3}$. The reciprocal of $-\frac{2}{5}$ is $-\frac{5}{2}$.
Distributive Property	
$a(b + c) = a \cdot b + a \cdot c$	$5(6 + 10) = 5 \cdot 6 + 5 \cdot 10$ $-2(3 + x) = -2 \cdot 3 + (-2)(x)$
Identities	
$a + 0 = a$ $\quad 0 + a = a$	$5 + 0 = 5$ $\quad 0 + (-2) = -2$
$a \cdot 1 = a$ $\quad 1 \cdot a = a$	$-14 \cdot 1 = -14$ $\quad 1 \cdot 27 = 27$
Inverses	
Additive or opposite: $a + (-a) = 0$	$7 + (-7) = 0$
Multiplicative or reciprocal: $b \cdot \frac{1}{b} = 1, \quad b \neq 0$	$3 \cdot \frac{1}{3} = 1$
Section 1.8 Simplifying Expressions	
The **numerical coefficient** of a **term** is its numerical factor.	**Term** / **Numerical Coefficient** $-7y$ / -7 x / 1 $\frac{1}{5}a^2b$ / $\frac{1}{5}$
Terms with the same variables raised to exactly the same powers are **like terms**.	**Like Terms** / **Unlike Terms** $12x, -x$ / $3y, 3y^2$ $-2xy, 5yx$ / $7a^2b, -2ab^2$
To combine like terms, add the numerical coefficients and multiply the result by the common variable factor.	$9y + 3y = 12y$ $-4z^2 + 5z^2 - 6z^2 = -5z^2$
To remove parentheses, apply the distributive property.	$-4(x + 7) + 10(3x - 1)$ $= -4x - 28 + 30x - 10$ $= 26x - 38$

Chapter 1 Review

(1.2) *Insert $<$, $>$, or $=$ in the appropriate space to make each statement true.*

1. $8 \quad 10$

2. $7 \quad 2$

3. $-4 \quad -5$

4. $\frac{12}{2} \quad -8$

5. $|-7| \quad |-8|$

6. $|-9| \quad -9$

7. $-|-1| \quad -1$

8. $|-14| \quad -(-14)$

9. $1.2 \quad 1.02$

10. $-\frac{3}{2} \quad -\frac{3}{4}$

Translate each statement into symbols.

11. Four is greater than or equal to negative three.

12. Six is not equal to five.

13. 0.03 is less than 0.3.

14. The United States is home to 1579 two-year colleges and 3004 four-year colleges. Write an inequality comparing the numbers 1579 and 3004. (*Source*: National Center for Education Statistics)

Given the sets of numbers below, list the numbers in each set that also belong to the set of:

a. Natural numbers
b. Whole numbers
c. Integers
d. Rational numbers
e. Irrational numbers
f. Real numbers

15. $\left\{-6, 0, 1, 1\frac{1}{2}, 3, \pi, 9.62\right\}$

16. $\left\{-3, -1.6, 2, 5, \frac{11}{2}, 15.1, \sqrt{5}, 2\pi\right\}$

The following chart shows the gains and losses in dollars of Density Oil and Gas stock for a particular week. Use this chart to answer Exercises 17 and 18.

Day	Gain or Loss (in dollars)
Monday	+1
Tuesday	−2
Wednesday	+5
Thursday	+1
Friday	−4

17. Which day showed the greatest loss?

18. Which day showed the greatest gain?

(1.3) *Choose the correct answer for each statement.*

19. The expression $6 \cdot 3^2 + 2 \cdot 8$ simplifies to
a. -52 **b.** 448 **c.** 70 **d.** 64

20. The expression $68 - 5 \cdot 2^3$ simplifies to
a. -232 **b.** 28 **c.** 38 **d.** 504

Simplify each expression.

21. $3(1 + 2 \cdot 5) + 4$

22. $8 + 3(2 \cdot 6 - 1)$

23. $\dfrac{4 + |6 - 2| + 8^2}{4 + 6 \cdot 4}$

24. $5[3(2 + 5) - 5]$

Translate each word statement to symbols.

25. The difference of twenty and twelve is equal to the product of two and four.

26. The quotient of nine and two is greater than negative five.

Evaluate each expression when $x = 6$, $y = 2$, and $z = 8$.

27. $2x + 3y$

28. $x(y + 2z)$

29. $\dfrac{x}{y} + \dfrac{z}{2y}$

30. $x^2 - 3y^2$

△ **31.** The expression $180 - a - b$ represents the measure of the unknown angle of the given triangle. Replace a with 37 and b with 80 to find the measure of the unknown angle.

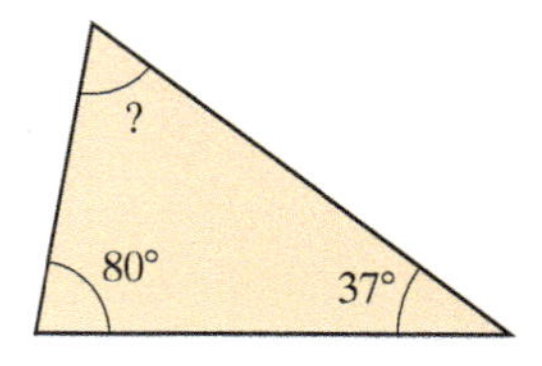

△ **32.** The expression $360 - a - b - c$ represents the measure of the unknown angle of the given quadrilateral. Replace a with 93, b with 80, and c with 82 to find the measure of the unknown angle.

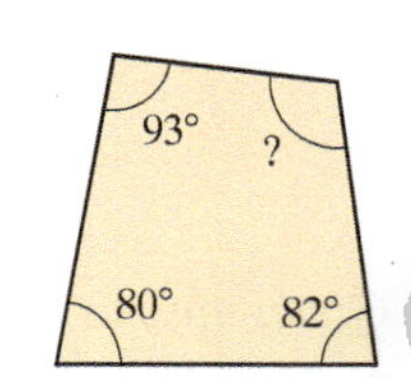

Decide whether the given number is a solution to the given equation.

33. $7x - 3 = 18$; 3

34. $3x + 4 = x - 1$; 1

(1.4) *Find the additive inverse or opposite of each number.*

35. -9

36. $\dfrac{2}{3}$

37. $|-2|$

38. $-|-7|$

Add.

39. $-15 + 4$

40. $-6 + (-11)$

41. $\dfrac{1}{16} + \left(-\dfrac{1}{4}\right)$

42. $-8 + |-3|$

43. $-4.6 + (-9.3)$

44. $-2.8 + 6.7$

(1.5) *Perform each indicated operation.*

45. $6 - 20$

46. $-3.1 - 8.4$

47. $-6 - (-11)$

48. $4 - 15$

49. $-21 - 16 + 3(8 - 2)$

50. $\dfrac{11 - (-9) + 6(8 - 2)}{2 + 3 \cdot 4}$

Evaluate each expression for $x = 3$, $y = -6$, and $z = -9$. Then choose the correct evaluation.

51. $2x^2 - y + z$
a. 15 **b.** 3 **c.** 27 **d.** -3

52. $\dfrac{|y - 4x|}{2x}$
a. 3 **b.** 1 **c.** -1 **d.** -3

*Use the chart for Exercises **17** and **18** to solve.*

53. At the beginning of the week the price of Density Oil and Gas stock is $50 per share. Find the price of a share of stock at the end of the week.

54. Find the price of a share of stock by the end of the day on Wednesday.

(1.6) *Find each multiplicative inverse or reciprocal.*

55. -6

56. $\dfrac{3}{5}$

Simplify each expression.

57. $6(-8)$

58. $(-2)(-14)$

59. $\dfrac{-18}{-6}$

60. $\dfrac{42}{-3}$

61. $-3(-6)(-2)$

62. $(-4)(-3)(0)(-6)$

63. $\dfrac{4(-3) + (-8)}{2 + (-2)}$

64. $\dfrac{3(-2)^2 - 5}{-14}$

(1.7) *Name the property illustrated in each equation.*

65. $-6 + 5 = 5 + (-6)$

66. $6 \cdot 1 = 6$

67. $3(8 - 5) = 3 \cdot 8 - 3 \cdot 5$

68. $4 + (-4) = 0$

69. $2 + (3 + 9) = (2 + 3) + 9$

70. $2 \cdot 8 = 8 \cdot 2$

71. $6(8 + 5) = 6 \cdot 8 + 6 \cdot 5$

72. $(3 \cdot 8) \cdot 4 = 3 \cdot (8 \cdot 4)$

73. $4 \cdot \frac{1}{4} = 1$

74. $8 + 0 = 8$

75. $4(8 + 3) = 4(3 + 8)$

76. $5(2 + 1) = 5 \cdot 2 + 5 \cdot 1$

(1.8) *Simplify each expression.*

77. $5x - x + 2x$

78. $0.2z - 4.6z - 7.4z$

79. $\frac{1}{2}x + 3 + \frac{7}{2}x - 5$

80. $\frac{4}{5}y + 1 + \frac{6}{5}y + 2$

81. $2(n - 4) + n - 10$

82. $3(w + 2) - (12 - w)$

83. Subtract $7x - 2$ from $x + 5$.

84. Subtract $1.4y - 3$ from $y - 0.7$.

Write each phrase as an algebraic expression. Simplify if possible.

85. Three times a number decreased by 7

86. Twice the sum of a number and 2.8, added to 3 times the number

Mixed Review

Insert $<$, $>$, or $=$ in the space between each pair of numbers.

87. $-|-11| \quad |11.4|$

88. $-1\frac{1}{2} \quad -2\frac{1}{2}$

Perform the indicated operations.

89. $-7.2 + (-8.1)$

90. $14 - 20$

91. $4(-20)$

92. $\frac{-20}{4}$

93. $-\frac{4}{5}\left(\frac{5}{16}\right)$

94. $-0.5(-0.3)$

95. $8 \div 2 \cdot 4$

96. $(-2)^4$

97. $\frac{-3 - 2(-9)}{-15 - 3(-4)}$

98. $5 + 2[(7 - 5)^2 + (1 - 3)]$

99. $-\frac{5}{8} \div \frac{3}{4}$

100. $\frac{-15 + (-4)^2 + |-9|}{10 - 2 \cdot 5}$

Remove parentheses and simplify each expression.

101. $7(3x - 3) - 5(x + 4)$

102. $8 + 2(9x - 10)$

Chapter 1 Getting Ready for the Test LC

*All the exercises below are **Multiple Choice**. Choose the correct letter(s). Also, letters may be used more than once. Select the given operation between the two numbers.*

1. For $-5 + (-3)$, the operation is

A. addition **B.** subtraction **C.** multiplication **D.** division

2. For $-5(-3)$, the operation is

A. addition **B.** subtraction **C.** multiplication **D.** division

Identify each as an

A. equation or an **B.** expression

3. $6x + 2 + 4x - 10$

4. $6x + 2 = 4x - 10$

For the exercises below, a and b are negative numbers. State whether each expression simplifies to

A. positive number **B.** negative number **C.** 0 **D.** not possible to determine

5. $a + b$

6. $a \cdot b$

7. $\frac{a}{b}$

8. $a - 0$

9. $0 \cdot b$

10. $a - b$

11. $0 + b$

12. $\frac{0}{a}$

The exercise statement and the correct answer are given. Select the correct directions.

A. Find the opposite. **B.** Find the reciprocal. **C.** Evaluate or simplify.

13. 5 Answer: $\frac{1}{5}$

14. 2^3 Answer: 8

15. -7 Answer: 7

MULTIPLE CHOICE *Exercises 16 through 18 below are given. Choose the best directions (choice **A**, **B**, or **C**) below for each exercise.*

A. Simplify. **B.** Identify the numerical coefficient. **C.** Are these like or unlike terms?

16. Given: $-3x^2$

17. Given: $5x^2$ and $4x$

18. Given: $4x - 5 + 2x + 3$

MULTIPLE CHOICE

19. Subtracting $100z$ from $8m$ translates to

A. $100z - 8m$ **B.** $8m - 100z$ **C.** $-800zm$ **D.** $92zm$

20. Subtracting $7x - 1$ from $9y$ translates to:

A. $7x - 1 - 9y$ **B.** $9y - 7x - 1$ **C.** $9y - (7x - 1)$ **D.** $7x - 1 - (9y)$

Chapter 1 Test MyLab Math

For additional practice go to your study plan in MyLab Math.

Answers

1. ______
2. ______
3. ______
4. ______
5. ______
6. ______
7. ______
8. ______
9. ______
10. ______
11. ______
12. ______
13. ______
14. ______
15. ______
16. ______
17. ______
18. ______
19. ______
20. ______
21. ______

Translate each statement into symbols.

1. The absolute value of negative seven is greater than five.

2. The sum of nine and five is greater than or equal to four.

Simplify each expression.

3. $-13 + 8$

4. $-13 - (-2)$

5. $6 \cdot 3 - 8 \cdot 4$

6. $13(-3)$

7. $(-6)(-2)$

8. $\frac{|-16|}{-8}$

9. $\frac{-8}{0}$

10. $\frac{|-6| + 2}{5 - 6}$

11. $\frac{1}{2} - \frac{5}{6}$

12. $-1\frac{1}{8} + 5\frac{3}{4}$

13. $-\frac{3}{5} + \frac{15}{8}$

14. $3(-4)^2 - 80$

15. $6[5 + 2(3 - 8) - 3]$

16. $\frac{-12 + 3 \cdot 8}{4}$

17. $\frac{(-2)(0)(-3)}{-6}$

Insert $<$, $>$, or $=$ in the appropriate space to make each statement true.

18. $-3 \quad -7$

19. $4 \quad -8$

20. $|-3| \quad 2$

21. $|-2| \quad -1 - (-3)$

22. Given $\left\{-5, -1, \frac{1}{4}, 0, 1, 7, 11.6, \sqrt{7}, 3\pi\right\}$, list the numbers in this set that also belong to the set of:

a. Natural numbers
b. Whole numbers
c. Integers
d. Rational numbers
e. Irrational numbers
f. Real numbers

Evaluate each expression when $x = 6$, $y = -2$, and $z = -3$.

23. $x^2 + y^2$

24. $x + yz$

25. $2 + 3x - y$

26. $\frac{y + z - 1}{x}$

Identify the property illustrated by each equation.

27. $8 + (9 + 3) = (8 + 9) + 3$

28. $6 \cdot 8 = 8 \cdot 6$

29. $-6(2 + 4) = -6 \cdot 2 + (-6) \cdot 4$

30. $\frac{1}{6}(6) = 1$

31. Find the opposite of -9.

32. Find the reciprocal of $-\frac{1}{3}$.

The New Orleans Saints were 22 yards from the goal when the series of gains and losses shown in the chart occurred. Use this chart to answer Exercises 33 and 34.

	Gains and Losses (in yards)
First down	5
Second down	−10
Third down	−2
Fourth down	29

33. During which down did the greatest loss of yardage occur?

34. Was a touchdown scored?

35. The temperature at the Winter Olympics was a frigid 14° below zero in the morning, but by noon it had risen 31°. What was the temperature at noon? (Temperatures are in degrees Fahrenheit.)

36. A stockbroker decided to sell 280 shares of stock, which decreased in value by \$1.50 per share yesterday. How much money did she lose?

Simplify each expression.

37. $2y - 6 - y - 4$

38. $2.7x + 6.1 + 3.2x - 4.9$

39. $4(x - 2) - 3(2x - 6)$

40. $-5(y + 1) + 2(3 - 5y)$

22. a. ______
b. ______
c. ______
d. ______
e. ______
f. ______
23. ______
24. ______
25. ______
26. ______
27. ______
28. ______
29. ______
30. ______
31. ______
32. ______
33. ______
34. ______
35. ______
36. ______
37. ______
38. ______
39. ______
40. ______

2 Equations, Inequalities, and Problem Solving

In this chapter, we solve equations and inequalities. Once we know how to solve equations and inequalities, we may solve word problems. Of course, problem solving is an integral topic in algebra, and its discussion is continued throughout this text.

Sections

2.1 The Addition Property of Equality

2.2 The Multiplication Property of Equality

2.3 Further Solving Linear Equations

Integrated Review—Solving Linear Equations

2.4 An Introduction to Problem Solving

2.5 Formulas and Problem Solving

2.6 Percent and Mixture Problem Solving

2.7 Linear Inequalities and Problem Solving

Check Your Progress

Vocabulary Check

Chapter Highlights

Chapter Review

Getting Ready for the Test

Chapter Test

Cumulative Review

Top 5 Countries by Numbers of Internet-Crime Complaints
1. United States
2. Canada
3. India
4. United Kingdom
5. Australia

Top 10 States by Number of Internet-Crime Complaints

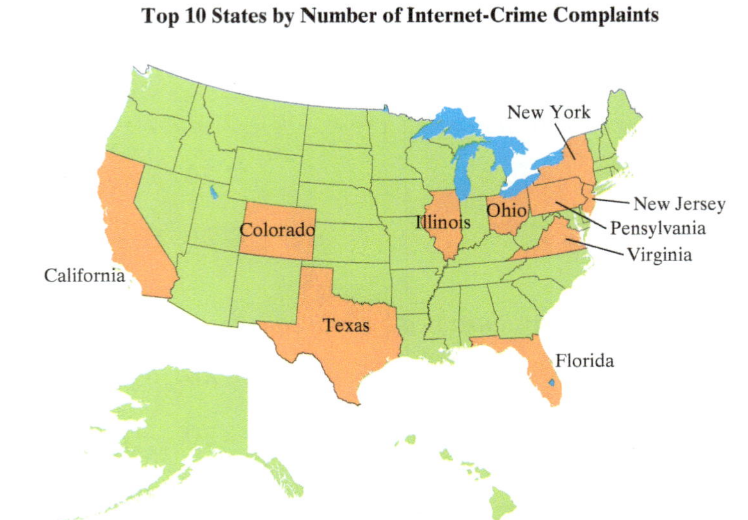

Internet Crime Continues

The Internet Crime Complaint Center (IC3) was established in May 2000. It is a joint operation between the FBI and the National White-Collar Crime Center. The IC3 receives and refers criminal complaints occurring on the Internet. Of course, nondelivery of merchandise or payment are the highest reported offenses.

In Section 2.6, Exercises 15 and 16, we analyze a bar graph on the yearly number of complaints received by the IC3.

Yearly Losses Reported to the IC3

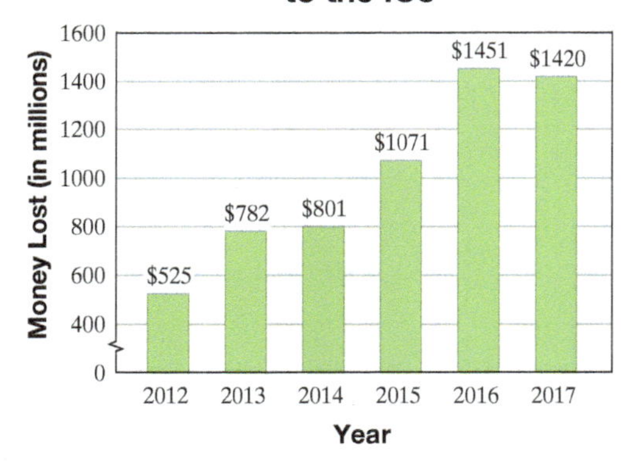

Ages of Persons Filing Complaints in 2017

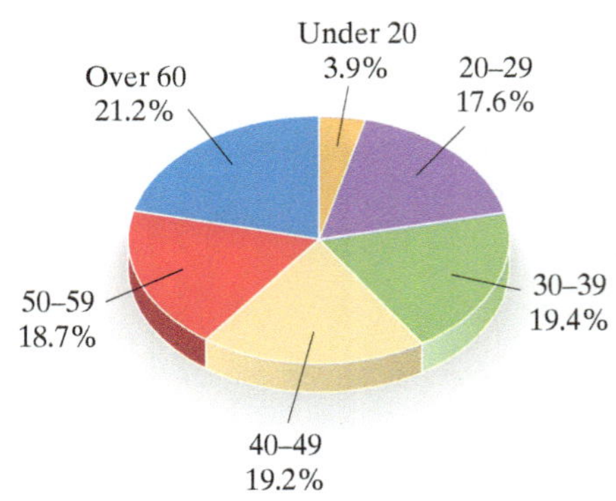

2.1 The Addition Property of Equality

Objectives

A Use the Addition Property of Equality to Solve Linear Equations.

B Simplify an Equation and Then Use the Addition Property of Equality.

C Write Word Phrases as Algebraic Expressions.

Let's recall from Section 1.3 the difference between an equation and an expression. A combination of operations on variables and numbers is an expression, and an equation is of the form "expression = expression."

Equations	Expressions
$3x - 1 = -17$	$3x - 1$
area = length · width	$5(20 - 3) + 10$
$8 + 16 = 16 + 8$	y^3
$-9a + 11b = 14b + 3$	$-x^2 + y - 2$

Now, let's concentrate on equations.

Objective A Using the Addition Property

A value of the variable that makes an equation a true statement is called a solution or root of the equation. The process of finding the solution of an equation is called **solving** the equation for the variable. In this section, we concentrate on solving *linear equations* in one variable.

Helpful Hint Simply stated, an equation contains "=" while an expression does not. Also, we *simplify* expressions and *solve* equations.

Linear Equation in One Variable

A **linear equation in one variable** can be written in the form

$$Ax + B = C$$

where A, B, and C are real numbers and $A \neq 0$.

Evaluating each side of a linear equation for a given value of the variable, as we did in Section 1.3, can tell us whether that value is a solution. But we can't rely on this as our method of solving it—with what value would we start?

Instead, to solve a linear equation in x, we write a series of simpler equations, all *equivalent* to the original equation, so that the final equation has the form

$$x = \text{number} \quad \text{or} \quad \text{number} = x$$

Equivalent equations are equations that have the same solution. This means that the "number" above is the solution to the original equation.

The first property of equality that helps us write simpler equivalent equations is the **addition property of equality.**

Addition Property of Equality

Let a, b, and c represent numbers. Then

$a = b$	Also, $a = b$
and $a + c = b + c$	and $a - c = b - c$
are equivalent equations.	are equivalent equations.

In other words, **the same number may be added to or subtracted from both sides** of an equation without changing the solution of the equation. (We may subtract the same number from both sides since subtraction is defined in terms of addition.)

Let's visualize how we use the addition property of equality to solve an equation. Picture the equation $x - 2 = 1$ as a balanced scale (see next page). The left side of the equation has the same value (weight) as the right side.

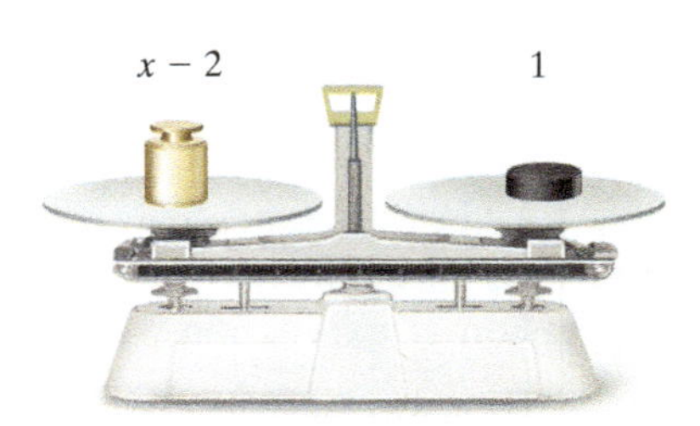

If the same weight is added to each side of a scale, the scale remains balanced. Likewise, if the same number is added to each side of an equation, the left side continues to have the same value as the right side.

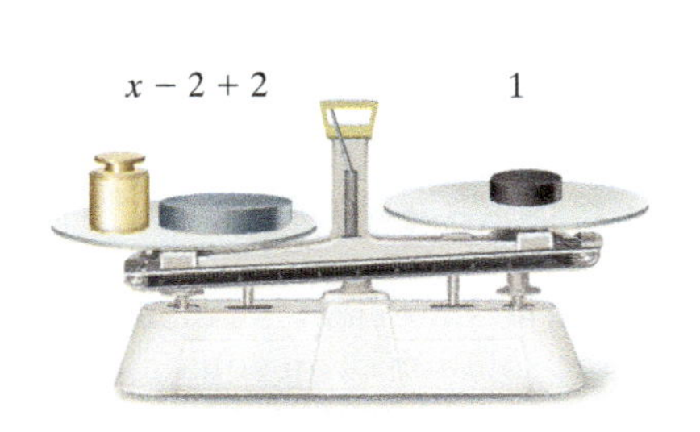

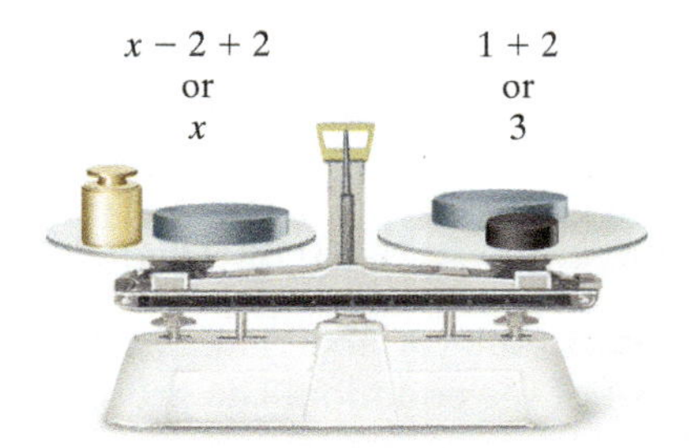

We use the addition property of equality to write equivalent equations until the variable is alone (by itself on one side of the equation) and the equation looks like "$x =$ number" or "number $= x$."

✓Concept Check Use the addition property to fill in the blanks so that the middle equation simplifies to the last equation.

$$x - 5 = 3$$
$$x - 5 + \_\_ = 3 + \_\_$$
$$x = 8$$

Practice 1

Solve: $x - 5 = 8$ for x.

Example 1 Solve $x - 7 = 10$ for x.

Solution: To solve for x, we first get x alone on one side of the equation. To do this, we add 7 to both sides of the equation.

$$x - 7 = 10$$
$$x - 7 + 7 = 10 + 7 \quad \text{Add 7 to both sides.}$$
$$x = 17 \quad \text{Simplify.}$$

The solution of the equation $x = 17$ is obviously 17.
Since we are writing equivalent equations, the solution of the equation $x - 7 = 10$ is also 17.

Check: To check, replace x with 17 in the original equation.

$$x - 7 = 10 \quad \text{Original equation.}$$
$$17 - 7 \stackrel{?}{=} 10 \quad \text{Replace } x \text{ with 17.}$$
$$10 = 10 \quad \text{True}$$

Since the statement is true, 17 is the solution.

Work Practice 1

Answer

1. $x = 13$

✓Concept Check Answer

5; 5

Example 2 Solve: $y + 0.6 = -1.0$

Solution: To solve for y (get y alone on one side of the equation), we subtract 0.6 from both sides of the equation.

$$y + 0.6 = -1.0$$
$$y + 0.6 - 0.6 = -1.0 - 0.6 \quad \text{Subtract 0.6 from both sides.}$$
$$y = -1.6 \quad \text{Combine like terms.}$$

Check:
$$y + 0.6 = -1.0 \quad \text{Original equation.}$$
$$-1.6 + 0.6 \stackrel{?}{=} -1.0 \quad \text{Replace } y \text{ with } -1.6.$$
$$-1.0 = -1.0 \quad \text{True}$$

The solution is -1.6.

Work Practice 2

Practice 2

Solve: $y + 1.7 = 0.3$

Example 3 Solve: $\frac{1}{2} = x - \frac{3}{4}$

Solution: To get x alone, we add $\frac{3}{4}$ to both sides.

$$\frac{1}{2} = x - \frac{3}{4}$$
$$\frac{1}{2} + \frac{3}{4} = x - \frac{3}{4} + \frac{3}{4} \quad \text{Add } \frac{3}{4} \text{ to both sides.}$$
$$\frac{1}{2} \cdot \frac{2}{2} + \frac{3}{4} = x \quad \text{The LCD is 4.}$$
$$\frac{2}{4} + \frac{3}{4} = x \quad \text{Add the fractions.}$$
$$\frac{5}{4} = x$$

Check:
$$\frac{1}{2} = x - \frac{3}{4} \quad \text{Original equation.}$$
$$\frac{1}{2} \stackrel{?}{=} \frac{5}{4} - \frac{3}{4} \quad \text{Replace } x \text{ with } \frac{5}{4}.$$
$$\frac{1}{2} \stackrel{?}{=} \frac{2}{4} \quad \text{Subtract.}$$
$$\frac{1}{2} = \frac{1}{2} \quad \text{True}$$

The solution is $\frac{5}{4}$.

Work Practice 3

Practice 3

Solve: $\frac{7}{8} = y - \frac{1}{3}$

Helpful Hint

We may solve an equation so that the variable is alone on *either* side of the equation. For example, $\frac{5}{4} = x$ is equivalent to $x = \frac{5}{4}$.

Example 4 Solve: $5t - 5 = 6t$

Solution: To solve for t, we first want all terms containing t on one side of the equation and numbers on the other side. Notice that if we subtract $5t$ from both sides of the equation, then variable terms will be on one side of the equation and the number -5 will be alone on the other side.

$$5t - 5 = 6t$$
$$5t - 5 - 5t = 6t - 5t \quad \text{Subtract } 5t \text{ from both sides.}$$
$$-5 = t \quad \text{Combine like terms.}$$

(*Continued on next page*)

Practice 4

Solve: $3x + 10 = 4x$

Answers

2. $y = -1.4$ **3.** $y = \frac{29}{24}$ **4.** $x = 10$

Helpful Hint For the equation from Example 4,

$$5t - 5 = 6t,$$

can we subtract $6t$ from both sides? Yes! The addition property allows us to do this, and we have the equivalent equation

$$-t - 5 = 0.$$

We are just no closer to our goal of having variable terms on one side of the equation and numbers on the other.

Check:

$$5t - 5 = 6t \quad \text{Original equation.}$$
$$5(-5) - 5 \stackrel{?}{=} 6(-5) \quad \text{Replace } t \text{ with } -5.$$
$$-25 - 5 \stackrel{?}{=} -30$$
$$-30 = -30 \quad \text{True}$$

The solution is -5.

Work Practice 4

Objective B Simplifying Equations

Many times, it is best to simplify one or both sides of an equation before applying the addition property of equality.

Practice 5

Solve:
$10w + 3 - 4w + 4 = -2w + 3 + 7w$

Example 5 Solve: $2x + 3x - 5 + 7 = 10x + 3 - 6x - 4$

Solution: First we simplify both sides of the equation.

$$2x + 3x - 5 + 7 = 10x + 3 - 6x - 4$$
$$5x + 2 = 4x - 1 \quad \text{Combine like terms on each side of the equation.}$$

Next, we want all terms with a variable on one side of the equation and all numbers on the other side.

$$5x + 2 - 4x = 4x - 1 - 4x \quad \text{Subtract } 4x \text{ from both sides.}$$
$$x + 2 = -1 \quad \text{Combine like terms.}$$
$$x + 2 - 2 = -1 - 2 \quad \text{Subtract 2 from both sides to get } x \text{ alone.}$$
$$x = -3 \quad \text{Combine like terms.}$$

Check:

$$2x + 3x - 5 + 7 = 10x + 3 - 6x - 4 \quad \text{Original equation.}$$
$$2(-3) + 3(-3) - 5 + 7 \stackrel{?}{=} 10(-3) + 3 - 6(-3) - 4 \quad \text{Replace } x \text{ with } -3.$$
$$-6 - 9 - 5 + 7 \stackrel{?}{=} -30 + 3 + 18 - 4 \quad \text{Multiply.}$$
$$-13 = -13 \quad \text{True}$$

The solution is -3.

Work Practice 5

If an equation contains parentheses, we use the distributive property to remove them, as before. Then we combine any like terms.

Practice 6

Solve:
$3(2w - 5) - (5w + 1) = -3$

Example 6 Solve: $6(2a - 1) - (11a + 6) = 7$

Solution:

$$6(2a - 1) - 1(11a + 6) = 7$$
$$6(2a) + 6(-1) - 1(11a) - 1(6) = 7 \quad \text{Apply the distributive property.}$$
$$12a - 6 - 11a - 6 = 7 \quad \text{Multiply.}$$
$$a - 12 = 7 \quad \text{Combine like terms.}$$
$$a - 12 + 12 = 7 + 12 \quad \text{Add 12 to both sides.}$$
$$a = 19 \quad \text{Simplify.}$$

Check: Check by replacing a with 19 in the original equation.

Work Practice 6

Answers

5. $w = -4$ 6. $w = 13$

Example 7 Solve: $3 - x = 7$

Solution: First we subtract 3 from both sides.

$$3 - x = 7$$
$$3 - x - 3 = 7 - 3 \quad \text{Subtract 3 from both sides.}$$
$$-x = 4 \quad \text{Simplify.}$$

We have not yet solved for x since x is not alone. However, this equation does say that the opposite of x is 4. If the opposite of x is 4, then x is the opposite of 4, or $x = -4$.

If $-x = 4$,

then $x = -4$.

Check:

$$3 - x = 7 \quad \text{Original equation.}$$
$$3 - (-4) \stackrel{?}{=} 7 \quad \text{Replace } x \text{ with } -4.$$
$$3 + 4 \stackrel{?}{=} 7 \quad \text{Add.}$$
$$7 = 7 \quad \text{True}$$

The solution is -4.

Work Practice 7

Practice 7

Solve: $12 - y = 9$

Objective C Writing Algebraic Expressions

In this section, we continue to practice writing algebraic expressions.

Example 8

a. The sum of two numbers is 8. If one number is 3, find the other number.

b. The sum of two numbers is 8. If one number is x, write an expression representing the other number.

Solution:

a. If the sum of two numbers is 8 and one number is 3, we find the other number by subtracting 3 from 8. The other number is $8 - 3$, or 5.

b. If the sum of two numbers is 8 and one number is x, we find the other number by subtracting x from 8. The other number is represented by $8 - x$.

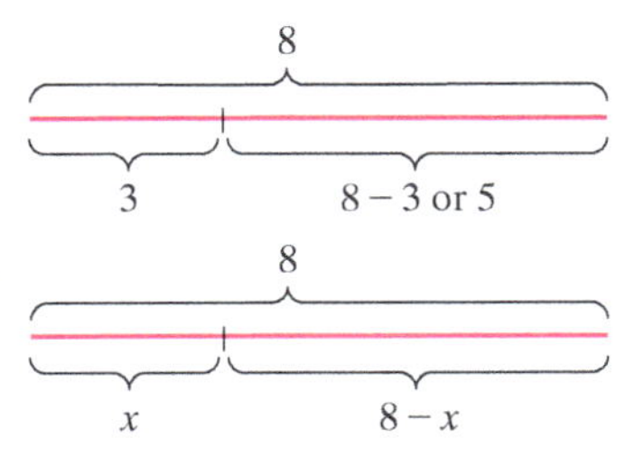

Work Practice 8

Practice 8

a. The sum of two numbers is 11. If one number is 4, find the other number.

b. The sum of two numbers is 11. If one number is x, write an expression representing the other number.

c. The sum of two numbers is 56. If one number is a, write an expression representing the other number.

Example 9 The Verrazano-Narrows Bridge in New York City is the longest suspension bridge in North America. The Golden Gate Bridge in San Francisco is 60 feet shorter than the Verrazano-Narrows Bridge. If the length of the Verrazano-Narrows Bridge is m feet, express the length of the Golden Gate Bridge as an algebraic expression in m. (*Source:* Survey of State Highway Engineers)

(*Continued on next page*)

Practice 9

In a recent year, the two top-selling Xbox 360 games were *Kinect Adventures* and *Grand Theft Auto V*. A price for *Grand Theft Auto V* is \$24 more than a price for *Kinect Adventures*. If the price of *Kinect Adventures* is n, what is the price for *Grand Theft Auto V*? (*Source:* Gamestop.com)

Answers

7. $y = 3$ **8. a.** $11 - 4$ or 7 **b.** $11 - x$ **c.** $56 - a$ **9.** $(n + 24)$ dollars

Solution: Since the Golden Gate Bridge is 60 feet shorter than the Verrazano-Narrows Bridge, we have that its length is

In words:	Length of Verrazano-Narrows Bridge	minus	60
Translate:	m	$-$	60

The Golden Gate Bridge is $(m - 60)$ feet long.

Work Practice 9

Vocabulary, Readiness & Video Check

Use the choices below to fill in each blank. Some choices may be used more than once or not at all.

equation	multiplication	addition
expression	solution	equivalent

1. A combination of operations on variables and numbers is called a(n) __________.

2. A statement of the form "expression = expression" is called a(n) __________.

3. A(n) __________ contains an equal sign (=).

4. A(n) __________ does not contain an equal sign (=).

5. A(n) __________ may be simplified and evaluated while a(n) __________ may be solved.

6. A(n) __________ of an equation is a number that when substituted for the variable makes the equation a true statement.

7. __________ equations have the same solution.

8. By the __________ property of equality, the same number may be added to or subtracted from both sides of an equation without changing the solution of the equation.

Solve each equation mentally. See Examples 1 and 2.

9. $x + 4 = 6$

10. $x + 7 = 17$

11. $n + 18 = 30$

12. $z + 22 = 40$

13. $b - 11 = 6$

14. $d - 16 = 5$

Martin-Gay Interactive Videos

See Video 2.1

Watch the section lecture video and answer the following questions.

Objective A **15.** Complete this statement based on the lecture given before Example 1. The addition property of equality means that if we have an equation, we can add the same real number to __________ of the equation and have an equivalent equation.

Objective B **16.** After both sides of Example 5 are simplified, write down the simplified equation.

Objective C **17.** Suppose we were to solve Example 8 again, this time letting the area of the Sahara Desert be x square miles. Use this to express the area of the Gobi Desert as an algebraic expression in x.

2.1 Exercise Set MyLab Math 

Objective A *Solve each equation. Check each solution. See Examples 1 through 4.*

1. $x + 7 = 10$
2. $x + 14 = 25$
3. $x - 2 = -4$
4. $y - 9 = 1$
5. $-11 = 3 + x$
6. $-8 = 8 + z$
7. $r - 8.6 = -8.1$
8. $t - 9.2 = -6.8$
9. $x - \frac{2}{5} = -\frac{3}{20}$
10. $y - \frac{4}{7} = -\frac{3}{14}$
11. $\frac{1}{3} + f = \frac{3}{4}$
12. $c + \frac{1}{6} = \frac{3}{8}$

Objective B *Solve each equation. Don't forget to first simplify each side of the equation, if possible. Check each solution. See Examples 5 through 7.*

13. $7x + 2x = 8x - 3$
14. $3n + 2n = 7 + 4n$
15. $\frac{5}{6}x + \frac{1}{6}x = -9$
16. $\frac{13}{11}y - \frac{2}{11}y = -3$
17. $2y + 10 = 5y - 4y$
18. $4x - 4 = 10x - 7x$
19. $-5(n - 2) = 8 - 4n$
20. $-4(z - 3) = 2 - 3z$
21. $\frac{3}{7}x + 2 = -\frac{4}{7}x - 5$
22. $\frac{1}{5}x - 1 = -\frac{4}{5}x - 13$
23. $5x - 6 = 6x - 5$
24. $2x + 7 = x - 10$
25. $8y + 2 - 6y = 3 + y - 10$
26. $4p - 11 - p = 2 + 2p - 20$
27. $-3(x - 4) = -4x$
28. $-2(x - 1) = -3x$
29. $\frac{3}{8}x - \frac{1}{6} = -\frac{5}{8}x - \frac{2}{3}$
30. $\frac{2}{5}x - \frac{1}{12} = -\frac{3}{5}x - \frac{3}{4}$
31. $2(x - 4) = x + 3$
32. $3(y + 7) = 2y - 5$
33. $3(n - 5) - (6 - 2n) = 4n$
34. $5(3 + z) - (8z + 9) = -4z$
35. $-2(x + 6) + 3(2x - 5) = 3(x - 4) + 10$
36. $-5(x + 1) + 4(2x - 3) = 2(x + 2) - 8$

Objectives A B Mixed Practice *Solve. See Examples 1 through 7.*

37. $13x - 3 = 14x$
38. $18x - 9 = 19x$
39. $5b - 0.7 = 6b$
40. $9x + 5.5 = 10x$
41. $3x - 6 = 2x + 5$
42. $7y + 2 = 6y + 2$
43. $13x - 9 + 2x - 5 = 12x - 1 + 2x$
44. $15x + 20 - 10x - 9 = 25x + 8 - 21x - 7$
45. $7(6 + w) = 6(2 + w)$
46. $6(5 + c) = 5(c - 4)$
47. $n + 4 = 3.6$
48. $m + 2 = 7.1$
49. $10 - (2x - 4) = 7 - 3x$
50. $15 - (6 - 7k) = 2 + 6k$
51. $\frac{1}{3} = x + \frac{2}{3}$
52. $\frac{1}{11} = y + \frac{10}{11}$
53. $-6.5 - 4x - 1.6 - 3x = -6x + 9.8$
54. $-1.4 - 7x - 3.6 - 2x = -8x + 4.4$

Objective C *Write each algebraic expression described. See Examples 8 and 9.*

55. Two numbers have a sum of 20. If one number is p, express the other number in terms of p.

56. Two numbers have a sum of 13. If one number is y, express the other number in terms of y.

57. A 10-foot board is cut into two pieces. If one piece is x feet long, express the other length in terms of x.

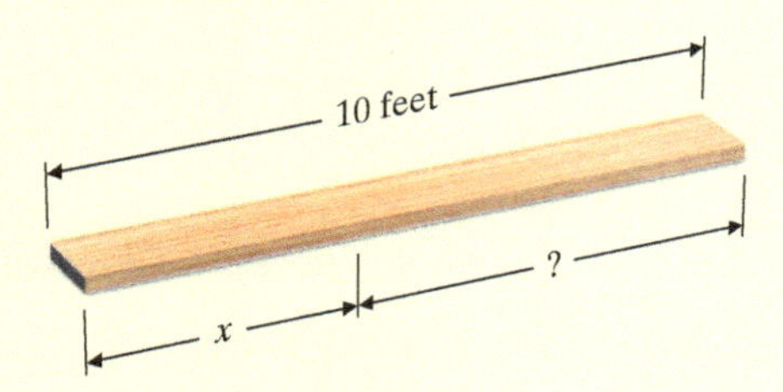

58. A 5-foot piece of string is cut into two pieces. If one piece is x feet long, express the other length in terms of x.

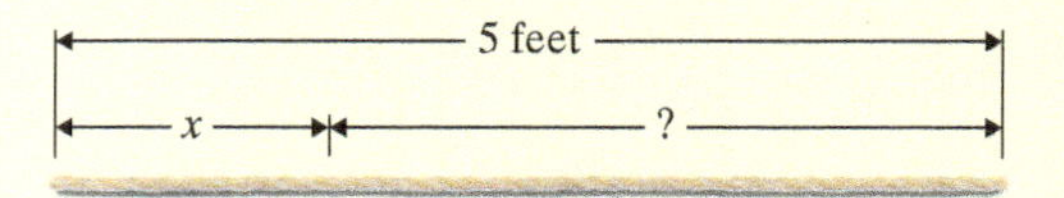

△ **59.** Recall that two angles are *supplementary* if their sum is 180°. If one angle measures $x°$, express the measure of its supplement in terms of x.

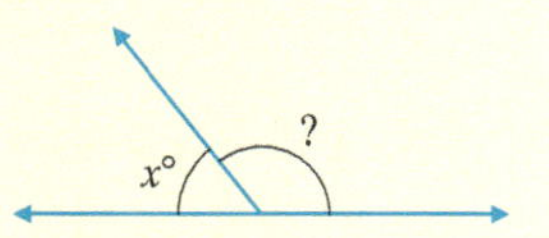

△ **60.** Recall that two angles are *complementary* if their sum is 90°. If one angle measures $x°$, express the measure of its complement in terms of x.

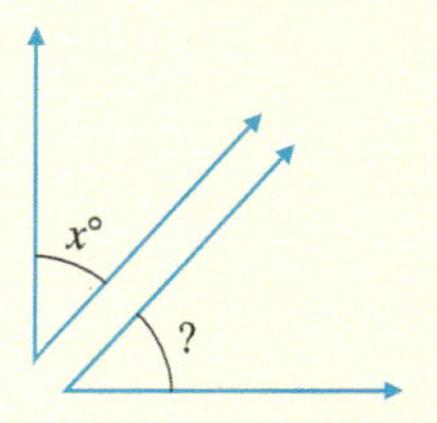

61. The longest north/south interstate highway is I-95, while the second longest north/south interstate highway is I-75. I-75 is about 139 miles shorter than I-95. If the length of I-75 is m miles, express the length of I-95 as an algebraic expression in m.

62. The longest interstate highway in the United States is I-90, which connects Seattle, Washington, and Boston, Massachusetts. The second longest interstate highway, I-80 (connecting San Francisco, California, and Teaneck, New Jersey), is 202 miles shorter than I-90. If the length of I-80 is m miles, express the length of I-90 as an algebraic expression in m.

63. The area of the Sahara Desert in Africa is 7 times the area of the Gobi Desert in Asia. If the area of the Gobi Desert is x square miles, express the area of the Sahara Desert as an algebraic expression in x.

64. The largest meteorite in the world is the Hoba West located in Namibia. Its weight is 3 times the weight of the Armanty meteorite located in Outer Mongolia. If the weight of the Armanty meteorite is y kilograms, express the weight of the Hoba West meteorite as an algebraic expression in y.

Review

Find each multiplicative inverse or reciprocal. See Section 1.6.

65. $\frac{5}{8}$ **66.** $\frac{7}{6}$ **67.** 2 **68.** 5 **69.** $-\frac{1}{9}$ **70.** $-\frac{3}{5}$

Perform each indicated operation and simplify. See Sections 1.6 and 1.7.

71. $\frac{3x}{3}$ **72.** $\frac{-2y}{-2}$ **73.** $-5\left(-\frac{1}{5}y\right)$ **74.** $7\left(\frac{1}{7}r\right)$ **75.** $\frac{3}{5}\left(\frac{5}{3}x\right)$ **76.** $\frac{9}{2}\left(\frac{2}{9}x\right)$

Concept Extensions

77. Write two terms whose sum is $-3x$.

78. Write four terms whose sum is $2y - 6$.

Use the addition property to fill in the blanks so that each second equation simplifies to the last equation. See the Concept Check in this section.

79.
$$x - 4 = -9$$
$$x - 4 + (\_\_\_) = -9 + (\_\_\_)$$
$$x = -5$$

80.
$$a + 9 = 15$$
$$a + 9 + (\_\_\_) = 15 + (\_\_\_)$$
$$a = 6$$

Fill in the blanks with numbers of your choice so that each equation has the given solution. Note: Each blank will be replaced with a different number.

81. _____ $+ x =$ _____; Solution: -3

82. $x -$ _____ $=$ _____; Solution: -10

Solve.

△ **83.** The sum of the angles of a triangle is 180°. If one angle of a triangle measures $x°$ and a second angle measures $(2x + 7)°$, express the measure of the third angle in terms of x. Simplify the expression.

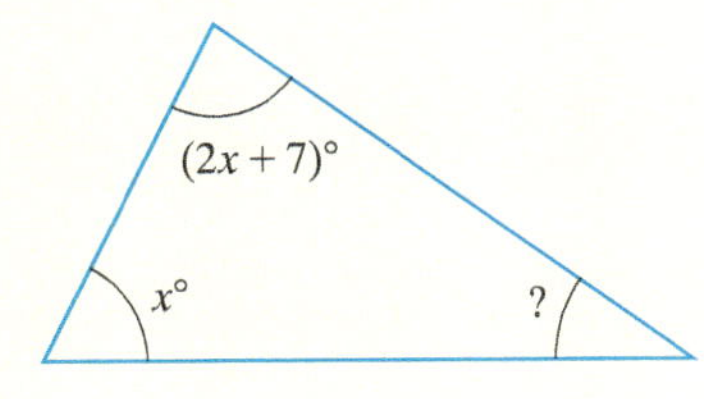

△ **84.** A quadrilateral is a four-sided figure (like the one shown in the figure) whose angle sum is 360°. If one angle measures $x°$, a second angle measures $3x°$, and a third angle measures $5x°$, express the measure of the fourth angle in terms of x. Simplify the expression.

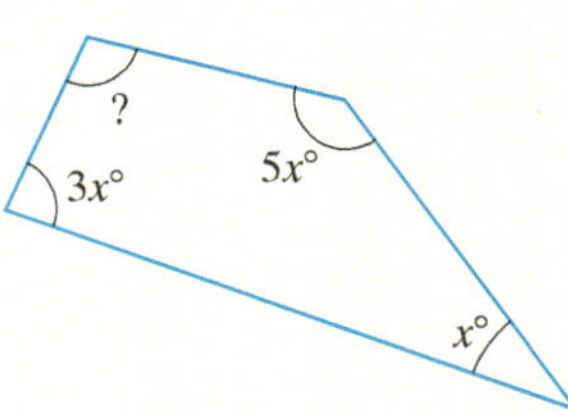

85. In your own words, explain what is meant by the solution of an equation.

86. In your own words, explain how to check a solution of an equation.

Use a calculator to determine the solution of each equation.

87. $36.766 + x = -108.712$

88. $-85.325 = x - 97.985$

2.2 The Multiplication Property of Equality 

Objectives

A Use the Multiplication Property of Equality to Solve Linear Equations.

B Use Both the Addition and Multiplication Properties of Equality to Solve Linear Equations.

C Write Word Phrases as Algebraic Expressions.

Objective A Using the Multiplication Property 

As useful as the addition property of equality is, it cannot help us solve every type of linear equation in one variable. For example, adding or subtracting a value on both sides of the equation does not help solve

$$\frac{5}{2}x = 15$$

because the variable x is being multiplied by a number (other than 1). Instead, we apply another important property of equality, the **multiplication property of equality.**

Multiplication Property of Equality

Let a, b, and c represent numbers and let $c \neq 0$. Then

$a = b$	Also, $a = b$
and $a \cdot c = b \cdot c$	and $\frac{a}{c} = \frac{b}{c}$
are equivalent equations.	are equivalent equations.

In other words, **both sides** of an equation **may be multiplied or divided by the same nonzero number** without changing the solution of the equation. (We may divide both sides by the same nonzero number since division is defined in terms of multiplication.)

Picturing again our balanced scale, if we multiply or divide the weight on each side by the same nonzero number, the scale (or equation) remains balanced.

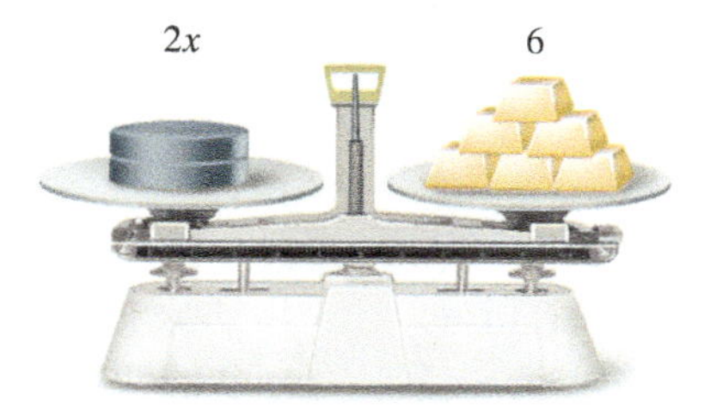

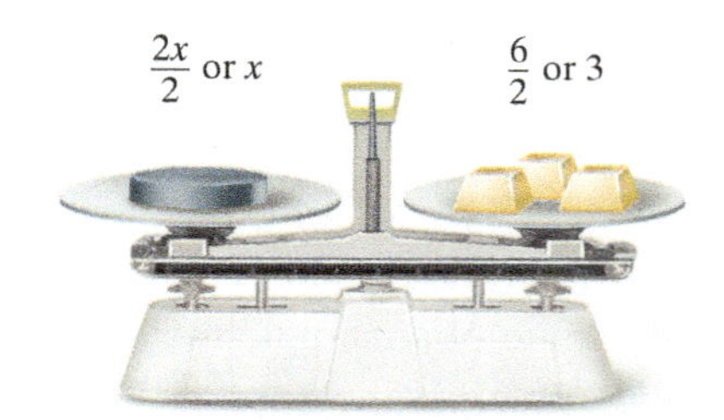

Practice 1

Solve: $\frac{3}{7}x = 9$

Example 1 Solve: $\frac{5}{2}x = 15$

Solution: To get x alone, we multiply both sides of the equation by the reciprocal (or multiplicative inverse) of $\frac{5}{2}$, which is $\frac{2}{5}$.

$$\frac{5}{2}x = 15$$

$$\frac{2}{5} \cdot \left(\frac{5}{2}x\right) = \frac{2}{5} \cdot 15 \quad \text{Multiply both sides by } \frac{2}{5}.$$

$$\left(\frac{2}{5} \cdot \frac{5}{2}\right)x = \frac{2}{5} \cdot 15 \quad \text{Apply the associative property.}$$

$$1x = 6 \quad \text{Simplify.}$$

or

$$x = 6$$

Answer

1. $x = 21$

Check: Replace x with 6 in the original equation.

$$\frac{5}{2}x = 15 \quad \text{Original equation.}$$
$$\frac{5}{2}(6) \stackrel{?}{=} 15 \quad \text{Replace } x \text{ with 6.}$$
$$15 = 15 \quad \text{True}$$

The solution is 6.

Work Practice 1

In the equation $\frac{5}{2}x = 15$, $\frac{5}{2}$ is the coefficient of x. When the coefficient of x is a *fraction,* we will get x alone by multiplying by the reciprocal. When the coefficient of x is an integer or a decimal, it is usually more convenient to divide both sides by the coefficient. (Dividing by a number is, of course, the same as multiplying by the reciprocal of the number.)

Example 2 Solve: $5x = 30$

Solution: To get x alone, we divide both sides of the equation by 5, the coefficient of x.

$$5x = 30$$
$$\frac{5x}{5} = \frac{30}{5} \quad \text{Divide both sides by 5.}$$
$$1 \cdot x = 6 \quad \text{Simplify.}$$
$$x = 6$$

Check:

$$5x = 30 \quad \text{Original equation.}$$
$$5 \cdot 6 \stackrel{?}{=} 30 \quad \text{Replace } x \text{ with 6.}$$
$$30 = 30 \quad \text{True}$$

The solution is 6.

Work Practice 2

Practice 2

Solve: $7x = 42$

Example 3 Solve: $-3x = 33$

Solution: Recall that $-3x$ means $-3 \cdot x$. To get x alone, we divide both sides by the coefficient of x, that is, -3.

$$-3x = 33$$
$$\frac{-3x}{-3} = \frac{33}{-3} \quad \text{Divide both sides by } -3.$$
$$1x = -11 \quad \text{Simplify.}$$
$$x = -11$$

Check:

$$-3x = 33 \quad \text{Original equation.}$$
$$-3(-11) \stackrel{?}{=} 33 \quad \text{Replace } x \text{ with } -11.$$
$$33 = 33 \quad \text{True}$$

The solution is -11.

Work Practice 3

Practice 3

Solve: $-4x = 52$

Answers

2. $x = 6$ **3.** $x = -13$

Practice 4

Solve: $\frac{y}{5} = 13$

Example 4 Solve: $\frac{y}{7} = 20$

Solution: Recall that $\frac{y}{7} = \frac{1}{7}y$. To get y alone, we multiply both sides of the equation by 7, the reciprocal of $\frac{1}{7}$.

$$\frac{y}{7} = 20$$

$$\frac{1}{7}y = 20$$

$$7 \cdot \frac{1}{7}y = 7 \cdot 20 \quad \text{Multiply both sides by 7.}$$

$$1y = 140 \quad \text{Simplify.}$$

$$y = 140$$

Check:

$$\frac{y}{7} = 20 \quad \text{Original equation.}$$

$$\frac{140}{7} \stackrel{?}{=} 20 \quad \text{Replace } y \text{ with 140.}$$

$$20 = 20 \quad \text{True}$$

The solution is 140.

Work Practice 4

Practice 5

Solve: $2.6x = 13.52$

Example 5 Solve: $3.1x = 4.96$

Solution:

$$3.1x = 4.96$$

$$\frac{3.1x}{3.1} = \frac{4.96}{3.1} \quad \text{Divide both sides by 3.1.}$$

$$1x = 1.6 \quad \text{Simplify.}$$

$$x = 1.6$$

Check: Check by replacing x with 1.6 in the original equation. The solution is 1.6.

Work Practice 5

Practice 6

Solve: $-\frac{5}{6}y = -\frac{3}{5}$

Example 6 Solve: $-\frac{2}{3}x = -\frac{5}{2}$

Solution: To get x alone, we multiply both sides of the equation by $-\frac{3}{2}$, the reciprocal of the coefficient of x.

$$-\frac{2}{3}x = -\frac{5}{2}$$

$$-\frac{3}{2} \cdot -\frac{2}{3}x = -\frac{3}{2} \cdot -\frac{5}{2} \quad \text{Multiply both sides by } -\frac{3}{2}\text{, the reciprocal of } -\frac{2}{3}.$$

$$x = \frac{15}{4} \quad \text{Simplify.}$$

Check: Check by replacing x with $\frac{15}{4}$ in the original equation. The solution is $\frac{15}{4}$.

Work Practice 6

Answers

4. $y = 65$ **5.** $x = 5.2$ **6.** $y = \frac{18}{25}$

Objective B Using Both the Addition and Multiplication Properties

We are now ready to combine the skills learned in the last section with the skills learned in this section to solve equations by applying more than one property.

Example 7 Solve: $-z - 4 = 6$

Solution: First, let's get $-z$, the term containing the variable, alone. To do so, we add 4 to both sides of the equation.

$$-z - 4 + 4 = 6 + 4 \quad \text{Add 4 to both sides.}$$
$$-z = 10 \quad \text{Simplify.}$$

Next, recall that $-z$ means $-1 \cdot z$. Thus to get z alone, we either multiply or divide both sides of the equation by -1. In this example, we divide.

$$-z = 10$$
$$\frac{-z}{-1} = \frac{10}{-1} \quad \text{Divide both sides by the coefficient } -1.$$
$$1z = -10 \quad \text{Simplify.}$$
$$z = -10$$

Check: $-z - 4 = 6$ Original equation.
$$-(-10) - 4 \stackrel{?}{=} 6 \quad \text{Replace } z \text{ with } -10.$$
$$10 - 4 \stackrel{?}{=} 6$$
$$6 = 6 \quad \text{True}$$

The solution is -10.

Work Practice 7

Practice 7

Solve: $-x + 7 = -12$

Don't forget to first simplify one or both sides of an equation, if possible.

Example 8 Solve: $a + a - 10 + 7 = -13$

Solution: First, we simplify the left side of the equation by combining like terms.

$$a + a - 10 + 7 = -13$$
$$2a - 3 = -13 \quad \text{Combine like terms.}$$
$$2a - 3 + 3 = -13 + 3 \quad \text{Add 3 to both sides.}$$
$$2a = -10 \quad \text{Simplify.}$$
$$\frac{2a}{2} = \frac{-10}{2} \quad \text{Divide both sides by 2.}$$
$$a = -5 \quad \text{Simplify.}$$

Check: To check, replace a with -5 in the original equation. The solution is -5.

Work Practice 8

Practice 8

Solve: $-7x + 2x + 3 - 20 = -2$

Example 9 Solve: $7x - 3 = 5x + 9$

Solution: To get x alone, let's first use the addition property to get variable terms on one side of the equation and numbers on the other side. One way to get variable terms on one side is to subtract $5x$ from both sides.

$$7x - 3 = 5x + 9$$
$$7x - 3 - 5x = 5x + 9 - 5x \quad \text{Subtract } 5x \text{ from both sides.}$$
$$2x - 3 = 9 \quad \text{Simplify.}$$

(Continued on next page)

Practice 9

Solve: $10x - 4 = 7x + 14$

Answers

7. $x = 19$ **8.** $x = -3$ **9.** $x = 6$

Now, to get numbers on the other side, let's add 3 to both sides.

$$2x - 3 + 3 = 9 + 3 \quad \text{Add 3 to both sides.}$$
$$2x = 12 \quad \text{Simplify.}$$

Use the multiplication property to get x alone.

$$\frac{2x}{2} = \frac{12}{2} \quad \text{Divide both sides by 2.}$$
$$x = 6 \quad \text{Simplify.}$$

Check: To check, replace x with 6 in the original equation to see that a true statement results. The solution is 6.

Work Practice 9

If an equation has parentheses, don't forget to use the distributive property to remove them. Then combine any like terms.

Practice 10

Solve: $4(3x - 2) = -1 + 4$

Example 10 Solve: $5(2x + 3) = -1 + 7$

Solution:

$$5(2x + 3) = -1 + 7$$
$$5(2x) + 5(3) = -1 + 7 \quad \text{Apply the distributive property.}$$
$$10x + 15 = 6 \quad \text{Multiply and write } -1 + 7 \text{ as 6.}$$
$$10x + 15 - 15 = 6 - 15 \quad \text{Subtract 15 from both sides.}$$
$$10x = -9 \quad \text{Simplify.}$$
$$\frac{10x}{10} = -\frac{9}{10} \quad \text{Divide both sides by 10.}$$
$$x = -\frac{9}{10} \quad \text{Simplify.}$$

Check: To check, replace x with $-\frac{9}{10}$ in the original equation to see that a true statement results. The solution is $-\frac{9}{10}$.

Work Practice 10

Practice 11

a. If x is the first of two consecutive integers, express the sum of the two integers in terms of x. Simplify if possible.

b. If x is the first of two consecutive odd integers (see next page), express the sum of the two integers in terms of x. Simplify if possible.

Objective C Writing Algebraic Expressions

We continue to sharpen our problem-solving skills by writing algebraic expressions.

Example 11 Writing an Expression for Consecutive Integers

If x is the first of three consecutive integers, express the sum of the three integers in terms of x. Simplify if possible.

Solution: An example of three consecutive integers is 7, 8, and 9.

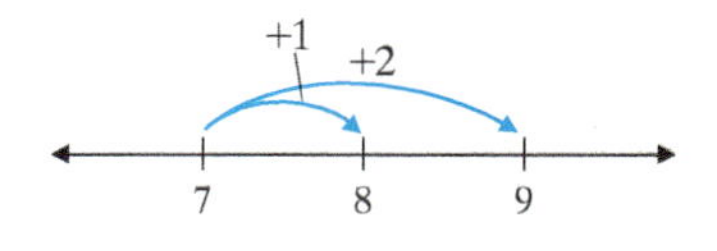

Answers

10. $x = \frac{11}{12}$ **11. a.** $2x + 1$ **b.** $2x + 2$

The second consecutive integer is always 1 more than the first, and the third consecutive integer is 2 more than the first. If x is the first of three consecutive integers, the three consecutive integers are x, $x + 1$, and $x + 2$.

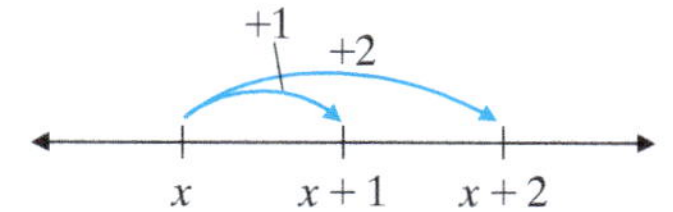

Their sum is shown below.

In words:	first integer	+	second integer	+	third integer
Translate:	x	+	$(x + 1)$	+	$(x + 2)$

This simplifies to $3x + 3$.

Work Practice 11

Study these examples of consecutive even and consecutive odd integers.

Consecutive even integers:

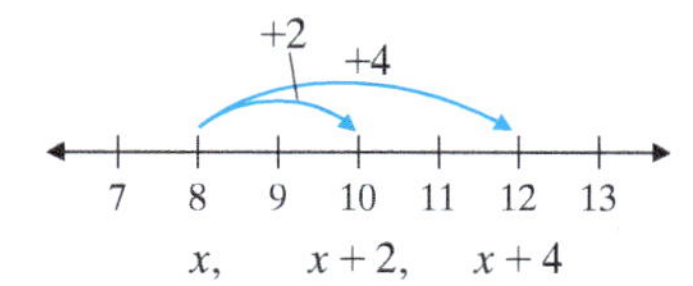

Consecutive odd integers:

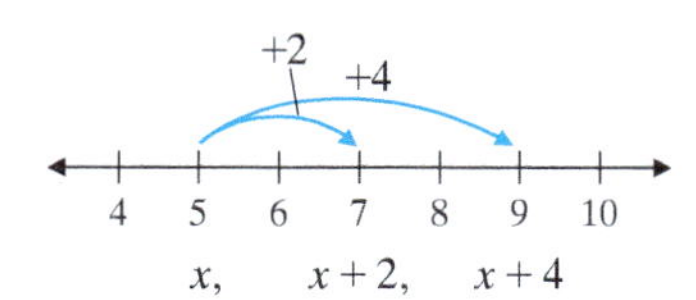

Helpful Hint

If x is an odd integer, then $x + 2$ is the next odd integer. This 2 simply means that odd integers are always 2 units from each other.

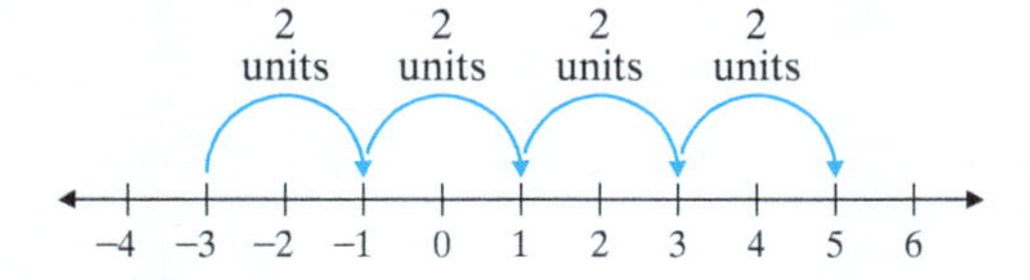

Vocabulary, Readiness & Video Check

Use the choices below to fill in each blank. Some choices may be used more than once. Many of these exercises contain an important review of Section 2.1 as well.

equation	multiplication	addition
expression	solution	equivalent

1. By the _______________ property of equality, both sides of an equation may be multiplied or divided by the same nonzero number without changing the solution of the equation.

2. By the _______________ property of equality, the same number may be added to or subtracted from both sides of an equation without changing the solution of the equation.

3. A(n)______________ may be solved while a(n) ______________ may be simplified and evaluated.

4. A(n)______________ contains an equal sign (=) while a(n) ______________ does not.

5. ______________ equations have the same solution.

6. A(n) ______________ of an equation is a number that when substituted for the variable makes the equation a true statement.

Solve each equation mentally. See Examples 2 and 3.

7. $3a = 27$ **8.** $9c = 54$ **9.** $5b = 10$ **10.** $7t = 14$ **11.** $6x = -30$ **12.** $8r = -64$

Martin-Gay Interactive Videos

See Video 2.2

Watch the section lecture video and answer the following questions.

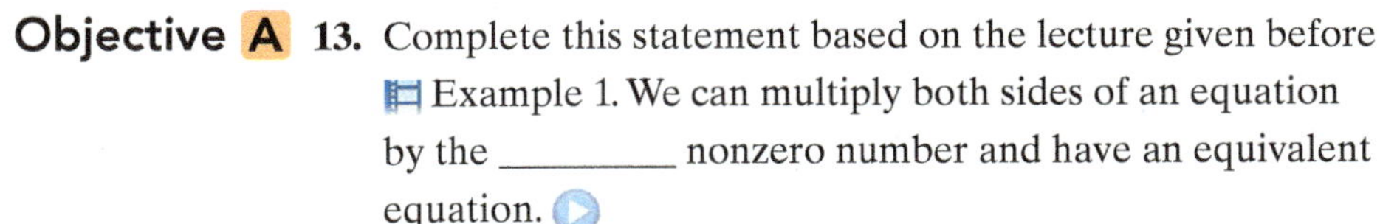

Objective A **13.** Complete this statement based on the lecture given before Example 1. We can multiply both sides of an equation by the ________ nonzero number and have an equivalent equation.

Objective B **14.** Both the addition and multiplication properties of equality are used to solve Examples 4–6. In each of these exercises, what property is applied first? What property is applied last? What conclusion, if any, can you make?

Objective C **15.** Let x be the first of four consecutive integers, as in Example 8. Now express the sum of the second integer and the fourth integer as an algebraic expression containing x.

2.2 Exercise Set MyLab Math

Objective A *Solve each equation. Check each solution. See Examples 1 through 6.*

1. $-5x = -20$ **2.** $-7x = -49$ **3.** $3x = 0$ **4.** $2x = 0$

5. $-x = -12$ **6.** $-y = 8$ **7.** $\frac{2}{3}x = -8$ **8.** $\frac{3}{4}n = -15$

9. $\frac{1}{6}d = \frac{1}{2}$ **10.** $\frac{1}{8}v = \frac{1}{4}$ **11.** $\frac{a}{2} = 1$ **12.** $\frac{d}{15} = 2$

13. $\frac{k}{-7} = 0$ **14.** $\frac{f}{-5} = 0$ **15.** $1.7x = 10.71$ **16.** $8.5y = 19.55$

Objective B *Solve each equation. Check each solution. See Examples 7 and 8.*

17. $2x - 4 = 16$ **18.** $3x - 1 = 26$ **19.** $-x + 2 = 22$ **20.** $-x + 4 = -24$

21. $6a + 3 = 3$ **22.** $8t + 5 = 5$ **23.** $\frac{x}{3} - 2 = -5$ **24.** $\frac{b}{4} - 1 = -7$

25. $6z - 8 - z + 3 = 0$ **26.** $4a + 1 + a - 11 = 0$ **27.** $1 = 0.4x - 0.6x - 5$ **28.** $19 = 0.4x - 0.9x - 6$

29. $\frac{2}{3}y - 11 = -9$ **30.** $\frac{3}{5}x - 14 = -8$ **31.** $\frac{3}{4}t - \frac{1}{2} = \frac{1}{3}$ **32.** $\frac{2}{7}z - \frac{1}{5} = \frac{1}{2}$

Solve each equation. See Examples 9 and 10.

33. $8x + 20 = 6x + 18$ **34.** $11x + 13 = 9x + 9$ **35.** $3(2x + 5) = -18 + 9$ **36.** $2(4x + 1) = -12 + 6$

37. $2x - 5 = 20x + 4$ **38.** $6x - 4 = -2x - 10$ **39.** $2 + 14 = -4(3x - 4)$ **40.** $8 + 4 = -6(5x - 2)$

41. $-6y - 3 = -5y - 7$ **42.** $-17z - 4 = -16z - 20$ **43.** $\frac{1}{2}(2x - 1) = -\frac{1}{7} - \frac{3}{7}$

44. $\frac{1}{3}(3x - 1) = -\frac{1}{10} - \frac{2}{10}$ **45.** $-10z - 0.5 = -20z + 1.6$ **46.** $-14y - 1.8 = -24y + 3.9$

47. $-4x + 20 = 4x - 20$ **48.** $-3x + 15 = 3x - 15$

Objectives A B Mixed Practice *See Examples 1 through 10.*

49. $42 = 7x$ **50.** $81 = 3x$ **51.** $4.4 = -0.8x$

52. $6.3 = -0.6x$ **53.** $6x + 10 = -20$ **54.** $10y + 15 = -5$

55. $5 - 0.3k = 5$ **56.** $2 - 0.4p = 2$ **57.** $13x - 5 = 11x - 11$

58. $20x - 20 = 16x - 40$ **59.** $9(3x + 1) = 4x - 5x$ **60.** $7(2x + 1) = 18x - 19x$

61. $-\frac{3}{7}p = -2$ **62.** $-\frac{4}{5}r = -5$ **63.** $-\frac{4}{3}x = 12$

64. $-\frac{10}{3}x = 30$ **65.** $-2x - \frac{1}{2} = \frac{7}{2}$ **66.** $-3n - \frac{1}{3} = \frac{8}{3}$

67. $10 = 2x - 1$ **68.** $12 = 3j - 4$ **69.** $10 - 3x - 6 - 9x = 7$

70. $12x + 30 + 8x - 6 = 10$ **71.** $z - 5z = 7z - 9 - z$ **72.** $t - 6t = -13 + t - 3t$

73. $-x - \frac{4}{5} = x + \frac{1}{2} + \frac{2}{5}$ **74.** $x + \frac{3}{7} = -x + \frac{1}{3} + \frac{4}{7}$

75. $-15 + 37 = -2(x + 5)$ **76.** $-19 + 74 = -5(x + 3)$

Objective C *Write each algebraic expression described. Simplify if possible. See Example 11.*

77. If x represents the first of two consecutive odd integers, express the sum of the two integers in terms of x.

78. If x is the first of three consecutive even integers, write their sum as an algebraic expression in x.

79. If x is the first of four consecutive integers, express the sum of the first integer and the third integer as an algebraic expression containing the variable x.

80. If x is the first of two consecutive integers, express the sum of 20 and the second consecutive integer as an algebraic expression containing the variable x.

81. Classrooms on one side of the science building are all numbered with consecutive even integers. If the first room on this side of the building is numbered x, write an expression in x for the sum of five classroom numbers in a row. Then simplify this expression.

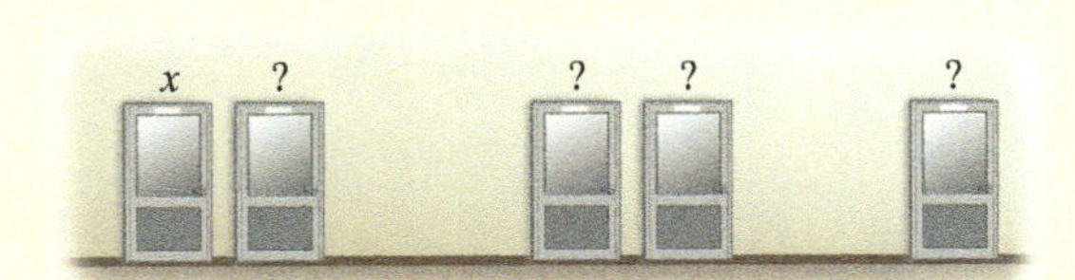

82. Two sides of a quadrilateral have the same length, x, while the other two sides have the same length, both being the next consecutive odd integer. Write the sum of these lengths. Then simplify this expression.

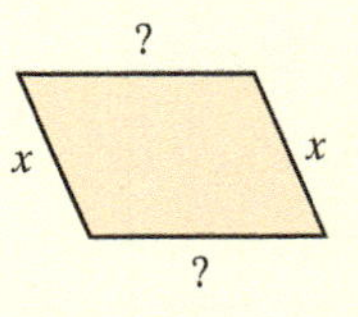

Review

Simplify each expression. See Section 1.8.

83. $5x + 2(x - 6)$

84. $-7y + 2y - 3(y + 1)$

85. $6(2z + 4) + 20$

86. $-(3a - 3) + 2a - 6$

87. $-(x - 1) + x$

88. $8(z - 6) + 7z - 1$

Concept Extensions

For Exercises 89 and 90, fill in the blank so that each equation has the given solution.

89. $6x =$ _____; solution: -8

90. _____ $x = 10$; solution: $\frac{1}{2}$

91. The equation $3x + 6 = 2x + 10 + x - 4$ is true for all real numbers. Substitute a few real numbers for x to see that this is so and then try solving the equation. Describe what happens.

92. The equation $6x + 2 - 2x = 4x + 1$ has no solution. Try solving this equation for x and describe what happens.

93. From the result of Exercise **91**, when do you think an equation has all real numbers as its solutions?

94. From the result of Exercise **92**, when do you think an equation has no solution?

Solve.

95. $0.07x - 5.06 = -4.92$

96. $0.06y + 2.63 = 2.5562$

2.3 Further Solving Linear Equations 

Objective A Solving Linear Equations 

Let's begin by restating the formal definition of a linear equation in one variable.

A **linear equation in one variable** can be written in the form

$$Ax + B = C$$

where A, B, and C are real numbers and $A \neq 0$.

We now combine our knowledge from the previous sections into a general strategy for solving linear equations.

To Solve Linear Equations in One Variable

Step 1: If an equation contains fractions or decimals, multiply both sides by the LCD to clear the equation of fractions or decimals.

Step 2: Use the distributive property to remove parentheses if they are present.

Step 3: Simplify each side of the equation by combining like terms.

Step 4: Get all variable terms on one side and all numbers on the other side by using the addition property of equality.

Step 5: Get the variable alone by using the multiplication property of equality.

Step 6: Check the solution by substituting it into the original equation.

Objectives

A Apply the General Strategy for Solving a Linear Equation.

B Solve Equations Containing Fractions or Decimals.

C Recognize Identities and Equations with No Solution.

Example 1 Solve: $4(2x - 3) + 7 = 3x + 5$

Solution: There are no fractions, so we begin with Step 2.

$$4(2x - 3) + 7 = 3x + 5$$

Step 2: $8x - 12 + 7 = 3x + 5$ Use the distributive property.

Step 3: $8x - 5 = 3x + 5$ Combine like terms.

Step 4: Get all variable terms on one side of the equation and all numbers on the other side. One way to do this is by subtracting $3x$ from both sides and then adding 5 to both sides.

$8x - 5 - 3x = 3x + 5 - 3x$ Subtract $3x$ from both sides.

$5x - 5 = 5$ Simplify.

$5x - 5 + 5 = 5 + 5$ Add 5 to both sides.

$5x = 10$ Simplify.

Step 5: Use the multiplication property of equality to get x alone.

$\frac{5x}{5} = \frac{10}{5}$ Divide both sides by 5.

$x = 2$ Simplify.

Step 6: Check.

$4(2x - 3) + 7 = 3x + 5$ Original equation

$4[2(2) - 3] + 7 \stackrel{?}{=} 3(2) + 5$ Replace x with 2.

$4(4 - 3) + 7 \stackrel{?}{=} 6 + 5$

$4(1) + 7 \stackrel{?}{=} 11$

$4 + 7 \stackrel{?}{=} 11$

$11 = 11$ True

The solution is 2.

Helpful Hint When checking solutions, use the original equation.

Work Practice 1

Practice 1

Solve:

$5(3x - 1) + 2 = 12x + 6$

Answer

1. $x = 3$

Practice 2

Solve: $9(5 - x) = -3x$

Example 2 Solve: $8(2 - t) = -5t$

Solution: First, we apply the distributive property.

$$8(2 - t) = -5t$$

Step 2: $16 - 8t = -5t$ — Use the distributive property.

Step 4: $16 - 8t + 8t = -5t + 8t$ — Add $8t$ to both sides.

$16 = 3t$ — Combine like terms.

Step 5: $\frac{16}{3} = \frac{3t}{3}$ — Divide both sides by 3.

$\frac{16}{3} = t$ — Simplify.

Step 6: Check.

$8(2 - t) = -5t$ — Original equation

$8\left(2 - \frac{16}{3}\right) \stackrel{?}{=} -5\left(\frac{16}{3}\right)$ — Replace t with $\frac{16}{3}$.

$8\left(\frac{6}{3} - \frac{16}{3}\right) \stackrel{?}{=} -\frac{80}{3}$ — The LCD is 3.

$8\left(-\frac{10}{3}\right) \stackrel{?}{=} -\frac{80}{3}$ — Subtract fractions.

$-\frac{80}{3} = -\frac{80}{3}$ — True

The solution is $\frac{16}{3}$.

Work Practice 2

Objective B Solving Equations Containing Fractions or Decimals

If an equation contains fractions, we can clear the equation of fractions by multiplying both sides by the LCD of all denominators. By doing this, we avoid working with time-consuming fractions.

Practice 3

Solve: $\frac{5}{2}x - 1 = \frac{3}{2}x - 4$

Example 3 Solve: $\frac{x}{2} - 1 = \frac{2}{3}x - 3$

Solution: We begin by clearing fractions. To do this, we multiply both sides of the equation by the LCD, which is 6.

$$\frac{x}{2} - 1 = \frac{2}{3}x - 3$$

Step 1: $6\left(\frac{x}{2} - 1\right) = 6\left(\frac{2}{3}x - 3\right)$ — Multiply both sides by the LCD, 6.

Step 2: $6\left(\frac{x}{2}\right) - 6(1) = 6\left(\frac{2}{3}x\right) - 6(3)$ — Use the distributive property.

Helpful Hint Don't forget to multiply *each* term by the LCD.

$3x - 6 = 4x - 18$ — Simplify.

There are no longer grouping symbols and no like terms on either side of the equation, so we continue with Step 4.

Answers

2. $x = \frac{15}{2}$ **3.** $x = -3$

$$3x - 6 = 4x - 18$$

Step 4: $3x - 6 - 3x = 4x - 18 - 3x$ Subtract $3x$ from both sides.

$-6 = x - 18$ Simplify.

$-6 + 18 = x - 18 + 18$ Add 18 to both sides.

$12 = x$ Simplify.

Step 5: The variable is now alone, so there is no need to apply the multiplication property of equality.

Step 6: Check.

$\frac{x}{2} - 1 = \frac{2}{3}x - 3$ Original equation

$\frac{12}{2} - 1 \stackrel{?}{=} \frac{2}{3} \cdot 12 - 3$ Replace x with 12.

$6 - 1 \stackrel{?}{=} 8 - 3$ Simplify.

$5 = 5$ True

The solution is 12.

Work Practice 3

Example 4 Solve: $\frac{2(a + 3)}{3} = 6a + 2$

Solution: We clear the equation of fractions first.

$$\frac{2(a + 3)}{3} = 6a + 2$$

Step 1: $3 \cdot \frac{2(a + 3)}{3} = 3(6a + 2)$ Clear the fraction by multiplying both sides by the LCD, 3.

$2(a + 3) = 3(6a + 2)$ Simplify.

Step 2: Next, we use the distributive property to remove parentheses.

$2a + 6 = 18a + 6$ Use the distributive property.

Step 4: $2a + 6 - 18a = 18a + 6 - 18a$ Subtract $18a$ from both sides.

$-16a + 6 = 6$ Simplify.

$-16a + 6 - 6 = 6 - 6$ Subtract 6 from both sides.

$-16a = 0$

Step 5: $\frac{-16a}{-16} = \frac{0}{-16}$ Divide both sides by -16.

$a = 0$ Simplify.

Step 6: To check, replace a with 0 in the original equation. The solution is 0.

Work Practice 4

Practice 4

Solve: $\frac{3(x - 2)}{5} = 3x + 6$

Helpful Hint

Remember: When solving an equation, it makes no difference on which side of the equation variable terms lie. Just make sure that constant terms lie on the other side.

When solving a problem about money, you may need to solve an equation containing decimals. If you choose, you may multiply to clear the equation of decimals.

Answer

4. $x = -3$

Practice 5

Solve:
$0.06x - 0.10(x - 2) = -0.16$

Helpful Hint If you have trouble with this step, try removing parentheses first.

$$0.25x + 0.10(x - 3) = 1.1$$
$$0.25x + 0.10x - 0.3 = 1.1$$
$$0.25x + 0.10x - 0.30 = 1.10$$
$$25x + 10x - 30 = 110$$

Then continue.

Example 5 Solve: $0.25x + 0.10(x - 3) = 1.1$

Solution: First we clear this equation of decimals by multiplying both sides of the equation by 100. Recall that multiplying a decimal number by 100 has the effect of moving the decimal point 2 places to the right.

$$0.25x + 0.10(x - 3) = 1.1$$

Step 1: $0.25x + 0.10(x - 3) = 1.10$ Multiply both sides by 100.
$$25x + 10(x - 3) = 110$$

Step 2: $25x + 10x - 30 = 110$ Apply the distributive property.

Step 3: $35x - 30 = 110$ Combine like terms.

Step 4: $35x - 30 + 30 = 110 + 30$ Add 30 to both sides.
$35x = 140$ Combine like terms.

Step 5: $\frac{35x}{35} = \frac{140}{35}$ Divide both sides by 35.
$$x = 4$$

Step 6: To check, replace x with 4 in the original equation. The solution is 4.

Work Practice 5

Objective C Recognizing Identities and Equations with No Solution

So far, each equation that we have solved has had a single solution. However, not every equation in one variable has a single solution. Some equations have no solution, while others have an infinite number of solutions. For example,

$$x + 5 = x + 7$$

has **no solution** since no matter which real number we replace x with, the equation is false.

real number + 5 = same real number + 7 FALSE

On the other hand,

$$x + 6 = x + 6$$

has infinitely many solutions since x can be replaced by any real number and the equation will always be true.

real number + 6 = same real number + 6 TRUE

The equation $x + 6 = x + 6$ is called an **identity.** The next two examples illustrate special equations like these.

Practice 6

Solve:
$5(2 - x) + 8x = 3(x - 6)$

Example 6 Solve: $-2(x - 5) + 10 = -3(x + 2) + x$

Solution:

$$-2(x - 5) + 10 = -3(x + 2) + x$$

$-2x + 10 + 10 = -3x - 6 + x$ Apply the distributive property on both sides.

$-2x + 20 = -2x - 6$ Combine like terms.

$-2x + 20 + 2x = -2x - 6 + 2x$ Add $2x$ to both sides.

$20 = -6$ Combine like terms.

Answers
5. $x = 9$
6. no solution

The final equation contains

- no variable terms, and
- the result is the false statement $20 = -6$.

There is no value for x that makes $20 = -6$ a true equation. Thus, we conclude that there is **no solution** to this equation.

Work Practice 6

Example 7 Solve: $3(x - 4) = 3x - 12$

Solution: $3(x - 4) = 3x - 12$

$3x - 12 = 3x - 12$ Apply the distributive property.

The left side of the equation is now identical to the right side. Every real number may be substituted for x and a true statement will result. We arrive at the same conclusion if we continue.

$$3x - 12 = 3x - 12$$

$$3x - 12 - 3x = 3x - 12 - 3x \quad \text{Subtract } 3x \text{ from both sides.}$$

$$-12 = -12 \quad \text{Combine like terms.}$$

Again, the final equation contains

- no variables, but this time
- the result is the true statement $-12 = -12$.

This means that one side of the equation is identical to the other side. Thus, $3(x - 4) = 3x - 12$ is an **identity** and **all real numbers** are solutions.

Work Practice 7

Practice 7

Solve:

$-6(2x + 1) - 14 = -10(x + 2) - 2x$

Answer

7. All real numbers are solutions.

✓**Concept Check** Suppose you have simplified several equations and obtained the following results. What can you conclude about the solutions to the original equation?

a. $7 = 7$ **b.** $x = 0$ **c.** $7 = -4$

✓**Concept Check Answer**

a. All real numbers are solutions.
b. The solution is 0.
c. There is no solution.

Calculator Explorations **Checking Equations**

We can use a calculator to check possible solutions of equations. To do this, replace the variable by the possible solution and evaluate each side of the equation separately.

Equation: $3x - 4 = 2(x + 6)$ Solution: $x = 16$

$3x - 4 = 2(x + 6)$ Original equation

$3(16) - 4 \stackrel{?}{=} 2(16 + 6)$ Replace x with 16.

Now evaluate each side with your calculator.

Evaluate left side: [3] [×] [16] [−] [4] [=]

or

[ENTER]

Display: [44]

Evaluate right side: [2] [(] [16] [+] [6] [)] [=]

or

[ENTER]

Display: [44]

Since the left side equals the right side, the equation checks.

Use a calculator to check the possible solutions to each equation.

1. $2x = 48 + 6x;\quad x = -12$
2. $-3x - 7 = 3x - 1;\quad x = -1$
3. $5x - 2.6 = 2(x + 0.8);\quad x = 4.4$
4. $-1.6x - 3.9 = -6.9x - 25.6;\quad x = 5$
5. $\dfrac{564x}{4} = 200x - 11(649);\quad x = 121$
6. $20(x - 39) = 5x - 432;\quad x = 23.2$

Vocabulary, Readiness & Video Check

Throughout algebra, it is important to be able to distinguish between equations and expressions.

Remember,
- an equation contains an equal sign and
- an expression does not.

Among other things,
- we solve equations and
- we simplify or perform operations on expressions.

Identify each as an equation or an expression.

1. $x = -7$ ________ **2.** $x - 7$ ________

3. $4y - 6 + 9y + 1$ ________ **4.** $4y - 6 = 9y + 1$ ________

5. $\frac{1}{x} - \frac{x-1}{8}$ ________ **6.** $\frac{1}{x} - \frac{x-1}{8} = 6$ ________

7. $0.1x + 9 = 0.2x$ ________ **8.** $0.1x^2 + 9y - 0.2x^2$ ________

Martin-Gay Interactive Videos *Watch the section lecture video and answer the following questions.*

See Video 2.3

Objective A **9.** The general strategy for solving linear equations in one variable is discussed after Example 1. How many properties (not steps) are mentioned in this strategy and what are they?

Objective B **10.** In the first step for solving Example 2, both sides of the equation are multiplied by the LCD. Why is the distributive property mentioned?

Objective B **11.** In Example 3, why is the number of decimal places in each term of the equation important?

Objective C **12.** Complete each statement based on Examples 4 and 5. When solving an equation and all variable terms subtract out:

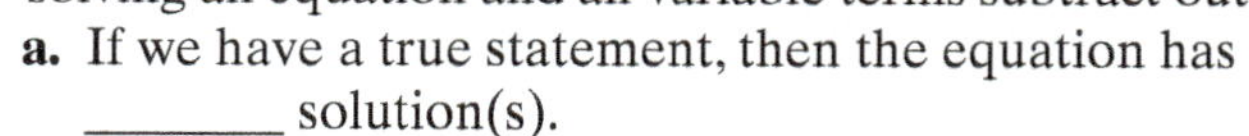

a. If we have a true statement, then the equation has ________ solution(s).

b. If we have a false statement, then the equation has ________ solution(s).

2.3 Exercise Set MyLab Math

Objective A *Solve each equation. See Examples 1 and 2.*

1. $-4y + 10 = -2(3y + 1)$ **2.** $-3x + 1 = -2(4x + 2)$

3. $15x - 8 = 10 + 9x$ **4.** $15x - 5 = 7 + 12x$

5. $-2(3x - 4) = 2x$ **6.** $-(5x - 10) = 5x$

7. $5(2x - 1) - 2(3x) = 1$ **8.** $3(2 - 5x) + 4(6x) = 12$

9. $-6(x-3)-26=-8$

10. $-4(n-4)-23=-7$

11. $8-2(a+1)=9+a$

12. $5-6(2+b)=b-14$

13. $4x+3=-3+2x+14$

14. $6y-8=-6+3y+13$

15. $-2y-10=5y+18$

16. $-7n+5=8n-10$

Objective B *Solve each equation. See Examples 3 through 5.*

17. $\frac{2}{3}x+\frac{4}{3}=-\frac{2}{3}$

18. $\frac{4}{5}x-\frac{8}{5}=-\frac{16}{5}$

19. $\frac{3}{4}x-\frac{1}{2}=1$

20. $\frac{2}{9}x-\frac{1}{3}=1$

21. $0.50x+0.15(70)=35.5$

22. $0.40x+0.06(30)=9.8$

23. $\frac{2(x+1)}{4}=3x-2$

24. $\frac{3(y+3)}{5}=2y+6$

25. $x+\frac{7}{6}=2x-\frac{7}{6}$

26. $\frac{5}{2}x-1=x+\frac{1}{4}$

27. $0.12(y-6)+0.06y=0.08y-0.7$

28. $0.60(z-300)+0.05z=0.70z-205$

Objective C *Solve each equation. See Examples 6 and 7.*

29. $4(3x+2)=12x+8$

30. $14x+7=7(2x+1)$

31. $\frac{x}{4}+1=\frac{x}{4}$

32. $\frac{x}{3}-2=\frac{x}{3}$

33. $3x-7=3(x+1)$

34. $2(x-5)=2x+10$

35. $-2(6x-5)+4=-12x+14$

36. $-5(4y-3)+2=-20y+17$

Objectives A B C Mixed Practice *Solve. See Examples 1 through 7.*

37. $\frac{6(3-z)}{5}=-z$

38. $\frac{4(5-w)}{3}=-w$

39. $-3(2t-5)+2t=5t-4$

40. $-(4a-7)-5a=10+a$

41. $5y+2(y-6)=4(y+1)-2$

42. $9x+3(x-4)=10(x-5)+7$

43. $\frac{3(x-5)}{2}=\frac{2(x+5)}{3}$

44. $\frac{5(x-1)}{4}=\frac{3(x+1)}{2}$

45. $0.7x-2.3=0.5$

46. $0.9x-4.1=0.4$

47. $5x-5=2(x+1)+3x-7$

48. $3(2x-1)+5=6x+2$

49. $4(2n+1)=3(6n+3)+1$

50. $4(4y+2)=2(1+6y)+8$

51. $x + \frac{5}{4} = \frac{3}{4}x$

52. $\frac{7}{8}x + \frac{1}{4} = \frac{3}{4}x$

53. $\frac{x}{2} - 1 = \frac{x}{5} + 2$

54. $\frac{x}{5} - 7 = \frac{x}{3} - 5$

55. $2(x + 3) - 5 = 5x - 3(1 + x)$

56. $4(2 + x) + 1 = 7x - 3(x - 2)$

57. $0.06 - 0.01(x + 1) = -0.02(2 - x)$

58. $-0.01(5x + 4) = 0.04 - 0.01(x + 4)$

59. $\frac{9}{2} + \frac{5}{2}y = 2y - 4$

60. $3 - \frac{1}{2}x = 5x - 8$

61. $\frac{3}{4}x - 1 + \frac{1}{2}x = \frac{5}{12}x + \frac{1}{6}$

62. $\frac{5}{9}x + 2 - \frac{1}{6}x = \frac{11}{18}x + \frac{1}{3}$

63. $3x + \frac{5}{16} = \frac{3}{4} - \frac{1}{8}x - \frac{1}{2}$

64. $2x - \frac{1}{10} = \frac{2}{5} - \frac{1}{4}x - \frac{17}{20}$

Review

Translating *Write each algebraic expression described. See Section 1.8. Recall that the perimeter of a figure is the total distance around the figure.*

△ 65. A plot of land is in the shape of a triangle. If one side is x meters, a second side is $(2x - 3)$ meters, and a third side is $(3x - 5)$ meters, express the perimeter of the lot as a simplified expression in x.

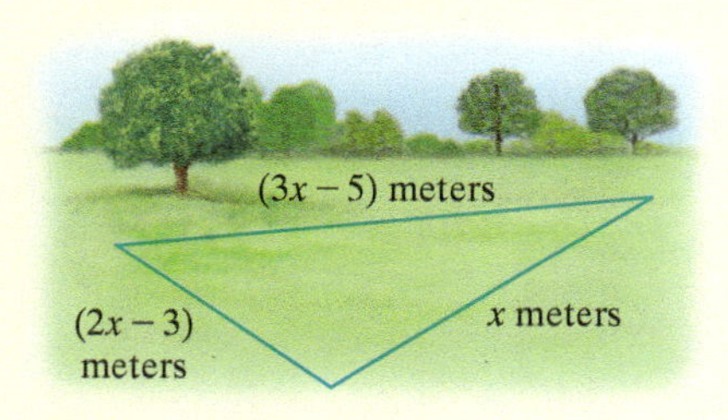

66. A portion of a board has length x feet. The other part has length $(7x - 9)$ feet. Express the total length of the board as a simplified expression in x.

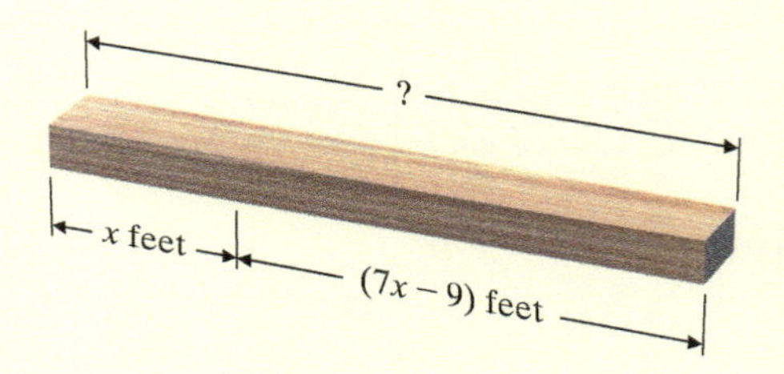

Translating *Write each phrase as an algebraic expression. Use x for the unknown number. See Section 1. 8.*

67. A number subtracted from −8

68. Three times a number

69. The sum of −3 and twice a number

70. The difference of 8 and twice a number

71. The product of 9 and the sum of a number and 20

72. The quotient of −12 and the difference of a number and 3

Concept Extensions

See the Concept Check in this section.

73. **a.** Solve: $x + 3 = x + 3$
b. If you simplify an equation (such as the one in part **a**) and get a true statement such as $3 = 3$ or $0 = 0$, what can you conclude about the solution(s) of the original equation?
c. On your own, construct an equation for which every real number is a solution.

74. **a.** Solve: $x + 3 = x + 5$
b. If you simplify an equation (such as the one in part **a**) and get a false statement such as $3 = 5$ or $10 = 17$, what can you conclude about the solution(s) of the original equation?
c. On your own, construct an equation that has no solution.

For Exercises 75 through 80, match each equation in the first column with its solution in the second column. Items in the second column may be used more than once.

75. $5x + 1 = 5x + 1$

76. $3x + 1 = 3x + 2$

77. $2x - 6x - 10 = -4x + 3 - 10$

78. $x - 11x - 3 = -10x - 1 - 2$

79. $9x - 20 = 8x - 20$

80. $-x + 15 = x + 15$

a. all real numbers
b. no solution
c. 0

81. Explain the difference between simplifying an expression and solving an equation.

82. On your own, write an expression and then an equation. Label each.

For Exercises 83 and 84, **a.** *Write an equation for the perimeter.* **b.** *Solve the equation in part (a).* **c.** *Find the length of each side.*

83. The perimeter of a geometric figure is the sum of the lengths of its sides. If the perimeter of the following pentagon (five-sided figure) is 28 centimeters, find the length of each side.

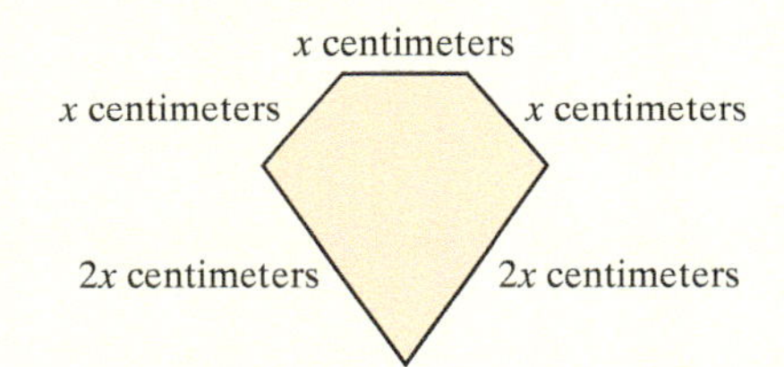

84. The perimeter of the following triangle is 35 meters. Find the length of each side.

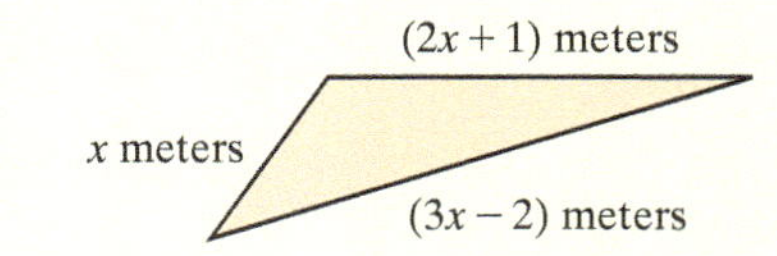

Fill in the blanks with numbers of your choice so that each equation has the given solution. Note: Each blank may be replaced by a different number.

85. $x +$ ____ $= 2x -$ ____; solution: 9

86. $-5x -$ ____ $=$ ____; solution: 2

Solve.

87. $1000(7x - 10) = 50(412 + 100x)$

88. $1000(x + 40) = 100(16 + 7x)$

89. $0.035x + 5.112 = 0.010x + 5.107$

90. $0.127x - 2.685 = 0.027x - 2.38$

Integrated Review Sections 2.1–2.3

Answers

1. ______
2. ______
3. ______
4. ______
5. ______
6. ______
7. ______
8. ______
9. ______
10. ______
11. ______
12. ______
13. ______
14. ______
15. ______
16. ______
17. ______
18. ______
19. ______
20. ______
21. ______
22. ______

Solving Linear Equations

Solve. Feel free to use the steps given in Section 2.3.

1. $x - 10 = -4$

2. $y + 14 = -3$

3. $9y = 108$

4. $-3x = 78$

5. $-6x + 7 = 25$

6. $5y - 42 = -47$

7. $\frac{2}{3}x = 9$

8. $\frac{4}{5}z = 10$

9. $\frac{r}{-4} = -2$

10. $\frac{y}{-8} = 8$

11. $6 - 2x + 8 = 10$

12. $-5 - 6y + 6 = 19$

13. $2x - 7 = 6x - 27$

14. $3 + 8y = 3y - 2$

15. $9(3x - 1) = -4 + 49$

16. $12(2x + 1) = -6 + 66$

17. $-3a + 6 + 5a = 7a - 8a$

18. $4b - 8 - b = 10b - 3b$

19. $-\frac{2}{3}x = \frac{5}{9}$

20. $-\frac{3}{8}y = -\frac{1}{16}$

21. $10 = -6n + 16$

22. $-5 = -2m + 7$

23. $3(5c - 1) - 2 = 13c + 3$

24. $4(3t + 4) - 20 = 3 + 5t$

25. $\frac{2(z + 3)}{3} = 5 - z$

26. $\frac{3(w + 2)}{4} = 2w + 3$

27. $-2(2x - 5) = -3x + 7 - x + 3$

28. $-4(5x - 2) = -12x + 4 - 8x + 4$

29. $0.02(6t - 3) = 0.04(t - 2) + 0.02$

30. $0.03(m + 7) = 0.02(5 - m) + 0.03$

31. $-3y = \frac{4(y - 1)}{5}$

32. $-4x = \frac{5(1 - x)}{6}$

33. $\frac{5}{3}x - \frac{7}{3} = x$

34. $\frac{7}{5}n + \frac{3}{5} = -n$

35. $\frac{1}{10}(3x - 7) = \frac{3}{10}x + 5$

36. $\frac{1}{7}(2x - 5) = \frac{2}{7}x + 1$

37. $5 + 2(3x - 6) = -4(6x - 7)$

38. $3 + 5(2x - 4) = -7(5x + 2)$

23. ______

24. ______

25. ______

26. ______

27. ______

28. ______

29. ______

30. ______

31. ______

32. ______

33. ______

34. ______

35. ______

36. ______

37. ______

38. ______

2.4 An Introduction to Problem Solving 

Objectives

A Solve Problems Involving Direct Translations.

B Solve Problems Involving Relationships Among Unknown Quantities.

C Solve Problems Involving Consecutive Integers.

First, let's review a list of key words and phrases from Section 1.3 to help us translate.

Helpful Hint

Order matters when subtracting and also dividing, so be especially careful with these translations.

Addition (+)	Subtraction (−)	Multiplication (·)	Division (÷)	Equality (=)
Sum	Difference of	Product	Quotient	Equals
Plus	Minus	Times	Divide	Gives
Added to	Subtracted from	Multiply	Into	Is/was/should be
More than	Less than	Twice	Ratio	Yields
Increased by	Decreased by	Of	Divided by	Amounts to
Total	Less			Represents
				Is the same as

We are now ready to put all our translating skills to practical use. To begin, we present a general strategy for problem solving.

General Strategy for Problem Solving

1. UNDERSTAND the problem. During this step, become comfortable with the problem. Some ways of doing this are:

   Read and reread the problem.

   Choose a variable to represent the unknown.

   Construct a drawing.

   Propose a solution and check. Pay careful attention to how you check your proposed solution. This will help when writing an equation to model the problem.
2. TRANSLATE the problem into an equation.
3. SOLVE the equation.
4. INTERPRET the results: *Check* the proposed solution in the stated problem and *state* your conclusion.

Objective A Solving Direct Translation Problems 

Much of problem solving involves a direct translation from a sentence to an equation.

Example 1 Finding an Unknown Number

Twice a number, added to seven, is the same as three subtracted from the number. Find the number.

Solution:

1. UNDERSTAND. Read and reread the problem. Let $x =$ the unknown number.
2. TRANSLATE.

twice a number	added to	seven	is the same as	three subtracted from the number
↓	↓	↓	↓	↓
$2x$	$+$	7	$=$	$x - 3$

Practice 1

Three times a number, minus 6, is the same as two times the number, plus 3. Find the number.

Answer

1. The number is 9.

3. SOLVE. Begin by subtracting x from both sides to isolate the variable term.

$$2x + 7 = x - 3$$
$$2x + 7 - x = x - 3 - x \quad \text{Subtract } x \text{ from both sides.}$$
$$x + 7 = -3 \quad \text{Combine like terms.}$$
$$x + 7 - 7 = -3 - 7 \quad \text{Subtract 7 from both sides.}$$
$$x = -10 \quad \text{Combine like terms.}$$

4. INTERPRET.

Check: Check the solution in the problem as it was originally stated. To do so, replace "number" in the sentence with −10. Twice "−10" added to 7 is the same as 3 subtracted from "−10."

$$2(-10) + 7 \stackrel{?}{=} -10 - 3$$
$$-13 = -13$$

State: The unknown number is −10.

Helpful Hint When checking solutions, go back to the original stated problem, rather than to your equation in case errors have been made in translating to an equation.

Work Practice 1

Example 2 Finding an Unknown Number

Twice the sum of a number and 4 is the same as four times the number decreased by 12. Find the number.

Solution:

1. UNDERSTAND. Read and reread the problem. If we let $x =$ the unknown number, then
"the sum of a number and 4" translates to "$x + 4$" and
"four times the number" translates to "$4x$"

2. TRANSLATE.

twice	the sum of a number and 4	is the same as	four times the number	decreased by	12
↓	↓	↓	↓	↓	↓
2	$(x + 4)$	$=$	$4x$	$-$	12

3. SOLVE.

$$2(x + 4) = 4x - 12$$
$$2x + 8 = 4x - 12 \quad \text{Apply the distributive property.}$$
$$2x + 8 - 4x = 4x - 12 - 4x \quad \text{Subtract } 4x \text{ from both sides.}$$
$$-2x + 8 = -12$$
$$-2x + 8 - 8 = -12 - 8 \quad \text{Subtract 8 from both sides.}$$
$$-2x = -20$$
$$\frac{-2x}{-2} = \frac{-20}{-2} \quad \text{Divide both sides by } -2.$$
$$x = 10$$

4. INTERPRET.

Check: Check this solution in the problem as it was originally stated. To do so, replace "number" with 10. Twice the sum of "10" and 4 is 28, which is the same as 4 times "10" decreased by 12.

State: The number is 10.

Work Practice 2

Practice 2

Three times the difference of a number and 5 is the same as twice the number decreased by 3. Find the number.

Answer

2. The number is 12.

Objective B Solving Problems Involving Relationships Among Unknown Quantities

Practice 3

An 18-foot wire is to be cut so that the length of the longer piece is 5 times the length of the shorter piece. Find the length of each piece.

Example 3 Finding the Length of a Board

A 10-foot board is to be cut into two pieces so that the length of the longer piece is 4 times the length of the shorter. Find the length of each piece.

Solution:

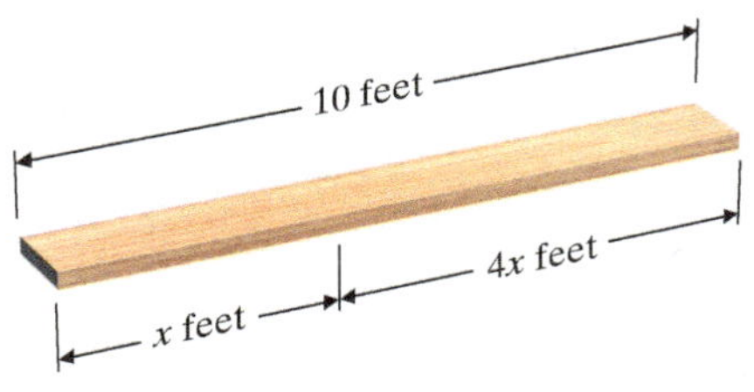

1. UNDERSTAND the problem. To do so, read and reread the problem. You may also want to propose a solution. For example, if 3 feet represents the length of the shorter piece, then $4(3) = 12$ feet is the length of the longer piece, since it is 4 times the length of the shorter piece. This guess gives a total board length of 3 feet + 12 feet = 15 feet, which is too long. However, the purpose of proposing a solution is not to guess correctly, but to help better understand the problem and how to model it.

   In general, if we let

   x = length of shorter piece, then

   $4x$ = length of longer piece

2. TRANSLATE the problem. First, we write the equation in words.

length of shorter piece	added to	length of longer piece	equals	total length of board
↓	↓	↓	↓	↓
x	$+$	$4x$	$=$	10

3. SOLVE.

$$x + 4x = 10$$
$$5x = 10 \quad \text{Combine like terms.}$$
$$\frac{5x}{5} = \frac{10}{5} \quad \text{Divide both sides by 5.}$$
$$x = 2$$

4. INTERPRET.

   Check: Check the solution in the stated problem. If the length of the shorter piece of board is 2 feet, the length of the longer piece is $4 \cdot (2 \text{ feet}) = 8$ feet and the sum of the lengths of the two pieces is 2 feet + 8 feet = 10 feet.

   State: The shorter piece of board is 2 feet and the longer piece of board is 8 feet.

Work Practice 3

Helpful Hint

Make sure that units are included in your answer, if appropriate.

Practice 4

Through the year 2020, the state of California will have 17 more electoral votes for president than the state of Texas. If the total electoral votes for these two states is 93, find the number of electoral votes for each state.

Example 4 Finding the Number of Republican and Democratic Representatives

As of January 2018, the total number of Democrats and Republicans in the U.S. House of Representatives was 435. There were 47 more Republican representatives than Democratic. Find the number of representatives from each party. (*Source:* Congressional Research Service)

Answers

3. shorter piece: 3 feet; longer piece: 15 feet 4. Texas: 38 electoral votes; California: 55 electoral votes

Solution:

1. UNDERSTAND the problem. Read and re-read the problem. Let's suppose that there are 200 Democratic representatives. Since there are 47 more Republicans than Democrats, there must be $200 + 47 = 247$ Republicans. The total number of Democrats and Republicans is then $200 + 247 = 447$. This is incorrect since the total should be 435, but we now have a better understanding of the problem.

   In general, if we let

   $x =$ number of Democrats, then

   $x + 47 =$ number of Republicans

2. TRANSLATE the problem. First, we write the equation in words.

number of Democrats	added to	number of Republicans	equals	435
↓	↓	↓	↓	↓
x	$+$	$(x + 47)$	$=$	435

3. SOLVE.

$$x + (x + 47) = 435$$
$$2x + 47 = 435 \quad \text{Combine like terms.}$$
$$2x + 47 - 47 = 435 - 47 \quad \text{Subtract 47 from both sides.}$$
$$2x = 388$$
$$\frac{2x}{2} = \frac{388}{2} \quad \text{Divide both sides by 2.}$$
$$x = 194$$

4. INTERPRET.

   Check: If there were 194 Democratic representatives, then there were $194 + 47 = 241$ Republican representatives. The total number of representatives is then $194 + 241 = 435$. The results check.

   State: There were 194 Democratic and 241 Republican representatives in Congress in January 2018.

Work Practice 4

Example 5 Calculating Hours on the Job

A computer science major at a local university has a part-time job working on computers for his clients. He charges \$20 to go to your home or office and then \$25 per hour. During one month he visited 10 homes or offices and his total income was \$575. How many hours did he spend working on computers?

Solution:

1. UNDERSTAND. Read and reread the problem. Let's propose that the student spent 20 hours working on computers. Pay careful attention as to how his income is calculated. For 20 hours and 10 visits, his income is $20(\$25) + 10(\$20) = \$700$, which is more than \$575. We now have a better understanding of the problem and know that the time spent working on computers was less than 20 hours.

   Let's let

   $x =$ hours working on computers. Then

   $25x =$ amount of money made while working on computers

(*Continued on next page*)

Practice 5

A car rental agency charges \$28 a day and \$0.15 a mile. If you rent a car for a day and your bill (before taxes) is \$52, how many miles did you drive?

Answer

5. 160 miles

2. TRANSLATE.

money made while working on computers	plus	money made for visits	is equal to	575
↓	↓	↓	↓	↓
$25x$	$+$	$10(20)$	$=$	575

3. SOLVE.

$$25x + 200 = 575$$

$$25x + 200 - 200 = 575 - 200 \quad \text{Subtract 200 from both sides.}$$

$$25x = 375 \quad \text{Simplify.}$$

$$\frac{25x}{25} = \frac{375}{25} \quad \text{Divide both sides by 25.}$$

$$x = 15 \quad \text{Simplify.}$$

4. INTERPRET.

Check: If the student works 15 hours and makes 10 visits, his income is $15(\$25) + 10(\$20) = \$575$.

State: The student spent 15 hours working on computers.

Work Practice 5

Practice 6

The measure of the second angle of a triangle is twice the measure of the smallest angle. The measure of the third angle of the triangle is three times the measure of the smallest angle. Find the measures of the angles.

Example 6 Finding Angle Measures

If the two walls of the Vietnam Veterans Memorial in Washington, D.C., were connected, an isosceles triangle would be formed. The measure of the third angle is 97.5° more than the measure of either of the two equal angles. Find the measure of the third angle. (*Source:* National Park Service)

Solution:

1. UNDERSTAND. Read and reread the problem. We then draw a diagram (recall that an isosceles triangle has two angles with the same measure) and let

x = degree measure of one angle

x = degree measure of the second, equal angle

$x + 97.5$ = degree measure of the third angle

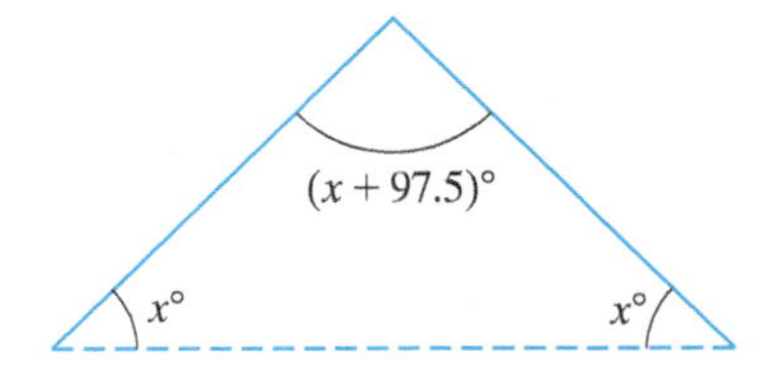

2. TRANSLATE. Recall that the sum of the measures of the angles of a triangle equals 180.

measure of first angle	+	measure of second angle	+	measure of third angle	equals	180
↓		↓		↓	↓	↓
x	$+$	x	$+$	$(x + 97.5)$	$=$	180

Answer

6. smallest: 30°; second: 60°; third: 90°

3. SOLVE.

$$x + x + (x + 97.5) = 180$$
$$3x + 97.5 = 180 \quad \text{Combine like terms.}$$
$$3x + 97.5 - 97.5 = 180 - 97.5 \quad \text{Subtract 97.5 from both sides.}$$
$$3x = 82.5$$
$$\frac{3x}{3} = \frac{82.5}{3} \quad \text{Divide both sides by 3.}$$
$$x = 27.5$$

4. INTERPRET.

Check: If $x = 27.5$, then the measure of the third angle is $x + 97.5 = 125$. The sum of the angles is then $27.5 + 27.5 + 125 = 180$, the correct sum.

State: The third angle measures 125°.*

Work Practice 6

Objective C Solving Consecutive Integer Problems 

The next example has to do with consecutive integers. (See Section 2.2 Objective C for further review of consecutive integers.)

	Example	*General Representation*	
Consecutive Integers	11, 12, 13 (+1, +1)	Let x be an integer.	x, $x+1$, $x+2$ (+1, +1)
Consecutive Even Integers	38, 40, 42 (+2, +2)	Let x be an even integer.	x, $x+2$, $x+4$ (+2, +2)
Consecutive Odd Integers	57, 59, 61 (+2, +2)	Let x be an odd integer.	x, $x+2$ $x+4$ (+2, +2)

Example 7 Finding Area Codes

Some states have a single area code for the entire state. Two such states have area codes that are consecutive odd integers. If the sum of these integers is 1208, find the two area codes. (*Source: World Almanac*)

Solution:

1. UNDERSTAND. Read and reread the problem. If we let

x = the first odd integer, then

$x + 2$ = the next odd integer

Helpful Hint Remember, the 2 here means that odd integers are 2 units apart—for example, the odd integers 13 and $13 + 2 = 15$.

2. TRANSLATE.

first odd integer	the sum of	next odd integer	is	1208
↓	↓	↓	↓	↓
x	$+$	$(x + 2)$	$=$	1208

(*Continued on next page*)

Practice 7

The sum of three consecutive even integers is 144. Find the integers.

*The two walls actually meet at an angle of 125 degrees 12 minutes. The measurement of 97.5° given in the problem is an approximation.

Answer

7. 46, 48, 50

3. SOLVE.

$$\begin{aligned} x + x + 2 &= 1208 \\ 2x + 2 &= 1208 \\ 2x + 2 - 2 &= 1208 - 2 \\ 2x &= 1206 \\ \frac{2x}{2} &= \frac{1206}{2} \\ x &= 603 \end{aligned}$$

4. INTERPRET.

Check: If $x = 603$, then the next odd integer $x + 2 = 603 + 2 = 605$. Notice their sum, $603 + 605 = 1208$, as needed.

State: The area codes are 603 and 605.

Note: New Hampshire's area code is 603 and South Dakota's area code is 605.

Work Practice 7

Vocabulary, Readiness & Video Check

Fill in the table.

1.	A number: x	→ Double the number:	→ Double the number, decreased by 31:
2.	A number: x	→ Three times the number:	→ Three times the number, increased by 17:
3.	A number: x	→ The sum of the number and 5:	→ Twice the sum of the number and 5:
4.	A number: x	→ The difference of the number and 11:	→ Seven times the difference of the number and 11:
5.	A number: y	→ The difference of 20 and the number:	→ The difference of 20 and the number, divided by 3:
6.	A number: y	→ The sum of −10 and the number:	→ The sum of −10 and the number, divided by 9:

Martin-Gay Interactive Videos

See Video 2.4

Watch the section lecture video and answer the following questions.

Objective A **7.** At the end of Example 1, where are we told is the best place to check the solution of an application problem?

Objective B **8.** The solution of the equation for Example 3 is $x = 43$. Why is this not the solution to the application?

Objective C **9.** What are two things that should be checked to make sure the solution of Example 4 is correct?

2.4 Exercise Set MyLab Math 

Objective A *Solve. See Examples 1 and 2. For Exercises 1 through 4, write each of the following as equations. Then solve.*

1. The sum of twice a number and 7 is equal to the sum of the number and 6. Find the number.

2. The difference of three times a number and 1 is the same as twice the number. Find the number.

3. Three times a number, minus 6, is equal to two times the number, plus 8. Find the number.

4. The sum of 4 times a number and -2 is equal to the sum of 5 times the number and -2. Find the number.

5. Twice the difference of a number and 8 is equal to three times the sum of the number and 3. Find the number.

6. Five times the sum of a number and -1 is the same as 6 times the number. Find the number.

7. The product of twice a number and 3 is the same as the difference of five times the number and $\frac{3}{4}$. Find the number.

8. If the difference of a number and 4 is doubled, the result is $\frac{1}{4}$ less than the number. Find the number.

Objective B *Solve. For Exercises 9 and 10, the solutions have been started for you. See Examples 3 and 4.*

9. A 25-inch piece of steel is cut into three pieces so that the second piece is twice as long as the first piece, and the third piece is one inch more than five times the length of the first piece. Find the lengths of the pieces.

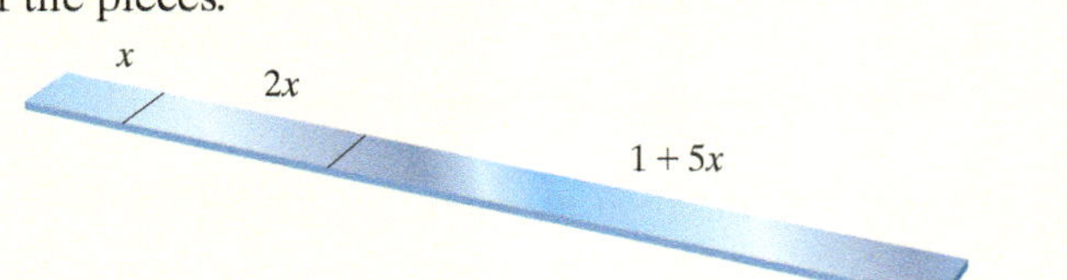

Start the solution:

1. UNDERSTAND the problem. Reread it as many times as needed.

2. TRANSLATE into an equation. (Fill in the blanks below.)

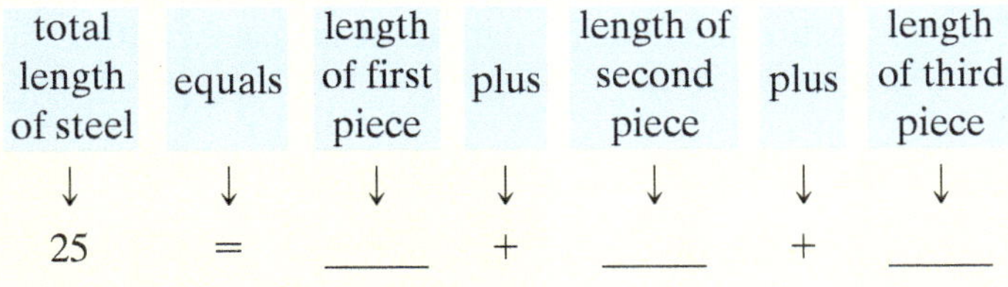

total length of steel	equals	length of first piece	plus	length of second piece	plus	length of third piece
↓	↓	↓	↓	↓	↓	↓
25	=	____	+	____	+	____

Finish with:

3. SOLVE and **4.** INTERPRET

10. A 46-foot piece of rope is cut into three pieces so that the second piece is three times as long as the first piece, and the third piece is two feet more than seven times the length of the first piece. Find the lengths of the pieces.

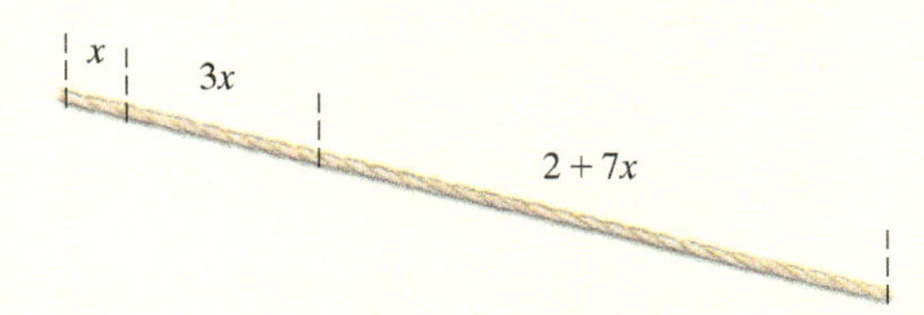

Start the solution:

1. UNDERSTAND the problem. Reread it as many times as needed.

2. TRANSLATE into an equation. (Fill in the blanks below.)

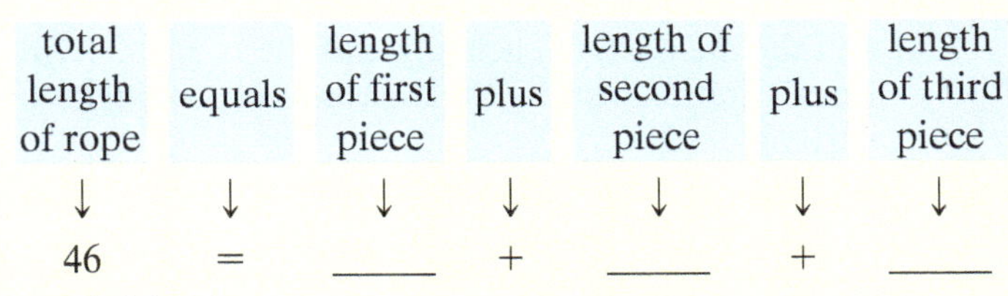

total length of rope	equals	length of first piece	plus	length of second piece	plus	length of third piece
↓	↓	↓	↓	↓	↓	↓
46	=	____	+	____	+	____

Finish with:

3. SOLVE and **4.** INTERPRET

11. A 40-inch board is to be cut into three pieces so that the second piece is twice as long as the first piece and the third piece is 5 times as long as the first piece. If x represents the length of the first piece, find the lengths of all three pieces.

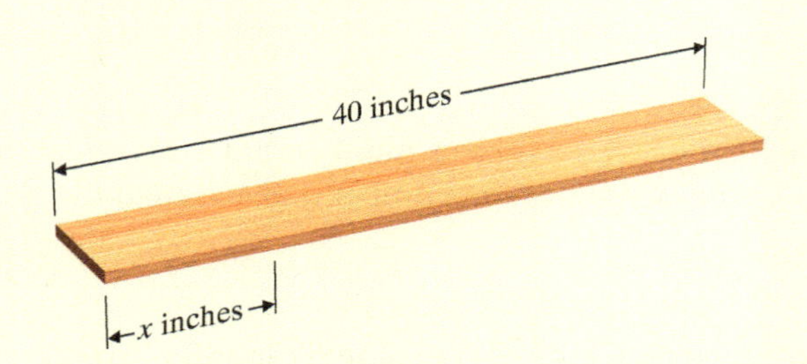

12. A 21-foot beam is to be divided so that the longer piece is 1 foot more than 3 times the length of the shorter piece. If x represents the length of the shorter piece, find the lengths of both pieces.

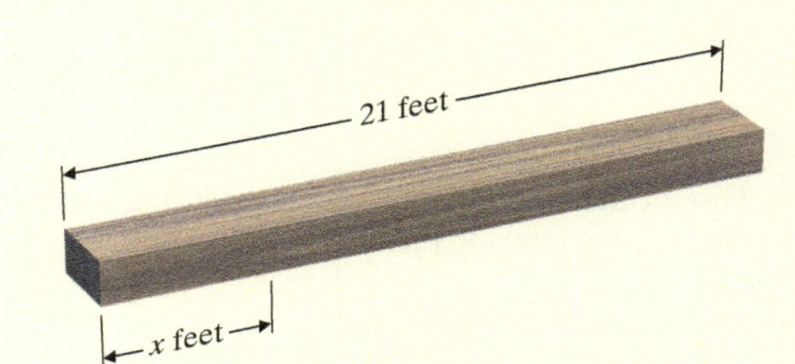

13. In 2017, New York produced 772 million pounds more apples than Pennsylvania. Together, the two states produced 1828 million pounds of apples. Find the amount of apples grown in New York and Pennsylvania in 2017. (*Source:* National Agriculture Statistics Service)

14. In the 2018 Winter Olympics, the team from Norway won 25 more medals than the team from Sweden. If the total number of medals won by both teams was 53, find the number of medals won by each team. (*Source:* NBC Sports)

Solve. See Example 5.

15. A car rental agency advertised renting a Buick Century for \$24.95 per day and \$0.29 per mile. If you rent this car for 2 days, how many whole miles can you drive on a \$100 budget?

16. A plumber gave an estimate for the renovation of a kitchen. Her hourly pay is \$27 per hour and the plumbing parts will cost \$80. If her total estimate is \$404, how many hours does she expect this job to take?

17. In one U.S. city, the taxi cost is \$3 plus \$0.80 per mile. If you are traveling from the airport, there is an additional charge of \$4.50 for tolls. How far can you travel from the airport by taxi for \$27.50?

18. A professional carpet cleaning service charges \$30 plus \$25.50 per hour to come to your home. If your total bill from this company is \$119.25 before taxes, for how many hours were you charged?

Solve. See Example 6.

△ **19.** The flag of Equatorial Guinea contains an isosceles triangle. (Recall that an isosceles triangle contains two angles with the same measure.) If the measure of the third angle of the triangle is 30° more than twice the measure of either of the other two angles, find the measure of each angle of the triangle. (*Hint:* Recall that the sum of the measures of the angles of a triangle is 180°.)

△ **20.** The flag of Brazil contains a parallelogram. One angle of the parallelogram is 15° less than twice the measure of the angle next to it. Find the measure of each angle of the parallelogram. (*Hint:* Recall that opposite angles of a parallelogram have the same measure and that the sum of the measures of the angles is 360°.)

21. The sum of the measures of the angles of a parallelogram is 360°. In the parallelogram below, angles A and D have the same measure and angles C and B have the same measure. If the measure of angle C is twice the measure of angle A, find the measure of each angle.

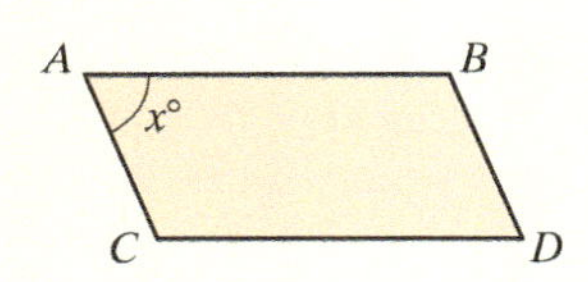

22. Recall that the sum of the measures of the angles of a triangle is 180°. In the triangle below, angle C has the same measure as angle B, and angle A measures 42° less than angle B. Find the measure of each angle.

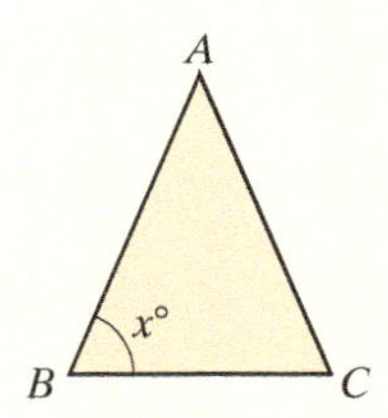

Objective C *Solve. See Example 7. Fill in the table. Most of the first row has been completed for you.*

	First Integer →	Next Integers		→	Indicated Sum
23. Three consecutive integers:	Integer: x	$x + 1$	$x + 2$		Sum of the three consecutive integers, simplified:
24. Three consecutive integers:	Integer: x				Sum of the second and third consecutive integers, simplified:
25. Three consecutive even integers:	Even integer: x				Sum of the first and third even consecutive integers, simplified:
26. Three consecutive odd integers:	Odd integer: x				Sum of the three consecutive odd integers, simplified:
27. Four consecutive integers:	Integer: x				Sum of the four consecutive integers, simplified:
28. Four consecutive integers:	Integer: x				Sum of the first and fourth consecutive integers, simplified:
29. Three consecutive odd integers:	Odd integer: x				Sum of the second and third consecutive odd integers, simplified:
30. Three consecutive even integers:	Even integer: x				Sum of the three consecutive even integers, simplified:

Solve. See Example 7.

31. The left and right page numbers of an open book are two consecutive integers whose sum is 469. Find these page numbers.

32. The room numbers of two adjacent classrooms are two consecutive even numbers. If their sum is 654, find the classroom numbers.

33. To make an international telephone call, you need the code for the country you are calling. The codes for Belgium, France, and Spain are three consecutive integers whose sum is 99. Find the code for each country. (*Source:* The World Almanac and Book of Facts)

34. The code to unlock a student's combination lock happens to be three consecutive odd integers whose sum is 51. Find the integers.

Objectives A B C Mixed Practice *Solve. See Examples 1 through 7.*

35. A 17-foot piece of string is cut into two pieces so that the longer piece is 2 feet longer than twice the length of the shorter piece. Find the lengths of both pieces.

36. A 25-foot wire is to be cut so that the longer piece is one foot longer than 5 times the length of the shorter piece. Find the length of each piece.

37. Currently, the two fastest trains in the world are China's CRH380A and Germany's Transrapid TR-09. The sum of their fastest speeds is 581 miles per hour. If the maximum speed of the CRH380A is 23 miles per hour faster than the maximum speed of the Transrapid TR-09, find the speeds of each. (*Note:* The Transrapid TR-09 is technically a monorail. *Source:* tiptoptens.com and telegraph.co.uk)

38. The Pentagon is the world's largest office building in terms of floor space. It has three times the amount of floor space as the Empire State Building. If the total floor space for these two buildings is approximately 8700 thousand square feet, find the floor space of each building.

39. Two angles are supplementary if their sum is 180°. The larger angle below measures eight degrees more than three times the measure of the smaller angle. If x represents the measure of the smaller angle and these two angles are supplementary, find the measure of each angle.

$x°$

40. Two angles are complementary if their sum is 90°. Given the measures of the complementary angles shown, find the measure of each angle.

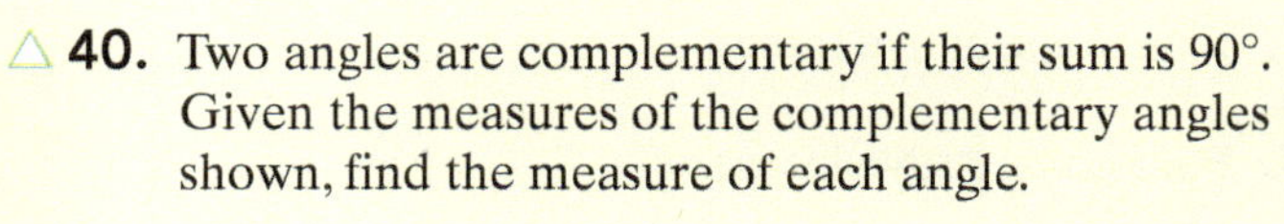

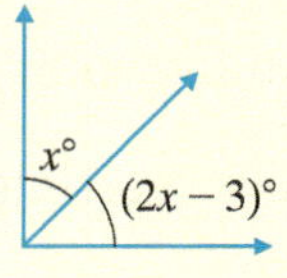

41. The measures of the angles of a triangle are 3 consecutive even integers. Find the measure of each angle.

42. A quadrilateral is a polygon with 4 sides. The sum of the measures of the 4 angles in a quadrilateral is 360°. If the measures of the angles of a quadrilateral are consecutive odd integers, find the measures.

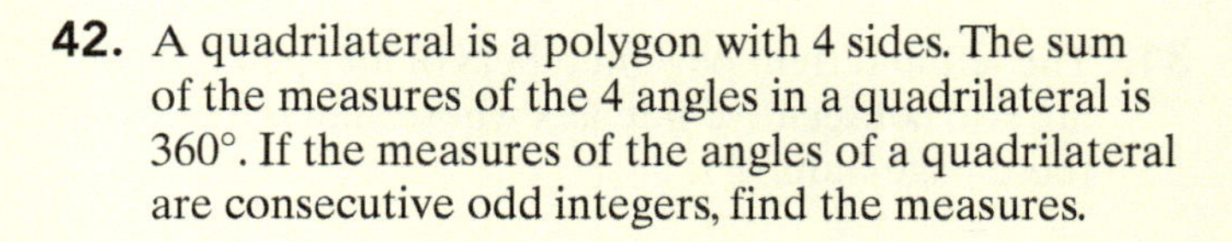

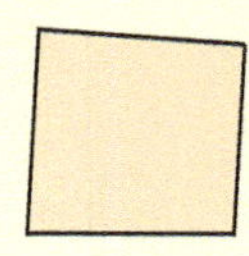

43. The sum of $\frac{1}{5}$ and twice a number is equal to $\frac{4}{5}$ subtracted from three times the number. Find the number.

44. The sum of $\frac{2}{3}$ and four times a number is equal to $\frac{5}{6}$ subtracted from five times the number. Find the number.

45. Hertz Car Rental charges a daily rate of \$39 plus \$0.20 per mile for a certain car. Suppose that you rent that car for a day and your bill (before taxes) is \$95. How many miles did you drive?

46. A woman's \$16,500 estate is to be divided so that her husband receives twice as much as her son. Find the amount of money that her husband receives and the amount of money that her son receives.

47. One of the biggest rivalries in college football is the University of Michigan Wolverines and the Ohio State University Buckeyes. During their match-up in 2018, Ohio beat Michigan by 23 points. If their combined scores totaled 101, find the individual team scores.

48. In January 2018 there was one independent governor in the United States, and there were 19 more Republican governors than Democratic governors. How many Democrats and how many Republicans held governor's offices at that time? (*Source:* Multistate.com)

49. The number of counties in California and the number of counties in Montana are consecutive even integers whose sum is 114. If California has more counties than Montana, how many counties does each state have? (*Source: The World Almanac and Book of Facts*)

50. A student is building a bookcase with stepped shelves for her dorm room. She buys a 48-inch board and wants to cut the board into three pieces with lengths equal to three consecutive even integers. Find the three board lengths.

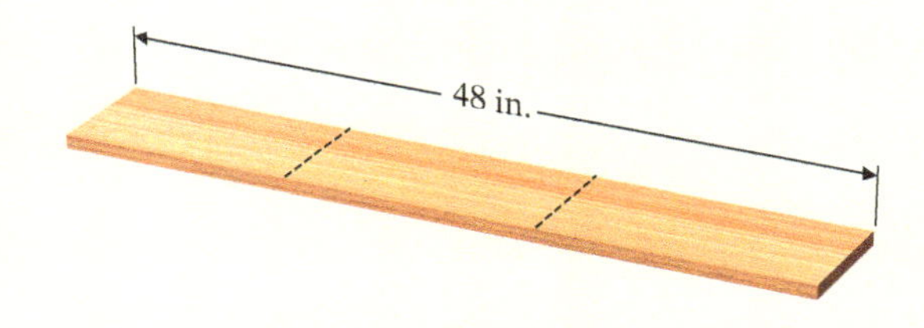

51. Scientists are continually updating information about the planets in our solar system, including the number of satellites orbiting each. Uranus is now believed to have 13 more satellites than Neptune. Also, Saturn is now believed to have 8 more than twice the number of satellites of Uranus. If the total number of satellites for these planets is 103, find the number of satellites for each planet. (*Source:* National Space Science Data Center)

52. The height of the Apple iPhone X is 0.07 inch more than twice its width. If the sum of the height and the width is 8.44 inches, find each dimension. (*Source:* Apple, Inc.)

53. If the sum of a number and five is tripled, the result is one less than twice the number. Find the number.

54. Twice the sum of a number and six equals three times the sum of the number and four. Find the number.

55. The area of the Sahara Desert is 7 times the area of the Gobi Desert. If the sum of their areas is 4,000,000 square miles, find the area of each desert.

56. The largest meteorite in the world is the Hoba West, located in Namibia. Its weight is 3 times the weight of the Armanty meteorite, located in Outer Mongolia. If the sum of their weights is 88 tons, find the weight of each.

57. In the 2018 Winter Olympic Games, France won more medals than Austria, which won more medals than Japan. If the numbers of medals won by these three countries are three consecutive integers whose sum is 42, find the number of medals won by each. (*Source:* NBS Sports)

58. To make an international telephone call, you need the code for the country you are calling. The codes for Mali Republic, Côte d'Ivoire, and Niger are three consecutive odd integers whose sum is 675. Find the code for each country.

59. In the fall of 2017, there were 48 thousand more female undergraduate students enrolled at CSU than male undergraduate students. If the total undergraduate enrollment was 430 thousand that fall, find the numbers of female undergraduate students and male undergraduate students who were enrolled. (*Source:* MIT)

60. In 2017, approximately 5 million fewer cars were sold in the United States than light trucks. If the total number of trucks and cars sold was 17.2 million, find the number of vehicles sold in each category. (*Source:* Alliance of Automobile Manufacturers)

61. A geodesic dome, based on the design by Buckminster Fuller, is composed of two different types of triangular panels. One of these is an isosceles triangle. In one geodesic dome, the measure of the third angle is 76.5° more than the measure of either of the two equal angles. Find the measure of the three angles. (*Source:* Buckminster Fuller Institute)

62. The measures of the angles of a particular triangle are such that the second and third angles are each four times the measure of the smallest angle. Find the measures of the angles of this triangle.

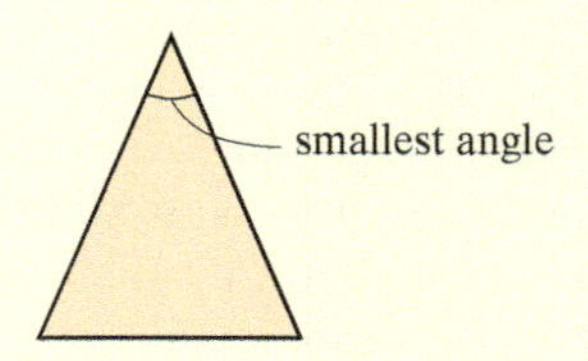

The graph below shows the states with the highest provisional tourism budgets in a recent year. Use the graph for Exercises 63 through 68.

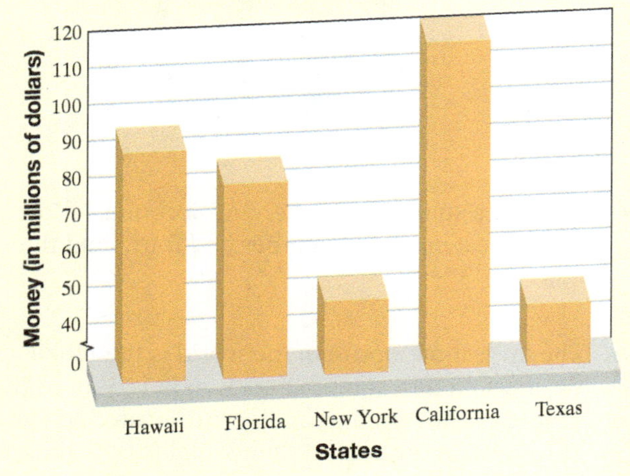

63. Which state spent the most money on tourism?

64. Which state shown spent less than $50 million on tourism?

65. The states of Florida and Hawaii spent a total of $175.9 million on tourism. The state of Florida spent $10.5 million less than the state of Hawaii. Find the amount that each state spent on tourism.

66. The states of California and Texas spent a total of $166.3 million on tourism. The state of Texas spent $73.3 million less than the state of California. Find the amount that each state spent on tourism.

Compare the heights of the bars in the graph with your results of the exercises below. Are your answers reasonable?

67. Exercise **65**

68. Exercise **66**

Review

Evaluate each expression for the given values. See Section 1.3.

69. $2W + 2L$; $W = 7$ and $L = 10$

70. $\frac{1}{2}Bh$; $B = 14$ and $h = 22$

71. πr^2; $r = 15$

72. $r \cdot t$; $r = 15$ and $t = 2$

Concept Extensions

73. A golden rectangle is a rectangle whose length is approximately 1.6 times its width. The early Greeks thought that a rectangle with these dimensions was the most pleasing to the eye, and examples of the golden rectangle are found in many early works of art. For example, the Parthenon in Athens contains many examples of golden rectangles.

Mike Hallahan would like to plant a rectangular garden in the shape of a golden rectangle. If he has 78 feet of fencing available, find the dimensions of the garden.

74. Dr. Dorothy Smith gave the students in her geometry class at the University of New Orleans the following question: Is it possible to construct a triangle such that the second angle of the triangle has a measure that is twice the measure of the first angle and the measure of the third angle is 5 times the measure of the first? If so, find the measure of each angle. (*Hint:* Recall that the sum of the measures of the angles of a triangle is 180°.)

75. Only male crickets chirp. They chirp at different rates depending on their species and the temperature of their environment. Suppose a certain species is currently chirping at a rate of 90 chirps per minute. At this rate, how many chirps occur in one hour? In one 24-hour day? In one year?

76. The human eye blinks once every 5 seconds on average. How many times does the average eye blink in one hour? In one 16-hour day while awake? In one year while awake?

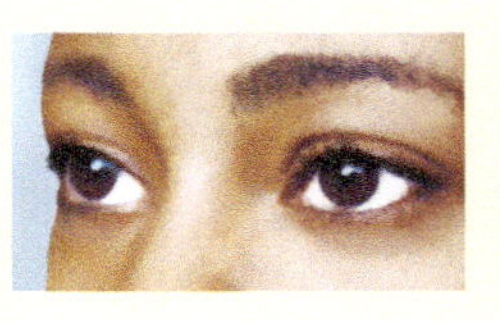

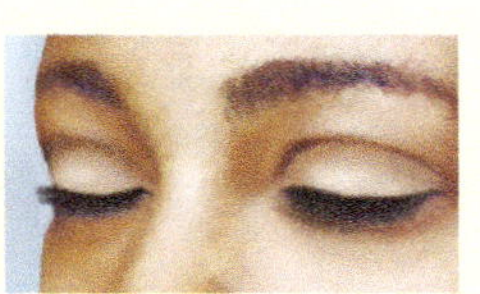

77. In your own words, explain why a solution of a word problem should be checked using the original wording of the problem and not the equation written from the wording.

78. Give an example of how you recently solved a problem using mathematics.

Recall from Exercise 73 that a golden rectangle is a rectangle whose length is approximately 1.6 times its width.

△ **79.** It is thought that for about 75% of adults, a rectangle in the shape of the golden rectangle is the most pleasing to the eye. Draw three rectangles, one in the shape of the golden rectangle, and poll your class. Do the results agree with the percentage given above?

△ **80.** Examples of golden rectangles can be found today in architecture and manufacturing packaging. Find an example of a golden rectangle in your home. A few suggestions: the front face of a book, the floor of a room, the front of a box of food.

For Exercises 81 and 82, measure the dimensions of each rectangle and decide which one best approximates the shape of a golden rectangle.

△ **81.**

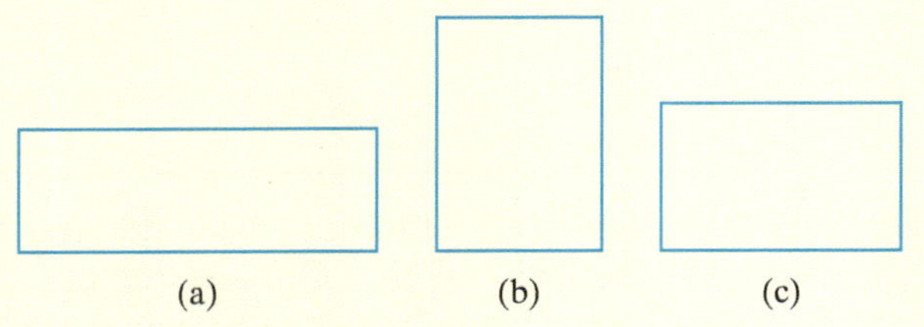

(a) (b) (c)

82.

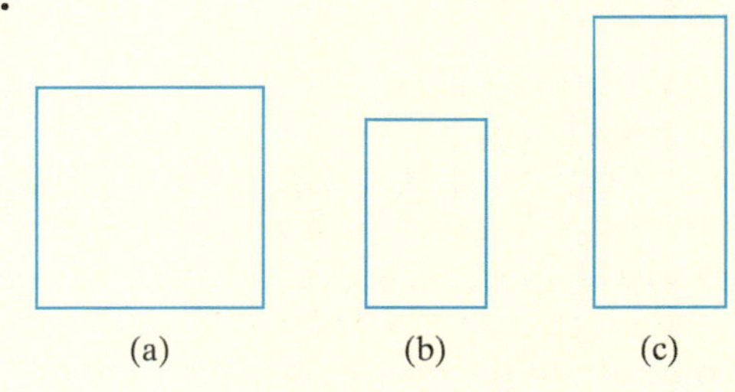

(a) (b) (c)

2.5 Formulas and Problem Solving 

Objectives

A Use Formulas to Solve Problems.

B Solve a Formula or Equation for One of Its Variables.

Objective A Using Formulas to Solve Problems 

A **formula** describes a known relationship among quantities. Many formulas are given as equations. For example, the formula

$$d = r \cdot t$$

stands for the relationship

$$\text{distance} = \text{rate} \cdot \text{time}$$

Let's look at one way that we can use this formula.

If we know we traveled a distance of 100 miles at a rate of 40 miles per hour, we can replace the variables d and r in the formula $d = rt$ and find our travel time, t.

$d = rt$ Formula

$100 = 40t$ Replace d with 100 and r with 40.

To solve for t, we divide both sides of the equation by 40.

$\frac{100}{40} = \frac{40t}{40}$ Divide both sides by 40.

$\frac{5}{2} = t$ Simplify.

The travel time was $\frac{5}{2}$ hours, or $2\frac{1}{2}$ hours, or 2.5 hours.

In this section, we solve problems that can be modeled by known formulas. We use the same problem-solving strategy that was used in the previous section.

Example 1 Finding Time Given Rate and Distance

A glacier is a giant mass of rocks and ice that flows downhill like a river. Portage Glacier in Alaska is about 6 miles, or 31,680 *feet*, long and moves 400 *feet* per year. Icebergs are created when the front end of the glacier flows into Portage Lake. How long does it take for ice at the head (beginning) of the glacier to reach the lake?

Solution:

1. UNDERSTAND. Read and reread the problem. The appropriate formula needed to solve this problem is the distance formula, $d = rt$. To become familiar with this formula, let's find the distance that ice traveling at a rate of 400 feet per year travels in 100 years. To do so, we let time t be 100 years and rate r be the given 400 feet per year, and substitute these values into the formula $d = rt$. We then have that distance $d = 400(100) = 40{,}000$ feet. Since we are interested in finding how long it takes ice to travel 31,680 feet, we now know that it is less than 100 years.

   Since we are using the formula $d = rt$, we let

   $t =$ the time in years for ice to reach the lake

   $r =$ rate or speed of ice

   $d =$ distance from beginning of glacier to lake

2. TRANSLATE. To translate to an equation, we use the formula $d = rt$ and let distance $d = 31{,}680$ feet and rate $r = 400$ feet per year.

$$d = r \cdot t$$

$$31{,}680 = 400 \cdot t \quad \text{Let } d = 31{,}680 \text{ and } r = 400.$$

3. SOLVE. Solve the equation for t. To solve for t, we divide both sides by 400.

$$\frac{31{,}680}{400} = \frac{400 \cdot t}{400} \quad \text{Divide both sides by 400.}$$

$$79.2 = t \quad \text{Simplify.}$$

4. INTERPRET.

   Check: To check, substitute 79.2 for t and 400 for r in the distance formula and check to see that the distance is 31,680 feet.

   State: It takes 79.2 years for the ice at the head of Portage Glacier to reach the lake.

Work Practice 1

Practice 1

A family is planning their vacation to visit relatives. They will drive from Cincinnati, Ohio, to Rapid City, South Dakota, a distance of 1180 miles. They plan to average a rate of 50 miles per hour. How much time will they spend driving?

Helpful Hint

Don't forget to include units, if appropriate.

Answer

1. 23.6 hours

Practice 2

A wood deck is being built behind a house. The width of the deck must be 18 feet because of the shape of the house. If there is 450 square feet of decking material, find the length of the deck.

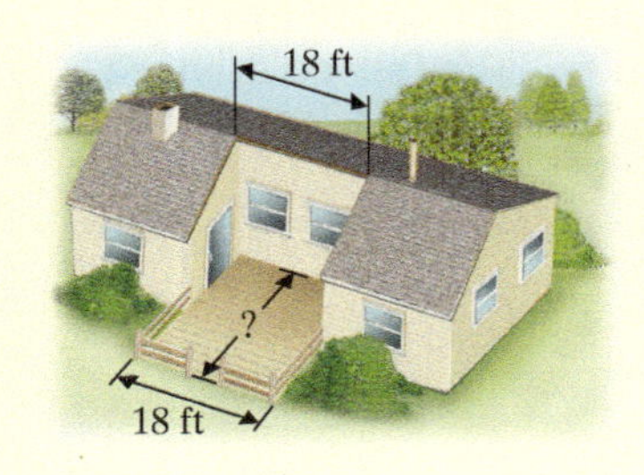

Example 2 Calculating the Length of a Garden

Charles Pecot can afford enough fencing to enclose a rectangular garden with a perimeter of 140 feet. If the width of his garden is to be 30 feet, find the length.

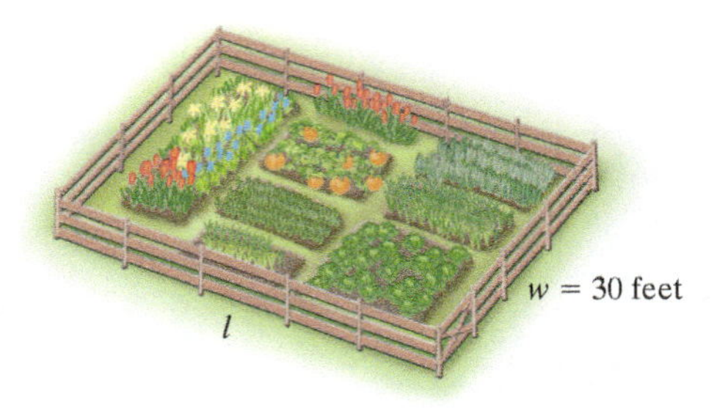

Solution:

1. UNDERSTAND. Read and reread the problem. The formula needed to solve this problem is the formula for the perimeter of a rectangle, $P = 2l + 2w$. Before continuing, let's become familar with this formula.

   l = the length of the rectangular garden
   w = the width of the rectangular garden
   P = perimeter of the garden

2. TRANSLATE. To translate to an equation, we use the formula $P = 2l + 2w$ and let perimeter $P = 140$ feet and width $w = 30$ feet.

   $$P = 2l + 2w \quad \text{Let } P = 140 \text{ and } w = 30.$$
   $$140 = 2l + 2(30)$$

3. SOLVE.

   $$140 = 2l + 2(30)$$
   $$140 = 2l + 60 \quad \text{Multiply 2(30).}$$
   $$140 - 60 = 2l + 60 - 60 \quad \text{Subtract 60 from both sides.}$$
   $$80 = 2l \quad \text{Combine like terms.}$$
   $$40 = l \quad \text{Divide both sides by 2.}$$

4. INTERPRET.

   Check: Substitute 40 for l and 30 for w in the perimeter formula and check to see that the perimeter is 140 feet.

   State: The length of the rectangular garden is 40 feet.

Work Practice 2

Practice 3

Convert the temperature 41°F to degrees Celsius.

Example 3 Finding an Equivalent Temperature

The average maximum temperature for January in Algiers, Algeria, is 59° Fahrenheit. Find the equivalent temperature in degrees Celsius.

Solution:

1. UNDERSTAND. Read and reread the problem. A formula that can be used to solve this problem is the formula for converting degrees Celsius to degrees Fahrenheit, $F = \frac{9}{5}C + 32$. Before continuing, become familiar with this formula. Using this formula, we let

   C = temperature in degrees Celsius, and
   F = temperature in degrees Fahrenheit.

Answers
2. 25 feet **3.** 5°C

2. TRANSLATE. To translate to an equation, we use the formula $F = \frac{9}{5}C + 32$ and let degrees Fahrenheit $F = 59$.

Formula: $F = \frac{9}{5}C + 32$

Substitute: $59 = \frac{9}{5}C + 32$ Let $F = 59$.

3. SOLVE.

$$59 = \frac{9}{5}C + 32$$

$$59 - 32 = \frac{9}{5}C + 32 - 32 \quad \text{Subtract 32 from both sides.}$$

$$27 = \frac{9}{5}C \quad \text{Combine like terms.}$$

$$\frac{5}{9} \cdot 27 = \frac{5}{9} \cdot \frac{9}{5}C \quad \text{Multiply both sides by } \frac{5}{9}.$$

$$15 = C \quad \text{Simplify.}$$

4. INTERPRET.

Check: To check, replace C with 15 and F with 59 in the formula and see that a true statement results.

State: Thus, 59° Fahrenheit is equivalent to 15° Celsius.

Work Practice 3

In the next example, we use the formula for perimeter of a rectangle as in Example 2. In Example 2, we knew the width of the rectangle. In this example, both the length and width are unknown.

Example 4 Finding Road Sign Dimensions

The length of a rectangular road sign is 2 feet less than three times its width. Find the dimensions if the perimeter is 28 feet.

Practice 4

The length of a rectangle is 1 meter more than 4 times its width. Find the dimensions if the perimeter is 52 meters.

Solution:

1. UNDERSTAND. Read and reread the problem. Recall that the formula for the perimeter of a rectangle is $P = 2l + 2w$. Draw a rectangle and guess the solution. If the width of the rectangular sign is 5 feet, its length is 2 feet less than 3 times the width, or 3(5 feet) − 2 feet = 13 feet. The perimeter P of the rectangle is then 2(13 feet) + 2(5 feet) = 36 feet, too large. We now know that the width is less than 5 feet.

Proposed rectangle:

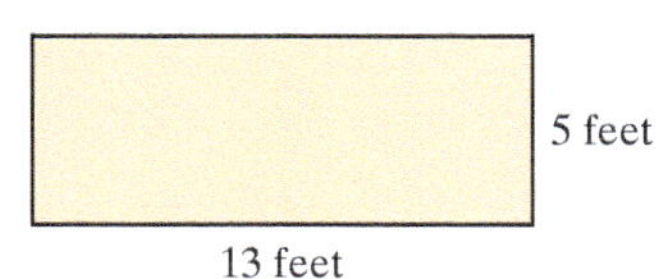

Answer

4. length: 21 m; width: 5 m

(*Continued on next page*)

Let

w = the width of the rectangular sign; then

$3w - 2$ = the length of the sign.

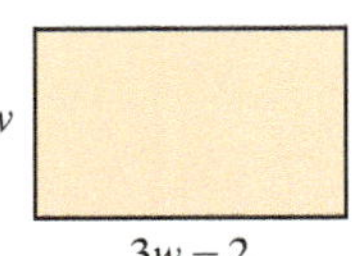

Draw a rectangle and label it with the assigned variables.

2. TRANSLATE.

Formula: $P = 2l + 2w$

Substitute: $28 = 2(3w - 2) + 2w$

3. SOLVE.

$$28 = 2(3w - 2) + 2w$$

$$28 = 6w - 4 + 2w \quad \text{Apply the distributive property.}$$

$$28 = 8w - 4$$

$$28 + 4 = 8w - 4 + 4 \quad \text{Add 4 to both sides.}$$

$$32 = 8w$$

$$\frac{32}{8} = \frac{8w}{8} \quad \text{Divide both sides by 8.}$$

$$4 = w$$

4. INTERPRET.

Check: If the width of the sign is 4 feet, the length of the sign is $3(4 \text{ feet}) - 2 \text{ feet} = 10 \text{ feet}$. This gives the rectangular sign a perimeter of $P = 2(4 \text{ feet}) + 2(10 \text{ feet}) = 28 \text{ feet}$, the correct perimeter.

State: The width of the sign is 4 feet and the length of the sign is 10 feet.

Work Practice 4

Objective B Solving a Formula for a Variable

We say that the formula

$$d = rt$$

is solved for d because d is alone on one side of the equation and the other side contains no d's. Suppose that we have a large number of problems to solve where we are given distance d and rate r and asked to find time t. In this case, it may be easier to first solve the formula $d = rt$ for t. To solve for t, we divide both sides of the equation by r.

$$d = rt$$

$$\frac{d}{r} = \frac{rt}{r} \quad \text{Divide both sides by } r.$$

$$\frac{d}{r} = t \quad \text{Simplify.}$$

To solve a formula or an equation for a specified variable, we use the same steps as for solving a linear equation except that we treat the specified variable as the only variable in the equation. These steps are listed next.

Solving Equations for a Specified Variable

Step 1: Multiply on both sides to clear the equation of fractions if they appear.

Step 2: Use the distributive property to remove parentheses if they appear.

Step 3: Simplify each side of the equation by combining like terms.

Step 4: Get all terms containing the specified variable on one side and all other terms on the other side by using the addition property of equality.

Step 5: Get the specified variable alone by using the multiplication property of equality.

Example 5 Solve $V = lwh$ for l.

Solution: This formula is used to find the volume of a box. To solve for l, we divide both sides by wh.

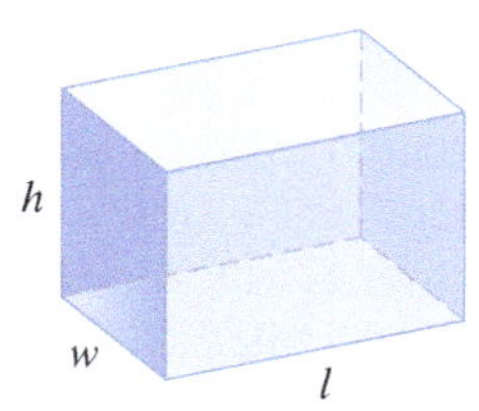

$$V = lwh$$

$$\frac{V}{wh} = \frac{lwh}{wh} \quad \text{Divide both sides by } wh.$$

$$\frac{V}{wh} = l \quad \text{Simplify.}$$

Since we have l alone on one side of the equation, we have solved for l in terms of V, w, and h. Remember that it does not matter on which side of the equation we get the variable alone.

Work Practice 5

Practice 5

Solve $C = 2\pi r$ for r. (This formula is used to find the circumference, C, of a circle given its radius, r.)

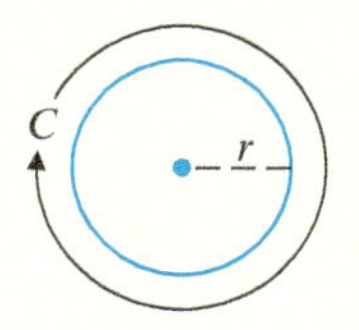

Example 6 Solve $y = mx + b$ for x.

Solution: First we get mx alone by subtracting b from both sides.

$$y = mx + b$$

$$y - b = mx + b - b \quad \text{Subtract } b \text{ from both sides.}$$

$$y - b = mx \quad \text{Combine like terms.}$$

Next we solve for x by dividing both sides by m.

$$\frac{y - b}{m} = \frac{mx}{m}$$

$$\frac{y - b}{m} = x \quad \text{Simplify.}$$

Work Practice 6

Practice 6

Solve $P = 2l + 2w$ for l.

✓Concept Check Solve:

a. ● = ■ − ■ for ■

b. ● = ■ · ▲ − ■ for ■

Example 7 Solve $P = 2l + 2w$ for w.

Solution: This formula relates the perimeter of a rectangle to its length and width. Find the term containing the variable w. To get this term, $2w$, alone, subtract $2l$ from both sides.

$$P = 2l + 2w$$

$$P - 2l = 2l + 2w - 2l \quad \text{Subtract } 2l \text{ from both sides.}$$

$$P - 2l = 2w \quad \text{Combine like terms.}$$

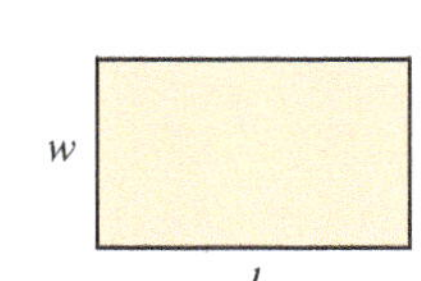

(*Continued on next page*)

Practice 7

Solve $P = 2a + b - c$ for a.

Answers

5. $r = \frac{C}{2\pi}$

6. $l = \frac{P - 2w}{2}$ **7.** $a = \frac{P - b + c}{2}$

✓Concept Check Answer

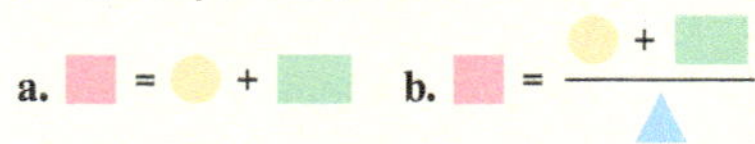

The 2s may *not* be divided out here. Although 2 is a factor of the denominator, 2 is *not* a factor of the numerator since it is not a factor of both terms in the numerator.

$$\frac{P - 2l}{2} = \frac{2w}{2}$$ Divide both sides by 2.

$$\frac{P - 2l}{2} = w$$ Simplify.

Work Practice 7

For the next example, we recall a formula used in Example 3 for converting between degrees Fahrenheit and degrees Celsius.

This formula below is solved for F, but we will solve it for C to give us an equivalent formula that may be more convenient to use. To do so, we will first clear the equation of fractions and then solve for the specified variable.

Practice 8

Solve $A = \frac{a + b}{2}$ for b.

Example 8 Solve $F = \frac{9}{5}C + 32$ for C.

Solution:

$$F = \frac{9}{5}C + 32$$

$$5(F) = 5\left(\frac{9}{5}C + 32\right)$$ Clear the fraction by multiplying both sides by the LCD.

$$5F = 9C + 160$$ Distribute the 5.

$$5F - 160 = 9C + 160 - 160$$ To get the term containing the variable C alone, subtract 160 from both sides.

$$5F - 160 = 9C$$ Combine like terms.

$$\frac{5F - 160}{9} = \frac{9C}{9}$$ Divide both sides by 9.

$$\frac{5F - 160}{9} = C$$ Simplify.

Helpful Hint

We now have two formulas for temperature conversion.

$$F = \frac{9}{5}C + 32$$

$$C = \frac{5F - 160}{9}$$

Use the one most convenient.

Answer

8. $b = 2A - a$

Work Practice 8

Vocabulary, Readiness & Video Check

Martin-Gay Interactive Videos

See Video 2.5

Watch the section lecture video and answer the following questions.

Objective A

1. Complete this statement based on the lecture given before Example 1: A formula is an equation that describes known ______ among quantities.

2. In Example 2, how are the units for the solution determined?

Objective B

3. In Example 4, what is the equation $5x = 30$ used to show?

2.5 Exercise Set MyLab Math

Objective A *Substitute the given values into each given formula and solve for the unknown variable. See Examples 1 through 4.*

△ 1. $A = bh$; $A = 45, b = 15$ (Area of a parallelogram)

2. $d = rt$; $d = 195, t = 3$ (Distance formula)

△ 3. $S = 4lw + 2wh$; $S = 102, l = 7, w = 3$ (Surface area of a special rectangular box)

△ 4. $V = lwh$; $l = 14, w = 8, h = 3$ (Volume of a rectangular box)

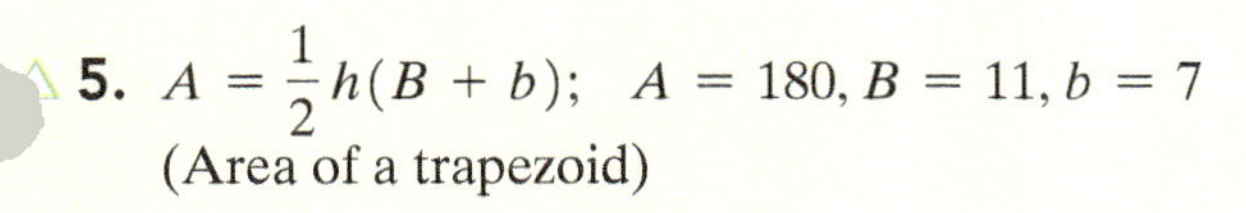

5. $A = \frac{1}{2}h(B + b)$; $A = 180, B = 11, b = 7$ (Area of a trapezoid)

6. $A = \frac{1}{2}h(B + b)$; $A = 60, B = 7, b = 3$ (Area of a trapezoid)

7. $P = a + b + c$; $P = 30, a = 8, b = 10$ (Perimeter of a triangle)

8. $V = \frac{1}{3}Ah$; $V = 45, h = 5$ (Volume of a pyramid)

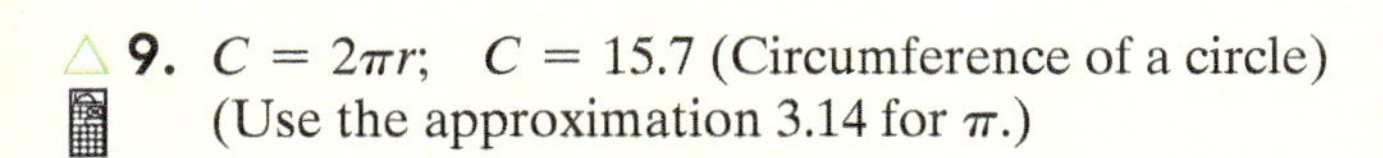

9. $C = 2\pi r$; $C = 15.7$ (Circumference of a circle) (Use the approximation 3.14 for π.)

10. $A = \pi r^2$; $r = 4$ (Area of a circle) (Use the approximation 3.14 for π.)

Objective B *Solve each formula for the specified variable. See Examples 5 through 8.*

11. $f = 5gh$ for h

12. $x = 4\pi y$ for y

13. $V = lwh$ for w

14. $T = mnr$ for n

15. $3x + y = 7$ for y

16. $-x + y = 13$ for y

17. $A = P + PRT$ for R

18. $A = P + PRT$ for T

19. $V = \frac{1}{3}Ah$ for A

20. $D = \frac{1}{4}fk$ for k

21. $P = a + b + c$ for a

22. $PR = x + y + z + w$ for z

23. $S = 2\pi rh + 2\pi r^2$ for h

24. $S = 4lw + 2wh$ for h

Objective A *Solve. For Exercises 25 and 26, the solutions have been started for you. See Examples 1 through 4.*

25. The iconic NASDAQ sign in New York's Times Square has a width of 84 feet and an area of 10,080 square feet. Find the height (or length) of the sign. (*Source:* livedesignonline.com)

Start the solution:

1. UNDERSTAND the problem. Reread it as many times as needed.
2. TRANSLATE into an equation. (Fill in the blanks below.)

Area	=	length	times	width
↓	↓	↓	↓	↓
____	=	x	·	____

Finish with:
3. SOLVE and **4.** INTERPRET

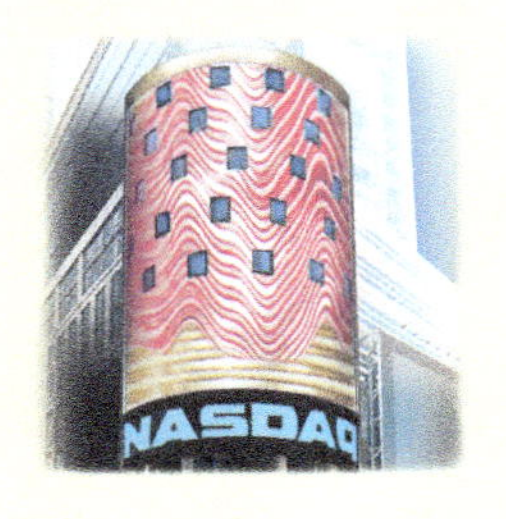

26. The world's largest sign for Coca-Cola is located in Arica, Chile. The rectangular sign has a length of 400 feet and an area of 52,400 square feet. Find the width of the sign. (*Source:* Fabulous Facts about Coca-Cola, Atlanta, GA)

Start the solution:

1. UNDERSTAND the problem. Reread it as many times as needed.
2. TRANSLATE into an equation. (Fill in the blanks below.)

Area	=	length	times	width
↓	↓	↓	↓	↓
____	=	____	·	w

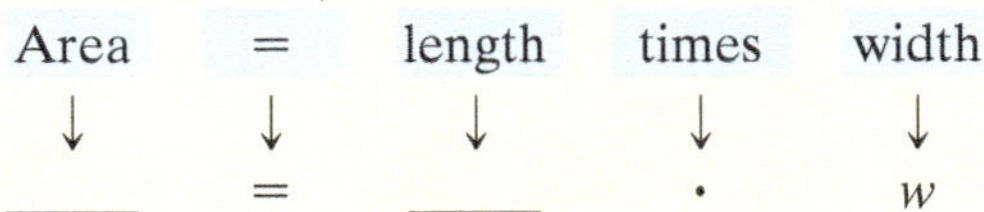

Finish with:
3. SOLVE and **4.** INTERPRET

27. A frame shop charges according to both the amount of framing needed to surround the picture and the amount of glass needed to cover the picture.

a. Find the area and perimeter of the picture below.

b. Identify whether the frame has to do with perimeter or area and the same with the glass.

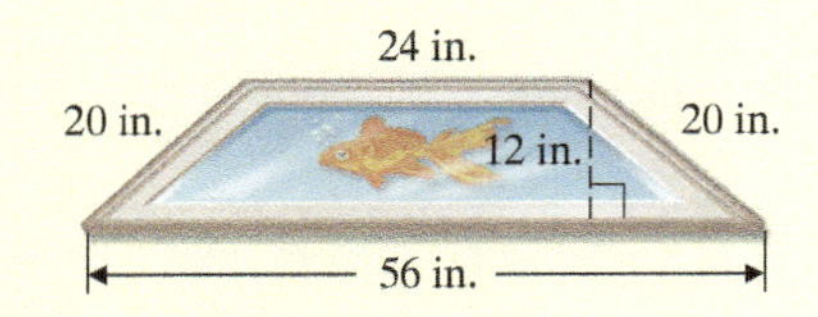

28. A decorator is painting and placing a border completely around a parallelogram-shaped wall.

a. Find the area and perimeter of the wall below.

b. Identify whether the border has to do with perimeter or area and the same with paint.

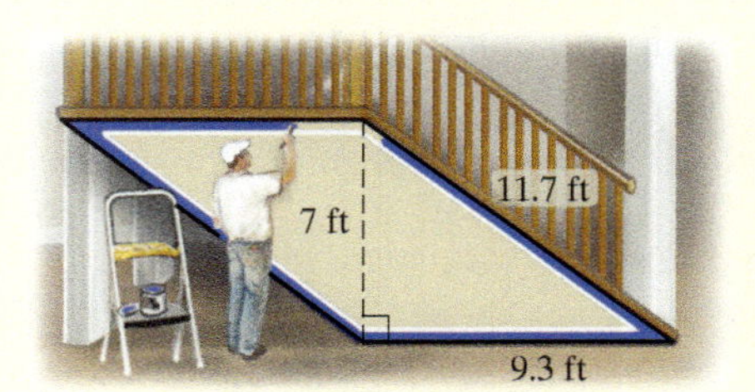

29. For the purpose of purchasing new baseboard and carpet,

a. Find the area and perimeter of the room below (neglecting doors).

b. Identify whether baseboard has to do with area or perimeter and the same with carpet.

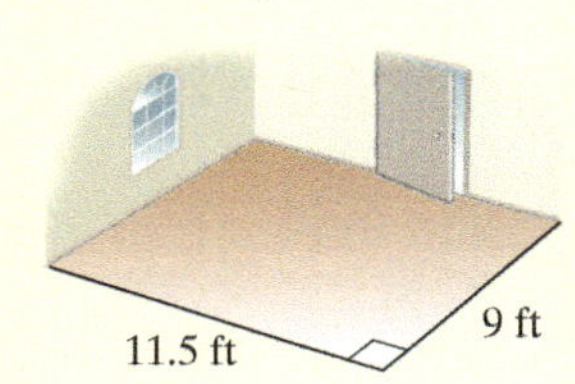

30. For the purpose of purchasing lumber for a new fence and seed to plant grass,

a. Find the area and perimeter of the yard below.

b. Identify whether a fence has to do with area or perimeter and the same with grass seed.

$\left(A = \frac{1}{2}bh\right)$

27 ft
45 ft
36 ft

For Exercises 31 and 32, use the temperature conversion formula of your choice. See Example 8 and the Helpful Hint beside the example.

31. Convert Nome, Alaska's 14°F high temperature to Celsius.

32. Convert Paris, France's low temperature of −5°C to Fahrenheit.

33. The X-30 is a "space plane" that skims the edge of space at 4000 miles per hour. Neglecting altitude, if the circumference of the Earth is approximately 25,000 miles, how long does it take for the X-30 to travel around the Earth?

34. In the United States, a notable hang glider flight was a 303-mile, $8\frac{1}{2}$-hour flight from New Mexico to Kansas. What was the average rate during this flight?

35. △ An architect designs a rectangular flower garden such that the width is exactly two-thirds of the length. If 260 feet of antique picket fencing is to be used to enclose the garden, find the dimensions of the garden.

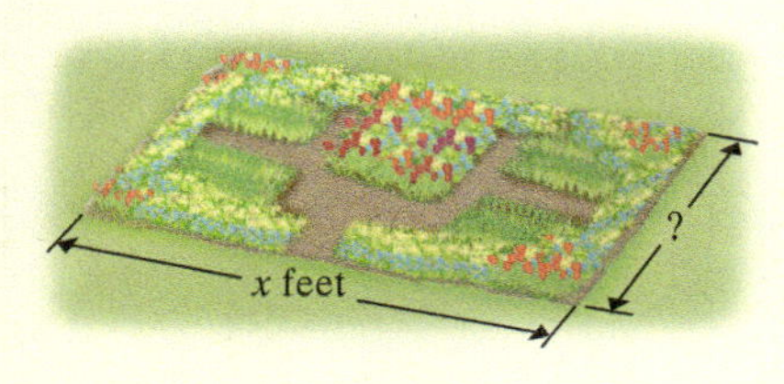

△ **36.** If the length of a rectangular parking lot is 10 meters less than twice its width, and the perimeter is 400 meters, find the length of the parking lot.

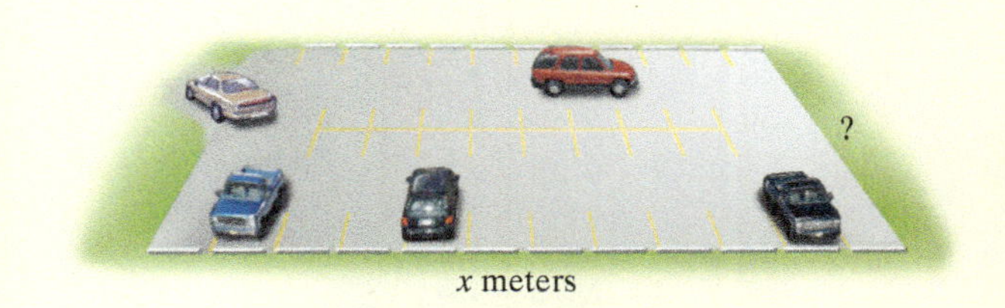

△ **37.** A flower bed is in the shape of a triangle with one side twice the length of the shortest side and the third side 30 feet more than the length of the shortest side. Find the dimensions if the perimeter is 102 feet.

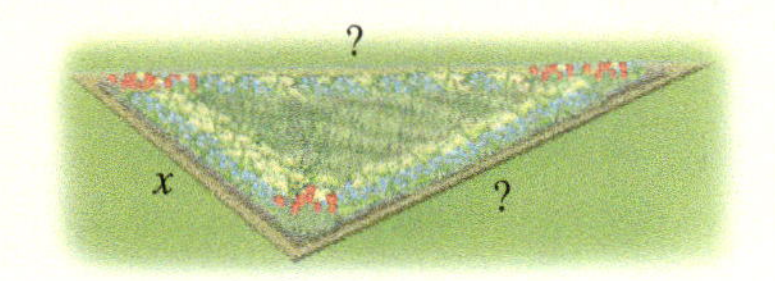

△ **38.** The perimeter of a yield sign in the shape of an isosceles triangle is 22 feet. If the shortest side is 2 feet less than the other two sides, find the length of the shortest side. (*Hint:* An isosceles triangle has two sides the same length.)

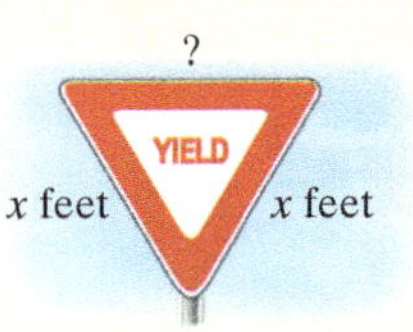

39. The Eurostar is a high-speed train that shuttles passengers between London, England, and Paris, France, via the Channel Tunnel. The Eurostar can make the trip in about $2\frac{1}{4}$ hours at an average speed of 136 mph. About how long is the Eurostar's route connecting London and Paris? (*Source:* Eurostar)

40. A family is planning their vacation to Disney World. They will drive from a small town outside New Orleans, Louisiana, to Orlando, Florida, a distance of 700 miles. They plan to average a rate of 56 mph. How long will this trip take?

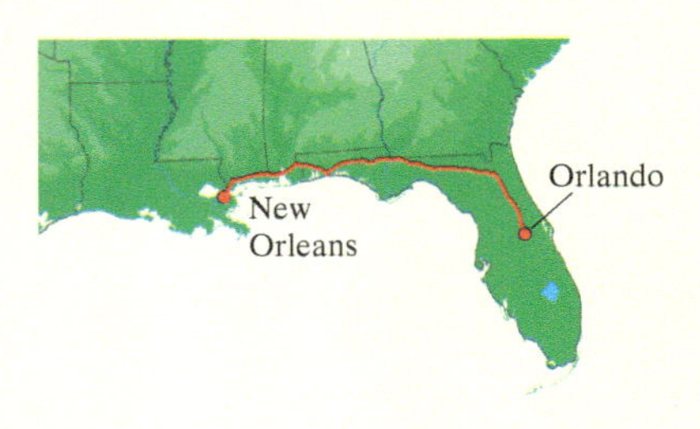

Dolbear's Law states the relationship between the rate at which Snowy Tree Crickets chirp and the air temperature of their environment. The formula is

$$T = 50 + \frac{N - 40}{4}, \text{ where } \begin{array}{l} T = \text{temperature in degrees Fahrenheit and} \\ N = \text{number of chirps per minute} \end{array}$$

41. If $N = 86$, find the temperature in degrees Fahrenheit, T.

42. If $N = 94$, find the temperature in degrees Fahrenheit, T.

43. If $T = 55°F$, find the number of chirps per minute.

44. If $T = 65°F$, find the number of chirps per minute.

Use the results of Exercises 41 through 44 to complete each sentence with "increases" or "decreases."

45. As the number of cricket chirps per minute increases, the air temperature of their environment ________.

46. As the air temperature of their environment decreases, the number of cricket chirps per minute ________.

Solve. See Examples 1 through 4.

△ **47.** Piranha fish require 1.5 cubic feet of water per fish to maintain a healthy environment. Find the maximum number of piranhas you could put in a tank measuring 8 feet by 3 feet by 6 feet.

△ **48.** Find the maximum number of goldfish you can put in a cylindrical tank whose diameter is 8 meters and whose height is 3 meters, if each goldfish needs 2 cubic meters of water. ($V = \pi r^2 h$)

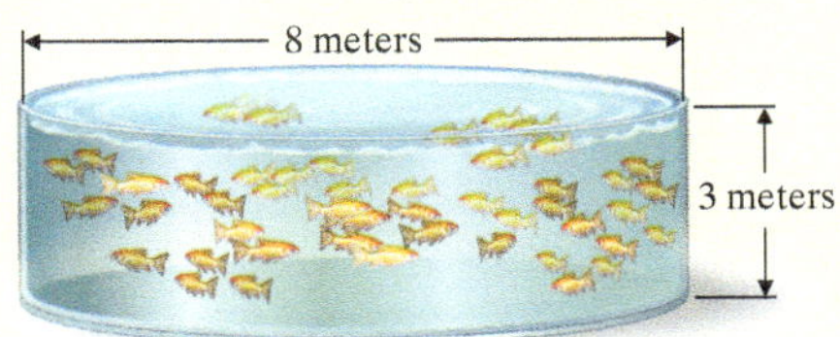

△ **49.** A lawn is in the shape of a trapezoid with a height of 60 feet and bases of 70 feet and 130 feet. How many bags of fertilizer must be purchased to cover the lawn if each bag covers 4000 square feet?

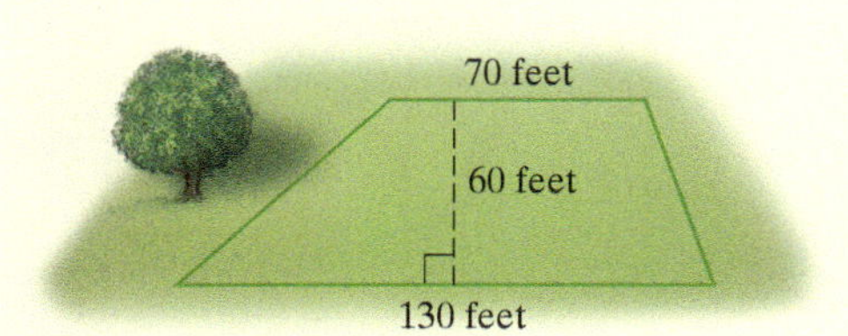

△ **50.** If the area of a right-triangularly shaped sail is 20 square feet and its base is 5 feet, find the height of the sail. $\left(A = \frac{1}{2}bh\right)$

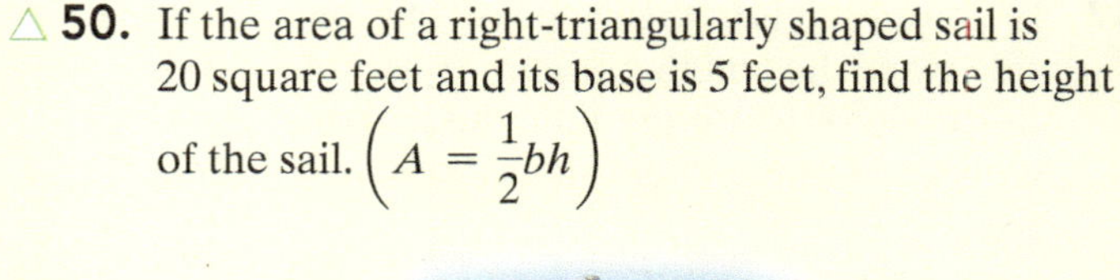

△ **51.** Maria's Pizza sells one 16-inch cheese pizza or two 10-inch cheese pizzas for $9.99. Determine which size gives more pizza.

△ **52.** Find how much rope is needed to wrap around the Earth at the equator, if the radius of the Earth is 4000 miles. (*Hint:* Use 3.14 for π and the formula for circumference.)

△ **53.** The perimeter of a geometric figure is the sum of the lengths of its sides. If the perimeter of the following pentagon (five-sided figure) is 48 meters, find the length of each side.

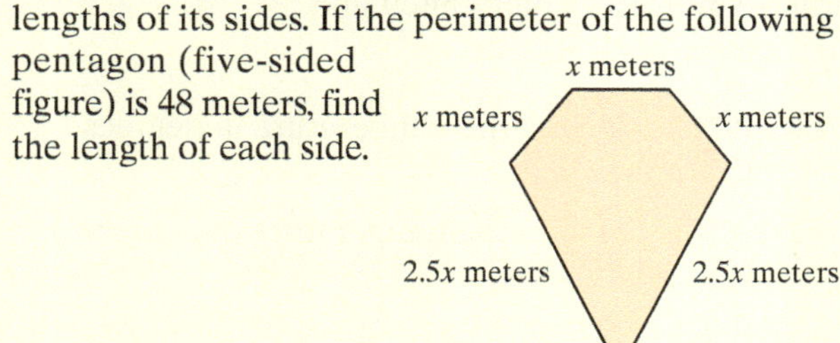

△ **54.** The perimeter of the following triangle is 82 feet. Find the length of each side.

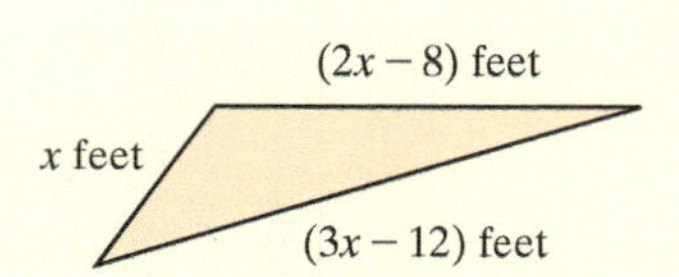

55. The Hawaiian volcano Kilauea is one of the world's most active volcanoes and has had continuous eruptive activity since 1983. Erupting lava flows through a tube system about 11 kilometers to the sea. Assume a lava flow speed of 0.5 kilometer per hour and calculate how long it takes to reach the sea.

56. The world's largest pink ribbon, the sign of the fight against breast cancer, was erected out of pink Post-it® notes on a billboard in New York City in October 2004. If the area of the rectangular billboard covered by the ribbon was approximately 3990 square feet, and the width of the billboard was approximately 57 feet, what was the height of this billboard?

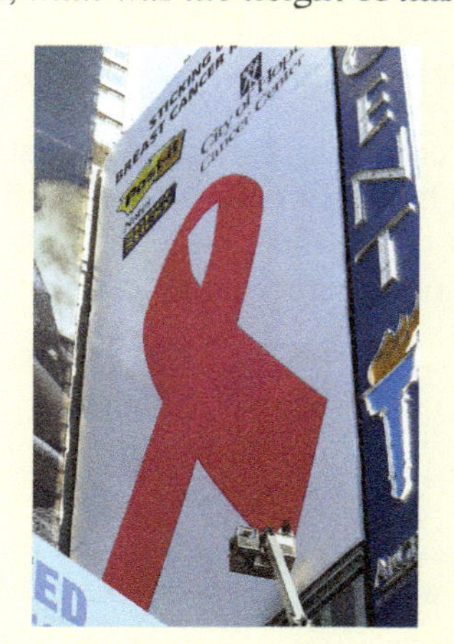

△ **57.** The perimeter of an equilateral triangle is 7 inches more than the perimeter of a square, and the sides of the triangle are 5 inches longer than the sides of the square. Find the side of the triangle. (*Hint:* An equilateral triangle has three sides the same length.)

△ **58.** A square animal pen and a pen shaped like an equilateral triangle have equal perimeters. Find the length of the sides of each pen if the sides of the triangular pen are 15 less than twice a side of the square pen. (*Hint:* An equilateral triangle has three sides the same length.)

59. Find how long it takes Tran Nguyen to drive 135 miles on I-10 if he merges onto I-10 at 10 a.m. and drives nonstop with his cruise control set on 60 mph.

60. Beaumont, Texas, is about 150 miles from Toledo Bend. If Leo Miller leaves Beaumont at 4 a.m. and averages 45 mph, when should he arrive at Toledo Bend?

△ **61.** The longest runway at Los Angeles International Airport has the shape of a rectangle and an area of 1,813,500 square feet. This runway is 150 feet wide. How long is the runway? (*Source:* Los Angeles World Airports)

62. Normal room temperature is about 78°F. Convert this temperature to Celsius.

63. The highest temperature ever recorded in Europe was 122°F in Seville, Spain, in August 1881. Convert this record high temperature to Celsius. (*Source:* National Climatic Data Center)

64. The lowest temperature ever recorded in Oceania was −10°C at the Haleakala Summit in Maui, Hawaii, in January 1961. Convert this record low temperature to Fahrenheit. (*Source:* National Climatic Data Center)

△ **65.** The IZOD IndyCar series is an open-wheeled race car competition based in the United States. An IndyCar car has a maximum length of 204 inches, a maximum width of 76.5 inches, and a maximum height of 44 inches. When the IZOD IndyCar series travels to another country for a grand prix, teams must ship their cars. Find the volume of the smallest shipping crate needed to ship an IndyCar car of maximum dimensions. (*Source:* Championship Auto Racing Teams, Inc.)

IndyCar Racing Car

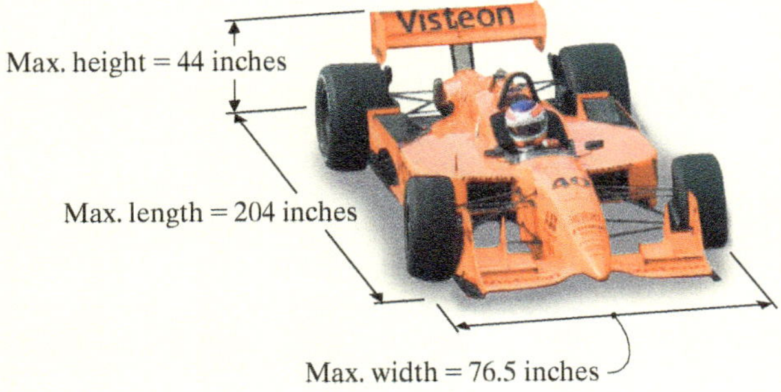

66. During a recent IndyCar road course race, the winner's average speed was 118 mph. Based on this speed, how long would it take an IndyCar driver to travel from Los Angeles to New York City, a distance of about 2810 miles by road, without stopping? Round to the nearest tenth of an hour.

△ **67.** The Hoberman Sphere is a toy ball that expands and contracts. When it is completely closed, it has a diameter of 9.5 inches. Find the volume of the Hoberman Sphere when it is completely closed. Use 3.14 for π. Round to the nearest whole cubic inch. (*Hint:* Volume of a sphere $= \frac{4}{3}\pi r^3$.) (*Source:* Hoberman Designs, Inc.)

△ **68.** When the Hoberman Sphere (see Exercise **67**) is completely expanded, its diameter is 30 inches. Find the volume of the Hoberman Sphere when it is completely expanded. Use 3.14 for π. (*Source:* Hoberman Designs, Inc.)

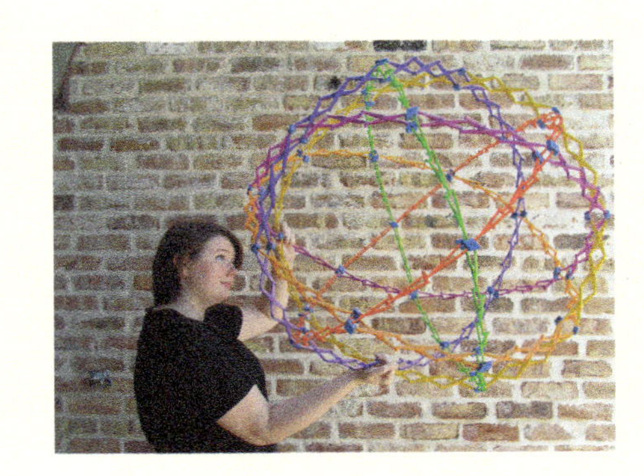

69. The average temperature on the planet Mercury is 167°C. Convert this temperature to degrees Fahrenheit. Round to the nearest degree. (*Source:* National Space Science Data Center)

70. The average temperature on the planet Jupiter is −227°F. Convert this temperature to degrees Celsius. Round to the nearest degree. (*Source:* National Space Science Data Center)

Review

Write each percent as a decimal. See Section R.3.

71. 32%

72. 8%

73. 200%

74. 0.5%

Write each decimal as a percent. See Section R.3.

75. 0.17

76. 0.03

77. 7.2

78. 5

Concept Extensions

Solve.

79. $N = R + \frac{V}{G}$ for V (Urban forestry: tree plantings per year)

80. $B = \frac{F}{P - V}$ for V (Business: break-even point)

81. The formula $V = lwh$ is used to find the volume of a box. If the length of a box is doubled, the width is doubled, and the height is doubled, how does this affect the volume? Explain your answer.

82. The formula $A = bh$ is used to find the area of a parallelogram. If the base of a parallelogram is doubled and its height is doubled, how does this affect the area? Explain your answer.

83. Use the Dolbear's Law formula from Exercises **41** through **46** and calculate when the number of cricket chirps per minute is the same as the temperature in degrees Fahrenheit. (*Hint:* Replace T with N and solve for N, or replace N with T and solve for T.)

84. Find the temperature at which the Celsius measurement and the Fahrenheit measurement are the same number.

Solve. See the Concept Check in this section.

85. ▲ − ● · ■ = ■ for ●

86. ⬟ · ■ + ▲ = ● for ■

87. Flying fish do not *actually* fly, but glide. They have been known to travel a distance of 1300 feet at a rate of 20 miles per hour. How many seconds would it take to travel this distance? (*Hint:* First convert miles per hour to feet per second. Recall that 1 mile = 5280 feet.) Round to the nearest tenth of a second.

88. A glacier is a giant mass of rocks and ice that flows downhill like a river. Exit Glacier, near Seward, Alaska, moves at a rate of 20 inches a day. Find the distance in feet the glacier moves in a year. (Assume 365 days a year.) Round to two decimal places.

Substitute the given values into each given formula and solve for the unknown variable. If necessary, round to one decimal place.

89. $I = PRT$; $I = 1{,}056{,}000, R = 0.055, T = 6$ (Simple interest formula)

90. $I = PRT$; $I = 3750, P = 25{,}000, R = 0.05$ (Simple interest formula)

91. $V = \frac{4}{3}\pi r^3$; $r = 3$ (Volume of a sphere) (Use a calculator approximation for π.)

92. $V = \frac{1}{3}\pi r^2 h$; $V = 565.2, r = 6$ (Volume of a cone) (Use a calculator approximation for π.)

2.6 Percent and Mixture Problem Solving

This section is devoted to solving problems in the categories listed. The same problem-solving steps used in previous sections are also followed in this section. They are listed below for review.

Objectives

A Solve Percent Equations.

B Solve Discount and Mark-Up Problems.

C Solve Percent of Increase and Percent of Decrease Problems.

D Solve Mixture Problems.

General Strategy for Problem Solving

1. UNDERSTAND the problem. During this step, become comfortable with the problem. Some ways of doing this are as follows:

   Read and reread the problem.

   Choose a variable to represent the unknown.

   Construct a drawing, whenever possible.

   Propose a solution and check. Pay careful attention to how you check your proposed solution. This will help writing an equation to model the problem.
2. TRANSLATE the problem into an equation.
3. SOLVE the equation.
4. INTERPRET the results: *Check* the proposed solution in the stated problem and *state* your conclusion.

Objective A Solving Percent Equations

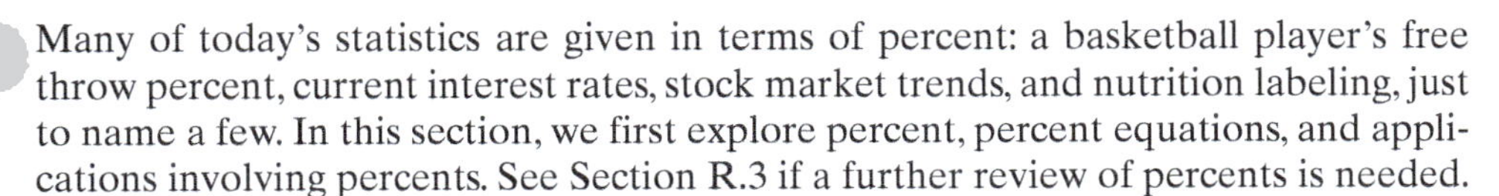

Many of today's statistics are given in terms of percent: a basketball player's free throw percent, current interest rates, stock market trends, and nutrition labeling, just to name a few. In this section, we first explore percent, percent equations, and applications involving percents. See Section R.3 if a further review of percents is needed.

Example 1 The number 63 is what percent of 72?

Solution:

1. UNDERSTAND. Read and reread the problem. Next, let's suppose that the percent is 80%. To check, we find 80% of 72.

   $$80\% \text{ of } 72 = 0.80(72) = 57.6$$

   This is close, but not 63. At this point, though, we have a better understanding of the problem; we know the correct answer is close to and greater than 80%, and we know how to check our proposed solution later.

   Let $x =$ the unknown percent.
2. TRANSLATE. Recall that "is" means "equals" and "of" signifies multiplying. Let's translate the sentence directly.

the number 63	is	what percent	of	72
↓	↓	↓	↓	↓
63	=	x	·	72
3. SOLVE.

   $63 = 72x$

   $0.875 = x$ Divide both sides by 72.

   $87.5\% = x$ Write as a percent.

(*Continued on next page*)

Practice 1

The number 22 is what percent of 40?

Answer

1. 55%

4. INTERPRET.

Check: Verify that 87.5% of 72 is 63.

State: The number 63 is 87.5% of 72.

Work Practice 1

Practice 2

The number 150 is 40% of what number?

Example 2

The number 120 is 15% of what number?

Solution:

1. UNDERSTAND. Read and reread the problem.

   Let $x =$ the unknown number.

2. TRANSLATE.

the number 120	is	15%	of	what number
↓	↓	↓	↓	↓
120	=	15%	·	x

3. SOLVE.

$120 = 0.15x$ Write 15% as 0.15.

$800 = x$ Divide both sides by 0.15.

4. INTERPRET.

Check: Check the proposed solution by finding 15% of 800 and verifying that the result is 120.

State: Thus, 120 is 15% of 800.

Work Practice 2

Practice 3

Use the circle graph to answer each question.

a. What percent of trips made by American travelers are solely for pleasure?

b. What percent of trips made by American travelers are for the purpose of pleasure or combined business/pleasure?

c. On an airplane flight of 250 Americans, how many of these people might we expect to be traveling solely for pleasure?

Example 3

The circle graph below shows the purpose of trips made by American travelers. Use this graph to answer the questions below.

Purpose of Trip

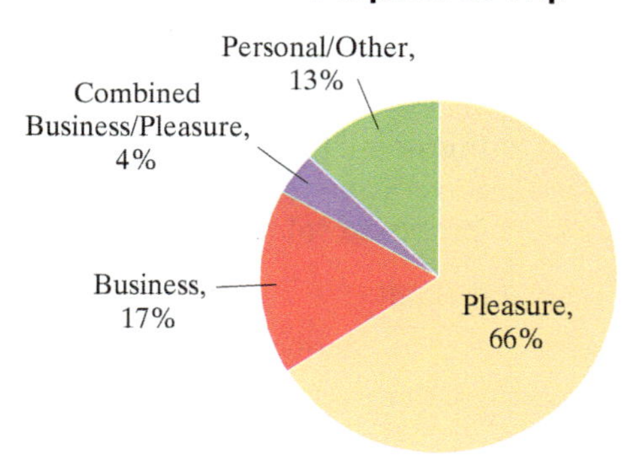

Source: Travel Industry Association of America

a. What percent of trips made by American travelers are solely for the purpose of business?

b. What percent of trips made by American travelers are for the purpose of business or combined business/pleasure?

c. On an airplane flight of 253 Americans, how many of these people might we expect to be traveling solely for business?

Answers

2. 375 3. a. 66% b. 70% c. 165 people

Solution:

a. From the circle graph, we see that 17% of trips made by American travelers are solely for the purpose of business.

b. From the circle graph, we know that 17% of trips are solely for business and 4% of trips are for combined business/pleasure. The sum 17% + 4% or 21% of trips made by American travelers are for the purpose of business or combined business/pleasure.

c. Since 17% of trips made by American travelers are for business, we find 17% of 253. Remember that "of" translates to "multiplication."

$$17\% \text{ of } 253 = 0.17(253) \quad \text{Replace "of" with the operation of multiplication.}$$
$$= 43.01$$

We might then expect that about 43 American travelers on the flight are traveling solely for business.

Work Practice 3

Objective B Solving Discount and Mark-Up Problems 

The next example has to do with discounting the price of a cell phone.

Example 4 Cell Phones Unlimited recently reduced the price of a \$140 phone by 20%. What is the discount and the new price?

Practice 4

A surfboard, originally purchased for \$400, was sold on eBay at a discount of 40% of the original price. What is the discount and the new price?

Solution:

1. UNDERSTAND. Read and reread the problem. Make sure you understand the meaning of the word "discount." Discount is the amount of money by which an item has been decreased. To find the discount, we simply find 20% of \$140. In other words, we have the formulas,

discount = percent · original price Then

new price = original price − discount

2, 3. TRANSLATE and SOLVE.

discount	=	percent	·	original price
	=	20%	·	\$140
	=	0.20	·	\$140
	= \$28			

Thus, the discount in price is \$28.

new price	=	original price	−	discount
	=	\$140	−	\$28
	= \$112			

4. INTERPRET.

Check: Check your calculations in the formulas, and also see if our results are reasonable. They are.

State: The discount in price is \$28 and the new price is \$112.

Work Practice 4

A concept similar to discount is mark-up. What is the difference between the two? A discount is subtracted from the original price while a mark-up is added to the original price. For mark-ups,

Answer

4. discount: \$160; new price: \$240

mark-up = percent · original price

new price = original price + mark-up

Mark-up exercises can be found in Exercise Set 2.6.

Objective C Solving Percent of Increase and Percent of Decrease Problems

Percent of increase or percent of decrease is a common way to describe how some measurement has increased or decreased. For example, crime increased by 8%, teachers received a 5.5% increase in salary, or a company decreased its employees by 10%. The next example is a review of percent of increase.

Practice 5

If a number increases from 120 to 200, find the percent of increase. Round to the nearest tenth of a percent.

Example 5 Calculating the Percent of Increase of Attending College

The average cost of tuition and fees for attending a four-year public college as a state resident rose from \$4845 during the 2000–2001 academic year to \$9650 during the 2016–2017 year. Find the percent of increase. (*Source*: *Forbes*)

Solution:

1. UNDERSTAND. Read and reread the problem. Notice that the new tuition, \$9650, is almost double the old tuition of \$4845. Because of that, we know that the percent of increase is close to 100%. To see this, let's guess that the percent of increase is 100%. To check, we find 100% of \$4845 to find the *increase* in cost. Then we add this increase to \$4845 to find the *new cost*. In other words, 100%(\$4845) = 1.00(\$4845) = \$4845, the increase in cost. The *new cost* would be old cost + increase = \$4845 + \$4845 = \$9690, close to the actual new cost of \$9650. We now know that the increase is close to, but less than, 100% and we know how to check our proposed solution.

   Let $x =$ the percent of increase.

2. TRANSLATE. First, find the **increase,** and then the **percent of increase.** The increase in cost is found by:

In words:	increase	=	new cost	−	old cost	or
Translate:	increase	=	\$9650	−	\$4845	
		=	\$4805			

   Next, find the percent of increase. The percent of increase or percent of decrease is always a percent of the original number or, in this case, the old cost.

In words:	increase	is	what percent	of	old cost
Translate:	\$4805	=	x	·	\$4845

3. SOLVE.

$$4805 = 4845x$$

$$0.992 \approx x \quad \text{Divide both sides by 4845 and round to 3 decimal places.}$$

$$99.2\% \approx x \quad \text{Write as a percent.}$$

4. INTERPRET.

   Check: Check the proposed solution

   State: The percent of increase in cost is approximately 99.2%.

Work Practice 5

Answer

5. 66.7%

Percent of decrease is found using a similar method. First find the decrease, then determine what percent of the original or first amount is that decrease.

Read the next example carefully. For Example 5, we were asked to find percent of increase. In Example 6, we are given the percent of increase and asked to find the number before the increase.

Example 6 Growth in digital 3-D theater screens is fastest in the Asia/Pacific entertainment market. Find the number of digital 3-D screens in Asia/Pacific in 2016 if, after a 24% increase, the number in 2017 was 58,259. Round to the nearest whole. (*Source:* MPAA)

Practice 6

Find the original price of a suit if the sale price is $46 after a 20% discount.

Solution:

1. UNDERSTAND. Read and reread the problem. Let's guess a solution and see how we would check our guess. If the number of digital 3-D screens in 2016 was 30,000, we would see if 30,000 plus the increase is 58,259; that is,

$$30{,}000 + 24\%(30{,}000) = 30{,}000 + 0.24(30{,}000) = 30{,}000 + 7200 = 37{,}200$$

Since 37,200 is too small, we know that our guess of 30,000 is too small. We also have a better understanding of the problem. Let

x = number of digital 3-D screens in 2016

2. TRANSLATE. To translate an equation, we remember that

In words:	number of digital 3-D screens in 2016	plus	increase	equals	number of digital 3-D screens in 2017
Translate:	x	$+$	$0.24x$	$=$	58,259

3. SOLVE.

$$1.24x = 58{,}259$$
$$x = \frac{58{,}259}{1.24}$$
$$x \approx 46{,}983$$

4. INTERPRET.

Check: Recall that x represents the number of digital 3-D screens in 2016. If this number is approximately 46,983, let's see if 46,983 plus the increase is close to 58,259. (We use the word "close" since 46,983 is rounded.)

$$46{,}983 + 24\%(46{,}983) = 46{,}983 + 0.24(46{,}983) = 46{,}983 + 11{,}275.92$$
$$= 58{,}258.92$$

which is close to 58,259.

State: There were approximately 46,983 digital 3-D screens in the Asia/Pacific region in 2016.

Work Practice 6

Objective D Solving Mixture Problems 

Mixture problems involve two or more different quantities being combined to form a new mixture. These applications range from Dow Chemical's need to form a chemical mixture of a required strength to Planter's Peanut Company's need to find the correct mixture of peanuts and cashews, given taste and price constraints.

Answer

6. $57.50

Practice 7

How much 20% dye solution and 50% dye solution should be mixed to obtain 6 liters of a 40% solution?

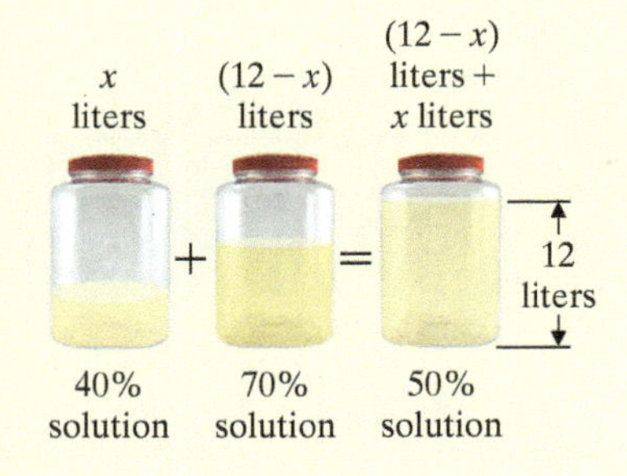

Example 7 Calculating Percent for a Lab Experiment

A chemist working on his doctoral degree at Massachusetts Institute of Technology needs 12 liters of a 50% acid solution for a lab experiment. The stockroom has only 40% and 70% solutions. How much of each solution should be mixed together to form 12 liters of a 50% solution?

Solution:

1. UNDERSTAND. First, read and reread the problem a few times. Next, guess a solution. Suppose that we need 7 liters of the 40% solution. Then we need $12 - 7 = 5$ liters of the 70% solution. To see if this is indeed the solution, find the amount of pure acid in 7 liters of the 40% solution, in 5 liters of the 70% solution, and in 12 liters of a 50% solution, the required amount and strength.

number of liters	×	acid strength	=	amount of pure acid
7 liters	×	40%	=	7(0.40) or 2.8 liters
5 liters	×	70%	=	5(0.70) or 3.5 liters
12 liters	×	50%	=	12(0.50) or 6 liters

Since 2.8 liters + 3.5 liters = 6.3 liters and not 6, our guess is incorrect, but we have gained some valuable insight into how to model and check this problem.

Let

x = number of liters of 40% solution; then

$12 - x$ = number of liters of 70% solution.

2. TRANSLATE. To help us translate to an equation, the following table summarizes the information given. Recall that the amount of acid in each solution is found by multiplying the acid strength of each solution by the number of liters.

	No. of Liters	· Acid Strength	= Amount of Acid
40% Solution	x	40%	$0.40x$
70% Solution	$12 - x$	70%	$0.70(12 - x)$
50% Solution Needed	12	50%	0.50(12)

The amount of acid in the final solution is the sum of the amounts of acid in the two beginning solutions.

In words:	acid in 40% solution	+	acid in 70% solution	=	acid in 50% mixture
Translate:	$0.40x$	+	$0.70(12 - x)$	=	$0.50(12)$

3. SOLVE.

$$0.40x + 0.70(12 - x) = 0.50(12)$$

$$0.4x + 8.4 - 0.7x = 6 \quad \text{Apply the distributive property.}$$

$$-0.3x + 8.4 = 6 \quad \text{Combine like terms.}$$

$$-0.3x = -2.4 \quad \text{Subtract 8.4 from both sides.}$$

$$x = 8 \quad \text{Divide both sides by } -0.3.$$

4. INTERPRET.

Check: To check, recall how we checked our guess.

State: If 8 liters of the 40% solution are mixed with $12 - 8$ or 4 liters of the 70% solution, the result is 12 liters of a 50% solution.

Work Practice 7

Answer

7. 2 liters of the 20% solution; 4 liters of the 50% solution

Vocabulary, Readiness & Video Check

Tell whether the percent labels in the circle graphs are correct.

1.
25%
40%
25%

2.
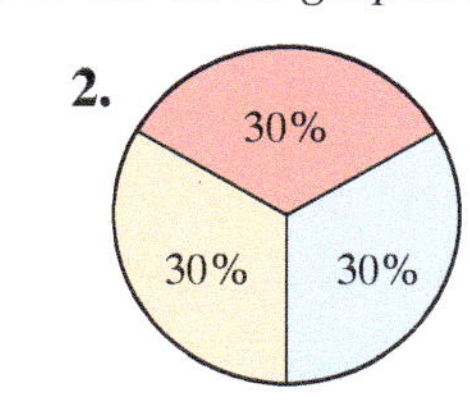

3.
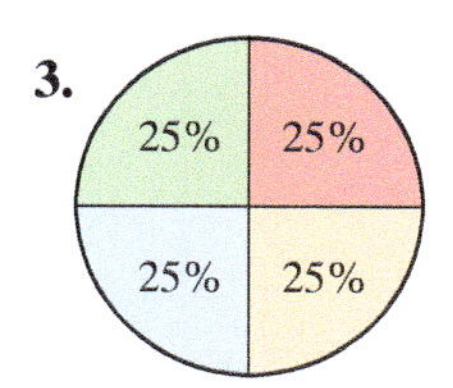

4.
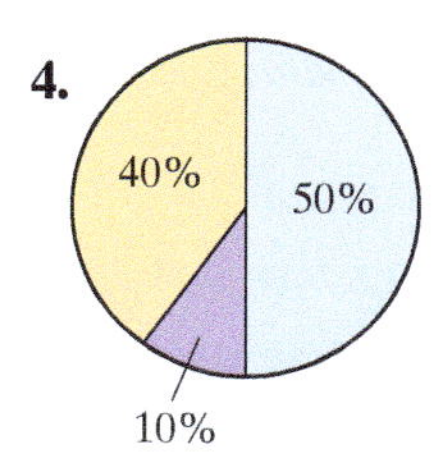

Martin-Gay Interactive Videos

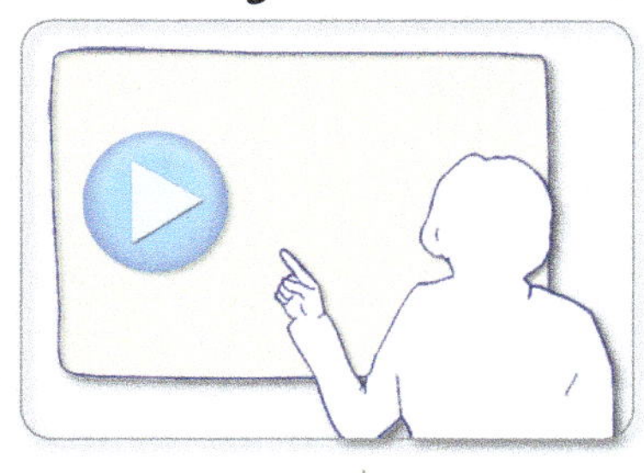

See Video 2.6

Watch the section lecture video and answer the following questions.

Objective A **5.** Answer these questions based on how Example 2 was translated to an equation.

a. What does "is" translate to?

b. What does "of" translate to?

c. How do you write a percent as an equivalent decimal?

Objective B **6.** At the end of Example 3 we are told that the process for finding discount is *almost* the same as finding mark-up.

a. How is discount similar?

b. How does discount differ?

Objective C **7.** According to Example 4, what amount must you find before you can find a percent of increase in price? How do you find this amount?

Objective D **8.** The following problem is worded like Example 6, but using different quantities.

How much of an alloy that is 10% copper should be mixed with 400 ounces of an alloy that is 30% copper in order to get an alloy that is 20% copper? Fill in the table and set up an equation that could be used to solve for the unknowns (do not actually solve). Use Example 6 as a model for your work.

Alloy	Ounces	Copper Percent	Amount of Copper

2.6 Exercise Set MyLab Math 

Objective A *Find each number described. For Exercises 1 and 2, the solutions have been started for you. See Examples 1 and 2.*

1. What number is 16% of 70?

Start the solution:

1. UNDERSTAND the problem. Reread it as many times as needed.

2. TRANSLATE into an equation. (Fill in the blanks below.)

what number	is	16%	of	70
↓	↓	↓	↓	↓
x	___	0.16	___	70

Finish with:

3. SOLVE and

4. INTERPRET

2. What number is 88% of 1000?

Start the solution:

1. UNDERSTAND the problem. Reread it as many times as needed.

2. TRANSLATE into an equation. (Fill in the blanks below.)

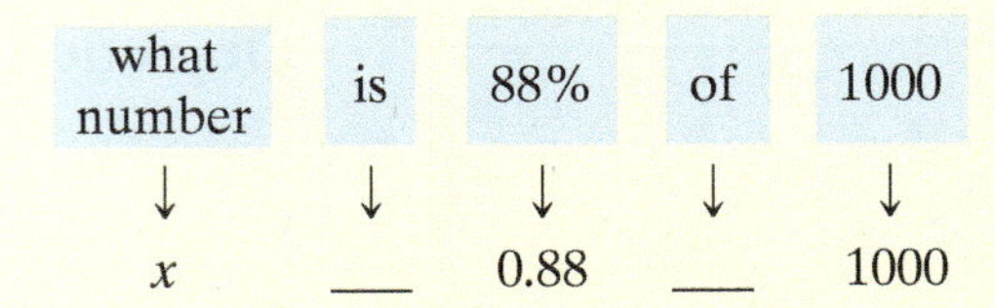

Finish with:

3. SOLVE and

4. INTERPRET

3. The number 28.6 is what percent of 52?

4. The number 87.2 is what percent of 436?

5. The number 45 is 25% of what number?

6. The number 126 is 35% of what number?

The circle graph below shows the types of accommodations that overnight visitors to national parks used. Use this graph for Exercises 7 through 10. See Example 3.

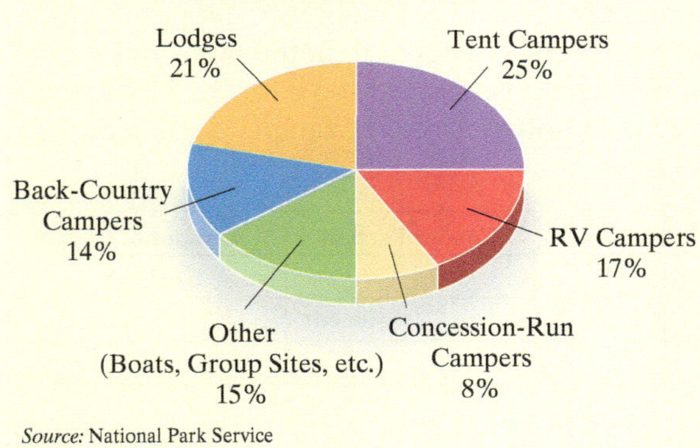

7. What percent of overnight stays were in RVs?

8. What percent of overnight stays involved tent camping?

9. In 2017, Yellowstone National Park reported approximately 1,448,000 overnight stays. How many of these stays might you expect were in lodges?

10. In 2017, Yosemite National Park reported approximately 1,771,000 overnight stays. How many of these stays might you expect involved back-country camping?

Objective B *Solve. If needed, round answers to the nearest cent. See Example 4.*

11. A used automobile dealership recently reduced the price of a used sports car by 8%. If the price of the car before discount was $18,500, find the discount and the new price.

12. A music store is advertising a 25%-off sale on all new releases. Find the discount and the sale price of a newly released CD that regularly sells for $12.50.

13. A birthday celebration meal is $40.50 including tax. Find the total cost if a 15% tip is added to the cost.

14. A retirement dinner for two is $65.40 including tax. Find the total cost if a 20% tip is added to the cost.

Objective C Solve. Round percents to the nearest tenth. See Example 5.

Use the graph below for Exercises 15 and 16.

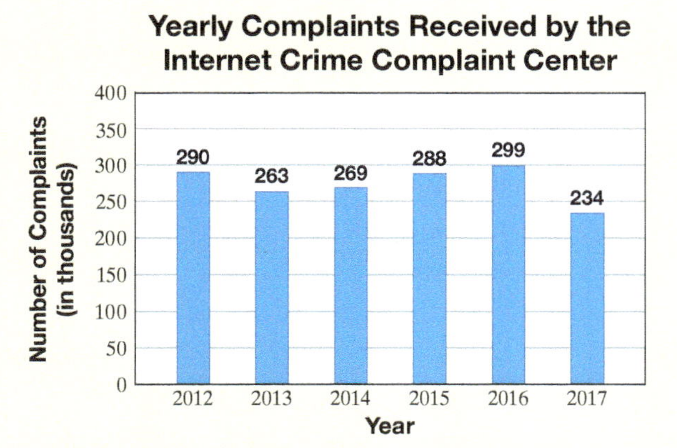

Source: Data from Internet Crime Complaint Center (www.ic3.gov)

15. The number of Internet-crime complaints decreased from 2016 to 2017. Find the percent of decrease.

16. The number of Internet-crime complaints increased from 2014 to 2015. Find the percent of increase.

17. By decreasing each dimension by 1 unit, the area of a rectangle decreased from 40 square feet (on the left) to 28 square feet (on the right). Find the percent of decrease in area.

8 ft, 5 ft — Area: 40 sq ft   7 ft, 4 ft — Area: 28 sq ft

18. By decreasing the length of the side by one unit, the area of a square decreased from 100 square meters to 81 square meters. Find the percent of decrease in area.

10 m — Area: 100 sq m   9 m — Area: 81 sq m

Solve. See Example 6.

19. Find the original price of a pair of shoes if the sale price is $78 after a 25% discount.

20. Find the original price of a popular pair of shoes if the increased price is $80 after a 25% increase.

21. Find last year's salary if after a 4% pay raise, this year's salary is $44,200.

22. Find last year's salary if after a 3% pay raise, this year's salary is $55,620.

Objective D Solve. For each exercise, a table is given for you to complete and use to write an equation that models the situation. See Example 7.

23. How much pure acid should be mixed with 2 gallons of a 40% acid solution in order to get a 70% acid solution?

	Number of Gallons ·	Acid Strength =	Amount of Acid
Pure Acid		100%	
40% Acid Solution			
70% Acid Solution Needed			

24. How many cubic centimeters (cc) of a 25% antibiotic solution should be added to 10 cubic centimeters of a 60% antibiotic solution in order to get a 30% antibiotic solution?

	Number of Cubic cm ·	Antibiotic Strength =	Amount of Antibiotic
25% Antibiotic Solution			
60% Antibiotic Solution			
30% Antibiotic Solution Needed			

25. Community Coffee Company wants a new flavor of Cajun coffee. How many pounds of coffee worth $7 a pound should be added to 14 pounds of coffee worth $4 a pound to get a mixture worth $5 a pound?

	Number of Pounds	· Cost per Pound	= Value
$7 per lb Coffee			
$4 per lb Coffee			
$5 per lb Coffee Wanted			

26. Planter's Peanut Company wants to mix 20 pounds of peanuts worth $3 a pound with cashews worth $5 a pound in order to make an experimental mix worth $3.50 a pound. How many pounds of cashews should be added to the peanuts?

	Number of Pounds	· Cost per Pound	= Value
$3 per lb Peanuts			
$5 per lb Cashews			
$3.50 per lb Mixture Wanted			

Objectives A B C D Mixed Practice *Solve. If needed, round money amounts to two decimal places and all other amounts to one decimal place. See Examples 1 through 7.*

27. Find 23% of 20.

28. Find 140% of 86.

29. The number 40 is 80% of what number?

30. The number 56.25 is 45% of what number?

31. The number 144 is what percent of 480?

32. The number 42 is what percent of 35?

The graph shows the percent of how frequently adult Americans shop online. Use the graph to answer Exercises 33 through 36.

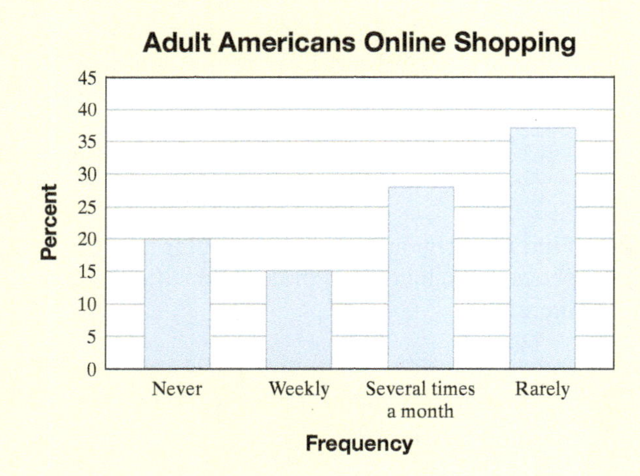

33. Estimate the percent of American adults who never shop online.

34. Estimate the percent of American adults who shop online several times a month.

35. According to the U.S. Census Bureau, in 2017, there were approximately 252 million adults in the United States. How many adults might we predict rarely shop online? Round to the nearest tenth.

36. According to the U.S. Census Bureau, in 2017, there were approximately 252 million adults in the United States. How many adults might we predict shop online weekly?

For Exercises 37 and 38, fill in the percent column in each table. Each table contains a worked-out example.

37.

Top Cranberry-Producing States in 2017 (in millions of pounds)

	Millions of Pounds	Percent of Total (rounded to nearest percent)
Wisconsin	560	
Oregon	48	
Massachusetts	220	
Washington	18	
New Jersey	59	Example: $\frac{59}{905} \approx 7\%$
Total	905	

Source: National Agricultural Statistics Service

38.

New Housing Starts in the United States by Region 2016 (in hundred thousands)

	Hundred-Thousand Units	Percent of Total (rounded to nearest percent)
Northeast	15	
Midwest	25	
South	75	
West	38	Example: $\frac{38}{153} \approx 25\%$
Total	153	

Source: U.S. Census Bureau

39. Iceberg lettuce is grown and shipped to stores for about 40 cents a head, and consumers purchase it for about 86 cents a head. Find the percent of increase.

40. U.S. macadamia nut production in 2016 was 21,000 tons, and in 2017 the production increased to 24,500 tons. Find the percent of increase. (*Source:* USDA)

41. A student at the University of New Orleans makes money by buying and selling used cars. Charles bought a used car and later sold it for a 20% profit. If he sold it for $4680, how much did Charles pay for the car?

42. From 2014 to 2024, the number of people employed as nurse practitioners in the United States is expected to increase by 35%. The number of people employed as nurse practitioners in 2014 was 170,400. Find the predicted number of nurse practitioners in 2024. (*Source:* Bureau of Labor Statistics)

43. By doubling each dimension, the area of a parallelogram increased from 36 square centimeters to 144 square centimeters. Find the percent of increase in area.

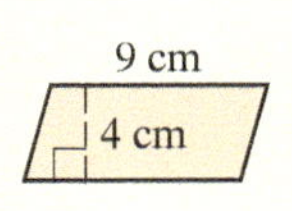

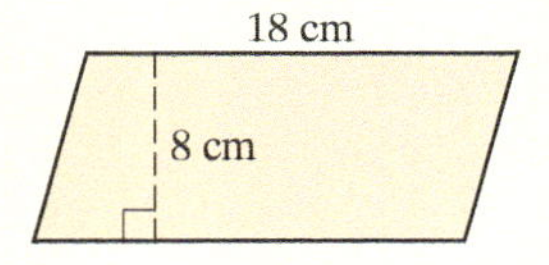

44. By doubling each dimension, the area of a triangle increased from 6 square miles to 24 square miles. Find the percent of increase in area.

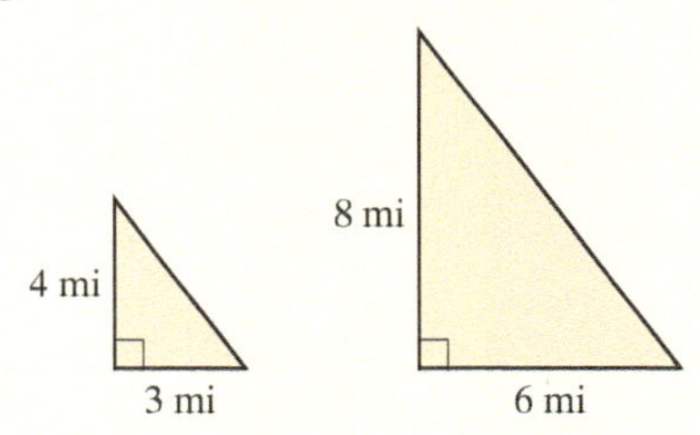

45. A gasoline station recently increased the price of one grade of gasoline by 5%. If this gasoline originally cost $2.20 per gallon, find the mark-up and the new price.

46. The price of a biology book recently increased by 10%. If this book originally cost $89.90, find the mark-up and the new price.

47. How much of an alloy that is 20% copper should be mixed with 200 ounces of an alloy that is 50% copper in order to get an alloy that is 30% copper?

48. How much water should be added to 30 gallons of a solution that is 70% antifreeze in order to get a mixture that is 60% antifreeze?

49. In 2016, there were approximately 71 million virtual reality devices in use worldwide. This is expected to grow to 337 million in 2020. What is the projected percent of increase? Round to the nearest tenth of a percent. *Source:* CTIA—The Wireless Association)

50. In 2010, the average size of a farm in the United States was 426 acres. In 2017, the average size of a farm in the United States had increased to 444 acres. What is this percent of increase? Round to the nearest tenth of a percent. (*Source:* National Agricultural Statistics Service)

51. A company recently downsized its number of employees by 35%. If there are still 78 employees, how many employees were there prior to the layoffs?

52. The average number of children born to each U.S. woman has decreased by 50% since 1960. If this average is now 1.8, find the average in 1960.

53. Nordstrom advertised a 25%-off sale. If a London Fog coat originally sold for $256, find the decrease in price and the sale price.

54. A gasoline station decreased the price of a $0.95 cola by 15%. Find the decrease in price and the new price.

55. Scoville units are used to measure the hotness of a pepper. Measuring 577 thousand Scoville units, the "Red Savina" habañero pepper was known as the hottest chili pepper. That has recently changed with the discovery of Naga Jolokia pepper from India. It measures 48% hotter than the habañero. Find the hotness of the Naga Jolokia pepper. Round to the nearest thousand units.

56. The number of cell phone tower sites in the United States was 253,086 in 2010. By 2017, the number of cell sites had increased by 27.8%. Find the number of cell towers in 2017. Round to the nearest whole number. (*Source:* CTIA—The Wireless Association)

57. In the beginning of 2018, about 53% of all households in the United States were wireless only, which means they had no landline telephone. There were roughly 127 million households in the United States at that time. How many U.S. households were wireless only in 2018? Round to the nearest tenth of a million. (*Source:* National Center for Health Statistics)

58. In 2017, there were 40,393 cinema screens in the United States. If about 38.4% of the total screens in the United States were digital 3-D screens, find the approximate number of digital 3-D screens. Round to the nearest whole number.

59. A new self-tanning lotion for everyday use is to be sold. First, an experimental lotion mixture is made by mixing 800 ounces of everyday moisturizing lotion worth $0.30 an ounce with self-tanning lotion worth $3 per ounce. If the experimental lotion is to cost $1.20 per ounce, how many ounces of the self-tanning lotion should be in the mixture?

60. The owner of a local chocolate shop wants to develop a new trail mix. How many pounds of chocolate-covered peanuts worth $5 a pound should be mixed with 10 pounds of granola bites worth $2 a pound to get a mixture worth $3 per pound?

Review

Place $<, >,$ or $=$ in the appropriate space to make each a true statement. See Sections 1.2, 1.3, and 1.6.

61. $-5 \quad -7$

62. $\frac{12}{3} \quad 2^2$

63. $|-5| \quad -(-5)$

64. $-3^3 \quad (-3)^3$

65. $(-3)^2 \quad -3^2$

66. $|-2| \quad -|-2|$

Concept Extensions

67. Is it possible to mix a 10% acid solution and a 40% acid solution to obtain a 60% acid solution? Why or why not?

68. Must the percents in a circle graph have a sum of 100%? Why or why not?

Standardized nutrition labels like the one below have been displayed on food items since 1994. The percent column on the right shows the percent of daily values (based on a 2000-calorie diet) shown at the bottom of the label. For example, a serving of this food contains 4 grams of total fat, where the recommended daily fat based on a 2000-calorie diet is less than 65 grams of fat. This means that $\frac{4}{65}$ *or approximately 6% (as shown) of your daily recommended fat is taken in by eating a serving of this food. Use this nutrition label to answer Exercises 69 through 71.*

Nutrition Facts

Serving Size	18 Crackers (31g)
Servings Per Container	About 9
Amount Per Serving	
Calories 130	Calories from Fat 35
	% Daily Value*
Total Fat 4g	6%
Saturated Fat 0.5g	3%
Polyunsaturated Fat 0g	
Monounsaturated Fat 1.5g	
Cholesterol 0mg	0%
Sodium 230mg	x
Total Carbohydrate 23g	y
Dietary Fiber 2g	8%
Sugars 3g	
Protein 2g	
Vitamin A 0% •	Vitamin C 0%
Calcium 2% •	Iron 6%

* Percent Daily Values are based on a 2,000 calorie diet. Your daily values may be higher or lower depending on your calorie needs.

	Calories	2,000	2,500
Total Fat	Less than	65g	80g
Sat. Fat	Less than	20g	25g
Cholesterol	Less than	300mg	300mg
Sodium	Less than	2400mg	2400mg
Total Carbohydrate		300g	375g
Dietary Fiber		25g	30g

69. Based on a 2000-calorie diet, what percent of daily value of sodium is contained in a serving of this food? In other words, find x in the label. (Round to the nearest tenth of a percent.)

70. Based on a 2000-calorie diet, what percent of daily value of total carbohydrate is contained in a serving of this food? In other words, find y in the label. (Round to the nearest tenth of a percent.)

71. Notice on the nutrition label that one serving of this food contains 130 calories and 35 of these calories are from fat. Find the percent of calories from fat. (Round to the nearest tenth of a percent.) It is recommended that no more than 30% of calorie intake come from fat. Does this food satisfy this recommendation?

Use the nutrition label below to answer Exercises 72 through 74.

NUTRITIONAL INFORMATION PER SERVING

Serving Size: 9.8 oz		**Servings Per Container: 1**	
Calories	280	Polyunsaturated Fat	1g
Protein	12g	Saturated Fat	3g
Carbohydrate	45g	Cholesterol	20mg
Fat	6g	Sodium	520mg
Percent of Calories from Fat	?	Potassium	220mg

72. If fat contains approximately 9 calories per gram, find the percent of calories from fat in one serving of this food. (Round to the nearest tenth of a percent.)

73. If protein contains approximately 4 calories per gram, find the percent of calories from protein from one serving of this food. (Round to the nearest tenth of a percent.)

74. Find a food that contains more than 30% of its calories per serving from fat. Analyze the nutrition label and verify that the percents shown are correct.

2.7 Linear Inequalities and Problem Solving 

Objectives

A Graph Inequalities on a Number Line.

B Use the Addition Property of Inequality to Solve Inequalities.

C Use the Multiplication Property of Inequality to Solve Inequalities.

D Use Both Properties to Solve Inequalities.

E Solve Problems Modeled by Inequalities.

In Chapter 1, we reviewed these inequality symbols and their meanings:

$<$ means "is less than" $\quad$ $\leq$ means "is less than or equal to"

$>$ means "is greater than" $\quad$ $\geq$ means "is greater than or equal to"

$\neq$ means "is not equal to"

An **inequality** is a statement that contains one of the symbols above.

Equations	Inequalities
$x = 3$	$x \leq 3$
$5n - 6 = 14$	$5n - 6 > 14$
$12 = 7 - 3y$	$12 \leq 7 - 3y$
$\frac{x}{4} - 6 = 1$	$\frac{x}{4} - 6 > 1$
$x = -2$	$x \neq -2$

Objective A Graphing Inequalities on a Number Line 

Recall that the single solution to the equation $x = 3$ is 3. The solutions of the inequality $x \leq 3$ include 3 and *all real numbers less than 3* (for example, -10, $\frac{1}{2}$, 2, and 2.9). Because we can't list all numbers less than 3, we instead show a picture of the solutions by graphing them on a number line.

To graph the solutions of $x \leq 3$, we shade the numbers to the left of 3 since they are less than 3. Then we place a closed circle on the point representing 3. The closed circle indicates that 3 *is* a solution: 3 *is* less than or equal to 3.

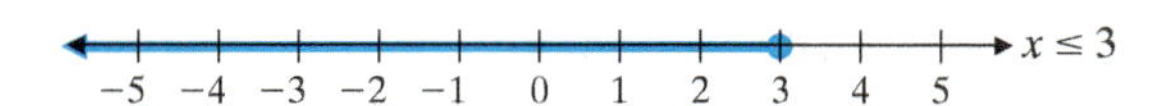

To graph the solutions of $x < 3$, we shade the numbers to the left of 3. Then we place an open circle on the point representing 3. The open circle indicates that 3 *is not* a solution: 3 *is not* less than 3.

Practice 1

Graph: $x \geq -2$

−5 −4 −3 −2 −1 0 1 2 3 4 5

Example 1 Graph: $x \geq -1$

Solution: To graph the solutions of $x \geq -1$, we place a closed circle at -1 since the inequality symbol is $\geq$ and -1 is greater than or equal to -1. Then we shade to the right of -1.

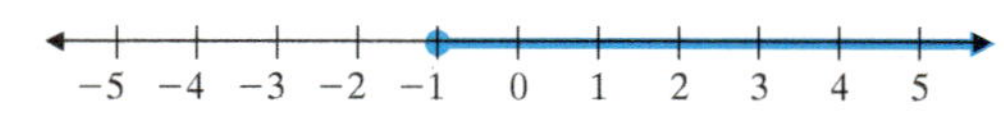

Work Practice 1

Practice 2

Graph: $5 > x$

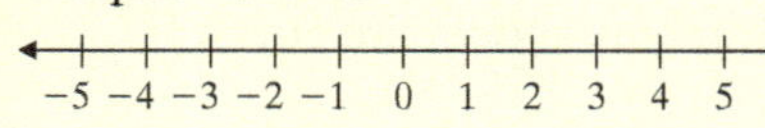

Example 2 Graph: $-1 > x$

Solution: Recall from Section 1.2 that $-1 > x$ means the same as $x < -1$. The graph of the solutions of $x < -1$ is shown below.

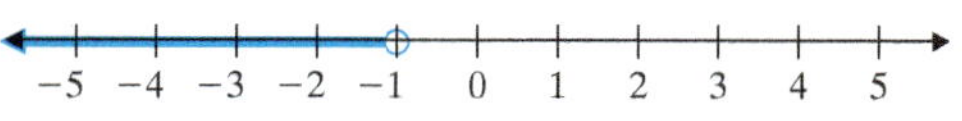

Work Practice 2

Answers

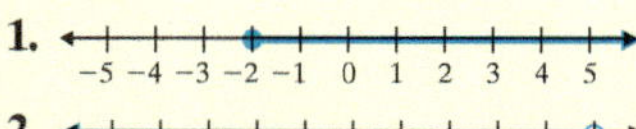

Example 3 Graph: $-4 < x \le 2$

Solution: We read $-4 < x \le 2$ as "-4 is less than x and x is less than or equal to 2," or as "x is greater than -4 and x is less than or equal to 2." To graph the solutions of this inequality, we place an open circle at -4 (-4 is not part of the graph), a closed circle at 2 (2 is part of the graph), and shade all numbers between -4 and 2. Why? All numbers between -4 and 2 are greater than -4 *and* also less than 2.

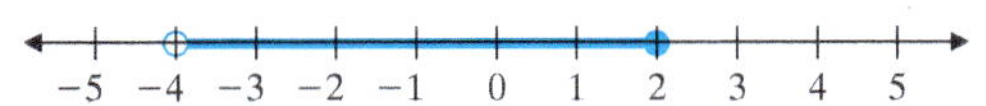

Work Practice 3

Objective B Using the Addition Property

When solutions of a linear inequality are not immediately obvious, they are found through a process similar to the one used to solve a linear equation. Our goal is to get the variable alone on one side of the inequality. We use properties of inequality similar to properties of equality.

Addition Property of Inequality

If a, b, and c are real numbers, then

$a < b$ and $a + c < b + c$

are equivalent inequalities.

This property also holds true for subtracting values, since subtraction is defined in terms of addition. In other words, adding or subtracting the same quantity from both sides of an inequality does not change the solutions of the inequality.

Example 4 Solve $x + 4 \le -6$. Graph the solutions.

Solution: To solve for x, subtract 4 from both sides of the inequality.

$$x + 4 \le -6 \quad \text{Original inequality}$$
$$x + 4 - 4 \le -6 - 4 \quad \text{Subtract 4 from both sides.}$$
$$x \le -10 \quad \text{Simplify.}$$

The graph of the solutions is shown below.

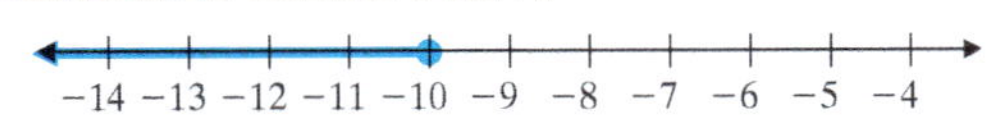

Work Practice 4

Helpful Hint

Notice that any number less than or equal to -10 is a solution to $x \le -10$. For example, solutions include

-10, -200, $-11\frac{1}{2}$, $-\sqrt{130}$, and -50.3

Objective C Using the Multiplication Property

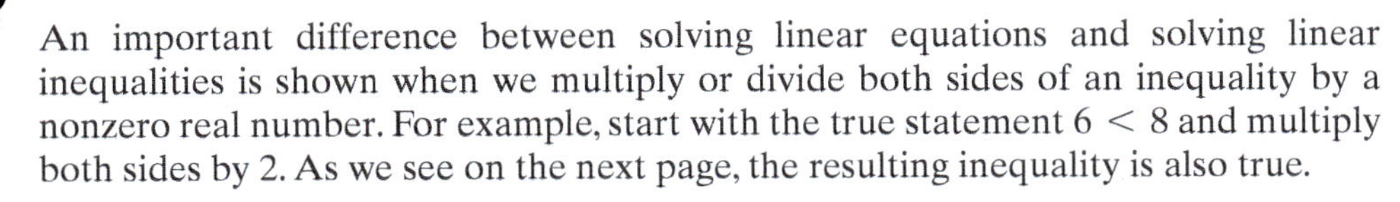

An important difference between solving linear equations and solving linear inequalities is shown when we multiply or divide both sides of an inequality by a nonzero real number. For example, start with the true statement $6 < 8$ and multiply both sides by 2. As we see on the next page, the resulting inequality is also true.

Practice 3

Graph: $-3 \le x < 1$

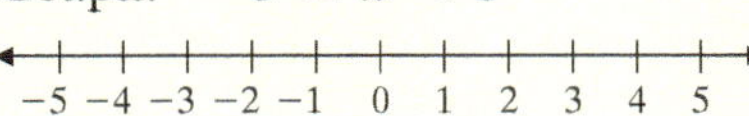

Practice 4

Solve $x - 6 \ge -11$. Graph the solutions.

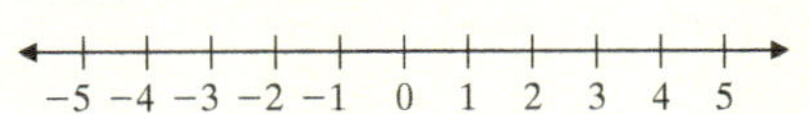

Answers

3.

4. $x \ge -5$

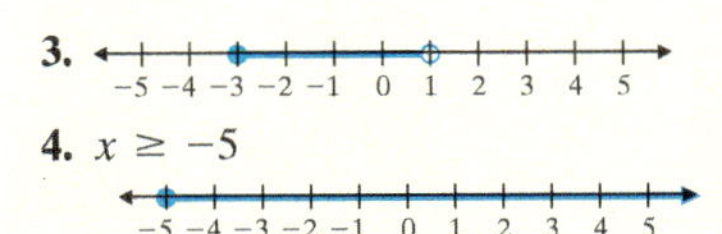

$$6 < 8 \quad \text{True}$$
$$2(6) < 2(8) \quad \text{Multiply both sides by 2.}$$
$$12 < 16 \quad \text{True}$$

But if we start with the same true statement $6 < 8$ and multiply both sides by -2, the resulting inequality is not a true statement.

$$6 < 8 \quad \text{True}$$
$$-2(6) < -2(8) \quad \text{Multiply both sides by } -2.$$
$$-12 < -16 \quad \text{False}$$

Notice, however, that if we reverse the direction of the inequality symbol, the resulting inequality is true.

$$-12 < -16 \quad \text{False}$$
$$-12 > -16 \quad \text{True}$$

This demonstrates the multiplication property of inequality.

Multiplication Property of Inequality

1. If a, b, and c are real numbers, and c is **positive,** then

$$a < b \quad \text{and} \quad ac < bc$$

are equivalent inequalities.

2. If a, b, and c are real numbers, and c is **negative,** then

$$a < b \quad \text{and} \quad ac > bc$$

are equivalent inequalities.

Because division is defined in terms of multiplication, this property also holds true when dividing both sides of an inequality by a nonzero number: If we multiply or divide both sides of an inequality by a negative number, **the direction of the inequality sign must be reversed for the inequalities to remain equivalent.**

✓Concept Check Fill in each box with $<$, $>$, $\leq$, or $\geq$.

a. Since $-8 < -4$, then $3(-8) \square 3(-4)$.

b. Since $5 \geq -2$, then $\frac{5}{-7} \square \frac{-2}{-7}$.

c. If $a < b$, then $2a \square 2b$.

d. If $a \geq b$, then $\frac{a}{-3} \square \frac{b}{-3}$.

Example 5 Solve $-2x \leq -4$. Graph the solutions.

Solution: Remember to reverse the direction of the inequality symbol when dividing by a negative number.

$$-2x \leq -4$$
$$\frac{-2x}{-2} \geq \frac{-4}{-2} \quad \text{Divide both sides by } -2 \text{ and reverse the inequality sign.}$$
$$x \geq 2 \quad \text{Simplify.}$$

The graph of the solutions is shown.

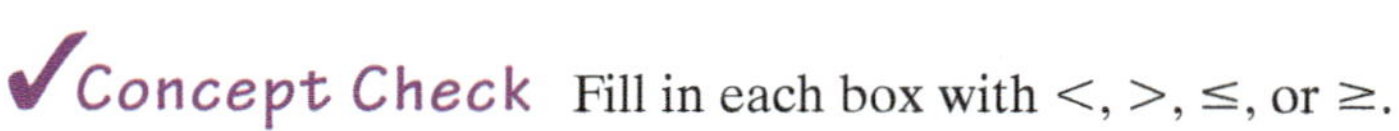

Work Practice 5

Practice 5

Solve $-3x \leq 12$. Graph the solutions.

−5 −4 −3 −2 −1 0 1 2 3 4 5

Answer

5. $x \geq -4$

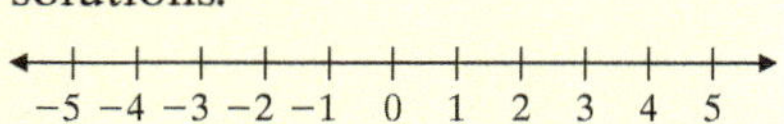

✓Concept Check Answers

a. $<$ **b.** $\leq$ **c.** $<$ **d.** $\leq$

Example 6 Solve $2x < -4$. Graph the solutions.

Solution:

$2x < -4$

$\frac{2x}{2} < \frac{-4}{2}$ Divide both sides by 2. Do not reverse the inequality sign.

$x < -2$ Simplify.

The graph of the solutions is shown.

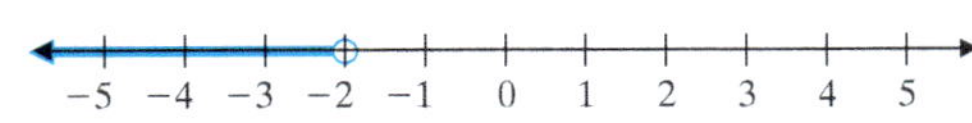

Work Practice 6

Practice 6

Solve $5x > -20$. Graph the solutions.

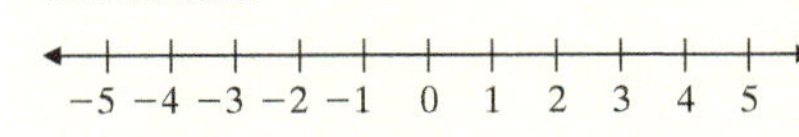

Since we cannot list all solutions to an inequality such as $x < -2$, we will use the set notation $\{x \mid x < -2\}$. Recall from Section 1.2 that this is read "the set of all numbers x such that x is less than -2." We will use this notation when solving inequalities.

Objective D Using Both Properties of Inequality

The following steps may be helpful when solving inequalities in one variable. Notice that these steps are similar to the ones given in Section 2.3 for solving equations.

To Solve Linear Inequalities in One Variable

Step 1: If an inequality contains fractions or decimals, multiply both sides by the LCD to clear the inequality of fractions or decimals.

Step 2: Use the distributive property to remove parentheses if they appear.

Step 3: Simplify each side of the inequality by combining like terms.

Step 4: Get all variable terms on one side and all numbers on the other side by using the addition property of inequality.

Step 5: Get the variable alone by using the multiplication property of inequality.

Helpful Hint

Don't forget that if both sides of an inequality are multiplied or divided by a negative number, the direction of the inequality sign must be reversed.

Example 7 Solve $-4x + 7 \geq -9$. Graph the solution set.

Solution:

$-4x + 7 \geq -9$

$-4x + 7 - 7 \geq -9 - 7$ Subtract 7 from both sides.

$-4x \geq -16$ Simplify.

$\frac{-4x}{-4} \leq \frac{-16}{-4}$ Divide both sides by -4 and reverse the direction of the inequality sign.

$x \leq 4$ Simplify.

The graph of the solution set $\{x \mid x \leq 4\}$ is shown.

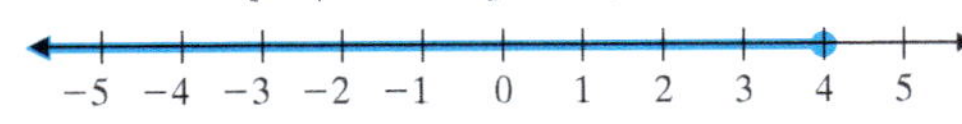

Work Practice 7

Practice 7

Solve $-3x + 11 \leq -13$. Graph the solution set.

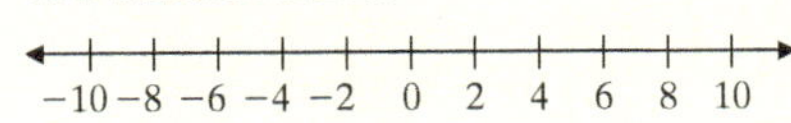

Answers

6. $x > -4$

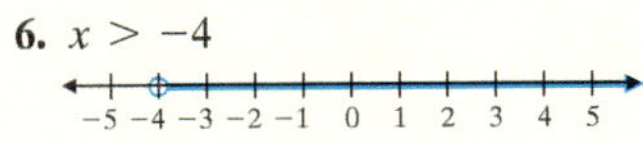

7. $\{x \mid x \geq 8\}$

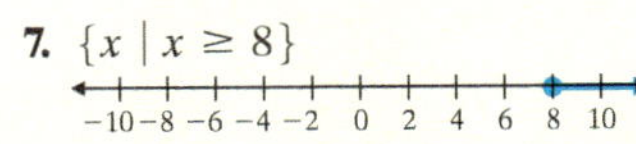

Practice 8

Solve $2x - 3 > 4(x - 1)$. Graph the solution set.

$-5\ -4\ -3\ -2\ -1\ 0\ 1\ 2\ 3\ 4\ 5$

Example 8 Solve $-5x + 7 < 2(x - 3)$. Graph the solution set.

Solution: $-5x + 7 < 2(x - 3)$

$$-5x + 7 < 2x - 6 \quad \text{Apply the distributive property.}$$
$$-5x + 7 - 2x < 2x - 6 - 2x \quad \text{Subtract } 2x \text{ from both sides.}$$
$$-7x + 7 < -6 \quad \text{Combine like terms.}$$
$$-7x + 7 - 7 < -6 - 7 \quad \text{Subtract 7 from both sides.}$$
$$-7x < -13 \quad \text{Combine like terms.}$$
$$\frac{-7x}{-7} > \frac{-13}{-7} \quad \text{Divide both sides by } -7 \text{ and reverse the direction of the inequality sign.}$$
$$x > \frac{13}{7} \quad \text{Simplify.}$$

The graph of the solution set $\left\{x \middle| x > \frac{13}{7}\right\}$ is shown.

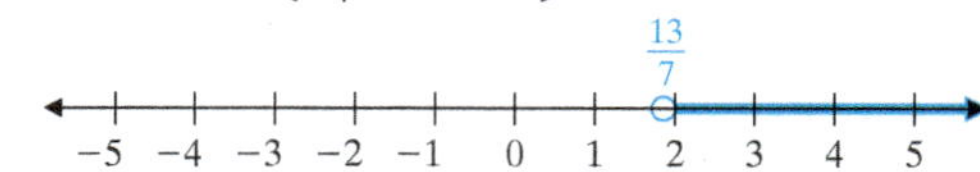

Work Practice 8

Practice 9

Solve:
$3(x + 5) - 1 \geq 5(x - 1) + 7$

Example 9 Solve: $2(x - 3) - 5 \leq 3(x + 2) - 18$

Solution: $2(x - 3) - 5 \leq 3(x + 2) - 18$

$$2x - 6 - 5 \leq 3x + 6 - 18 \quad \text{Apply the distributive property.}$$
$$2x - 11 \leq 3x - 12 \quad \text{Combine like terms.}$$
$$-x - 11 \leq -12 \quad \text{Subtract } 3x \text{ from both sides.}$$
$$-x \leq -1 \quad \text{Add 11 to both sides.}$$
$$\frac{-x}{-1} \geq \frac{-1}{-1} \quad \text{Divide both sides by } -1 \text{ and reverse the direction of the inequality sign.}$$
$$x \geq 1 \quad \text{Simplify.}$$

The solution set is $\{x \mid x \geq 1\}$.

Work Practice 9

Objective E Solving Problems Modeled by Inequalities

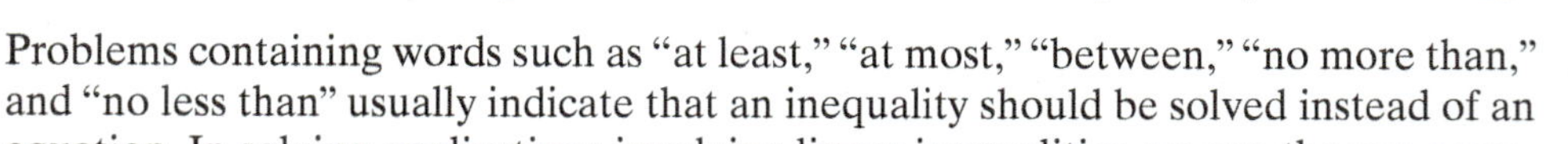

Problems containing words such as "at least," "at most," "between," "no more than," and "no less than" usually indicate that an inequality should be solved instead of an equation. In solving applications involving linear inequalities, we use the same procedure we used to solve applications involving linear equations.

Some Inequality Translations

≥	≤	<	>
at least	at most	is less than	is greater than
no less than	no more than		

Practice 10

Twice a number, subtracted from 35 is greater than 15. Find all numbers that make this true.

Answers

8. $\left\{x \middle| x < \frac{1}{2}\right\}$

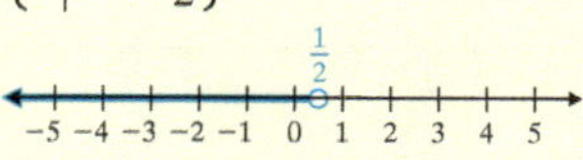

9. $\{x \mid x \leq 6\}$
10. all numbers less than 10

Example 10 12 subtracted from 3 times a number is less than 21. Find all numbers that make this statement true.

Solution:

1. UNDERSTAND. Read and reread the problem. This is a direct translation problem, and let's let

   $x =$ the unknown number

2. TRANSLATE.

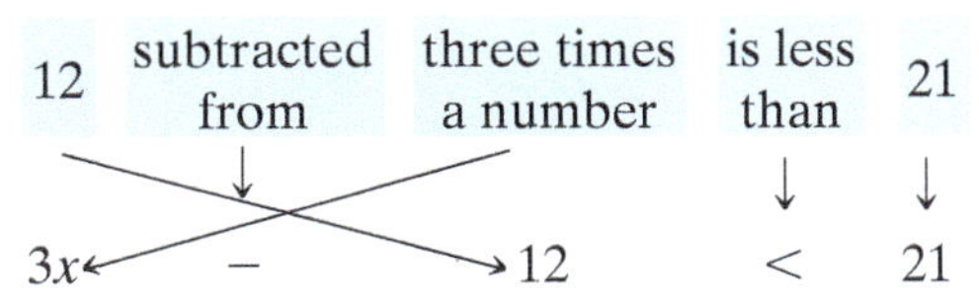

3. SOLVE. $3x - 12 < 21$

$$3x < 33 \quad \text{Add 12 to both sides.}$$

$$\frac{3x}{3} < \frac{33}{3} \quad \text{Divide both sides by 3 and do not reverse the direction of the inequality sign.}$$

$$x < 11 \quad \text{Simplify.}$$

4. INTERPRET.

Check: Check the translation; then let's choose a number less than 11 to see if it checks. For example, let's check 10. 12 subtracted from 3 times 10 is 12 subtracted from 30, or 18. Since 18 is less than 21, the number 10 checks.

State: All numbers less than 11 make the original statement true.

Work Practice 10

Example 11 Budgeting for a Wedding

A couple is having their wedding reception at the Gallery reception hall. They may spend at most \$2000 for the reception. If the reception hall charges a \$100 cleanup fee plus \$36 per person, find the greatest number of people that they can invite and still stay within their budget.

Solution:

1. UNDERSTAND. Read and reread the problem. Suppose that 40 people attend the reception. The cost is then $\$100 + \$36(40) = \$100 + \$1440 = \$1540$. Let $x =$ the number of people who attend the reception.

2. TRANSLATE.

cleanup fee	+	cost per person	times	number of people	must be less than or equal to	\$2000
↓		↓	↓	↓	↓	↓
100	+	36	$\cdot$	x	$\leq$	2000

3. SOLVE.

$$100 + 36x \leq 2000$$

$$36x \leq 1900 \quad \text{Subtract 100 from both sides.}$$

$$x \leq 52\frac{7}{9} \quad \text{Divide both sides by 36.}$$

4. INTERPRET.

Check: Since x represents the number of people, we round down to the nearest whole, or 52. Notice that if 52 people attend, the cost is $\$100 + \$36(52) = \$1972$. If 53 people attend, the cost is $\$100 + \$36(53) = \$2008$, which is more than the given \$2000.

State: The couple can invite at most 52 people to the reception.

Work Practice 11

Practice 11

Alex earns \$600 per month plus 4% of all his sales. Find the minimum sales that will allow Alex to earn at least \$3000 per month.

Answer

11. \$60,000

Vocabulary, Readiness & Video Check

Identify each as an equation, expression, or inequality.

1. $6x - 7(x + 9)$ ________

2. $6x = 7(x + 9)$ ________

3. $6x < 7(x + 9)$ ________

4. $5y - 2 \geq -38$ ________

5. $\frac{9}{7} = \frac{x + 2}{14}$ ________

6. $\frac{9}{7} - \frac{x + 2}{14}$ ________

Decide which number listed is not a solution to each given inequality.

7. $x \geq -3$; $-3, 0, -5, \pi$ ________

8. $x < 6$; $-6, |-6|, 0, -3.2$ ________

9. $x < 4.01$; $4, -4.01, 4.1, -4.1$ ________

10. $x \geq -3$; $-4, -3, -2, -(-2)$ ________

Martin-Gay Interactive Videos

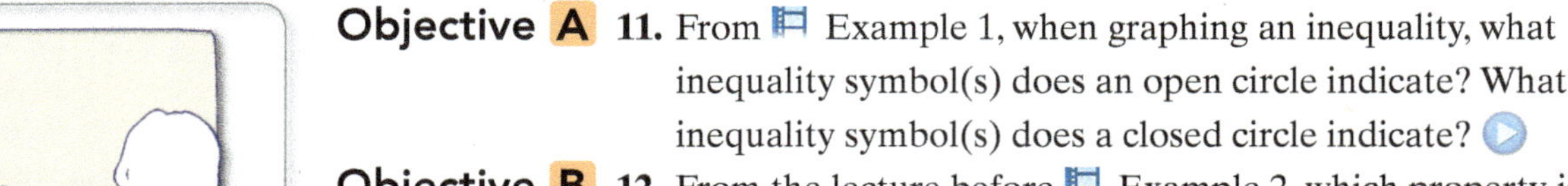

See Video 2.7

Watch the section lecture video and answer the following questions.

Objective A **11.** From Example 1, when graphing an inequality, what inequality symbol(s) does an open circle indicate? What inequality symbol(s) does a closed circle indicate?

Objective B **12.** From the lecture before Example 2, which property is the addition property of inequality very similar to?

Objective C **13.** What is the answer to Example 3, written in solution set notation?

Objective D **14.** When solving Example 4, why is special attention given to the coefficient of x in the last step?

Objective E **15.** What is the phrase in Example 5 that tells us to translate to an inequality? What does this phrase translate to?

2.7 Exercise Set MyLab Math

Objective A *Graph each inequality on the number line. See Examples 1 and 2.*

1. $x \leq -1$

2. $y < 0$

3. $x > \frac{1}{2}$

4. $z \geq -\frac{2}{3}$

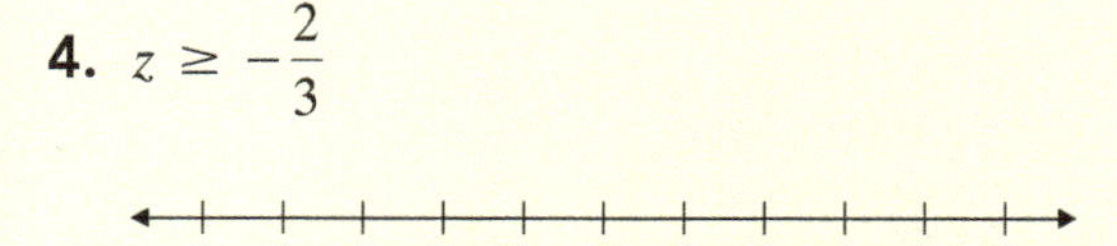

5. $y < 4$

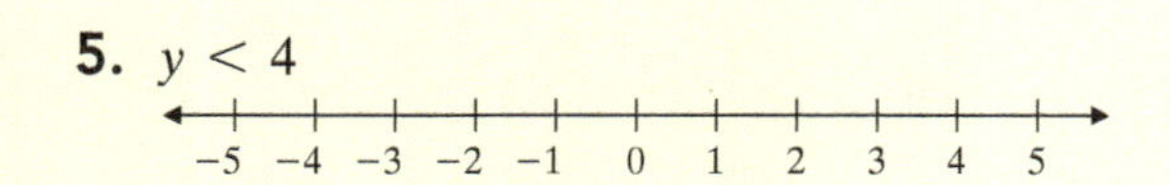

6. $x > 3$

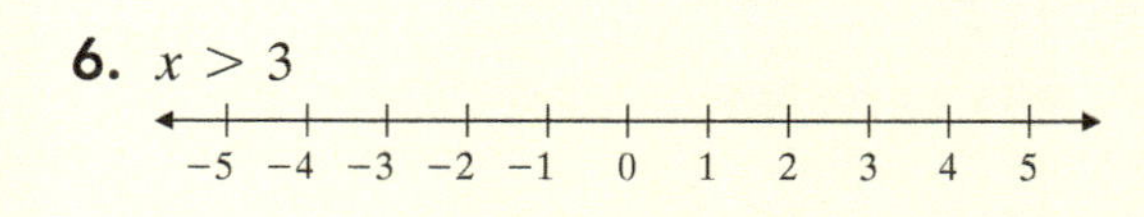

7. $-2 \leq m$

8. $-5 \geq x$

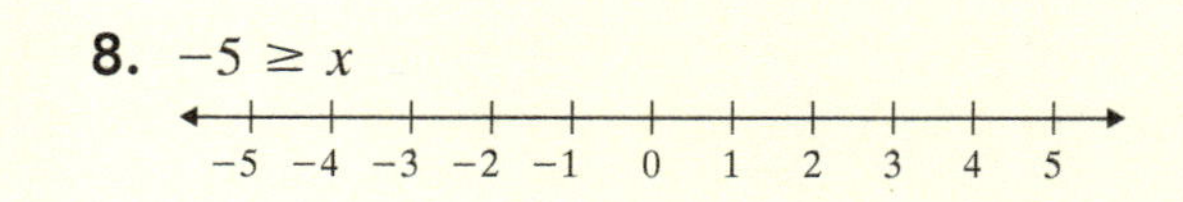

Graph each inequality on the number line. See Example 3.

9. $-1 < x < 3$

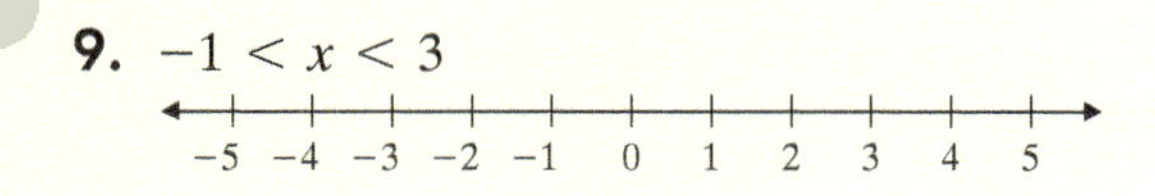

10. $-2 \leq x \leq 3$

11. $0 \leq y < 2$

12. $-4 < x \leq 0$

Objective B *Solve each inequality. Graph the solution set. Write each answer using set notation. See Example 4.*

13. $x - 2 \geq -7$

14. $x + 4 \leq 1$

15. $-9 + y < 0$

16. $-3 + m > 5$

17. $3x - 5 > 2x - 8$

18. $3 - 7x \geq 10 - 8x$

19. $4x - 1 \leq 5x - 2x$

20. $7x + 3 < 9x - 3x$

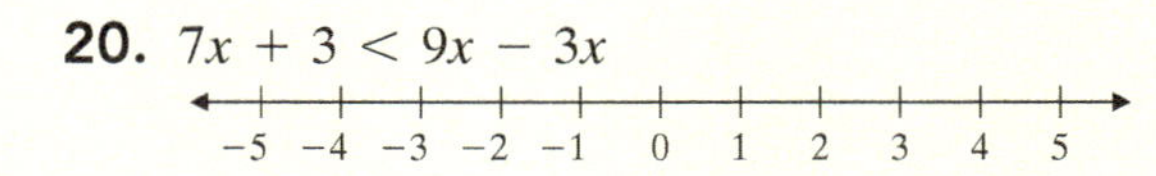

Objective C *Solve each inequality. Graph the solution set. Write each answer using set notation. See Examples 5 and 6.*

21. $2x < -6$

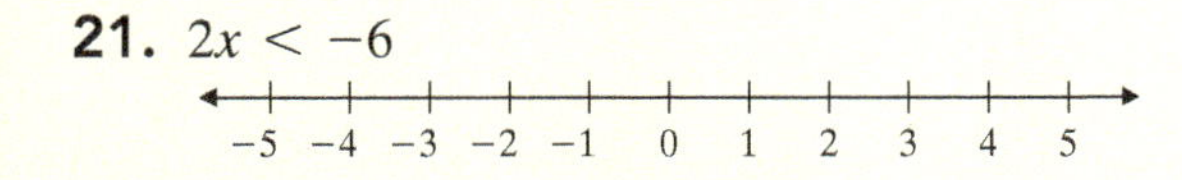

22. $3x > -9$

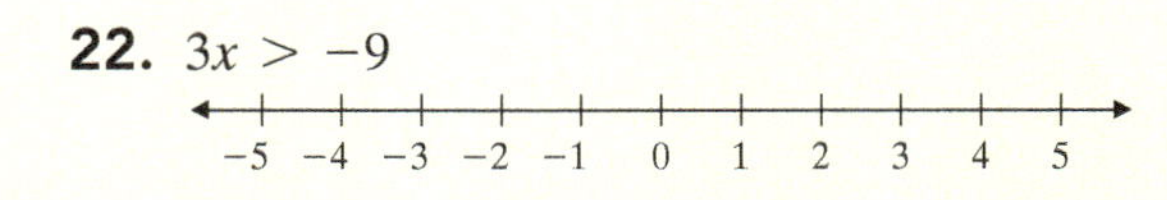

23. $-8x \leq 16$

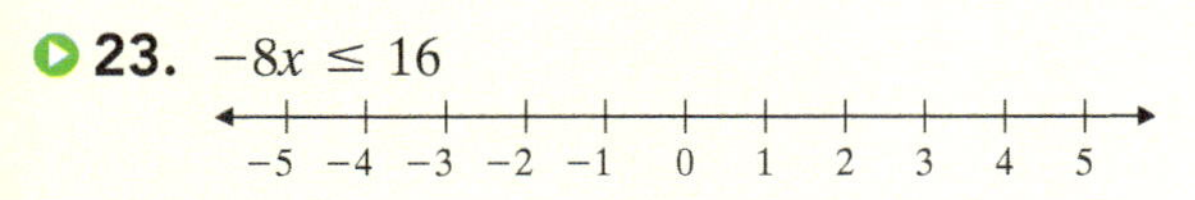

24. $-5x < 20$

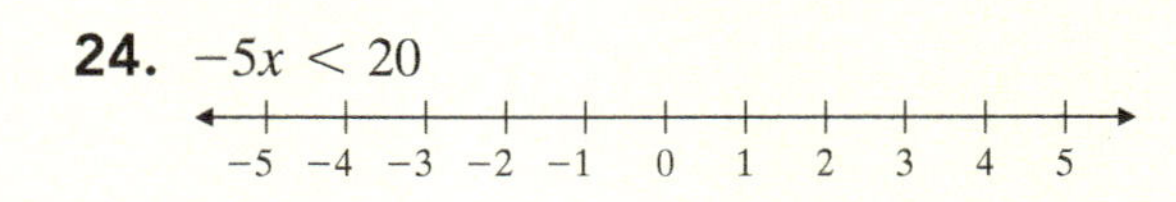

25. $-x > 0$

26. $-y \geq 0$

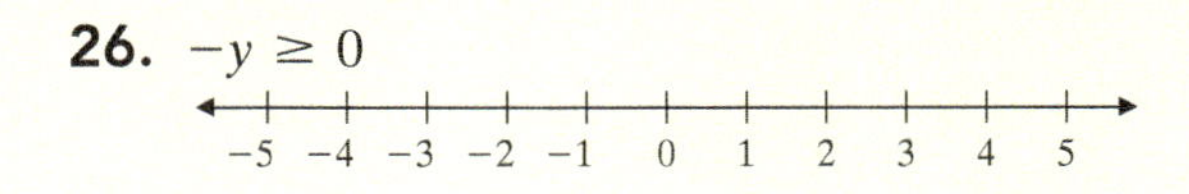

27. $\frac{3}{4}y \geq -2$

28. $\frac{5}{6}x \leq -8$

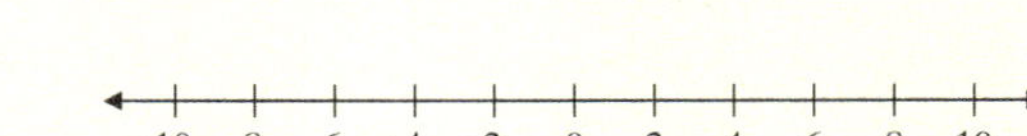

29. $-0.6y < -1.8$

30. $-0.3x > -2.4$

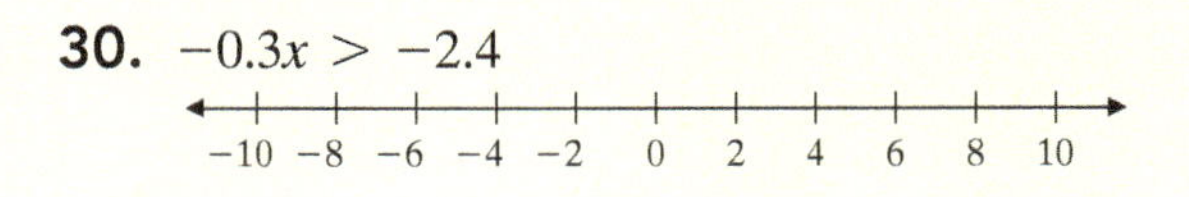

Objectives B C D Mixed Practice *Solve each inequality. Write each answer using set notation. See Examples 4 through 9.*

31. $-8 < x + 7$

32. $-11 > x + 4$

33. $7(x + 1) - 6x \geq -4$

34. $10(x + 2) - 9x \leq -1$

35. $4x > 1$

36. $6x < 5$

37. $-\frac{2}{3}y \leq 8$

38. $-\frac{3}{4}y \geq 9$

39. $4(2z + 1) < 4$

40. $6(2 - z) \geq 12$

41. $3x - 7 < 6x + 2$

42. $2x - 1 \geq 4x - 5$

43. $5x - 7x \leq x + 2$

44. $4 - x < 8x + 2x$

45. $-6x + 2 \geq 2(5 - x)$

46. $-7x + 4 > 3(4 - x)$

47. $3(x - 5) < 2(2x - 1)$

48. $5(x - 2) \leq 3(2x - 1)$

49. $4(3x - 1) \leq 5(2x - 4)$

50. $3(5x - 4) \leq 4(3x - 2)$

51. $3(x + 2) - 6 > -2(x - 3) + 14$

52. $7(x - 2) + x \leq -4(5 - x) - 12$

53. $-5(1 - x) + x \leq -(6 - 2x) + 6$

54. $-2(x - 4) - 3x < -(4x + 1) + 2x$

55. $\frac{1}{4}(x + 4) < \frac{1}{5}(2x + 3)$

56. $\frac{1}{2}(x - 5) < \frac{1}{3}(2x - 1)$

57. $-5x + 4 \leq -4(x - 1)$

58. $-6x + 2 < -3(x + 4)$

Objective E *Solve the following. For Exercises 61 and 62, the solutions have been started for you. See Examples 10 and 11.*

59. Six more than twice a number is greater than negative fourteen. Find all numbers that make this statement true.

60. One more than five times a number is less than or equal to ten. Find all such numbers.

△ **61.** The perimeter of a rectangle is to be no greater than 100 centimeters and the width must be 15 centimeters. Find the maximum length of the rectangle.

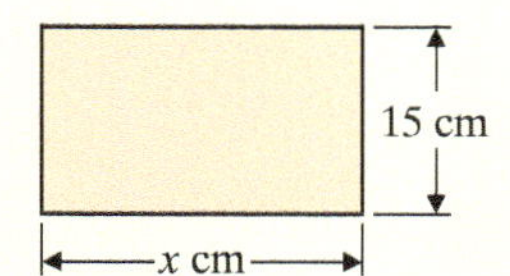

Start the solution:

1. UNDERSTAND the problem. Reread it as many times as needed.

2. TRANSLATE into an equation. (Fill in the blank below.)

the perimeter of the rectangle	is no greater than	100
↓	↓	↓
$x + 15 + x + 15$	______	100

Finish with:

3. SOLVE and **4.** INTERPRET

△ **62.** One side of a triangle is four times as long as another side, and the third side is 12 inches long. If the perimeter can be no longer than 87 inches, find the maximum lengths of the other two sides.

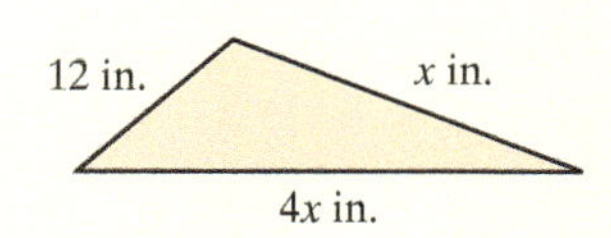

Start the solution:

1. UNDERSTAND the problem. Reread it as many times as needed.

2. TRANSLATE into an equation. (Fill in the blank below.)

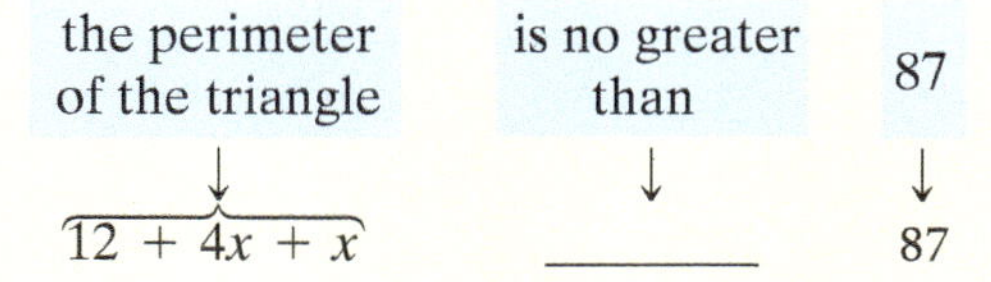

Finish with:

3. SOLVE and **4.** INTERPRET

63. Ben Holladay bowled 146 and 201 in his first two games. What must he bowl in his third game to have an average of at least 180? (*Hint:* The average of a list of numbers is their sum divided by the number of numbers in the list.)

64. On an NBA team the two forwards measure 6′8″ and 6′6″ tall and the two guards measure 6′0″ and 5′9″ tall. How tall should the center be if they wish to have a starting team average height of at least 6′5″?

65. A couple are celebrating their 30th wedding anniversary by having a reception at Tiffany Oaks reception hall. They have budgeted \$3000 for their reception. If the reception hall charges a \$50.00 cleanup fee plus \$34 per person, find the greatest number of people that they may invite and still stay within their budget.

66. A surprise retirement party is being planned for Pratap Puri. A total of \$860 has been collected for the event, which is to be held at a local reception hall. This reception hall charges a cleanup fee of \$40 and \$15 per person for drinks and light snacks. Find the greatest number of people that may be invited and still stay within the \$860 budget.

67. A 150-pound person uses 5.8 calories per minute when walking at a speed of 4 mph. How long must a person walk at this speed to use at least 300 calories? Round up to the nearest minute. (*Source: Home & Garden Bulletin* No. 72)

68. A 170-pound person uses 5.3 calories per minute when bicycling at a speed of 5.5 mph. How long must a person ride a bike at this speed in order to use at least 200 calories? Round up to the nearest minute. (*Source:* Same as Exercise **67**)

Review

Evaluate each expression. See Section 1.3.

69. 3^4 **70.** 4^3 **71.** 1^8 **72.** 0^7 **73.** $\left(\frac{7}{8}\right)^2$ **74.** $\left(\frac{2}{3}\right)^3$

The graph shows the number of Starbucks locations worldwide from 2010 to 2017. The height of the graph for each year shown corresponds to the number of Starbucks locations worldwide. Use this graph to answer Exercises 75 through 80. See Section R.4.

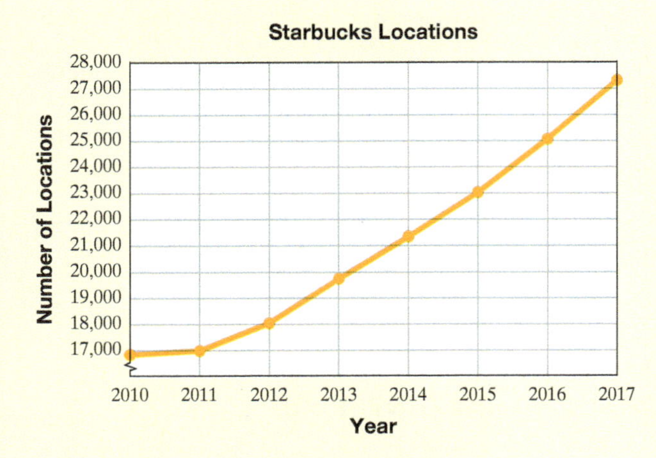

75. Approximate the number of Starbucks locations in 2013.

76. Approximate the number of Starbucks locations in 2016.

77. Between which two years did the greatest increase in the number of Starbucks locations occur?

78. Between which two years did the number of Starbucks locations appear to remain about the same?

79. During which year did the number of Starbucks locations rise above 18,000?

80. During which year did the number of Starbucks locations rise above 25,000?

Concept Extensions

Fill in the box with $<$, $>$, $\leq$, or $\geq$. See the Concept Check in this section.

81. Since $3 < 5$, then $3(-4) \square 5(-4)$.

82. If $m \leq n$, then $2m \square 2n$.

83. If $m \leq n$, then $-2m \square -2n$.

84. If $-x < y$, then $x \square -y$.

85. When solving an inequality, when must you reverse the direction of the inequality symbol?

86. If both sides of the inequality $-3x < -30$ are divided by 3, do you reverse the direction of the inequality symbol? Why or why not?

Solve.

87. A history major has scores of 75, 83, and 85 on his history tests. Use an inequality to find the scores he can make on his final exam to receive a B in the class. The final exam counts as **two** tests, and a B is received if the final course average is greater than or equal to 80.

88. A mathematics major has scores of 85, 95, and 92 on her college geometry tests. Use an inequality to find the scores she can make on her final exam to receive an A in the course. The final exam counts as **three** tests, and an A is received if the final course average is greater than or equal to 90. Round to one decimal place.

89. Explain how solving a linear inequality is similar to solving a linear equation.

90. Explain how solving a linear inequality is different from solving a linear equation.

Chapter 2 Group Activity

Investigating Averages

Sections 2.1–2.6

Materials:

- small rubber ball or crumpled paper ball
- bucket or waste can

This activity may be completed by working in groups or individually.

1. Try shooting the ball into the bucket or waste can 5 times. Record your results below.

   Shots Made **Shots Missed**

2. Find your shooting percent for the 5 shots (that is, the percent of the shots you actually made out of the number you tried).
3. Suppose you are going to try an additional 5 shots. How many of the next 5 shots will you have to make to have a 50% shooting percent for all 10 shots? An 80% shooting percent?
4. Did you solve an equation in Question **3**? If so, explain what you did. If not, explain how you could use an equation to find the answers.
5. Now suppose you are going to try an additional 22 shots. How many of the next 22 shots will you have to make to have at least a 50% shooting percent for all 27 shots? At least a 70% shooting percent?
6. Choose one of the sports played at your college that is currently in season. How many regular-season games are scheduled? What is the team's current percent of games won?
7. Suppose the team has a goal of finishing the season with a winning percent better than 110% of their current wins. At least how many of the remaining games must they win to achieve their goal?

Chapter 2 Vocabulary Check

Fill in each blank with one of the words or phrases listed below.

no solution	all real numbers	linear equation in one variable
equivalent equations	formula	reversed
inequality	the same	

1. A(n) ______________ can be written in the form $Ax + B = C$.
2. Equations that have the same solution are called ______________.
3. An equation that describes a known relationship among quantities is called a(n) ______________.
4. $3x - 6 < 5$ is an example of a(n) ______________.
5. The solution(s) to the equation $x + 5 = x + 5$ is/are ______________.
6. The solution(s) to the equation $x + 5 = x + 4$ is/are ______________.
7. If both sides of an inequality are multiplied or divided by the same positive number, the direction of the inequality symbol is ______________.
8. If both sides of an inequality are multiplied or divided by the same negative number, the direction of the inequality symbol is ______________.

Helpful Hint Are you preparing for your test?
To help, don't forget to take these:
- Chapter 2 Getting Ready for the Test on page 182
- Chapter 2 Test on page 183

Then check all of your answers at the back of this text. For further review, the step-by-step video solutions to any of these exercises are located in MyLab Math.

2 Chapter Highlights

Definitions and Concepts	Examples
Section 2.1 The Addition Property of Equality	
A **linear equation in one variable** can be written in the form $Ax + B = C$ where A, B, and C are real numbers and $A \neq 0$.	$-3x + 7 = 2$ $3(x - 1) = -8(x + 5) + 4$
Equivalent equations are equations that have the same solution.	$x - 7 = 10$ and $x = 17$ are equivalent equations.
Addition Property of Equality Adding the same number to or subtracting the same number from both sides of an equation does not change its solution.	$y + 9 = 3$ $y + 9 - 9 = 3 - 9$ $y = -6$
Section 2.2 The Multiplication Property of Equality	
Multiplication Property of Equality Multiplying both sides or dividing both sides of an equation by the same nonzero number does not change its solution.	$\frac{2}{3}a = 18$ $\frac{3}{2}\left(\frac{2}{3}a\right) = \frac{3}{2}(18)$ $a = 27$
Section 2.3 Further Solving Linear Equations	
To Solve Linear Equations	*Solve:* $\frac{5(-2x + 9)}{6} + 3 = \frac{1}{2}$
1. Clear the equation of fractions.	**1.** $6 \cdot \frac{5(-2x + 9)}{6} + 6 \cdot 3 = 6 \cdot \frac{1}{2}$
2. Remove any grouping symbols such as parentheses.	**2.** $5(-2x + 9) + 18 = 3$ $-10x + 45 + 18 = 3$ Apply the distributive property.
3. Simplify each side by combining like terms.	**3.** $-10x + 63 = 3$ Combine like terms.
4. Get all variable terms on one side and all numbers on the other side by using the addition property of equality.	**4.** $-10x + 63 - 63 = 3 - 63$ Subtract 63. $-10x = -60$
5. Get the variable alone by using the multiplication property of equality.	**5.** $\frac{-10x}{-10} = \frac{-60}{-10}$ Divide by -10. $x = 6$
6. Check the solution by substituting it into the original equation.	

Definitions and Concepts	Examples
Section 2.4 An Introduction to Problem Solving	
Problem-Solving Steps	The height of the Hudson volcano in Chile is twice the height of the Kiska volcano in the Aleutian Islands. If the sum of their heights is 12,870 feet, find the height of each.
1. UNDERSTAND the problem.	**1.** Read and reread the problem. Guess a solution and check your guess. Let x be the height of the Kiska volcano. Then $2x$ is the height of the Hudson volcano. 
2. TRANSLATE the problem.	**2.** height of Kiska → x; added to → $+$; height of Hudson → $2x$; is → $=$; 12,870 → 12,870
3. SOLVE the equation.	**3.** $x + 2x = 12{,}870$ $3x = 12{,}870$ $x = 4290$
4. INTERPRET the results.	**4. Check:** If x is 4290, then $2x$ is 2(4290) or 8580. Their sum is 4290 + 8580 or 12,870, the required amount. **State:** The Kiska volcano is 4290 feet tall, and the Hudson volcano is 8580 feet tall.
Section 2.5 Formulas and Problem Solving	
An equation that describes a known relationship among quantities is called a **formula.**	$A = lw$ (area of a rectangle) $I = PRT$ (simple interest)
To solve a formula for a specified variable, use the same steps as for solving a linear equation. Treat the specified variable as the only variable of the equation.	*Solve:* $P = 2l + 2w$ for l. $P = 2l + 2w$ $P - 2w = 2l + 2w - 2w$ Subtract $2w$. $P - 2w = 2l$ $\frac{P - 2w}{2} = \frac{2l}{2}$ Divide by 2. $\frac{P - 2w}{2} = l$
Section 2.6 Percent and Mixture Problem Solving	
Let's use our problem-solving steps to solve a problem about percents.	32% of what number is 36.8?
1. UNDERSTAND.	**1.** Read and reread. Propose a solution and check. Let $x =$ the unknown number.
2. TRANSLATE.	**2.** 32% → 32%; of → $\cdot$; what number → x; is → $=$; 36.8 → 36.8

(continued)

Definitions and Concepts	Examples

Section 2.6 Percent and Mixture Problem Solving (*continued*)

3. SOLVE.

3. *Solve:*

$$32\% \cdot x = 36.8$$
$$0.32x = 36.8$$
$$\frac{0.32x}{0.32} = \frac{36.8}{0.32} \quad \text{Divide by 0.32.}$$
$$x = 115 \quad \text{Simplify.}$$

4. INTERPRET.

4. *Check, then state:* 32% of 115 is 36.8.

Let's use our problem-solving steps to solve an application about mixtures.

How many liters of a 20% acid solution must be mixed with a 50% acid solution in order to obtain 12 liters of a 30% solution?

1. UNDERSTAND.

1. Read and reread. Guess a solution and check.
Let x = number of liters of 20% solution.
Then $12 - x$ = number of liters of 50% solution.

2. TRANSLATE.

2.

	No. of Liters ·	Acid Strength =	Amount of Acid
20% Solution	x	20%	$0.20x$
50% Solution	$12 - x$	50%	$0.50(12 - x)$
30% Solution Needed	12	30%	$0.30(12)$

In words: acid in 20% solution + acid in 50% solution = acid in 30% solution

Translate: $0.20x + 0.50(12 - x) = 0.30(12)$

3. SOLVE.

3. *Solve:*

$$0.20x + 0.50(12 - x) = 0.30(12)$$
$$0.20x + 6 - 0.50x = 3.6 \quad \text{Apply the distributive property.}$$
$$-0.30x + 6 = 3.6 \quad \text{Combine like terms.}$$
$$-0.30x = -2.4 \quad \text{Subtract 6.}$$
$$x = 8 \quad \text{Divide by } -0.30.$$

4. INTERPRET.

4. *Check, then state:*

If 8 liters of a 20% acid solution are mixed with 12 − 8 or 4 liters of a 50% acid solution, the result is 12 liters of a 30% solution.

Definitions and Concepts	Examples

Section 2.7 Linear Inequalities and Problem Solving

Properties of inequalities are similar to properties of equations. However, if you multiply or divide both sides of an inequality by the same *negative* number, you must reverse the direction of the inequality symbol.

$-2x \le 4$

$\frac{-2x}{-2} \ge \frac{4}{-2}$ Divide by -2; reverse the inequality symbol.

$x \ge -2$

To Solve Linear Inequalities

1. Clear the inequality of fractions.
2. Remove grouping symbols.
3. Simplify each side by combining like terms.
4. Write all variable terms on one side and all numbers on the other side using the addition property of inequality.
5. Get the variable alone by using the multiplication property of inequality.

Solve: $3(x + 2) \le -2 + 8$

1. $3(x + 2) \le -2 + 8$ No fractions to clear.
2. $3x + 6 \le -2 + 8$ Apply the distributive property.
3. $3x + 6 \le 6$ Combine like terms.
4. $3x + 6 - 6 \le 6 - 6$ Subtract 6.

   $3x \le 0$
5. $\frac{3x}{3} \le \frac{0}{3}$ Divide by 3.

   $x \le 0$

The solution set is $\{x \mid x \le 0\}$.

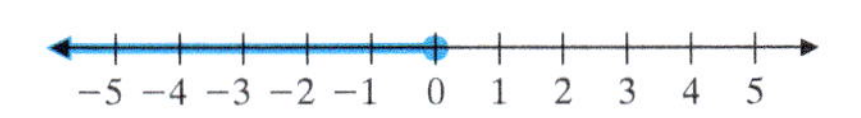

Chapter 2 Review

(2.1) *Solve each equation.*

1. $8x + 4 = 9x$
2. $5y - 3 = 6y$
3. $\frac{2}{7}x + \frac{5}{7}x = 6$
4. $3x - 5 = 4x + 1$
5. $2x - 6 = x - 6$
6. $4(x + 3) = 3(1 + x)$
7. $6(3 + n) = 5(n - 1)$
8. $5(2 + x) - 3(3x + 2) = -5(x - 6) + 2$

Choose the correct algebraic expression.

9. The sum of two numbers is 10. If one number is x, express the other number in terms of x.
   a. $x - 10$
   b. $10 - x$
   c. $10 + x$
   d. $10x$

10. Mandy is 5 inches taller than Melissa. If x inches represents the height of Mandy, express Melissa's height in terms of x.
    a. $x - 5$
    b. $5 - x$
    c. $5 + x$
    d. $5x$

△ **11.** If one angle measures $x°$, express the measure of its complement in terms of x.
a. $(180 - x)°$
b. $(90 - x)°$
c. $(x - 180)°$
d. $(x - 90)°$

△ **12.** If one angle measures $(x + 5)°$, express the measure of its supplement in terms of x.
a. $(185 + x)°$
b. $(95 + x)°$
c. $(175 - x)°$
d. $(x - 170)°$

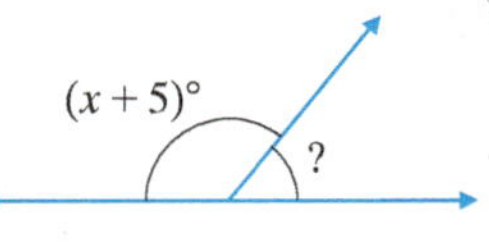

(2.2) *Solve each equation.*

13. $\frac{3}{4}x = -9$

14. $\frac{x}{6} = \frac{2}{3}$

15. $-5x = 0$

16. $-y = 7$

17. $0.2x = 0.15$

18. $\frac{-x}{3} = 1$

19. $-3x + 1 = 19$

20. $5x + 25 = 20$

21. $7(x - 1) + 9 = 5x$

22. $7x - 6 = 5x - 3$

23. $-5x + \frac{3}{7} = \frac{10}{7}$

24. $5x + x = 9 + 4x - 1 + 6$

25. Write the sum of three consecutive integers as an expression in x. Let x be the first integer.

26. Write the sum of the first and fourth of four consecutive even integers. Let x be the first even integer.

(2.3) *Solve each equation.*

27. $\frac{5}{3}x + 4 = \frac{2}{3}x$

28. $\frac{7}{8}x + 1 = \frac{5}{8}x$

29. $-(5x + 1) = -7x + 3$

30. $-4(2x + 1) = -5x + 5$

31. $-6(2x - 5) = -3(9 + 4x)$

32. $3(8y - 1) = 6(5 + 4y)$

33. $\frac{3(2 - z)}{5} = z$

34. $\frac{4(n + 2)}{5} = -n$

35. $0.5(2n - 3) - 0.1 = 0.4(6 + 2n)$

36. $-9 - 5a = 3(6a - 1)$

37. $\frac{5(c + 1)}{6} = 2c - 3$

38. $\frac{2(8 - a)}{3} = 4 - 4a$

39. $200(70x - 3560) = -179(150x - 19{,}300)$

40. $1.72y - 0.04y = 0.42$

(2.4) *Solve each of the following.*

41. The height of the Washington Monument is 50.5 inches more than 10 times the length of a side of its square base. If the sum of these two dimensions is 7327 inches, find the height of the Washington Monument. (*Source:* National Park Service)

42. A 12-foot board is to be divided into two pieces so that one piece is twice as long as the other. If x represents the length of the shorter piece, find the length of each piece.

43. The national park system in the United States includes a variety of park unit types. In 2018, there were a total of 41 national battlefields and national memorials. The number of national memorials was three less than three times the number of national battlefields. How many of each park unit were there? (*Source:* National Park System)

44. Find three consecutive integers whose sum is -114.

45. The quotient of a number and 3 is the same as the difference of the number and two. Find the number.

46. Double the sum of a number and 6 is the opposite of the number. Find the number.

(2.5) *Substitute the given values into the given formulas and solve for the unknown variable.*

47. $P = 2l + 2w;\quad P = 46, l = 14$

48. $V = lwh;\quad V = 192, l = 8, w = 6$

Solve each equation for the indicated variable or constant.

49. $y = mx + b$ for m

50. $r = vst - 5$ for s

51. $2y - 5x = 7$ for x

52. $3x - 6y = -2$ for y

△ **53.** $C = \pi d$ for π

△ **54.** $C = 2\pi r$ for π

△ **55.** A swimming pool holds 900 cubic meters of water. If its length is 20 meters and its height is 3 meters, find its width.

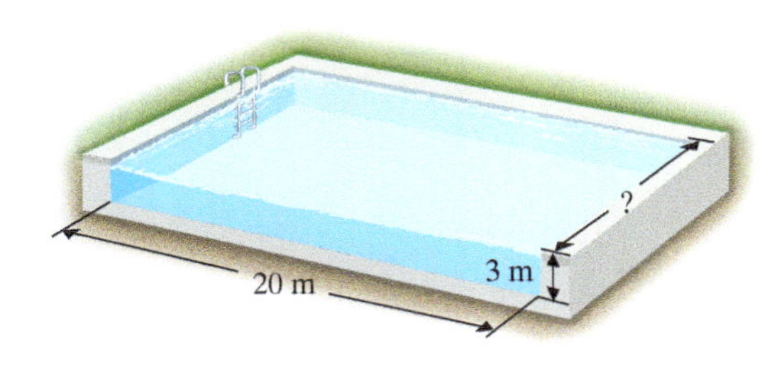

56. The perimeter of a rectangular billboard is 60 feet and the billboard has a length 6 feet longer than its width. Find the dimensions of the billboard.

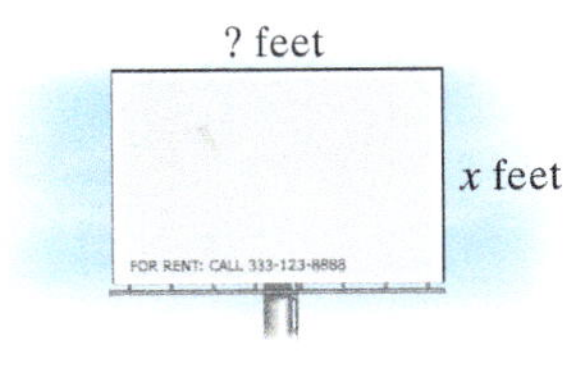

57. A charity 10K race is given annually to benefit a local hospice organization. How long will it take to run/walk a 10K race (10 kilometers or 10,000 meters) if your average pace is 125 **meters** per minute? Give your time in hours and minutes.

58. On July 10, 1913, the highest temperature ever recorded in the United States was 134°F, which occurred in Death Valley, California. Convert this temperature to degrees Celsius. (*Source:* National Weather Service)

(2.6) *Find each of the following.*

59. The number 9 is what percent of 45?

60. The number 59.5 is what percent of 85?

61. The number 137.5 is 125% of what number?

62. The number 768 is 60% of what number?

63. The price of a small diamond ring was recently increased by 11%. If the ring originally cost $1900, find the mark-up and the new price of the ring.

64. The U.S. motion picture and television industry is made up of over 108,000 businesses. About 85% of these are small businesses with fewer than 10 employees. How many motion picture and television industry businesses have fewer than 10 employees? (*Source:* Motion Picture Association of America)

65. Thirty gallons of a 20% acid solution are needed for an experiment. Only 40% and 10% acid solutions are available. How much of each should be mixed to form the needed solution?

66. In 2010, the average price of a cinema ticket was $7.89. By 2017, this price had increased to $8.97. What was the percent of increase? Round to the nearest tenth of a percent. (*Source:* MPAA)

The graph below shows the percent(s) of cell phone users who have engaged in various behaviors while driving and talking on their cell phones. Use this graph to answer Exercises 67 through 70.

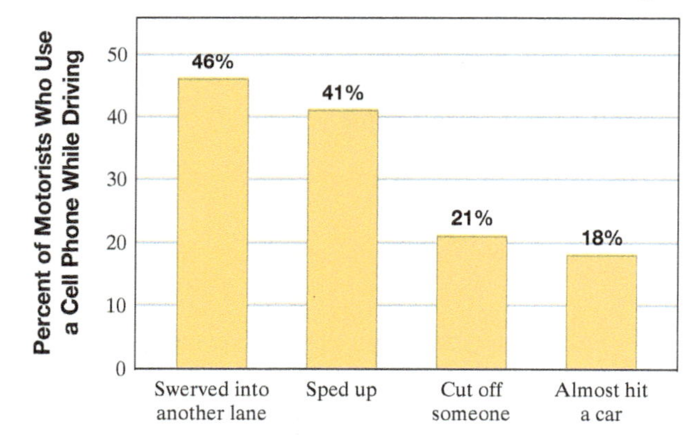

Source: Progressive Insurance

67. What percent of motorists who use a cell phone while driving have almost hit another car?

68. What is the most common effect of cell phone use on driving?

69. If a cell phone service has an estimated 4600 customers who use their cell phones while driving, how many of these customers would you expect to have cut someone off while driving and talking on their cell phones?

70. Do the percents in the graph to the left have a sum of 100%? Why or why not?

(2.7) *Graph on a number line.*

71. $x \leq -2$

−5 −4 −3 −2 −1 0 1 2 3 4 5

72. $0 < x \leq 5$

−5 −4 −3 −2 −1 0 1 2 3 4 5

Solve each inequality.

73. $x - 5 \leq -4$

74. $x + 7 > 2$

75. $-2x \geq -20$

76. $-3x > 12$

77. $5x - 7 > 8x + 5$

78. $x + 4 \geq 6x - 16$

79. $\frac{2}{3}y > 6$

80. $-0.5y \leq 7.5$

81. $-2(x - 5) > 2(3x - 2)$

82. $4(2x - 5) \leq 5x - 1$

83. Carol Abolafia earns \$175 per week plus a 5% commission on all her sales. Find the minimum amount of sales she must make to ensure that she earns at least \$300 per week.

84. Joseph Barrow shot rounds of 76, 82, and 79 golfing. What must he shoot on his next round so that his average will be below 80?

Mixed Review

Solve each equation.

85. $6x + 2x - 1 = 5x + 11$

86. $2(3y - 4) = 6 + 7y$

87. $4(3 - a) - (6a + 9) = -12a$

88. $\frac{x}{3} - 2 = 5$

89. $2(y + 5) = 2y + 10$

90. $7x - 3x + 2 = 2(2x - 1)$

Solve.

91. The sum of six and twice a number is equal to seven less than the number. Find the number.

92. A 23-inch piece of string is to be cut into two pieces so that the length of the longer piece is three more than four times the shorter piece. Find the lengths of both pieces.

93. Solve $V = \frac{1}{3}Ah$ for h.

94. What number is 26% of 85?

95. The number 72 is 45% of what number?

96. A company recently increased its number of employees from 235 to 282. Find the percent of increase.

Solve each inequality. Graph the solution set.

97. $4x - 7 > 3x + 2$

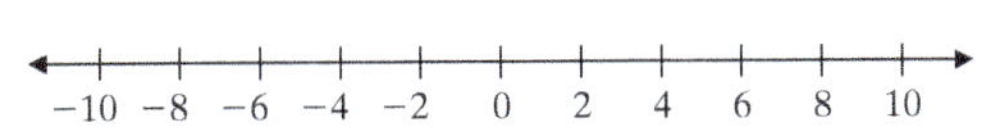

98. $-5x < 20$

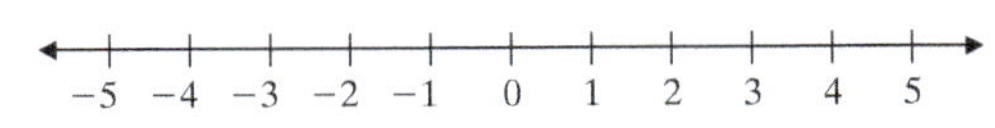

99. $-3(1 + 2x) + x \geq -(3 - x)$

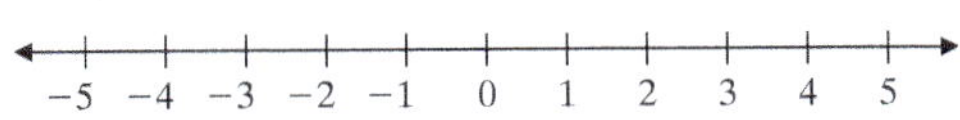

Chapter 2 Getting Ready for the Test LC

MULTIPLE CHOICE *Exercises 1 through 4 below are given. Choose the best directions (choice **A**, **B**, **C**, or **D**) below for each exercise.*

A. Solve for x.

B. Simplify.

C. Identify the numerical coefficient.

D. Are these like or unlike terms?

1. Given: $-3x^2$
2. Given: $4x - 5 = 2x + 3$
3. Given: $5x^2$ and $4x$
4. Given: $4x - 5 + 2x + 3$

MULTIPLE CHOICE

5. Subtracting $100z$ from $8m$ translates to

A. $100z - 8m$ **B.** $8m - 100z$ **C.** $-800zm$ **D.** $92zm$

6. Subtracting $7x - 1$ from $9y$ translates to:

A. $7x - 1 - 9y$ **B.** $9y - 7x - 1$ **C.** $9y - (7x - 1)$ **D.** $7x - 1 - (9y)$

MATCHING *Match each equation in Exercises 7 through 10 with its solution in choices **A.**, **B.**, or **C** below. Choices **A.**, **B.**, or **C** may be used more than once.*

A. all real numbers **B.** no solution **C.** the solution is 0

7. $7x + 6 = 7x + 9$
8. $2y - 5 = 2y - 5$
9. $11x - 13 = 10x - 13$
10. $x + 15 = -x + 15$

MULTIPLE CHOICE

11. To solve $5(3x - 2) = -(x + 20)$, we first use the distributive property and remove parentheses by multiplying. Once this is done, the equation is

A. $15x - 2 = -x + 20$ **B.** $15x - 10 = -x - 20$ **C.** $15x - 10 = -x + 20$ **D.** $15x - 7 = -x - 20$

12. To solve $\frac{8x}{3} + 1 = \frac{x - 2}{10}$ we multiply through by the LCD, 30. Once this is done, the simplified equation is

A. $80x + 1 = 3x - 6$ **B.** $80x + 6 = 3x - 6$ **C.** $8x + 1 = x - 2$ **D.** $80x + 30 = 3x - 6$

Solve each equation.

1. $-\frac{4}{5}x = 4$

2. $4(n - 5) = -(4 - 2n)$

3. $5y - 7 + y = -(y + 3y)$

4. $4z + 1 - z = 1 + z$

5. $\frac{2(x + 6)}{3} = x - 5$

6. $\frac{4(y - 1)}{5} = 2y + 3$

7. $\frac{1}{2} - x + \frac{3}{2} = x - 4$

8. $\frac{1}{3}(y + 3) = 4y$

9. $-0.3(x - 4) + x = 0.5(3 - x)$

10. $-4(a + 1) - 3a = -7(2a - 3)$

11. $-2(x - 3) = x + 5 - 3x$

Solve each application.

12. A number increased by two-thirds of the number is 35. Find the number.

13. A gallon of water seal covers 200 square feet. How many gallons are needed to paint two coats of water seal on a deck that measures 20 feet by 35 feet?

14. Find the value of x if $y = -14$, $m = -2$, and $b = -2$ in the formula $y = mx + b$.

Solve each equation for the indicated variable.

15. $V = \pi r^2 h$ for h

16. $3x - 4y = 10$ for y

Answers

1. ____________

2. ____________

3. ____________

4. ____________

5. ____________

6. ____________

7. ____________

8. ____________

9. ____________

10. ____________

11. ____________

12. ____________

13. ____________

14. ____________

15. ____________

16. ____________

17. ______

18. ______

19. ______

20. ______

21. ______

22. ______

23. ______

24. ______

25. ______

Solve each inequality. Graph the solution set.

17. $3x - 5 \geq 7x + 3$

−5 −4 −3 −2 −1 0 1 2 3 4 5

18. $x + 6 > 4x - 6$

−5 −4 −3 −2 −1 0 1 2 3 4 5

Solve each inequality.

19. $-0.3x \geq 2.4$

20. $-5(x - 1) + 6 \leq -3(x + 4) + 1$

21. $\dfrac{2(5x + 1)}{3} > 2$

The following graph shows the breakdown of tornadoes occurring in the United States by strength. The corresponding Fujita Tornado Scale categories are shown in parentheses. Use this graph to answer Exercise 22.

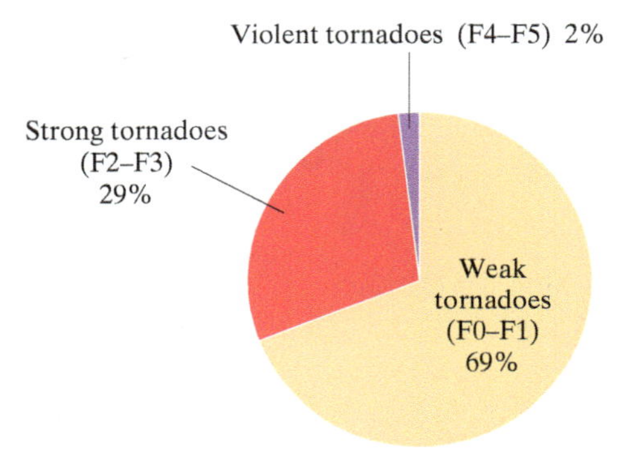

Source: National Climatic Data Center

22. According to the National Climatic Data Center, in an average year, about 800 tornadoes are reported in the United States. How many of these would you expect to be classified as "weak" tornadoes?

23. The number 72 is what percent of 180?

24. Some states have a single area code for the entire state. Two such states have area codes where one is double the other. If the sum of these integers is 1203, find the two area codes.

25. California has more public libraries than any other state. It has 387 more public libraries than Ohio. If the total number of public libraries for these states is 1827, find the number of public libraries in California and the number in Ohio. (*Source:* Institute of Museum and Library Services)

Cumulative Review — Chapters 1–2

Determine whether each statement is true or false.

1. $8 \geq 8$

2. $-4 < -6$

3. $8 \leq 8$

4. $3 > -3$

5. $23 \leq 0$

6. $-8 \geq -8$

7. $0 \leq 23$

8. $-8 \leq -8$

9. Insert $<$, $>$, or $=$ in the appropriate space to make each statement true.

a. $|0| \quad 2$

b. $|-5| \quad 5$

c. $|-3| \quad |-2|$

d. $|-9| \quad |-9.7|$

e. $\left|-7\frac{1}{6}\right| \quad |7|$

10. Find the absolute value of each number.

a. $|5|$

b. $|-8|$

c. $\left|-\frac{2}{3}\right|$

Simplify.

11. $\dfrac{3 + |4 - 3| + 2^2}{6 - 3}$

12. $1 + 2(9 - 7)^3 + 4^2$

Add without using number lines.

13. $(-8) + (-11)$

14. $-2 + (-8)$

15. $(-2) + 10$

16. $-10 + 20$

17. $0.2 + (-0.5)$

18. $1.2 + (-1.2)$

Answers

1. ____
2. ____
3. ____
4. ____
5. ____
6. ____
7. ____
8. ____
9. a. ____
   b. ____
   c. ____
   d. ____
   e. ____
10. a. ____
   b. ____
   c. ____
11. ____
12. ____
13. ____
14. ____
15. ____
16. ____
17. ____
18. ____

19. a. ______
b. ______
20. a. ______
b. ______
c. ______
d. ______
21. a. ______
b. ______
c. ______
22. a. ______
b. ______
c. ______
23. a. ______
b. ______
c. ______
24. a. ______
b. ______
25. ______
26. ______
27. ______
28. ______
29. a. ______
b. ______
c. ______
d. ______
e. ______

19. Simplify each expression.

a. $-3 + [(-2 - 5) - 2]$

b. $2^3 - 10 + [-6 - (-5)]$

20. Simplify each expression.

a. $-(-5)$

b. $-\left(-\frac{2}{3}\right)$

c. $-(-a)$

d. $-|-3|$

21. Multiply.

a. $7(0)(-6)$

b. $(-2)(-3)(-4)$

c. $(-1)(-5)(-9)(-2)$

22. Subtract.

a. $-2.7 - 8.4$

b. $-\frac{4}{5} - \left(-\frac{3}{5}\right)$

c. $\frac{1}{4} - \left(-\frac{1}{2}\right)$

23. Find each quotient.

a. $-18 \div 3$

b. $\frac{-14}{-2}$

c. $\frac{20}{-4}$

24. Find each product.

a. $(4.5)(-0.08)$

b. $-\frac{3}{4} \cdot -\frac{8}{17}$

Use the distributive property to write each expression without parentheses. Then simplify the result.

25. $-5(-3 + 2z)$

26. $2(y - 3x + 4)$

27. $\frac{1}{2}(6x + 14) + 10$

28. $-(x + 4) + 3(x + 4)$

29. Determine whether the terms are like or unlike.

a. $2x, 3x^2$

b. $4x^2y, x^2y, -2x^2y$

c. $-2yz, -3zy$

d. $-x^4, x^4$

e. $-8a^5, 8a^5$

30. Divide.

a. $\frac{-32}{8}$ b. $\frac{-108}{-12}$

c. $-\frac{5}{7} \div \left(-\frac{9}{2}\right)$

31. Subtract $4x - 2$ from $2x - 3$.

32. Subtract $10x + 3$ from $-5x + 1$.

33. Solve: $x - 7 = 10$

Solve.

34. $\frac{5}{6} + x = \frac{2}{3}$

35. $-z - 4 = 6$

36. $-3x + 1 - (-4x - 6) = 10$

37. $\frac{2(a + 3)}{3} = 6a + 2$

38. $\frac{x}{4} = 18$

39. As of January 2018, the total number of Democrats and Republicans in the U.S. House of Representatives was 435. There were 47 more Republican representatives than Democratic. Find the number of representatives from each party. (*Source:* Congressional Research Service)

40. $6x + 5 = 4(x + 4) - 1$

41. A glacier is a giant mass of rocks and ice that flows downhill like a river. Portage Glacier in Alaska is about 6 miles, or 31,680 feet, long and moves 400 feet per year. Icebergs are created when the front end of the glacier flows into Portage Lake. How long does it take for ice at the head (beginning) of the glacier to reach the lake?

42. A number increased by 4 is the same as 3 times the number decreased by 8. Find the number.

43. The number 63 is what percent of 72?

44. Solve: $C = 2\pi r$ for r.

45. Solve: $5(2x + 3) = -1 + 7$

46. Solve: $x - 3 > 2$

47. Graph $-1 > x$.

−5 −4 −3 −2 −1 0 1 2 3 4 5

48. Solve: $3x - 4 \leq 2x - 14$

49. Solve: $2(x - 3) - 5 \leq 3(x + 2) - 18$

50. Solve: $-3x \geq 9$

30. a. ______

b. ______

c. ______

31. ______

32. ______

33. ______

34. ______

35. ______

36. ______

37. ______

38. ______

39. ______

40. ______

41. ______

42. ______

43. ______

44. ______

45. ______

46. ______

47. ______

48. ______

49. ______

50. ______

3 Exponents and Polynomials

Recall from Chapter 1 that an exponent is a shorthand notation for repeated factors. This chapter explores additional concepts about exponents and exponential expressions. An especially useful type of exponential expression is a polynomial. Polynomials model many real-world phenomena. Our goal in this chapter is to become proficient with operations on polynomials.

Sections

3.1 Exponents

3.2 Negative Exponents and Scientific Notation

3.3 Introduction to Polynomials

3.4 Adding and Subtracting Polynomials

3.5 Multiplying Polynomials

3.6 Special Products

Integrated Review—Exponents and Operations on Polynomials

3.7 Dividing Polynomials

Check Your Progress

Vocabulary Check

Chapter Highlights

Chapter Review

Getting Ready for the Test

Chapter Test

Cumulative Review

How Do You Listen to Music? Downloading? Streaming? A Physical CD or LP/Vinyl?

No matter how you listen to music, this industry is booming. In the United States, the music industry has grown to an estimated $15 billion. The number of paid subscribers to subscription streaming is increasing, and the number of digital music downloads is expected to decline. Interestingly enough, LP/vinyl albums are making a comeback. The bar graph below shows the increase in LP/vinyl album sales in the United States, but also study the circle graph to its right. Notice that although LP/vinyl album sales are increasing, they still represent a small part of the "total album sales pie." These sales are mostly through digital and CDs.

In Section 3.3, Exercises 25 and 26, we use the data below to predict future sales of LP/vinyl albums. (*Source:* IFPI.org)

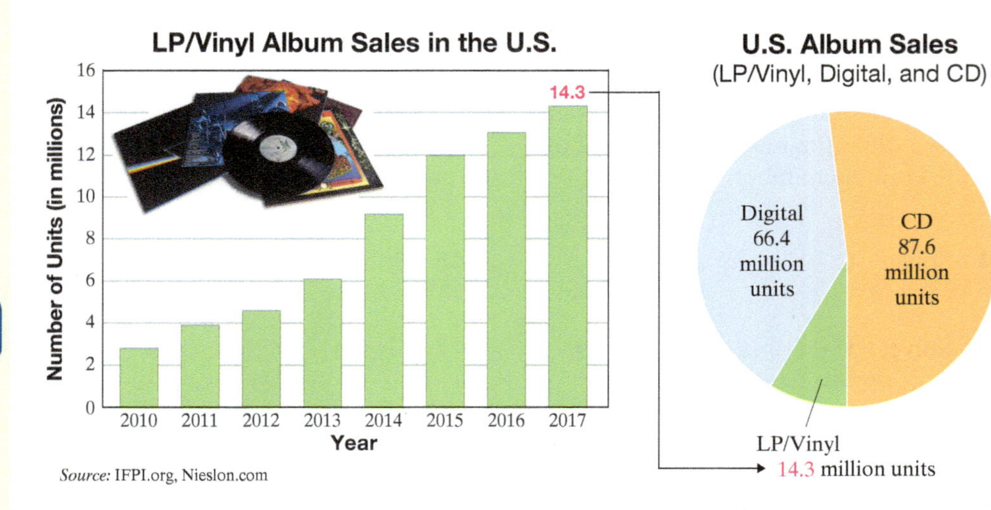

Source: IFPI.org, Nieslon.com

3.1 Exponents

Objective A Evaluating Exponential Expressions

Objectives

A Evaluate Exponential Expressions.

B Use the Product Rule for Exponents.

C Use the Power Rule for Exponents.

D Use the Power Rules for Products and Quotients.

E Use the Quotient Rule for Exponents, and Define a Number Raised to the 0 Power.

F Decide Which Rule(s) to Use to Simplify an Expression.

In this section, we continue our work with integer exponents. Recall from Section 1.3 that repeated multiplication of the same factor can be written using exponents. For example,

$$2 \cdot 2 \cdot 2 \cdot 2 \cdot 2 = 2^5$$

The exponent 5 tells us how many times 2 is a factor. The expression 2^5 is called an **exponential expression.** It is also called the fifth **power** of 2, or we can say that 2 is **raised** to the fifth power.

$$5^6 = \underbrace{5 \cdot 5 \cdot 5 \cdot 5 \cdot 5 \cdot 5}_{\text{6 factors; each factor is 5}} \quad \text{and} \quad (-3)^4 = \underbrace{(-3) \cdot (-3) \cdot (-3) \cdot (-3)}_{\text{4 factors; each factor is } -3}$$

The base of an exponential expression is the repeated factor. The exponent is the number of times that the base is used as a factor.

$$a^n = \underbrace{a \cdot a \cdot a \cdots a}_{n \text{ factors; each factor is } a}$$

exponent or power (points to n); base (points to a)

Examples Evaluate each expression.

Practice 1–6

Evaluate each expression.

1. 3^4 **2.** 7^1 **3.** $(-2)^3$

4. -2^3 **5.** $\left(\frac{2}{3}\right)^2$ **6.** $5 \cdot 6^2$

1. $2^3 = 2 \cdot 2 \cdot 2 = 8$

2. $3^1 = 3$. To raise 3 to the first power means to use 3 as a factor only once. When no exponent is shown, the exponent is assumed to be 1.

3. $(-4)^2 = (-4)(-4) = 16$

4. $-4^2 = -(4 \cdot 4) = -16$

5. $\left(\frac{1}{2}\right)^4 = \frac{1}{2} \cdot \frac{1}{2} \cdot \frac{1}{2} \cdot \frac{1}{2} = \frac{1}{16}$

6. $4 \cdot 3^2 = 4 \cdot 9 = 36$

Work Practice 1–6

Notice how similar -4^2 is to $(-4)^2$ in the preceding examples. The difference between the two is the parentheses. In $(-4)^2$, the parentheses tell us that the base, or the repeated factor, is -4. In -4^2, only 4 is the base.

Helpful Hint

Be careful when identifying the base of an exponential expression. Pay close attention to the use of parentheses.

$(-3)^2$	-3^2	$2 \cdot 3^2$
The base is -3.	The base is 3.	The base is 3.
$(-3)^2 = (-3)(-3) = 9$	$-3^2 = -(3 \cdot 3) = -9$	$2 \cdot 3^2 = 2 \cdot 3 \cdot 3 = 18$

An exponent has the same meaning whether the base is a number or a variable. If x is a real number and n is a positive integer, then x^n is the product of n factors, each of which is x.

$$x^n = \underbrace{x \cdot x \cdot x \cdot x \cdot x \cdots x}_{n \text{ factors; each factor is } x}$$

Answers

1. 81 **2.** 7 **3.** -8 **4.** -8 **5.** $\frac{4}{9}$ **6.** 180

Practice 7

Evaluate each expression for the given value of x.

a. $3x^2$ when x is 4

b. $\frac{x^4}{-8}$ when x is -2

Example 7 Evaluate each expression for the given value of x.

a. $2x^3$ when x is 5 **b.** $\frac{9}{x^2}$ when x is -3

Solution:

a. When x is 5, $2x^3 = 2 \cdot 5^3$

$$= 2 \cdot (5 \cdot 5 \cdot 5)$$
$$= 2 \cdot 125$$
$$= 250$$

b. When x is -3, $\frac{9}{x^2} = \frac{9}{(-3)^2}$

$$= \frac{9}{(-3)(-3)}$$
$$= \frac{9}{9} = 1$$

Work Practice 7

Objective B Using the Product Rule

Exponential expressions can be multiplied, divided, added, subtracted, and themselves raised to powers. Let's see if we can discover a shortcut method for multiplying exponential expressions with the same base. By our definition of an exponent,

$$5^4 \cdot 5^3 = \underbrace{(5 \cdot 5 \cdot 5 \cdot 5)}_{4 \text{ factors of } 5} \cdot \underbrace{(5 \cdot 5 \cdot 5)}_{3 \text{ factors of } 5}$$
$$= \underbrace{5 \cdot 5 \cdot 5 \cdot 5 \cdot 5 \cdot 5 \cdot 5}_{7 \text{ factors of } 5}$$
$$= 5^7$$

Also,

$$x^2 \cdot x^3 = (x \cdot x) \cdot (x \cdot x \cdot x)$$
$$= x \cdot x \cdot x \cdot x \cdot x$$
$$= x^5$$

In both cases, notice that the result is exactly the same if the exponents are added.

$$5^4 \cdot 5^3 = 5^{4+3} = 5^7 \quad \text{and} \quad x^2 \cdot x^3 = x^{2+3} = x^5$$

This suggests the following rule.

Product Rule for Exponents

If m and n are positive integers and a is a real number, then

$$a^m \cdot a^n = a^{m+n}$$ ← Add exponents. (Keep common base.)

For example,

$$3^5 \cdot 3^7 = 3^{5+7} = 3^{12}$$ ← Add exponents. (Keep common base.)

Answers
7. a. 48 b. -2

Helpful Hint

Don't forget that

$$3^5 \cdot 3^7 \neq 9^{12}$$ ← Add exponents.
↑ Common base *not* kept.

$$3^5 \cdot 3^7 = \underbrace{3 \cdot 3 \cdot 3 \cdot 3 \cdot 3}_{5 \text{ factors of } 3} \cdot \underbrace{3 \cdot 3 \cdot 3 \cdot 3 \cdot 3 \cdot 3 \cdot 3}_{7 \text{ factors of } 3}$$

$$= 3^{12}$$ 12 factors of 3, *not* 9

In other words, to multiply two exponential expressions with the **same base,** we keep the base and add the exponents. We call this **simplifying** the exponential expression.

Examples Use the product rule to simplify each expression.

8. $4^2 \cdot 4^5 = 4^{2+5} = 4^7$ ← Add exponents.
↑ Keep common base.

9. $x^2 \cdot x^5 = x^{2+5} = x^7$

10. $y^3 \cdot y = y^3 \cdot y^1$
$= y^{3+1}$
$= y^4$

Helpful Hint Don't forget that if no exponent is written, it is assumed to be 1.

11. $y^3 \cdot y^2 \cdot y^7 = y^{3+2+7} = y^{12}$

12. $(-5)^7 \cdot (-5)^8 = (-5)^{7+8} = (-5)^{15}$

Work Practice 8–12

Practice 8–12

Use the product rule to simplify each expression.

8. $7^3 \cdot 7^2$ **9.** $x^4 \cdot x^9$

10. $r^5 \cdot r$ **11.** $s^6 \cdot s^2 \cdot s^3$

12. $(-3)^9 \cdot (-3)$

✓**Concept Check** Where possible, use the product rule to simplify the expression.

a. $z^2 \cdot z^{14}$ **b.** $x^2 \cdot z^{14}$ **c.** $9^8 \cdot 9^3$ **d.** $9^8 \cdot 2^7$

Example 13 Use the product rule to simplify $(2x^2)(-3x^5)$.

Solution: Recall that $2x^2$ means $2 \cdot x^2$ and $-3x^5$ means $-3 \cdot x^5$.

$(2x^2)(-3x^5) = (2 \cdot x^2) \cdot (-3 \cdot x^5)$

$= (2 \cdot -3) \cdot (x^2 \cdot x^5)$ Group factors with common bases (using commutative and associative properties).

$= -6x^7$ Simplify.

Work Practice 13

Practice 13

Use the product rule to simplify $(6x^3)(-2x^9)$.

Examples Simplify.

14. $(x^2y)(x^3y^2) = (x^2 \cdot x^3) \cdot (y^1 \cdot y^2)$ Group like bases and write y as y^1.
$= x^5 \cdot y^3$ or x^5y^3 Multiply.

15. $(-a^7b^4)(3ab^9) = (-1 \cdot 3) \cdot (a^7 \cdot a^1) \cdot (b^4 \cdot b^9)$
$= -3a^8b^{13}$

Work Practice 14–15

Practice 14–15

Simplify.

14. $(m^5n^{10})(mn^8)$

15. $(-x^9y)(4x^2y^{11})$

Answers

8. 7^5 **9.** x^{13} **10.** r^6 **11.** s^{11} **12.** $(-3)^{10}$ **13.** $-12x^{12}$ **14.** m^6n^{18} **15.** $-4x^{11}y^{12}$

✓**Concept Check Answers**

a. z^{16} **b.** cannot be simplified
c. 9^{11} **d.** cannot be simplified

Helpful Hint

These examples will remind you of the difference between adding and multiplying terms.

Addition

$5x^3 + 3x^3 = (5 + 3)x^3 = 8x^3$ — By the distributive property

$7x + 4x^2 = 7x + 4x^2$ — Cannot be combined

Multiplication

$(5x^3)(3x^3) = 5 \cdot 3 \cdot x^3 \cdot x^3 = 15x^{3+3} = 15x^6$ — By the product rule

$(7x)(4x^2) = 7 \cdot 4 \cdot x \cdot x^2 = 28x^{1+2} = 28x^3$ — By the product rule

Objective C Using the Power Rule

Exponential expressions can themselves be raised to powers. Let's try to discover a rule that simplifies an expression like $(x^2)^3$. By the definition of a^n,

$$(x^2)^3 = (x^2)(x^2)(x^2) \quad (x^2)^3 \text{ means 3 factors of } (x^2).$$

which can be simplified by the product rule for exponents.

$$(x^2)^3 = (x^2)(x^2)(x^2) = x^{2+2+2} = x^6$$

Notice that the result is exactly the same if we multiply the exponents.

$$(x^2)^3 = x^{2 \cdot 3} = x^6$$

The following rule states this result.

Power Rule for Exponents

If m and n are positive integers and a is a real number, then

$$(a^m)^n = a^{mn}$$
← Multiply exponents.
Keep the base.

For example,

$$(7^2)^5 = 7^{2 \cdot 5} = 7^{10}$$
← Multiply exponents.
Keep the base.

$$[(-5)^3]^7 = (-5)^{3 \cdot 7} = (-5)^{21}$$
← Multiply exponents.
Keep the base.

In other words, to raise an exponential expression to a power, we keep the base and multiply the exponents.

Examples Use the power rule to simplify each expression.

16. $(5^3)^6 = 5^{3 \cdot 6} = 5^{18}$ **17.** $(y^8)^2 = y^{8 \cdot 2} = y^{16}$

Work Practice 16–17

Practice 16–17

Use the power rule to simplify each expression.

16. $(9^4)^{10}$ **17.** $(z^6)^3$

Helpful Hint

Take a moment to make sure that you understand when to apply the product rule and when to apply the power rule.

Product Rule → Add Exponents
$x^5 \cdot x^7 = x^{5+7} = x^{12}$
$y^6 \cdot y^2 = y^{6+2} = y^8$

Power Rule → Multiply Exponents
$(x^5)^7 = x^{5 \cdot 7} = x^{35}$
$(y^6)^2 = y^{6 \cdot 2} = y^{12}$

Answers
16. 9^{40} **17.** z^{18}

Objective D Using the Power Rules for Products and Quotients

When the base of an exponential expression is a product, the definition of a^n still applies. For example, simplify $(xy)^3$ as follows.

$$(xy)^3 = (xy)(xy)(xy) \quad (xy)^3 \text{ means 3 factors of } (xy).$$
$$= x \cdot x \cdot x \cdot y \cdot y \cdot y \quad \text{Group factors with common bases.}$$
$$= x^3y^3 \quad \text{Simplify.}$$

Notice that to simplify the expression $(xy)^3$, we raise each factor within the parentheses to a power of 3.

$$(xy)^3 = x^3y^3$$

In general, we have the following rule.

Power of a Product Rule

If n is a positive integer and a and b are real numbers, then

$$(ab)^n = a^nb^n$$

For example,

$$(3x)^5 = 3^5x^5$$

In other words, to raise a product to a power, we raise each factor to the power.

Examples Simplify each expression.

18. $(st)^4 = s^4 \cdot t^4 = s^4t^4$ Use the power of a product rule.

19. $(2a)^3 = 2^3 \cdot a^3 = 8a^3$ Use the power of a product rule.

20. $(-5x^2y^3z)^2 = (-5)^2 \cdot (x^2)^2 \cdot (y^3)^2 \cdot (z^1)^2 = 25x^4y^6z^2$ Use the power of a product rule.

21. $(-xy^3)^5 = (-1xy^3)^5 = (-1)^5 \cdot x^5 \cdot (y^3)^5 = -1x^5y^{15}$ or $-x^5y^{15}$ Use the power of a product rule.

Work Practice 18–21

Practice 18–21

Simplify each expression.

18. $(xy)^7$ **19.** $(3y)^4$

20. $(-2p^4q^2r)^3$ **21.** $(-a^4b)^7$

Let's see what happens when we raise a quotient to a power. For example, we simplify $\left(\frac{x}{y}\right)^3$ as follows.

$$\left(\frac{x}{y}\right)^3 = \left(\frac{x}{y}\right)\left(\frac{x}{y}\right)\left(\frac{x}{y}\right) \quad \left(\frac{x}{y}\right)^3 \text{ means 3 factors of } \left(\frac{x}{y}\right).$$
$$= \frac{x \cdot x \cdot x}{y \cdot y \cdot y} \quad \text{Multiply fractions.}$$
$$= \frac{x^3}{y^3} \quad \text{Simplify.}$$

Notice that to simplify the expression $\left(\frac{x}{y}\right)^3$, we raise both the numerator and the denominator to a power of 3.

$$\left(\frac{x}{y}\right)^3 = \frac{x^3}{y^3}$$

Answers

18. x^7y^7 **19.** $81y^4$ **20.** $-8p^{12}q^6r^3$ **21.** $-a^{28}b^7$

In general, we have the following rule.

Power of a Quotient Rule

If n is a positive integer and a and c are real numbers, then

$$\left(\frac{a}{c}\right)^n = \frac{a^n}{c^n}, \quad c \neq 0$$

For example,

$$\left(\frac{y}{7}\right)^3 = \frac{y^3}{7^3}$$

In other words, to raise a quotient to a power, we raise both the numerator and the denominator to the power.

Practice 22–23

Simplify each expression.

22. $\left(\frac{r}{s}\right)^6$ **23.** $\left(\frac{5x^6}{9y^3}\right)^2$

Examples Simplify each expression.

22. $\left(\frac{m}{n}\right)^7 = \frac{m^7}{n^7}, \quad n \neq 0$ Use the power of a quotient rule.

23. $\left(\frac{2x^4}{3y^5}\right)^4 = \frac{2^4 \cdot (x^4)^4}{3^4 \cdot (y^5)^4}$ Use the power of a quotient rule.

$= \frac{16x^{16}}{81y^{20}}, \quad y \neq 0$ Use the power rule for exponents.

Work Practice 22–23

Objective E Using the Quotient Rule and Defining the Zero Exponent

Another pattern for simplifying exponential expressions involves quotients.

$$\frac{x^5}{x^3} = \frac{x \cdot x \cdot x \cdot x \cdot x}{x \cdot x \cdot x}$$
$$= \frac{x \cdot x \cdot x \cdot x \cdot x}{x \cdot x \cdot x}$$
$$= 1 \cdot 1 \cdot 1 \cdot x \cdot x$$
$$= x \cdot x$$
$$= x^2$$

Notice that the result is exactly the same if we subtract exponents of the common bases.

$$\frac{x^5}{x^3} = x^{5-3} = x^2$$

The following rule states this result in a general way.

Quotient Rule for Exponents

If m and n are positive integers and a is a real number, then

$$\frac{a^m}{a^n} = a^{m-n}, \quad a \neq 0$$

For example,

$$\frac{x^6}{x^2} = x^{6-2} = x^4, \quad x \neq 0$$

Answers

22. $\frac{r^6}{s^6}, \ s \neq 0$ **23.** $\frac{25x^{12}}{81y^6}, \ y \neq 0$

In other words, to divide one exponential expression by another with a common base, we keep the base and subtract the exponents.

Examples Simplify each quotient.

24. $\frac{x^5}{x^2} = x^{5-2} = x^3$ Use the quotient rule.

25. $\frac{4^7}{4^3} = 4^{7-3} = 4^4 = 256$ Use the quotient rule.

26. $\frac{(-3)^5}{(-3)^2} = (-3)^3 = -27$ Use the quotient rule.

27. $\frac{2x^5y^2}{xy} = 2 \cdot \frac{x^5}{x^1} \cdot \frac{y^2}{y^1}$

$= 2 \cdot (x^{5-1}) \cdot (y^{2-1})$ Use the quotient rule.

$= 2x^4y^1$ or $2x^4y$

Work Practice 24–27

Practice 24–27

Simplify each quotient.

24. $\frac{y^7}{y^3}$ **25.** $\frac{5^9}{5^6}$

26. $\frac{(-2)^{14}}{(-2)^{10}}$ **27.** $\frac{7a^4b^{11}}{ab}$

Let's now give meaning to an expression such as x^0. To do so, we will simplify $\frac{x^3}{x^3}$ in two ways and compare the results.

$\frac{x^3}{x^3} = x^{3-3} = x^0$ Apply the quotient rule.

$\frac{x^3}{x^3} = \frac{x \cdot x \cdot x}{x \cdot x \cdot x} = 1$ Apply the fundamental principle for fractions.

Since $\frac{x^3}{x^3} = x^0$ and $\frac{x^3}{x^3} = 1$, we define that $x^0 = 1$ as long as x is not 0.

Zero Exponent

$a^0 = 1$, as long as a is not 0.

For example, $5^0 = 1$.

In other words, a base raised to the 0 power is 1, as long as the base is not 0.

Examples Simplify each expression.

28. $3^0 = 1$

29. $(5x^3y^2)^0 = 1$

30. $(-4)^0 = 1$

31. $-4^0 = -1 \cdot 4^0 = -1 \cdot 1 = -1$

32. $5x^0 = 5 \cdot x^0 = 5 \cdot 1 = 5$

Work Practice 28–32

Practice 28–32

Simplify each expression.

28. 8^0 **29.** $(2r^2s)^0$

30. $(-7)^0$ **31.** -7^0

32. $7y^0$

Answers

24. y^4 **25.** 125 **26.** 16 **27.** $7a^3b^{10}$

28. 1 **29.** 1 **30.** 1 **31.** -1 **32.** 7

✓Concept Check Suppose you are simplifying each expression. Tell whether you would *add* the exponents, *subtract* the exponents, *multiply* the exponents, *divide* the exponents, or *none of these.*

a. $(x^{63})^{21}$ **b.** $\frac{y^{15}}{y^3}$ **c.** $z^{16} + z^8$ **d.** $w^{45} \cdot w^9$

Objective F Deciding Which Rule to Use

Let's practice deciding which rule to use to simplify an expression. We will continue this discussion with more examples in the next section.

Practice 33

Simplify each expression.

a. $\frac{x^7}{x^4}$ **b.** $(3y^4)^4$ **c.** $\left(\frac{x}{4}\right)^3$

Example 33 Simplify each expression.

a. $x^7 \cdot x^4$ **b.** $\left(\frac{t}{2}\right)^4$ **c.** $(9y^5)^2$

Solution:

a. Here, we have a product, so we use the product rule to simplify.

$x^7 \cdot x^4 = x^{7+4} = x^{11}$

b. This is a quotient raised to a power, so we use the power of a quotient rule.

$\left(\frac{t}{2}\right)^4 = \frac{t^4}{2^4} = \frac{t^4}{16}$

c. This is a product raised to a power, so we use the power of a product rule.

$(9y^5)^2 = 9^2(y^5)^2 = 81y^{10}$

Work Practice 33

Answers

33. a. x^3 **b.** $81y^{16}$ **c.** $\frac{x^3}{64}$

✓Concept Check Answers

a. multiply **b.** subtract
c. none of these **d.** add

Vocabulary, Readiness & Video Check

Use the choices below to fill in each blank. Some choices may be used more than once.

0	base	add
1	exponent	multiply

1. Repeated multiplication of the same factor can be written using a(n) ________.

2. In 5^2, the 2 is called the ________ and the 5 is called the ________.

3. To simplify $x^2 \cdot x^7$, keep the base and ________ the exponents.

4. To simplify $(x^3)^6$, keep the base and ________ the exponents.

5. The understood exponent on the term y is ________.

6. If $x^{\square} = 1$, the exponent is ________.

Each expression contains an exponent of 2. For each exercise, name the base for this exponent of 2.

7. 3^2 ________
8. $(-3)^2$ ________
9. -4^2 ________
10. $5 \cdot 3^2$ ________
11. $5x^2$ ________
12. $(5x)^2$ ________

Martin-Gay Interactive Videos

See Video 3.1

Watch the section lecture video and answer the following questions.

Objective A 13. Examples 3 and 4 illustrate how to find the base of an exponential expression both with and without parentheses. Explain how identifying the base of Example 7 is similar to identifying the base of Example 4.

Objective B 14. Why were the commutative and associative properties applied in Example 12?

Objective C 15. What point is made at the end of Example 15?

Objective D 16. Although it's not especially emphasized in Example 20, what is helpful to remind ourself about the -2 in the problem?

Objective E 17. In Example 24, which exponent rule is used to show that any nonzero base raised to the power of zero is 1?

Objective F 18. When simplifying an exponential expression that's a fraction, will we always use the quotient rule? Refer to Example 30 to support your answer.

3.1 Exercise Set MyLab Math

Objective A *Evaluate each expression. See Examples 1 through 6.*

1. 7^2
2. -3^2
3. $(-5)^1$
4. $(-3)^2$
5. -2^4
6. -4^3
7. $(-2)^4$
8. $(-4)^3$
9. $\left(\frac{1}{3}\right)^3$
10. $\left(-\frac{1}{9}\right)^2$
11. $7 \cdot 2^4$
12. $9 \cdot 2^2$

Evaluate each expression for the given replacement values. See Example 7.

13. x^2 when $x = -2$
14. x^3 when $x = -2$
15. $5x^3$ when $x = 3$
16. $4x^2$ when $x = 5$
17. $2xy^2$ when $x = 3$ and $y = -5$
18. $-4x^2y^3$ when $x = 2$ and $y = -1$
19. $\frac{2z^4}{5}$ when $z = -2$
20. $\frac{10}{3y^3}$ when $y = -3$

Objective B *Use the product rule to simplify each expression. Write the results using exponents. See Examples 8 through 15.*

21. $x^2 \cdot x^5$
22. $y^2 \cdot y$
23. $(-3)^3 \cdot (-3)^9$
24. $(-5)^7 \cdot (-5)^6$
25. $(5y^4)(3y)$
26. $(-2z^3)(-2z^2)$
27. $(x^9y)(x^{10}y^5)$
28. $(a^2b)(a^{13}b^{17})$
29. $(-8mn^6)(9m^2n^2)$
30. $(-7a^3b^3)(7a^{19}b)$
31. $(4z^{10})(-6z^7)(z^3)$
32. $(12x^5)(-x^6)(x^4)$

33. The rectangle below has width $4x^2$ feet and length $5x^3$ feet. Find its area as an expression in x.

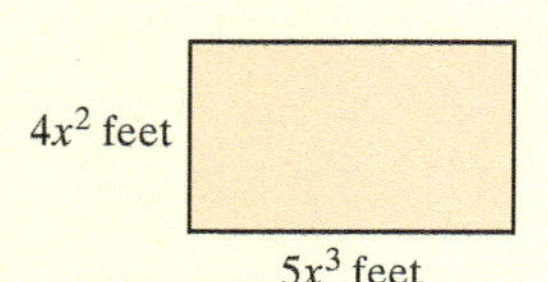

△ **34.** The parallelogram below has base length $9y^7$ meters and height $2y^{10}$ meters. Find its area as an expression in y.

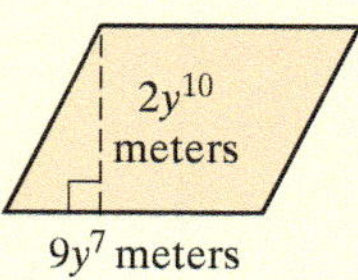

Objectives C D Mixed Practice *Use the power rule and the power of a product or quotient rule to simplify each expression. See Examples 16 through 23.*

35. $(x^9)^4$

36. $(y^7)^5$

37. $(pq)^8$

38. $(ab)^6$

39. $(2a^5)^3$

40. $(4x^6)^2$

41. $(x^2y^3)^5$

42. $(a^4b)^7$

43. $(-7a^2b^5c)^2$

44. $(-3x^7yz^2)^3$

45. $\left(\frac{r}{s}\right)^9$

46. $\left(\frac{q}{t}\right)^{11}$

47. $\left(\frac{mp}{n}\right)^9$

48. $\left(\frac{xy}{7}\right)^2$

49. $\left(\frac{-2xz}{y^5}\right)^2$

50. $\left(\frac{xy^4}{-3z^3}\right)^3$

△ **51.** The square shown has sides of length $8z^5$ decimeters. Find its area.

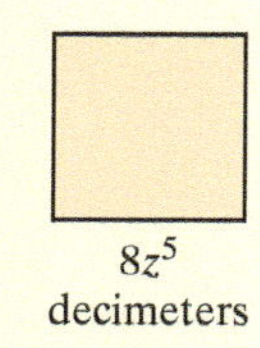

△ **52.** Given the circle below with radius $5y$ centimeters, find its area. Do not approximate π.

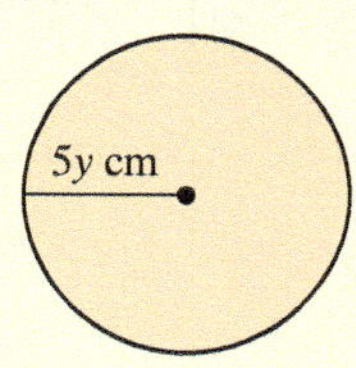

53. The vault below is in the shape of a cube. If each side is $3y^4$ feet, find its volume.

54. The silo shown is in the shape of a cylinder. If its radius is $4x$ meters and its height is $5x^3$ meters, find its volume. Do not approximate π.

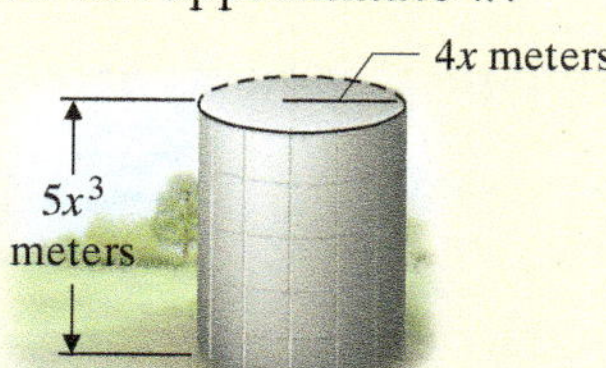

Objective E *Use the quotient rule and simplify each expression. See Examples 24 through 27.*

55. $\frac{x^3}{x}$

56. $\frac{y^{10}}{y^9}$

57. $\frac{(-4)^6}{(-4)^3}$

58. $\frac{(-6)^{13}}{(-6)^{11}}$

59. $\frac{p^7q^{20}}{pq^{15}}$

60. $\frac{x^8y^6}{xy^5}$

61. $\frac{7x^2y^6}{14xy^3}$

62. $\frac{9a^4b^7}{27ab^2}$

Simplify each expression. See Examples 28 through 32.

63. 7^0

64. 23^0

65. $(2x)^0$

66. $(4y)^0$

67. $-7x^0$

68. $-2x^0$

69. $5^0 + y^0$

70. $-3^0 + 4^0$

Objectives A B C D E F Mixed Practice *Simplify each expression. See Examples 1 through 6 and 8 through 33.*

71. -9^2

72. $(-9)^2$

73. $\left(\frac{1}{4}\right)^3$

74. $\left(\frac{2}{3}\right)^3$

75. b^4b^2

76. y^4y

77. $a^2a^3a^4$

78. $x^2x^{15}x^9$

79. $(2x^3)(-8x^4)$

80. $(3y^4)(-5y)$

81. $(a^7b^{12})(a^4b^8)$

82. $(y^2z^2)(y^{15}z^{13})$

83. $(-2mn^6)(-13m^8n)$

84. $(-3s^5t)(-7st^{10})$

85. $(z^4)^{10}$

86. $(t^5)^{11}$

87. $(4ab)^3$

88. $(2ab)^4$

89. $(-6xyz^3)^2$

90. $(-3xy^2a^3)^3$

91. $\frac{z^{12}}{z^4}$

92. $\frac{b^6}{b^3}$

93. $\frac{3x^5}{x}$

94. $\frac{5x^9}{x}$

95. $(6b)^0$

96. $(5ab)^0$

97. $(9xy)^2$

98. $(2ab)^5$

99. $2^3 + 2^5$

100. $7^2 - 7^0$

101. $\left(\frac{3y^5}{6x^4}\right)^3$

102. $\left(\frac{2ab}{6yz}\right)^4$

103. $\frac{2x^3y^2z}{xyz}$

104. $\frac{5x^{12}y^{13}z}{x^5y^7z}$

Review

Subtract. See Section 1.5.

105. $5 - 7$

106. $9 - 12$

107. $3 - (-2)$

108. $5 - (-10)$

109. $-11 - (-4)$

110. $-15 - (-21)$

Concept Extensions

Solve. See the Concept Checks in this section. For Exercises 111 through 114, match the expression with the operation needed to simplify each. A letter may be used more than once and a letter may not be used at all.

111. $(x^{14})^{23}$

112. $x^{14} \cdot x^{23}$

113. $x^{14} + x^{23}$

114. $\frac{x^{35}}{x^{17}}$

a. Add the exponents.
b. Subtract the exponents.
c. Multiply the exponents.
d. Divide the exponents.
e. None of these

Fill in the boxes so that each statement is true. (More than one answer is possible for each exercise.)

115. $x^{\square} \cdot x^{\square} = x^{12}$

116. $(x^{\square})^{\square} = x^{20}$

117. $\dfrac{y^{\square}}{y^{\square}} = y^{7}$

118. $(y^{\square})^{\square} \cdot (y^{\square})^{\square} = y^{30}$

△ **119.** The formula $V = x^3$ can be used to find the volume V of a cube with side length x. Find the volume of a cube with side length 7 meters. (Volume is the capacity of a solid such as a cube and is measured in cubic units.)

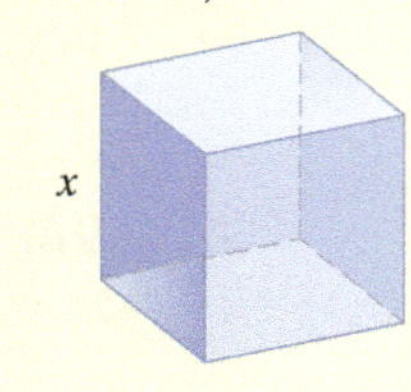

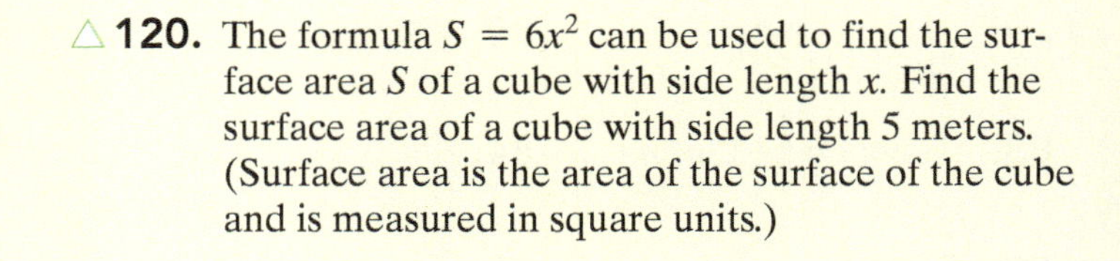

△ **120.** The formula $S = 6x^2$ can be used to find the surface area S of a cube with side length x. Find the surface area of a cube with side length 5 meters. (Surface area is the area of the surface of the cube and is measured in square units.)

△ **121.** To find the amount of water that a swimming pool in the shape of a cube can hold, do we use the formula for volume of the cube or surface area of the cube? (See Exercises **119** and **120**.)

△ **122.** To find the amount of material needed to cover an ottoman in the shape of a cube, do we use the formula for volume of the cube or surface area of the cube? (See Exercises **119** and **120**.)

123. Explain why $(-5)^4 = 625$, while $-5^4 = -625$.

124. Explain why $5 \cdot 4^2 = 80$, while $(5 \cdot 4)^2 = 400$.

125. In your own words, explain why $5^0 = 1$.

126. In your own words, explain when $(-3)^n$ is positive and when it is negative.

Simplify each expression. Assume that variables represent positive integers.

127. $x^{5a}x^{4a}$

128. $b^{9a}b^{4a}$

129. $(a^b)^5$

130. $(2a^{4b})^4$

131. $\dfrac{x^{9a}}{x^{4a}}$

132. $\dfrac{y^{15b}}{y^{6b}}$

3.2 Negative Exponents and Scientific Notation

Objective A Simplifying Expressions Containing Negative Exponents

Our work with exponential expressions so far has been limited to exponents that are positive integers or 0. Here we will also give meaning to an expression like x^{-3}.

Suppose that we wish to simplify the expression $\frac{x^2}{x^5}$. If we use the quotient rule for exponents, we subtract exponents:

$$\frac{x^2}{x^5} = x^{2-5} = x^{-3}, \quad x \neq 0$$

But what does x^{-3} mean? Let's simplify $\frac{x^2}{x^5}$ using the definition of a^n.

$$\frac{x^2}{x^5} = \frac{x \cdot x}{x \cdot x \cdot x \cdot x \cdot x}$$

$$= \frac{x \cdot x}{x \cdot x \cdot x \cdot x \cdot x}$$ Divide numerator and denominator by common factors by applying the fundamental principle for fractions.

$$= \frac{1}{x^3}$$

If the quotient rule is to hold true for negative exponents, then x^{-3} must equal $\frac{1}{x^3}$.

From this example, we state the definition for negative exponents.

Negative Exponents

If a is a real number other than 0 and n is an integer, then

$$a^{-n} = \frac{1}{a^n}$$

For example,

$$x^{-3} = \frac{1}{x^3}$$

In other words, another way to write a^{-n} is to take its reciprocal and change the sign of its exponent.

Examples Simplify by writing each expression with positive exponents only.

1. $3^{-2} = \frac{1}{3^2} = \frac{1}{9}$ Use the definition of negative exponents.
2. $2x^{-3} = 2^1 \cdot \frac{1}{x^3} = \frac{2^1}{x^3}$ or $\frac{2}{x^3}$ Use the definition of negative exponents.
3. $2^{-1} + 4^{-1} = \frac{1}{2} + \frac{1}{4} = \frac{2}{4} + \frac{1}{4} = \frac{3}{4}$
4. $(-2)^{-4} = \frac{1}{(-2)^4} = \frac{1}{(-2)(-2)(-2)(-2)} = \frac{1}{16}$

Helpful Hint Don't forget that since there are no parentheses, only x is the base for the exponent -3.

Work Practice 1–4

Objectives

A Simplify Expressions Containing Negative Exponents.

B Use the Rules and Definitions for Exponents to Simplify Exponential Expressions.

C Write Numbers in Scientific Notation.

D Convert Numbers in Scientific Notation to Standard Form.

E Perform Operations on Numbers Written in Scientific Notation.

Practice 1–4

Simplify by writing each expression with positive exponents only.

1. 5^{-3}
2. $7x^{-4}$
3. $5^{-1} + 3^{-1}$
4. $(-3)^{-4}$

Answers

1. $\frac{1}{125}$ 2. $\frac{7}{x^4}$ 3. $\frac{8}{15}$ 4. $\frac{1}{81}$

Helpful Hint

A negative exponent *does not affect* the sign of its base.
Remember: Another way to write a^{-n} is to take its reciprocal and change the sign of its exponent: $a^{-n} = \frac{1}{a^n}$. For example,

$$x^{-2} = \frac{1}{x^2}, \qquad 2^{-3} = \frac{1}{2^3} \text{ or } \frac{1}{8}$$

$$\frac{1}{y^{-4}} = \frac{1}{\frac{1}{y^4}} = y^4, \qquad \frac{1}{5^{-2}} = 5^2 \text{ or } 25$$

From the preceding Helpful Hint, we know that $x^{-2} = \frac{1}{x^2}$ and $\frac{1}{y^{-4}} = y^4$. We can use this to include another statement in our definition of negative exponents.

Negative Exponents

If a is a real number other than 0 and n is an integer, then

$$a^{-n} = \frac{1}{a^n} \quad \text{and} \quad \frac{1}{a^{-n}} = a^n$$

Practice 5–8

Simplify each expression. Write each result using positive exponents only.

5. $\left(\frac{6}{7}\right)^{-2}$
6. $\frac{x}{x^{-4}}$
7. $\frac{y^{-9}}{z^{-5}}$
8. $\frac{y^{-4}}{y^6}$

Examples

Simplify each expression. Write each result using positive exponents only.

5. $\left(\frac{2}{x}\right)^{-3} = \frac{2^{-3}}{x^{-3}} = \frac{2^{-3}}{1} \cdot \frac{1}{x^{-3}} = \frac{1}{2^3} \cdot \frac{x^3}{1} = \frac{x^3}{2^3} = \frac{x^3}{8}$ Use the negative exponent rule.

6. $\frac{y}{y^{-2}} = \frac{y^1}{y^{-2}} = y^{1-(-2)} = y^3$ Use the quotient rule.

7. $\frac{p^{-4}}{q^{-9}} = p^{-4} \cdot \frac{1}{q^{-9}} = \frac{1}{p^4} \cdot q^9 = \frac{q^9}{p^4}$ Use the negative exponents rule.

8. $\frac{x^{-5}}{x^7} = x^{-5-7} = x^{-12} = \frac{1}{x^{12}}$

Work Practice 5–8

Objective B Simplifying Exponential Expressions 

All the previously stated rules for exponents apply for negative exponents also. Here is a summary of the rules and definitions for exponents.

Summary of Exponent Rules

If m and n are integers and a, b, and c are real numbers, then

Product rule for exponents:	$a^m \cdot a^n = a^{m+n}$
Power rule for exponents:	$(a^m)^n = a^{m \cdot n}$
Power of a product:	$(ab)^n = a^n b^n$
Power of a quotient:	$\left(\frac{a}{c}\right)^n = \frac{a^n}{c^n}, \quad c \neq 0$
Quotient rule for exponents:	$\frac{a^m}{a^n} = a^{m-n}, \quad a \neq 0$
Zero exponent:	$a^0 = 1, \quad a \neq 0$
Negative exponent:	$a^{-n} = \frac{1}{a^n}, \quad a \neq 0$

Answers

5. $\frac{49}{36}$ 6. x^5 7. $\frac{z^5}{y^9}$ 8. $\frac{1}{y^{10}}$

Examples Simplify each expression. Write each result using positive exponents only.

9. $\dfrac{(x^3)^4 x}{x^7} = \dfrac{x^{12} \cdot x}{x^7} = \dfrac{x^{12+1}}{x^7} = \dfrac{x^{13}}{x^7} = x^{13-7} = x^6$ Use the power rule.

10. $\left(\dfrac{3a^2}{b}\right)^{-3} = \dfrac{3^{-3}(a^2)^{-3}}{b^{-3}}$ Raise each factor in the numerator and the denominator to the −3 power.

$= \dfrac{3^{-3}a^{-6}}{b^{-3}}$ Use the power rule.

$= \dfrac{b^3}{3^3a^6}$ Use the negative exponent rule.

$= \dfrac{b^3}{27a^6}$ Write 3^3 as 27.

11. $(y^{-3}z^6)^{-6} = (y^{-3})^{-6}(z^6)^{-6}$ Raise each factor to the −6 power.

$= y^{18}z^{-36} = \dfrac{y^{18}}{z^{36}}$

12. $\dfrac{(2x)^5}{x^3} = \dfrac{2^5 \cdot x^5}{x^3} = 2^5 \cdot x^{5-3} = 32x^2$ Raise each factor in the numerator to the fifth power.

13. $\dfrac{x^{-7}}{(x^4)^3} = \dfrac{x^{-7}}{x^{12}} = x^{-7-12} = x^{-19} = \dfrac{1}{x^{19}}$

14. $(5y^3)^{-2} = 5^{-2}(y^3)^{-2} = 5^{-2}y^{-6} = \dfrac{1}{5^2y^6} = \dfrac{1}{25y^6}$

15. $-\dfrac{22a^7b^{-5}}{11a^{-2}b^3} = -\dfrac{22}{11} \cdot a^{7-(-2)}b^{-5-3} = -2a^9b^{-8} = -\dfrac{2a^9}{b^8}$

16. $\dfrac{(2xy)^{-3}}{(x^2y^3)^2} = \dfrac{2^{-3}x^{-3}y^{-3}}{(x^2)^2(y^3)^2} = \dfrac{2^{-3}x^{-3}y^{-3}}{x^4y^6} = 2^{-3}x^{-3-4}y^{-3-6}$

$= 2^{-3}x^{-7}y^{-9} = \dfrac{1}{2^3x^7y^9}$ or $\dfrac{1}{8x^7y^9}$

Work Practice 9–16

Practice 9–16

Simplify each expression. Write each result using positive exponents only.

9. $\dfrac{(x^5)^3 x}{x^4}$

10. $\left(\dfrac{9x^3}{y}\right)^{-2}$

11. $(a^{-4}b^7)^{-5}$

12. $\dfrac{(2x)^4}{x^8}$

13. $\dfrac{y^{-10}}{(y^5)^4}$

14. $(4a^2)^{-3}$

15. $-\dfrac{32x^{-3}y^{-6}}{8x^{-5}y^{-2}}$

16. $\dfrac{(3x^{-2}y)^{-2}}{(2x^7y)^3}$

Objective C Writing Numbers in Scientific Notation

Both very large and very small numbers frequently occur in many fields of science. For example, the distance between the sun and the dwarf planet Pluto is approximately 5,906,000,000 kilometers, and the mass of a proton is approximately 0.000 000 000 000 000 000 000 001 65 gram. It can be tedious to write these numbers in this standard decimal notation, so **scientific notation** is used as a convenient shorthand for expressing very large and very small numbers.

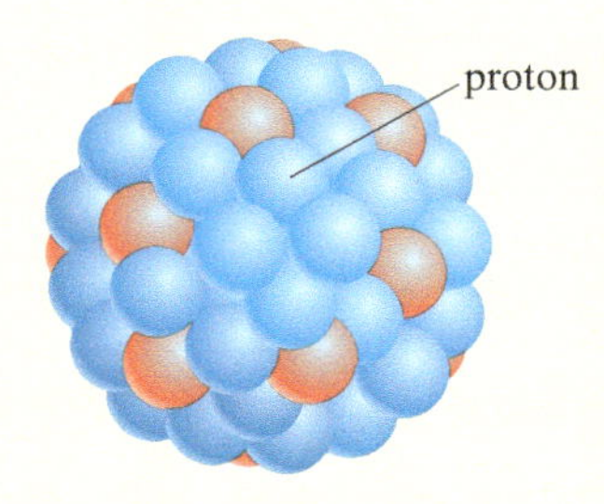

Mass of proton is approximately 0.000 000 000 000 000 000 000 001 65 gram

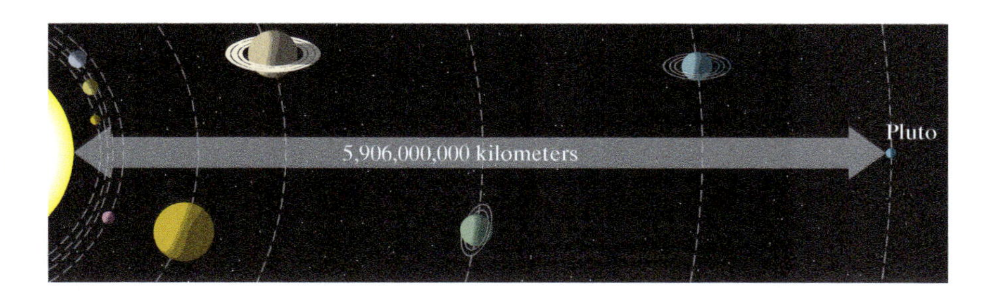

Answers

9. x^{12} 10. $\dfrac{y^2}{81x^6}$ 11. $\dfrac{a^{20}}{b^{35}}$ 12. $\dfrac{16}{x^4}$ 13. $\dfrac{1}{y^{30}}$ 14. $\dfrac{1}{64a^6}$ 15. $-\dfrac{4x^2}{y^4}$ 16. $\dfrac{1}{72x^{17}y^5}$

Scientific Notation

A positive number is written in scientific notation if it is written as the product of a number a, where $1 \le a < 10$, and an integer power r of 10: $a \times 10^r$.

The following numbers are written in scientific notation. The $\times$ sign for multiplication is used as part of the notation.

2.03×10^2 7.362×10^7 5.906×10^9 (Distance between the sun and Pluto)

1×10^{-3} 8.1×10^{-5} 1.65×10^{-24} (Mass of a proton)

The following steps are useful when writing positive numbers in scientific notation.

To Write a Number in Scientific Notation

Step 1: Move the decimal point in the original number so that the new number has a value between 1 and 10.

Step 2: Count the number of decimal places the decimal point is moved in Step 1. If the original number is 10 or greater, the count is positive. If the original number is less than 1, the count is negative.

Step 3: Multiply the new number in Step 1 by 10 raised to an exponent equal to the count found in Step 2.

Practice 17

Write each number in scientific notation.

a. 420,000 **b.** 0.00017
c. 9,060,000,000 **d.** 0.000007

Example 17 Write each number in scientific notation.

a. 367,000,000 **b.** 0.000003
c. 20,520,000,000 **d.** 0.00085

Solution:

a. Step 1: Move the decimal point until the number is between 1 and 10.
367,000,000.
8 places

Step 2: The decimal point is moved 8 places and the original number is 10 or greater, so the count is positive 8.

Step 3: $367{,}000{,}000 = 3.67 \times 10^8$

b. Step 1: Move the decimal point until the number is between 1 and 10.
0.000003
6 places

Step 2: The decimal point is moved 6 places and the original number is less than 1, so the count is -6.

Step 3: $0.000003 = 3.0 \times 10^{-6}$

c. $20{,}520{,}000{,}000 = 2.052 \times 10^{10}$

d. $0.00085 = 8.5 \times 10^{-4}$

Work Practice 17

Objective D Converting Numbers to Standard Form

A number written in scientific notation can be rewritten in standard form. For example, to write 8.63×10^3 in standard form, recall that $10^3 = 1000$.

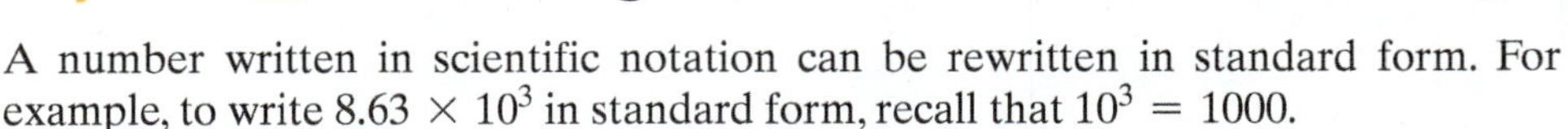

$$8.63 \times 10^3 = 8.63(1000) = 8630$$

Answers

17. a. 4.2×10^5 **b.** 1.7×10^{-4} **c.** 9.06×10^9 **d.** 7×10^{-6}

Notice that the exponent on the 10 is positive 3, and we moved the decimal point 3 places to the right.

To write 7.29×10^{-3} in standard form, recall that $10^{-3} = \frac{1}{10^3} = \frac{1}{1000}$.

$$7.29 \times 10^{-3} = 7.29\left(\frac{1}{1000}\right) = \frac{7.29}{1000} = 0.00729$$

The exponent on the 10 is negative 3, and we moved the decimal to the left 3 places.

In general, **to write a scientific notation number in standard form,** move the decimal point the same number of places as the exponent on 10. If the exponent is positive, move the decimal point to the right; if the exponent is negative, move the decimal point to the left.

Example 18 Write each number in standard form, without exponents.

a. 1.02×10^5 **b.** 7.358×10^{-3}

c. 8.4×10^7 **d.** 3.007×10^{-5}

Solution:

a. Move the decimal point 5 places to the right.
$1.02 \times 10^5 = 102{,}000.$

b. Move the decimal point 3 places to the left.
$7.358 \times 10^{-3} = 0.007358$

c. $8.4 \times 10^7 = 84{,}000{,}000.$ 7 places to the right

d. $3.007 \times 10^{-5} = 0.00003007$ 5 places to the left

Work Practice 18

Practice 18

Write the numbers in standard form, without exponents.

a. 3.062×10^{-4}

b. 5.21×10^4

c. 9.6×10^{-5}

d. 6.002×10^6

✓**Concept Check** Which number in each pair is larger?

a. 7.8×10^3 or 2.1×10^5
b. 9.2×10^{-2} or 2.7×10^4
c. 5.6×10^{-4} or 6.3×10^{-5}

Objective E Performing Operations with Scientific Notation

Performing operations on numbers written in scientific notation makes use of the rules and definitions for exponents.

Example 19 Perform each indicated operation. Write each result in standard decimal form.

a. $(8 \times 10^{-6})(7 \times 10^3)$

b. $\frac{12 \times 10^2}{6 \times 10^{-3}}$

Solution:

a. $(8 \times 10^{-6})(7 \times 10^3) = 8 \cdot 7 \cdot 10^{-6} \cdot 10^3$
$= 56 \times 10^{-3}$
$= 0.056$

b. $\frac{12 \times 10^2}{6 \times 10^{-3}} = \frac{12}{6} \times 10^{2-(-3)} = 2 \times 10^5 = 200{,}000$

Work Practice 19

Practice 19

Perform each indicated operation. Write each result in standard decimal form.

a. $(9 \times 10^7)(4 \times 10^{-9})$

b. $\frac{8 \times 10^4}{2 \times 10^{-3}}$

Answers

18. a. 0.0003062 **b.** 52,100 **c.** 0.000096 **d.** 6,002,000

19. a. 0.36 **b.** 40,000,000

✓**Concept Check Answers**

a. 2.1×10^5 **b.** 2.7×10^4 **c.** 5.6×10^{-4}

Calculator Explorations Scientific Notation

To enter a number written in scientific notation on a scientific calculator, locate the scientific notation key, which may be marked [EE] or [EXP]. To enter 3.1×10^7, press [3.1] [EE] [7]. The display should read [3.1 07].

Enter each number written in scientific notation on your calculator.

1. 5.31×10^3
2. 4.8×10^{14}
3. 6.6×10^{-9}
4. 9.9811×10^{-2}

Multiply each of the following on your calculator. Notice the form of the result.

5. $3{,}000{,}000 \times 5{,}000{,}000$
6. $230{,}000 \times 1000$

Multiply each of the following on your calculator. Write the product in scientific notation.

7. $(3.26 \times 10^6)(2.5 \times 10^{13})$
8. $(8.76 \times 10^{-4})(1.237 \times 10^9)$

Vocabulary, Readiness & Video Check

Fill in each blank with the correct choice.

1. The expression x^{-3} equals ________.
   a. $-x^3$ b. $\frac{1}{x^3}$ c. $\frac{-1}{x^3}$ d. $\frac{1}{x^{-3}}$

2. The expression 5^{-4} equals ________.
   a. -20 b. -625 c. $\frac{1}{20}$ d. $\frac{1}{625}$

3. The number 3.021×10^{-3} is written in ________.
   a. standard form b. expanded form
   c. scientific notation

4. The number 0.0261 is written in ________.
   a. standard form b. expanded form
   c. scientific notation

Write each expression using positive exponents only.

5. $5x^{-2}$
6. $3x^{-3}$
7. $\frac{1}{y^{-6}}$
8. $\frac{1}{x^{-3}}$
9. $\frac{4}{y^{-3}}$
10. $\frac{16}{y^{-7}}$

Martin-Gay Interactive Videos

See Video 3.2

Watch the section lecture video and answer the following questions.

Objective A 11. What important reminder is given at the end of Example 1?

Objective B 12. Name all the rules and definitions used to simplify Example 8.

Objective C 13. From Examples 9 and 10, explain how the movement of the decimal point in Step 1 suggests the sign of the exponent on the number 10.

Objective D 14. From Example 11, what part of a number written in scientific notation is key in telling us how to write the number in standard form?

Objective E 15. In Example 13, what exponent rules were needed to evaluate the expression?

3.2 Exercise Set MyLab Math 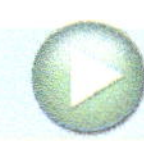

Objective A *Simplify each expression. Write each result using positive exponents only. See Examples 1 through 8.*

1. 4^{-3} **2.** 6^{-2} **3.** $7x^{-3}$ **4.** $(7x)^{-3}$ **5.** $\left(-\frac{1}{4}\right)^{-3}$ **6.** $\left(-\frac{1}{8}\right)^{-2}$

7. $3^{-1} + 2^{-1}$ **8.** $4^{-1} + 4^{-2}$ **9.** $\frac{1}{p^{-3}}$ **10.** $\frac{1}{q^{-5}}$ **11.** $\frac{p^{-5}}{q^{-4}}$ **12.** $\frac{r^{-5}}{s^{-2}}$

13. $\frac{x^{-2}}{x}$ **14.** $\frac{y}{y^{-3}}$ **15.** $\frac{z^{-4}}{z^{-7}}$ **16.** $\frac{x^{-4}}{x^{-1}}$ **17.** $3^{-2} + 3^{-1}$ **18.** $4^{-2} - 4^{-3}$

19. $(-3)^{-2}$ **20.** $(-2)^{-6}$ **21.** $\frac{-1}{p^{-4}}$ **22.** $\frac{-1}{y^{-6}}$ **23.** $-2^0 - 3^0$ **24.** $5^0 + (-5)^0$

Objective B *Simplify each expression. Write each result using positive exponents only. See Examples 9 through 16.*

25. $\frac{x^2x^5}{x^3}$ **26.** $\frac{y^4y^5}{y^6}$ **27.** $\frac{p^2p}{p^{-1}}$ **28.** $\frac{y^3y}{y^{-2}}$ **29.** $\frac{(m^5)^4m}{m^{10}}$ **30.** $\frac{(x^2)^8x}{x^9}$

31. $\frac{r}{r^{-3}r^{-2}}$ **32.** $\frac{p}{p^{-3}p^{-5}}$ **33.** $(x^5y^3)^{-3}$ **34.** $(z^5x^5)^{-3}$ **35.** $\frac{(x^2)^3}{x^{10}}$ **36.** $\frac{(y^4)^2}{y^{12}}$

37. $\frac{(a^5)^2}{(a^3)^4}$ **38.** $\frac{(x^2)^5}{(x^4)^3}$ **39.** $\frac{8k^4}{2k}$ **40.** $\frac{27r^6}{3r^4}$ **41.** $\frac{-6m^4}{-2m^3}$ **42.** $\frac{15a^4}{-15a^5}$

43. $\frac{-24a^6b}{6ab^2}$ **44.** $\frac{-5x^4y^5}{15x^4y^2}$ **45.** $\frac{6x^2y^3}{-7x^2y^5}$ **46.** $\frac{-8xa^2b}{-5xa^5b}$ **47.** $(3a^2b^{-4})^3$ **48.** $(5x^3y^{-2})^2$

49. $(a^{-5}b^2)^{-6}$ **50.** $(4^{-1}x^5)^{-2}$ **51.** $\left(\frac{x^{-2}y^4}{x^3y^7}\right)^{-2}$ **52.** $\left(\frac{a^5b}{a^7b^{-2}}\right)^{-3}$ **53.** $\frac{4^2z^{-3}}{4^3z^{-5}}$ **54.** $\frac{5^{-1}z^7}{5^{-2}z^9}$

55. $\frac{3^{-1}x^4}{3^3x^{-7}}$ **56.** $\frac{2^{-3}x^{-4}}{2^2x}$ **57.** $\frac{7ab^{-4}}{7^{-1}a^{-3}b^2}$ **58.** $\frac{6^{-5}x^{-1}y^2}{6^{-2}x^{-4}y^4}$ **59.** $\frac{-12m^5n^{-7}}{4m^{-2}n^{-3}}$ **60.** $\frac{-15r^{-6}s}{5r^{-4}s^{-3}}$

61. $\left(\frac{a^{-5}b}{ab^3}\right)^{-4}$ **62.** $\left(\frac{r^{-2}s^{-3}}{r^{-4}s^{-3}}\right)^{-3}$ **63.** $(5^2)(8)(2^0)$ **64.** $(3^4)(7^0)(2)$ **65.** $\frac{(xy^3)^5}{(xy)^{-4}}$ **66.** $\frac{(rs)^{-3}}{(r^2s^3)^2}$

67. $\frac{(-2xy^{-3})^{-3}}{(xy^{-1})^{-1}}$ **68.** $\frac{(-3x^2y^2)^{-2}}{(xyz)^{-2}}$ **69.** $\frac{(a^4b^{-7})^{-5}}{(5a^2b^{-1})^{-2}}$ **70.** $\frac{(a^6b^{-2})^4}{(4a^{-3}b^{-3})^3}$

71. Find the volume of the cube.

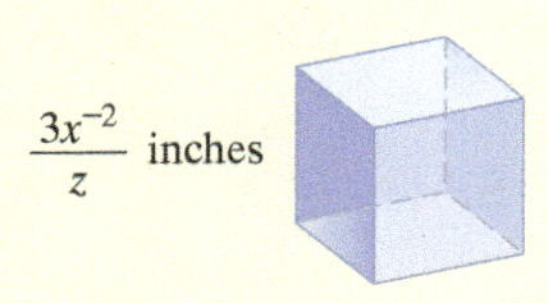

72. Find the area of the triangle.

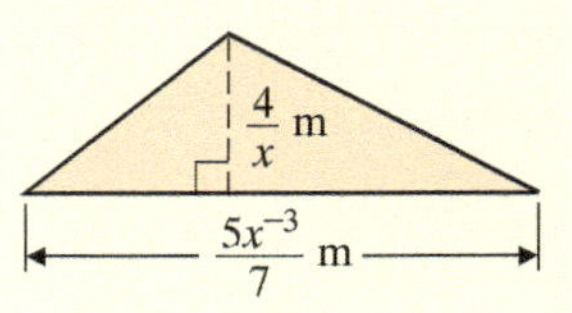

Objective C *Write each number in scientific notation. See Example 17.*

73. 78,000

74. 9,300,000,000

75. 0.00000167

76. 0.00000017

77. 0.00635

78. 0.00194

79. 1,160,000

80. 700,000

81. When it is completed in 2022, the Thirty Meter Telescope is expected to be the world's largest optical telescope. Located in an observatory complex at the summit of Mauna Kea in Hawaii, the elevation of the Thirty Meter Telescope will be roughly 4200 meters above sea level. Write 4200 in scientific notation.

82. The Thirty Meter Telescope (see Exercise **81**) will have the ability to view objects 13,000,000,000 light-years away. Write 13,000,000,000 in scientific notation.

Objective D *Write each number in standard form. See Example 18.*

83. 8.673×10^{-10}

84. 9.056×10^{-4}

85. 3.3×10^{-2}

86. 4.8×10^{-6}

87. 2.032×10^{4}

88. 9.07×10^{10}

89. Each second, the Sun converts 7.0×10^8 tons of hydrogen into helium and energy in the form of gamma rays. Write this number in standard form. (*Source:* Students for the Exploration and Development of Space)

90. In chemistry, Avogadro's number is the number of atoms in one mole of an element. Avogadro's number is $6.02214199 \times 10^{23}$. Write this number in standard form. (*Source:* National Institute of Standards and Technology)

Objectives C D Mixed Practice *See Examples 17 and 18. If a number is written in standard form, write it in scientific notation. If a number is written in scientific notation, write it in standard form. The bar graph below shows estimates of the top six national debts as of the beginning of 2017.* (*Source:* The Economist)

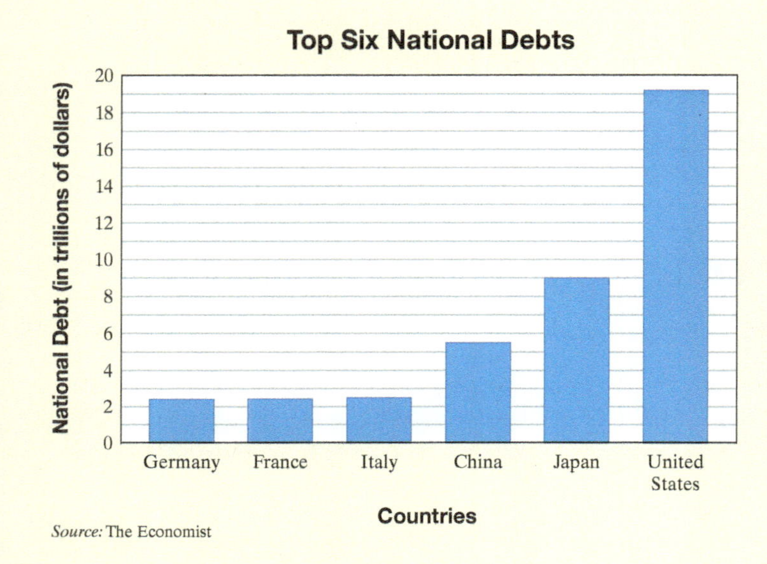

91. Germany's national debt as of the beginning of 2017 was \$2,415,000,000,000.

92. Italy's national debt as of the beginning of 2017 was \$2,500,000,000,000.

93. China's national debt as of the beginning of 2017 was $\$5.5 \times 10^{12}$.

94. France's national debt as of the beginning of 2017 was $\$2.435 \times 10^{12}$.

95. Use the bar graph to estimate the national debt of Japan and then express it in both standard form and scientific notation.

96. Use the bar graph to estimate (to the nearest trillion) the national debt of the United States and then express it in both standard form and scientific notation.

Objective E *Evaluate each expression using exponential rules. Write each result in standard form. See Example 19.*

97. $(1.2 \times 10^{-3})(3 \times 10^{-2})$

98. $(2.5 \times 10^{6})(2 \times 10^{-6})$

99. $(4 \times 10^{-10})(7 \times 10^{-9})$

100. $(5 \times 10^{6})(4 \times 10^{-8})$

101. $\dfrac{8 \times 10^{-1}}{16 \times 10^{5}}$

102. $\dfrac{25 \times 10^{-4}}{5 \times 10^{-9}}$

103. $\dfrac{1.4 \times 10^{-2}}{7 \times 10^{-8}}$

104. $\dfrac{0.4 \times 10^{5}}{0.2 \times 10^{11}}$

105. Although the actual amount varies by season and time of day, the average volume of water that flows over Niagara Falls (the American and Canadian falls combined) each second is 7.5×10^5 gallons. How much water flows over Niagara Falls in an hour? Write the result in scientific notation. (*Hint:* 1 hour equals 3600 seconds.) (*Source:* http://niagarafallslive.com)

106. A beam of light travels 9.460×10^{12} kilometers per year. How far does light travel in 10,000 years? Write the result in scientific notation.

Review

Simplify each expression by combining any like terms. See Section 1.8.

107. $3x - 5x + 7$

108. $7w + w - 2w$

109. $y - 10 + y$

110. $-6z + 20 - 3z$

111. $7x + 2 - 8x - 6$

112. $10y - 14 - y - 14$

Concept Extensions

For Exercises 113 through 120, write each number in standard form. Then write the number in scientific notation.

113. The wireless subscriber connections in the United States at year's beginning 2018 were 435.3 million. (*Source:* CTIA—The Wireless Association)

114. Google estimates approximately 1.2 trillion searches for 2018. (*Source:* Google.com)

115. The surface of the Arctic Ocean encompasses 14.056 million square kilometers of water. (*Source:* CIA World Factbook)

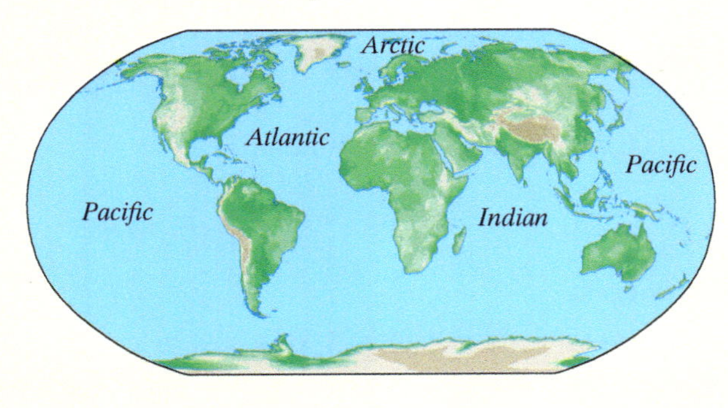

116. The surface of the Pacific Ocean encompasses 155.557 million square kilometers of water. (*Source:* CIA World Factbook)

Solve.

117. A nanometer is one-billionth, or 10^{-9}, of a meter. A strand of DNA is about 2.5 nanometers in diameter. Use scientific notation to write the diameter of a DNA strand in terms of meters. (*Source:* United States National Nanotechnology Initiative)

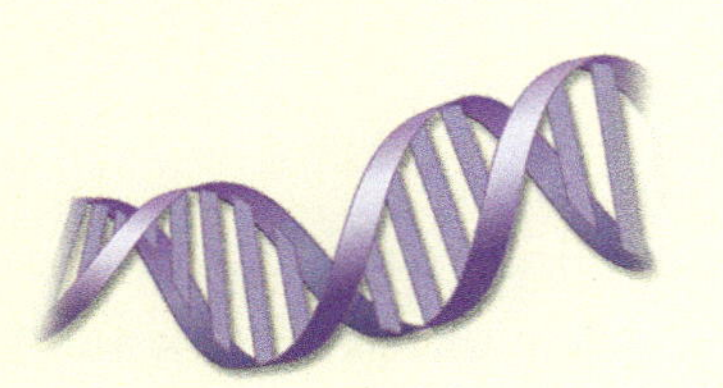

118. A micrometer (sometimes referred to as a micron) is one-millionth, or 10^{-6}, of a meter. A single red blood cell is about 7 micrometers in diameter. Use scientific notation to write the diameter of a red blood cell in terms of meters. (*Source:* National Institute of Standards and Technology)

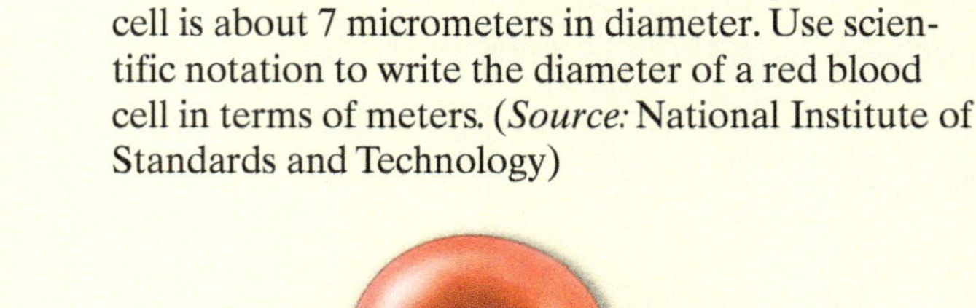

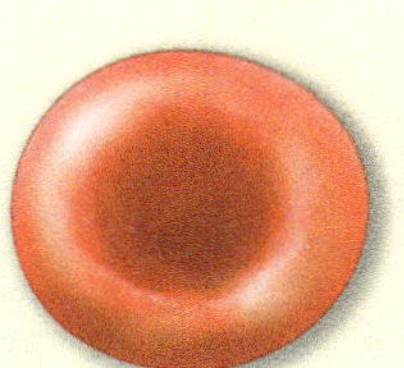

119. The Thirty Meter Telescope, described in Exercises **81** and **82**, will be capable of observing ultraviolet wavelengths measuring 310 nanometers. Express this wavelength in terms of meters using both standard form and scientific notation. (See Exercise **117** for a definition of nanometer.) (*Source:* TMT Observatory Corporation)

120. The Thirty Meter Telescope, described in Exercises **81** and **82**, will be capable of observing infrared wavelengths measuring 28 micrometers. Express this wavelength in terms of meters using both standard form and scientific notation. (See Exercise **118** for a definition of micrometer.) (*Source:* TMT Observatory Corporation)

Simplify.

121. $(2a^3)^3a^4 + a^5a^8$

122. $(2a^3)^3a^{-3} + a^{11}a^{-5}$

Fill in the boxes so that each statement is true. (More than one answer is possible for some of these exercises.)

123. $x^{\square} = \frac{1}{x^5}$

124. $7^{\square} = \frac{1}{49}$

125. $z^{\square} \cdot z^{\square} = z^{-10}$

126. $(x^{\square})^{\square} = x^{-15}$

127. Which is larger? See the Concept Check in this section.

a. 9.7×10^{-2} or 1.3×10^1

b. 8.6×10^5 or 4.4×10^7

c. 6.1×10^{-2} or 5.6×10^{-4}

128. Determine whether each statement is true or false.

a. $5^{-1} < 5^{-2}$

b. $\left(\frac{1}{5}\right)^{-1} < \left(\frac{1}{5}\right)^{-2}$

c. $a^{-1} < a^{-2}$ for all nonzero numbers.

129. It was stated earlier that for an integer n,

$$x^{-n} = \frac{1}{x^n}, \quad x \neq 0.$$

Explain why x may not equal 0.

130. The quotient rule states that

$$\frac{a^m}{a^n} = a^{m-n}, \; a \neq 0.$$

Explain why a may not equal 0.

Simplify each expression. Assume that variables represent positive integers.

131. $(x^{-3s})^3$

132. $a^{-4m} \cdot a^{5m}$

133. $a^{4m+1} \cdot a^4$

134. $(3y^{2z})^3$

3.3 Introduction to Polynomials 

Objective A Defining Term and Coefficient

In this section, we introduce a special algebraic expression called a polynomial. Let's first review some definitions presented in Section 1.8.

Recall that a term is a number or the product of a number and variables raised to powers. The terms of an expression are separated by plus signs. The terms of the expression $4x^2 + 3x$ are $4x^2$ and $3x$. The terms of the expression $9x^4 - 7x - 1$, or $9x^4 + (-7x) + (-1)$, are $9x^4$, $-7x$, and -1.

Expression	Terms
$4x^2 + 3x$	$4x^2, 3x$
$9x^4 - 7x - 1$	$9x^4, -7x, -1$
$7y^3$	$7y^3$

The **numerical coefficient** of a term, or simply the **coefficient**, is the numerical factor of each term. If no numerical factor appears in the term, then the coefficient is understood to be 1. If the term is a number only, it is called a **constant term** or simply a **constant**.

Term	Coefficient
x^5	1
$3x^2$	3
$-4x$	-4
$-x^2y$	-1
3 (constant)	3

Objectives

A Define Term and Coefficient of a Term.

B Define Polynomial, Monomial, Binomial, Trinomial, and Degree.

C Evaluate a Polynomial for Given Replacement Values.

D Simplify a Polynomial by Combining Like Terms.

E Simplify a Polynomial in Several Variables.

F Write a Polynomial in Descending Powers of the Variable and with No Missing Powers of the Variable.

Example 1

Complete the table for the expression $7x^5 - 8x^4 + x^2 - 3x + 5$.

Term	Coefficient
$7x^5$	
	-8
x^2	
	-3
5	

Solution: The completed table is shown below.

Term	Coefficient
$7x^5$	7
$-8x^4$	-8
x^2	1
$-3x$	-3
5	5

Work Practice 1

Practice 1

Complete the table for the expression $-6x^6 + 4x^5 + 7x^3 - 9x^2 - 1$.

Term	Coefficient
$-6x^6$	
	4
$7x^3$	
	-9
-1	

Answer

1. term: $4x^5$; $-9x^2$, coefficient: $-6, 7, -1$

Objective B Defining Polynomial, Monomial, Binomial, Trinomial, and Degree

Now we are ready to define what we mean by a polynomial.

Polynomial

A **polynomial in *x*** is a finite sum of terms of the form ax^n, where a is a real number and n is a whole number.

For example,

$$x^5 - 3x^3 + 2x^2 - 5x + 1$$

is a polynomial in x. Notice that this polynomial is written in **descending powers** of x, because the powers of x decrease from left to right. (Recall that the term 1 can be thought of as $1x^0$.)

On the other hand,

$$x^{-5} + 2x - 3$$

is **not** a polynomial because one of its terms contains a variable with an exponent, -5, that is not a whole number.

Types of Polynomials

A **monomial** is a polynomial with exactly one term.
A **binomial** is a polynomial with exactly two terms.
A **trinomial** is a polynomial with exactly three terms.

The following are examples of monomials, binomials, and trinomials. Each of these examples is also a polynomial.

Polynomials			
Monomials	**Binomials**	**Trinomials**	**More Than Three Terms**
ax^2	$x + y$	$x^2 + 4xy + y^2$	$5x^3 - 6x^2 + 3x - 6$
$-3z$	$3p + 2$	$x^5 + 7x^2 - x$	$-y^5 + y^4 - 3y^3 - y^2 + y$
4	$4x^2 - 7$	$-q^4 + q^3 - 2q$	$x^6 + x^4 - x^3 + 1$

Each term of a polynomial has a degree. The **degree of a term in one variable** is the exponent on the variable.

Practice 2

Identify the degree of each term of the trinomial

$-15x^3 + 2x^2 - 5$.

Example 2 Identify the degree of each term of the trinomial $12x^4 - 7x + 3$.

Solution: The term $12x^4$ has degree 4.

The term $-7x$ has degree 1 since $-7x$ is $-7x^1$.

The term 3 has degree 0 since 3 is $3x^0$.

Work Practice 2

Each polynomial also has a degree.

Degree of a Polynomial

The **degree of a polynomial** is the greatest degree of any term of the polynomial.

Answer

2. 3; 2; 0

Example 3 Find the degree of each polynomial and tell whether the polynomial is a monomial, binomial, trinomial, or none of these.

a. $-2t^2 + 3t + 6$ **b.** $15x - 10$ **c.** $7x + 3x^3 + 2x^2 - 1$

Solution:

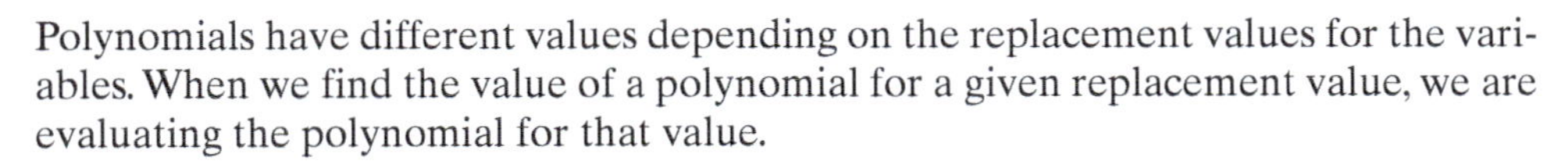

a. The degree of the trinomial $-2t^2 + 3t + 6$ is 2, the greatest degree of any of its terms.

b. The degree of the binomial $15x - 10$ or $15x^1 - 10$ is 1.

c. The degree of the polynomial $7x + 3x^3 + 2x^2 - 1$ is 3. The polynomial is neither a monomial, binomial, nor trinomial.

Work Practice 3

Practice 3

Find the degree of each polynomial and tell whether the polynomial is a monomial, binomial, trinomial, or none of these.

a. $-6x + 14$

b. $9x - 3x^6 + 5x^4 + 2$

c. $10x^2 - 6x - 6$

Objective C Evaluating Polynomials

Polynomials have different values depending on the replacement values for the variables. When we find the value of a polynomial for a given replacement value, we are evaluating the polynomial for that value.

Example 4 Evaluate each polynomial when $x = -2$.

a. $-5x + 6$ **b.** $3x^2 - 2x + 1$

Solution:

a. $-5x + 6 = -5(-2) + 6$ Replace x with -2.
$= 10 + 6$
$= 16$

b. $3x^2 - 2x + 1 = 3(-2)^2 - 2(-2) + 1$ Replace x with -2.
$= 3(4) - 2(-2) + 1$
$= 12 + 4 + 1$
$= 17$

Work Practice 4

Practice 4

Evaluate each polynomial when $x = -1$.

a. $-2x + 10$

b. $6x^2 + 11x - 20$

Many physical phenomena can be modeled by polynomials.

Example 5 Finding Free-Fall Time

The Swiss Re Building in London is a unique building. Londoners often refer to it as the "pickle building." The building is 592.1 feet tall. An object is dropped from the highest point of this building. Neglecting air resistance, the height in feet of the object above ground at time t seconds is given by the polynomial $-16t^2 + 592.1$. Find the height of the object when $t = 1$ second and when $t = 6$ seconds. (See next page for illustration.)

Solution: To find each height, we evaluate the polynomial when $t = 1$ and when $t = 6$.

$-16t^2 + 592.1 = -16(1)^2 + 592.1$ Replace t with 1.
$= -16(1) + 592.1$
$= -16 + 592.1$
$= 576.1$

The height of the object at 1 second is 576.1 feet.

$-16t^2 + 592.1 = -16(6)^2 + 592.1$ Replace t with 6.
$= -16(36) + 592.1$
$= -576 + 592.1 = 16.1$

(Continued on next page)

Practice 5

Find the height of the object in Example 5 when $t = 2$ seconds and $t = 4$ seconds.

Answers

3. a. binomial, 1 **b.** none of these, 6 **c.** trinomial, 2 **4. a.** 12 **b.** -25 **5.** 528.1 feet; 336.1 feet

The height of the object at 6 seconds is 16.1 feet.

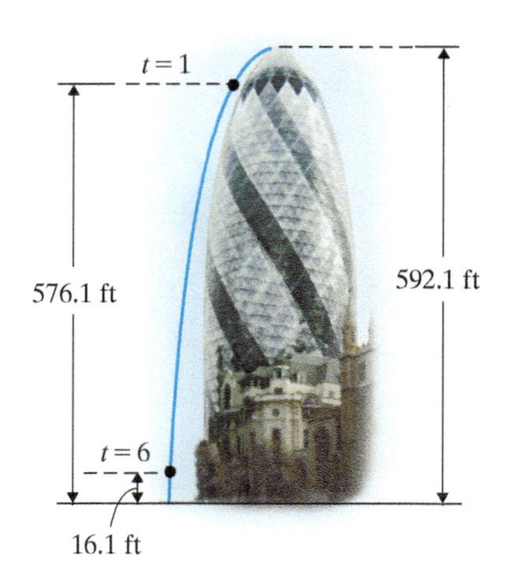

Work Practice 5

Objective D Simplifying Polynomials by Combining Like Terms

We can simplify polynomials with like terms by combining the like terms. Recall from Section 1.8 that like terms are terms that contain exactly the same variables raised to exactly the same powers.

Like Terms	Unlike Terms
$5x^2, -7x^2$	$3x, 3y$
$y, 2y$	$-2x^2, -5x$
$\frac{1}{2}a^2b, -a^2b$	$6st^2, 4s^2t$

Only like terms can be combined. We combine like terms by applying the distributive property.

Examples Simplify each polynomial by combining any like terms.

6. $-3x + 7x = (-3 + 7)x = 4x$

7. $11x^2 + 5 + 2x^2 - 7 = 11x^2 + 2x^2 + 5 - 7$
$= 13x^2 - 2$

8. $9x^3 + x^3 = 9x^3 + 1x^3$ Write x^3 as $1x^3$.
$= 10x^3$

9. $5x^2 + 6x - 9x - 3 = 5x^2 - 3x - 3$ Combine like terms $6x$ and $-9x$.

10. $\frac{2}{5}x^4 + \frac{2}{3}x^3 - x^2 + \frac{1}{10}x^4 - \frac{1}{6}x^3$
$= \left(\frac{2}{5} + \frac{1}{10}\right)x^4 + \left(\frac{2}{3} - \frac{1}{6}\right)x^3 - x^2$
$= \left(\frac{4}{10} + \frac{1}{10}\right)x^4 + \left(\frac{4}{6} - \frac{1}{6}\right)x^3 - x^2$
$= \frac{5}{10}x^4 + \frac{3}{6}x^3 - x^2$
$= \frac{1}{2}x^4 + \frac{1}{2}x^3 - x^2$

Work Practice 6–10

Practice 6–10

Simplify each polynomial by combining any like terms.

6. $-6y + 8y$
7. $14y^2 + 3 - 10y^2 - 9$
8. $7x^3 + x^3$
9. $23x^2 - 6x - x - 15$
10. $\frac{2}{7}x^3 - \frac{1}{4}x + 2 - \frac{1}{2}x^3 + \frac{3}{8}x$

Answers

6. $2y$ 7. $4y^2 - 6$ 8. $8x^3$
9. $23x^2 - 7x - 15$
10. $-\frac{3}{14}x^3 + \frac{1}{8}x + 2$

Example 11 Write a polynomial that describes the total area of the squares and rectangles shown below. Then simplify the polynomial.

Solution: Recall that the area of a rectangle is length times width.

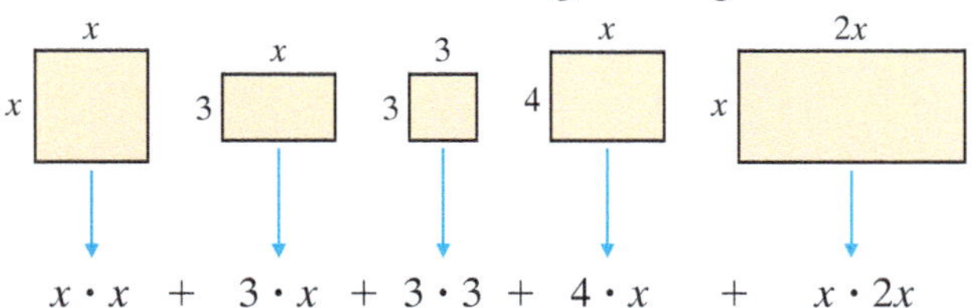

Area: $x \cdot x + 3 \cdot x + 3 \cdot 3 + 4 \cdot x + x \cdot 2x$

$= x^2 + 3x + 9 + 4x + 2x^2$

$= 3x^2 + 7x + 9$ Combine like terms.

Work Practice 11

Practice 11

Write a polynomial that describes the total area of the squares and rectangles shown below. Then simplify the polynomial.

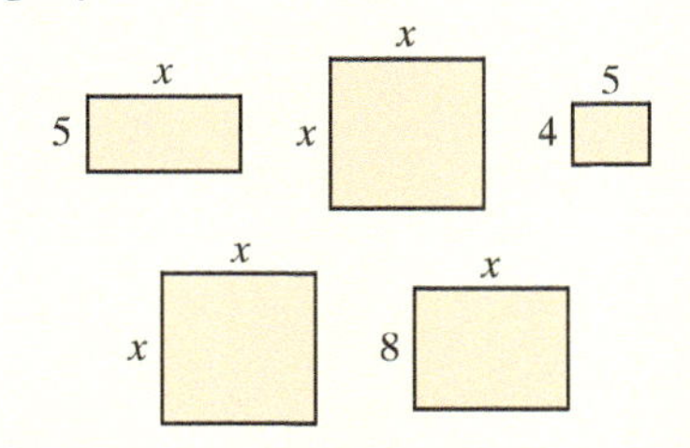

Objective E Simplifying Polynomials Containing Several Variables

A polynomial may contain more than one variable. One example is

$$5x + 3xy^2 - 6x^2y^2 + x^2y - 2y + 1$$

We call this expression a polynomial in several variables.

The **degree of a term** with more than one variable is the sum of the exponents on the variables. The **degree of a polynomial** in several variables is still the greatest degree of the terms of the polynomial.

Example 12 Identify the degrees of the terms and the degree of the polynomial $5x + 3xy^2 - 6x^2y^2 + x^2y - 2y + 1$.

Solution: To organize our work, we use a table.

Terms of Polynomial	Degree of Term	Degree of Polynomial
$5x$	1	
$3xy^2$	1 + 2, or 3	
$-6x^2y^2$	2 + 2, or 4	4 (greatest degree)
x^2y	2 + 1, or 3	
$-2y$	1	
1	0	

Work Practice 12

Practice 12

Identify the degrees of the terms and the degree of the polynomial $-2x^3y^2 + 4 - 8xy + 3x^3y + 5xy^2$.

To simplify a polynomial containing several variables, we combine any like terms.

Examples Simplify each polynomial by combining any like terms.

13. $3xy - 5y^2 + 7yx - 9x^2 = (3 + 7)xy - 5y^2 - 9x^2$

$= 10xy - 5y^2 - 9x^2$

14. $9a^2b - 6a^2 + 5b^2 + a^2b - 11a^2 + 2b^2$

$= 10a^2b - 17a^2 + 7b^2$

Helpful Hint This term can be written as $7yx$ or $7xy$.

Work Practice 13–14

Practice 13–14

Simplify each polynomial by combining any like terms.

13. $11ab - 6a^2 - ba + 8b^2$

14. $7x^2y^2 + 2y^2 - 4y^2x^2 + x^2 - y^2 + 5x^2$

Answers

11. $5x + x^2 + 20 + x^2 + 8x$; $2x^2 + 13x + 20$

12. 5, 0, 2, 4, 3; 5

13. $10ab - 6a^2 + 8b^2$

14. $3x^2y^2 + y^2 + 6x^2$

Objective F Inserting "Missing" Terms

To prepare for dividing polynomials in Section 3.7, let's practice writing a polynomial in descending powers of the variable and with no "missing" powers.

Recall from objective B that a polynomial such as

$$x^5 - 3x^3 + 2x^2 - 5x + 1$$

is written in descending powers of x because the powers of x decrease from left to right. Study the decreasing powers of x and notice that there is a "missing" power of x. This missing power is x^4. Writing a polynomial in decreasing powers of the variable helps you immediately determine important features of the polynomial, such as its degree. It is also sometimes helpful to write a polynomial so that there are no "missing" powers of x. For our polynomial above, if we simply insert a term of $0x^4$, which equals 0, we have an equivalent polynomial with no missing powers of x.

$$x^5 - 3x^3 + 2x^2 - 5x + 1 = x^5 + 0x^4 - 3x^3 + 2x^2 - 5x + 1$$

Practice 15

Write each polynomial in descending powers of the variable with no missing powers.

a. $x^2 + 9$
b. $9m^3 + m^2 - 5$
c. $-3a^3 + a^4$

Example 15 Write each polynomial in descending powers of the variable with no missing powers.

a. $x^2 - 4$ **b.** $3m^3 - m + 1$ **c.** $2x + x^4$

Solution:

a. $x^2 - 4 = x^2 + 0x^1 - 4$ or $x^2 + 0x - 4$ — Insert a missing term of $0x^1$ or $0x$.

b. $3m^3 - m + 1 = 3m^3 + 0m^2 - m + 1$ — Insert a missing term of $0m^2$.

c. $2x + x^4 = x^4 + 2x$ — Write in descending powers of variable.

$= x^4 + 0x^3 + 0x^2 + 2x + 0x^0$ — Insert missing terms of $0x^3$, $0x^2$, and $0x^0$ (or 0).

Work Practice 15

Helpful Hint

Since there is no constant as a last term, we insert a $0x^0$. This $0x^0$ (or 0) is the final power of x in our polynomial.

Answers

15. a. $x^2 + 0x + 9$ **b.** $9m^3 + m^2 + 0m - 5$ **c.** $a^4 - 3a^3 + 0a^2 + 0a + 0a^0$

Vocabulary, Readiness & Video Check

Use the choices below to fill in each blank. Not all choices will be used.

least	monomial	trinomial	coefficient
greatest	binomial	constant	

1. A ______________ is a polynomial with exactly two terms.
2. A ______________ is a polynomial with exactly one term.
3. A ______________ is a polynomial with exactly three terms.
4. The numerical factor of a term is called the ______________.
5. A number term is also called a ______________.
6. The degree of a polynomial is the ______________ degree of any term of the polynomial.

Martin-Gay Interactive Videos

See Video 3.3 

Watch the section lecture video and answer the following questions.

Objective A 7. How many terms does the polynomial in Example 1 have? What are they?

Objective B 8. For Example 2, why is the degree of each **term** found when the example asks for the degree of the **polynomial** only?

Objective C 9. From Example 3, what does the value of a polynomial depend on?

Objective D 10. When combining any like terms in a polynomial, as in Example 5, what are we doing to the polynomial?

Objective E 11. In Example 6, after combining like terms what is the degree of the binomial? Which term determines this?

Objective F 12. In Example 7, what power is "missing" from the original polynomial? What term is inserted to replace this missing power?

3.3 Exercise Set MyLab Math

Objective A *Complete the table for each polynomial. See Example 1.*

1. $x^2 - 3x + 5$

Term	Coefficient
x^2	
	-3
5	

2. $2x^3 - x + 4$

Term	Coefficient
	2
$-x$	
4	

3. $-5x^4 + 3.2x^2 + x - 5$

Term	Coefficient
$-5x^4$	
$3.2x^2$	
x	
-5	

4. $9.7x^7 - 3x^5 + x^3 - \frac{1}{4}x^2$

Term	Coefficient
$9.7x^7$	
$-3x^5$	
x^3	
$-\frac{1}{4}x^2$	

Objective B *Find the degree of each polynomial and determine whether it is a monomial, binomial, trinomial, or none of these. See Examples 2 and 3.*

5. $x + 2$

6. $-6y + 4$

7. $9m^3 - 5m^2 + 4m - 8$

8. $a + 5a^2 + 3a^3 - 4a^4$

9. $12x^4 - x^6 - 12x^2$

10. $7r^2 + 2r - 3r^5$

11. $3z - 5z^4$

12. $5y^6 + 2$

Objective C *Evaluate each polynomial when* **a.** $x = 0$ *and* **b.** $x = -1$. *See Examples 4 and 5.*

13. $5x - 6$

14. $2x - 10$

15. $x^2 - 5x - 2$

16. $x^2 + 3x - 4$

17. $-x^3 + 4x^2 - 15$

18. $-2x^3 + 3x^2 - 6$

A rocket is fired upward from the ground with an initial velocity of 200 feet per second. Neglecting air resistance, the height of the rocket in feet at any time t can be described by the polynomial $-16t^2 + 200t$. Find the height of the rocket at the times given in Exercises 19 through 22. See Example 5.

	Time, t (in seconds)	Height $-16t^2 + 200t$
19.	1	
20.	5	
21.	7.6	
22.	10.3	

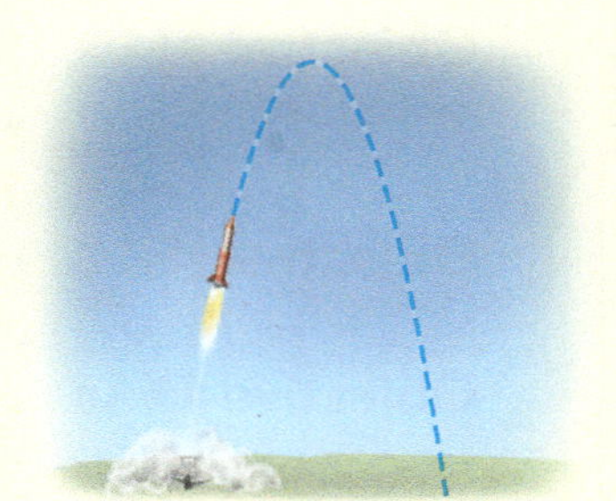

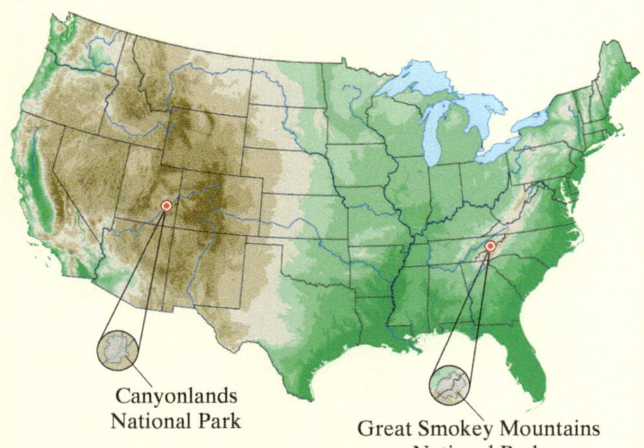

23. The polynomial $12x^2 - 26x + 454$ models the yearly number of visitors (in thousands) x years after 2010 at Canyonlands National Park. Use this polynomial to estimate the number of visitors to the park in 2020 ($x = 10$).

24. The polynomial $100x^2 - 238x + 9398$ models the yearly number of visitors (in thousands) x years after 2010 at Great Smoky Mountains National Park. Use this polynomial to estimate the number of visitors to the park in 2022 ($x = 12$).

25. The number of vinyl album sales (in millions) in the United States x years after 2010 is given by the polynomial $0.06x^2 + 1.4x + 2.4$ for 2010 through 2017. Use this model to predict the number of vinyl album sales in the United States in the year 2024 ($x = 14$). (See the Chapter Opener.)

26. Use the model in Exercise **25** to predict the number of vinyl album sales in the United States in the year 2022. (See the Chapter Opener.)

Objective D *Simplify each expression by combining like terms. See Examples 6 through 10.*

27. $9x - 20x$

28. $14y - 30y$

29. $14x^3 + 9x^3$

30. $18x^3 + 4x^3$

31. $7x^2 + 3 + 9x^2 - 10$

32. $8x^2 + 4 + 11x^2 - 20$

33. $15x^2 - 3x^2 - 13$

34. $12k^3 - 9k^3 + 11$

35. $8s - 5s + 4s$

36. $5y + 7y - 6y$

37. $0.1y^2 - 1.2y^2 + 6.7 - 1.9$

38. $7.6y + 3.2y^2 - 8y - 2.5y^2$

39. $\frac{2}{3}x^4 + 12x^3 + \frac{1}{6}x^4 - 19x^3 - 19$

40. $\frac{2}{5}x^4 - 23x^2 + \frac{1}{15}x^4 + 5x^2 - 5$

41. $\frac{3}{20}x^3 + \frac{1}{10} - \frac{3}{10}x - \frac{1}{5} - \frac{7}{20}x + 6x^2$

42. $\frac{5}{16}x^3 - \frac{1}{8} + \frac{3}{8}x + \frac{1}{4} - \frac{9}{16}x - 14x^2$

Write a polynomial that describes the total area of each set of rectangles and squares shown in Exercises 43 and 44. Then simplify the polynomial. See Example 11.

△ **43.**

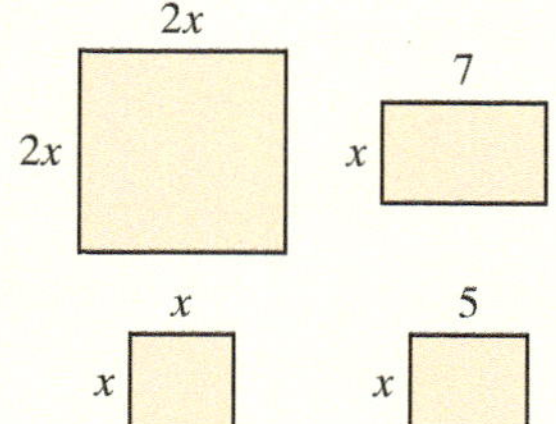

△ **44.**

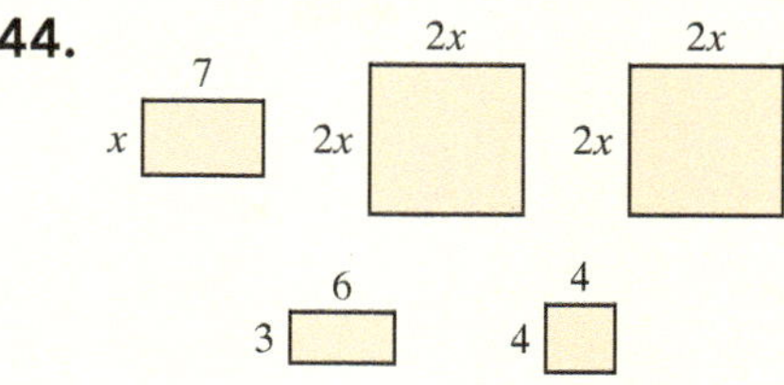

Recall that the perimeter of a figure such as the ones shown in Exercises 45 and 46 is the sum of the lengths of its sides. Write each perimeter as a polynomial. Then simplify the polynomial.

△ **45.**

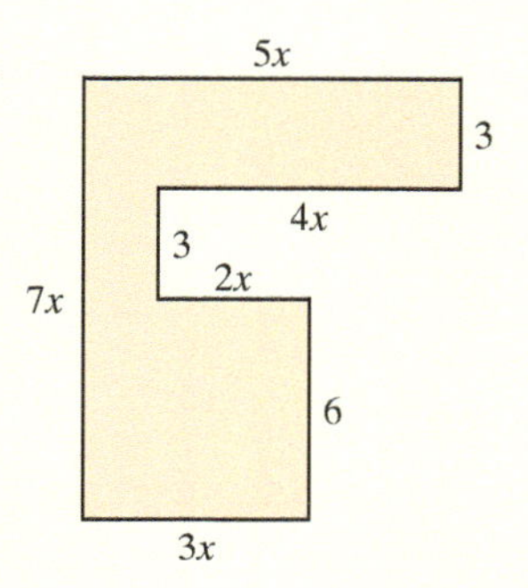

△ **46.**

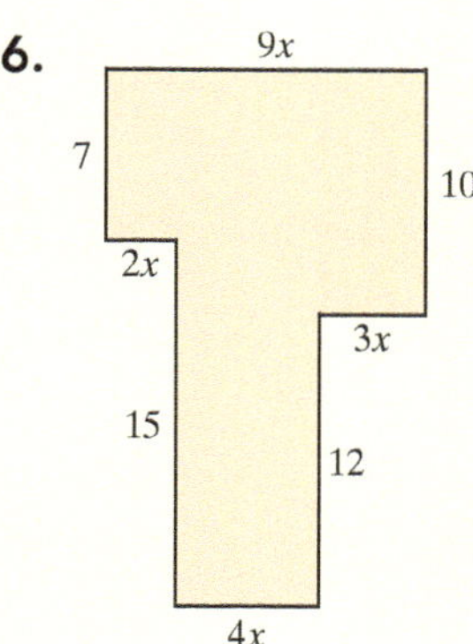

Objective E

Identify the degrees of the terms and the degree of the polynomial. See Example 12.

47. $9ab - 6a + 5b - 3$

48. $y^4 - 6y^3x + 2x^2y^2 - 5y^2 + 3$

49. $x^3y - 6 + 2x^2y^2 + 5y^3$

50. $2a^2b + 10a^4b - 9ab + 6$

Simplify each polynomial by combining any like terms. See Examples 13 and 14.

51. $3ab - 4a + 6ab - 7a$

52. $-9xy + 7y - xy - 6y$

53. $4x^2 - 6xy + 3y^2 - xy$

54. $3a^2 - 9ab + 4b^2 - 7ab$

55. $5x^2y + 6xy^2 - 5yx^2 + 4 - 9y^2x$

56. $17a^2b - 16ab^2 + 3a^3 + 4ba^3 - b^2a$

57. $14y^3 - 9 + 3a^2b^2 - 10 - 19b^2a^2$

58. $18x^4 + 2x^3y^3 - 1 - 2y^3x^3 - 17x^4$

Objective F

Write each polynomial in descending powers of the variable and with no missing powers. See Example 15.

59. $7x^2 + 3$

60. $5x^2 - 2$

61. $x^3 - 64$

62. $x^3 - 8$

63. $5y^3 + 2y - 10$

64. $6m^3 - 3m + 4$

65. $8y + 2y^4$

66. $11z + 4z^4$

67. $6x^5 + x^3 - 3x + 15$

68. $9y^5 - y^2 + 2y - 11$

Review

Simplify each expression. See Section 1.8.

69. $4 + 5(2x + 3)$

70. $9 - 6(5x + 1)$

71. $2(x - 5) + 3(5 - x)$

72. $-3(w + 7) + 5(w + 1)$

Concept Extensions

73. Describe how to find the degree of a term.

74. Describe how to find the degree of a polynomial.

75. Explain why xyz is a monomial while $x + y + z$ is a trinomial.

76. Explain why the degree of the term $5y^3$ is 3 and the degree of the polynomial $2y + y + 2y$ is 1.

Simplify, if possible.

77. $x^4 \cdot x^9$

78. $x^4 + x^9$

79. $a \cdot b^3 \cdot a^2 \cdot b^7$

80. $a + b^3 + a^2 + b^7$

81. $(y^5)^4 + (y^2)^{10}$

82. $x^5y^2 + y^2x^5$

Fill in the boxes so that the terms in each expression can be combined. Then simplify. Each exercise has more than one solution.

83. $7x^{\square} + 2x^{\square}$

84. $(3y^2)^{\square} + (4y^3)^{\square}$

85. Explain why the height of the rocket in Exercises **19** through **22** increases and then decreases as time passes.

86. Approximate (to the nearest tenth of a second) how long before the rocket in Exercises **19** through **22** hits the ground.

Simplify each polynomial by combining like terms.

87. $1.85x^2 - 3.76x + 9.25x^2 + 10.76 - 4.21x$

88. $7.75x + 9.16x^2 - 1.27 - 14.58x^2 - 18.34$

3.4 Adding and Subtracting Polynomials 

Objective A Adding Polynomials 

To add polynomials, we use commutative and associative properties and then combine like terms. To see if you are ready to add polynomials, try the Concept Check.

Concept Check When combining like terms in the expression $5x - 8x^2 - 8x$, which of the following is the proper result?

a. $-11x^2$ **b.** $-3x - 8x^2$ **c.** $-11x$ **d.** $-11x^4$

To Add Polynomials

To add polynomials, combine all like terms.

Examples Add.

1. $(4x^3 - 6x^2 + 2x + 7) + (5x^2 - 2x)$

$= 4x^3 - 6x^2 + 2x + 7 + 5x^2 - 2x$ Remove parentheses.

$= 4x^3 + (-6x^2 + 5x^2) + (2x - 2x) + 7$ Group like terms.

$= 4x^3 - x^2 + 7$ Simplify.

2. $(-2x^2 + 5x - 1) + (-2x^2 + x + 3)$

$= -2x^2 + 5x - 1 - 2x^2 + x + 3$ Remove parentheses.

$= (-2x^2 - 2x^2) + (5x + 1x) + (-1 + 3)$ Group like terms.

$= -4x^2 + 6x + 2$ Simplify.

Work Practice 1–2

Just as we can add numbers vertically, polynomials can be added vertically if we line up like terms underneath one another.

Example 3 Add $(7y^3 - 2y^2 + 7)$ and $(6y^2 + 1)$ using a vertical format.

Solution: Vertically line up like terms and add.

$$\begin{array}{r} 7y^3 - 2y^2 + 7 \\ 6y^2 + 1 \\ \hline 7y^3 + 4y^2 + 8 \end{array}$$

Work Practice 3

Objective B Subtracting Polynomials 

To subtract one polynomial from another, recall the definition of subtraction. To subtract a number, we add its opposite: $a - b = a + (-b)$. To subtract a polynomial, we also add its opposite. Just as $-b$ is the opposite of b, $-(x^2 + 5)$ is the opposite of $(x^2 + 5)$.

To Subtract Polynomials

To subtract two polynomials, change the signs of the terms of the polynomial being subtracted and then add.

Objectives

A Add Polynomials.

B Subtract Polynomials.

C Add or Subtract Polynomials in One Variable.

D Add or Subtract Polynomials in Several Variables.

Practice 1–2

Add.

1. $(3x^5 - 7x^3 + 2x - 1) + (3x^3 - 2x)$

2. $(5x^2 - 2x + 1) + (-6x^2 + x - 1)$

Practice 3

Add $(9y^2 - 6y + 5)$ and $(4y + 3)$ using a vertical format.

Answers

1. $3x^5 - 4x^3 - 1$ **2.** $-x^2 - x$

3. $9y^2 - 2y + 8$

Concept Check Answer

b

Practice 4

Subtract:
$(9x + 5) - (4x - 3)$

Example 4 Subtract: $(5x - 3) - (2x - 11)$

Solution: From the definition of subtraction, we have

$$(5x - 3) - (2x - 11) = (5x - 3) + [-(2x - 11)] \quad \text{Add the opposite.}$$
$$= (5x - 3) + (-2x + 11) \quad \text{Apply the distributive property.}$$
$$= 5x - 3 - 2x + 11 \quad \text{Remove parentheses.}$$
$$= 3x + 8 \quad \text{Combine like terms.}$$

Work Practice 4

Practice 5

Subtract:
$(4x^3 - 10x^2 + 1)$
$-(-4x^3 + x^2 - 11)$

Example 5 Subtract: $(2x^3 + 8x^2 - 6x) - (2x^3 - x^2 + 1)$

Solution: First, we change the sign of each term of the second polynomial; then we add.

$$(2x^3 + 8x^2 - 6x) - (2x^3 - x^2 + 1)$$
$$= (2x^3 + 8x^2 - 6x) + (-2x^3 + x^2 - 1)$$
$$= 2x^3 + 8x^2 - 6x - 2x^3 + x^2 - 1$$
$$= 2x^3 - 2x^3 + 8x^2 + x^2 - 6x - 1$$
$$= 9x^2 - 6x - 1 \quad \text{Combine like terms.}$$

Work Practice 5

Just as polynomials can be added vertically, so can they be subtracted vertically.

Practice 6

Subtract $(6y^2 - 3y + 2)$ from $(2y^2 - 2y + 7)$ using a vertical format.

Example 6 Subtract $(5y^2 + 2y - 6)$ from $(-3y^2 - 2y + 11)$ using a vertical format.

Solution: Arrange the polynomials in a vertical format, lining up like terms.

$$\begin{array}{r} -3y^2 - 2y + 11 \\ \underline{-(5y^2 + 2y - 6)} \end{array} \qquad \begin{array}{r} -3y^2 - 2y + 11 \\ \underline{-5y^2 - 2y + 6} \\ -8y^2 - 4y + 17 \end{array}$$

Work Practice 6

Helpful Hint

Don't forget to change the sign of each term in the polynomial being subtracted.

Objective C Adding and Subtracting Polynomials in One Variable

Let's practice adding and subtracting polynomials in one variable.

Practice 7

Subtract $(3x + 1)$ from the sum of $(4x - 3)$ and $(12x - 5)$.

Example 7 Subtract $(5z - 7)$ from the sum of $(8z + 11)$ and $(9z - 2)$.

Solution: Notice that $(5z - 7)$ is to be subtracted **from** a sum. The translation is

$$[(8z + 11) + (9z - 2)] - (5z - 7)$$
$$= 8z + 11 + 9z - 2 - 5z + 7 \quad \text{Remove grouping symbols.}$$
$$= 8z + 9z - 5z + 11 - 2 + 7 \quad \text{Group like terms.}$$
$$= 12z + 16 \quad \text{Combine like terms.}$$

Work Practice 7

Answers

4. $5x + 8$ **5.** $8x^3 - 11x^2 + 12$ **6.** $-4y^2 + y + 5$ **7.** $13x - 9$

Objective D Adding and Subtracting Polynomials in Several Variables

Now that we know how to add or subtract polynomials in one variable, we can also add and subtract polynomials in several variables.

Examples Add or subtract as indicated.

8. $(3x^2 - 6xy + 5y^2) + (-2x^2 + 8xy - y^2)$
$= 3x^2 - 6xy + 5y^2 - 2x^2 + 8xy - y^2$
$= x^2 + 2xy + 4y^2$ Combine like terms.

9. $(9a^2b^2 + 6ab - 3ab^2) - (5b^2a + 2ab - 3 - 9b^2)$
$= 9a^2b^2 + 6ab - 3ab^2 - 5b^2a - 2ab + 3 + 9b^2$
$= 9a^2b^2 + 4ab - 8ab^2 + 9b^2 + 3$ Combine like terms.

Work Practice 8–9

Practice 8–9

Add or subtract as indicated.

8. $(2a^2 - ab + 6b^2) + (-3a^2 + 4ab - 7b^2)$

9. $(5x^2y^2 + 3 - 9x^2y + y^2) - (-x^2y^2 + 7 - 8xy^2 + 2y^2)$

✓**Concept Check** If possible, simplify each expression by performing the indicated operation.

a. $2y + y$
b. $2y \cdot y$
c. $-2y - y$
d. $(-2y)(-y)$
e. $2x + y$

Answers
8. $-a^2 + 3ab - b^2$
9. $6x^2y^2 - 4 - 9x^2y + 8xy^2 - y^2$

✓**Concept Check Answers**
a. $3y$ **b.** $2y^2$ **c.** $-3y$ **d.** $2y^2$
e. cannot be simplified

Vocabulary, Readiness & Video Check

Simplify by combining like terms if possible.

1. $-9y - 5y$
2. $6m^5 + 7m^5$
3. $x + 6x$
4. $7z - z$
5. $5m^2 + 2m$
6. $8p^3 + 3p^2$

Martin-Gay Interactive Videos

See Video 3.4

Watch the section lecture video and answer the following questions.

Objective A **7.** In Example 1, like terms are combined when adding the polynomials. What are the two sets of like terms?

Objective B **8.** In Example 2, why can't parentheses just be removed as they were in Example 1?

Objective C **9.** In Example 3, why are we told to be careful when translating to an expression?

Objective D **10.** In Example 5, why aren't any signs changed when parentheses are removed?

3.4 Exercise Set MyLab Math 

Objective A *Add. See Examples 1 and 2.*

1. $(3x + 7) + (9x + 5)$
2. $(-y - 2) + (3y + 5)$
3. $(-7x + 5) + (-3x^2 + 7x + 5)$
4. $(3x - 8) + (4x^2 - 3x + 3)$
5. $(-5x^2 + 3) + (2x^2 + 1)$
6. $(3x^2 + 7) + (3x^2 + 9)$
7. $(-3y^2 - 4y) + (2y^2 + y - 1)$
8. $(7x^2 + 2x - 9) + (-3x^2 + 5)$
9. $(1.2x^3 - 3.4x + 7.9) + (6.7x^3 + 4.4x^2 - 10.9)$
10. $(9.6y^3 + 2.7y^2 - 8.6) + (1.1y^3 - 8.8y + 11.6)$
11. $\left(\frac{3}{4}m^2 - \frac{2}{5}m + \frac{1}{8}\right) + \left(-\frac{1}{4}m^2 - \frac{3}{10}m + \frac{11}{16}\right)$
12. $\left(-\frac{4}{7}n^2 + \frac{5}{6}n - \frac{1}{20}\right) + \left(\frac{3}{7}n^2 - \frac{5}{12}n - \frac{3}{10}\right)$

Add using a vertical format. See Example 3.

13. $\begin{array}{r} 3t^2 + 4 \\ \underline{5t^2 - 8} \end{array}$
14. $\begin{array}{r} 7x^3 + 3 \\ \underline{2x^3 - 7} \end{array}$
15. $\begin{array}{r} 10a^3 - 8a^2 + 4a + 9 \\ \underline{5a^3 + 9a^2 - 7a + 7} \end{array}$
16. $\begin{array}{r} 2x^3 - 3x^2 + x - 4 \\ \underline{5x^3 + 2x^2 - 3x + 2} \end{array}$

Objective B *Subtract. See Examples 4 and 5.*

17. $(2x + 5) - (3x - 9)$
18. $(4 + 5a) - (-a - 5)$
19. $(5x^2 + 4) - (-2x^2 + 6)$
20. $(-7y^2 + 5) - (-8y^2 + 12)$
21. $3x - (5x - 9)$
22. $4 - (-y - 4)$
23. $(2x^2 + 3x - 9) - (-4x + 7)$
24. $(-7x^2 + 4x + 7) - (-8x + 2)$
25. $(5x + 8) - (-2x^2 - 6x + 8)$
26. $(-6y^2 + 3y - 4) - (9y^2 - 4)$
27. $(0.7x^2 + 0.2x - 0.8) - (0.9x^2 + 1.4)$
28. $(-0.3y^2 + 0.6y - 0.3) - (0.5y^2 + 0.3)$
29. $\left(\frac{1}{4}z^2 - \frac{1}{5}z\right) - \left(-\frac{3}{20}z^2 + \frac{1}{10}z - \frac{7}{20}\right)$
30. $\left(\frac{1}{3}x^2 - \frac{2}{7}x\right) - \left(\frac{4}{21}x^2 + \frac{1}{21}x - \frac{2}{3}\right)$

Subtract using a vertical format. See Example 6.

31. $\begin{array}{r} 4z^2 - 8z + 3 \\ \underline{-(6z^2 + 8z - 3)} \end{array}$

32. $\begin{array}{r} 7a^2 - 9a + 6 \\ \underline{-(11a^2 - 4a + 2)} \end{array}$

33. $\begin{array}{r} 5u^5 - 4u^2 + 3u - 7 \\ \underline{-(3u^5 + 6u^2 - 8u + 2)} \end{array}$

34. $\begin{array}{r} 5x^3 - 4x^2 + 6x - 2 \\ \underline{-(3x^3 - 2x^2 - x - 4)} \end{array}$

Objectives A B C Mixed Practice *Add or subtract as indicated. See Examples 1 through 7.*

35. $(3x + 5) + (2x - 14)$

36. $(2y + 20) + (5y - 30)$

37. $(9x - 1) - (5x + 2)$

38. $(7y + 7) - (y - 6)$

39. $(14y + 12) + (-3y - 5)$

40. $(26y + 17) + (-20y - 10)$

41. $(x^2 + 2x + 1) - (3x^2 - 6x + 2)$

42. $(5y^2 - 3y - 1) - (2y^2 + y + 1)$

43. $(3x^2 + 5x - 8) + (5x^2 + 9x + 12) - (8x^2 - 14)$

44. $(2x^2 + 7x - 9) + (x^2 - x + 10) - (3x^2 - 30)$

45. $(-a^2 + 1) - (a^2 - 3) + (5a^2 - 6a + 7)$

46. $(-m^2 + 3) - (m^2 - 13) + (6m^2 - m + 1)$

Translating *Perform each indicated operation. See Examples 3, 6, and 7.*

47. Subtract $4x$ from $(7x - 3)$.

48. Subtract y from $(y^2 - 4y + 1)$.

49. Add $(4x^2 - 6x + 1)$ and $(3x^2 + 2x + 1)$.

50. Add $(-3x^2 - 5x + 2)$ and $(x^2 - 6x + 9)$.

51. Subtract $(5x + 7)$ from $(7x^2 + 3x + 9)$.

52. Subtract $(5y^2 + 8y + 2)$ from $(7y^2 + 9y - 8)$.

53. Subtract $(4y^2 - 6y - 3)$ from the sum of $(8y^2 + 7)$ and $(6y + 9)$.

54. Subtract $(4x^2 - 2x + 2)$ from the sum of $(x^2 + 7x + 1)$ and $(7x + 5)$.

55. Subtract $(3x^2 - 4)$ from the sum of $(x^2 - 9x + 2)$ and $(2x^2 - 6x + 1)$.

56. Subtract $(y^2 - 9)$ from the sum of $(3y^2 + y + 4)$ and $(-2y^2 - 6y - 10)$.

Objective D *Add or subtract as indicated. See Examples 8 and 9.*

57. $(9a + 6b - 5) + (-11a - 7b + 6)$

58. $(3x - 2 + 6y) + (7x - 2 - y)$

59. $(4x^2 + y^2 + 3) - (x^2 + y^2 - 2)$

60. $(7a^2 - 3b^2 + 10) - (-2a^2 + b^2 - 12)$

61. $(x^2 + 2xy - y^2) + (5x^2 - 4xy + 20y^2)$

62. $(a^2 - ab + 4b^2) + (6a^2 + 8ab - b^2)$

63. $(11r^2s + 16rs - 3 - 2r^2s^2) - (3sr^2 + 5 - 9r^2s^2)$

64. $(3x^2y - 6xy + x^2y^2 - 5) - (11x^2y^2 - 1 + 5yx^2)$

Objectives A B C Mixed Practice

For Exercises 65 through 68, find the perimeter of each figure.

65.

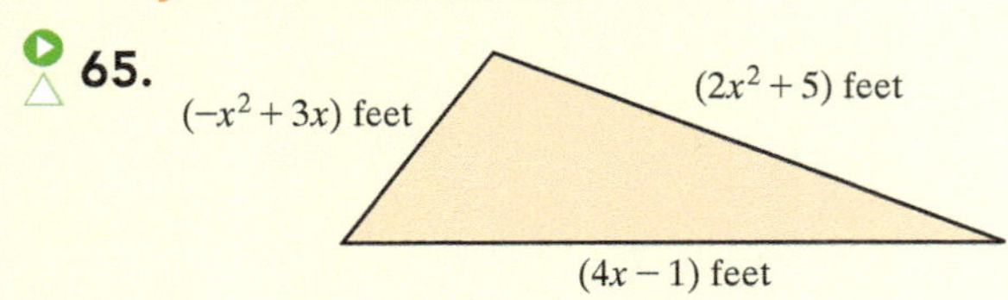

66.

$(-x + 4)$ centimeters
x^2 centimeters
$5x$ centimeters
$(x^2 - 6x - 2)$ centimeters

67.

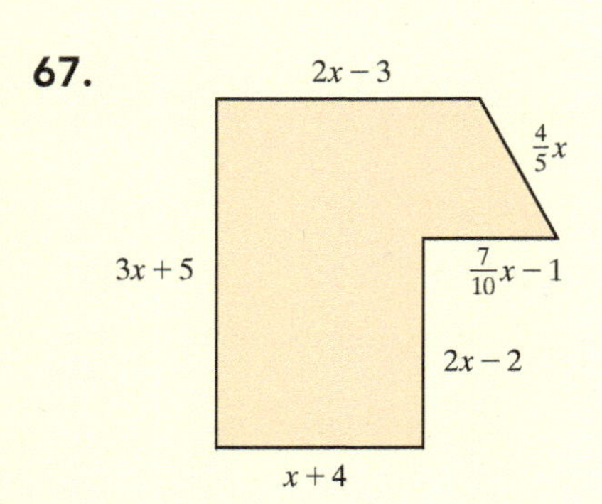

68.

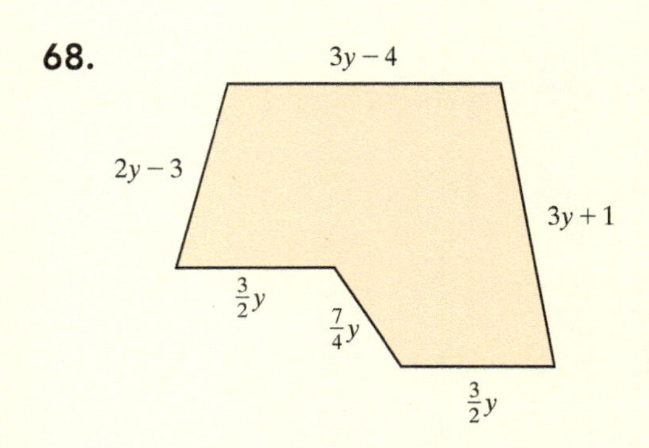

69. A wooden beam is $(4y^2 + 4y + 1)$ meters long. If a piece $(y^2 - 10)$ meters is cut off, express the length of the remaining piece of beam as a polynomial in y.

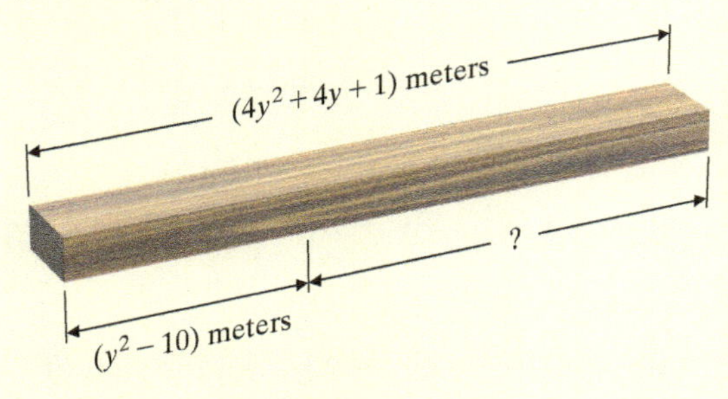

70. A piece of quarter-round molding is $(13x - 7)$ inches long. If a piece $(2x + 2)$ inches long is removed, express the length of the remaining piece of molding as a polynomial in x.

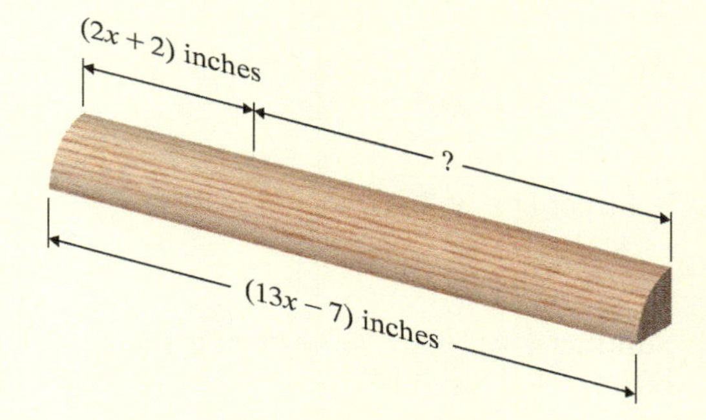

Perform each indicated operation.

71. $[(1.2x^2 - 3x + 9.1) - (7.8x^2 - 3.1 + 8)] + (1.2x - 6)$

72. $[(7.9y^4 - 6.8y^3 + 3.3y) + (6.1y^3 - 5)] - (4.2y^4 + 1.1y - 1)$

Review

Multiply. See Section 3.1.

73. $3x(2x)$

74. $-7x(x)$

75. $(12x^3)(-x^5)$

76. $6r^3(7r^{10})$

77. $10x^2(20xy^2)$

78. $-z^2y(11zy)$

Concept Extensions

Fill in the squares so that each is a true statement.

79. $3x^{\square} + 4x^2 = 7x^{\square}$

80. $9y^7 + 3y^{\square} = 12y^7$

81. $2x^{\square} + 3x^{\square} - 5x^{\square} + 4x^{\square} = 6x^4 - 2x^3$

82. $3y^{\square} + 7y^{\square} - 2y^{\square} - y^{\square} = 10y^5 - 3y^2$

Match each expression on the left with its simplification on the right. Not all letters on the right must be used and a letter may be used more than once. See the first Concept Check in this section.

83. $10y - 6y^2 - y$

84. $5x + 5x$

85. $(5x - 3) + (5x - 3)$

86. $(15x - 3) - (5x - 3)$

a. $3y$
b. $9y - 6y^2$
c. $10x$
d. $25x^2$
e. $10x - 6$
f. none of these

Simplify each expression by performing the indicated operation. Explain how you arrived at each answer. See the second Concept Check in this section.

87. **a.** $z + 3z$
b. $z \cdot 3z$
c. $-z - 3z$
d. $(-z)(-3z)$

88. **a.** $2y + y$
b. $2y \cdot y$
c. $-2y - y$
d. $(-2y)(-y)$

89. **a.** $m \cdot m \cdot m$
b. $m + m + m$
c. $(-m)(-m)(-m)$
d. $-m - m - m$

90. **a.** $x + x$
b. $x \cdot x$
c. $-x - x$
d. $(-x)(-x)$

91. The polynomial $437x^2 + 4868x + 3292$ represents the electricity generated (in thousand megawatthours) by photovoltaic solar sources in the United States during 2012–2015. The polynomial $-4489x^2 + 29{,}816x + 141{,}244$ represents the electricity generated (in thousand megawatthours) by wind power in the United States during 2012–2015. In both polynomials, x represents the number of years after 2012. Find a polynomial for the total electricity generated by both solar and wind power during 2012–2015. (*Source:* Based on information from the Energy Information Administration)

92. The polynomial $-43x^2 + 264x + 15{,}565$ represents the electricity generated (in thousand megawatt hours) by geothermal sources in the United States during 2012–2015. The polynomial $-603x^2 - 7199x + 276{,}272$ represents the electricity generated (in thousand megawatt hours) by conventional hydropower in the United States during 2012–2015. In both polynomials, x represents the number of years after 2012. Find a polynomial for the total electricity generated by both geothermal and hydropower during 2012–2015. (*Source:* based on data from the Energy Information Administration)

3.5 Multiplying Polynomials

Objectives

A Multiply Monomials.

B Multiply a Monomial by a Polynomial.

C Multiply Two Polynomials.

D Multiply Polynomials Vertically.

Objective A Multiplying Monomials

Recall from Section 3.1 that to multiply two monomials such as $(-5x^3)$ and $(-2x^4)$, we use the associative and commutative properties and regroup. Remember also that to multiply exponential expressions with a common base, we use the product rule for exponents and add exponents.

$$(-5x^3)(-2x^4) = (-5)(-2)(x^3 \cdot x^4)$$ Use the commutative and associative properties.

$$= 10x^7$$ Multiply.

Practice 1–3

Multiply.

1. $10x \cdot 9x$
2. $8x^3(-11x^7)$
3. $(-5x^4)(-x)$

Examples Multiply.

1. $6x \cdot 4x = (6 \cdot 4)(x \cdot x)$ Use the commutative and associative properties.
   $= 24x^2$ Multiply.
2. $-7x^2 \cdot 2x^5 = (-7 \cdot 2)(x^2 \cdot x^5)$
   $= -14x^7$
3. $(-12x^5)(-x) = (-12x^5)(-1x)$
   $= (-12)(-1)(x^5 \cdot x)$
   $= 12x^6$

Work Practice 1–3

✓**Concept Check** Simplify.

a. $3x \cdot 2x$ b. $3x + 2x$

Objective B Multiplying Monomials by Polynomials

To multiply a monomial such as $7x$ by a trinomial such as $x^2 + 2x + 5$, we use the distributive property.

Practice 4–6

Multiply.

4. $4x(x^2 + 4x + 3)$
5. $8x(7x^4 + 1)$
6. $-2x^3(3x^2 - x + 2)$

Examples Multiply.

4. $7x(x^2 + 2x + 5) = 7x(x^2) + 7x(2x) + 7x(5)$ Apply the distributive property.
   $= 7x^3 + 14x^2 + 35x$ Multiply.
5. $5x(2x^3 + 6) = 5x(2x^3) + 5x(6)$ Apply the distributive property.
   $= 10x^4 + 30x$ Multiply.
6. $-3x^2(5x^2 + 6x - 1)$
   $= (-3x^2)(5x^2) + (-3x^2)(6x) + (-3x^2)(-1)$ Apply the distributive property.
   $= -15x^4 - 18x^3 + 3x^2$ Multiply.

Work Practice 4–6

Answers

1. $90x^2$ 2. $-88x^{10}$ 3. $5x^5$
4. $4x^3 + 16x^2 + 12x$ 5. $56x^5 + 8x$
6. $-6x^5 + 2x^4 - 4x^3$

✓**Concept Check Answers**

a. $6x^2$ b. $5x$

Objective C Multiplying Two Polynomials

We also use the distributive property to multiply two binomials.

Example 7 Multiply.

a. $(m + 4)(m + 6)$ **b.** $(3x + 2)(2x - 5)$

Solution:

a. $(m + 4)(m + 6) = m(m + 6) + 4(m + 6)$ Use the distributive property.

$= m \cdot m + m \cdot 6 + 4 \cdot m + 4 \cdot 6$ Use the distributive property.

$= m^2 + 6m + 4m + 24$ Multiply.

$= m^2 + 10m + 24$ Combine like terms.

b. $(3x + 2)(2x - 5) = 3x(2x - 5) + 2(2x - 5)$ Use the distributive property.

$= 3x(2x) + 3x(-5) + 2(2x) + 2(-5)$

$= 6x^2 - 15x + 4x - 10$ Multiply.

$= 6x^2 - 11x - 10$ Combine like terms.

Work Practice 7

Practice 7

Multiply:

a. $(x + 5)(x + 10)$

b. $(4x + 5)(3x - 4)$

This idea can be expanded so that we can multiply any two polynomials.

To Multiply Two Polynomials

Multiply each term of the first polynomial by each term of the second polynomial, and then combine like terms.

Examples Multiply.

8. $(2x - y)^2$

$= (2x - y)(2x - y)$ Using the meaning of an exponent, we have 2 factors of $(2x - y)$.

$= 2x(2x) + 2x(-y) + (-y)(2x) + (-y)(-y)$

$= 4x^2 - 2xy - 2xy + y^2$ Multiply.

$= 4x^2 - 4xy + y^2$ Combine like terms.

9. $(t + 2)(3t^2 - 4t + 2)$

$= t(3t^2) + t(-4t) + t(2) + 2(3t^2) + 2(-4t) + 2(2)$

$= 3t^3 - 4t^2 + 2t + 6t^2 - 8t + 4$

$= 3t^3 + 2t^2 - 6t + 4$ Combine like terms.

Work Practice 8–9

Practice 8–9

Multiply.

8. $(3x - 2y)^2$

9. $(x + 3)(2x^2 - 5x + 4)$

✓**Concept Check** Square where indicated. Simplify if possible.

a. $(4a)^2 + (3b)^2$ **b.** $(4a + 3b)^2$

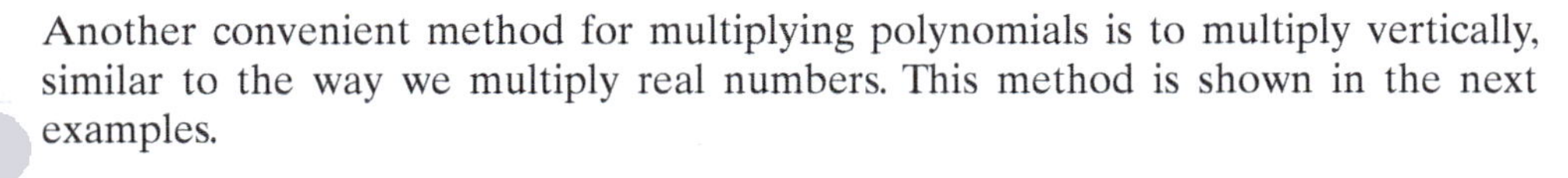

Objective D Multiplying Polynomials Vertically

Another convenient method for multiplying polynomials is to multiply vertically, similar to the way we multiply real numbers. This method is shown in the next examples.

Answers

7. a. $x^2 + 15x + 50$ **b.** $12x^2 - x - 20$ **8.** $9x^2 - 12xy + 4y^2$ **9.** $2x^3 + x^2 - 11x + 12$

✓**Concept Check Answers**

a. $16a^2 + 9b^2$ **b.** $16a^2 + 24ab + 9b^2$

Practice 10

Multiply vertically:
$(3y^2 + 1)(y^2 - 4y + 5)$

Example 10 Multiply vertically: $(2y^2 + 5)(y^2 - 3y + 4)$

Solution:

$$\begin{array}{r} y^2 - 3y + 4 \\ 2y^2 + 5 \\ \hline 5y^2 - 15y + 20 \\ 2y^4 - 6y^3 + 8y^2 \qquad\quad \\ \hline 2y^4 - 6y^3 + 13y^2 - 15y + 20 \end{array}$$

Multiply $y^2 - 3y + 4$ by 5.
Multiply $y^2 - 3y + 4$ by $2y^2$.
Combine like terms.

Work Practice 10

Practice 11

Find the product of $(4x^2 - x - 1)$ and $(3x^2 + 6x - 2)$ using a vertical format.

Example 11 Find the product of $(2x^2 - 3x + 4)$ and $(x^2 + 5x - 2)$ using a vertical format.

Solution: First, we arrange the polynomials in a vertical format. Then we multiply each term of the first polynomial by each term of the second polynomial.

$$\begin{array}{r} 2x^2 - 3x + 4 \\ x^2 + 5x - 2 \\ \hline -4x^2 + 6x - 8 \\ 10x^3 - 15x^2 + 20x \qquad \\ 2x^4 - 3x^3 + 4x^2 \qquad\qquad\quad \\ \hline 2x^4 + 7x^3 - 15x^2 + 26x - 8 \end{array}$$

Multiply $2x^2 - 3x + 4$ by -2.
Multiply $2x^2 - 3x + 4$ by $5x$.
Multiply $2x^2 - 3x + 4$ by x^2.
Combine like terms.

Work Practice 11

Answers

10. $3y^4 - 12y^3 + 16y^2 - 4y + 5$
11. $12x^4 + 21x^3 - 17x^2 - 4x + 2$

Vocabulary, Readiness & Video Check

Fill in each blank with the correct choice.

1. The expression $5x(3x + 2)$ equals $5x \cdot 3x + 5x \cdot 2$ by the ______________ property.
a. commutative **b.** associative **c.** distributive

2. The expression $(x + 4)(7x - 1)$ equals $x(7x - 1) + 4(7x - 1)$ by the ______________ property.
a. commutative **b.** associative **c.** distributive

3. The expression $(5y - 1)^2$ equals ______________.
a. $2(5y - 1)$ **b.** $(5y - 1)(5y + 1)$ **c.** $(5y - 1)(5y - 1)$

4. The expression $9x \cdot 3x$ equals ______________.
a. $27x$ **b.** $27x^2$ **c.** $12x$ **d.** $12x^2$

Perform the indicated operation, if possible.

5. $x^3 \cdot x^5$
6. $x^2 \cdot x^6$
7. $x^3 + x^5$
8. $x^2 + x^6$
9. $x^7 \cdot x^7$
10. $x^{11} \cdot x^{11}$
11. $x^7 + x^7$
12. $x^{11} + x^{11}$

Martin-Gay Interactive Videos

See Video 3.5

Watch the section lecture video and answer the following questions.

Objective A **13.** For Example 1, we use the product property to multiply the monomials. Is it possible to add the same two monomials? Why or why not?

Objective B **14.** What property and what exponent rule are used in Examples 3 and 4?

Objective C **15.** In Example 5, how many times is the distributive property actually applied? Explain.

Objective D **16.** Would you say the vertical format used in Example 8 also applies the distributive property? Explain.

3.5 Exercise Set MyLab Math

Objective A *Multiply. See Examples 1 through 3.*

1. $8x^2 \cdot 3x$

2. $6x \cdot 3x^2$

3. $(-x^3)(-x)$

4. $(-x^6)(-x)$

5. $-4n^3 \cdot 7n^7$

6. $9t^6(-3t^5)$

7. $(-3.1x^3)(4x^9)$

8. $(-5.2x^4)(3x^4)$

9. $\left(-\frac{1}{3}y^2\right)\left(\frac{2}{5}y\right)$

10. $\left(-\frac{3}{4}y^7\right)\left(\frac{1}{7}y^4\right)$

11. $(2x)(-3x^2)(4x^5)$

12. $(x)(5x^4)(-6x^7)$

Objective B *Multiply. See Examples 4 through 6.*

13. $3x(2x + 5)$

14. $2x(6x + 3)$

15. $7x(x^2 + 2x - 1)$

16. $5y(y^2 + y - 10)$

17. $-2a(a + 4)$

18. $-3a(2a + 7)$

19. $3x(2x^2 - 3x + 4)$

20. $4x(5x^2 - 6x - 10)$

21. $3a^2(4a^3 + 15)$

22. $9x^3(5x^2 + 12)$

23. $-2a^2(3a^2 - 2a + 3)$

24. $-4b^2(3b^3 - 12b^2 - 6)$

25. $3x^2y(2x^3 - x^2y^2 + 8y^3)$

26. $4xy^2(7x^3 + 3x^2y^2 - 9y^3)$

27. $-y(4x^3 - 7x^2y + xy^2 + 3y^3)$

28. $-x(6y^3 - 5xy^2 + x^2y - 5x^3)$

29. $\frac{1}{2}x^2(8x^2 - 6x + 1)$

30. $\frac{1}{3}y^2(9y^2 - 6y + 1)$

Objective C *Multiply. See Examples 7 through 9.*

31. $(x + 4)(x + 3)$

32. $(x + 2)(x + 9)$

33. $(a + 7)(a - 2)$

34. $(y - 10)(y + 11)$

35. $\left(x + \frac{2}{3}\right)\left(x - \frac{1}{3}\right)$

36. $\left(x + \frac{3}{5}\right)\left(x - \frac{2}{5}\right)$

37. $(3x^2 + 1)(4x^2 + 7)$

38. $(5x^2 + 2)(6x^2 + 2)$

39. $(4x - 3)(3x - 5)$

40. $(8x - 3)(2x - 4)$

41. $(1 - 3a)(1 - 4a)$

42. $(3 - 2a)(2 - a)$

43. $(2y - 4)^2$

44. $(6x - 7)^2$

45. $(x - 2)(x^2 - 3x + 7)$

46. $(x + 3)(x^2 + 5x - 8)$

47. $(x + 5)(x^3 - 3x + 4)$

48. $(a + 2)(a^3 - 3a^2 + 7)$

49. $(2a - 3)(5a^2 - 6a + 4)$

50. $(3 + b)(2 - 5b - 3b^2)$

51. $(7xy - y)^2$

52. $(x^2 - 4)^2$

Objective D *Multiply vertically. See Examples 10 and 11.*

53. $(2x - 11)(6x + 1)$

54. $(4x - 7)(5x + 1)$

55. $(x + 3)(2x^2 + 4x - 1)$

56. $(4x - 5)(8x^2 + 2x - 4)$

57. $(x^2 + 5x - 7)(2x^2 - 7x - 9)$

58. $(3x^2 - x + 2)(x^2 + 2x + 1)$

Objectives A B C D Mixed Practice *Multiply. See Examples 1 through 11.*

59. $-1.2y(-7y^6)$

60. $-4.2x(-2x^5)$

61. $-3x(x^2 + 2x - 8)$

62. $-5x(x^2 - 3x + 10)$

63. $(x + 19)(2x + 1)$

64. $(3y + 4)(y + 11)$

65. $\left(x + \frac{1}{7}\right)\left(x - \frac{3}{7}\right)$

66. $\left(m + \frac{2}{9}\right)\left(m - \frac{1}{9}\right)$

67. $(3y + 5)^2$

68. $(7y + 2)^2$

69. $(a + 4)(a^2 - 6a + 6)$

70. $(t + 3)(t^2 - 5t + 5)$

Express as the product of polynomials. Then multiply.

71. Find the area of the rectangle.

$(2x + 5)$ yards

$(2x - 5)$ yards

72. Find the area of the square field.

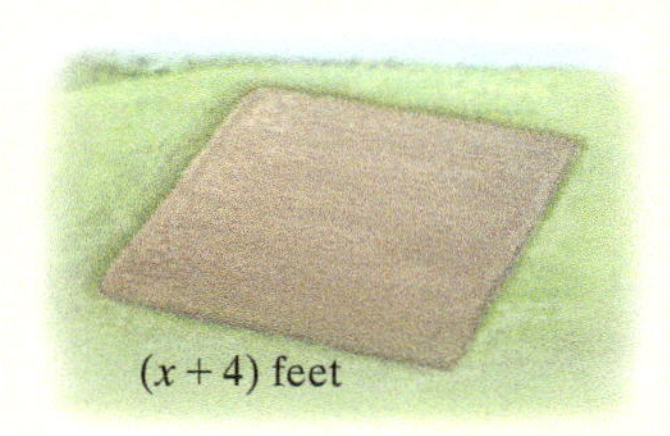

73. Find the area of the triangle.

$4x$ inches

$(3x - 2)$ inches

△ **74.** Find the volume of the cube-shaped glass block.

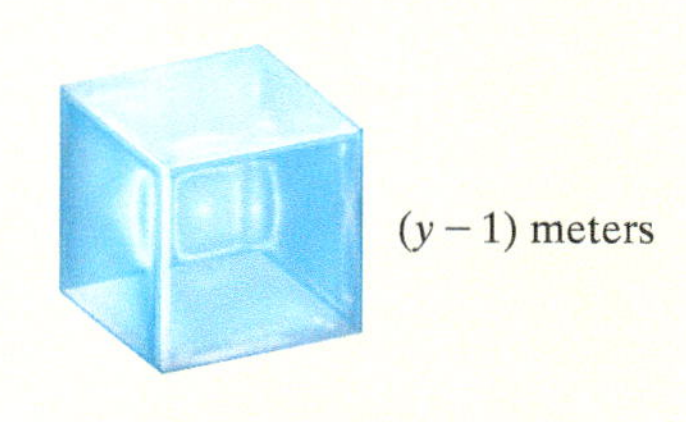

Review

In this section, we review operations on monomials. Study the box below, and then proceed. See Sections 1.8, 3.1, and 3.2.

Operations on Monomials	
Multiply	Review the product rule for exponents.
Divide	Review the quotient rule for exponents.
Add or Subtract	Remember, we may only combine like terms.

Perform the operations on the monomials, if possible. The first two rows have been completed for you.

	Monomials	Add	Subtract	Multiply	Divide
	$6x, 3x$	$6x + 3x = 9x$	$6x - 3x = 3x$	$6x \cdot 3x = 18x^2$	$\frac{6x}{3x} = 2$
	$-12x^2, 2x$	$-12x^2 + 2x$; can't be simplified	$-12x^2 - 2x$; can't be simplified	$-12x^2 \cdot 2x = -24x^3$	$\frac{-12x^2}{2x} = -6x$
75.	$5a, 15a$				
76.	$4y^3, 4y^7$				
77.	$-3y^5, 9y^4$				
78.	$-14x^2, 2x^2$				

Concept Extensions

79. Perform each indicated operation. Explain the difference between the two expressions.

a. $(3x + 5) + (3x + 7)$

b. $(3x + 5)(3x + 7)$

80. Perform each indicated operation. Explain the difference between the two expressions.

a. $(8x - 3) - (5x - 2)$

b. $(8x - 3)(5x - 2)$

Mixed Practice (*Sections 3.4 and 3.5*) *Perform the indicated operations.*

81. $(3x - 1) + (10x - 6)$

82. $(2x - 1) + (10x - 7)$

83. $(3x - 1)(10x - 6)$

84. $(2x - 1)(10x - 7)$

85. $(3x - 1) - (10x - 6)$

86. $(2x - 1) - (10x - 7)$

△ **87.** The area of the largest rectangle below is $x(x + 3)$. Find another expression for this area by finding the sum of the areas of the smaller rectangles.

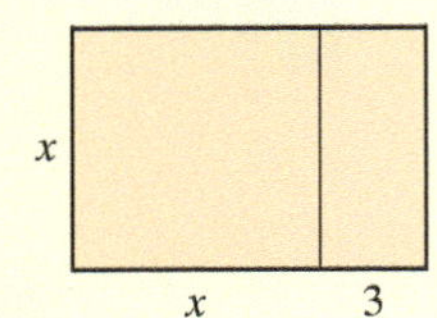

△ **88.** The area of the figure below is $(x + 2)(x + 3)$. Find another expression for this area by finding the sum of the areas of the smaller rectangles.

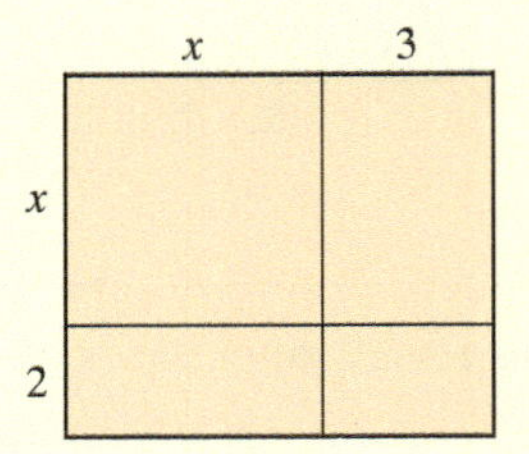

89. Write an expression for the area of the largest rectangle below in two different ways.

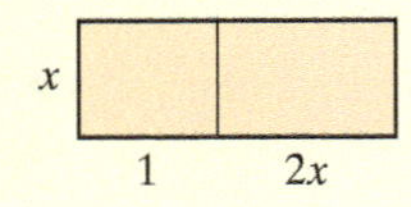

90. Write an expression for the area of the figure below in two different ways.

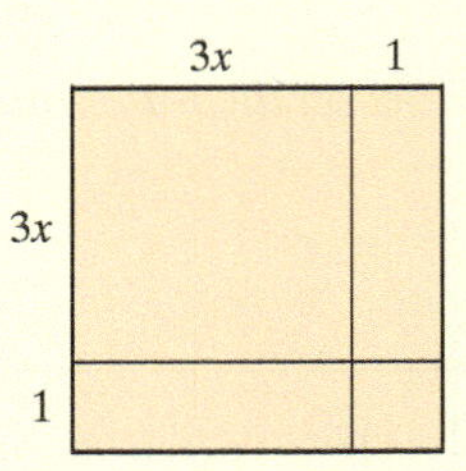

Simplify. See the Concept Checks in this section.

91. $5a + 6a$

92. $5a \cdot 6a$

Square where indicated. Simplify if possible.

93. $(5x)^2 + (2y)^2$

94. $(5x + 2y)^2$

95. Multiply each of the following polynomials.

a. $(a + b)(a - b)$

b. $(2x + 3y)(2x - 3y)$

c. $(4x + 7)(4x - 7)$

d. Can you make a general statement about all products of the form $(x + y)(x - y)$?

96. Evaluate each of the following.

a. $(2 + 3)^2$; $2^2 + 3^2$

b. $(8 + 10)^2$; $8^2 + 10^2$

c. Does $(a + b)^2 = a^2 + b^2$ no matter what the values of a and b are? Why or why not?

3.6 Special Products

Objective A Using the FOIL Method

Objectives

A Multiply Two Binomials Using the FOIL Method.

B Square a Binomial.

C Multiply the Sum and Difference of Two Terms.

D Use Special Products to Multiply Binomials.

In this section, we multiply binomials using special products. First, we introduce a special order for multiplying binomials called the FOIL order or method. This order, or pattern, is a result of the distributive property. We demonstrate by multiplying $(3x + 1)$ by $(2x + 5)$.

The FOIL Method

F stands for the product of the **First** terms. $(3x + 1)(2x + 5)$

$(3x)(2x) = 6x^2$ F

O stands for the product of the **Outer** terms. $(3x + 1)(2x + 5)$

$(3x)(5) = 15x$ O

I stands for the product of the **Inner** terms. $(3x + 1)(2x + 5)$

$(1)(2x) = 2x$ I

L stands for the product of the **Last** terms. $(3x + 1)(2x + 5)$

$(1)(5) = 5$ L

$$(3x + 1)(2x + 5) = \overset{F}{6x^2} + \overset{O}{15x} + \overset{I}{2x} + \overset{L}{5}$$
$$= 6x^2 + 17x + 5 \quad \text{Combine like terms.}$$

Let's practice multiplying binomials using the FOIL method.

Example 1 Multiply: $(x - 3)(x + 4)$

Solution:

$$(x - 3)(x + 4) = \overset{F}{(x)(x)} + \overset{O}{(x)(4)} + \overset{I}{(-3)(x)} + \overset{L}{(-3)(4)}$$
$$= x^2 + 4x - 3x - 12$$
$$= x^2 + x - 12 \quad \text{Combine like terms.}$$

Work Practice 1

Practice 1

Multiply: $(x + 7)(x - 5)$

Helpful Hint Remember that the FOIL order for multiplying can be used only for the product of 2 binomials.

Example 2 Multiply: $(5x - 7)(x - 2)$

Solution:

$$(5x - 7)(x - 2) = \overset{F}{5x(x)} + \overset{O}{5x(-2)} + \overset{I}{(-7)(x)} + \overset{L}{(-7)(-2)}$$
$$= 5x^2 - 10x - 7x + 14$$
$$= 5x^2 - 17x + 14 \quad \text{Combine like terms.}$$

Work Practice 2

Practice 2

Multiply: $(6x - 1)(x - 4)$

Answers

1. $x^2 + 2x - 35$ **2.** $6x^2 - 25x + 4$

Practice 3

Multiply: $(2y^2 + 3)(y - 4)$

Example 3 Multiply: $(y^2 + 6)(2y - 1)$

Solution:

$$(y^2 + 6)(2y - 1) = \overset{F}{2y^3} - \overset{O}{1y^2} + \overset{I}{12y} - \overset{L}{6}$$

Notice in this example that there are no like terms that can be combined, so the product is $2y^3 - y^2 + 12y - 6$.

Work Practice 3

Objective B Squaring Binomials

An expression such as $(3y + 1)^2$ is called the square of a binomial. Since $(3y + 1)^2 = (3y + 1)(3y + 1)$, we can use the FOIL method to find this product.

Practice 4

Multiply: $(2x + 9)^2$

Example 4 Multiply: $(3y + 1)^2$

Solution:

$$\begin{aligned}(3y + 1)^2 &= (3y + 1)(3y + 1) \\ &= \overset{F}{(3y)(3y)} + \overset{O}{(3y)(1)} + \overset{I}{1(3y)} + \overset{L}{1(1)} \\ &= 9y^2 + 3y + 3y + 1 \\ &= 9y^2 + 6y + 1\end{aligned}$$

Work Practice 4

Notice the pattern that appears in Example 4.

$$(3y + 1)^2 = 9y^2 + 6y + 1$$

$9y^2$ is the first term of the binomial squared: $(3y)^2 = 9y^2$.

$6y$ is 2 times the product of both terms of the binomial: $(2)(3y)(1) = 6y$.

1 is the second term of the binomial squared: $(1)^2 = 1$.

This pattern leads to the formulas below, which can be used when squaring a binomial. We call these **special products**.

Squaring a Binomial

A binomial squared is equal to the square of the first term plus or minus twice the product of both terms plus the square of the second term.

$$(a + b)^2 = a^2 + 2ab + b^2$$

$$(a - b)^2 = a^2 - 2ab + b^2$$

This product can be visualized geometrically.

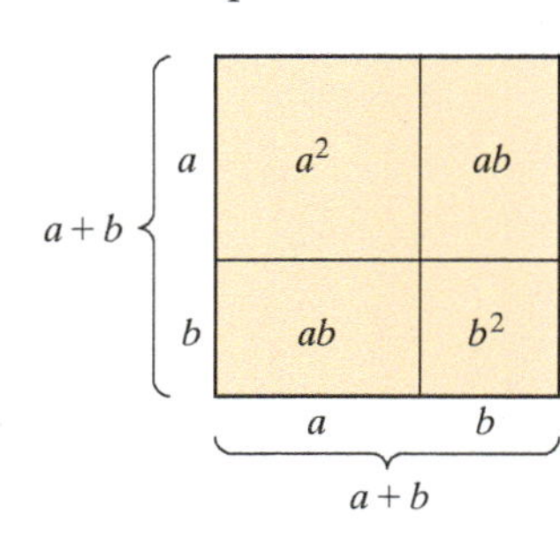

The area of the large square is side · side.

Area $= (a + b)(a + b) = (a + b)^2$

The area of the large square is also the sum of the areas of the smaller rectangles.

Area $= a^2 + ab + ab + b^2 = a^2 + 2ab + b^2$

Thus, $(a + b)^2 = a^2 + 2ab + b^2$.

Answers

3. $2y^3 - 8y^2 + 3y - 12$
4. $4x^2 + 36x + 81$

Examples Use a special product to square each binomial.

	first term squared	plus or minus	twice the product of the terms	plus	second term squared
5. $(t+2)^2 =$	t^2	$+$	$2(t)(2)$	$+$	$2^2 = t^2 + 4t + 4$
6. $(p-q)^2 =$	p^2	$-$	$2(p)(q)$	$+$	$q^2 = p^2 - 2pq + q^2$
7. $(2x+5)^2 =$	$(2x)^2$	$+$	$2(2x)(5)$	$+$	$5^2 = 4x^2 + 20x + 25$
8. $(x^2-7y)^2 =$	$(x^2)^2$	$-$	$2(x^2)(7y)$	$+$	$(7y)^2 = x^4 - 14x^2y + 49y^2$

Work Practice 5–8

Practice 5–8

Use a special product to square each binomial.

5. $(y+3)^2$
6. $(r-s)^2$
7. $(6x+5)^2$
8. $(x^2-3y)^2$

Helpful Hint

Notice that

$(a+b)^2 \neq a^2 + b^2$ The middle term, $2ab$, is missing.

$(a+b)^2 = (a+b)(a+b) = a^2 + 2ab + b^2$

Likewise,

$(a-b)^2 \neq a^2 - b^2$

$(a-b)^2 = (a-b)(a-b) = a^2 - 2ab + b^2$

Objective C Multiplying the Sum and Difference of Two Terms

Another special product is the product of the sum and difference of the same two terms, such as $(x+y)(x-y)$. Finding this product by the FOIL method, we see a pattern emerge.

$$(x+y)(x-y) = \overset{F}{x^2} - \overset{O}{xy} + \overset{I}{xy} - \overset{L}{y^2}$$
$$= x^2 - y^2$$

Notice that the two middle terms subtract out. This is because the **O**uter product is the opposite of the **I**nner product. Only the **difference of squares** remains.

Multiplying the Sum and Difference of Two Terms

The product of the sum and difference of two terms is the square of the first term minus the square of the second term.

$$(a+b)(a-b) = a^2 - b^2$$

Answers

5. $y^2 + 6y + 9$
6. $r^2 - 2rs + s^2$
7. $36x^2 + 60x + 25$
8. $x^4 - 6x^2y + 9y^2$

Practice 9–13

Use a special product to multiply.

9. $(x+9)(x-9)$
10. $(5+4y)(5-4y)$
11. $\left(x-\frac{1}{3}\right)\left(x+\frac{1}{3}\right)$
12. $(3a-b)(3a+b)$
13. $(2x^2-6y)(2x^2+6y)$

Examples Use a special product to multiply.

first term squared — minus — second term squared

9. $(x+4)(x-4) = x^2 - 4^2 = x^2 - 16$
10. $(6t+7)(6t-7) = (6t)^2 - 7^2 = 36t^2 - 49$
11. $\left(x-\frac{1}{4}\right)\left(x+\frac{1}{4}\right) = x^2 - \left(\frac{1}{4}\right)^2 = x^2 - \frac{1}{16}$
12. $(2p-q)(2p+q) = (2p)^2 - q^2 = 4p^2 - q^2$
13. $(3x^2-5y)(3x^2+5y) = (3x^2)^2 - (5y)^2 = 9x^4 - 25y^2$

Work Practice 9–13

✓**Concept Check** Match each expression on the left to the equivalent expression or expressions in the list on the right.

$(a+b)^2$

$(a+b)(a-b)$

a. $(a+b)(a+b)$
b. a^2-b^2
c. a^2+b^2
d. $a^2-2ab+b^2$
e. $a^2+2ab+b^2$

Objective D Using Special Products 

Let's now practice using our special products on a variety of multiplication problems. This practice will help us recognize when to apply what special product formula.

Practice 14–17

Use a special product to multiply, if possible.

14. $(7x-1)^2$
15. $(5y+3)(2y-5)$
16. $(2a-1)(2a+1)$
17. $\left(5y-\frac{1}{9}\right)^2$

Examples Use a special product to multiply, if possible.

14. $(4x-9)(4x+9)$
$= (4x)^2 - 9^2 = 16x^2 - 81$ — This is the sum and difference of the same two terms.

15. $(3y+2)^2$ — This is a binomial squared.
$= (3y)^2 + 2(3y)(2) + 2^2$
$= 9y^2 + 12y + 4$

16. $(6a+1)(a-7)$ — No special product applies. Use the FOIL method.
F O I L
$= 6a \cdot a + 6a(-7) + 1 \cdot a + 1(-7)$
$= 6a^2 - 42a + a - 7$
$= 6a^2 - 41a - 7$

17. $\left(4x-\frac{1}{11}\right)^2$ — This is a binomial squared.
$= (4x)^2 - 2(4x)\left(\frac{1}{11}\right) + \left(\frac{1}{11}\right)^2$
$= 16x^2 - \frac{8}{11}x + \frac{1}{121}$

Work Practice 14–17

Answers

9. x^2-81 10. $25-16y^2$ 11. $x^2-\frac{1}{9}$ 12. $9a^2-b^2$ 13. $4x^4-36y^2$ 14. $49x^2-14x+1$ 15. $10y^2-19y-15$ 16. $4a^2-1$ 17. $25y^2-\frac{10}{9}y+\frac{1}{81}$

✓**Concept Check Answer**

a and e, b

Helpful Hint

- When multiplying two binomials, you may always use the FOIL order or method.
- When multiplying any two polynomials, you may always use the distributive property to find the product.

Vocabulary, Readiness & Video Check

Answer each exercise true or false.

1. $(x + 4)^2 = x^2 + 16$ ______

2. For $(x + 6)(2x - 1)$, the product of the first terms is $2x^2$. ______

3. $(x + 4)(x - 4) = x^2 + 16$ ______

4. The product $(x - 1)(x^3 + 3x - 1)$ is a polynomial of degree 5. ______

Martin-Gay Interactive Videos

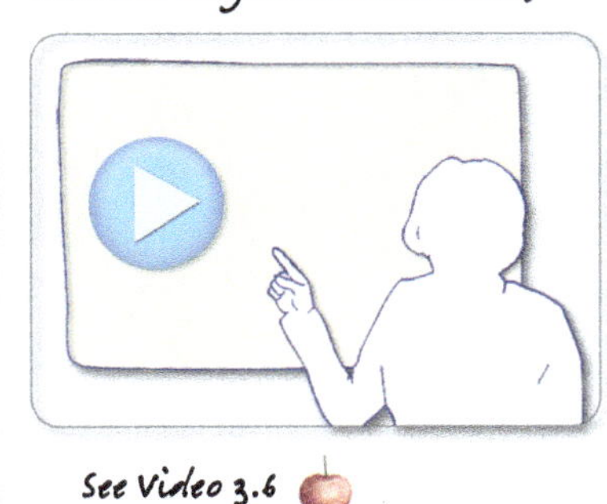

See Video 3.6 

Watch the section lecture video and answer the following questions.

Objective A **5.** From Examples 1–3, for what type of multiplication problem is the FOIL order of multiplication used?

Objective B **6.** Name at least one other method we can use to multiply Example 4.

Objective C **7.** From Example 5, why does multiplying the sum and difference of the same two terms always give us a binomial answer?

Objective D **8.** At the end of Example 8, what three special products for multiplying binomials are summarized?

3.6 Exercise Set MyLab Math

Objective A *Multiply using the FOIL method. See Examples 1 through 3.*

1. $(x + 3)(x + 4)$

2. $(x + 5)(x + 1)$

3. $(x - 5)(x + 10)$

4. $(y - 12)(y + 4)$

5. $(5x - 6)(x + 2)$

6. $(3y - 5)(2y + 7)$

7. $(y - 6)(4y - 1)$

8. $(2x - 9)(x - 11)$

9. $(2x + 5)(3x - 1)$

10. $(6x + 2)(x - 2)$

11. $(y^2 + 7)(6y + 4)$

12. $(y^2 + 3)(5y + 6)$

13. $\left(x - \frac{1}{3}\right)\left(x + \frac{2}{3}\right)$

14. $\left(x - \frac{2}{5}\right)\left(x + \frac{1}{5}\right)$

15. $(0.4 - 3a)(0.2 - 5a)$

16. $(0.3 - 2a)(0.6 - 5a)$

17. $(x + 5y)(2x - y)$

18. $(x + 4y)(3x - y)$

Objective B *Multiply. See Examples 4 through 8.*

19. $(x + 2)^2$ **20.** $(x + 7)^2$ **21.** $(2a - 3)^2$ **22.** $(7x - 3)^2$

23. $(3a - 5)^2$ **24.** $(5a - 2)^2$ **25.** $(x^2 + 0.5)^2$ **26.** $(x^2 + 0.3)^2$

27. $\left(y - \frac{2}{7}\right)^2$ **28.** $\left(y - \frac{3}{4}\right)^2$ **29.** $(2x - 1)^2$ **30.** $(5b - 4)^2$

31. $(5x + 9)^2$ **32.** $(6s + 2)^2$ **33.** $(3x - 7y)^2$ **34.** $(4s - 2y)^2$

35. $(4m + 5n)^2$ **36.** $(3n + 5m)^2$ **37.** $(5x^4 - 3)^2$ **38.** $(7x^3 - 6)^2$

Objective C *Multiply. See Examples 9 through 13.*

39. $(a - 7)(a + 7)$ **40.** $(b + 3)(b - 3)$ **41.** $(x + 6)(x - 6)$ **42.** $(x - 8)(x + 8)$

43. $(3x - 1)(3x + 1)$ **44.** $(7x - 5)(7x + 5)$ **45.** $(x^2 + 5)(x^2 - 5)$ **46.** $(a^2 + 6)(a^2 - 6)$

47. $(2y^2 - 1)(2y^2 + 1)$ **48.** $(3x^2 + 1)(3x^2 - 1)$ **49.** $(4 - 7x)(4 + 7x)$ **50.** $(8 - 7x)(8 + 7x)$

51. $\left(3x - \frac{1}{2}\right)\left(3x + \frac{1}{2}\right)$ **52.** $\left(10x + \frac{2}{7}\right)\left(10x - \frac{2}{7}\right)$ **53.** $(9x + y)(9x - y)$ **54.** $(2x - y)(2x + y)$

55. $(2m + 5n)(2m - 5n)$ **56.** $(5m + 4n)(5m - 4n)$

Objective D Mixed Practice *Multiply. See Examples 14 through 17.*

57. $(a + 5)(a + 4)$ **58.** $(a + 5)(a + 7)$ **59.** $(a - 7)^2$ **60.** $(b - 2)^2$

61. $(4a + 1)(3a - 1)$ **62.** $(6a + 7)(6a + 5)$ **63.** $(x + 2)(x - 2)$ **64.** $(x - 10)(x + 10)$

65. $(3a + 1)^2$ **66.** $(4a + 2)^2$ **67.** $(x + y)(4x - y)$ **68.** $(3x + 2)(4x - 2)$

69. $\left(\frac{1}{3}a^2 - 7\right)\left(\frac{1}{3}a^2 + 7\right)$

70. $\left(\frac{a}{2} + 4y\right)\left(\frac{a}{2} - 4y\right)$

71. $(3b + 7)(2b - 5)$

72. $(3y - 13)(y - 3)$

73. $(x^2 + 10)(x^2 - 10)$

74. $(x^2 + 8)(x^2 - 8)$

75. $(4x + 5)(4x - 5)$

76. $(3x + 5)(3x - 5)$

77. $(5x - 6y)^2$

78. $(4x - 9y)^2$

79. $(2r - 3s)(2r + 3s)$

80. $(6r - 2x)(6r + 2x)$

Express each as a product of polynomials in x. Then multiply and simplify.

81. Find the area of the square rug if its side is $(2x + 1)$ feet.

82. Find the area of the rectangular canvas if its length is $(3x - 2)$ inches and its width is $(x - 4)$ inches.

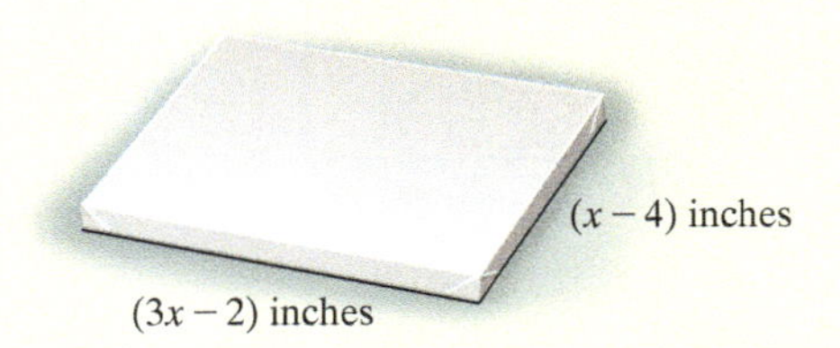

Review

Simplify each expression. See Sections 3.1 and 3.2.

83. $\frac{50b^{10}}{70b^5}$

84. $\frac{60y^6}{80y^2}$

85. $\frac{8a^{17}b^5}{-4a^7b^{10}}$

86. $\frac{-6a^8y}{3a^4y}$

87. $\frac{2x^4y^{12}}{3x^4y^4}$

88. $\frac{-48ab^6}{32ab^3}$

Concept Extensions

Match each expression on the left to the equivalent expression on the right. See the Concept Check in this section. (Not all choices will be used.)

89. $(a - b)^2$

90. $(a - b)(a + b)$

91. $(a + b)^2$

92. $(a + b)^2(a - b)^2$

a. $a^2 - b^2$

b. $a^2 + b^2$

c. $a^2 - 2ab + b^2$

d. $a^2 + 2ab + b^2$

e. none of these

Fill in the squares so that a true statement forms.

93. $(x^{\square} + 7)(x^{\square} + 3) = x^4 + 10x^2 + 21$

94. $(5x^{\square} - 2)^2 = 25x^6 - 20x^3 + 4$

Find the area of the shaded figure. To do so, subtract the area of the smaller square(s) from the area of the larger geometric figure.

△ **95.**

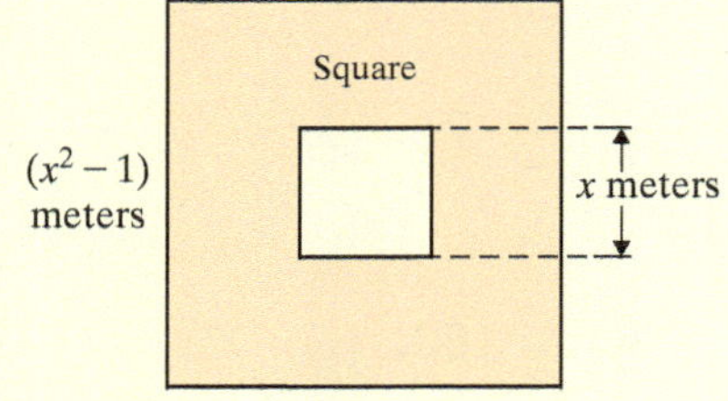

△ **96.**

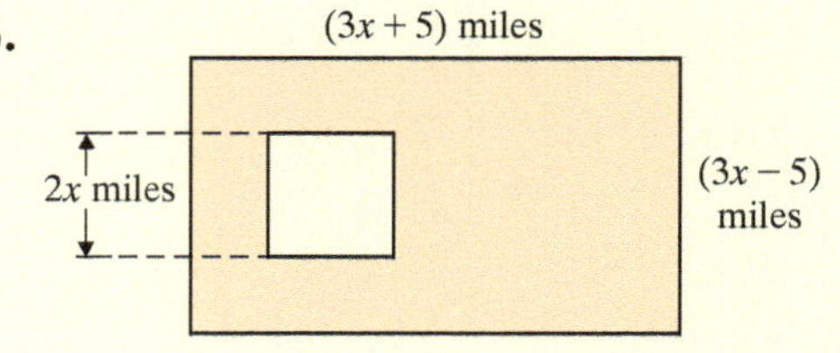

△ **97.**

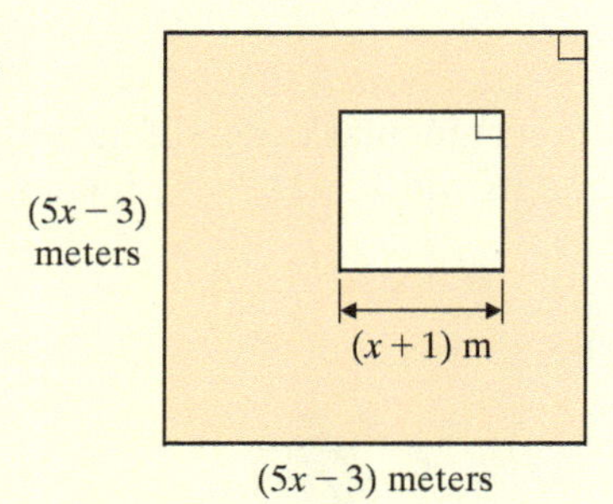

△ **98.**

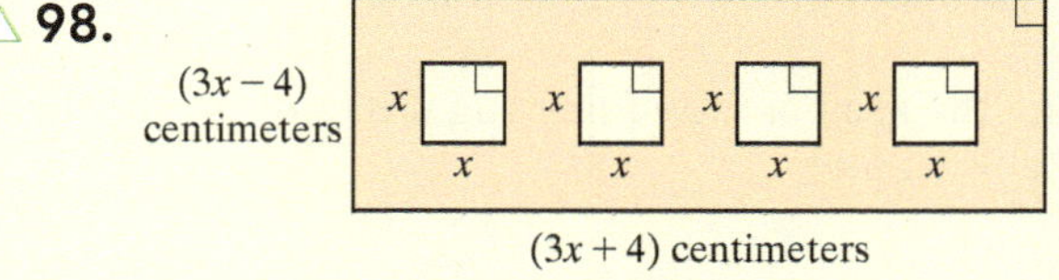

99. In your own words, describe the different methods that can be used to find the product: $(2x - 5)(3x + 1)$.

100. In your own words, describe the different methods that can be used to find the product: $(5x + 1)^2$.

101. Suppose that a classmate asked you why $(2x + 1)^2$ is **not** $(4x^2 + 1)$. Write down your response to this classmate.

102. Suppose that a classmate asked you why $(2x + 1)^2$ **is** $(4x^2 + 4x + 1)$. Write down your response to this classmate.

Exponents and Operations on Polynomials

Perform operations and simplify.

1. $(5x^2)(7x^3)$
2. $(4y^2)(-8y^7)$
3. -4^2
4. $(-4)^2$
5. $(x - 5)(2x + 1)$
6. $(3x - 2)(x + 5)$
7. $(x - 5) + (2x + 1)$
8. $(3x - 2) + (x + 5)$
9. $\dfrac{7x^9y^{12}}{x^3y^{10}}$
10. $\dfrac{20a^2b^8}{14a^2b^2}$
11. $(12m^7n^6)^2$
12. $(4y^9z^{10})^3$
13. $(4y - 3)(4y + 3)$
14. $(7x - 1)(7x + 1)$
15. $(x^{-7}y^5)^9$
16. 8^{-2}
17. $(3^{-1}x^9)^3$
18. $\dfrac{(r^7s^{-5})^6}{(2r^{-4}s^{-4})^4}$
19. $(7x^2 - 2x + 3) - (5x^2 + 9)$
20. $(10x^2 + 7x - 9) - (4x^2 - 6x + 2)$

Answers

1. ______
2. ______
3. ______
4. ______
5. ______
6. ______
7. ______
8. ______
9. ______
10. ______
11. ______
12. ______
13. ______
14. ______
15. ______
16. ______
17. ______
18. ______
19. ______
20. ______

21. ______
22. ______
23. ______
24. ______
25. ______
26. ______
27. ______
28. ______
29. ______
30. ______
31. ______
32. ______
33. ______
34. ______
35. ______
36. ______
37. ______
38. ______

21. $0.7y^2 - 1.2 + 1.8y^2 - 6y + 1$

22. $7.8x^2 - 6.8x - 3.3 + 0.6x^2 - 0.9$

23. Subtract $(y^2 + 2)$ from $(3y^2 - 6y + 1)$.

24. $(z^2 + 5) - (3z^2 - 1) + \left(8z^2 + 2z - \frac{1}{2}\right)$

25. $(x + 4)^2$

26. $(y - 9)^2$

27. $(x + 4) + (x + 4)$

28. $(y - 9) + (y - 9)$

29. $7x^2 - 6xy + 4(y^2 - xy)$

30. $5a^2 - 3ab + 6(b^2 - a^2)$

31. $(x - 3)(x^2 + 5x - 1)$

32. $(x + 1)(x^2 - 3x - 2)$

33. $(2x - 7)(3x + 10)$

34. $(5x - 1)(4x + 5)$

35. $(2x - 7)(x^2 - 6x + 1)$

36. $(5x - 1)(x^2 + 2x - 3)$

37. $\left(2x + \frac{5}{9}\right)\left(2x - \frac{5}{9}\right)$

38. $\left(12y + \frac{3}{7}\right)\left(12y - \frac{3}{7}\right)$

3.7 Dividing Polynomials

Objective A Dividing by a Monomial

To divide a polynomial by a monomial, recall addition of fractions. Fractions that have a common denominator are added by adding the numerators:

$$\frac{a}{c} + \frac{b}{c} = \frac{a + b}{c}$$

If we read this equation from right to left and let a, b, and c be monomials, $c \neq 0$, we have the following.

To Divide a Polynomial by a Monomial

Divide each term of the polynomial by the monomial.

$$\frac{a + b}{c} = \frac{a}{c} + \frac{b}{c}, \quad c \neq 0$$

Throughout this section, we assume that denominators are not 0.

Example 1 Divide: $(6m^2 + 2m) \div 2m$

Solution: We begin by writing the quotient in fraction form. Then we divide each term of the polynomial $6m^2 + 2m$ by the monomial $2m$ and use the quotient rule for exponents to simplify.

$$\frac{6m^2 + 2m}{2m} = \frac{6m^2}{2m} + \frac{2m}{2m}$$

$$= 3m + 1 \qquad \text{Simplify.}$$

Check: To check, we multiply.

$$2m(3m + 1) = 2m(3m) + 2m(1) = 6m^2 + 2m$$

The quotient $3m + 1$ checks.

Work Practice 1

Concept Check In which of the following is $\frac{x + 5}{5}$ simplified correctly?

a. $\frac{x}{5} + 1$ **b.** x **c.** $x + 1$

Example 2 Divide: $\frac{9x^5 - 12x^2 + 3x}{3x^2}$

Solution:

$$\frac{9x^5 - 12x^2 + 3x}{3x^2} = \frac{9x^5}{3x^2} - \frac{12x^2}{3x^2} + \frac{3x}{3x^2} \qquad \text{Divide each term by } 3x^2.$$

$$= 3x^3 - 4 + \frac{1}{x} \qquad \text{Simplify.}$$

Notice that the quotient is not a polynomial because of the term $\frac{1}{x}$. This expression is called a rational expression—we will study rational expressions in Chapter 5. Although the quotient of two polynomials is not always a polynomial, we may still check by multiplying.

(*Continued on next page*)

Objectives

A Divide a Polynomial by a Monomial.

B Use Long Division to Divide a Polynomial by a Polynomial Other Than a Monomial.

Practice 1

Divide: $(25x^3 + 5x^2) \div 5x^2$

Practice 2

Divide: $\frac{24x^7 + 12x^2 - 4x}{4x^2}$

Answers

1. $5x + 1$ **2.** $6x^5 + 3 - \frac{1}{x}$

Concept Check Answer

a

Check: $3x^2\left(3x^3 - 4 + \frac{1}{x}\right) = 3x^2(3x^3) - 3x^2(4) + 3x^2\left(\frac{1}{x}\right)$

$= 9x^5 - 12x^2 + 3x$

Work Practice 2

Practice 3

Divide: $\frac{12x^3y^3 - 18xy + 6y}{3xy}$

Example 3 Divide: $\frac{8x^2y^2 - 16xy + 2x}{4xy}$

Solution: $\frac{8x^2y^2 - 16xy + 2x}{4xy} = \frac{8x^2y^2}{4xy} - \frac{16xy}{4xy} + \frac{2x}{4xy}$ Divide each term by $4xy$.

$= 2xy - 4 + \frac{1}{2y}$ Simplify.

Check: $4xy\left(2xy - 4 + \frac{1}{2y}\right) = 4xy(2xy) - 4xy(4) + 4xy\left(\frac{1}{2y}\right)$

$= 8x^2y^2 - 16xy + 2x$

Work Practice 3

Objective B Dividing by a Polynomial Other Than a Monomial

To divide a polynomial by a polynomial other than a monomial, we use a process known as long division. Polynomial long division is similar to number long division, so we review long division by dividing 13 into 3660.

Helpful Hint Recall that 3660 is called the dividend.

```
     281
13)3660
   -26        2·13 = 26
    106       Subtract and bring down the next digit in the dividend.
   -104       8·13 = 104
     20       Subtract and bring down the next digit in the dividend.
    -13       1·13 = 13
      7       Subtract. There are no more digits to bring down, so the remainder is 7.
```

The quotient is 281 R 7, which can be written as $281\frac{7}{13}$. ← remainder ← divisor

Recall that division can be checked by multiplication. To check this division problem, we see that

$13 \cdot 281 + 7 = 3660$, the dividend.

Now we demonstrate long division of polynomials.

Practice 4

Divide $x^2 + 12x + 35$ by $x + 5$ using long division.

Example 4 Divide $x^2 + 7x + 12$ by $x + 3$ using long division.

Solution:

To subtract, change the signs of these terms and add.

```
           x
x + 3)x² + 7x + 12
     -x² ∓ 3x         ↓
           4x + 12
```

How many times does x divide x^2? $\frac{x^2}{x} = x$.

Multiply: $x(x + 3)$

Subtract and bring down the next term.

Answers

3. $4x^2y^2 - 6 + \frac{2}{x}$ 4. $x + 7$

Now we repeat this process.

$$\begin{array}{r} x + 4 \\ x + 3\overline{)x^2 + 7x + 12} \\ \underline{x^2 + 3x} \\ 4x + 12 \\ \underline{-4x \mp 12} \\ 0 \end{array}$$

How many times does x divide $4x$? $\frac{4x}{x} = 4$.

Multiply: $4(x + 3)$

To subtract, change the signs of these terms and add.

Subtract. The remainder is 0.

The quotient is $x + 4$.

Check: We check by multiplying.

divisor	·	quotient	+	remainder	=	dividend
↓		↓		↓		↓
$(x + 3)$	·	$(x + 4)$	+	0	=	$x^2 + 7x + 12$

or

The quotient checks.

■ **Work Practice 4**

Example 5 Divide $6x^2 + 10x - 5$ by $3x - 1$ using long division.

Solution:

$$\begin{array}{r} 2x + 4 \\ 3x - 1\overline{)6x^2 + 10x - 5} \\ \underline{-6x^2 \pm 2x} \\ 12x - 5 \\ \underline{-12x \pm 4} \\ -1 \end{array}$$

$\frac{6x^2}{3x} = 2x$, so $2x$ is a term of the quotient.

Multiply: $2x(3x - 1)$

Subtract and bring down the next term.

$\frac{12x}{3x} = 4$. Multiply: $4(3x - 1)$

Subtract. The remainder is -1.

Thus $(6x^2 + 10x - 5)$ divided by $(3x - 1)$ is $(2x + 4)$ with a remainder of -1. This can be written as follows.

$$\frac{6x^2 + 10x - 5}{3x - 1} = 2x + 4 + \frac{-1}{3x - 1} \quad \begin{array}{l} \leftarrow \text{remainder} \\ \leftarrow \text{divisor} \end{array}$$

$$\text{or } 2x + 4 - \frac{1}{3x - 1}$$

Check: To check, we multiply $(3x - 1)(2x + 4)$. Then we add the remainder, -1, to this product.

$$(3x - 1)(2x + 4) + (-1) = (6x^2 + 12x - 2x - 4) - 1$$
$$= 6x^2 + 10x - 5$$

The quotient checks.

■ **Work Practice 5**

Practice 5

Divide: $8x^2 + 2x - 7$ by $2x - 1$

Notice that the division process is continued until the degree of the remainder polynomial is less than the degree of the divisor polynomial.

Recall that in Section 3.3 we practiced writing polynomials in descending order of powers and with no missing terms. For example, $2 - 4x^2$ written in this form is $-4x^2 + 0x + 2$. Writing the dividend and divisor in this form is helpful when dividing polynomials.

Answer

5. $4x + 3 + \frac{-4}{2x - 1}$ or $4x + 3 - \frac{4}{2x - 1}$

Practice 6

Divide: $(15 - 2x^2) \div (x - 3)$

Example 6

Divide: $(2 - 4x^2) \div (x + 1)$

Solution: We use the rewritten form of $2 - 4x^2$ from the previous page.

$$
\begin{array}{r l}
-4x + 4 & \dfrac{-4x^2}{x} = -4x\text{, so } -4x \text{ is a term of the quotient.} \\
x + 1\overline{)-4x^2 + 0x + 2} & \\
\underline{\overset{+}{-}4x^2 \overset{+}{-} 4x} & \text{Multiply: } -4x(x + 1) \\
4x + 2 & \text{Subtract and bring down the next term.} \\
\underline{\overset{-}{}4x \overset{-}{+} 4} & \dfrac{4x}{x} = 4\text{. Multiply: } 4(x + 1) \\
-2 & \text{Remainder}
\end{array}
$$

Thus, $\dfrac{-4x^2 + 0x + 2}{x + 1}$ or $\dfrac{2 - 4x^2}{x + 1} = -4x + 4 + \dfrac{-2}{x + 1}$ or $-4x + 4 - \dfrac{2}{x + 1}$.

Check: To check, see that $(x + 1)(-4x + 4) + (-2) = 2 - 4x^2$.

Work Practice 6

Practice 7

Divide: $\dfrac{5 - x + 9x^3}{3x + 2}$

Example 7

Divide: $\dfrac{4x^2 + 7 + 8x^3}{2x + 3}$

Solution: Before we begin the division process, we rewrite $4x^2 + 7 + 8x^3$ as $8x^3 + 4x^2 + 0x + 7$. Notice that we have written the polynomial in descending order and have represented the missing x-term by $0x$.

$$
\begin{array}{r l}
4x^2 - 4x + 6 & \\
2x + 3\overline{)8x^3 + 4x^2 + 0x + 7} & \\
\underline{\overset{-}{}8x^3 \overset{-}{+} 12x^2} & \\
-8x^2 + 0x & \\
\underline{\overset{+}{-}8x^2 \overset{+}{-} 12x} & \\
12x + 7 & \\
\underline{\overset{-}{}12x \overset{-}{+} 18} & \\
-11 & \text{Remainder}
\end{array}
$$

Thus, $\dfrac{4x^2 + 7 + 8x^3}{2x + 3} = 4x^2 - 4x + 6 + \dfrac{-11}{2x + 3}$ or $4x^2 - 4x + 6 - \dfrac{11}{2x + 3}$.

Work Practice 7

Practice 8

Divide: $x^3 - 1$ by $x - 1$

Example 8

Divide $x^3 - 8$ by $x - 2$.

Solution: Notice that the polynomial $x^3 - 8$ is missing an x^2-term and an x-term. We'll represent these terms by inserting $0x^2$ and $0x$.

$$
\begin{array}{r}
x^2 + 2x + 4 \\
x - 2\overline{)x^3 + 0x^2 + 0x - 8} \\
\underline{\overset{-}{}x^3 \overset{+}{-} 2x^2} \\
2x^2 + 0x \\
\underline{\overset{-}{}2x^2 \overset{+}{-} 4x} \\
4x - 8 \\
\underline{\overset{-}{}4x \overset{+}{-} 8} \\
0
\end{array}
$$

Thus, $\dfrac{x^3 - 8}{x - 2} = x^2 + 2x + 4$.

Check: To check, see that $(x^2 + 2x + 4)(x - 2) = x^3 - 8$.

Work Practice 8

Answers

6. $-2x - 6 + \dfrac{-3}{x - 3}$ or $-2x - 6 - \dfrac{3}{x - 3}$
7. $3x^2 - 2x + 1 + \dfrac{3}{3x + 2}$
8. $x^2 + x + 1$

Vocabulary, Readiness & Video Check

Use the choices below to fill in each blank. Choices may be used more than once.

dividend divisor quotient

1. In $6\overline{)18}$ with quotient 3, the 18 is the ________, the 3 is the ________, and the 6 is the ________.

2. In $x + 1\overline{)x^2 + 3x + 2}$ with quotient $x + 2$, the $x + 1$ is the ________, the $x^2 + 3x + 2$ is the ________, and the $x + 2$ is the ________.

Simplify each expression mentally.

3. $\frac{a^6}{a^4}$ **4.** $\frac{p^8}{p^3}$ **5.** $\frac{y^2}{y}$ **6.** $\frac{a^3}{a}$

Martin-Gay Interactive Videos

See Video 3.7

Watch the section lecture video and answer the following questions.

Objective A **7.** The lecture before Example 1 begins with adding two fractions with the same denominator. From there, the lecture continues to a method for dividing a polynomial by a monomial. What role does the monomial play in the fraction example?

Objective B **8.** In Example 5, we're told that although we don't have to fill in missing powers in the divisor and the dividend, it really is a good idea to do so. Why?

3.7 Exercise Set MyLab Math

Objective A *Perform each division. See Examples 1 through 3.*

1. $\frac{12x^4 + 3x^2}{x}$ **2.** $\frac{15x^2 - 9x^5}{x}$ **3.** $\frac{20x^3 - 30x^2 + 5x + 5}{5}$ **4.** $\frac{8x^3 - 4x^2 + 6x + 2}{2}$

5. $\frac{15p^3 + 18p^2}{3p}$ **6.** $\frac{6x^5 + 3x^4}{3x^4}$ **7.** $\frac{-9x^4 + 18x^5}{6x^5}$ **8.** $\frac{14m^2 - 27m^3}{7m}$

9. $\frac{-9x^5 + 3x^4 - 12}{3x^3}$ **10.** $\frac{6a^2 - 4a + 12}{-2a^2}$ **11.** $\frac{4x^4 - 6x^3 + 7}{-4x^4}$ **12.** $\frac{-12a^3 + 36a - 15}{3a}$

Objective B *Find each quotient using long division. See Examples 4 and 5.*

13. $\dfrac{x^2 + 4x + 3}{x + 3}$

14. $\dfrac{x^2 + 7x + 10}{x + 5}$

15. $\dfrac{2x^2 + 13x + 15}{x + 5}$

16. $\dfrac{3x^2 + 8x + 4}{x + 2}$

17. $\dfrac{2x^2 - 7x + 3}{x - 4}$

18. $\dfrac{3x^2 - x - 4}{x - 1}$

19. $\dfrac{9a^3 - 3a^2 - 3a + 4}{3a + 2}$

20. $\dfrac{4x^3 + 12x^2 + x - 14}{2x + 3}$

21. $\dfrac{8x^2 + 10x + 1}{2x + 1}$

22. $\dfrac{3x^2 + 17x + 7}{3x + 2}$

23. $\dfrac{2x^3 + 2x^2 - 17x + 8}{x - 2}$

24. $\dfrac{4x^3 + 11x^2 - 8x - 10}{x + 3}$

Find each quotient using long division. Don't forget to write the polynomials in descending order and fill in any missing terms. See Examples 6 through 8.

25. $\dfrac{x^2 - 36}{x - 6}$

26. $\dfrac{a^2 - 49}{a - 7}$

27. $\dfrac{x^3 - 27}{x - 3}$

28. $\dfrac{x^3 + 64}{x + 4}$

29. $\dfrac{1 - 3x^2}{x + 2}$

30. $\dfrac{7 - 5x^2}{x + 3}$

31. $\dfrac{-4b + 4b^2 - 5}{2b - 1}$

32. $\dfrac{-3y + 2y^2 - 15}{2y + 5}$

Objectives A B Mixed Practice *Divide. If the divisor contains 2 or more terms, use long division. See Examples 1 through 8.*

33. $\dfrac{a^2b^2 - ab^3}{ab}$

34. $\dfrac{m^3n^2 - mn^4}{mn}$

35. $\dfrac{8x^2 + 6x - 27}{2x - 3}$

36. $\dfrac{18w^2 + 18w - 8}{3w + 4}$

37. $\dfrac{2x^2y + 8x^2y^2 - xy^2}{2xy}$

38. $\dfrac{11x^3y^3 - 33xy + x^2y^2}{11xy}$

39. $\dfrac{2b^3 + 9b^2 + 6b - 4}{b + 4}$

40. $\dfrac{2x^3 + 3x^2 - 3x + 4}{x + 2}$

41. $\dfrac{y^3 + 3y^2 + 4}{y - 2}$

42. $\dfrac{3x^3 + 11x + 12}{x + 4}$

43. $\dfrac{5 - 6x^2}{x - 2}$

44. $\dfrac{3 - 7x^2}{x - 3}$

45. $\dfrac{x^5 + x^2}{x^2 + x}$

46. $\dfrac{x^6 - x^3}{x^3 - x^2}$

Review

Fill in each blank. See Section 3.1.

47. $12 = 4 \cdot$ ___

48. $12 = 2 \cdot$ ___

49. $20 = -5 \cdot$ ___

50. $20 = -4 \cdot$ ___

51. $9x^2 = 3x \cdot$ ___

52. $9x^2 = 9x \cdot$ ___

53. $36x^2 = 4x \cdot$ ___

54. $36x^2 = 2x \cdot$ ___

Concept Extensions

Solve.

55. The perimeter of a square is $(12x^3 + 4x - 16)$ feet. Find the length of its side.

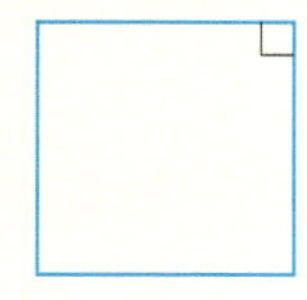

Perimeter is $(12x^3 + 4x - 16)$ feet

△ **56.** The volume of the swimming pool shown is $(36x^5 - 12x^3 + 6x^2)$ cubic feet. If its depth is $2x$ feet and its width is $3x$ feet, find its length.

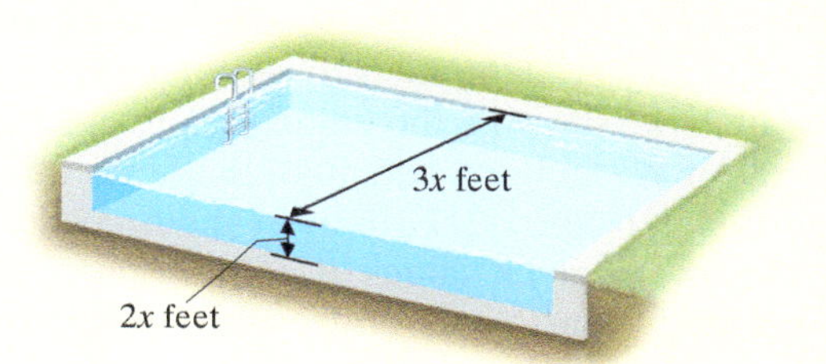

57. The area of the parallelogram shown is $(10x^2 + 31x + 15)$ square meters. If its base is $(5x + 3)$ meters, find its height.

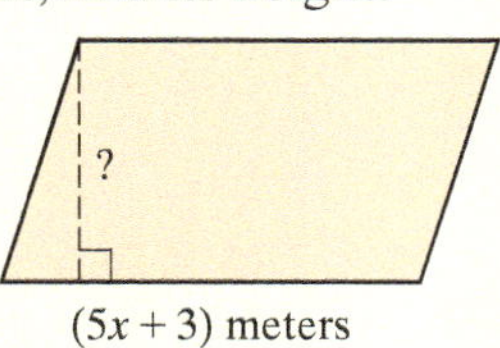

58. The area of the top of the Ping-Pong table shown is $(49x^2 + 70x - 200)$ square inches. If its length is $(7x + 20)$ inches, find its width.

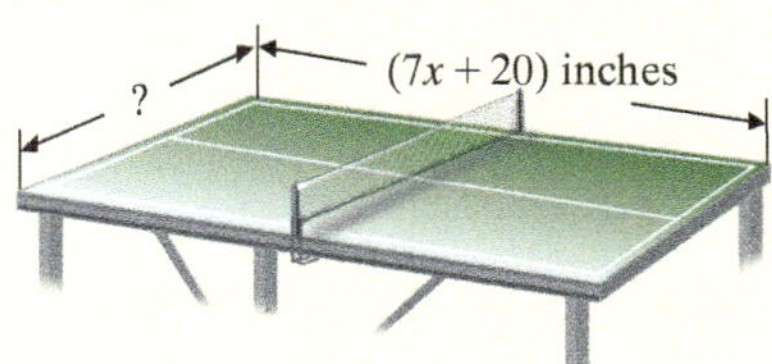

59. Explain how to check a polynomial long division result when the remainder is 0.

60. Explain how to check a polynomial long division result when the remainder is not 0.

61. In which of the following is $\frac{a+7}{7}$ simplified correctly? See the Concept Check in this section.

a. $a + 1$
b. a
c. $\frac{a}{7} + 1$

62. In which of the following is $\frac{5x+15}{5}$ simplified correctly? See the Concept Check in this section.

a. $x + 15$
b. $x + 3$
c. $x + 1$

Chapter 3 Group Activity

Modeling with Polynomials

Materials:

- calculator

This activity may be completed by working in groups or individually.

Washington state is the leading producer of apples in the United States. The polynomial model $-312x^2 + 903x + 6227$ gives Washington's annual apple production (in million pounds) for the period 2013–2016. The polynomial model $-287x^2 + 730x + 10{,}531$ gives the total U.S. annual apple production (in million pounds) for the same period. In both models, x is the number of years after 2013. (*Source:* Based on data from the National Agricultural Statistics Service)

1. Use the given polynomials to complete the following table showing the annual apple production (both for Washington and all of the United States) over the period 2013–2016 by evaluating each polynomial at the given values of x. Then subtract each value in the fourth column from the corresponding value in the third column. Record the result in the last column, labeled "Difference." What do you think these values represent?

Year	x	Total U.S. Annual Apple Production (million pounds)	Washington's Annual Apple Production (million pounds)	Difference
2013	0			
2014	1			
2015	2			
2016	3			

2. Use the polynomial models to find a new polynomial model representing the annual apple production of *all other* U.S. states, excluding Washington. Then evaluate your new polynomial model to complete the accompanying table.

Year	x	Other Annual Apple Production (million pounds)
2013	0	
2014	1	
2015	2	
2016	3	

3. Compare the values in the last column of the table in Question **1** to the values in the last column of the table in Question **2.** What do you notice? What can you conclude?

4. Make a bar graph of the data in the table in Question **2.** Describe what you see.

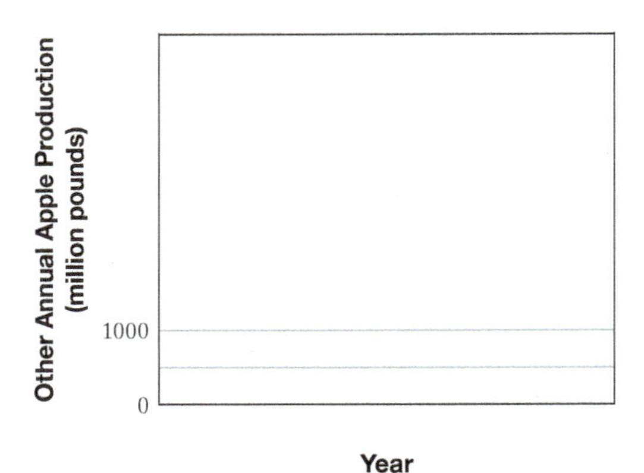

Chapter 3 Vocabulary Check

Fill in each blank with one of the words or phrases listed below.

term	coefficient	monomial	binomial	trinomial
polynomials	degree of a term	degree of a polynomial	distributive	FOIL

1. A ______________ is a number or the product of a number and variables raised to powers.
2. The ______________ method may be used when multiplying two binomials.
3. A polynomial with exactly 3 terms is called a ______________.
4. The ______________ is the greatest degree of any term of the polynomial.
5. A polynomial with exactly 2 terms is called a ______________.
6. The ______________ of a term is its numerical factor.

7. The ________________ is the sum of the exponents on the variables in the term.
8. A polynomial with exactly 1 term is called a ________________.
9. Monomials, binomials, and trinomials are all examples of ________________.
10. The ________________ property is used to multiply $2x(x - 4)$.

Helpful Hint Are you preparing for your test? To help, don't forget to take these:

- Chapter 3 Getting Ready for the Test on page 261
- Chapter 3 Test on page 262

Then check all of your answers at the back of this text. For further review, the step-by-step video solutions to any of these exercises are located in MyLab Math.

3 Chapter Highlights

Definitions and Concepts	Examples
Section 3.1 Exponents	
a^n means the product of n factors, each of which is a.	$3^2 = 3 \cdot 3 = 9$ $(-5)^3 = (-5)(-5)(-5) = -125$ $\left(\frac{1}{2}\right)^4 = \frac{1}{2} \cdot \frac{1}{2} \cdot \frac{1}{2} \cdot \frac{1}{2} = \frac{1}{16}$
Let m and n be integers and no denominators be 0. **Product Rule:** $a^m \cdot a^n = a^{m+n}$ **Power Rule:** $(a^m)^n = a^{mn}$ **Power of a Product Rule:** $(ab)^n = a^n b^n$ **Power of a Quotient Rule:** $\left(\frac{a}{b}\right)^n = \frac{a^n}{b^n}$ **Quotient Rule:** $\frac{a^m}{a^n} = a^{m-n}, a \neq 0$ **Zero Exponent:** $a^0 = 1, a \neq 0$	$x^2 \cdot x^7 = x^{2+7} = x^9$ $(5^3)^8 = 5^{3 \cdot 8} = 5^{24}$ $(7y)^4 = 7^4 y^4$ $\left(\frac{x}{8}\right)^3 = \frac{x^3}{8^3}$ $\frac{x^9}{x^4} = x^{9-4} = x^5, x \neq 0$ $5^0 = 1; x^0 = 1, x \neq 0$
Section 3.2 Negative Exponents and Scientific Notation	
If $a \neq 0$ and n is an integer, $a^{-n} = \frac{1}{a^n}$	$3^{-2} = \frac{1}{3^2} = \frac{1}{9}; 5x^{-2} = \frac{5}{x^2}$ Simplify: $\left(\frac{x^{-2}y}{x^5}\right)^{-2} = \frac{x^4y^{-2}}{x^{-10}} = x^{4-(-10)}y^{-2} = \frac{x^{14}}{y^2}$
A positive number is written in scientific notation if it is written as the product of a number a, where $1 \leq a < 10$, and an integer power r of 10. $a \times 10^r$	$1200 = 1.2 \times 10^3$ $0.000000568 = 5.68 \times 10^{-7}$

Definitions and Concepts	Examples
Section 3.3 Introduction to Polynomials	
A **term** is a number or the product of a number and variables raised to powers.	$-5x, 7a^2b, \frac{1}{4}y^4, 0.2$
The **numerical coefficient,** or **coefficient,** of a term is its numerical factor.	**Term** / **Coefficient** $7x^2$ — 7 y — 1 $-a^2b$ — −1
A **polynomial** is a finite sum of terms of the form ax^n where a is a real number and n is a whole number.	$5x^3 - 6x^2 + 3x - 6$ (Polynomial)
A **monomial** is a polynomial with exactly 1 term.	$\frac{5}{6}y^3$ (Monomial)
A **binomial** is a polynomial with exactly 2 terms.	$-0.2a^2b - 5b^2$ (Binomial)
A **trinomial** is a polynomial with exactly 3 terms.	$3x^2 - 2x + 1$ (Trinomial)
The **degree of a term with one variable** is the exponent on the variable. The **degree of a term with more than one variable** is the sum of the exponents on the variables.	**Term** / **Degree** $3x^5$ — 5 $7x^4y^2$ — $4 + 2 = 5$
The **degree of a polynomial** is the greatest degree of any term of the polynomial.	**Polynomial** / **Degree** $5x^2 - 3x + 2$ — 2 $7y + 8y^2z^3 - 12$ — $2 + 3 = 5$
Section 3.4 Adding and Subtracting Polynomials	
To add polynomials, combine like terms.	Add. $(7x^2 - 3x + 2) + (-5x - 6)$ $= 7x^2 - 3x + 2 - 5x - 6$ $= 7x^2 - 8x - 4$
To subtract two polynomials, change the signs of the terms of the second polynomial, and then add.	Subtract. $(17y^2 - 2y + 1) - (-3y^3 + 5y - 6)$ $= (17y^2 - 2y + 1) + (3y^3 - 5y + 6)$ $= 17y^2 - 2y + 1 + 3y^3 - 5y + 6$ $= 3y^3 + 17y^2 - 7y + 7$
Section 3.5 Multiplying Polynomials	
To multiply two polynomials, multiply each term of one polynomial by each term of the other polynomial, and then combine like terms.	Multiply. $(2x + 1)(5x^2 - 6x + 2)$ $= 2x(5x^2 - 6x + 2) + 1(5x^2 - 6x + 2)$ $= 10x^3 - 12x^2 + 4x + 5x^2 - 6x + 2$ $= 10x^3 - 7x^2 - 2x + 2$

Definitions and Concepts	Examples
Section 3.6 Special Products	
The **FOIL method** may be used when multiplying two binomials.	Multiply: $(5x-3)(2x+3)$ First, Last, Inner, Outer: $(5x-3)(2x+3)$ $= (5x)(2x) + (5x)(3) + (-3)(2x) + (-3)(3)$ (F, O, I, L) $= 10x^2 + 15x - 6x - 9$ $= 10x^2 + 9x - 9$
Squaring a Binomial $(a+b)^2 = a^2 + 2ab + b^2$ $(a-b)^2 = a^2 - 2ab + b^2$	Square each binomial. $(x+5)^2 = x^2 + 2(x)(5) + 5^2$ $= x^2 + 10x + 25$ $(3x-2y)^2 = (3x)^2 - 2(3x)(2y) + (2y)^2$ $= 9x^2 - 12xy + 4y^2$
Multiplying the Sum and Difference of Two Terms $(a+b)(a-b) = a^2 - b^2$	Multiply. $(6y+5)(6y-5) = (6y)^2 - 5^2$ $= 36y^2 - 25$
Section 3.7 Dividing Polynomials	
To divide a polynomial by a monomial, $\frac{a+b}{c} = \frac{a}{c} + \frac{b}{c}, c \neq 0$	Divide. $\frac{15x^5 - 10x^3 + 5x^2 - 2x}{5x^2}$ $= \frac{15x^5}{5x^2} - \frac{10x^3}{5x^2} + \frac{5x^2}{5x^2} - \frac{2x}{5x^2}$ $= 3x^3 - 2x + 1 - \frac{2}{5x}$
To divide a polynomial by a polynomial other than a monomial, use long division.	$2x+3 \overline{)10x^2 + 13x - 7}$ with quotient $5x - 1 + \frac{-4}{2x+3}$ $\underline{-10x^2 \mp 15x}$ $-2x - 7$ $\underline{\pm 2x \pm 3}$ -4 or $5x - 1 - \frac{4}{2x+3}$

Chapter 3 Review

(3.1) *Each expression contains an exponent of 4. For each exercise, name the base for this exponent of 4.*

1. 3^4 **2.** $(-5)^4$ **3.** -5^4 **4.** x^4

Evaluate each expression.

5. 8^3 **6.** $(-6)^2$ **7.** -6^2 **8.** $-4^3 - 4^0$ **9.** $(3b)^0$ **10.** $\frac{8b}{8b}$

Simplify each expression.

11. $y^2 \cdot y^7$ **12.** $x^9 \cdot x^5$ **13.** $(2x^5)(-3x^6)$ **14.** $(-5y^3)(4y^4)$

15. $(x^4)^2$ **16.** $(y^3)^5$ **17.** $(3y^6)^4$ **18.** $(2x^3)^3$

19. $\frac{x^9}{x^4}$ **20.** $\frac{z^{12}}{z^5}$ **21.** $\frac{a^5b^4}{ab}$ **22.** $\frac{x^4y^6}{xy}$

23. $\frac{3x^4y^{10}}{12xy^6}$ **24.** $\frac{2x^7y^8}{8xy^2}$ **25.** $5a^7(2a^4)^3$ **26.** $(2x)^2(9x)$

27. $(-5a)^0 + 7^0 + 8^0$ **28.** $8x^0 + 9^0$

Simplify the given expression and choose the correct result.

29. $\left(\frac{3x^4}{4y}\right)^3$

a. $\frac{27x^{64}}{64y^3}$ b. $\frac{27x^{12}}{64y^3}$

c. $\frac{9x^{12}}{12y^3}$ d. $\frac{3x^{12}}{4y^3}$

30. $\left(\frac{5a^6}{b^3}\right)^2$

a. $\frac{10a^{12}}{b^6}$ b. $\frac{25a^{36}}{b^9}$

c. $\frac{25a^{12}}{b^6}$ d. $25a^{12}b^6$

(3.2) *Simplify each expression.*

31. 7^{-2} **32.** -7^{-2} **33.** $2x^{-4}$ **34.** $(2x)^{-4}$

35. $\left(\frac{1}{5}\right)^{-3}$ **36.** $\left(\frac{-2}{3}\right)^{-2}$ **37.** $2^0 + 2^{-4}$ **38.** $6^{-1} - 7^{-1}$

Simplify each expression. Write each answer using positive exponents only.

39. $\frac{x^5}{x^{-3}}$

40. $\frac{z^4}{z^{-4}}$

41. $\frac{r^{-3}}{r^{-4}}$

42. $\frac{y^{-2}}{y^{-5}}$

43. $\left(\frac{bc^{-2}}{bc^{-3}}\right)^4$

44. $\left(\frac{x^{-3}y^{-4}}{x^{-2}y^{-5}}\right)^{-3}$

45. $\frac{x^{-4}y^{-6}}{x^2y^7}$

46. $\frac{a^5b^{-5}}{a^{-5}b^5}$

Write each number in scientific notation.

47. 0.00027

48. 0.8868

49. 80,800,000

50. 868,000

51. In November 2016, approximately 137,000,000 people cast ballots in the U.S. presidential election. Write this number in scientific notation. (*Source:* U.S. election atlas)

52. The approximate diameter of the Milky Way galaxy is 150,000 light-years. Write this number in scientific notation. (*Source:* NASA IMAGE/POETRY Education and Public Outreach Program)

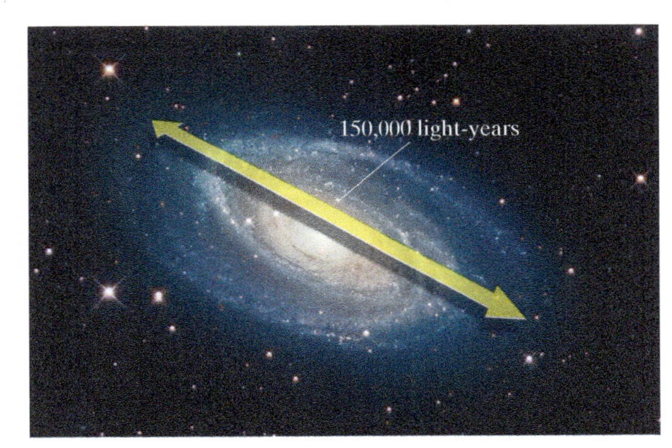

Write each number in standard form.

53. 8.67×10^5

54. 3.86×10^{-3}

55. 8.6×10^{-4}

56. 8.936×10^5

57. The volume of the planet Jupiter is 1.43128×10^{15} cubic kilometers. Write this number in standard form. (*Source:* National Space Science Data Center)

58. An angstrom is a unit of measure, equal to 1×10^{-10} meter, used for measuring wavelengths or the diameters of atoms. Write this number in standard form. (*Source:* National Institute of Standards and Technology)

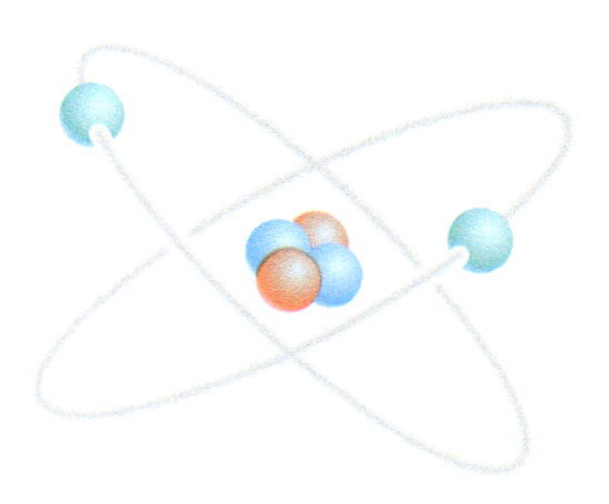

Simplify. Express each result in standard form.

59. $(8 \times 10^4)(2 \times 10^{-7})$

60. $\dfrac{8 \times 10^4}{2 \times 10^{-7}}$

(3.3) *Find the degree of each polynomial.*

61. $y^5 + 7x - 8x^4$

62. $9y^2 + 30y + 25$

63. $-14x^2y - 28x^2y^3 - 42x^2y^2$

64. $6x^2y^2z^2 + 5x^2y^3 - 12xyz$

65. The Glass Bridge Skywalk is suspended 4000 feet over the Colorado River at the very edge of the Grand Canyon. Neglecting air resistance, the height of an object dropped from the Skywalk at time t seconds is given by the polynomial $-16t^2 + 4000$. Find the height of the object at the given times below.

t	0 seconds	1 second	3 seconds	5 seconds
$-16t^2 + 4000$				

△ **66.** The surface area of a box with a square base and a height of 5 units is given by the polynomial $2x^2 + 20x$. Fill in the table below by evaluating $2x^2 + 20x$ for the given values of x.

x	1	3	5.1	10
$2x^2 + 20x$				

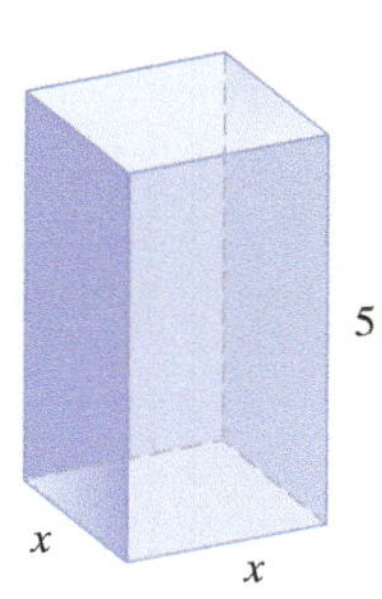

Combine like terms in each expression.

67. $7a^2 - 4a^2 - a^2$

68. $9y + y - 14y$

69. $6a^2 + 4a + 9a^2$

70. $21x^2 + 3x + x^2 + 6$

71. $4a^2b - 3b^2 - 8q^2 - 10a^2b + 7q^2$

72. $2s^{14} + 3s^{13} + 12s^{12} - s^{10}$

(3.4) *Add or subtract as indicated.*

73. $(3x^2 + 2x + 6) + (5x^2 + x)$

74. $(2x^5 + 3x^4 + 4x^3 + 5x^2) + (4x^2 + 7x + 6)$

75. $(-5y^2 + 3) - (2y^2 + 4)$

76. $(2m^7 + 3x^4 + 7m^6) - (8m^7 + 4m^2 + 6x^4)$

77. $(3x^2 - 7xy + 7y^2) - (4x^2 - xy + 9y^2)$

78. $(8x^6 - 5xy - 10y^2) - (7x^6 - 9xy - 12y^2)$

Translating *Perform the indicated operations.*

79. Add $(-9x^2 + 6x + 2)$ and $(4x^2 - x - 1)$.

80. Subtract $(4x^2 + 8x - 7)$ from the sum of $(x^2 + 7x + 9)$ and $(x^2 + 4)$.

(3.5) *Multiply each expression.*

81. $6(x + 5)$

82. $9(x - 7)$

83. $4(2a + 7)$

84. $9(6a - 3)$

85. $-7x(x^2 + 5)$

86. $-8y(4y^2 - 6)$

87. $-2(x^3 - 9x^2 + x)$

88. $-3a(a^2b + ab + b^2)$

89. $(3a^3 - 4a + 1)(-2a)$

90. $(6b^3 - 4b + 2)(7b)$

91. $(2x + 2)(x - 7)$

92. $(2x - 5)(3x + 2)$

93. $(4a - 1)(a + 7)$

94. $(6a - 1)(7a + 3)$

95. $(x + 7)(x^3 + 4x - 5)$

96. $(x + 2)(x^5 + x + 1)$

97. $(x^2 + 2x + 4)(x^2 + 2x - 4)$

98. $(x^3 + 4x + 4)(x^3 + 4x - 4)$

99. $(x + 7)^3$

100. $(2x - 5)^3$

(3.6) *Use special products to multiply each of the following.*

101. $(x + 7)^2$

102. $(x - 5)^2$

103. $(3x - 7)^2$

104. $(4x + 2)^2$

105. $(5x - 9)^2$

106. $(5x + 1)(5x - 1)$

107. $(7x + 4)(7x - 4)$

108. $(a + 2b)(a - 2b)$

109. $(2x - 6)(2x + 6)$

110. $(4a^2 - 2b)(4a^2 + 2b)$

Express each as a product of polynomials in x. Then multiply and simplify.

111. Find the area of the square if its side is $(3x - 1)$ meters.

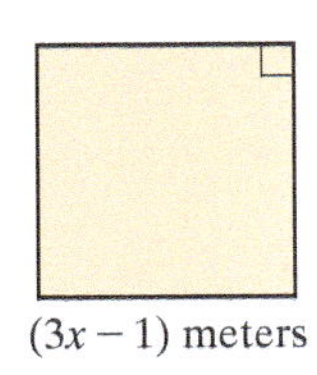

112. Find the area of the rectangle.

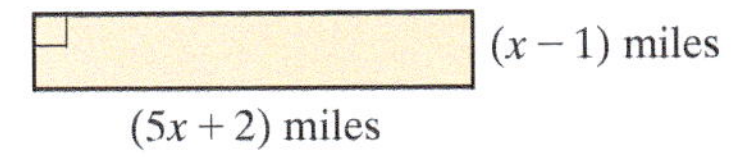

(3.7) *Divide.*

113. $\dfrac{x^2 + 21x + 49}{7x^2}$

114. $\dfrac{5a^3b - 15ab^2 + 20ab}{-5ab}$

115. $(a^2 - a + 4) \div (a - 2)$

116. $(4x^2 + 20x + 7) \div (x + 5)$

117. $\dfrac{a^3 + a^2 + 2a + 6}{a - 2}$

118. $\dfrac{9b^3 - 18b^2 + 8b - 1}{3b - 2}$

119. $\dfrac{4x^4 - 4x^3 + x^2 + 4x - 3}{2x - 1}$

120. $\dfrac{-10x^2 - x^3 - 21x + 18}{x - 6}$

△ **121.** The area of the rectangle below is $(15x^3 - 3x^2 + 60)$ square feet. If its length is $3x^2$ feet, find its width.

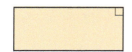

Area is $(15x^3 - 3x^2 + 60)$ sq feet

122. The perimeter of the equilateral triangle below is $(21a^3b^6 + 3a - 3)$ units. Find the length of a side.

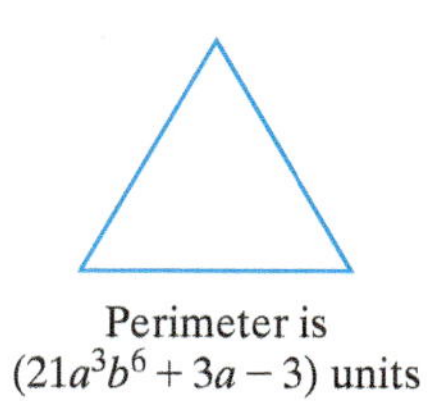

Perimeter is $(21a^3b^6 + 3a - 3)$ units

Mixed Review

Evaluate.

123. 3^3

124. $\left(-\frac{1}{2}\right)^3$

Simplify each expression. Write each answer using positive exponents only.

125. $(4xy^2)(x^3y^5)$

126. $\dfrac{18x^9}{27x^3}$

127. $\left(\dfrac{3a^4}{b^2}\right)^3$

128. $(2x^{-4}y^3)^{-4}$

129. $\dfrac{a^{-3}b^6}{9^{-1}a^{-5}b^{-2}}$

Perform the indicated operations and simplify.

130. $(-y^2 - 4) + (3y^2 - 6)$

131. $(6x + 2) + (5x - 7)$

132. $(5x^2 + 2x - 6) - (-x - 4)$

133. $(8y^2 - 3y + 1) - (3y^2 + 2)$

134. $(2x + 5)(3x - 2)$

135. $4x(7x^2 + 3)$

136. $(7x - 2)(4x - 9)$

137. $(x - 3)(x^2 + 4x - 6)$

Use special products to multiply.

138. $(5x + 4)^2$

139. $(6x + 3)(6x - 3)$

Divide.

140. $\dfrac{8a^4 - 2a^3 + 4a - 5}{2a^3}$

141. $\dfrac{x^2 + 2x + 10}{x + 5}$

142. $\dfrac{4x^3 + 8x^2 - 11x + 4}{2x - 3}$

Chapter 3 Getting Ready for the Test 

MATCHING *For Exercises 1 through 4, match the expression in the left column with the exponent operation needed to simplify in the right column. Letters may be used more than once or not at all.*

1. $x^2 \cdot x^5$
2. $(x^2)^5$
3. $x^2 + x^5$
4. $\frac{x^5}{x^2}$

A. multiply the exponents
B. divide the exponents
C. add the exponents
D. subtract the exponents
E. this expression will not simplify

MATCHING *For Exercises 5 through 8, match the operation in the left column with the result when the operation is performed on the given terms in the right columns. Letters may be used more than once or not at all.*

Given Terms: 20y and 4y

5. Add the terms
6. Subtract the terms
7. Multiply the terms
8. Divide the terms.

A. $80y$
B. $24y^2$
C. $16y$
D. 16
E. $80y^2$
F. $24y$
G. $16y^2$
H. $5y$
I. 5

9. MULTIPLE CHOICE *The expression* 5^{-1} *is equivalent to*

A. -5 **B.** 4 **C.** $\frac{1}{5}$ **D.** $-\frac{1}{5}$

10. MULTIPLE CHOICE *The expression* 2^{-3} *is equivalent to*

A. -6 **B.** -1 **C.** $-\frac{1}{6}$ **D.** $\frac{1}{8}$

MATCHING *For Exercises 11 through 14, match each expression in the left column with its simplified form in the right columns. Letters may be used more than once or not at all.*

11. $y + y + y$
12. $y \cdot y \cdot y$
13. $(-y)(-y)(-y)$
14. $-y - y - y$

A. $3y^3$
B. y^3
C. $3y$
D. $-3y$
E. $-3y^3$
F. $-y^3$

Chapter 3 Test MyLab Math

For additional practice go to your study plan in MyLab Math.

Answers

1. ______
2. ______
3. ______
4. ______
5. ______
6. ______
7. ______
8. ______
9. ______
10. ______
11. ______
12. ______
13. ______
14. ______
15. a. ______
    b. ______
16. ______
17. ______
18. ______
19. ______

Evaluate each expression.

1. 2^5 **2.** $(-3)^4$ **3.** -3^4 **4.** 4^{-3}

Simplify each expression. Write the result using only positive exponents.

5. $(3x^2)(-5x^9)$ **6.** $\frac{y^7}{y^2}$ **7.** $\frac{r^{-8}}{r^{-3}}$

8. $\left(\frac{4x^2y^3}{x^3y^{-4}}\right)^2$ **9.** $\frac{6^2x^{-4}y^{-1}}{6^3x^{-3}y^7}$

Express each number in scientific notation.

10. 563,000 **11.** 0.0000863

Write each number in standard form.

12. 1.5×10^{-3} **13.** 6.23×10^4

14. Simplify. Write the answer in standard form.
$(1.2 \times 10^5)(3 \times 10^{-7})$

15. a. Complete the table for the polynomial $4xy^2 + 7xyz + x^3y - 2$.

Term	Numerical Coefficient	Degree of Term
$4xy^2$		
$7xyz$		
x^3y		
-2		

b. What is the degree of the polynomial?

16. Simplify by combining like terms.
$5x^2 + 4x - 7x^2 + 11 + 8x$

Perform each indicated operation.

17. $(8x^3 + 7x^2 + 4x - 7) + (8x^3 - 7x - 6)$

18.

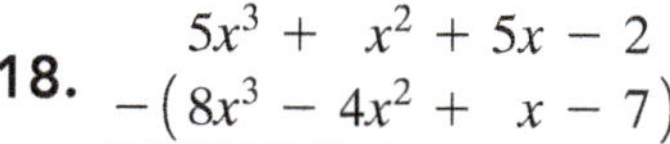

$$\begin{array}{r} 5x^3 + x^2 + 5x - 2 \\ -(8x^3 - 4x^2 + x - 7) \\ \hline \end{array}$$

19. Subtract $(4x + 2)$ from the sum of $(8x^2 + 7x + 5)$ and $(x^3 - 8)$.

Multiply.

20. $(3x + 7)(x^2 + 5x + 2)$

21. $3x^2(2x^2 - 3x + 7)$

22. $(x + 7)(3x - 5)$

23. $\left(3x - \frac{1}{5}\right)\left(3x + \frac{1}{5}\right)$

24. $(4x - 2)^2$

25. $(8x + 3)^2$

26. $(x^2 - 9b)(x^2 + 9b)$

27. The height of the Bank of China in Hong Kong is 1001 feet. Neglecting air resistance, the height of an object dropped from this building at time t seconds is given by the polynomial $-16t^2 + 1001$. Find the height of the object at the given times below.

t	0 seconds	1 second	3 seconds	5 seconds
$-16t^2 + 1001$				

28. Find the area of the top of the table. Express the area as a product, then multiply and simplify.

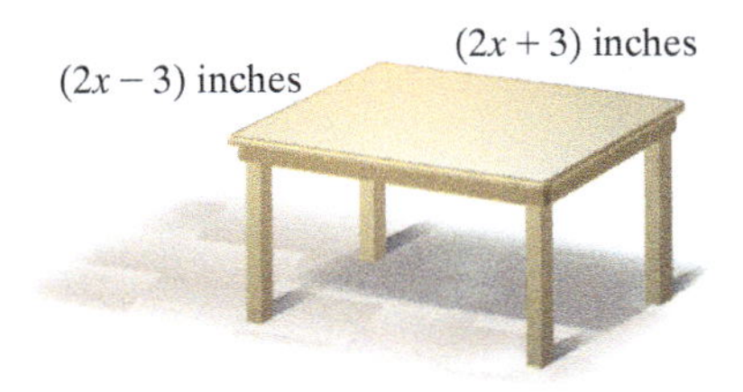

Divide.

29. $\dfrac{4x^2 + 2xy - 7x}{8xy}$

30. $(x^2 + 7x + 10) \div (x + 5)$

31. $\dfrac{27x^3 - 8}{3x + 2}$

20. ____________

21. ____________

22. ____________

23. ____________

24. ____________

25. ____________

26. ____________

27. ____________

28. ____________

29. ____________

30. ____________

31. ____________

Chapters 1–3 Cumulative Review

Answers

1. a. ______
   b. ______
   c. ______
   d. ______
   e. ______
   f. ______
2. a. ______
   b. ______
   c. ______
3. a. ______
   b. ______
   c. ______
   d. ______
   e. ______
   f. ______
4. a. ______
   b. ______
5. ______
6. ______
7. a. ______
   b. ______
   c. ______
   d. ______
   e. ______

1. Given the set $\left\{-2, 0, \frac{1}{4}, 112, -3, 11, \sqrt{2}\right\}$, list the numbers in this set that belong to the set of:

a. Natural numbers
b. Whole numbers
c. Integers
d. Rational numbers
e. Irrational numbers
f. Real numbers

2. Find the absolute value of each number.

a. $|-7.2|$
b. $|0|$
c. $\left|-\frac{1}{2}\right|$

3. Evaluate (find the value of) the following:

a. 3^2
b. 5^3
c. 2^4
d. 7^1
e. $\left(\frac{3}{7}\right)^2$
f. $(0.6)^2$

4. Multiply. Write products in lowest terms.

a. $\frac{3}{4} \cdot \frac{7}{21}$
b. $\frac{1}{2} \cdot 4\frac{5}{6}$

5. Simplify: $\frac{3}{2} \cdot \frac{1}{2} - \frac{1}{2}$

6. Evaluate $\frac{2x - 7y}{x^2}$ for $x = 5$ and $y = 1$.

7. Write an algebraic expression that represents each phrase. Let the variable x represent the unknown number.

a. The sum of a number and 3
b. The product of 3 and a number
c. The quotient of 7.3 and a number
d. 10 decreased by a number
e. 5 times a number, increased by 7

8. Simplify: $8 + 3(2 \cdot 6 - 1)$

9. Add: $11.4 + (-4.7)$

10. Is $x = 1$ a solution of $5x + 2 = x - 8$?

11. Find the value of each expression when $x = 2$ and $y = -5$.

a. $\dfrac{x - y}{12 + x}$

b. $x^2 - y$

12. Subtract:

a. $7 - 40$

b. $-5 - (-10)$

Divide.

13. $\dfrac{-30}{-10}$

14. $\dfrac{-48}{6}$

15. $\dfrac{42}{-0.6}$

16. $\dfrac{-30}{-0.2}$

Find each product by using the distributive property to remove parentheses.

17. $5(3x + 2)$

18. $-3(2x - 3)$

19. $-2(y + 0.3z - 1)$

20. $4(-x^2 + 6x - 1)$

21. $-(9x + y - 2z + 6)$

22. $-(-4xy + 6y - 2)$

23. Solve: $6(2a - 1) - (11a + 6) = 7$

24. Solve: $2x + \dfrac{1}{8} = x - \dfrac{3}{8}$

25. Solve: $\dfrac{y}{7} = 20$

26. Solve: $10 = 5j - 2$

27. Solve: $0.25x + 0.10(x - 3) = 1.1$

28. Solve: $\dfrac{7x + 5}{3} = x + 3$

29. Twice the sum of a number and 4 is the same as four times the number decreased by 12. Find the number.

30. Write the phrase as an algebraic expression and simplify if possible. Double a number, subtracted from the sum of the number and seven.

8. ______
9. ______
10. ______
11. a. ______
b. ______
12. a. ______
b. ______
13. ______
14. ______
15. ______
16. ______
17. ______
18. ______
19. ______
20. ______
21. ______
22. ______
23. ______
24. ______
25. ______
26. ______
27. ______
28. ______
29. ______
30. ______

31. ____
32. ____
33. ____
34. ____
35. ____
36. a. ____
b. ____
c. ____
37. a. ____
b. ____
c. ____
38. ____
39. ____
40. ____
41. ____
42. ____
43. ____
44. ____
45. ____
46. ____
47. ____
48. ____
49. ____

△ **31.** Charles Pecot can afford enough fencing to enclose a rectangular garden with a perimeter of 140 feet. If the width of his garden is to be 30 feet, find the length.

32. Simplify: $\frac{4(-3) + (-8)}{5 + (-5)}$

33. The number 120 is 15% of what number?

34. Graph $x < 5$.

−5 −4 −3 −2 −1 0 1 2 3 4 5

35. Solve: $-4x + 7 \geq -9$. Graph the solution set.

−5 −4 −3 −2 −1 0 1 2 3 4 5

36. Evaluate.

a. $(-5)^2$ **b.** -5^2 **c.** $2 \cdot 5^2$

37. Simplify each expression.

a. $x^7 \cdot x^4$

b. $\left(\frac{t}{2}\right)^4$

c. $(9y^5)^2$

38. Simplify: $\frac{(z^2)^3 \cdot z^7}{z^9}$

Simplify the following expressions. Write each result using positive exponents only.

39. $\left(\frac{3a^2}{b}\right)^{-3}$ **40.** $(5x^7)(-3x^9)$ **41.** $(5y^3)^{-2}$ **42.** $(-3)^{-2}$

Simplify each polynomial.

43. $9x^3 + x^3$ **44.** $(5y^2 - 6) - (y^2 + 2)$ **45.** $5x^2 + 6x - 9x - 3$

46. Multiply: $(10x^2 - 3)(10x^2 + 3)$.

47. Multiply: $7x(x^2 + 2x + 5)$

48. Multiply: $(10x^2 + 3)^2$.

49. Divide: $\frac{9x^5 - 12x^2 + 3x}{3x^2}$

Factoring Polynomials

4

In Chapter 3, we multiplied polynomials. Now we will learn the reverse operation of multiplying—factoring. Factoring allows us to write a sum as a product. As we will see in this chapter, factoring can be used to solve equations other than linear equations. In Chapter 5, we will also use factoring to perform operations on rational expressions.

Sections

4.1 The Greatest Common Factor and Factoring by Grouping

4.2 Factoring Trinomials of the Form $x^2 + bx + c$

4.3 Factoring Trinomials of the Form $ax^2 + bx + c$

4.4 Factoring Trinomials of the Form $ax^2 + bx + c$ by Grouping

4.5 Factoring Perfect Square Trinomials and the Difference of Two Squares

Integrated Review—Choosing a Factoring Strategy

4.6 Solving Quadratic Equations by Factoring

4.7 Quadratic Equations and Problem Solving

Check Your Progress

Vocabulary Check

Chapter Highlights

Chapter Review

Getting Ready for the Test

Chapter Test

Cumulative Review

Why Are You in College?

There are probably as many answers as there are students. It may help you to know that college graduates have higher earnings and lower rates of unemployment. The double line graph below shows the increasing number of associate and bachelor degrees awarded over the years. It is also enlightening to know that an increasing number of high school graduates are interested in higher education.

In Exercise 99 of Section 4.1, we will explore how many students graduate from U.S. high schools each year and how many of those may expect to go to college.

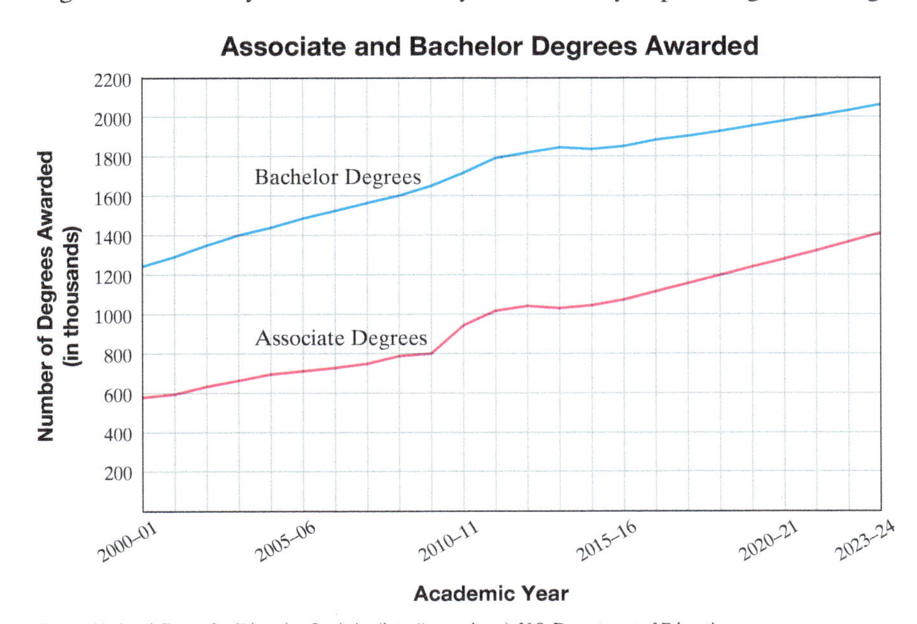

Source: National Center for Education Statistics (http://nces.ed.gov); U.S. Department of Education

Note: Some years are projected.

4.1 The Greatest Common Factor and Factoring by Grouping

Objectives

A Find the Greatest Common Factor of a List of Numbers.

B Find the Greatest Common Factor of a List of Terms.

C Factor Out the Greatest Common Factor from the Terms of a Polynomial.

D Factor a Polynomial by Grouping.

In the product $2 \cdot 3 = 6$, the numbers 2 and 3 are called **factors** of 6 and $2 \cdot 3$ is a **factored form** of 6. This is true of polynomials also. Since $(x + 2)(x + 3) = x^2 + 5x + 6$, then $(x + 2)$ and $(x + 3)$ are factors of $x^2 + 5x + 6$, and $(x + 2)(x + 3)$ is a factored form of the polynomial.

a factored form of 6

$$\underbrace{2 \cdot 3}_{\text{factor} \quad \text{factor}} = \underbrace{6}_{\text{product}}$$

a factored form of x^5

$$\underbrace{x^2 \cdot x^3}_{\text{factor} \quad \text{factor}} = \underbrace{x^5}_{\text{product}}$$

a factored form of $x^2 + 5x + 6$

$$\underbrace{(x + 2)(x + 3)}_{\text{factor} \quad \text{factor}} = \underbrace{x^2 + 5x + 6}_{\text{product}}$$

The process of writing a polynomial as a product is called **factoring** the polynomial.

Study the examples below and look for a pattern.

Multiplying: $5(x^2 + 3) = 5x^2 + 15$ | $2x(x - 7) = 2x^2 - 14x$

↓

Factoring: $5x^2 + 15 = 5(x^2 + 3)$ | $2x^2 - 14x = 2x(x - 7)$

Do you see that factoring is the reverse process of multiplying?

$$x^2 + 5x + 6 = (x + 2)(x + 3)$$

factoring →

← multiplying

✓**Concept Check** Multiply: $2(x - 4)$

What do you think the result of factoring $2x - 8$ would be? Why?

Objective A Finding the Greatest Common Factor of a List of Numbers

The first step in factoring a polynomial is to see whether the terms of the polynomial have a common factor. If there is one, we can write the polynomial as a product by **factoring out** the common factor. We will usually factor out the *greatest* common factor (GCF).

The GCF of a list of integers is the largest integer that is a factor of all the integers in the list. For example, the GCF of 12 and 20 is 4 because 4 is the largest integer that is a factor of both 12 and 20. With large integers, the GCF may not be easily found by inspection. When this happens, use the following steps.

✓**Concept Check Answer**

$2x - 8$; the result would be $2(x - 4)$ because factoring is the reverse process of multiplying.

Finding the GCF of a List of Integers

Step 1: Write each number as a product of prime numbers.

Step 2: Identify the common prime factors.

Step 3: The product of all common prime factors found in Step 2 is the greatest common factor. If there are no common prime factors, the greatest common factor is 1.

Recall from Section R.1 that a prime number is a whole number other than 1 whose only factors are 1 and itself.

Example 1 Find the GCF of each list of numbers.

a. 28 and 40 **b.** 55 and 21 **c.** 15, 18, and 66

Solution:

a. Write each number as a product of primes.

$$28 = 2 \cdot 2 \cdot 7 = 2^2 \cdot 7$$
$$40 = 2 \cdot 2 \cdot 2 \cdot 5 = 2^3 \cdot 5$$

There are two common factors, each of which is 2, so the GCF is

$$\text{GCF} = 2 \cdot 2 = 4$$

b. $55 = 5 \cdot 11$
$21 = 3 \cdot 7$

There are no common prime factors; thus, the GCF is 1.

c. $15 = 3 \cdot 5$
$18 = 2 \cdot 3 \cdot 3 = 2 \cdot 3^2$
$66 = 2 \cdot 3 \cdot 11$

The only prime factor common to all three numbers is 3, so the GCF is

$\text{GCF} = 3$

Work Practice 1

Practice 1

Find the GCF of each list of numbers.

a. 45 and 75

b. 32 and 33

c. 14, 24, and 60

Objective B Finding the Greatest Common Factor of a List of Terms

The greatest common factor of a list of variables raised to powers is found in a similar way. For example, the GCF of x^2, x^3, and x^5 is x^2 because each term contains a factor of x^2 and no higher power of x is a factor of each term.

$$x^2 = x \cdot x$$
$$x^3 = x \cdot x \cdot x$$
$$x^5 = x \cdot x \cdot x \cdot x \cdot x$$

There are two common factors, each of which is x, so the GCF $= x \cdot x$ or x^2. From this example, we see that **the GCF of a list of common variables raised to powers is the variable raised to the smallest exponent in the list.**

Example 2 Find the GCF of each list of terms.

a. x^3, x^7, and x^5

b. y, y^4, and y^7

(Continued on next page)

Practice 2

Find the GCF of each list of terms.

a. y^4, y^5, and y^8

b. x and x^{10}

Answers

1. a. 15 **b.** 1 **c.** 2

2. a. y^4 **b.** x

Solution:

a. The GCF is x^3, since 3 is the smallest exponent to which x is raised.

b. The GCF is y^1 or y, since 1 is the smallest exponent on y.

Work Practice 2

The **greatest common factor (GCF) of a list of terms** is the product of the GCF of the numerical coefficients and the GCF of the variable factors.

$$20x^2y^2 = 2 \cdot 2 \cdot 5 \cdot x \cdot x \cdot y \cdot y$$
$$6xy^3 = 2 \cdot 3 \cdot x \cdot y \cdot y \cdot y$$
$$\text{GCF} = 2 \cdot x \cdot y \cdot y = 2xy^2$$

Helpful Hint

Remember that the GCF of a list of terms contains the smallest exponent on each common variable.

The GCF of x^5y^6, x^2y^7, and x^3y^4 is x^2y^4. (Smallest exponent on x; Smallest exponent on y)

Practice 3

Find the greatest common factor of each list of terms.

a. $6x^2$, $9x^4$, and $-12x^5$

b. $-16y$, $-20y^6$, and $40y^4$

c. a^5b^4, ab^3, and a^3b^2

Example 3 Find the greatest common factor of each list of terms.

a. $6x^2$, $10x^3$, and $-8x$

b. $-18y^2$, $-63y^3$, and $27y^4$

c. a^3b^2, a^5b, and a^6b^2

Solution:

a.
$$6x^2 = 2 \cdot 3 \cdot x^2$$
$$10x^3 = 2 \cdot 5 \cdot x^3$$
$$-8x = -1 \cdot 2 \cdot 2 \cdot 2 \cdot x^1$$
The GCF of x^2, x^3, and x^1 is x^1 or x.
$$\text{GCF} = 2 \cdot x^1 \quad \text{or} \quad 2x$$

b.
$$-18y^2 = -1 \cdot 2 \cdot 3 \cdot 3 \cdot y^2$$
$$-63y^3 = -1 \cdot 3 \cdot 3 \cdot 7 \cdot y^3$$
$$27y^4 = 3 \cdot 3 \cdot 3 \cdot y^4$$
The GCF of y^2, y^3, and y^4 is y^2.
$$\text{GCF} = 3 \cdot 3 \cdot y^2 \quad \text{or} \quad 9y^2$$

c. The GCF of a^3, a^5, and a^6 is a^3.
The GCF of b^2, b, and b^2 is b. Thus,
the GCF of a^3b^2, a^5b, and a^6b^2 is a^3b.

Work Practice 3

Objective C Factoring Out the Greatest Common Factor

To factor a polynomial such as $8x + 14$, we first see whether the terms have a greatest common factor other than 1. In this case, they do: The GCF of $8x$ and 14 is 2.

We factor out 2 from each term by writing each term as the product of 2 and the term's remaining factors.

$$8x + 14 = 2 \cdot 4x + 2 \cdot 7$$

Using the distributive property, we can write

$$8x + 14 = 2 \cdot 4x + 2 \cdot 7$$
$$= 2(4x + 7)$$

Answers

3. a. $3x^2$ **b.** $4y$ **c.** ab^2

Thus, a factored form of $8x + 14$ is $2(4x + 7)$. We can check by multiplying:

$$2(4x + 7) = 2 \cdot 4x + 2 \cdot 7 = 8x + 14$$

Helpful Hint

A factored form of $8x + 14$ is *not*

$$2 \cdot 4x + 2 \cdot 7$$

Although the *terms* have been factored (written as products), the *polynomial* $8x + 14$ has not been factored. A factored form of $8x + 14$ is the *product* $2(4x + 7)$.

✓Concept Check Which of the following is/are factored form(s) of $6t + 18$?

a. 6 **b.** $6 \cdot t + 6 \cdot 3$ **c.** $6(t + 3)$ **d.** $3(t + 6)$

Example 4

Factor each polynomial by factoring out the greatest common factor (GCF).

a. $5ab + 10a$ **b.** $y^5 - y^{12}$

Solution:

a. The GCF of terms $5ab$ and $10a$ is $5a$. Thus,

$$5ab + 10a = 5a \cdot b + 5a \cdot 2$$
$$= 5a(b + 2) \quad \text{Apply the distributive property.}$$

We can check our work by multiplying $5a$ and $(b + 2)$.
$5a(b + 2) = 5a \cdot b + 5a \cdot 2 = 5ab + 10a$, the original polynomial.

b. The GCF of y^5 and y^{12} is y^5. Thus,

$$y^5 - y^{12} = y^5(1) - y^5(y^7)$$
$$= y^5(1 - y^7)$$

Helpful Hint
Don't forget the 1.

Work Practice 4

Practice 4

Factor each polynomial by factoring out the greatest common factor (GCF).

a. $10y + 25$

b. $x^4 - x^9$

Example 5

Factor: $-9a^5 + 18a^2 - 3a$

Solution:

$$-9a^5 + 18a^2 - 3a = 3a(-3a^4) + 3a(6a) + 3a(-1)$$
$$= 3a(-3a^4 + 6a - 1)$$

Helpful Hint
Don't forget the -1.

Work Practice 5

Practice 5

Factor: $-10x^3 + 8x^2 - 2x$

In Example 5, we could have chosen to factor out $-3a$ instead of $3a$. If we factor out $-3a$, we have

$$-9a^5 + 18a^2 - 3a = (-3a)(3a^4) + (-3a)(-6a) + (-3a)(1)$$
$$= -3a(3a^4 - 6a + 1)$$

Helpful Hint
Notice the changes in signs when factoring out $-3a$.

Answers
4. a. $5(2y + 5)$ **b.** $x^4(1 - x^5)$
5. $2x(-5x^2 + 4x - 1)$

✓Concept Check Answer
c

Practice 6–8

Factor.

6. $4x^3 + 12x$

7. $\frac{2}{5}a^5 - \frac{4}{5}a^3 + \frac{1}{5}a^2$

8. $6a^3b + 3a^3b^2 + 9a^2b^4$

Examples Factor.

6. $6a^4 - 12a = 6a(a^3 - 2)$

7. $\frac{3}{7}x^4 + \frac{1}{7}x^3 - \frac{5}{7}x^2 = \frac{1}{7}x^2(3x^2 + x - 5)$

8. $15p^2q^4 + 20p^3q^5 + 5p^3q^3 = 5p^2q^3(3q + 4pq^2 + p)$

Work Practice 6–8

Practice 9

Factor: $7(p + 2) + q(p + 2)$

Example 9 Factor: $5(x + 3) + y(x + 3)$

Solution: The binomial $(x + 3)$ is present in both terms and is the greatest common factor. We use the distributive property to factor out $(x + 3)$.

$$5(x + 3) + y(x + 3) = (x + 3)(5 + y)$$

Work Practice 9

Practice 10

Factor $7xy^3(p + q) - (p + q)$

Example 10 Factor: $3m^2n(a + b) - (a + b)$

Solution: The greatest common factor is $(a + b)$.

$$3m^2n(a + b) - 1(a + b) = (a + b)(3m^2n - 1)$$

Work Practice 10

Objective D Factoring by Grouping

Once the GCF is factored out, we can sometimes continue to factor the polynomial using a variety of techniques. We discuss here a technique called **factoring by grouping.** This technique can be used to factor some polynomials with four terms.

Practice 11

Factor $ab + 7a + 2b + 14$ by grouping.

Helpful Hint Notice that this form, $x(y + 2) + 3(y + 2)$, is *not* a factored form of the original polynomial. It is a sum, not a product.

Example 11 Factor $xy + 2x + 3y + 6$ by grouping.

Solution: Notice that the first two terms of this polynomial have a common factor of x and that the second two terms have a common factor of 3. Because of this, group the first two terms, then the last two terms, and then factor out these common factors.

$$xy + 2x + 3y + 6 = (xy + 2x) + (3y + 6) \quad \text{Group terms.}$$
$$= x(y + 2) + 3(y + 2) \quad \text{Factor out GCF from each grouping.}$$

Next we factor out the common binomial factor, $(y + 2)$.

$$x(y + 2) + 3(y + 2) = (y + 2)(x + 3)$$

Now the result is a factored form because it is a product. We were able to write the polynomial as a product because of the common binomial factor, $(y + 2)$, that appeared. If this does not happen, try rearranging the terms of the original polynomial.

Check: Multiply $(y + 2)$ by $(x + 3)$.

$$(y + 2)(x + 3) = xy + 2x + 3y + 6,$$

the original polynomial.

Thus, a factored form of $xy + 2x + 3y + 6$ is the product $(y + 2)(x + 3)$.

Work Practice 11

You may want to try these steps on the next page when factoring by grouping.

Answers

6. $4x(x^2 + 3)$

7. $\frac{1}{5}a^2(2a^3 - 4a + 1)$

8. $3a^2b(2a + ab + 3b^3)$

9. $(p + 2)(7 + q)$

10. $(p + q)(7xy^3 - 1)$

11. $(b + 7)(a + 2)$

To Factor a Four-Term Polynomial by Grouping

Step 1: Group the terms in two groups of two terms so that each group has a common factor.

Step 2: Factor out the GCF from each group.

Step 3: If there is a common binomial factor, factor it out.

Step 4: If not, rearrange the terms and try these steps again.

Examples Factor by grouping.

12. $15x^3 - 10x^2 + 6x - 4$

$= (15x^3 - 10x^2) + (6x - 4)$ Group the terms.

$= 5x^2(3x - 2) + 2(3x - 2)$ Factor each group.

$= (3x - 2)(5x^2 + 2)$ Factor out the common factor, $(3x - 2)$.

13. $3x^2 + 4xy - 3x - 4y$

$= (3x^2 + 4xy) + (-3x - 4y)$

$= x(3x + 4y) - 1(3x + 4y)$ Factor each group. A -1 is factored from the second pair of terms so that there is a common factor, $(3x + 4y)$.

$= (3x + 4y)(x - 1)$ Factor out the common factor, $(3x + 4y)$.

14. $2a^2 + 5ab + 2a + 5b$

$= (2a^2 + 5ab) + (2a + 5b)$

$= a(2a + 5b) + 1(2a + 5b)$ Factor each group. An understood 1 is written before $(2a + 5b)$ to help remember that $(2a + 5b)$ is $1(2a + 5b)$.

$= (2a + 5b)(a + 1)$ Factor out the common factor, $(2a + 5b)$.

Work Practice 12–14

Practice 12–14

Factor by grouping.

12. $28x^3 - 7x^2 + 12x - 3$

13. $2xy + 5y^2 - 4x - 10y$

14. $3x^2 + 4xy + 3x + 4y$

Helpful Hint Notice that the factor of 1 is written when $(2a + 5b)$ is factored out.

Examples Factor by grouping.

15. $3x^3 - 2x - 9x^2 + 6$

$= x(3x^2 - 2) - 3(3x^2 - 2)$ Factor each group. A -3 is factored from the second pair of terms so that there is a common factor, $(3x^2 - 2)$.

$= (3x^2 - 2)(x - 3)$ Factor out the common factor, $(3x^2 - 2)$.

16. $3xy + 2 - 3x - 2y$

Notice that the first two terms have no common factor other than 1. However, if we rearrange these terms, a grouping emerges that does lead to a common factor.

$3xy + 2 - 3x - 2y$

$= (3xy - 3x) + (-2y + 2)$

$= 3x(y - 1) - 2(y - 1)$ Factor -2 from the second group.

$= (y - 1)(3x - 2)$ Factor out the common factor, $(y - 1)$.

17. $5x - 10 + x^3 - x^2 = 5(x - 2) + x^2(x - 1)$

There is no common binomial factor that can now be factored out. No matter how we rearrange the terms, no grouping will lead to a common factor. Thus, this polynomial is not factorable by grouping.

Work Practice 15–17

Practice 15–17

Factor by grouping.

15. $4x^3 + x - 20x^2 - 5$

16. $3xy - 4 + x - 12y$

17. $2x - 2 + x^3 - 3x^2$

Helpful Hint Throughout this chapter, we will be factoring polynomials. Even when the instructions do not so state, it is always a good idea to check your answers by multiplying.

Helpful Hint

One more reminder: When **factoring** a polynomial, make sure the polynomial is written as a **product.** For Example 13, it is true that

$$3x^2 + 4xy - 3x - 4y = \underbrace{x(3x + 4y) - 1(3x + 4y)}_{\text{but this is not a factored form}},$$

since it is a **sum (difference),** not a **product.**

A factored form of $3x^2 + 4xy - 3x - 4y$ is the **product** $(3x + 4y)(x - 1)$.

Answers

12. $(4x - 1)(7x^2 + 3)$
13. $(2x + 5y)(y - 2)$
14. $(3x + 4y)(x + 1)$
15. $(4x^2 + 1)(x - 5)$
16. $(3y + 1)(x - 4)$
17. cannot be factored by grouping

Vocabulary, Readiness & Video Check

Use the choices below to fill in each blank. Some choices may be used more than once and some may not be used at all.

greatest common factor   factors   factoring   true   false   least   greatest

1. Since $5 \cdot 4 = 20$, the numbers 5 and 4 are called ______________ of 20.
2. The ______________ of a list of integers is the largest integer that is a factor of all the integers in the list.
3. The greatest common factor of a list of common variables raised to powers is the variable raised to the ______________ exponent in the list.
4. The process of writing a polynomial as a product is called ______________.
5. True or false? A factored form of $7x + 21 + xy + 3y$ is $7(x + 3) + y(x + 3)$. ______________
6. True or false? A factored form of $3x^3 + 6x + x^2 + 2$ is $3x(x^2 + 2)$. ______________

Write the prime factorization of the following integers.

7. 14   8. 15

Write the GCF of the following pairs of integers.

9. 18, 3   10. 7, 35   11. 20, 15   12. 6, 15

Martin-Gay Interactive Videos

See Video 4.1

Watch the section lecture video and answer the following questions.

Objective A 13. Based on Example 1, give a general definition for the greatest common factor (GCF) of a list of numbers.

Objective B 14. When finding the GCF of the terms in Example 3, why are the numerical parts of the terms factored out, but not the variable parts?

Objective C 15. From Example 5, once we factor out the GCF, how can the number of terms in the other factor help us determine if our factorization is correct?

Objective D 16. In Examples 7 and 8, what are we reminded to always do first when factoring a polynomial? Also, a polynomial with how many terms suggests it might be factored by grouping?

4.1 Exercise Set MyLab Math 

Objectives A B Mixed Practice *Find the GCF for each list. See Examples 1 through 3.*

1. 32, 36   2. 36, 90   3. 18, 42, 84   4. 30, 75, 135

5. 24, 14, 21   6. 15, 25, 27   7. y^2, y^4, y^7 8. x^3, x^2, x^5

9. z^7, z^9, z^{11} 10. y^8, y^{10}, y^{12} 11. $x^{10}y^2, xy^2, x^3y^3$ 12. p^7q, p^8q^2, p^9q^3

13. $14x, 21$ 14. $20y, 15$ 15. $12y^4, 20y^3$ 16. $32x^5, 18x^2$

17. $-10x^2, 15x^3$ **18.** $-21x^3, 14x$ **19.** $12x^3, -6x^4, 3x^5$ **20.** $15y^2, 5y^7, -20y^3$

21. $-18x^2y, 9x^3y^3, 36x^3y$ **22.** $7x^3y^3, -21x^2y^2, 14xy^4$ **23.** $20a^6b^2c^8, 50a^7b$ **24.** $40x^7y^2z, 64x^9y$

Objective C *Factor out the GCF from each polynomial. See Examples 4 through 10.*

25. $3a + 6$ **26.** $18a + 12$ **27.** $30x - 15$ **28.** $42x - 7$ **29.** $x^3 + 5x^2$

30. $y^5 + 6y^4$ **31.** $6y^4 + 2y^3$ **32.** $5x^2 + 10x^6$ **33.** $32xy - 18x^2$ **34.** $10xy - 15x^2$

35. $4x - 8y + 4$ **36.** $7x + 21y - 7$ **37.** $6x^3 - 9x^2 + 12x$ **38.** $12x^3 + 16x^2 - 8x$

39. $a^7b^6 - a^3b^2 + a^2b^5 - a^2b^2$ **40.** $x^9y^6 + x^3y^5 - x^4y^3 + x^3y^3$ **41.** $5x^3y - 15x^2y + 10xy$

42. $14x^3y + 7x^2y - 7xy$ **43.** $8x^5 + 16x^4 - 20x^3 + 12$ **44.** $9y^6 - 27y^4 + 18y^2 + 6$

45. $\frac{1}{3}x^4 + \frac{2}{3}x^3 - \frac{4}{3}x^5 + \frac{1}{3}x$ **46.** $\frac{2}{5}y^7 - \frac{4}{5}y^5 + \frac{3}{5}y^2 - \frac{2}{5}y$ **47.** $y(x^2 + 2) + 3(x^2 + 2)$

48. $x(y^2 + 1) - 3(y^2 + 1)$ **49.** $z(y + 4) + 3(y + 4)$ **50.** $8(x + 2) - y(x + 2)$

51. $r(z^2 - 6) + (z^2 - 6)$ **52.** $q(b^3 - 5) + (b^3 - 5)$

Factor a negative number or a GCF with a negative coefficient from each polynomial. See directly after Example 5.

53. $-2x - 14$ **54.** $-7y - 21$ **55.** $-2x^5 + x^7$

56. $-5y^3 + y^6$ **57.** $-6a^4 + 9a^3 - 3a^2$ **58.** $-5m^6 + 10m^5 - 5m^3$

Objective D *Factor each four-term polynomial by grouping. If this is not possible, write "not factorable by grouping." See Examples 11 through 17.*

59. $x^3 + 2x^2 + 5x + 10$ **60.** $x^3 + 4x^2 + 3x + 12$ **61.** $5x + 15 + xy + 3y$

62. $xy + y + 2x + 2$ **63.** $6x^3 - 4x^2 + 15x - 10$ **64.** $16x^3 - 28x^2 + 12x - 21$

65. $5m^3 + 6mn + 5m^2 + 6n$ **66.** $8w^2 + 7wv + 8w + 7v$ **67.** $2y - 8 + xy - 4x$

68. $6x - 42 + xy - 7y$ **69.** $2x^3 + x^2 + 8x + 4$ **70.** $2x^3 - x^2 - 10x + 5$

71. $3x - 3 + x^3 - 4x^2$ **72.** $7x - 21 + x^3 - 2x^2$ **73.** $4x^2 - 8xy - 3x + 6y$

74. $5xy - 15x - 6y + 18$ **75.** $5q^2 - 4pq - 5q + 4p$ **76.** $6m^2 - 5mn - 6m + 5n$

Objectives C D Mixed Practice *Factor out the GCF from each polynomial. Then factor by grouping.*

77. $12x^2y - 42x^2 - 4y + 14$

78. $90 + 15y^2 - 18x - 3xy^2$

79. $6a^2 + 9ab^2 + 6ab + 9b^3$

80. $16x^2 + 4xy^2 + 8xy + 2y^3$

Review

Multiply. See Section 3.5.

81. $(x + 2)(x + 5)$

82. $(y + 3)(y + 6)$

83. $(b + 1)(b - 4)$

84. $(x - 5)(x + 10)$

Fill in the chart by finding two numbers that have the given product and sum. The first column is filled in for you.

		85.	86.	87.	88.	89.	90.	91.	92.
Two Numbers	4, 7								
Their Product	28	12	20	8	16	−10	−9	−24	−36
Their Sum	11	8	9	−9	−10	3	0	−5	−5

Concept Extensions

See the Concept Checks in this section.

93. Which of the following is/are factored form(s) of $-2x + 14$?

a. $-2(x + 7)$

b. $-2 \cdot x + 14$

c. $-2(x - 14)$

d. $-2(x - 7)$

94. Which of the following is/are factored form(s) of $8a - 24$?

a. $8 \cdot a - 24$

b. $8(a - 3)$

c. $4(2a - 12)$

d. $8 \cdot a - 2 \cdot 12$

Which of the following expressions are factored?

95. $(a + 6)(a + 2)$

96. $(x + 5)(x + y)$

97. $5(2y + z) - b(2y + z)$

98. $3x(a + 2b) + 2(a + 2b)$

Solve.

99. The number (in thousands) of students who graduated from U.S. high schools, both public and private, each year during 2010 through 2019 can be modeled by $\frac{16}{10}x^2 + \frac{67}{10}x + \frac{34{,}382}{10}$, where x is the number of years since 2010. (*Source:* National Center for Educational Statistics)

a. Find the number of students who graduated from U.S. high schools in 2018. To do so, let $x = 8$ and evaluate $\frac{16}{10}x^2 + \frac{67}{10}x + \frac{34{,}382}{10}$.

b. Use this expression to predict the number of students who will graduate from U.S. high schools in 2024.

c. Factor the polynomial $\frac{16}{10}x^2 + \frac{67}{10}x + \frac{34{,}382}{10}$ by factoring $\frac{1}{10}$ from each term.

d. For the year 2018, the National Center for Higher Education determined that about 70% of U.S. high school graduates went on to higher education. Using your answer from part **a**, approximate how many of those graduating in 2018 pursued higher education.

100. The amount of bottled water consumed in the United States, in gallons per person, for the period 2010–2017 can be approximated by the polynomial $\frac{16}{100}x^2 + \frac{91}{100}x + \frac{2825}{100}$, where x is the number of years after 2010. (*Source:* Bottledwater.org and *Fortune*)

a. Find the approximate U.S. annual per capita consumption of bottled water in 2016. To do so, let $x = 6$ and evaluate $\frac{16}{100}x^2 + \frac{91}{100}x + \frac{2825}{100}$

b. Find the approximate U.S. annual per capita consumption of bottled water in 2017.

c. Suppose the annual per capita consumption of bottled water continues to be approximated by the polynomial $\frac{16}{100}x^2 + \frac{91}{100}x + \frac{2825}{100}$. Use this polynomial to predict the per capita consumption of bottled water in 2022.

d. Factor out the GCF $\frac{1}{100}$ from the polynomial $\frac{16}{100}x^2 + \frac{91}{100}x + \frac{2825}{100}$.

101. The annual orange production (in thousand tons) in the United States for the period 2011–2016 can be approximated by the polynomial $87x^2 - 1131x + 9048$, where x is the number of years after 2011. (*Source:* Based on data from the National Agricultural Statistics Service and bloomberg.com)

a. Find the approximate U.S. orange production in 2015. To do so, let $x = 4$ and evaluate $87x^2 - 1131x + 9048$.

b. Find the approximate U.S. orange production in 2016.

c. Factor out the GCF from the polynomial $87x^2 - 1131x + 9048$.

102. The polynomial $-6x^2 + 72x + 384$ represents the approximate number of visitors (in thousands) per year to Redwoods National Park in California, during 2013–2016. In this polynomial, x represents the years since 2013. (*Source:* Based on data from the National Park Service)

a. Find the approximate number of visitors to Redwoods National Park in 2014. To do so, let $x = 1$ and evaluate $-6x^2 + 72x + 384$.

b. Find the approximate number of visitors to Redwoods National Park in 2016.

c. Factor out a common factor of -6 from the polynomial $-6x^2 + 72x + 384$.

Write an expression for the area of each shaded region. Then write the expression as a factored polynomial.

△ **103.**

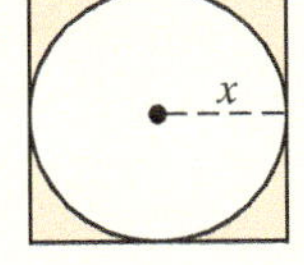

△ **104.**

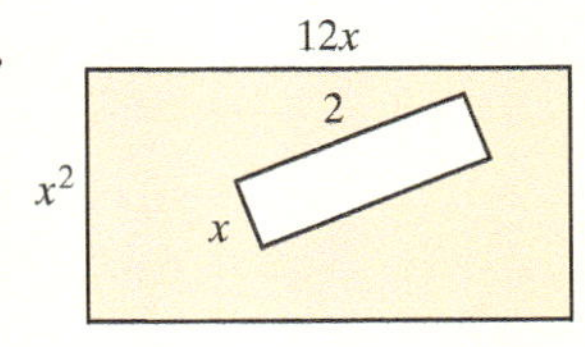

Write an expression for the length of each rectangle. (Hint: Factor the area binomial and recall that Area = width · length.)

△ **105.**

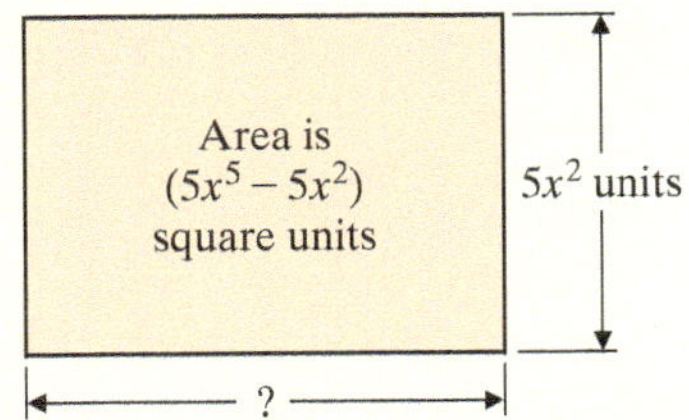

△ **106.**

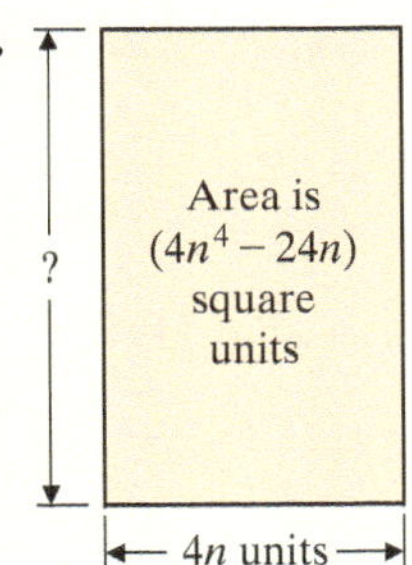

107. Construct a binomial whose greatest common factor is $5a^3$. (*Hint:* Multiply $5a^3$ by a binomial whose terms contain no common factor other than 1: $5a^3(\square + \square)$.)

108. Construct a trinomial whose greatest common factor is $2x^2$. See the hint for Exercise **107**.

109. Explain how you can tell whether a polynomial is written in factored form.

110. Construct a four-term polynomial that can be factored by grouping. Explain how you constructed the polynomial.

4.2 Factoring Trinomials of the Form $x^2 + bx + c$

Objectives

A Factor Trinomials of the Form $x^2 + bx + c$.

B Factor Out the Greatest Common Factor and Then Factor a Trinomial of the Form $x^2 + bx + c$.

Objective A Factoring Trinomials of the Form $x^2 + bx + c$

In this section, we factor trinomials of the form $x^2 + bx + c$, such as

$$x^2 + 7x + 12, \quad x^2 - 12x + 35, \quad x^2 + 4x - 12, \quad \text{and} \quad r^2 - r - 42$$

Notice that for these trinomials, the coefficient of the squared variable is 1.

Recall that factoring means to write as a product and that factoring and multiplying are reverse processes. Using the FOIL method of multiplying binomials, we have the following.

$$\begin{aligned} & \quad\;\; \text{F} \quad \text{O} \quad \text{I} \quad \text{L} \\ (x + 3)(x + 1) &= x^2 + 1x + 3x + 3 \\ &= x^2 + 4x + 3 \end{aligned}$$

Thus, a factored form of $x^2 + 4x + 3$ is $(x + 3)(x + 1)$.

Notice that the product of the first terms of the binomials is $x \cdot x = x^2$, the first term of the trinomial. Also, the product of the last two terms of the binomials is $3 \cdot 1 = 3$, the third term of the trinomial. The sum of these same terms is $3 + 1 = 4$, the coefficient of the middle, x, term of the trinomial.

The product of these numbers is 3.

$$x^2 + 4x + 3 = (x + 3)(x + 1)$$

The sum of these numbers is 4.

Many trinomials, such as the one above, factor into two binomials. To factor $x^2 + 7x + 10$, let's assume that it factors into two binomials and begin by writing two pairs of parentheses. The first term of the trinomial is x^2, so we use x and x as the first terms of the binomial factors.

$$x^2 + 7x + 10 = (x + \square)(x + \square)$$

To determine the last term of each binomial factor, we look for two integers whose product is 10 and whose sum is 7. The integers are 2 and 5. Thus,

$$x^2 + 7x + 10 = (x + 2)(x + 5)$$

Check: To see if we have factored correctly, we multiply.

$$\begin{aligned} (x + 2)(x + 5) &= x^2 + 5x + 2x + 10 \\ &= x^2 + 7x + 10 \qquad \text{Combine like terms.} \end{aligned}$$

Helpful Hint

Since multiplication is commutative, the factored form of $x^2 + 7x + 10$ can be written as either $(x + 2)(x + 5)$ or $(x + 5)(x + 2)$.

To Factor a Trinomial of the Form $x^2 + bx + c$

The product of these numbers is c.

$$x^2 + bx + c = (x + \square)(x + \square)$$

The sum of these numbers is b.

Example 1 Factor: $x^2 + 7x + 12$

Solution: We begin by writing the first terms of the binomial factors.

$(x + \square)(x + \square)$

Next we look for two numbers whose product is 12 and whose sum is 7. Since our numbers must have a positive product and a positive sum, we look at pairs of positive factors of 12 only.

Factors of 12	Sum of Factors
1, 12	13
2, 6	8
3, 4	7

Correct sum, so the numbers are 3 and 4.

Thus, $x^2 + 7x + 12 = (x + 3)(x + 4)$

Check: $(x + 3)(x + 4) = x^2 + 4x + 3x + 12 = x^2 + 7x + 12$

Work Practice 1

Practice 1

Factor: $x^2 + 12x + 20$

Example 2 Factor: $x^2 - 12x + 35$

Solution: Again, we begin by writing the first terms of the binomials.

$(x + \square)(x + \square)$

Now we look for two numbers whose product is 35 and whose sum is -12. Since our numbers must have a positive product and a negative sum, we look at pairs of negative factors of 35 only.

Factors of 35	Sum of Factors
$-1, -35$	-36
$-5, -7$	-12

Correct sum, so the numbers are -5 and -7.

$x^2 - 12x + 35 = (x - 5)(x - 7)$

Check: To check, multiply $(x - 5)(x - 7)$.

Work Practice 2

Practice 2

Factor each trinomial.

a. $x^2 - 23x + 22$

b. $x^2 - 27x + 50$

Example 3 Factor: $x^2 + 4x - 12$

Solution: $x^2 + 4x - 12 = (x + \square)(x + \square)$

We look for two numbers whose product is -12 and whose sum is 4. Since our numbers must have a negative product, we look at pairs of factors with opposite signs.

Factors of -12	Sum of Factors
$-1, 12$	11
$1, -12$	-11
$-2, 6$	4
$2, -6$	-4
$-3, 4$	1
$3, -4$	-1

Correct sum, so the numbers are -2 and 6.

$x^2 + 4x - 12 = (x - 2)(x + 6)$

Work Practice 3

Practice 3

Factor: $x^2 + 5x - 36$

Answers

1. $(x + 10)(x + 2)$

2. a. $(x - 1)(x - 22)$

b. $(x - 2)(x - 25)$

3. $(x + 9)(x - 4)$

Practice 4

Factor each trinomial.

a. $q^2 - 3q - 40$

b. $y^2 + 2y - 48$

Example 4 Factor: $r^2 - r - 42$

Solution: Because the variable in this trinomial is r, the first term of each binomial factor is r.

$$r^2 - r - 42 = (r + \square)(r + \square)$$

Now we look for two numbers whose product is -42 and whose sum is -1, the numerical coefficient of r. The numbers are 6 and -7. Therefore,

$$r^2 - r - 42 = (r + 6)(r - 7)$$

Work Practice 4

Practice 5

Factor: $x^2 + 6x + 15$

Example 5 Factor: $a^2 + 2a + 10$

Solution: Look for two numbers whose product is 10 and whose sum is 2. Neither 1 and 10 nor 2 and 5 give the required sum, 2. We conclude that $a^2 + 2a + 10$ is not factorable with integers. A polynomial such as $a^2 + 2a + 10$ is called a **prime polynomial.**

Work Practice 5

Practice 6

Factor each trinomial.

a. $x^2 + 9xy + 14y^2$

b. $a^2 - 13ab + 30b^2$

Example 6 Factor: $x^2 + 5xy + 6y^2$

Solution: $x^2 + 5xy + 6y^2 = (x + \square)(x + \square)$

Recall that the middle term, $5xy$, is the same as $5yx$. Thus, we can see that $5y$ is the "coefficient" of x. We then look for two terms whose product is $6y^2$ and whose sum is $5y$. The terms are $2y$ and $3y$ because $2y \cdot 3y = 6y^2$ and $2y + 3y = 5y$. Therefore,

$$x^2 + 5xy + 6y^2 = (x + 2y)(x + 3y)$$

Work Practice 6

Practice 7

Factor: $x^4 + 8x^2 + 12$

Example 7 Factor: $x^4 + 5x^2 + 6$

Solution: As usual, we begin by writing the first terms of the binomials. Since the greatest power of x in this polynomial is x^4, we write

$$(x^2 + \square)(x^2 + \square) \quad \text{Since } x^2 \cdot x^2 = x^4$$

Now we look for two factors of 6 whose sum is 5. The numbers are 2 and 3. Thus,

$$x^4 + 5x^2 + 6 = (x^2 + 2)(x^2 + 3)$$

Work Practice 7

If the terms of a polynomial are not written in descending powers of the variable, you may want to rearrange the terms before factoring.

Practice 8

Factor: $48 - 14x + x^2$

Example 8 Factor: $40 - 13t + t^2$

Solution: First, we rearrange terms so that the trinomial is written in descending powers of t.

$$40 - 13t + t^2 = t^2 - 13t + 40$$

Next, try to factor.

$$t^2 - 13t + 40 = (t + \square)(t + \square)$$

Now we look for two factors of 40 whose sum is -13. The numbers are -8 and -5. Thus,

$$t^2 - 13t + 40 = (t - 8)(t - 5)$$

Work Practice 8

Answers

4. a. $(q - 8)(q + 5)$ **b.** $(y + 8)(y - 6)$

5. prime polynomial

6. a. $(x + 2y)(x + 7y)$ **b.** $(a - 3b)(a - 10b)$

7. $(x^2 + 6)(x^2 + 2)$

8. $(x - 6)(x - 8)$

The following sign patterns may be useful when factoring trinomials.

Helpful Hint

A positive constant in a trinomial tells us to look for two numbers with the same sign. The sign of the coefficient of the middle term tells us whether the signs are both positive or both negative.

both positive ↓ same sign ↓

$$x^2 + 10x + 16 = (x + 2)(x + 8)$$

both negative ↓ same sign ↓

$$x^2 - 10x + 16 = (x - 2)(x - 8)$$

A negative constant in a trinomial tells us to look for two numbers with opposite signs.

opposite signs ↓

$$x^2 + 6x - 16 = (x + 8)(x - 2)$$

opposite signs ↓

$$x^2 - 6x - 16 = (x - 8)(x + 2)$$

Objective B Factoring Out the Greatest Common Factor

Remember that the first step in factoring any polynomial is to factor out the greatest common factor (if there is one other than 1 or −1).

Example 9 Factor: $3m^2 - 24m - 60$

Solution: First we factor out the greatest common factor, 3, from each term.

$$3m^2 - 24m - 60 = 3(m^2 - 8m - 20)$$

Now we factor $m^2 - 8m - 20$ by looking for two factors of −20 whose sum is −8. The factors are −10 and 2. Therefore, the complete factored form is

$$3m^2 - 24m - 60 = 3(m + 2)(m - 10)$$

Work Practice 9

Helpful Hint

Remember to write the common factor, 3, as part of the factored form.

Example 10 Factor: $2x^4 - 26x^3 + 84x^2$

Solution:

$$2x^4 - 26x^3 + 84x^2 = 2x^2(x^2 - 13x + 42) \quad \text{Factor out common factor, } 2x^2.$$
$$= 2x^2(x - 6)(x - 7) \quad \text{Factor } x^2 - 13x + 42.$$

Work Practice 10

Practice 9

Factor each trinomial.

a. $4x^2 - 24x + 36$

b. $x^3 + 3x^2 - 4x$

Practice 10

Factor: $5x^5 - 25x^4 - 30x^3$

Answers

9. a. $4(x - 3)(x - 3)$ or $4(x - 3)^2$
b. $x(x + 4)(x - 1)$
10. $5x^3(x + 1)(x - 6)$

Vocabulary, Readiness & Video Check

Fill in each blank with "true" or "false."

1. To factor $x^2 + 7x + 6$, we look for two numbers whose product is 6 and whose sum is 7. ____
2. We can write the factorization $(y + 2)(y + 4)$ also as $(y + 4)(y + 2)$. ____
3. The factorization $(4x - 12)(x - 5)$ is completely factored. ____
4. The factorization $(x + 2y)(x + y)$ may also be written as $(x + 2y)^2$. ____

Complete each factored form.

5. $x^2 + 9x + 20 = (x + 4)(x \quad)$
6. $x^2 + 12x + 35 = (x + 5)(x \quad)$
7. $x^2 - 7x + 12 = (x - 4)(x \quad)$
8. $x^2 - 13x + 22 = (x - 2)(x \quad)$
9. $x^2 + 4x + 4 = (x + 2)(x \quad)$
10. $x^2 + 10x + 24 = (x + 6)(x \quad)$

Martin-Gay Interactive Videos

See Video 4.2

Watch the section lecture video and answer the following questions.

Objective A 11. In Example 2, why are only negative factors of 15 considered?

Objective B 12. In Example 5, we know we need a positive and a negative factor of −10. How do we determine which factor is negative?

4.2 Exercise Set MyLab Math

Objective A *Factor each trinomial completely. If a polynomial can't be factored, write "prime." See Examples 1 through 8.*

1. $x^2 + 7x + 6$
2. $x^2 + 6x + 8$
3. $y^2 - 10y + 9$
4. $y^2 - 12y + 11$
5. $x^2 - 6x + 9$
6. $x^2 - 10x + 25$
7. $x^2 - 3x - 18$
8. $x^2 - x - 30$
9. $x^2 + 3x - 70$
10. $x^2 + 4x - 32$
11. $x^2 + 5x + 2$
12. $x^2 - 7x + 5$
13. $x^2 + 8xy + 15y^2$
14. $x^2 + 6xy + 8y^2$
15. $a^4 - 2a^2 - 15$
16. $y^4 - 3y^2 - 70$
17. $13 + 14m + m^2$
18. $17 + 18n + n^2$
19. $10t - 24 + t^2$
20. $6q - 27 + q^2$
21. $a^2 - 10ab + 16b^2$
22. $a^2 - 9ab + 18b^2$

Objectives A B Mixed Practice *Factor each trinomial completely. Some of these trinomials contain a greatest common factor (other than 1). Don't forget to factor out the GCF first. See Examples 1 through 10.*

23. $2z^2 + 20z + 32$

24. $3x^2 + 30x + 63$

25. $2x^3 - 18x^2 + 40x$

26. $3x^3 - 12x^2 - 36x$

27. $x^2 - 3xy - 4y^2$

28. $x^2 - 4xy - 77y^2$

29. $x^2 + 15x + 36$

30. $x^2 + 19x + 60$

31. $x^4 - x^2 - 2$

32. $x^4 - 5x^2 - 14$

33. $r^2 - 16r + 48$

34. $r^2 - 10r + 21$

35. $x^2 + xy - 2y^2$

36. $x^2 - xy - 6y^2$

37. $3x^2 + 9x - 30$

38. $4x^2 - 4x - 48$

39. $3x^4 - 60x^2 + 108$

40. $2x^4 - 24x^2 + 70$

41. $x^2 - 18x - 144$

42. $x^2 + x - 42$

43. $r^2 - 3r + 6$

44. $x^2 + 4x - 10$

45. $x^2 - 8x + 15$

46. $x^2 - 9x + 14$

47. $6x^3 + 54x^2 + 120x$

48. $3x^3 + 3x^2 - 126x$

49. $4x^2y + 4xy - 12y$

50. $3x^2y - 9xy + 45y$

51. $x^2 - 4x - 21$

52. $x^2 - 4x - 32$

53. $x^2 + 7xy + 10y^2$

54. $x^2 - 3xy - 4y^2$

55. $64 + 24t + 2t^2$

56. $50 + 20t + 2t^2$

57. $x^3 - 2x^2 - 24x$

58. $x^3 - 3x^2 - 28x$

59. $2t^5 - 14t^4 + 24t^3$

60. $3x^6 + 30x^5 + 72x^4$

61. $5x^3y - 25x^2y^2 - 120xy^3$

62. $7a^3b - 35a^2b^2 + 42ab^3$

63. $162 - 45m + 3m^2$

64. $48 - 20n + 2n^2$

65. $-x^2 + 12x - 11$ (Factor out -1 first.)

66. $-x^2 + 8x - 7$ (Factor out -1 first.)

67. $\frac{1}{2}y^2 - \frac{9}{2}y - 11$ (Factor out $\frac{1}{2}$ first.)

68. $\frac{1}{3}y^2 - \frac{5}{3}y - 8$ (Factor out $\frac{1}{3}$ first.)

69. $x^3y^2 + x^2y - 20x$

70. $a^2b^3 + ab^2 - 30b$

Review

Multiply. See Sections 3.5 and 3.6.

71. $(2x + 1)(x + 5)$

72. $(3x + 2)(x + 4)$

73. $(5y - 4)(3y - 1)$

74. $(4z - 7)(7z - 1)$

75. $(a + 3b)(9a - 4b)$

76. $(y - 5x)(6y + 5x)$

Concept Extensions

77. Write a polynomial that factors as $(x - 3)(x + 8)$.

78. To factor $x^2 + 13x + 42$, think of two numbers whose _____ is 42 and whose _____ is 13.

Complete each sentence in your own words.

79. If $x^2 + bx + c$ is factorable and c is negative, then the signs of the last-term factors of the binomials are opposite because …

80. If $x^2 + bx + c$ is factorable and c is positive, then the signs of the last-term factors of the binomials are the same because …

Remember that perimeter means distance around. Write the perimeter of each rectangle as a simplified polynomial. Then factor the polynomial completely.

△ **81.**

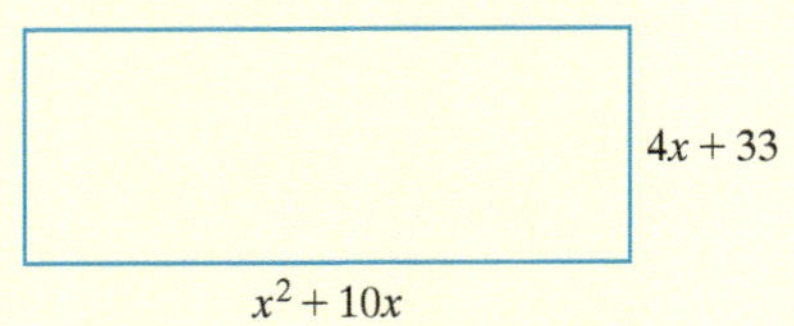

△ **82.**

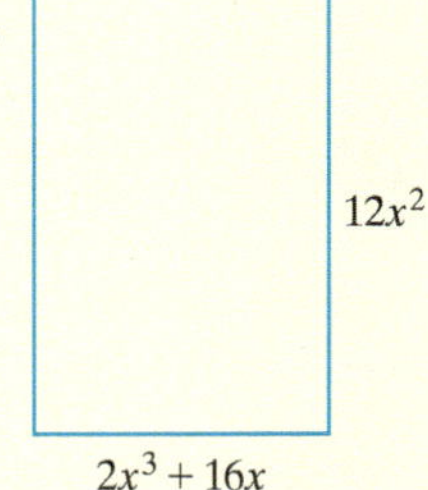

83. An object is thrown upward from the top of an 80-foot building with an initial velocity of 64 feet per second. Neglecting air resistance, the height of the object after t seconds is given by $-16t^2 + 64t + 80$. Factor this polynomial.

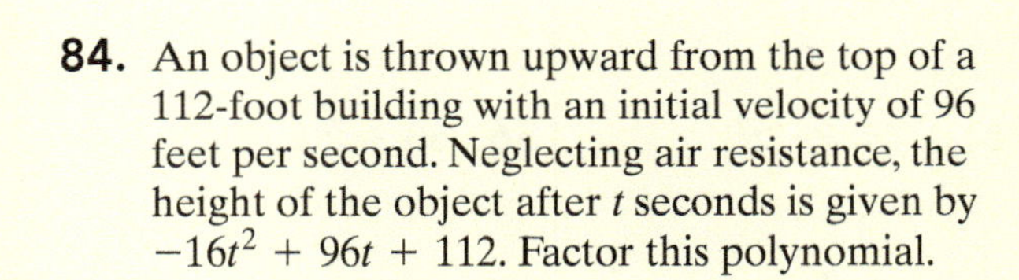

84. An object is thrown upward from the top of a 112-foot building with an initial velocity of 96 feet per second. Neglecting air resistance, the height of the object after t seconds is given by $-16t^2 + 96t + 112$. Factor this polynomial.

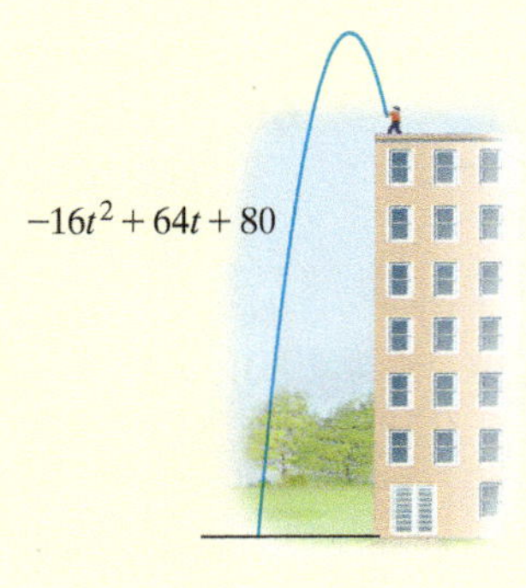

Factor each trinomial completely.

85. $x^2 + \frac{1}{2}x + \frac{1}{16}$

86. $x^2 + x + \frac{1}{4}$

87. $z^2(x + 1) - 3z(x + 1) - 70(x + 1)$

88. $y^2(x + 1) - 2y(x + 1) - 15(x + 1)$

Find a positive value of c so that each trinomial is factorable.

89. $n^2 - 16n + c$

90. $y^2 - 4y + c$

Find a positive value of b so that each trinomial is factorable.

91. $y^2 + by + 20$

92. $x^2 + bx + 15$

Factor each trinomial. (Hint: Notice that $x^{2n} + 4x^n + 3$ factors as $(x^n + 1)(x^n + 3)$. Remember: $x^n \cdot x^n = x^{n+n}$ or x^{2n}.)

93. $x^{2n} + 8x^n - 20$

94. $x^{2n} + 5x^n + 6$

4.3 Factoring Trinomials of the Form $ax^2 + bx + c$

Objectives

A Factor Trinomials of the Form $ax^2 + bx + c$, Where $a \neq 1$.

B Factor Out the GCF Before Factoring a Trinomial of the Form $ax^2 + bx + c$.

Objective A Factoring Trinomials of the Form $ax^2 + bx + c$

In this section, we factor trinomials of the form $ax^2 + bx + c$, such as

$$3x^2 + 11x + 6, \quad 8x^2 - 22x + 5, \quad \text{and} \quad 2x^2 + 13x - 7$$

Notice that the coefficient of the squared variable in these trinomials is a number other than 1. We will factor these trinomials using a trial-and-check method based on our work in the last section.

To begin, let's review the relationship between the numerical coefficients of the trinomial and the numerical coefficients of its factored form. For example, since

$$(2x + 1)(x + 6) = 2x^2 + 13x + 6,$$

a factored form of $2x^2 + 13x + 6$ is $(2x + 1)(x + 6)$.

Notice that $2x$ and x are factors of $2x^2$, the first term of the trinomial. Also, 6 and 1 are factors of 6, the last term of the trinomial, as shown:

$$2x^2 + 13x + 6 = (2x + 1)(x + 6)$$

$2x \cdot x$

$1 \cdot 6$

Also notice that $13x$, the middle term, is the sum of the following products:

$$2x^2 + 13x + 6 = (2x + 1)(x + 6)$$

$$\begin{array}{r} 1x \\ +12x \\ \hline 13x \end{array}$$

Middle term

Let's use this pattern to factor $5x^2 + 7x + 2$. First, we find factors of $5x^2$. Since all numerical coefficients in this trinomial are positive, we will use factors with positive numerical coefficients only. Thus, the factors of $5x^2$ are $5x$ and x. Let's try these factors as first terms of the binomials. Thus far, we have

$$5x^2 + 7x + 2 = (5x + \square)(x + \square)$$

Next, we need to find positive factors of 2. Positive factors of 2 are 1 and 2. Now we try possible combinations of these factors as second terms of the binomials until we obtain a middle term of $7x$.

$$(5x + 1)(x + 2) = 5x^2 + 11x + 2$$

$1x$
$+10x$
$11x$ ⟶ **Incorrect** middle term

Let's try switching factors 2 and 1.

$$(5x + 2)(x + 1) = 5x^2 + 7x + 2$$

$2x$
$+5x$
$7x$ ⟶ **Correct** middle term

Thus a factored form of $5x^2 + 7x + 2$ is $(5x + 2)(x + 1)$. To check, we multiply $(5x + 2)$ and $(x + 1)$. The product is $5x^2 + 7x + 2$.

Practice 1

Factor each trinomial.

a. $5x^2 + 27x + 10$

b. $4x^2 + 12x + 5$

Example 1

Factor: $3x^2 + 11x + 6$

Solution: Since all numerical coefficients are positive, we use factors with positive numerical coefficients. We first find factors of $3x^2$.

Factors of $3x^2$: $3x^2 = 3x \cdot x$

If factorable, the trinomial will be of the form

$$3x^2 + 11x + 6 = (3x + \square)(x + \square)$$

Next we factor 6.

Factors of 6: $6 = 1 \cdot 6, \quad 6 = 2 \cdot 3$

Now we try combinations of factors of 6 until a middle term of $11x$ is obtained. Let's try 1 and 6 first.

$$(3x + 1)(x + 6) = 3x^2 + 19x + 6$$

$1x$
$+18x$
$19x$ ⟶ **Incorrect** middle term

Now let's next try 6 and 1.

$$(3x + 6)(x + 1)$$

Before multiplying, notice that the terms of the factor $3x + 6$ have a common factor of 3. The terms of the original trinomial $3x^2 + 11x + 6$ have no common factor other than 1, so the terms of its factors will also contain no common factor other than 1. This means that $(3x + 6)(x + 1)$ is not a factored form.

Next let's try 2 and 3 as last terms.

$$(3x + 2)(x + 3) = 3x^2 + 11x + 6$$

$2x$
$+9x$
$11x$ ⟶ **Correct** middle term

Thus a factored form of $3x^2 + 11x + 6$ is $(3x + 2)(x + 3)$.

Helpful Hint This is true in general: If the terms of a trinomial have no common factor (other than 1), then the terms of each of its binomial factors will contain no common factor (other than 1).

Work Practice 1

Answers

1. a. $(5x + 2)(x + 5)$
b. $(2x + 5)(2x + 1)$

✓**Concept Check** Do the terms of $3x^2 + 29x + 18$ have a common factor? Without multiplying, decide which of the following factored forms could not be a factored form of $3x^2 + 29x + 18$.

a. $(3x + 18)(x + 1)$ **b.** $(3x + 2)(x + 9)$
c. $(3x + 6)(x + 3)$ **d.** $(3x + 9)(x + 2)$

Example 2

Factor: $8x^2 - 22x + 5$

Solution: Factors of $8x^2$: $8x^2 = 8x \cdot x$, $8x^2 = 4x \cdot 2x$

We'll try $8x$ and x.

$$8x^2 - 22x + 5 = (8x + \square)(x + \square)$$

Since the middle term, $-22x$, has a negative numerical coefficient, we factor 5 into negative factors.

Factors of 5: $5 = -1 \cdot -5$

Let's try -1 and -5.

$$(8x - 1)(x - 5) = 8x^2 - 41x + 5$$

$-1x$
$+(-40x)$
$-41x$ → **Incorrect** middle term

Now let's try -5 and -1.

$$(8x - 5)(x - 1) = 8x^2 - 13x + 5$$

$-5x$
$+(-8x)$
$-13x$ → **Incorrect** middle term

Don't give up yet! We can still try other factors of $8x^2$. Let's try $4x$ and $2x$ with -1 and -5.

$$(4x - 1)(2x - 5) = 8x^2 - 22x + 5$$

$-2x$
$+(-20x)$
$-22x$ → **Correct** middle term

A factored form of $8x^2 - 22x + 5$ is $(4x - 1)(2x - 5)$.

Work Practice 2

Practice 2

Factor each trinomial.

a. $2x^2 - 11x + 12$
b. $6x^2 - 5x + 1$

Example 3

Factor: $2x^2 + 13x - 7$

Solution: Factors of $2x^2$: $2x^2 = 2x \cdot x$

Factors of -7: $-7 = 1 \cdot -7$, $-7 = -1 \cdot 7$

We try possible combinations of these factors:

$(2x + 1)(x - 7) = 2x^2 - 13x - 7$ **Incorrect** middle term

$(2x - 1)(x + 7) = 2x^2 + 13x - 7$ **Correct** middle term

A factored form of $2x^2 + 13x - 7$ is $(2x - 1)(x + 7)$.

Work Practice 3

Practice 3

Factor each trinomial.

a. $3x^2 + 14x - 5$
b. $35x^2 + 4x - 4$

Answers

2. a. $(2x - 3)(x - 4)$ **b.** $(3x - 1)(2x - 1)$
3. a. $(3x - 1)(x + 5)$ **b.** $(5x + 2)(7x - 2)$

✓**Concept Check Answer**

no; **a, c, d**

Practice 4

Factor each trinomial.

a. $14x^2 - 3xy - 2y^2$

b. $12a^2 - 16ab - 3b^2$

Example 4 Factor: $10x^2 - 13xy - 3y^2$

Solution: Factors of $10x^2$: $10x^2 = 10x \cdot x$, $10x^2 = 2x \cdot 5x$

Factors of $-3y^2$: $-3y^2 = -3y \cdot y$, $-3y^2 = 3y \cdot -y$

We try some combinations of these factors:

Correct ↓ Correct ↓

$$(10x - 3y)(x + y) = 10x^2 + 7xy - 3y^2$$
$$(x + 3y)(10x - y) = 10x^2 + 29xy - 3y^2$$
$$(5x + 3y)(2x - y) = 10x^2 + xy - 3y^2$$
$$(2x - 3y)(5x + y) = 10x^2 - 13xy - 3y^2$$ **Correct** middle term

A factored form of $10x^2 - 13xy - 3y^2$ is $(2x - 3y)(5x + y)$.

Work Practice 4

Practice 5

Factor: $2x^4 - 5x^2 - 7$

Example 5 Factor: $3x^4 - 5x^2 - 8$

Solution: Factors of $3x^4$: $3x^4 = 3x^2 \cdot x^2$

Factors of -8: $-8 = -2 \cdot 4, 2 \cdot -4, -1 \cdot 8, 1 \cdot -8$

Try combinations of these factors:

Correct ↓ Correct ↓

$$(3x^2 - 2)(x^2 + 4) = 3x^4 + 10x^2 - 8$$
$$(3x^2 + 4)(x^2 - 2) = 3x^4 - 2x^2 - 8$$
$$(3x^2 + 8)(x^2 - 1) = 3x^4 + 5x^2 - 8$$ **Incorrect sign** on middle term, so switch signs in binomial factors.
$$(3x^2 - 8)(x^2 + 1) = 3x^4 - 5x^2 - 8$$ **Correct** middle term

Work Practice 5

Helpful Hint

Study the last two lines of Example 5. If a factoring attempt gives you a middle term whose numerical coefficient is the opposite of the desired numerical coefficient, try switching the signs of the last terms in the binomials.

Switched signs
$$(3x^2 + 8)(x^2 - 1) = 3x^4 + 5x^2 - 8$$ Middle term: $+5x$
$$(3x^2 - 8)(x^2 + 1) = 3x^4 - 5x^2 - 8$$ Middle term: $-5x$

Objective B Factoring Out the Greatest Common Factor

Don't forget that the first step in factoring any polynomial is to look for a common factor to factor out.

Practice 6

Factor each trinomial.

a. $3x^3 + 17x^2 + 10x$

b. $6xy^2 + 33xy - 18x$

Example 6 Factor: $24x^4 + 40x^3 + 6x^2$

Solution: Notice that all three terms have a common factor of $2x^2$. Thus we factor out $2x^2$ first.

$$24x^4 + 40x^3 + 6x^2 = 2x^2(12x^2 + 20x + 3)$$

Answers

4. a. $(7x + 2y)(2x - y)$ b. $(6a + b)(2a - 3b)$

5. $(2x^2 - 7)(x^2 + 1)$

6. a. $x(3x + 2)(x + 5)$ b. $3x(2y - 1)(y + 6)$

Next we factor $12x^2 + 20x + 3$.

Factors of $12x^2$: $12x^2 = 4x \cdot 3x$, $\quad 12x^2 = 12x \cdot x$, $\quad 12x^2 = 6x \cdot 2x$

Since all terms in the trinomial have positive numerical coefficients, we factor 3 using positive factors only.

Factors of 3: $3 = 1 \cdot 3$

We try some combinations of the factors.

$$2x^2(4x + 3)(3x + 1) = 2x^2(12x^2 + 13x + 3)$$
$$2x^2(12x + 1)(x + 3) = 2x^2(12x^2 + 37x + 3)$$
$$2x^2(2x + 3)(6x + 1) = 2x^2(12x^2 + 20x + 3) \quad \text{Correct middle term}$$

A factored form of $24x^4 + 40x^3 + 6x^2$ is $2x^2(2x + 3)(6x + 1)$.

Helpful Hint Don't forget to include the common factor in the factored form.

Work Practice 6

When the term containing the squared variable has a negative coefficient, you may want to first factor out a common factor of -1.

Example 7 Factor: $-6x^2 - 13x + 5$

Solution: We begin by factoring out a common factor of -1.

$$-6x^2 - 13x + 5 = -1(6x^2 + 13x - 5) \quad \text{Factor out } -1.$$
$$= -1(3x - 1)(2x + 5) \quad \text{Factor } 6x^2 + 13x - 5.$$

Work Practice 7

Practice 7

Factor: $-5x^2 - 19x + 4$

Answer

7. $-1(x + 4)(5x - 1)$

Vocabulary, Readiness & Video Check

Complete each factorization.

1. $2x^2 + 5x + 3$ factors as $(2x + 3)($ _____ $)$.
   a. $(x + 3)$ b. $(2x + 1)$ c. $(3x + 4)$ d. $(x + 1)$
2. $7x^2 + 9x + 2$ factors as $(7x + 2)($ _____ $)$.
   a. $(3x + 1)$ b. $(x + 1)$ c. $(x + 2)$ d. $(7x + 1)$
3. $3x^2 + 31x + 10$ factors as _____.
   a. $(3x + 2)(x + 5)$ b. $(3x + 5)(x + 2)$ c. $(3x + 1)(x + 10)$
4. $5x^2 + 61x + 12$ factors as _____.
   a. $(5x + 1)(x + 12)$ b. $(5x + 3)(x + 4)$ c. $(5x + 2)(x + 6)$

Martin-Gay Interactive Videos

See Video 4.3

Watch the section lecture video and answer the following questions.

Objective A 5. From Example 1, explain in general terms how we would go about factoring a trinomial with a first-term coefficient $\neq 1$.

Objective B 6. From Examples 3 and 5, how can factoring the GCF from a trinomial help us save time when trying to factor the remaining trinomial?

4.3 Exercise Set MyLab Math 

Objective A *Complete each factored form. See Examples 1 through 5.*

1. $5x^2 + 22x + 8 = (5x + 2)$
2. $2y^2 + 15y + 25 = (2y + 5)$
3. $50x^2 + 15x - 2 = (5x + 2)$
4. $6y^2 + 11y - 10 = (2y + 5)$
5. $20x^2 - 7x - 6 = (5x + 2)$
6. $8y^2 - 2y - 55 = (2y + 5)$

Factor each trinomial completely. If a polynomial can't be factored, write "prime." See Examples 1 through 5.

7. $2x^2 + 13x + 15$
8. $3x^2 + 8x + 4$
9. $8y^2 - 17y + 9$
10. $21x^2 - 41x + 10$
11. $2x^2 - 9x - 5$
12. $36r^2 - 5r - 24$
13. $20r^2 + 27r - 8$
14. $3x^2 + 20x - 63$
15. $10x^2 + 31x + 3$
16. $12x^2 + 17x + 5$
17. $x + 3x^2 - 2$
18. $y + 8y^2 - 9$
19. $6x^2 - 13xy + 5y^2$
20. $8x^2 - 14xy + 3y^2$
21. $15m^2 - 16m - 15$
22. $25n^2 - 5n - 6$
23. $-9x + 20 + x^2$
24. $-7x + 12 + x^2$
25. $2x^2 - 7x - 99$
26. $2x^2 + 7x - 72$
27. $-27t + 7t^2 - 4$
28. $-3t + 4t^2 - 7$
29. $3a^2 + 10ab + 3b^2$
30. $2a^2 + 11ab + 5b^2$
31. $49p^2 - 7p - 2$
32. $3r^2 + 10r - 8$
33. $18x^2 - 9x - 14$
34. $42a^2 - 43a + 6$
35. $2m^2 + 17m + 10$
36. $3n^2 + 20n + 5$
37. $24x^2 + 41x + 12$
38. $24x^2 - 49x + 15$

Objectives A B Mixed Practice *Factor each trinomial completely. If a polynomial can't be factored, write "prime." See Examples 1 through 7.*

39. $12x^3 + 11x^2 + 2x$
40. $8a^3 + 14a^2 + 3a$
41. $21b^2 - 48b - 45$
42. $12x^2 - 14x - 10$
43. $7z + 12z^2 - 12$
44. $16t + 15t^2 - 15$
45. $6x^2y^2 - 2xy^2 - 60y^2$
46. $8x^2y + 34xy - 84y$
47. $4x^2 - 8x - 21$
48. $6x^2 - 11x - 10$
49. $3x^2 - 42x + 63$
50. $5x^2 - 75x + 60$
51. $8x^2 + 6xy - 27y^2$
52. $54a^2 + 39ab - 8b^2$
53. $-x^2 + 2x + 24$
54. $-x^2 + 4x + 21$
55. $4x^3 - 9x^2 - 9x$
56. $6x^3 - 31x^2 + 5x$
57. $24x^2 - 58x + 9$
58. $36x^2 + 55x - 14$

59. $40a^2b + 9ab - 9b$

60. $24y^2x + 7yx - 5x$

61. $30x^3 + 38x^2 + 12x$

62. $6x^3 - 28x^2 + 16x$

63. $6y^3 - 8y^2 - 30y$

64. $12x^3 - 34x^2 + 24x$

65. $10x^4 + 25x^3y - 15x^2y^2$

66. $42x^4 - 99x^3y - 15x^2y^2$

67. $-14x^2 + 39x - 10$

68. $-15x^2 + 26x - 8$

69. $16p^4 - 40p^3 + 25p^2$

70. $9q^4 - 42q^3 + 49q^2$

71. $-2x^2 + 9x + 5$

72. $-3x^2 + 8x + 16$

73. $-4 + 52x - 48x^2$

74. $-5 + 55x - 50x^2$

75. $2t^4 + 3t^2 - 27$

76. $4r^4 - 17r^2 - 15$

77. $5x^2y^2 + 20xy + 1$

78. $3a^2b^2 + 12ab + 1$

79. $6a^5 + 37a^3b^2 + 6ab^4$

80. $5m^5 + 26m^3h^2 + 5mh^4$

Review

Multiply. See Section 3.6.

81. $(x - 4)(x + 4)$

82. $(2x - 9)(2x + 9)$

83. $(x + 2)^2$

84. $(x + 3)^2$

85. $(2x - 1)^2$

86. $(3x - 5)^2$

The following graph shows the average number of texts sent per day by text message users in each age group. See Section R.4. Source: Experion Marketing)

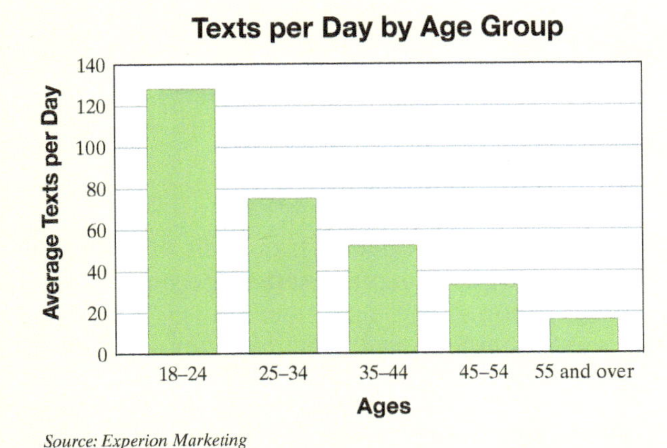

Source: Experion Marketing

87. What range of ages sends and receives the greatest number of texts per day?

88. What range of ages sends and receives the fewest number of texts per day?

89. Describe any trend you see.

90. What do you think this graph would look like in 10 years? Explain your reasoning.

Concept Extensions

See the Concept Check in this section.

91. Do the terms of $4x^2 + 19x + 12$ have a common factor (other than 1)?

92. Without multiplying, decide which of the following factored forms is not a factored form of $4x^2 + 19x + 12$.

a. $(2x + 4)(2x + 3)$

b. $(4x + 4)(x + 3)$

c. $(4x + 3)(x + 4)$

d. $(2x + 2)(2x + 6)$

Write the perimeter of each figure as a simplified polynomial. Then factor the polynomial completely.

93.

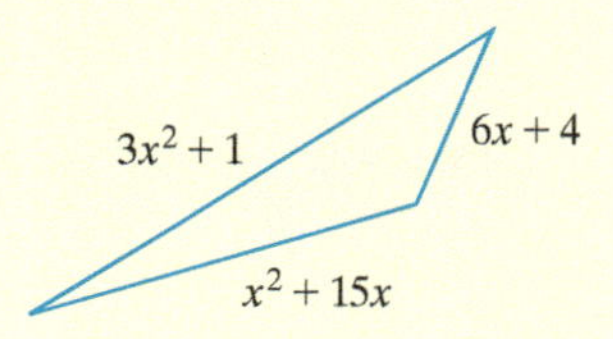

94.

$3y^2$

$-22y + 7$

Factor each trinomial completely.

95. $4x^2 + 2x + \frac{1}{4}$

96. $27x^2 + 2x - \frac{1}{9}$

97. $4x^2(y - 1)^2 + 25x(y - 1)^2 + 25(y - 1)^2$

98. $3x^2(a + 3)^3 - 28x(a + 3)^3 + 25(a + 3)^3$

Find a positive value of b so that each trinomial is factorable.

99. $3x^2 + bx - 5$

100. $2z^2 + bz - 7$

Find a positive value of c so that each trinomial is factorable.

101. $5x^2 + 7x + c$

102. $3x^2 - 8x + c$

103. In your own words, describe the steps you use to factor a trinomial.

104. A student in your class factored $6x^2 + 7x + 1$ as $(3x + 1)(2x + 1)$. Write down how you would explain the student's error.

4.4 Factoring Trinomials of the Form $ax^2 + bx + c$ by Grouping

Objective

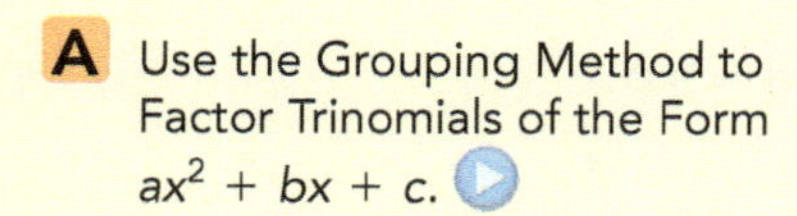

A Use the Grouping Method to Factor Trinomials of the Form $ax^2 + bx + c$.

Objective A Using the Grouping Method

There is an alternative method that can be used to factor trinomials of the form $ax^2 + bx + c, a \neq 1$. This method is called the **grouping method** because it uses factoring by grouping as we learned in Section 4.1.

To see how this method works, recall from Section 4.2 that to factor a trinomial such as $x^2 + 11x + 30$, we find two numbers such that

Product is 30.

$$x^2 + 11x + 30$$

Sum is 11.

To factor a trinomial such as $2x^2 + 11x + 12$ by grouping, we use an extension of the method in Section 4.1. Here we look for two numbers such that

Product is $2 \cdot 12 = 24$.

$$2x^2 + 11x + 12$$

Sum is 11.

This time, we use the two numbers to write

$2x^2 + 11x + 12$ as

$= 2x^2 + \square x + \square x + 12$

Then we factor by grouping. Since we want a positive product, 24, and a positive sum, 11, we consider pairs of positive factors of 24 only.

Factors of 24	Sum of Factors
1, 24	25
2, 12	14
3, 8	11

Correct sum

The factors are 3 and 8. Now we use these factors to write the middle term, $11x$, as $3x + 8x$ (or $8x + 3x$). We replace $11x$ with $3x + 8x$ in the original trinomial and then we can factor by grouping.

$$\begin{aligned}2x^2 + 11x + 12 &= 2x^2 + 3x + 8x + 12 \\ &= (2x^2 + 3x) + (8x + 12) && \text{Group the terms.} \\ &= x(2x + 3) + 4(2x + 3) && \text{Factor each group.} \\ &= (2x + 3)(x + 4) && \text{Factor out } (2x + 3).\end{aligned}$$

In general, we have the following procedure.

To Factor Trinomials by Grouping

Step 1: Factor out a greatest common factor, if there is one other than 1.

Step 2: For the resulting trinomial $ax^2 + bx + c$, find two numbers whose product is $a \cdot c$ and whose sum is b.

Step 3: Write the middle term, bx, using the factors found in Step 2.

Step 4: Factor by grouping.

Example 1

Factor $8x^2 - 14x + 5$ by grouping.

Solution:

Step 1: The terms of this trinomial contain no greatest common factor other than 1.

Step 2: This trinomial is of the form $ax^2 + bx + c$, with $a = 8$, $b = -14$, and $c = 5$. Find two numbers whose product is $a \cdot c$ or $8 \cdot 5 = 40$ and whose sum is b or -14.

The numbers are -4 and -10.

Factors of 40	Sum of Factors
−40, −1	−41
−20, −2	−22
−10, −4	−14 Correct sum

Step 3: Write $-14x$ as $-4x - 10x$ so that

$8x^2 - 14x + 5 = 8x^2 - 4x - 10x + 5$

Step 4: Factor by grouping.

$$\begin{aligned}8x^2 - 4x - 10x + 5 &= 4x(2x - 1) - 5(2x - 1) \\ &= (2x - 1)(4x - 5)\end{aligned}$$

Work Practice 1

Example 2

Factor $6x^2 - 2x - 20$ by grouping.

Solution:

Step 1: First factor out the greatest common factor, 2.

$6x^2 - 2x - 20 = 2(3x^2 - x - 10)$

(Continued on next page)

Practice 1

Factor each trinomial by grouping.

a. $3x^2 + 14x + 8$

b. $12x^2 + 19x + 5$

Practice 2

Factor each trinomial by grouping.

a. $30x^2 - 26x + 4$

b. $6x^2y - 7xy - 5y$

Answers

1. a. $(x + 4)(3x + 2)$ **b.** $(4x + 5)(3x + 1)$

2. a. $2(5x - 1)(3x - 2)$ **b.** $y(2x + 1)(3x - 5)$

Step 2: Next notice that $a = 3$, $b = -1$, and $c = -10$ in the resulting trinomial. Find two numbers whose product is $a \cdot c$ or $3(-10) = -30$ and whose sum is b, -1. The numbers are -6 and 5.

Step 3: $3x^2 - x - 10 = 3x^2 - 6x + 5x - 10$

Step 4: $3x^2 - 6x + 5x - 10 = 3x(x - 2) + 5(x - 2)$
$= (x - 2)(3x + 5)$

A factored form of $6x^2 - 2x - 20 = 2(x - 2)(3x + 5)$.

Don't forget to include the common factor of 2.

Work Practice 2

Practice 3

Factor $12y^5 + 10y^4 - 42y^3$ by grouping.

Example 3

Factor $18y^4 + 21y^3 - 60y^2$ by grouping.

Solution:

Step 1: First factor out the greatest common factor, $3y^2$.

$$18y^4 + 21y^3 - 60y^2 = 3y^2(6y^2 + 7y - 20)$$

Step 2: Notice that $a = 6$, $b = 7$, and $c = -20$ in the resulting trinomial. Find two numbers whose product is $a \cdot c$ or $6(-20) = -120$ and whose sum is 7. It may help to factor -120 as a product of primes and -1.

$$-120 = 2 \cdot 2 \cdot 2 \cdot 3 \cdot 5 \cdot (-1)$$

Then choose pairings of factors until you have two pairings whose sum is 7.

$$\underbrace{2 \cdot 2 \cdot 2}_{} \cdot \underbrace{3 \cdot 5}_{15} \cdot (-1) \quad (2 \cdot 2 \cdot 2 \cdot (-1) = -8)$$

The numbers are -8 and 15.

Step 3: $6y^2 + 7y - 20 = 6y^2 - 8y + 15y - 20$

Step 4: $6y^2 - 8y + 15y - 20 = 2y(3y - 4) + 5(3y - 4)$
$= (3y - 4)(2y + 5)$

A factored form of $18y^4 + 21y^3 - 60y^2$ is $3y^2(3y - 4)(2y + 5)$.

Don't forget to include the common factor of $3y^2$.

Work Practice 3

Answer

3. $2y^3(3y + 7)(2y - 3)$

Vocabulary, Readiness & Video Check

For each trinomial $ax^2 + bx + c$, choose two numbers whose product is $a \cdot c$ and whose sum is b.

1. $x^2 + 6x + 8$
a. 4, 2 **b.** 7, 1 **c.** 6, 2 **d.** 6, 8

2. $x^2 + 11x + 24$
a. 6, 4 **b.** 12, 2 **c.** 8, 3 **d.** 5, 6

3. $2x^2 + 13x + 6$
a. 2, 6 **b.** 12, 1 **c.** 13, 1 **d.** 3, 4

4. $4x^2 + 8x + 3$
a. 4, 3 **b.** 4, 4 **c.** 12, 1 **d.** 2, 6

Martin-Gay Interactive Videos

See Video 4.4

Watch the section lecture video and answer the following questions.

Objective A 5. In the lecture following Example 1, why does writing a term as the sum or difference of two terms suggest we'd then try to factor by grouping?

4.4 Exercise Set MyLab Math

Objective A *Factor each polynomial by grouping. Notice that Step 3 has already been done in these exercises. See Examples 1 through 3.*

1. $x^2 + 3x + 2x + 6$

2. $x^2 + 5x + 3x + 15$

3. $y^2 + 8y - 2y - 16$

4. $z^2 + 10z - 7z - 70$

5. $8x^2 - 5x - 24x + 15$

6. $4x^2 - 9x - 32x + 72$

7. $5x^4 - 3x^2 + 25x^2 - 15$

8. $2y^4 - 10y^2 + 7y^2 - 35$

Factor each trinomial by grouping. Exercises 9 through 12 are broken into parts to help you get started. See Examples 1 through 3.

9. $6x^2 + 11x + 3$

- **a.** Find two numbers whose product is $6 \cdot 3 = 18$ and whose sum is 11.
- **b.** Write $11x$ using the factors from part **a.**
- **c.** Factor by grouping.

10. $8x^2 + 14x + 3$

- **a.** Find two numbers whose product is $8 \cdot 3 = 24$ and whose sum is 14.
- **b.** Write $14x$ using the factors from part **a.**
- **c.** Factor by grouping.

11. $15x^2 - 23x + 4$

- **a.** Find two numbers whose product is $15 \cdot 4 = 60$ and whose sum is -23.
- **b.** Write $-23x$ using the factors from part **a.**
- **c.** Factor by grouping.

12. $6x^2 - 13x + 5$

- **a.** Find two numbers whose product is $6 \cdot 5 = 30$ and whose sum is -13.
- **b.** Write $-13x$ using the factors from part **a.**
- **c.** Factor by grouping.

13. $21y^2 + 17y + 2$

14. $15x^2 + 11x + 2$

15. $7x^2 - 4x - 11$

16. $8x^2 - x - 9$

17. $10x^2 - 9x + 2$

18. $30x^2 - 23x + 3$

19. $2x^2 - 7x + 5$

20. $2x^2 - 7x + 3$

21. $12x + 4x^2 + 9$

22. $20x + 25x^2 + 4$

23. $4x^2 - 8x - 21$

24. $6x^2 - 11x - 10$

25. $10x^2 - 23x + 12$

26. $21x^2 - 13x + 2$

27. $2x^3 + 13x^2 + 15x$

28. $3x^3 + 8x^2 + 4x$

29. $16y^2 - 34y + 18$

30. $4y^2 - 2y - 12$

31. $-13x + 6 + 6x^2$

32. $-25x + 12 + 12x^2$

33. $54a^2 - 9a - 30$

34. $30a^2 + 38a - 20$

35. $20a^3 + 37a^2 + 8a$

36. $10a^3 + 17a^2 + 3a$

37. $12x^3 - 27x^2 - 27x$

38. $30x^3 - 155x^2 + 25x$

39. $3x^2y + 4xy^2 + y^3$

40. $6r^2t + 7rt^2 + t^3$

41. $20z^2 + 7z + 1$

42. $36z^2 + 6z + 1$

43. $24a^2 - 6ab - 30b^2$

44. $30a^2 + 5ab - 25b^2$

45. $15p^4 + 31p^3q + 2p^2q^2$

46. $20s^4 + 61s^3t + 3s^2t^2$

47. $35 + 12x + x^2$

48. $33 + 14x + x^2$

49. $6 - 11x + 5x^2$

50. $5 - 12x + 7x^2$

Review

Multiply. See Section 3.6.

51. $(x - 2)(x + 2)$

52. $(y - 5)(y + 5)$

53. $(y + 4)(y + 4)$

54. $(x + 7)(x + 7)$

55. $(9z + 5)(9z - 5)$

56. $(8y + 9)(8y - 9)$

57. $(4x - 3)^2$

58. $(2z - 1)^2$

Concept Extensions

Write the perimeter of each figure as a simplified polynomial. Then factor the polynomial.

59.

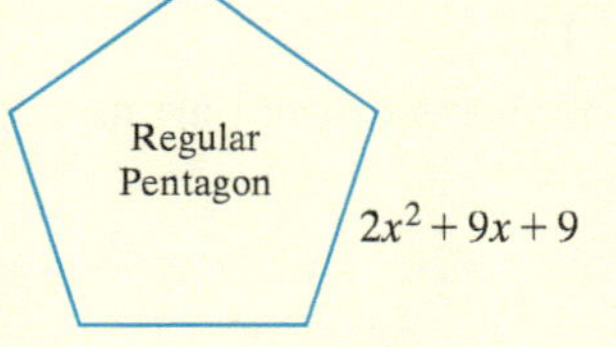

60.

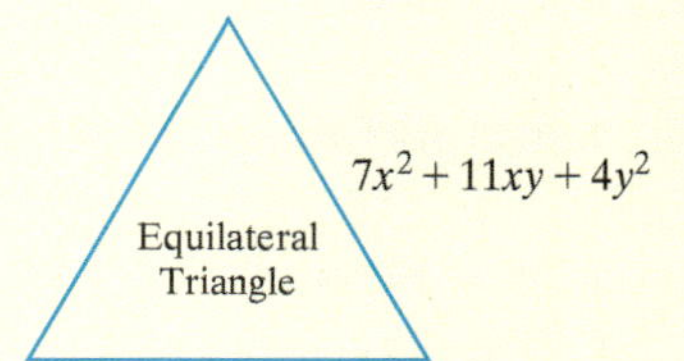

Factor each polynomial by grouping.

61. $x^{2n} + 2x^n + 3x^n + 6$
(*Hint:* Don't forget that $x^{2n} = x^n \cdot x^n$.)

62. $x^{2n} + 6x^n + 10x^n + 60$

63. $3x^{2n} + 16x^n - 35$

64. $12x^{2n} - 40x^n + 25$

65. In your own words, explain how to factor a trinomial by grouping.

66. In your own words, explain how to factor a 4-form polynomial by grouping.

4.5 Factoring Perfect Square Trinomials and the Difference of Two Squares 

Objective A Recognizing Perfect Square Trinomials 

Objectives

A Recognize Perfect Square Trinomials.

B Factor Perfect Square Trinomials.

C Factor the Difference of Two Squares.

A trinomial that is the square of a binomial is called a **perfect square trinomial.** For example,

$$(x + 3)^2 = (x + 3)(x + 3)$$
$$= x^2 + 6x + 9$$

Thus $x^2 + 6x + 9$ is a perfect square trinomial.

In Chapter 3, we discovered special product formulas for squaring binomials.

$$(a + b)^2 = a^2 + 2ab + b^2 \quad \text{and} \quad (a - b)^2 = a^2 - 2ab + b^2$$

Because multiplication and factoring are reverse processes, we can now use these special products to help us factor perfect square trinomials. If we reverse these equations, we have the following.

Factoring Perfect Square Trinomials

$$a^2 + 2ab + b^2 = (a + b)^2$$
$$a^2 - 2ab + b^2 = (a - b)^2$$

Helpful Hint

Notice that for both given forms of a perfect square trinomial, the last term is positive. This is because the last term is a square.

To use these equations to help us factor, we must first be able to recognize a perfect square trinomial. A trinomial is a perfect square when

1. two terms, a^2 and b^2, are squares and
2. another term is $2 \cdot a \cdot b$ or $-2 \cdot a \cdot b$. That is, this term is twice the product of a and b, or its opposite.

Example 1 Decide whether $x^2 + 8x + 16$ is a perfect square trinomial.

Solution:

1. Two terms, x^2 and 16, are squares $(16 = 4^2)$.
2. Twice the product of x and 4 is the other term of the trinomial.

$$2 \cdot x \cdot 4 = 8x$$

Thus, $x^2 + 8x + 16$ is a perfect square trinomial.

Work Practice 1

Practice 1

Decide whether each trinomial is a perfect square trinomial.

a. $x^2 + 12x + 36$

b. $x^2 + 20x + 100$

Example 2 Decide whether $4x^2 + 10x + 9$ is a perfect square trinomial.

Solution:

1. Two terms, $4x^2$ and 9, are squares.

$$4x^2 = (2x)^2 \quad \text{and} \quad 9 = 3^2$$

2. Twice the product of $2x$ and 3 is *not* the other term of the trinomial.

$$2 \cdot 2x \cdot 3 = 12x, \textit{ not } 10x$$

The trinomial is *not* a perfect square trinomial.

Work Practice 2

Practice 2

Decide whether each trinomial is a perfect square trinomial.

a. $9x^2 + 20x + 25$

b. $4x^2 + 8x + 11$

Answers

1. a. yes **b.** yes

2. a. no **b.** no

Practice 3

Decide whether each trinomial is a perfect square trinomial.

a. $25x^2 - 10x + 1$

b. $9x^2 - 42x + 49$

Example 3 Decide whether $9x^2 - 12xy + 4y^2$ is a perfect square trinomial.

Solution:

1. Two terms, $9x^2$ and $4y^2$, are squares.

$$9x^2 = (3x)^2 \quad \text{and} \quad 4y^2 = (2y)^2$$

2. Twice the product of $3x$ and $2y$ is the opposite of the other term of the trinomial.

$$2 \cdot 3x \cdot 2y = 12xy, \text{ the opposite of } -12xy$$

Thus, $9x^2 - 12xy + 4y^2$ is a perfect square trinomial.

Work Practice 3

Objective B Factoring Perfect Square Trinomials

Now that we can recognize perfect square trinomials, we are ready to factor them.

Practice 4

Factor: $x^2 + 16x + 64$

Example 4 Factor: $x^2 + 12x + 36$

Solution:

$$x^2 + 12x + 36 = x^2 + 2 \cdot x \cdot 6 + 6^2 \qquad 36 = 6^2 \text{ and } 12x = 2 \cdot x \cdot 6$$

$$a^2 + 2 \cdot a \cdot b + b^2$$

$$= (x + 6)^2$$

$$(a + b)^2$$

Work Practice 4

Practice 5

Factor: $9r^2 + 24rs + 16s^2$

Example 5 Factor: $25x^2 + 20xy + 4y^2$

Solution:

$$25x^2 + 20xy + 4y^2 = (5x)^2 + 2 \cdot 5x \cdot 2y + (2y)^2$$

$$= (5x + 2y)^2$$

Work Practice 5

Practice 6

Factor: $9n^4 - 6n^2 + 1$

Example 6 Factor: $4m^4 - 4m^2 + 1$

Solution:

$$4m^4 - 4m^2 + 1 = (2m^2)^2 - 2 \cdot 2m^2 \cdot 1 + 1^2$$

$$a^2 \quad - 2 \cdot a \cdot b + b^2$$

$$= (2m^2 - 1)^2$$

$$(a - b)^2$$

Work Practice 6

Answers

3. a. yes b. yes 4. $(x + 8)^2$
5. $(3r + 4s)^2$ 6. $(3n^2 - 1)^2$

Example 7 Factor: $25x^2 + 50x + 9$

Solution: Notice that this trinomial is not a perfect square trinomial.

$25x^2 = (5x)^2, 9 = 3^2$

but

$2 \cdot 5x \cdot 3 = 30x$

and $30x$ is not the middle term, $50x$.

Although $25x^2 + 50x + 9$ is not a perfect square trinomial, it is factorable. Using techniques we learned in Section 4.3 or 4.4, we find that

$$25x^2 + 50x + 9 = (5x + 9)(5x + 1)$$

Work Practice 7

Example 8 Factor: $162x^3 - 144x^2 + 32x$

Solution: Don't forget to first look for a common factor. There is a greatest common factor of $2x$ in this trinomial.

$$\begin{aligned} 162x^3 - 144x^2 + 32x &= 2x(81x^2 - 72x + 16) \\ &= 2x[(9x)^2 - 2 \cdot 9x \cdot 4 + 4^2] \\ &= 2x(9x - 4)^2 \end{aligned}$$

Work Practice 8

Objective C Factoring the Difference of Two Squares 

In Chapter 3, we discovered another special product, the product of the sum and difference of two terms a and b:

$$(a + b)(a - b) = a^2 - b^2$$

Reversing this equation gives us another factoring pattern, which we use to factor the difference of two squares.

Factoring the Difference of Two Squares

$$a^2 - b^2 = (a + b)(a - b)$$

To use this equation to help us factor, we must first be able to recognize the difference of two squares. A binomial is a difference of two squares if

1. both terms are squares and
2. the signs of the terms are different.

Let's practice using this pattern.

Examples Factor each binomial.

9. $z^2 - 4 = z^2 - 2^2 = (z + 2)(z - 2)$

$a^2 - b^2 = (a + b)(a - b)$

10. $y^2 - 25 = y^2 - 5^2 = (y + 5)(y - 5)$

11. $y^2 - \frac{4}{9} = y^2 - \left(\frac{2}{3}\right)^2 = \left(y + \frac{2}{3}\right)\left(y - \frac{2}{3}\right)$

12. $x^2 + 4$

(Continued on next page)

Practice 7

Factor: $9x^2 + 15x + 4$

Helpful Hint A perfect square trinomial can also be factored by the methods found in Sections 4.2 through 4.4.

Practice 8

Factor:

a. $8n^2 + 40n + 50$

b. $12x^3 - 84x^2 + 147x$

Practice 9–12

Factor each binomial.

9. $x^2 - 9$ **10.** $a^2 - 16$

11. $c^2 - \frac{9}{25}$ **12.** $s^2 + 9$

Answers

7. $(3x + 1)(3x + 4)$

8. a. $2(2n + 5)^2$ **b.** $3x(2x - 7)^2$

9. $(x - 3)(x + 3)$

10. $(a - 4)(a + 4)$

11. $\left(c - \frac{3}{5}\right)\left(c + \frac{3}{5}\right)$

12. prime polynomial

Helpful Hint After the greatest common factor has been removed, the *sum* of two squares cannot be factored further using real numbers.

Note that the binomial $x^2 + 4$ is the *sum* of two squares since we can write $x^2 + 4$ as $x^2 + 2^2$. We might try to factor using $(x + 2)(x + 2)$ or $(x - 2)(x - 2)$. But when we multiply to check, we find that neither factoring is correct.

$$(x + 2)(x + 2) = x^2 + 4x + 4$$
$$(x - 2)(x - 2) = x^2 - 4x + 4$$

In both cases, the product is a trinomial, not the required binomial. In fact, $x^2 + 4$ is a prime polynomial.

Work Practice 9–12

Practice 13–15

Factor each difference of two squares.

13. $9s^2 - 1$
14. $16x^2 - 49y^2$
15. $p^4 - 81$

Examples

Factor each difference of two squares.

13. $4x^2 - 1 = (2x)^2 - 1^2 = (2x + 1)(2x - 1)$

14. $25a^2 - 9b^2 = (5a)^2 - (3b)^2 = (5a + 3b)(5a - 3b)$

15. $y^4 - 16 = (y^2)^2 - 4^2$

$= (y^2 + 4)(y^2 - 4)$ Factor the difference of two squares.

$= (y^2 + 4)(y + 2)(y - 2)$ Factor the difference of two squares.

Work Practice 13–15

Helpful Hint

1. Don't forget to first see whether there's a greatest common factor (other than 1) that can be factored out.
2. Factor completely. In other words, check to see whether any factors can be factored further (as in Example 15).

Practice 16–18

Factor each binomial.

16. $9x^3 - 25x$ **17.** $48x^4 - 3$
18. $-9x^2 + 100$

Examples

Factor each binomial.

16. $4x^3 - 49x = x(4x^2 - 49)$ Factor out the common factor, x.

$= x[(2x)^2 - 7^2]$

$= x(2x + 7)(2x - 7)$ Factor the difference of two squares.

17. $162x^4 - 2 = 2(81x^4 - 1)$ Factor out the common factor, 2.

$= 2(9x^2 + 1)(9x^2 - 1)$ Factor the difference of two squares.

$= 2(9x^2 + 1)(3x + 1)(3x - 1)$ Factor the difference of two squares.

18. $-49x^2 + 16 = -1(49x^2 - 16)$ Factor out -1.

$= -1(7x + 4)(7x - 4)$ Factor the difference of two squares.

Work Practice 16–18

Practice 19

Factor: $121 - m^2$

Example 19

Factor: $36 - x^2$

Solution: This is the difference of two squares. Factor as is or, if you like, first write the binomial with the variable term first.

Factor as is: $36 - x^2 = 6^2 - x^2 = (6 + x)(6 - x)$

Rewrite binomial: $36 - x^2 = -x^2 + 36 = -1(x^2 - 36)$

$= -1(x + 6)(x - 6)$

Both factorizations are correct and are equal. To see this, factor -1 from $(6 - x)$ in the first factorization.

Work Practice 19

Helpful Hint

When rearranging terms, keep in mind that the sign of a term is in front of the term.

Answers

13. $(3s - 1)(3s + 1)$
14. $(4x - 7y)(4x + 7y)$
15. $(p^2 + 9)(p + 3)(p - 3)$
16. $x(3x - 5)(3x + 5)$
17. $3(4x^2 + 1)(2x + 1)(2x - 1)$
18. $-1(3x - 10)(3x + 10)$
19. $(11 + m)(11 - m)$ or $-1(m + 11)(m - 11)$

Calculator Explorations Graphing

A graphing calculator is a convenient tool for evaluating an expression at a given replacement value. For example, let's evaluate $x^2 - 6x$ when $x = 2$. To do so, store the value 2 in the variable x and then enter and evaluate the algebraic expression.

```
2→X
                 2
X²-6X
                -8
```

The value of $x^2 - 6x$ when $x = 2$ is -8. You may want to use this method for evaluating expressions as you explore the following.

We can use a graphing calculator to explore factoring patterns numerically. Use your calculator to evaluate $x^2 - 2x + 1$, $x^2 - 2x - 1$, and $(x - 1)^2$ for each value of x given in the table. What do you observe?

	$x^2 - 2x + 1$	$x^2 - 2x - 1$	$(x - 1)^2$
$x = 5$			
$x = -3$			
$x = 2.7$			
$x = -12.1$			
$x = 0$			

Notice in each case that $x^2 - 2x - 1 \neq (x - 1)^2$. Because for each x in the table the value of $x^2 - 2x + 1$ and the value of $(x - 1)^2$ are the same, we might guess that $x^2 - 2x + 1 = (x - 1)^2$. We can verify our guess algebraically with multiplication:

$$(x - 1)(x - 1) = x^2 - x - x + 1 = x^2 - 2x + 1$$

Vocabulary, Readiness & Video Check

Use the choices below to fill in each blank. Some choices may be used more than once and some choices may not be used at all.

perfect square trinomial	true	$(5y)^2$	$(x + 5y)^2$
difference of two squares	false	$(x - 5y)^2$	$5y^2$

1. A ______________ is a trinomial that is the square of a binomial.
2. The term $25y^2$ written as a square is ______________.
3. The expression $x^2 + 10xy + 25y^2$ is called a ______________.
4. The expression $x^2 - 49$ is called a ______________.
5. The factorization $(x + 5y)(x + 5y)$ may also be written as ______________.
6. True or false? The factorization $(x - 5y)(x + 5y)$ may also be written as $(x - 5y)^2$. ______________
7. The trinomial $x^2 - 6x - 9$ is a perfect square trinomial. ______________
8. The binomial $y^2 + 9$ factors as $(y + 3)^2$. ______________

Write each number or term as a square. For example, 16 written as a square is 4^2.

9. 64 **10.** 9 **11.** $121a^2$

12. $81b^2$ **13.** $36p^4$ **14.** $4q^4$

Martin-Gay Interactive Videos

See Video 4.5

Watch the section lecture video and answer the following questions.

Objective A **15.** The polynomial in Example 2 is shown to *not* be a perfect square trinomial. Does this necessarily mean it can't be factored?

Objective B **16.** Describe in words the special patterns that the trinomials in Examples 3 and 4 have that identify them as perfect square trinomials.

Objective C **17.** In Examples 5 and 6, what are two reasons the original binomial is rewritten so that each term is a square?

Objective C **18.** For Example 7, what is a prime polynomial?

4.5 Exercise Set MyLab Math

Objective A *Determine whether each trinomial is a perfect square trinomial. See Examples 1 through 3.*

1. $x^2 + 16x + 64$

2. $x^2 + 22x + 121$

3. $y^2 + 5y + 25$

4. $y^2 + 4y + 16$

5. $m^2 - 2m + 1$

6. $p^2 - 4p + 4$

7. $a^2 - 16a + 49$

8. $n^2 - 20n + 144$

9. $4x^2 + 12xy + 8y^2$

10. $25x^2 + 20xy + 2y^2$

11. $25a^2 - 40ab + 16b^2$

12. $36a^2 - 12ab + b^2$

Objective B *Factor each trinomial completely. See Examples 4 through 8.*

13. $x^2 + 22x + 121$

14. $x^2 + 18x + 81$

15. $x^2 - 16x + 64$

16. $x^2 - 12x + 36$

17. $16a^2 - 24a + 9$

18. $25x^2 - 20x + 4$

19. $x^4 + 4x^2 + 4$

20. $m^4 + 10m^2 + 25$

21. $2n^2 - 28n + 98$

22. $3y^2 - 6y + 3$

23. $16y^2 + 40y + 25$

24. $9y^2 + 48y + 64$

25. $x^2y^2 - 10xy + 25$

26. $4x^2y^2 - 28xy + 49$

27. $m^3 + 18m^2 + 81m$

28. $y^3 + 12y^2 + 36y$

29. $1 + 6x^2 + x^4$

30. $1 + 16x^2 + x^4$

31. $9x^2 - 24xy + 16y^2$

32. $25x^2 - 60xy + 36y^2$

Objective C *Factor each binomial completely. See Examples 9 through 19.*

33. $x^2 - 4$ **34.** $x^2 - 36$ **35.** $81 - p^2$ **36.** $100 - t^2$

37. $-4r^2 + 1$ **38.** $-9t^2 + 1$ **39.** $9x^2 - 16$ **40.** $36y^2 - 25$

41. $16r^2 + 1$ **42.** $49y^2 + 1$ **43.** $-36 + x^2$ **44.** $-1 + y^2$

45. $m^4 - 1$ **46.** $n^4 - 16$ **47.** $x^2 - 169y^2$ **48.** $x^2 - 225y^2$

49. $18r^2 - 8$ **50.** $32t^2 - 50$ **51.** $9xy^2 - 4x$ **52.** $36x^2y - 25y$

53. $16x^4 - 64x^2$ **54.** $25y^4 - 100y^2$ **55.** $xy^3 - 9xyz^2$ **56.** $x^3y - 4xy^3$

57. $36x^2 - 64y^2$ **58.** $225a^2 - 81b^2$ **59.** $144 - 81x^2$ **60.** $12x^2 - 27$

61. $25y^2 - 9$ **62.** $49a^2 - 16$ **63.** $121m^2 - 100n^2$ **64.** $169a^2 - 49b^2$

65. $x^2y^2 - 1$ **66.** $a^2b^2 - 16$ **67.** $x^2 - \frac{1}{4}$

68. $y^2 - \frac{1}{16}$ **69.** $49 - \frac{9}{25}m^2$ **70.** $100 - \frac{4}{81}n^2$

Objectives B C Mixed Practice *Factor each binomial or trinomial completely. See Examples 4 through 19.*

71. $81a^2 - 25b^2$ **72.** $49y^2 - 100z^2$ **73.** $x^2 + 14xy + 49y^2$ **74.** $x^2 + 10xy + 25y^2$

75. $32n^4 - 112n^2 + 98$ **76.** $162a^4 - 72a^2 + 8$ **77.** $x^6 - 81x^2$

78. $n^9 - n^5$ **79.** $64p^3q - 81pq^3$ **80.** $100x^3y - 49xy^3$

Review

Solve each equation. See Section 2.3.

81. $x - 6 = 0$

82. $y + 5 = 0$

83. $2m + 4 = 0$

84. $3x - 9 = 0$

85. $5z - 1 = 0$

86. $4a + 2 = 0$

Concept Extensions

Factor each expression completely.

87. $x^2 - \frac{2}{3}x + \frac{1}{9}$

88. $x^2 - \frac{1}{25}$

89. $(x + 2)^2 - y^2$

90. $(y - 6)^2 - z^2$

91. $a^2(b - 4) - 16(b - 4)$

92. $m^2(n + 8) - 9(n + 8)$

93. $(x^2 + 6x + 9) - 4y^2$ (*Hint:* Factor the trinomial in parentheses first.)

94. $(x^2 + 2x + 1) - 36y^2$ (See the hint for Exercise **93.**)

95. $x^{2n} - 100$

96. $x^{2n} - 81$

Solve.

97. Fill in the blank so that $x^2 +$ ______$x + 16$ is a perfect square trinomial.

98. Fill in the blank so that $9x^2 +$ ______$x + 25$ is a perfect square trinomial.

99. Describe a perfect square trinomial.

100. Write a perfect square trinomial that factors as $(x + 3y)^2$.

101. What binomial multiplied by $(x - 6)$ gives the difference of two squares?

102. What binomial multiplied by $(5 + y)$ gives the difference of two squares?

The area of the largest square in the figure is $(a + b)^2$. Use this figure to answer Exercises 103 and 104.

103. Write the area of the largest square as the sum of the areas of the smaller squares and rectangles.

104. What factoring formula from this section is visually represented by this square?

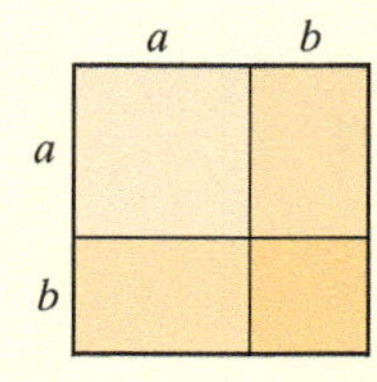

105. The Toroweap Overlook, on the North Rim of the Grand Canyon, lies 3000 vertical feet above the Colorado River. The view is spectacular, and the sheer drop is dramatic. A film crew creating a documentary about the Grand Canyon has suspended a camera platform 296 feet below the Overlook. A camera filter comes loose and falls to the river below. The height of the filter above the river, after t seconds, is given by the expression $2704 - 16t^2$.

a. Find the height of the filter above the river after 3 seconds.

b. Find the height of the filter above the river after 7 seconds.

c. To the nearest whole second, estimate when the filter lands in the river.

d. Factor $2704 - 16t^2$.

106. An object is dropped from the top of Pittsburgh's U.S. Steel Tower, which is 841 feet tall. (*Source: World Almanac* research) The height of the object after t seconds is given by the expression $841 - 16t^2$.

a. Find the height of the object after 2 seconds.

b. Find the height of the object after 5 seconds.

c. To the nearest whole second, estimate when the object hits the ground.

d. Factor $841 - 16t^2$.

107. The world's tallest building is the Burj Khalifa in Dubai, United Arab Emirates, at a height of 2717 feet. (*Source:* Council on Tall Buildings and Urban Habitat) Suppose a worker is suspended 413 feet below the tip of the building, at a height of 2304 feet above the ground. If the worker accidentally drops a bolt, the height of the bolt after t seconds is given by the expression $2304 - 16t^2$.

a. Find the height of the bolt after 3 seconds.

b. Find the height of the bolt after 7 seconds.

c. To the nearest whole second, estimate when the bolt hits the ground.

d. Factor $2304 - 16t^2$.

108. A performer with the Moscow Circus is planning a stunt involving a free fall from the top of the MV Lomonosov State University building, which is 784 feet tall. (*Source:* Council on Tall Buildings and Urban Habitat) Neglecting air resistance, the performer's height above gigantic cushions positioned at ground level after t seconds is given by the expression $784 - 16t^2$.

a. Find the performer's height after 2 seconds.

b. Find the performer's height after 5 seconds.

c. To the nearest whole second, estimate when the performer reaches the cushions positioned at ground level.

d. Factor $784 - 16t^2$.

Integrated Review Sections 4.1–4.5

Answers

1. ______
2. ______
3. ______
4. ______
5. ______
6. ______
7. ______
8. ______
9. ______
10. ______
11. ______
12. ______
13. ______
14. ______
15. ______
16. ______
17. ______
18. ______
19. ______
20. ______
21. ______
22. ______
23. ______
24. ______
25. ______
26. ______
27. ______
28. ______
29. ______
30. ______
31. ______
32. ______
33. ______
34. ______
35. ______
36. ______

Choosing a Factoring Strategy

The following steps may be helpful when factoring polynomials.

To Factor a Polynomial

Step 1: Are there any common factors? If so, factor out the GCF.

Step 2: How many terms are in the polynomial?

a. Two terms: Is it the difference of two squares? $a^2 - b^2 = (a - b)(a + b)$

b. Three terms: Try one of the following.

i. Perfect square trinomial: $a^2 + 2ab + b^2 = (a + b)^2$
$a^2 - 2ab + b^2 = (a - b)^2$

ii. If not a perfect square trinomial, factor using the methods presented in Sections 4.2 through 4.4.

c. Four terms: Try factoring by grouping.

Step 3: See if any factors in the factored polynomial can be factored further.

Step 4: Check by multiplying.

Factor each polynomial completely.

1. $x^2 + x - 12$

2. $x^2 - 10x + 16$

3. $x^2 + 2x + 1$

4. $x^2 - 6x + 9$

5. $x^2 - x - 6$

6. $x^2 + x - 2$

7. $x^2 + x - 6$

8. $x^2 + 7x + 12$

9. $x^2 - 7x + 10$

10. $x^2 - x - 30$

11. $2x^2 - 98$

12. $3x^2 - 75$

13. $x^2 + 3x + 5x + 15$

14. $3y - 21 + xy - 7x$

15. $x^2 + 6x - 16$

16. $x^2 - 3x - 28$

17. $4x^3 + 20x^2 - 56x$

18. $6x^3 - 6x^2 - 120x$

19. $12x^2 + 34x + 24$

20. $24a^2 + 18ab - 15b^2$

21. $4a^2 - b^2$

22. $x^2 - 25y^2$

23. $28 - 13x - 6x^2$

24. $20 - 3x - 2x^2$

25. $4 - 2x + x^2$

26. $a + a^2 - 3$

27. $6y^2 + y - 15$

28. $4x^2 - x - 5$

29. $18x^3 - 63x^2 + 9x$

30. $12a^3 - 24a^2 + 4a$

31. $16a^2 - 56a + 49$

32. $25p^2 - 70p + 49$

33. $14 + 5x - x^2$

34. $3 - 2x - x^2$

35. $3x^4y + 6x^3y - 72x^2y$

36. $2x^3y + 8x^2y^2 - 10xy^3$

37. $12x^3y + 243xy$

38. $6x^3y^2 + 8xy^2$

39. $2xy - 72x^3y$

40. $2x^3 - 18x$

41. $x^3 + 6x^2 - 4x - 24$

42. $x^3 - 2x^2 - 36x + 72$

43. $6a^3 + 10a^2$

44. $4n^2 - 6n$

45. $3x^3 - x^2 + 12x - 4$

46. $x^3 - 2x^2 + 3x - 6$

47. $6x^2 + 18xy + 12y^2$

48. $12x^2 + 46xy - 8y^2$

49. $5(x + y) + x(x + y)$

50. $7(x - y) + y(x - y)$

51. $14t^2 - 9t + 1$

52. $3t^2 - 5t + 1$

53. $-3x^2 - 2x + 5$

54. $-7x^2 - 19x + 6$

55. $1 - 8a - 20a^2$

56. $1 - 7a - 60a^2$

57. $x^4 - 10x^2 + 9$

58. $x^4 - 13x^2 + 36$

59. $x^2 - 23x + 120$

60. $y^2 + 22y + 96$

61. $25p^2 - 70pq + 49q^2$

62. $16a^2 - 56ab + 49b^2$

63. $x^2 - 14x - 48$

64. $7x^2 + 24xy + 9y^2$

65. $-x^2 - x + 30$

66. $-x^2 + 6x - 8$

67. $3rs - s + 12r - 4$

68. $x^3 - 2x^2 + x - 2$

69. $4x^2 - 8xy - 3x + 6y$

70. $4x^2 - 2xy - 7yz + 14xz$

71. $x^2 + 9xy - 36y^2$

72. $3x^2 + 10xy - 8y^2$

73. $x^4 - 14x^2 - 32$

74. $x^4 - 22x^2 - 75$

75. Explain why it makes good sense to factor out the GCF first, before using other methods of factoring.

76. The sum of two squares usually does not factor. Is the sum of two squares $9x^2 + 81y^2$ factorable? If so, factor the polynomial.

37. ______
38. ______
39. ______
40. ______
41. ______
42. ______
43. ______
44. ______
45. ______
46. ______
47. ______
48. ______
49. ______
50. ______
51. ______
52. ______
53. ______
54. ______
55. ______
56. ______
57. ______
58. ______
59. ______
60. ______
61. ______
62. ______
63. ______
64. ______
65. ______
66. ______
67. ______
68. ______
69. ______
70. ______
71. ______
72. ______
73. ______
74. ______
75. ______
76. ______

4.6 Solving Quadratic Equations by Factoring

Objectives

A Solve Quadratic Equations by Factoring.

B Solve Equations with Degree Greater Than Two by Factoring.

In this section, we introduce a new type of equation—the **quadratic equation.**

Quadratic Equation

A quadratic equation is one that can be written in the form

$$ax^2 + bx + c = 0$$

where a, b, and c are real numbers and $a \neq 0$.

Some examples of quadratic equations are shown below.

$$x^2 - 9x - 22 = 0 \qquad 4x^2 - 28 = -49 \qquad x(2x - 7) = 4$$

The form $ax^2 + bx + c = 0$ is called the **standard form** of a quadratic equation. The quadratic equation $x^2 - 9x - 22 = 0$ is the only equation above that is in standard form.

Quadratic equations model many real-life situations. For example, let's suppose we want to know how long before a person diving from a 144-foot cliff reaches the ocean. The answer to this question is found by solving the quadratic equation $-16t^2 + 144 = 0$. (See Example 1 in Section 4.7.)

Objective A Solving Quadratic Equations by Factoring

Some quadratic equations can be solved by making use of factoring and the **zero-factor property.**

Zero-Factor Property

If a and b are real numbers and if $ab = 0$, then $a = 0$ or $b = 0$.

In other words, if the product of two numbers is 0, then at least one of the numbers must be 0.

Practice 1

Solve: $(x - 7)(x + 2) = 0$

Example 1 Solve: $(x - 3)(x + 1) = 0$

Solution: If this equation is to be a true statement, then either the factor $x - 3$ must be 0 or the factor $x + 1$ must be 0. In other words, either

$$x - 3 = 0 \qquad \text{or} \qquad x + 1 = 0$$

If we solve these two linear equations, we have

$$x = 3 \qquad \text{or} \qquad x = -1$$

Answer

1. 7 and −2

Thus, 3 and -1 are both solutions of the equation $(x-3)(x+1)=0$. To check, we replace x with 3 in the original equation. Then we replace x with -1 in the original equation.

Check:

$(x-3)(x+1)=0$
$(3-3)(3+1)\stackrel{?}{=}0$ Replace x with 3.
$0(4)=0$ True

$(x-3)(x+1)=0$
$(-1-3)(-1+1)\stackrel{?}{=}0$ Replace x with -1.
$(-4)(0)=0$ True

The solutions are 3 and -1.

Work Practice 1

Helpful Hint

The zero-factor property says that *if a product is 0, then a factor is 0.*

If $a \cdot b=0$, then $a=0$ or $b=0$.
If $x(x+5)=0$, then $x=0$ or $x+5=0$.
If $(x+7)(2x-3)=0$, then $x+7=0$ or $2x-3=0$.

Use this property only when the product is 0. For example, if $a \cdot b=8$, we do not know the value of a or b. The values may be $a=2, b=4$ or $a=8, b=1$, or any other two numbers whose product is 8.

Example 2 Solve: $(x-5)(2x+7)=0$

Solution: The product is 0. By the zero-factor property, this is true only when a factor is 0. To solve, we set each factor equal to 0 and solve the resulting linear equations.

$(x-5)(2x+7)=0$

$x-5=0$ or $2x+7=0$
$x=5$ $\quad 2x=-7$
$x=-\frac{7}{2}$

Check: Let $x=5$.

$(x-5)(2x+7)=0$
$(5-5)(2\cdot 5+7)\stackrel{?}{=}0$ Replace x with 5.
$0\cdot 17\stackrel{?}{=}0$
$0=0$ True

Let $x=-\frac{7}{2}$.

$(x-5)(2x+7)=0$
$\left(-\frac{7}{2}-5\right)\left(2\left(-\frac{7}{2}\right)+7\right)\stackrel{?}{=}0$ Replace x with $-\frac{7}{2}$.
$\left(-\frac{17}{2}\right)(-7+7)\stackrel{?}{=}0$
$\left(-\frac{17}{2}\right)\cdot 0\stackrel{?}{=}0$
$0=0$ True

The solutions are 5 and $-\frac{7}{2}$.

Work Practice 2

Practice 2

Solve: $(x-10)(3x+1)=0$

Answer

2. 10 and $-\frac{1}{3}$

Practice 3

Solve each equation.

a. $y(y+3)=0$

b. $x(4x-3)=0$

Example 3

Solve: $x(5x-2)=0$

Solution:

$$x(5x-2)=0$$

$$x=0 \quad \text{or} \quad 5x-2=0 \qquad \text{Use the zero-factor property.}$$

$$5x=2$$

$$x=\frac{2}{5}$$

Check these solutions in the original equation. The solutions are 0 and $\frac{2}{5}$.

Work Practice 3

Practice 4

Solve: $x^2-3x-18=0$

Example 4

Solve: $x^2-9x-22=0$

Solution: One side of the equation is 0. However, to use the zero-factor property, one side of the equation must be 0 *and* the other side must be written as a product (must be factored). Thus, we must first factor this polynomial.

$$x^2-9x-22=0$$

$$(x-11)(x+2)=0 \qquad \text{Factor.}$$

Now we can apply the zero-factor property.

$$x-11=0 \quad \text{or} \quad x+2=0$$

$$x=11 \qquad x=-2$$

Check: Let $x=11$.

$$x^2-9x-22=0$$

$$11^2-9\cdot 11-22 \stackrel{?}{=} 0$$

$$121-99-22 \stackrel{?}{=} 0$$

$$22-22 \stackrel{?}{=} 0$$

$$0=0 \qquad \text{True}$$

Let $x=-2$.

$$x^2-9x-22=0$$

$$(-2)^2-9(-2)-22 \stackrel{?}{=} 0$$

$$4+18-22 \stackrel{?}{=} 0$$

$$22-22 \stackrel{?}{=} 0$$

$$0=0 \qquad \text{True}$$

The solutions are 11 and -2.

Work Practice 4

Practice 5

Solve: $9x^2-24x=-16$

Example 5

Solve: $4x^2-28x=-49$

Solution: First we rewrite the equation in standard form so that one side is 0. Then we factor the polynomial.

$$4x^2-28x=-49$$

$$4x^2-28x+49=0 \qquad \text{Write in standard form by adding 49 to both sides.}$$

$$(2x-7)(2x-7)=0 \qquad \text{Factor.}$$

Next we use the zero-factor property and set each factor equal to 0. Since the factors are the same, the related equations will give the same solution.

$$2x-7=0 \quad \text{or} \quad 2x-7=0 \qquad \text{Set each factor equal to 0.}$$

$$2x=7 \qquad 2x=7 \qquad \text{Solve.}$$

$$x=\frac{7}{2} \qquad x=\frac{7}{2}$$

Check this solution in the original equation. The solution is $\frac{7}{2}$.

Work Practice 5

Answers

3. a. 0 and -3 **b.** 0 and $\frac{3}{4}$

4. 6 and -3 **5.** $\frac{4}{3}$

The following steps may be used to solve a quadratic equation by factoring.

To Solve Quadratic Equations by Factoring

Step 1: Write the equation in standard form so that one side of the equation is 0.

Step 2: Factor the quadratic equation completely.

Step 3: Set each factor containing a variable equal to 0.

Step 4: Solve the resulting equations.

Step 5: Check each solution in the original equation.

Since it is not always possible to factor a quadratic polynomial, not all quadratic equations can be solved by factoring. Other methods of solving quadratic equations are presented in Chapter 9.

Example 6 Solve: $x(2x - 7) = 4$

Solution: First we write the equation in standard form; then we factor.

$$x(2x - 7) = 4$$

$$2x^2 - 7x = 4 \quad \text{Multiply.}$$

$$2x^2 - 7x - 4 = 0 \quad \text{Write in standard form.}$$

$$(2x + 1)(x - 4) = 0 \quad \text{Factor.}$$

$$2x + 1 = 0 \quad \text{or} \quad x - 4 = 0 \quad \text{Set each factor equal to zero.}$$

$$2x = -1 \qquad x = 4 \quad \text{Solve.}$$

$$x = -\frac{1}{2}$$

Check the solutions in the original equation. The solutions are $-\frac{1}{2}$ and 4.

Work Practice 6

Practice 6

Solve each equation.

a. $x(x - 4) = 5$

b. $x(3x + 7) = 6$

Helpful Hint

To solve the equation $x(2x - 7) = 4$, do **not** set each factor equal to 4. Remember that to apply the zero-factor property, one side of the equation must be 0 and the other side of the equation must be in factored form.

✓**Concept Check** Explain the error and solve the equation correctly.

$$(x - 3)(x + 1) = 5$$

$$x - 3 = 0 \quad \text{or} \quad x + 1 = 0$$

$$x = 3 \quad \text{or} \quad x = -1$$

Answers

6. a. 5 and -1 **b.** $\frac{2}{3}$ and -3

✓**Concept Check Answer**

To use the zero-factor property, one side of the equation must be 0, not 5. Correctly, $(x - 3)(x + 1) = 5$, $x^2 - 2x - 3 = 5$, $x^2 - 2x - 8 = 0$, $(x - 4)(x + 2) = 0$, $x - 4 = 0$ or $x + 2 = 0$, $x = 4$ or $x = -2$.

Objective B Solving Equations with Degree Greater Than Two by Factoring

Some equations with degree greater than 2 can be solved by factoring and then using the zero-factor property.

Practice 7

Solve: $2x^3 - 18x = 0$

Example 7 Solve: $3x^3 - 12x = 0$

Solution: To factor the left side of the equation, we begin by factoring out the greatest common factor, $3x$.

$$3x^3 - 12x = 0$$

$$3x(x^2 - 4) = 0 \quad \text{Factor out the GCF, } 3x.$$

$$3x(x + 2)(x - 2) = 0 \quad \text{Factor } x^2 - 4 \text{, a difference of two squares.}$$

$$3x = 0 \quad \text{or} \quad x + 2 = 0 \quad \text{or} \quad x - 2 = 0 \quad \text{Set each factor equal to 0.}$$

$$x = 0 \qquad x = -2 \qquad x = 2 \quad \text{Solve.}$$

Thus, the equation $3x^3 - 12x = 0$ has three solutions: 0, −2, and 2.

Check: Replace x with each solution in the original equation.
Let $x = 0$.

$$3(0)^3 - 12(0) \stackrel{?}{=} 0$$

$$0 = 0 \quad \text{True}$$

Let $x = -2$.

$$3(-2)^3 - 12(-2) \stackrel{?}{=} 0$$

$$3(-8) + 24 \stackrel{?}{=} 0$$

$$0 = 0 \quad \text{True}$$

Let $x = 2$.

$$3(2)^3 - 12(2) \stackrel{?}{=} 0$$

$$3(8) - 24 \stackrel{?}{=} 0$$

$$0 = 0 \quad \text{True}$$

The solutions are 0, −2, and 2.

Work Practice 7

Practice 8

Solve:
$(x + 3)(3x^2 - 20x - 7) = 0$

Example 8 Solve: $(5x - 1)(2x^2 + 15x + 18) = 0$

Solution:

$$(5x - 1)(2x^2 + 15x + 18) = 0$$

$$(5x - 1)(2x + 3)(x + 6) = 0 \quad \text{Factor the trinomial.}$$

$$5x - 1 = 0 \quad \text{or} \quad 2x + 3 = 0 \quad \text{or} \quad x + 6 = 0 \quad \text{Set each factor equal to 0.}$$

$$5x = 1 \qquad 2x = -3 \qquad x = -6 \quad \text{Solve.}$$

$$x = \frac{1}{5} \qquad 3x = -\frac{3}{2}$$

Check each solution in the original equation. The solutions are $\frac{1}{5}, -\frac{3}{2}$, and −6.

Work Practice 8

Answers

7. 0, 3, and −3 8. −3, $-\frac{1}{3}$, and 7

Vocabulary, Readiness & Video Check

Use the choices below to fill in each blank. Not all choices will be used.

−3, 5	$a = 0$ or $b = 0$	0	linear
3, −5	quadratic	1	

1. An equation that can be written in the form $ax^2 + bx + c = 0$ (with $a \neq 0$) is called a ______________ equation.
2. If the product of two numbers is 0, then at least one of the numbers must be ______________.

3. The solutions to $(x - 3)(x + 5) = 0$ are ______________.

4. If $a \cdot b = 0$, then ______________.

Martin-Gay Interactive Videos

See Video 4.6

Watch the section lecture video and answer the following questions.

Objective A **5.** As shown in Examples 1–3, what two things have to be true in order to use the zero-factor theorem?

Objective B **6.** Example 4 implies that the zero-factor theorem can be used with any number of factors on one side of the equation as long as the other side of the equation is zero. Why do you think this is true?

4.6 Exercise Set MyLab Math

Objective A *Solve each equation. See Examples 1 through 3.*

1. $(x - 2)(x + 1) = 0$

2. $(x + 3)(x + 2) = 0$

3. $(x - 6)(x - 7) = 0$

4. $(x + 4)(x - 10) = 0$

5. $(x + 9)(x + 17) = 0$

6. $(x - 11)(x - 1) = 0$

7. $x(x + 6) = 0$

8. $x(x - 7) = 0$

9. $3x(x - 8) = 0$

10. $2x(x + 12) = 0$

11. $(2x + 3)(4x - 5) = 0$

12. $(3x - 2)(5x + 1) = 0$

13. $(2x - 7)(7x + 2) = 0$

14. $(9x + 1)(4x - 3) = 0$

15. $\left(x - \frac{1}{2}\right)\left(x + \frac{1}{3}\right) = 0$

16. $\left(x + \frac{2}{9}\right)\left(x - \frac{1}{4}\right) = 0$

17. $(x + 0.2)(x + 1.5) = 0$

18. $(x + 1.7)(x + 2.3) = 0$

Solve. See Examples 4 through 6.

19. $x^2 - 13x + 36 = 0$

20. $x^2 + 2x - 63 = 0$

21. $x^2 + 2x - 8 = 0$

22. $x^2 - 5x + 6 = 0$

23. $x^2 - 7x = 0$

24. $x^2 - 3x = 0$

25. $x^2 + 20x = 0$

26. $x^2 + 15x = 0$

27. $x^2 = 16$

28. $x^2 = 9$

29. $x^2 - 4x = 32$

30. $x^2 - 5x = 24$

31. $(x + 4)(x - 9) = 4x$

32. $(x + 3)(x + 8) = x$

33. $x(3x - 1) = 14$

34. $x(4x - 11) = 3$

35. $3x^2 + 19x - 72 = 0$

36. $36x^2 + x - 21 = 0$

Objectives A B and Section 2.3 Mixed Practice

Solve each equation. See Examples 1 through 8. (A few exercises are linear equations.)

37. $4x^3 - x = 0$

38. $4y^3 - 36y = 0$

39. $4(x - 7) = 6$

40. $5(3 - 4x) = 9$

41. $(4x - 3)(16x^2 - 24x + 9) = 0$

42. $(2x + 5)(4x^2 + 20x + 25) = 0$

43. $4y^2 - 1 = 0$

44. $4y^2 - 81 = 0$

45. $(2x + 3)(2x^2 - 5x - 3) = 0$

46. $(2x - 9)(x^2 + 5x - 36) = 0$

47. $x^2 - 15 = -2x$

48. $x^2 - 26 = -11x$

49. $30x^2 - 11x = 30$

50. $9x^2 + 7x = 2$

51. $5x^2 - 6x - 8 = 0$

52. $12x^2 + 7x - 12 = 0$

53. $6y^2 - 22y - 40 = 0$

54. $3x^2 - 6x - 9 = 0$

55. $(y - 2)(y + 3) = 6$

56. $(y - 5)(y - 2) = 28$

57. $x^3 - 12x^2 + 32x = 0$

58. $x^3 - 14x^2 + 49x = 0$

59. $x^2 + 14x + 49 = 0$

60. $x^2 + 22x + 121 = 0$

61. $12y = 8y^2$

62. $9y = 6y^2$

63. $7x^3 - 7x = 0$

64. $3x^3 - 27x = 0$

65. $3x^2 + 8x - 11 = 13 - 6x$

66. $2x^2 + 12x - 1 = 4 + 3x$

67. $3x^2 - 20x = -4x^2 - 7x - 6$

68. $4x^2 - 20x = -5x^2 - 6x - 5$

Review

Perform each indicated operation. Write all results in lowest terms. See Section R.2.

69. $\frac{3}{5} + \frac{4}{9}$

70. $\frac{2}{3} + \frac{3}{7}$

71. $\frac{7}{10} - \frac{5}{12}$

72. $\frac{5}{9} - \frac{5}{12}$

73. $\frac{4}{5} \cdot \frac{7}{8}$

74. $\frac{3}{7} \cdot \frac{12}{17}$

Concept Extensions

For Exercises 75 and 76, see the Concept Check in this section.

75. Explain the error and solve correctly:

$x(x - 2) = 8$

$x = 8$ or $x - 2 = 8$

$x = 10$

76. Explain the error and solve correctly:

$(x - 4)(x + 2) = 0$

$x = -4$ or $x = 2$

Solve.

77. Write a quadratic equation that has two solutions, 6 or -1. Leave the polynomial in the equation in factored form.

78. Write a quadratic equation that has two solutions, 0 or -2. Leave the polynomial in the equation in factored form.

79. Write a quadratic equation in standard form that has two solutions, 5 and 7.

80. Write an equation that has three solutions, 0, 1, and 2.

81. A compass is accidentally thrown upward and out of a hot-air balloon at a height of 300 feet. The height, y, of the compass at time x is given by the equation $y = -16x^2 + 20x + 300$.

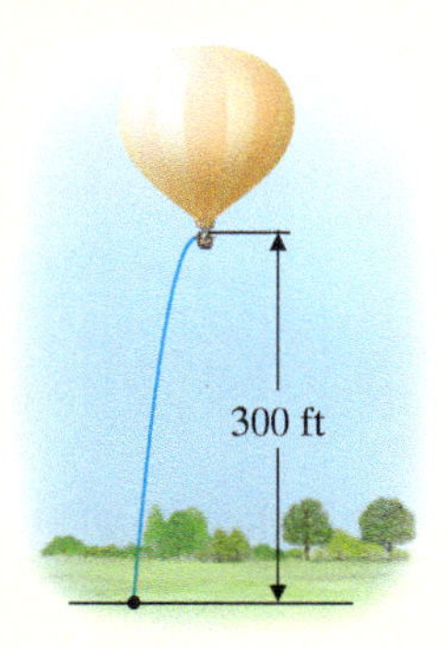

a. Find the height of the compass at the given times by filling in the table below.

Time, x (in seconds)	0	1	2	3	4	5	6
Height, y (in feet)							

b. Use the table to determine when the compass strikes the ground.

c. Use the table to approximate the maximum height of the compass.

82. A rocket is fired upward from the ground with an initial velocity of 100 feet per second. The height, y, of the rocket at any time x is given by the equation $y = -16x^2 + 100x$.

a. Find the height of the rocket at the given times by filling in the table below.

Time, x (in seconds)	0	1	2	3	4	5	6	7
Height, y (in feet)								

b. Use the table to determine between what two whole numbered seconds the rocket strikes the ground.

c. Use the table to approximate the maximum height of the rocket.

Solve each equation.

83. $(x - 3)(3x + 4) = (x + 2)(x - 6)$

84. $(2x - 3)(x + 6) = (x - 9)(x + 2)$

85. $(2x - 3)(x + 8) = (x - 6)(x + 4)$

86. $(x + 6)(x - 6) = (2x - 9)(x + 4)$

4.7 Quadratic Equations and Problem Solving

Objective

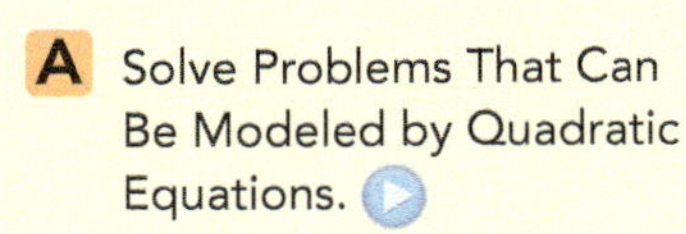

A Solve Problems That Can Be Modeled by Quadratic Equations.

Objective A Solving Problems Modeled by Quadratic Equations

Some problems may be modeled by quadratic equations. To solve these problems, we use the same problem-solving steps that were introduced in Section 2.4. When solving these problems, keep in mind that a solution of an equation that models a problem may not be a solution to the problem. For example, a person's age or the length of a rectangle is always a positive number. Thus we discard solutions that do not make sense as solutions of the problem.

Practice 1

Cliff divers also frequent the falls at Waimea Falls Park in Oahu, Hawaii. Here, a diver can jump from a ledge 64 feet up the waterfall into a rocky pool below. Neglecting air resistance, the height of a diver above the pool after t seconds is $h = -16t^2 + 64$. Find how long it takes the diver to reach the pool.

Example 1 Finding Free-Fall Time

Since the 1940s, one of the top tourist attractions in Acapulco, Mexico, is watching the cliff divers off La Quebrada. The divers' platform is about 144 feet above the sea. These divers must time their descent just right, since they land in the crashing Pacific, in an inlet that is at most $9\frac{1}{2}$ feet deep. Neglecting air resistance, the height h in feet of a cliff diver above the ocean after t seconds is given by the quadratic equation $h = -16t^2 + 144$.

Find out how long it takes the diver to reach the ocean.

Solution:

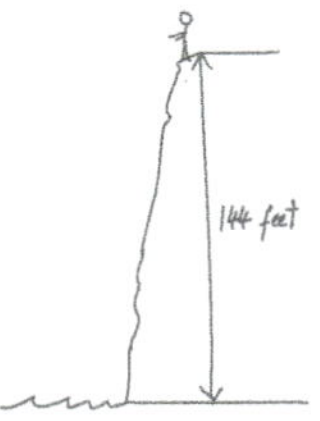

1. UNDERSTAND. Read and reread the problem. Then draw a picture of the problem.

   The equation $h = -16t^2 + 144$ models the height of the falling diver at time t. Familiarize yourself with this equation by finding the height of the diver at time $t = 1$ second and $t = 2$ seconds.

   When $t = 1$ second, the height of the diver is $h = -16(1)^2 + 144 = 128$ feet.
   When $t = 2$ seconds, the height of the diver is $h = -16(2)^2 + 144 = 80$ feet.

2. TRANSLATE. To find out how long it takes the diver to reach the ocean, we want to know the value of t for which $h = 0$.

3. SOLVE. Solve the equation.

$$0 = -16t^2 + 144$$

$$0 = -16(t^2 - 9) \quad \text{Factor out } -16.$$

$$0 = -16(t - 3)(t + 3) \quad \text{Factor completely.}$$

$$t - 3 = 0 \quad \text{or} \quad t + 3 = 0 \quad \text{Set each factor containing a variable equal to 0.}$$

$$t = 3 \quad \text{or} \quad t = -3 \quad \text{Solve.}$$

4. INTERPRET. Since the time t cannot be negative, the proposed solution is 3 seconds.

   Check: Verify that the height of the diver when t is 3 seconds is 0.

   When $t = 3$ seconds, $h = -16(3)^2 + 144 = -144 + 144 = 0$.

Work Practice 1

Answer
1. 2 sec

Example 2 Finding a Number

The square of a number plus three times the number is 70. Find the number.

Solution:

1. UNDERSTAND. Read and reread the problem. Suppose that the number is 5. The square of 5 is 5^2 or 25. Three times 5 is 15. Then $25 + 15 = 40$, not 70, so the number must be greater than 5. Remember, the purpose of proposing a number, such as 5, is to better understand the problem. Now that we do, we will let $x =$ the number.
2. TRANSLATE.

the square of a number	plus	three times the number	is	70
↓	↓	↓	↓	↓
x^2	$+$	$3x$	$=$	70

3. SOLVE.

$$x^2 + 3x = 70$$
$$x^2 + 3x - 70 = 0 \quad \text{Subtract 70 from both sides.}$$
$$(x + 10)(x - 7) = 0 \quad \text{Factor.}$$
$$x + 10 = 0 \quad \text{or} \quad x - 7 = 0 \quad \text{Set each factor equal to 0.}$$
$$x = -10 \qquad x = 7 \quad \text{Solve.}$$

4. INTERPRET.

Check: The square of -10 is $(-10)^2$, or 100. Three times -10 is $3(-10)$ or -30. Then $100 + (-30) = 70$, the correct sum, so -10 checks.

The square of 7 is 7^2 or 49. Three times 7 is 3(7), or 21. Then $49 + 21 = 70$, the correct sum, so 7 checks.

State: There are two numbers. They are -10 and 7.

Work Practice 2

Practice 2

The square of a number minus twice the number is 63. Find the number.

Example 3 Finding the Dimensions of a Sail

The height of a triangular sail is 2 meters less than twice the length of the base. If the sail has an area of 30 square meters, find the length of its base and the height.

Solution:

1. UNDERSTAND. Read and reread the problem. Since we are finding the length of the base and the height, we let

$x =$ the length of the base

Since the height is 2 meters less than twice the length of the base,

$2x - 2 =$ the height

An illustration is shown in the margin.

2. TRANSLATE. We are given that the area of the triangle is 30 square meters, so we use the formula for area of a triangle.

area of triangle	$=$	$\frac{1}{2}$	$\cdot$	base	$\cdot$	height
↓		↓		↓		↓
30	$=$	$\frac{1}{2}$	$\cdot$	x	$\cdot$	$(2x - 2)$

(*Continued on next page*)

Practice 3

The length of a rectangular garden is 5 feet more than its width. The area of the garden is 176 square feet. Find the length and the width of the garden.

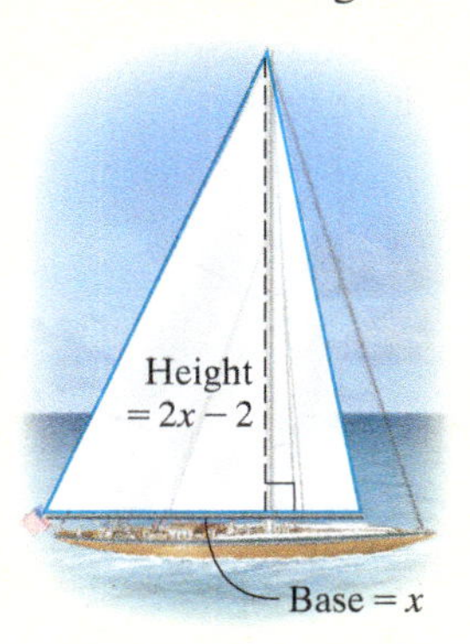

Answers

2. 9 and -7
3. length: 16 ft; width: 11 ft

3. SOLVE. Now we solve the quadratic equation.

$$30 = \frac{1}{2}x(2x - 2)$$

$$30 = x^2 - x \quad \text{Multiply.}$$

$$0 = x^2 - x - 30 \quad \text{Write in standard form.}$$

$$0 = (x - 6)(x + 5) \quad \text{Factor.}$$

$$x - 6 = 0 \quad \text{or} \quad x + 5 = 0 \quad \text{Set each factor equal to 0.}$$

$$x = 6 \qquad\qquad x = -5$$

4. INTERPRET. Since x represents the length of the base, we discard the solution -5. The base of a triangle cannot be negative. The base is then 6 meters and the height is $2(6) - 2 = 10$ meters.

Check: To check this problem, we recall that

$$\text{area} = \frac{1}{2}\text{base} \cdot \text{height or}$$

$$30 \stackrel{?}{=} \frac{1}{2}(6)(10)$$

$$30 = 30 \quad \text{True}$$

State: The base of the triangular sail is 6 meters and the height is 10 meters.

Work Practice 3

The next example has to do with consecutive integers. Study the following diagrams for a review of consecutive integers.

Examples

If x is the first integer, then consecutive integers are $x, x + 1, x + 2, \ldots.$

If x is the first even integer, then consecutive even integers are $x, x + 2, x + 4, \ldots.$

If x is the first odd integer, then consecutive odd integers are $x, x + 2, x + 4, \ldots.$

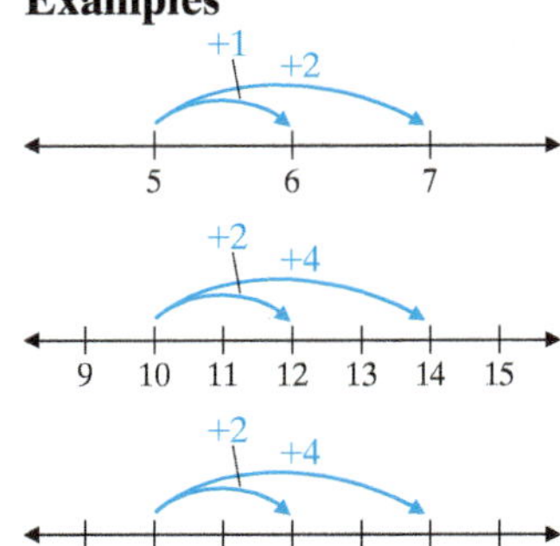

Practice 4

Find two consecutive odd integers whose product is 23 more than their sum.

Example 4 Finding Consecutive Even Integers

Find two consecutive even integers whose product is 34 more than their sum.

Solution:

1. UNDERSTAND. Read and reread the problem. Let's just choose two consecutive even integers to help us better understand the problem. Let's choose 10 and 12. Their product is $10(12) = 120$ and their sum is $10 + 12 = 22$. The product is $120 - 22$, or 98 greater than the sum. Thus our guess is incorrect, but we have a better understanding of this example.

   Let's let x and $x + 2$ be the consecutive even integers.

2. TRANSLATE.

Product of integers	is	34	more than	sum of integers
↓	↓			
$x(x + 2)$	$=$	$x + (x + 2) + 34$		

Answer

4. 5 and 7 or -5 and -3

3. SOLVE. Now we solve the equation.

$$x(x+2) = x + (x+2) + 34$$
$$x^2 + 2x = x + x + 2 + 34 \quad \text{Multiply.}$$
$$x^2 + 2x = 2x + 36 \quad \text{Combine like terms.}$$
$$x^2 - 36 = 0 \quad \text{Write in standard form.}$$
$$(x+6)(x-6) = 0 \quad \text{Factor.}$$
$$x + 6 = 0 \quad \text{or} \quad x - 6 = 0 \quad \text{Set each factor equal to 0.}$$
$$x = -6 \qquad\qquad x = 6 \quad \text{Solve.}$$

4. INTERPRET. If $x = -6$, then $x + 2 = -6 + 2$, or -4.
If $x = 6$, then $x + 2 = 6 + 2$, or 8.

Check: $-6, -4$ $\qquad$ $6, 8$

$$-6(-4) \stackrel{?}{=} -6 + (-4) + 34 \qquad 6(8) \stackrel{?}{=} 6 + 8 + 34$$
$$24 \stackrel{?}{=} -10 + 34 \qquad 48 \stackrel{?}{=} 14 + 34$$
$$24 = 24 \quad \text{True} \qquad 48 = 48 \quad \text{True}$$

State: The two consecutive even integers are -6 and -4 or 6 and 8.

Work Practice 4

The next example makes use of the **Pythagorean theorem.** Before we review this theorem, recall that a **right triangle** is a triangle that contains a 90° or right angle. The **hypotenuse** of a right triangle is the side opposite the right angle and is the longest side of the triangle. The **legs** of a right triangle are the other sides of the triangle.

Pythagorean Theorem

In a right triangle, the sum of the squares of the lengths of the two legs is equal to the square of the length of the hypotenuse.

$$(\text{leg})^2 + (\text{leg})^2 = (\text{hypotenuse})^2 \quad \text{or} \quad a^2 + b^2 = c^2$$

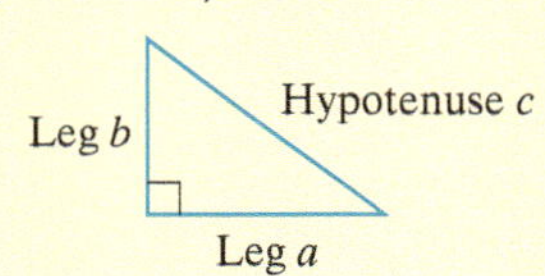

Helpful Hint If you use this formula, don't forget that c represents the length of the hypotenuse.

Example 5 Finding the Dimensions of a Triangle

Find the lengths of the sides of a right triangle if the lengths can be expressed as three consecutive even integers.

Solution:

1. UNDERSTAND. Read and reread the problem. Let's suppose that the length of one leg of the right triangle is 4 units. Then the other leg is the next even integer, or 6 units, and the hypotenuse of the triangle is the next even integer, or 8 units. Remember that the hypotenuse is the longest side. Let's see if a triangle with sides of these lengths forms a right triangle. To do this, we check to see whether the Pythagorean theorem holds true.

$$4^2 + 6^2 \stackrel{?}{=} 8^2$$
$$16 + 36 \stackrel{?}{=} 64$$
$$52 = 64 \quad \text{False}$$

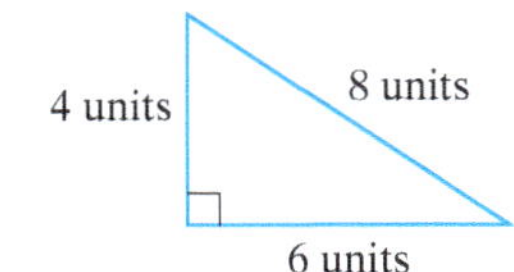

Our proposed numbers do not check, but we now have a better understanding of the problem.

(*Continued on next page*)

Practice 5

The length of one leg of a right triangle is 7 meters less than the length of the other leg. The length of the hypotenuse is 13 meters. Find the lengths of the legs.

Answer

5. 5 meters, 12 meters

We let x, $x + 2$, and $x + 4$ be three consecutive even integers. Since these integers represent lengths of the sides of a right triangle, we have the following.

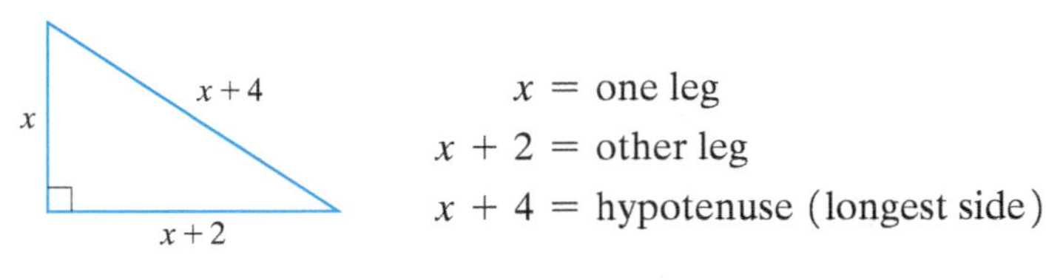

x = one leg
$x + 2$ = other leg
$x + 4$ = hypotenuse (longest side)

2. TRANSLATE. By the Pythagorean theorem, we have that

$$(\text{leg})^2 + (\text{leg})^2 = (\text{hypotenuse})^2$$
$$(x)^2 + (x + 2)^2 = (x + 4)^2$$

3. SOLVE. Now we solve the equation.

$$x^2 + (x + 2)^2 = (x + 4)^2$$
$$x^2 + x^2 + 4x + 4 = x^2 + 8x + 16 \quad \text{Multiply.}$$
$$2x^2 + 4x + 4 = x^2 + 8x + 16 \quad \text{Combine like terms.}$$
$$x^2 - 4x - 12 = 0 \quad \text{Write in standard form.}$$
$$(x - 6)(x + 2) = 0 \quad \text{Factor.}$$
$$x - 6 = 0 \quad \text{or} \quad x + 2 = 0 \quad \text{Set each factor equal to 0.}$$
$$x = 6 \qquad x = -2$$

4. INTERPRET. We discard $x = -2$ since length cannot be negative. If $x = 6$, then $x + 2 = 8$ and $x + 4 = 10$.

Check: Verify that

$$(\text{leg})^2 + (\text{leg})^2 = (\text{hypotenuse})^2$$
$$6^2 + 8^2 \stackrel{?}{=} 10^2$$
$$36 + 64 \stackrel{?}{=} 100$$
$$100 = 100 \quad \text{True}$$

State: The sides of the right triangle have lengths 6 units, 8 units, and 10 units.

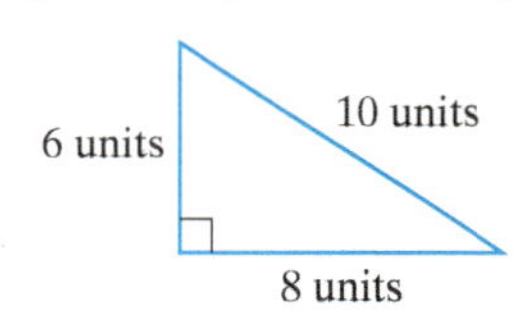

Work Practice 5

Vocabulary, Readiness & Video Check

Martin-Gay Interactive Videos

See Video 4.7

Watch the section lecture video and answer the following question.

Objective A **1.** In each of Examples 1–3, why aren't both solutions of the translated equation accepted as solutions to the application?

4.7 Exercise Set MyLab Math 

Objective A *See Examples 1 through 5 for all exercises.*

Translating *For Exercises 1 through 6, represent each given condition using a single variable, x.*

△ **1.** The length and width of a rectangle whose length is 4 centimeters more than its width

△ **2.** The length and width of a rectangle whose length is twice its width

3. Two consecutive odd integers

4. Two consecutive even integers

△ **5.** The base and height of a triangle whose height is one more than four times its base

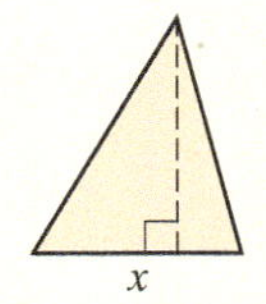

△ **6.** The base and height of a trapezoid whose base is three less than five times its height

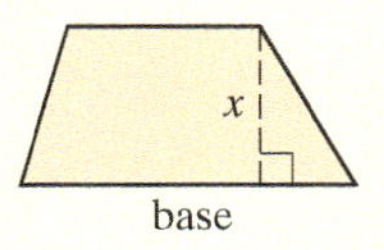

Use the information given to find the dimensions of each figure.

△ **7.** The *area* of the square is 121 square units. Find the length of its sides.

△ **8.** The *area* of the rectangle is 84 square inches. Find its length and width.

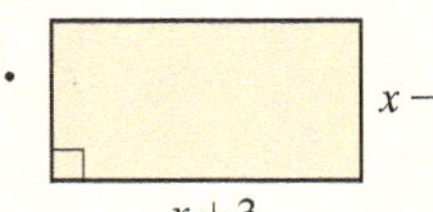

△ **9.** The *perimeter* of the quadrilateral is 120 centimeters. Find the lengths of its sides.

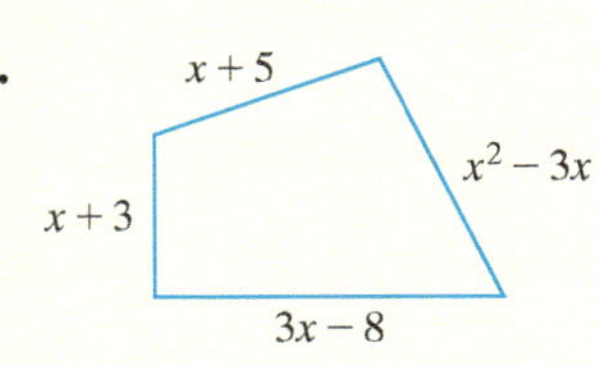

10. The *perimeter* of the triangle is 29 feet. Find the lengths of its sides.

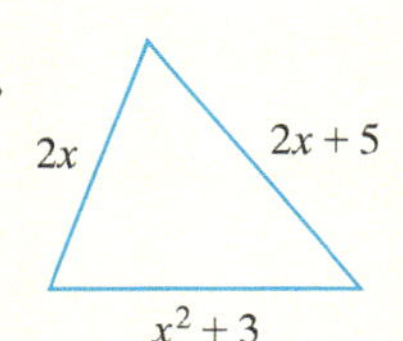

△ **11.** The *area* of the parallelogram is 96 square miles. Find its base and height.

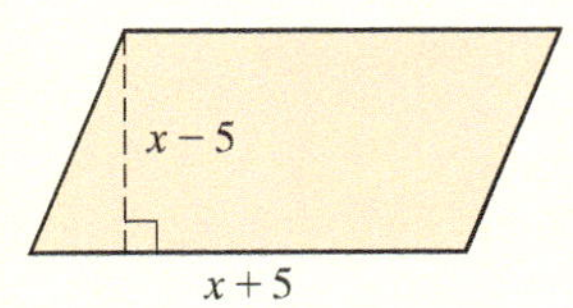

△ **12.** The *area* of the circle is 25π square kilometers. Find its radius.

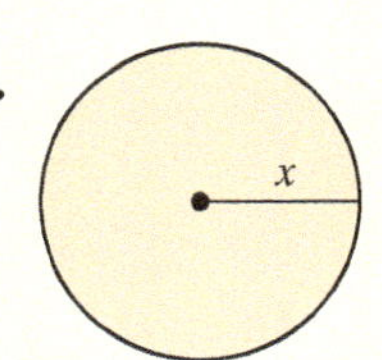

Solve.

13. An object is thrown upward from the top of an 80-foot building with an initial velocity of 64 feet per second. The height h of the object after t seconds is given by the quadratic equation $h = -16t^2 + 64t + 80$. When will the object hit the ground?

14. A hang glider accidentally drops her compass from the top of a 400-foot cliff. The height h of the compass after t seconds is given by the quadratic equation $h = -16t^2 + 400$. When will the compass hit the ground?

15. The width of a rectangle is 7 centimeters less than twice its length. Its area is 30 square centimeters. Find the dimensions of the rectangle.

16. The length of a rectangle is 9 inches more than its width. Its area is 112 square inches. Find the dimensions of the rectangle.

△ *The equation* $D = \frac{1}{2}n(n - 3)$ *gives the number of diagonals D for a polygon with n sides. For example, a polygon with 6 sides has* $D = \frac{1}{2} \cdot 6(6 - 3)$ *or* $D = 9$ *diagonals. (See if you can count all 9 diagonals. Some are shown in the figure.)*

Use this equation, $D = \frac{1}{2}n(n - 3)$, *for Exercises 17 through 20.*

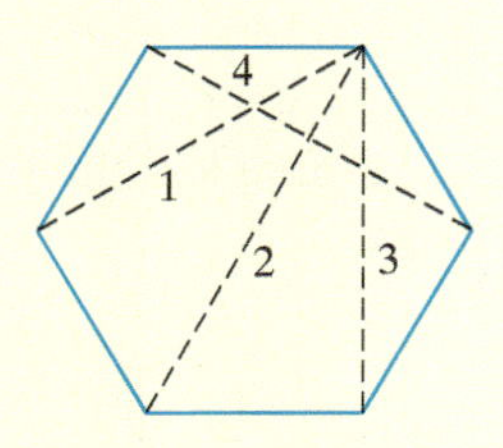

17. Find the number of diagonals for a polygon that has 12 sides.

18. Find the number of diagonals for a polygon that has 15 sides.

19. Find the number of sides n for a polygon that has 35 diagonals.

20. Find the number of sides n for a polygon that has 14 diagonals.

21. The sum of a number and its square is 132. Find the number.

22. The sum of a number and its square is 182. Find the number.

23. The product of two consecutive room numbers is 210. Find the room numbers.

24. The product of two consecutive page numbers is 420. Find the page numbers.

25. A ladder is leaning against a building so that the distance from the ground to the top of the ladder is one foot less than the length of the ladder. Find the length of the ladder if the distance from the bottom of the ladder to the building is 5 feet.

26. Use the given figure to find the length of the guy wire.

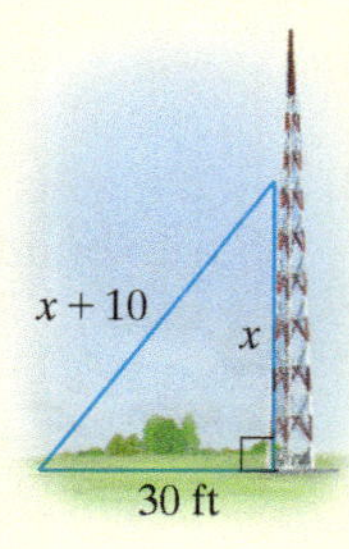

△ **27.** If the sides of a square are increased by 3 inches, the area becomes 64 square inches. Find the length of the sides of the original square.

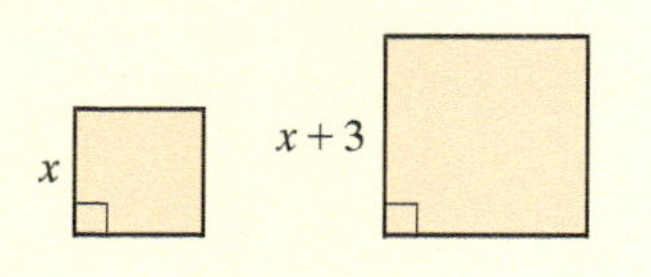

△ **28.** If the sides of a square are increased by 5 meters, the area becomes 100 square meters. Find the length of the sides of the original square.

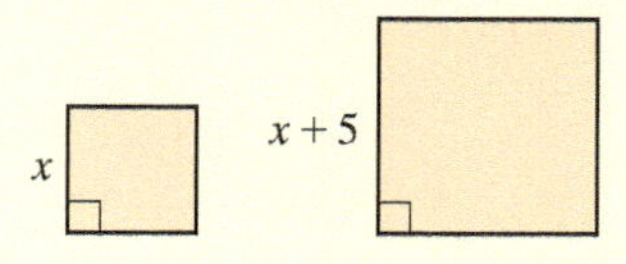

△ **29.** One leg of a right triangle is 4 millimeters longer than the shorter leg and the hypotenuse is 8 millimeters longer than the shorter leg. Find the lengths of the sides of the triangle.

△ **30.** One leg of a right triangle is 9 centimeters longer than the other leg and the hypotenuse is 45 centimeters. Find the lengths of the legs of the triangle.

△ **31.** The length of the base of a triangle is twice its height. If the area of the triangle is 100 square kilometers, find the height.

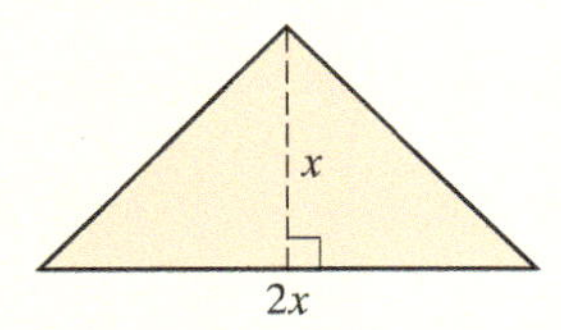

△ **32.** The height of a triangle is 2 millimeters less than the base. If the area is 60 square millimeters, find the base.

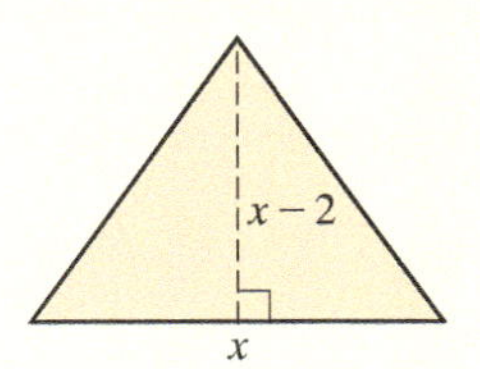

△ **33.** Find the length of the shorter leg of a right triangle if the longer leg is 12 feet more than the shorter leg and the hypotenuse is 12 feet less than twice the shorter leg.

△ **34.** Find the length of the shorter leg of a right triangle if the longer leg is 10 miles more than the shorter leg and the hypotenuse is 10 miles less than twice the shorter leg.

35. An object is dropped from 39 feet below the tip of the pinnacle atop one of the 1483-foot-tall Petronas Twin Towers in Kuala Lumpur, Malaysia. (*Source:* Council on Tall Buildings and Urban Habitat) The height h of the object after t seconds is given by the equation $h = -16t^2 + 1444$. Find how many seconds pass before the object reaches the ground.

36. An object is dropped from the top of 311 South Wacker Drive, a 961-foot-tall office building in Chicago. (*Source:* Council on Tall Buildings and Urban Habitat) The height h of the object after t seconds is given by the equation $h = -16t^2 + 961$. Find how many seconds pass before the object reaches the ground.

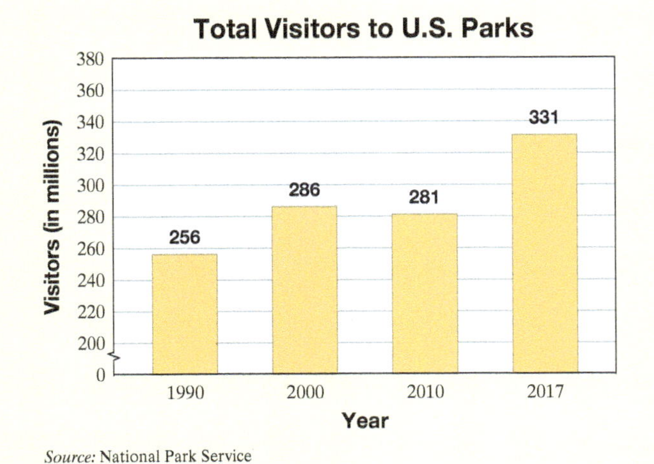

Source: National Park Service

37. In 2017, two parks in the US had visitors (in the millions) that happened to be two consecutive integers. To find the number of visitors for each park, find two positive consecutive integers who product is 209 more than their sum.

Let x = 2017 Golden Gate National Recreation Area visitors (in millions)

Let $(x + 1)$ = 2017 Blue Ridge Parkway visitors (in millions)

38. In 2017, two parks in the US had visitors (in the millions) that happened to be two consecutive even integers. To find the number of visitors for each park, find two positive consecutive even integers who product is 79 more than their sum.

Let x = 2017 Gateway National Recreation Area visitors (in millions)

Let $(x + 2)$ = 2017 Great Smoky Mountains National Park visitors (in millions)

△ **39.** Find the dimensions of a rectangle whose width is 7 miles less than its length and whose area is 120 square miles.

△ **40.** Find the dimensions of a rectangle whose width is 2 inches less than half its length and whose area is 160 square inches.

41. If the cost, C, for manufacturing x units of a certain product is given by $C = x^2 - 15x + 50$, find the number of units manufactured at a cost of \$9500.

42. If a switchboard handles n telephones, the number C of telephone connections it can make simultaneously is given by the equation $C = \frac{n(n-1)}{2}$. Find how many telephones are handled by a switchboard making 120 telephone connections simultaneously.

Review

The following double line graph shows a comparison of the number of annual visitors (in millions) to Acadia National Park and Grand Teton National Park for the years shown. Use this graph to answer Exercises 43 through 50. See Section R.4.

Annual Visitors to Acadia and Grand Teton National Parks

Visitors (in millions): 0, 2.3, 2.4, 2.5, 2.6, 2.7, 2.8, 2.9, 3.0, 3.1, 3.2, 3.3, 3.4, 3.5

Year: 2011, 2012, 2013, 2014, 2015, 2016, 2017

Grand Teton

Acadia

Source: National Park Service

43. Approximate the number of visitors to Acadia National Park in 2015.

44. Approximate the number of visitors to Grand Teton National Park in 2015.

45. Approximate the number of visitors to Acadia National Park in 2017.

46. Approximate the number of visitors to Grand Teton National Park in 2017.

47. Determine the year that the colored lines in this graph intersect.

48. For what year(s) on the graph is the number of visitors to Grand Teton National Park greater than 3 million?

49. In your own words, explain the meaning of the point of intersection in the graph.

50. Describe the trends shown in this graph and speculate as to why these trends have occurred.

Write each fraction in simplest form. See Section R.2.

51. $\frac{20}{35}$ **52.** $\frac{24}{32}$ **53.** $\frac{27}{18}$ **54.** $\frac{15}{27}$ **55.** $\frac{14}{42}$ **56.** $\frac{45}{50}$

Concept Extensions

△ **57.** The side of a square equals the width of a rectangle. The length of the rectangle is 6 meters longer than its width. The sum of the areas of the square and the rectangle is 176 square meters. Find the side of the square.

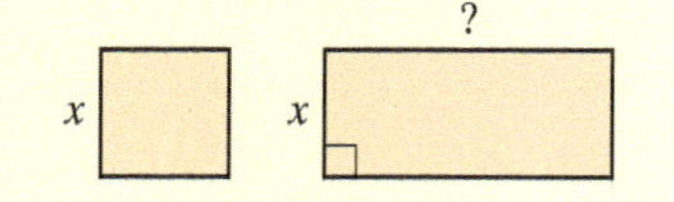

△ **58.** Two boats travel at right angles to each other after leaving the same dock at the same time. One hour later the boats are 17 miles apart. If one boat travels 7 miles per hour faster than the other boat, find the rate of each boat.

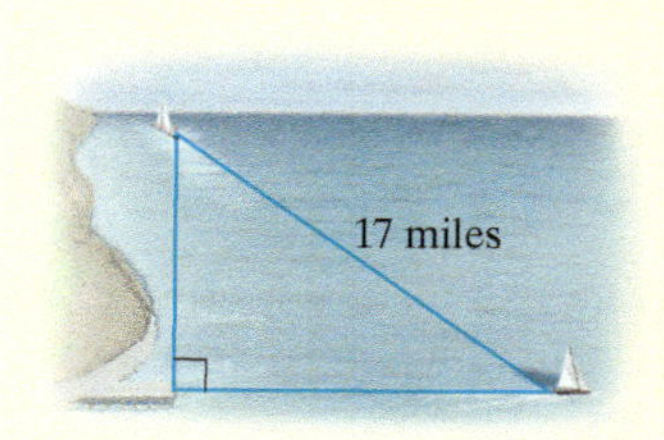

59. The sum of two numbers is 25, and the sum of their squares is 325. Find the numbers.

60. The sum of two numbers is 20, and the sum of their squares is 218. Find the numbers.

△ **61.** A rectangular pool is surrounded by a walk 4 meters wide. The pool is 6 meters longer than its width. If the total area of the pool and walk is 576 square meters more than the area of the pool, find the dimensions of the pool.

△ **62.** A rectangular garden is surrounded by a walk of uniform width. The area of the garden is 180 square yards. If the dimensions of the garden plus the walk are 16 yards by 24 yards, find the width of the walk.

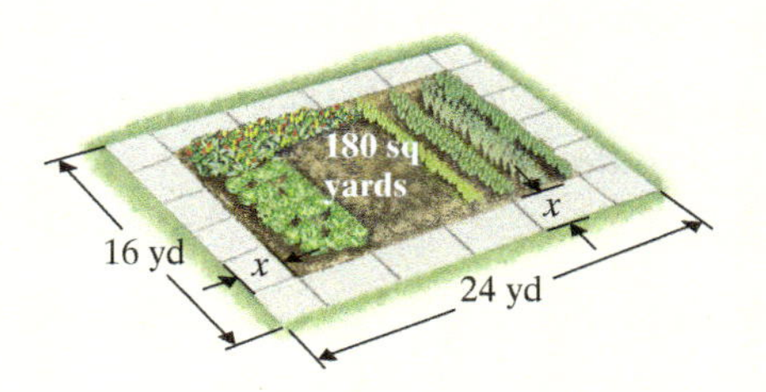

63. Write down two numbers whose sum is 10. Square each number and find the sum of the squares. Use this work to write a word problem like Exercise **59**. Then give the word problem to a classmate to solve.

64. Write down two numbers whose sum is 12. Square each number and find the sum of the squares. Use this work to write a word problem like Exercise **60**. Then give the word problem to a classmate to solve.

Chapter 4 Group Activity

Factoring polynomials can be visualized using areas of rectangles. To see this, let's first find the areas of the following squares and rectangles. (Recall that Area = Length · Width.)

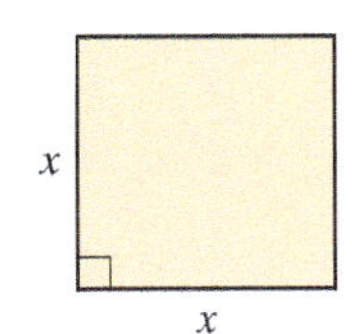

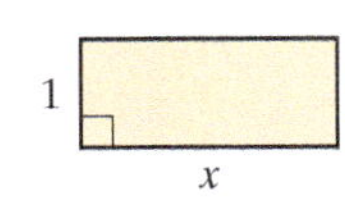

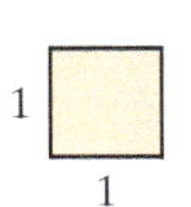

To use these areas to visualize factoring the polynomial $x^2 + 3x + 2$, for example, use the shapes below to form a rectangle. The factored form is found by reading the length and the width of the rectangle, as shown below.

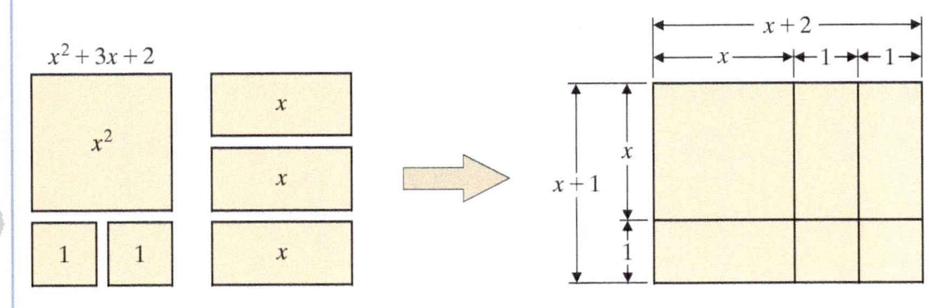

Thus, $x^2 + 3x + 2 = (x + 2)(x + 1)$.

Try using this method to visualize the factored form of each polynomial below.

Work in a group and use tiles to find a factored form for the polynomials below. (Tiles can be handmade from index cards.)

1. $x^2 + 6x + 5$
2. $x^2 + 5x + 6$
3. $x^2 + 5x + 4$
4. $x^2 + 4x + 3$
5. $x^2 + 6x + 9$
6. $x^2 + 4x + 4$

Chapter 4 Vocabulary Check

Fill in each blank with one of the words or phrases listed below. Some words or phrases may be used more than once.

factoring	hypotenuse	quadratic equation
greatest common factor	leg	perfect square trinomial

1. An equation that can be written in the form $ax^2 + bx + c = 0$ (with a not 0) is called a ________________.
2. ________________ is the process of writing an expression as a product.
3. The ________________ of a list of terms is the product of all common factors.
4. A trinomial that is the square of some binomial is called a ________________.
5. In a right triangle, the side opposite the right angle is called the ________________.
6. In a right triangle, each side adjacent to the right angle is called a ________________.
7. The Pythagorean theorem states that $(\text{leg})^2 + (\text{leg})^2 = (\_\_\_\_\_\_\_\_)^2$.

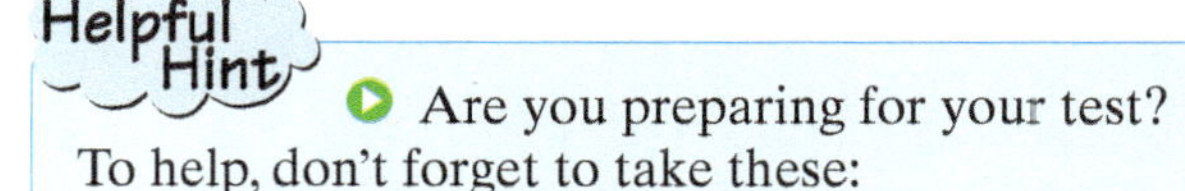

Helpful Hint Are you preparing for your test? To help, don't forget to take these:

- Chapter 4 Getting Ready for the Test on page 333
- Chapter 4 Test on page 334

Then check all of your answers at the back of this text. For further review, the step-by-step video solutions to any of these exercises are located in MyLab Math.

4 Chapter Highlights

Definitions and Concepts	Examples
Section 4.1 The Greatest Common Factor and Factoring by Grouping	
Factoring is the process of writing an expression as a product.	Factor: $6 = 2 \cdot 3$ Factor: $x^2 + 5x + 6 = (x + 2)(x + 3)$
The GCF of a list of variable terms contains the smallest exponent on each common variable.	The GCF of z^5, z^3, and z^{10} is z^3.
The GCF of a list of terms is the product of all common factors.	Find the GCF of $8x^2y$, $10x^3y^2$, and $50x^2y^3$. $8x^2y = 2 \cdot 2 \cdot 2 \cdot x^2 \cdot y$ $10x^3y^2 = 2 \cdot 5 \cdot x^3 \cdot y^2$ $50x^2y^3 = 2 \cdot 5 \cdot 5 \cdot x^2 \cdot y^3$ $\text{GCF} = 2 \cdot x^2 \cdot y$ or $2x^2y$
To Factor by Grouping **Step 1:** Group the terms in two groups so that each group has a common factor. **Step 2:** Factor out the GCF from each group. **Step 3:** If there is a common binomial factor, factor it out. **Step 4:** If not, rearrange the terms and try these steps again.	Factor: $10ax + 15a - 6xy - 9y$ **Step 1:** $(10ax + 15a) + (-6xy - 9y)$ **Step 2:** $5a(2x + 3) - 3y(2x + 3)$ **Step 3:** $(2x + 3)(5a - 3y)$

Definitions and Concepts	Examples
Section 4.2 Factoring Trinomials of the Form $x^2 + bx + c$	
The product of these numbers is c. $x^2 + bx + c = (x + \square)(x + \square)$ The sum of these numbers is b.	Factor: $x^2 + 7x + 12$ $3 + 4 = 7 \quad 3 \cdot 4 = 12$ $x^2 + 7x + 12 = (x + 3)(x + 4)$
Section 4.3 Factoring Trinomials of the Form $ax^2 + bx + c$	
To factor $ax^2 + bx + c$, try various combinations of factors of ax^2 and c until a middle term of bx is obtained when checking.	Factor: $3x^2 + 14x - 5$ Factors of $3x^2$: $3x, x$ Factors of -5: $-1, 5$ and $1, -5$ $(3x - 1)(x + 5)$ $-1x$ $+15x$ $14x$ Correct middle term
Section 4.4 Factoring Trinomials of the Form $ax^2 + bx + c$ by Grouping	
To Factor $ax^2 + bx + c$ by Grouping **Step 1:** Factor out a greatest common factor, if there is one other than 1. **Step 2:** For the resulting trinomial $ax^2 + bx + c$, find two numbers whose product is $a \cdot c$ and whose sum is b. **Step 3:** Write the middle term, bx, using the factors found in Step 2. **Step 4:** Factor by grouping.	Factor: $3x^2 + 14x - 5$ **Step 1:** There is no common factor other than 1. **Step 2:** Find two numbers whose product is $3 \cdot (-5)$ or -15 and whose sum is 14. They are 15 and -1. **Step 3:** $3x^2 + 14x - 5$ $= 3x^2 + 15x - 1x - 5$ **Step 4:** $= 3x(x + 5) - 1(x + 5)$ $= (x + 5)(3x - 1)$
Section 4.5 Factoring Perfect Square Trinomials and the Difference of Two Squares	
A **perfect square trinomial** is a trinomial that is the square of some binomial.	**Perfect Square Trinomial = Square of Binomial** $x^2 + 4x + 4 = (x + 2)^2$ $25x^2 - 10x + 1 = (5x - 1)^2$
Factoring Perfect Square Trinomials $a^2 + 2ab + b^2 = (a + b)^2$ $a^2 - 2ab + b^2 = (a - b)^2$	Factor. $x^2 + 6x + 9 = x^2 + 2 \cdot x \cdot 3 + 3^2 = (x + 3)^2$ $4x^2 - 12x + 9 = (2x)^2 - 2 \cdot 2x \cdot 3 + 3^2$ $= (2x - 3)^2$
Difference of Two Squares $a^2 - b^2 = (a + b)(a - b)$	Factor. $x^2 - 9 = x^2 - 3^2 = (x + 3)(x - 3)$

Definitions and Concepts	Examples
Section 4.6 Solving Quadratic Equations by Factoring	
A **quadratic equation** is an equation that can be written in the form $ax^2 + bx + c = 0$ with a not 0. The form $ax^2 + bx + c = 0$ is called the **standard form** of a quadratic equation.	**Quadratic Equation** / **Standard Form** $x^2 = 16$ → $x^2 - 16 = 0$ $y = -2y^2 + 5$ → $2y^2 + y - 5 = 0$
Zero-Factor Property If a and b are real numbers and if $ab = 0$, then $a = 0$ or $b = 0$.	If $(x + 3)(x - 1) = 0$, then $x + 3 = 0$ or $x - 1 = 0$.
To Solve Quadratic Equations by Factoring **Step 1:** Write the equation in standard form so that one side of the equation is 0. **Step 2:** Factor completely. **Step 3:** Set each factor containing a variable equal to 0. **Step 4:** Solve the resulting equations. **Step 5:** Check solutions in the original equation.	Solve: $3x^2 = 13x - 4$ **Step 1:** $3x^2 - 13x + 4 = 0$ **Step 2:** $(3x - 1)(x - 4) = 0$ **Step 3:** $3x - 1 = 0$ or $x - 4 = 0$ **Step 4:** $3x = 1$ $\quad x = 4$ $x = \frac{1}{3}$ **Step 5:** Check both $\frac{1}{3}$ and 4 in the original equation.
Section 4.7 Quadratic Equations and Problem Solving	
Problem-Solving Steps	A garden is in the shape of a rectangle whose length is two feet more than its width. If the area of the garden is 35 square feet, find its dimensions.
1. UNDERSTAND the problem.	**1.** Read and reread the problem. Guess a solution and check your guess. Draw a diagram. Let x be the width of the rectangular garden. Then $x + 2$ is the length. 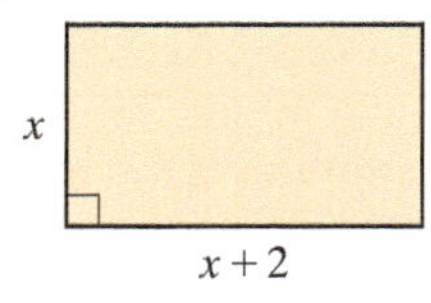
2. TRANSLATE.	**2.** length · width = area $(x + 2) \cdot x = 35$
3. SOLVE.	**3.** $(x + 2)x = 35$ $x^2 + 2x - 35 = 0$ $(x - 5)(x + 7) = 0$ $x - 5 = 0$ or $x + 7 = 0$ $x = 5$ $\quad x = -7$
4. INTERPRET.	**4.** Discard the solution $x = -7$ since x represents width. *Check:* If x is 5 feet, then $x + 2 = 5 + 2 = 7$ feet. The area of a rectangle whose width is 5 feet and whose length is 7 feet is (5 feet)(7 feet) or 35 square feet. *State:* The garden is 5 feet by 7 feet.

Chapter 4 Review

(4.1) *Factor out the GCF from each polynomial.*

1. $5m + 30$

2. $6x^2 - 15x$

3. $4x^5 + 2x - 10x^4$

4. $20x^3 + 12x^2 + 24x$

5. $3x(2x + 3) - 5(2x + 3)$

6. $5x(x + 1) - (x + 1)$

Factor each polynomial by grouping.

7. $3x^2 - 3x + 2x - 2$

8. $3a^2 + 9ab + 3b^2 + ab$

9. $10a^2 + 5ab + 7b^2 + 14ab$

10. $6x^2 + 10x - 3x - 5$

(4.2) *Factor each trinomial.*

11. $x^2 + 6x + 8$

12. $x^2 - 11x + 24$

13. $x^2 + x + 2$

14. $x^2 - 5x - 6$

15. $x^2 + 2x - 8$

16. $x^2 + 4xy - 12y^2$

17. $x^2 + 8xy + 15y^2$

18. $72 - 18x - 2x^2$

19. $32 + 12x - 4x^2$

20. $5y^3 - 50y^2 + 120y$

21. To factor $x^2 + 2x - 48$, think of two numbers whose product is _____ and whose sum is _____.

22. What is the first step in factoring $3x^2 + 15x + 30$?

(4.3) or (4.4) *Factor each trinomial.*

23. $2x^2 + 13x + 6$

24. $4x^2 + 4x - 3$

25. $6x^2 + 5xy - 4y^2$

26. $x^2 - x + 2$

27. $2x^2 - 23x - 39$

28. $18x^2 - 9xy - 20y^2$

29. $10y^3 + 25y^2 - 60y$

30. $60y^3 - 39y^2 + 6y$

Write the perimeter of each figure as a simplified polynomial. Then factor each polynomial completely.

△ **31.**

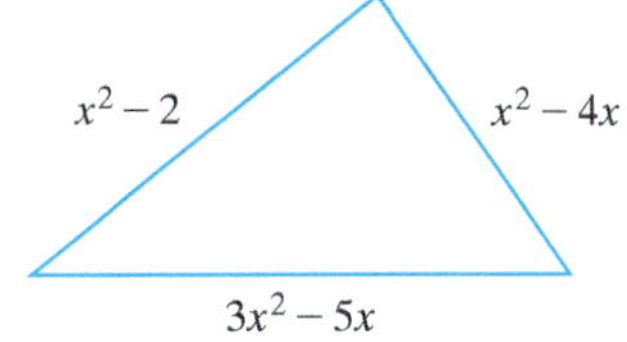

△ **32.**

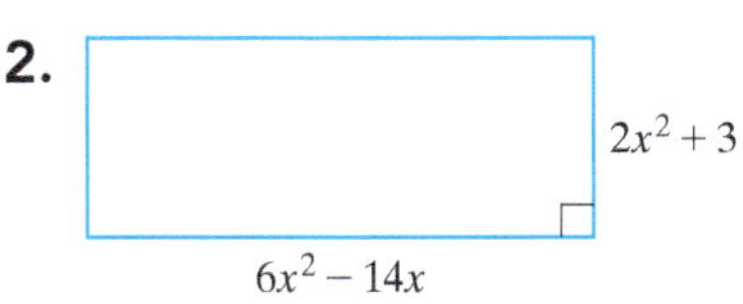

(4.5) *Determine whether each polynomial is a perfect square trinomial.*

33. $x^2 + 6x + 9$

34. $x^2 + 8x + 64$

35. $9m^2 - 12m + 16$

36. $4y^2 - 28y + 49$

Determine whether each binomial is a difference of two squares.

37. $x^2 - 9$

38. $x^2 + 16$

39. $4x^2 - 25y^2$

40. $9a^3 - 1$

Factor each polynomial completely.

41. $x^2 - 81$

42. $x^2 + 12x + 36$

43. $4x^2 - 9$

44. $9t^2 - 25s^2$

45. $16x^2 + y^2$

46. $n^2 - 18n + 81$

47. $3r^2 + 36r + 108$

48. $9y^2 - 42y + 49$

49. $5m^8 - 5m^6$

50. $4x^2 - 28xy + 49y^2$

51. $3x^2y + 6xy^2 + 3y^3$

52. $16x^4 - 1$

(4.6) *Solve each equation.*

53. $(x + 6)(x - 2) = 0$

54. $(x - 7)(x + 11) = 0$

55. $3x(x + 1)(7x - 2) = 0$

56. $4(5x + 1)(x + 3) = 0$

57. $x^2 + 8x + 7 = 0$

58. $x^2 - 2x - 24 = 0$

59. $x^2 + 10x = -25$

60. $x(x - 10) = -16$

61. $(3x - 1)(9x^2 - 6x + 1) = 0$

62. $56x^2 - 5x - 6 = 0$

63. $m^2 = 6m$

64. $r^2 = 25$

65. Write a quadratic equation in standard form that has the two solutions 4 and 5.

66. Write a quadratic equation in standard form that has two solutions, both -1.

(4.7) *Use the given information to choose the correct dimensions.*

△ **67.** The perimeter of a rectangle is 24 inches. The length is twice the width. Find the dimensions of the rectangle.

a. 5 inches by 7 inches **b.** 5 inches by 10 inches
c. 4 inches by 8 inches **d.** 2 inches by 10 inches

△ **68.** The area of a rectangle is 80 meters. The length is one more than three times the width. Find the dimensions of the rectangle.

a. 8 meters by 10 meters **b.** 4 meters by 13 meters
c. 4 meters by 20 meters **d.** 5 meters by 16 meters

Use the given information to find the dimensions of each figure.

69. The *area* of the square is 81 square units. Find the length of a side.

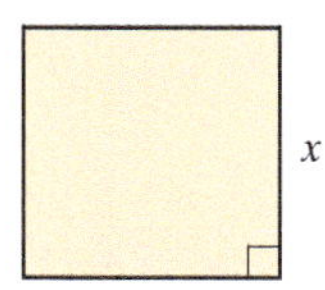

70. The *perimeter* of the quadrilateral is 47 units. Find the lengths of the sides.

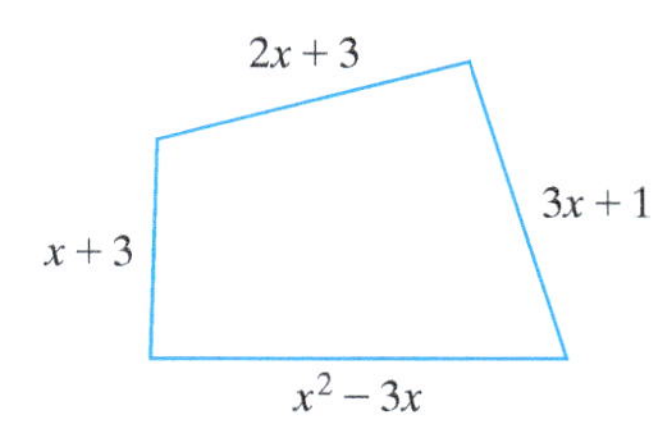

Solve.

71. A flag for a local organization is in the shape of a rectangle whose length is 15 inches less than twice its width. If the area of the flag is 500 square inches, find its dimensions.

72. The base of a triangular sail is four times its height. If the area of the triangle is 162 square yards, find the base.

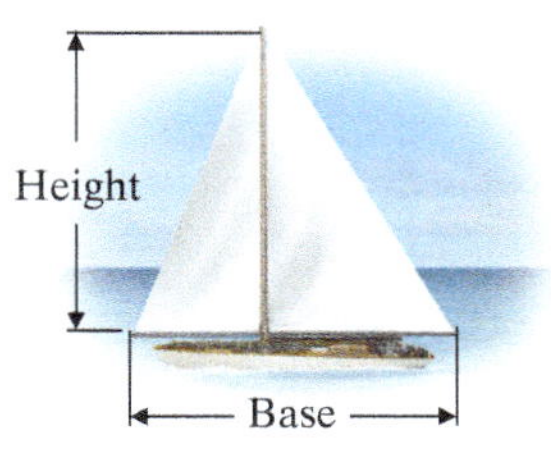

73. Find two consecutive positive integers whose product is 380.

74. Find two consecutive even positive integers whose product is 440.

75. A rocket is fired from the ground with an initial velocity of 440 feet per second. Its height h after t seconds is given by the equation $h = -16t^2 + 440t$.

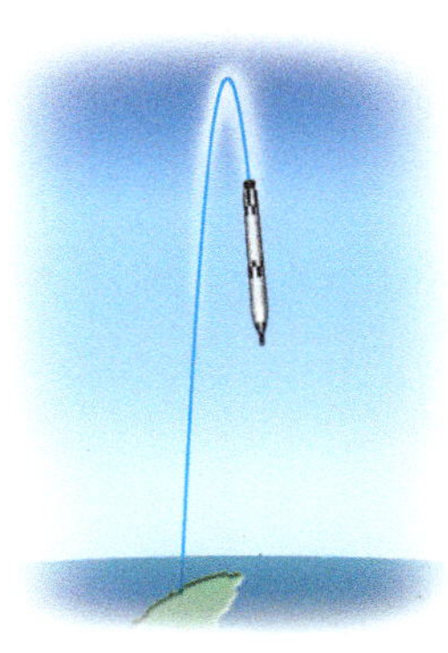

a. Find how many seconds pass before the rocket reaches a height of 2800 feet. Explain why two answers are obtained.

b. Find how many seconds pass before the rocket reaches the ground again.

76. An architect's squaring instrument is in the shape of a right triangle. Find the length of the longer leg of the right triangle if the hypotenuse is 8 centimeters longer than the longer leg and the shorter leg is 8 centimeters shorter than the longer leg.

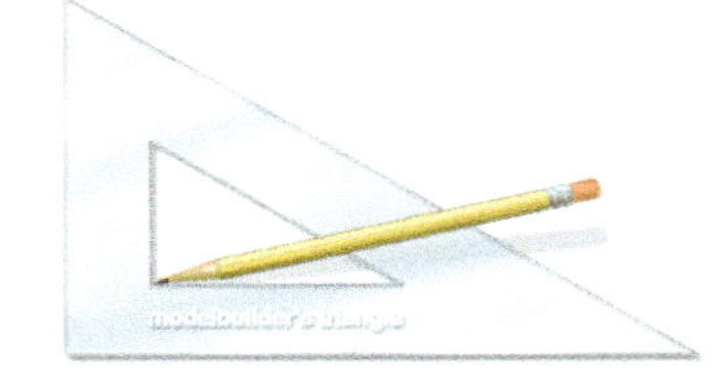

Mixed Review

Factor completely.

77. $6x + 24$

78. $7x - 63$

79. $11x(4x - 3) - 6(4x - 3)$

80. $2x(x - 5) - (x - 5)$

81. $3x^3 - 4x^2 + 6x - 8$

82. $xy + 2x - y - 2$

83. $2x^2 + 2x - 24$

84. $3x^3 - 30x^2 + 27x$

85. $4x^2 - 81$

86. $2x^2 - 18$

87. $16x^2 - 24x + 9$

88. $5x^2 + 20x + 20$

Solve.

89. $2x^2 - x - 28 = 0$

90. $x^2 - 2x = 15$

91. $2x(x + 7)(x + 4) = 0$

92. $x(x - 5) = -6$

93. $x^2 = 16x$

94. The perimeter of the following triangle is 48 inches. Find the lengths of its sides.

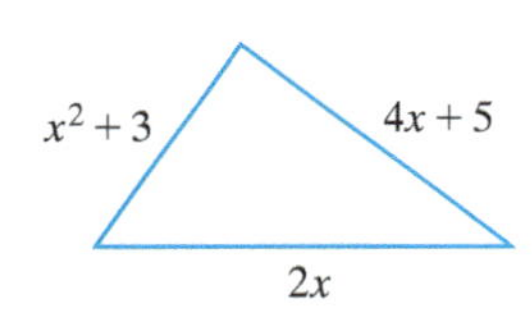

95. The width of a rectangle is 4 inches less than its length. Its area is 12 square inches. Find the dimensions of the rectangle.

Chapter 4 Getting Ready for the Test 

MULTIPLE CHOICE All the exercises below are **Multiple Choice.** Choose the correct letter. Also, letters may be used more than once.

1. The greatest common factor of the terms of $10x^4 - 70x^3 + 2x^2 - 14x$ is

A. $2x^2$ **B.** $2x$ **C.** $7x^2$ **D.** $7x$

2. Choose the expression that is NOT a factored form of $9y^3 - 18y^2$.

A. $9(y^3 - 2y^2)$ **B.** $9y(y^2 - 2y)$ **C.** $9y^2(y - 2)$ **D.** $9 \cdot y^3 - 18 \cdot y^2$

For Exercises 3 through 6, identify each expression as:

A. A factored expression or **B.** Not a factored expression

3. $(x - 1)(x + 5)$

4. $z(z + 12)(z - 12)$

5. $y(x - 6) + 1(x - 6)$

6. $m \cdot m - 5 \cdot 5$

For Exercises 7 through 9, choose the correct letter.

7. Choose the correct factored form for $4x^2 + 16$ or select "can't be factored."

A. can't be factored **B.** $4(x^2 + 4)$ **C.** $4(x + 2)^2$ **D.** $4(x + 2)(x - 2)$

8. Which of the binomials can't be factored using real numbers?

A. $x^2 + 64$ **B.** $x^2 - 64$

9. To solve $x(x + 2) = 15$, which is an incorrect next step?

A. $x^2 + 2x = 15$ **B.** $x(x + 2) - 15 = 0$ **C.** $x = 15$ or $x + 2 = 15$

Chapter 4 Test MyLab Math

For additional practice go to your study plan in MyLab Math.

Answers

1. ______
2. ______
3. ______
4. ______
5. ______
6. ______
7. ______
8. ______
9. ______
10. ______
11. ______
12. ______
13. ______
14. ______
15. ______
16. ______
17. ______
18. ______
19. ______
20. ______
21. ______
22. ______
23. ______

Factor each polynomial completely. If a polynomial cannot be factored, write "prime."

1. $9x^2 - 3x$

2. $x^2 + 11x + 28$

3. $49 - m^2$

4. $y^2 + 22y + 121$

5. $x^4 - 16$

6. $4(a + 3) - y(a + 3)$

7. $x^2 + 4$

8. $y^2 - 8y - 48$

9. $3a^2 + 3ab - 7a - 7b$

10. $3x^2 - 5x + 2$

11. $180 - 5x^2$

12. $3x^3 - 21x^2 + 30x$

13. $6t^2 - t - 5$

14. $xy^2 - 7y^2 - 4x + 28$

15. $x - x^5$

16. $x^2 + 14xy + 24y^2$

Solve each equation.

17. $(x - 3)(x + 9) = 0$

18. $x^2 + 5x = 14$

19. $x(x + 6) = 7$

20. $3x(2x - 3)(3x + 4) = 0$

21. $5t^3 - 45t = 0$

22. $t^2 - 2t - 15 = 0$

23. $6x^2 = 15x$

Solve.

24. A deck for a home is in the shape of a triangle. The length of the base of the triangle is 9 feet longer than its height. If the area of the triangle is 68 square feet, find the length of the base.

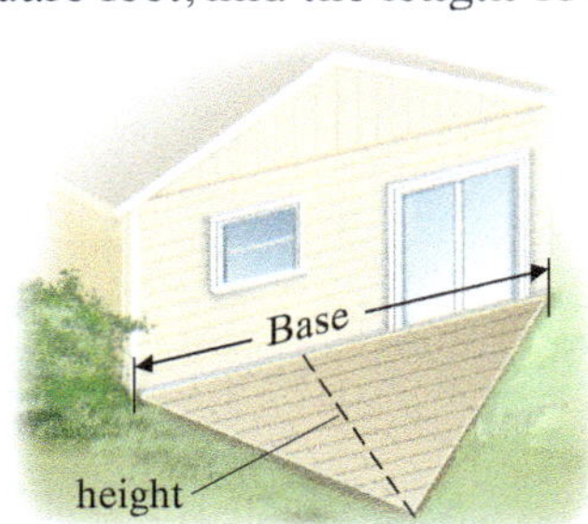

25. The *area* of the rectangle is 54 square units. Find the dimensions of the rectangle.

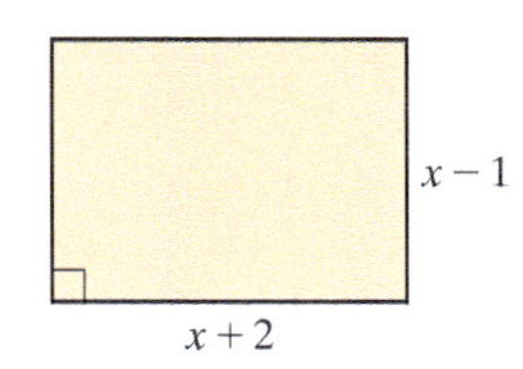

26. An object is dropped from the top of the Woolworth Building on Broadway in New York City. The height h of the object after t seconds is given by the equation

$h = -16t^2 + 784$

Find how many seconds pass before the object reaches the ground.

27. Find the lengths of the sides of a right triangle if the hypotenuse is 10 centimeters longer than the shorter leg and 5 centimeters longer than the longer leg.

28. A window washer is suspended 38 feet below the roof of the 1127-foot-tall John Hancock Center in Chicago. (*Source:* Council on Tall Buildings and Urban Habitat) If the window washer drops an object from this height, the object's height h after t seconds is given by the equation $h = -16t^2 + 1089$. Find how many seconds pass before the object reaches the ground.

24. ______

25. ______

26. ______

27. ______

28. ______

Chapters 1–4 Cumulative Review

Answers

1. a. ____
   b. ____
   c. ____
2. a. ____
   b. ____
3. ____
4. ____
5. ____
6. ____
7. a. ____
   b. ____
   c. ____
8. ____
9. ____
10. ____
11. ____
12. ____
13. ____
14. ____
15. ____
16. ____
17. ____
18. ____
19. ____
20. ____
21. ____
22. ____
23. ____
24. ____

1. Translate each sentence into a mathematical statement.

a. Nine is less than or equal to eleven.
b. Eight is greater than one.
c. Three is not equal to four.

2. Insert $<$ or $>$ in the space to make each statement true.

a. $|-5| \quad |-3|$
b. $|0| \quad |-2|$

3. Decide whether 2 is a solution of $3x + 10 = 8x$.

4. Evaluate $\frac{x}{y} + 5x$ if $x = 20$ and $y = 10$.

5. Subtract 8 from -4.

6. Evaluate $\frac{x}{y} + 5x$ if $x = -20$ and $y = 10$.

7. Evaluate each expression when $x = -2$ and $y = -4$.

a. $\frac{3x}{2y}$ **b.** $x^3 - y^2$ **c.** $\frac{x - y}{-x}$

8. Evaluate $\frac{x}{y} + 5x$ if $x = -20$ and $y = -10$.

Simplify each expression by combining like terms.

9. $2x + 3x + 5 + 2$

10. $5 - 2(3x - 7)$

11. $-5a - 3 + a + 2$

12. $5(x - 6) + 9(-2x + 1)$

13. $2.3x + 5x - 6$

Solve each equation.

14. $0.8y + 0.2(y - 1) = 1.8$

15. $-3x = 33$

16. $\frac{x}{-7} = -4$

17. $3(x - 4) = 3x - 12$

18. $-\frac{2}{3}x = -22$

19. Solve $V = lwh$ for l.

20. Solve for y: $3x + 2y = -7$

Simplify each expression.

21. $(5^3)^6$

22. $5^2 + 5^1$

23. $(y^8)^2$

24. $y^8 \cdot y^2$

Simplify the following expressions. Write each result using positive exponents only.

25. $\dfrac{(x^3)^4 x}{x^7}$

26. 3^{-2}

27. $(y^{-3}z^6)^{-6}$

28. $\dfrac{x^{-3}}{x^{-7}}$

29. $\dfrac{x^{-7}}{(x^4)^3}$

30. $\dfrac{(5a^7)^2}{a^5}$

Simplify each polynomial by combining any like terms.

31. $-3x + 7x$

32. $\dfrac{2}{3}x + 23 + \dfrac{1}{6}x - 100$

33. $11x^2 + 5 + 2x^2 - 7$

34. $0.2x - 1.1 + 2.3 - 0.7x$

35. Multiply: $(2x - y)^2$

36. Multiply: $(3x - 7y)^2$

Use a special product to square each binomial.

37. $(t + 2)^2$

38. $(x - 13)^2$

39. $(x^2 - 7y)^2$

40. $(7x + y)^2$

41. Divide: $\dfrac{8x^2y^2 - 16xy + 2x}{4xy}$

Factor each polynomial. For Exercise 47, factor by grouping.

42. $z^3 + 7z + z^2 + 7$

43. $5(x + 3) + y(x + 3)$

44. $2x^3 + 2x^2 - 84x$

45. $x^4 + 5x^2 + 6$

46. $-4x^2 - 23x + 6$

47. $6x^2 - 2x - 20$

48. $9xy^2 - 16x$

49. The platform for the cliff divers in Acapulco, Mexico, is about 144 feet above the sea. Neglecting air resistance, the height h in feet of a cliff diver above the ocean after t seconds is given by the quadratic equation $h = -16t^2 + 144$. Find how long it takes the diver to reach the ocean.

50. Solve $x^2 - 13x = -36$.

25. ____________
26. ____________
27. ____________
28. ____________
29. ____________
30. ____________
31. ____________
32. ____________
33. ____________
34. ____________
35. ____________
36. ____________
37. ____________
38. ____________
39. ____________
40. ____________
41. ____________
42. ____________
43. ____________
44. ____________
45. ____________
46. ____________
47. ____________
48. ____________
49. ____________
50. ____________

5 Rational Expressions

In this chapter, we expand our knowledge of algebraic expressions to include algebraic fractions, called *rational expressions*. We explore the operations of addition, subtraction, multiplication, and division using principles similar to the principles for numerical fractions.

Sections

5.1 Simplifying Rational Expressions

5.2 Multiplying and Dividing Rational Expressions

5.3 Adding and Subtracting Rational Expressions with the Same Denominator and Least Common Denominator

5.4 Adding and Subtracting Rational Expressions with Different Denominators

5.5 Solving Equations Containing Rational Expressions

Integrated Review—Summary on Rational Expressions

5.6 Proportions and Problem Solving with Rational Equations

5.7 Simplifying Complex Fractions

Check Your Progress

Vocabulary Check

Chapter Highlights

Chapter Review

Getting Ready for the Test

Chapter Test

Cumulative Review

Most Popular Sports in the World
(by estimated number of fans)

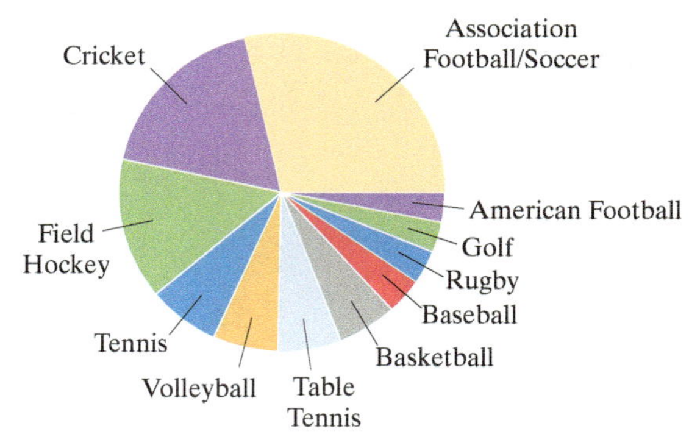

Source: worldatlas

What Do the Sports Above Have in Common?

All sports, from little league to professional, have numerous statistics kept on them. Although the final score of a match (or game) is a relatively easy statistic, determining the best player in a particular sport usually requires one or more agreed-upon formulas.

Let's focus on baseball and specifically what is called a player's slugging percentage. This is a popular measure of the hitting power of a player. Slugging percentage is different from batting average, for example, as it only deals with hits and not walks or being hit by pitches.

The bar graph below shows the all-time leaders in baseball slugging percentage. In Section 5.1, Exercises 75 through 82, we use the slugging percentage formula to calculate this particular statistic.

Baseball Slugging Percentage—Top Ten All-Time Leaders

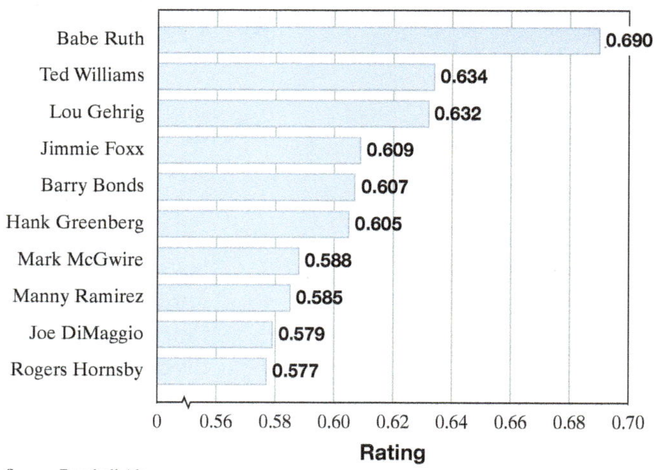

Source: Baseball Almanac

5.1 Simplifying Rational Expressions 

Objective A Evaluating Rational Expressions

A rational number is a number that can be written as a quotient of integers. A *rational expression* is also a quotient; it is a quotient of polynomials. Examples are

$$\frac{2}{3}, \quad \frac{3y^3}{8}, \quad \frac{-4p}{p^3 + 2p + 1}, \quad \text{and} \quad \frac{5x^2 - 3x + 2}{3x + 7}$$

Rational Expression

A **rational expression** is an expression that can be written in the form

$$\frac{P}{Q}$$

where P and Q are polynomials and $Q \neq 0$.

Rational expressions have different numerical values depending on what values replace the variables.

Example 1 Find the numerical value of $\frac{x + 4}{2x - 3}$ for each replacement value.

a. $x = 5$ **b.** $x = -2$

Solution:

a. We replace each x in the expression with 5 and then simplify.

$$\frac{x + 4}{2x - 3} = \frac{5 + 4}{2(5) - 3} = \frac{9}{10 - 3} = \frac{9}{7}$$

b. We replace each x in the expression with -2 and then simplify.

$$\frac{x + 4}{2x - 3} = \frac{-2 + 4}{2(-2) - 3} = \frac{2}{-7} \quad \text{or} \quad -\frac{2}{7}$$

Work Practice 1

In the example above, we wrote $\frac{2}{-7}$ as $-\frac{2}{7}$. For a negative fraction such as $\frac{2}{-7}$, recall from Section 1.6 that

$$\frac{2}{-7} = \frac{-2}{7} = -\frac{2}{7}$$

In general, for any fraction,

$$\frac{-a}{b} = \frac{a}{-b} = -\frac{a}{b}, \qquad b \neq 0$$

This is also true for rational expressions. For example,

$$\frac{-(x + 2)}{x} = \frac{x + 2}{-x} = -\frac{x + 2}{x}$$

Notice the parentheses.

Objectives

A Find the Value of a Rational Expression Given a Replacement Number.

B Identify Values for Which a Rational Expression Is Undefined.

C Simplify, or Write Rational Expressions in Lowest Terms.

D Write Equivalent Rational Expressions of the Forms $-\frac{a}{b} = \frac{-a}{b} = \frac{a}{-b}$.

Practice 1

Find the value of $\frac{x - 3}{5x + 1}$ for each replacement value.

a. $x = 4$

b. $x = -3$

Answers

1. a. $\frac{1}{21}$ **b.** $\frac{3}{7}$

Objective B Identifying When a Rational Expression Is Undefined

Do you recall why division by 0 is not defined? Remember, for example, that

$\frac{8}{4} = 2$ because $2 \cdot 4 = 8$.

Thus, if $\frac{8}{0} = $ *a number*,

then *the number* $\cdot\ 0 = 8$. There is no number that when multiplied by 0 equals 8; thus $\frac{8}{0}$ is undefined. This is true in general for fractions and rational expressions.

In the definition of rational expression (first "box" in this section), notice that we wrote $Q \neq 0$ for the denominator Q. The denominator of a rational expression must not equal 0 since division by 0 is not defined. (See the Helpful Hint to the left.) This means we must be careful when replacing the variable in a rational expression by a number. For example, suppose we replace x with 5 in the rational expression $\frac{3+x}{x-5}$. The expression becomes

$$\frac{3+x}{x-5} = \frac{3+5}{5-5} = \frac{8}{0}$$

But division by 0 is undefined. Therefore, in this expression we can allow x to be any real number *except* 5. **A rational expression is undefined for values that make the denominator 0.** Thus,

> To find values for which a rational expression is undefined, find values for which the denominator is 0.

Example 2

Are there any values for x for which each expression is undefined?

a. $\frac{x}{x-3}$ **b.** $\frac{x^2+2}{x^2-3x+2}$ **c.** $\frac{x^3-6x^2-10x}{3}$

Solution: To find values for which a rational expression is undefined, we find values that make the denominator 0.

a. The denominator of $\frac{x}{x-3}$ is 0 when $x - 3 = 0$ or when $x = 3$. Thus, when $x = 3$, the expression $\frac{x}{x-3}$ is undefined.

b. We set the denominator equal to 0.

$$x^2 - 3x + 2 = 0$$
$$(x-2)(x-1) = 0 \quad \text{Factor.}$$
$$x - 2 = 0 \quad \text{or} \quad x - 1 = 0 \quad \text{Set each factor equal to 0.}$$
$$x = 2 \qquad\qquad x = 1 \quad \text{Solve.}$$

Thus, when $x = 2$ or $x = 1$, the denominator $x^2 - 3x + 2$ is 0. So the rational expression $\frac{x^2+2}{x^2-3x+2}$ is undefined when $x = 2$ or when $x = 1$.

c. The denominator of $\frac{x^3-6x^2-10x}{3}$ is never 0, so there are no values of x for which this expression is undefined.

Work Practice 2

Practice 2

Are there any values for x for which each rational expression is undefined?

a. $\frac{x}{x+8}$

b. $\frac{x-3}{x^2+5x+4}$

c. $\frac{x^2-3x+2}{5}$

Note: For Sections 5.1 through 5.4, unless otherwise stated, we will now assume that variables in rational expressions are replaced only by values for which the expressions are defined.

Answers

2. a. $x = -8$ **b.** $x = -4, x = -1$ **c.** no

Objective C Simplifying Rational Expressions 

A fraction is said to be written in lowest terms or simplest form when the numerator and denominator have no common factors other than 1 (or -1). For example, the fraction $\frac{7}{10}$ is written in lowest terms since the numerator and denominator have no common factors other than 1 (or -1).

The process of writing a rational expression in lowest terms or simplest form is called **simplifying** the rational expression.

Simplifying a rational expression is similar to simplifying a fraction. Recall from Section R.2 that to simplify a fraction, we essentially "remove factors of 1." Our ability to do this comes from these facts:

- Any nonzero number over itself simplifies to 1 $\left(\frac{5}{5} = 1, \frac{-7.26}{-7.26} = 1, \text{ and } \frac{c}{c} = 1\right.$ as long as c is not $\left.0\right)$, and
- The product of any number and 1 is that number $\left(19 \cdot 1 = 19, -8.9 \cdot 1 = -8.9, \frac{a}{b} \cdot 1 = \frac{a}{b}\right)$.

In other words, we have the following:

$$\frac{a \cdot c}{b \cdot c} = \frac{a}{b} \cdot \frac{c}{c} = \frac{a}{b}$$

Since $\frac{a}{b} \cdot 1 = \frac{a}{b}$

Simplify: $\frac{15}{20}$

$$\frac{15}{20} = \frac{3 \cdot 5}{2 \cdot 2 \cdot 5} \quad \text{Factor the numerator and the denominator.}$$

$$= \frac{3 \cdot 5}{2 \cdot 2 \cdot 5} \quad \text{Look for common factors.}$$

$$= \frac{3}{2 \cdot 2} \cdot \frac{5}{5} \quad \text{Common factors in the numerator and denominator form factors of 1.}$$

$$= \frac{3}{2 \cdot 2} \cdot 1 \quad \text{Write } \frac{5}{5} \text{ as 1.}$$

$$= \frac{3}{2 \cdot 2} = \frac{3}{4} \quad \text{Multiply to remove a factor of 1.}$$

Before we use the same technique to simplify a rational expression, remember that as long as the denominator is not 0, $\frac{a^3b}{a^3b} = 1$, $\frac{x+3}{x+3} = 1$, and $\frac{7x^2 + 5x - 100}{7x^2 + 5x - 100} = 1$.

Simplify: $\frac{x^2 - 9}{x^2 + x - 6}$

$$\frac{x^2 - 9}{x^2 + x - 6} = \frac{(x-3)(x+3)}{(x-2)(x+3)} \quad \text{Factor the numerator and the denominator.}$$

$$= \frac{(x-3)(x+3)}{(x-2)(x+3)} \quad \text{Look for common factors.}$$

$$= \frac{x-3}{x-2} \cdot \frac{x+3}{x+3}$$

$$= \frac{x-3}{x-2} \cdot 1 \quad \text{Write } \frac{x+3}{x+3} \text{ as 1.}$$

$$= \frac{x-3}{x-2} \quad \text{Multiply to remove a factor of 1.}$$

Helpful Hint

We use the Fundamental Principle of Fractions to simplify rational expressions. This process is also sometimes called

- Dividing out common factors

or

- Removing a factor of 1

(See Section R.2 for a review.)

Just as for numerical fractions, we can use a shortcut notation. Remember that as long as exact factors in both the numerator and denominator are divided out, we are "removing a factor of 1." We will use the following notation to show this:

$$\frac{x^2 - 9}{x^2 + x - 6} = \frac{(x - 3)(x + 3)}{(x - 2)(x + 3)}$$ A factor of 1 is identified by the shading.

$$= \frac{x - 3}{x - 2}$$ Remove a factor of 1.

Thus, the rational expression $\frac{x^2 - 9}{x^2 + x - 6}$ has the same value as the rational expression $\frac{x - 3}{x - 2}$ for all values of x except 2 and -3. (Remember that when x is 2, the denominator of both rational expressions is 0 and that when x is -3, the original rational expression has a denominator of 0.)

As we simplify rational expressions, we will assume that the simplified rational expression is equal to the original rational expression for all real numbers except those for which either denominator is 0. The following steps may be used to simplify rational expressions.

To Simplify a Rational Expression

Step 1: Completely factor the numerator and denominator.

Step 2: Divide out factors common to the numerator and denominator. (This is the same as "removing a factor of 1.")

Practice 3

Simplify: $\frac{x^4 + x^3}{5x + 5}$

Example 3 Simplify: $\frac{5x - 5}{x^3 - x^2}$

Solution: To begin, we factor the numerator and denominator if possible. Then we look for common factors.

$$\frac{5x - 5}{x^3 - x^2} = \frac{5(x - 1)}{x^2(x - 1)} = \frac{5}{x^2}$$

Work Practice 3

Practice 4

Simplify: $\frac{x^2 + 11x + 18}{x^2 + x - 2}$

Example 4 Simplify: $\frac{x^2 + 8x + 7}{x^2 - 4x - 5}$

Solution: We factor the numerator and denominator and then look for common factors.

$$\frac{x^2 + 8x + 7}{x^2 - 4x - 5} = \frac{(x + 7)(x + 1)}{(x - 5)(x + 1)} = \frac{x + 7}{x - 5}$$

Work Practice 4

Practice 5

Simplify: $\frac{x^2 + 10x + 25}{x^2 + 5x}$

Example 5 Simplify: $\frac{x^2 + 4x + 4}{x^2 + 2x}$

Solution: We factor the numerator and denominator and then look for common factors.

$$\frac{x^2 + 4x + 4}{x^2 + 2x} = \frac{(x + 2)(x + 2)}{x(x + 2)} = \frac{x + 2}{x}$$

Work Practice 5

Answers

3. $\frac{x^3}{5}$ 4. $\frac{x + 9}{x - 1}$ 5. $\frac{x + 5}{x}$

Helpful Hint

When simplifying a rational expression, we look for **common *factors,* not common *terms.***

$$\frac{x \cdot (x+2)}{x \cdot x} = \frac{x+2}{x}$$

Common factors. These can be divided out.

$$\frac{x+2}{x}$$

Common terms. There is no factor of 1 that can be generated.

✓**Concept Check** Recall that we can remove only *factors* of 1. Which of the following are *not* true? Explain why.

a. $\frac{3-1}{3+5}$ simplifies to $-\frac{1}{5}$.

b. $\frac{2x+10}{2}$ simplifies to $x+5$.

c. $\frac{37}{72}$ simplifies to $\frac{3}{2}$.

d. $\frac{2x+3}{2}$ simplifies to $x+3$.

Example 6 Simplify: $\frac{x+9}{x^2-81}$

Solution: We factor and then apply the fundamental principle. Remember that this principle allows us to divide the numerator and denominator by all common factors.

$$\frac{x+9}{x^2-81} = \frac{x+9}{(x+9)(x-9)} = \frac{1}{x-9}$$

Work Practice 6

Practice 6

Simplify: $\frac{x+5}{x^2-25}$

Example 7 Simplify each rational expression.

a. $\frac{x+y}{y+x}$ **b.** $\frac{x-y}{y-x}$

Solution:

a. The expression $\frac{x+y}{y+x}$ can be simplified by using the commutative property of addition to rewrite the denominator $y+x$ as $x+y$.

$$\frac{x+y}{y+x} = \frac{x+y}{x+y} = 1$$

b. The expression $\frac{x-y}{y-x}$ can be simplified by recognizing that $y-x$ and $x-y$ are opposites. In other words, $y-x=-1(x-y)$. We proceed as follows:

$$\frac{x-y}{y-x} = \frac{1 \cdot (x-y)}{(-1)(x-y)} = \frac{1}{-1} = -1$$

Work Practice 7

Practice 7

Simplify each rational expression.

a. $\frac{x+4}{4+x}$ **b.** $\frac{x-4}{4-x}$

Answers

6. $\frac{1}{x-5}$ **7. a.** 1 **b.** -1

✓**Concept Check Answer**

a, c, d

Practice 8

Simplify: $\dfrac{2x^2 - 5x - 12}{16 - x^2}$

Example 8 Simplify: $\dfrac{4 - x^2}{3x^2 - 5x - 2}$

Solution:

$$\frac{4 - x^2}{3x^2 - 5x - 2} = \frac{(2 - x)(2 + x)}{(x - 2)(3x + 1)} \quad \text{Factor.}$$

$$= \frac{(-1)(x - 2)(2 + x)}{(x - 2)(3x + 1)} \quad \text{Write } 2 - x \text{ as } -1(x - 2).$$

$$= \frac{(-1)(2 + x)}{3x + 1} \text{ or } \frac{-2 - x}{3x + 1} \quad \text{Simplify.}$$

Work Practice 8

Objective D Writing Equivalent Forms of Rational Expressions

From Example 7(a), we have $y + x = x + y$. ($y + x$ and $x + y$ are equivalent.)

From Example 7(b), we have $y - x = -1(x - y)$. ($y - x$ and $x - y$ are opposites.)

Thus, $\dfrac{x + y}{y + x} = \dfrac{x + y}{x + y} = 1$ and $\dfrac{x - y}{y - x} = \dfrac{x - y}{-1(x - y)} = \dfrac{1}{-1} = -1$.

When performing operations on rational expressions, equivalent forms of answers often result. For this reason, it is very important to be able to recognize equivalent answers.

Practice 9

List 4 equivalent forms of $-\dfrac{3x + 7}{x - 6}$.

Remember, a negative sign in front of a fraction or rational expression may be moved to the numerator or the denominator, but *not* both.

Example 9 List some equivalent forms of $-\dfrac{5x - 1}{x + 9}$.

Solution: To do so, recall that $-\dfrac{a}{b} = \dfrac{-a}{b} = \dfrac{a}{-b}$. Thus

$$-\frac{5x - 1}{x + 9} = \frac{-(5x - 1)}{x + 9} = \frac{-5x + 1}{x + 9} \text{ or } \frac{1 - 5x}{x + 9}$$

Also,

$$-\frac{5x - 1}{x + 9} = \frac{5x - 1}{-(x + 9)} = \frac{5x - 1}{-x - 9} \text{ or } \frac{5x - 1}{-9 - x}$$

Thus $-\dfrac{5x - 1}{x + 9} = \dfrac{-(5x - 1)}{x + 9} = \dfrac{-5x + 1}{x + 9} = \dfrac{5x - 1}{-(x + 9)} = \dfrac{5x - 1}{-x - 9}$

Work Practice 9

Keep in mind that many rational expressions may look different but in fact are equivalent.

Answers

8. $-\dfrac{2x + 3}{x + 4}$ or $\dfrac{-2x - 3}{x + 4}$

9. $\dfrac{-(3x + 7)}{x - 6}$; $\dfrac{-3x - 7}{x - 6}$; $\dfrac{3x + 7}{-(x - 6)}$; $\dfrac{3x + 7}{-x + 6}$

Vocabulary, Readiness & Video Check

Use the choices below to fill in each blank. Not all choices will be used.

-1	0	simplifying	$\frac{-a}{-b}$	$\frac{-a}{b}$	$\frac{a}{-b}$
1	2	rational expression			

1. A ______________ is an expression that can be written in the form $\frac{P}{Q}$, where P and Q are polynomials and $Q \neq 0$.

2. The expression $\frac{x+3}{3+x}$ simplifies to ______________.

3. The expression $\frac{x-3}{3-x}$ simplifies to ______________.

4. A rational expression is undefined for values that make the denominator ______________.

5. The expression $\frac{7x}{x-2}$ is undefined for $x =$ ______________.

6. The process of writing a rational expression in lowest terms is called ______________.

7. For a rational expression, $-\frac{a}{b} =$ ______________ $=$ ______________.

Decide which rational expression(s) can be simplified. (Do not actually simplify.)

8. $\frac{x}{x+7}$ **9.** $\frac{3+x}{x+3}$ **10.** $\frac{5-x}{x-5}$ **11.** $\frac{x+2}{x+8}$

Martin-Gay Interactive Videos

See Video 5.1

Watch the section lecture video and answer the following questions.

Objective A **12.** From the lecture before Example 1, what do the different values of a rational expression depend on? How are these different values found?

Objective B **13.** Why can't the denominators of rational expressions be zero? How can we find the numbers for which a rational expression is undefined?

Objective C **14.** In Example 7, why isn't a factor of x divided out of the expression at the end?

Objective D **15.** From Example 9, if we move a negative sign from in front of a rational expression to either the numerator or the denominator, when would we need to use parentheses and why?

5.1 Exercise Set MyLab Math

Objective A *Find the value of the following expressions when $x = 2$, $y = -2$, and $z = -5$. See Example 1.*

1. $\frac{x+5}{x+2}$ **2.** $\frac{x+8}{x+1}$ **3.** $\frac{4z-1}{z-2}$ **4.** $\frac{7y-1}{y-1}$

5. $\frac{y^3}{y^2-1}$ **6.** $\frac{z}{z^2-5}$ **7.** $\frac{x^2+8x+2}{x^2-x-6}$ **8.** $\frac{x+5}{x^2+4x-8}$

Objective B *Find any numbers for which each rational expression is undefined. See Example 2.*

9. $\frac{7}{2x}$

10. $\frac{3}{5x}$

11. $\frac{x+3}{x+2}$

12. $\frac{5x+1}{x-9}$

13. $\frac{x-4}{2x-5}$

14. $\frac{x+1}{5x-2}$

15. $\frac{9x^3+4}{15x^2+30x}$

16. $\frac{19x^3+2}{x^2-x}$

17. $\frac{x^2-5x-2}{4}$

18. $\frac{9y^5+y^3}{9}$

19. $\frac{3x^2+9}{x^2-5x-6}$

20. $\frac{11x^2+1}{x^2-5x-14}$

21. $\frac{x}{3x^2+13x+14}$

22. $\frac{x}{2x^2+15x+27}$

Objective C *Simplify each expression. See Examples 3 through 8.*

23. $\frac{x+7}{7+x}$

24. $\frac{y+9}{9+y}$

25. $\frac{x-7}{7-x}$

26. $\frac{y-9}{9-y}$

27. $\frac{2}{8x+16}$

28. $\frac{3}{9x+18}$

29. $\frac{x-2}{x^2-4}$

30. $\frac{x-5}{x^2-25}$

31. $\frac{2x-10}{3x-30}$

32. $\frac{3x-9}{4x-16}$

33. $\frac{-5a-5b}{a+b}$

34. $\frac{-4x-4y}{x+y}$

35. $\frac{7x+35}{x^2+5x}$

36. $\frac{9x+99}{x^2+11x}$

37. $\frac{x+5}{x^2-4x-45}$

38. $\frac{x-3}{x^2-6x+9}$

39. $\frac{5x^2+11x+2}{x+2}$

40. $\frac{12x^2+4x-1}{2x+1}$

41. $\frac{x^3+7x^2}{x^2+5x-14}$

42. $\frac{x^4-10x^3}{x^2-17x+70}$

43. $\frac{14x^2-21x}{2x-3}$

44. $\frac{4x^2+24x}{x+6}$

45. $\frac{x^2+7x+10}{x^2-3x-10}$

46. $\frac{2x^2+7x-4}{x^2+3x-4}$

47. $\frac{3x^2+7x+2}{3x^2+13x+4}$

48. $\frac{4x^2-4x+1}{2x^2+9x-5}$

49. $\frac{2x^2-8}{4x-8}$

50. $\frac{5x^2 - 500}{35x + 350}$

51. $\frac{4 - x^2}{x - 2}$

52. $\frac{49 - y^2}{y - 7}$

53. $\frac{x^2 - 1}{x^2 - 2x + 1}$

54. $\frac{x^2 - 16}{x^2 - 8x + 16}$

Simplify each expression. Each exercise contains a four-term polynomial that should be factored by grouping. See Examples 3 through 8.

55. $\frac{x^2 + xy + 2x + 2y}{x + 2}$

56. $\frac{ab + ac + b^2 + bc}{b + c}$

57. $\frac{5x + 15 - xy - 3y}{2x + 6}$

58. $\frac{xy - 6x + 2y - 12}{y^2 - 6y}$

59. $\frac{2xy + 5x - 2y - 5}{3xy + 4x - 3y - 4}$

60. $\frac{2xy + 2x - 3y - 3}{2xy + 4x - 3y - 6}$

Objective D

Study Example 9. Then list four equivalent forms for each rational expression.

61. $-\frac{x - 10}{x + 8}$

62. $-\frac{x + 11}{x - 4}$

63. $-\frac{5y - 3}{y - 12}$

64. $-\frac{8y - 1}{y - 15}$

Objectives C D Mixed Practice

Simplify each expression. Then determine whether the given answer is correct. See Examples 3 through 9.

65. $\frac{9 - x^2}{x - 3}$; Answer: $-3 - x$?

66. $\frac{100 - x^2}{x - 10}$; Answer: $-10 - x$?

67. $\frac{7 - 34x - 5x^2}{25x^2 - 1}$; Answer: $\frac{x + 7}{-5x - 1}$?

68. $\frac{2 - 15x - 8x^2}{64x^2 - 1}$; Answer: $\frac{x + 2}{-8x - 1}$?

Review

Perform each indicated operation. See Section R.2.

69. $\frac{1}{3} \cdot \frac{9}{11}$

70. $\frac{5}{27} \cdot \frac{2}{5}$

71. $\frac{1}{3} \div \frac{1}{4}$

72. $\frac{7}{8} \div \frac{1}{2}$

73. $\frac{13}{20} \div \frac{2}{9}$

74. $\frac{8}{15} \div \frac{5}{8}$

Concept Extensions

A baseball player's slugging percentage S can be calculated with the following formula: $S = \dfrac{H + D + 2T + 3R}{B}$, *where* H = number of hits, D = number of doubles, T = number of triples, R = number of home runs, *and* B = number of times at bat. *Use this formula to complete the table below and then rank players by their slugging percentage in 2018. Round answers to 3 decimal places.*

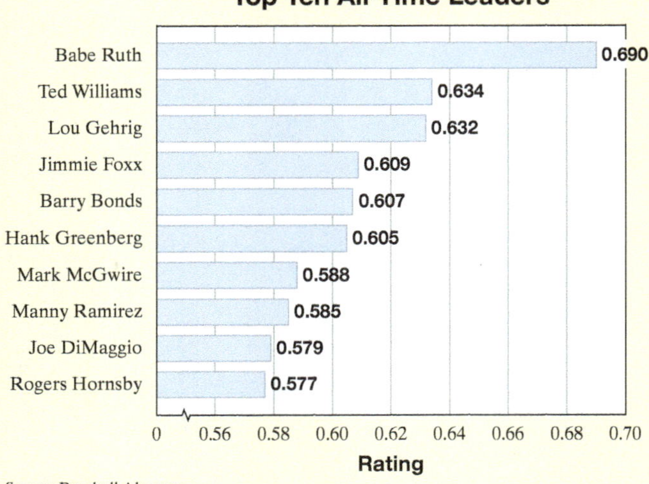

Source: Baseball Almanac

	Player Name (2018 Data)	B	H	D	T	HR	$S = \dfrac{H + D + 2T + 3R}{B}$
75.	Trevor Story (Colorado Rockies)	598	174	42	6	37	
76.	Mike Trout (Los Angeles Angels)	471	147	24	4	39	
77.	Mookie Betts (Boston Red Sox)	520	180	47	5	32	
78.	Nolan Arenado (Colorado Rockies)	590	175	38	2	38	
79.	J. D. Martinez (Boston Red Sox)	569	188	37	2	43	
80.	Christian Yelich (Milwaukee Brewers)	574	187	34	7	36	

81. Use your calculations above to name the player with the greatest slugging percentage.

82. Use your calculations above to name the player with the second-greatest slugging percentage.

Which of the following are incorrect and why? See the Concept Check in this section.

83. $\dfrac{5a - 15}{5}$ simplifies to $a - 3$?

84. $\dfrac{7m - 9}{7}$ simplifies to $m - 9$?

85. $\dfrac{1 + 2}{1 + 3}$ simplifies to $\dfrac{2}{3}$?

86. $\dfrac{46}{54}$ simplifies to $\dfrac{6}{5}$?

87. Explain how to write a fraction in lowest terms.

88. Explain how to write a rational expression in lowest terms.

89. Explain why the denominator of a fraction or a rational expression must not equal 0.

90. Does $\dfrac{(x - 3)(x + 3)}{x - 3}$ have the same value as $x + 3$ for all real numbers? Explain why or why not.

91. The average cost per DVD, in dollars, for a company to produce x DVDs on exercising is given by the formula $A = \frac{3x + 400}{x}$, where A is the average cost per DVD and x is the number of DVDs produced.

a. Find the cost for producing 1 DVD.

b. Find the average cost for producing 100 DVDs.

c. Does the cost per DVD decrease or increase when more DVDs are produced? Explain your answer.

92. Fax usage in large businesses is increasing. (See the bar graph.) For a certain model of fax machine, the manufacturing cost C per machine is given by the equation

$$C = \frac{250x + 10{,}000}{x}$$

where x is the number of fax machines manufactured and cost C is in dollars per machine.

a. Find the cost per fax machine when manufacturing 100 fax machines.

b. Find the cost per fax machine when manufacturing 1000 fax machines.

c. Does the cost per machine decrease or increase when more machines are manufactured? Explain why this is so.

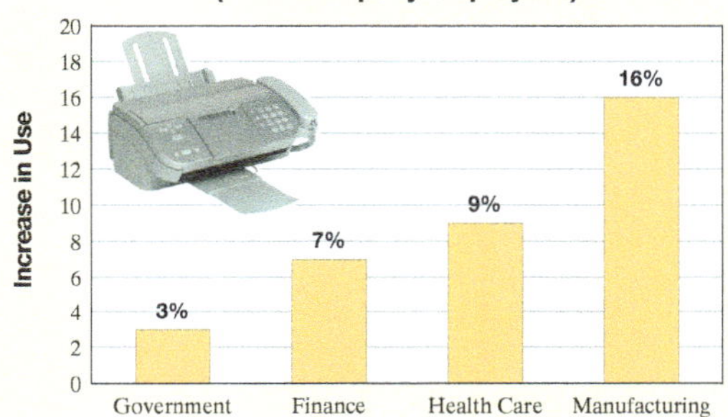

Source: Progressive Insurance

93. The dose of medicine prescribed for a child depends on the child's age A in years and the adult dose D for the medication. Young's Rule is a formula used by pediatricians that gives a child's dose C as

$$C = \frac{DA}{A + 12}$$

Suppose that an 8-year-old child needs medication, and the normal adult dose is 1000 mg. What size dose should the child receive?

94. Calculating body-mass index is a way to gauge whether a person should lose weight. Doctors recommend that body-mass index values fall between 19 and 25. The formula for body-mass index B is

$$B = \frac{705w}{h^2}$$

where w is weight in pounds and h is height in inches. Should a 148-pound person who is 5 feet 6 inches tall lose weight?

95. Anthropologists and forensic scientists use a measure called the cephalic index to help classify skulls. The cephalic index of a skull with width W and length L from front to back is given by the formula

$$C = \frac{100W}{L}$$

A long skull has an index value less than 75, a medium skull has an index value between 75 and 85, and a broad skull has an index value over 85. Find the cephalic index of a skull that is 5 inches wide and 6.4 inches long. Classify the skull.

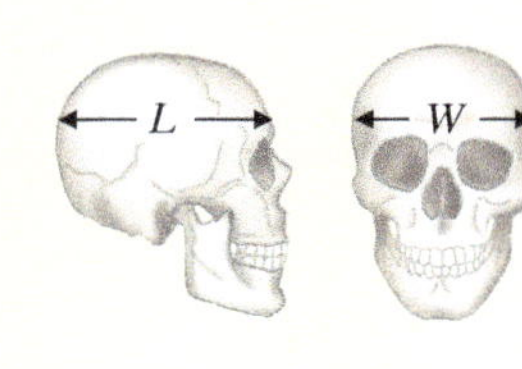

96. A company's gross profit margin P can be computed with the formula $P = \frac{R - C}{R}$, where $R =$ the company's revenue and $C =$ cost of goods sold. For the fiscal year 2017, Tesla, Inc. had revenues of \$11.76 billion and cost of goods sold \$9.5 billion. (*Source:* Tesla, Inc.) What was Tesla's gross profit margin in 2017? Express the answer as a percent, rounded to the nearest whole percent.

5.2 Multiplying and Dividing Rational Expressions

Objectives

A Multiply Rational Expressions.

B Divide Rational Expressions.

C Multiply and Divide Rational Expressions.

D Convert Between Units of Measure.

Objective A Multiplying Rational Expressions

Just as simplifying rational expressions is similar to simplifying number fractions, multiplying and dividing rational expressions is similar to multiplying and dividing number fractions.

Fractions	Rational Expressions
Multiply: $\frac{3}{5} \cdot \frac{10}{11}$	Multiply: $\frac{x-3}{x+5} \cdot \frac{2x+10}{x^2-9}$

Multiply numerators and then multiply denominators.

$\frac{3}{5} \cdot \frac{10}{11} = \frac{3 \cdot 10}{5 \cdot 11}$	$\frac{x-3}{x+5} \cdot \frac{2x+10}{x^2-9} = \frac{(x-3) \cdot (2x+10)}{(x+5) \cdot (x^2-9)}$

Simplify by factoring numerators and denominators.

$= \frac{3 \cdot 2 \cdot 5}{5 \cdot 11}$	$= \frac{(x-3) \cdot 2(x+5)}{(x+5)(x+3)(x-3)}$

Apply the fundamental principle.

$= \frac{3 \cdot 2}{11}$ or $\frac{6}{11}$	$= \frac{2}{x+3}$

Multiplying Rational Expressions

If $\frac{P}{Q}$ and $\frac{R}{S}$ are rational expressions, then

$$\frac{P}{Q} \cdot \frac{R}{S} = \frac{PR}{QS}$$

To multiply rational expressions, multiply the numerators and then multiply the denominators.

Note: Recall that for Sections 5.1 through 5.4, we assume variables in rational expressions have only those replacement values for which the expressions are defined.

Practice 1

Multiply.

a. $\frac{16y}{3} \cdot \frac{1}{x^2}$

b. $\frac{-5a^3}{3b^3} \cdot \frac{2b^2}{15a}$

Example 1

Multiply.

a. $\frac{25x}{2} \cdot \frac{1}{y^3}$ **b.** $\frac{-7x^2}{5y} \cdot \frac{3y^5}{14x^2}$

Solution: To multiply rational expressions, we multiply the numerators and then multiply the denominators of both expressions. Then we write the product in lowest terms.

a. $\frac{25x}{2} \cdot \frac{1}{y^3} = \frac{25x \cdot 1}{2 \cdot y^3} = \frac{25x}{2y^3}$

The expression $\frac{25x}{2y^3}$ is in lowest terms.

b. $\frac{-7x^2}{5y} \cdot \frac{3y^5}{14x^2} = \frac{-7x^2 \cdot 3y^5}{5y \cdot 14x^2}$ Multiply.

Answers

1. a. $\frac{16y}{3x^2}$ **b.** $-\frac{2a^2}{9b}$

The expression $\frac{-7x^2 \cdot 3y^5}{5y \cdot 14x^2}$ is not in lowest terms, so we factor the numerator and the denominator and apply the fundamental principle to "remove factors of 1."

$$= \frac{-1 \cdot 7 \cdot 3 \cdot x^2 \cdot y \cdot y^4}{5 \cdot 2 \cdot 7 \cdot x^2 \cdot y}$$ Common factors in the numerator and denominator form factors of 1.

$$= -\frac{3y^4}{10}$$ Divide out common factors. (This is the same as "removing a factor of 1.")

Work Practice 1

Helpful Hint It is the Fundamental Principle of Fractions that allows us to simplify.

When multiplying rational expressions, it is usually best to factor each numerator and denominator first. This will help us when we apply the fundamental principle to write the product in lowest terms.

Example 2 Multiply: $\frac{x^2 + x}{3x} \cdot \frac{6}{5x + 5}$

Solution:

$$\frac{x^2 + x}{3x} \cdot \frac{6}{5x + 5} = \frac{x(x + 1)}{3x} \cdot \frac{2 \cdot 3}{5(x + 1)}$$ Factor numerators and denominators.

$$= \frac{x(x + 1) \cdot 2 \cdot 3}{3x \cdot 5(x + 1)}$$ Multiply.

$$= \frac{2}{5}$$ Divide out common factors.

Work Practice 2

Practice 2

Multiply: $\frac{3x + 6}{14} \cdot \frac{7x^2}{x^3 + 2x^2}$

The following steps may be used to multiply rational expressions.

To Multiply Rational Expressions

Step 1: Completely factor numerators and denominators.

Step 2: Multiply numerators and multiply denominators.

Step 3: Simplify or write the product in lowest terms by dividing out common factors.

✓Concept Check Which of the following is a true statement?

a. $\frac{1}{3} \cdot \frac{1}{2} = \frac{1}{5}$ **b.** $\frac{2}{x} \cdot \frac{5}{x} = \frac{10}{x}$ **c.** $\frac{3}{x} \cdot \frac{1}{2} = \frac{3}{2x}$ **d.** $\frac{x}{7} \cdot \frac{x + 5}{4} = \frac{2x + 5}{28}$

Example 3 Multiply: $\frac{3x + 3}{5x^2 - 5x} \cdot \frac{2x^2 + x - 3}{4x^2 - 9}$

Solution:

$$\frac{3x + 3}{5x^2 - 5x} \cdot \frac{2x^2 + x - 3}{4x^2 - 9} = \frac{3(x + 1)}{5x(x - 1)} \cdot \frac{(2x + 3)(x - 1)}{(2x - 3)(2x + 3)}$$ Factor.

$$= \frac{3(x + 1)(2x + 3)(x - 1)}{5x(x - 1)(2x - 3)(2x + 3)}$$ Multiply.

$$= \frac{3(x + 1)}{5x(2x - 3)}$$ Simplify.

Work Practice 3

Practice 3

Multiply:

$\frac{4x + 8}{7x^2 - 14x} \cdot \frac{3x^2 - 5x - 2}{9x^2 - 1}$

Answers

2. $\frac{3}{2}$ **3.** $\frac{4(x + 2)}{7x(3x - 1)}$

✓Concept Check Answer

c

Objective B Dividing Rational Expressions 

We can divide by a rational expression in the same way we divide by a number fraction. Recall that to divide by a fraction, we multiply by its reciprocal.

For example, to divide $\frac{3}{2}$ by $\frac{7}{8}$, we multiply $\frac{3}{2}$ by $\frac{8}{7}$.

$$\frac{3}{2} \div \frac{7}{8} = \frac{3}{2} \cdot \frac{8}{7} = \frac{3 \cdot 4 \cdot 2}{2 \cdot 7} = \frac{12}{7}$$

Helpful Hint

Don't forget how to find reciprocals. The reciprocal of $\frac{a}{b}$ is $\frac{b}{a}$, $a \neq 0, b \neq 0$.

Dividing Rational Expressions

If $\frac{P}{Q}$ and $\frac{R}{S}$ are rational expressions and $\frac{R}{S}$ is not 0, then

$$\frac{P}{Q} \div \frac{R}{S} = \frac{P}{Q} \cdot \frac{S}{R} = \frac{PS}{QR}$$

To divide two rational expressions, multiply the first rational expression by the reciprocal of the second rational expression.

Practice 4

Divide: $\frac{7x^2}{6} \div \frac{x}{2y}$

Helpful Hint Remember, **to divide by a rational expression,** multiply by its reciprocal.

Example 4 Divide: $\frac{3x^3}{40} \div \frac{4x^3}{y^2}$

Solution:

$$\frac{3x^3}{40} \div \frac{4x^3}{y^2} = \frac{3x^3}{40} \cdot \frac{y^2}{4x^3} \quad \text{Multiply by the reciprocal of } \frac{4x^3}{y^2}.$$

$$= \frac{3x^3 \cdot y^2}{160x^3}$$

$$= \frac{3y^2}{160} \quad \text{Simplify.}$$

Work Practice 4

Practice 5

Divide: $\frac{(x-4)^2}{6} \div \frac{3x-12}{2}$

Example 5 Divide: $\frac{(x+2)^2}{10} \div \frac{2x+4}{5}$

Solution:

$$\frac{(x+2)^2}{10} \div \frac{2x+4}{5} = \frac{(x+2)^2}{10} \cdot \frac{5}{2x+4} \quad \text{Multiply by the reciprocal of } \frac{2x+4}{5}.$$

$$= \frac{(x+2)(x+2) \cdot 5}{5 \cdot 2 \cdot 2 \cdot (x+2)} \quad \text{Factor and multiply.}$$

$$= \frac{x+2}{4} \quad \text{Simplify.}$$

Work Practice 5

Answers

4. $\frac{7xy}{3}$ 5. $\frac{x-4}{9}$

Example 6 Divide: $\frac{6x + 2}{x^2 - 1} \div \frac{3x^2 + x}{x - 1}$

Solution:

$$\frac{6x + 2}{x^2 - 1} \div \frac{3x^2 + x}{x - 1} = \frac{6x + 2}{x^2 - 1} \cdot \frac{x - 1}{3x^2 + x}$$ Multiply by the reciprocal.

$$= \frac{2(3x + 1)(x - 1)}{(x + 1)(x - 1) \cdot x(3x + 1)}$$ Factor and multiply.

$$= \frac{2}{x(x + 1)}$$ Simplify.

Work Practice 6

Practice 6

Divide: $\frac{10x + 4}{x^2 - 4} \div \frac{5x^3 + 2x^2}{x + 2}$

Example 7 Divide: $\frac{2x^2 - 11x + 5}{5x - 25} \div \frac{4x - 2}{10}$

Solution:

$$\frac{2x^2 - 11x + 5}{5x - 25} \div \frac{4x - 2}{10} = \frac{2x^2 - 11x + 5}{5x - 25} \cdot \frac{10}{4x - 2}$$ Multiply by the reciprocal.

$$= \frac{(2x - 1)(x - 5) \cdot 2 \cdot 5}{5(x - 5) \cdot 2(2x - 1)}$$ Factor and multiply.

$$= \frac{1}{1} \text{ or } 1$$ Simplify.

Work Practice 7

Practice 7

Divide:

$\frac{3x^2 - 10x + 8}{7x - 14} \div \frac{9x - 12}{21}$

Objective C Multiplying and Dividing Rational Expressions

Let's make sure that we understand the difference between multiplying and dividing rational expressions.

Rational Expressions	
Multiplication	Multiply the numerators and multiply the denominators.
Division	Multiply by the reciprocal of the divisor.

Example 8 Multiply or divide as indicated.

a. $\frac{x - 4}{5} \cdot \frac{x}{x - 4}$

b. $\frac{x - 4}{5} \div \frac{x}{x - 4}$

c. $\frac{x^2 - 4}{2x + 6} \cdot \frac{x^2 + 4x + 3}{2 - x}$

Solution:

a. $\frac{x - 4}{5} \cdot \frac{x}{x - 4} = \frac{(x - 4) \cdot x}{5 \cdot (x - 4)} = \frac{x}{5}$

b. $\frac{x - 4}{5} \div \frac{x}{x - 4} = \frac{x - 4}{5} \cdot \frac{x - 4}{x} = \frac{(x - 4)^2}{5x}$

c. $\frac{x^2 - 4}{2x + 6} \cdot \frac{x^2 + 4x + 3}{2 - x} = \frac{(x - 2)(x + 2) \cdot (x + 1)(x + 3)}{2(x + 3) \cdot (2 - x)}$ Factor and multiply.

(*Continued on next page*)

Practice 8

Multiply or divide as indicated.

a. $\frac{x + 3}{x} \cdot \frac{7}{x + 3}$

b. $\frac{x + 3}{x} \div \frac{7}{x + 3}$

c. $\frac{3 - x}{x^2 + 6x + 5} \cdot \frac{2x + 10}{x^2 - 7x + 12}$

Answers

6. $\frac{2}{x^2(x - 2)}$ **7.** 1

8. a. $\frac{7}{x}$ **b.** $\frac{(x + 3)^2}{7x}$

c. $-\frac{2}{(x + 1)(x - 4)}$

Recall from Section 5.1 that $x - 2$ and $2 - x$ are opposites. This means that $\frac{x-2}{2-x} = -1$. Thus,

$$\frac{(x-2)(x+2)\cdot(x+1)(x+3)}{2(x+3)\cdot(2-x)} = \frac{-1(x+2)(x+1)}{2}$$

$$= -\frac{(x+2)(x+1)}{2}$$

Work Practice 8

Objective D Converting Between Units of Measure

How many square inches are in 1 square foot?
How many cubic feet are in a cubic yard?

If you have trouble answering these questions, this section will be helpful to you.

Now that we know how to multiply fractions and rational expressions, we can use this knowledge to help us convert between units of measure. To do so, we will use **unit fractions.** A unit fraction is a fraction that equals 1. For example, since 12 in. = 1 ft, we have the unit fractions

$$\frac{12 \text{ in.}}{1 \text{ ft}} = 1 \quad \text{and} \quad \frac{1 \text{ ft}}{12 \text{ in.}} = 1$$

Practice 9

288 square inches = _____ square feet

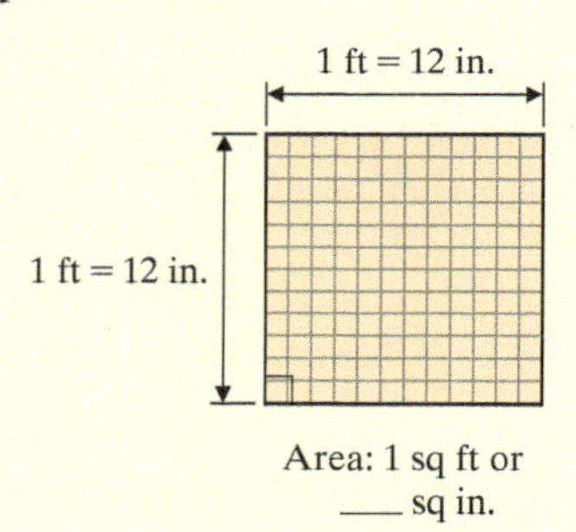

Example 9 18 square feet = _____ square yards

Solution: Let's multiply 18 square feet by a unit fraction that has square feet in the denominator and square yards in the numerator. From the diagram, you can see that

1 square yard = 9 square feet

Thus,

$$18 \text{ sq ft} = \frac{18 \text{ sq ft}}{1}\cdot 1 = \frac{\overset{2}{\cancel{18}}\ \cancel{\text{sq ft}}}{1}\cdot\frac{1 \text{ sq yd}}{\underset{1}{\cancel{9}}\ \cancel{\text{sq ft}}}$$

$$= \frac{2\cdot 1}{1\cdot 1}\text{ sq yd} = 2 \text{ sq yd}$$

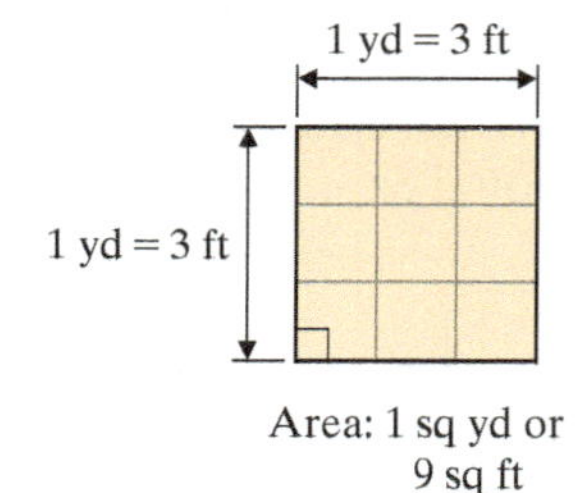

Thus, 18 sq ft = 2 sq yd.

Draw a diagram of 18 sq ft to help you see that this is reasonable.

Work Practice 9

Practice 10

3.5 square feet = _____ square inches

Example 10 5.2 square yards = _____ square feet

Solution:

$$5.2 \text{ sq yd} = \frac{5.2 \text{ sq yd}}{1}\cdot 1 = \frac{5.2\ \cancel{\text{sq yd}}}{1}\cdot\frac{9 \text{ sq ft}}{1\ \cancel{\text{sq yd}}} \begin{array}{l}\leftarrow \text{Units converting to}\\ \leftarrow \text{Units given}\end{array}$$

$$= \frac{5.2\cdot 9}{1\cdot 1}\text{ sq ft}$$

$$= 46.8 \text{ sq ft}$$

Thus, 5.2 sq yd = 46.8 sq ft.

Draw a diagram to see that this is reasonable.

Work Practice 10

Answers
9. 2 sq ft **10.** 504 sq in.

Example 11 Converting from Cubic Feet to Cubic Yards

The largest building in the world by volume is still The Boeing Company's Everett, Washington, factory complex, where Boeing's wide-body jetliners, the 747, 767, and 777, are built. The volume of this factory complex is 472,370,319 cubic feet. Find the volume of this Boeing facility in cubic yards. (*Source:* The Boeing Company)

Solution: There are 27 cubic feet in 1 cubic yard. (See the diagram.)

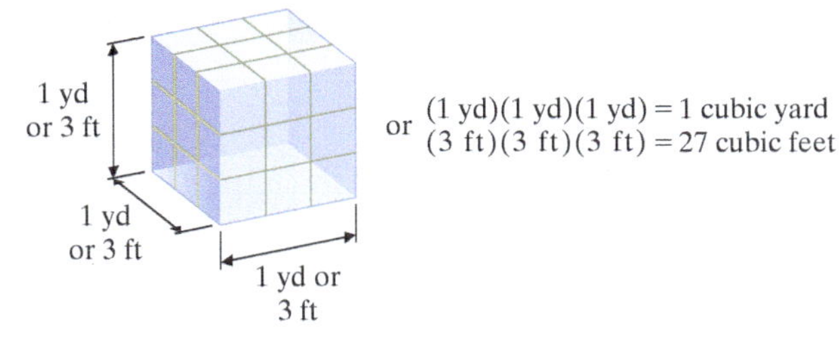

$$472{,}370{,}319 \text{ cu ft} = 472{,}370{,}319 \text{ cu ft} \cdot \frac{1 \text{ cu yd}}{27 \text{ cu ft}}$$

$$= \frac{472{,}370{,}319}{27} \text{ cu yd}$$

$$= 17{,}495{,}197 \text{ cu yd}$$

Work Practice 11

Practice 11

The largest casino in the world is the Venetian, in Macau, on the southern tip of China. The gaming area for this casino is approximately 61,000 *square yards*. Find the size of the gaming area in *square feet*. (*Source: USA Today*)

Helpful Hint

When converting between units of measurement, if possible, write the unit fraction so that **the numerator contains the units you are converting to** and **the denominator contains the original units.**

$$48 \text{ in.} = \frac{48 \text{ in.}}{1} \cdot \underbrace{\frac{1 \text{ ft}}{12 \text{ in.}}}_{\text{Unit fraction}} \begin{array}{l} \leftarrow \text{Units converting to} \\ \leftarrow \text{Original units} \end{array}$$

$$= \frac{48}{12} \text{ ft} = 4 \text{ ft}$$

Example 12

At the 2016 Summer Olympics, Jamaican athlete Usain Bolt won the gold medal in the men's 100-meter track event. He ran the distance at an average speed of 33.4 feet per second. Convert this speed to miles per hour. (*Source:* International Olympic Committee)

Solution: Recall that 1 mile = 5280 feet and 1 hour = 3600 seconds (60 · 60).

$$33.4 \text{ feet/second} = \frac{33.4 \text{ feet}}{1 \text{ second}} \cdot \underbrace{\frac{3600 \text{ seconds}}{1 \text{ hour}} \cdot \frac{1 \text{ mile}}{5280 \text{ feet}}}_{\text{Unit fractions}}$$

$$= \frac{33.4 \cdot 3600}{5280} \text{ miles/hour}$$

$$= 22.77 \text{ miles/hour}$$

Work Practice 12

Practice 12

The cheetah is the fastest land animal, being clocked at about 102.7 feet per second. Convert this to miles per hour. Round to the nearest tenth. (Source: *World Almanac and Book of Facts*)

Answers

11. 549,000 sq ft
12. 70.0 miles per hour

Vocabulary, Readiness & Video Check

Use the choices below to fill in each blank. Not all choices will be used.

opposites $\frac{a \cdot d}{b \cdot c}$ $\frac{a \cdot c}{b \cdot d}$ $\frac{x}{42}$ $\frac{x^2}{42}$ $\frac{2x}{42}$ $\frac{6}{7}$ $\frac{7}{6}$

reciprocals

1. The expressions $\frac{x}{2y}$ and $\frac{2y}{x}$ are called ________.

2. $\frac{a}{b} \cdot \frac{c}{d} =$ ________

3. $\frac{a}{b} \div \frac{c}{d} =$ ________

4. $\frac{x}{7} \cdot \frac{x}{6} =$ ________

5. $\frac{x}{7} \div \frac{x}{6} =$ ________

Martin-Gay Interactive Videos

See Video 5.2 

Watch the section lecture video and answer the following questions.

Objective A **6.** Would you say a person needs to be quite comfortable with factoring polynomials in order to be successful with multiplying rational expressions? Explain, referencing Example 2 in your answer.

Objective B **7.** Based on the lecture before Example 3, complete the following statements: Dividing rational expressions is exactly like dividing ________. Therefore, to divide by a rational expression, multiply by its ________.

Objective C **8.** In Examples 4 and 5, determining the operation is the first step in deciding how to perform the operation. Why is this so?

Objective D **9.** In Example 6, why is the unit fraction $\frac{27 \text{ cu ft}}{1 \text{ cu yd}}$ used?

5.2 Exercise Set MyLab Math

Objective A *Find each product and simplify if possible. See Examples 1 through 3.*

1. $\frac{3x}{y^2} \cdot \frac{7y}{4x}$

2. $\frac{9x^2}{y} \cdot \frac{4y}{3x^3}$

3. $\frac{8x}{2} \cdot \frac{x^5}{4x^2}$

4. $\frac{6x^2}{10x^3} \cdot \frac{5x}{12}$

5. $-\frac{5a^2b}{30a^2b^2} \cdot b^3$

6. $-\frac{9x^3y^2}{18xy^5} \cdot y^3$

7. $\frac{x}{2x - 14} \cdot \frac{x^2 - 7x}{5}$

8. $\frac{4x - 24}{20x} \cdot \frac{5}{x - 6}$

9. $\frac{6x + 6}{5} \cdot \frac{10}{36x + 36}$

10. $\frac{x^2 + x}{8} \cdot \frac{16}{x + 1}$

11. $\frac{(m + n)^2}{m - n} \cdot \frac{m}{m^2 + mn}$

12. $\frac{(m - n)^2}{m + n} \cdot \frac{m}{m^2 - mn}$

13. $\frac{x^2 - 25}{x^2 - 3x - 10} \cdot \frac{x + 2}{x}$

14. $\frac{a^2 - 4a + 4}{a^2 - 4} \cdot \frac{a + 3}{a - 2}$

15. $\frac{x^2 + 6x + 8}{x^2 + x - 20} \cdot \frac{x^2 + 2x - 15}{x^2 + 8x + 16}$

16. $\frac{x^2 + 9x + 20}{x^2 - 15x + 44} \cdot \frac{x^2 - 11x + 28}{x^2 + 12x + 35}$

Objective B *Find each quotient and simplify. See Examples 4 through 7.*

17. $\frac{5x^7}{2x^5} \div \frac{15x}{4x^3}$

18. $\frac{9y^4}{6y} \div \frac{y^2}{3}$

19. $\frac{8x^2}{y^3} \div \frac{4x^2y^3}{6}$

20. $\frac{7a^2b}{3ab^2} \div \frac{21a^2b^2}{14ab}$

21. $\frac{(x - 6)(x + 4)}{4x} \div \frac{2x - 12}{8x^2}$

22. $\frac{(x + 3)^2}{5} \div \frac{5x + 15}{25}$

23. $\frac{3x^2}{x^2 - 1} \div \frac{x^5}{(x + 1)^2}$

24. $\frac{9x^5}{a^2 - b^2} \div \frac{27x^2}{3b - 3a}$

25. $\frac{m^2 - n^2}{m + n} \div \frac{m}{m^2 + nm}$

26. $\frac{(m - n)^2}{m + n} \div \frac{m^2 - mn}{m}$

27. $\frac{x + 2}{7 - x} \div \frac{x^2 - 5x + 6}{x^2 - 9x + 14}$

28. $\frac{x - 3}{2 - x} \div \frac{x^2 + 3x - 18}{x^2 + 2x - 8}$

29. $\frac{x^2 + 7x + 10}{x - 1} \div \frac{x^2 + 2x - 15}{x - 1}$

30. $\frac{x + 1}{2x^2 + 5x + 3} \div \frac{20x + 100}{2x + 3}$

Objective C Mixed Practice *Multiply or divide as indicated. See Example 8.*

31. $\frac{5x - 10}{12} \div \frac{4x - 8}{8}$

32. $\frac{6x + 6}{5} \div \frac{9x + 9}{10}$

33. $\frac{x^2 + 5x}{8} \cdot \frac{9}{3x + 15}$

34. $\frac{3x^2 + 12x}{6} \cdot \frac{9}{2x + 8}$

35. $\frac{7}{6p^2 + q} \div \frac{14}{18p^2 + 3q}$

36. $\frac{3x + 6}{20} \div \frac{4x + 8}{8}$

37. $\frac{3x + 4y}{x^2 + 4xy + 4y^2} \cdot \frac{x + 2y}{2}$

38. $\frac{x^2 - y^2}{3x^2 + 3xy} \cdot \frac{3x^2 + 6x}{3x^2 - 2xy - y^2}$

39. $\frac{(x + 2)^2}{x - 2} \div \frac{x^2 - 4}{2x - 4}$

40. $\frac{x + 3}{x^2 - 9} \div \frac{5x + 15}{(x - 3)^2}$

41. $\frac{x^2 - 4}{24x} \div \frac{2 - x}{6xy}$

42. $\frac{3y}{3 - x} \div \frac{12xy}{x^2 - 9}$

43. $\frac{a^2 + 7a + 12}{a^2 + 5a + 6} \cdot \frac{a^2 + 8a + 15}{a^2 + 5a + 4}$

44. $\frac{b^2 + 2b - 3}{b^2 + b - 2} \cdot \frac{b^2 - 4}{b^2 + 6b + 8}$

45. $\frac{5x - 20}{3x^2 + x} \cdot \frac{3x^2 + 13x + 4}{x^2 - 16}$

46. $\frac{9x + 18}{4x^2 - 3x} \cdot \frac{4x^2 - 11x + 6}{x^2 - 4}$

47. $\frac{8n^2 - 18}{2n^2 - 5n + 3} \div \frac{6n^2 + 7n - 3}{n^2 - 9n + 8}$

48. $\frac{36n^2 - 64}{3n^2 - 10n + 8} \div \frac{3n^2 - 5n - 12}{n^2 - 9n + 14}$

Objective D *Convert as indicated. See Examples 9 through 12.*

49. 10 square feet = ________ square inches.

50. 1008 square inches = _____ square feet.

51. 45 square feet = _____ square yards.

52. 2 square yards = ________ square inches.

53. 3 cubic yards = ______ cubic feet.

54. 2 cubic yards = _________ cubic inches.

55. 50 miles per hour = ______ feet per second (round to the nearest whole).

56. 10 feet per second = ______ miles per hour (round to the nearest tenth).

57. 6.3 square yards = _______ square feet.

58. 3.6 square yards = _______ square feet.

59. In January 2010, the Burj Khalifa Tower officially became the tallest building in the world. This tower has a curtain wall (the exterior skin of the building) that is approximately 133,500 square yards. Convert this to square feet. (*Source:* Burj Khalifa)

60. The Pentagon, headquarters for the Department of Defense, contains 3,705,793 square feet of office and storage space. Convert this to square yards. Round to the nearest square yard. (*Source:* U.S. Department of Defense)

61. On July 24, 2014, the Sunswift eVe solar-powered car set a new solar-powered-car land speed record of 91.1 feet per second. This car was built by a student team at the University of New South Wales, Australia. Convert this speed to miles per hour. Round to the nearest tenth. (*Source:* University of New South Wales)

62. Peregrine falcons are among the fastest birds in the world. When engaged in a high-speed dive for prey, a peregrine falcon can reach speeds over 200 miles per hour. Find this speed in feet per second. Round to the nearest tenth. (*Source:* Ohio Department of Natural Resources)

Review

Perform each indicated operation. See Section R.2.

63. $\frac{1}{5} + \frac{4}{5}$

64. $\frac{3}{15} + \frac{6}{15}$

65. $\frac{9}{9} - \frac{19}{9}$

66. $\frac{4}{3} - \frac{8}{3}$

67. $\frac{6}{5} + \left(\frac{1}{5} - \frac{8}{5}\right)$

68. $-\frac{3}{2} + \left(\frac{1}{2} - \frac{3}{2}\right)$

Concept Extensions

Identify each statement as true or false. If false, correct the multiplication. See the Concept Check in this section.

69. $\frac{4}{a} \cdot \frac{1}{b} = \frac{4}{ab}$

70. $\frac{2}{3} \cdot \frac{2}{4} = \frac{2}{7}$

71. $\frac{x}{5} \cdot \frac{x+3}{4} = \frac{2x+3}{20}$

72. $\frac{7}{a} \cdot \frac{3}{a} = \frac{21}{a}$

△ **73.** Find the area of the rectangle.

$\frac{2x}{x^2-25}$ feet

$\frac{x+5}{9x}$ feet

△ **74.** Find the area of the square.

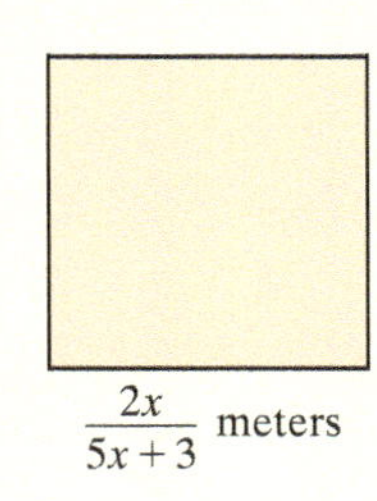

$\frac{2x}{5x+3}$ meters

Multiply or divide as indicated.

75. $\left(\frac{x^2 - y^2}{x^2 + y^2} \div \frac{x^2 - y^2}{3x}\right) \cdot \frac{x^2 + y^2}{6}$

76. $\left(\frac{x^2 - 9}{x^2 - 1} \cdot \frac{x^2 + 2x + 1}{2x^2 + 9x + 9}\right) \div \frac{2x + 3}{1 - x}$

77. $\left(\frac{2a + b}{b^2} \cdot \frac{3a^2 - 2ab}{ab + 2b^2}\right) \div \frac{a^2 - 3ab + 2b^2}{5ab - 10b^2}$

78. $\left(\frac{x^2y^2 - xy}{4x - 4y} \div \frac{3y - 3x}{8x - 8y}\right) \cdot \frac{y - x}{8}$

79. In your own words, explain how you multiply rational expressions.

80. Explain how dividing rational expressions is similar to dividing rational numbers.

81. On a day in January 2018, 1 euro was equivalent to 1.237 American dollars. If you had wanted to exchange \$2000 U.S. for euros on that day for a European vacation, how many would you have received? Round to the nearest hundredth. (*Source:* Barclay's Bank)

82. An environmental technician finds that warm water from an industrial process is being discharged into a nearby pond at a rate of 30 gallons per minute. Plant regulations state that the flow rate should be no more than 0.1 cubic foot per second. Is the flow rate of 30 gallons per minute in violation of the plant regulations? (*Hint:* 1 cubic foot is equivalent to 7.48 gallons.)

5.3 Adding and Subtracting Rational Expressions with the Same Denominator and Least Common Denominator

Objectives

A Add and Subtract Rational Expressions with Common Denominators.

B Find the Least Common Denominator of a List of Rational Expressions.

C Write a Rational Expression as an Equivalent Expression Whose Denominator Is Given.

Objective A Adding and Subtracting Rational Expressions with the Same Denominator

Like multiplication and division, addition and subtraction of rational expressions are similar to addition and subtraction of rational numbers. In this section, we add and subtract rational expressions with a common denominator.

Add: $\frac{6}{5} + \frac{2}{5}$	Add: $\frac{9}{x+2} + \frac{3}{x+2}$

Add the numerators and place the sum over the common denominator.

$\frac{6}{5} + \frac{2}{5} = \frac{6+2}{5}$	$\frac{9}{x+2} + \frac{3}{x+2} = \frac{9+3}{x+2}$
$= \frac{8}{5}$ Simplify.	$= \frac{12}{x+2}$ Simplify.

Adding and Subtracting Rational Expressions with Common Denominators

If $\frac{P}{R}$ and $\frac{Q}{R}$ are rational expressions, then

$$\frac{P}{R} + \frac{Q}{R} = \frac{P+Q}{R} \quad \text{and} \quad \frac{P}{R} - \frac{Q}{R} = \frac{P-Q}{R}$$

To add or subtract rational expressions, add or subtract numerators and place the sum or difference over the common denominator.

Practice 1

Add: $\frac{8x}{3y} + \frac{x}{3y}$

Example 1 Add: $\frac{5m}{2n} + \frac{m}{2n}$

Solution:

$\frac{5m}{2n} + \frac{m}{2n} = \frac{5m+m}{2n}$ Add the numerators.

$= \frac{6m}{2n}$ Simplify the numerator by combining like terms.

$= \frac{3m}{n}$ Simplify by applying the fundamental principle.

Work Practice 1

Practice 2

Subtract: $\frac{3x}{3x-7} - \frac{7}{3x-7}$

Example 2 Subtract: $\frac{2y}{2y-7} - \frac{7}{2y-7}$

Solution:

$\frac{2y}{2y-7} - \frac{7}{2y-7} = \frac{2y-7}{2y-7}$ Subtract the numerators.

$= \frac{1}{1}$ or 1 Simplify.

Work Practice 2

Answers

1. $\frac{3x}{y}$ 2. 1

Example 3 Subtract: $\frac{3x^2 + 2x}{x - 1} - \frac{10x - 5}{x - 1}$

Solution:

$$\frac{3x^2 + 2x}{x - 1} - \frac{10x - 5}{x - 1} = \frac{3x^2 + 2x - (10x - 5)}{x - 1}$$ Subtract the numerators. Notice the parentheses.

$$= \frac{3x^2 + 2x - 10x + 5}{x - 1}$$ Use the distributive property.

$$= \frac{3x^2 - 8x + 5}{x - 1}$$ Combine like terms.

$$= \frac{(x - 1)(3x - 5)}{x - 1}$$ Factor.

$$= 3x - 5$$ Simplify.

Work Practice 3

Practice 3

Subtract: $\frac{2x^2 + 5x}{x + 2} - \frac{4x + 6}{x + 2}$

Helpful Hint

Parentheses are inserted so that the entire numerator, $10x - 5$, is subtracted.

Helpful Hint

Notice how the numerator $10x - 5$ was subtracted in Example 3.

This − sign applies to the entire numerator $10x - 5$.

So parentheses are inserted here to indicate this.

$$\frac{3x^2 + 2x}{x - 1} - \frac{10x - 5}{x - 1} = \frac{3x^2 + 2x - (10x - 5)}{x - 1}$$

Objective B Finding the Least Common Denominator 

Recall from Section R.2 that to add and subtract fractions with different denominators, we first find the least common denominator (LCD). Then we write all fractions as equivalent fractions with the LCD.

For example, suppose we want to add $\frac{3}{8}$ and $\frac{1}{6}$. To find the LCD of the denominators, factor 8 and 6. Remember, the LCD is the same as the least common multiple, LCM. It is the smallest number that is a multiple of 6 and also 8.

$$8 = 2 \cdot 2 \cdot 2$$

$$6 = 2 \cdot 3$$

The LCM is a multiple of 6.

$$\text{LCM} = 2 \cdot 2 \cdot 2 \cdot 3 = 24$$

The LCM is a multiple of 8.

In the next section, we will find the sum $\frac{3}{8} + \frac{1}{6}$, but for now, let's concentrate on the LCD.

To add or subtract rational expressions with different denominators, we also first find the LCD and then write all rational expressions as equivalent expressions with the LCD. The **least common denominator (LCD) of a list of rational expressions** is a polynomial of least degree whose factors include all the factors of the denominators in the list.

To Find the Least Common Denominator (LCD)

Step 1: Factor each denominator completely.

Step 2: The least common denominator (LCD) is the product of all unique factors found in Step 1, each raised to a power equal to the greatest number of times that the factor appears in any one factored denominator.

Answer

3. $2x - 3$

Practice 4

Find the LCD for each pair.

a. $\frac{2}{9}, \frac{7}{15}$

b. $\frac{5}{6x^3}, \frac{11}{18x^5}$

Example 4

Find the LCD for each pair.

a. $\frac{1}{8}, \frac{3}{22}$ **b.** $\frac{7}{5x}, \frac{6}{15x^2}$

Solution:

a. We start by finding the prime factorization of each denominator.

$$8 = 2^3 \quad \text{and}$$
$$22 = 2 \cdot 11$$

Next we write the product of all the unique factors, each raised to a power equal to the greatest number of times that the factor appears.

The greatest number of times that the factor 2 appears is 3.

The greatest number of times that the factor 11 appears is 1.

$$\text{LCD} = 2^3 \cdot 11^1 = 8 \cdot 11 = 88$$

b. We factor each denominator.

$$5x = 5 \cdot x \quad \text{and}$$
$$15x^2 = 3 \cdot 5 \cdot x^2$$

The greatest number of times that the factor 5 appears is 1.
The greatest number of times that the factor 3 appears is 1.
The greatest number of times that the factor x appears is 2.

$$\text{LCD} = 3^1 \cdot 5^1 \cdot x^2 = 15x^2$$

Work Practice 4

Practice 5

Find the LCD of $\frac{3a}{a+5}$ and $\frac{7a}{a-5}$.

Example 5

Find the LCD of $\frac{7x}{x+2}$ and $\frac{5x^2}{x-2}$.

Solution: The denominators $x + 2$ and $x - 2$ are completely factored already. The factor $x + 2$ appears once and the factor $x - 2$ appears once.

$$\text{LCD} = (x + 2)(x - 2)$$

Work Practice 5

Practice 6

Find the LCD of $\frac{7x^2}{(x-4)^2}$ and $\frac{5x}{3x-12}$.

Example 6

Find the LCD of $\frac{6m^2}{3m+15}$ and $\frac{2}{(m+5)^2}$.

Solution: We factor each denominator.

$$3m + 15 = 3(m + 5)$$
$$(m + 5)^2 = (m + 5)^2 \quad \text{This denominator is already factored.}$$

The greatest number of times that the factor 3 appears is 1.
The greatest number of times that the factor $m + 5$ appears *in any one denominator* is 2.

$$\text{LCD} = 3(m + 5)^2$$

Work Practice 6

✓Concept Check Choose the correct LCD of $\frac{x}{(x+1)^2}$ and $\frac{5}{x+1}$.

a. $x + 1$ **b.** $(x + 1)^2$ **c.** $(x + 1)^3$ **d.** $5x(x + 1)^2$

Answers
4. a. 45 **b.** $18x^5$
5. $(a + 5)(a - 5)$ **6.** $3(x - 4)^2$

✓Concept Check Answer
b

Example 7 Find the LCD of $\frac{t-10}{2t^2+t-6}$ and $\frac{t+5}{t^2+3t+2}$.

Solution:

$$2t^2+t-6 = (2t-3)(t+2)$$
$$t^2+3t+2 = (t+1)(t+2)$$
$$\text{LCD} = (2t-3)(t+2)(t+1)$$

Work Practice 7

Practice 7

Find the LCD of $\frac{y+5}{y^2+2y-3}$ and $\frac{y+4}{y^2-3y+2}$.

Example 8 Find the LCD of $\frac{2}{x-2}$ and $\frac{10}{2-x}$.

Solution: The denominators $x-2$ and $2-x$ are opposites. That is, $2-x=-1(x-2)$. We can use either $x-2$ or $2-x$ as the LCD.

$$\text{LCD} = x-2 \quad \text{or} \quad \text{LCD} = 2-x$$

Work Practice 8

Practice 8

Find the LCD of $\frac{6}{x-4}$ and $\frac{9}{4-x}$.

Objective C Writing Equivalent Rational Expressions

Next we practice writing a rational expression as an equivalent rational expression with a given denominator. To do this, we multiply by a form of 1. Recall that multiplying an expression by 1 produces an equivalent expression. In other words,

$$\frac{P}{Q} = \frac{P}{Q}\cdot 1 = \frac{P}{Q}\cdot\frac{R}{R} = \frac{PR}{QR}$$

Example 9 Write each rational expression as an equivalent rational expression with the given denominator.

a. $\frac{4b}{9a} = \frac{\quad}{27a^2b}$ **b.** $\frac{7x}{2x+5} = \frac{\quad}{6x+15}$

Solution:

a. We can ask ourselves: "What do we multiply $9a$ by to get $27a^2b$?" The answer is $3ab$, since $9a(3ab) = 27a^2b$. So we multiply by 1 in the form of $\frac{3ab}{3ab}$.

$$\frac{4b}{9a} = \frac{4b}{9a}\cdot 1 = \frac{4b}{9a}\cdot\frac{3ab}{3ab}$$
$$= \frac{4b(3ab)}{9a(3ab)} = \frac{12ab^2}{27a^2b}$$

b. First, factor the denominator on the right.

$$\frac{7x}{2x+5} = \frac{\quad}{3(2x+5)}$$

To obtain the denominator on the right from the denominator on the left, we multiply by 1 in the form of $\frac{3}{3}$.

$$\frac{7x}{2x+5} = \frac{7x}{2x+5}\cdot\frac{3}{3} = \frac{7x\cdot 3}{(2x+5)\cdot 3} = \frac{21x}{3(2x+5)}$$

Work Practice 9

Practice 9

Write the rational expression as an equivalent rational expression with the given denominator.

$$\frac{2x}{5y} = \frac{\quad}{20x^2y^2}$$

Answers

7. $(y+3)(y-1)(y-2)$
8. $x-4$ or $4-x$
9. $\frac{8x^3y}{20x^2y^2}$

Practice 10

Write the rational expression as an equivalent rational expression with the given denominator.

$$\frac{3}{x^2-25} = \frac{\quad}{(x+5)(x-5)(x-3)}$$

Example 10 Write the rational expression as an equivalent rational expression with the given denominator.

$$\frac{5}{x^2-4} = \frac{\quad}{(x-2)(x+2)(x-4)}$$

Solution: First we factor the denominator $x^2 - 4$ as $(x-2)(x+2)$. If we multiply the original denominator $(x-2)(x+2)$ by $x-4$, the result is the new denominator $(x-2)(x+2)(x-4)$. Thus, we multiply by 1 in the form of $\frac{x-4}{x-4}$.

$$\begin{aligned}\frac{5}{x^2-4} = \frac{5}{(x-2)(x+2)} &= \frac{5}{(x-2)(x+2)} \cdot \frac{x-4}{x-4} \\ &= \frac{5(x-4)}{(x-2)(x+2)(x-4)} \\ &= \frac{5x-20}{(x-2)(x+2)(x-4)}\end{aligned}$$

Work Practice 10

Answer

10. $\frac{3x-9}{(x+5)(x-5)(x-3)}$

Vocabulary, Readiness & Video Check

Use the choices below to fill in each blank. Not all choices will be used.

$\frac{9}{22}$ $\quad\frac{5}{22}$ $\quad\frac{9}{11}$ $\quad\frac{5}{11}$ $\quad\frac{ac}{b}$ $\quad\frac{a-c}{b}$ $\quad\frac{a+c}{b}$ $\quad\frac{5-6+x}{x}$ $\quad\frac{5-(6+x)}{x}$

1. $\frac{7}{11} + \frac{2}{11} =$ ________

2. $\frac{7}{11} - \frac{2}{11} =$ ________

3. $\frac{a}{b} + \frac{c}{b} =$ ________

4. $\frac{a}{b} - \frac{c}{b} =$ ________

5. $\frac{5}{x} - \frac{6+x}{x} =$ ________

Martin-Gay Interactive Videos

See Video 5.3

Watch the section lecture video and answer the following questions.

Objective A **6.** In Example 3, why is it important to place parentheses around the second numerator when writing as one expression?

7. In Examples 4 and 5, we factor the denominators completely. How does this help determine the LCD?

8. Based on Example 6, complete the following statements: To write an equivalent rational expression, we multiply the ________ of a rational expression by the same expression as the denominator. This means we're multiplying the original rational expression by a factor of ________ and therefore not changing the ________ of the original expression.

5.3 Exercise Set MyLab Math 

Objective A *Add or subtract as indicated. Simplify the result if possible. See Examples 1 through 3.*

1. $\frac{a}{13} + \frac{9}{13}$

2. $\frac{x+1}{7} + \frac{6}{7}$

3. $\frac{4m}{3n} + \frac{5m}{3n}$

4. $\frac{3p}{2q} + \frac{11p}{2q}$

5. $\frac{4m}{m-6} - \frac{24}{m-6}$

6. $\frac{8y}{y-2} - \frac{16}{y-2}$

7. $\frac{9}{3+y} + \frac{y+1}{3+y}$

8. $\frac{9}{y+9} + \frac{y-5}{y+9}$

9. $\frac{5x^2+4x}{x-1} - \frac{6x+3}{x-1}$

10. $\frac{x^2+9x}{x+7} - \frac{4x+14}{x+7}$

11. $\frac{4a}{a^2+2a-15} - \frac{12}{a^2+2a-15}$

12. $\frac{3y}{y^2+3y-10} - \frac{6}{y^2+3y-10}$

13. $\frac{2x+3}{x^2-x-30} - \frac{x-2}{x^2-x-30}$

14. $\frac{3x-1}{x^2+5x-6} - \frac{2x-7}{x^2+5x-6}$

15. $\frac{2x+1}{x-3} + \frac{3x+6}{x-3}$

16. $\frac{4p-3}{2p+7} + \frac{3p+8}{2p+7}$

17. $\frac{2x^2}{x-5} - \frac{25+x^2}{x-5}$

18. $\frac{6x^2}{2x-5} - \frac{25+2x^2}{2x-5}$

19. $\frac{5x+4}{x-1} - \frac{2x+7}{x-1}$

20. $\frac{7x+1}{x-4} - \frac{2x+21}{x-4}$

Objective B *Find the LCD for each list of rational expressions. See Examples 4 through 8.*

21. $\frac{19}{2x}, \frac{5}{4x^3}$

22. $\frac{17x}{4y^5}, \frac{2}{8y}$

23. $\frac{9}{8x}, \frac{3}{2x+4}$

24. $\frac{1}{6y}, \frac{3x}{4y+12}$

25. $\frac{2}{x+3}, \frac{5}{x-2}$

26. $\frac{-6}{x-1}, \frac{4}{x+5}$

27. $\frac{x}{x+6}, \frac{10}{3x+18}$

28. $\frac{12}{x+5}, \frac{x}{4x+20}$

29. $\frac{8x^2}{(x-6)^2}, \frac{13x}{5x-30}$

30. $\frac{9x^2}{7x-14}, \frac{6x}{(x-2)^2}$

31. $\frac{1}{3x+3}, \frac{8}{2x^2+4x+2}$

32. $\frac{19x+5}{4x-12}, \frac{3}{2x^2-12x+18}$

33. $\frac{5}{x-8}, \frac{3}{8-x}$

34. $\frac{2x+5}{3x-7}, \frac{5}{7-3x}$

35. $\frac{5x+1}{x^2+3x-4}, \frac{3x}{x^2+2x-3}$

36. $\frac{4}{x^2+4x+3}, \frac{4x-2}{x^2+10x+21}$

37. $\frac{2x}{3x^2+4x+1}, \frac{7}{2x^2-x-1}$

38. $\frac{3x}{4x^2+5x+1}, \frac{5}{3x^2-2x-1}$

39. $\frac{1}{x^2-16}, \frac{x+6}{2x^3-8x^2}$

40. $\frac{5}{x^2-25}, \frac{x+9}{3x^3-15x^2}$

Objective C *Rewrite each rational expression as an equivalent rational expression with the given denominator. See Examples 9 and 10.*

41. $\frac{3}{2x} = \frac{}{4x^2}$

42. $\frac{3}{9y^5} = \frac{}{72y^9}$

43. $\frac{6}{3a} = \frac{}{12ab^2}$

44. $\frac{5}{4y^2x} = \frac{}{32y^3x^2}$

45. $\frac{9}{2x + 6} = \frac{}{2y(x + 3)}$

46. $\frac{4x + 1}{3x + 6} = \frac{}{3y(x + 2)}$

47. $\frac{9a + 2}{5a + 10} = \frac{}{5b(a + 2)}$

48. $\frac{5 + y}{2x^2 + 10} = \frac{}{4(x^2 + 5)}$

49. $\frac{x}{x^3 + 6x^2 + 8x} = \frac{}{x(x + 4)(x + 2)(x + 1)}$

50. $\frac{5x}{x^3 + 2x^2 - 3x} = \frac{}{x(x - 1)(x - 5)(x + 3)}$

51. $\frac{9y - 1}{15x^2 - 30} = \frac{}{30x^2 - 60}$

52. $\frac{6m - 5}{3x^2 - 9} = \frac{}{12x^2 - 36}$

Mixed Practice (*Sections 5.2 and 5.3*) *Perform the indicated operations.*

53. $\frac{5x}{7} + \frac{9x}{7}$

54. $\frac{5x}{7} \cdot \frac{9x}{7}$

55. $\frac{x + 3}{4} \div \frac{2x - 1}{4}$

56. $\frac{x + 3}{4} - \frac{2x - 1}{4}$

57. $\frac{x^2}{x - 6} - \frac{5x + 6}{x - 6}$

58. $\frac{-2x}{x^3 - 8x} + \frac{3x}{x^3 - 8x}$

59. $\frac{x^2 + 5x}{x^2 - 25} \cdot \frac{3x - 15}{x^2}$

60. $\frac{-2x}{x^3 - 8x} \div \frac{3x}{x^3 - 8x}$

61. $\frac{x^3 + 7x^2}{3x^3 - x^2} \div \frac{5x^2 + 36x + 7}{9x^2 - 1}$

62. $\frac{12x - 6}{x^2 + 3x} \cdot \frac{4x^2 + 13x + 3}{4x^2 - 1}$

Review

Perform each indicated operation. See Section R.2.

63. $\frac{2}{3} + \frac{5}{7}$

64. $\frac{9}{10} - \frac{3}{5}$

65. $\frac{1}{3} - \frac{3}{4}$

66. $\frac{11}{15} + \frac{5}{9}$

67. $\frac{1}{12} + \frac{3}{20}$

68. $\frac{7}{30} + \frac{1}{6}$

Concept Extensions

For Exercises 69 and 70, see the Concept Check in this section.

69. Choose the correct LCD of $\frac{11a^3}{4a - 20}$ and $\frac{15a^3}{(a - 5)^2}$.

a. $4a(a - 5)(a + 5)$
b. $a - 5$
c. $(a - 5)^2$
d. $4(a - 5)^2$
e. $(4a - 20)(a - 5)^2$

70. Choose the correct LCD of $\frac{5}{14x^2}$ and $\frac{y}{6x^3}$.

a. $84x^5$
b. $84x^3$
c. $42x^3$
d. $42x^5$

For Exercises 71 and 72, an algebra student approaches you with each incorrect solution. Find the error and correct the work shown below.

71. $\frac{2x - 6}{x - 5} - \frac{x + 4}{x - 5}$

$= \frac{2x - 6 - x + 4}{x - 5}$

$= \frac{x - 2}{x - 5}$

72. $\frac{x}{x + 3} + \frac{2}{x + 3}$

$= \frac{x + 2}{x + 3}$

$= \frac{2}{3}$

Multiple choice. Select the correct result.

73. $\frac{3}{x} + \frac{y}{x} =$

a. $\frac{3 + y}{x^2}$ b. $\frac{3 + y}{2x}$ c. $\frac{3 + y}{x}$

74. $\frac{3}{x} - \frac{y}{x} =$

a. $\frac{3 - y}{x^2}$ b. $\frac{3 - y}{2x}$ c. $\frac{3 - y}{x}$

75. $\frac{3}{x} \cdot \frac{y}{x} =$

a. $\frac{3y}{x}$ b. $\frac{3y}{x^2}$ c. $3y$

76. $\frac{3}{x} \div \frac{y}{x} =$

a. $\frac{3}{y}$ b. $\frac{y}{3}$ c. $\frac{3}{x^2y}$

Write each rational expression as an equivalent expression with a denominator of $x - 2$.

77. $\frac{5}{2 - x}$

78. $\frac{8y}{2 - x}$

79. $-\frac{7 + x}{2 - x}$

80. $\frac{x - 3}{-(x - 2)}$

△ **81.** A square has a side of length $\frac{5}{x - 2}$ meters. Express its perimeter as a rational expression.

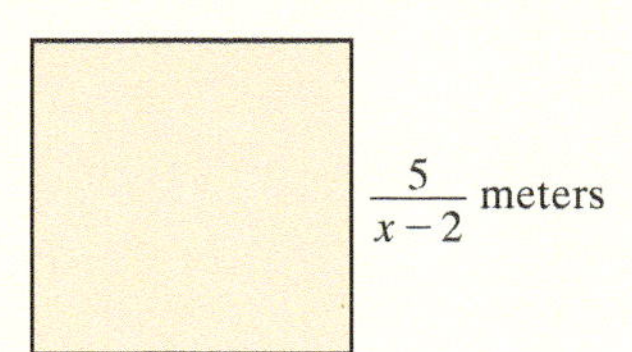

△ **82.** A trapezoid has sides of the indicated lengths. Find its perimeter.

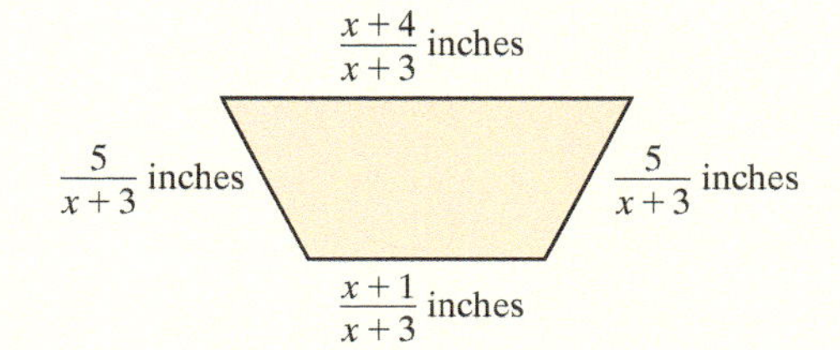

83. Write two rational expressions with the same denominator whose sum is $\frac{5}{3x - 1}$.

84. Write two rational expressions with the same denominator whose difference is $\frac{x - 7}{x^2 + 1}$.

85. The planet Mercury revolves around the Sun in 88 Earth days. It takes Jupiter 4332 Earth days to make one revolution around the Sun. (*Source:* National Space Science Data Center) If the two planets are aligned as shown in the figure, how long will it take for them to align again?

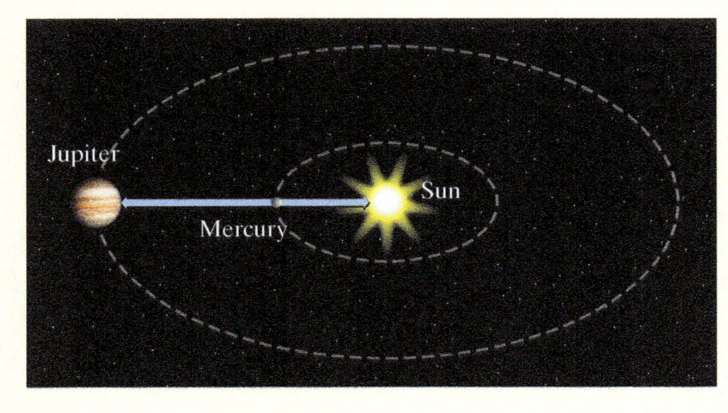

86. You are throwing a barbecue and you want to make sure that you purchase the same number of hot dogs as hot dog buns. Hot dogs come 8 to a package and hot dog buns come 12 to a package. What is the least number of each type of package you should buy?

87. Write some instructions to help a friend who is having difficulty finding the LCD of two rational expressions.

88. In your own words, describe how to add or subtract two rational expressions with the same denominator.

89. Explain why the LCD of the rational expressions $\frac{7}{x + 1}$ and $\frac{9x}{(x + 1)^2}$ is $(x + 1)^2$ and not $(x + 1)^3$.

90. Explain the similarities between subtracting $\frac{3}{8}$ from $\frac{7}{8}$ and subtracting $\frac{6}{x + 3}$ from $\frac{9}{x + 3}$.

5.4 Adding and Subtracting Rational Expressions with Different Denominators

Objective

Add and Subtract Rational Expressions with Different Denominators.

Objective A Adding and Subtracting Rational Expressions with Different Denominators

Let's add $\frac{3}{8}$ and $\frac{1}{6}$. In the previous section, we found the LCD of 8 and 6 to be 24. Now let's write equivalent fractions with denominator 24 by multiplying by different forms of 1.

$$\frac{3}{8} = \frac{3}{8} \cdot 1 = \frac{3}{8} \cdot \frac{3}{3} = \frac{3 \cdot 3}{8 \cdot 3} = \frac{9}{24}$$

$$\frac{1}{6} = \frac{1}{6} \cdot 1 = \frac{1}{6} \cdot \frac{4}{4} = \frac{1 \cdot 4}{6 \cdot 4} = \frac{4}{24}$$

Now that the denominators are the same, we may add.

$$\frac{3}{8} + \frac{1}{6} = \frac{9}{24} + \frac{4}{24} = \frac{9 + 4}{24} = \frac{13}{24}$$

We add or subtract rational expressions the same way. You may want to use the steps below.

To Add or Subtract Rational Expressions with Different Denominators

Step 1: Find the LCD of the rational expressions.

Step 2: Rewrite each rational expression as an equivalent expression whose denominator is the LCD found in Step 1.

Step 3: Add or subtract numerators and write the sum or difference over the common denominator.

Step 4: Simplify or write the rational expression in lowest terms.

Practice 1

Perform each indicated operation.

a. $\frac{y}{5} - \frac{3y}{15}$ **b.** $\frac{5}{8x} + \frac{11}{10x^2}$

Example 1

Perform each indicated operation.

a. $\frac{a}{4} - \frac{2a}{8}$ **b.** $\frac{3}{10x^2} + \frac{7}{25x}$

Solution:

a. First, we must find the LCD. Since $4 = 2^2$ and $8 = 2^3$, the LCD $= 2^3 = 8$. Next we write each fraction as an equivalent fraction with the denominator 8, and then we subtract.

$$\frac{a}{4} = \frac{a}{4} \cdot 1 = \frac{a}{4} \cdot \frac{2}{2} = \frac{a \cdot 2}{4 \cdot 2} = \frac{2a}{8}$$

$$\frac{a}{4} - \frac{2a}{8} = \frac{2a}{8} - \frac{2a}{8} = \frac{2a - 2a}{8} = \frac{0}{8} = 0$$

Notice that we wrote $\frac{a}{4}$ as the equivalent expression $\frac{2a}{8}$. Multiplying by a form of 1 means we multiply the numerator and the denominator by the same number. Since this is so, we will start using the shorthand notation on the next page.

Answers

1. a. 0 **b.** $\frac{25x + 44}{40x^2}$

$$\frac{a}{4} = \frac{a(2)}{4(2)} = \frac{2a}{8}$$

Multiplying the numerator and denominator by 2 is the same as multiplying by $\frac{2}{2}$ or 1.

b. Since $10x^2 = 2 \cdot 5 \cdot x \cdot x$ and $25x = 5 \cdot 5 \cdot x$, the LCD $= 2 \cdot 5^2 \cdot x^2 = 50x^2$. We write each fraction as an equivalent fraction with a denominator of $50x^2$.

$$\frac{3}{10x^2} + \frac{7}{25x} = \frac{3(5)}{10x^2(5)} + \frac{7(2x)}{25x(2x)}$$

$$= \frac{15}{50x^2} + \frac{14x}{50x^2}$$

$$= \frac{15 + 14x}{50x^2}$$ Add numerators. Write the sum over the common denominator.

Work Practice 1

Example 2 Subtract: $\frac{6x}{x^2 - 4} - \frac{3}{x + 2}$

Solution: Since $x^2 - 4 = (x + 2)(x - 2)$, the LCD $= (x + 2)(x - 2)$. We write equivalent expressions with the LCD as denominators.

$$\frac{6x}{x^2 - 4} - \frac{3}{x + 2} = \frac{6x}{(x + 2)(x - 2)} - \frac{3(x - 2)}{(x + 2)(x - 2)}$$

$$= \frac{6x - 3(x - 2)}{(x + 2)(x - 2)}$$ Subtract numerators. Write the difference over the common denominator.

$$= \frac{6x - 3x + 6}{(x + 2)(x - 2)}$$ Apply the distributive property in the numerator.

$$= \frac{3x + 6}{(x + 2)(x - 2)}$$ Combine like terms in the numerator.

Next we factor the numerator to see if this rational expression can be simplified.

$$\frac{3x + 6}{(x + 2)(x - 2)} = \frac{3\,(x + 2)}{(x + 2)\,(x - 2)}$$ Factor.

$$= \frac{3}{x - 2}$$ Apply the fundamental principle to simplify.

Work Practice 2

Practice 2

Subtract: $\frac{10x}{x^2 - 9} - \frac{5}{x + 3}$

Example 3 Add: $\frac{2}{3t} + \frac{5}{t + 1}$

Solution: The LCD is $3t(t + 1)$. We write each rational expression as an equivalent rational expression with a denominator of $3t(t + 1)$.

$$\frac{2}{3t} + \frac{5}{t + 1} = \frac{2(t + 1)}{3t(t + 1)} + \frac{5(3t)}{(t + 1)(3t)}$$

$$= \frac{2(t + 1) + 5(3t)}{3t(t + 1)}$$ Add numerators. Write the sum over the common denominator.

$$= \frac{2t + 2 + 15t}{3t(t + 1)}$$ Apply the distributive property in the numerator.

$$= \frac{17t + 2}{3t(t + 1)}$$ Combine like terms in the numerator.

Work Practice 3

Practice 3

Add: $\frac{5}{7x} + \frac{2}{x + 1}$

Answers

2. $\frac{5}{x - 3}$ **3.** $\frac{19x + 5}{7x(x + 1)}$

Practice 4

Subtract: $\frac{10}{x-6} - \frac{15}{6-x}$

Example 4 Subtract: $\frac{7}{x-3} - \frac{9}{3-x}$

Solution: To find a common denominator, we notice that $x - 3$ and $3 - x$ are opposites. That is, $3 - x = -(x - 3)$. We write the denominator $3 - x$ as $-(x - 3)$ and simplify.

$$\frac{7}{x-3} - \frac{9}{3-x} = \frac{7}{x-3} - \frac{9}{-(x-3)}$$

$$= \frac{7}{x-3} - \frac{-9}{x-3} \quad \text{Apply } \frac{a}{-b} = \frac{-a}{b}.$$

$$= \frac{7-(-9)}{x-3} \quad \text{Subtract numerators. Write the difference over the common denominator.}$$

$$= \frac{16}{x-3}$$

Work Practice 4

Practice 5

Add: $2 + \frac{x}{x+5}$

Example 5 Add: $1 + \frac{m}{m+1}$

Solution: Recall that 1 is the same as $\frac{1}{1}$. The LCD of $\frac{1}{1}$ and $\frac{m}{m+1}$ is $m + 1$.

$$1 + \frac{m}{m+1} = \frac{1}{1} + \frac{m}{m+1} \quad \text{Write 1 as } \frac{1}{1}.$$

$$= \frac{1(m+1)}{1(m+1)} + \frac{m}{m+1} \quad \text{Multiply both the numerator and the denominator of } \frac{1}{1} \text{ by } m+1.$$

$$= \frac{m+1+m}{m+1} \quad \text{Add numerators. Write the sum over the common denominator.}$$

$$= \frac{2m+1}{m+1} \quad \text{Combine like terms in the numerator.}$$

Work Practice 5

Practice 6

Subtract: $\frac{4}{3x^2+2x} - \frac{3x}{12x+8}$

Example 6 Subtract: $\frac{3}{2x^2+x} - \frac{2x}{6x+3}$

Solution: First, we factor the denominators.

$$\frac{3}{2x^2+x} - \frac{2x}{6x+3} = \frac{3}{x(2x+1)} - \frac{2x}{3(2x+1)}$$

The LCD is $3x(2x + 1)$. We write equivalent expressions with denominator $3x(2x + 1)$.

$$\frac{3}{x(2x+1)} - \frac{2x}{3(2x+1)} = \frac{3(3)}{x(2x+1)(3)} - \frac{2x(x)}{3(2x+1)(x)}$$

$$= \frac{9-2x^2}{3x(2x+1)} \quad \text{Subtract numerators. Write the difference over the common denominator.}$$

Work Practice 6

Answers

4. $\frac{25}{x-6}$ 5. $\frac{3x+10}{x+5}$ 6. $\frac{16-3x^2}{4x(3x+2)}$

Example 7 Add: $\frac{2x}{x^2 + 2x + 1} + \frac{x}{x^2 - 1}$

Solution: First we factor the denominators.

$$\frac{2x}{x^2 + 2x + 1} + \frac{x}{x^2 - 1}$$

$$= \frac{2x}{(x+1)(x+1)} + \frac{x}{(x+1)(x-1)}$$ Rewrite each expression with LCD $(x+1)(x+1)(x-1)$.

$$= \frac{2x(x-1)}{(x+1)(x+1)(x-1)} + \frac{x(x+1)}{(x+1)(x-1)(x+1)}$$

$$= \frac{2x(x-1) + x(x+1)}{(x+1)^2(x-1)}$$ Add numerators. Write the sum over the common denominator.

$$= \frac{2x^2 - 2x + x^2 + x}{(x+1)^2(x-1)}$$ Apply the distributive property in the numerator.

$$= \frac{3x^2 - x}{(x+1)^2(x-1)} \text{ or } \frac{x(3x-1)}{(x+1)^2(x-1)}$$

The numerator was factored as a last step to see if the rational expression could be simplified further. Since there are no factors common to the numerator and the denominator, we can't simplify further.

Work Practice 7

Practice 7

Add: $\frac{6x}{x^2 + 4x + 4} + \frac{x}{x^2 - 4}$

Answer

7. $\frac{x(7x-10)}{(x+2)^2(x-2)}$

Vocabulary, Readiness & Video Check

Multiple choice. Choose the correct response.

1. $\frac{3}{7x} + \frac{5}{7} =$

a. $\frac{3}{7x} + \frac{5}{7x} = \frac{8}{7x}$ **b.** $\frac{3}{7x} + \frac{5}{7} \cdot \frac{x}{x} = \frac{3 + 5x}{7x}$ **c.** $\frac{3}{7x} + \frac{5}{7} \cdot \frac{x}{x} = \frac{8x}{7x} \text{ or } \frac{8}{7}$

2. $\frac{1}{x} + \frac{2}{x^2} =$

a. $\frac{1}{x} \cdot \frac{x}{x} + \frac{2}{x^2} = \frac{x+2}{x^2}$ **b.** $\frac{3}{x^3}$ **c.** $\frac{1}{x} \cdot \frac{x}{x} + \frac{2}{x^2} = \frac{3x}{x^2} \text{ or } \frac{3}{x}$

Martin-Gay Interactive Videos

See Video 5.4

Watch the section lecture video and answer the following question.

Objective A **3.** What special case is shown in Example 2, and what's the purpose of presenting it?

5.4 Exercise Set MyLab Math 

Objective A *Perform each indicated operation. Simplify if possible. See Examples 1 through 7.*

1. $\frac{4}{2x} + \frac{9}{3x}$

2. $\frac{15}{7a} + \frac{8}{6a}$

3. $\frac{15a}{b} + \frac{6b}{5}$

4. $\frac{4c}{d} - \frac{8d}{5}$

5. $\frac{3}{x} + \frac{5}{2x^2}$

6. $\frac{14}{3x^2} + \frac{6}{x}$

7. $\frac{6}{x+1} + \frac{10}{2x+2}$

8. $\frac{8}{x+4} - \frac{3}{3x+12}$

9. $\frac{3}{x+2} - \frac{2x}{x^2-4}$

10. $\frac{5}{x-4} + \frac{4x}{x^2-16}$

11. $\frac{3}{4x} + \frac{8}{x-2}$

12. $\frac{5}{y^2} - \frac{y}{2y+1}$

13. $\frac{6}{x-3} + \frac{8}{3-x}$

14. $\frac{15}{y-4} + \frac{20}{4-y}$

15. $\frac{9}{x-3} + \frac{9}{3-x}$

16. $\frac{5}{a-7} + \frac{5}{7-a}$

17. $\frac{-8}{x^2-1} - \frac{7}{1-x^2}$

18. $\frac{-9}{25x^2-1} + \frac{7}{1-25x^2}$

19. $\frac{5}{x} + 2$

20. $\frac{7}{x^2} - 5x$

21. $\frac{5}{x-2} + 6$

22. $\frac{6y}{y+5} + 1$

23. $\frac{y+2}{y+3} - 2$

24. $\frac{7}{2x-3} - 3$

25. $\frac{-x+2}{x} - \frac{x-6}{4x}$

26. $\frac{-y+1}{y} - \frac{2y-5}{3y}$

27. $\frac{5x}{x+2} - \frac{3x-4}{x+2}$

28. $\frac{7x}{x-3} - \frac{4x+9}{x-3}$

29. $\frac{3x^4}{7} - \frac{4x^2}{21}$

30. $\frac{5x}{6} + \frac{11x^2}{2}$

31. $\frac{1}{x+3} - \frac{1}{(x+3)^2}$

32. $\frac{5x}{(x-2)^2} - \frac{3}{x-2}$

33. $\frac{4}{5b} + \frac{1}{b-1}$

34. $\frac{1}{y+5} + \frac{2}{3y}$

35. $\frac{2}{m} + 1$

36. $\frac{6}{x} - 1$

37. $\frac{2x}{x-7} - \frac{x}{x-2}$

38. $\frac{9x}{x-10} - \frac{x}{x-3}$

39. $\frac{6}{1-2x} - \frac{4}{2x-1}$

40. $\frac{10}{3n-4} - \frac{5}{4-3n}$

41. $\frac{7}{(x+1)(x-1)} + \frac{8}{(x+1)^2}$

42. $\frac{5}{(x+1)(x+5)} - \frac{2}{(x+5)^2}$

43. $\frac{x}{x^2-1} - \frac{2}{x^2-2x+1}$

44. $\frac{x}{x^2-4} - \frac{5}{x^2-4x+4}$

45. $\frac{3a}{2a+6} - \frac{a-1}{a+3}$

46. $\frac{5y}{2x+2y} - \frac{y-1}{x+y}$

47. $\frac{y-1}{2y+3} + \frac{3}{(2y+3)^2}$

48. $\frac{x-6}{5x+1} + \frac{6}{(5x+1)^2}$

49. $\frac{5}{2-x} + \frac{x}{2x-4}$

50. $\frac{-1}{a-2} + \frac{4}{4-2a}$

51. $\frac{15}{x^2+6x+9} + \frac{2}{x+3}$

52. $\frac{2}{x^2+4x+4} + \frac{1}{x+2}$

53. $\frac{13}{x^2-5x+6} - \frac{5}{x-3}$

54. $\frac{-7}{y^2-3y+2} - \frac{2}{y-1}$

55. $\frac{70}{m^2-100} + \frac{7}{2(m+10)}$

56. $\frac{27}{y^2-81} + \frac{3}{2(y+9)}$

57. $\frac{x+8}{x^2-5x-6} + \frac{x+1}{x^2-4x-5}$

58. $\frac{x+4}{x^2+12x+20} + \frac{x+1}{x^2+8x-20}$

59. $\frac{5}{4n^2-12n+8} - \frac{3}{3n^2-6n}$

60. $\frac{6}{5y^2-25y+30} - \frac{2}{4y^2-8y}$

Mixed Practice (*Sections 5.2, 5.3, and 5.4*) *Perform the indicated operations. Addition, subtraction, multiplication, and division of rational expressions are included here.*

61. $\frac{15x}{x+8} \cdot \frac{2x+16}{3x}$

62. $\frac{9z+5}{15} \cdot \frac{5z}{81z^2-25}$

63. $\frac{8x+7}{3x+5} - \frac{2x-3}{3x+5}$

64. $\frac{2z^2}{4z-1} - \frac{z-2z^2}{4z-1}$

65. $\frac{5a+10}{18} \div \frac{a^2-4}{10a}$

66. $\frac{9}{x^2-1} \div \frac{12}{3x+3}$

67. $\frac{5}{x^2-3x+2} + \frac{1}{x-2}$

68. $\frac{4}{2x^2+5x-3} + \frac{2}{x+3}$

Review

Solve each linear or quadratic equation. See Sections 2.3 and 4.6.

69. $3x + 5 = 7$

70. $5x - 1 = 8$

71. $2x^2 - x - 1 = 0$

72. $4x^2 - 9 = 0$

73. $4(x + 6) + 3 = -3$

74. $2(3x + 1) + 15 = -7$

Concept Extensions

Perform each indicated operation.

75. $\frac{3}{x} - \frac{2x}{x^2 - 1} + \frac{5}{x + 1}$

76. $\frac{5}{x - 2} + \frac{7x}{x^2 - 4} - \frac{11}{x}$

77. $\frac{5}{x^2 - 4} + \frac{2}{x^2 - 4x + 4} - \frac{3}{x^2 - x - 6}$

78. $\frac{8}{x^2 + 6x + 5} - \frac{3x}{x^2 + 4x - 5} + \frac{2}{x^2 - 1}$

79. $\frac{9}{x^2 + 9x + 14} - \frac{3x}{x^2 + 10x + 21} + \frac{x + 4}{x^2 + 5x + 6}$

80. $\frac{x + 10}{x^2 - 3x - 4} - \frac{8}{x^2 + 6x + 5} - \frac{9}{x^2 + x - 20}$

81. A board of length $\frac{3}{x + 4}$ inches was cut into two pieces. If one piece is $\frac{1}{x - 4}$ inches, express the length of the other piece as a rational expression.

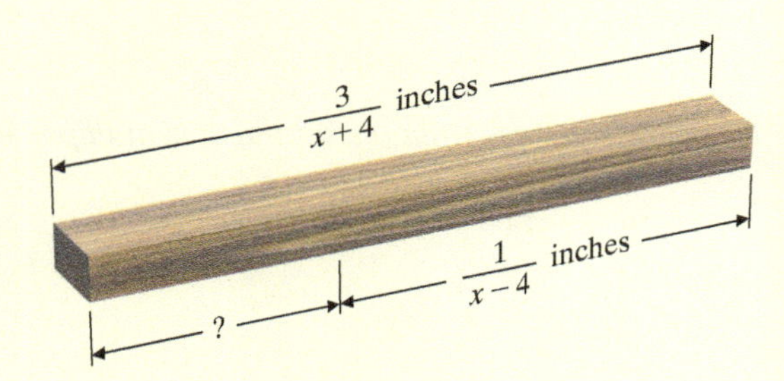

△ **82.** The length of a rectangle is $\frac{3}{y - 5}$ feet, while its width is $\frac{2}{y}$ feet. Find its perimeter and then find its area.

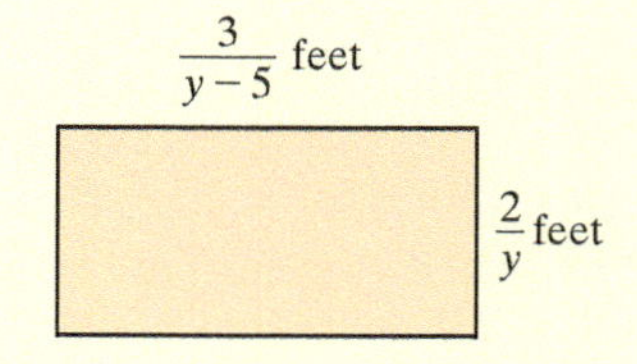

83. In ice hockey, penalty killing percentage is a statistic calculated as $1 - \frac{G}{P}$, where G = opponent's power play goals and P = opponent's power play opportunities. Simplify this expression.

84. The dose of medicine prescribed for a child depends on the child's age A in years and the adult dose D for the medication. Two expressions that give a child's dose are Young's Rule, $\frac{DA}{A + 12}$, and Cowling's Rule, $\frac{D(A + 1)}{24}$. Find an expression for the difference in the doses given by these expressions.

85. Explain when the LCD of the rational expressions in a sum is the product of the denominators.

86. Explain when the LCD is the same as one of the denominators of a rational expression to be added or subtracted.

87. Two angles are said to be complementary if the sum of their measures is 90°. If one angle measures $\frac{40}{x}$ degrees, find the measure of its complement.

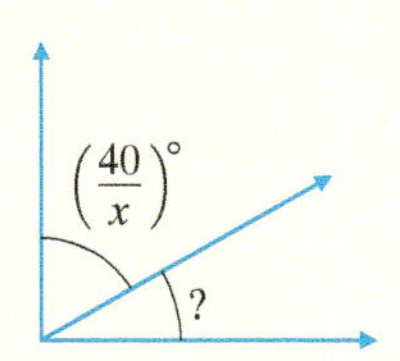

88. Two angles are said to be supplementary if the sum of their measures is 180°. If one angle measures $\frac{x+2}{x}$ degrees, find the measure of its supplement.

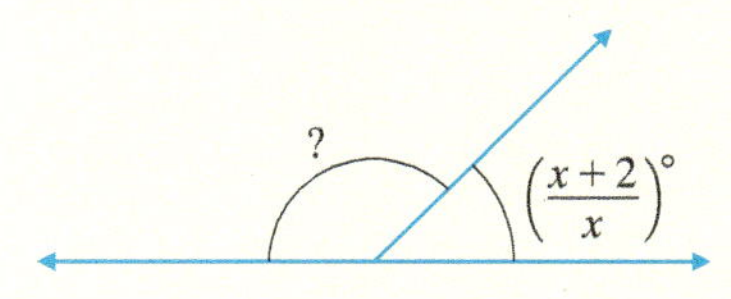

89. In your own words, explain how to add two rational expressions with different denominators.

90. In your own words, explain how to subtract two rational expressions with different denominators.

5.5 Solving Equations Containing Rational Expressions

Objective A Solving Equations Containing Rational Expressions

Objectives

A Solve Equations Containing Rational Expressions.

B Solve Equations Containing Rational Expressions for a Specified Variable.

In Chapter 2, we solved equations containing fractions. In this section, we continue the work we began in that chapter by solving equations containing rational expressions. For example,

$$\frac{x}{2}+\frac{8}{3}=\frac{1}{6} \quad \text{and} \quad \frac{4x}{x^2+x-30}+\frac{2}{x-5}=\frac{1}{x+6}$$

are equations containing rational expressions. To solve equations such as these, we use the multiplication property of equality to clear the equation of fractions by multiplying both sides of the equation by the LCD.

Example 1 Solve: $\frac{x}{2}+\frac{8}{3}=\frac{1}{6}$

Solution: The LCD of denominators 2, 3, and 6 is 6, so we multiply both sides of the equation by 6.

$$6\left(\frac{x}{2}+\frac{8}{3}\right)=6\left(\frac{1}{6}\right)$$

$$6\left(\frac{x}{2}\right)+6\left(\frac{8}{3}\right)=6\left(\frac{1}{6}\right) \quad \text{Apply the distributive property.}$$

$$3\cdot x+16=1 \quad \text{Multiply and simplify.}$$

$$3x=-15 \quad \text{Subtract 16 from both sides.}$$

$$x=-5 \quad \text{Divide both sides by 3.}$$

(Continued on next page)

Practice 1

Solve: $\frac{x}{4}+\frac{4}{5}=\frac{1}{20}$

Make sure that *each* term is multiplied by the LCD.

Answer

1. $x=-3$

Check: To check, we replace x with -5 in the original equation.

$$\frac{-5}{2}+\frac{8}{3}\stackrel{?}{=}\frac{1}{6} \quad \text{Replace } x \text{ with } -5.$$

$$\frac{1}{6}=\frac{1}{6} \quad \text{True}$$

This number checks, so the solution is -5.

Work Practice 1

Practice 2

Solve: $\frac{x+2}{3}-\frac{x-1}{5}=\frac{1}{15}$

Example 2 Solve: $\frac{t-4}{2}-\frac{t-3}{9}=\frac{5}{18}$

Solution: The LCD of denominators 2, 9, and 18 is 18, so we multiply both sides of the equation by 18.

$$18\left(\frac{t-4}{2}-\frac{t-3}{9}\right)=18\left(\frac{5}{18}\right)$$

$$18\left(\frac{t-4}{2}\right)-18\left(\frac{t-3}{9}\right)=18\left(\frac{5}{18}\right) \quad \text{Apply the distributive property.}$$

Multiply *each* term by 18.

$$9(t-4)-2(t-3)=5 \quad \text{Simplify.}$$

$$9t-36-2t+6=5 \quad \text{Use the distributive property.}$$

$$7t-30=5 \quad \text{Combine like terms.}$$

$$7t=35$$

$$t=5 \quad \text{Solve for } t.$$

Check:

$$\frac{t-4}{2}-\frac{t-3}{9}=\frac{5}{18}$$

$$\frac{5-4}{2}-\frac{5-3}{9}\stackrel{?}{=}\frac{5}{18} \quad \text{Replace } t \text{ with } 5.$$

$$\frac{1}{2}-\frac{2}{9}\stackrel{?}{=}\frac{5}{18} \quad \text{Simplify.}$$

$$\frac{5}{18}=\frac{5}{18} \quad \text{True}$$

The solution is 5.

Work Practice 2

Recall from Section 5.1 that a rational expression is defined for all real numbers except those that make the denominator of the expression 0. This means that if an equation contains *rational expressions with variables in the denominator,* we must be certain that the proposed solution does not make the denominator 0. If replacing the variable with the proposed solution makes the denominator 0, the rational expression is undefined and this proposed solution must be rejected.

Answer

2. $x=-6$

Example 3 Solve: $3 - \frac{6}{x} = x + 8$

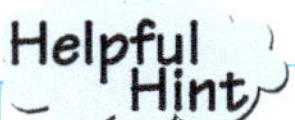

Notice that Example 3 contains our first equation with a variable in the denominator.

Solution: In this equation, 0 cannot be a solution because if x is 0, the rational expression $\frac{6}{x}$ is undefined. The LCD is x, so we multiply both sides of the equation by x.

$$x\left(3 - \frac{6}{x}\right) = x(x + 8)$$

$$x(3) - x\left(\frac{6}{x}\right) = x \cdot x + x \cdot 8 \quad \text{Apply the distributive property.}$$

$$3x - 6 = x^2 + 8x \quad \text{Simplify.}$$

Helpful Hint: Multiply *each* term by x.

Now we write the quadratic equation in standard form and solve for x.

$$0 = x^2 + 5x + 6$$

$$0 = (x + 3)(x + 2) \quad \text{Factor.}$$

$$x + 3 = 0 \quad \text{or} \quad x + 2 = 0 \quad \text{Set each factor equal to 0 and solve.}$$

$$x = -3 \qquad\qquad x = -2$$

Notice that neither -3 nor -2 makes the denominator in the original equation equal to 0.

Check: To check these solutions, we replace x in the original equation by -3, and then by -2.

If $x = -3$:

$$3 - \frac{6}{x} = x + 8$$

$$3 - \frac{6}{-3} \stackrel{?}{=} -3 + 8$$

$$3 - (-2) \stackrel{?}{=} 5$$

$$5 = 5 \quad \text{True}$$

If $x = -2$:

$$3 - \frac{6}{x} = x + 8$$

$$3 - \frac{6}{-2} \stackrel{?}{=} -2 + 8$$

$$3 - (-3) \stackrel{?}{=} 6$$

$$6 = 6 \quad \text{True}$$

Both -3 and -2 are solutions.

Work Practice 3

Practice 3

Solve: $2 + \frac{6}{x} = x + 7$

The following steps may be used to solve an equation containing rational expressions.

To Solve an Equation Containing Rational Expressions

Step 1: Multiply both sides of the equation by the LCD of all rational expressions in the equation.

Step 2: Remove any grouping symbols and solve the resulting equation.

Step 3: Check the solution in the original equation.

Answer

3. $x = -6, x = 1$

Practice 4

Solve:

$$\frac{2}{x+3} + \frac{3}{x-3} = \frac{-2}{x^2-9}$$

Example 4 Solve: $\frac{4x}{x^2+x-30} + \frac{2}{x-5} = \frac{1}{x+6}$

Solution: The denominator $x^2 + x - 30$ factors as $(x+6)(x-5)$. The LCD is then $(x+6)(x-5)$, so we multiply both sides of the equation by this LCD.

$$(x+6)(x-5)\left(\frac{4x}{x^2+x-30} + \frac{2}{x-5}\right) = (x+6)(x-5)\left(\frac{1}{x+6}\right) \quad \text{Multiply by the LCD.}$$

$$(x+6)(x-5)\cdot\frac{4x}{x^2+x-30} + (x+6)(x-5)\cdot\frac{2}{x-5} = (x+6)(x-5)\cdot\frac{1}{x+6} \quad \text{Apply the distributive property.}$$

$$4x + 2(x+6) = x - 5 \quad \text{Simplify.}$$

$$4x + 2x + 12 = x - 5 \quad \text{Apply the distributive property.}$$

$$6x + 12 = x - 5 \quad \text{Combine like terms.}$$

$$5x = -17$$

$$x = -\frac{17}{5} \quad \text{Divide both sides by 5.}$$

Check: Check by replacing x with $-\frac{17}{5}$ in the original equation. The solution is $-\frac{17}{5}$.

Work Practice 4

Practice 5

Solve: $\frac{5x}{x-1} = \frac{5}{x-1} + 3$

Example 5 Solve: $\frac{2x}{x-4} = \frac{8}{x-4} + 1$

Solution: Multiply both sides by the LCD, $x - 4$.

$$(x-4)\left(\frac{2x}{x-4}\right) = (x-4)\left(\frac{8}{x-4} + 1\right) \quad \text{Multiply by the LCD.}$$

$$(x-4)\cdot\frac{2x}{x-4} = (x-4)\cdot\frac{8}{x-4} + (x-4)\cdot 1 \quad \text{Use the distributive property.}$$

$$2x = 8 + (x-4) \quad \text{Simplify.}$$

$$2x = 4 + x$$

$$x = 4$$

Notice that 4 makes a denominator 0 in the original equation. Therefore, 4 is *not* a solution and this equation has *no solution*.

Helpful Hint

As we can see from Example 5, it is important to check the proposed solution(s) in the original equation.

Work Practice 5

✓Concept Check When can we clear fractions by multiplying through by the LCD?

a. When adding or subtracting rational expressions
b. When solving an equation containing rational expressions
c. Both of these
d. Neither of these

Answers

4. $x = -1$ **5.** no solution

✓Concept Check Answer

b

Example 6 Solve: $x + \frac{14}{x-2} = \frac{7x}{x-2} + 1$

Solution: Notice the denominators in this equation. We can see that 2 can't be a solution. The LCD is $x - 2$, so we multiply both sides of the equation by $x - 2$.

$$(x-2)\left(x + \frac{14}{x-2}\right) = (x-2)\left(\frac{7x}{x-2} + 1\right)$$

$$(x-2)(x) + (x-2)\left(\frac{14}{x-2}\right) = (x-2)\left(\frac{7x}{x-2}\right) + (x-2)(1)$$

$x^2 - 2x + 14 = 7x + x - 2$ Simplify.

$x^2 - 2x + 14 = 8x - 2$ Combine like terms.

$x^2 - 10x + 16 = 0$ Write the quadratic equation in standard form.

$(x-8)(x-2) = 0$ Factor.

$x - 8 = 0$ or $x - 2 = 0$ Set each factor equal to 0.

$x = 8$ $\quad x = 2$ Solve.

As we have already noted, 2 can't be a solution of the original equation. So we need replace x only with 8 in the original equation. We find that 8 is a solution; the only solution is 8.

Work Practice 6

Practice 6

Solve:

$$x - \frac{6}{x+3} = \frac{2x}{x+3} + 2$$

Objective B Solving Equations for a Specified Variable

The last example in this section is an equation containing several variables, and we are directed to solve for one of the variables. The steps used in the preceding examples can be applied to solve equations for a specified variable as well.

Example 7 Solve $\frac{1}{a} + \frac{1}{b} = \frac{1}{x}$ for x.

Solution: (This type of equation often models a work problem, as we shall see in the next section.) The LCD is abx, so we multiply both sides by abx.

$$abx\left(\frac{1}{a} + \frac{1}{b}\right) = abx\left(\frac{1}{x}\right)$$

$$abx\left(\frac{1}{a}\right) + abx\left(\frac{1}{b}\right) = abx \cdot \frac{1}{x}$$

$bx + ax = ab$ Simplify.

$x(b + a) = ab$ Factor out x from each term on the left side.

$\frac{x(b+a)}{b+a} = \frac{ab}{b+a}$ Divide both sides by $b + a$.

$x = \frac{ab}{b+a}$ Simplify.

This equation is now solved for x.

Work Practice 7

Practice 7

Solve $\frac{1}{a} + \frac{1}{b} = \frac{1}{x}$ for a.

Answers

6. $x = 4$ **7.** $a = \frac{bx}{b-x}$

Vocabulary, Readiness & Video Check

Multiple choice. Choose the correct response.

1. Multiply both sides of the equation $\frac{3x}{2} + 5 = \frac{1}{4}$ by 4. The result is:

a. $3x + 5 = 1$ **b.** $6x + 5 = 1$ **c.** $6x + 20 = 1$ **d.** $6x + 9 = 1$

2. Multiply both sides of the equation $\frac{1}{x} - \frac{3}{5x} = 2$ by $5x$. The result is:

a. $1 - 3 = 10x$ **b.** $5 - 3 = 10x$ **c.** $5x - 3 = 10x$ **d.** $5 - 3 = 7x$

Choose the correct LCD for the fractions in each equation.

3. Equation: $\frac{9}{x} + \frac{3}{4} = \frac{1}{12}$; LCD: ______

a. $4x$ **b.** $12x$ **c.** $48x$ **d.** x

4. Equation: $\frac{8}{3x} - \frac{1}{x} = \frac{7}{9}$; LCD: ______

a. x **b.** $3x$ **c.** $27x$ **d.** $9x$

5. Equation: $\frac{9}{x-1} = \frac{7}{(x-1)^2}$; LCD: ______

a. $(x - 1)^2$ **b.** $(x - 1)$ **c.** $(x - 1)^3$ **d.** 63

6. Equation: $\frac{1}{x-2} - \frac{3}{x^2-4} = 8$; LCD: ______

a. $(x - 2)$ **b.** $(x + 2)$ **c.** $(x^2 - 4)$ **d.** $(x - 2)(x^2 - 4)$

Martin-Gay Interactive Videos

See Video 5.5

Watch the section lecture video and answer the following questions.

Objective A 7. After multiplying through by the LCD and then simplifying, why is it important to take a moment and determine whether we have a linear or a quadratic equation before we finish solving the problem?

Objective A 8. From Examples 2–5, what extra step is needed when checking solutions to an equation containing rational expressions?

Objective B 9. The steps for solving Example 6 for a specified variable are the same as what other steps? How do we treat this specified variable?

5.5 Exercise Set MyLab Math 

Objective A *Solve each equation and check each solution. See Examples 1 through 3.*

1. $\frac{x}{5} + 3 = 9$

2. $\frac{x}{5} - 2 = 9$

3. $\frac{x}{2} + \frac{5x}{4} = \frac{x}{12}$

4. $\frac{x}{6} + \frac{4x}{3} = \frac{x}{18}$

5. $2 - \frac{8}{x} = 6$

6. $5 + \frac{4}{x} = 1$

7. $2 + \frac{10}{x} = x + 5$

8. $6 + \frac{5}{y} = y - \frac{2}{y}$

9. $\frac{a}{5} = \frac{a - 3}{2}$

10. $\frac{b}{5} = \frac{b + 2}{6}$

11. $\frac{x - 3}{5} + \frac{x - 2}{2} = \frac{1}{2}$

12. $\frac{a + 5}{4} + \frac{a + 5}{2} = \frac{a}{8}$

Solve each equation and check each proposed solution. See Examples 4 through 6.

13. $\frac{3}{2a - 5} = -1$

14. $\frac{6}{4 - 3x} = -3$

15. $\frac{4y}{y - 4} + 5 = \frac{5y}{y - 4}$

16. $\frac{2a}{a + 2} - 5 = \frac{7a}{a + 2}$

17. $2 + \frac{3}{a - 3} = \frac{a}{a - 3}$

18. $\frac{2y}{y - 2} - \frac{4}{y - 2} = 4$

19. $\frac{1}{x + 3} + \frac{6}{x^2 - 9} = 1$

20. $\frac{1}{x + 2} + \frac{4}{x^2 - 4} = 1$

21. $\frac{2y}{y + 4} + \frac{4}{y + 4} = 3$

22. $\frac{5y}{y + 1} - \frac{3}{y + 1} = 4$

23. $\frac{2x}{x + 2} - 2 = \frac{x - 8}{x - 2}$

24. $\frac{4y}{y - 3} - 3 = \frac{3y - 1}{y + 3}$

Solve each equation. See Examples 1 through 6.

25. $\frac{2}{y} + \frac{1}{2} = \frac{5}{2y}$

26. $\frac{6}{3y} + \frac{3}{y} = 1$

27. $\frac{a}{a - 6} = \frac{-2}{a - 1}$

28. $\frac{5}{x - 6} = \frac{x}{x - 2}$

29. $\frac{11}{2x} + \frac{2}{3} = \frac{7}{2x}$

30. $\frac{5}{3} - \frac{3}{2x} = \frac{3}{2}$

31. $\frac{2}{x - 2} + 1 = \frac{x}{x + 2}$

32. $1 + \frac{3}{x + 1} = \frac{x}{x - 1}$

33. $\frac{x + 1}{3} - \frac{x - 1}{6} = \frac{1}{6}$

34. $\frac{3x}{5} - \frac{x - 6}{3} = -\frac{2}{5}$

35. $\frac{t}{t - 4} = \frac{t + 4}{6}$

36. $\frac{15}{x + 4} = \frac{x - 4}{x}$

37. $\frac{y}{2y + 2} + \frac{2y - 16}{4y + 4} = \frac{2y - 3}{y + 1}$

38. $\frac{1}{x + 2} = \frac{4}{x^2 - 4} - \frac{1}{x - 2}$

39. $\frac{4r - 4}{r^2 + 5r - 14} + \frac{2}{r + 7} = \frac{1}{r - 2}$

40. $\frac{3}{x + 3} = \frac{12x + 19}{x^2 + 7x + 12} - \frac{5}{x + 4}$

41. $\frac{x + 1}{x + 3} = \frac{x^2 - 11x}{x^2 + x - 6} - \frac{x - 3}{x - 2}$

42. $\frac{2t + 3}{t - 1} - \frac{2}{t + 3} = \frac{5 - 6t}{t^2 + 2t - 3}$

Objective B *Solve each equation for the indicated variable. See Example 7.*

43. $R = \frac{E}{I}$ for I (Electronics: resistance of a circuit)

44. $T = \frac{V}{Q}$ for Q (Water purification: settling time)

45. $T = \frac{2U}{B + E}$ for B (Merchandising: stock turnover rate)

46. $i = \frac{A}{t + B}$ for t (Hydrology: rainfall intensity)

47. $B = \frac{705w}{h^2}$ for w (Health: body-mass index)

△ 48. $\frac{A}{W} = L$ for W (Geometry: area of a rectangle)

49. $N = R + \frac{V}{G}$ for G (Urban forestry: tree plantings per year)

50. $C = \frac{D(A + 1)}{24}$ for A (Medicine: Cowling's Rule for child's dose)

△ 51. $\frac{C}{\pi r} = 2$ for r (Geometry: circumference of a circle)

52. $W = \frac{CE^2}{2}$ for C (Electronics: energy stored in a capacitor)

53. $\frac{1}{y} + \frac{1}{3} = \frac{1}{x}$ for x

54. $\frac{1}{5} + \frac{2}{y} = \frac{1}{x}$ for x

Review

Translating *Write each phrase as an expression. See Sections R.2 and 1.8.*

55. The reciprocal of x

56. The reciprocal of $x + 1$

57. The reciprocal of x, added to the reciprocal of 2

58. The reciprocal of x, subtracted from the reciprocal of 5

Answer each question.

59. If a tank is filled in 3 hours, what part of the tank is filled in 1 hour?

60. If a strip of beach is cleaned in 4 hours, what part of the beach is cleaned in 1 hour?

Concept Extensions

61. Explain the difference between solving an equation such as $\frac{x}{2} + \frac{3}{4} = \frac{x}{4}$ for x and performing an operation such as adding $\frac{x}{2} + \frac{3}{4}$.

62. When solving an equation such as $\frac{y}{4} = \frac{y}{2} - \frac{1}{4}$, we may multiply all terms by 4. When subtracting two rational expressions such as $\frac{y}{2} - \frac{1}{4}$, we may not. Explain why.

Determine whether each of the following is an equation or an expression. If it is an equation, then solve it for its variable. If it is an expression, perform the indicated operation.

63. $\frac{1}{x} + \frac{5}{9}$

64. $\frac{1}{x} + \frac{5}{9} = \frac{2}{3}$

65. $\frac{5}{x-1} - \frac{2}{x} = \frac{5}{x(x-1)}$

66. $\frac{5}{x-1} - \frac{2}{x}$

Recall that two angles are supplementary if the sum of their measures is 180°. Find the measures of the supplementary angles.

67.

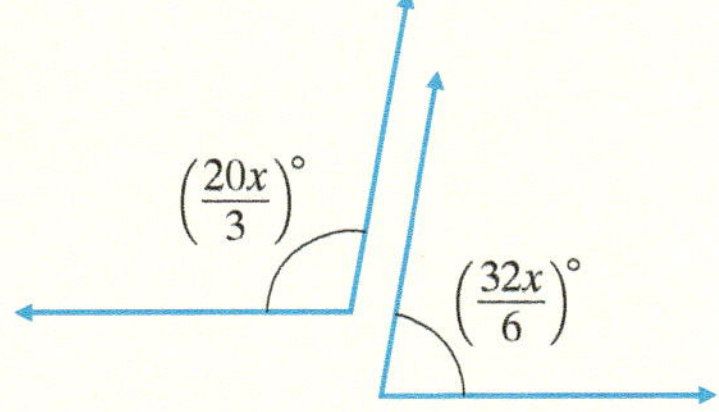

68.

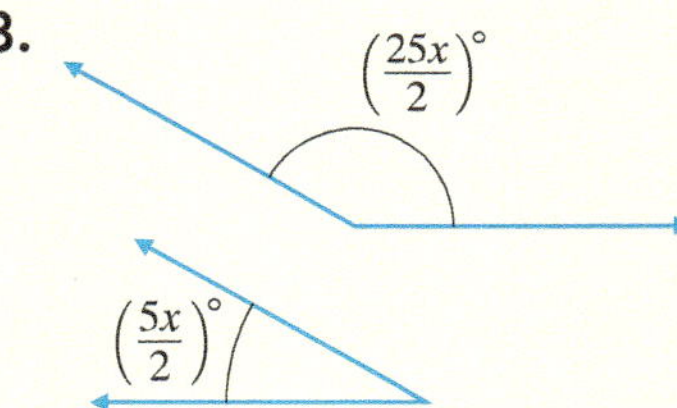

Recall that two angles are complementary if the sum of their measures is 90°. Find the measures of the complementary angles.

69.

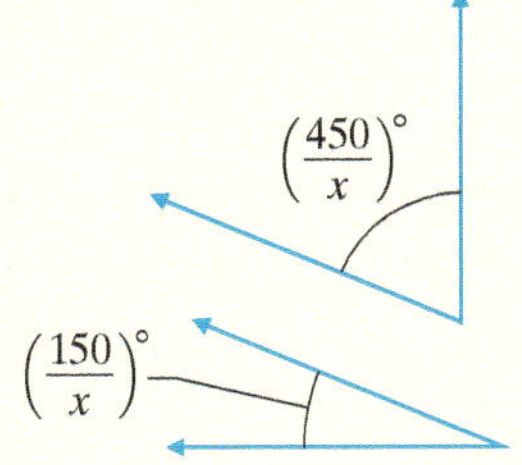

70.

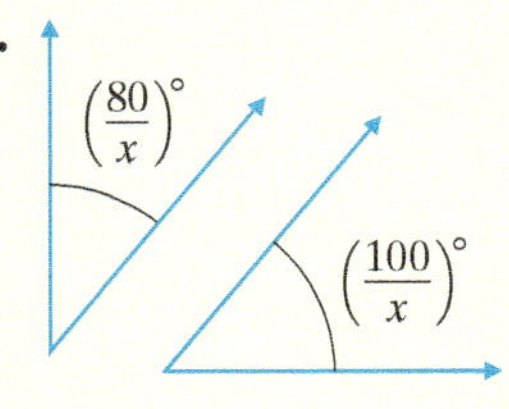

Solve each equation.

71. $\frac{4}{a^2 + 4a + 3} + \frac{2}{a^2 + a - 6} - \frac{3}{a^2 - a - 2} = 0$

72. $\frac{-4}{a^2 + 2a - 8} + \frac{1}{a^2 + 9a + 20} = \frac{-4}{a^2 + 3a - 10}$

Integrated Review Sections 5.1–5.5

Answers

1. ______

2. ______

3. ______

4. ______

5. ______

6. ______

7. ______

8. ______

9. ______

10. ______

Summary on Rational Expressions

It is important to know the difference between performing operations with rational expressions and solving an equation containing rational expressions. Study the examples below.

Performing Operations with Rational Expressions

Adding: $\frac{1}{x} + \frac{1}{x+5} = \frac{1 \cdot (x+5)}{x(x+5)} + \frac{1 \cdot x}{x(x+5)} = \frac{x+5+x}{x(x+5)} = \frac{2x+5}{x(x+5)}$

Subtracting: $\frac{3}{x} - \frac{5}{x^2y} = \frac{3 \cdot xy}{x \cdot xy} - \frac{5}{x^2y} = \frac{3xy-5}{x^2y}$

Multiplying: $\frac{2}{x} \cdot \frac{5}{x-1} = \frac{2 \cdot 5}{x(x-1)} = \frac{10}{x(x-1)}$

Dividing: $\frac{4}{2x+1} \div \frac{x-3}{x} = \frac{4}{2x+1} \cdot \frac{x}{x-3} = \frac{4x}{(2x+1)(x-3)}$

Solving an Equation Containing Rational Expressions

To solve an equation containing rational expressions, we clear the equation of fractions by multiplying both sides by the LCD.

$$\frac{3}{x} - \frac{5}{x-1} = \frac{1}{x(x-1)} \quad \text{Note that } x \text{ can't be 0 or 1.}$$

$$x(x-1)\left(\frac{3}{x}\right) - x(x-1)\left(\frac{5}{x-1}\right) = x(x-1) \cdot \frac{1}{x(x-1)} \quad \text{Multiply both sides by the LCD.}$$

$$3(x-1) - 5x = 1 \quad \text{Simplify.}$$

$$3x - 3 - 5x = 1 \quad \text{Use the distributive property.}$$

$$-2x - 3 = 1 \quad \text{Combine like terms.}$$

$$-2x = 4 \quad \text{Add 3 to both sides.}$$

$$x = -2 \quad \text{Divide both sides by } -2.$$

Don't forget to check to make sure our proposed solution of -2 does not make any denominators 0. If it does, this proposed solution is *not* a solution of the equation. -2 checks and is the solution.

Determine whether each of the following is an equation or an expression. If it is an equation, solve it for its variable. If it is an expression, perform the indicated operation.

1. $\frac{1}{x} + \frac{2}{3}$

2. $\frac{3}{a} + \frac{5}{6}$

3. $\frac{1}{x} + \frac{2}{3} = \frac{3}{x}$

4. $\frac{3}{a} + \frac{5}{6} = 1$

5. $\frac{2}{x+1} - \frac{1}{x}$

6. $\frac{4}{x-3} - \frac{1}{x}$

7. $\frac{2}{x+1} - \frac{1}{x} = 1$

8. $\frac{4}{x-3} - \frac{1}{x} = \frac{6}{x(x-3)}$

9. $\frac{15x}{x+8} \cdot \frac{2x+16}{3x}$

10. $\frac{9z+5}{15} \cdot \frac{5z}{81z^2-25}$

11. $\dfrac{2x+1}{x-3}+\dfrac{3x+6}{x-3}$

12. $\dfrac{4p-3}{2p+7}+\dfrac{3p+8}{2p+7}$

13. $\dfrac{x+5}{7}=\dfrac{8}{2}$

14. $\dfrac{1}{2}=\dfrac{x+1}{8}$

15. $\dfrac{5a+10}{18}\div\dfrac{a^2-4}{10a}$

16. $\dfrac{9}{x^2-1}\div\dfrac{12}{3x+3}$

17. $\dfrac{x+2}{3x-1}+\dfrac{5}{(3x-1)^2}$

18. $\dfrac{4}{(2x-5)^2}+\dfrac{x+1}{2x-5}$

19. $\dfrac{x-7}{x}-\dfrac{x+2}{5x}$

20. $\dfrac{10x-9}{x}-\dfrac{x-4}{3x}$

21. $\dfrac{3}{x+3}=\dfrac{5}{x^2-9}-\dfrac{2}{x-3}$

22. $\dfrac{9}{x^2-4}+\dfrac{2}{x+2}=\dfrac{-1}{x-2}$

23. Explain the difference between solving an equation such as $\dfrac{x}{3}+\dfrac{1}{6}=\dfrac{x}{6}$ for x and performing an operation such as adding $\dfrac{x}{3}+\dfrac{1}{6}$.

24. When solving an equation such as $\dfrac{y}{6}=\dfrac{y}{3}-\dfrac{1}{6}$, we may multiply all terms by 6. When subtracting two rational expressions such as $\dfrac{y}{3}-\dfrac{1}{6}$, we may not. Explain why.

11. ____________

12. ____________

13. ____________

14. ____________

15. ____________

16. ____________

17. ____________

18. ____________

19. ____________

20. ____________

21. ____________

22. ____________

23. ____________

24. ____________

5.6 Proportions and Problem Solving with Rational Equations

Objectives

A Solve Proportions.

B Use Proportions to Solve Problems, Including Similar Triangle Problems.

C Solve Problems About Numbers.

D Solve Problems About Work.

E Solve Problems About Distance.

Objective A Solving Proportions

A **ratio** is the quotient of two numbers or two quantities. For example, the ratio of 2 to 5 can be written in fraction form as $\frac{2}{5}$, the quotient of 2 and 5.

A **rate** is a special type of ratio with different kinds of measurement. For example, the ratio "110 miles in 2 hours" written as a fraction in simplest form is $\frac{110 \text{ miles}}{2 \text{ hours}} = \frac{55 \text{ mi}}{1 \text{ hr}}$ or 55 mph.

If two ratios are equal, we say the ratios are **in proportion** to each other. A **proportion** is a mathematical statement that two ratios are equal.

For example, the equation $\frac{1}{2} = \frac{4}{8}$ is a proportion, as is $\frac{x}{5} = \frac{8}{10}$, because both sides of the equations are ratios. When we want to emphasize the equation as a proportion, we

read the proportion $\frac{1}{2} = \frac{4}{8}$ as "one is to two as four is to eight"

In a proportion, cross products are equal. To understand cross products, let's start with the proportion

$$\frac{a}{b} = \frac{c}{d}$$

and multiply both sides by the LCD, bd.

$$bd\left(\frac{a}{b}\right) = bd\left(\frac{c}{d}\right) \quad \text{Multiply both sides by the LCD, } bd.$$

$$\underbrace{ad}_{\text{Cross product}} = \underbrace{bc}_{\text{Cross product}} \quad \text{Simplify.}$$

Notice why ad and bc are called cross products.

$$ad \qquad \frac{a}{b} = \frac{c}{d} \qquad bc$$

Cross Products

If $\frac{a}{b} = \frac{c}{d}$, then $ad = bc$.

For example, if

$$\frac{1}{2} = \frac{4}{8}, \quad \text{then} \quad 1 \cdot 8 = 2 \cdot 4 \quad \text{or} \quad 8 = 8$$

Notice that a proportion contains four numbers (or expressions). If any three numbers are known, we can solve and find the fourth number.

Example 1 Solve for x: $\frac{45}{x} = \frac{5}{7}$

Solution: This is an equation with rational expressions, and also a proportion. Below are two ways to solve.

Since this is a rational equation, we can use the methods of the previous section.

$$\frac{45}{x} = \frac{5}{7}$$

$7x \cdot \frac{45}{x} = 7x \cdot \frac{5}{7}$ Multiply both sides by LCD, $7x$.

$7 \cdot 45 = x \cdot 5$ Divide out common factors.

$315 = 5x$ Multiply.

$\frac{315}{5} = \frac{5x}{5}$ Divide both sides by 5.

$63 = x$ Simplify.

Since this is also a proportion, we may set cross products equal.

$$\frac{45}{x} = \frac{5}{7}$$

$45 \cdot 7 = x \cdot 5$ Set cross products equal.

$315 = 5x$ Multiply.

$\frac{315}{5} = \frac{5x}{5}$ Divide both sides by 5.

$63 = x$ Simplify.

Check: Both methods give us a solution of 63. To check, substitute 63 for x in the original proportion. The solution is 63.

Work Practice 1

Practice 1

Solve for x: $\frac{3}{8} = \frac{63}{x}$

In this section, if the rational equation is a proportion, we will use cross products to solve.

Example 2 Solve for x: $\frac{x - 5}{3} = \frac{x + 2}{5}$

Solution:

$$\frac{x - 5}{3} = \frac{x + 2}{5}$$

$5(x - 5) = 3(x + 2)$ Set cross products equal.

$5x - 25 = 3x + 6$ Multiply.

$5x = 3x + 31$ Add 25 to both sides.

$2x = 31$ Subtract $3x$ from both sides.

$\frac{2x}{2} = \frac{31}{2}$ Divide both sides by 2.

$x = \frac{31}{2}$

Check: Verify that $\frac{31}{2}$ is the solution.

Work Practice 2

Practice 2

Solve for x: $\frac{2x + 1}{7} = \frac{x - 3}{5}$

Objective B Using Proportions to Solve Problems

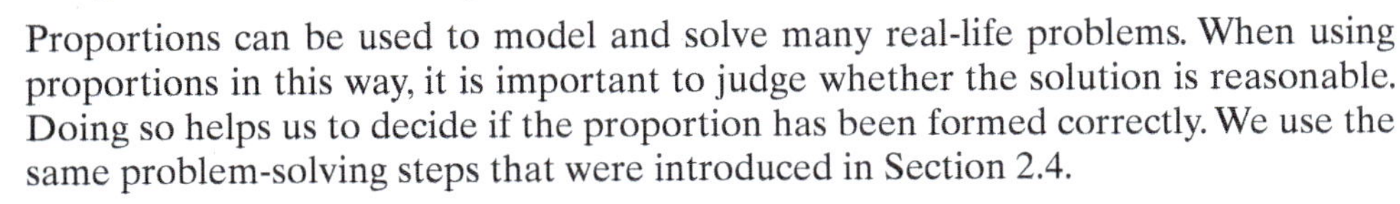

Proportions can be used to model and solve many real-life problems. When using proportions in this way, it is important to judge whether the solution is reasonable. Doing so helps us to decide if the proportion has been formed correctly. We use the same problem-solving steps that were introduced in Section 2.4.

Answers

1. $x = 168$ **2.** $x = -\frac{26}{3}$

Practice 3

To estimate the number of people in Jackson, population 50,000, who have a flu shot, 250 people were polled. Of those polled, 26 had a flu shot. How many people in the city might we expect to have a flu shot?

Example 3 Calculating the Cost of a Redbox Rental

Not everyone is streaming movies. Many people still rent movies and video games from Redbox. If renting 3 movies costs \$4.50, how much should 5 movies cost?

Solution

1. UNDERSTAND. Read and reread the problem. We know that the cost of renting 5 movies is more than the cost of renting three movies, or \$4.50, and less than the cost of 6 rentals, which is double the cost of 3 rentals, or $2(\$4.50) = \$9.00$. Let's suppose that 5 rentals cost \$7.00. To check, we see if 3 rentals is to 5 rentals as the *price* of three rentals is to the *price* of 5 rentals. In other words, we see if

$$\frac{3\text{ rentals}}{5\text{ rentals}} = \frac{\textit{price}\text{ of 3 rentals}}{\textit{price}\text{ of 5 rentals}}$$

or

$$\frac{3}{5} = \frac{4.50}{7}$$

$$3(7) = 5(4.50) \quad \text{Set cross products equal}$$

or

$$21 = 22.5 \quad \text{Not a true statement.}$$

Thus, \$7 is not correct, but we now have a better understanding of the problem.
Let x = price of renting 5 videos.

2. TRANSLATE.

$$\frac{3\text{ rentals}}{5\text{ rentals}} = \frac{\textit{price}\text{ of 3 rentals}}{\textit{price}\text{ of 5 rentals}}$$

$$\frac{3}{5} = \frac{4.50}{x}$$

3. SOLVE.

$$\frac{3}{5} = \frac{4.50}{x}$$

$$3x = 5(4.50) \quad \text{Set cross products equal.}$$

$$3x = 22.50$$

$$x = 7.50 \quad \text{Divide both sides by 3.}$$

4. INTERPRET.

Check: Verify that 3 rentals is to 5 rentals as \$4.50 is to \$7.50. Also, notice that our solution is a reasonable one as discussed in Step 1.

State: Five rentals from Redbox cost \$7.50.

Work Practice 3

Helpful Hint

The proportion $\dfrac{5\text{ rentals}}{3\text{ rentals}} = \dfrac{\textit{price}\text{ of 5 rentals}}{\textit{price}\text{ of 3 rentals}}$ could also have been used to solve Example 3. Notice that the cross products are the same.

Answer

3. 5200 people

Similar triangles have the same shape but not necessarily the same size. In similar triangles, the measures of corresponding angles are equal, and corresponding sides are in proportion.

If triangle ABC and triangle XYZ shown are similar, then we know that the measure of angle A = the measure of angle X, the measure of angle B = the measure of angle Y, and the measure of angle C = the measure of angle Z. We also know that corresponding sides are in proportion: $\frac{a}{x} = \frac{b}{y} = \frac{c}{z}$.

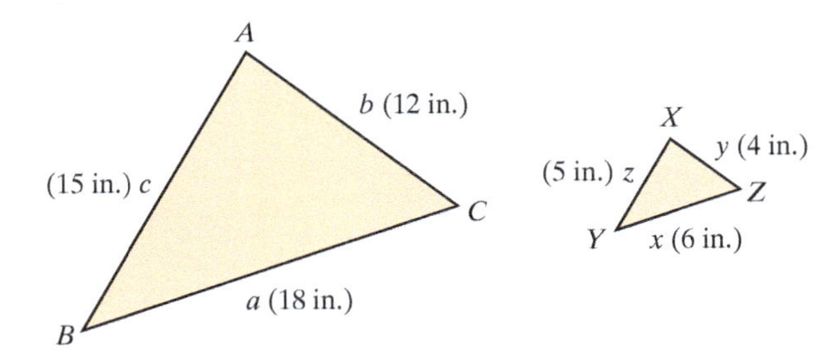

In this section, we will position similar triangles so that they have the same orientation.

To show that corresponding sides are in proportion for the triangles above, we write the ratios of the corresponding sides.

$$\frac{a}{x} = \frac{18}{6} = 3 \quad \frac{b}{y} = \frac{12}{4} = 3 \quad \frac{c}{z} = \frac{15}{5} = 3$$

Example 4 Finding the Length of a Side of a Triangle

If the following two triangles are similar, find the missing length x.

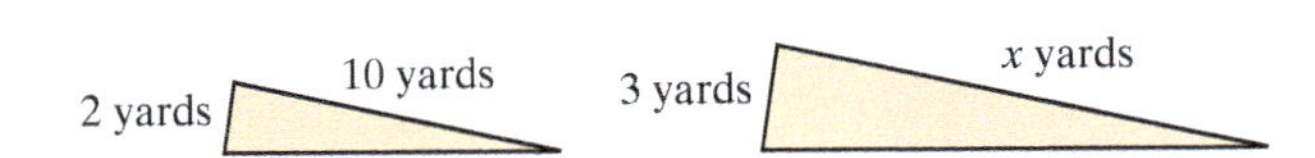

Solution:

1. UNDERSTAND. Read the problem and study the figure.
2. TRANSLATE. Since the triangles are similar, their corresponding sides are in proportion and we have

$$\frac{2}{3} = \frac{10}{x}$$

3. SOLVE. To solve, we multiply both sides by the LCD, $3x$, or cross multiply.

$$2x = 30$$
$$x = 15 \quad \text{Divide both sides by 2.}$$

4. INTERPRET.

Check: To check, replace x with 15 in the original proportion and see that a true statement results.

State: The missing length is 15 yards.

Work Practice 4

Practice 4

For the similar triangles, find x.

12 units
9 units
x units
15 units

Answer

4. 20 units

Objective C Solving Problems About Numbers

Let's continue to solve problems. The remaining problems are all modeled by rational equations.

Practice 5

The quotient of a number and 2, minus $\frac{1}{3}$, is the quotient of the number and 6. Find the number.

Example 5 Finding an Unknown Number

The quotient of a number and 6, minus $\frac{5}{3}$, is the quotient of the number and 2. Find the number.

Solution:

1. UNDERSTAND. Read and reread the problem. Suppose that the unknown number is 2; then we see if the quotient of 2 and 6, or $\frac{2}{6}$, minus $\frac{5}{3}$ is equal to the quotient of 2 and 2, or $\frac{2}{2}$.

$$\frac{2}{6} - \frac{5}{3} = \frac{1}{3} - \frac{5}{3} = -\frac{4}{3}, \text{ not } \frac{2}{2}$$

Don't forget that the purpose of a proposed solution is to better understand the problem.
Let x = the unknown number.

2. TRANSLATE.

In words:	the quotient of x and 6	minus	$\frac{5}{3}$	is	the quotient of x and 2
	↓	↓	↓	↓	↓
Translate:	$\frac{x}{6}$	$-$	$\frac{5}{3}$	$=$	$\frac{x}{2}$

3. SOLVE. Here, we solve the equation $\frac{x}{6} - \frac{5}{3} = \frac{x}{2}$. We begin by multiplying both sides of the equation by the LCD, 6.

$$6\left(\frac{x}{6} - \frac{5}{3}\right) = 6\left(\frac{x}{2}\right)$$

$$6\left(\frac{x}{6}\right) - 6\left(\frac{5}{3}\right) = 6\left(\frac{x}{2}\right) \quad \text{Apply the distributive property.}$$

$$x - 10 = 3x \quad \text{Simplify.}$$

$$-10 = 2x \quad \text{Subtract } x \text{ from both sides.}$$

$$\frac{-10}{2} = \frac{2x}{2} \quad \text{Divide both sides by 2.}$$

$$-5 = x \quad \text{Simplify.}$$

4. INTERPRET.

Check: To check, we verify that "the quotient of -5 and 6 minus $\frac{5}{3}$ is the quotient of -5 and 2," or $-\frac{5}{6} - \frac{5}{3} = -\frac{5}{2}$.

State: The unknown number is -5.

Work Practice 5

Answer
5. 1

Objective D Solving Problems About Work 

The next example is often called a work problem. Work problems usually involve people or machines doing a certain task.

Example 6 Finding Work Rates

Sam Waterton and Frank Schaffer work in a plant that manufactures automobiles. Sam can complete a quality control tour of the plant in 3 hours while his assistant, Frank, needs 7 hours to complete the same job. The regional manager is coming to inspect the plant facilities, so both Sam and Frank are directed to complete a quality control tour together. How long will this take?

Solution:

1. UNDERSTAND. Read and reread the problem. The key idea here is the relationship between the **time** (hours) it takes to complete the job and the **part of the job** completed in 1 unit of time (hour). For example, if the **time** it takes Sam to complete the job is 3 hours, the **part of the job** he can complete in 1 hour is $\frac{1}{3}$. Similarly, Frank can complete $\frac{1}{7}$ of the job in 1 hour.

   Let $x =$ the **time** in hours it takes Sam and Frank to complete the job together.

   Then $\frac{1}{x} =$ the **part of the job** they complete in 1 hour.

	Hours to Complete Total Job	Part of Job Completed in 1 Hour
Sam	3	$\frac{1}{3}$
Frank	7	$\frac{1}{7}$
Together	x	$\frac{1}{x}$

2. TRANSLATE.

In words:	part of job job Sam completes in 1 hour	added to	part of job Frank completes in 1 hour	is equal to	part of job they complete together in in 1 hour
	↓	↓	↓	↓	↓
Translate:	$\frac{1}{3}$	+	$\frac{1}{7}$	=	$\frac{1}{x}$

3. SOLVE. Here, we solve the equation $\frac{1}{3} + \frac{1}{7} = \frac{1}{x}$. We begin by multiplying both sides of the equation by the LCD, $21x$.

$$21x\left(\frac{1}{3}\right) + 21x\left(\frac{1}{7}\right) = 21x\left(\frac{1}{x}\right)$$

$$7x + 3x = 21 \quad \text{Simplify.}$$

$$10x = 21$$

$$x = \frac{21}{10} \text{ or } 2\frac{1}{10} \text{ hours}$$

(*Continued on next page*)

Practice 6

Guillaume Beauchesne and Greg Langacker volunteer at a local recycling plant. Guillaume can sort a batch of recyclables in 2 hours alone while his friend Greg needs 3 hours to complete the same job. If they work together, how long will it take them to sort one batch?

Answer

6. $1\frac{1}{5}$ hours

4. INTERPRET.

Check: Our proposed solution is $2\frac{1}{10}$ hours. This proposed solution is reasonable since $2\frac{1}{10}$ hours is more than half of Sam's time and less than half of Frank's time. Check this solution in the originally *stated* problem.

State: Sam and Frank can complete the quality control tour in $2\frac{1}{10}$ hours.

Work Practice 6

✓Concept Check Solve $E = mc^2$

a. for m **b.** for c^2

Objective E Solving Problems About Distance

Next we look at a problem solved by the distance formula,

$d = r \cdot t$

Example 7 Finding Speeds of Vehicles

A car travels 180 miles in the same time that a truck travels 120 miles. If the car's speed is 20 miles per hour faster than the truck's, find the car's speed and the truck's speed.

Solution:

1. UNDERSTAND. Read and reread the problem. Suppose that the truck's speed is 45 miles per hour. Then the car's speed is 20 miles per hour faster, or 65 miles per hour.

We are given that the car travels 180 miles in the same time that the truck travels 120 miles. To find the time it takes the car to travel 180 miles, remember that since $d = rt$, we know that $\frac{d}{r} = t$.

Car's Time

$$t = \frac{d}{r} = \frac{180}{65} = 2\frac{50}{65} = 2\frac{10}{13} \text{ hours}$$

Truck's Time

$$t = \frac{d}{r} = \frac{120}{45} = 2\frac{30}{45} = 2\frac{2}{3} \text{ hours}$$

Since the times are not the same, our proposed solution is not correct. But we have a better understanding of the problem.

Let $x =$ the speed of the truck.

Since the car's speed is 20 miles per hour faster than the truck's, then

$x + 20 =$ the speed of the car

Use the formula $d = r \cdot t$ or **d**istance = **r**ate · **t**ime. Prepare a chart to organize the information in the problem.

	Distance	=	Rate	·	Time
Truck	120		x		$\frac{120}{x}$ ← distance / ← rate
Car	180		$x + 20$		$\frac{180}{x+20}$ ← distance / ← rate

Practice 7

A car travels 600 miles in the same time that a motorcycle travels 450 miles. If the car's speed is 15 miles per hour faster than the motorcycle's, find the speed of the car and the speed of the motorcycle.

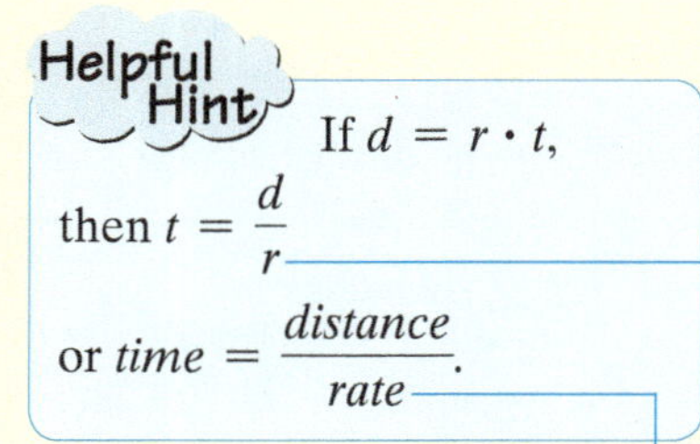

If $d = r \cdot t$,

then $t = \frac{d}{r}$

or $time = \frac{distance}{rate}$.

Answer

7. car: 60 mph; motorcycle: 45 mph

✓Concept Check Answers

a. $m = \frac{E}{c^2}$ **b.** $c^2 = \frac{E}{m}$

2. TRANSLATE. Since the car and the truck traveled the same amount of time, we have that

In words: car's time = truck's time

Translate: $\frac{180}{x + 20} = \frac{120}{x}$

3. SOLVE. We begin by multiplying both sides of the equation by the LCD, $x(x + 20)$, or cross multiplying.

$$\frac{180}{x + 20} = \frac{120}{x}$$

$$180x = 120(x + 20)$$

$$180x = 120x + 2400 \quad \text{Use the distributive property.}$$

$$60x = 2400 \quad \text{Subtract } 120x \text{ from both sides.}$$

$$x = 40 \quad \text{Divide both sides by 60.}$$

4. INTERPRET. The speed of the truck is 40 miles per hour. The speed of the car must then be $x + 20$ or 60 miles per hour.

Check: Find the time it takes the car to travel 180 miles and the time it takes the truck to travel 120 miles.

Car's Time

$$t = \frac{d}{r} = \frac{180}{60} = 3 \text{ hours}$$

Truck's Time

$$t = \frac{d}{r} = \frac{120}{40} = 3 \text{ hours}$$

Since both travel the same amount of time, the proposed solution is correct.

State: The car's speed is 60 miles per hour and the truck's speed is 40 miles per hour.

Work Practice 7

Vocabulary, Readiness & Video Check

Without solving algebraically, select the best choice for each exercise.

1. One person can complete a job in 7 hours. A second person can complete the same job in 5 hours. How long will it take them to complete the job if they work together?

a. more than 7 hours
b. between 5 and 7 hours
c. less than 5 hours

2. One inlet pipe can fill a pond in 30 hours. A second inlet pipe can fill the same pond in 25 hours. How long before the pond is filled if both inlet pipes are on?

a. less than 25 hours
b. between 25 and 30 hours
c. more than 30 hours

Fill in a Table *Given the variable in the first column, use the phrase in the second column to translate to an expression, and then continue to the phrase in the third column to translate to another expression.*

3.	A number: x	The reciprocal of the number:	The reciprocal of the number, decreased by 3:
4.	A number: y	The reciprocal of the number:	The reciprocal of the number, increased by 2:
5.	A number: z	The sum of the number and 5:	The reciprocal of the sum of the number and 5:
6.	A number: x	The difference of the number and 1:	The reciprocal of the difference of the number and 1:
7.	A number: y	Twice the number:	Eleven divided by twice the number:
8.	A number: z	Triple the number:	Negative ten divided by triple the number:

Martin-Gay Interactive Videos

See Video 5.6

Watch the section lecture video and answer the following questions.

Objective A **9.** Based on Examples 1 and 2, can proportions only be solved by using cross products? Explain.

Objective B **10.** In Example 3 we are told there are many ways to set up a correct proportion. Why does this fact make it even more important to check that our solution is reasonable?

Objective C **11.** What words or phrases in Example 5 told us to translate to an equation containing rational expressions?

Objective D **12.** From Example 6, how can we determine a somewhat reasonable answer to a work problem before we even begin to solve it?

Objective E **13.** The following problem is worded like Example 7, but using different quantities.

A car travels 325 miles in the same time that a motorcycle travels 290 miles. If the car's speed is 7 miles per hour more than the motorcycle's, find the speed of the car and the speed of the motorcycle. Fill in the table and set up an equation based on this problem (do not solve). Use Example 7 as a model for our work.

	d	$= r$	$\cdot\ t$
car			
motorcycle			

5.6 Exercise Set MyLab Math

Objective A *Solve each proportion. See Examples 1 and 2.*

1. $\frac{2}{3} = \frac{x}{6}$

2. $\frac{x}{2} = \frac{16}{6}$

3. $\frac{x}{10} = \frac{5}{9}$

4. $\frac{9}{4x} = \frac{6}{2}$

5. $\frac{x+1}{2x+3} = \frac{2}{3}$

6. $\frac{x+1}{x+2} = \frac{5}{3}$

7. $\frac{9}{5} = \frac{12}{3x+2}$

8. $\frac{6}{11} = \frac{27}{3x-2}$

Objective B *Solve. See Example 3.*

9. The ratio of the weight of an object on Earth to the weight of the same object on Pluto is 100 to 3. If an elephant weighs 4100 pounds on Earth, find the elephant's weight on Pluto.

10. If a 170-pound person weighs approximately 65 pounds on Mars, about how much does a 9000-pound satellite weigh? Round your answer to the nearest pound.

11. There are 110 calories per 28.4 grams of Crispy Rice cereal. Find how many calories are in 42.6 grams of this cereal.

12. On an architect's blueprint, 1 inch corresponds to 4 feet. Find the length of a wall represented by a line that is $3\frac{7}{8}$ inches long on the blueprint.

Find the unknown length x or y in the following pairs of similar triangles. See Example 4.

△ **13.**

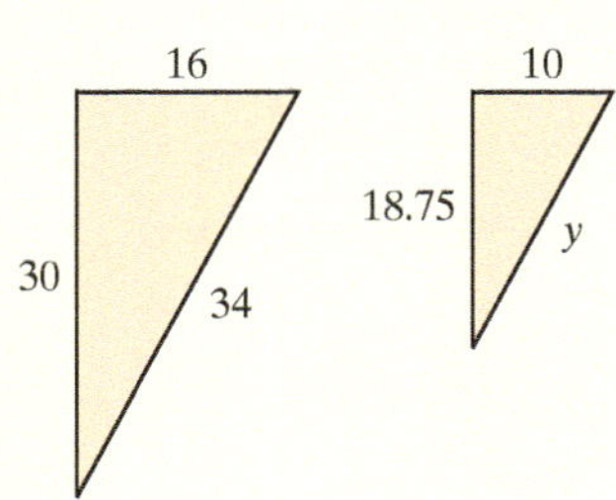

△ **14.**

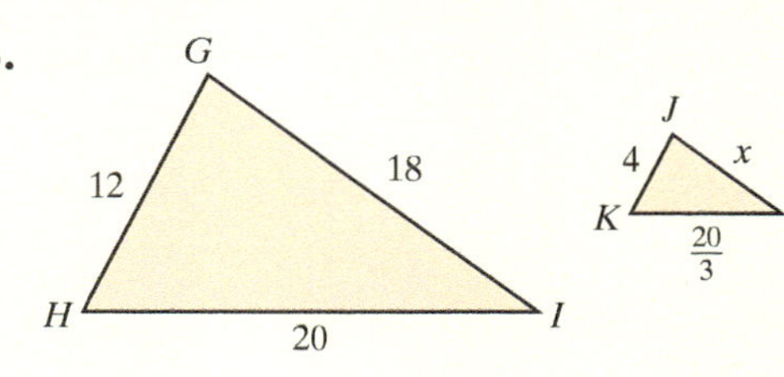

△ **15.**

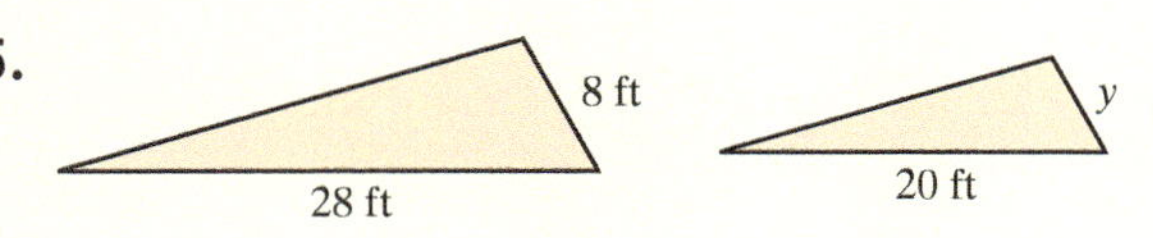

△ **16.**

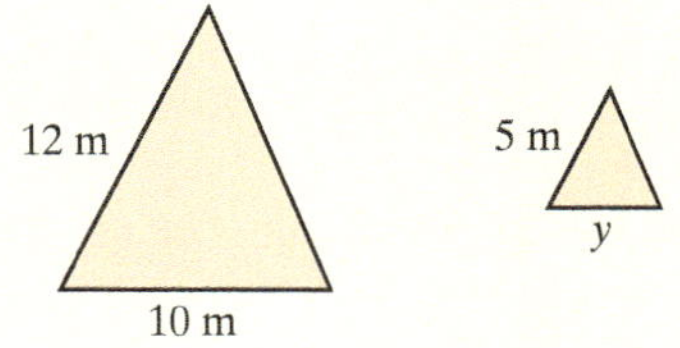

Objective C *Solve the following. See Example 5.*

17. If twice a number added to 3 is divided by the number plus 1, the result is three halves. Find the number.

18. Twelve divided by the sum of x and 2 equals the quotient of 4 and the difference of x and 2. Find x.

19. A number added to the product of 14 and the reciprocal of the number equals 9. Find the number.

20. A number added to the product of 6 and the reciprocal of the number equals -5. Find the number.

Objective D *See Example 6.*

21. Smith Engineering found that an experienced surveyor surveys a roadbed in 4 hours. An apprentice surveyor needs 5 hours to survey the same stretch of road. If the two work together, find how long it takes them to complete the job.

22. An experienced bricklayer constructs a small wall in 3 hours. The apprentice completes the job in 6 hours. Find how long it takes if they work together.

23. In 2 minutes, a conveyor belt moves 300 pounds of recyclable aluminum from the delivery truck to a storage area. A smaller belt moves the same quantity of cans the same distance in 6 minutes. If both belts are used, find how long it takes to move the cans to the storage area.

24. Find how long it takes the conveyor belts described in Exercise **23** to move 1200 pounds of cans. (*Hint:* Think of 1200 pounds as four 300-pound jobs.)

Objective E See Example 7.

25. A jogging enthusiast begins her workout by jogging to the park, a distance of 12 miles. She then jogs home at the same speed but along a different route. This return trip is 18 miles and her time is one hour longer. Find her jogging speed. Complete the accompanying chart and use it to find her jogging speed.

	Distance	=	Rate	·	Time
Trip to Park	12				
Return Trip	18				

26. A boat can travel 9 miles upstream in the same amount of time it takes to travel 11 miles downstream. If the current of the river is 3 miles per hour, complete the chart below and use it to find the speed of the boat in still water.

	Distance	=	Rate	·	Time
Upstream	9		$r - 3$		
Downstream	11		$r + 3$		

27. A cyclist rode the first 20-mile portion of his workout at a constant speed. For the 16-mile cooldown portion of his workout, he reduced his speed by 2 miles per hour. Each portion of the workout took the same time. Find the cyclist's speed during the first portion and find his speed during the cooldown portion.

28. A semi-truck travels 300 miles through the flatland in the same amount of time that it travels 180 miles through mountains. The rate of the truck is 20 miles per hour slower in the mountains than in the flatland. Find both the flatland rate and the mountain rate.

Objectives A B C D E Mixed Practice

Solve the following. See Examples 1 through 7. (Note: Some exercises can be modeled by equations without rational expressions.)

29. A human factors expert recommends that there be at least 9 square feet of floor space in a college classroom for every student in the class. Find the minimum floor space that 40 students need.

30. Due to space problems at a local university, a 20-foot by 12-foot conference room is converted into a classroom. Find the maximum number of students the room can accommodate. (See Exercise **29.**)

31. One-fourth equals the quotient of a number and 8. Find the number.

32. Four times a number added to 5 is divided by 6. The result is $\frac{7}{2}$. Find the number.

33. Marcus and Tony work for Lombardo's Pipe and Concrete. Mr. Lombardo is preparing an estimate for a customer. He knows that Marcus lays a slab of concrete in 6 hours. Tony lays the same size slab in 4 hours. If both work on the job and the cost of labor is $45.00 per hour, decide what the labor estimate should be.

34. Mr. Dodson can paint his house by himself in 5 days. His son needs an additional day to complete the job if he works by himself. If they work together, find how long it takes to paint the house.

35. A pilot can travel 400 miles with the wind in the same amount of time as 336 miles against the wind. Find the speed of the wind if the pilot's speed in still air is 230 miles per hour.

36. A fisherman on Pearl River rows 9 miles downstream in the same amount of time he rows 3 miles upstream. If the current is 6 miles per hour, find how long it takes him to cover the 12 miles.

37. Find the unknown length y.

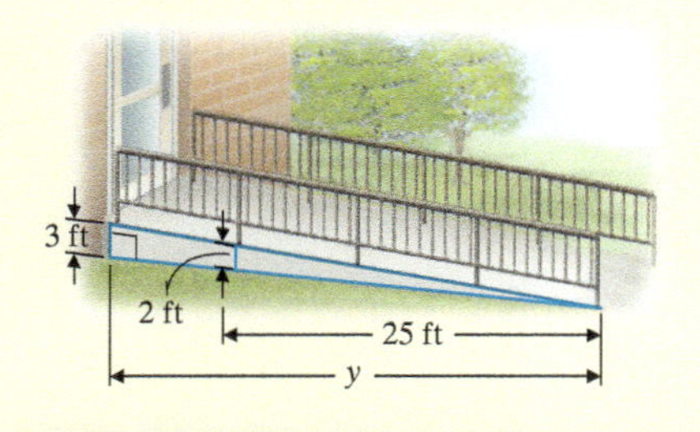

△ **38.** Find the unknown length y.

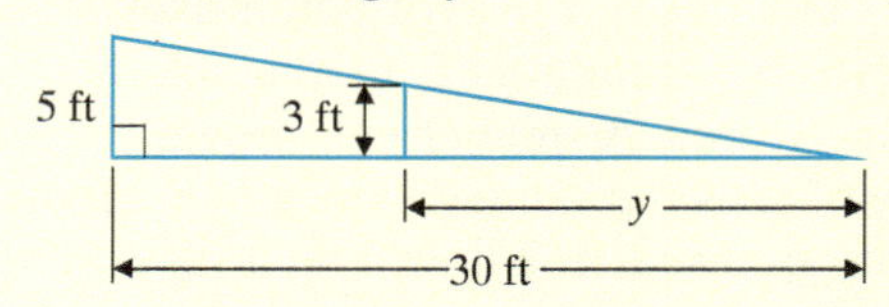

39. Suppose two trains leave Holbrook, Arizona, at the same time, traveling in opposite directions. One train travels 10 mph faster than the other. In 3.5 hours, the trains are 322 miles apart. Find the speed of each train.

40. Suppose two cars leave Brinkley, Arkansas, at the same time, traveling in opposite directions. One car travels 8 mph faster than the other car. In 2.5 hours, the cars are 280 miles apart. Find the speed of each car.

41. Two divided by the difference of a number and 3 minus 4 divided by the number plus 3, equals 8 times the reciprocal of the difference of the number squared and 9. What is the number?

42. If 15 times the reciprocal of a number is added to the ratio of 9 times the number minus 7 and the number plus 2, the result is 9. What is the number?

43. A pilot flies 630 miles with a tailwind of 35 miles per hour. Against the wind, he flies only 455 miles in the same amount of time. Find the rate of the plane in still air.

44. A marketing manager travels 1080 miles in a corporate jet and then an additional 240 miles by car. If the car ride takes one hour longer than the jet ride takes, and if the rate of the jet is 6 times the rate of the car, find the time the manager travels by jet and find the time the manager travels by car.

45. To mix weed killer with water correctly, it is necessary to mix 8 teaspoons of weed killer with 2 gallons of water. Find how many gallons of water are needed to mix with the entire box if it contains 36 teaspoons of weed killer.

46. The directions for a certain bug spray concentrate are to mix 3 ounces of concentrate with 2 gallons of water. How many ounces of concentrate are needed to mix with 5 gallons of water?

47. A boater travels 16 miles per hour on the water on a still day. During one particularly windy day, he finds that he travels 48 miles with the wind behind him in the same amount of time that he travels 16 miles into the wind. Find the rate of the wind.

Let x be the rate of the wind.

	r · t = d	
with wind	$16 + x$	48
into wind	$16 - x$	16

48. The current on a portion of the Mississippi River is 3 miles per hour. A barge can go 6 miles upstream in the same amount of time it takes to go 10 miles downstream. Find the speed of the boat in still water.

Let x be the speed of the boat in still water.

	r · t = d	
upstream	$x - 3$	6
downstream	$x + 3$	10

49. Two hikers are 11 miles apart and walking toward each other. They meet in 2 hours. Find the rate of each hiker if one hiker walks 1.1 mph faster than the other.

50. On a 255-mile trip, Gary Alessandrini traveled at an average speed of 70 mph, got a speeding ticket, and then traveled at 60 mph for the remainder of the trip. If the entire trip took 4.5 hours and the speeding ticket stop took 30 minutes, how long did Gary speed before getting stopped?

51. One custodian cleans a suite of offices in 3 hours. When a second worker is asked to join the regular custodian, the job takes only $1\frac{1}{2}$ hours. How long does it take the second worker to do the same job alone?

52. One person proofreads copy for a small newspaper in 4 hours. If a second proofreader is also employed, the job can be done in $2\frac{1}{2}$ hours. How long does it take for the second proofreader to do the same job alone?

△ **53.** An architect is completing the plans for a triangular deck. Use the diagram below to find the missing dimension.

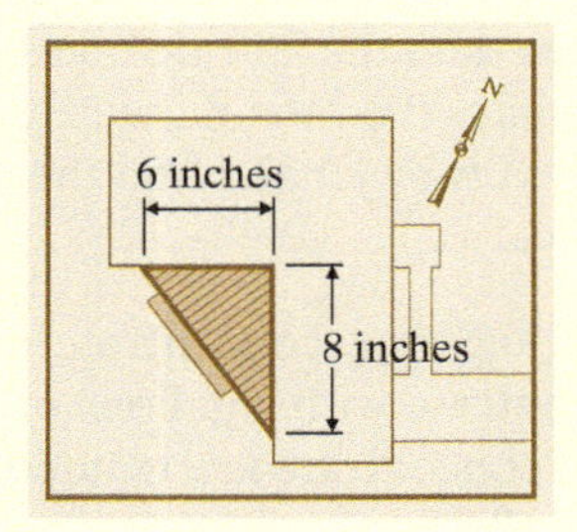

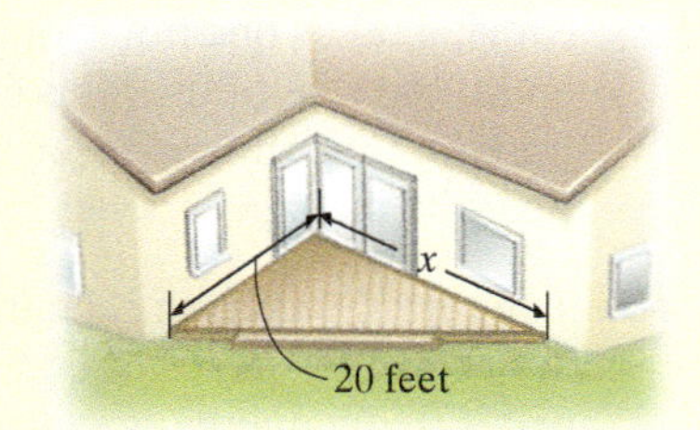

△ **54.** A student wishes to make a small model of a triangular mainsail in order to study the effects of wind on the sail. The smaller model will be the same shape as a regular-size sailboat's mainsail. Use the following diagram to find the missing dimensions.

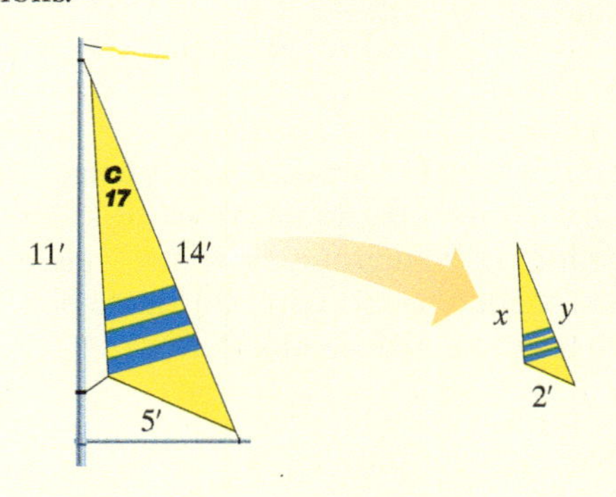

55. A manufacturer of cans of salted mixed nuts states that the ratio of peanuts to other nuts is 3 to 2. If 324 peanuts are in a can, find how many other nuts should also be in the can.

56. There are 1280 calories in a 14-ounce portion of Eagle Brand Milk. Find how many calories are in 2 ounces of Eagle Brand Milk.

57. A jet plane traveling at 500 mph overtakes a propeller plane traveling at 200 mph that had a 2-hour head start. How far from the starting point are the planes?

58. How long will it take a bus traveling at 60 miles per hour to overtake a car traveling at 40 mph if the car had a 1.5-hour head start?

59. One pipe fills a storage pool in 20 hours. A second pipe fills the same pool in 15 hours. When a third pipe is added and all three are used to fill the pool, it takes only 6 hours. Find how long it takes the third pipe to do the job.

60. One pump fills a tank in 9 hours. A second pump fills the same tank in 6 hours. When a third pump is added and all three are used to fill the tank, it takes only 3 hours. Find how long it takes the third pump to fill the tank.

61. A car travels 280 miles in the same time that a motorcycle travels 240 miles. If the car's speed is 10 miles per hour more than the motorcycle's, find the speed of the car and the speed of the motorcycle.

62. A bus traveled on a level road for 3 hours at an average speed 20 miles per hour faster than it traveled on a winding road. The time spent on the winding road was 4 hours. Find the average speed on the level road if the entire trip was 305 miles.

63. In 6 hours, an experienced cook prepares enough pies to supply a local restaurant's daily order. Another cook prepares the same number of pies in 7 hours. Together with a third cook, they prepare the pies in 2 hours. Find how long it takes the third cook to prepare the pies alone.

64. Mrs. Smith balances the company books in 8 hours. It takes her assistant 12 hours to do the same job. Together with a third person, they balance the company books in 2 hours. Find how long it takes the third person to balance the books alone.

65. The quotient of a number and 3, minus 1, equals $\frac{5}{3}$. Find the number.

66. The quotient of a number and 5, minus 1, equals $\frac{7}{5}$. Find the number.

67. Currently, the Toyota Corolla is the best-selling car in the world. A driver of this car took a day trip around the Maine coastline driving at two different speeds. He drove 70 miles at a slower speed and 300 miles at a speed 40 miles per hour faster. If the time spent driving at the faster speed was twice that spent driving at the slower speed, find the two speeds during the trip. (*Source: Forbes*)

68. The second best-selling car in the world is the Volkswagen Golf. Suppose that during a test drive of two Golfs, one car travels 224 miles in the same time that the second car travels 175 miles. If the speed of the first car is 14 miles per hour faster than the speed of the second car, find the speed of both cars. (*Source: Forbes*)

69. A pilot can fly an MD-11 2160 miles with the wind in the same time she can fly 1920 miles against the wind. If the speed of the wind is 30 mph, find the speed of the plane in still air. (*Source:* Air Transport Association of America)

70. A pilot can fly a DC-10 1365 miles against the wind in the same time he can fly 1575 miles with the wind. If the speed of the plane in still air is 490 miles per hour, find the speed of the wind. (*Source:* Air Transport Association of America)

Given that the following pairs of triangles are similar, find each missing length.

△**71.**

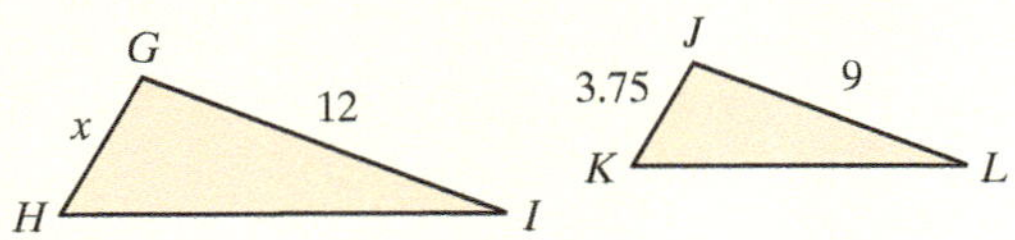

△**72.**

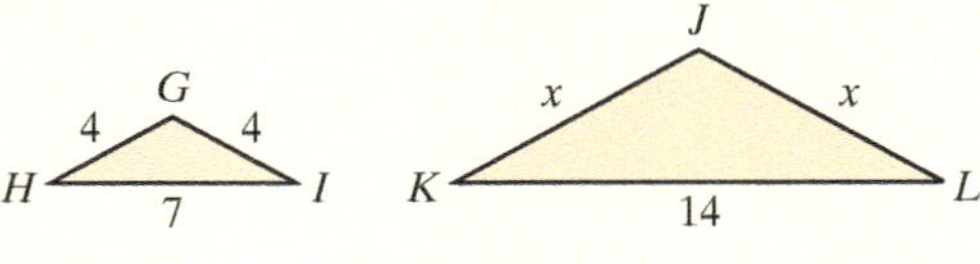

△**73.**

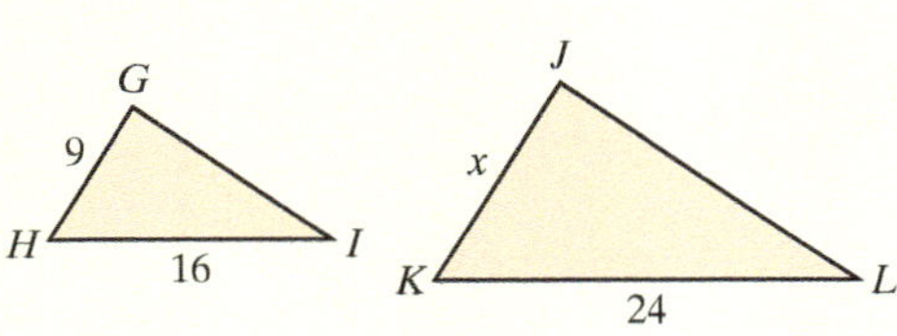

△**74.**

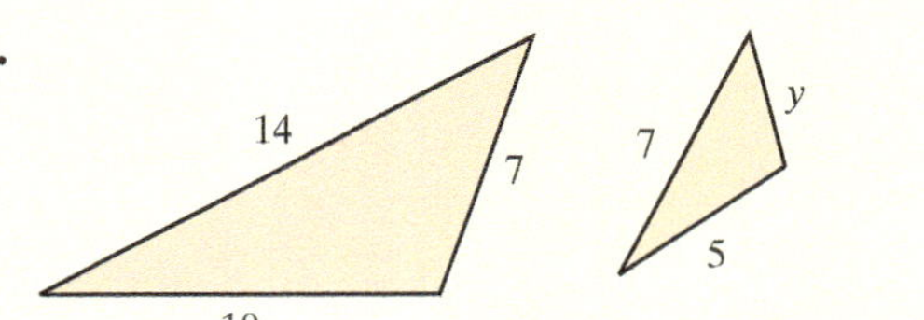

Review

Simplify. Follow the circled steps in the order shown. See Section R.2.

75. $\dfrac{\frac{3}{4}+\frac{1}{4}}{\frac{3}{8}+\frac{13}{8}}$ ← ① Add. (numerator) ← ② Add. (denominator)

76. $\dfrac{\frac{9}{5}+\frac{6}{5}}{\frac{17}{6}+\frac{7}{6}}$ ← ① Add. (numerator) ← ② Add. (denominator)

77. $\dfrac{\frac{2}{5}+\frac{1}{5}}{\frac{7}{10}+\frac{7}{10}}$ ① Add. ← ③ Divide. ② Add.

78. $\dfrac{\frac{1}{4}+\frac{5}{4}}{\frac{3}{8}+\frac{7}{8}}$ ① Add. ← ③ Divide. ② Add.

Concept Extensions

79. One pump fills a tank 3 times as fast as another pump. If the pumps work together, they fill the tank in 21 minutes. How long does it take each pump to fill the tank alone?

80. It takes 9 hours for pump A to fill a tank alone. Pump C fills a tank 3 times as fast as Pump B. If pumps A, B, and C are used, the tank fills in 5 hours. How long does it take pump B to fill the tank alone and pump C to fill the tank alone?

81. Person A can complete a job in 5 hours, and person B can complete the same job in 3 hours. Without solving algebraically, discuss reasonable and unreasonable answers for how long it would take them to complete the job together.

82. For which of the following equations can we immediately use cross products to solve for x?

a. $\frac{2-x}{5}=\frac{1+x}{3}$ **b.** $\frac{2}{5}-x=\frac{1+x}{3}$

Solve. See the Concept Check in this section.

Solve $D = RT$

83. for R

84. for T

85. A hyena spots a giraffe 0.5 mile away and begins running toward it. The giraffe starts running away from the hyena just as the hyena begins running toward it. A hyena can run at a speed of 40 mph and a giraffe can run at 32 mph. How long will it take the hyena to overtake the giraffe? (*Source: The World Almanac and Book of Facts*)

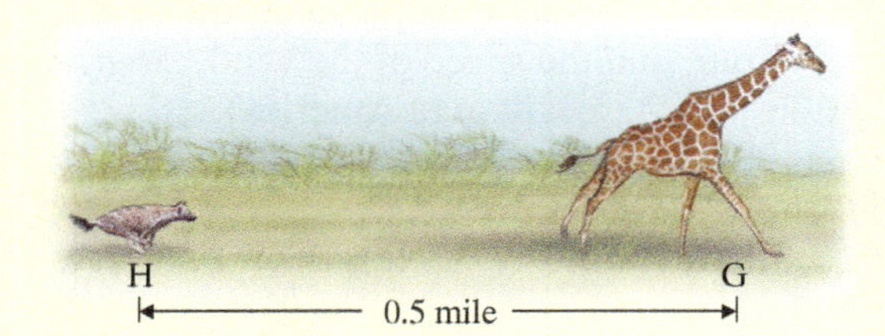

86. The two fastest cars in the world are the Hennessey Venom GT and the Bugatti Veyron 16.4 Super Sport. At an international auto demonstration, the Hennessey traveled 390 miles in the same time the Bugatti traveled 363 miles. If the speed of the Hennessey was 18 mph faster than the speed of the Bugatti, find the speed of both cars. (*Source: Top Ten of Everything*)

5.7 Simplifying Complex Fractions

Objectives

A Simplify Complex Fractions Using Method 1.

B Simplify Complex Fractions Using Method 2.

A rational expression whose numerator or denominator or both numerator and denominator contain fractions is called a **complex rational expression** or a **complex fraction.** Some examples are

$$\frac{4}{2-\frac{1}{2}} \qquad \frac{\frac{3}{2}}{\frac{4}{7}-x} \qquad \frac{\frac{1}{x+2}}{x+2-\frac{1}{x}}$$

← Numerator of complex fraction
← Main fraction bar
← Denominator of complex fraction

Our goal in this section is to write complex fractions in simplest form. A complex fraction is in simplest form when it is in the form $\frac{P}{Q}$, where P and Q are polynomials that have no common factors.

In this section, two methods of simplifying complex fractions are presented.

Objective A Simplifying Complex Fractions—Method 1

The first method presented uses the fact that the main fraction bar indicates division.

Method 1: To Simplify a Complex Fraction

Step 1: Add or subtract fractions in the numerator or denominator so that the numerator is a single fraction and the denominator is a single fraction.

Step 2: Perform the indicated division by multiplying the numerator of the complex fraction by the reciprocal of the denominator of the complex fraction.

Step 3: Write the rational expression in lowest terms.

Example 1 Simplify the complex fraction $\dfrac{\frac{5}{8}}{\frac{2}{3}}$.

Solution: Since the numerator and denominator of the complex fraction are already single fractions, we proceed to Step 2: Perform the indicated division by multiplying the numerator $\frac{5}{8}$ by the reciprocal of the denominator $\frac{2}{3}$.

$$\frac{\frac{5}{8}}{\frac{2}{3}} = \frac{5}{8} \div \frac{2}{3} = \frac{5}{8} \cdot \frac{3}{2} = \frac{15}{16}$$

The reciprocal of $\frac{2}{3}$ is $\frac{3}{2}$.

Work Practice 1

Practice 1

Simplify the complex fraction $\dfrac{\frac{3}{7}}{\frac{5}{9}}$.

Example 2 Simplify: $\dfrac{\frac{2}{3} + \frac{1}{5}}{\frac{2}{3} - \frac{2}{9}}$

Solution: We simplify the numerator and denominator of the complex fraction separately. First we add $\frac{2}{3}$ and $\frac{1}{5}$ to obtain a single fraction in the numerator. Then we subtract $\frac{2}{9}$ from $\frac{2}{3}$ to obtain a single fraction in the denominator.

$$\frac{\frac{2}{3} + \frac{1}{5}}{\frac{2}{3} - \frac{2}{9}} = \frac{\frac{2(5)}{3(5)} + \frac{1(3)}{5(3)}}{\frac{2(3)}{3(3)} - \frac{2}{9}}$$

The LCD of the numerator's fractions is 15.
The LCD of the denominator's fractions is 9.

$$= \frac{\frac{10}{15} + \frac{3}{15}}{\frac{6}{9} - \frac{2}{9}}$$

Simplify.

$$= \frac{\frac{13}{15}}{\frac{4}{9}}$$

Add the numerator's fractions.
Subtract the denominator's fractions.

(*Continued on next page*)

Practice 2

Simplify: $\dfrac{\frac{3}{4} - \frac{2}{3}}{\frac{1}{2} + \frac{3}{8}}$

Answers

1. $\frac{27}{35}$ **2.** $\frac{2}{21}$

Next we perform the indicated division by multiplying the numerator of the complex fraction by the reciprocal of the denominator of the complex fraction.

$$\frac{\frac{13}{15}}{\frac{4}{9}} = \frac{13}{15} \cdot \frac{9}{4} \quad \text{The reciprocal of } \frac{4}{9} \text{ is } \frac{9}{4}.$$

$$= \frac{13 \cdot 3 \cdot 3}{3 \cdot 5 \cdot 4} = \frac{39}{20}$$

Work Practice 2

Practice 3

Simplify: $\dfrac{\frac{2}{5} - \frac{1}{x}}{\frac{2x}{15} - \frac{1}{3}}$

Example 3 Simplify: $\dfrac{\frac{1}{z} - \frac{1}{2}}{\frac{1}{3} - \frac{z}{6}}$

Solution: Subtract to get a single fraction in the numerator and a single fraction in the denominator of the complex fraction.

$$\frac{\frac{1}{z} - \frac{1}{2}}{\frac{1}{3} - \frac{z}{6}} = \frac{\frac{2}{2z} - \frac{z}{2z}}{\frac{2}{6} - \frac{z}{6}} \quad \begin{array}{l}\text{The LCD of the numerator's fractions is } 2z.\\ \text{The LCD of the denominator's fractions is } 6.\end{array}$$

$$= \frac{\frac{2-z}{2z}}{\frac{2-z}{6}}$$

$$= \frac{2-z}{2z} \cdot \frac{6}{2-z} \quad \text{Multiply by the reciprocal of } \frac{2-z}{6}.$$

$$= \frac{2 \cdot 3 \cdot (2-z)}{2 \cdot z \cdot (2-z)} \quad \text{Factor.}$$

$$= \frac{3}{z} \quad \text{Write in lowest terms.}$$

Work Practice 3

Objective B Simplifying Complex Fractions—Method 2

Next we study a second method for simplifying complex fractions. In this method, we multiply the numerator and the denominator of the complex fraction by the LCD of all fractions in the complex fraction.

Method 2: To Simplify a Complex Fraction

Step 1: Find the LCD of all the fractions in the complex fraction.

Step 2: Multiply both the numerator and the denominator of the complex fraction by the LCD from Step 1.

Step 3: Perform the indicated operations and write the result in lowest terms.

We use Method 2 to rework Example 2.

Answer

3. $\frac{3}{x}$

Example 4 Simplify: $\dfrac{\frac{2}{3}+\frac{1}{5}}{\frac{2}{3}-\frac{2}{9}}$

Solution: The LCD of $\frac{2}{3}, \frac{1}{5}, \frac{2}{3}$, and $\frac{2}{9}$ is 45, so we multiply the numerator and the denominator of the complex fraction by 45. Then we perform the indicated operations, and write in lowest terms.

$$\frac{\frac{2}{3}+\frac{1}{5}}{\frac{2}{3}-\frac{2}{9}} = \frac{45\left(\frac{2}{3}+\frac{1}{5}\right)}{45\left(\frac{2}{3}-\frac{2}{9}\right)}$$

$$= \frac{45\left(\frac{2}{3}\right)+45\left(\frac{1}{5}\right)}{45\left(\frac{2}{3}\right)-45\left(\frac{2}{9}\right)} \quad \text{Apply the distributive property.}$$

$$= \frac{30+9}{30-10} = \frac{39}{20} \quad \text{Simplify.}$$

Work Practice 4

Practice 4

Use Method 2 to simplify the complex fraction in Practice 2:

$$\frac{\frac{3}{4}-\frac{2}{3}}{\frac{1}{2}+\frac{3}{8}}$$

Helpful Hint

The same complex fraction was simplified using two different methods in Examples 2 and 4. Notice that the simplified results are the same.

Example 5 Simplify: $\dfrac{\frac{x+1}{y}}{\frac{x}{y}+2}$

Solution: The LCD of $\frac{x+1}{y}, \frac{x}{y}$ and $\frac{2}{1}$ is y, so we multiply the numerator and the denominator of the complex fraction by y.

$$\frac{\frac{x+1}{y}}{\frac{x}{y}+2} = \frac{y\left(\frac{x+1}{y}\right)}{y\left(\frac{x}{y}+2\right)}$$

$$= \frac{y\left(\frac{x+1}{y}\right)}{y\left(\frac{x}{y}\right)+y\cdot 2} \quad \text{Apply the distributive property in the denominator.}$$

$$= \frac{x+1}{x+2y} \quad \text{Simplify.}$$

Work Practice 5

Practice 5

Simplify: $\dfrac{1+\frac{x}{y}}{\frac{2x+1}{y}}$

Answers

4. $\frac{2}{21}$ **5.** $\frac{y+x}{2x+1}$

Practice 6

Simplify: $\dfrac{\dfrac{5}{6y} + \dfrac{y}{x}}{\dfrac{y}{3} - x}$

Example 6 Simplify: $\dfrac{\dfrac{x}{y} + \dfrac{3}{2x}}{\dfrac{x}{2} + y}$

Solution: The LCD of $\frac{x}{y}, \frac{3}{2x}, \frac{x}{2}$, and $\frac{y}{1}$ is $2xy$, so we multiply both the numerator and the denominator of the complex fraction by $2xy$.

$$\frac{\dfrac{x}{y} + \dfrac{3}{2x}}{\dfrac{x}{2} + y} = \frac{2xy\left(\dfrac{x}{y} + \dfrac{3}{2x}\right)}{2xy\left(\dfrac{x}{2} + y\right)}$$

$$= \frac{2xy\left(\dfrac{x}{y}\right) + 2xy\left(\dfrac{3}{2x}\right)}{2xy\left(\dfrac{x}{2}\right) + 2xy(y)} \quad \text{Apply the distributive property.}$$

$$= \frac{2x^2 + 3y}{x^2y + 2xy^2}$$

$$\text{or } \frac{2x^2 + 3y}{xy(x + 2y)}$$

Work Practice 6

Answer

6. $\dfrac{5x + 6y^2}{2xy^2 - 6x^2y}$ or $\dfrac{5x + 6y^2}{2xy(y - 3x)}$

Vocabulary, Readiness & Video Check

Complete the steps by writing the simplified complex fraction.

1. $\dfrac{\dfrac{y}{2}}{\dfrac{5x}{2}} = \dfrac{2\left(\dfrac{y}{2}\right)}{2\left(\dfrac{5x}{2}\right)} = \dfrac{?}{?}$

2. $\dfrac{\dfrac{10}{x}}{\dfrac{z}{x}} = \dfrac{x\left(\dfrac{10}{x}\right)}{x\left(\dfrac{z}{x}\right)} = \dfrac{?}{?}$

3. $\dfrac{\dfrac{3}{x}}{\dfrac{5}{x^2}} = \dfrac{x^2\left(\dfrac{3}{x}\right)}{x^2\left(\dfrac{5}{x^2}\right)} = \dfrac{?}{?}$

4. $\dfrac{\dfrac{a}{10}}{\dfrac{b}{20}} = \dfrac{20\left(\dfrac{a}{10}\right)}{20\left(\dfrac{b}{20}\right)} = \dfrac{?}{?}$

One method for simplifying a complex fraction is to multiply the fraction's numerator and denominator by the LCD of all fractions in the complex fraction. For each complex fraction, choose the LCD of its fractions.

5. $\dfrac{\dfrac{1}{4} + \dfrac{1}{2}}{\dfrac{1}{3} + \dfrac{1}{2}}$ The LCD of $\frac{1}{4}, \frac{1}{2}, \frac{1}{3}$, and $\frac{1}{2}$ is

a. 4 **b.** 2 **c.** 12 **d.** 6

6. $\dfrac{\dfrac{3}{5} + \dfrac{2}{3}}{\dfrac{1}{10} + \dfrac{1}{6}}$ The LCD of $\frac{3}{5}, \frac{2}{3}, \frac{1}{10}$, and $\frac{1}{6}$ is

a. 15 **b.** 30 **c.** 60 **d.** 180

7. $\dfrac{\dfrac{5}{2x^2} + \dfrac{3}{16x}}{\dfrac{x}{8} + \dfrac{3}{4x}}$ The LCD of $\frac{5}{2x^2}, \frac{3}{16x}, \frac{x}{8}$, and $\frac{3}{4x}$ is

a. $16x^2$ **b.** $32x^3$ **c.** $16x$ **d.** $16x^3$

8. $\dfrac{\dfrac{11}{6} + \dfrac{10}{x^2}}{\dfrac{7}{9} + \dfrac{5}{x}}$ The LCD of $\frac{11}{6}, \frac{10}{x^2}, \frac{7}{9}$, and $\frac{5}{x}$ is

a. 18 **b.** x^2 **c.** $18x^2$ **d.** $54x^3$

Martin-Gay Interactive Videos

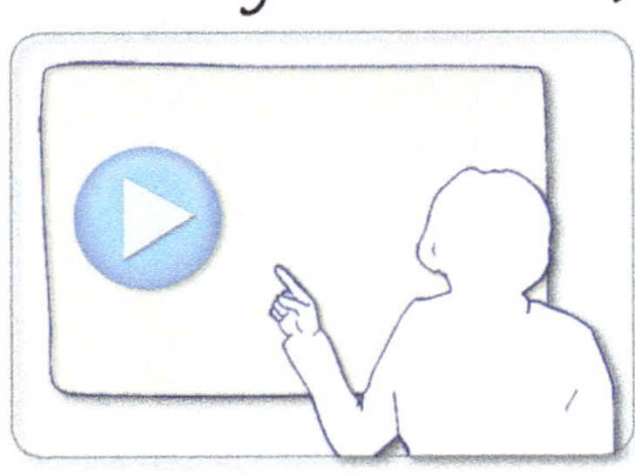

See Video 5.7

Watch the section lecture video and answer the following questions.

Objective A 9. From Example 2, before we can rewrite the complex fraction as a division problem, what must we make sure we have?

Objective B 10. How does finding an LCD in Method 2, as in Examples 4 and 5, differ from finding an LCD in Method 1? Mention the purpose of the LCD in each method.

5.7 Exercise Set MyLab Math

Objectives A B Mixed Practice *Simplify each complex fraction. See Examples 1 through 6.*

1. $\dfrac{\frac{1}{2}}{\frac{3}{4}}$

2. $\dfrac{\frac{1}{8}}{-\frac{5}{12}}$

3. $\dfrac{-\frac{4x}{9}}{-\frac{2x}{3}}$

4. $\dfrac{-\frac{6y}{11}}{\frac{4y}{9}}$

5. $\dfrac{\frac{1+x}{6}}{\frac{1+x}{3}}$

6. $\dfrac{\frac{6x-3}{5x^2}}{\frac{2x-1}{10x}}$

7. $\dfrac{\frac{1}{2}+\frac{2}{3}}{\frac{5}{9}-\frac{5}{6}}$

8. $\dfrac{\frac{3}{4}-\frac{1}{2}}{\frac{3}{8}+\frac{1}{6}}$

9. $\dfrac{2+\frac{7}{10}}{1+\frac{3}{5}}$

10. $\dfrac{4-\frac{11}{12}}{5+\frac{1}{4}}$

11. $\dfrac{\frac{1}{3}}{\frac{1}{2}-\frac{1}{4}}$

12. $\dfrac{\frac{7}{10}-\frac{3}{5}}{\frac{1}{2}}$

13. $\dfrac{-\frac{2}{9}}{-\frac{14}{3}}$

14. $\dfrac{\frac{3}{8}}{\frac{4}{15}}$

15. $\dfrac{-\frac{5}{12x^2}}{\frac{25}{16x^3}}$

16. $\dfrac{-\frac{7}{8y}}{\frac{21}{4y}}$

17. $\dfrac{\frac{m}{n}-1}{\frac{m}{n}+1}$

18. $\dfrac{\frac{x}{2}+2}{\frac{x}{2}-2}$

19. $\dfrac{\frac{1}{5}-\frac{1}{x}}{\frac{7}{10}+\frac{1}{x^2}}$

20. $\dfrac{\frac{1}{y^2}+\frac{2}{3}}{\frac{1}{y}-\frac{5}{6}}$

21. $\dfrac{1+\frac{1}{y-2}}{y+\frac{1}{y-2}}$

22. $\dfrac{x-\frac{1}{2x+1}}{1-\frac{x}{2x+1}}$

23. $\dfrac{\frac{4y-8}{16}}{\frac{6y-12}{4}}$

24. $\dfrac{\frac{7y+21}{3}}{\frac{3y+9}{8}}$

25. $\dfrac{\frac{x}{y}+1}{\frac{x}{y}-1}$

26. $\dfrac{\frac{3}{5y}+8}{\frac{3}{5y}-8}$

27. $\dfrac{1}{2+\frac{1}{3}}$

28. $\dfrac{3}{1-\frac{4}{3}}$

29. $\dfrac{\frac{ax+ab}{x^2-b^2}}{\frac{x+b}{x-b}}$

30. $\dfrac{\frac{m+2}{m-2}}{\frac{2m+4}{m^2-4}}$

31. $\dfrac{\frac{-3+y}{4}}{\frac{8+y}{28}}$

32. $\dfrac{\frac{-x+2}{18}}{\frac{8}{9}}$

33. $\dfrac{3+\frac{12}{x}}{1-\frac{16}{x^2}}$

34. $\dfrac{2+\frac{6}{x}}{1-\frac{9}{x^2}}$

35. $\dfrac{\frac{8}{x+4}+2}{\frac{12}{x+4}-2}$

36. $\dfrac{\frac{25}{x+5}+5}{\frac{3}{x+5}-5}$

37. $\dfrac{\frac{s}{r}+\frac{r}{s}}{\frac{s}{r}-\frac{r}{s}}$

38. $\dfrac{\frac{2}{x}+\frac{x}{2}}{\frac{2}{x}-\frac{x}{2}}$

39. $\dfrac{\frac{6}{x-5}+\frac{x}{x-2}}{\frac{3}{x-6}-\frac{2}{x-5}}$

40. $\dfrac{\frac{4}{x}+\frac{x}{x+1}}{\frac{1}{2x}+\frac{1}{x+6}}$

Review

Use the bar graph below to answer Exercises 41 through 44. See Section R.4. Note: Some of these players are still competing; thus, their total prize money may increase.

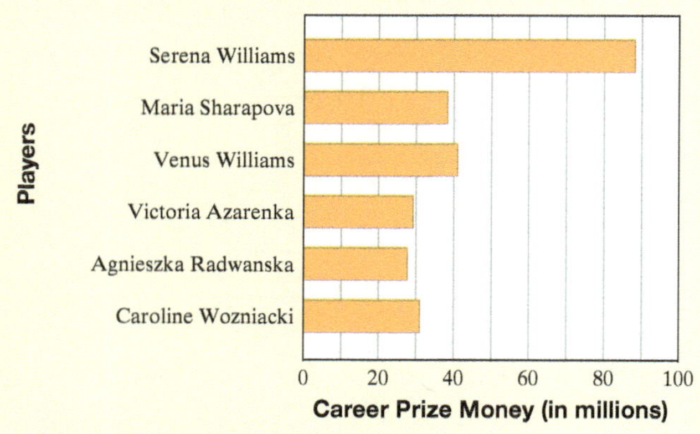

Source: Women's Tennis Association, 2018

41. Which women's tennis player has earned the most prize money in her career?

42. Estimate how much more prize money Maria Sharapova has earned in her career than Victoria Azarenka.

43. What is the approximate spread in lifetime prize money between Agnieszka Radwanska and Venus Williams?

44. To date in her career, Serena Williams has won 97 doubles and singles tournament titles. Assuming her prize money is earned only for tournament titles, how much prize money has she earned, on average, per tournament title?

Concept Extensions

45. Explain how to simplify a complex fraction using Method 1.

46. Explain how to simplify a complex fraction using Method 2.

To find the average of two numbers, we find their sum and divide by 2. For example, the average of 65 and 81 is found by simplifying $\frac{65+81}{2}$. *This simplifies to* $\frac{146}{2} = 73$. *Use this for Exercises 47 through 50.*

47. Find the average of $\frac{1}{3}$ and $\frac{3}{4}$.

48. Write the average of $\frac{3}{n}$ and $\frac{5}{n^2}$ as a simplified rational expression.

49. A carpenter needs to drill a hole halfway between the two marked points. An intersecting board keeps him from measuring between the marked points, but he does have earlier measurements as shown. How far from the left side of the marked board should he drill?

50. Use the same diagram as for Exercise **49**. Suppose the measurements are 7.2 inches and 10.3 inches. How far from the left side of the marked board should he drill?

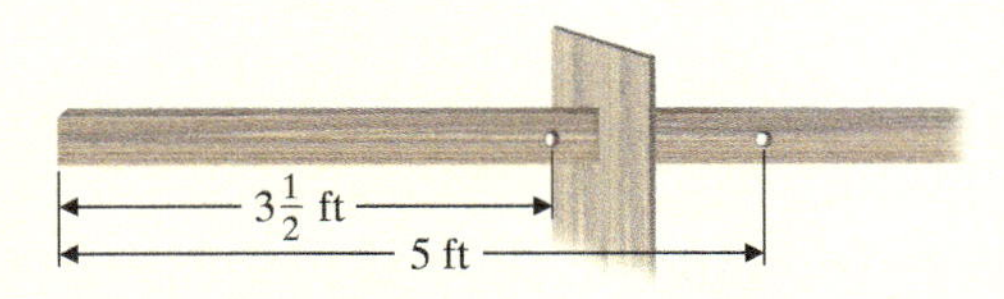

Solve.

51. In electronics, when two resistors R_1 (read R sub 1) and R_2 (read R sub 2) are connected in parallel, the total resistance is given by the complex fraction

$$\frac{1}{\frac{1}{R_1} + \frac{1}{R_2}}.$$

Simplify this expression.

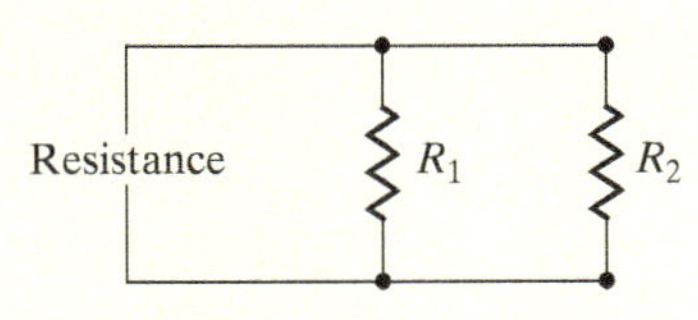

52. Astronomers occasionally need to know the day of the week a particular date fell on. The complex fraction

$$\frac{J + \frac{3}{2}}{7}$$

where J is the *Julian day number*, is used to make this calculation. Simplify this expression.

Simplify each of the following. First, write each expression with positive exponents. Then simplify the complex fraction. The first step has been completed for Exercise 53.

53. $\frac{x^{-1} + 2^{-1}}{x^{-2} - 4^{-1}} = \frac{\frac{1}{x} + \frac{1}{2}}{\frac{1}{x^2} - \frac{1}{4}}$

54. $\frac{3^{-1} - x^{-1}}{9^{-1} - x^{-2}}$

55. $\frac{y^{-2}}{1 - y^{-2}}$

56. $\frac{4 + x^{-1}}{3 + x^{-1}}$

57. If the distance formula $d = r \cdot t$ is solved for t, then $t = \frac{d}{r}$. Use this formula to find t if distance d is $\frac{20x}{3}$ miles and rate r is $\frac{5x}{9}$ miles per hour. Write t in simplified form.

△ **58.** If the formula for the area of a rectangle, $A = l \cdot w$, is solved for w, then $w = \frac{A}{l}$. Use this formula to find w if area A is $\frac{4x - 2}{3}$ square meters and length l is $\frac{6x - 3}{5}$ meters. Write w in simplified form.

Chapter 5 Group Activity

Fast-Growing Careers

According to U.S. Bureau of Labor Statistics projections, the careers listed below will have the largest job growth in the years shown.

Occupation	Employment (number in thousands)		
	2014	2024	Change
1. Home health aides	913.5	1260.6	+347.1
2. Physical therapists	210.9	282.6	+71.7
3. Nurse practitioners	170.4	230.0	+59.6
4. Physical therapist assistants and aides	128.7	180.2	+51.5
5. Physician assistants	94.4	122.7	+28.3
6. Operations research analyst	91.3	118.7	+27.4
7. Occupational therapy assistants and aides	41.9	58.7	+16.8
8. Statisticians	30.0	40.2	+10.2
9. Ambulance drivers and attendants but not EMTs	20.0	26.5	+6.5
10. Wind turbine technician	4.4	9.2	+4.8

What do all of these in-demand occupations have in common? They all require a knowledge of math! For some careers, like nurse practitioners, statisticians, and operations research analysts, the ways math is used on the job may be obvious. For other occupations, the use of math may not be quite as obvious. However, tasks common to many jobs, such as filling in a time sheet or a medication log, writing up an expense report, planning a budget, figuring a bill, ordering supplies, and even making a work schedule, all require math.

Activity

Suppose that your college placement office is planning to publish an occupational handbook on math in popular occupations. Choose one of the occupations from the given list that interests you. Research the occupation. Then write a brief entry for the occupational handbook that describes how a person in that career would use math in his or her job. Include an example if possible.

Chapter 5 Vocabulary Check

Fill in each blank with one of the words or phrases listed below. Not all choices will be used.

least common denominator	simplifying	reciprocals	numerator	$\frac{-a}{b}$		rate
rational expression	proportion	complex fraction	denominator	$\frac{-a}{-b}$	$\frac{a}{-b}$	unit

1. A ______________, is an expression that can be written in the form $\frac{P}{Q}$, where P and Q are polynomials and Q is not 0.

2. In a ______________ the numerator or denominator or both may contain fractions.

3. For a rational expression, $-\frac{a}{b}$ = ______________ = ______________.

4. A rational expression is undefined when the ______________ is 0.

5. The process of writing a rational expression in lowest terms is called ______________.

6. The expressions $\frac{2x}{7}$ and $\frac{7}{2x}$ are called ______________.

7. A ______________ is a mathematical statement that two ratios are equal.

8. The ______________________ of a list of rational expressions is a polynomial of least degree whose factors include all factors of the denominators in the list.

9. A ______________________ fraction is a fraction that equals 1.

10. A ______________________ is a special type of ratio with different kinds of measurement.

Helpful Hint Are you preparing for your test? To help, don't forget to take these:

- Chapter 5 Getting Ready for the Test on page 417
- Chapter 5 Test on page 418

Then check all of your answers at the back of this text. For further review, the step-by-step video solutions to any of these exercises are located in MyLab Math.

5 Chapter Highlights

Definitions and Concepts	Examples
Section 5.1 Simplifying Rational Expressions	
A **rational expression** is an expression that can be written in the form $\frac{P}{Q}$, where P and Q are polynomials and Q does not equal 0.	$\frac{7y^3}{4}, \frac{x^2+6x+1}{x-3}, \frac{-5}{s^3+8}$
To find values for which a rational expression is undefined, find values for which the denominator is 0.	Find any values for which the expression $\frac{5y}{y^2-4y+3}$ is undefined. $y^2-4y+3=0$ Set the denominator equal to 0. $(y-3)(y-1)=0$ Factor. $y-3=0$ or $y-1=0$ Set each factor equal to 0. $y=3$ $\quad y=1$ Solve. The expression is undefined when y is 3 and when y is 1.
To Simplify a Rational Expression **Step 1:** Factor the numerator and denominator. **Step 2:** Divide out factors common to the numerator and denominator. (This is the same as removing a factor of 1.)	Simplify: $\frac{4x+20}{x^2-25}$ $\frac{4x+20}{x^2-25}=\frac{4(x+5)}{(x+5)(x-5)}=\frac{4}{x-5}$
Section 5.2 Multiplying and Dividing Rational Expressions	
$\frac{P}{Q}\cdot\frac{R}{S}=\frac{PR}{QS}$ **To Multiply Rational Expressions** **Step 1:** Completely factor numerators and denominators. **Step 2:** Multiply numerators and multiply denominators. **Step 3:** Write the product in lowest terms.	Multiply: $\frac{4x+4}{2x-3}\cdot\frac{2x^2+x-6}{x^2-1}$ $\frac{4x+4}{2x-3}\cdot\frac{2x^2+x-6}{x^2-1}$ $=\frac{4(x+1)}{2x-3}\cdot\frac{(2x-3)(x+2)}{(x+1)(x-1)}$ $=\frac{4(x+1)(2x-3)(x+2)}{(2x-3)(x+1)(x-1)}$ $=\frac{4(x+2)}{x-1}$

(*Continued*)

Definitions and Concepts	Examples
Section 5.2 Multiplying and Dividing Rational Expressions (*continued*)	
To divide by a rational expression, multiply by the reciprocal. $\frac{P}{Q} \div \frac{R}{S} = \frac{P}{Q} \cdot \frac{S}{R} = \frac{PS}{QR}$	Divide: $\frac{15x + 5}{3x^2 - 14x - 5} \div \frac{15}{3x - 12}$ $\frac{15x + 5}{3x^2 - 14x - 5} \div \frac{15}{3x - 12}$ $= \frac{5(3x + 1)}{(3x + 1)(x - 5)} \cdot \frac{3(x - 4)}{3 \cdot 5}$ $= \frac{x - 4}{x - 5}$
Section 5.3 Adding and Subtracting Rational Expressions with the Same Denominator and Least Common Denominator	
To add or subtract rational expressions with the same denominator, add or subtract numerators, and place the sum or difference over the common denominator. $\frac{P}{R} + \frac{Q}{R} = \frac{P + Q}{R}$ $\frac{P}{R} - \frac{Q}{R} = \frac{P - Q}{R}$	Perform each indicated operation. $\frac{5}{x + 1} + \frac{x}{x + 1} = \frac{5 + x}{x + 1}$ $\frac{2y + 7}{y^2 - 9} - \frac{y + 4}{y^2 - 9}$ $= \frac{(2y + 7) - (y + 4)}{y^2 - 9}$ $= \frac{2y + 7 - y - 4}{y^2 - 9}$ $= \frac{y + 3}{(y + 3)(y - 3)}$ $= \frac{1}{y - 3}$
To Find the Least Common Denominator (LCD) **Step 1:** Factor each denominator completely. **Step 2:** The LCD is the product of all unique factors, each raised to a power equal to the greatest number of times that it appears in any one factored denominator.	Find the LCD for $\frac{7x}{x^2 + 10x + 25}$ and $\frac{11}{3x^2 + 15x}$ $x^2 + 10x + 25 = (x + 5)(x + 5)$ $3x^2 + 15x = 3x(x + 5)$ LCD $= 3x(x + 5)(x + 5)$ or $3x(x + 5)^2$
Section 5.4 Adding and Subtracting Rational Expressions with Different Denominators	
To Add or Subtract Rational Expressions with Different Denominators **Step 1:** Find the LCD. **Step 2:** Rewrite each rational expression as an equivalent expression whose denominator is the LCD. **Step 3:** Add or subtract numerators and place the sum or difference over the common denominator.	Perform the indicated operation. $\frac{9x + 3}{x^2 - 9} - \frac{5}{x - 3}$ $= \frac{9x + 3}{(x + 3)(x - 3)} - \frac{5}{x - 3}$

Definitions and Concepts	Examples
Section 5.4 Adding and Subtracting Rational Expressions with Different Denominators (*continued*)	
Step 4: Write the result in lowest terms.	LCD is $(x+3)(x-3)$. $= \frac{9x+3}{(x+3)(x-3)} - \frac{5(x+3)}{(x-3)(x+3)}$ $= \frac{9x+3-5(x+3)}{(x+3)(x-3)}$ $= \frac{9x+3-5x-15}{(x+3)(x-3)}$ $= \frac{4x-12}{(x+3)(x-3)}$ $= \frac{4(x-3)}{(x+3)(x-3)} = \frac{4}{x+3}$
Section 5.5 Solving Equations Containing Rational Expressions	
To Solve an Equation Containing Rational Expressions **Step 1:** Multiply both sides of the equation by the LCD of all rational expressions in the equation. **Step 2:** Remove any grouping symbols and solve the resulting equation. **Step 3:** Check the solution in the original equation.	Solve: $\frac{5x}{x+2} + 3 = \frac{4x-6}{x+2}$ The LCD is $x+2$. $(x+2)\left(\frac{5x}{x+2}+3\right) = (x+2)\left(\frac{4x-6}{x+2}\right)$ $(x+2)\left(\frac{5x}{x+2}\right) + (x+2)(3) = (x+2)\left(\frac{4x-6}{x+2}\right)$ $5x + 3x + 6 = 4x - 6$ $4x = -12$ $x = -3$ The solution checks; the solution is -3.
Section 5.6 Proportions and problem solving with Rational Equations	
A **ratio** is the quotient of two numbers or two quantities. A **proportion** is a mathematical statement that two ratios are equal. **Cross products:** If $\frac{a}{b} = \frac{c}{d}$, then $ad = bc$.	**Proportions** $\frac{2}{3} = \frac{8}{12}$ $\frac{x}{7} = \frac{15}{35}$ **Cross Products** $2 \cdot 12$ or 24 $\frac{2}{3} = \frac{8}{12}$ $3 \cdot 8$ or 24 Solve: $\frac{3}{4} = \frac{x}{x-1}$ $\frac{3}{4} = \frac{x}{x-1}$ $3(x-1) = 4x$ Set cross products equal. $3x - 3 = 4x$ $-3 = x$
Problem-Solving Steps **1.** UNDERSTAND. Read and reread the problem.	A small plane and a car leave Kansas City, Missouri, and head for Minneapolis, Minnesota, a distance of 450 miles. The speed of the plane is 3 times the speed of the car, and the plane arrives 6 hours ahead of the car. Find the speed of the car.

(*Continued*)

Definitions and Concepts	Examples

Section 5.6 Proportions and problem solving with Rational Equations (*continued*)

Let x = the speed of the car.
Then $3x$ = the speed of the plane.

	Distance =	Rate	· Time
Car	450	x	$\frac{450}{x}\left(\frac{\text{distance}}{\text{rate}}\right)$
Plane	450	$3x$	$\frac{450}{3x}\left(\frac{\text{distance}}{\text{rate}}\right)$

2. TRANSLATE.

In words: plane's time + 6 hours = car's time

Translate: $\frac{450}{3x} + 6 = \frac{450}{x}$

3. SOLVE.

$$\frac{450}{3x} + 6 = \frac{450}{x}$$

$$3x\left(\frac{450}{3x}\right) + 3x(6) = 3x\left(\frac{450}{x}\right)$$

$$450 + 18x = 1350$$

$$18x = 900$$

$$x = 50$$

4. INTERPRET.

Check this solution in the originally stated problem.
State the conclusion: The speed of the car is 50 miles per hour.

Section 5.7 Simplifying Complex Fractions

Method 1: To Simplify a Complex Fraction

Step 1: Add or subtract fractions in the numerator and the denominator of the complex fraction.

Step 2: Perform the indicated division.

Step 3: Write the result in lowest terms.

Simplify:

$$\frac{\frac{1}{x} + 2}{\frac{1}{x} - \frac{1}{y}} = \frac{\frac{1}{x} + \frac{2x}{x}}{\frac{y}{xy} - \frac{x}{xy}} = \frac{\frac{1 + 2x}{x}}{\frac{y - x}{xy}} = \frac{1 + 2x}{x} \cdot \frac{xy}{y - x} = \frac{y(1 + 2x)}{y - x}$$

Method 2: To Simplify a Complex Fraction

Step 1: Find the LCD of all fractions in the complex fraction.

Step 2: Multiply the numerator and the denominator of the complex fraction by the LCD.

Step 3: Perform the indicated operations and write the result in lowest terms.

$$\frac{\frac{1}{x} + 2}{\frac{1}{x} - \frac{1}{y}} = \frac{xy\left(\frac{1}{x} + 2\right)}{xy\left(\frac{1}{x} - \frac{1}{y}\right)} = \frac{xy\left(\frac{1}{x}\right) + xy(2)}{xy\left(\frac{1}{x}\right) - xy\left(\frac{1}{y}\right)} = \frac{y + 2xy}{y - x} \text{ or } \frac{y(1 + 2x)}{y - x}$$

Chapter 5 Review

(5.1) *Find any real number(s) for which each rational expression is undefined.*

1. $\frac{x + 5}{x^2 - 4}$

2. $\frac{5x + 9}{4x^2 - 4x - 15}$

Find the value of each rational expression when $x = 5$, $y = 7$, and $z = -2$.

3. $\frac{2 - z}{z + 5}$

4. $\frac{x^2 + xy - y^2}{x + y}$

Simplify each rational expression.

5. $\frac{2x + 6}{x^2 + 3x}$

6. $\frac{3x - 12}{x^2 - 4x}$

7. $\frac{x + 2}{x^2 - 3x - 10}$

8. $\frac{x + 4}{x^2 + 5x + 4}$

9. $\frac{x^3 - 4x}{x^2 + 3x + 2}$

10. $\frac{5x^2 - 125}{x^2 + 2x - 15}$

11. $\frac{x^2 - x - 6}{x^2 - 3x - 10}$

12. $\frac{x^2 - 2x}{x^2 + 2x - 8}$

Simplify each expression. First, factor the four-term polynomials by grouping.

13. $\frac{x^2 + xa + xb + ab}{x^2 - xc + bx - bc}$

14. $\frac{x^2 + 5x - 2x - 10}{x^2 - 3x - 2x + 6}$

(5.2) *Perform each indicated operation and simplify.*

15. $\frac{15x^3y^2}{z} \cdot \frac{z}{5xy^3}$

16. $\frac{-y^3}{8} \cdot \frac{9x^2}{y^3}$

17. $\frac{x^2 - 9}{x^2 - 4} \cdot \frac{x - 2}{x + 3}$

18. $\frac{2x + 5}{x - 6} \cdot \frac{2x}{-x + 6}$

19. $\frac{x^2 - 5x - 24}{x^2 - x - 12} \div \frac{x^2 - 10x + 16}{x^2 + x - 6}$

20. $\frac{4x + 4y}{xy^2} \div \frac{3x + 3y}{x^2y}$

21. $\frac{x^2 + x - 42}{x - 3} \cdot \frac{(x - 3)^2}{x + 7}$

22. $\frac{2a + 2b}{3} \cdot \frac{a - b}{a^2 - b^2}$

23. $\frac{2x^2 - 9x + 9}{8x - 12} \div \frac{x^2 - 3x}{2x}$

24. $\frac{x^2 - y^2}{x^2 + xy} \div \frac{3x^2 - 2xy - y^2}{3x^2 + 6x}$

(5.3) *Perform each indicated operation and simplify.*

25. $\frac{x}{x^2 + 9x + 14} + \frac{7}{x^2 + 9x + 14}$

26. $\frac{x}{x^2 + 2x - 15} + \frac{5}{x^2 + 2x - 15}$

27. $\frac{4x - 5}{3x^2} - \frac{2x + 5}{3x^2}$

28. $\frac{9x + 7}{6x^2} - \frac{3x + 4}{6x^2}$

Find the LCD of each pair of rational expressions.

29. $\frac{x + 4}{2x}, \frac{3}{7x}$

30. $\frac{x - 2}{x^2 - 5x - 24}, \frac{3}{x^2 + 11x + 24}$

Rewrite each rational expression as an equivalent expression whose denominator is the given polynomial.

31. $\frac{5}{7x} = \frac{}{14x^3y}$

32. $\frac{9}{4y} = \frac{}{16y^3x}$

33. $\frac{x + 2}{x^2 + 11x + 18} = \frac{}{(x + 2)(x - 5)(x + 9)}$

34. $\frac{3x - 5}{x^2 + 4x + 4} = \frac{}{(x + 2)^2(x + 3)}$

(5.4) *Perform each indicated operation and simplify.*

35. $\frac{4}{5x^2} + \frac{6}{y}$

36. $\frac{2}{x - 3} - \frac{4}{x - 1}$

37. $\frac{4}{x + 3} - 2$

38. $\frac{3}{x^2 + 2x - 8} + \frac{2}{x^2 - 3x + 2}$

39. $\frac{2x - 5}{6x + 9} - \frac{4}{2x^2 + 3x}$

40. $\frac{x - 1}{x^2 - 2x + 1} - \frac{x + 1}{x - 1}$

(5.5) *Solve each equation.*

41. $\frac{n}{10} = 9 - \frac{n}{5}$

42. $\frac{2}{x + 1} - \frac{1}{x - 2} = -\frac{1}{2}$

43. $\frac{y}{2y + 2} + \frac{2y - 16}{4y + 4} = \frac{y - 3}{y + 1}$

44. $\frac{2}{x - 3} - \frac{4}{x + 3} = \frac{8}{x^2 - 9}$

45. $\frac{x - 3}{x + 1} - \frac{x - 6}{x + 5} = 0$

46. $x + 5 = \frac{6}{x}$

(5.6) *Solve each proportion.*

47. $\dfrac{2}{x-1} = \dfrac{3}{x+3}$

48. $\dfrac{4}{y-3} = \dfrac{2}{y-3}$

Solve.

49. A machine can process 300 parts in 20 minutes. Find how many parts can be processed in 45 minutes.

50. As his consulting fee, Mr. Visconti charges \$90.00 per day. Find how much he charges for 3 hours of consulting. Assume an 8-hour workday.

51. Five times the reciprocal of a number equals the sum of $\frac{3}{2}$ the reciprocal of the number and $\frac{7}{6}$. What is the number?

52. The reciprocal of a number equals the reciprocal of the difference of 4 and the number. Find the number.

53. A car travels 90 miles in the same time that a car traveling 10 miles per hour slower travels 60 miles. Find the speed of each car.

54. The current in a bayou near Lafayette, Louisiana, is 4 miles per hour. A paddleboat travels 48 miles upstream in the same amount of time it takes to travel 72 miles downstream. Find the speed of the boat in still water.

55. When Mark and Maria manicure Mr. Stergeon's lawn, it takes them 5 hours. If Mark works alone, it takes 7 hours. Find how long it takes Maria alone.

56. It takes pipe A 20 days to fill a fish pond. Pipe B takes 15 days. Find how long it takes both pipes together to fill the pond.

Given that the pairs of triangles are similar, find each missing length x.

△ **57.**

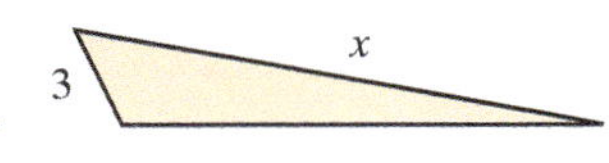

△ **58.**

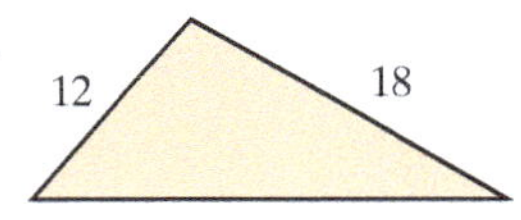

4 x

(5.7) *Simplify each complex fraction.*

59. $\dfrac{\dfrac{5x}{27}}{-\dfrac{10xy}{21}}$

60. $\dfrac{\dfrac{3}{5}+\dfrac{2}{7}}{\dfrac{1}{5}+\dfrac{5}{6}}$

61. $\dfrac{3-\dfrac{1}{y}}{2-\dfrac{1}{y}}$

62. $\dfrac{\dfrac{6}{x+2}+4}{\dfrac{8}{x+2}-4}$

Mixed Review

Simplify each rational expression.

63. $\dfrac{4x+12}{8x^2+24x}$

64. $\dfrac{x^3-6x^2+9x}{x^2+4x-21}$

Perform the indicated operations and simplify.

65. $\dfrac{x^2+9x+20}{x^2-25} \cdot \dfrac{x^2-9x+20}{x^2+8x+16}$

66. $\dfrac{x^2-x-72}{x^2-x-30} \div \dfrac{x^2+6x-27}{x^2-9x+18}$

67. $\frac{x}{x^2 - 36} + \frac{6}{x^2 - 36}$

68. $\frac{5x - 1}{4x} - \frac{3x - 2}{4x}$

69. $\frac{4}{3x^2 + 8x - 3} + \frac{2}{3x^2 - 7x + 2}$

70. $\frac{3x}{x^2 + 9x + 14} - \frac{6x}{x^2 + 4x - 21}$

Solve.

71. $\frac{4}{a - 1} + 2 = \frac{3}{a - 1}$

72. $\frac{x}{x + 3} + 4 = \frac{x}{x + 3}$

Solve.

73. The quotient of twice a number and 3, minus one-sixth, is the quotient of the number and 2. Find the number.

74. Mr. Crocker can paint his shed by himself in three days. His son will need an additional day to complete the job if he works alone. If they work together, find how long it takes to paint the shed.

Given that the following pairs of triangles are similar, find each missing length.

75.

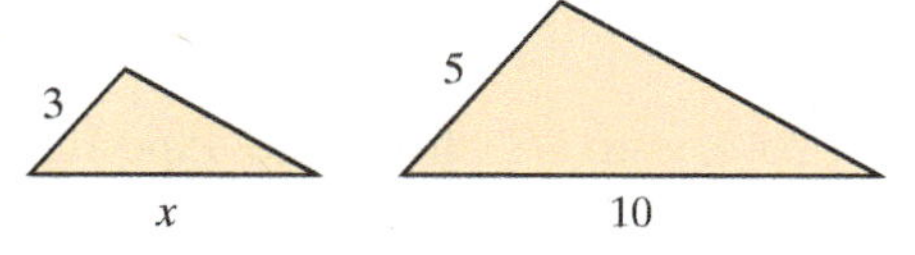

76.

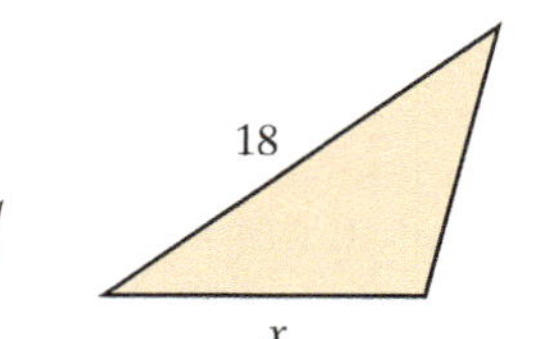

Simplify each complex fraction.

77. $\frac{\frac{1}{4}}{\frac{1}{3} + \frac{1}{2}}$

78. $\frac{4 + \frac{2}{x}}{6 + \frac{3}{x}}$

Convert as indicated.

79. 1.8 square yards = ____ square feet

80. 135 cubic feet = ____ cubic yards

Chapter 5 Getting Ready for the Test LC

MULTIPLE CHOICE Exercises 1 through 12 are **Multiple Choice**. Select the correct choice.

1. $\frac{x-8}{8-x}$ simplifies to

A. 1 **B.** -1 **C.** -2 **D.** -8

2. $\frac{8}{x^2} \cdot \frac{4}{x^2} =$

A. $\frac{32}{x^2}$ **B.** $\frac{2}{x^2}$ **C.** $\frac{32}{x^4}$ **D.** 2 **E.** $\frac{1}{2}$

3. $\frac{8}{x^2} \div \frac{4}{x^2} =$

A. $\frac{32}{x^2}$ **B.** $\frac{2}{x^2}$ **C.** $\frac{32}{x^4}$ **D.** 2 **E.** $\frac{1}{2}$

4. $\frac{8}{x^2} + \frac{4}{x^2} =$

A. $\frac{32}{x^2}$ **B.** $\frac{2}{x^2}$ **C.** $\frac{12}{x^4}$ **D.** $\frac{12}{x^2}$

5. $\frac{7x}{x-1} - \frac{5+2x}{x-1} =$

A. 5 **B.** $\frac{9x-5}{x-1}$ **C.** $\frac{5}{x-1}$ **D.** $\frac{14}{x-1}$

6. The LCD of $\frac{9}{25x}$ and $\frac{z}{10x^3}$ is

A. $250x^4$ **B.** $250x$ **C.** $50x^4$ **D.** $50x^3$

For Exercises 7 through 10, identify each as an

A. expression or **B.** equation.

7. $\frac{5}{x} + \frac{1}{3}$

8. $\frac{5}{x} + \frac{1}{3} = \frac{2}{x}$

9. $\frac{a+5}{11} = 9$

10. $\frac{a+5}{11} \cdot 9$

For Exercises 11 and 12, select the correct choice.

11. Multiply the given equation through by the LCD of its terms. Choose the correct equivalent equation once this is done and terms are simplified. Given Equation: $\frac{x+3}{4} + \frac{5}{6} = 3$

A. $(x+3) + 5 = 3$ **B.** $3(x+3) + 2 \cdot 5 = 3$ **C.** $3(x+3) + 2 \cdot 5 = 12 \cdot 3$ **D.** $6(x+3) + 4 \cdot 5 = 3$

12. Translate to an equation. Let x be the unknown number. "The quotient of a number and 5 equals the sum of that number and 12."

A. $\frac{x}{5} = x + 12$ **B.** $\frac{5}{x} = x + 12$ **C.** $\frac{x}{5} = x \cdot 12$ **D.** $\frac{x}{5} \cdot (x + 12)$

Chapter 5 Test

Answers

1. ______
2. a. ______
   b. ______
3. ______
4. ______
5. ______
6. ______
7. ______
8. ______
9. ______
10. ______
11. ______
12. ______
13. ______
14. ______
15. ______
16. ______
17. ______

1. Find any real numbers for which the following expression is undefined.

$$\frac{x+5}{x^2+4x+3}$$

2. For a certain computer desk, the average manufacturing cost C per desk (in dollars) is

$$C = \frac{100x + 3000}{x}$$

where x is the number of desks manufactured.

a. Find the average cost per desk when manufacturing 200 computer desks.

b. Find the average cost per desk when manufacturing 1000 computer desks.

Simplify each rational expression.

3. $\frac{3x-6}{5x-10}$

4. $\frac{x+6}{x^2+12x+36}$

5. $\frac{7-x}{x-7}$

6. $\frac{y-x}{x^2-y^2}$

7. $\frac{2m^3-2m^2-12m}{m^2-5m+6}$

8. $\frac{ay+3a+2y+6}{ay+3a+5y+15}$

Perform each indicated operation and simplify if possible.

9. $\frac{x^2-13x+42}{x^2+10x+21} \div \frac{x^2-4}{x^2+x-6}$

10. $\frac{3}{x-1} \cdot (5x-5)$

11. $\frac{y^2-5y+6}{2y+4} \cdot \frac{y+2}{2y-6}$

12. $\frac{5}{2x+5} - \frac{6}{2x+5}$

13. $\frac{5a}{a^2-a-6} - \frac{2}{a-3}$

14. $\frac{6}{x^2-1} + \frac{3}{x+1}$

15. $\frac{x^2-9}{x^2-3x} \div \frac{x^2+4x+1}{2x+10}$

16. $\frac{x+2}{x^2+11x+18} + \frac{5}{x^2-3x-10}$

17. $\frac{4y}{y^2+6y+5} - \frac{3}{y^2+5y+4}$

Solve each equation.

18. $\frac{4}{y} - \frac{5}{3} = -\frac{1}{5}$

19. $\frac{5}{y + 1} = \frac{4}{y + 2}$

20. $\frac{a}{a - 3} = \frac{3}{a - 3} - \frac{3}{2}$

21. $\frac{10}{x^2 - 25} = \frac{3}{x + 5} + \frac{1}{x - 5}$

22. $x - \frac{14}{x - 1} = 4 - \frac{2x}{x - 1}$

Simplify each complex fraction.

23. $\dfrac{\frac{5x^2}{yz^2}}{\frac{10x}{z^3}}$

24. $\dfrac{\frac{b}{a} - \frac{a}{b}}{\frac{1}{b} + \frac{1}{a}}$

25. $\dfrac{5 - \frac{1}{y^2}}{\frac{1}{y} + \frac{2}{y^2}}$

26. One number plus five times its reciprocal is equal to six. Find the number.

27. A pleasure boat traveling down the Red River takes the same time to go 14 miles upstream as it takes to go 16 miles downstream. If the current of the river is 2 miles per hour, find the speed of the boat in still water.

28. An inlet pipe can fill a tank in 12 hours. A second pipe can fill the tank in 15 hours. If both pipes are used, find how long it takes to fill the tank.

△ **29.** Given that the two triangles are similar, find x.

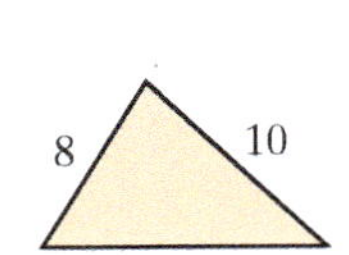

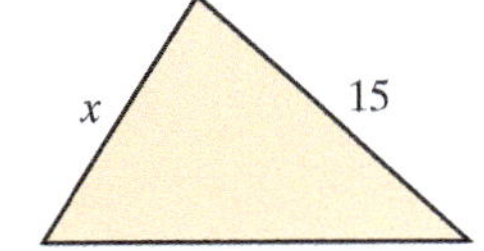

30. In a sample of 85 fluorescent bulbs, 3 were found to be defective. At this rate, how many defective bulbs should be found in 510 bulbs?

18. ______________

19. ______________

20. ______________

21. ______________

22. ______________

23. ______________

24. ______________

25. ______________

26. ______________

27. ______________

28. ______________

29. ______________

30. ______________

Chapters 1–5 Cumulative Review

Answers

1. a. ______
   b. ______
   c. ______
2. a. ______
   b. ______
   c. ______
3. a. ______
   b. ______
4. a. ______
   b. ______
5. ______
6. ______
7. ______
8. ______
9. ______
10. ______
11. ______
12. ______
13. ______
14. ______

1. Write each sentence as an equation. Let x represent the unknown number.

a. The quotient of 15 and a number is 4.

b. Three subtracted from 12 is a number.

c. 17 added to four times a number is 21.

2. Write each sentence as an equation. Let x represent the unknown number.

a. The difference of 12 and a number is -45.

b. The product of 12 and a number is -45.

c. A number less 10 is twice the number.

3. Find each sum.

a. $3 + (-7) + (-8)$

b. $[7 + (-10)] + [-2 + (-4)]$

4. Simplify each expression.

a. $28 - 6 - 30$

b. $7 - 2 - 22$

For Exercises 5 through 8, name the property illustrated by each true statement.

5. $3(x + y) = 3 \cdot x + 3 \cdot y$

6. $3 + y = y + 3$

7. $(x + 7) + 9 = x + (7 + 9)$

8. $(x \cdot 7) \cdot 9 = x \cdot (7 \cdot 9)$

9. Solve: $3 - x = 7$

10. Solve: $7x - 6 = 6x - 6$

11. A 10-foot board is to be cut into two pieces so that the length of the longer piece is 4 times the length of the shorter. Find the length of each piece.

12. Find two consecutive even integers whose sum is 382.

13. Solve $y = mx + b$ for x.

14. Solve $3x - 2y = 6$ for x.

15. Solve $x + 4 \le -6$. Graph the solutions.

16. Solve: $-3x + 7 > -x + 9$

Simplify.

17. $\frac{x^5}{x^2}$

18. $\frac{y^{14}}{y^{14}}$

19. $\frac{4^7}{4^3}$

20. $(x^5y^2)^3$

21. $\frac{(-3)^5}{(-3)^2}$

22. $\frac{x^{19}y^5}{xy}$

23. $\frac{2x^5y^2}{xy}$

24. $(-3a^2b)(5a^3b)$

Simplify by writing each expression with positive exponents only.

25. $2x^{-3}$

26. 7^{-2}

27. $(-2)^{-4}$

28. $5z^{-7}$

Multiply.

29. $5x(2x^3 + 6)$

30. $(x + 9)^2$

31. $-3x^2(5x^2 + 6x - 1)$

32. $(2x + 1)(2x - 1)$

Perform the indicated operations.

33. Divide: $\frac{4x^2 + 7 + 8x^3}{2x + 3}$

34. Divide $(4x^3 - 9x + 2)$ by $(x - 4)$.

35. Factor: $x^2 + 7x + 12$

36. Factor: $-2a^2 + 10a + 12$

15. ____

16. ____

17. ____

18. ____

19. ____

20. ____

21. ____

22. ____

23. ____

24. ____

25. ____

26. ____

27. ____

28. ____

29. ____

30. ____

31. ____

32. ____

33. ____

34. ____

35. ____

36. ____

37. ______

38. ______

39. ______

40. ______

41. ______

42. ______

43. ______

44. ______

45. ______

46. ______

47. ______

48. ______

49. ______

50. ______

37. Factor: $25x^2 + 20xy + 4y^2$

38. Factor: $x^2 - 4$

39. Solve: $x^2 - 9x - 22 = 0$

40. Solve: $3x^2 + 5x = 2$

41. Multiply: $\frac{x^2 + x}{3x} \cdot \frac{6}{5x + 5}$

42. Simplify: $\frac{2x^2 - 50}{4x^4 - 20x^3}$

43. Subtract: $\frac{3x^2 + 2x}{x - 1} - \frac{10x - 5}{x - 1}$

44. Factor: $7x^6 - 7x^5 + 7x^4$

45. Subtract: $\frac{6x}{x^2 - 4} - \frac{3}{x + 2}$

46. Factor: $4x^2 + 12x + 9$

47. Solve: $\frac{t - 4}{2} - \frac{t - 3}{9} = \frac{5}{18}$

48. Multiply: $\frac{6x^2 - 18x}{3x^2 - 2x} \cdot \frac{15x - 10}{x^2 - 9}$

49. Sam Waterton and Frank Schaffer work in a plant that manufactures automobiles. Sam can complete a quality control tour of the plant in 3 hours while his assistant, Frank, needs 7 hours to complete the same job. The regional manager is coming to inspect the plant facilities, so both Sam and Frank are directed to complete a quality control tour together. How long will this take?

50. Simplify: $\dfrac{\frac{m}{3} + \frac{n}{6}}{\frac{m + n}{12}}$

Graphing Equations and Inequalities

In Chapter 2 we learned to solve and graph the solutions of linear equations and inequalities in one variable on number lines. Now we define and present techniques for solving and graphing linear equations and inequalities in two variables on grids. Two-variable equations lead directly to the concept of *function,* perhaps the most important concept in all of mathematics. Functions are introduced in Section 6.6.

International Tourist Arrivals Forecast for 2020–2030 (numbers shown in millions)

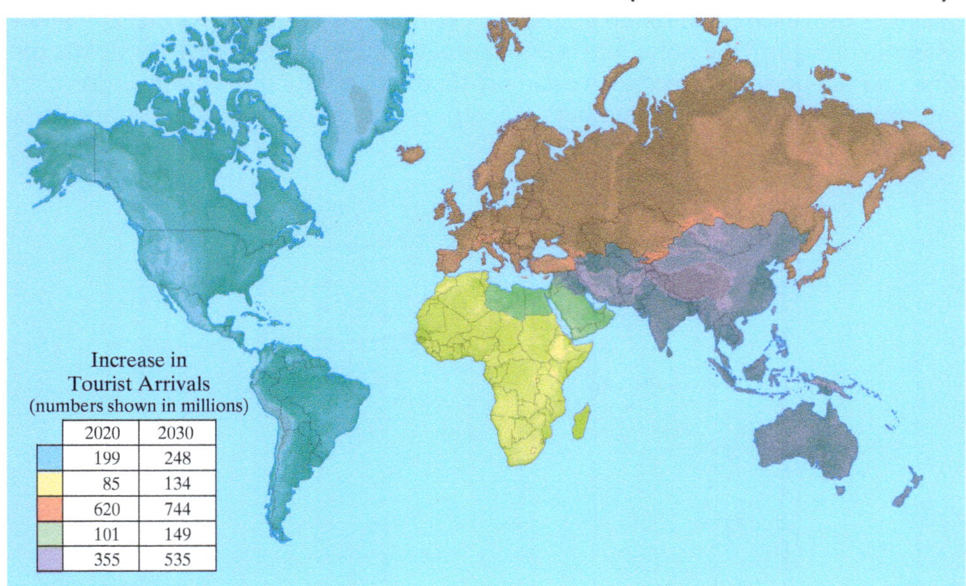

	2020	2030
(blue)	199	248
(yellow)	85	134
(orange)	620	744
(green)	101	149
(purple)	355	535

What Is *Tourism Toward 2030*?

Tourism 2020 Vision was the World Tourism Organization's long-term forecast of world tourism through 2020. *Tourism Toward 2030* is its new program title for longer-term forecasts to 2030. The broken-line graph below shows the forecast for number of tourists, which is extremely important as these numbers greatly affect a country's economy. In Section 6.1, Exercises 45 through 50, we read a bar graph showing the top tourist destinations by country.

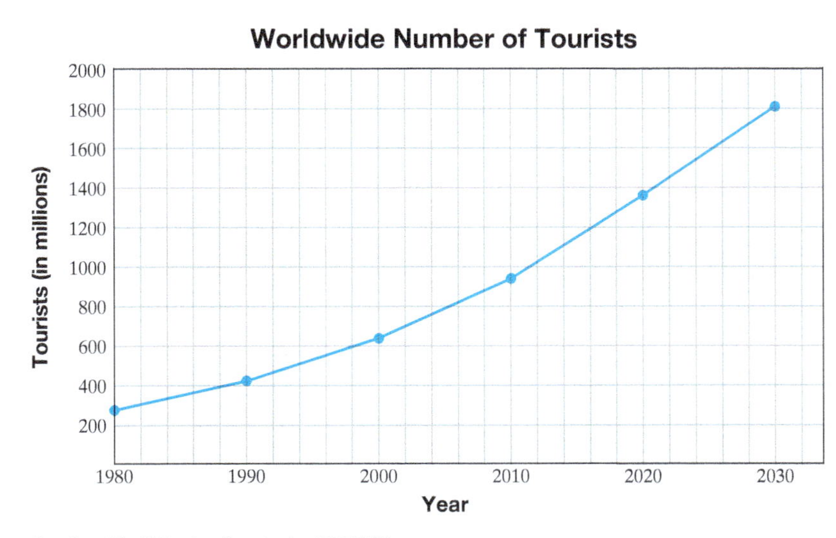

Data from World Tourism Organization (UNWTO)

Sections

6.1 The Rectangular Coordinate System

6.2 Graphing Linear Equations

6.3 Intercepts

6.4 Slope and Rate of Change

6.5 Equations of Lines

Integrated Review—Summary on Linear Equations

6.6 Introduction to Functions

6.7 Graphing Linear Inequalities in Two Variables

Check Your Progress

Vocabulary Check

Chapter Highlights

Chapter Review

Getting Ready for the Test

Chapter Test

Cumulative Review

6.1 The Rectangular Coordinate System 

Objectives

A Plot Ordered Pairs of Numbers on the Rectangular Coordinate System.

B Graph Paired Data to Create a Scatter Diagram.

C Find the Missing Coordinate of an Ordered Pair Solution, Given One Coordinate of the Pair.

In Section R.4, we learned how to read graphs. The broken line graph below shows the relationship between the time before and after smoking a cigarette and pulse rate. The horizontal line or axis shows time in minutes and the vertical line or axis shows the pulse rate in heartbeats per minute. Notice that there are two numbers associated with each point of the graph. For example, the graph shows that 15 minutes after "lighting up," the pulse rate is 80 beats per minute. If we agree to write the time first and the pulse rate second, we can say there is a point on the graph corresponding to the **ordered pair** of numbers (15, 80). A few more ordered pairs are shown alongside their corresponding points.

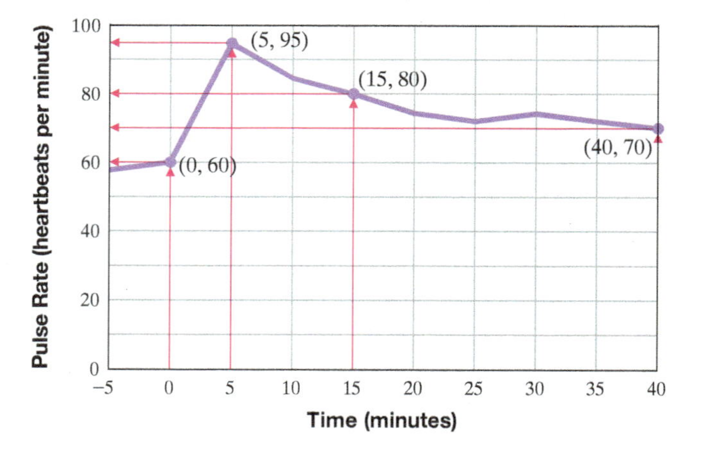

Objective A Plotting Ordered Pairs of Numbers 

In general, we use the idea of ordered pairs to describe the location of a point in a plane (such as a piece of paper). We start with a horizontal and a vertical axis. Each axis is a number line, and for the sake of consistency we construct our axes to intersect at the 0 coordinate of both. This point of intersection is called the **origin.** Notice that these two number lines or axes divide the plane into four regions called **quadrants.** The quadrants are usually numbered with Roman numerals as shown. The axes are not considered to be in any quadrant.

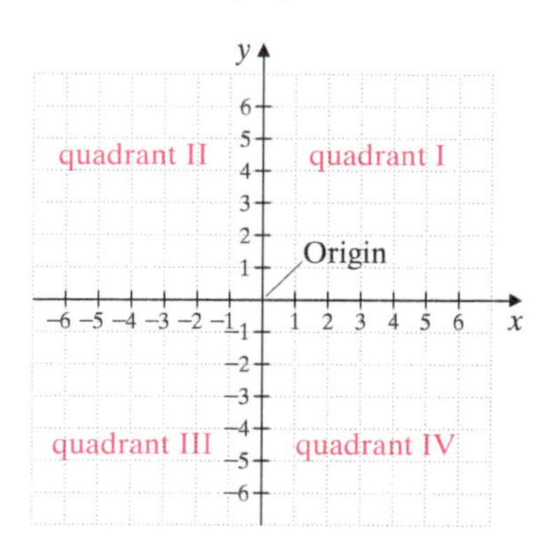

It is helpful to label axes, so we label the horizontal axis the ***x*-axis** and the vertical axis the ***y*-axis.** We call the system described above the **rectangular coordinate system,** or the **coordinate plane.** Just as with other graphs shown, we can then describe the locations of points by ordered pairs of numbers. We list the horizontal, *x*-axis measurement first and the vertical, *y*-axis measurement second.

To plot or graph the point corresponding to the ordered pair (a, b) we start at the origin. We then move a units left or right (right if a is positive, left if a is negative). From there, we move b units up or down (up if b is positive, down if b is negative). For example, to plot the point corresponding to the ordered pair $(3, 2)$, we start at the origin, move 3 units right, and from there move 2 units up. (See the figure below.) The x-value, 3, is also called the **x-coordinate** and the y-value, 2, is also called the **y-coordinate.** From now on, we will call the point with coordinates $(3, 2)$ simply the point $(3, 2)$. The point $(-2, 5)$ is also graphed below.

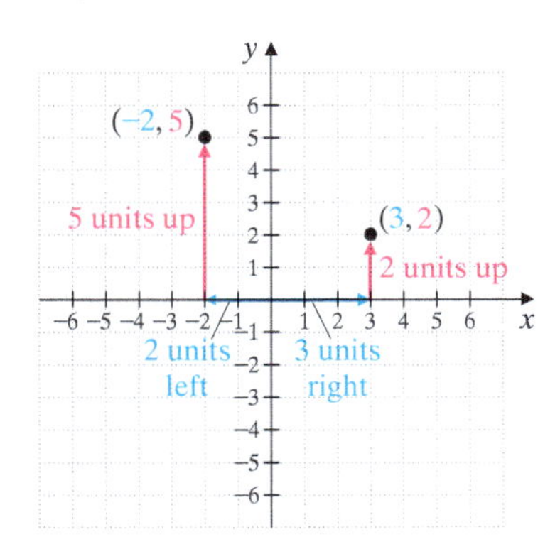

Helpful Hint

Don't forget that **each ordered pair corresponds to exactly one point in the plane and that each point in the plane corresponds to exactly one ordered pair.**

✓Concept Check Is the graph of the point $(-5, 1)$ in the same location as the graph of the point $(1, -5)$? Explain.

Example 1 On a single coordinate system, plot each ordered pair. State in which quadrant, or on which axis, each point lies.

a. $(5, 3)$ **b.** $(-2, -4)$ **c.** $(1, -2)$ **d.** $(-5, 3)$ **e.** $(0, 0)$

f. $(0, 2)$ **g.** $(-5, 0)$ **h.** $\left(0, -5\frac{1}{2}\right)$ **i.** $\left(4\frac{2}{3}, -3\right)$

Solution:

a. Point $(5, 3)$ lies in quadrant I.
b. Point $(-2, -4)$ lies in quadrant III.
c. Point $(1, -2)$ lies in quadrant IV.
d. Point $(-5, 3)$ lies in quadrant II.
e–h. Points $(0, 0)$, $(0, 2)$, and $\left(0, -5\frac{1}{2}\right)$ lie on the y-axis. Points $(0, 0)$ and $(-5, 0)$ lie on the x-axis.
i. Point $\left(4\frac{2}{3}, -3\right)$ lies in quadrant IV.

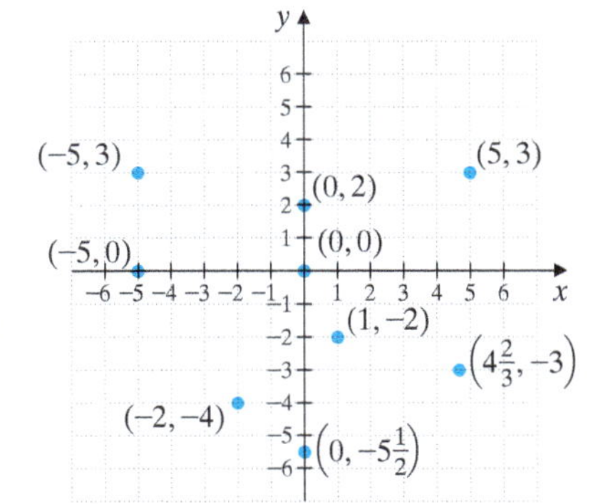

Work Practice 1

Helpful Hint

In Example 1, notice that the point $(0, 0)$ lies on both the x-axis and the y-axis. It is the only point in the entire rectangular coordinate system that has this feature. Why? It is the only point of intersection of the x-axis and the y-axis.

Practice 1

On a single coordinate system, plot each ordered pair. State in which quadrant, or on which axis, each point lies.

a. $(4, 2)$ **b.** $(-1, -3)$
c. $(2, -2)$ **d.** $(-5, 1)$
e. $(0, 3)$ **f.** $(3, 0)$
g. $(0, -4)$ **h.** $\left(-2\frac{1}{2}, 0\right)$
i. $\left(1, -3\frac{3}{4}\right)$

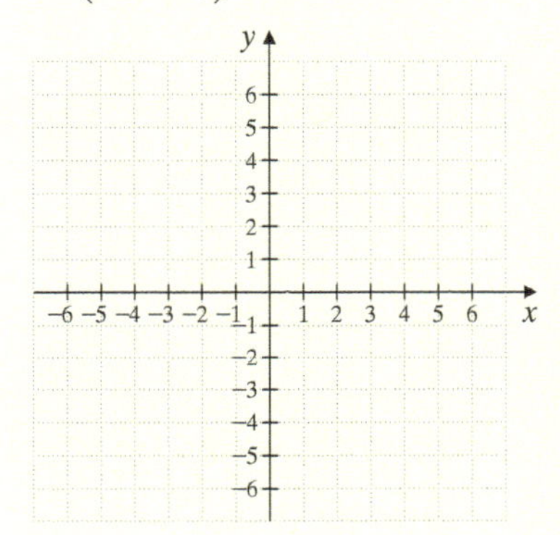

Answers

1.

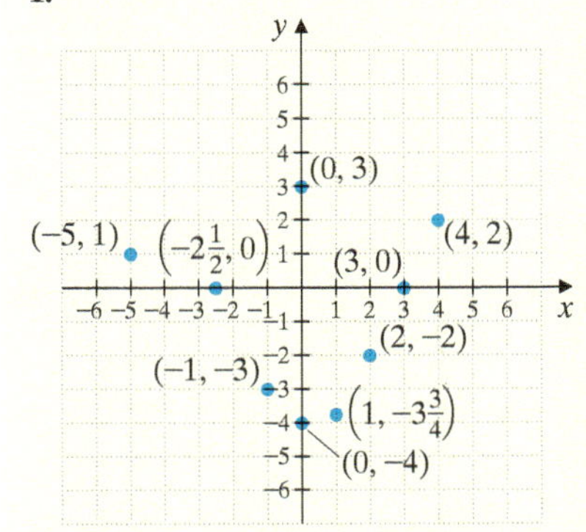

a. Point $(4, 2)$ lies in quadrant I.
b. Point $(-1, -3)$ lies in quadrant III.
c. Point $(2, -2)$ lies in quadrant IV.
d. Point $(-5, 1)$ lies in quadrant II.
e–h. Points $(3, 0)$ and $\left(-2\frac{1}{2}, 0\right)$ lie on the x-axis. Points $(0, 3)$ and $(0, -4)$ lie on the y-axis.
i. Point $\left(1, -3\frac{3}{4}\right)$ lies in quadrant IV.

✓Concept Check Answer

The graph of point $(-5, 1)$ lies in quadrant II and the graph of point $(1, -5)$ lies in quadrant IV. They are *not* in the same location.

Practice 2

The table gives the number of tornadoes that occurred in the United States for the years shown. (*Source:* Storm Prediction Center, National Oceanic and Atmospheric Administration)

Year	Tornadoes
2010	1525
2011	1897
2012	1116
2013	943
2014	1055
2015	1257
2016	1060
2017	1406

a. Write this paired data as a set of ordered pairs of the form (year, number of tornadoes).

b. Create a scatter diagram of the paired data.

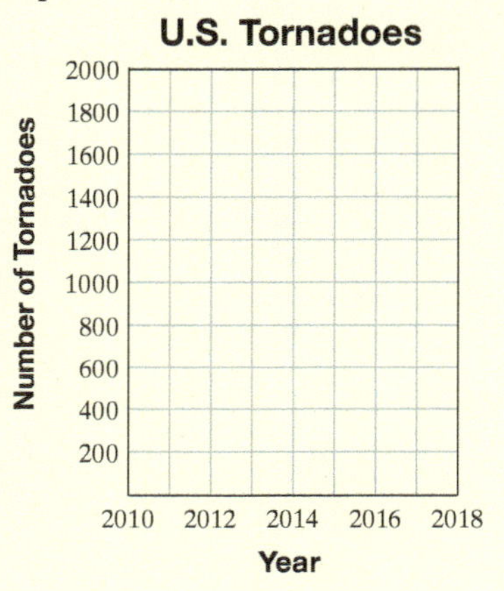

c. What trend in the paired data, if any, does the scatter diagram show?

Answers

2. a. (2010, 1525), (2011, 1897), (2012, 1116), (2013, 943), (2014, 1055), (2015, 1257), (2016, 1060), (2017, 1406)

b.

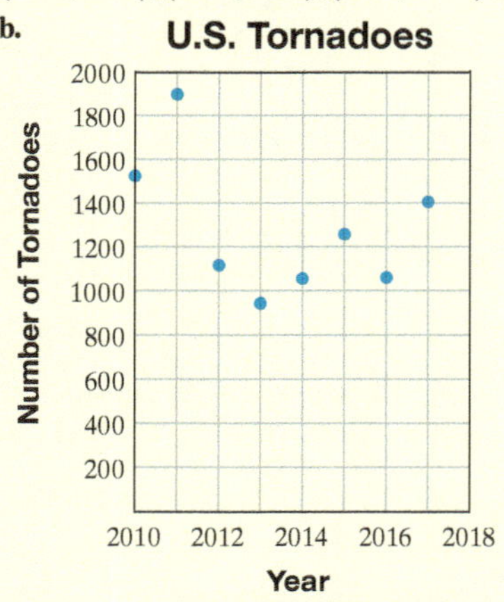

c. The number of tornadoes varies greatly from year to year.

✓**Concept Check Answers**

a. $(-3, 5)$ **b.** $(0, -6)$

✓**Concept Check** For each description of a point in the rectangular coordinate system, write an ordered pair that represents it.

a. Point A is located three units to the left of the y-axis and five units above the x-axis.

b. Point B is located six units below the origin.

From Example 1, notice that the y-coordinate of any point on the x-axis is 0. For example, the point $(-5, 0)$ lies on the x-axis. Also, the x-coordinate of any point on the y-axis is 0. For example, the point $(0, 2)$ lies on the y-axis.

Objective B Creating Scatter Diagrams 

Data that can be represented as ordered pairs are called **paired data.** Many types of data collected from the real world are paired data. For instance, the annual measurements of a child's height can be written as ordered pairs of the form (year, height in inches) and are paired data. The graph of paired data as points in a rectangular coordinate system is called a **scatter diagram.** Scatter diagrams can be used to look for patterns and trends in paired data.

Example 2

The table gives the annual net sales for Dick's Sporting Goods for the years shown. (*Source:* Dick's Sporting Goods)

a. Write this paired data as a set of ordered pairs of the form (year, net sales in billions of dollars).

b. Create a scatter diagram of the paired data.

c. What trend in the paired data does the scatter diagram show?

Year	Dick's Sporting Goods Net Sales (in billions of dollars)
2010	4.9
2011	5.2
2012	5.8
2013	6.2
2014	6.8
2015	7.3
2016	7.9
2017	8.6

Solution:

a. The ordered pairs are (2010, 4.9), (2011, 5.2), (2012, 5.8), (2013, 6.2), (2014, 6.8), (2015, 7.3), (2016, 7.9), and (2017, 8.6).

b. We begin by plotting the ordered pairs. Because the x-coordinate in each ordered pair is a year, we label the x-axis "Year" and mark the horizontal axis with the years given. Then we label the y-axis or vertical axis "Net Sales (in billions of dollars)." In this case, it is convenient to mark the vertical axis in increments of 1, starting with 0.

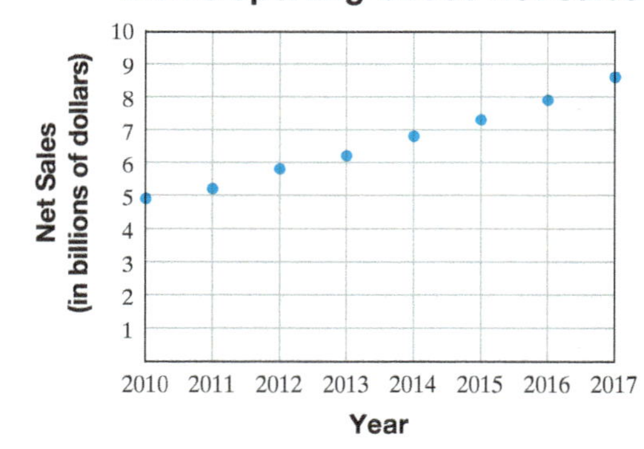

c. The scatter diagram shows that Dick's Sporting Goods net sales steadily increased over the years 2010–2017.

Work Practice 2

Objective C Completing Ordered Pair Solutions 

Let's see how we can use ordered pairs to record solutions of equations containing two variables. An equation in one variable such as $x + 1 = 5$ has one solution, 4: The number 4 is the value of the variable x that makes the equation true.

An equation in two variables, such as $2x + y = 8$, has solutions consisting of two values, one for x and one for y. For example, $x = 3$ and $y = 2$ is a solution of $2x + y = 8$ because, if x is replaced with 3 and y with 2, we get a true statement.

$$2x + y = 8$$
$$2(3) + 2 \stackrel{?}{=} 8 \quad \text{Replace } x \text{ with 3 and } y \text{ with 2.}$$
$$8 = 8 \quad \text{True}$$

The solution $x = 3$ and $y = 2$ can be written as $(3, 2)$, an ordered pair of numbers.

In general, an ordered pair is a **solution** of an equation in two variables if replacing the variables by the values of the ordered pair results in a *true statement.*

For example, another ordered pair solution of $2x + y = 8$ is $(5, -2)$. Replacing x with 5 and y with -2 results in a true statement.

$$2x + y = 8$$
$$2(5) + (-2) \stackrel{?}{=} 8 \quad \text{Replace } x \text{ with 5 and } y \text{ with } -2.$$
$$10 - 2 \stackrel{?}{=} 8$$
$$8 = 8 \quad \text{True}$$

Example 3 Complete each ordered pair so that it is a solution to the equation $3x + y = 12$.

a. $(0, \ \ )$ **b.** $(\ \ , 6)$ **c.** $(-1, \ \ )$

Solution:

a. In the ordered pair $(0, \ \ )$, the x-value is 0. We let $x = 0$ in the equation and solve for y.

$$3x + y = 12$$
$$3(0) + y = 12 \quad \text{Replace } x \text{ with 0.}$$
$$0 + y = 12$$
$$y = 12$$

The completed ordered pair is $(0, 12)$.

b. In the ordered pair $(\ \ , 6)$, the y-value is 6. We let $y = 6$ in the equation and solve for x.

$$3x + y = 12$$
$$3x + 6 = 12 \quad \text{Replace } y \text{ with 6.}$$
$$3x = 6 \quad \text{Subtract 6 from both sides.}$$
$$x = 2 \quad \text{Divide both sides by 3.}$$

The ordered pair is $(2, 6)$.

c. In the ordered pair $(-1, \ \ )$, the x-value is -1. We let $x = -1$ in the equation and solve for y.

$$3x + y = 12$$
$$3(-1) + y = 12 \quad \text{Replace } x \text{ with } -1.$$
$$-3 + y = 12$$
$$y = 15 \quad \text{Add 3 to both sides.}$$

The ordered pair is $(-1, 15)$.

Work Practice 3

Practice 3

Complete each ordered pair so that it is a solution to the equation $x + 2y = 8$.

a. $(0, \ \ )$

b. $(\ \ , 3)$

c. $(-4, \ \ )$

Answers

3. a. $(0, 4)$ **b.** $(2, 3)$ **c.** $(-4, 6)$

Solutions of equations in two variables can also be recorded in a **table of paired values,** as shown in the next example.

Practice 4

Complete the table for the equation $y = -2x$.

	x	y
a.	-3	
b.		0
c.		10

Example 4

Complete the table for the equation $y = 3x$.

	x	y
a.	-1	
b.		0
c.		-9

Solution:

a. We replace x with -1 in the equation and solve for y.

$y = 3x$

$y = 3(-1)$ Let $x = -1$.

$y = -3$

The ordered pair is $(-1, -3)$.

b. We replace y with 0 in the equation and solve for x.

$y = 3x$

$0 = 3x$ Let $y = 0$.

$0 = x$ Divide both sides by 3.

The ordered pair is $(0, 0)$.

c. We replace y with -9 in the equation and solve for x.

$y = 3x$

$-9 = 3x$ Let $y = -9$.

$-3 = x$ Divide both sides by 3.

The ordered pair is $(-3, -9)$.

The completed table is shown to the right.

x	y
-1	-3
0	0
-3	-9

Work Practice 4

Practice 5

Complete the table for the equation $y = \frac{1}{3}x - 1$.

	x	y
a.	-3	
b.	0	
c.		0

Example 5

Complete the table for the equation

$$y = \frac{1}{2}x - 5.$$

	x	y
a.	-2	
b.	0	
c.		0

Solution:

a. Let $x = -2$.

$y = \frac{1}{2}x - 5$

$y = \frac{1}{2}(-2) - 5$

$y = -1 - 5$

$y = -6$

b. Let $x = 0$.

$y = \frac{1}{2}x - 5$

$y = \frac{1}{2}(0) - 5$

$y = 0 - 5$

$y = -5$

c. Let $y = 0$.

$y = \frac{1}{2}x - 5$

$0 = \frac{1}{2}x - 5$ Now, solve for x.

$5 = \frac{1}{2}x$ Add 5.

$10 = x$ Multiply by 2.

Ordered pairs: $(-2, -6)$ $(0, -5)$ $(10, 0)$

The completed table is

x	-2	0	10
y	-6	-5	0

Work Practice 5

Answers

4.

	x	y
a.	-3	6
b.	0	0
c.	-5	10

5.

	x	y
a.	-3	-2
b.	0	-1
c.	3	0

Notice in the previous example, Example 5, that a table showing ordered pair solutions may be written vertically or horizontally.

By now, you may also have noticed that equations in two variables often have more than one solution. We discuss this more in the next section.

Example 6 A small business purchased a computer for \$2000. The business predicts that the computer will be used for 5 years and the value in dollars y of the computer in x years is $y = -300x + 2000$. Complete the table.

x	0	1	2	3	4	5
y						

Solution:

To find the value of y when x is 0, we replace x with 0 in the equation. We use this same procedure to find y when x is 1 and when x is 2.

When $x = 0$,
$y = -300x + 2000$
$y = -300 \cdot 0 + 2000$
$y = 0 + 2000$
$y = 2000$

When $x = 1$,
$y = -300x + 2000$
$y = -300 \cdot 1 + 2000$
$y = -300 + 2000$
$y = 1700$

When $x = 2$,
$y = -300x + 2000$
$y = -300 \cdot 2 + 2000$
$y = -600 + 2000$
$y = 1400$

We have the ordered pairs (0, 2000), (1, 1700), and (2, 1400). This means that in 0 years the value of the computer is \$2000, in 1 year the value of the computer is \$1700, and in 2 years the value is \$1400. To complete the table of values, we continue the procedure for $x = 3$, $x = 4$, and $x = 5$.

When $x = 3$,
$y = -300x + 2000$
$y = -300 \cdot 3 + 2000$
$y = -900 + 2000$
$y = 1100$

When $x = 4$,
$y = -300x + 2000$
$y = -300 \cdot 4 + 2000$
$y = -1200 + 2000$
$y = 800$

When $x = 5$,
$y = -300x + 2000$
$y = -300 \cdot 5 + 2000$
$y = -1500 + 2000$
$y = 500$

The completed table is shown below.

x	0	1	2	3	4	5
y	2000	1700	1400	1100	800	500

Work Practice 6

Practice 6

A company purchased a portable scanner for \$250. The business manager of the company predicts that the scanner will be used for 4 years and the value in dollars y of the machine in x years is $y = -50x + 250$. Complete the table.

x	1	2	3	4
y				

The ordered pair solutions recorded in the completed table for Example 6 are another set of paired data. They are graphed next. Notice that this scatter diagram gives a visual picture of the decrease in value of the computer.

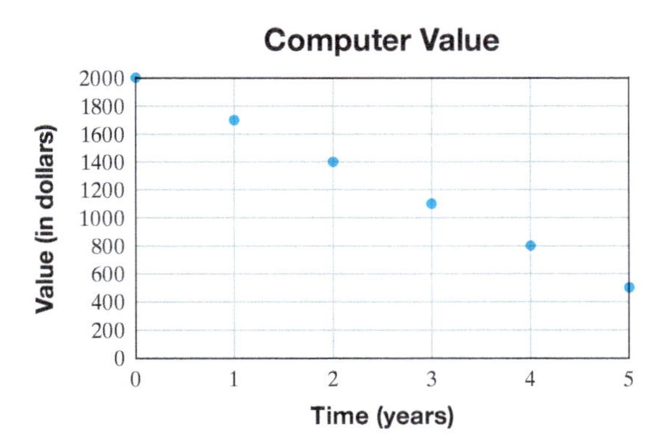

Answer

6.

x	1	2	3	4
y	200	150	100	50

Vocabulary, Readiness & Video Check

Use the choices below to fill in each blank. The exercises below all have to do with the rectangular coordinate system.

origin	x-coordinate	x-axis	scatter diagram	four
quadrants	y-coordinate	y-axis	solution	one

1. The horizontal axis is called the ______________.

2. The vertical axis is called the ______________.

3. The intersection of the horizontal axis and the vertical axis is a point called the ______________.

4. The axes divide the plane into regions, called ______________. There are ______________ of these regions.

5. In the ordered pair of numbers $(-2, 5)$, the number -2 is called the ______________ and the number 5 is called the ______________.

6. Each ordered pair of numbers corresponds to ______________ point in the plane.

7. An ordered pair is a(n) ______________ of an equation in two variables if replacing the variables by the coordinates of the ordered pair results in a true statement.

8. The graph of paired data as points in a rectangular coordinate system is called a(n) ______________.

Martin-Gay Interactive Videos

See Video 6.1 

Watch the section lecture video and answer the following questions.

Objective A **9.** Several points are plotted in Examples 1–6. Where do we always start when plotting a point? How does the first coordinate tell us to move? How does the second coordinate tell us to move?

Objective B **10.** From Example 7, what kind of data can be graphed in a scatter diagram?

Objective C **11.** In Example 8, when finding the missing value in an ordered pair solution of a linear equation in two variables, how can we check our solution?

6.1 Exercise Set MyLab Math

Objective A *Plot each ordered pair. State in which quadrant or on which axis each point lies. See Example 1.*

1. a. $(1, 5)$ **b.** $(-5, -2)$ **c.** $(-3, 0)$ **d.** $(0, -1)$ **e.** $(2, -4)$ **f.** $\left(-1, 4\frac{1}{2}\right)$ **g.** $(3.7, 2.2)$ **h.** $\left(\frac{1}{2}, -3\right)$

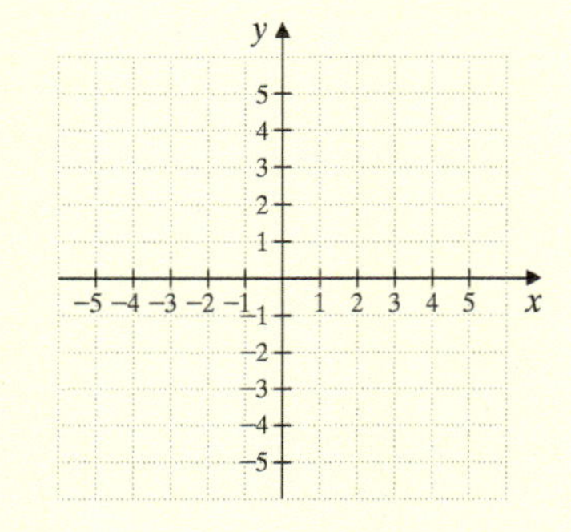

2. a. $(2, 4)$ **b.** $(0, 2)$ **c.** $(-2, 1)$ **d.** $(-3, -3)$ **e.** $\left(3\frac{3}{4}, 0\right)$ **f.** $(5, -4)$ **g.** $(-3.4, 4.8)$ **h.** $\left(\frac{1}{3}, -5\right)$

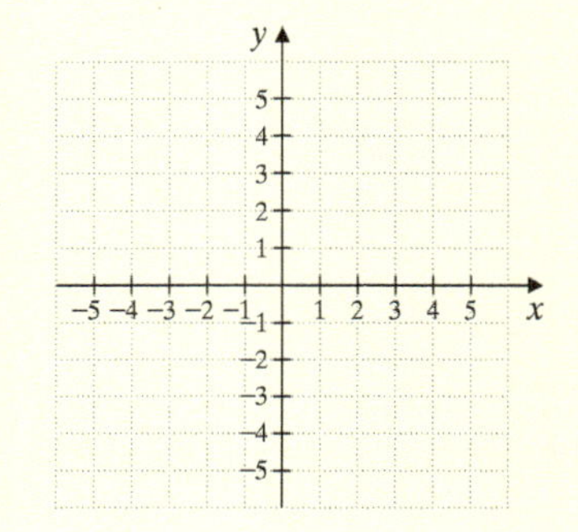

Find the x- and y-coordinates of each labeled point. See Example 1.

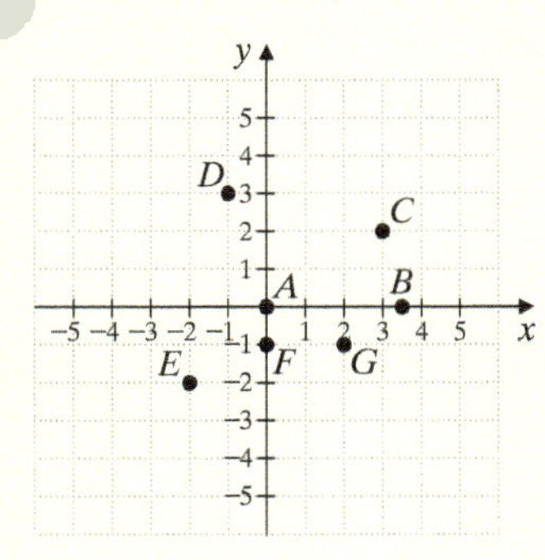

3. *A* **4.** *B* **5.** *C*

6. *D* **7.** *E* **8.** *F*

9. *G*

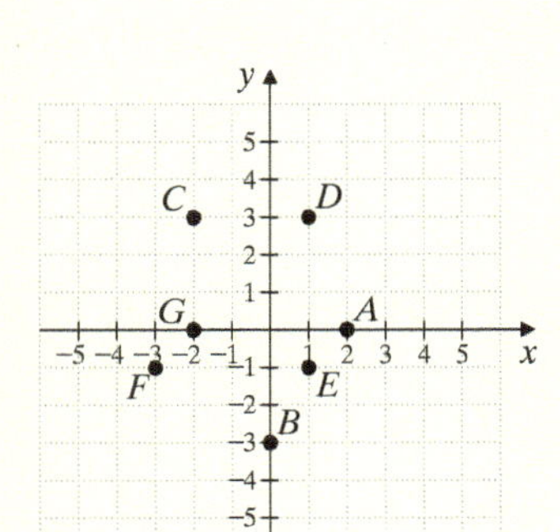

10. *A* **11.** *B* **12.** *C*

13. *D* **14.** *E* **15.** *F*

16. *G*

Objective B *Solve. See Example 2.*

17. The table shows the domestic box office (in billions of dollars) for the U.S. and Canadian movie industry during the years shown. (*Source:* Motion Picture Association of America)

Year	Box Office (in billions of dollars)
2010	10.6
2011	10.2
2012	10.8
2013	10.9
2014	10.4
2015	11.1
2016	11.4
2017	11.1

a. Write this paired data as a set of ordered pairs of the form (year, box office).

b. In your own words, write the meaning of the ordered pair (2017, 11.1).

c. Create a scatter diagram of the paired data. Be sure to label the axes appropriately.

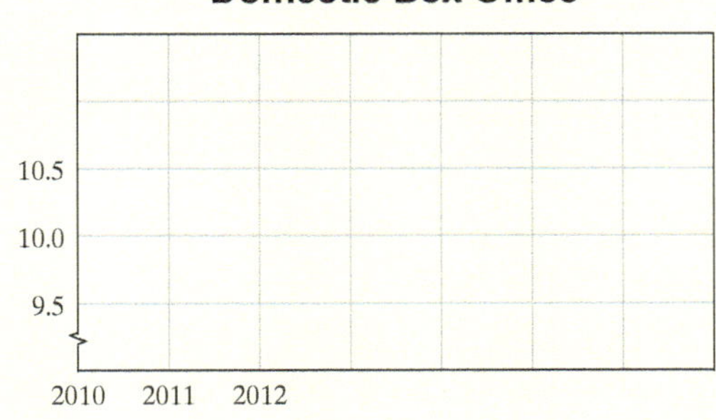

Note: ⸗ along the vertical axis indicates that some numbers are missing from the scale.

d. What trend in the paired data does the scatter diagram show?

18. The table shows the amount of money (in billions of dollars) that Americans spent on their pets for the years shown. (*Source:* American Pet Products Association, Inc.)

Year	Pet-Related Expenditures (in billions of dollars)
2011	51
2012	53
2013	56
2014	58
2015	60
2016	67
2017	70
2018	72

a. Write this paired data as a set of ordered pairs of the form (year, pet-related expenditures).

b. In your own words, write the meaning of the ordered pair (2016, 67).

c. Create a scatter diagram of the paired data. Be sure to label the axes appropriately.

Pet-Related Expenditures

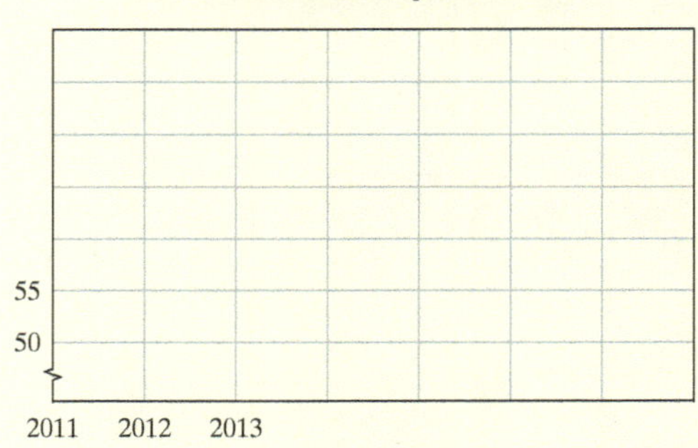

d. What trend in the paired data does the scatter diagram show?

19. Minh, a psychology student, kept a record of how much time she spent studying for each of her 20-point psychology quizzes and her score on each quiz.

Hours Spent Studying	0.50	0.75	1.00	1.25	1.50	1.50	1.75	2.00
Quiz Score	10	12	15	16	18	19	19	20

a. Write the data as ordered pairs of the form (hours spent studying, quiz score).

b. In your own words, write the meaning of the ordered pair (1.25, 16).

c. Create a scatter diagram of the paired data. Be sure to label the axes appropriately.

d. What might Minh conclude from the scatter diagram?

Minh's Chart for Psychology

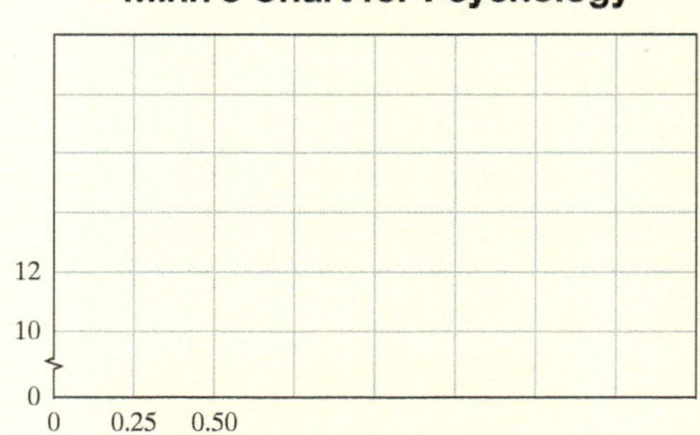

20. A local lumberyard uses quantity pricing. The table shows the price per board for different amounts of lumber purchased.

Price per Board (in dollars)	Number of Boards Purchased
8.00	1
7.50	10
6.50	25
5.00	50
2.00	100

Lumberyard Board Pricing

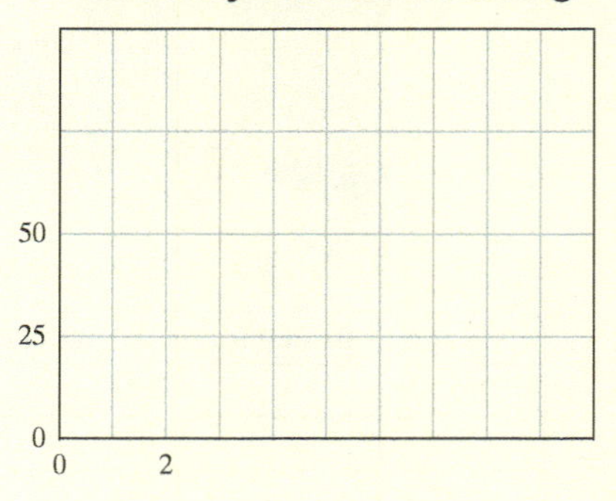

a. Write the data as ordered pairs of the form (price per board, number of boards purchased).

b. In your own words, write the meaning of the ordered pair (2.00, 100).

c. Create a scatter diagram of the paired data. Be sure to label the axes appropriately.

d. What trend in the paired data does the scatter diagram show?

Objective C *Complete each ordered pair so that it is a solution to the given linear equation. See Example 3.*

21. $x - 4y = 4$; (, −2), (4,)

22. $x - 5y = -1$; (, −2), (4,)

23. $y = \frac{1}{4}x - 3$; (−8,), (, 1)

24. $y = \frac{1}{5}x - 2$; (−10,), (, 1)

Complete the table of ordered pairs for each linear equation. See Examples 4 and 5.

25. $y = -7x$

x	y
0	
−1	
	2

26. $y = -9x$

x	y
	0
−3	
	2

27. $x = -y + 2$

x	y
0	
	0
−3	

28. $x = -y + 4$

x	y
	0
0	
	−3

29. $y = \frac{1}{2}x$

x	y
0	
−6	
	1

30. $y = \frac{1}{3}x$

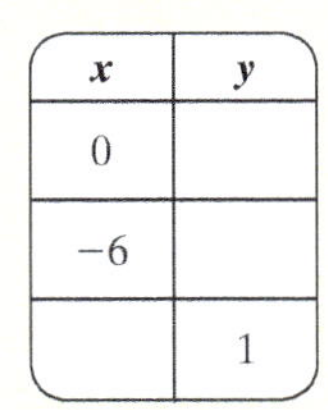

x	y
0	
−6	
	1

31. $x + 3y = 6$

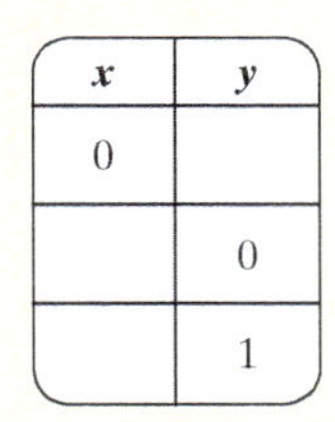

x	y
0	
	0
	1

32. $2x + y = 4$

x	y
0	
	0
	2

33. $y = 2x - 12$

x	y
0	
	−2
3	

34. $y = 5x + 10$

x	y
	0
	5
0	

35. $2x + 7y = 5$

x	y
0	
	0
	1

36. $x - 6y = 3$

x	y
0	
1	
	−1

Objectives A B C Mixed Practice *Complete the table of ordered pairs for each equation. Then plot the ordered pair solutions. See Examples 1 through 5.*

37. $x = -5y$

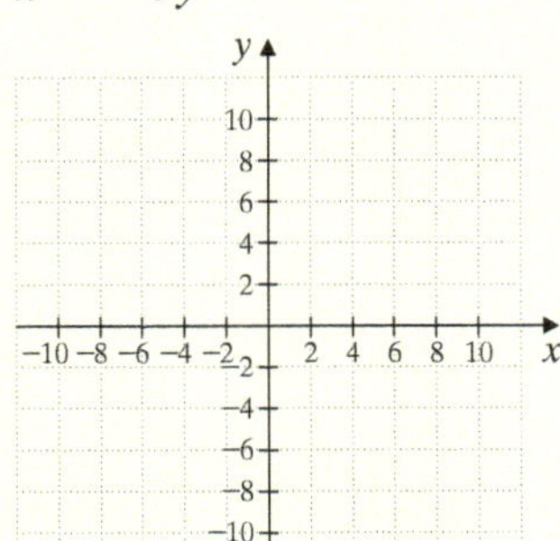

x	y
	0
	1
10	

38. $y = -3x$

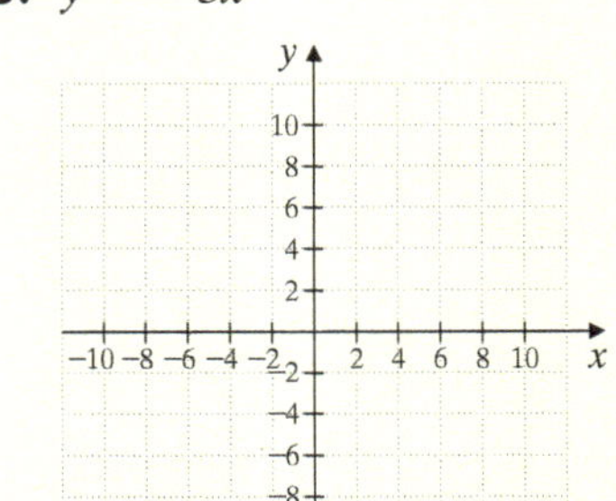

x	y
0	
−2	
	9

39. $y = \frac{1}{3}x + 2$

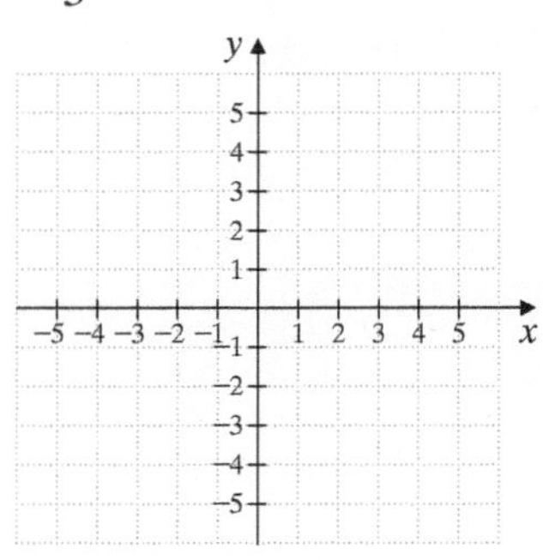

x	y
0	
−3	
	0

40. $y = \frac{1}{2}x + 3$

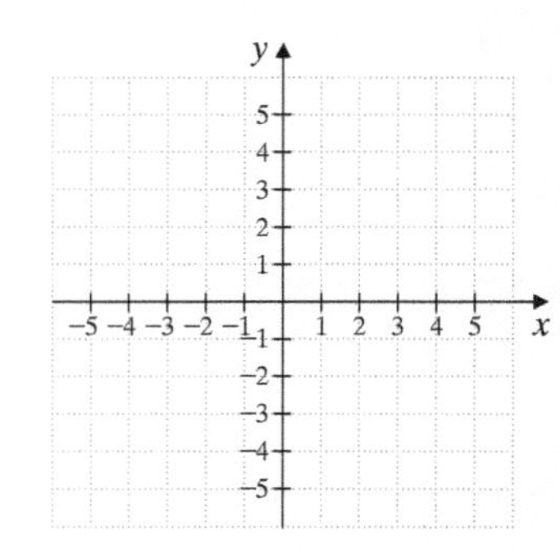

x	y
0	
−4	
	0

Solve. See Example 6.

41. The cost in dollars y of producing x computer desks is given by $y = 80x + 5000$.

a. Complete the table.

x	100	200	300
y			

b. Find the number of computer desks that can be produced for $8600. (*Hint:* Find x when $y = 8600$.)

42. The hourly wage y of an employee at a certain production company is given by $y = 0.25x + 9$, where x is the number of units produced by the employee in an hour.

a. Complete the table.

x	0	1	5	10
y				

b. Find the number of units that the employee must produce each hour to earn an hourly wage of $12.25. (*Hint:* Find x when $y = 12.25$.)

43. The average annual cinema admission price y (in dollars) from 2010 through 2017 is given by $y = 0.15x + 7.74$. In this equation, x represents the number of years after 2010. (*Source:* Motion Picture Association of America)

a. Complete the table.

x	0	1	7
y			

b. Find the year in which the average cinema admission price was approximately $8.00. (*Hint:* Find x when $y = 8.00$ and round to the nearest whole number.)

c. Use the given equation to predict when the cinema admission price might be $10.00. (Use the hint for part **b.**)

44. The number of farms y in the United States from 2010 through 2017 is given by $y = -14{,}000x + 2{,}145{,}000$. In the equation, x represents the number of years after 2010. (*Source:* Based on data from the National Agricultural Statistics Service)

a. Complete the table.

x	2	4	7
y			

b. Find the year in which there were approximately 2,030,000 farms. (*Hint:* Find x when $y = 2{,}030{,}000$ and round to the nearest whole number.)

c. Use the given equation to predict when the number of farms might be 2,000,000. (Use the hint for part **b.**)

Review

The following bar graph shows the top 10 tourist destinations and the number of tourists that visit each destination per year forecasted for 2020. Use this graph to answer Exercises 45 through 50. See Section R.4.

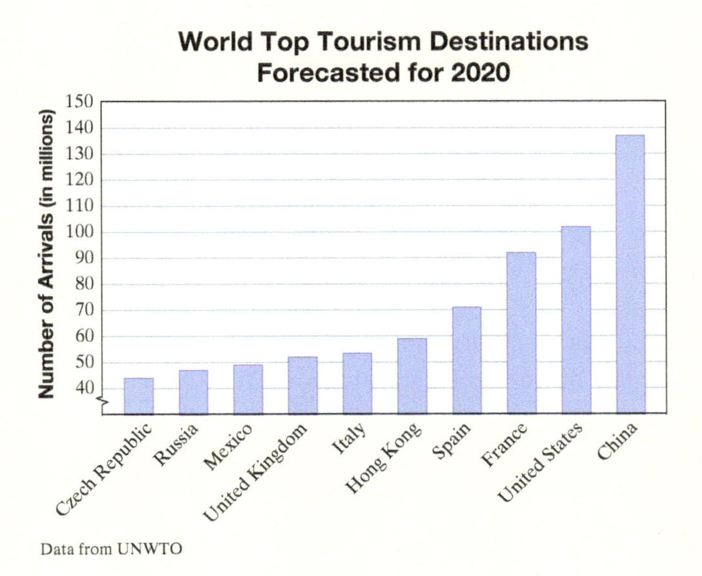

Data from UNWTO

45. Which location shown is predicted to be the most popular tourist destination?

46. Which location shown is predicted to be the least popular tourist destination?

47. Which locations shown are predicted to have more than 70 million tourists per year?

48. Which locations shown are predicted to have more than 100 million tourists per year?

49. Estimate the predicted number of tourists per year whose destination is Italy.

50. Estimate the predicted number of tourists per year whose destination is Mexico.

Solve each equation for y. See Section 2.5.

51. $x + y = 5$

52. $x - y = 3$

53. $2x + 4y = 5$

54. $5x + 2y = 7$

55. $10x = -5y$

56. $4y = -8x$

Concept Extensions

Answer each exercise with true or false.

57. Point $(-1, 5)$ lies in quadrant IV.

58. Point $(3, 0)$ lies on the y-axis.

59. For the point $\left(-\frac{1}{2}, 1.5\right)$, the first value, $-\frac{1}{2}$, is the x-coordinate and the second value, 1.5, is the y-coordinate.

60. The ordered pair $\left(2, \frac{2}{3}\right)$ is a solution of $2x - 3y = 6$.

For Exercises 61 through 65, fill in each blank with "0," "positive," or "negative." For Exercises 66 and 67, fill in each blank with "x" or "y."

	Point	Location
61.	(______, ______)	quadrant III
62.	(______, ______)	quadrant I
63.	(______, ______)	quadrant IV
64.	(______, ______)	quadrant II
65.	(______, ______)	origin
66.	(number, 0)	__-axis
67.	(0, number)	__-axis

68. Give an example of an ordered pair whose location is in (or on)

a. quadrant I **b.** quadrant II **c.** quadrant III
d. quadrant IV **e.** x-axis **f.** y-axis

Solve. See the Concept Checks in this section.

69. Is the graph of $(3, 0)$ in the same location as the graph of $(0, 3)$? Explain why or why not.

70. Give the coordinates of a point such that if the coordinates are reversed, their location is the same.

71. In general, what points can have coordinates reversed and still have the same location?

72. In your own words, describe how to plot or graph an ordered pair of numbers.

73. Discuss any similarities in the graphs of the ordered pair solutions for Exercises **37** through **40**.

74. Discuss any differences in the graphs of the ordered pair solutions for Exercises **37** through **40**.

Write an ordered pair for each point described.

75. Point C is four units to the right of the y-axis and seven units below the x-axis.

76. Point D is three units to the left of the origin.

△ **77.** Find the perimeter of the rectangle whose vertices are the points with coordinates $(-1, 5)$, $(3, 5)$, $(3, -4)$, and $(-1, -4)$.

△ **78.** Find the area of the rectangle whose vertices are the points with coordinates $(5, 2)$, $(5, -6)$, $(0, -6)$, and $(0, 2)$.

The scatter diagram below shows the annual number of people enrolled as Gold Star Members at Costco Wholesale. The horizontal axis represents the number of years after 2012.

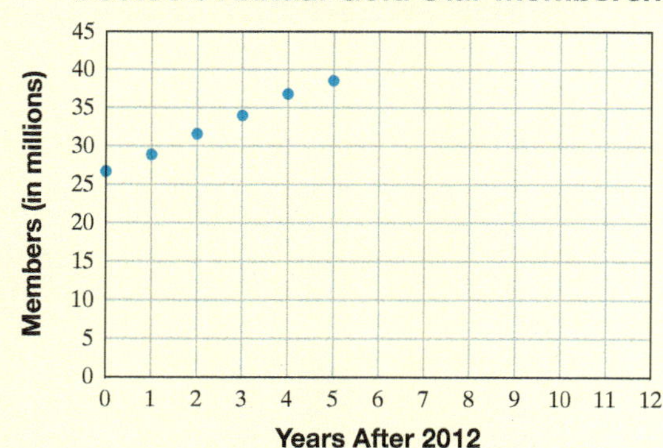

79. Estimate the annual Gold Star Membership for years 1, 2, 3, and 4.

80. Use a straightedge or ruler and this scatter diagram to predict Costco's Gold Star Membership in the year 2020.

6.2 Graphing Linear Equations 

Objective

A Graph a Linear Equation by Finding and Plotting Ordered Pair Solutions.

In the previous section, we found that equations in two variables may have more than one solution. For example, both $(2, 2)$ and $(0, 4)$ are solutions of the equation $x + y = 4$. In fact, this equation has an infinite number of solutions. Other solutions include $(-2, 6)$, $(4, 0)$, and $(6, -2)$. Notice the pattern that appears in the graph of these solutions.

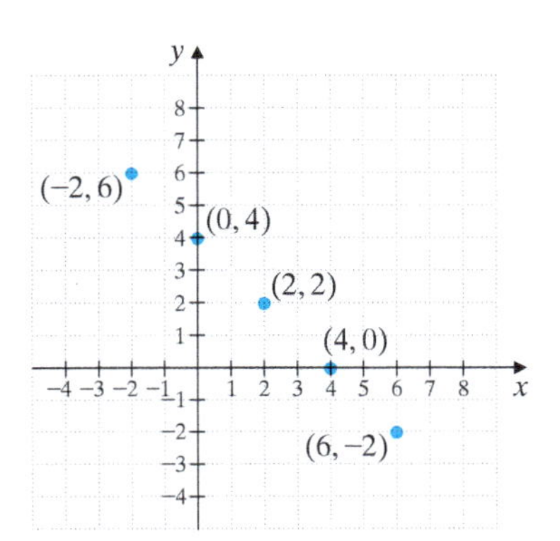

These solutions all appear to lie on the same line, as seen in the graph below. It can be shown that every ordered pair solution of the equation corresponds to a point on this line and that every point on this line corresponds to an ordered pair solution. Thus, we say that this line is the **graph of the equation** $x + y = 4$.

Helpful Hint Notice that we can show only a part of a line on a graph. The arrowheads on each end of the line remind us that the line actually extends indefinitely in both directions.

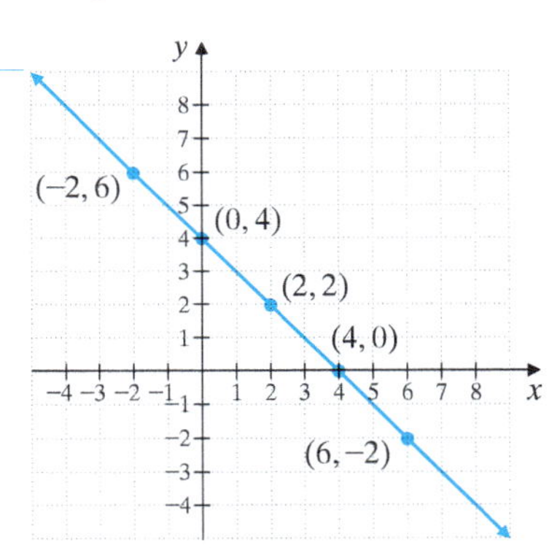

The equation $x + y = 4$ is called a *linear equation in two variables* and *the graph of every linear equation in two variables is a straight line.*

Linear Equation in Two Variables

A **linear equation in two variables** is an equation that can be written in the form

$$Ax + By = C$$

where A, B, and C are real numbers and A and B are not both 0. This form is called **standard form. The graph of a linear equation in two variables is a straight line.**

Helpful Hint

Notice from above that the form $Ax + By = C$

- is called standard form, and
- has an understood exponent of 1 on both x and y.

A linear equation in two variables may be written in many forms. Standard form, $Ax + By = C$, is just one of these many forms.

Following are examples of linear equations in two variables.

$2x + y = 8$ (Standard form) $\quad -2x = 7y \quad y = \frac{1}{3}x + 2 \quad y = 7$

Objective A Graphing Linear Equations

From geometry, we know that a straight line is determined by just two points. Thus, to graph a linear equation in two variables, we need to find just two of its infinitely many solutions. Once we do so, we plot the solution points and draw the line connecting the points. Usually, we find a third solution as well, as a check.

Practice 1

Graph the linear equation $x + 3y = 6$.

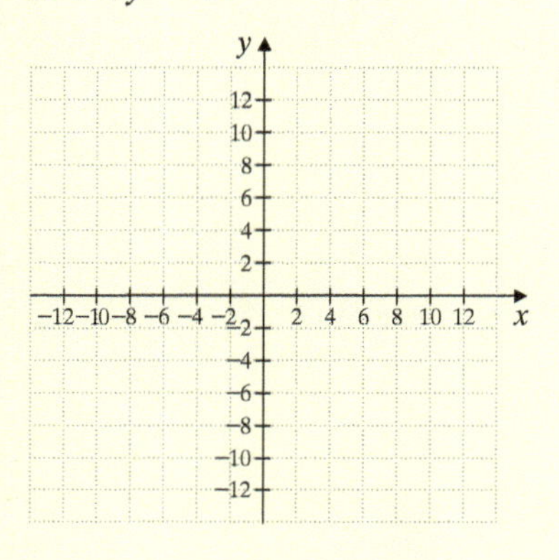

Example 1

Graph the linear equation $2x + y = 5$.

Solution: To graph this equation, we find three ordered pair solutions of $2x + y = 5$. To do this, we choose a value for one variable, x or y, and solve for the other variable. For example, if we let $x = 1$, then $2x + y = 5$ becomes

$$2x + y = 5$$
$$2(1) + y = 5 \quad \text{Replace } x \text{ with 1.}$$
$$2 + y = 5 \quad \text{Multiply.}$$
$$y = 3 \quad \text{Subtract 2 from both sides.}$$

Since $y = 3$ when $x = 1$, the ordered pair $(1, 3)$ is a solution of $2x + y = 5$. Next, we let $x = 0$.

$$2x + y = 5$$
$$2(0) + y = 5 \quad \text{Replace } x \text{ with 0.}$$
$$0 + y = 5$$
$$y = 5$$

The ordered pair $(0, 5)$ is a second solution.

The two solutions found so far allow us to draw the straight line that is the graph of all solutions of $2x + y = 5$. However, we will find a third ordered pair as a check. Let $y = -1$.

$$2x + y = 5$$
$$2x + (-1) = 5 \quad \text{Replace } y \text{ with } -1.$$
$$2x - 1 = 5$$
$$2x = 6 \quad \text{Add 1 to both sides.}$$
$$x = 3 \quad \text{Divide both sides by 2.}$$

Helpful Hint All three points should fall on the same straight line. If not, check your ordered pair solutions for a mistake.

The third solution is $(3, -1)$. These three ordered pair solutions are listed in the table and plotted on the coordinate plane. The graph of $2x + y = 5$ is the line through the three points.

x	y
1	3
0	5
3	−1

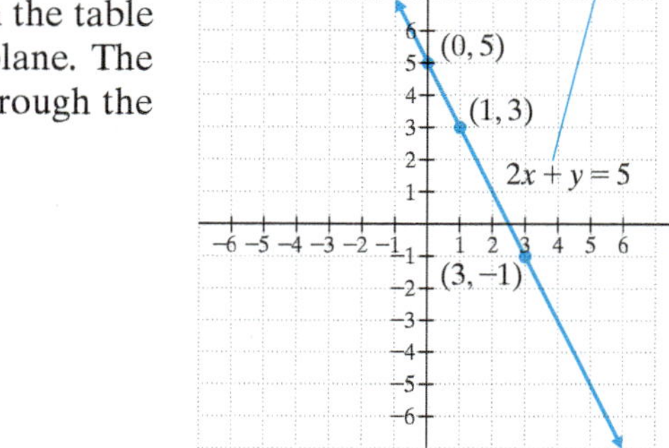

Work Practice 1

Answer

1.

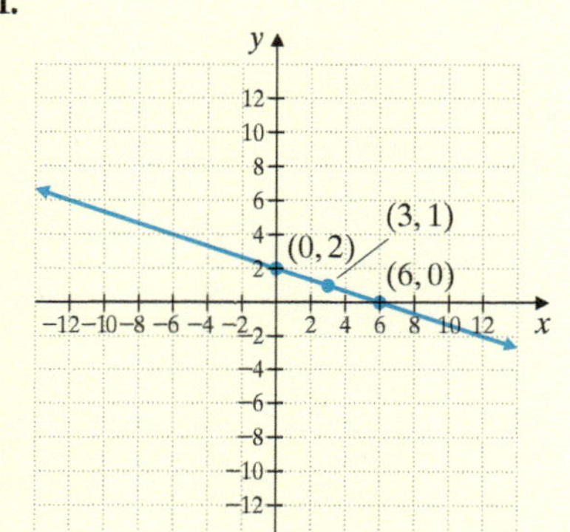

Example 2 Graph the linear equation $-5x + 3y = 15$.

Solution: We find three ordered pair solutions of $-5x + 3y = 15$.

Let $x = 0$.

$$-5x + 3y = 15$$
$$-5 \cdot 0 + 3y = 15$$
$$0 + 3y = 15$$
$$3y = 15$$
$$y = 5$$

Let $y = 0$.

$$-5x + 3y = 15$$
$$-5x + 3 \cdot 0 = 15$$
$$-5x + 0 = 15$$
$$-5x = 15$$
$$x = -3$$

Let $x = -2$.

$$-5x + 3y = 15$$
$$-5 \cdot -2 + 3y = 15$$
$$10 + 3y = 15$$
$$3y = 5$$
$$y = \frac{5}{3} \text{ or } 1\frac{2}{3}$$

The ordered pairs are $(0, 5)$, $(-3, 0)$, and $\left(-2, 1\frac{2}{3}\right)$. The graph of $-5x + 3y = 15$ is the line through the three points.

x	y
0	5
−3	0
−2	$1\frac{2}{3}$

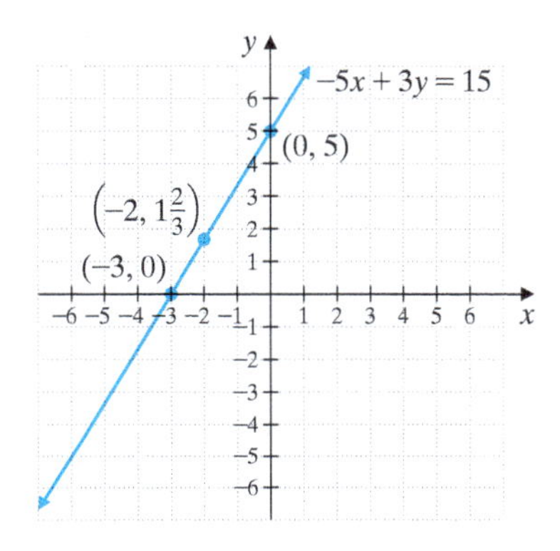

Work Practice 2

Example 3 Graph the linear equation $y = 3x$.

Solution: We find three ordered pair solutions. Since this equation is solved for y, we'll choose three x-values.

If $x = 2$, $y = 3 \cdot 2 = 6$.
If $x = 0$, $y = 3 \cdot 0 = 0$.
If $x = -1$, $y = 3 \cdot -1 = -3$.

Next, we plot the ordered pair solutions and draw a line through the plotted points. The line is the graph of $y = 3x$.

Think about the following for a moment: A line is made up of an infinite number of points. Every point on the line defined by $y = 3x$ represents an ordered pair solution of the equation, and every ordered pair solution is a point on this line.

x	y
2	6
0	0
−1	−3

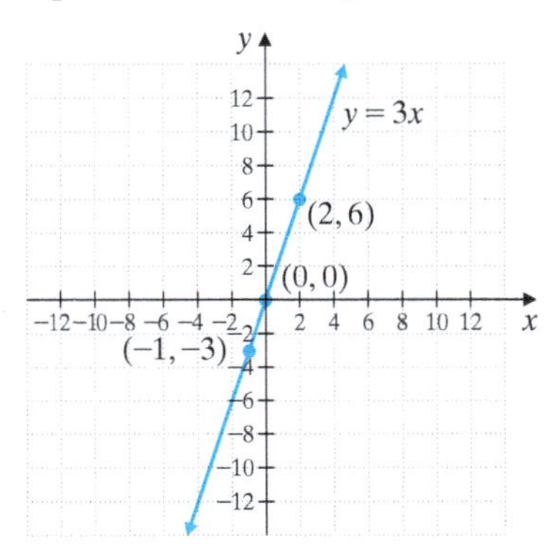

Work Practice 3

Practice 2

Graph the linear equation $-2x + 4y = 8$.

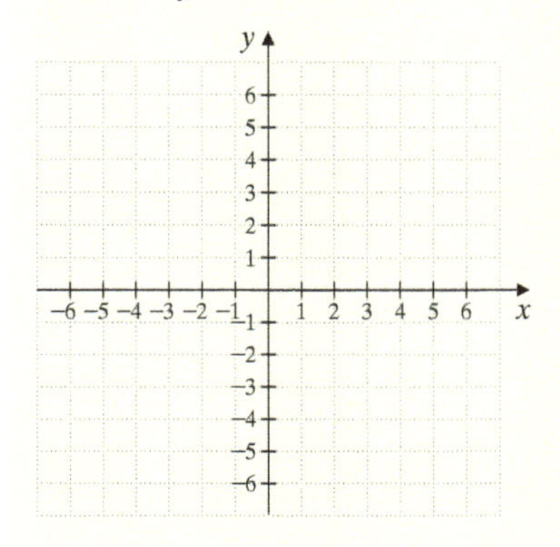

Practice 3

Graph the linear equation $y = 2x$.

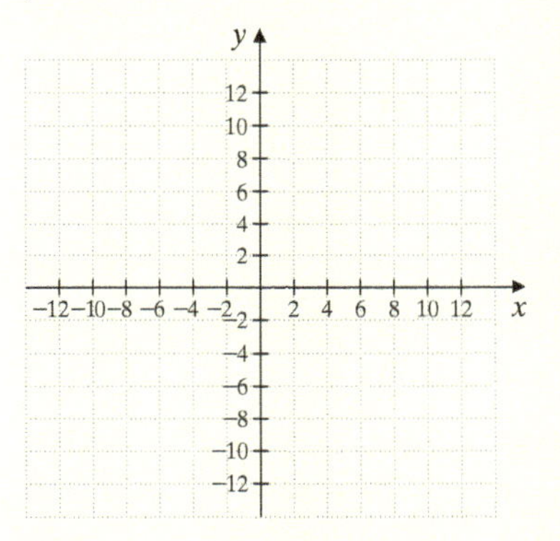

Answers

2.

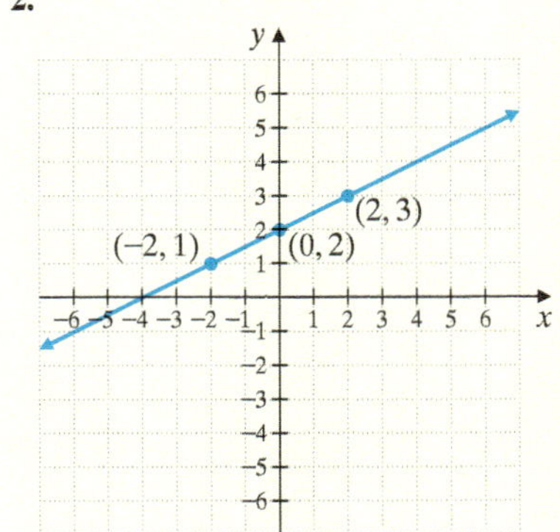

3.

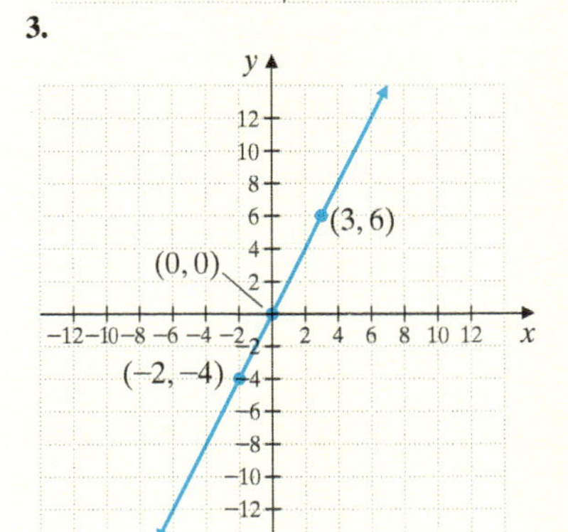

Practice 4

Graph the linear equation $y = -\frac{1}{2}x + 4$.

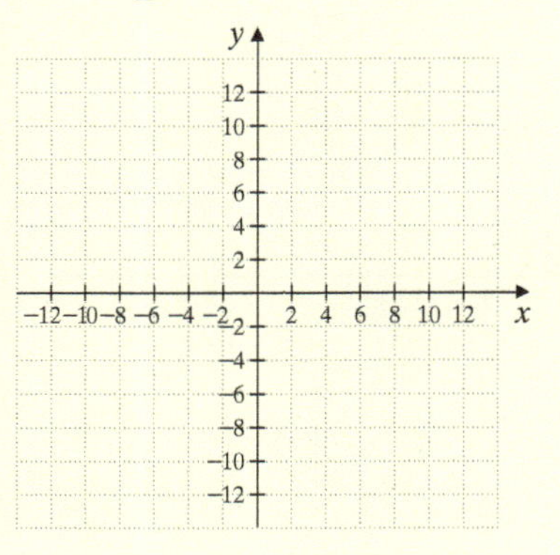

Practice 5

Graph the linear equation $x = 3$.

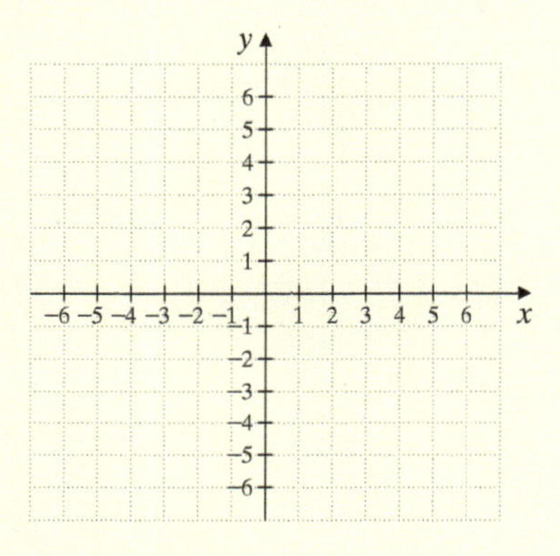

Answers

4.

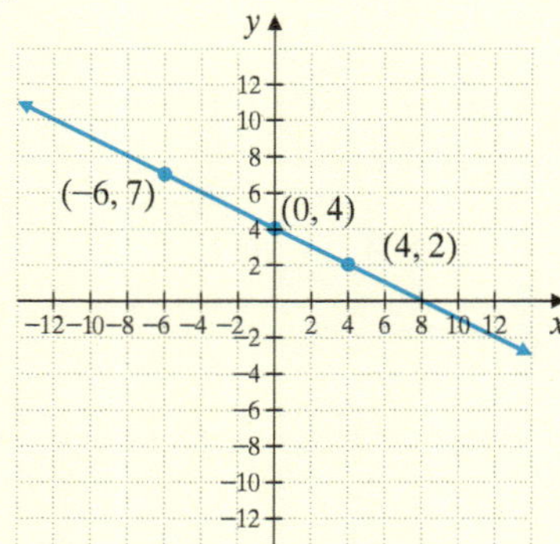

5.

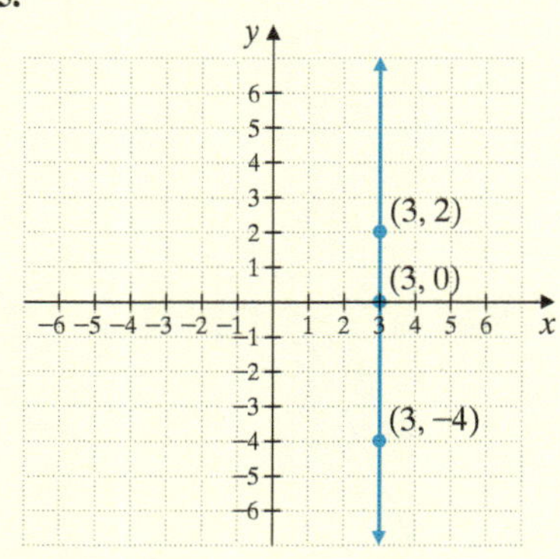

Helpful Hint

When graphing a linear equation in two variables, if it is

- solved for y, it may be easier to find ordered pair solutions by choosing x-values.
- solved for x, it may be easier to find ordered pair solutions by choosing y-values.

Example 4

Graph the linear equation $y = -\frac{1}{3}x + 2$.

Solution: We find three ordered pair solutions, plot the solutions, and draw a line through the plotted solutions. To avoid fractions, we'll choose x-values that are multiples of 3 to substitute into the equation.

If $x = 6$, then $y = -\frac{1}{3}\cdot 6 + 2 = -2 + 2 = 0$.

If $x = 0$, then $y = -\frac{1}{3}\cdot 0 + 2 = 0 + 2 = 2$.

If $x = -3$, then $y = -\frac{1}{3}\cdot -3 + 2 = 1 + 2 = 3$.

x	y
6	0
0	2
−3	3

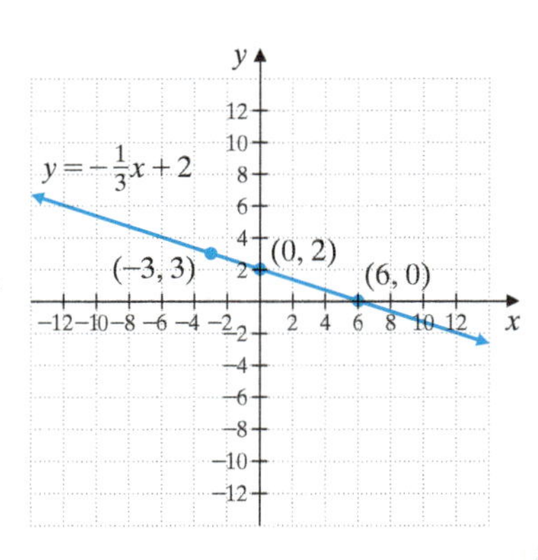

Work Practice 4

Let's take a moment and compare the graphs in Examples 3 and 4. The graph of $y = 3x$ tilts upward (as we follow the line from left to right) and the graph of $y = -\frac{1}{3}x + 2$ tilts downward (as we follow the line from left to right). We will learn more about the tilt, or slope, of a line in Section 6.4.

Example 5

Graph the linear equation $y = -2$.

Solution: The equation $y = -2$ can be written in standard form as $0x + y = -2$. No matter what value we replace x with, y is always -2.

x	y
0	−2
3	−2
−2	−2

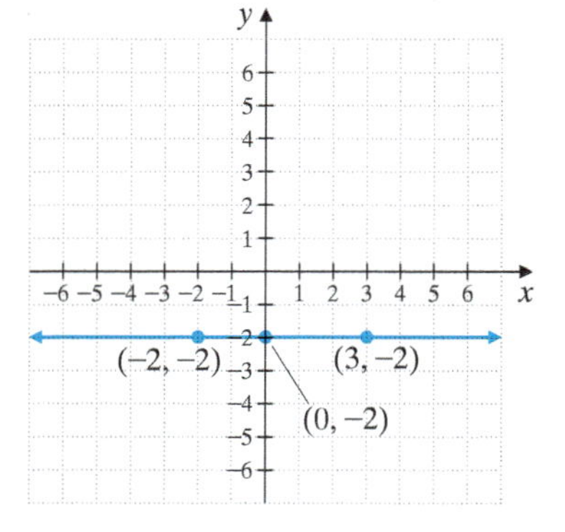

Helpful Hint

From Example 5, we learned that equations such as $y = -2$ are linear equations since $y = -2$ can be written as $0x + y = -2$.

Notice that the graph of $y = -2$ is a horizontal line.

Work Practice 5

Linear equations are often used to model real data, as seen in the next example.

Example 6 Estimating the Number of Registered Nurses

One of the occupations expected to have the most growth in the next few years is registered nurse. The number of people y (in thousands) employed as registered nurses in the United States can be estimated by the linear equation $y = 43.9x + 2751$, where x is the number of years after the year 2014. (*Source:* Based on data from the Bureau of Labor Statistics)

a. Graph the equation.

b. Use the graph to predict the number of registered nurses in the year 2025.

Solution:

a. To graph $y = 43.9x + 2751$, choose x-values and substitute into the equation.

If $x = 0$, then $y = 43.9(0) + 2751 = 2751$.

If $x = 2$, then $y = 43.9(2) + 2751 = 2838.8$.

If $x = 5$, then $y = 43.9(5) + 2751 = 2970.5$.

x	y
0	2751
2	2838.8
5	2970.5

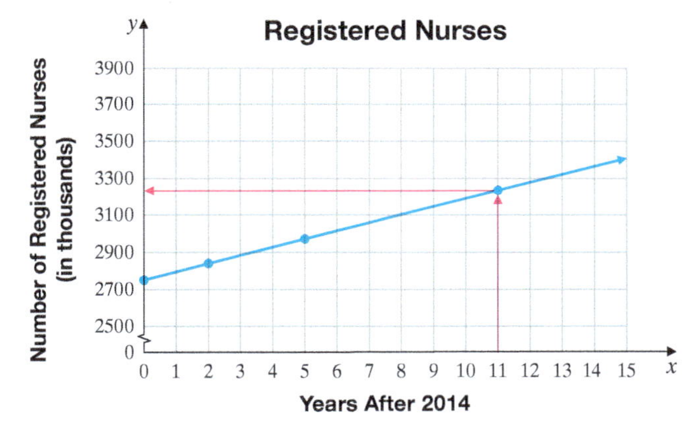

b. To use the graph to *predict* the number of registered nurses in the year 2025, we need to find the y-coordinate that corresponds to $x = 11$. (11 years after 2014 is the year 2025.) To do so, find 11 on the x-axis. Move vertically upward to the graphed line and then horizontally to the left. We approximate the number on the y-axis to be 3230. Thus, in the year 2025, we predict that there will be 3230 thousand registered nurses. (The value found by substituting 11 for x in the equation is 3223.9.)

Work Practice 6

Practice 6

Use the graph in Example 6 to predict the number of registered nurses in 2026.

Helpful Hint

Make sure you understand that models are mathematical approximations of the data for the known years. (For example, see the model in Example 6.) Any number of unknown factors can affect future years, so be cautious when using models to make predictions.

Answer

6. 3275 thousand

Calculator Explorations Graphing

In this section, we begin an optional study of graphing calculators and graphing software packages for computers. These graphers use the same point plotting technique that was introduced in this section. The advantage of this graphing technology is, of course, that graphing calculators and computers can find and plot ordered pair solutions much faster than we can. Note, however, that the features described in these boxes may not be available on all graphing calculators.

The rectangular screen where a portion of the rectangular coordinate system is displayed is called a **window.** We call it a **standard window** for graphing when both the x- and y-axes show coordinates between -10 and 10. This information is often displayed in the window menu on a graphing calculator as follows.

Xmin = −10
Xmax = 10
Xscl = 1
Ymin = −10
Ymax = 10
Yscl = 1

To use a graphing calculator to graph the equation $y = 2x + 3$, press the [Y =] key and enter the keystrokes [2] [x] [+] [3]. The top row should now read $Y_1 = 2x + 3$. Next press the [GRAPH] key, and the display should look like this:

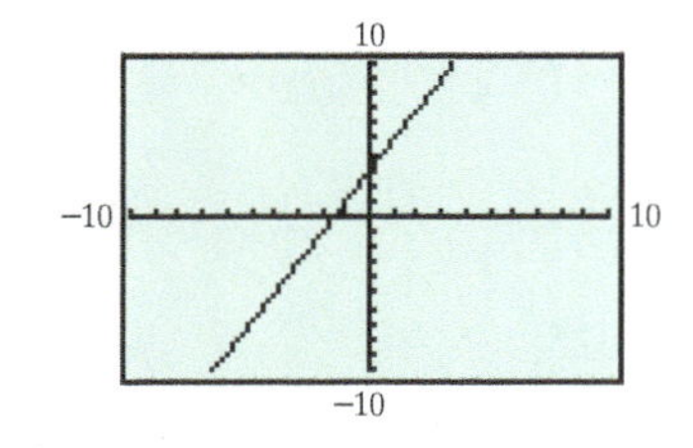

Graph the following linear equations. (Unless otherwise stated, use a standard window when graphing.)

1. $y = -3x + 7$

2. $y = -x + 5$

3. $y = 2.5x - 7.9$

4. $y = -1.3x + 5.2$

5. $y = -\frac{3}{10}x + \frac{32}{5}$

6. $y = \frac{2}{9}x - \frac{22}{3}$

Vocabulary, Readiness & Video Check

Martin-Gay Interactive Videos

See Video 6.2

Watch the section lecture video and answer the following questions.

Objective A 1. In the lecture before Example 1, it's mentioned that we need only two points to determine a line. Why, then, are three ordered pair solutions found in Examples 1–3?

Objective A 2. What does a graphed line represent, as discussed at the end of Examples 1 and 3?

6.2 Exercise Set MyLab Math 

Objective A *For each equation, find three ordered pair solutions by completing the table. Then use the ordered pairs to graph the equation. See Examples 1 through 5.*

1. $x - y = 6$

x	y
	0
4	
	−1

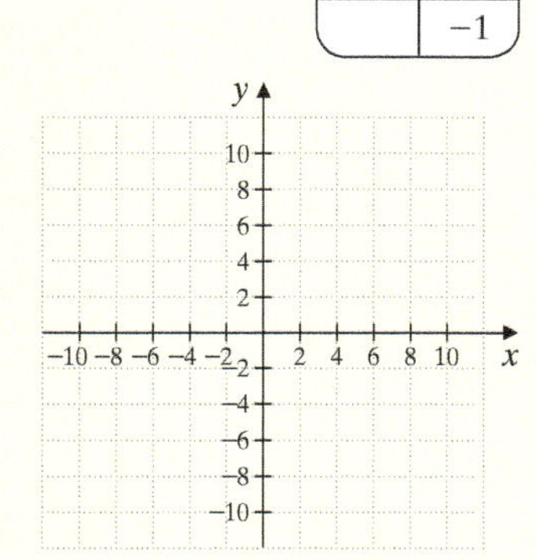

2. $x - y = 4$

x	y
0	
	2
−1	

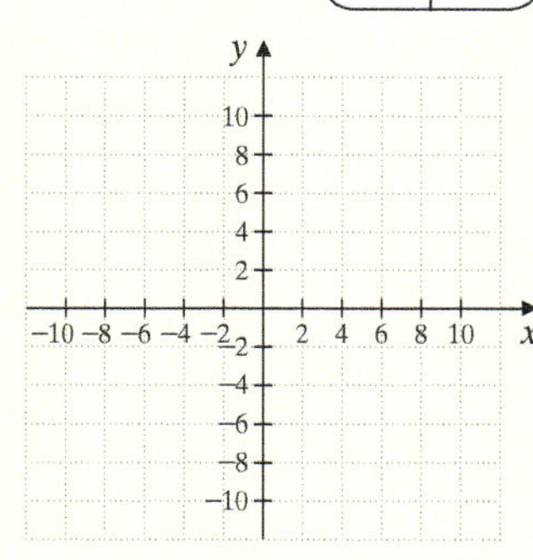

3. $y = -4x$

x	y
1	
0	
−1	

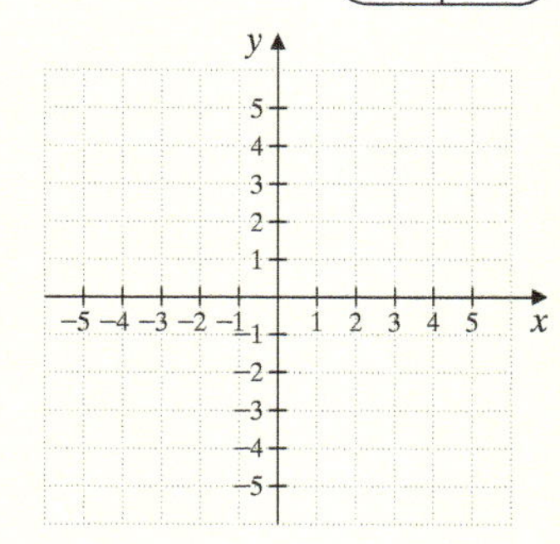

4. $y = -5x$

x	y
1	
0	
−1	

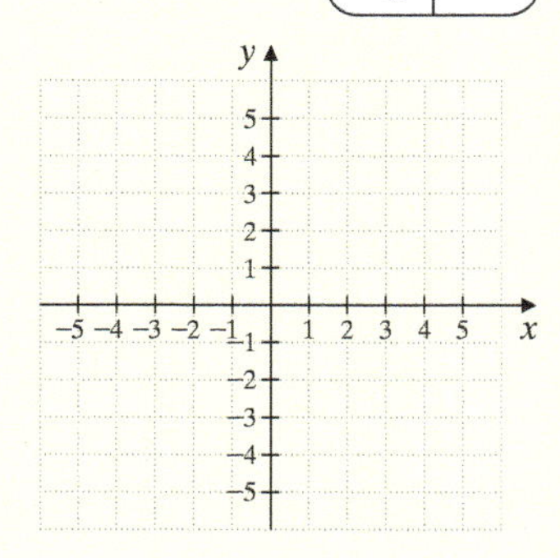

5. $y = \frac{1}{3}x$

x	y
0	
6	
−3	

6. $y = \frac{1}{2}x$

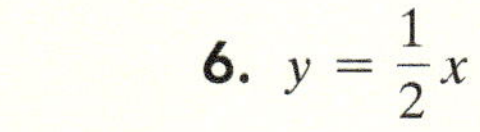

x	y
0	
−4	
2	

7. $y = -4x + 3$

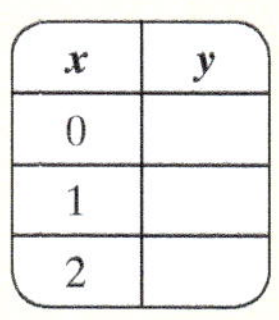

x	y
0	
1	
2	

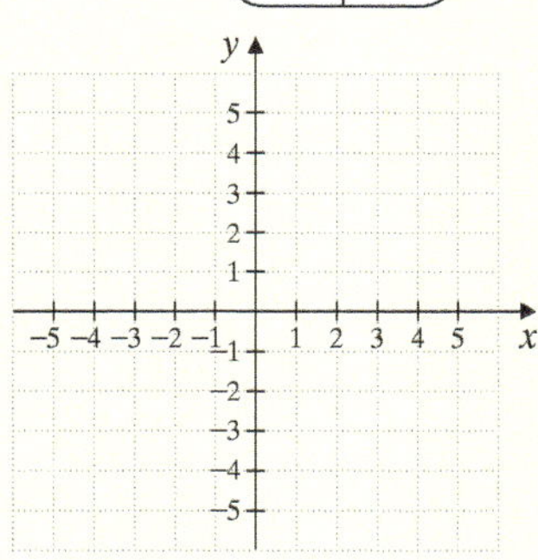

8. $y = -5x + 2$

x	y
0	
1	
2	

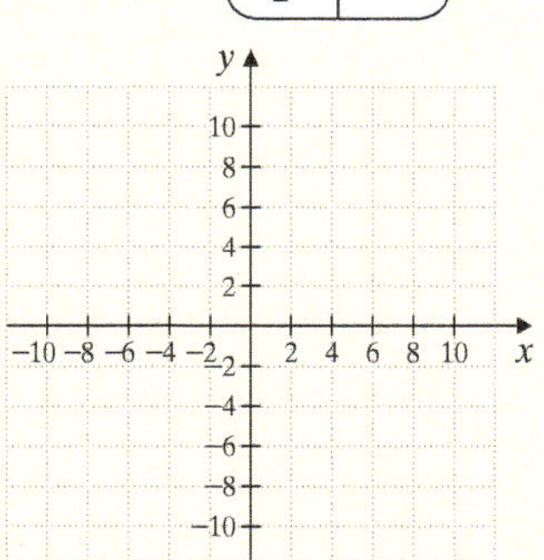

Graph each linear equation. See Examples 1 through 5.

9. $x + y = 1$

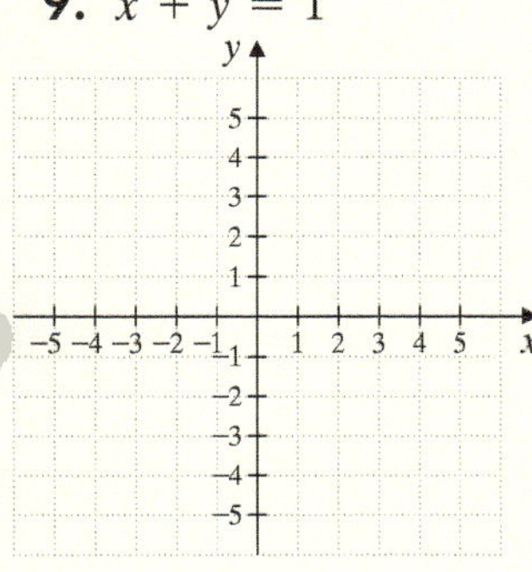

10. $x + y = 7$

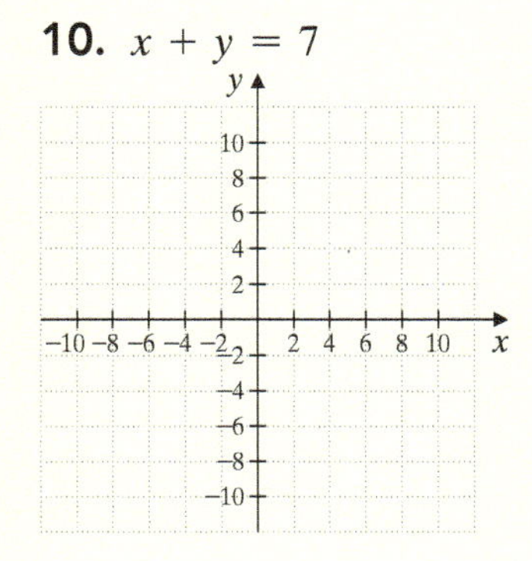

11. $x - y = -2$

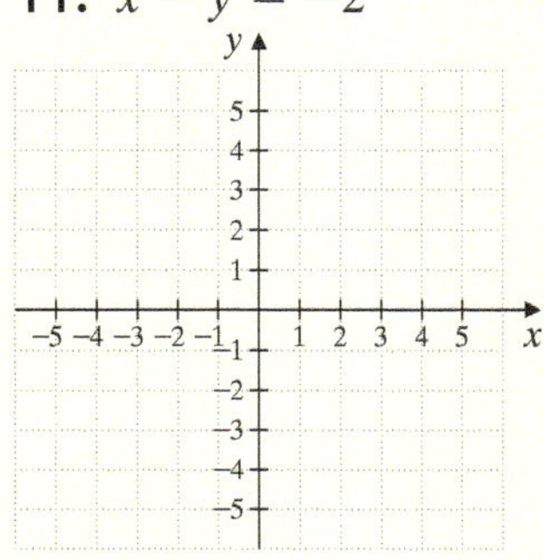

12. $-x + y = 6$

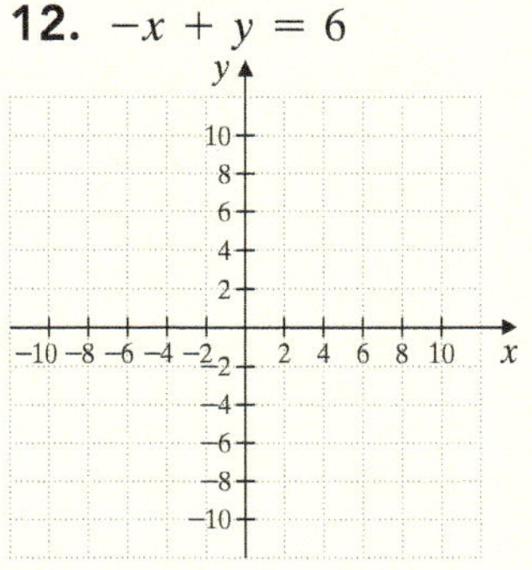

13. $x - 2y = 6$

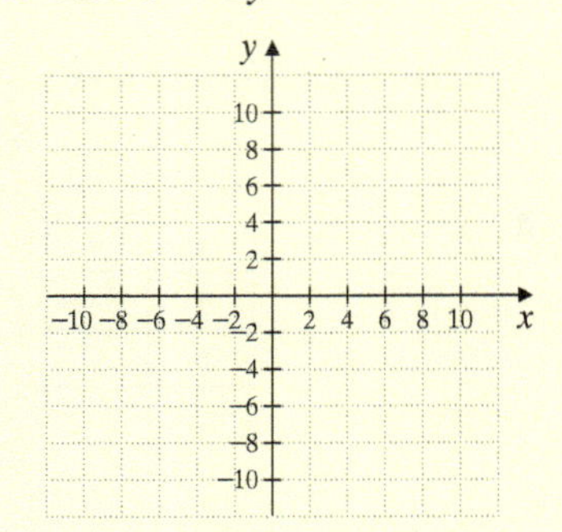

14. $-x + 5y = 5$

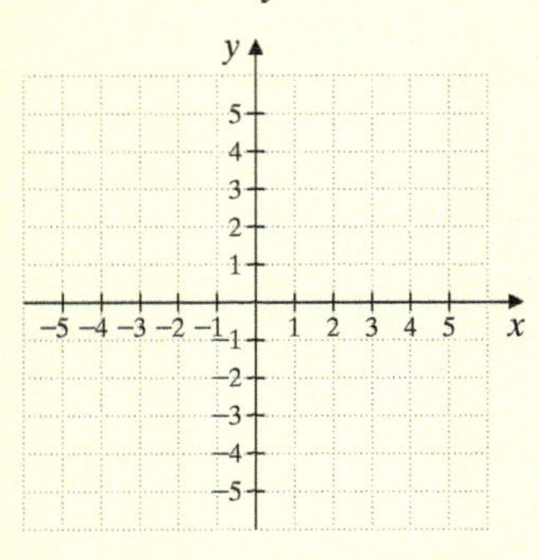

15. $y = 6x + 3$

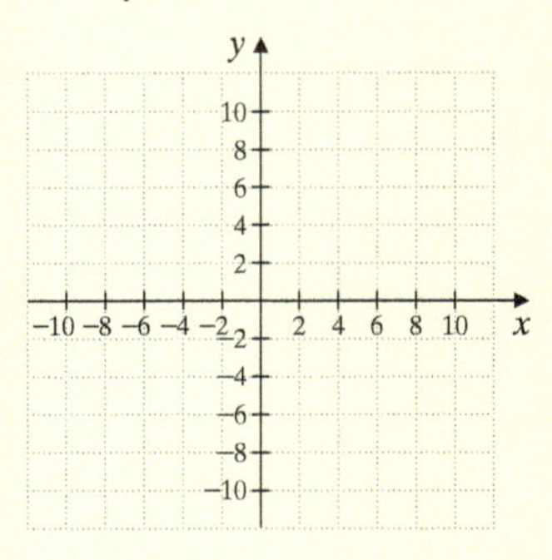

16. $y = -2x + 7$

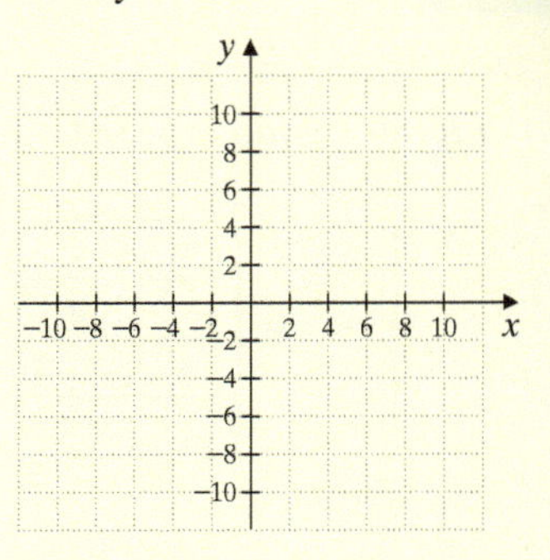

17. $x = -4$

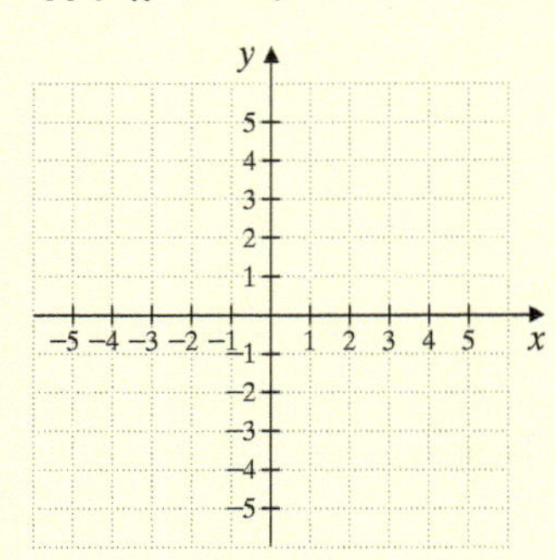

18. $y = 5$

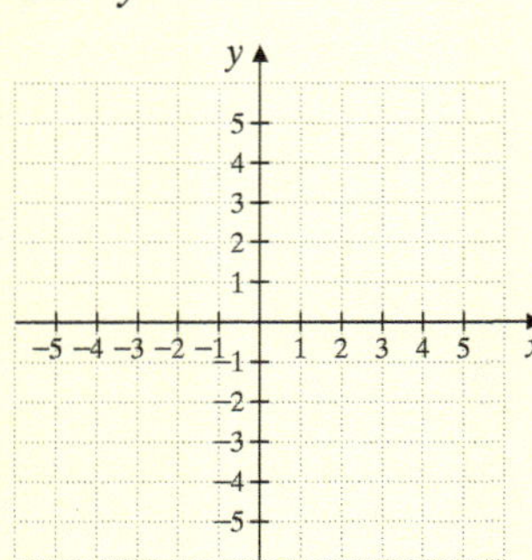

19. $y = 3$

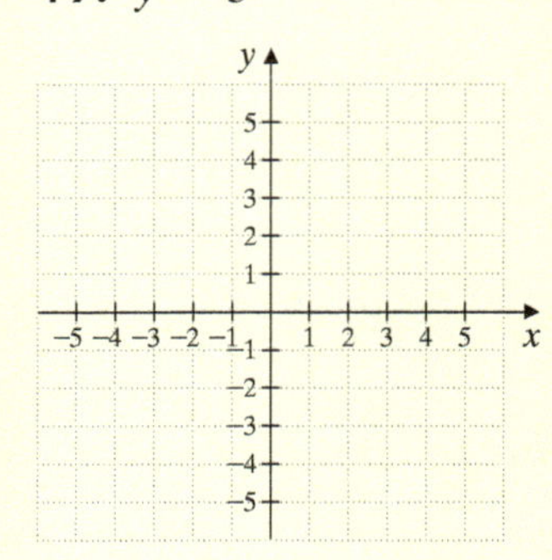

20. $x = -1$

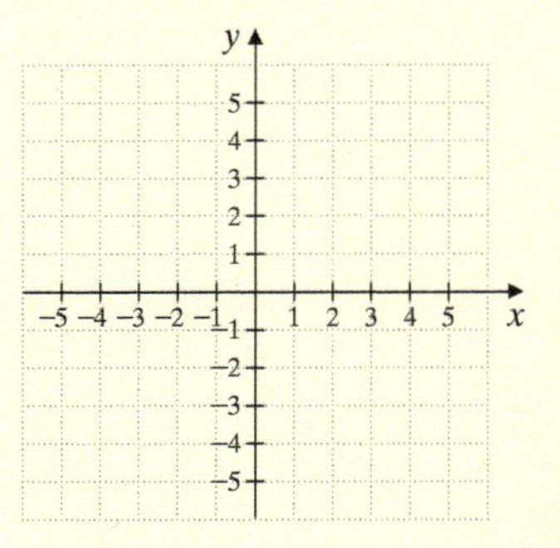

21. $y = x$

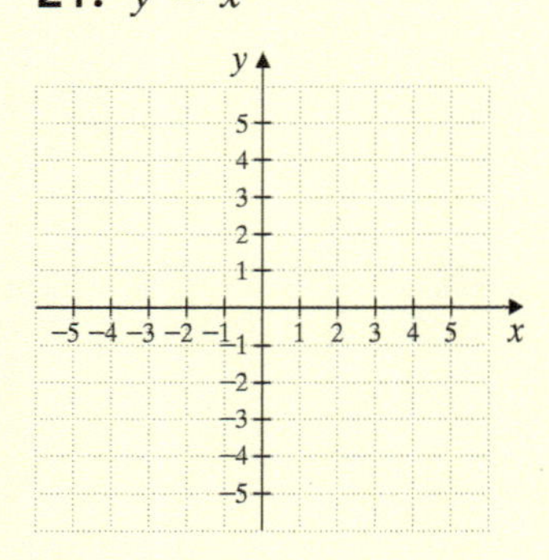

22. $y = -x$

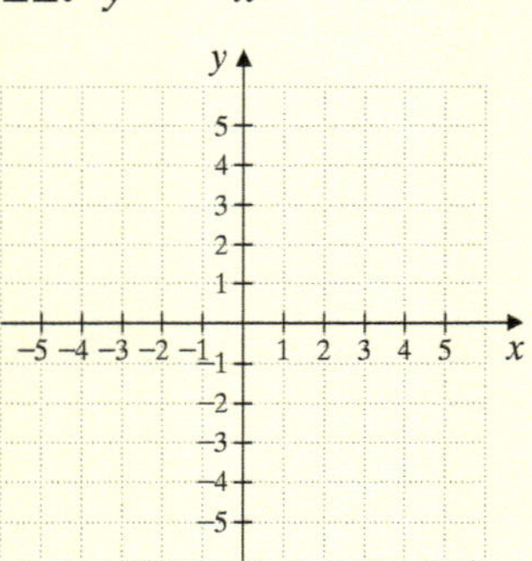

23. $x = -3y$

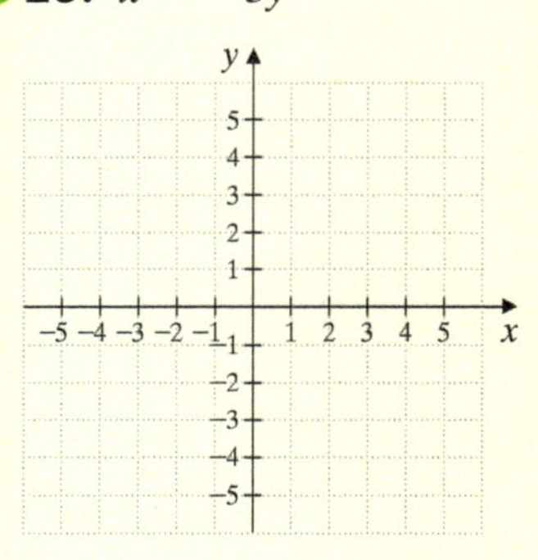

24. $x = 4y$

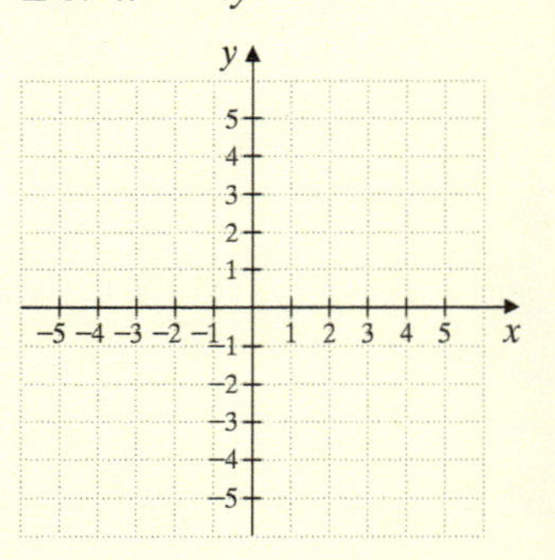

25. $x + 3y = 9$

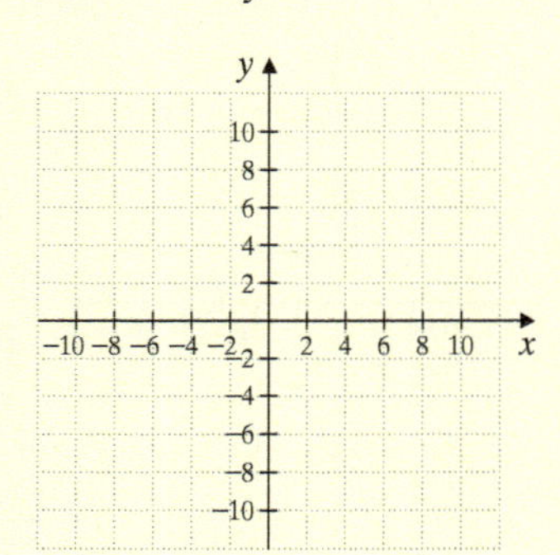

26. $2x + y = 2$

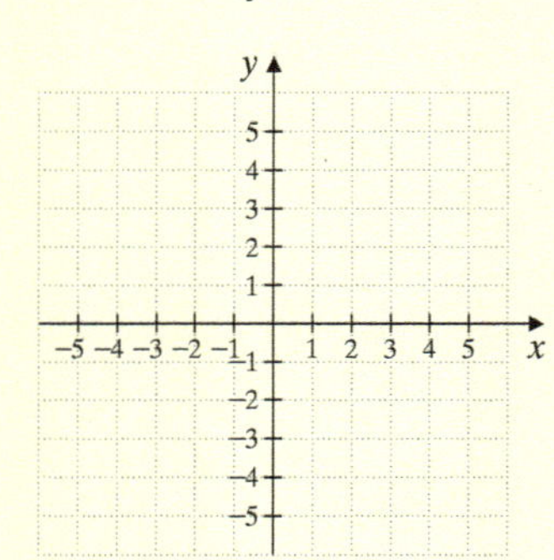

27. $y = \frac{1}{2}x + 2$

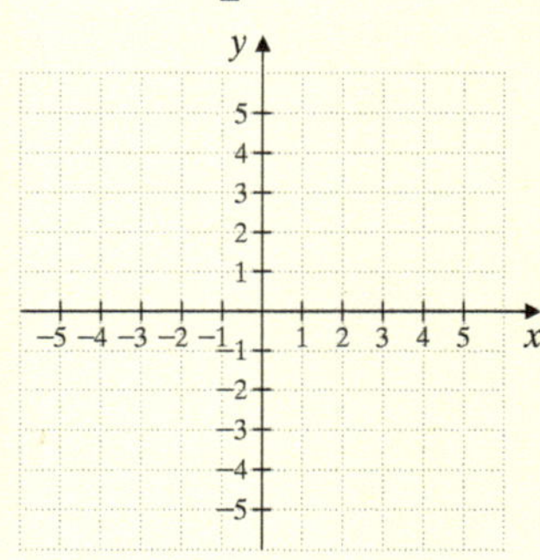

28. $y = \frac{1}{4}x + 3$

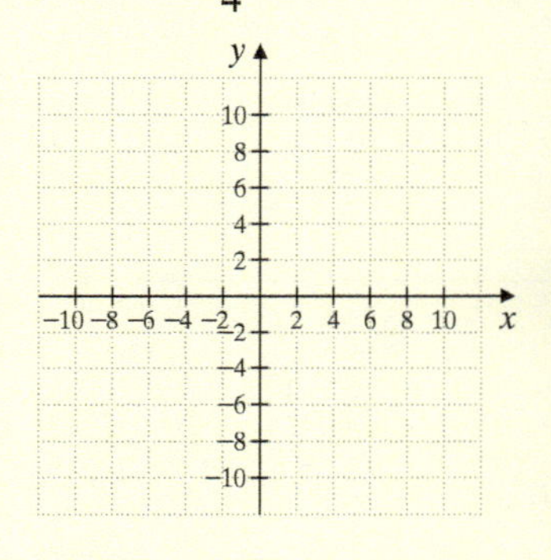

29. $3x - 2y = 12$

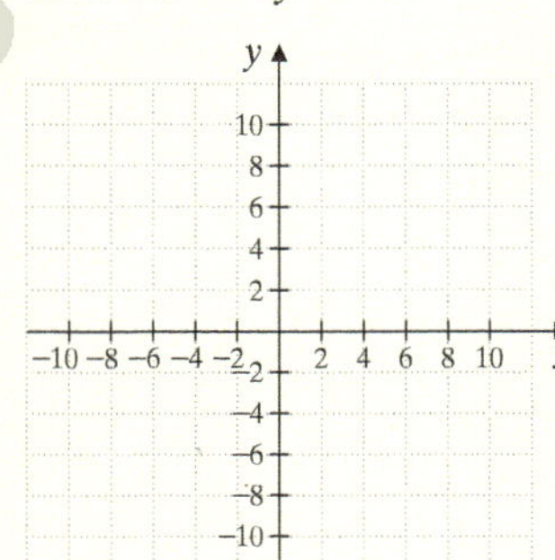

30. $2x - 7y = 14$

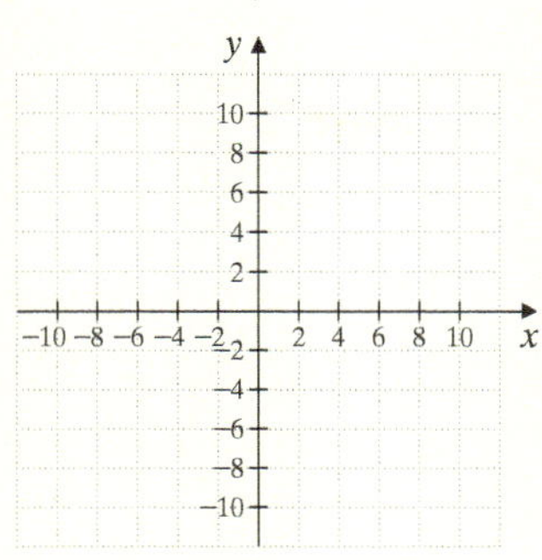

31. $y = -3.5x + 4$

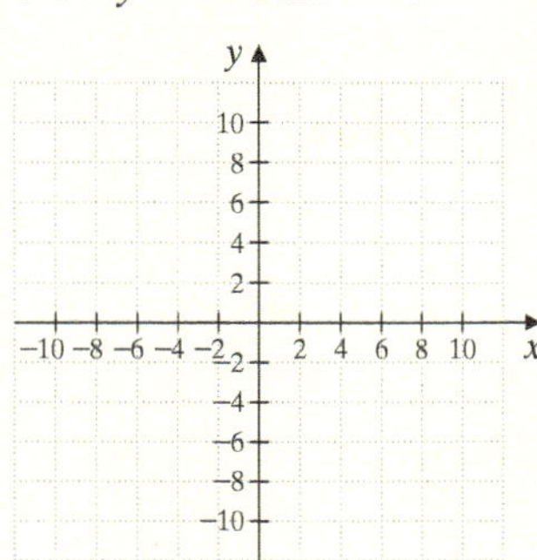

32. $y = -1.5x - 3$

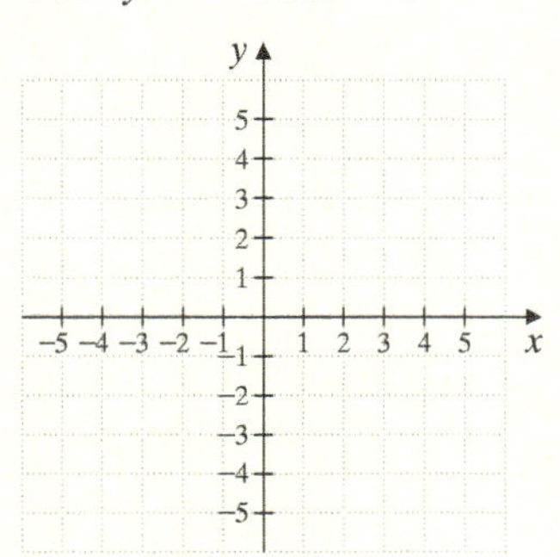

Solve. See Example 6.

33. The number of students y (in thousands) taking the SAT college entrance exam each year from 2010 through 2017 can be approximated by the linear equation $y = 15x + 1604$, where x is the number of years after 2010. (*Source:* Based on data from the College Board)

a. Graph the linear equation.

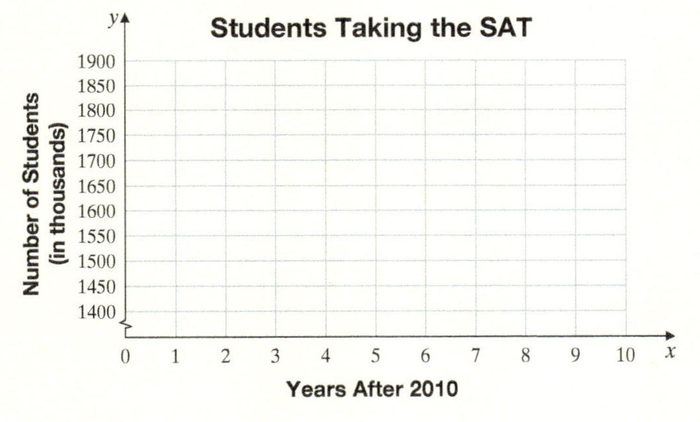

b. Does the point (7, 1709) lie on the line? If so, what does this ordered pair mean?

34. College is getting more expensive every year. The average cost for tuition and fees at a public two-year college y from 2001 through 2016 can be approximated by the linear equation $y = 90x + 2211$, where x is the number of years after 2001. (*Source:* The College Board: Trends in College Pricing 2016)

a. Graph the linear equation.

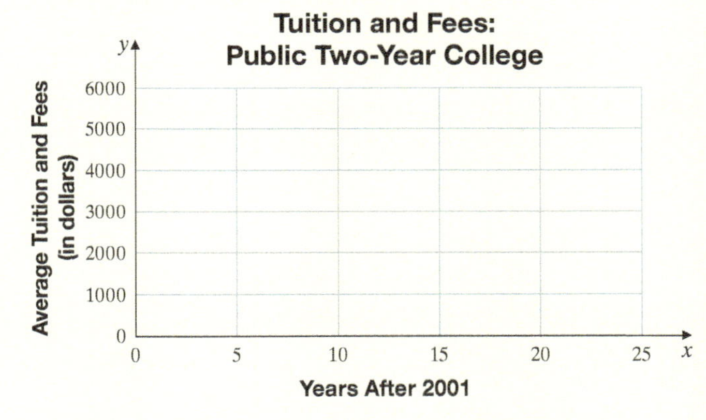

b. Does the point (20, 4011) lie on the line? If so, what does this ordered pair mean?

35. The total annual revenue y (in billions of euros) for IKEA from 2010 through 2016 can be approximated by the equation $y = 1.7x + 23.5$, where x is the number of years after 2010. (*Source:* Based on data from IKEA Group)

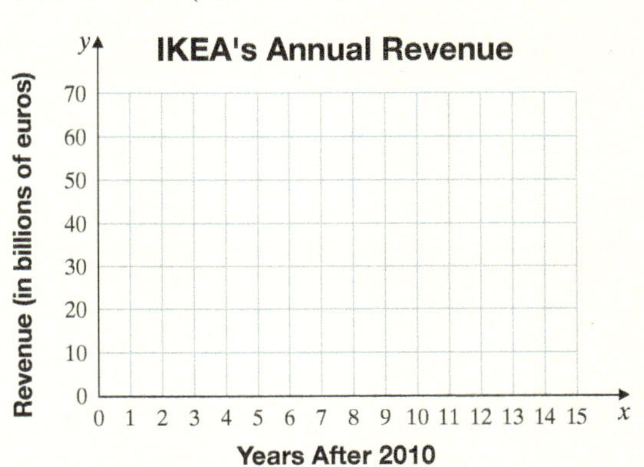

a. Graph the linear equation.

b. Complete the ordered pair (6,).

c. Write a sentence explaining the meaning of the ordered pair found in part **b**.

36. For the period 1970 through 2016, the annual food-and-drink sales for restaurants in the United States can be estimated by $y = 15.7x - 23.1$, where x is the number of years after 1970 and y is the food-and-drink sales in billions of dollars. (*Source:* Based on data from the National Restaurant Association)

a. Graph the linear equation.

b. Complete the ordered pair (43,).

c. Write a sentence explaining the meaning of the ordered pair found in part **b**.

Review

Solve. See Section 6.1.

△ **37.** The coordinates of three vertices of a rectangle are $(-2, 5)$, $(4, 5)$, and $(-2, -1)$. Find the coordinates of the fourth vertex.

△ **38.** The coordinates of two vertices of a square are $(-3, -1)$ and $(2, -1)$. Find the coordinates of two pairs of possible points for the third and fourth vertices.

Complete each table. See Section 6.1.

39. $x - y = -3$

x	y
0	
	0

40. $y - x = 5$

x	y
0	
	0

41. $y = 2x$

x	y
0	
	0

42. $x = -3y$

x	y
0	
	0

Concept Extensions

Graph each pair of linear equations on the same set of axes. Discuss how the graphs are similar and how they are different.

43. $y = 5x$
$y = 5x + 4$

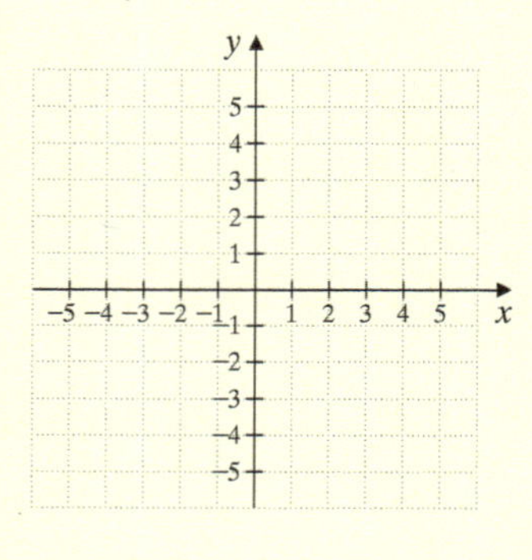

44. $y = 2x$
$y = 2x + 5$

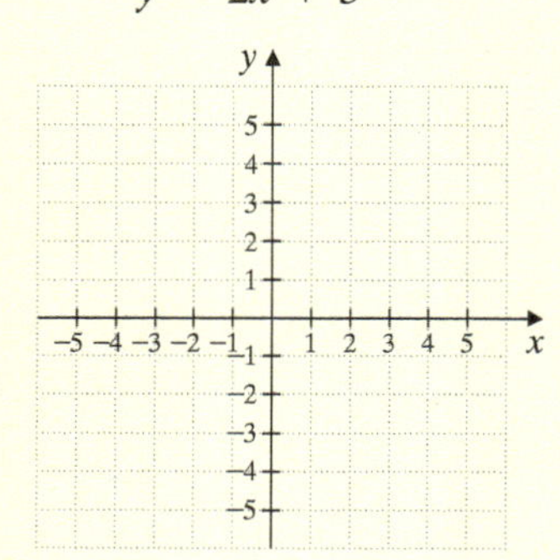

45. $y = -2x$
$y = -2x - 3$

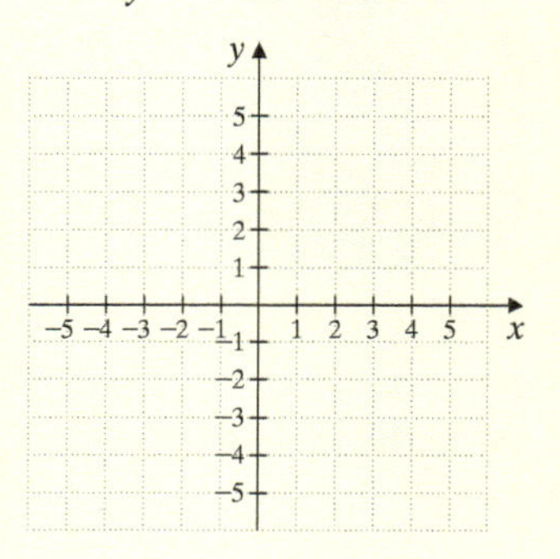

46. $y = x$
$y = x - 7$

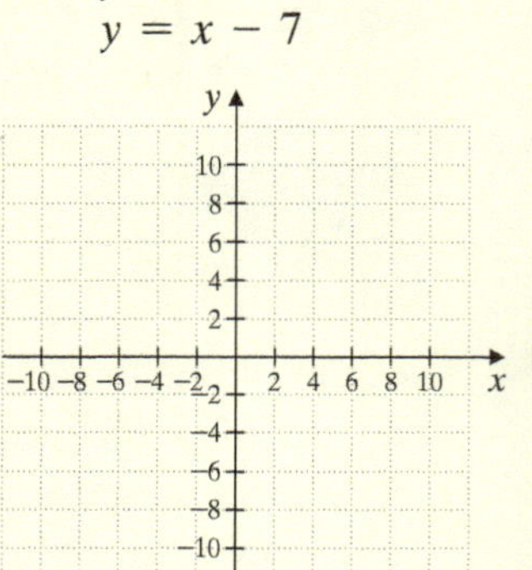

47. Graph the nonlinear equation $y = x^2$ by completing the table shown. Plot the ordered pairs and connect them with a smooth curve.

x	y
0	
1	
−1	
2	
−2	

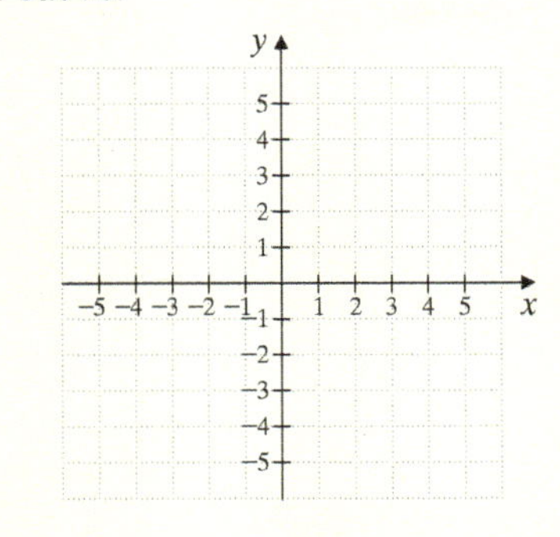

48. Graph the nonlinear equation $y = |x|$ by completing the table shown. Plot the ordered pairs and connect them. This curve is "V" shaped.

x	y
0	
1	
−1	
2	
−2	

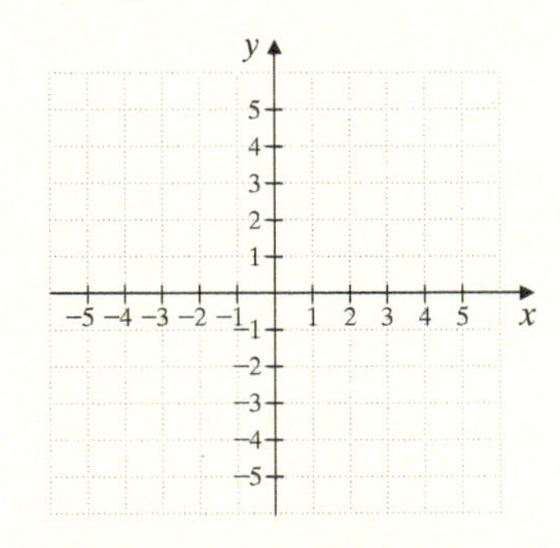

△ **49.** The perimeter of the trapezoid below is 22 centimeters. Write a linear equation in two variables for the perimeter. Find y if x is 3 centimeters.

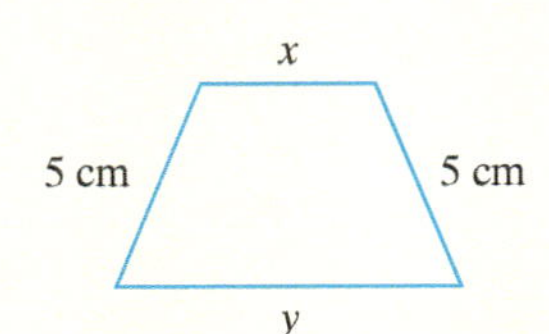

△ **50.** The perimeter of the rectangle below is 50 miles. Write a linear equation in two variables for the perimeter. Use this equation to find x when y is 20 miles.

51. If (a, b) is an ordered pair solution of $x + y = 5$, is (b, a) also a solution? Explain why or why not.

52. If (a, b) is an ordered pair solution of $x - y = 5$, is (b, a) also a solution? Explain why or why not.

6.3 Intercepts

Objective A Identifying Intercepts

The graph of $y = 4x - 8$ is shown below. Notice that this graph crosses the y-axis at the point $(0, -8)$. This point is called the ***y*-intercept.** Likewise the graph crosses the x-axis at $(2, 0)$. This point is called the ***x*-intercept.**

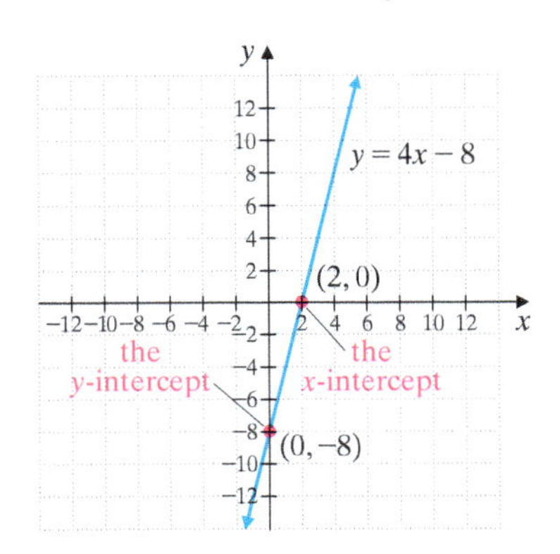

The intercepts are $(2, 0)$ and $(0, -8)$.

Objectives

A Identify Intercepts of a Graph.

B Graph a Linear Equation by Finding and Plotting Intercept Points.

C Identify and Graph Vertical and Horizontal Lines.

Helpful Hint

If a graph crosses the x-axis at $(2, 0)$ and the y-axis at $(0, -8)$, then

$(2, 0)$ ← x-intercept $(0, -8)$ ← y-intercept

Notice that for the x-intercept, the y-value is 0 and that for the y-intercept, the x-value is 0.

Note: Sometimes in mathematics, you may see just the number -8 stated as the y-intercept, and 2 stated as the x-intercept.

Examples

Identify the x- and y-intercepts.

1.

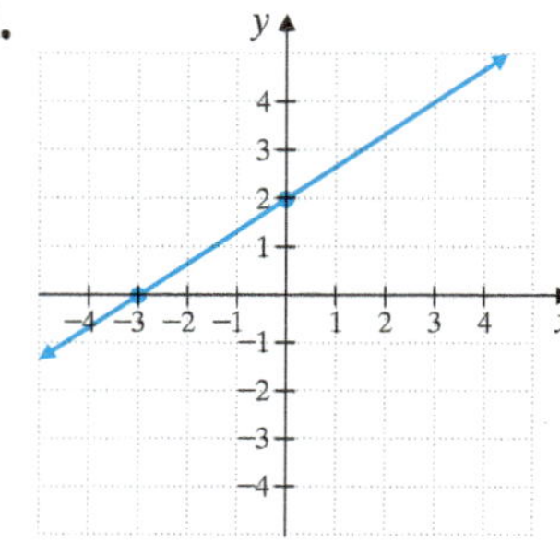

Solution:

x-intercept: $(-3, 0)$
y-interceipt: $(0, 2)$

2.

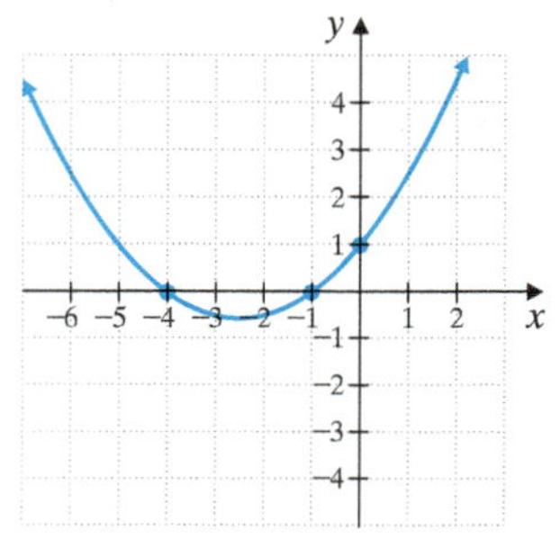

Solution:

x-intercepts: $(-4, 0)$, $(-1, 0)$
y-intercept: $(0, 1)$

3.

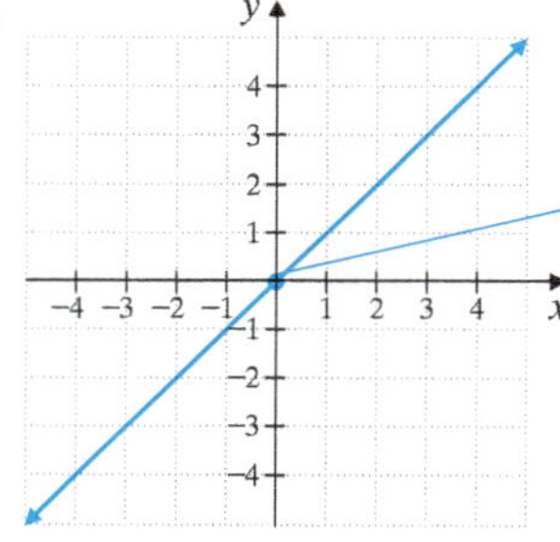

Helpful Hint

Notice that any time $(0, 0)$ is a point of a graph, then it is an x-intercept and a y-intercept. Why? It is the *only* point that lies on both axes.

Solution:

x-intercept: $(0, 0)$
y-intercept: $(0, 0)$

Here, the x- and y-intercepts happen to be the same point.

4.

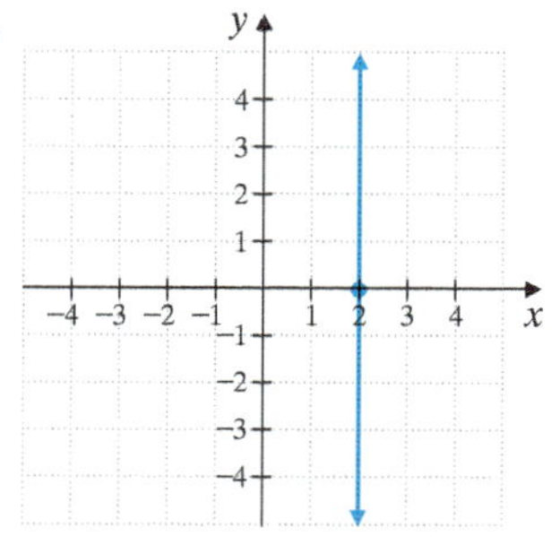

Solution:

x-intercept: $(2, 0)$
y-intercept: none

5.

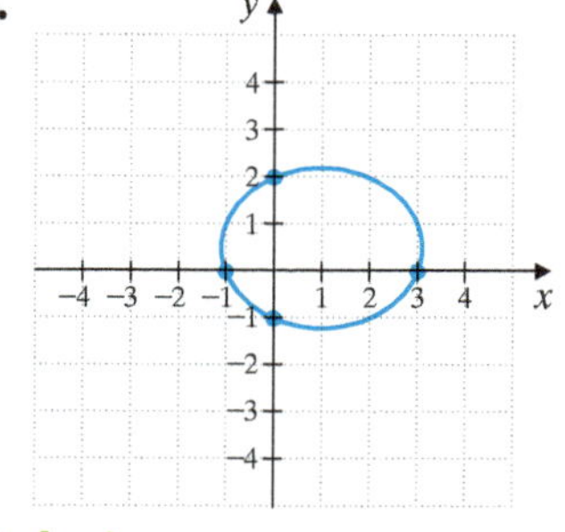

Solution:

x-intercepts: $(-1, 0)$, $(3, 0)$
y-intercepts: $(0, 2)$, $(0, -1)$

Work Practice 1–5

Practice 1–5

Identify the x- and y-intercepts.

1.

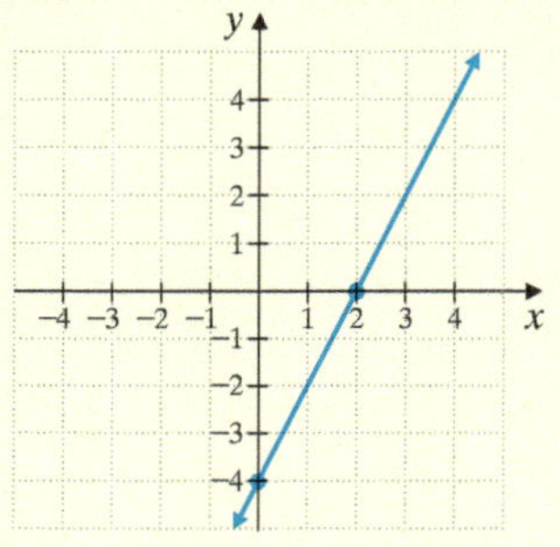

2.

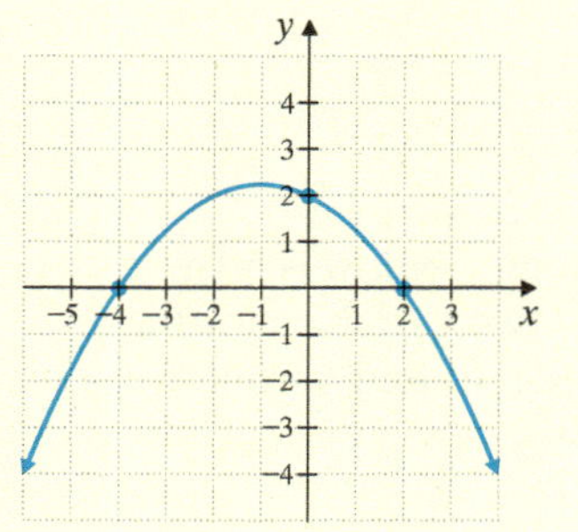

3.

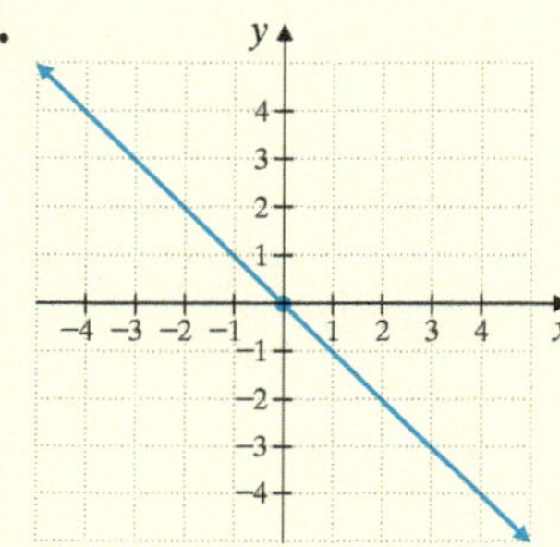

4.

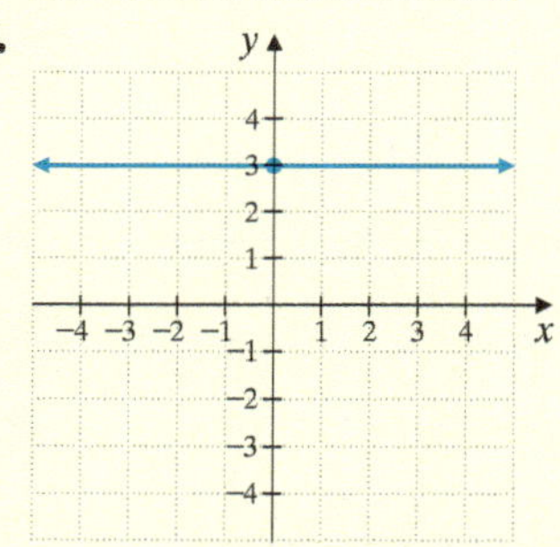

5.

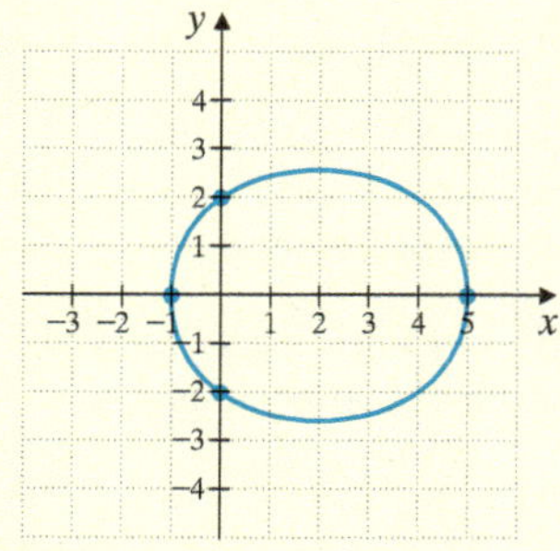

Answers

1. x-intercept: $(2, 0)$; y-intercept: $(0, -4)$
2. x-intercepts: $(-4, 0)$, $(2, 0)$; y-intercept: $(0, 2)$
3. x-intercept and y-intercept: $(0, 0)$
4. no x-intercept; y-intercept: $(0, 3)$
5. x-intercepts: $(-1, 0)$, $(5, 0)$; y-intercepts: $(0, 2)$, $(0, -2)$

Objective B Finding and Plotting Intercepts

Given an equation of a line, we can usually find intercepts easily since one coordinate is 0.

To find the x-intercept of a line from its equation, let $y = 0$, since a point on the x-axis has a y-coordinate of 0. To find the y-intercept of a line from its equation, let $x = 0$, since a point on the y-axis has an x-coordinate of 0.

Finding x- and y-Intercepts

To find the x-intercept, let $y = 0$ and solve for x.

To find the y-intercept, let $x = 0$ and solve for y.

Example 6 Graph $x - 3y = 6$ by finding and plotting its intercepts.

Solution: We let $y = 0$ to find the x-intercept and $x = 0$ to find the y-intercept.

Let $y = 0$.	Let $x = 0$.
$x - 3y = 6$	$x - 3y = 6$
$x - 3(0) = 6$	$0 - 3y = 6$
$x - 0 = 6$	$-3y = 6$
$x = 6$	$y = -2$

The x-intercept is $(6, 0)$ and the y-intercept is $(0, -2)$. We find a third ordered pair solution to check our work. If we let $y = -1$, then $x = 3$. We plot the points $(6, 0)$, $(0, -2)$, and $(3, -1)$. The graph of $x - 3y = 6$ is the line drawn through these points, as shown.

x	y
6	0
0	−2
3	−1

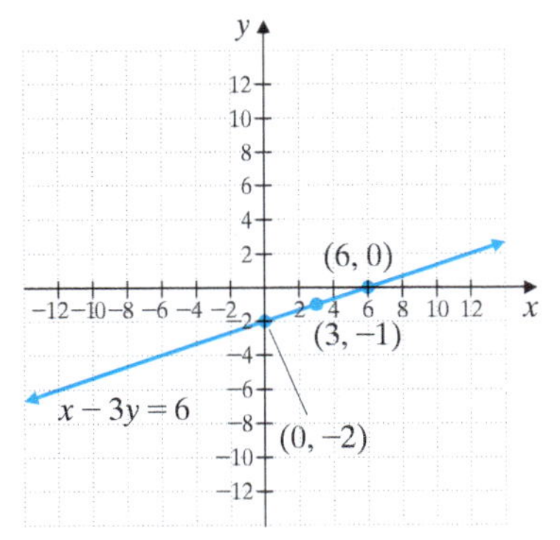

Work Practice 6

Example 7 Graph $x = -2y$ by finding and plotting its intercepts.

Solution: We let $y = 0$ to find the x-intercept and $x = 0$ to find the y-intercept.

Let $y = 0$.	Let $x = 0$.
$x = -2y$	$x = -2y$
$x = -2(0)$	$0 = -2y$
$x = 0$	$0 = y$

Both the x-intercept and y-intercept are $(0, 0)$. In other words, when $x = 0$, then $y = 0$, which gives the ordered pair $(0, 0)$. Also, when $y = 0$, then $x = 0$, which gives the same ordered pair, $(0, 0)$. This happens when the graph passes through the origin. Since two points are needed to determine a line, we must find at least one more ordered pair that satisfies $x = -2y$. Since the equation is solved for x, we

(Continued on next page)

Practice 6

Graph $2x - y = 4$ by finding and plotting its intercepts.

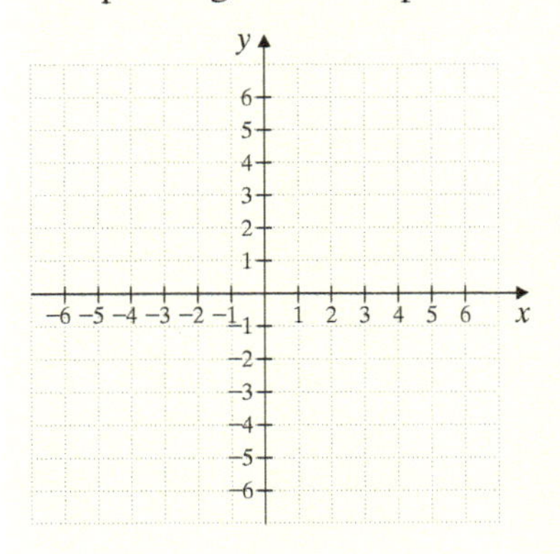

Practice 7

Graph $y = 3x$ by finding and plotting its intercepts.

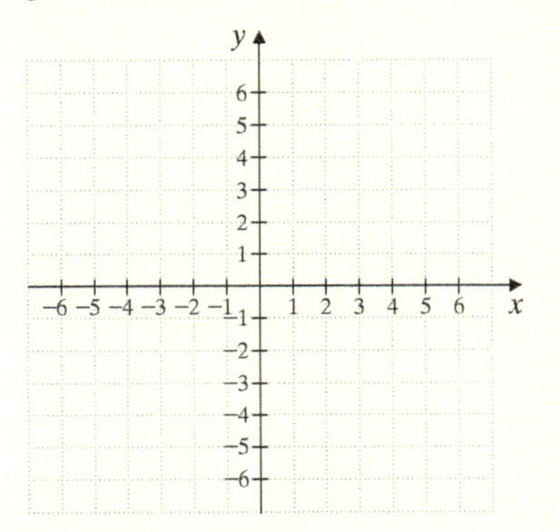

Answers

6.

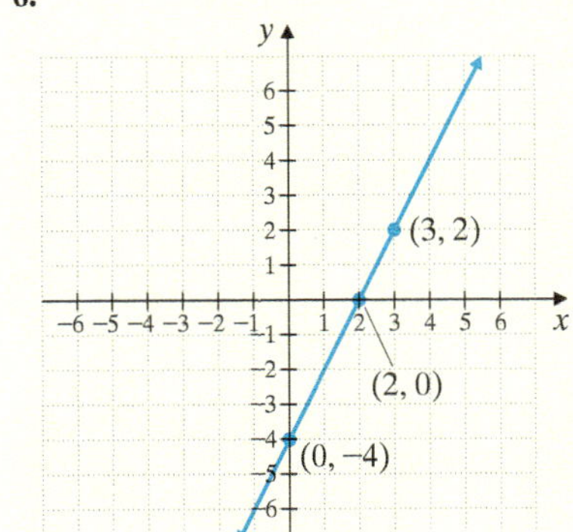

7.

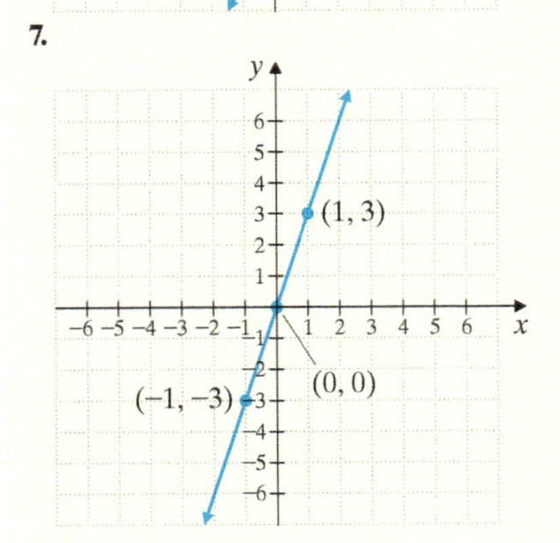

choose y-values so that there is no need to solve to find the corresponding x-value. We let $y = -1$ to find a second ordered pair solution and let $y = 1$ as a check point.

Let $y = -1$.
$x = -2(-1)$
$x = 2$ Multiply.

Let $y = 1$.
$x = -2(1)$
$x = -2$ Multiply.

The ordered pairs are $(0, 0)$, $(2, -1)$, and $(-2, 1)$. We plot these points to graph $x = -2y$.

x	y
0	0
2	−1
−2	1

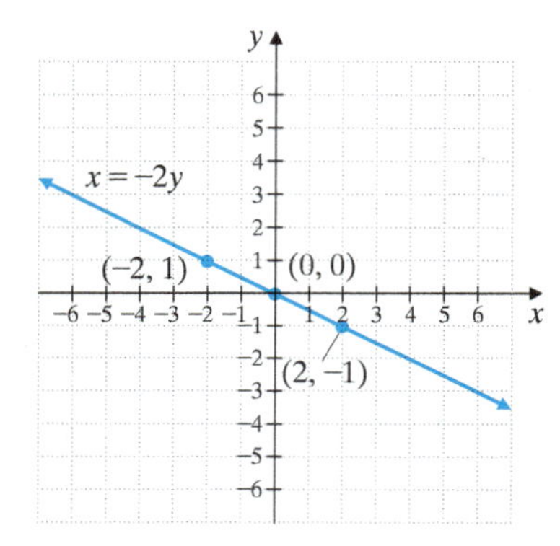

Work Practice 7

Practice 8

Graph: $3x = 2y + 4$

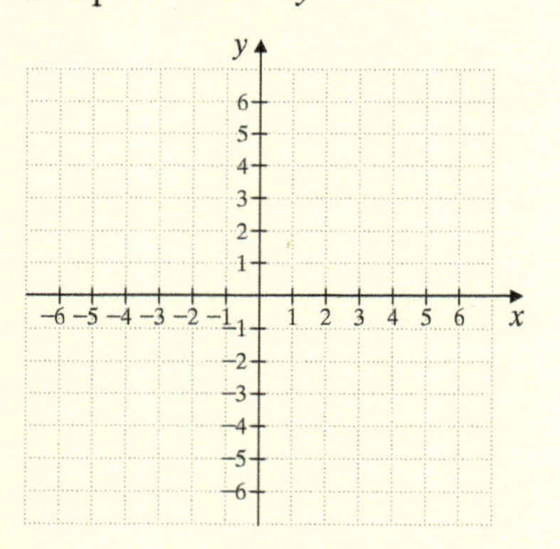

Answer

8.

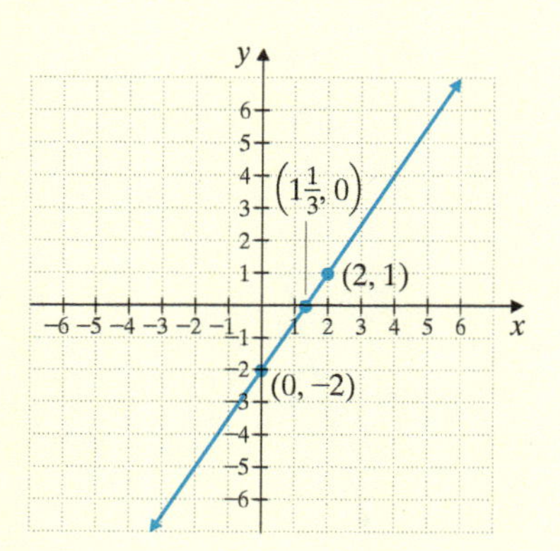

Example 8

Graph: $4x = 3y - 9$

Solution: Find the x- and y-intercepts, and then choose $x = 2$ to find a check point.

Let $y = 0$
$4x = 3(0) - 9$
$4x = -9$
Solve for x.
$x = -\frac{9}{4}$ or $-2\frac{1}{4}$

Let $x = 0$
$4 \cdot 0 = 3y - 9$
$9 = 3y$
Solve for y.
$3 = y$

Let $x = 2$
$4(2) = 3y - 9$
$8 = 3y - 9$
Solve for y.
$17 = 3y$
$\frac{17}{3} = y$ or $y = 5\frac{2}{3}$

The ordered pairs are $\left(-2\frac{1}{4}, 0\right)$, $(0, 3)$, and $\left(2, 5\frac{2}{3}\right)$. The equation $4x = 3y - 9$ is graphed as follows.

x	y
$-2\frac{1}{4}$	0
0	3
2	$5\frac{2}{3}$

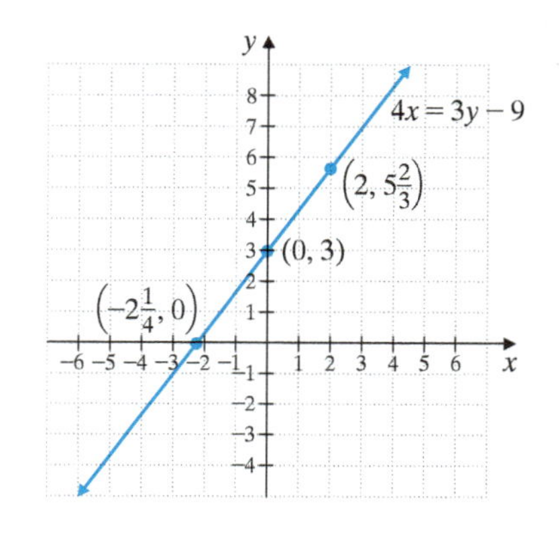

Work Practice 8

Objective C Graphing Vertical and Horizontal Lines 

From Section 6.2, recall that the equation $x = 2$, for example, is a linear equation in two variables because it can be written in the form $x + 0y = 2$. The graph of this equation is a vertical line, as reviewed in the next example.

Example 9 Graph: $x = 2$

Solution: The equation $x = 2$ can be written as $x + 0y = 2$. For any y-value chosen, notice that x is 2. No other value for x satisfies $x + 0y = 2$. Any ordered pair whose x-coordinate is 2 is a solution of $x + 0y = 2$. We will use the ordered pair solutions $(2, 3)$, $(2, 0)$, and $(2, -3)$ to graph $x = 2$.

x	y
2	3
2	0
2	−3

The graph is a vertical line with x-intercept (2, 0). Note that this graph has no y-intercept because x is never 0.

Work Practice 9

In general, we have the following.

Vertical Lines

The graph of $x = c$, where c is a real number, is a **vertical line** with x-intercept $(c, 0)$.

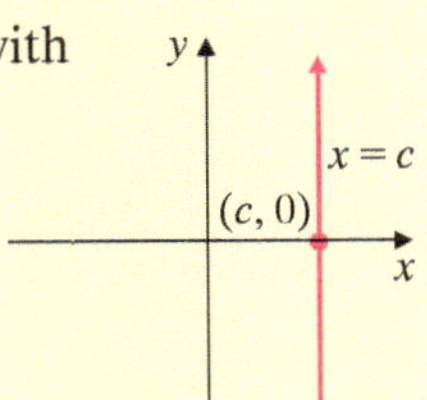

Example 10 Graph: $y = -3$

Solution: The equation $y = -3$ can be written as $0x + y = -3$. For any x-value chosen, y is -3. If we choose 4, 1, and -2 as x-values, the ordered pair solutions are $(4, -3)$, $(1, -3)$, and $(-2, -3)$. We use these ordered pairs to graph $y = -3$. The graph is a horizontal line with y-intercept $(0, -3)$ and no x-intercept.

x	y
4	−3
1	−3
−2	−3

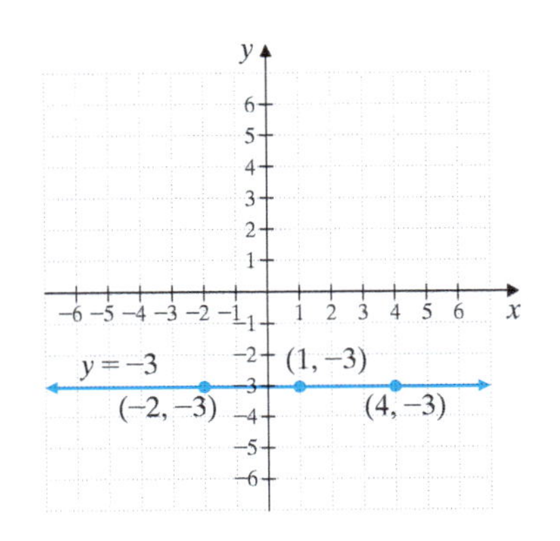

Work Practice 10

Practice 9

Graph: $x = -3$

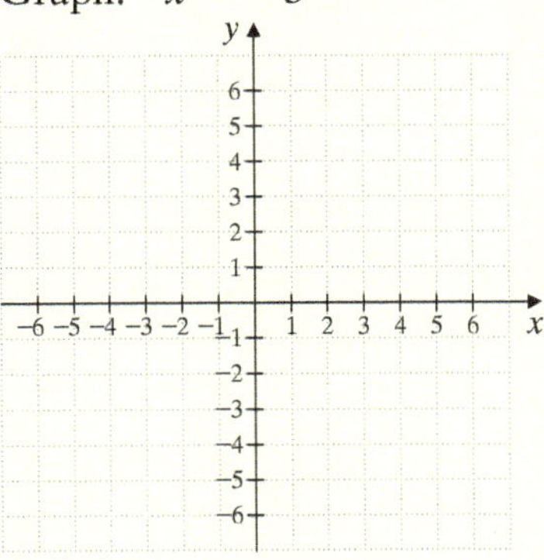

Practice 10

Graph: $y = 4$

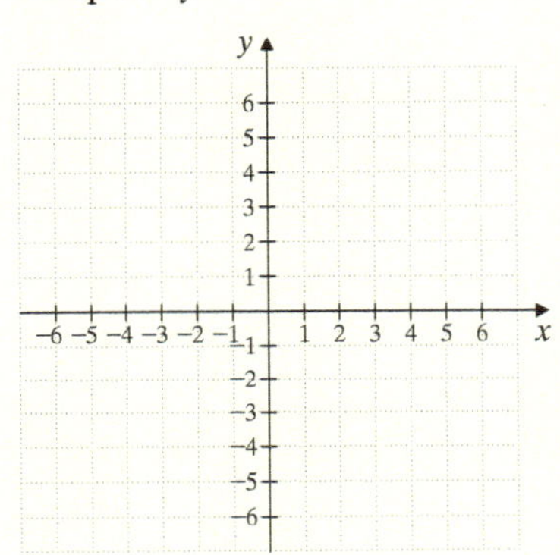

Answers

9.

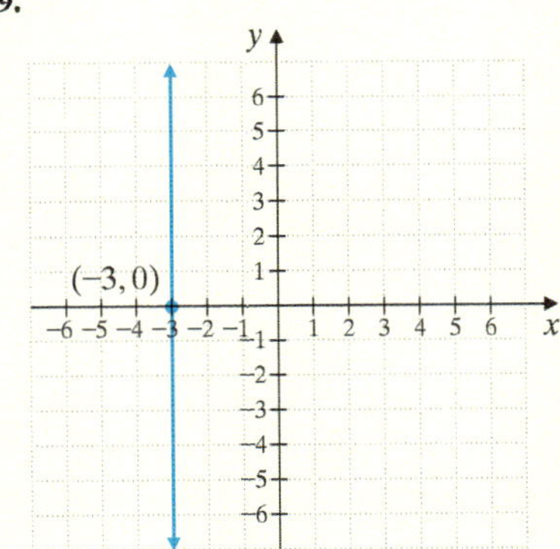

10.

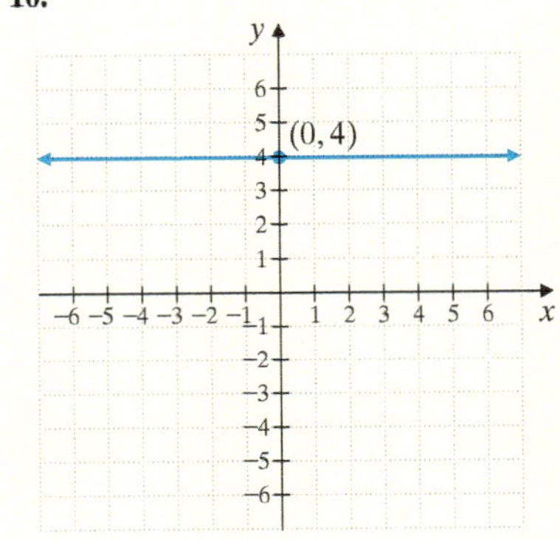

In general, we have the following.

Horizontal Lines

The graph of $y = c$, where c is a real number, is a **horizontal line** with y-intercept $(0, c)$.

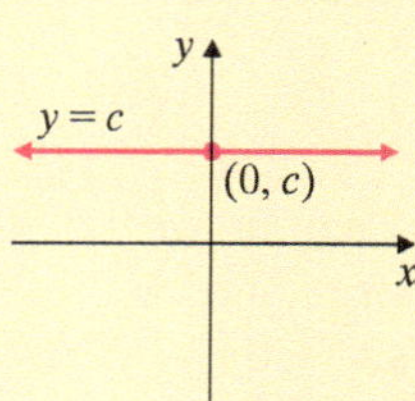

Calculator Explorations Graphing

You may have noticed that to use the Y = key on a graphing calculator to graph an equation, the equation must be solved for y. For example, to graph $2x + 3y = 7$, we solve the equation for y.

$$2x + 3y = 7$$

$$3y = -2x + 7 \quad \text{Subtract } 2x \text{ from both sides.}$$

$$\frac{3y}{3} = -\frac{2x}{3} + \frac{7}{3} \quad \text{Divide both sides by 3.}$$

$$y = -\frac{2}{3}x + \frac{7}{3} \quad \text{Simplify.}$$

To graph $2x + 3y = 7$ or $y = -\frac{2}{3}x + \frac{7}{3}$, press the Y = key and enter

$$Y_1 = -\frac{2}{3}x + \frac{7}{3}$$

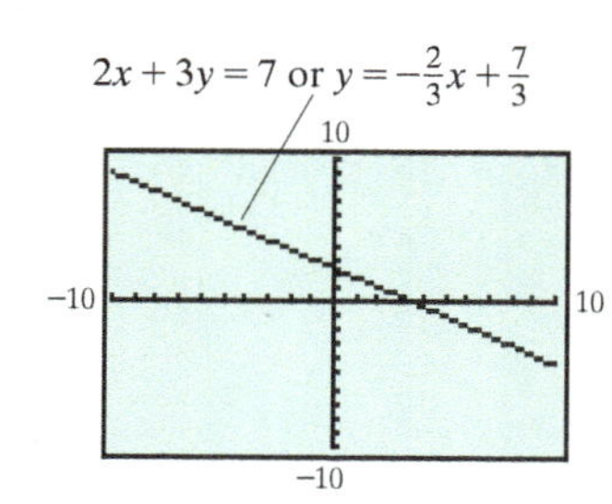

Graph each linear equation.

1. $x = 3.78y$

2. $-2.61y = x$

3. $3x + 7y = 21$

4. $-4x + 6y = 12$

5. $-2.2x + 6.8y = 15.5$

6. $5.9x - 0.8y = -10.4$

Vocabulary, Readiness & Video Check

Use the choices below to fill in each blank. Some choices may be used more than once. Exercises 1 and 2 come from Section 6.2.

x	vertical	x-intercept	linear
y	horizontal	y-intercept	standard

1. An equation that can be written in the form $Ax + By = C$ is called a(n) ____________ equation in two variables.
2. The form $Ax + By = C$ is called ____________ form.
3. The graph of the equation $y = -1$ is a(n) ____________ line.
4. The graph of the equation $x = 5$ is a(n) ____________ line.
5. A point where a graph crosses the y-axis is called a(n) ____________.
6. A point where a graph crosses the x-axis is called a(n) ____________.
7. Given an equation of a line, to find the x-intercept (if there is one), let ____________ $= 0$ and solve for ____________.
8. Given an equation of a line, to find the y-intercept (if there is one), let ____________ $= 0$ and solve for ____________.

Answer the following true or false.

9. All lines have an x-intercept *and* a y-intercept. ____________
10. The graph of $y = 4x$ contains the point $(0, 0)$. ____________
11. The graph of $x + y = 5$ has an x-intercept of $(5, 0)$ and a y-intercept of $(0, 5)$. ____________
12. The graph of $y = 5x$ contains the point $(5, 1)$. ____________

Martin-Gay Interactive Videos

See Video 6.3

Watch the section lecture video and answer the following questions.

Objective A 13. At the end of Example 2, patterns are discussed. What reason is given for why x-intercepts have y-values of 0? For why y-intercepts have x-values of 0?

Objective B 14. In Example 3, the goal is to use the x- and y-intercepts to graph a line. Yet once the two intercepts are found, a third point is also found before the line is graphed. Why do you think this practice of finding a third point is continued?

Objective C 15. From Examples 5 and 6, what is the coefficient of x when the equation of a horizontal line is written as $Ax + By = C$? What is the coefficient of y when the equation of a vertical line is written as $Ax + By = C$?

6.3 Exercise Set MyLab Math

Objective A *Identify the intercepts. See Examples 1 through 5.*

1.

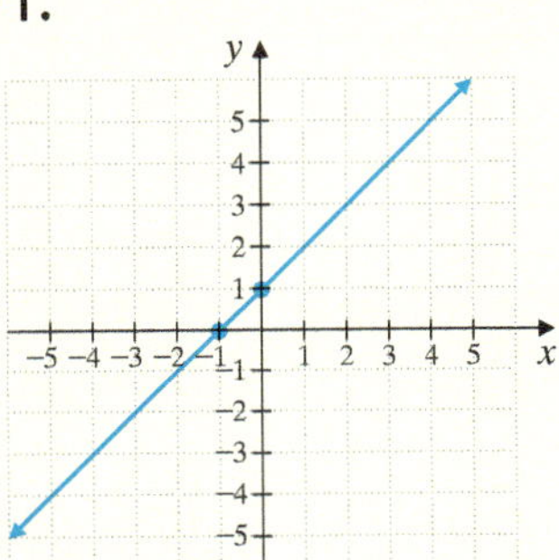

2.

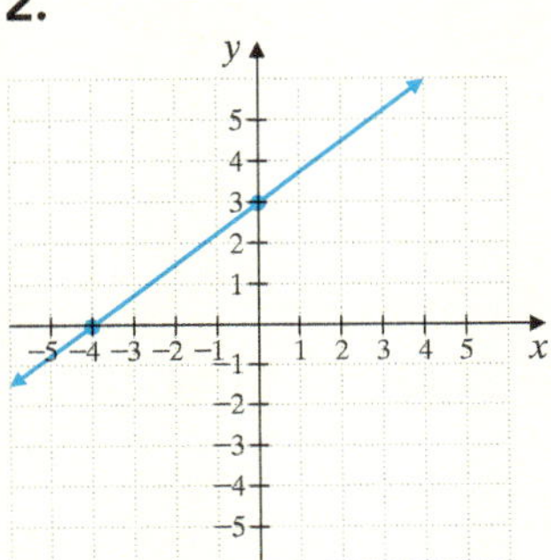

3.

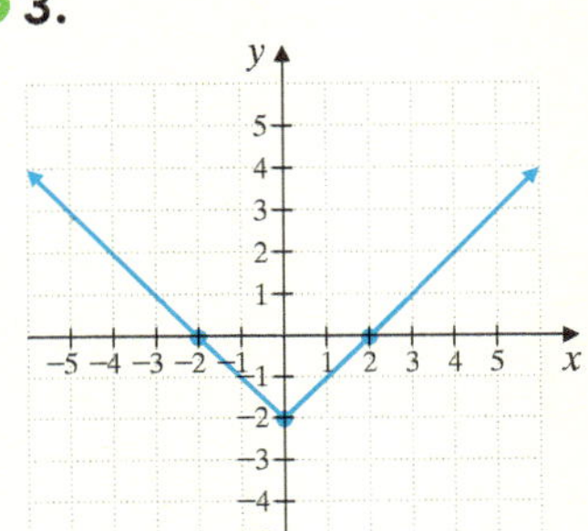

4.

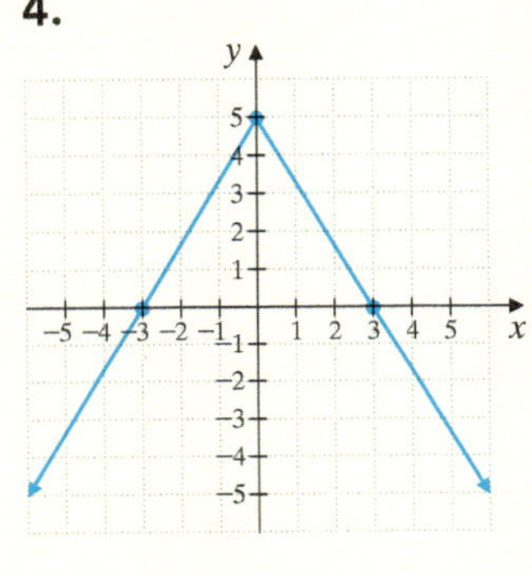

5.

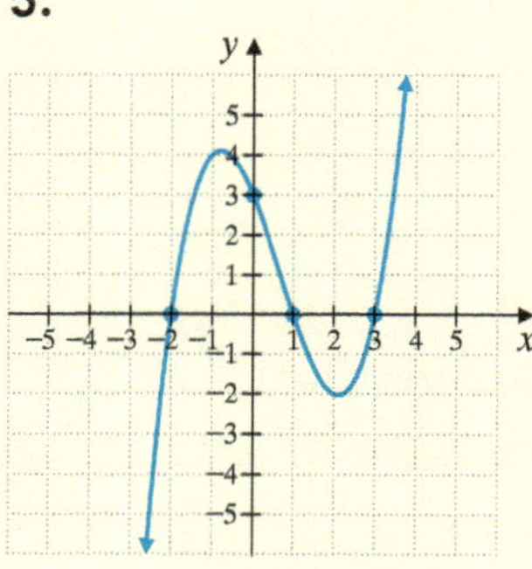

6.

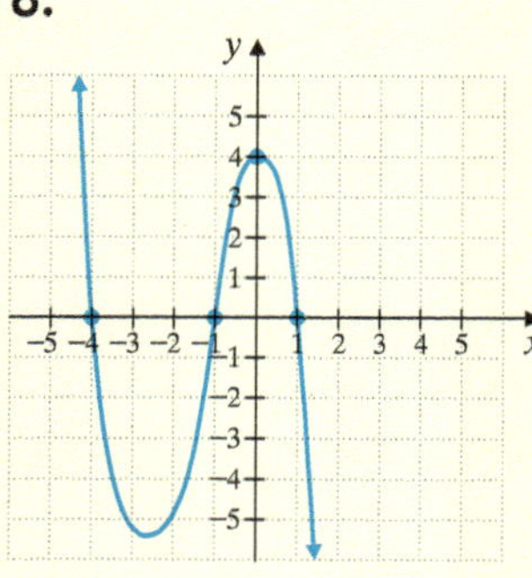

7.

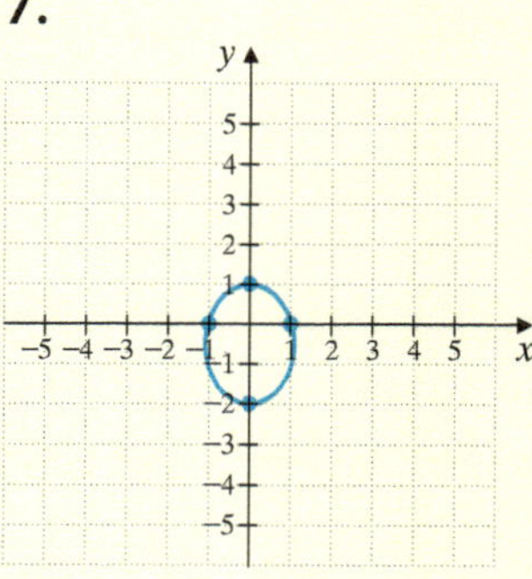

8.

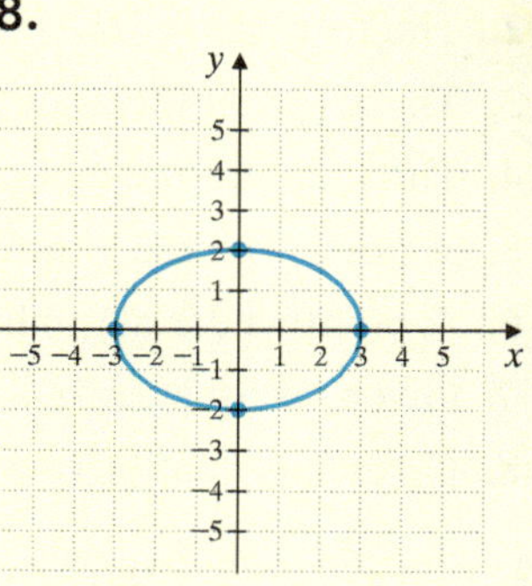

Objective B *Graph each linear equation by finding and plotting its intercepts. See Examples 6 through 8.*

9. $x - y = 3$

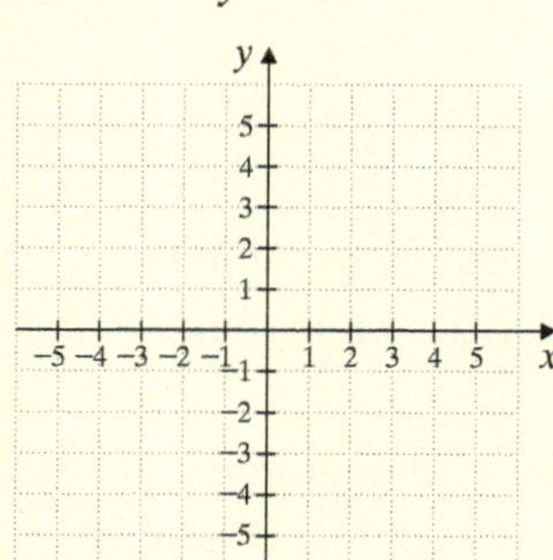

10. $x - y = -4$

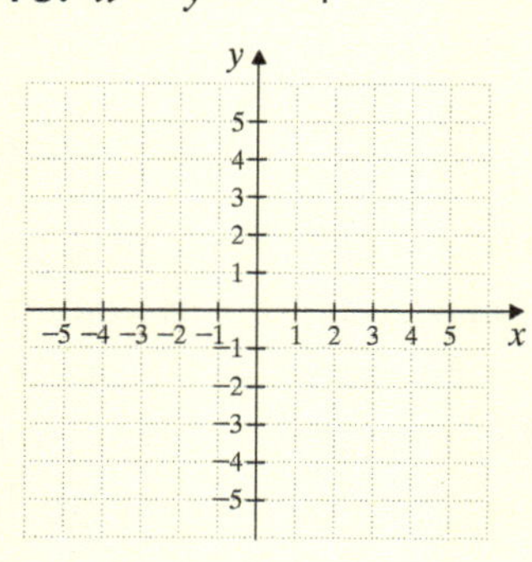

11. $x = 5y$

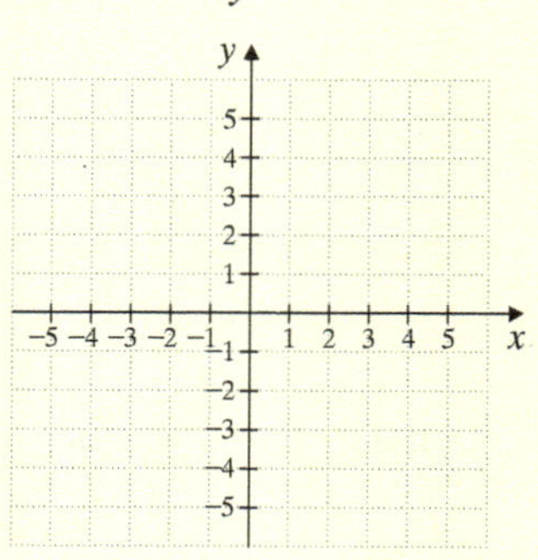

12. $x = 2y$

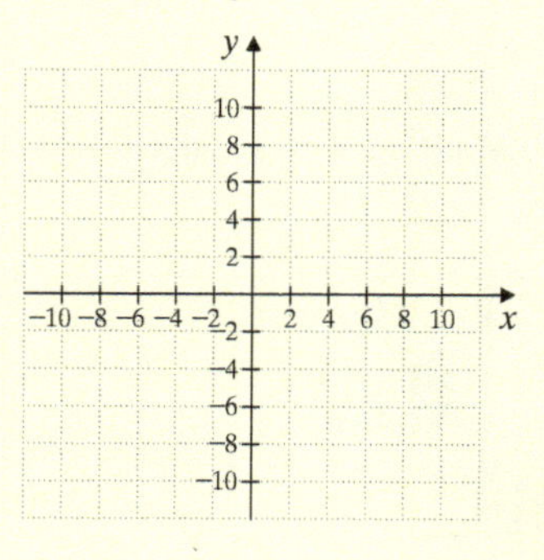

13. $-x + 2y = 6$

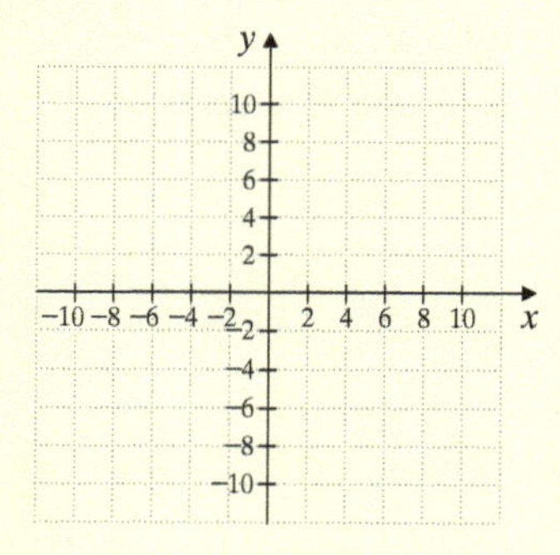

14. $x - 2y = -8$

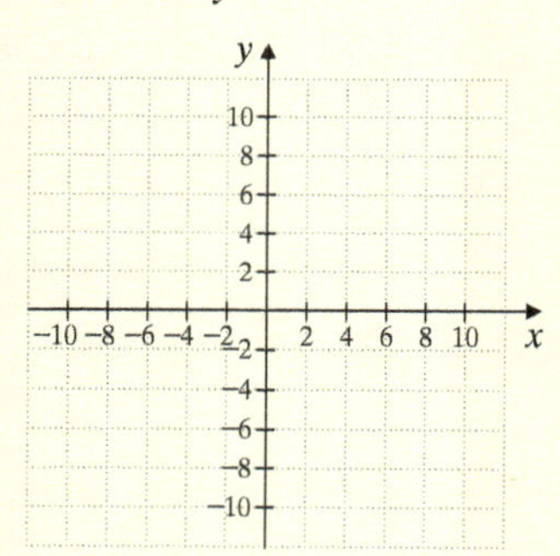

15. $2x - 4y = 8$

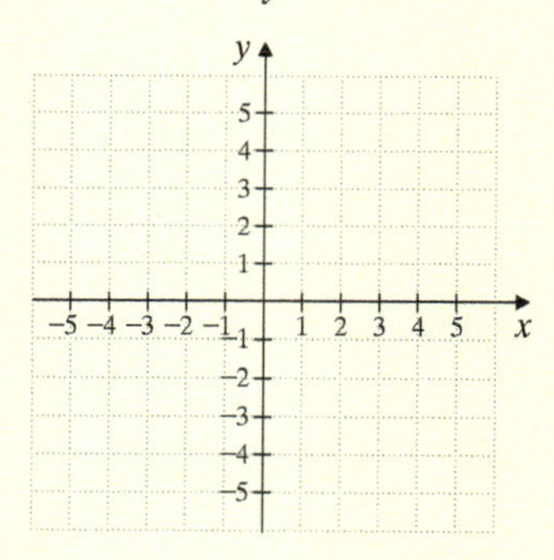

16. $2x + 3y = 6$

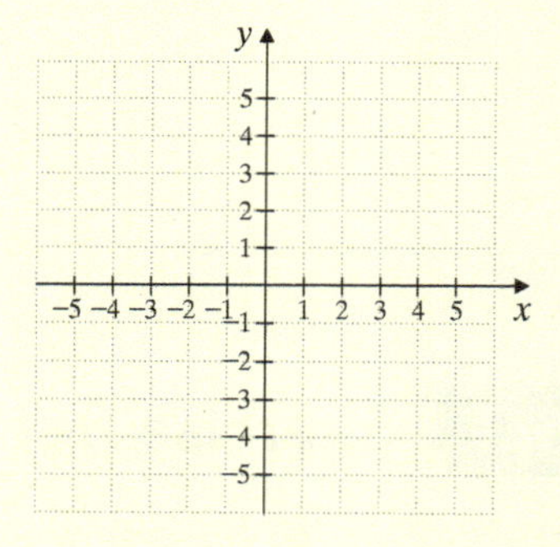

17. $y = 2x$

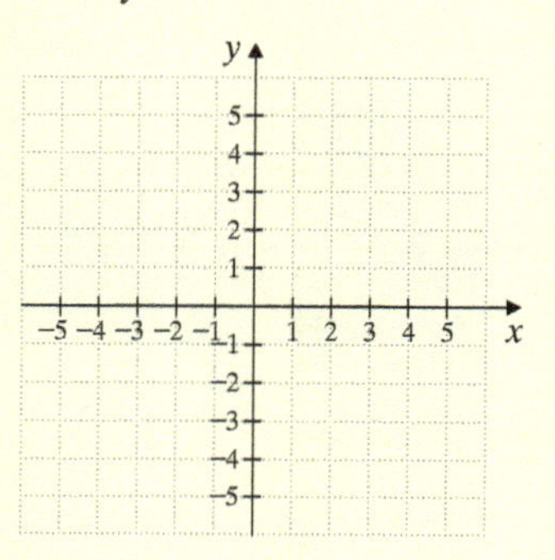

18. $y = -2x$

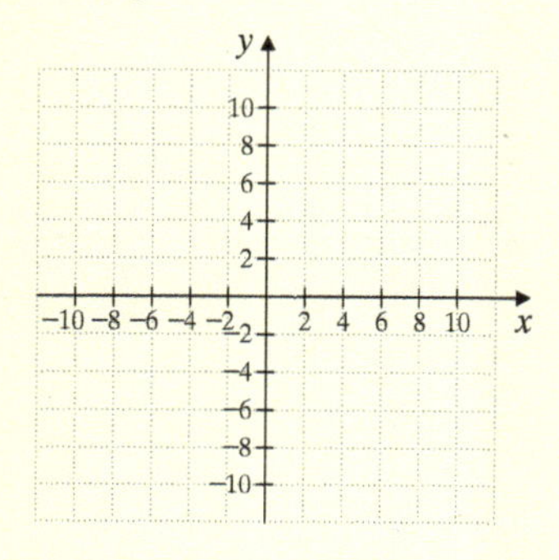

19. $y = 3x + 6$

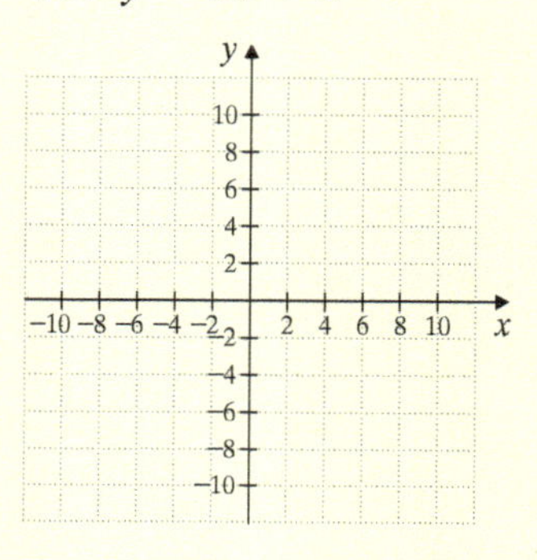

20. $y = 2x + 10$

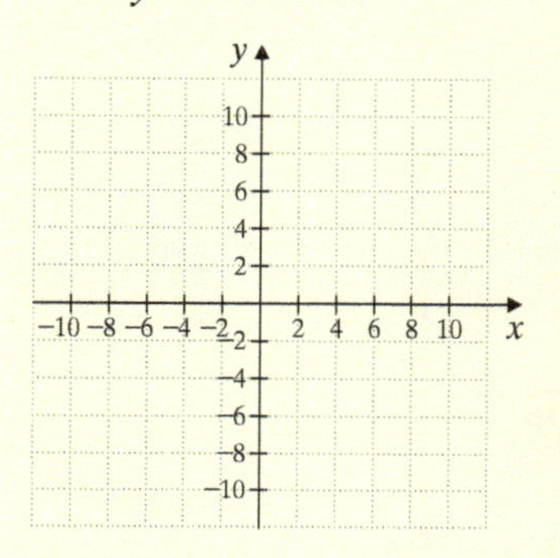

Objective C *Graph each linear equation. See Examples 9 and 10.*

21. $x = -1$

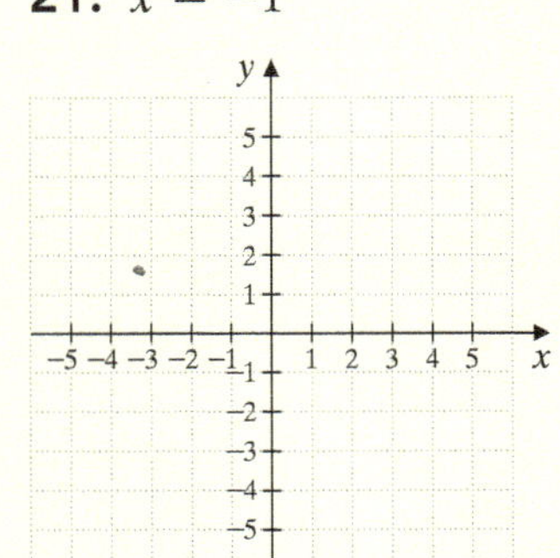

22. $y = 5$

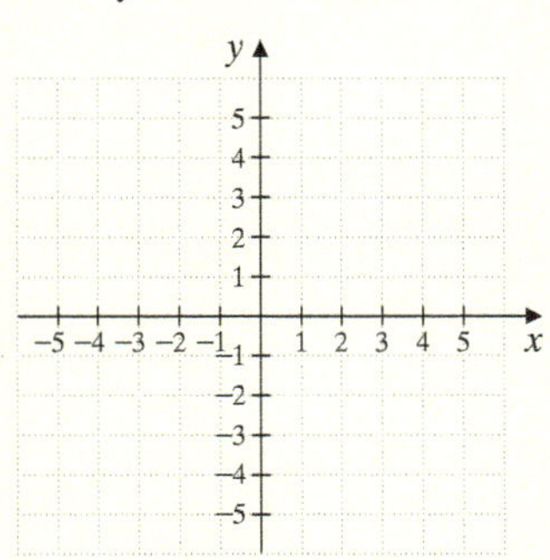

23. $y = 0$

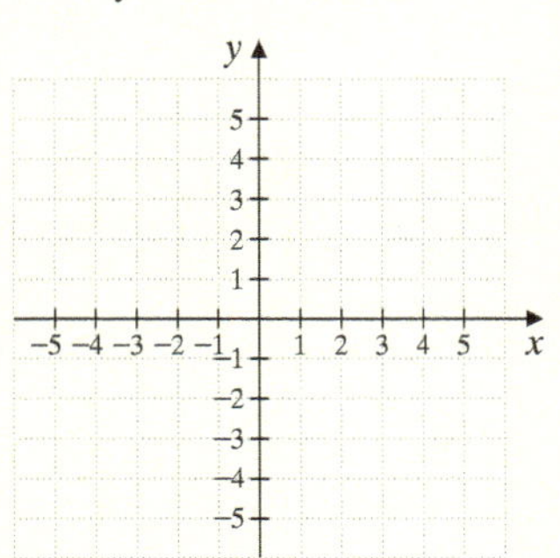

24. $x = 0$

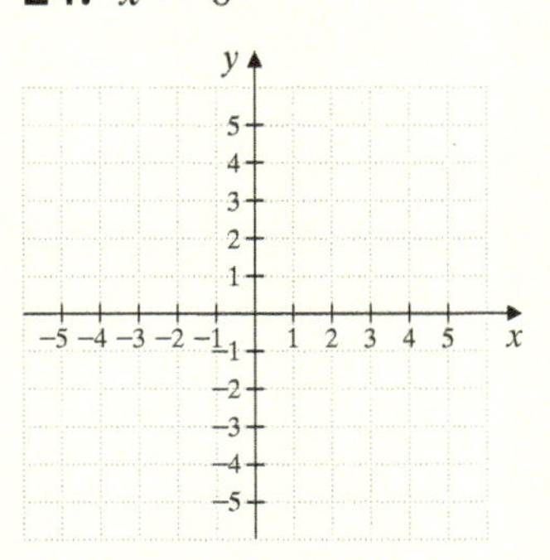

25. $y + 7 = 0$

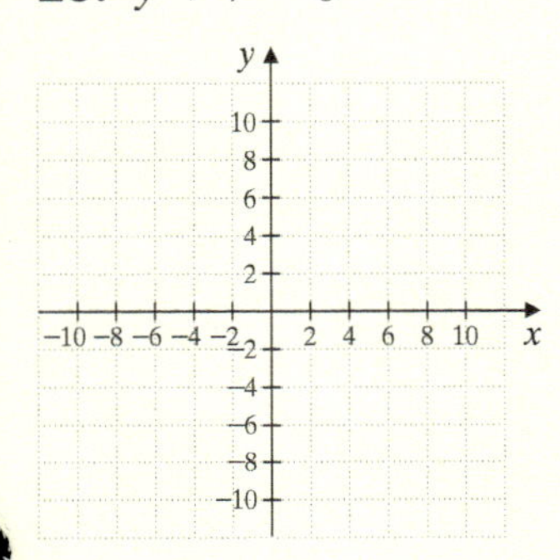

26. $x - 2 = 0$

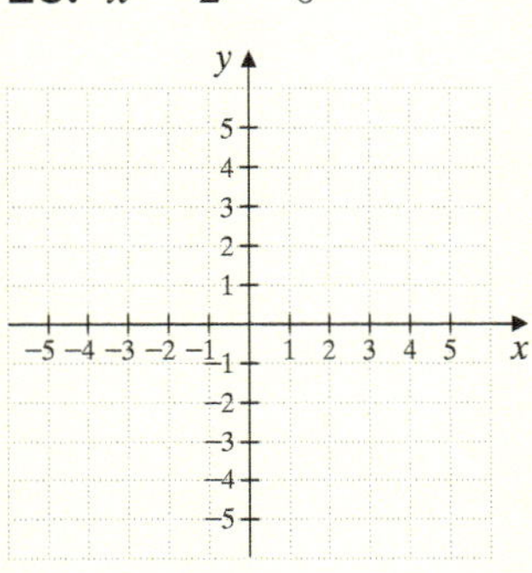

27. $x + 3 = 0$

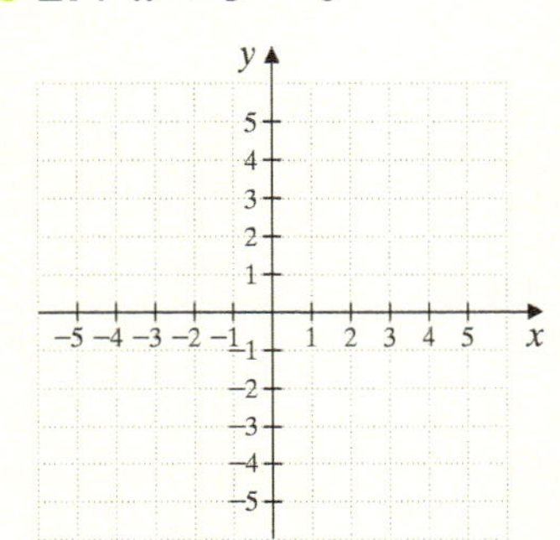

28. $y - 6 = 0$

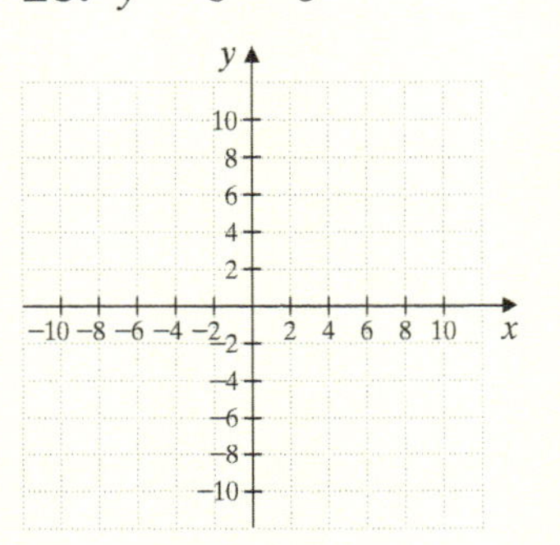

Objectives B C Mixed Practice *Graph each linear equation. See Examples 6 through 10.*

29. $x = y$

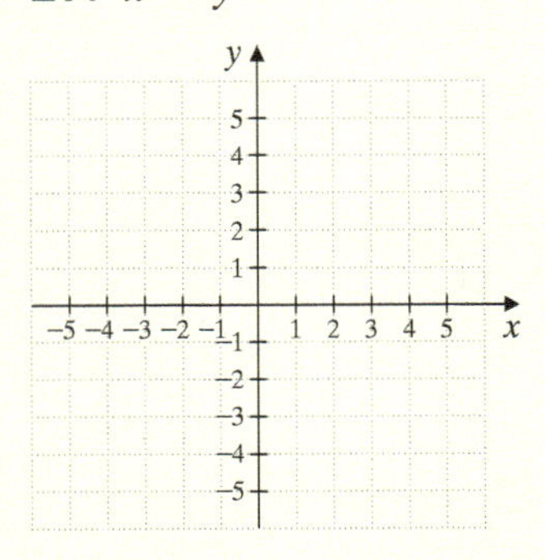

30. $x = -y$

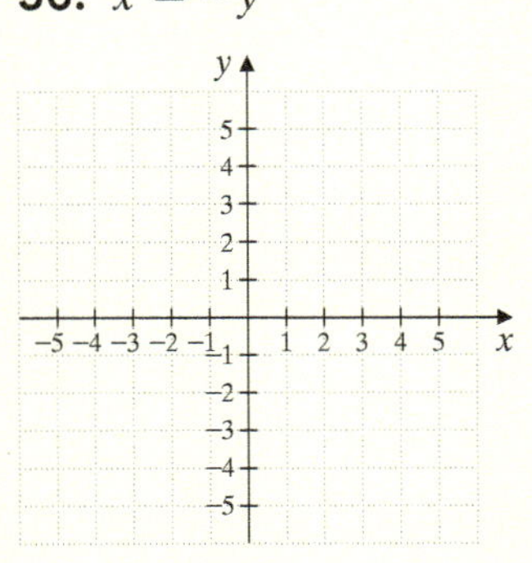

31. $x + 8y = 8$

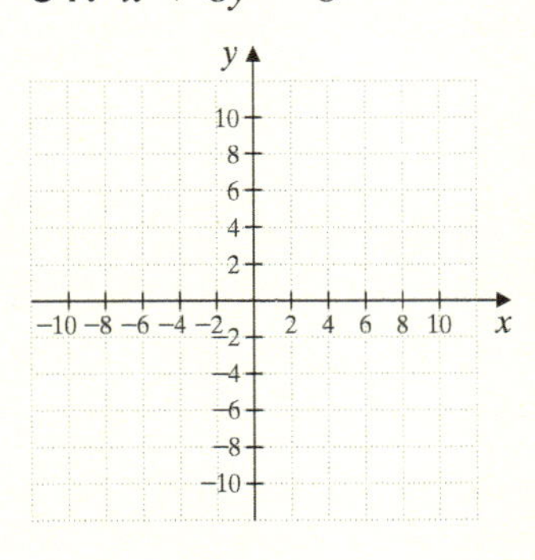

32. $x + 3y = 9$

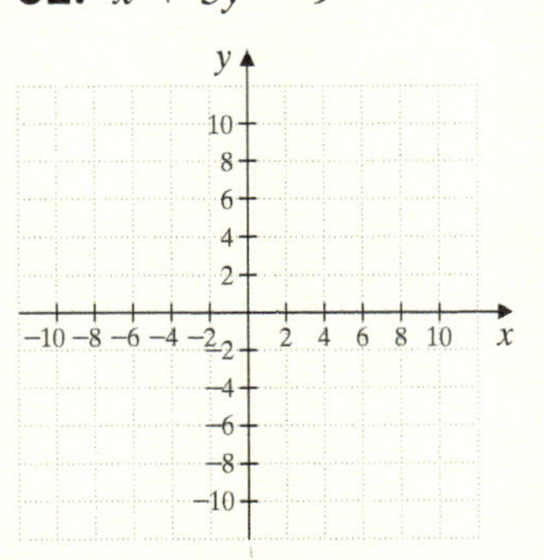

33. $5 = 6x - y$

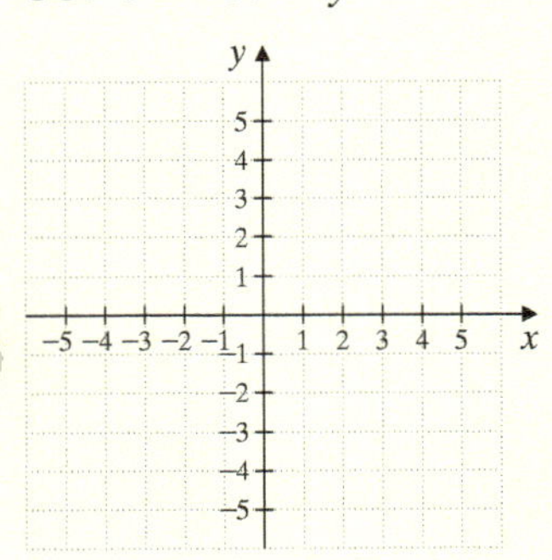

34. $4 = x - 3y$

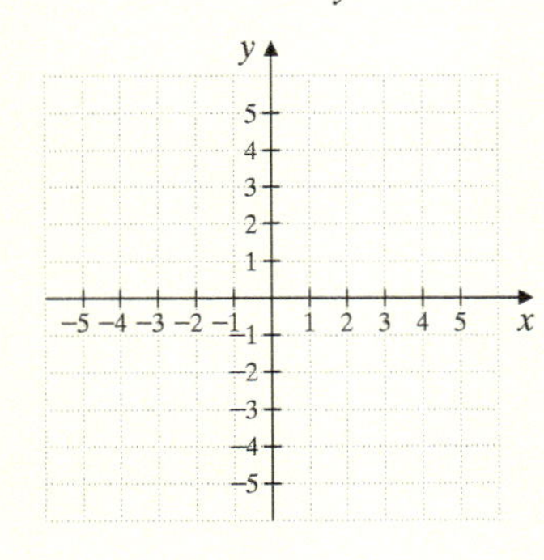

35. $-x + 10y = 11$

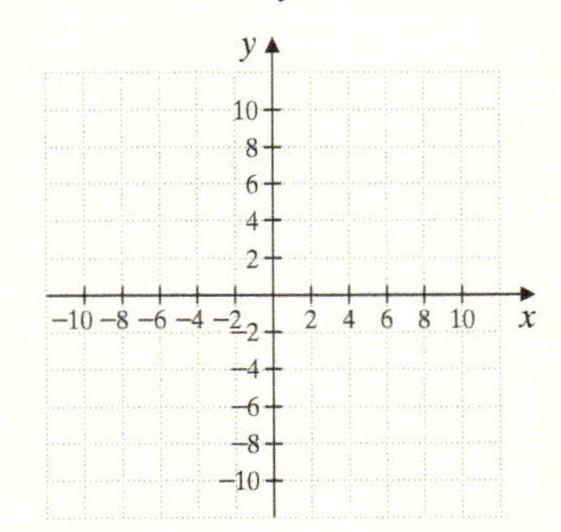

36. $-x + 9y = 10$

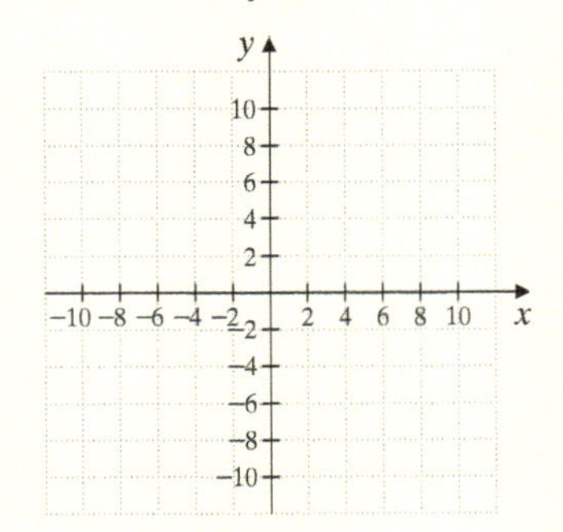

37. $x = -4\frac{1}{2}$

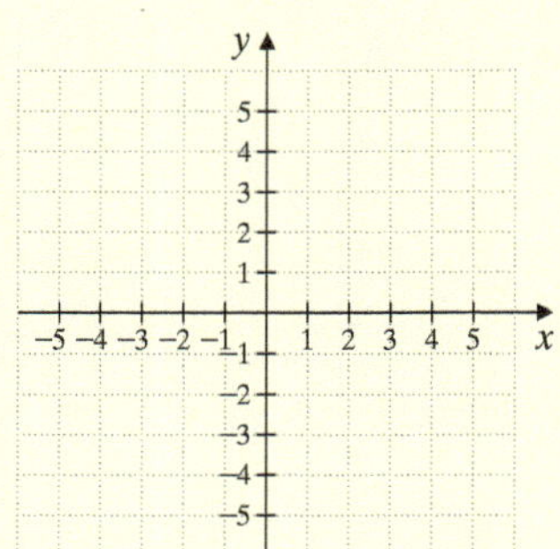

38. $x = -1\frac{3}{4}$

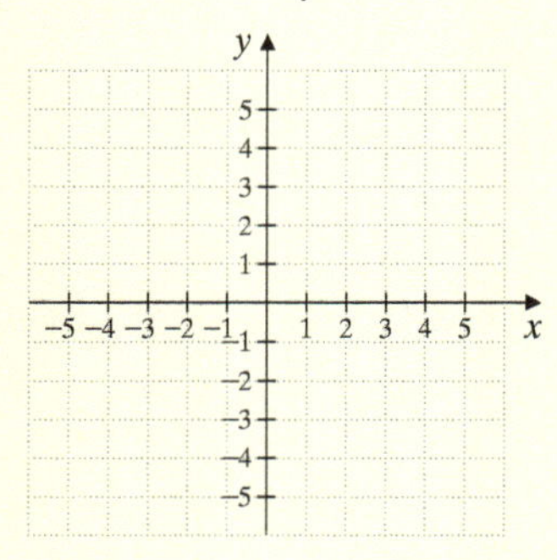

39. $y = 3\frac{1}{4}$

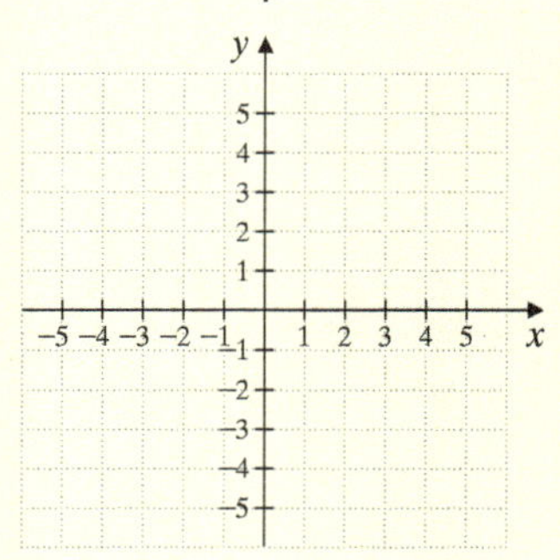

40. $y = 2\frac{1}{2}$

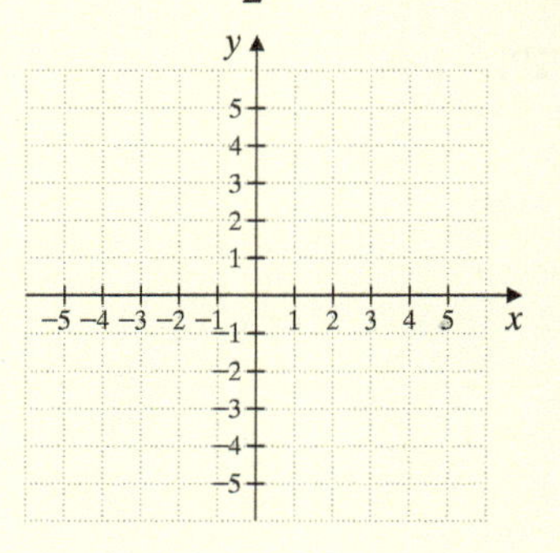

41. $y = -\frac{2}{3}x + 1$

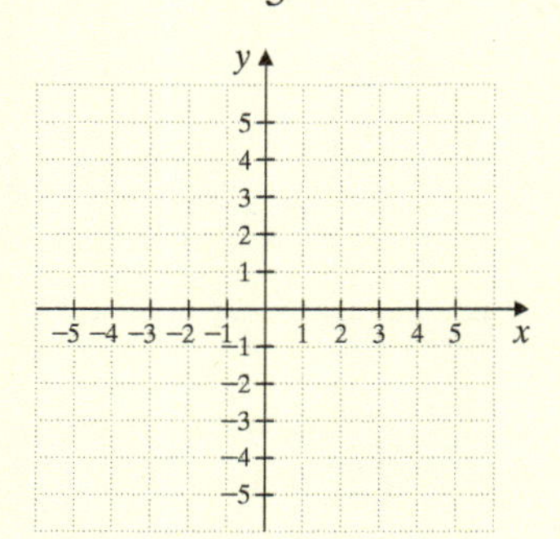

42. $y = -\frac{3}{5}x + 3$

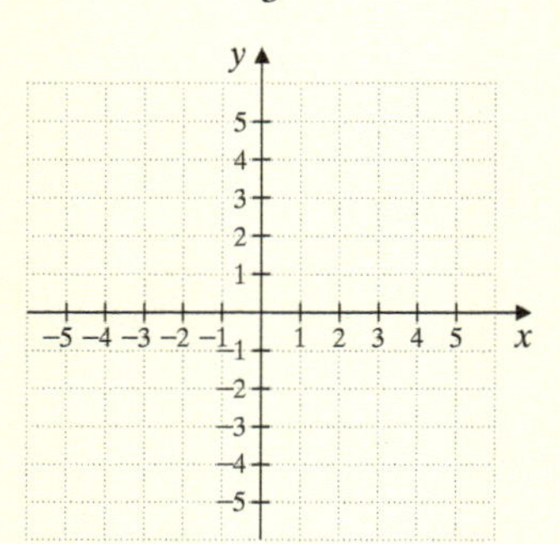

43. $4x - 6y + 2 = 0$

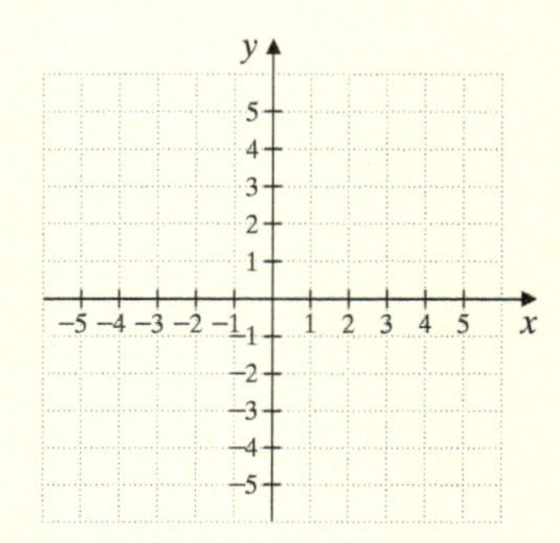

44. $9x - 6y + 3 = 0$

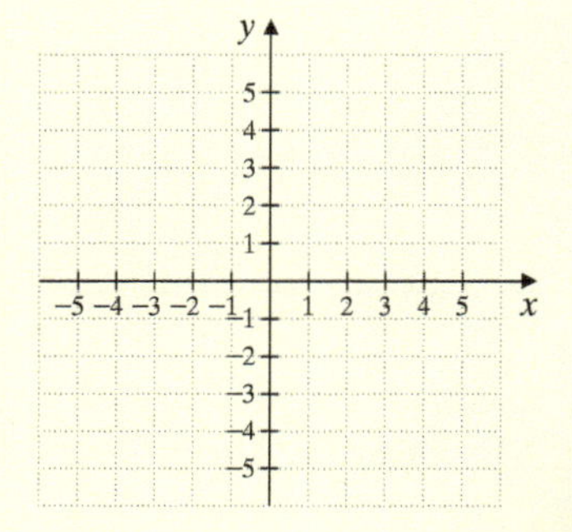

Review

Simplify. See Sections 1.5 and 1.6.

45. $\frac{-6 - 3}{2 - 8}$

46. $\frac{4 - 5}{-1 - 0}$

47. $\frac{-8 - (-2)}{-3 - (-2)}$

48. $\frac{12 - 3}{10 - 9}$

49. $\frac{0 - 6}{5 - 0}$

50. $\frac{2 - 2}{3 - 5}$

Concept Extensions

Match each equation with its graph.

51. $y = 3$

52. $y = 2x + 2$

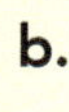

53. $x = 3$

54. $y = 2x + 3$

a.

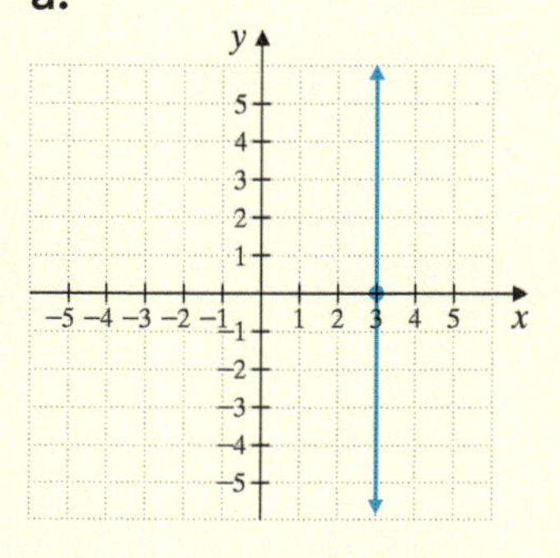

b.

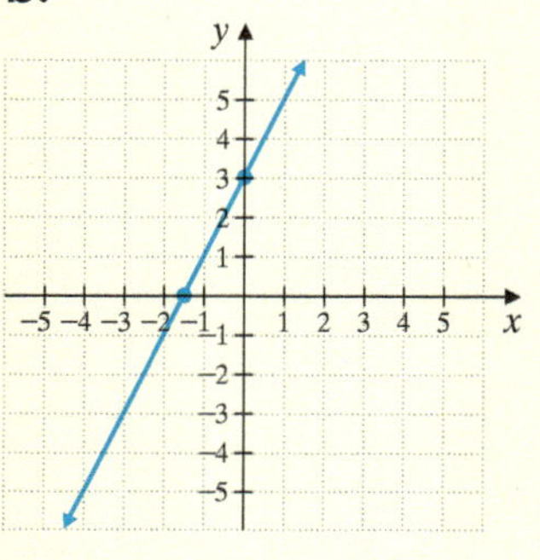

c.

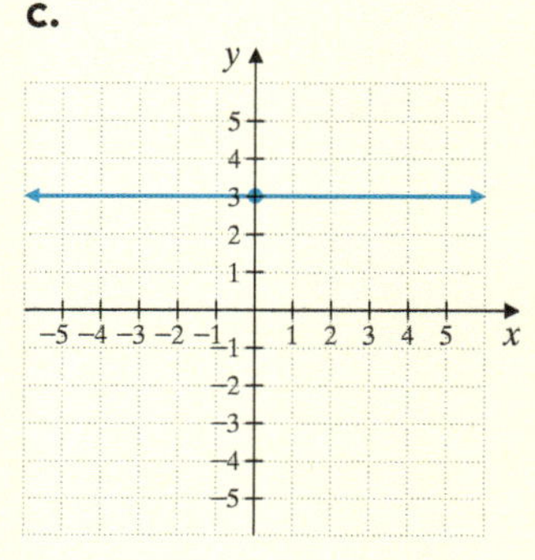

d.

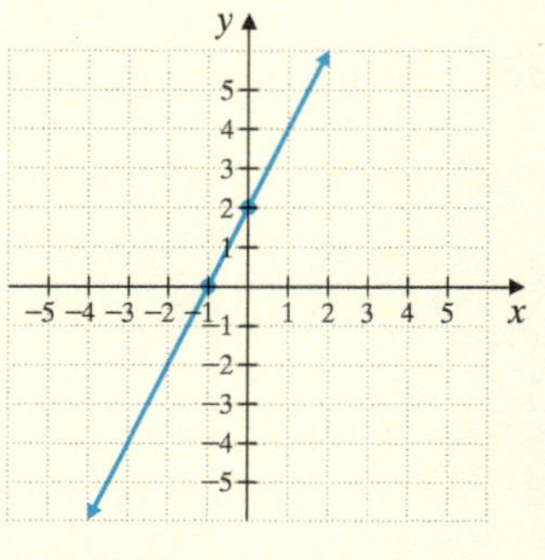

55. What is the greatest number of x- and y-intercepts that a line can have?

56. What is the smallest number of x- and y-intercepts that a line can have?

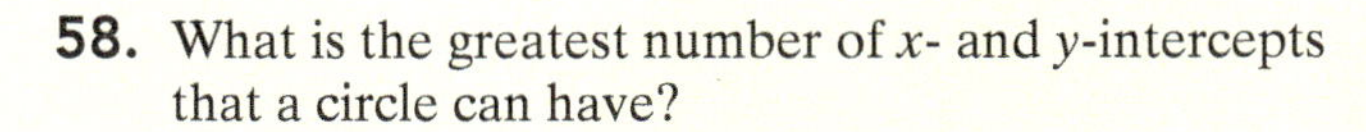

57. What is the smallest number of x- and y-intercepts that a circle can have?

58. What is the greatest number of x- and y-intercepts that a circle can have?

59. Discuss whether a vertical line ever has a y-intercept.

60. Discuss whether a horizontal line ever has an x-intercept.

The production supervisor at Alexandra's Office Products finds that it takes 3 hours to manufacture a particular office chair and 6 hours to manufacture an office desk. A total of 1200 hours is available to produce office chairs and desks of this style. The linear equation that models this situation is $3x + 6y = 1200$, *where x represents the number of chairs produced and y represents the number of desks manufactured.*

61. Complete the ordered pair solution $(0, \quad)$ of this equation. Describe the manufacturing situation that corresponds to this solution.

62. Complete the ordered pair solution $(\quad, 0)$ of this equation. Describe the manufacturing situation that corresponds to this solution.

63. If 50 desks are manufactured, find the greatest number of chairs that can be made.

64. If 50 chairs are manufactured, find the greatest number of desks that can be made.

Two lines in the same plane that do not intersect are called ***parallel lines.***

65. Use your own graph paper to draw a line parallel to the line $y = -1$ that intersects the y-axis at -4. What is the equation of this line?

66. Use your own graph paper to draw a line parallel to the line $x = 5$ that intersects the x-axis at 1. What is the equation of this line?

Solve.

67. As print newspaper sales decline, the number of employees in the print newspaper business is also declining. Employment in the print newspaper industry from 2010 to 2017 can be modeled by the equation $y = -2320x + 55{,}839$, where x represents the number of years after 2010. (*Source:* American Society of News Editors)

a. Find the x-intercept of this equation (round to the nearest tenth).

b. What does this x-intercept mean?

68. The number y of Barnes & Noble retail stores in operation for the years 2012 to 2016 can be modeled by the equation $y = -12.9x + 689$, where x represents the number of years after 2012. (*Source:* Based on data from Barnes & Noble, Inc.)

a. Find the y-intercept of this equation.

b. What does this y-intercept mean?

6.4 Slope and Rate of Change

Objective A Finding the Slope of a Line Given Two Points

Thus far, much of this chapter has been devoted to graphing lines. You have probably noticed by now that a key feature of a line is its slant or steepness. In mathematics, the slant or steepness of a line is formally known as its **slope.** We measure the slope of a line by the ratio of vertical change (rise) to the corresponding horizontal change (run) as we move along the line.

On the line at the top of the next page, for example, suppose that we begin at the point $(1, 2)$ and move to the point $(4, 6)$. The vertical change is the change in y-coordinates: $6 - 2$ or 4 units. The corresponding horizontal change is the change in x-coordinates: $4 - 1 = 3$ units. The ratio of these changes is

Objectives

A Find the Slope of a Line Given Two Points of the Line.

B Find the Slope of a Line Given Its Equation.

C Find the Slopes of Horizontal and Vertical Lines.

D Compare the Slopes of Parallel and Perpendicular Lines.

E Interpret Slope as a Rate of Change.

$$\text{slope} = \frac{\text{change in } y \text{ (vertical change or rise)}}{\text{change in } x \text{ (horizontal change or run)}} = \frac{4}{3}$$

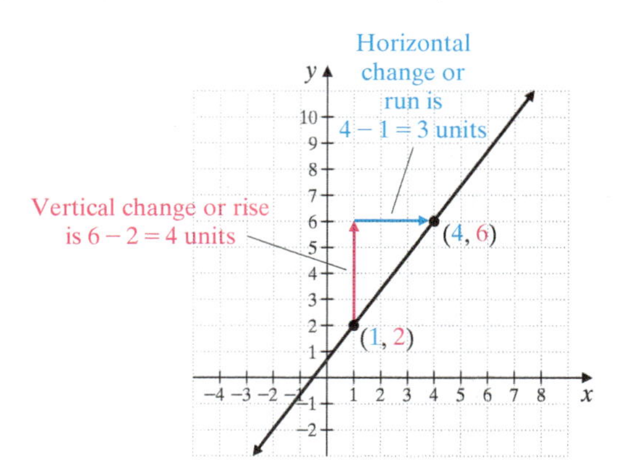

The slope of this line, then, is $\frac{4}{3}$. This means that for every 4 units of change in y-coordinates, there is a corresponding change of 3 units in x-coordinates.

Helpful Hint

It makes no difference what two points of a line are chosen to find its slope. The slope of a line is the same everywhere on the line.

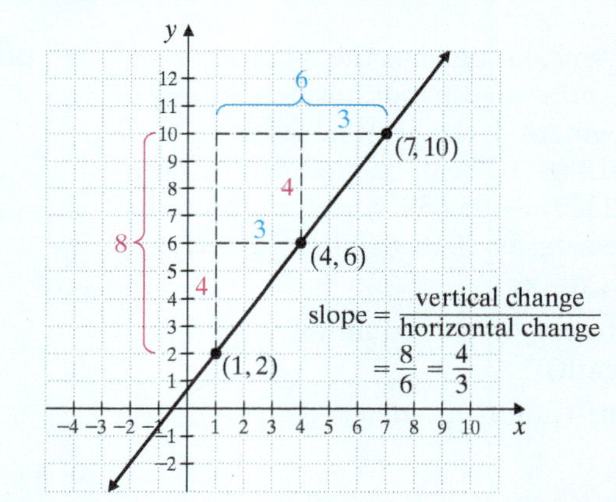

To find the slope of a line, then, choose two points of the line. Label the two x-coordinates of the two points x_1 and x_2 (read "x sub one" and "x sub two"), and label the corresponding y-coordinates y_1 and y_2.

The vertical change or **rise** between these points is the difference in the y-coordinates: $y_2 - y_1$. The horizontal change or **run** between the points is the difference of the x-coordinates: $x_2 - x_1$. The slope of the line is the ratio of $y_2 - y_1$ to $x_2 - x_1$, and we traditionally use the letter m to denote slope: $m = \frac{y_2 - y_1}{x_2 - x_1}$.

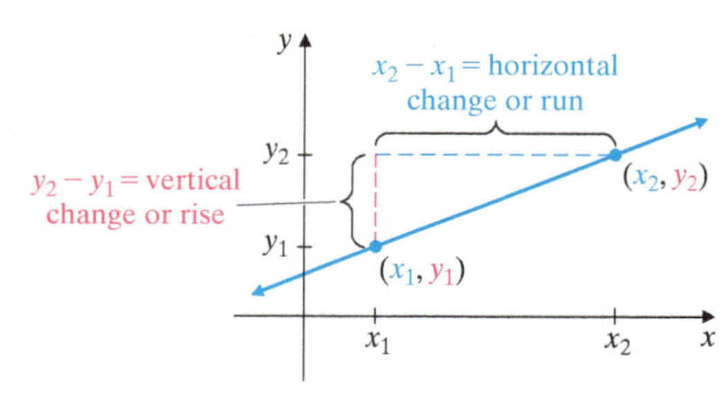

Slope of a Line

The slope m of the line containing the points (x_1, y_1) and (x_2, y_2) is given by

$$m = \frac{\text{rise}}{\text{run}} = \frac{\text{change in } y}{\text{change in } x} = \frac{y_2 - y_1}{x_2 - x_1}, \qquad \text{as long as } x_2 \neq x_1$$

Example 1 Find the slope of the line through $(-1, 5)$ and $(2, -3)$. Graph the line.

Solution: Let (x_1, y_1) be $(-1, 5)$ and (x_2, y_2) be $(2, -3)$. Then, by the definition of slope, we have the following.

$$m = \frac{y_2 - y_1}{x_2 - x_1} = \frac{-3 - 5}{2 - (-1)} = \frac{-8}{3} = -\frac{8}{3}$$

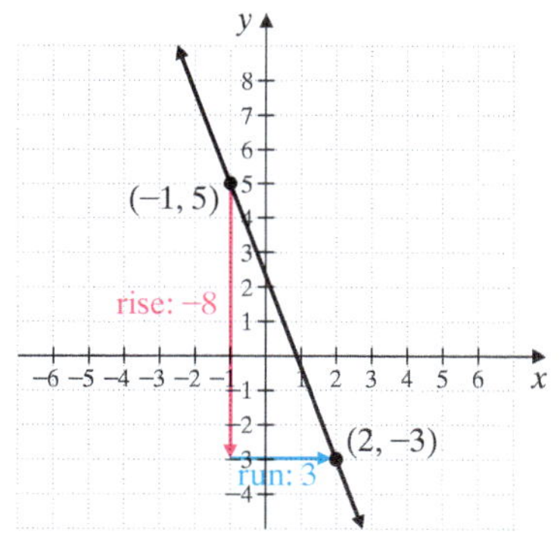

The slope of the line is $-\frac{8}{3}$.

Work Practice 1

Practice 1

Find the slope of the line through $(-2, 3)$ and $(4, -1)$. Graph the line.

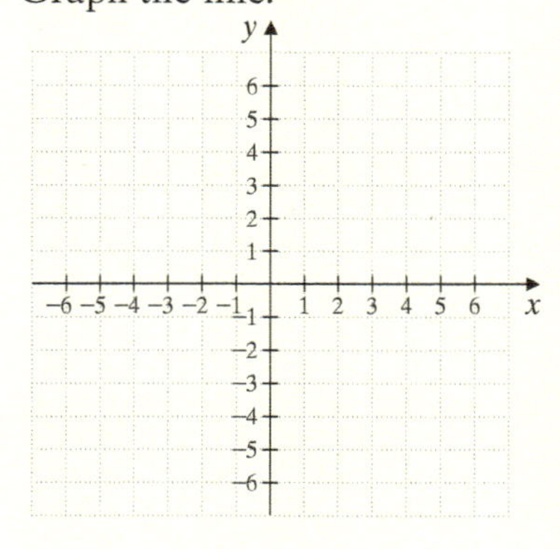

Helpful Hint

When finding slope, it makes no difference which point is identified as (x_1, y_1) and which is identified as (x_2, y_2). Just remember that whatever y-value is first in the numerator, its corresponding x-value is first in the denominator. Another way to calculate the slope in Example 1 is

$$m = \frac{y_2 - y_1}{x_2 - x_1} = \frac{5 - (-3)}{-1 - 2} = \frac{8}{-3} \text{ or } -\frac{8}{3} \leftarrow \text{Same slope as found in Example 1}$$

Practice 2

Find the slope of the line through $(-2, 1)$ and $(3, 5)$. Graph the line. (Answer on following page.)

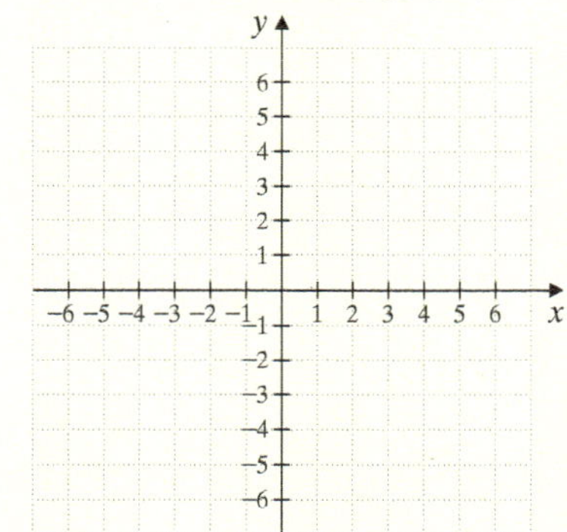

✓**Concept Check** The points $(-2, -5)$, $(0, -2)$, $(4, 4)$, and $(10, 13)$ all lie on the same line. Work with a partner and verify that the slope is the same no matter which points are used to find slope.

Example 2 Find the slope of the line through $(-1, -2)$ and $(2, 4)$. Graph the line.

Solution: Let (x_1, y_1) be $(2, 4)$ and (x_2, y_2) be $(-1, -2)$.

$$m = \frac{y_2 - y_1}{x_2 - x_1} = \frac{-2 - 4}{-1 - 2} \quad \begin{matrix} y\text{-value} \\ \text{corresponding } x\text{-value} \end{matrix}$$

$$= \frac{-6}{-3} = 2$$

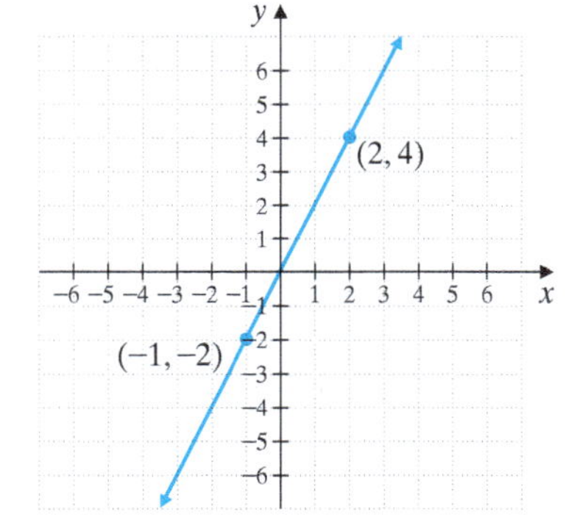

The slope is 2.

Work Practice 2

Answer

1. $m = -\frac{2}{3}$

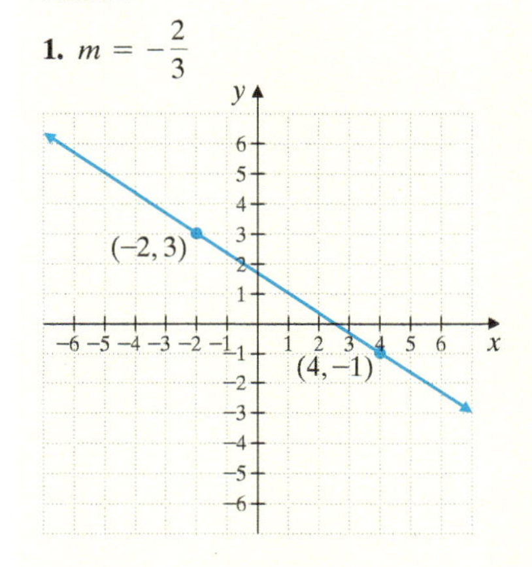

✓**Concept Check Answer**

$m = \frac{3}{2}$

✓**Concept Check** What is wrong with the following slope calculation for the points (3, 5) and (−2, 6)?

$$m = \frac{5-6}{-2-3} = \frac{-1}{-5} = \frac{1}{5}$$

Notice that the slope of the line in Example 1 is negative and that the slope of the line in Example 2 is positive. Let your eye follow the line with negative slope from left to right and notice that the line "goes down." If you follow the line with positive slope from left to right, you will notice that the line "goes up." This is true in general.

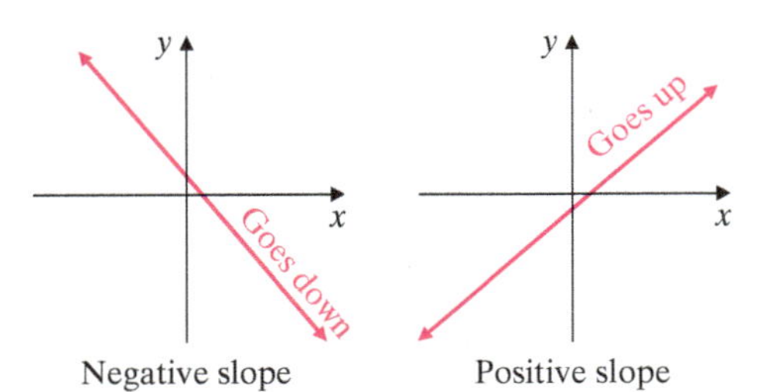

Helpful Hint To decide whether a line "goes up" or "goes down," always **follow the line from left to right.**

Objective B Finding the Slope of a Line Given Its Equation

As we have seen, the slope of a line is defined by two points on the line. Thus, if we know the equation of a line, we can find its slope by finding two of its points. For example, let's find the slope of the line

$$y = 3x - 2$$

To find two points, we can choose two values for x and substitute to find corresponding y-values. If $x = 0$, for example, $y = 3 \cdot 0 - 2$ or $y = -2$. If $x = 1$, $y = 3 \cdot 1 - 2$ or $y = 1$. This gives the ordered pairs $(0, -2)$ and $(1, 1)$. Using the definition for slope, we have

$$m = \frac{1-(-2)}{1-0} = \frac{3}{1} = 3 \quad \text{The slope is 3.}$$

Notice that the slope, 3, is the same as the coefficient of x in the equation $y = 3x - 2$. This is true in general.

If a linear equation is solved for y, the coefficient of x is the line's slope. In other words, the slope of the line given by $y = mx + b$ is m, the coefficient of x.

$$y = mx + b$$
(slope: m)

Example 3 Find the slope of the line $-2x + 3y = 11$.

Solution: When we solve for y, the coefficient of x is the slope.

$$-2x + 3y = 11$$
$$3y = 2x + 11 \quad \text{Add } 2x \text{ to both sides.}$$
$$y = \frac{2}{3}x + \frac{11}{3} \quad \text{Divide both sides by 3.}$$

The slope is $\frac{2}{3}$.

Work Practice 3

Practice 3

Find the slope of the line $5x + 4y = 10$.

Answers

2. $m = \frac{4}{5}$

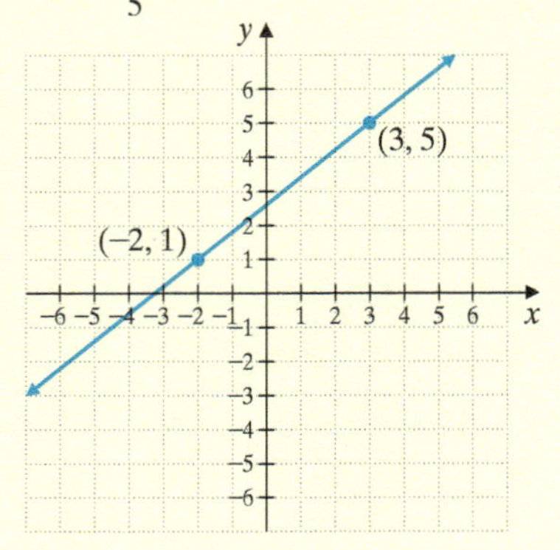

3. $m = -\frac{5}{4}$

✓**Concept Check Answer**

$$m = \frac{5-6}{3-(-2)} = \frac{-1}{5} = -\frac{1}{5}$$

Example 4 Find the slope of the line $-y = 5x - 2$.

Solution: Remember, the equation must be solved for y (not $-y$) in order for the coefficient of x to be the slope.

To solve for y, let's divide both sides of the equation by -1.

$$-y = 5x - 2$$

$$\frac{-y}{-1} = \frac{5x}{-1} - \frac{2}{-1} \quad \text{Divide both sides by } -1.$$

$$y = -5x + 2 \quad \text{Simplify.}$$

The slope is -5.

Work Practice 4

Practice 4

Find the slope of the line $-y = -2x + 7$.

Objective C Finding Slopes of Horizontal and Vertical Lines

Example 5 Find the slope of the line $y = -1$.

Solution: Recall that $y = -1$ is a horizontal line with y-intercept -1. To find the slope, we find two ordered pair solutions of $y = -1$, knowing that solutions of $y = -1$ must have a y-value of -1. We will use $(2, -1)$ and $(-3, -1)$. We let (x_1, y_1) be $(2, -1)$ and (x_2, y_2) be $(-3, -1)$.

$$m = \frac{y_2 - y_1}{x_2 - x_1} = \frac{-1 - (-1)}{-3 - 2} = \frac{0}{-5} = 0$$

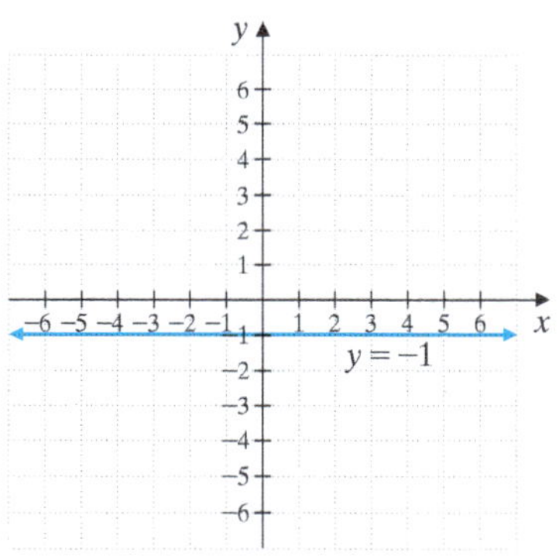

The slope of the line $y = -1$ is 0. Since the y-values will have a difference of 0 for every horizontal line, we can say that all **horizontal lines have a slope of 0.**

Work Practice 5

Practice 5

Find the slope of $y = 3$.

Example 6 Find the slope of the line $x = 5$.

Solution: Recall that the graph of $x = 5$ is a vertical line with x-intercept 5. To find the slope, we find two ordered pair solutions of $x = 5$. Ordered pair solutions of $x = 5$ must have an x-value of 5. We will use $(5, 0)$ and $(5, 4)$. We let $(x_1, y_1) = (5, 0)$ and $(x_2, y_2) = (5, 4)$.

$$m = \frac{y_2 - y_1}{x_2 - x_1} = \frac{4 - 0}{5 - 5} = \frac{4}{0}$$

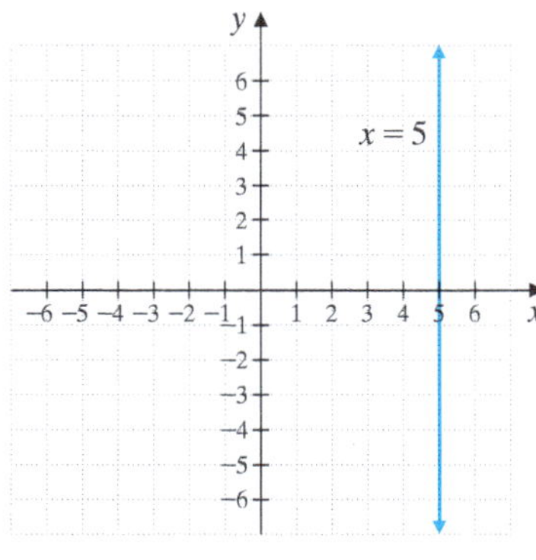

(*Continued on next page*)

Practice 6

Find the slope of the line $x = -2$.

Answers

4. $m = 2$ **5.** $m = 0$

6. undefined slope

Helpful Hint Slope of 0 and undefined slope are not the same. Vertical lines have undefined slope, while horizontal lines have a slope of 0.

Since $\frac{4}{0}$ is undefined, we say that the slope of the vertical line $x = 5$ is undefined.

Since the x-values will have a difference of 0 for every vertical line, we can say that all **vertical lines have undefined slope.**

Work Practice 6

Here is a general review of slope.

Summary of Slope

Slope m of the line through (x_1, y_1) and (x_2, y_2) is given by the equation

$$m = \frac{y_2 - y_1}{x_2 - x_1}.$$

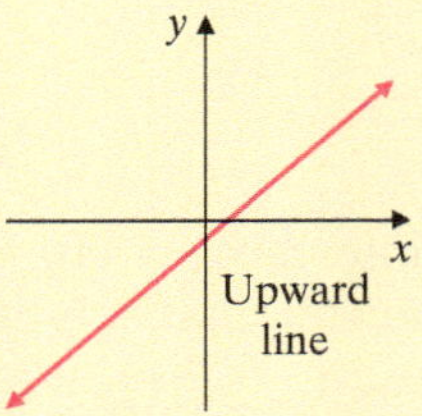

Positive slope: $m > 0$

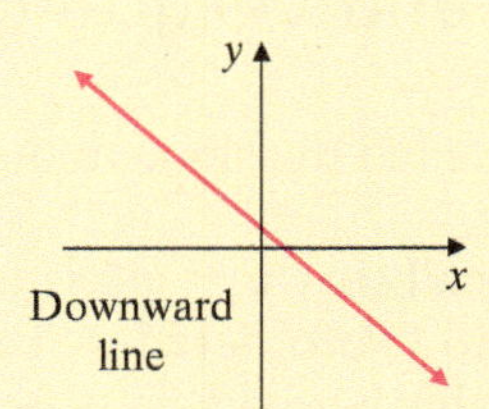

Negative slope: $m < 0$

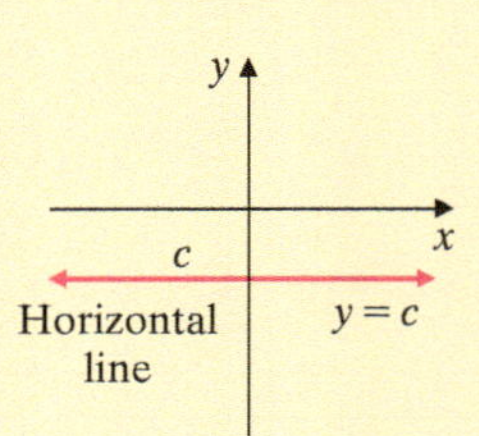

Zero slope: $m = 0$

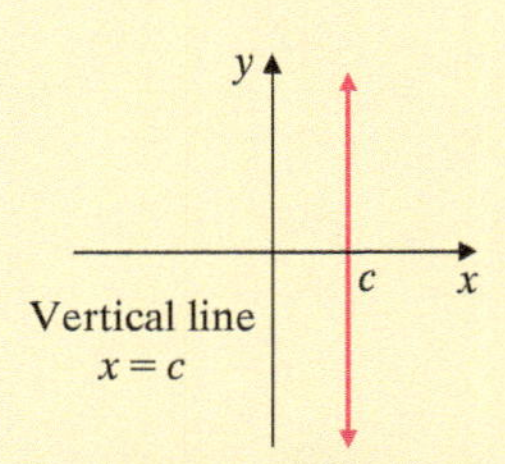

No slope or undefined slope

Objective D Comparing Slopes of Parallel and Perpendicular Lines

Two lines in the same plane are **parallel** if they do not intersect. Slopes of lines can help us determine whether lines are parallel. Since parallel lines have the same steepness, it follows that they have the same slope.

For example, the graphs of

$$y = -2x + 4$$

and

$$y = -2x - 3$$

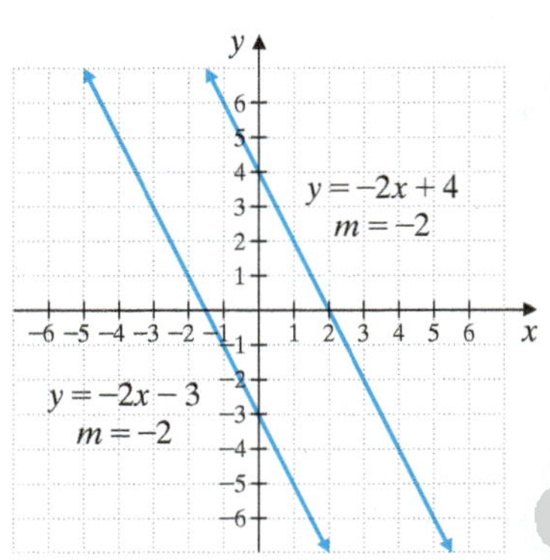

are shown. These lines have the same slope, −2. They also have different y-intercepts, so the lines are parallel. (If the y-intercepts were the same also, the lines would be the same.)

Parallel Lines

Nonvertical parallel lines have the same slope and different y-intercepts.

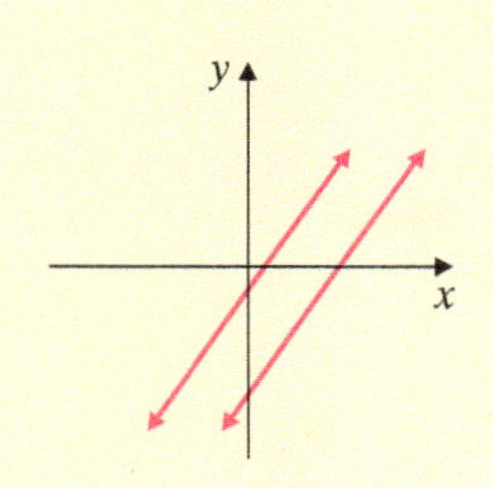

Two lines are **perpendicular** if they lie in the same plane and meet at a 90° (right) angle. How do the slopes of perpendicular lines compare? The product of the slopes of two perpendicular lines is -1.

For example, the graphs of

$$y = 4x + 1$$

and

$$y = -\frac{1}{4}x - 3$$

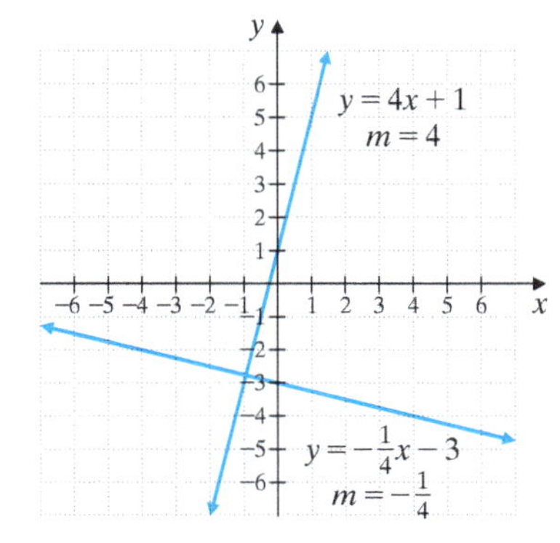

are shown. The slopes of the lines are 4 and $-\frac{1}{4}$. Their product is $4\left(-\frac{1}{4}\right) = -1$, so the lines are perpendicular.

Perpendicular Lines

If the product of the slopes of two lines is -1, then the lines are perpendicular.

(Two nonvertical lines are perpendicular if the slope of one is the negative reciprocal of the slope of the other.)

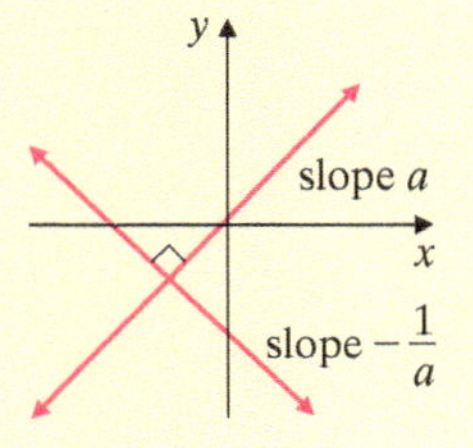

Helpful Hint

Here are examples of numbers that are negative (opposite) reciprocals.

Number	Negative Reciprocal	Their product is -1.
$\frac{2}{3}$	$-\frac{3}{2}$	$\frac{2}{3} \cdot -\frac{3}{2} = -\frac{6}{6} = -1$
-5 or $-\frac{5}{1}$	$\frac{1}{5}$	$-5 \cdot \frac{1}{5} = -\frac{5}{5} = -1$

Here are a few important points about vertical and horizontal lines.

- Two distinct vertical lines are parallel.
- Two distinct horizontal lines are parallel.
- A horizontal line and a vertical line are always perpendicular.

Practice 7

Determine whether each pair of lines is parallel, perpendicular, or neither.

a. $x + y = 5$
$2x + y = 5$

b. $5y = 2x - 3$
$5x + 2y = 1$

c. $y = 2x + 1$
$4x - 2y = 8$

Example 7

Determine whether each pair of lines is parallel, perpendicular, or neither.

a. $y = -\frac{1}{5}x + 1$
$2x + 10y = 3$

b. $x + y = 3$
$-x + y = 4$

c. $3x + y = 5$
$2x + 3y = 6$

Solution:

a. The slope of the line $y = -\frac{1}{5}x + 1$ is $-\frac{1}{5}$. We find the slope of the second line by solving its equation for y.

$$2x + 10y = 3$$
$$10y = -2x + 3 \quad \text{Subtract } 2x \text{ from both sides.}$$
$$y = \frac{-2}{10}x + \frac{3}{10} \quad \text{Divide both sides by 10.}$$
$$y = -\frac{1}{5}x + \frac{3}{10} \quad \text{Simplify.}$$

The slope of this line is $-\frac{1}{5}$ also. Since the lines have the same slope and different y-intercepts, they are parallel, as shown below on the left graph.

b. To find each slope, we solve each equation for y.

$$x + y = 3 \qquad\qquad -x + y = 4$$
$$y = -x + 3 \qquad\qquad y = x + 4$$

The slope is -1. The slope is 1.

The slopes are not the same, so the lines are not parallel. Next we check the product of the slopes: $(-1)(1) = -1$. Since the product is -1, the lines are perpendicular, as shown below on the right.

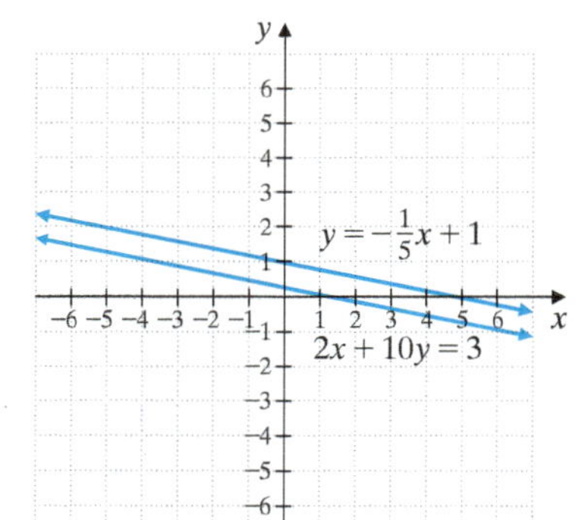

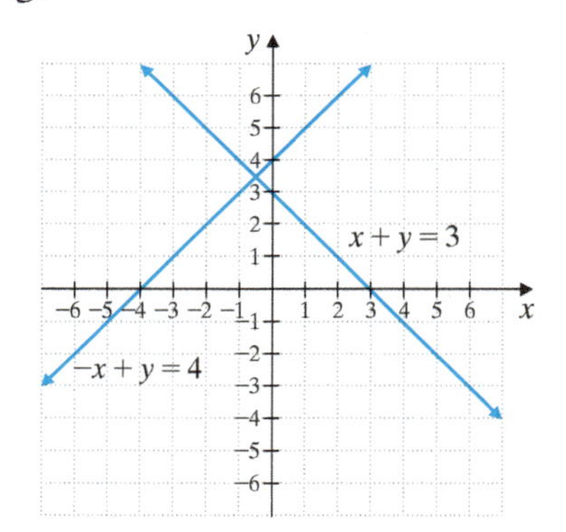

c. We solve each equation for y to find each slope. The slopes are -3 and $-\frac{2}{3}$. The slopes are not the same and their product is not -1. Thus, the lines are neither parallel nor perpendicular.

Work Practice 7

> **Helpful Hint**
> Note: To find the y-intercept of a line, let $x = 0$.
> - For $y = -\frac{1}{5}x + 1$, the y-intercept is $(0, 1)$.
> - For $y = -\frac{1}{5}x + \frac{3}{10}$, the y-intercept is $\left(0, \frac{3}{10}\right)$.
>
> Thus, the y-intercepts are different.

✓Concept Check Consider the line $-6x + 2y = 1$.

a. Write the equations of two lines parallel to this line.

b. Write the equations of two lines perpendicular to this line.

Answers

7. a. neither **b.** perpendicular **c.** parallel

✓Concept Check Answers

Answers may vary; for example,

a. $y = 3x - 3$, $y = 3x - 1$

b. $y = -\frac{1}{3}x$, $y = -\frac{1}{3}x + 1$

Objective E Interpreting Slope as a Rate of Change 

Slope can also be interpreted as a rate of change. In other words, slope tells us how fast y is changing with respect to x. To see this, let's look at a few of the many real-world applications of slope. For example, the pitch of a roof, used by builders and architects, is its slope. The pitch of the roof on the right is $\frac{7}{10}\left(\frac{\text{rise}}{\text{run}}\right)$. This means that the roof rises vertically 7 feet for every horizontal 10 feet. The rate of change for the roof is 7 vertical feet (y) per 10 horizontal feet (x).

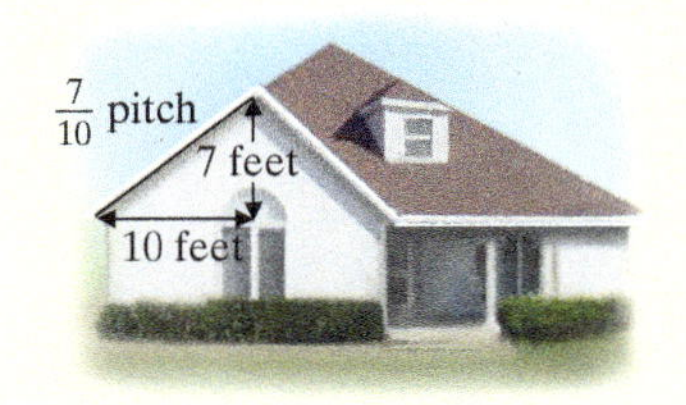

The grade of a road is its slope written as a percent. A 7% grade, as shown below, means that the road rises (or falls) 7 feet for every horizontal 100 feet. $\left(\text{Recall that } 7\% = \frac{7}{100}.\right)$ Here, the slope of $\frac{7}{100}$ gives us the rate of change. The road rises (in our diagram) 7 vertical feet (y) for every 100 horizontal feet (x).

$\frac{7}{100}$ = 7% grade

7 feet

100 feet

Example 8 Finding the Grade of a Road

At one part of the road to the summit of Pike's Peak, the road rises 15 feet for a horizontal distance of 250 feet. Find the grade of the road.

Solution: Recall that the grade of a road is its slope written as a percent.

$$\text{grade} = \frac{\text{rise}}{\text{run}} = \frac{15}{250} = 0.06 = 6\%$$

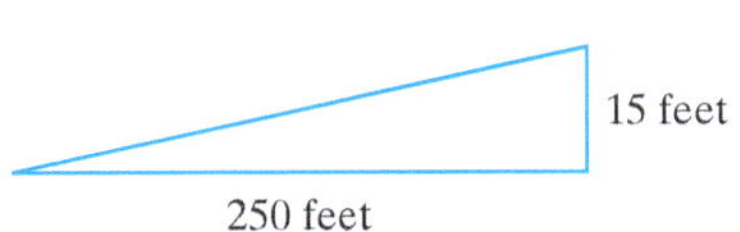

The grade is 6%.

Work Practice 8

Practice 8

Find the grade of the road shown.

Example 9 Finding the Slope of a Line

The following graph shows the number y of movie screens in the United States, where x is the number of years after 2010. Find the slope of the line and attach the proper units for the rate of change. Then write a sentence explaining the meaning of slope in this application. (*Source:* Motion Picture Association of America)

Solution: Use (2, 39.736) and (7, 40.336) to calculate slope.

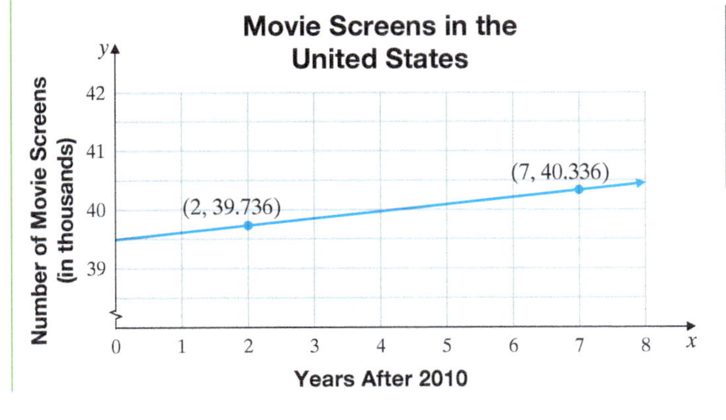

(*Continued on next page*)

Practice 9

Find the slope of the line and write the slope as a rate of change. This graph represents annual restaurant-industry employment y (in billions of workers) for year x. Write a sentence explaining the meaning of slope in this application.

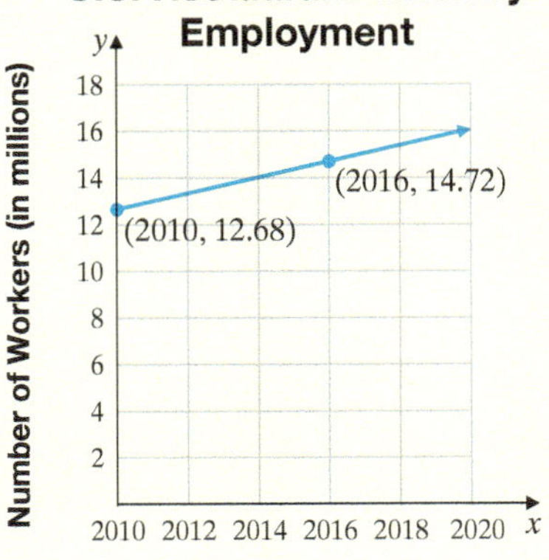

Source: National Restaurant Association

Answers

8. 15% **9.** $m = 0.34$; Each year the number of workers employed in the U.S. restaurant industry increases by 0.34 million, or 340,000, workers per year.

$$m = \frac{40.336 - 39.736}{7 - 2} = \frac{0.600}{5} = \frac{0.120 \text{ thousand screens}}{1 \text{ year}}$$

This means that the rate of change of the number of movie screens is 0.120 thousand screens per year, or each year 120 movie screens are added in the United States.

Work Practice 9

Calculator Explorations Graphing

It is possible to use a graphing calculator to sketch the graph of more than one equation on the same set of axes. This feature can be used to see parallel lines with the same slope. For example, graph the equations $y = \frac{2}{5}x$, $y = \frac{2}{5}x + 7$, and $y = \frac{2}{5}x - 4$ on the same set of axes. To do so, press the $\boxed{Y =}$ key and enter the equations on the first three lines.

$$Y_1 = \frac{2}{5}x$$

$$Y_2 = \frac{2}{5}x + 7$$

$$Y_3 = \frac{2}{5}x - 4$$

The displayed equations should look like this:

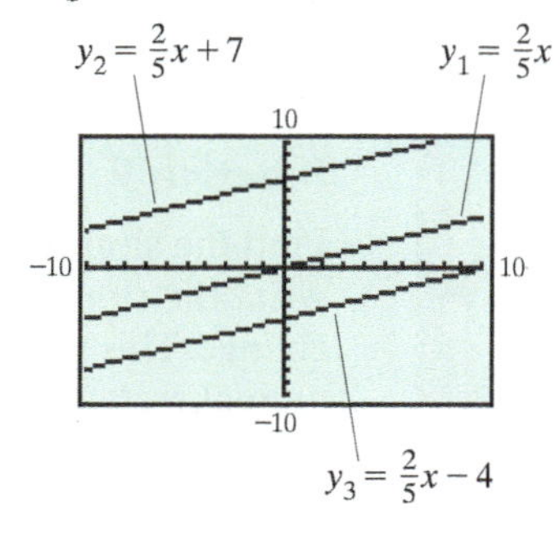

These lines are parallel, as expected, since they all have a slope of $\frac{2}{5}$. The graph of $y = \frac{2}{5}x + 7$ is the graph of $y = \frac{2}{5}x$ moved 7 units upward with a y-intercept of 7. Also, the graph of $y = \frac{2}{5}x - 4$ is the graph of $y = \frac{2}{5}x$ moved 4 units downward with a y-intercept of -4.

Graph the parallel lines on the same set of axes. Describe the similarities and differences in their graphs.

1. $y = 3.8x$, $y = 3.8x - 3$, $y = 3.8x + 9$

2. $y = -4.9x$, $y = -4.9x + 1$, $y = -4.9x + 8$

3. $y = \frac{1}{4}x$, $y = \frac{1}{4}x + 5$, $y = \frac{1}{4}x - 8$

4. $y = -\frac{3}{4}x$, $y = -\frac{3}{4}x - 5$, $y = -\frac{3}{4}x + 6$

Vocabulary, Readiness & Video Check

Use the choices below to fill in each blank. Not all choices will be used.

m	x	0	positive	undefined
b	y	slope	negative	

1. The measure of the steepness or tilt of a line is called ________.
2. If an equation is written in the form $y = mx + b$, the value of the letter ________ is the value of the slope of the graph.
3. The slope of a horizontal line is ________.
4. The slope of a vertical line is ________.
5. If the graph of a line moves upward from left to right, the line has ________ slope.
6. If the graph of a line moves downward from left to right, the line has ________ slope.
7. Given two points of a line, slope $= \dfrac{\text{change in \_\_\_\_\_\_\_\_}}{\text{change in \_\_\_\_\_\_\_\_}}$.

Decide whether a line with the given slope slants upward or downward or is horizontal or vertical.

8. $m = -\frac{2}{3}$ ________ **9.** $m = 5$ ________ **10.** m is undefined. ________ **11.** $m = 0$ ________

Martin-Gay Interactive Videos

See Video 6.4

Watch the section lecture video and answer the following questions.

Objective A **12.** What important point is made during Example 1 having to do with the order of the points in the slope formula?

Objective B **13.** From Example 5, how do we write an equation in "slope-intercept form"? Once the equation is in slope-intercept form, how do we identify the slope?

Objective C **14.** In the lecture after Example 8, different slopes are summarized. What is the difference between zero slope and undefined slope? What does "no slope" mean?

Objective D **15.** From Example 10, what form of the equation is best to determine if two lines are parallel or perpendicular? Why?

Objective E **16.** Writing the slope as a rate of change in Example 11 gave real-life meaning to the slope. What step in the general strategy for problem solving does this correspond to?

6.4 Exercise Set MyLab Math

Objective A *Find the slope of the line that passes through the given points. See Examples 1 and 2.*

1. $(-1, 5)$ and $(6, -2)$ **2.** $(-1, 16)$ and $(3, 4)$ **3.** $(1, 4)$ and $(5, 3)$ **4.** $(3, 1)$ and $(2, 6)$

5. $(5, 1)$ and $(-2, 1)$ **6.** $(-8, 3)$ and $(-2, 3)$ **7.** $(-4, 3)$ and $(-4, 5)$ **8.** $(-2, -3)$ and $(-2, 5)$

Use the points shown on each graph to find the slope of each line. See Examples 1 and 2.

9.

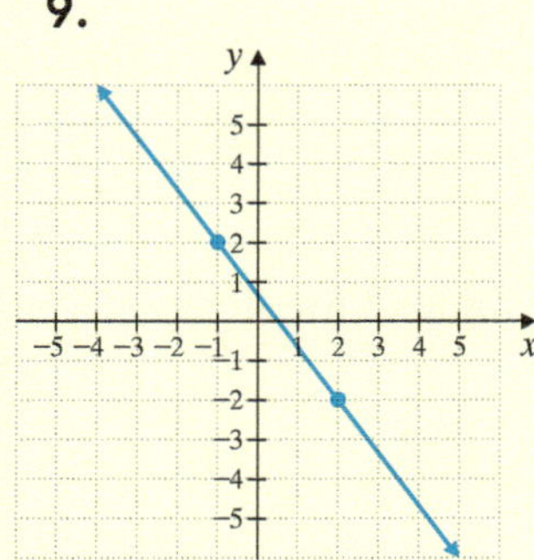

10.

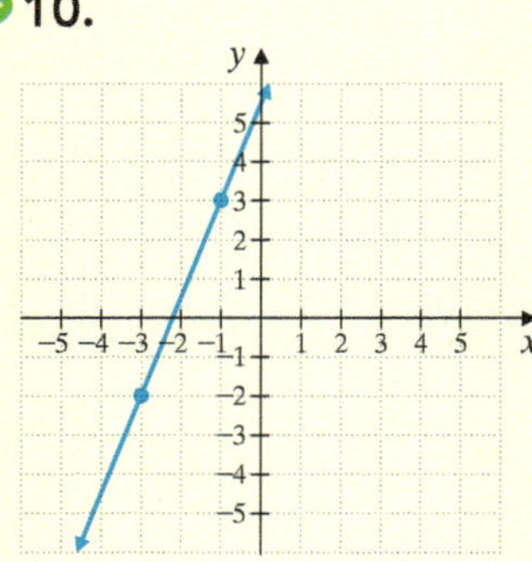

11.

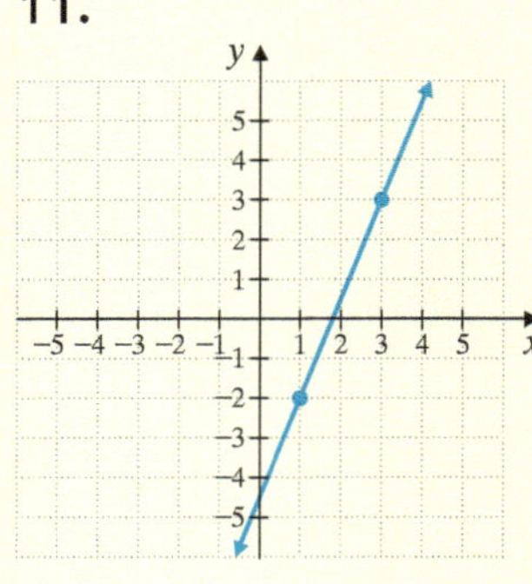

12.

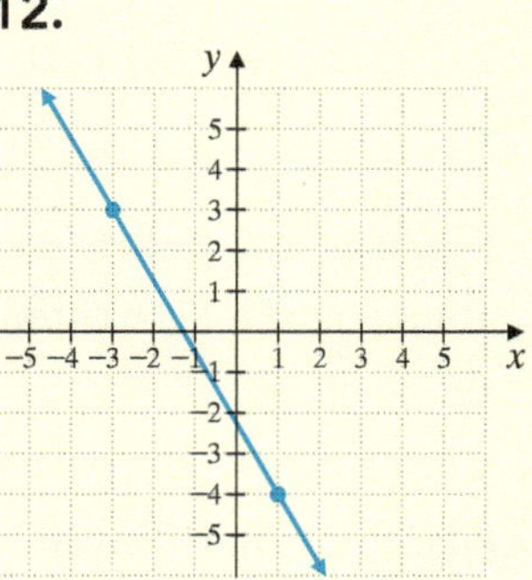

Objectives A C Mixed Practice

State whether the slope of the line is positive, negative, 0, or is undefined. See the box on page 462.

13.

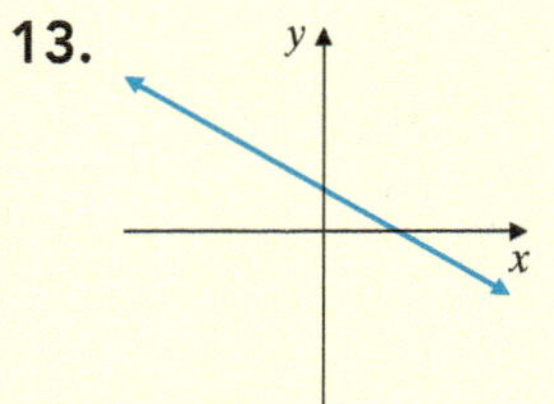

14.

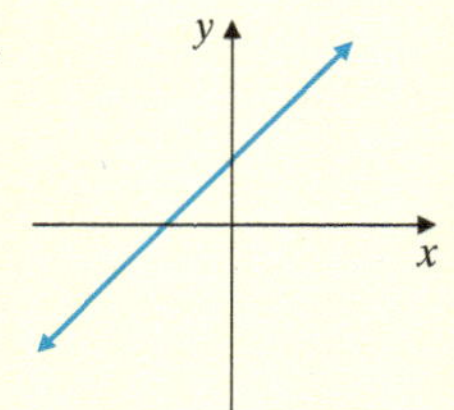

15.

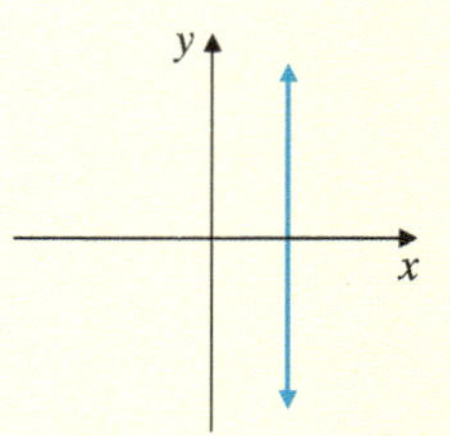

16.

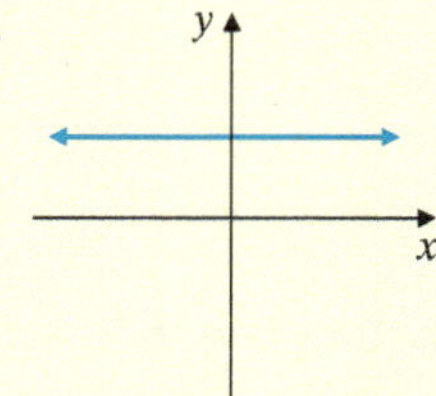

Decide whether a line with the given slope slants upward or downward, or is horizontal or vertical. See the box on page 462.

17. $m = \frac{7}{6}$ ________

18. $m = -3$ ________

19. $m = 0$ ________

20. m is undefined. ________

For each graph, determine which line has the greater slope.

21.

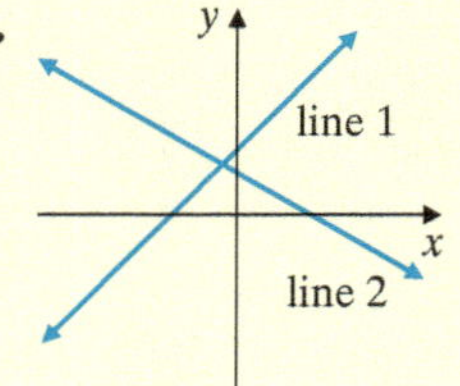

22.

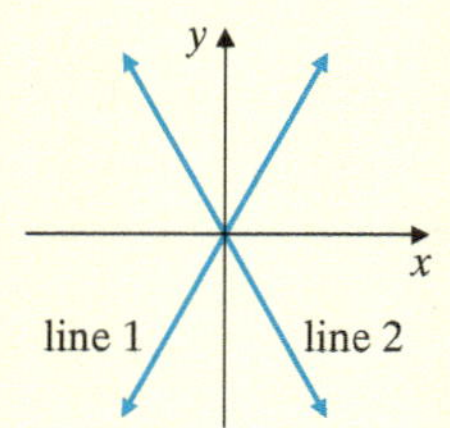

23.

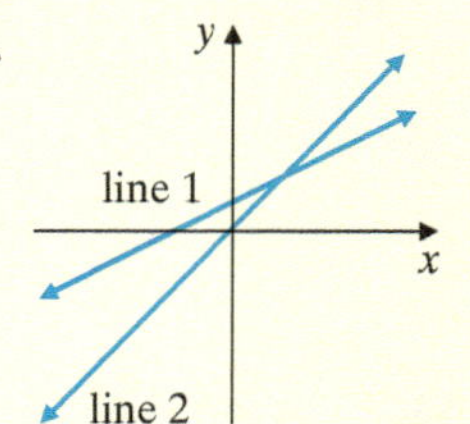

24.

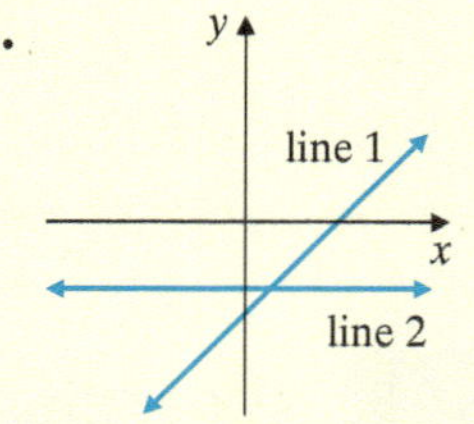

Objectives B C Mixed Practice

Find the slope of each line. See Examples 3 through 6.

25. $y = 5x - 2$

26. $y = -2x + 6$

27. $y = -0.3x + 2.5$

28. $y = -7.6x - 0.1$

29. $2x + y = 7$

30. $-5x + y = 10$

31.

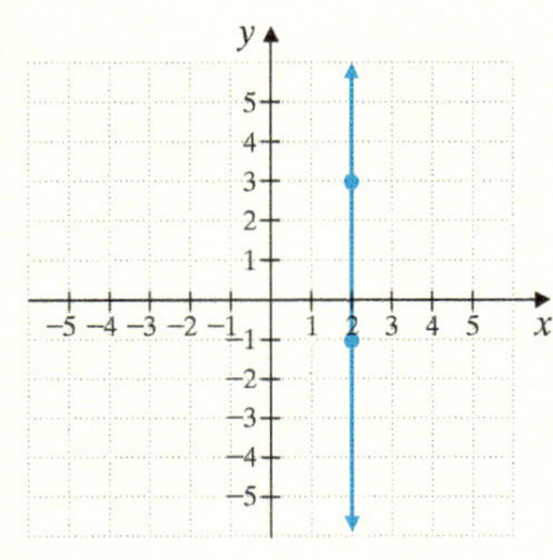

32.

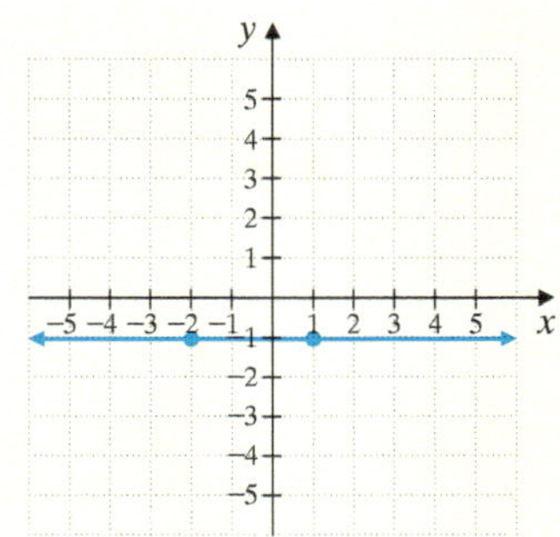

33. $2x - 3y = 10$

34. $3x - 5y = 1$

35. $x = 1$

36. $y = -2$

37. $x = 2y$

38. $x = -4y$

39. $y = -3$

40. $x = 5$

41. $-3x - 4y = 6$

42. $-4x - 7y = 9$

43. $20x - 5y = 1.2$

44. $24x - 3y = 5.7$

Objective D *Find the slope of a line that is (a) parallel and (b) perpendicular to the line through each pair of points. See Example 7.*

45. $(-3, -3)$ and $(0, 0)$

46. $(6, -2)$ and $(1, 4)$

47. $(-8, -4)$ and $(3, 5)$

48. $(6, -1)$ and $(-4, -10)$

△ *Determine whether each pair of lines is parallel, perpendicular, or neither. See Example 7.*

49. $y = \frac{2}{9}x + 3$
$y = -\frac{2}{9}x$

50. $y = \frac{1}{5}x + 20$
$y = -\frac{1}{5}x$

51. $x - 3y = -6$
$y = 3x - 9$

52. $y = 4x - 2$
$4x + y = 5$

53. $6x = 5y + 1$
$-12x + 10y = 1$

54. $-x + 2y = -2$
$2x = 4y + 3$

55. $6 + 4x = 3y$
$3x + 4y = 8$

56. $10 + 3x = 5y$
$5x + 3y = 1$

Objective E *The pitch of a roof is its slope. Find the pitch of each roof shown. See Example 8.*

57.

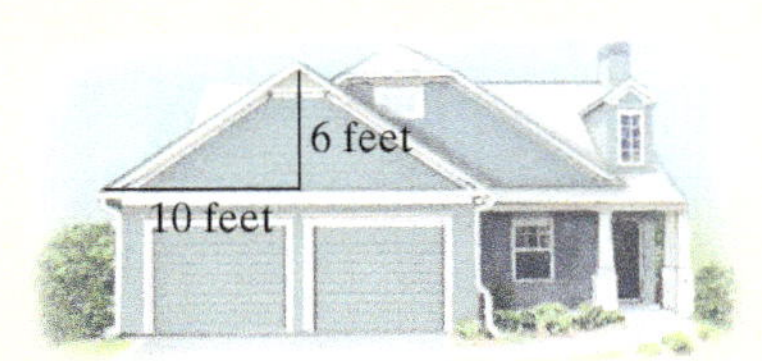

58.

The grade of a road is its slope written as a percent. Find the grade of each road shown. See Example 8.

59.

60.

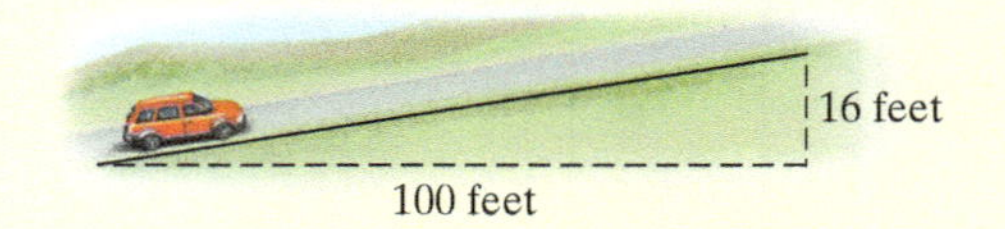

61. One of Japan's superconducting "bullet" trains is researched and tested at the Yamanashi Maglev Test Line near Otsuki City. The steepest section of the track has a rise of 2580 meters for a horizontal distance of 6450 meters. What is the grade (slope written as a percent) of this section of track? (*Source:* Japan Railways Central Co.)

62. Professional plumbers suggest that a sewer pipe rise 0.25 inch for every horizontal foot. Find the recommended slope for a sewer pipe and write the slope as a grade, or percent. Round to the nearest percent.

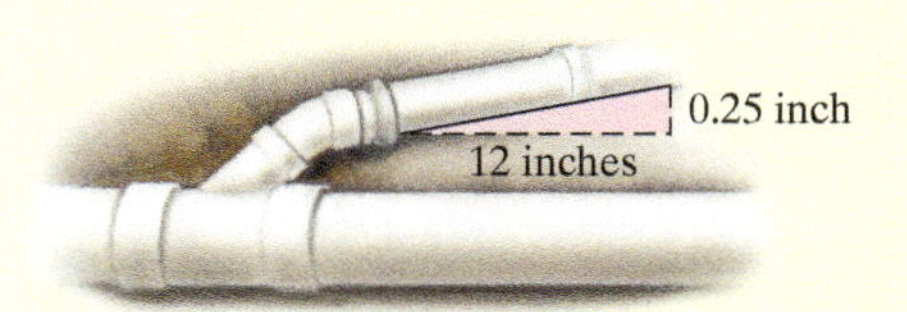

63. There has been controversy over the past few years about the world's steepest street. *The Guinness Book of Records* listed Baldwin Street, in Dunedin, New Zealand, as the world's steepest street, but Canton Avenue in the Pittsburgh neighborhood of Beechview may actually be steeper. Calculate each grade to the nearest percent.

Canton Avenue	For every 30 meters of horizontal distance, the vertical change is 11 meters.	
Baldwin Street	For every 2.86 meters of horizontal distance, the vertical change is 1 meter.	

64. According to federal regulations, a wheelchair ramp should rise no more than 1 foot for a horizontal distance of 12 feet. Write the slope as a grade. Round to the nearest tenth of a percent.

Find the slope of each line and write a sentence using the slope as a rate of change. Don't forget to attach the proper units. See Example 9.

65. This graph approximates the number of U.S. households that have televisions y (in millions) for year x.

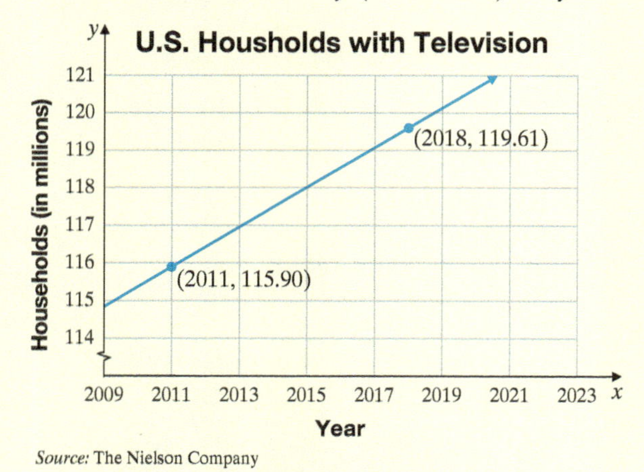

Source: The Nielson Company

66. The graph approximates the amount of money y (in billions of dollars) spent worldwide on tourism for year x.

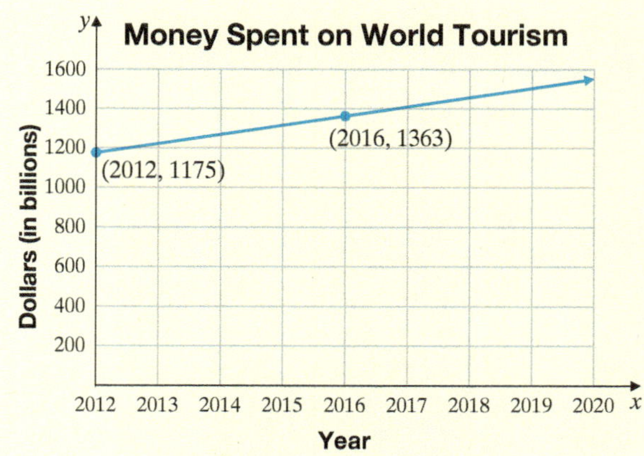

Source: World Tourism Organization

67. Americans are keeping their cars longer. The graph below shows the median age y (in years) of automobiles in the United States for the years shown.

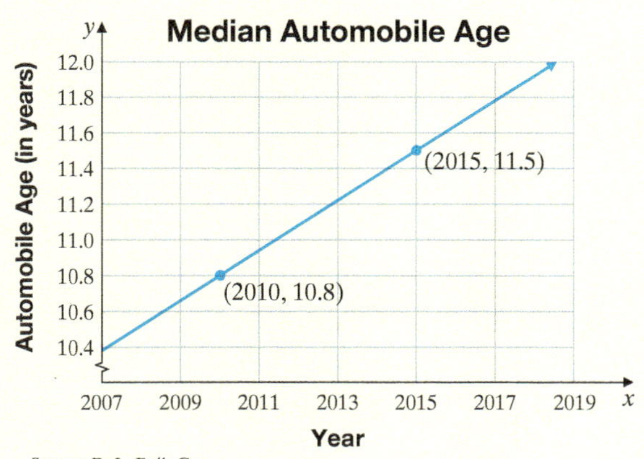

Source: R. L. Polk Co.

68. The graph below shows the approximate total cost y (in dollars) of owning and operating a small SUV in the United States in 2017, where x is the annual number of miles driven.

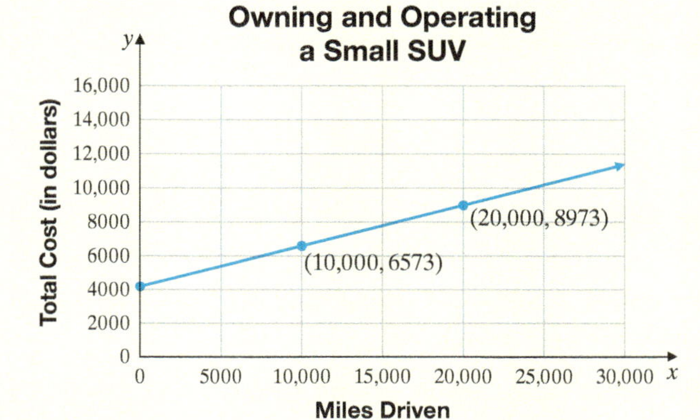

Source: AAA

Review

Solve each equation for y. See Section 2.5.

69. $y - (-6) = 2(x - 4)$

70. $y - 7 = -9(x - 6)$

71. $y - 1 = -6(x - (-2))$

72. $y - (-3) = 4(x - (-5))$

Concept Extensions

Match each line with its slope. (Exercises 76 through 78 are on the next page.)

A. $m = 0$

B. undefined slope

C. $m = 3$

D. $m = 1$

E. $m = -\frac{1}{2}$

F. $m = -\frac{3}{4}$

73.

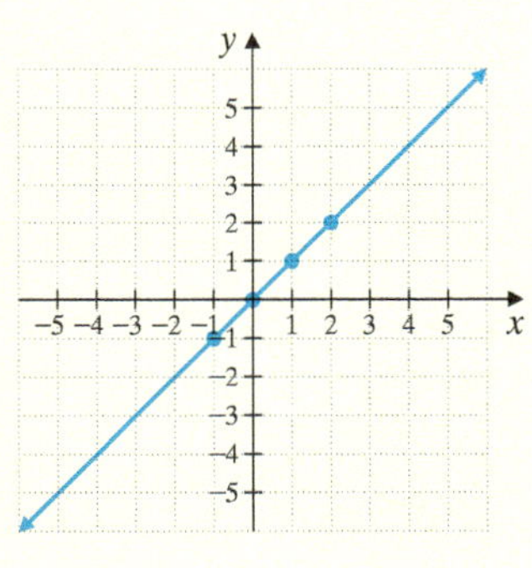

74.

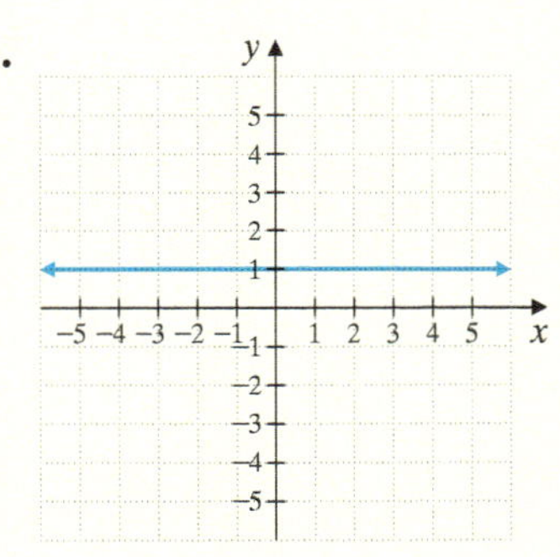

75.

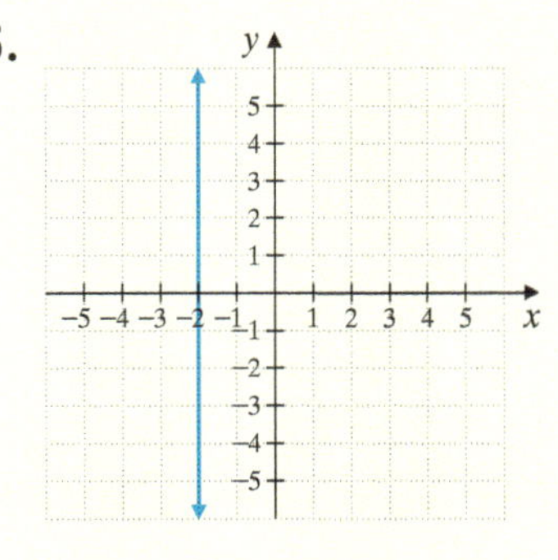

76.

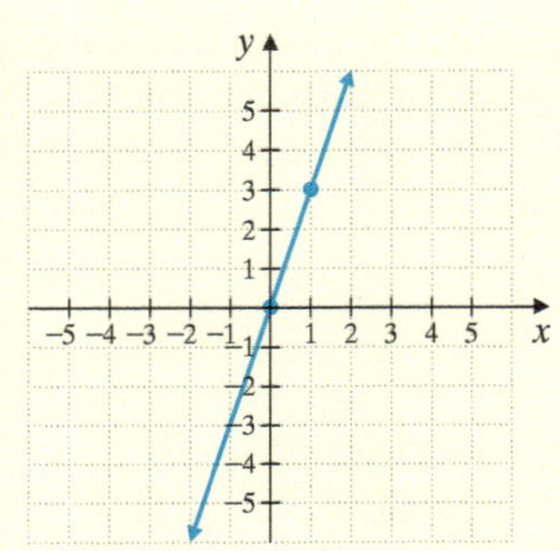

77.

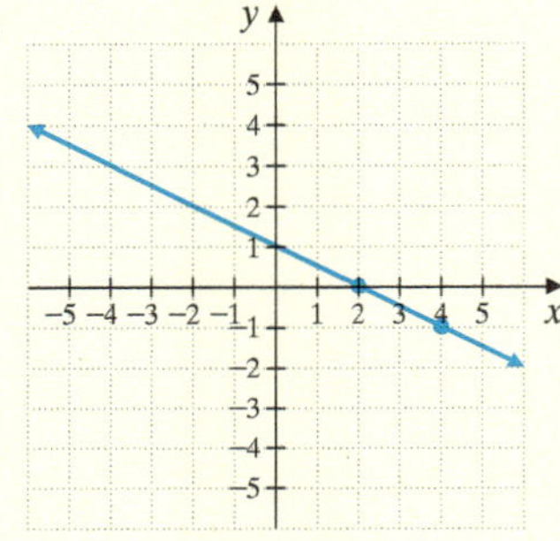

78.

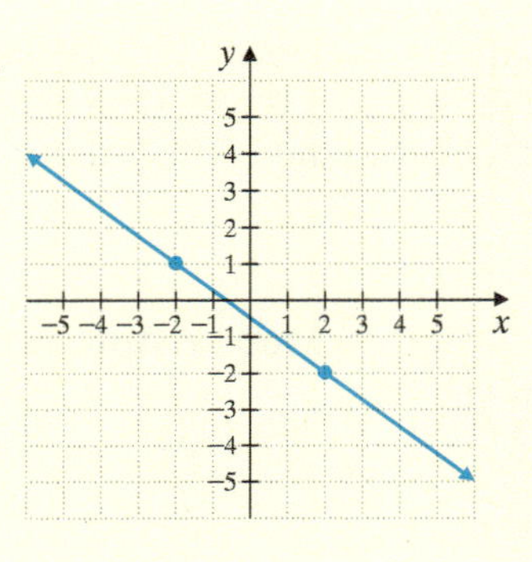

Solve. See a Concept Check in this section.

79. Verify that the points $(2, 1)$, $(0, 0)$, $(-2, -1)$, and $(-4, -2)$ are all on the same line by computing the slope between each pair of points. (See the first Concept Check.)

80. Given the points $(2, 3)$ and $(-5, 1)$, can the slope of the line through these points be calculated by $\frac{1 - 3}{2 - (-5)}$? Why or why not? (See the second Concept Check.)

81. Write the equations of three lines parallel to $10x - 5y = -7$. (See the third Concept Check.)

82. Write the equations of two lines perpendicular to $10x - 5y = -7$. (See the third Concept Check.)

The following line graph shows the average fuel economy (in miles per gallon) of passenger automobiles produced during each of the model years shown. Use this graph to answer Exercises 83 through 88.

83. What was the average fuel economy (in miles per gallon) for automobiles produced during 2008?

84. Find the decrease in average fuel economy for automobiles between the years 2010 and 2011.

85. During which of the model years shown was average fuel economy the lowest?
What was the average fuel economy for that year?

86. During which of the model years shown was average fuel economy the highest?
What was the average fuel economy for that year?

Average Fuel Economy for Autos

Average Miles per Gallon: 30, 31, 32, 33, 34, 35, 36, 37, 38, 39

Model Year: 2007, 2009, 2011, 2013, 2015, 2017

Source: Bureau of Transportation Statistics

87. Of the following line segments, which has the greatest slope: from 2007 to 2008, from 2009 to 2010, or from 2011 to 2012?

88. What line segment has a slope of 1?

89. Find x so that the pitch of the roof is $\frac{2}{5}$.

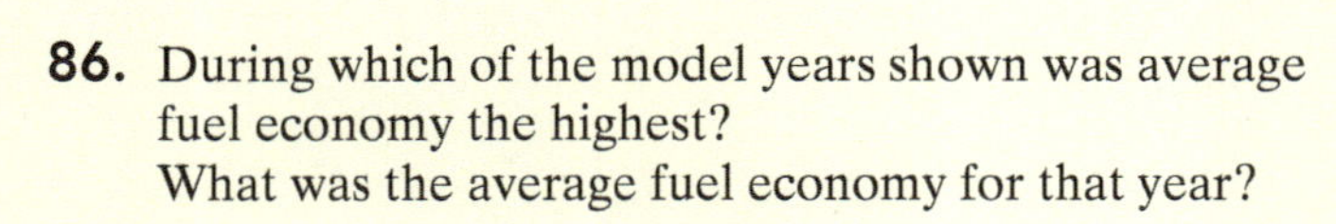

90. Find x so that the pitch of the roof is $\frac{1}{3}$.

91. There were 2378 heart transplants performed in the United States in 2012. In 2017, the number of heart transplants in the United States rose to 3244. (*Source:* Organ Procurement and Transplantation Network)

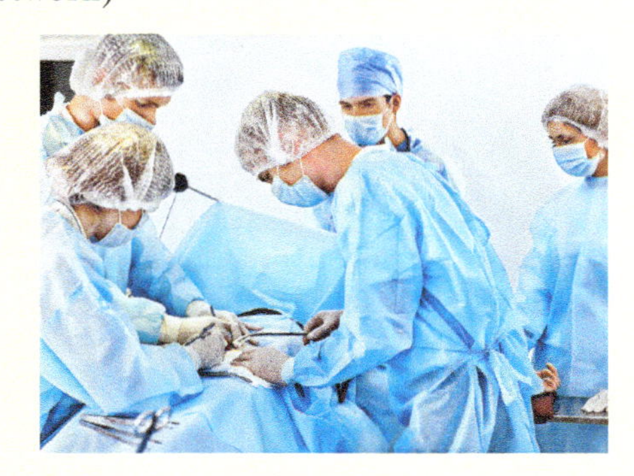

a. Write two ordered pairs of the form (year, number of heart transplants).

b. Find the slope of the line between the two points.

c. Write a sentence explaining the meaning of the slope as a rate of change.

92. The average price of an acre of U.S. cropland was \$2980 in 2011. In 2017, the price of an acre rose to \$4090. (*Source:* National Agricultural Statistics Service)

a. Write two ordered pairs of the form (year, price per acre).

b. Find the slope of the line through the two points.

c. Write a sentence explaining the meaning of the slope as a rate of change.

93. Show that the quadrilateral with vertices $(1, 3)$, $(2, 1)$, $(-4, 0)$, and $(-3, -2)$ is a parallelogram.

94. Show that a triangle with vertices at the points $(1, 1)$, $(-4, 4)$, and $(-3, 0)$ is a right triangle.

Find the slope of the line through the given points.

95. $(-3.8, 1.2)$ and $(-2.2, 4.5)$

96. $(2.1, 6.7)$ and $(-8.3, 9.3)$

97. $(14.3, -10.1)$ and $(9.8, -2.9)$

98. $(2.3, 0.2)$ and $(7.9, 5.1)$

99. The graph of $y = \frac{1}{2}x$ has a slope of $\frac{1}{2}$. The graph of $y = 3x$ has a slope of 3. The graph of $y = 5x$ has a slope of 5. Graph all three equations on a single coordinate system. As the slope becomes larger, how does the steepness of the line change?

100. The graph of $y = -\frac{1}{3}x + 2$ has a slope of $-\frac{1}{3}$. The graph of $y = -2x + 2$ has a slope of -2. The graph of $y = -4x + 2$ has a slope of -4. Graph all three equations on a single coordinate system. As the absolute value of the slope becomes larger, how does the steepness of the line change?

6.5 Equations of Lines 

Objectives

A Use the Slope-Intercept Form to Graph a Linear Equation.

B Use the Slope-Intercept Form to Write an Equation of a Line.

C Use the Point-Slope Form to Find an Equation of a Line Given Its Slope and a Point of the Line.

D Use the Point-Slope Form to Find an Equation of a Line Given Two Points of the Line.

E Find Equations of Vertical and Horizontal Lines.

F Use the Point-Slope Form to Solve Problems.

We know that when a linear equation is solved for y, the coefficient of x is the slope of the line. For example, the slope of the line whose equation is $y = 3x + 1$ is 3. In the equation $y = 3x + 1$, what does 1 represent? To find out, let $x = 0$ and watch what happens.

$$y = 3x + 1$$
$$y = 3 \cdot 0 + 1 \quad \text{Let } x = 0.$$
$$y = 1$$

We now have the ordered pair $(0, 1)$, which means that 1 is the y-intercept.

This is true in general. To see this, let $x = 0$ and solve for y in $y = mx + b$.

$$y = m \cdot 0 + b \quad \text{Let } x = 0.$$
$$y = b$$

We obtain the ordered pair $(0, b)$, which means that point is the y-intercept.

The form $y = mx + b$ is appropriately called the *slope-intercept form* of a linear equation. (m: slope; b: y-intercept is $(0, b)$.)

Slope-Intercept Form

When a linear equation in two variables is written in **slope-intercept form,**

$$y = mx + b$$

(m: slope; b: $(0, b)$, y-intercept)

then m is the slope of the line and $(0, b)$ is the y-intercept of the line.

Objective A Using the Slope-Intercept Form to Graph an Equation

We can use the slope-intercept form of the equation of a line to graph a linear equation.

Practice 1

Use the slope-intercept form to graph the equation $y = \frac{2}{3}x - 4$.

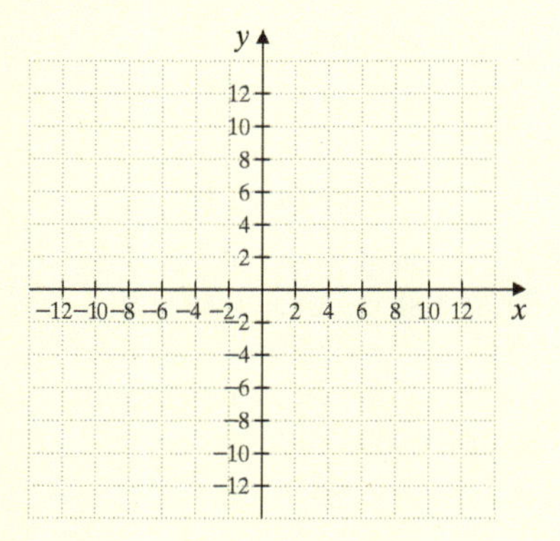

Example 1

Use the slope-intercept form to graph the equation $y = \frac{3}{5}x - 2$.

Solution: Since the equation $y = \frac{3}{5}x - 2$ is written in slope-intercept form $y = mx + b$, the slope of its graph is $\frac{3}{5}$ and the y-intercept is $(0, -2)$. To graph this equation, we begin by plotting the point $(0, -2)$. From this point, we can find another point of the graph by using the slope $\frac{3}{5}$ and recalling that slope is $\frac{\text{rise}}{\text{run}}$. We start at the y-intercept and move 3 units up since the numerator of the slope is 3; then we move 5 units to the right since the denominator of the slope is 5. We stop at the point $(5, 1)$. The line through $(0, -2)$ and $(5, 1)$ is the graph of $y = \frac{3}{5}x - 2$.

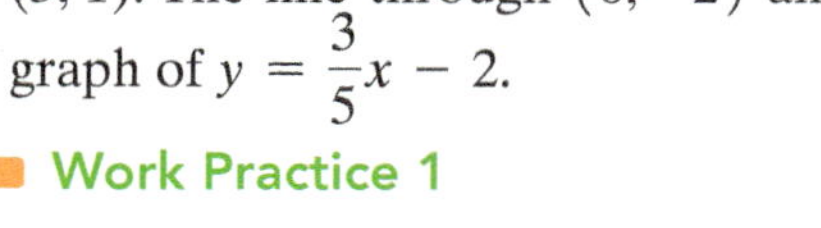

Work Practice 1

Example 2 Use the slope-intercept form to graph the equation $4x + y = 1$.

Solution: First we write the given equation in slope-intercept form.

$$4x + y = 1$$
$$y = -4x + 1$$

The graph of this equation will have slope -4 and y-intercept $(0, 1)$. To graph this line, we first plot the point $(0, 1)$. To find another point of the graph, we use the slope -4, which can be written as $\frac{-4}{1}$ $\left(\frac{4}{-1} \text{ could also be used}\right)$. We start at the point $(0, 1)$ and move 4 units down (since the numerator of the slope is -4), and then 1 unit to the right (since the denominator of the slope is 1). We arrive at the point $(1, -3)$. The line through $(0, 1)$ and $(1, -3)$ is the graph of $4x + y = 1$.

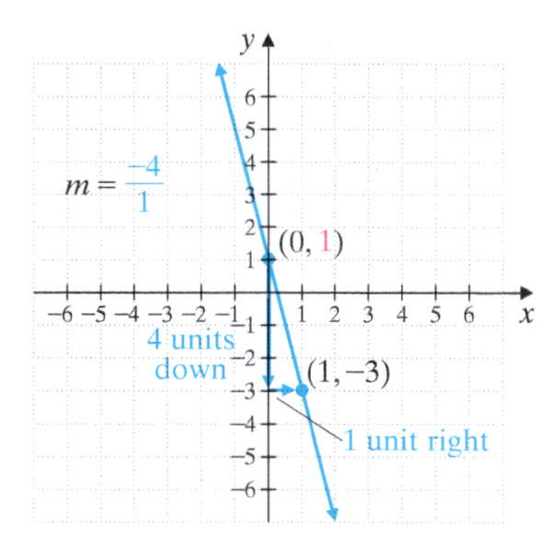

Work Practice 2

Practice 2

Use the slope-intercept form to graph $3x + y = 2$.

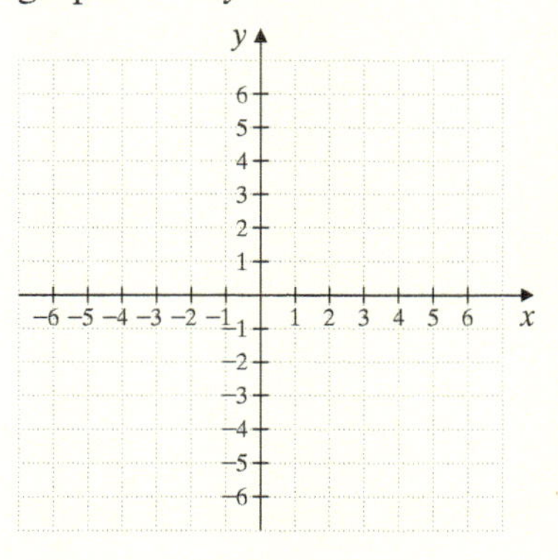

Helpful Hint

In Example 2, if we interpret the slope of -4 as $\frac{4}{-1}$, we arrive at $(-1, 5)$ for a second point. Notice that this point is also on the line.

Objective B Using the Slope-Intercept Form to Write an Equation

The slope-intercept form can also be used to write the equation of a line when we know its slope and y-intercept.

Example 3 Find an equation of the line with y-intercept $(0, -3)$ and slope of $\frac{1}{4}$.

Solution: We are given the slope and the y-intercept. We let $m = \frac{1}{4}$ and $b = -3$ and write the equation in slope-intercept form, $y = mx + b$.

$$y = mx + b$$
$$y = \frac{1}{4}x + (-3) \quad \text{Let } m = \frac{1}{4} \text{ and } b = -3.$$
$$y = \frac{1}{4}x - 3 \quad \text{Simplify.}$$

Work Practice 3

Practice 3

Find an equation of the line with y-intercept $(0, -2)$ and slope of $\frac{3}{5}$.

Objective C Writing an Equation Given Its Slope and a Point

Thus far, we have seen that we can write an equation of a line if we know its slope and y-intercept. We can also write an equation of a line if we know its slope and any

Answers

1.

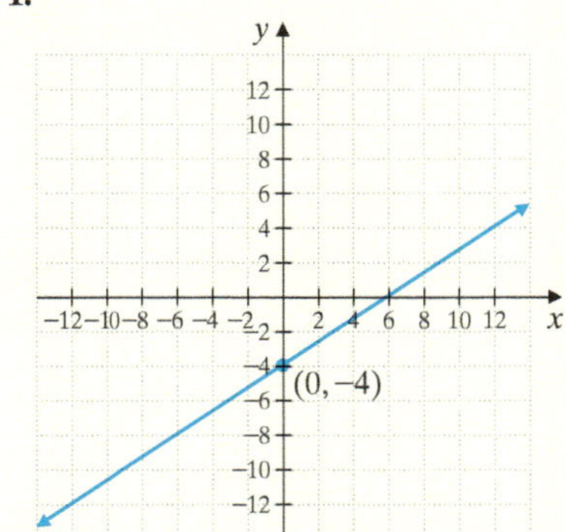

2.

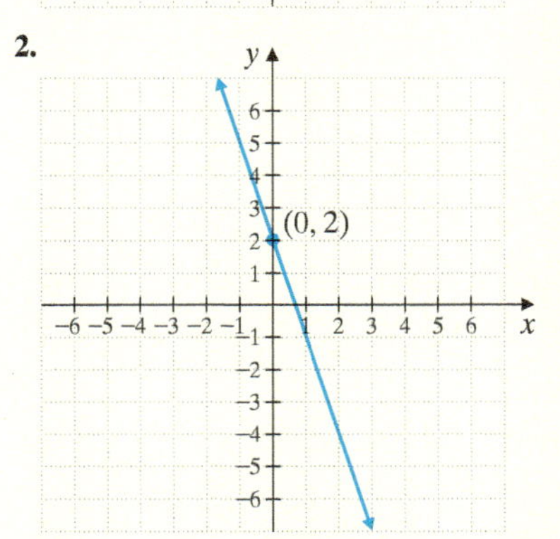

3. $y = \frac{3}{5}x - 2$

point on the line. To see how we do this, let m represent slope and (x_1, y_1) represent a point on the line. Then if (x, y) is any other point on the line, we have that

$$\frac{y - y_1}{x - x_1} = m$$

$$y - y_1 = \underset{\text{slope}}{\underset{\uparrow}{m}}(x - x_1) \quad \text{Multiply both sides by } (x - x_1).$$

This is the *point-slope form* of the equation of a line.

Point-Slope Form of the Equation of a Line

The **point-slope form** of the equation of a line is

$$y - y_1 = m(x - x_1)$$

(slope: m; (x_1, y_1) point on the line)

where m is the slope of the line and (x_1, y_1) is a point on the line.

Practice 4

Find an equation of the line with slope -3 that passes through $(2, -4)$. Write the equation in slope-intercept form, $y = mx + b$.

Example 4 Find an equation of the line with slope -2 that passes through $(-1, 5)$. Write the equation in slope-intercept form, $y = mx + b$, and in standard form, $Ax + By = C$.

Solution: Since the slope and a point on the line are given, we use point-slope form, $y - y_1 = m(x - x_1)$, to write the equation. Let $m = -2$ and $(-1, 5) = (x_1, y_1)$.

$$\begin{aligned} y - y_1 &= m(x - x_1) \\ y - 5 &= -2[x - (-1)] && \text{Let } m = -2 \text{ and } (x_1, y_1) = (-1, 5). \\ y - 5 &= -2(x + 1) && \text{Simplify.} \\ y - 5 &= -2x - 2 && \text{Use the distributive property.} \end{aligned}$$

To write the equation in slope-intercept form, $y = mx + b$, we simply solve the equation for y. To do this, we add 5 to both sides.

$$\begin{aligned} y - 5 &= -2x - 2 \\ y &= -2x + 3 && \text{Slope-intercept form} \\ 2x + y &= 3 && \text{Add } 2x \text{ to both sides and we have standard form.} \end{aligned}$$

Work Practice 4

Objective D Writing an Equation Given Two Points

We can also find an equation of a line when we are given any two points of the line.

Practice 5

Find an equation of the line through $(1, 3)$ and $(5, -2)$. Write the equation in the form $Ax + By = C$.

Example 5 Find an equation of the line through $(2, 5)$ and $(-3, 4)$. Write the equation in the form $Ax + By = C$.

Solution: First, use the two given points to find the slope of the line.

$$m = \frac{4 - 5}{-3 - 2} = \frac{-1}{-5} = \frac{1}{5}$$

Next we use the slope $\frac{1}{5}$ and either one of the given points to write the equation in point-slope form. We use $(2, 5)$. Let $x_1 = 2$, $y_1 = 5$, and $m = \frac{1}{5}$.

$$\begin{aligned} y - y_1 &= m(x - x_1) && \text{Use point-slope form.} \\ y - 5 &= \frac{1}{5}(x - 2) && \text{Let } x_1 = 2,\ y_1 = 5, \text{ and } m = \frac{1}{5}. \end{aligned}$$

Answers

4. $y = -3x + 2$ **5.** $5x + 4y = 17$

$$5(y - 5) = 5 \cdot \frac{1}{5}(x - 2) \quad \text{Multiply both sides by 5 to clear fractions.}$$
$$5y - 25 = x - 2 \quad \text{Use the distributive property and simplify.}$$
$$-x + 5y - 25 = -2 \quad \text{Subtract } x \text{ from both sides.}$$
$$-x + 5y = 23 \quad \text{Add 25 to both sides.}$$

Work Practice 5

Helpful Hint

When you multiply both sides of the equation from Example 5, $-x + 5y = 23$, by -1, it becomes $x - 5y = -23$.

Both $-x + 5y = 23$ and $x - 5y = -23$ are in the form $Ax + By = C$ and both are equations of the same line.

Objective E Finding Equations of Vertical and Horizontal Lines

Recall from Section 6.3 that:

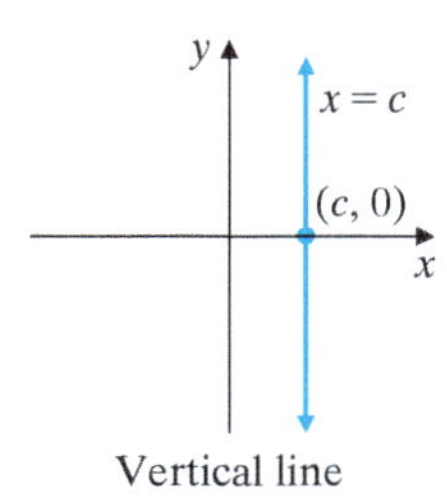

Vertical line

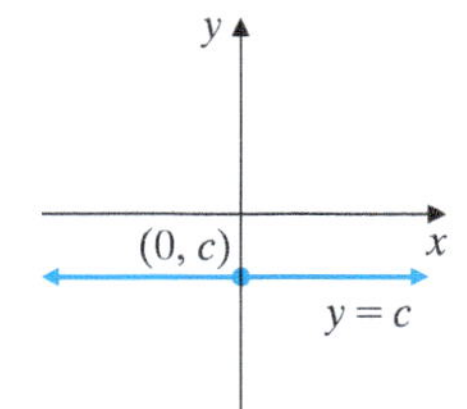

Horizontal line

Example 6 Find an equation of the vertical line through $(-1, 5)$.

Solution: The equation of a vertical line can be written in the form $x = c$, so an equation for the vertical line passing through $(-1, 5)$ is $x = -1$.

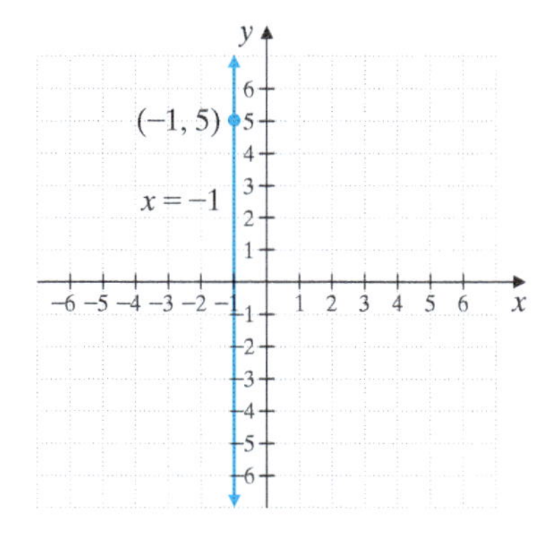

Work Practice 6

Practice 6

Find an equation of the vertical line through $(3, -2)$.

Example 7 Find an equation of the line parallel to the line $y = 5$ and passing through $(-2, -3)$.

Solution: Since the graph of $y = 5$ is a horizontal line, any line parallel to it is also horizontal. The equation of a horizontal line can be written in the form $y = c$. An equation for the horizontal line passing through $(-2, -3)$ is $y = -3$.

(Continued on next page)

Practice 7

Find an equation of the line parallel to the line $y = -2$ and passing through $(4, 3)$.

Answers

6. $x = 3$ 7. $y = 3$

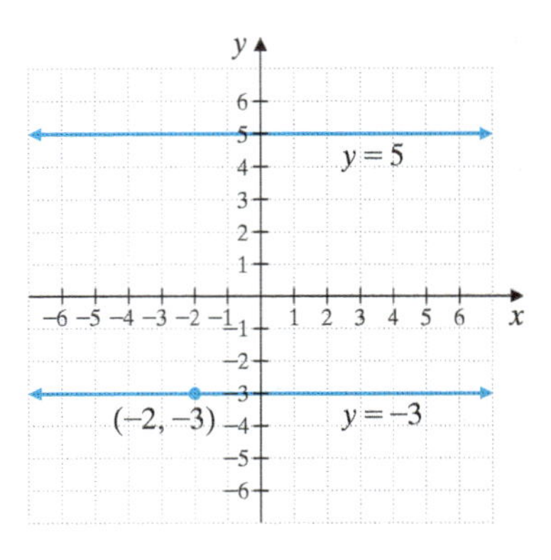

Work Practice 7

Objective F Using the Point-Slope Form to Solve Problems

Problems occurring in many fields can be modeled by linear equations in two variables. The next example is from the field of marketing and shows how consumer demand for a product depends on the price of the product.

Practice 8

The Pool Entertainment Company learned that by pricing a new pool toy at \$10, local sales will reach 200 a week. Lowering the price to \$9 will cause sales to rise to 250 a week.

a. Assume that the relationship between sales price and number of toys sold is linear, and write an equation describing this relationship. Write the equation in slope-intercept form. Use ordered pairs of the form (sales price, number sold).

b. Predict the weekly sales of the toy if the price is \$7.50.

Example 8 The Whammo Company has learned that by pricing a newly released Frisbee at \$6, sales will reach 2000 Frisbees per day. Raising the price to \$8 will cause the sales to fall to 1500 Frisbees per day.

a. Assume that the relationship between sales price and number of Frisbees sold is linear, and write an equation describing this relationship. Write the equation in slope-intercept form. Use ordered pairs of the form (sales price, number sold).

b. Predict the daily sales of Frisbees if the price is \$7.50.

Solution:

a. We use the given information and write two ordered pairs. Our ordered pairs are (6, 2000) and (8, 1500). To use the point-slope form to write an equation, we find the slope of the line that contains these points.

$$m = \frac{2000 - 1500}{6 - 8} = \frac{500}{-2} = -250$$

Next we use the slope and either one of the points to write the equation in point-slope form. We use (6, 2000).

$$y - y_1 = m(x - x_1) \quad \text{Use point-slope form.}$$
$$y - 2000 = -250(x - 6) \quad \text{Let } x_1 = 6,\ y_1 = 2000, \text{ and } m = -250.$$
$$y - 2000 = -250x + 1500 \quad \text{Use the distributive property.}$$
$$y = -250x + 3500 \quad \text{Write in slope-intercept form.}$$

Answer

8. a. $y = -50x + 700$ **b.** 325

b. To predict the sales if the price is \$7.50, we find y when $x = 7.50$.

$$y = -250x + 3500$$
$$y = -250(7.50) + 3500 \quad \text{Let } x = 7.50.$$
$$y = -1875 + 3500$$
$$y = 1625$$

If the price is \$7.50, sales will reach 1625 Frisbees per day.

Work Practice 8

We also could have solved Example 8 by using ordered pairs of the form (number sold, sales price).

Here is a summary of our discussion on linear equations thus far.

Forms of Linear Equations

$Ax + By = C$	**Standard form** of a linear equation. A and B are not both 0.
$y = mx + b$	**Slope-intercept form** of a linear equation. The slope is m and the y-intercept is $(0, b)$.
$y - y_1 = m(x - x_1)$	**Point-slope form** of a linear equation. The slope is m and (x_1, y_1) is a point on the line.
$y = c$	**Horizontal line** The slope is 0 and the y-intercept is $(0, c)$.
$x = c$	**Vertical line** The slope is undefined and the x-intercept is $(c, 0)$.

Parallel and Perpendicular Lines

Nonvertical parallel lines have the same slope.
The product of the slopes of two nonvertical perpendicular lines is -1.

Calculator Explorations Graphing

A graphing calculator is a very useful tool for discovering patterns. To discover the change in the graph of a linear equation caused by a change in slope, try the following: Use a standard window and graph a linear equation in the form $y = mx + b$. Recall that the graph of such an equation will have slope m and y-intercept $(0, b)$.

First graph $y = x + 3$. To do so, press the $\boxed{Y =}$ key and enter $Y_1 = x + 3$. Notice that this graph has slope 1 and that the y-intercept is 3. Next, on the same set of axes, graph $y = 2x + 3$ and $y = 3x + 3$ by pressing $\boxed{Y =}$ and entering $Y_2 = 2x + 3$ and $Y_3 = 3x + 3$.

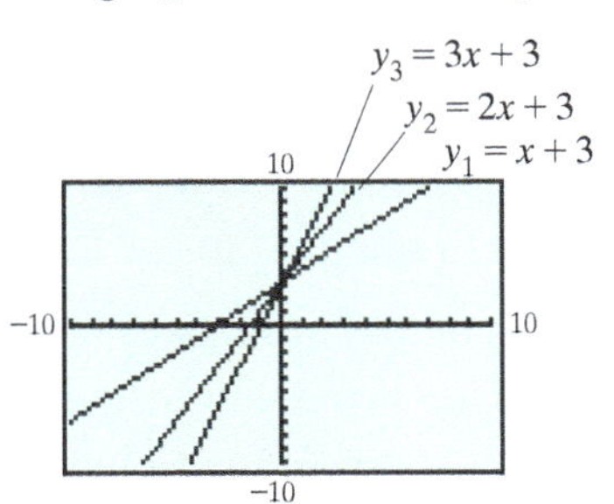

Notice the difference in the graph of each equation as the slope changes from 1 to 2 to 3. How would the graph of $y = 5x + 3$ appear? To see the change in the graph caused by a change to negative slope, try graphing $y = -x + 3$, $y = -2x + 3$, and $y = -3x + 3$ on the same set of axes.

Use a graphing calculator to graph the following equations. For each exercise, graph the first equation and use its graph to predict the appearance of the other equations. Then graph the other equations on the same set of axes and check your prediction.

1. $y = x$; $y = 6x$, $y = -6x$

2. $y = -x$; $y = -5x$, $y = -10x$

(Continued on next page)

3. $y = \frac{1}{2}x + 2; y = \frac{3}{4}x + 2, y = x + 2$

4. $y = x + 1; y = \frac{5}{4}x + 1, y = \frac{5}{2}x + 1$

Vocabulary, Readiness & Video Check

Use the choices below to fill in each blank. Some choices may be used more than once and some not at all.

b	(y_1, x_1)	point-slope	vertical	standard
m	(x_1, y_1)	slope-intercept	horizontal	

1. The form $y = mx + b$ is called ______________ form. When a linear equation in two variables is written in this form, ______________ is the slope of its graph and (0, ______________) is its y-intercept.

2. The form $y - y_1 = m(x - x_1)$ is called ______________ form. When a linear equation in two variables is written in this form, ______________ is the slope of its graph and ______________ is a point on the graph.

For Exercises 3 through 6, identify the form that the linear equation in two variables is written in. For Exercises 7 and 8, identify the appearance of the graph of the equation.

3. $y - 7 = 4(x + 3)$; ______________ form

4. $5x - 9y = 11$; ______________ form

5. $y = \frac{3}{4}x - \frac{1}{3}$; ______________ form

6. $y + 2 = \frac{-1}{3}(x - 2)$ ______________ form

7. $y = \frac{1}{2}$; ______________ line

8. $x = -17$; ______________ line

Martin-Gay Interactive Videos

See Video 6.5

Watch the section lecture video and answer the following questions.

Objective A **9.** We can use the slope-intercept form to graph a line. Complete these statements based on Example 1: Start by graphing the ______________. From this point, find another point by applying the slope–if necessary, rewrite the slope as a(n) ______________

Objective B **10.** In Example 3, what is the y-intercept?

Objective C **11.** In Example 4, we use the point-slope form to find an equation of a line given the slope and a point. How do we then write this equation in standard form?

Objective D **12.** The lecture before Example 5 discusses how to find the equation of a line given two points. Is there any circumstance when we might want to use the slope–intercept form to find the equation of a line given two points? If so, when?

Objective E 13. Solve Examples 6 and 7 again, this time using the point $(-2, -3)$ in each exercise.

Objective F 14. In Example 8, we are told to use ordered pairs of the form (time, speed). Explain why it is important to keep track of how we define our ordered pairs and/or our variables.

6.5 Exercise Set MyLab Math 

Objective A *Use the slope-intercept form to graph each equation. See Examples 1 and 2.*

1. $y = 2x + 1$

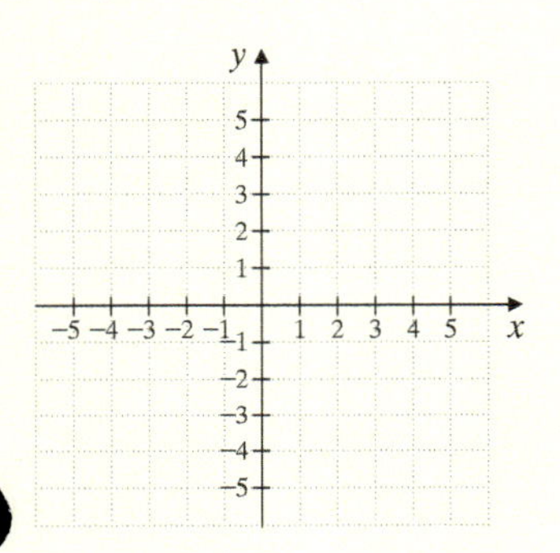

2. $y = -4x - 1$

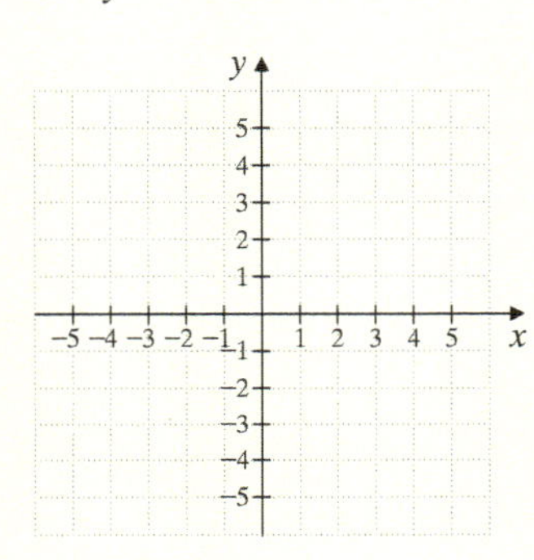

3. $y = \frac{2}{3}x + 5$

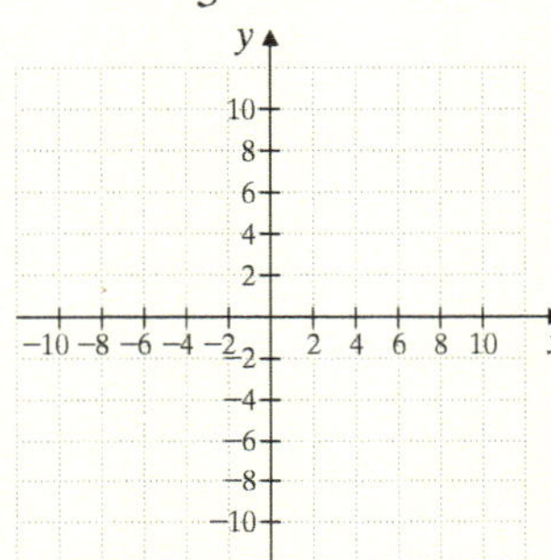

4. $y = \frac{1}{4}x - 3$

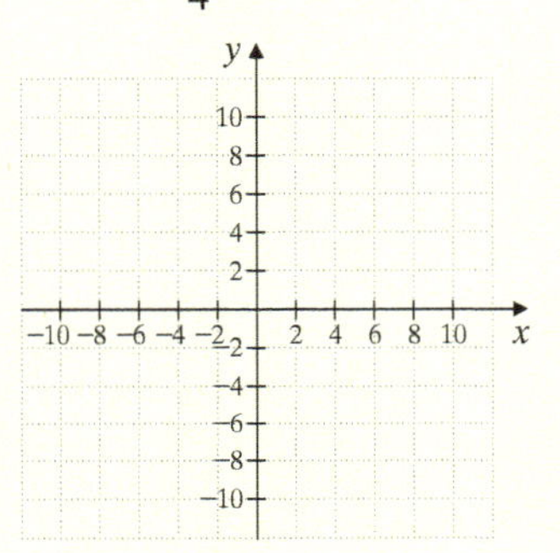

5. $y = -5x$

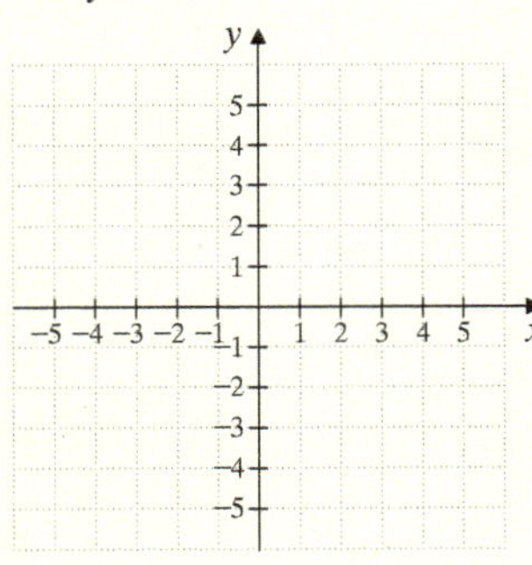

6. $y = -6x$

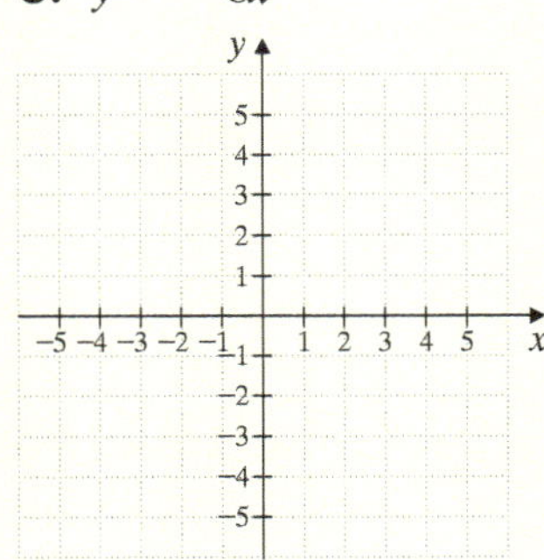

7. $4x + y = 6$

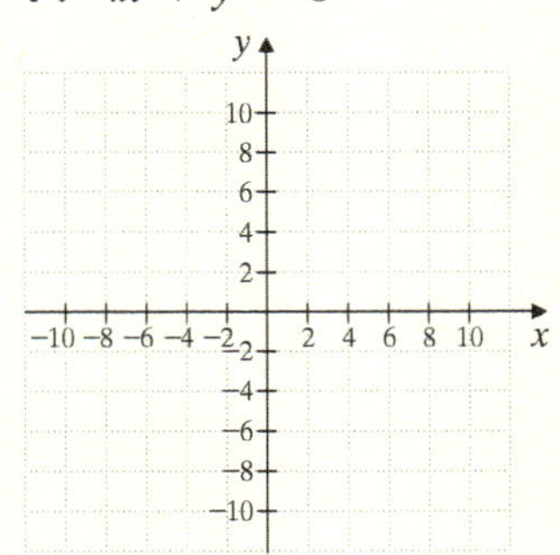

8. $-3x + y = 2$

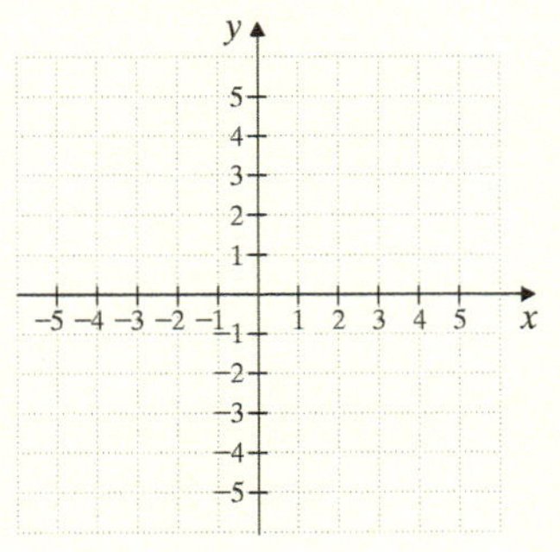

9. $4x - 7y = -14$

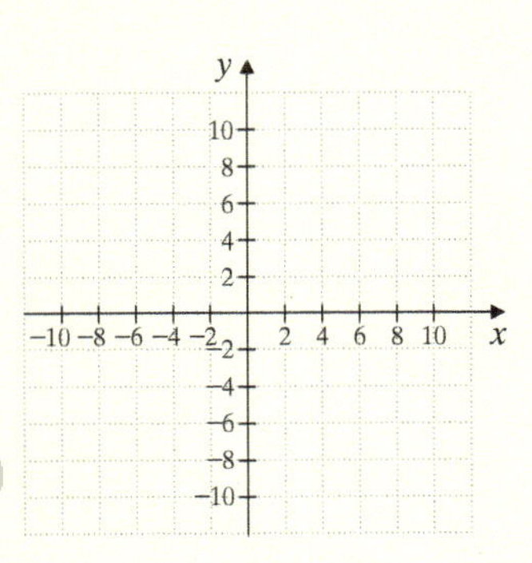

10. $3x - 4y = 4$

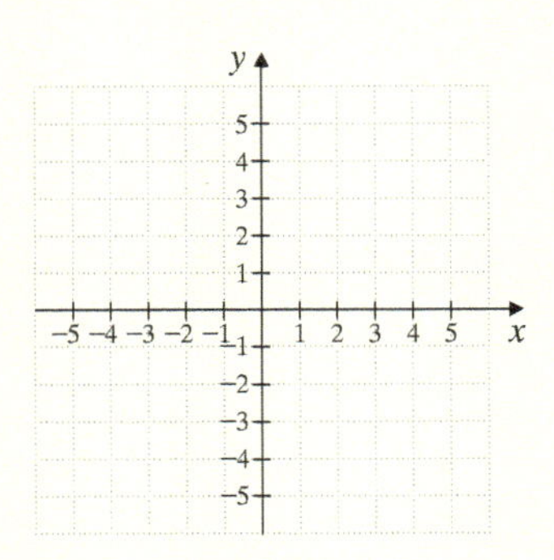

11. $x = \frac{5}{4}y$

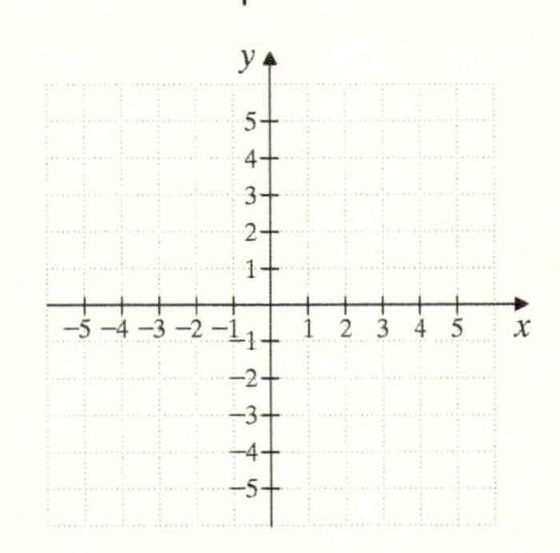

12. $x = \frac{3}{2}y$

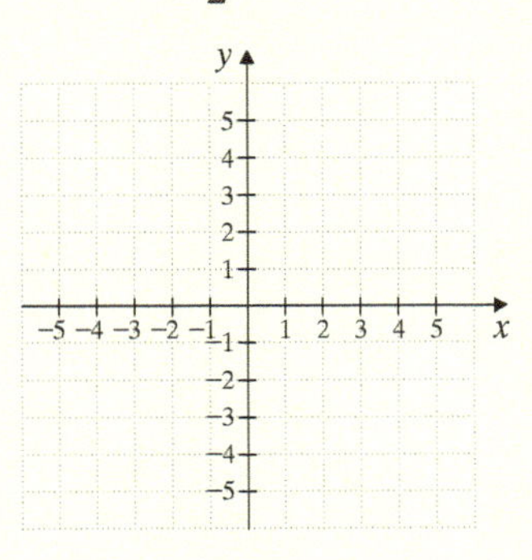

Objective B *Write an equation of the line with each given slope, m, and y-intercept, (0, b). See Example 3.*

13. $m = 5, b = 3$

14. $m = -3, b = -3$

15. $m = -4, b = -\frac{1}{6}$

16. $m = 2, b = \frac{3}{4}$

17. $m = \frac{2}{3}, b = 0$

18. $m = -\frac{4}{5}, b = 0$

19. $m = 0, b = -8$

20. $m = 0, b = -2$

21. $m = -\frac{1}{5}, b = \frac{1}{9}$

22. $m = \frac{1}{2}, b = -\frac{1}{3}$

Objective C *Find an equation of each line with the given slope that passes through the given point. Write the equation in the form $Ax + By = C$. See Example 4.*

23. $m = 6$; $(2, 2)$

24. $m = 4$; $(1, 3)$

25. $m = -8$; $(-1, -5)$

26. $m = -2$; $(-11, -12)$

27. $m = \frac{3}{2}$; $(5, -6)$

28. $m = \frac{2}{3}$; $(-8, 9)$

29. $m = -\frac{1}{2}$; $(-3, 0)$

30. $m = -\frac{1}{5}$; $(4, 0)$

Objective D *Find an equation of the line passing through each pair of points. Write the equation in the form $Ax + By = C$. See Example 5.*

31. $(3, 2)$ and $(5, 6)$

32. $(6, 2)$ and $(8, 8)$

33. $(-1, 3)$ and $(-2, -5)$

34. $(-4, 0)$ and $(6, -1)$

35. $(2, 3)$ and $(-1, -1)$

36. $(7, 10)$ and $(-1, -1)$

37. $(0, 0)$ and $\left(-\frac{1}{8}, \frac{1}{13}\right)$

38. $(0, 0)$ and $\left(-\frac{1}{2}, \frac{1}{3}\right)$

Objective E *Find an equation of each line. See Example 6.*

39. Vertical line through $(0, 2)$

40. Horizontal line through $(1, 4)$

41. Horizontal line through $(-1, 3)$

42. Vertical line through $(-1, 3)$

43. Vertical line through $\left(-\frac{7}{3}, -\frac{2}{5}\right)$

44. Horizontal line through $\left(\frac{2}{7}, 0\right)$

Find an equation of each line. See Example 7.

△ **45.** Parallel to $y = 5$, through $(1, 2)$

△ **46.** Perpendicular to $y = 5$, through $(1, 2)$

△ **47.** Perpendicular to $x = -3$, through $(-2, 5)$

△ **48.** Parallel to $y = -4$, through $(0, -3)$

△ **49.** Parallel to $x = 0$, through $(6, -8)$

△ **50.** Perpendicular to $x = 7$, through $(-5, 0)$

Objectives A C D E Mixed Practice *See Examples 1 and 4 through 7. Find an equation of each line described. Write each equation in slope-intercept form when possible.*

51. With slope $-\frac{1}{2}$, through $\left(0, \frac{5}{3}\right)$

52. With slope $\frac{5}{7}$, through $(0, -3)$

53. Through $(10, 7)$ and $(7, 10)$

54. Through $(5, -6)$ and $(-6, 5)$

55. With undefined slope, through $\left(-\frac{3}{4}, 1\right)$

56. With slope 0, through $(6.7, 12.1)$

57. Slope 1, through $(-7, 9)$

58. Slope 5, through $(6, -8)$

59. Slope -5, y-intercept $(0, 7)$

60. Slope -2, y-intercept $(0, -4)$

61. Through $(-8, 11)$, parallel to $x = 5$

62. Through $(9, -12)$, parallel to the y-axis

63. Through $(2, 3)$ and $(0, 0)$

64. Through $(4, 7)$ and $(0, 0)$

65. Through $(-2, -3)$, perpendicular to the y-axis

66. Through $(0, 12)$, perpendicular to the x-axis

67. Slope $-\frac{4}{7}$, through $(-1, -2)$

68. Slope $-\frac{3}{5}$, through $(4, 4)$

Objective F *Solve. Assume each exercise describes a linear relationship. Write the equations in slope-intercept form. See Example 8.*

69. In 2010, a total of about 7162 different magazines were in print in the United States. By 2016, this number was 7216. (*Source:* MPA—The Association of Magazine Media)

a. Write two ordered pairs of the form (years after 2010, number of magazines in print) for this situation.

b. Assume the relationship between years after 2010 and number of magazines in print is linear over this period. Use the ordered pairs from part **a** to write an equation for the line relating year after 2010 to number of magazines in print.

c. Use the linear equation in part **b** to estimate the number of magazines in print in 2015.

70. The number of print consumer magazines published in the United States is starting to decline in the magazine market. In 2014, there were approximately 7289 print magazines. By 2017, this had decreased to about 7175 magazines. (*Source:* MPA—The Association of Magazine Media)

a. Write two ordered pairs of the form (years after 2014, number of print magazines).

b. Assume the relationship between years after 2014 and number of print magazines is linear over this period. Use the ordered pairs from part **a** to write an equation for the line relating years after 2014 to number of print magazines.

c. Use the linear equation from part **b** to estimate the number of print magazines in 2025 if this trend were to continue.

71. A rock is dropped from the top of a 400-foot cliff. After 1 second, the rock is traveling 32 feet per second. After 3 seconds, the rock is traveling 96 feet per second.

a. Assume that the relationship between time and speed is linear and write an equation describing this relationship. Use ordered pairs of the form (time, speed).

b. Use this equation to determine the speed of the rock 4 seconds after it is dropped.

72. A Hawaiian fruit company is studying the sales of a pineapple sauce to see if this product is to be continued. At the end of its first year, profits on this product amounted to \$30,000. At the end of the fourth year, profits were \$66,000.

a. Assume that the relationship between years on the market and profit is linear and write an equation describing this relationship. Use ordered pairs of the form (years on the market, profit).

b. Use this equation to predict the profit at the end of 7 years.

73. In 2012 there were approximately 434,000 hybrid vehicles sold in the United States. In 2017, due to an overall decline in the market, there were approximately 368,000 such vehicles sold. (*Source:* Oak Ridge National Laboratory)

a. Write an equation describing the relationship between time and the number of hybrid vehicles sold. Use ordered pairs of the form (years past 2012, number of vehicles sold).

b. Use this equation to estimate the number of hybrid sales in 2016.

74. In 2012, there were approximately 980 thousand restaurants in the United States. In 2016, there were 1046 thousand restaurants. (*Source:* National Restaurant Association)

a. Write an equation describing the relationship between time and the number of restaurants. Use ordered pairs of the form (years past 2012, number of restaurants in thousands).

b. Use this equation to predict the number of eating establishments in 2018.

75. In 2012 there were 5320 indoor cinema sites in the United States. In 2018, there were 5482 indoor cinema sites. (*Source:* National Association of Theater Owners)

a. Write an equation describing this relationship. Use ordered pairs of the form (years past 2012, number of indoor cinema sites).

b. Use this equation to predict the number of indoor cinema sites in 2023.

76. In 2017, the U.S. population per square mile of land area was approximately 92.23. In 2010, this person-per-square-mile population was 87.4. (*Source:* World Bank)

a. Write an equation describing the relationship between year and persons per square mile. Use ordered pairs of the form (years past 2010, persons per square mile).

b. Use this equation to predict the person-per-square-mile population in 2024.

77. The Pool Fun Company has learned that, by pricing a newly released Fun Noodle at \$3, sales will reach 10,000 Fun Noodles per day during the summer. Raising the price to \$5 will cause the sales to fall to 8000 Fun Noodles per day.

a. Assume that the relationship between sales price and number of Fun Noodles sold is linear and write an equation describing this relationship. Use ordered pairs of the form (sales price, number sold).

b. Predict the daily sales of Fun Noodles if the price is \$3.50.

78. The value of a building bought in 2005 may be depreciated (or decreased) as time passes for income tax purposes. Seven years after the building was bought, this value was \$225,000 and 12 years after it was bought, this value was \$195,000.

a. If the relationship between number of years past 2005 and the depreciated value of the building is linear, write an equation describing this relationship. Use ordered pairs of the form (years past 2005, value of building).

b. Use this equation to estimate the depreciated value of the building in 2023.

Review

Find the value of $x^2 - 3x + 1$ *for each given value of x. See Section 1.3.*

79. 2 **80.** 5 **81.** -1 **82.** -3

Concept Extensions

Match each linear equation with its graph.

83. $y = 2x + 1$ **84.** $y = -x + 1$ **85.** $y = -3x - 2$ **86.** $y = \frac{5}{3}x - 2$

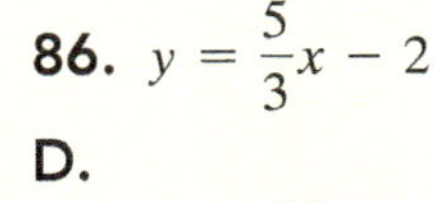

A.

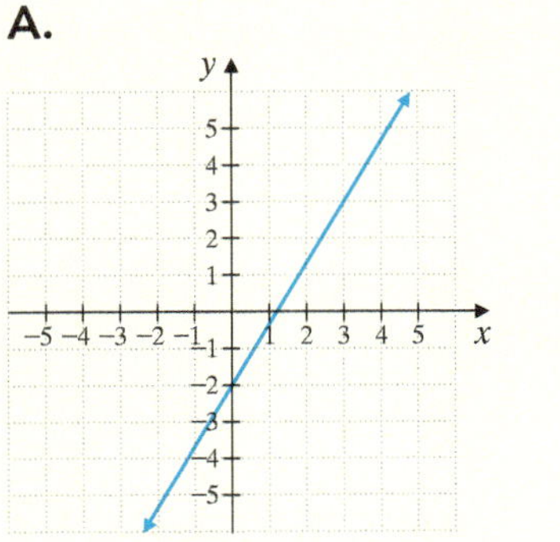

B.

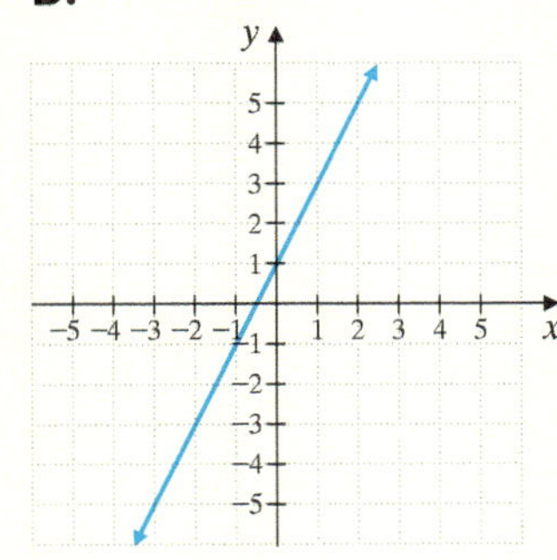

C.

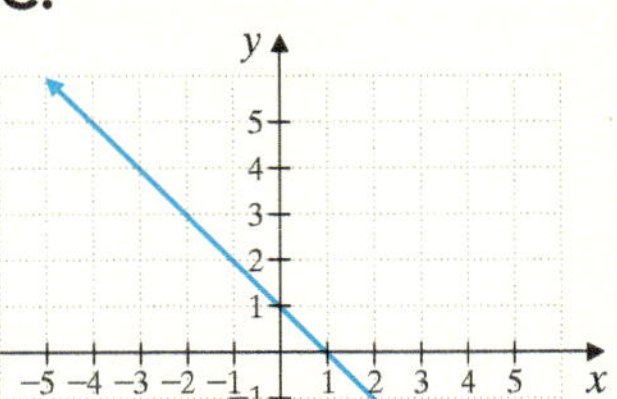

D.

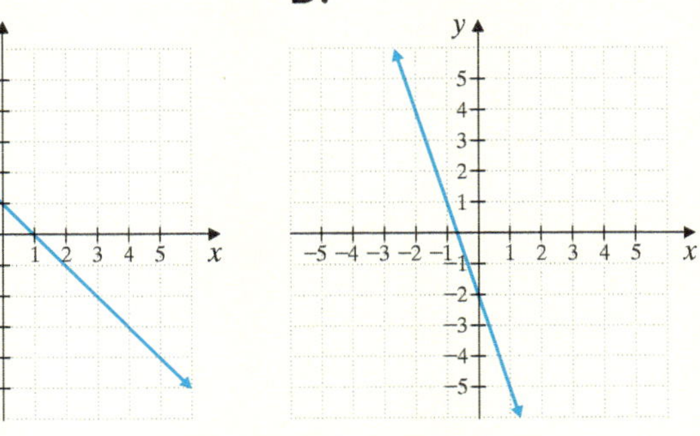

87. Write an equation in standard form of the line that contains the point $(-2, 4)$ and has the same slope as the line $y = 2x + 5$.

88. Write an equation in standard form of the line that contains the point $(3, 0)$ and has the same slope as the line $y = -3x - 1$.

89. Write an equation in standard form of the line that contains the point $(-1, 2)$ and is

a. parallel to the line $y = 3x - 1$.

b. perpendicular to the line $y = 3x - 1$.

△ **90.** Write an equation in standard form of the line that contains the point $(4, 0)$ and is

a. parallel to the line $y = -2x + 3$.

b. perpendicular to the line $y = -2x + 3$.

Answers

1. ______

2. ______

3. ______

4. ______

5. ______

6. ______

7. ______

8. ______

Summary on Linear Equations

Find the slope of each line.

1.

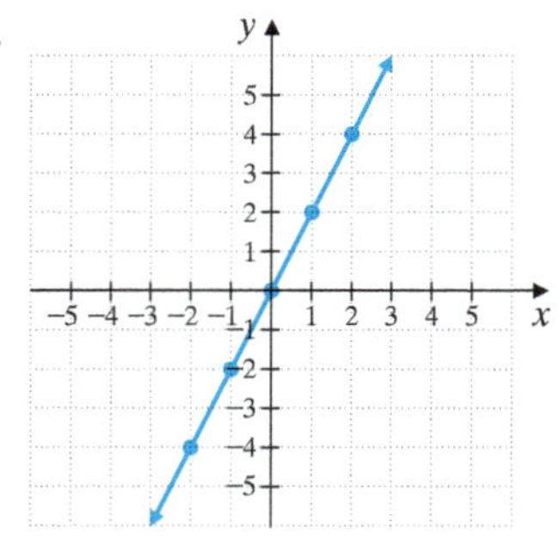

2.

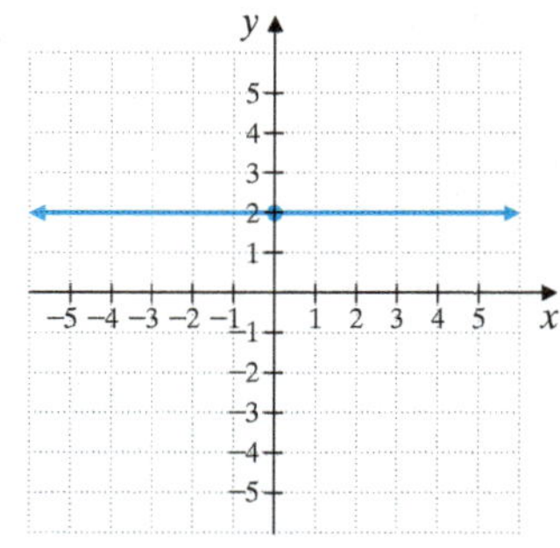

3.

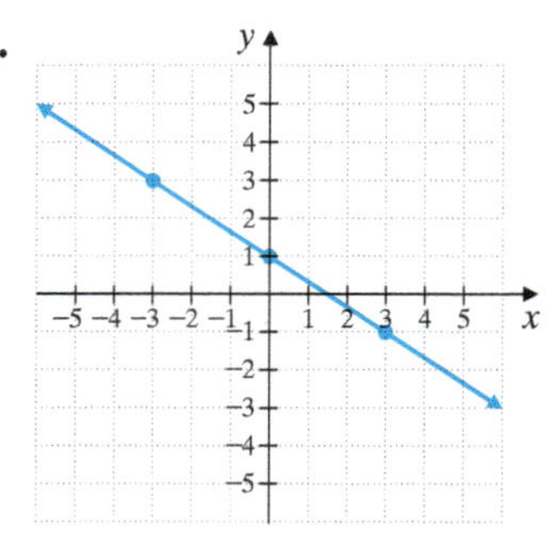

4.

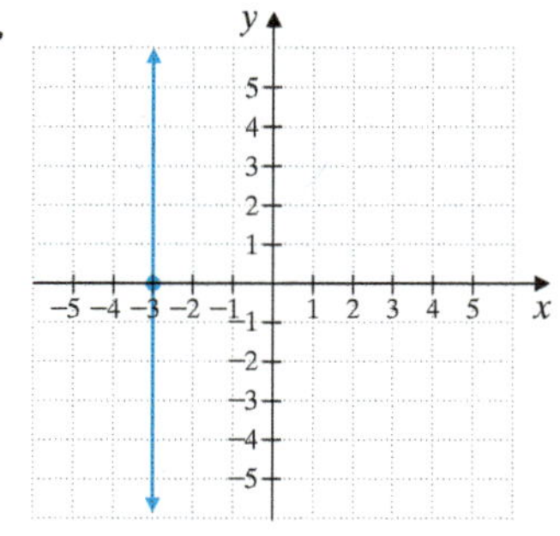

Graph each linear equation. For Exercises 11 and 12, label the intercepts.

5. $y = -2x$

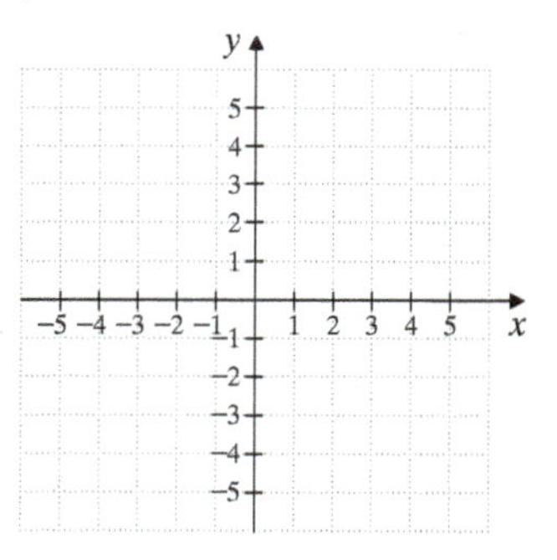

6. $x + y = 3$

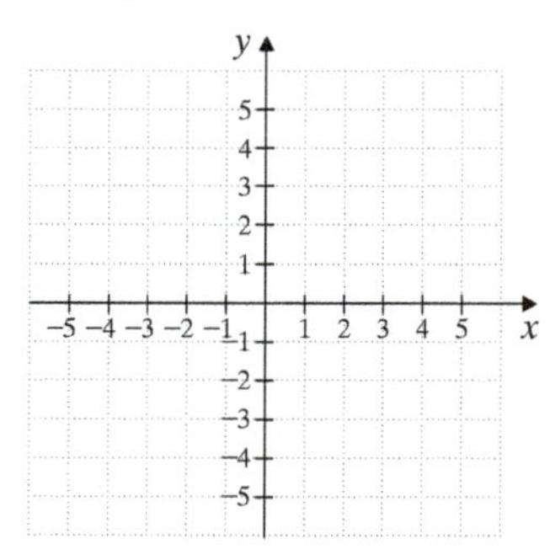

7. $x = -1$

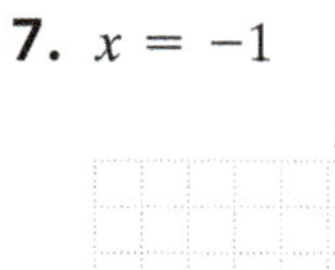

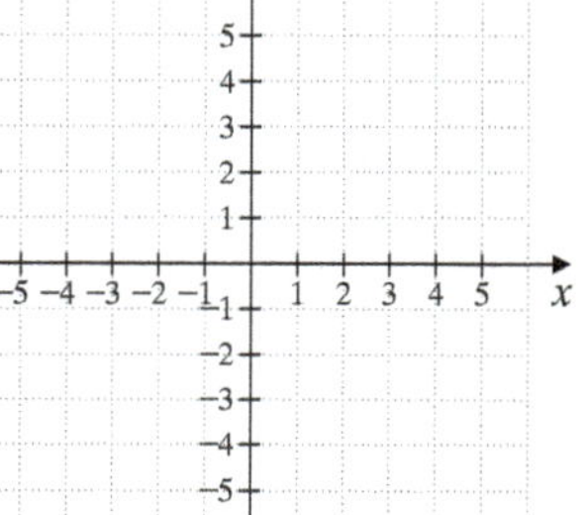

8. $y = 4$

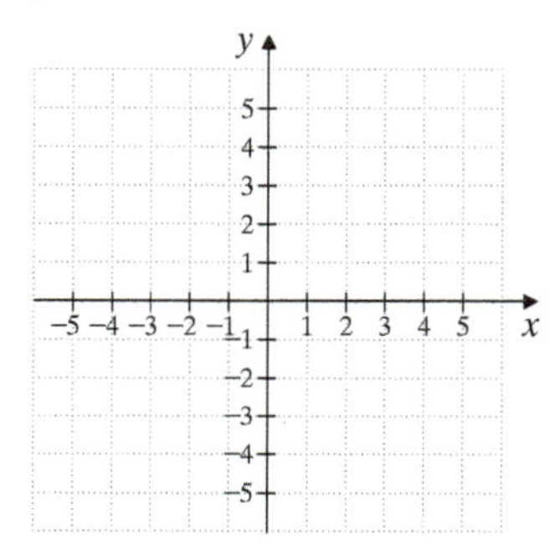

9. $x - 2y = 6$

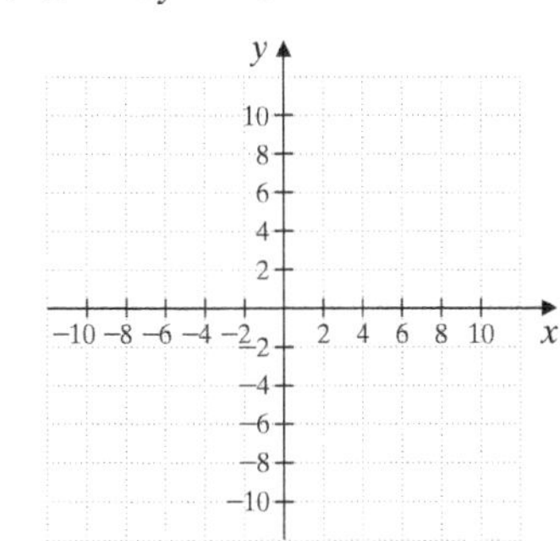

10. $y = 3x + 2$

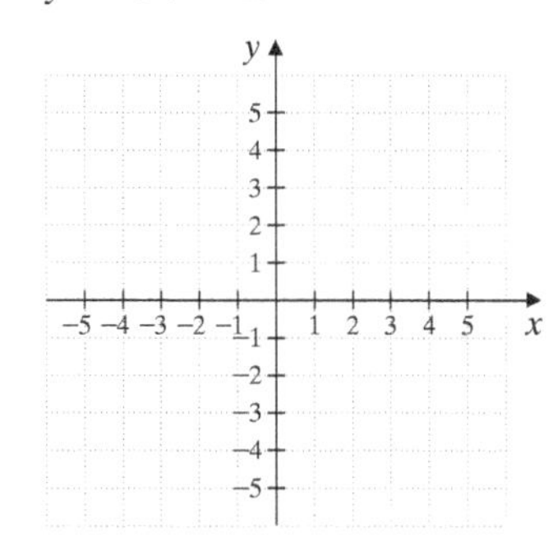

11. $y = -\frac{3}{4}x + 3$

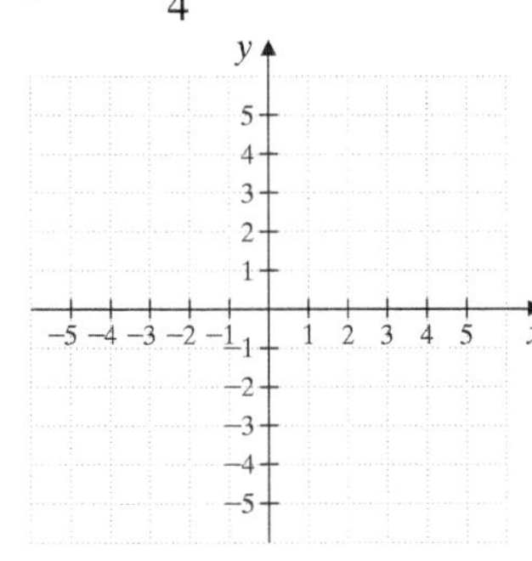

12. $5x - 2y = 8$

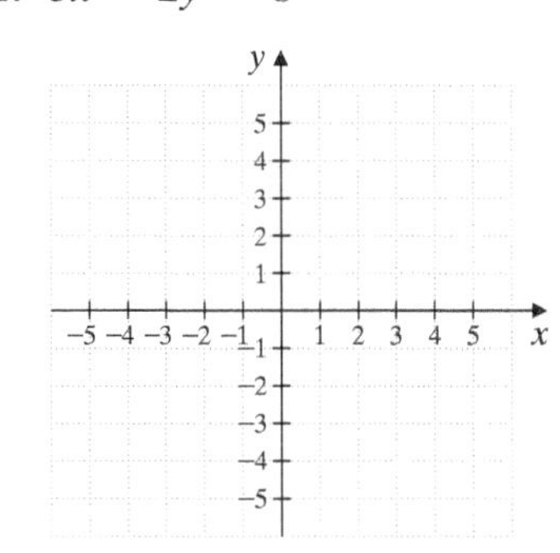

Find the slope of each line by writing the equation in slope-intercept form.

13. $y = 3x - 1$ **14.** $y = -6x + 2$ **15.** $7x + 2y = 11$ **16.** $2x - y = 0$

Find the slope of each line.

17. $x = 2$

18. $y = -4$

19. Write an equation of the line with slope $m = 2$ and y-intercept $\left(0, -\frac{1}{3}\right)$. Write the equation in the form $y = mx + b$.

20. Find an equation of the line with slope $m = -4$ that passes through the point $(-1, 3)$. Write the equation in the form $y = mx + b$.

21. Find an equation of the line that passes through the points $(2, 0)$ and $(-1, -3)$. Write the equation in the form $Ax + By = C$.

Determine whether each pair of lines is parallel, perpendicular, or neither.

22. $6x - y = 7$
$2x + 3y = 4$

23. $3x - 6y = 4$
$y = -2x$

24. Yogurt is an ever more popular food item. In 2012, U.S. production of yogurt stood at approximately 4418 million pounds. In 2017, this number rose to 4478 million pounds of yogurt. (*Source:* United States Department of Agriculture)

a. Write two ordered pairs of the form (year, millions of pounds of yogurt produced).

b. Find the slope of the line between these two points.

c. Write a sentence explaining the meaning of the slope as a rate of change.

9. _____
10. _____
11. _____
12. _____
13. _____
14. _____
15. _____
16. _____
17. _____
18. _____
19. _____
20. _____
21. _____
22. _____
23. _____
24. a. _____
b. _____
c. _____

6.6 Introduction to Functions

Objectives

A Identify Relations, Domains, and Ranges.

B Identify Functions.

C Use the Vertical Line Test.

D Use Function Notation.

Objective A Identifying Relations, Domains, and Ranges

In previous sections, we have discussed the relationships between two quantities. For example, the relationship between the length of the side of a square x and its area y is described by the equation $y = x^2$. Ordered pairs can be used to write down solutions of this equation. For example, (2, 4) is a solution of $y = x^2$, and this notation tells us that the x-value 2 is related to the y-value 4 for this equation. In other words, when the length of the side of a square is 2 units, its area is 4 square units.

Examples of Relationships Between Two Quantities

Area of Square: $y = x^2$

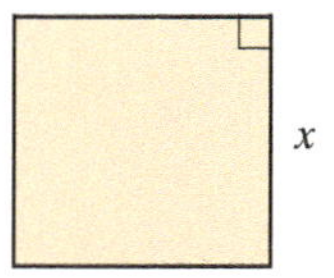

Some Ordered Pairs

x	y
2	4
5	25
7	49
12	144

Equation of Line: $y = x + 2$

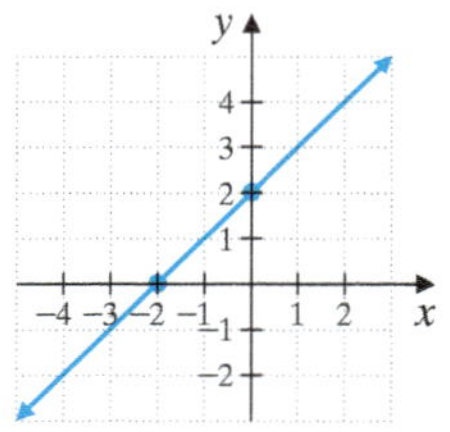

Some Ordered Pairs

x	y
−3	−1
0	2
2	4
9	11

Online Advertising Spending

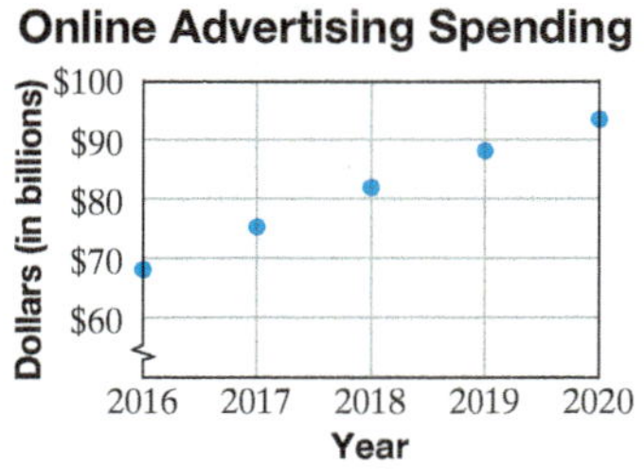

Ordered Pairs

Year	*Billions of Dollars*
2016	68.1
2017	75.3
2018	81.9
2019	88.1
2020	93.5

A set of ordered pairs is called a **relation.** The set of all x-coordinates is called the **domain** of a relation, and the set of all y-coordinates is called the **range** of a relation. Equations such as $y = x^2$ are also called relations since equations in two variables define a set of ordered pair solutions.

Practice 1

Find the domain and range of the relation $\{(-3, 5), (-3, 1), (4, 6), (7, 0)\}$.

Example 1 Find the domain and the range of the relation $\{(0, 2), (3, 3), (-1, 0), (3, -2)\}$.

Solution: The domain is the set of all x-coordinates, or $\{-1, 0, 3\}$, and the range is the set of all y-coordinates, or $\{-2, 0, 2, 3\}$.

Work Practice 1

Objective B Identifying Functions

Paired data occur often in real-life applications. Some special sets of paired data, or ordered pairs, are called *functions*.

Answer

1. domain: $\{-3, 4, 7\}$; range: $\{0, 1, 5, 6\}$

Function

A **function** is a set of ordered pairs in which each x-coordinate has exactly one y-coordinate.

In other words, a function cannot have two ordered pairs with the same x-coordinate but different y-coordinates.

Example 2 Determine whether each relation is also a function.

a. $\{(-1, 1), (2, 3), (7, 3), (8, 6)\}$

b. $\{(0, -2), (1, 5), (0, 3), (7, 7)\}$

Solution:

a. Although the ordered pairs $(2, 3)$ and $(7, 3)$ have the same y-value, each x-value is assigned to only one y-value, so this set of ordered pairs is a function.

b. The x-value 0 is paired with two y-values, -2 and 3, so this set of ordered pairs is not a function.

Work Practice 2

Practice 2

Are the following relations also functions?

a. $\{(2, 5), (-3, 7), (4, 5), (0, -1)\}$

b. $\{(1, 4), (6, 6), (1, -3), (7, 5)\}$

Relations and functions can be described by graphs of their ordered pairs.

Example 3 Which graph is the graph of a function?

a.

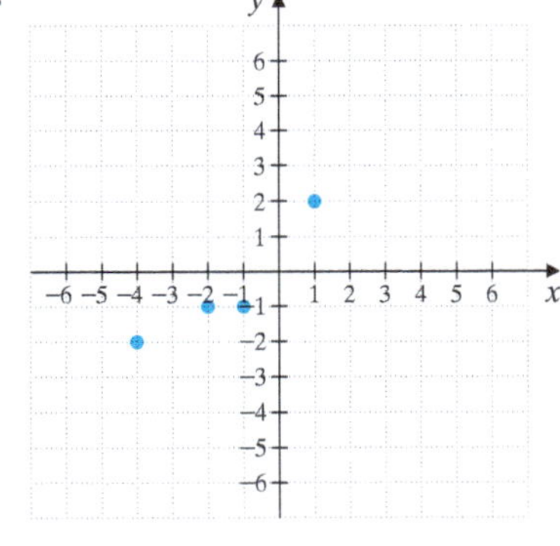

b.

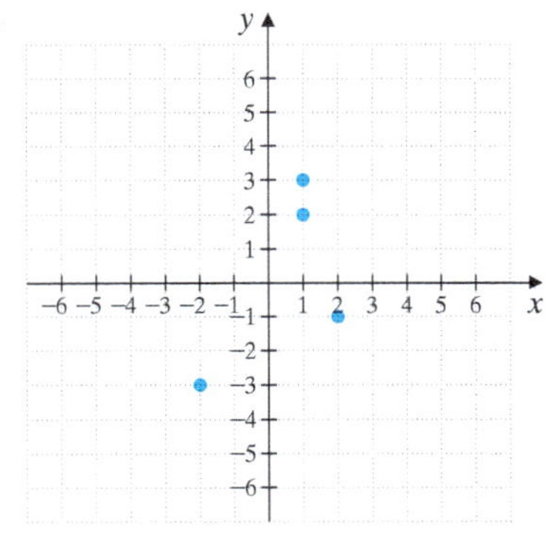

Solution:

a. This is the graph of the relation $\{(-4, -2), (-2, -1), (-1, -1), (1, 2)\}$. Each x-coordinate has exactly one y-coordinate, so this is the graph of a function.

b. This is the graph of the relation $\{(-2, -3), (1, 2), (1, 3), (2, -1)\}$. The x-coordinate 1 is paired with two y-coordinates, 2 and 3, so this is not the graph of a function.

Work Practice 3

Practice 3

Is each graph the graph of a function?

a.

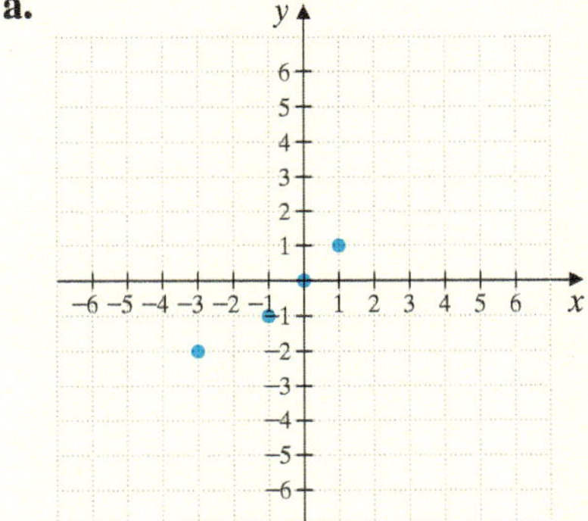

b.

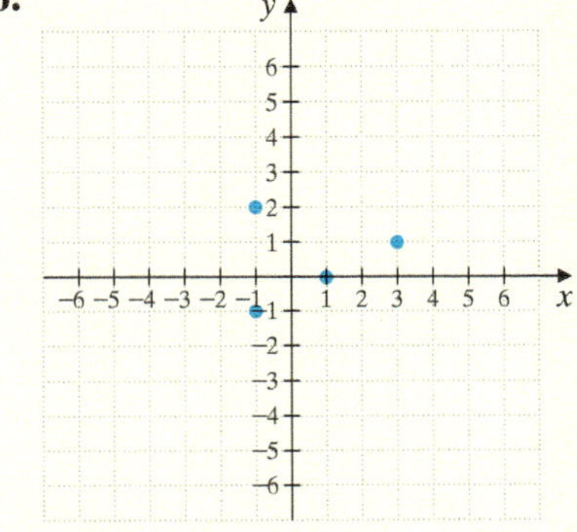

Objective C Using the Vertical Line Test

The graph in Example 3(b) was not the graph of a function because the x-coordinate 1 was paired with two y-coordinates, 2 and 3. Notice that when an x-coordinate is paired with more than one y-coordinate, a vertical line can be drawn that

Answers

2. a. a function **b.** not a function

3. a. a function **b.** not a function

will intersect the graph at more than one point. We can use this fact to determine whether a relation is also a function. We call this the vertical line test.

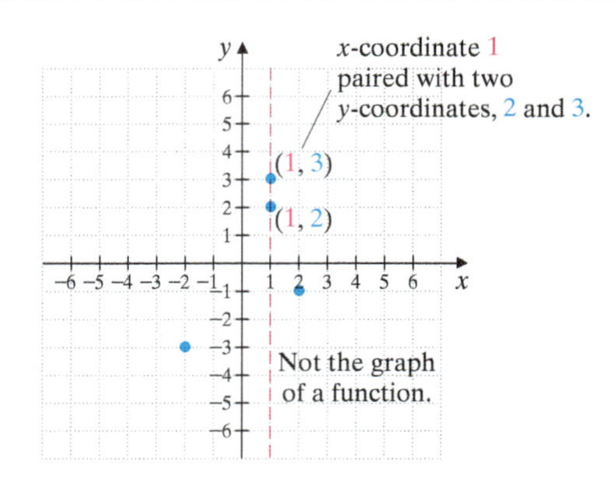

Practice 4

Determine whether each graph is the graph of a function.

a.

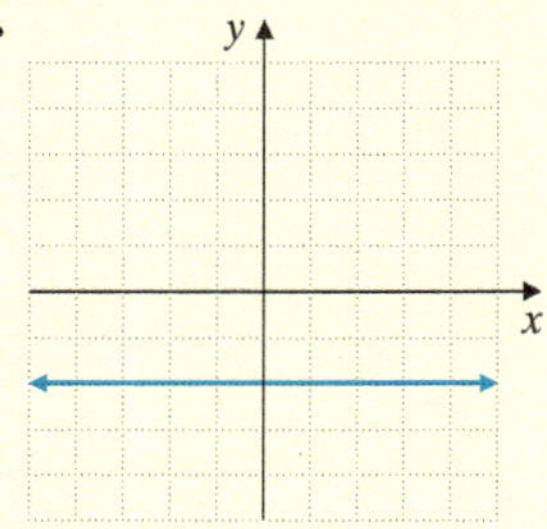

b.

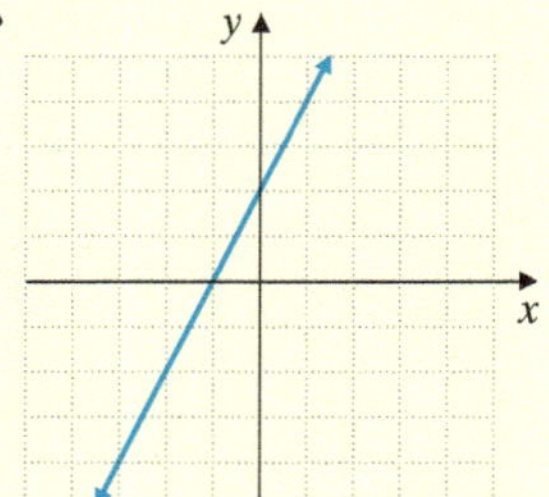

c.

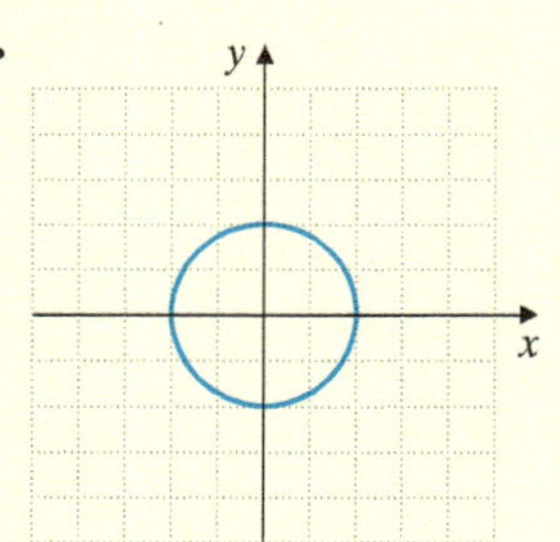

d.

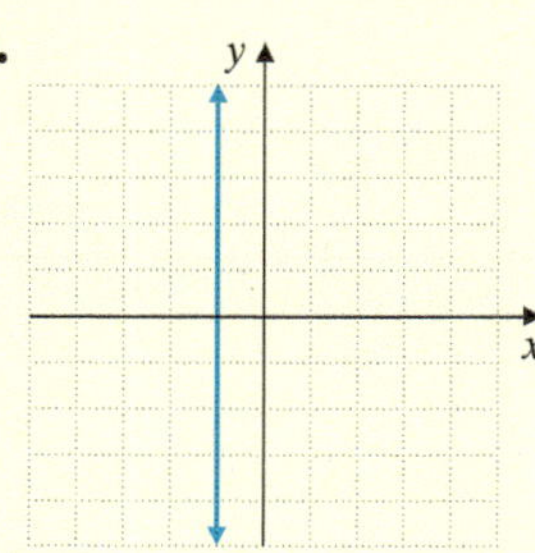

Vertical Line Test

If a vertical line can be drawn so that it intersects a graph more than once, the graph is not the graph of a function. (If no such vertical line can be drawn, the graph is that of a function.)

This vertical line test works for all types of graphs on the rectangular coordinate system.

Example 4

Use the vertical line test to determine whether each graph is the graph of a function.

a. b. c. d.

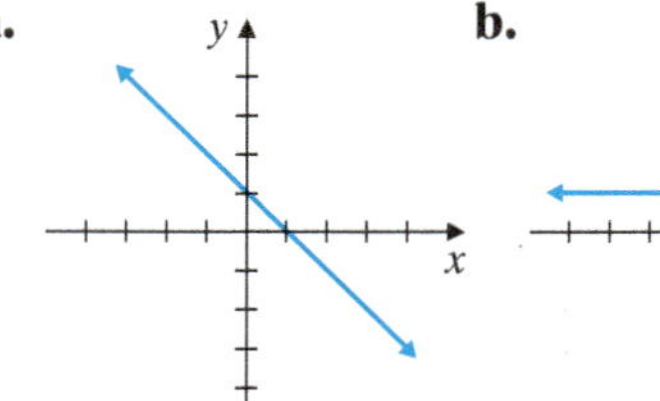

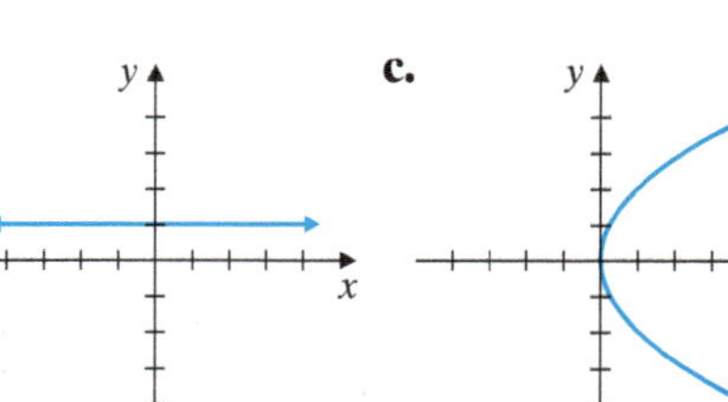

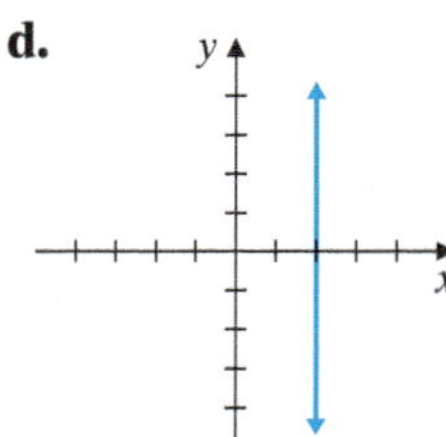

Solution:

a. This graph is the graph of a function since no vertical line will intersect this graph more than once.

b. This graph is also the graph of a function; no vertical line will intersect it more than once.

c. This graph is not the graph of a function. Vertical lines can be drawn that intersect the graph in two points. An example of one such line is shown.

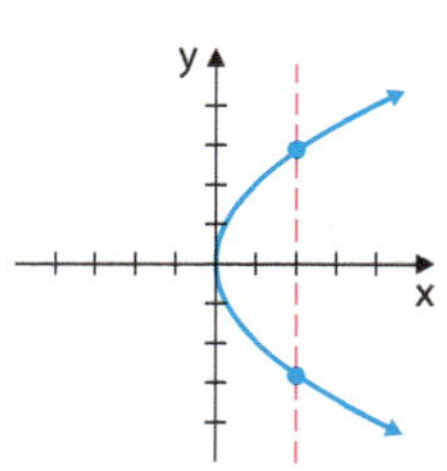

d. This graph is not the graph of a function. A vertical line can be drawn that intersects this line at every point.

Work Practice 4

Answers

4. a. a function **b.** a function **c.** not a function **d.** not a function

Recall that the graph of a linear equation is a line, and a line that is not vertical will pass the vertical line test. **Thus, all linear equations are functions except those of the form $x = c$, which are vertical lines.**

Example 5 Which of the following linear equations are functions?

a. $y = x$ **b.** $y = 2x + 1$ **c.** $y = 5$ **d.** $x = -1$

Solution: **a, b,** and **c** are functions because their graphs are nonvertical lines. **d** is not a function because its graph is a vertical line.

Work Practice 5

Practice 5

Which of the following linear equations are functions?

a. $y = 2x$ **b.** $y = -3x - 1$

c. $y = 8$ **d.** $x = 2$

Examples of functions can often be found in magazines, newspapers, books, and other printed material in the form of tables or graphs such as that in Example 6.

Example 6 The graph shows the sunrise time for Indianapolis, Indiana, for the first of each month for one year. Use this graph to answer the questions.

a. Approximate the time of sunrise on February 1.

b. Approximate the date(s) when the sun rises at 5 a.m.

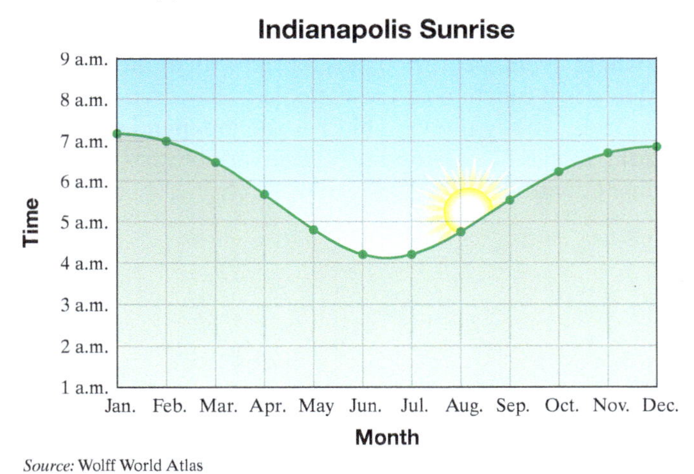

Source: Wolff World Atlas

c. Is this the graph of a function?

Solution:

a. As shown on the next page, to approximate the time of sunrise on February 1, we find the mark on the horizontal axis that corresponds to February 1. From this mark, we move vertically upward (shown in blue) until the graph is reached. From that point on the graph, we move horizontally to the left until the vertical axis is reached. The vertical axis there reads 7 a.m.

b. To approximate the date(s) when the sun rises at 5 a.m., we find 5 a.m. on the time axis and move horizontally to the right (shown in red). Notice that we will hit the graph at two points, corresponding to two dates for which the sun rises at 5 a.m. We follow both points on the graph vertically downward until the horizontal axis is reached. The sun rises at 5 a.m. at approximately the end of the month of April and early in the month of August.

(*Continued on next page*)

Practice 6

Use the graph in Example 6 to answer the questions.

a. Approximate the time of sunrise on March 1.

b. Approximate the date(s) when the sun rises at 6 a.m.

Answers

5. a, b, and **c** are functions.

6. a. 6:30 a.m. **b.** middle of March and middle of September

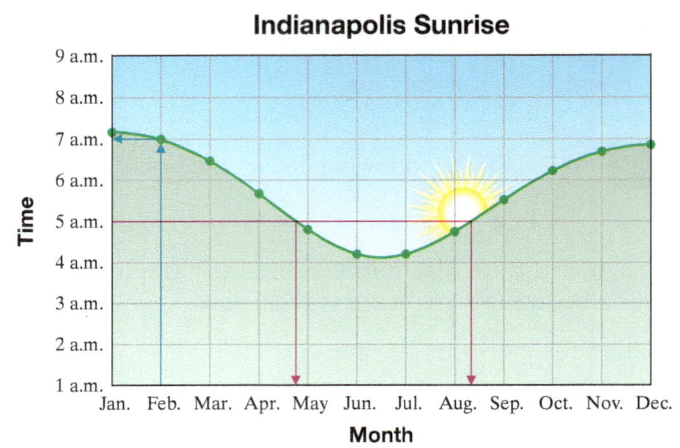

Source: Wolff World Atlas

c. The graph is the graph of a function since it passes the vertical line test. In other words, for every day of the year in Indianapolis, there is exactly one sunrise time.

Work Practice 6

Objective D Using Function Notation 

The graph of the linear equation $y = 2x + 1$ passes the vertical line test, so we say that $y = 2x + 1$ is a function. In other words, $y = 2x + 1$ gives us a rule for writing ordered pairs where every x-coordinate is paired with at most one y-coordinate.

The variable y is a function of the variable x. For each value of x, there is only one value of y. Thus, we say the variable x is the **independent variable** because any value in the domain can be assigned to x. The variable y is the **dependent variable** because its value depends on x.

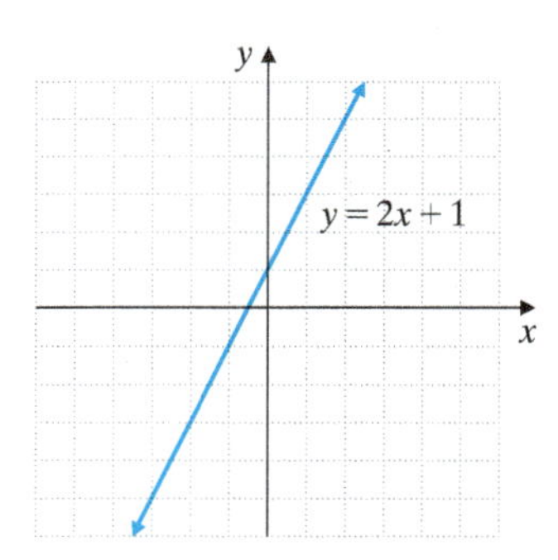

We often use letters such as f, g, and h to name functions. For example, the symbol $f(x)$ means *function of* x and is read "f of x." This notation is called **function notation.** The equation $y = 2x + 1$ can be written as $f(x) = 2x + 1$ using function notation, and these equations mean the same thing. In other words, $y = f(x)$.

The notation $f(1)$ means to replace x with 1 and find the resulting y or function value. Since

$$f(x) = 2x + 1$$

then

$$f(1) = 2(1) + 1 = 3$$

This means that, when $x = 1$, y or $f(x) = 3$, and we have the ordered pair (1, 3). Now let's find $f(2)$, $f(0)$, and $f(-1)$.

$$\begin{aligned} f(x) &= 2x + 1 \\ f(2) &= 2(2) + 1 \\ &= 4 + 1 \\ &= 5 \end{aligned} \qquad \begin{aligned} f(x) &= 2x + 1 \\ f(0) &= 2(0) + 1 \\ &= 0 + 1 \\ &= 1 \end{aligned} \qquad \begin{aligned} f(x) &= 2x + 1 \\ f(-1) &= 2(-1) + 1 \\ &= -2 + 1 \\ &= -1 \end{aligned}$$

Ordered Pairs: (2, 5) (0, 1) (−1, −1)

Helpful Hint Note that, for example, if $f(2) = 5$, the corresponding ordered pair is (2, 5).

Helpful Hint

Note that $f(x)$ is a special symbol in mathematics used to denote a function. The symbol $f(x)$ is read "f of x." It does **not** mean $f \cdot x$ (f times x).

Example 7

Given $g(x) = x^2 - 3$, find the following and list the corresponding ordered pairs generated.

a. $g(2)$ **b.** $g(-2)$ **c.** $g(0)$

Solution:

a. $$\begin{aligned} g(x) &= x^2 - 3 \\ g(2) &= 2^2 - 3 \\ &= 4 - 3 \\ &= 1 \end{aligned}$$

b. $$\begin{aligned} g(x) &= x^2 - 3 \\ g(-2) &= (-2)^2 - 3 \\ &= 4 - 3 \\ &= 1 \end{aligned}$$

c. $$\begin{aligned} g(x) &= x^2 - 3 \\ g(0) &= 0^2 - 3 \\ &= 0 - 3 \\ &= -3 \end{aligned}$$

Ordered Pairs:	$g(2) = 1$ gives (2, 1)	$g(-2) = 1$ gives (−2, 1)	$g(0) = -3$ gives (0, −3)

Work Practice 7

Practice 7

Given $f(x) = x^2 + 1$, find the following and list the corresponding ordered pairs.

a. $f(1)$ **b.** $f(-3)$ **c.** $f(0)$

We now practice finding the domain and the range of a function. The domain of our functions will be the set of all possible real numbers that x can be replaced by. The range is the set of corresponding y-values.

Example 8

Find the domain of each function.

a. $g(x) = \dfrac{1}{x}$ **b.** $f(x) = 2x + 1$

Solution:

a. Recall that we cannot divide by 0, so the domain of $g(x)$ is the set of all real numbers except 0.

b. In this function, x can be any real number. The domain of $f(x)$ is the set of all real numbers.

Work Practice 8

Practice 8

Find the domain of each function.

a. $h(x) = 6x + 3$

b. $f(x) = \dfrac{1}{x^2}$

Concept Check Suppose that the value of f is −7 when the function is evaluated at 2. Write this situation in function notation.

Answers

7. a. 2; (1, 2) **b.** 10; (−3, 10) **c.** 1; (0, 1)

8. a. Domain: all real numbers
b. Domain: all real numbers except 0

Concept Check Answer

$f(2) = -7$

Practice 9

Find the domain and the range of each function graphed.

a.

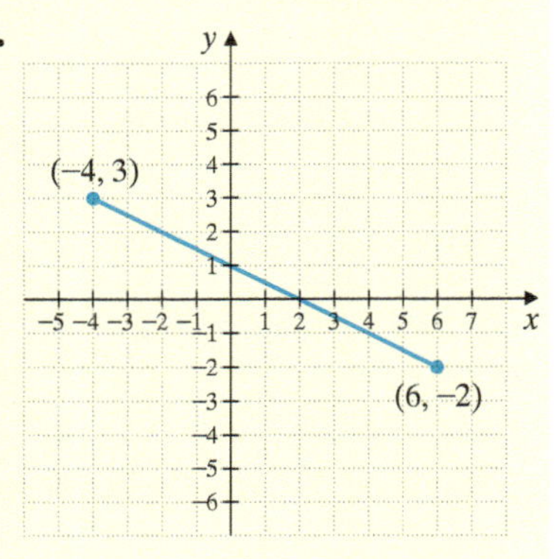

b.

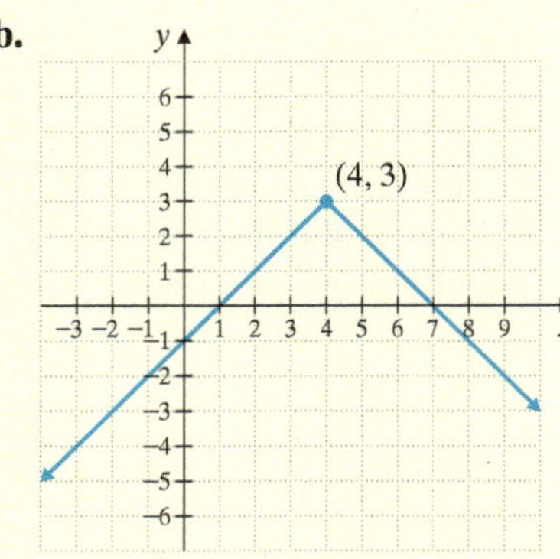

Answers

9. a. Domain: $-4 \leq x \leq 6$
Range: $-2 \leq y \leq 3$
b. Domain: all real numbers
Range: $y \leq 3$

Example 9

Find the domain and the range of each function graphed.

a.

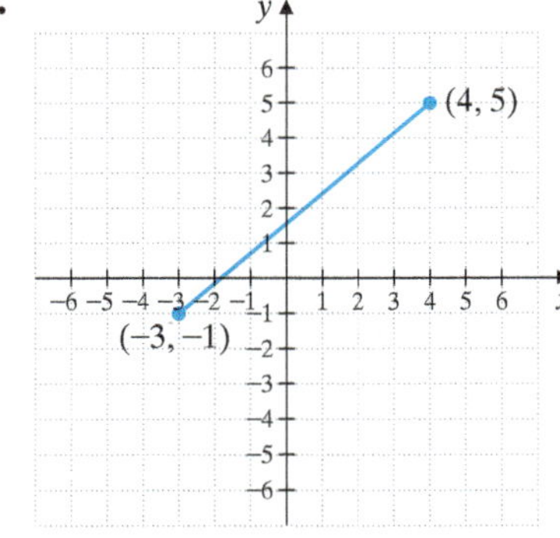

b.

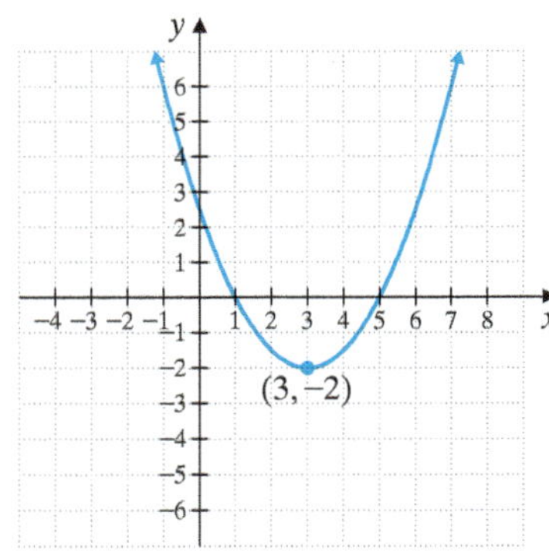

Solution:

a.

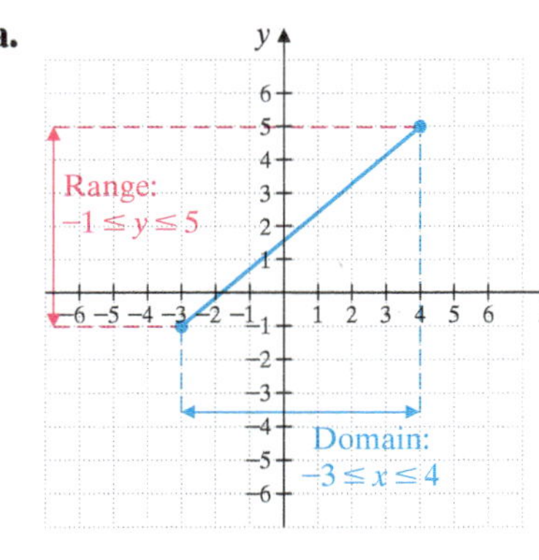

b.

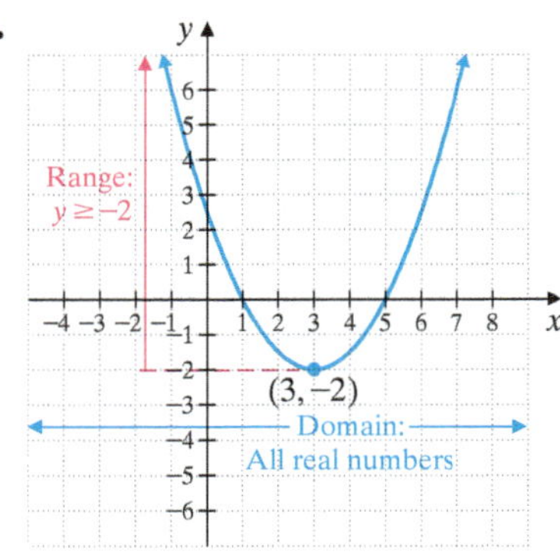

Work Practice 9

Vocabulary, Readiness & Video Check

Use the choices below to fill in each blank. Not all choices will be used.

$x = c$	horizontal	domain	relation	(7, 3)	x	$\{x \mid x \leq 5\}$
$y = c$	vertical	range	function	(3, 7)	y	all real numbers

1. A set of ordered pairs is called a(n) ______________.

2. A set of ordered pairs that assigns to each x-value exactly one y-value is called a(n) ______________.

3. The set of all y-coordinates of a relation is called the ______________.

4. The set of all x-coordinates of a relation is called the ______________.

5. All linear equations are functions except those whose graphs are ______________ lines.

6. All linear equations are functions except those whose equations are of the form ______________.

7. If $f(3) = 7$, the corresponding ordered pair is ______________.

8. The domain of $f(x) = x + 5$ is ______________.

9. For the function $y = mx + b$, the dependent variable is ______________ and the independent variable is ______________.

Martin-Gay Interactive Videos

See Video 6.6

Watch the section lecture video and answer the following questions.

Objective A **10.** In the lecture before Example 1, relations are discussed. Why can an equation in two variables define a relation?

Objective B **11.** Based on Examples 2 and 3, can a set of ordered pairs with no repeated x-values, but with repeated y-values, be a function? For example: $\{(0, 4), (-3, 4), (2, 4)\}$.

Objective C **12.** After reviewing Example 8, explain why the vertical line test works.

Objective D **13.** In Example 10, three function values are found along with their corresponding ordered pairs. One of the three is: $f(0) = 2$ corresponds to $(0, 2)$. Write down the other two.

6.6 Exercise Set MyLab Math 

Objective A *Find the domain and the range of each relation. See Example 1.*

1. $\{(2, 4), (0, 0), (-7, 10), (10, -7)\}$

2. $\{(3, -6), (1, 4), (-2, -2)\}$

3. $\{(0, -2), (1, -2), (5, -2)\}$

4. $\{(5, 0), (5, -3), (5, 4), (5, 3)\}$

Objective B *Determine whether each relation is also a function. See Example 2.*

5. $\{(1, 1), (2, 2), (-3, -3), (0, 0)\}$

6. $\{(11, 6), (-1, -2), (0, 0), (3, -2)\}$

7. $\{(-1, 0), (-1, 6), (-1, 8)\}$

8. $\{(1, 2), (3, 2), (1, 4)\}$

Objectives B C Mixed Practice *Determine whether each graph is the graph of a function. For Exercises 9 through 12, either write down the ordered pairs or use the vertical line test. See Examples 3 and 4.*

 **9.**

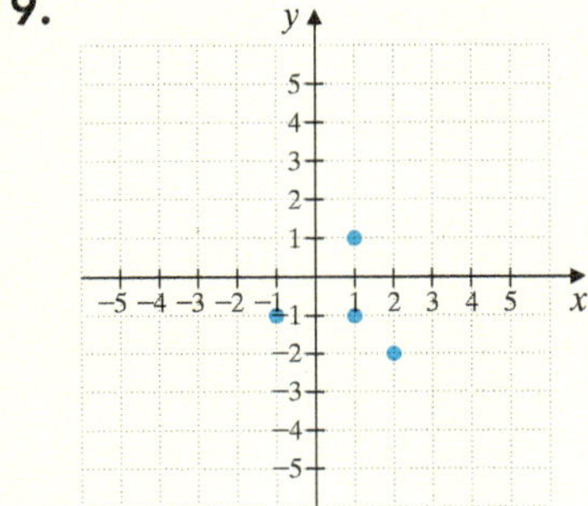

10.

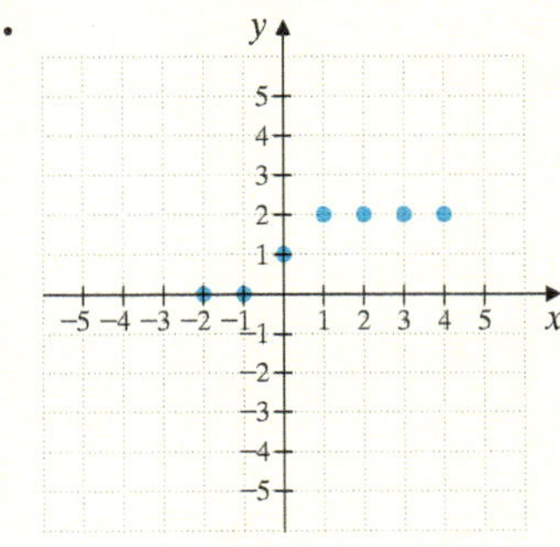

11.

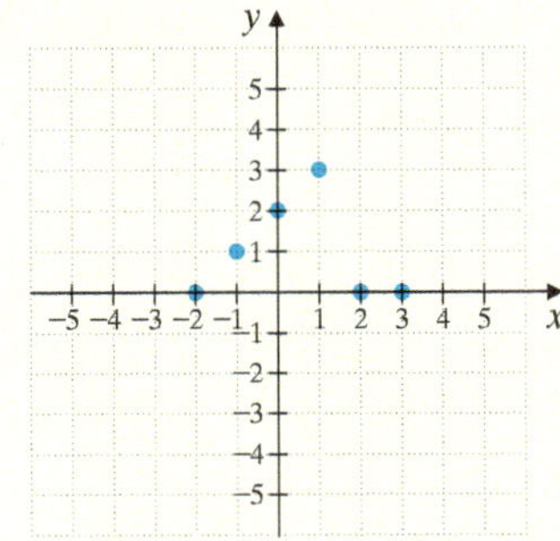

12.

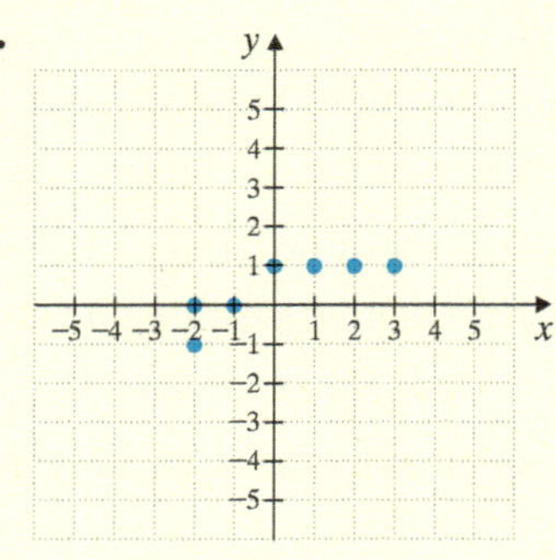

13.

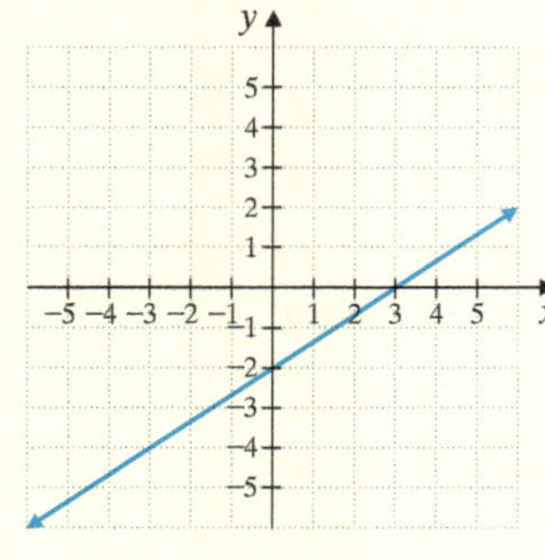

14.

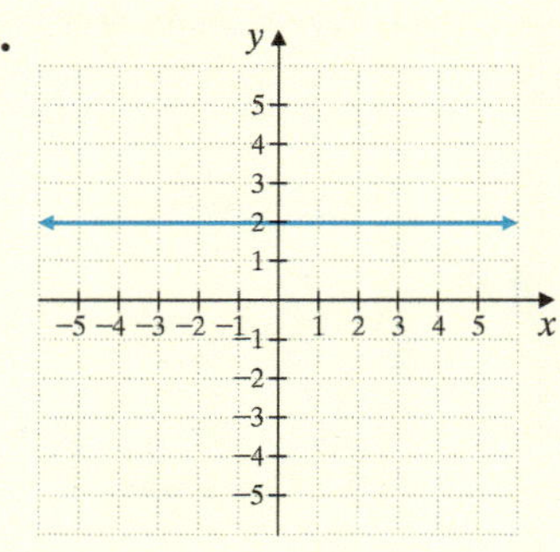

15.

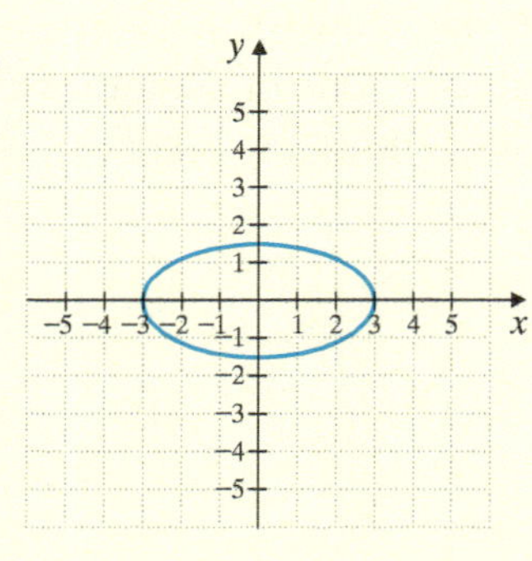

16.

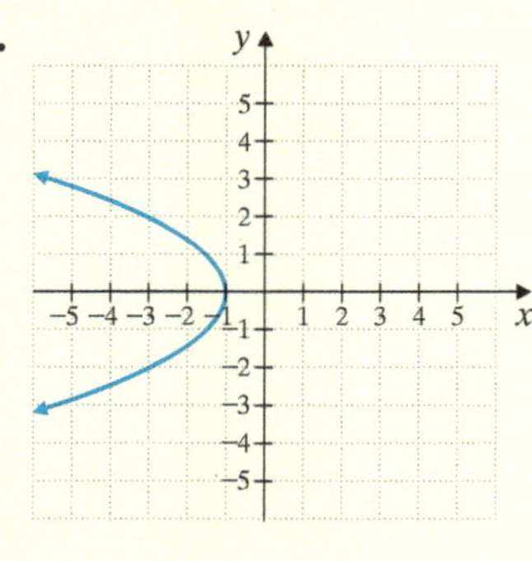

For each exercise, choose the value of x so that the relation is NOT also a function.

17. $\{(2, 3), (-1, 7), (x, 9)\}$
a. -1 **b.** 1 **c.** 9 **d.** 7

18. $\{(-8, 0), (x, 1), (5, -3)\}$
a. 0 **b.** -3 **c.** -5 **d.** 5

Decide whether the equation describes a function. See Example 5.

19. $y - x = 7$

20. $2x - 3y = 9$

21. $y = 6$

22. $x = 3$

23. $x = -2$

24. $y = -9$

25. $x = y^2$

26. $y = x^2 - 3$

(*Hint:* For Exercises **25** and **26**, check to see whether each x-value pairs with exactly one y-value.)

The graph shows the sunset times for Seward, Alaska for the first of each month for one year. Use this graph to answer Exercises 27 through 32. See Example 6.

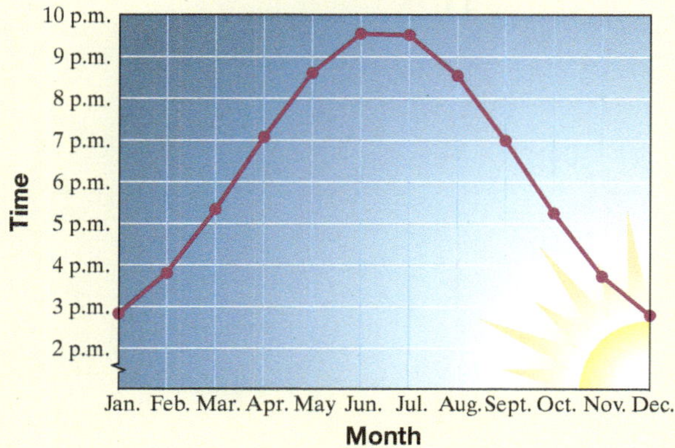

27. Approximate the time of sunset on June 1.

28. Approximate the time of sunset on November 1.

29. Approximate the date(s) when the sunset is at 3 p.m.

30. Approximate the date(s) when the sunset is at 9 p.m.

31. Is this graph the graph of a function? Why or why not?

32. Do you think a graph of sunset times for any location will always be a function? Why or why not?

This graph shows the U.S. hourly minimum wage for each year shown. Use this graph to answer Exercises 33 through 38. See Example 6.

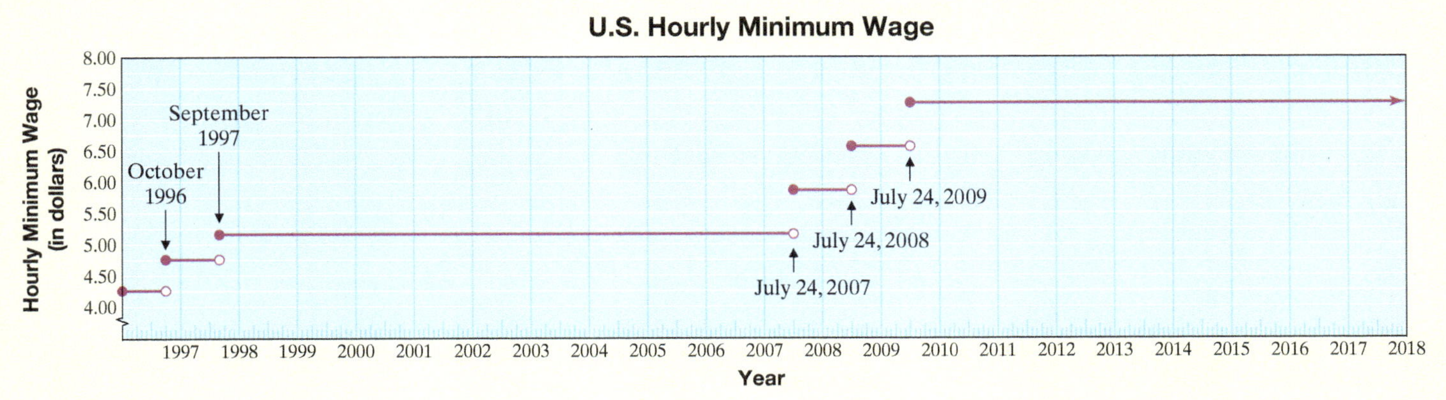

Source: U.S. Department of Labor

33. Approximate the minimum wage before October 1996.

34. Approximate the minimum wage in 2006.

35. Find the year when the minimum wage increased to over $7.00 per hour.

36. According to the graph, what hourly wage was in effect for the greatest number of years?

37. Is this graph the graph of a function? Why or why not?

38. Do you think that a similar graph of your hourly wage on January 1 of every year (whether you are working or not) would be the graph of a function? Why or why not?

This graph shows the cost of mailing a large envelope through the U.S. Postal Service by weight. Use this graph to answer Exercises 39 through 44. See Example 6.

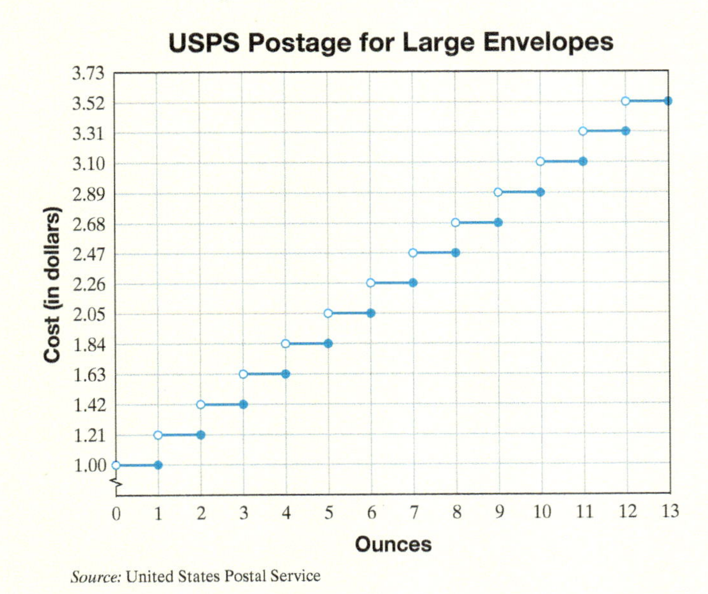

Source: United States Postal Service

39. Approximate the postage to mail a large envelope weighing more than 4 ounces but not more than 5 ounces.

40. Approximate the postage to mail a large envelope weighing more than 7 ounces but not more than 8 ounces.

41. Give the weight of a large envelope that costs $1.21 to mail.

42. If you have $3.00, what is the weight of the largest envelope you can mail for that amount of money?

43. Is this graph a function? Why or why not?

44. Do you think that a similar graph of postage to mail a first-class letter would be the graph of a function? Why or why not?

Objective D *Find $f(-2)$, $f(0)$, and $f(3)$ for each function. See Example 7.*

45. $f(x) = 2x - 5$

46. $f(x) = 3 - 7x$

47. $f(x) = x^2 + 2$

48. $f(x) = x^2 - 4$

49. $f(x) = 3x$ **50.** $f(x) = -3x$ **51.** $f(x) = |x|$ **52.** $f(x) = |2 - x|$

Find $h(-1)$, $h(0)$, and $h(4)$ for each function. See Example 7.

53. $h(x) = -5x$ **54.** $h(x) = -3x$ **55.** $h(x) = 2x^2 + 3$ **56.** $h(x) = 3x^2$

For each given function value, write a corresponding ordered pair.

57. $f(3) = 6$ **58.** $f(7) = -2$ **59.** $g(0) = -\frac{1}{2}$

60. $g(0) = -\frac{7}{8}$ **61.** $h(-2) = 9$ **62.** $h(-10) = 1$

Objectives A D Mixed Practice

Find the domain of each function. See Example 8.

63. $f(x) = 3x - 7$ **64.** $g(x) = 5 - 2x$ **65.** $h(x) = \frac{1}{x + 5}$ **66.** $f(x) = \frac{1}{x - 6}$

Find the domain and the range of each relation graphed. See Example 9.

67.

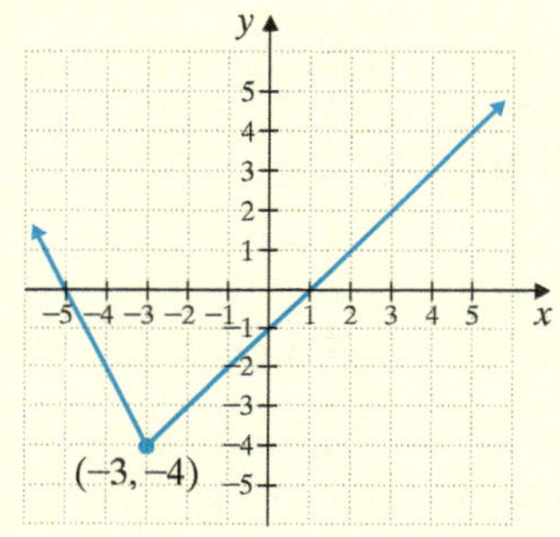

68.

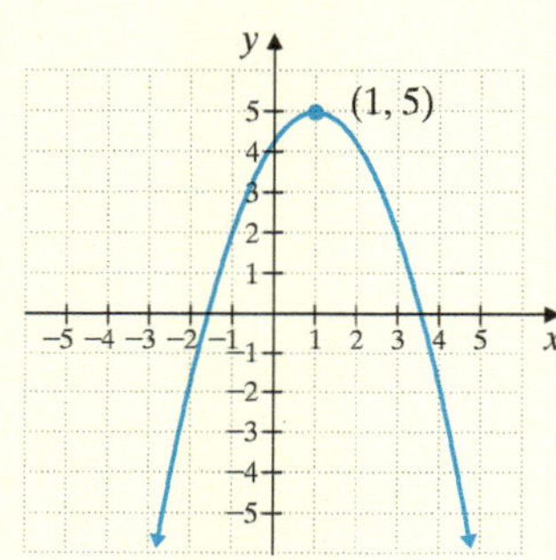

69.

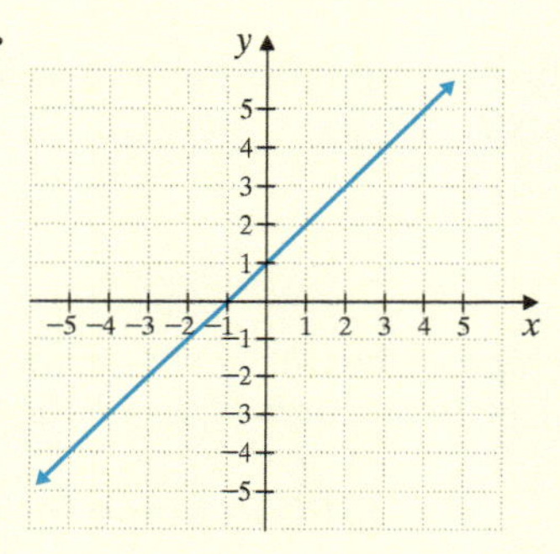

70.

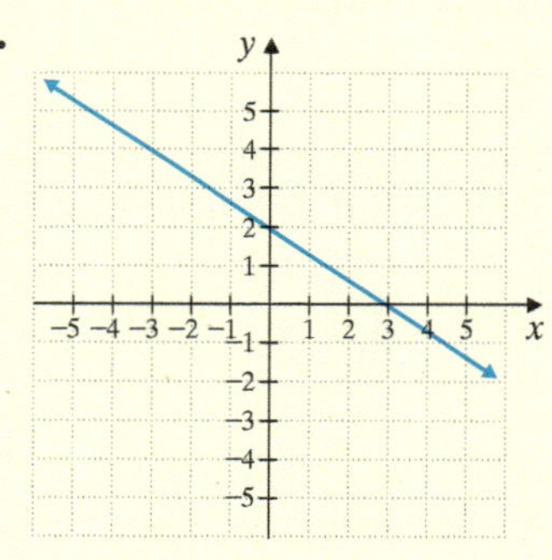

71.

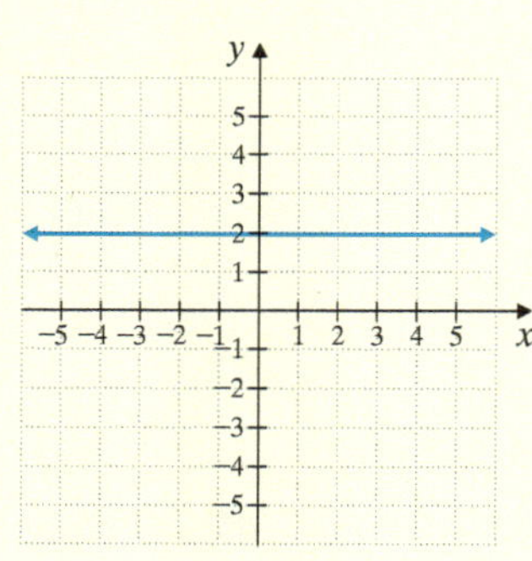

72.

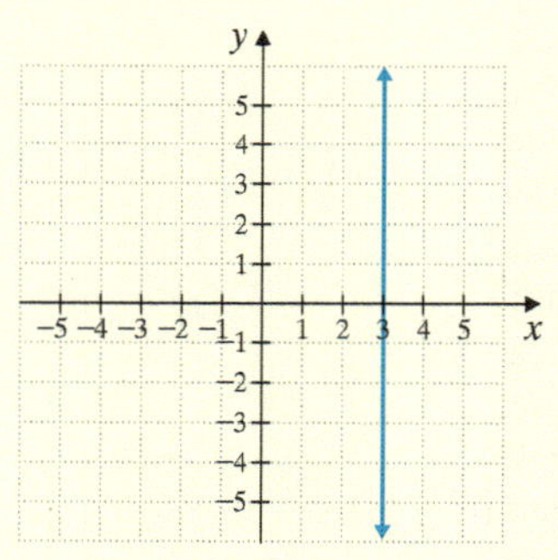

Use the graph of f below to answer Exercises 73 through 78.

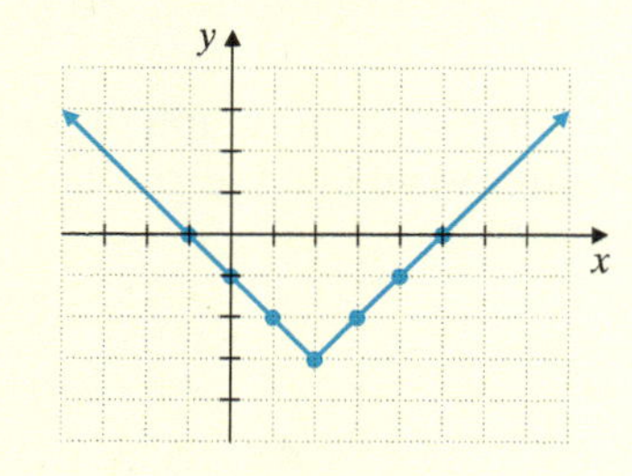

73. Complete the ordered pair solution for f. (0,)

74. Complete the ordered pair solution for f. (3,)

75. $f(0) =$ _____?

76. $f(3) =$ _____?

77. If $f(x) = 0$, find the value(s) of x.

78. If $f(x) = -1$, find the value(s) of x.

Review

Solve each inequality. See Section 2.7.

79. $2x + 5 < 7$

80. $3x - 1 \geq 11$

81. $-x + 6 \leq 9$

82. $-2x + 3 > 3$

Find the perimeter of each figure. See Section 5.4.

△ **83.**

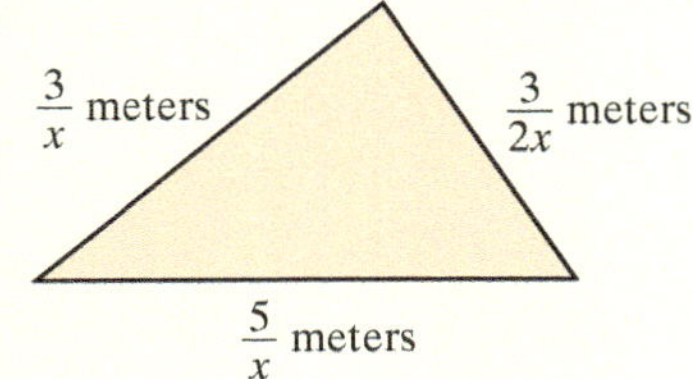

△ **84.**

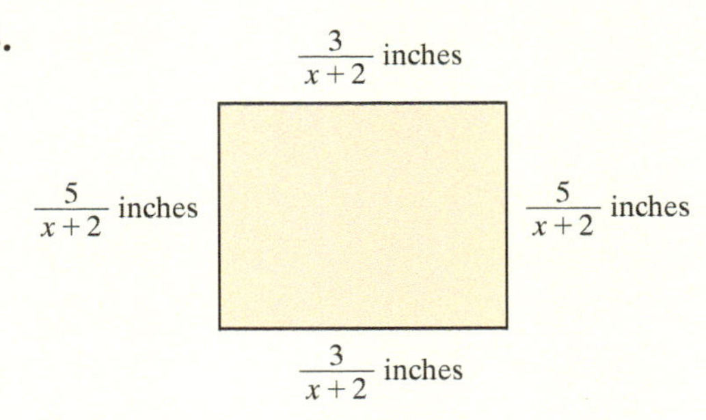

Concept Extensions

Solve. See the Concept Check in this section.

85. If a function f is evaluated at -5, the value of the function is 12. Write this situation using function notation.

86. Suppose $(9, 20)$ is an ordered pair solution for the function g. Write this situation using function notation.

The graph of the function f is below. Use this graph to answer Exercises 87 through 90.

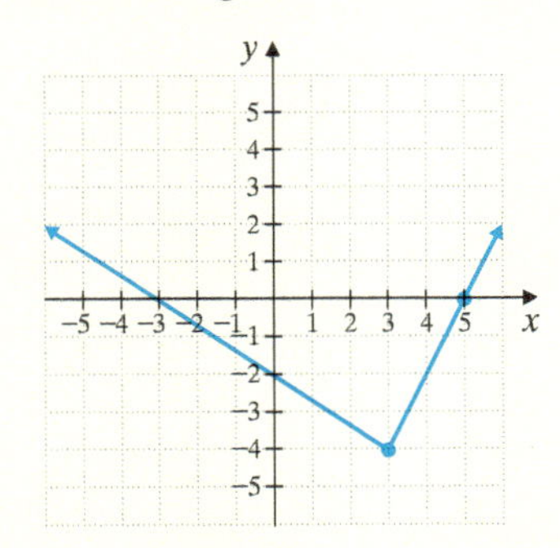

87. Write the coordinates of the lowest point of the graph.

88. Write the answer to Exercise **87** in function notation.

89. An x-intercept of this graph is $(5, 0)$. Write this using function notation.

90. Write the other x-intercept of this graph (see Exercise **89**) using function notation.

91. In your own words define (a) function; (b) domain; (c) range.

92. Explain the vertical line test and how it is used.

93. Since $y = x + 7$ is a function, rewrite the equation using function notation.

94. Since $y = 3$ is a function, rewrite the equation using function notation.

95. The dosage in milligrams of Ivermectin, a heartworm preventive for a dog who weighs x pounds, is given by the function

$$f(x) = \frac{136}{25}x$$

a. Find the proper dosage for a dog who weighs 35 pounds.

b. Find the proper dosage for a dog who weighs 70 pounds.

96. Forensic scientists use the function

$$f(x) = 2.59x + 47.24$$

to estimate the height of a woman, in centimeters, given the length x of her femur bone in centimeters.

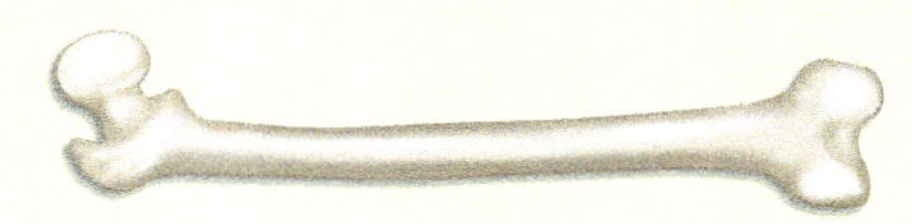

a. Estimate the height of a woman whose femur measures 46 centimeters.

b. Estimate the height of a woman whose femur measures 39 centimeters.

6.7 Graphing Linear Inequalities in Two Variables

Objectives

A Determine Whether an Ordered Pair Is a Solution of a Linear Inequality in Two Variables.

B Graph a Linear Inequality in Two Variables.

Recall that a linear equation in two variables is an equation that can be written in the form $Ax + By = C$, where A, B, and C are real numbers and A and B are not both 0. A **linear inequality in two variables** is an inequality that can be written in one of the forms

$$Ax + By < C \qquad Ax + By \le C$$
$$Ax + By > C \qquad Ax + By \ge C$$

where A, B, and C are real numbers and A and B are not both 0.

Objective A Determining Solutions of Linear Inequalities in Two Variables

Just as for linear equations in x and y, an ordered pair is a **solution** of an inequality in x and y if replacing the variables with the coordinates of the ordered pair results in a true statement.

Practice 1

Determine whether each ordered pair is a solution of $x - 4y > 8$.

a. $(-3, 2)$ **b.** $(9, 0)$

Example 1 Determine whether each ordered pair is a solution of the inequality $2x - y < 6$.

a. $(5, -1)$ **b.** $(2, 7)$

Solution:

a. We replace x with 5 and y with -1 and see if a true statement results.

$$2x - y < 6$$
$$2(5) - (-1) < 6 \quad \text{Replace } x \text{ with 5 and } y \text{ with } -1.$$
$$10 + 1 < 6$$
$$11 < 6 \quad \text{False}$$

The ordered pair $(5, -1)$ is not a solution since $11 < 6$ is a false statement.

b. We replace x with 2 and y with 7 and see if a true statement results.

$$2x - y < 6$$
$$2(2) - (7) < 6 \quad \text{Replace } x \text{ with 2 and } y \text{ with 7.}$$
$$4 - 7 < 6$$
$$-3 < 6 \quad \text{True}$$

The ordered pair $(2, 7)$ is a solution since $-3 < 6$ is a true statement.

Work Practice 1

Objective B Graphing Linear Inequalities in Two Variables

The linear equation $x - y = 1$ is graphed next. Recall that all points on the line correspond to ordered pairs that satisfy the equation $x - y = 1$.

Notice that the line defined by $x - y = 1$ divides the rectangular coordinate system plane into 2 sides. All points on one side of the line satisfy the inequality $x - y < 1$, and all points on the other side satisfy the inequality $x - y > 1$. The graph on the next page shows a few examples of this.

Answers

1. a. no **b.** yes

$x - y < 1$
$1 - 3 < 1$ True
$-2 - 1 < 1$ True
$-4 - (-4) < 1$ True

$x - y > 1$
$4 - 1 > 1$ True
$2 - (-2) > 1$ True
$0 - (-4) > 1$ True

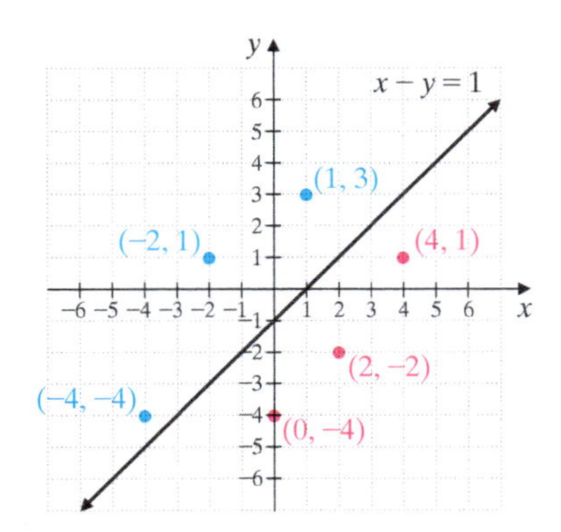

The graph of $x - y < 1$ is the region shaded blue and the graph of $x - y > 1$ is the region shaded red below.

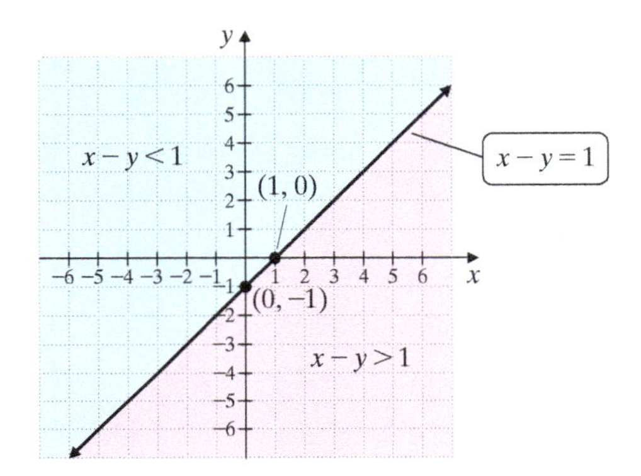

The region to the left of the line and the region to the right of the line are called **half-planes.** Every line divides the plane (similar to a sheet of paper extending indefinitely in all directions) into two half-planes; the line is called the **boundary.**

Recall that the inequality $x - y \le 1$ means

$$x - y = 1 \quad \text{or} \quad x - y < 1$$

Thus, the graph of $x - y \le 1$ is the blue half-plane $x - y < 1$ along with the boundary line $x - y = 1$.

To Graph a Linear Inequality in Two Variables

Step 1: Graph the boundary line found by replacing the inequality sign with an equal sign. If the inequality sign is $>$ or $<$, graph a dashed boundary line (indicating that the points on the line are not solutions of the inequality). If the inequality sign is $\ge$ or $\le$, graph a solid boundary line (indicating that the points on the line are solutions of the inequality).

Step 2: Choose a point *not* on the boundary line as a test point. Substitute the coordinates of this test point into the *original* inequality.

Step 3: If a true statement is obtained in Step 2, shade the half-plane that contains the test point. If a false statement is obtained, shade the half-plane that does not contain the test point.

Practice 2

Graph: $x - y > 3$

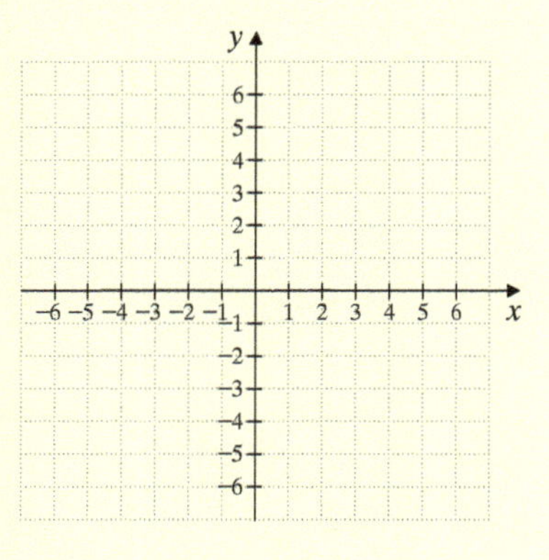

Practice 3

Graph: $x - 4y \le 4$

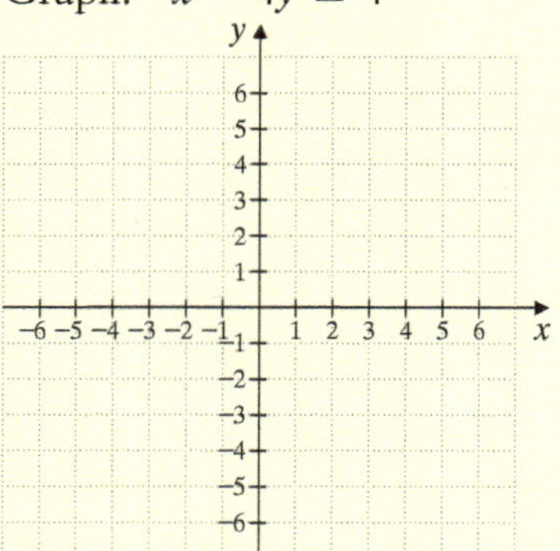

Answers

2.

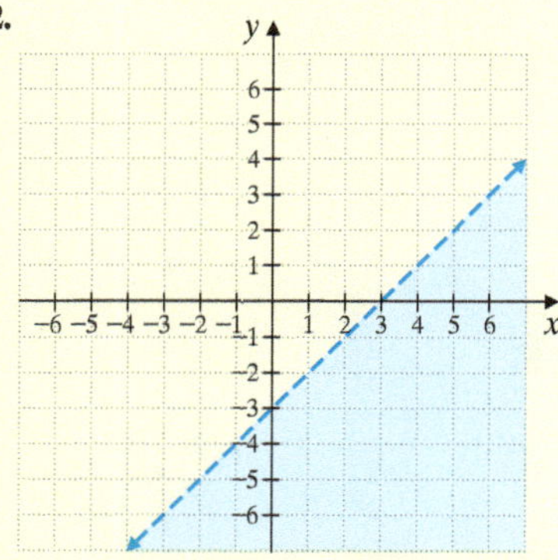

3.

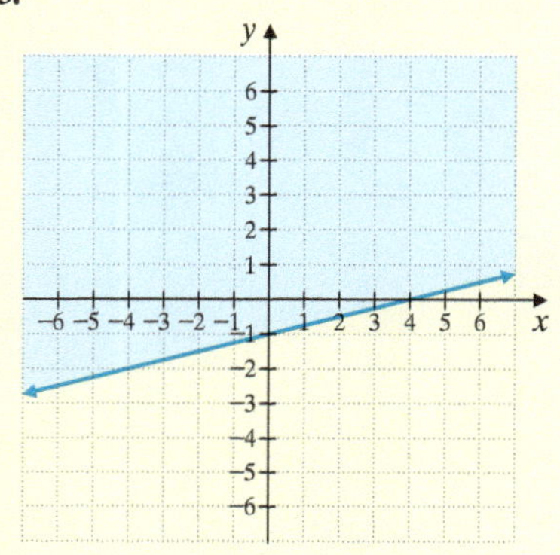

✓**Concept Check Answers**

a. no **b.** yes **c.** yes

Example 2

Graph: $x + y < 7$

Solution:

Step 1: First we graph the boundary line by graphing the equation $x + y = 7$. We graph this boundary as a *dashed line* because the inequality sign is $<$, and thus the points on the line are not solutions of the inequality $x + y < 7$.

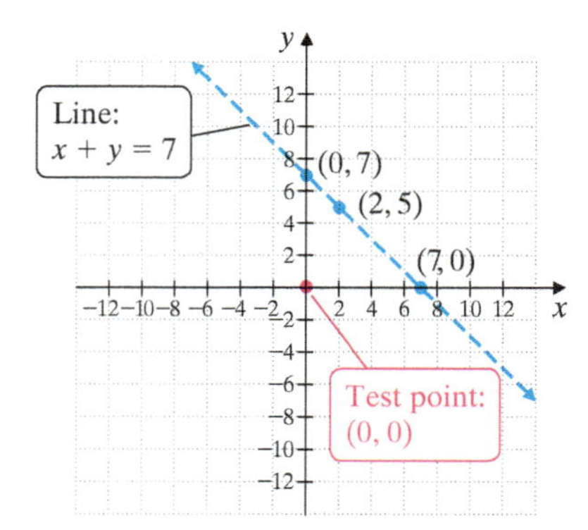

Step 2: Next we choose a test point, being careful *not* to choose a point on the boundary line. We choose (0, 0), and substitute the coordinates of (0, 0) into $x + y < 7$.

$x + y < 7$ Original inequality

$0 + 0 < 7$ Replace x with 0 and y with 0.

$0 < 7$ True

Step 3: Since the result is a true statement, (0, 0) is a solution of $x + y < 7$, and every point in the same half-plane as (0, 0) is also a solution. To indicate this, we shade the entire half-plane containing (0, 0), as shown.

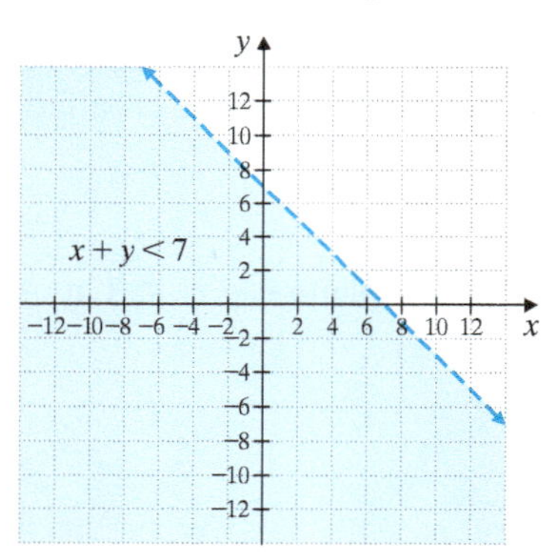

Graph of $x + y < 7$

Work Practice 2

✓**Concept Check** Determine whether (0, 0) is included in the graph of

a. $y \ge 2x + 3$ **b.** $x < 7$ **c.** $2x - 3y < 6$

Example 3

Graph: $2x - y \ge 3$

Solution:

Step 1: We graph the boundary line by graphing $2x - y = 3$. We draw this line as a solid line because the inequality sign is $\ge$, and thus the points on the line are solutions of $2x - y \ge 3$.

Step 2: Once again, (0, 0) is a convenient test point since it is not on the boundary line.

We substitute 0 for x and 0 for y into the original inequality.

$$2x - y \geq 3$$
$$2(0) - 0 \geq 3 \quad \text{Let } x = 0 \text{ and } y = 0.$$
$$0 \geq 3 \quad \text{False}$$

Step 3: Since the statement is false, no point in the half-plane containing (0, 0) is a solution. Therefore, we shade the half-plane that does not contain (0, 0). Every point in the shaded half-plane and every point on the boundary line is a solution of $2x - y \geq 3$.

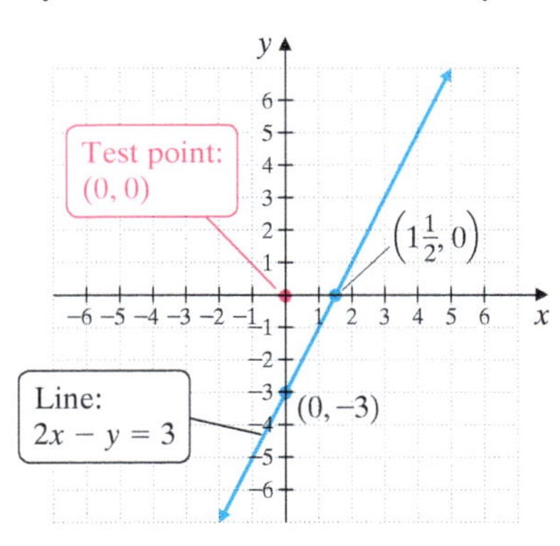

Step 1 and Step 2 above

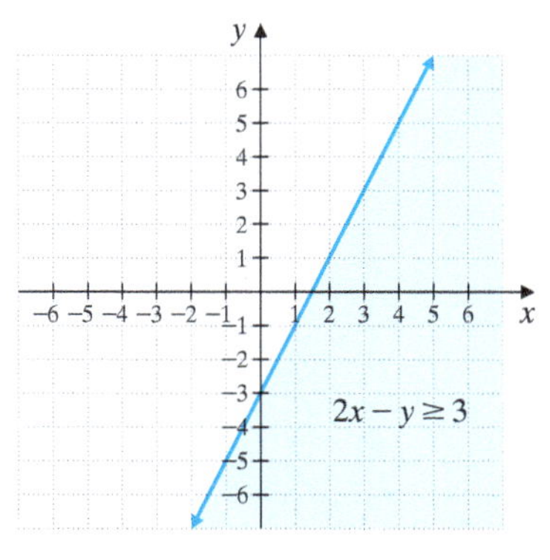

Graph of $2x - y \geq 3$

Work Practice 3

Helpful Hint

When graphing an inequality, make sure the test point is substituted into the **original inequality.** For Example 3, we substituted the test point (0, 0) into the **original inequality** $2x - y \geq 3$, *not* $2x - y = 3$.

Example 4

Graph: $x > 2y$

Solution:

Step 1: We find the boundary line by graphing $x = 2y$. The boundary line is a dashed line since the inequality symbol is $>$.

Step 2: We cannot use (0, 0) as a test point because it is a point on the boundary line. We choose instead (0, 2).

$$x > 2y$$
$$0 > 2(2) \quad \text{Let } x = 0 \text{ and } y = 2.$$
$$0 > 4 \quad \text{False}$$

Step 3: Since the statement is false, we shade the half-plane that does not contain the test point (0, 2), as shown.

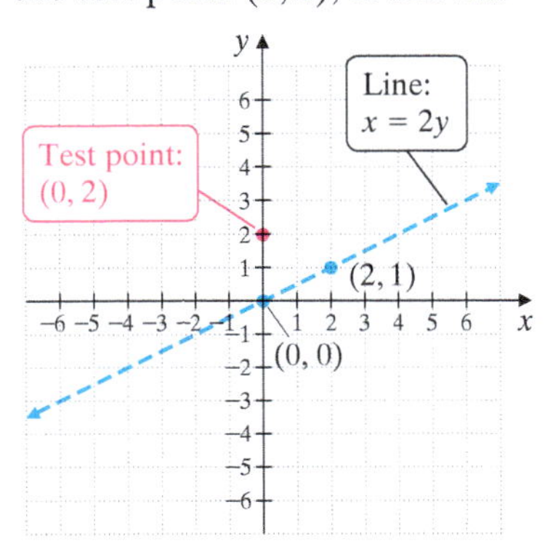

Step 1 and Step 2 above

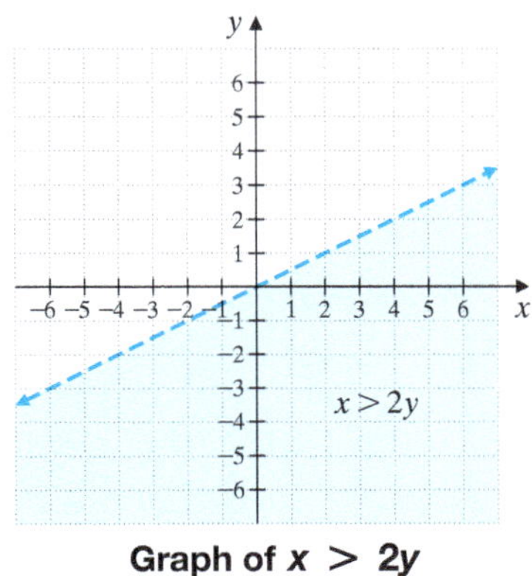

Graph of $x > 2y$

Work Practice 4

Practice 4

Graph: $y < 3x$

Answer

4.

Practice 5

Graph: $3x + 2y \geq 12$

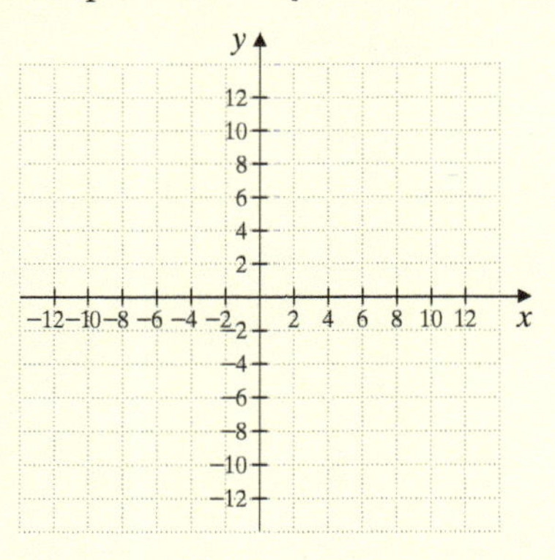

Practice 6

Graph: $x < 2$

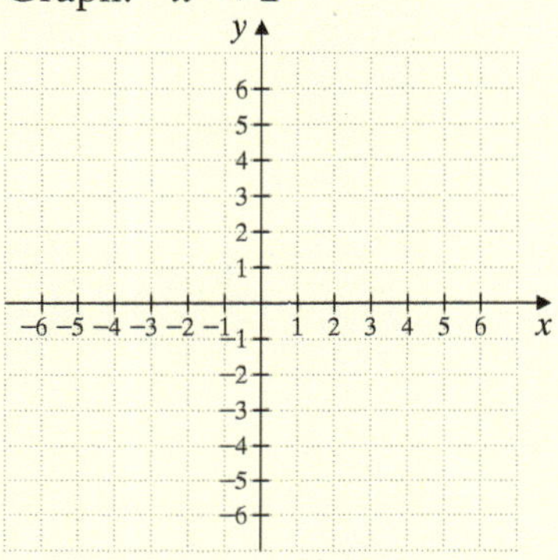

Answers

5.

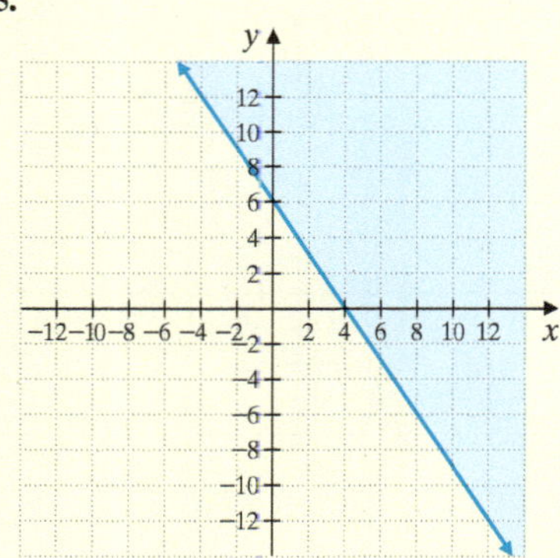

6.

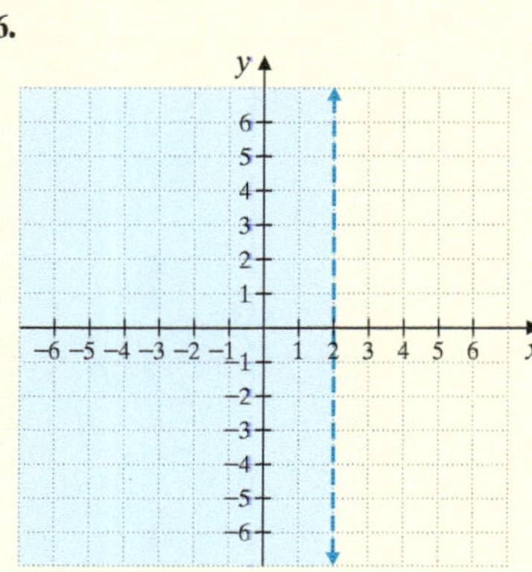

Example 5

Graph: $5x + 4y \leq 20$

Solution: We graph the solid boundary line $5x + 4y = 20$ and choose $(0, 0)$ as the test point.

$$5x + 4y \leq 20$$

$$5(0) + 4(0) \leq 20 \quad \text{Let } x = 0 \text{ and } y = 0.$$

$$0 \leq 20 \quad \text{True}$$

We shade the half-plane that contains $(0, 0)$, as shown.

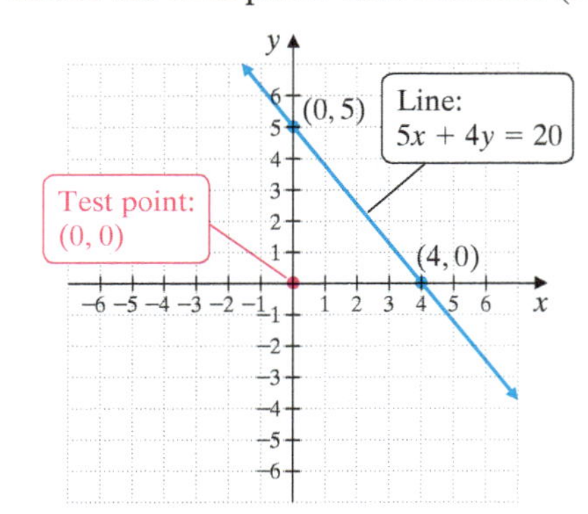

Steps 1 and 2 to graph $5x + 4y \leq 20$

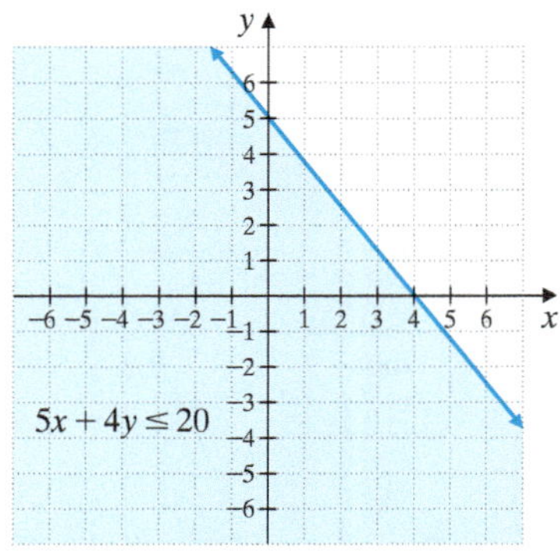

Graph of $5x + 4y \leq 20$

Work Practice 5

Example 6

Graph: $y > 3$

Solution: We graph the dashed boundary line $y = 3$ and choose $(0, 0)$ as the test point. (Recall that the graph of $y = 3$ is a horizontal line with y-intercept 3.)

$$y > 3$$

$$0 > 3 \quad \text{Let } y = 0.$$

$$0 > 3 \quad \text{False}$$

We shade the half-plane that does not contain $(0, 0)$, as shown.

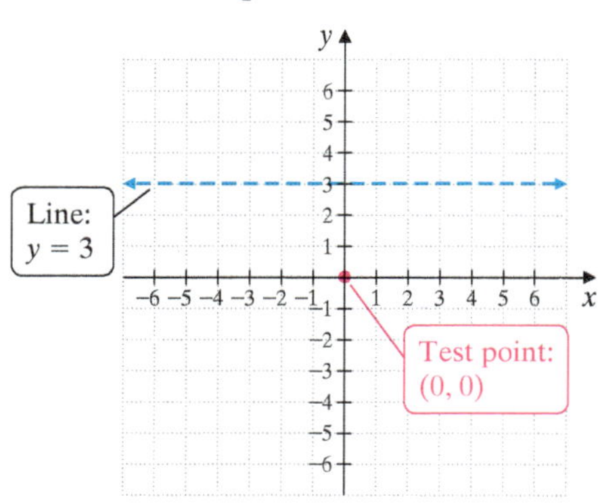

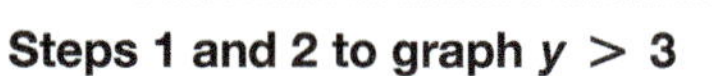

Steps 1 and 2 to graph $y > 3$

Graph of $y > 3$

Work Practice 6

Example 7 Graph: $y \le \frac{2}{3}x - 4$

Solution: Graph the solid boundary line $y = \frac{2}{3}x - 4$. This equation is in slope-intercept form, with slope $\frac{2}{3}$ and y-intercept -4.

We use this information to graph the line. Then we choose $(0, 0)$ as our test point.

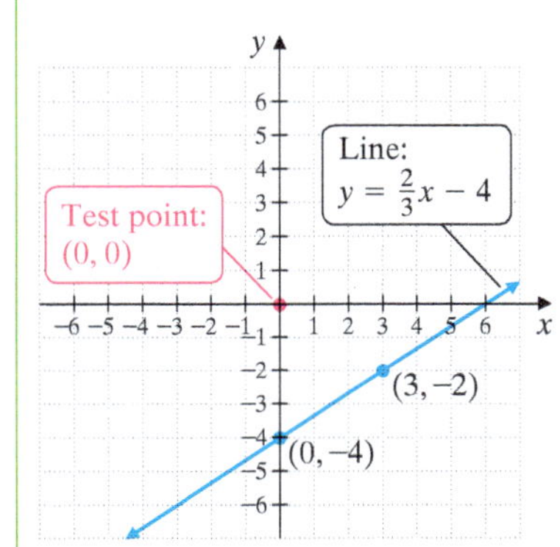

Steps 1 and 2 to graph $y \le \frac{2}{3}x - 4$

Check the test point.

$y \le \frac{2}{3}x - 4$

$0 \overset{?}{\le} \frac{2}{3} \cdot 0 - 4$

$0 \le -4$ False

Since false, we shade the half-plane that does not contain $(0, 0)$, as shown.

$y \le \frac{2}{3}x - 4$

Graph of $y \le \frac{2}{3}x - 4$

Work Practice 7

Practice 7

Graph: $y \ge \frac{1}{4}x + 3$

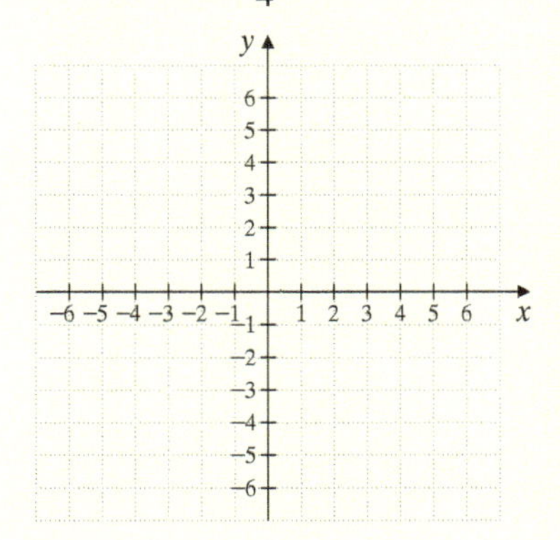

Answer

7.

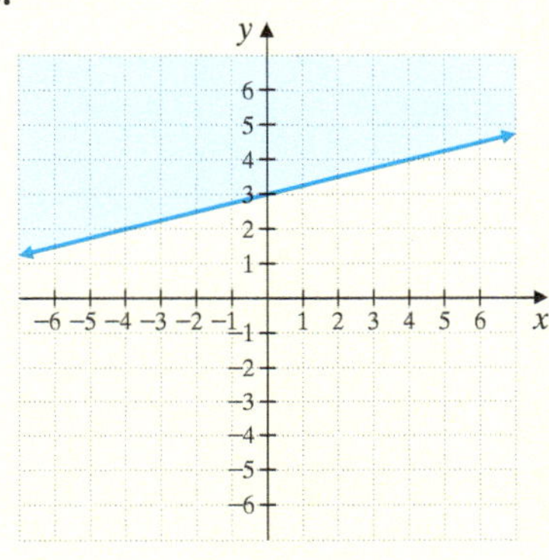

Vocabulary, Readiness & Video Check

Use the choices below to fill in each blank. Some choices may be used more than once, and some not at all.

true	$x < 2$	$y < 2$	half-planes
false	$x \le 2$	$y \le 2$	linear inequality in two variables

1. The statement $5x - 6y < 7$ is an example of a ______________.
2. A boundary line divides a plane into two regions called ______________.
3. True or false? The graph of $5x - 6y < 7$ includes its corresponding boundary line. ______________
4. True or false? When graphing a linear inequality, to determine which side of the boundary line to shade, choose a point *not* on the boundary line. ______________
5. True or false? The boundary line for the inequality $5x - 6y < 7$ is the graph of $5x - 6y = 7$.

____

6. The graph of ______________ is

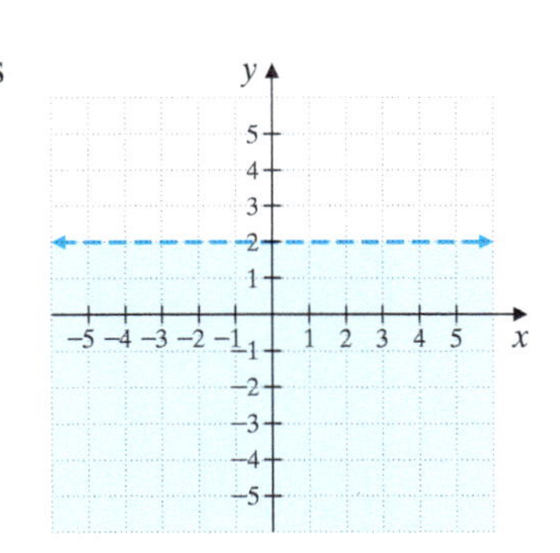

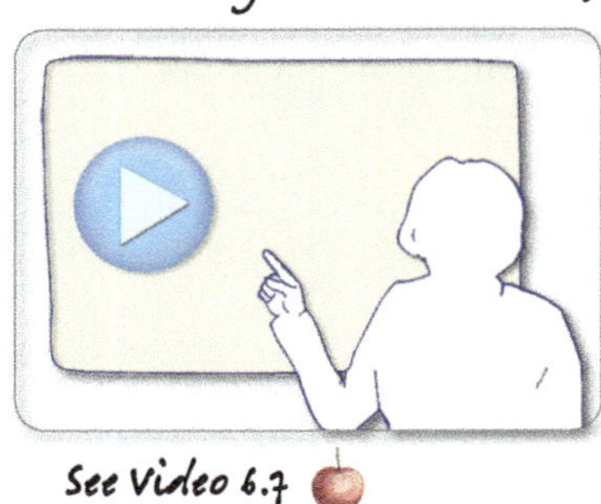

Watch the section lecture video and answer the following questions.

Objective A 7. From Example 1, how do we determine whether an ordered pair in x and y is a solution of an inequality in x and y?

Objective B 8. From Example 3, how do we find the equation of the boundary line? How do we determine if the points on the boundary line are solutions of the inequality?

6.7 Exercise Set MyLab Math

Objective A *Determine whether the ordered pairs given are solutions of the linear inequality in two variables. See Example 1.*

1. $x - y > 3$; $(0, 3)$, $(2, -1)$

2. $y - x < -2$; $(2, 1)$, $(5, -1)$

3. $3x - 5y \leq -4$; $(2, 3)$, $(-1, -1)$

4. $2x + y \geq 10$; $(0, 11)$, $(5, 0)$

5. $x < -y$; $(0, 2)$, $(-5, 1)$

6. $y > 3x$; $(0, 0)$, $(1, 4)$

Objective B *Graph each inequality. See Examples 2 through 7.*

7. $x + y \leq 1$

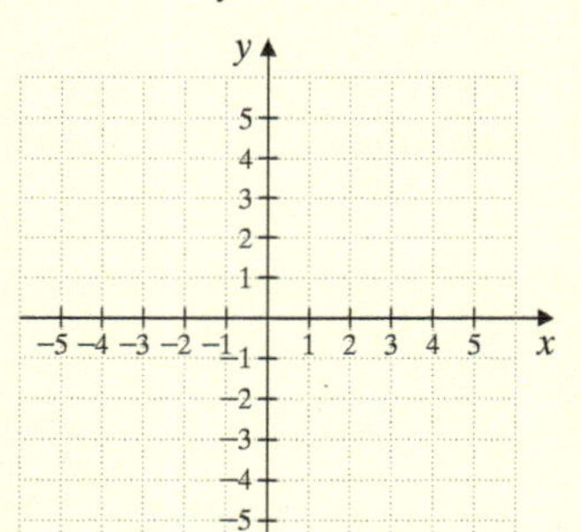

8. $x + y \geq -2$

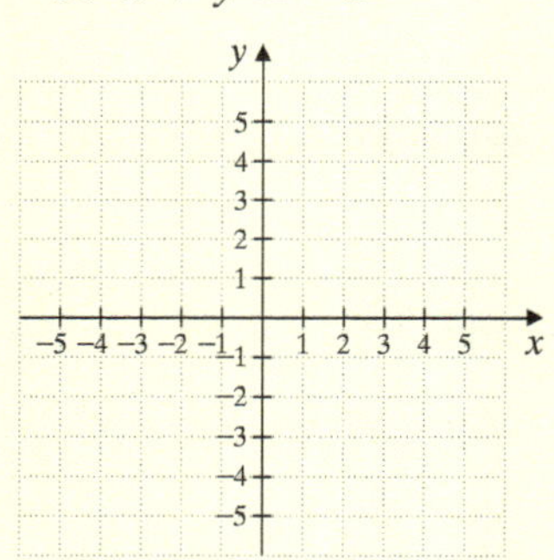

9. $2x - y > -4$

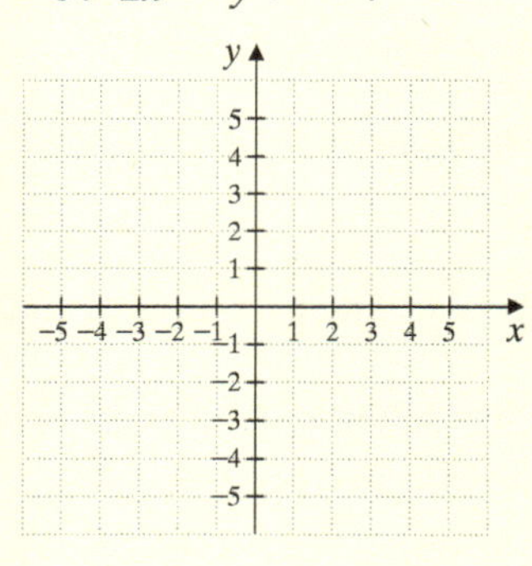

10. $x - 3y < 3$

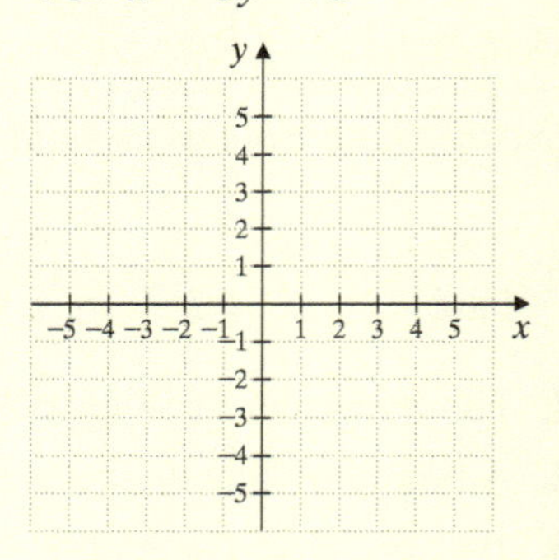

11. $y \geq 2x$

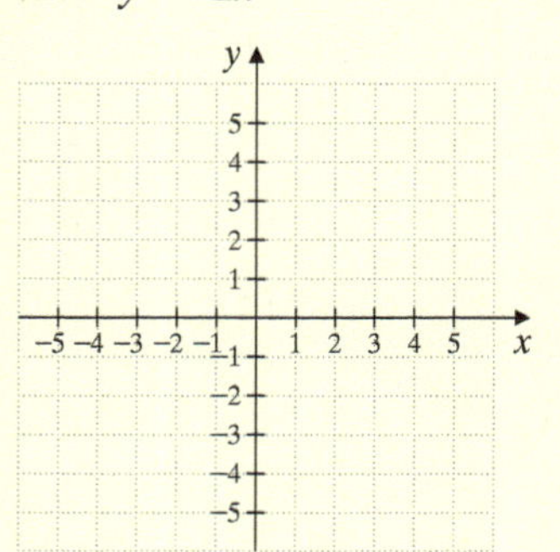

12. $y \leq 3x$

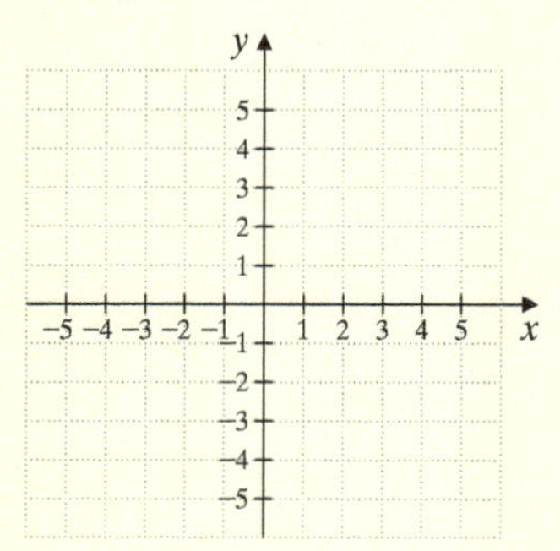

13. $x < -3y$

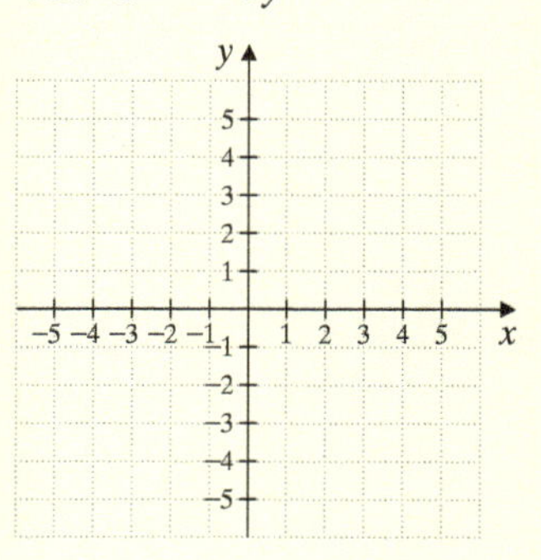

14. $x > -2y$

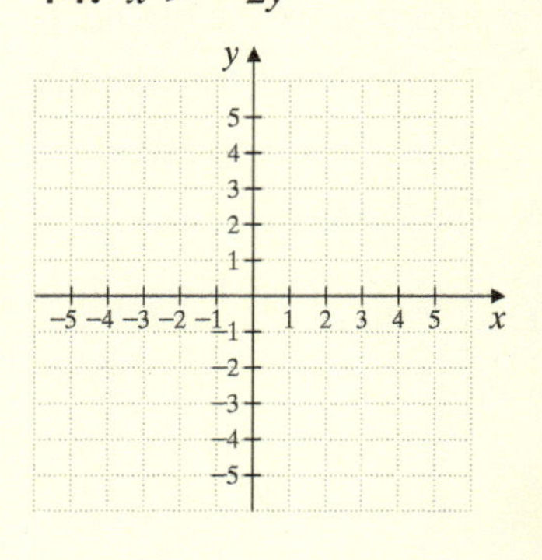

15. $y \geq x + 5$

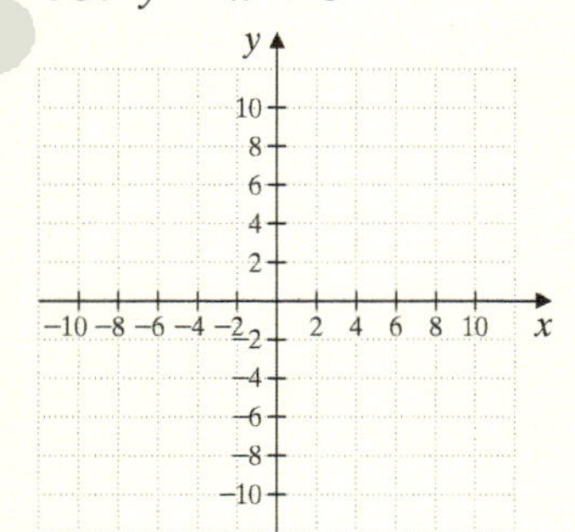

16. $y \leq x + 1$

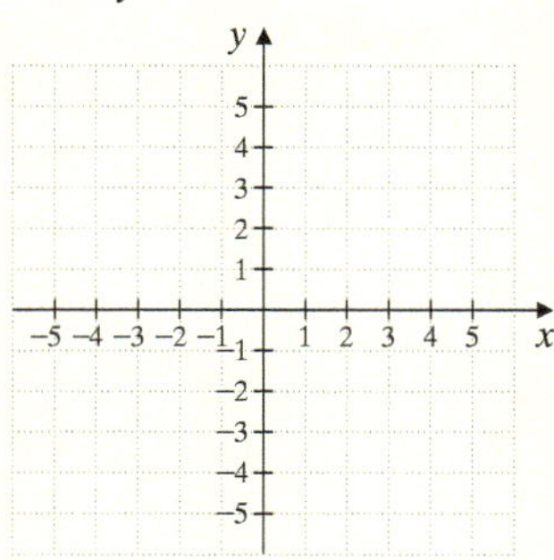

17. $y < 4$

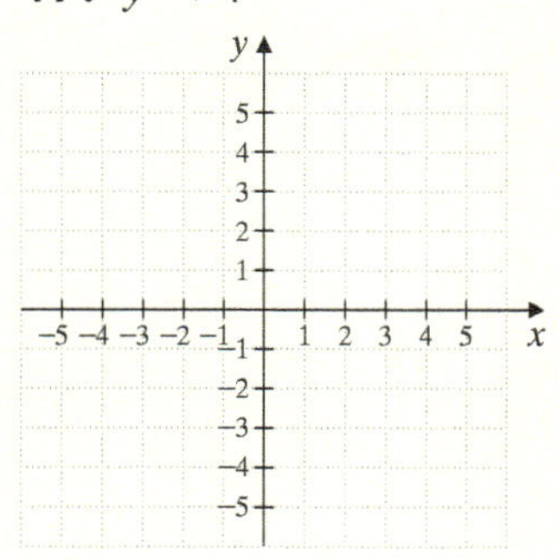

18. $y > 2$

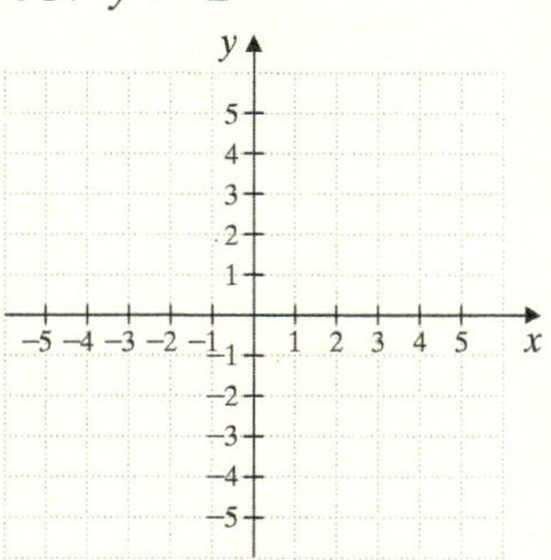

19. $x \geq -3$

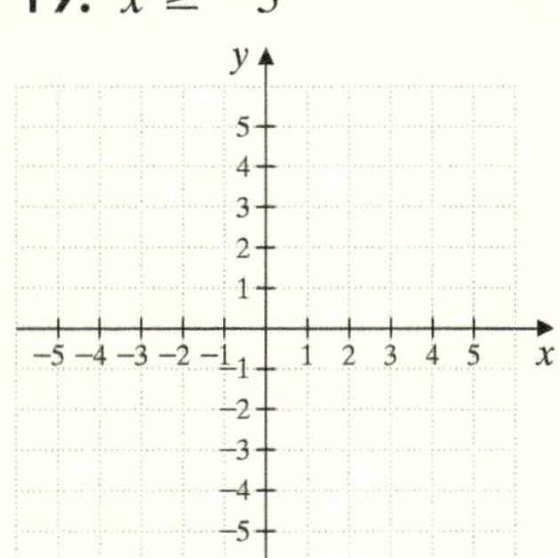

20. $x \leq -1$

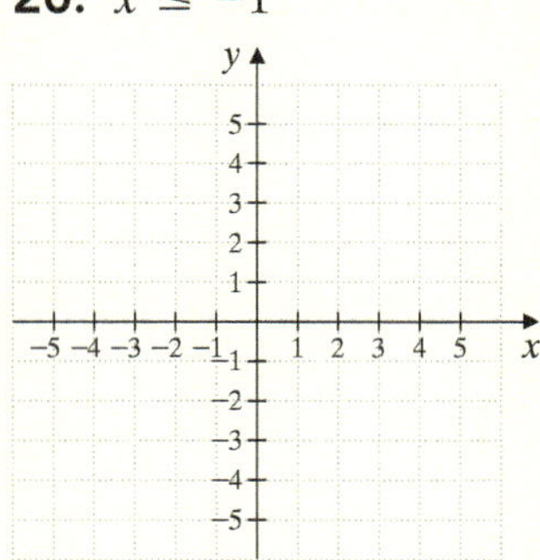

21. $5x + 2y \leq 10$

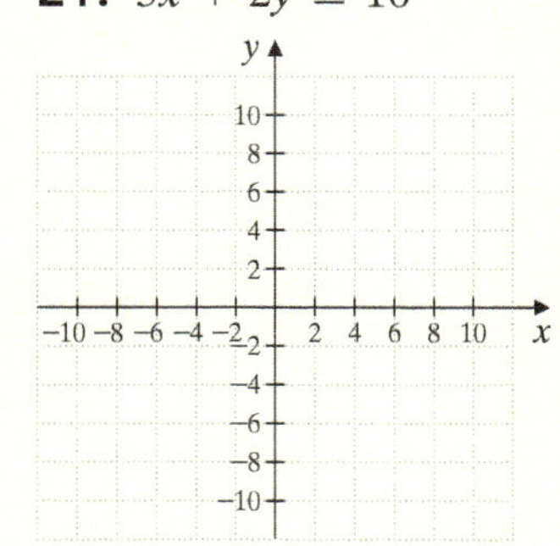

22. $4x + 3y \geq 12$

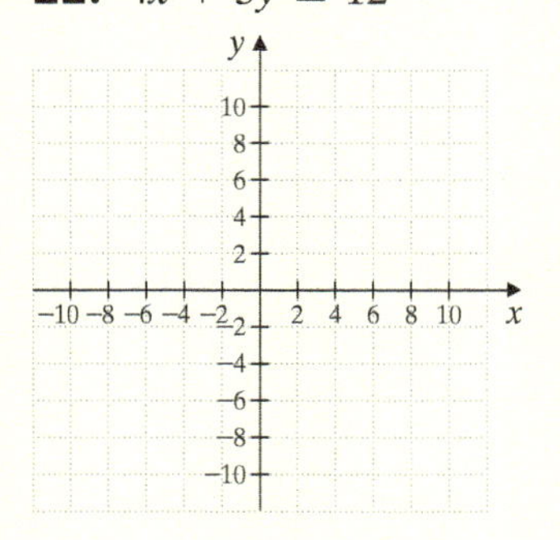

23. $x > y$

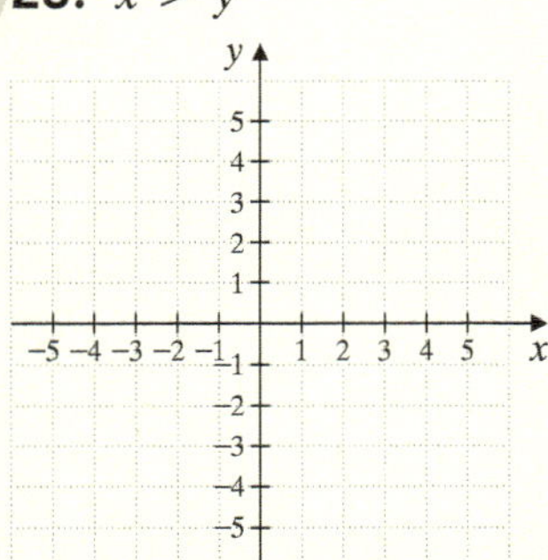

24. $x \leq -y$

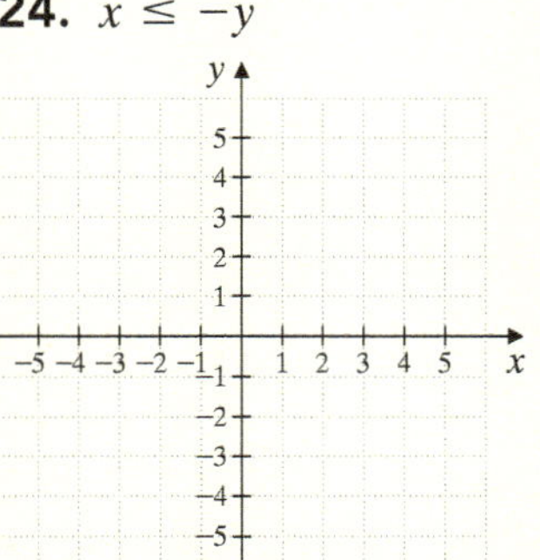

25. $x - y \leq 6$

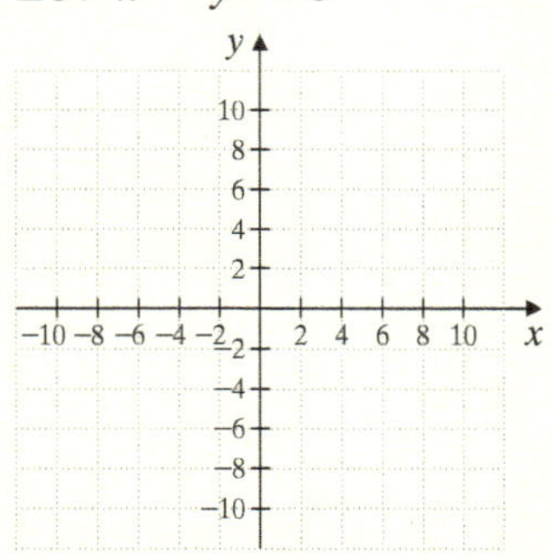

26. $x - y > 10$

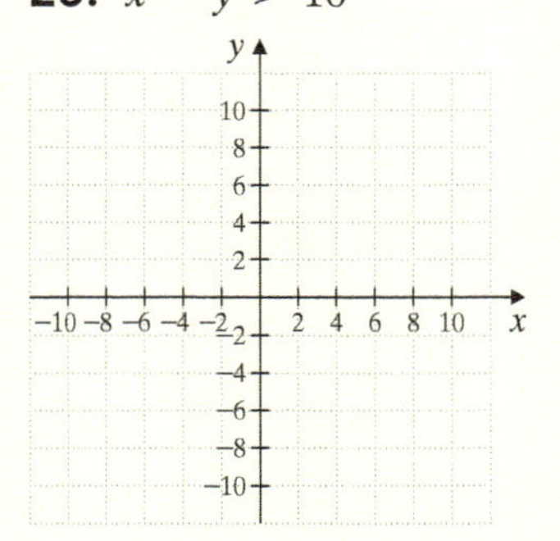

27. $x \geq 0$

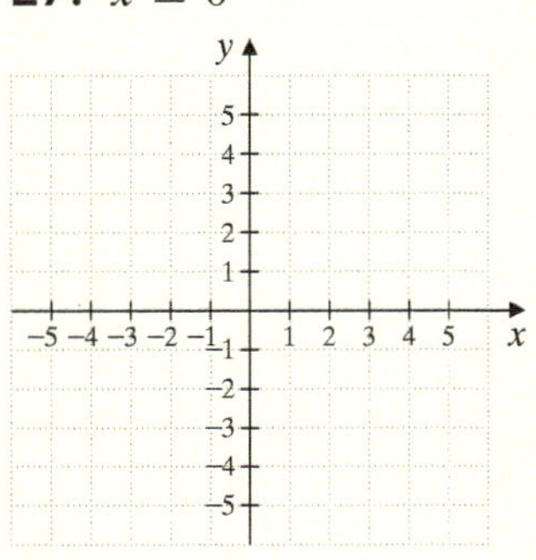

28. $y \leq 0$

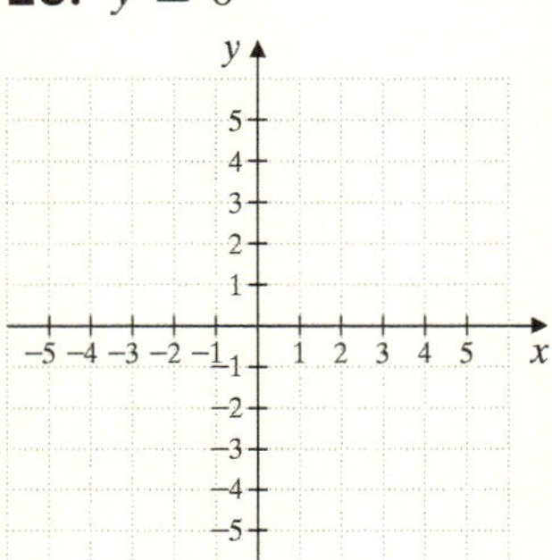

29. $2x + 7y > 5$

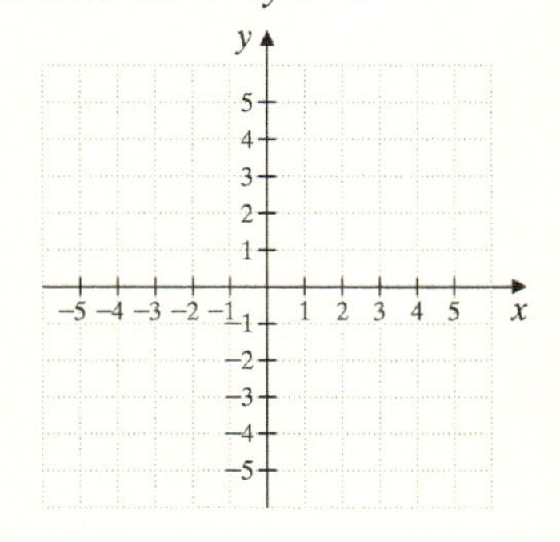

30. $3x + 5y \leq -2$

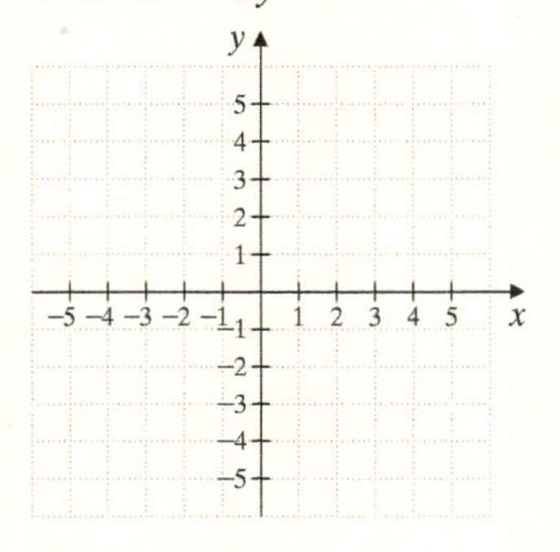

31. $y \geq \frac{1}{2}x - 4$

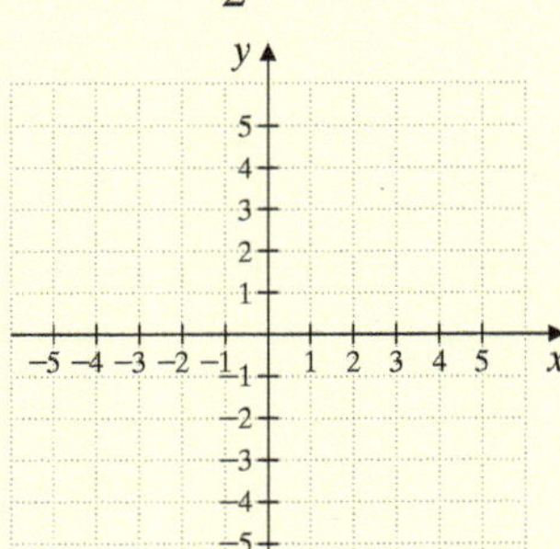

32. $y < \frac{2}{5}x - 3$

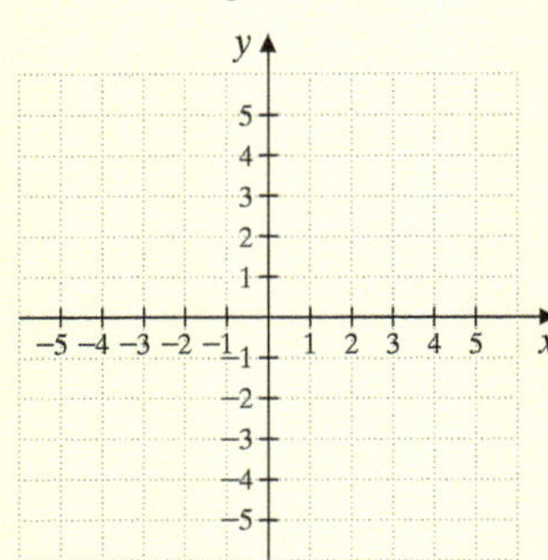

33. $y < -\frac{3}{4}x + 2$

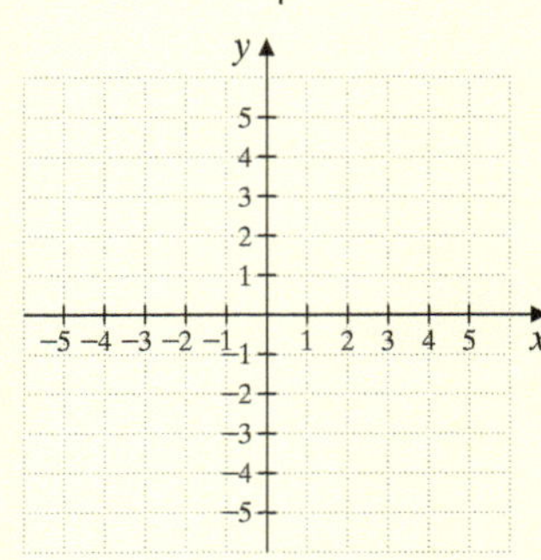

34. $y > -\frac{1}{3}x + 4$

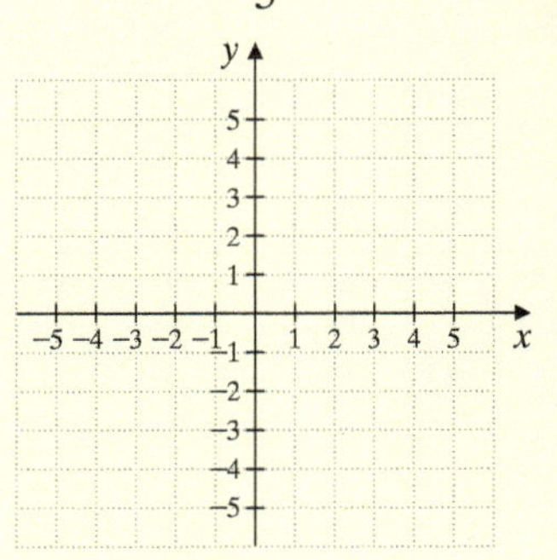

Review

Approximate the coordinates of each point of intersection. See Section 6.1.

35.

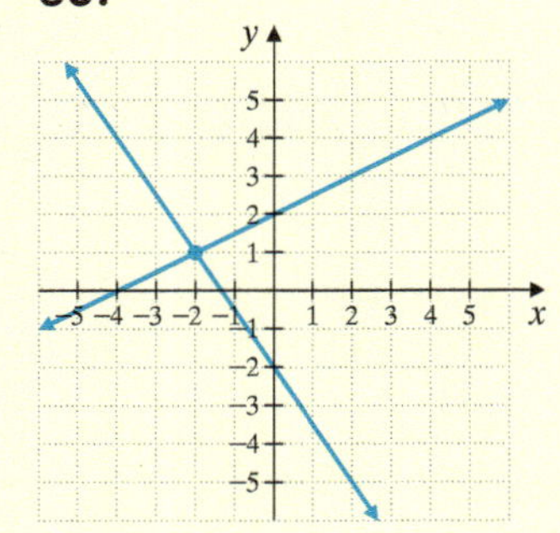

36.

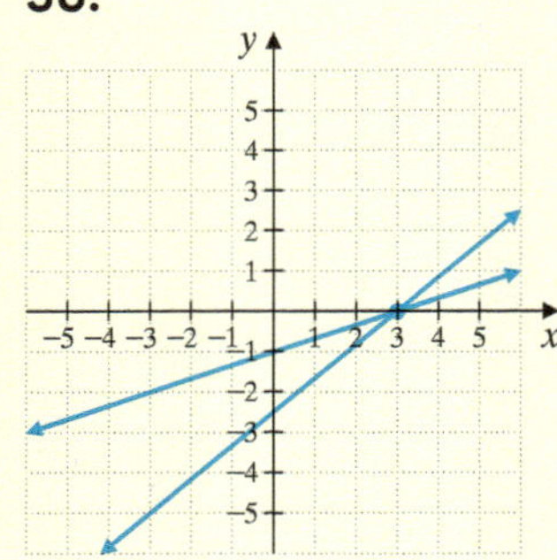

37.

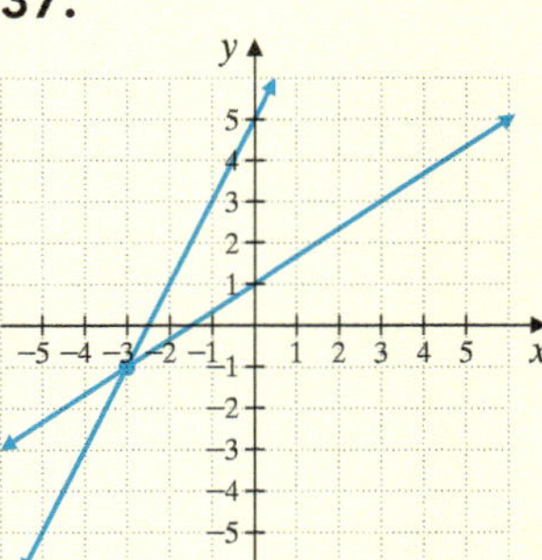

38.

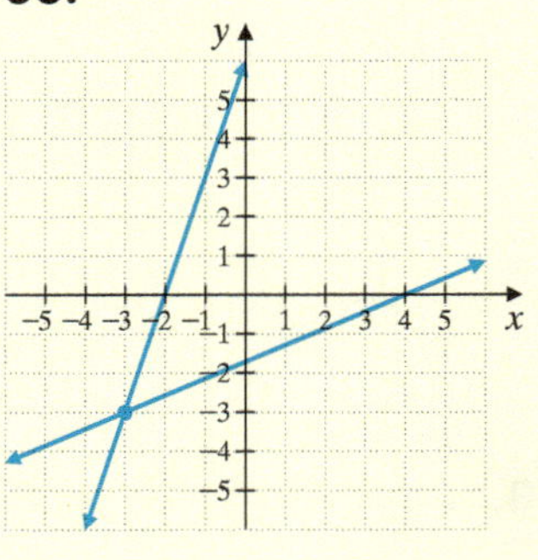

Concept Extensions

Match each inequality with its graph.

A. $x > 2$ **B.** $y < 2$ **C.** $y \leq 2x$ **D.** $y \leq -3x$

39.

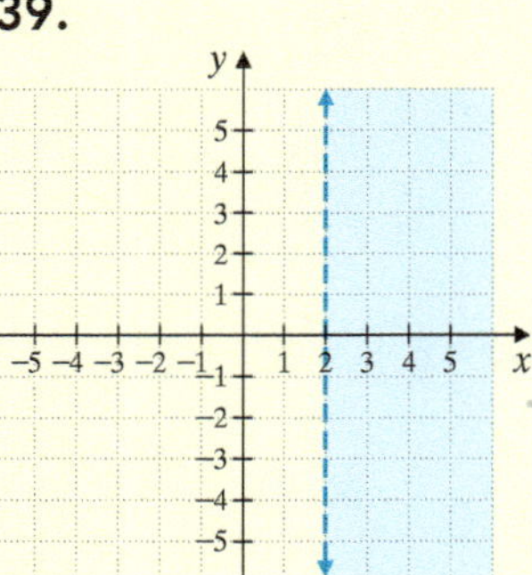

40.

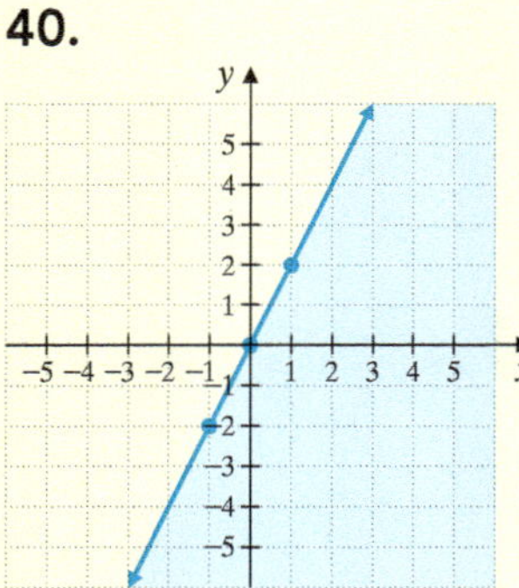

41.

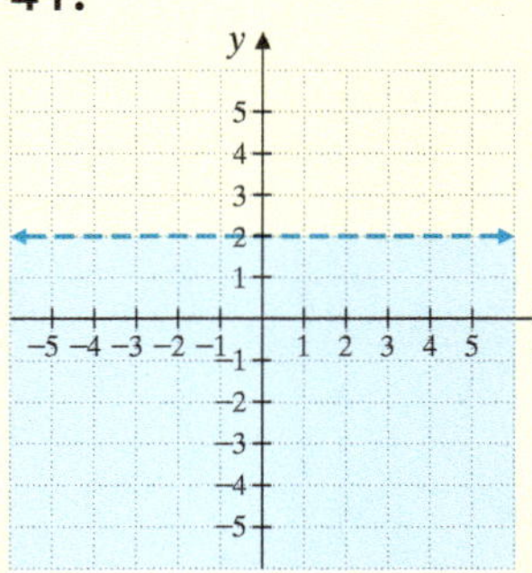

42.

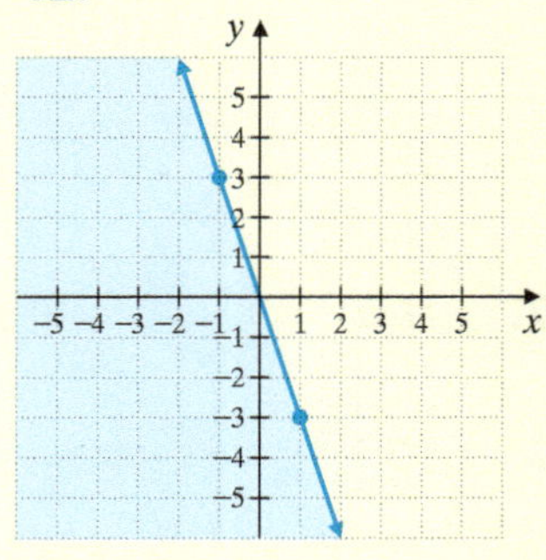

43. Explain why a point on the boundary line should not be chosen as the test point.

44. Write an inequality whose solutions are all points with coordinates whose sum is at least 13.

Determine whether (1, 1) is included in each graph. See the Concept Check in this section.

45. $3x + 4y < 8$

46. $y > 5x$

47. $y \geq -\frac{1}{2}x$

48. $x > 3$

49. It's the end of the budgeting period for Dennis Fernandes and he has $500 left in his budget for car rental expenses. He plans to spend this budget on a sales trip throughout southern Texas. He will rent a car that costs $30 per day and $0.15 per mile and he can spend no more than $500.

a. Write an inequality describing this situation. Let x = number of days and let y = number of miles.

b. Graph this inequality below.

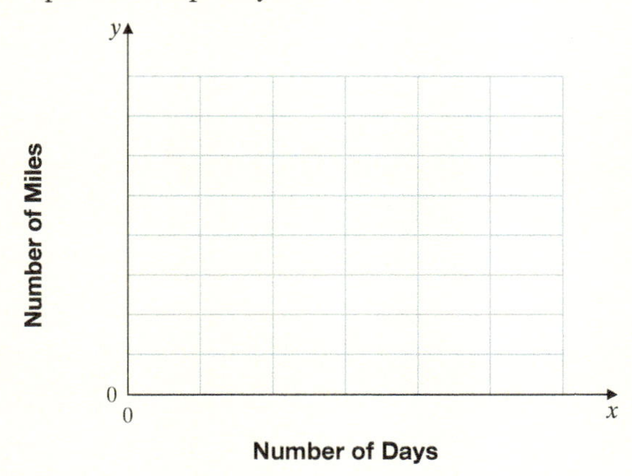

c. Why is the grid showing quadrant I only?

50. A couple are planning their wedding. They have calculated that they want the cost of their wedding ceremony x plus the cost of their reception y to be no more than $5000.

a. Write an inequality describing this relationship.

b. Graph this inequality below.

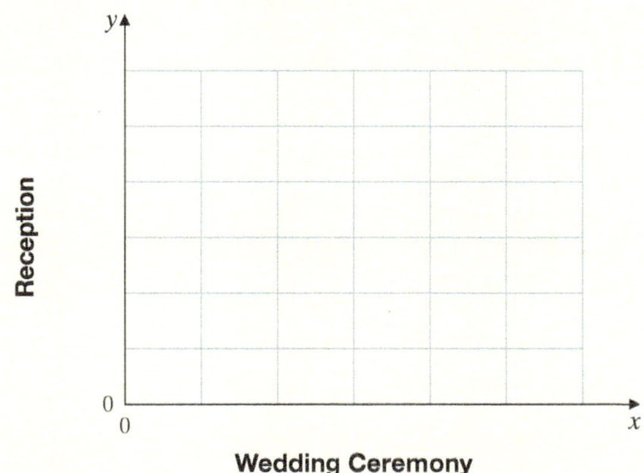

c. Why is the grid showing quadrant I only?

Chapter 6 Group Activity

Finding a Linear Model

This activity may be completed by working in groups or individually.

The following table shows the actual number of international tourist arrivals to the United States for the years 2012 through 2017.

Year	International Tourist Arrivals to the United States (in millions)
2012	67
2013	70
2014	75
2015	79
2016	77
2017	77

Source: World Tourism Organization

1. Make a scatter diagram of the paired data in the table.
2. Use what you have learned in this chapter to write an equation of the line representing the paired data in the table. Explain how you found the equation, and what each variable represents.
3. What is the slope of your line? What does the slope mean in this context?
4. Use your linear equation to predict the number of international tourist arrivals to the United States in 2024.
5. Compare your linear equation to that found by other students or groups. Is it the same, similar, or different? How?
6. Compare your prediction from question **4** to that of other students or groups. Describe what you find.
7. Suppose that the number of international tourist arrivals to the United States for 2024 was estimated to be 95 million. If this data point is added to the chart, how does it affect your results?

Chapter 6 Vocabulary Check

Fill in each blank with one of the words listed below.

y-axis	x-axis	solution	x	slope-intercept	function
x-intercept	y-intercept	y	standard	point-slope	
domain	range	linear	slope	relation	

1. An ordered pair is a(n) ______________ of an equation in two variables if replacing the variables by the coordinates of the ordered pair results in a true statement.
2. The vertical number line in the rectangular coordinate system is called the ______________.
3. A(n) ______________ equation can be written in the form $Ax + By = C$.
4. A(n) ______________ is a point of a graph where the graph crosses the x-axis.
5. The form $Ax + By = C$ is called ______________ form.
6. A(n) ______________ is a point of a graph where the graph crosses the y-axis.
7. A set of ordered pairs that assigns to each x-value exactly one y-value is called a(n) ______________.
8. The equation $y = 7x - 5$ is written in ______________ form.
9. The set of all x-coordinates of a relation is called the ______________ of the relation.
10. The set of all y-coordinates of a relation is called the ______________ of the relation.
11. A set of ordered pairs is called a(n) ______________.
12. The equation $y + 1 = 7(x - 2)$ is written in ______________ form.
13. To find an x-intercept of a graph, let ______________ $= 0$.
14. The horizontal number line in the rectangular coordinate system is called the ______________.
15. To find a y-intercept of a graph, let ______________ $= 0$.
16. The ______________ of a line measures the steepness or tilt of the line.

Helpful Hint

Are you preparing for your test? To help, don't forget to take these:

- Chapter 6 Getting Ready for the Test on page 521
- Chapter 6 Test on page 522

Then check all of your answers at the back of this text. For further review, the step-by-step video solutions to any of these exercises are located in MyLab Math.

6 Chapter Highlights

Definitions and Concepts	Examples
Section 6.1 The Rectangular Coordinate System	

The **rectangular coordinate system** consists of a plane and a vertical and a horizontal number line intersecting at their 0 coordinates. The vertical number line is called the ***y*-axis** and the horizontal number line is called the ***x*-axis.** The point of intersection of the axes is called the **origin.**

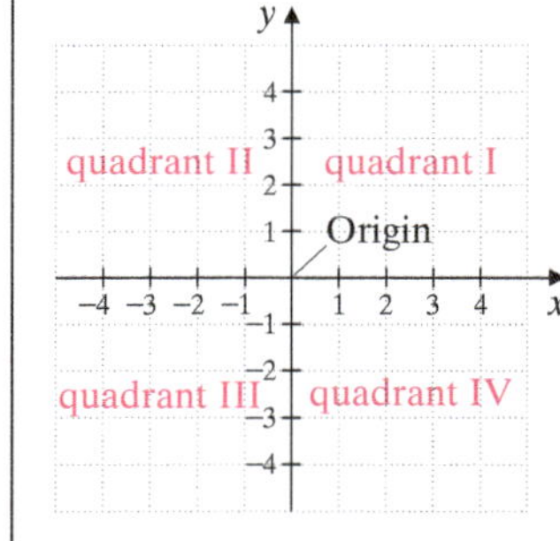

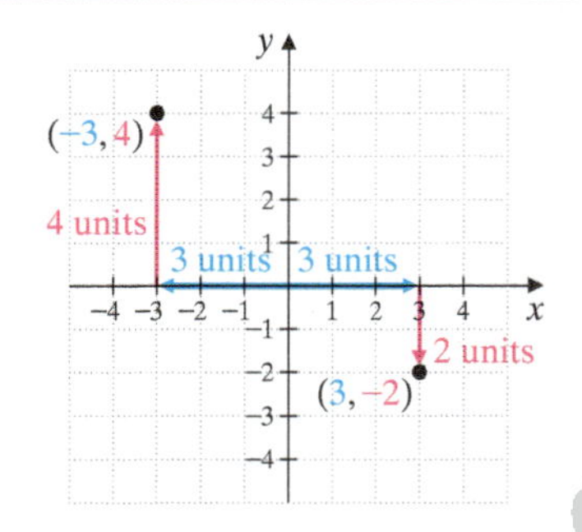

To **plot** or **graph** an ordered pair means to find its corresponding point on a rectangular coordinate system.

To plot or graph an ordered pair such as $(3, -2)$, start at the origin. Move 3 units to the right and from there, 2 units down.

To plot or graph $(-3, 4)$, start at the origin. Move 3 units to the left and from there, 4 units up.

An ordered pair is a **solution** of an equation in two variables if replacing the variables with the coordinates of the ordered pair results in a true statement.

Definitions and Concepts	Examples

Section 6.1 The Rectangular Coordinate System (*continued*)

If one coordinate of an ordered pair solution of an equation is known, the other value can be determined by substitution.

Complete the ordered pair (0,) for the equation $x - 6y = 12$.

$$x - 6y = 12$$
$$0 - 6y = 12 \quad \text{Let } x = 0.$$
$$\frac{-6y}{-6} = \frac{12}{-6} \quad \text{Divide by } -6.$$
$$y = -2$$

The ordered pair solution is $(0, -2)$.

Section 6.2 Graphing Linear Equations

A **linear equation in two variables** is an equation that can be written in the form $Ax + By = C$, where A and B are not both 0. The form $Ax + By = C$ is called **standard form.**

$$3x + 2y = -6 \qquad x = -5$$
$$y = 3 \qquad y = -x + 10$$

$3x + 2y = -6$ is in standard form.

To graph a linear equation in two variables, find three ordered pair solutions. Plot the solution points and draw the line connecting the points.

Graph: $x - 2y = 5$

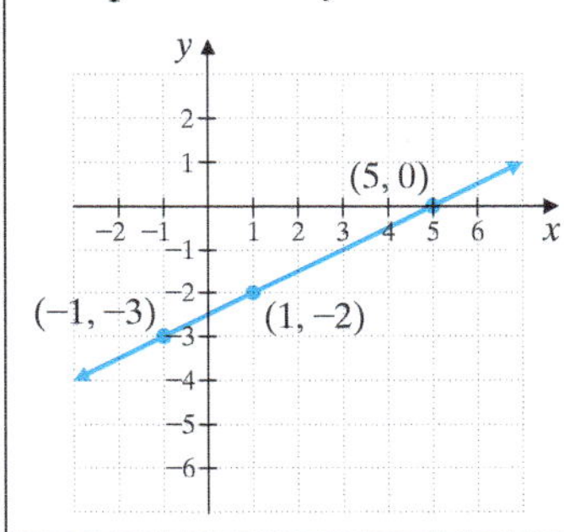

x	y
5	0
1	−2
−1	−3

Section 6.3 Intercepts

An **intercept** of a graph is a point where the graph intersects an axis. If a graph intersects the x-axis at a, then $(a, 0)$ is an ***x*-intercept.** If a graph intersects the y-axis at b, then $(0, b)$ is a ***y*-intercept.**

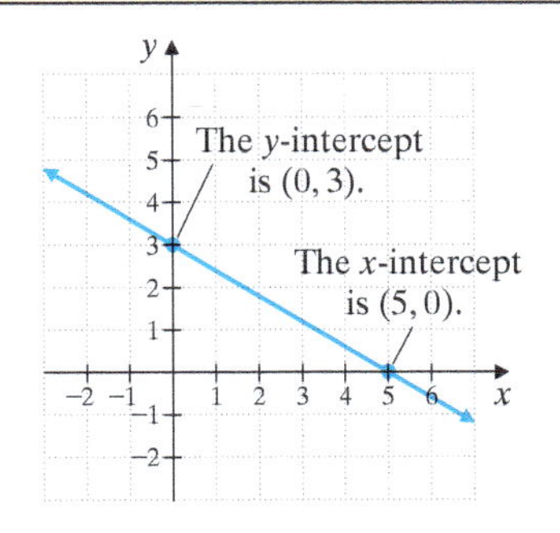

To find the *x*-intercept(s), let $y = 0$ and solve for x.

To find the *y*-intercept(s), let $x = 0$ and solve for y.

Find the intercepts for $2x - 5y = -10$.

If $y = 0$, then

$$2x - 5 \cdot 0 = -10$$
$$2x = -10$$
$$\frac{2x}{2} = \frac{-10}{2}$$
$$x = -5$$

If $x = 0$, then

$$2 \cdot 0 - 5y = -10$$
$$-5y = -10$$
$$\frac{-5y}{-5} = \frac{-10}{-5}$$
$$y = 2$$

(continued)

Definitions and Concepts	Examples
Section 6.3 Intercepts (*continued*)	
	The x-intercept is $(-5, 0)$. The y-intercept is $(0, 2)$. 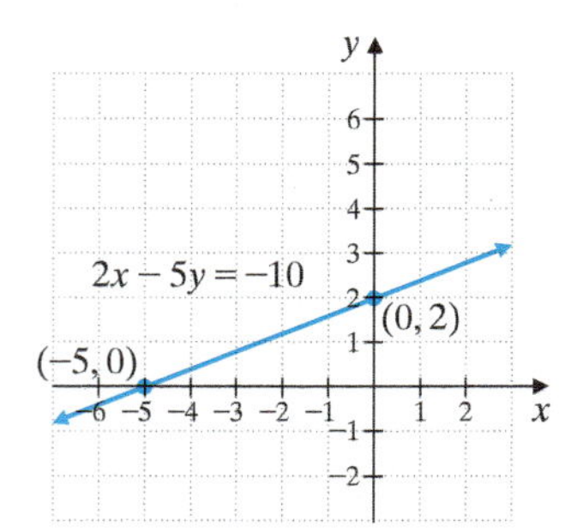
The graph of $x = c$ is a vertical line with x-intercept $(c, 0)$. The graph of $y = c$ is a horizontal line with y-intercept $(0, c)$.	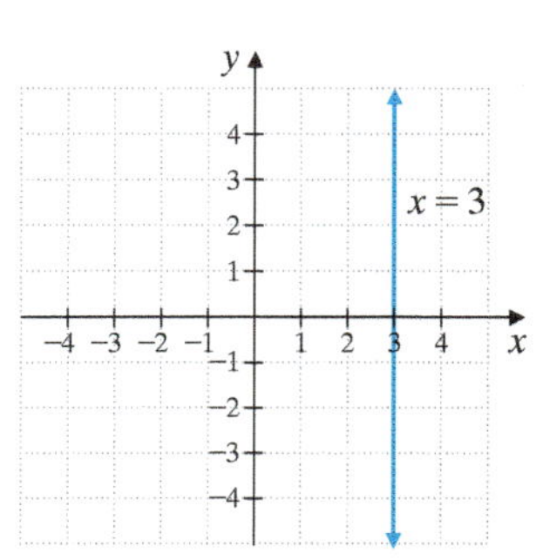 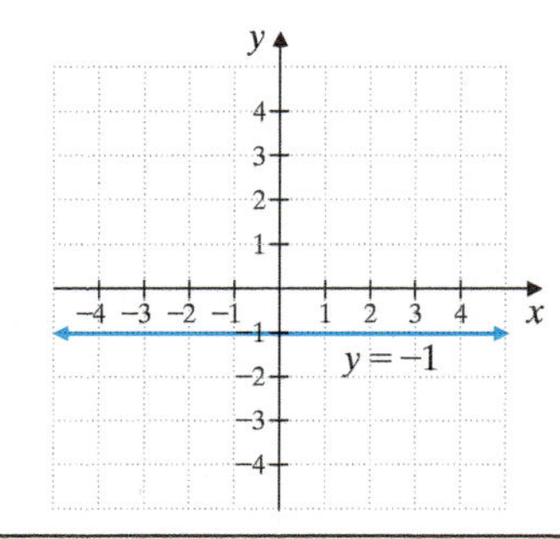
Section 6.4 Slope and Rate of Change	
The **slope m** of the line through points (x_1, y_1) and (x_2, y_2) is given by $$m = \frac{y_2 - y_1}{x_2 - x_1} \quad \text{as long as } x_2 \neq x_1$$ A horizontal line has slope 0. The slope of a vertical line is undefined. Nonvertical parallel lines have the same slope. Two nonvertical lines are perpendicular if the slope of one is the negative reciprocal of the slope of the other.	The slope of the line through points $(-1, 6)$ and $(-5, 8)$ is $$m = \frac{y_2 - y_1}{x_2 - x_1} = \frac{8 - 6}{-5 - (-1)} = \frac{2}{-4} = -\frac{1}{2}$$ The slope of the line $y = -5$ is 0. The line $x = 3$ has undefined slope. 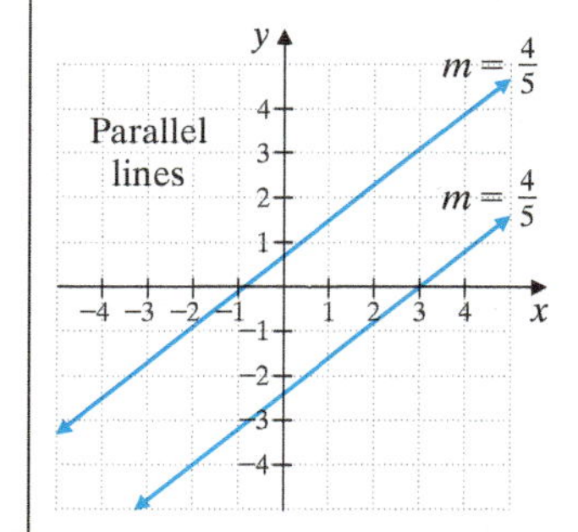 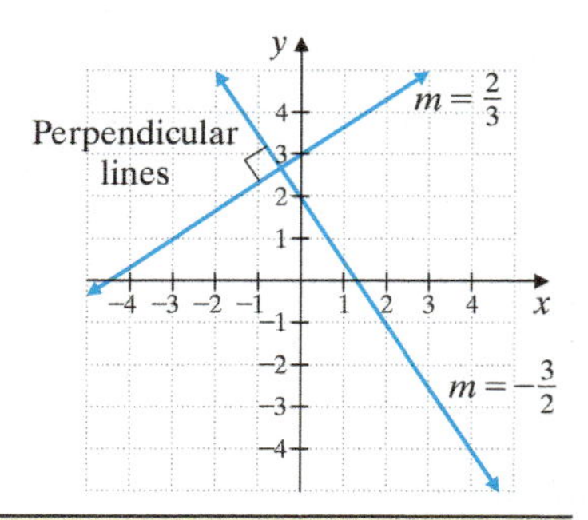
Section 6.5 Equations of Lines	
Slope-Intercept Form $$y = mx + b$$ m is the slope of the line. $(0, b)$ is the y-intercept.	Find the slope and the y-intercept of the line $2x + 3y = 6$. Solve for y: $$2x + 3y = 6$$ $$3y = -2x + 6 \quad \text{Subtract } 2x.$$ $$y = -\frac{2}{3}x + 2 \quad \text{Divide by 3.}$$ The slope of the line is $-\frac{2}{3}$ and the y-intercept is $(0, 2)$.

Definitions and Concepts	Examples
Section 6.5 Equations of Lines (*continued*)	
Point-Slope Form $y - y_1 = m(x - x_1)$ m is the slope. (x_1, y_1) is a point of the line.	Find an equation of the line with slope $\frac{3}{4}$ that contains the point $(-1, 5)$. $y - 5 = \frac{3}{4}[x - (-1)]$ $4(y - 5) = 3(x + 1)$ Multiply by 4. $4y - 20 = 3x + 3$ Distribute. $-3x + 4y = 23$ Subtract $3x$ and add 20.
Section 6.6 Introduction to Functions	
A set of ordered pairs is a **relation.** The set of all x-coordinates is called the **domain** of the relation and the set of all y-coordinates is called the **range** of the relation.	The domain of the relation $\{(0, 5), (2, 5), (4, 5), (5, -2)\}$ is $\{0, 2, 4, 5\}$. The range is $\{-2, 5\}$.
A **function** is a set of ordered pairs that assigns to each x-value exactly one y-value. **Vertical Line Test** If a vertical line can be drawn so that it intersects a graph more than once, the graph is not the graph of a function. (If no such line can be drawn, the graph is that of a function.)	Which are graphs of functions? 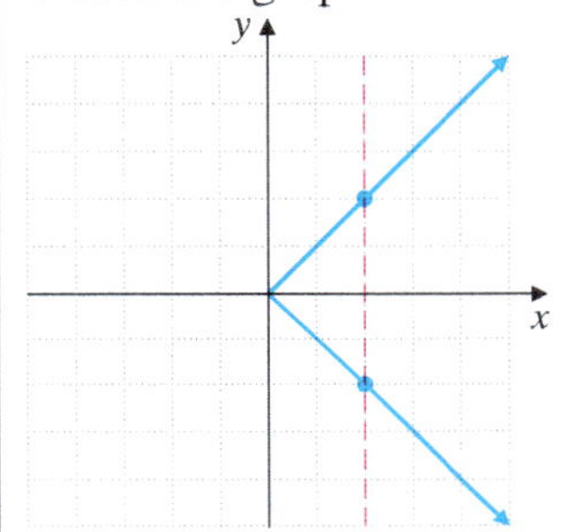 This graph is not the graph of a function. 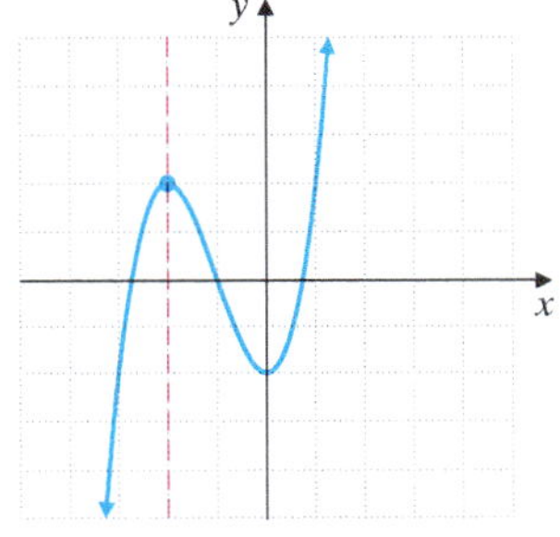 This graph is the graph of a function.
The symbol $f(x)$ means **function of x.** This notation is called **function notation.**	If $f(x) = 3x - 7$, then $f(-1) = 3(-1) - 7$ $= -3 - 7$ $= -10$
Section 6.7 Graphing Linear Inequalities in Two Variables	
A **linear inequality in two variables** is an inequality that can be written in one of these forms: $Ax + By < C$ $Ax + By \le C$ $Ax + By > C$ $Ax + By \ge C$ where A and B are not both 0.	$2x - 5y < 6$ $x \ge -5$ $y > -8x$ $y \le 2$

(*continued*)

Definitions and Concepts	Examples
Section 6.7 Graphing Linear Inequalities in Two Variables (*continued*)	
To Graph a Linear Inequality 1. Graph the boundary line by graphing the related equation. Draw the line solid if the inequality symbol is $\le$ or $\ge$. Draw the line dashed if the inequality symbol is $<$ or $>$. 2. Choose a test point not on the line. Substitute its coordinates into the original inequality. 3. If the resulting inequality is true, shade the half-plane that contains the test point. If the inequality is not true, shade the half-plane that does not contain the test point.	Graph: $2x - y \le 4$ 1. Graph $2x - y = 4$. Draw a solid line because the inequality symbol is $\le$. 2. Check the test point (0, 0) in the original inequality, $2x - y \le 4$. $2 \cdot 0 - 0 \le 4$ Let $x = 0$ and $y = 0$. $0 \le 4$ True 3. The inequality is true, so shade the half-plane containing (0, 0), as shown. 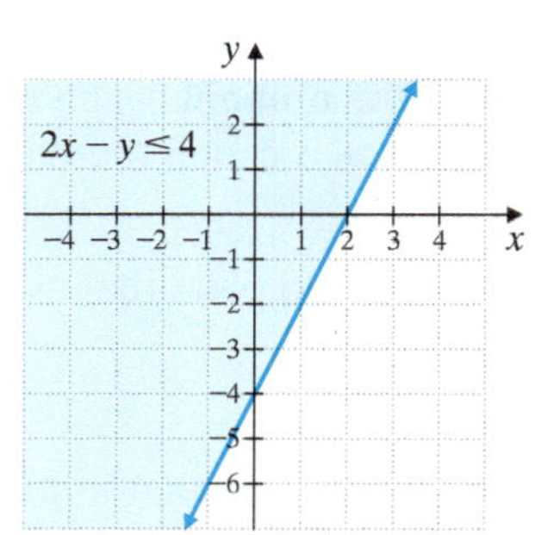

Chapter 6 Review

(6.1) *Plot each point on the same rectangular coordinate system.*

1. $(-7, 0)$
2. $\left(0, 4\frac{4}{5}\right)$
3. $(-2, -5)$
4. $(1, -3)$
5. $(0.7, 0.7)$
6. $(-6, 4)$

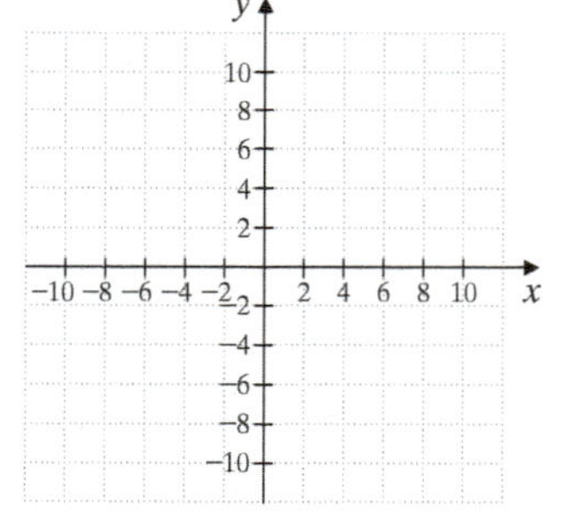

Complete each ordered pair so that it is a solution of the given equation.

7. $-2 + y = 6x$; $(7, \quad)$
8. $y = 3x + 5$; $(\quad, -8)$

Complete the table of values for each given equation.

9. $9 = -3x + 4y$

x	y
	0
	3
9	

10. $y = 5$

x	y
7	
−7	
0	

11. $x = 2y$

x	y
	0
	5
	−5

12. The cost in dollars of producing x compact disc holders is given by $y = 5x + 2000$.

a. Complete the table.

x	1	100	1000
y			

b. Find the number of compact disc holders that can be produced for \$6430.

(6.2) *Graph each linear equation.*

13. $x - y = 1$

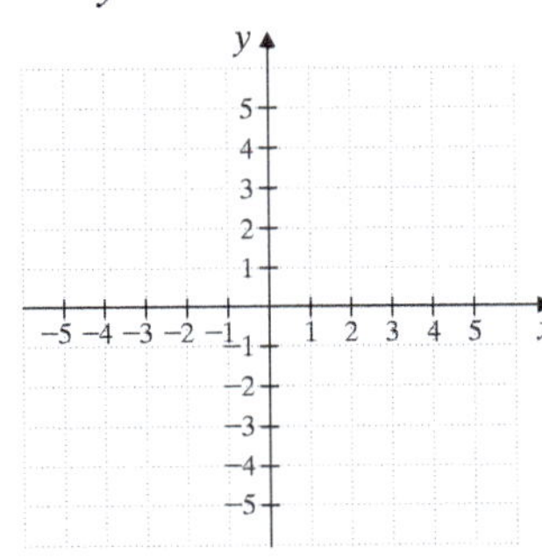

14. $x + y = 6$

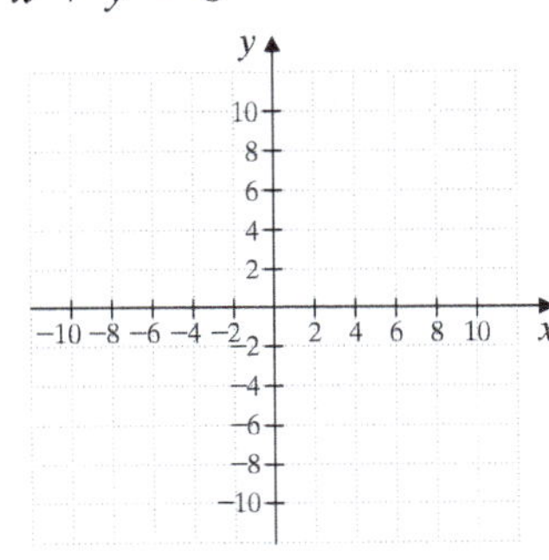

15. $x - 3y = 12$

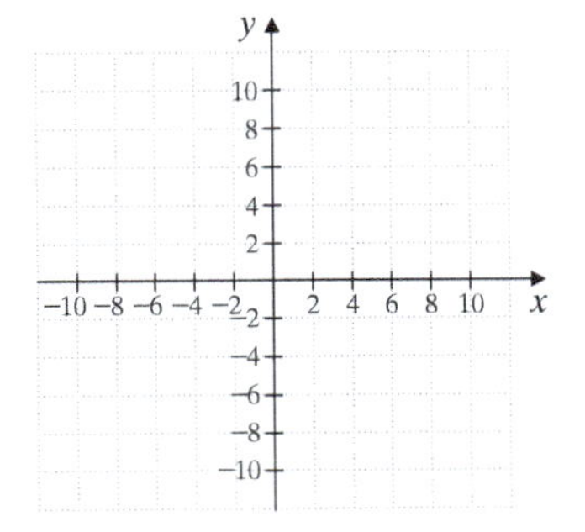

16. $5x - y = -8$

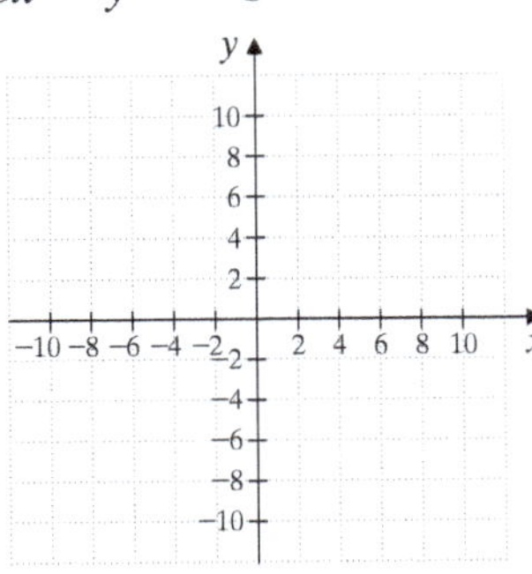

17. $x = 3y$

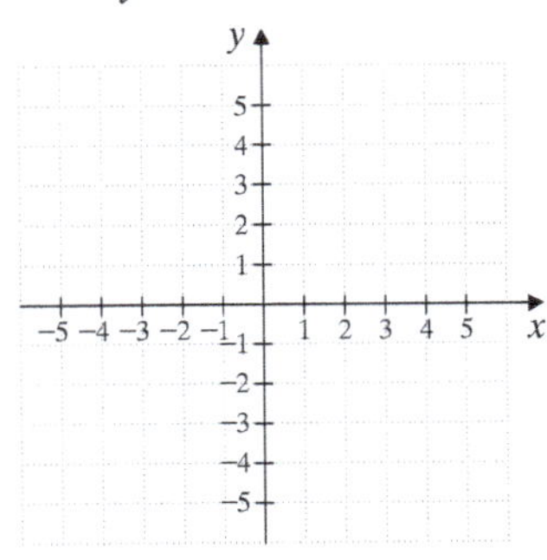

18. $y = -2x$

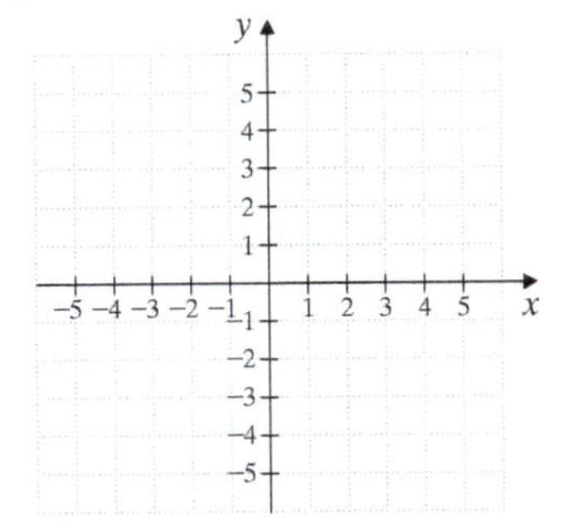

(6.3) *Identify the intercepts in each graph.*

19.

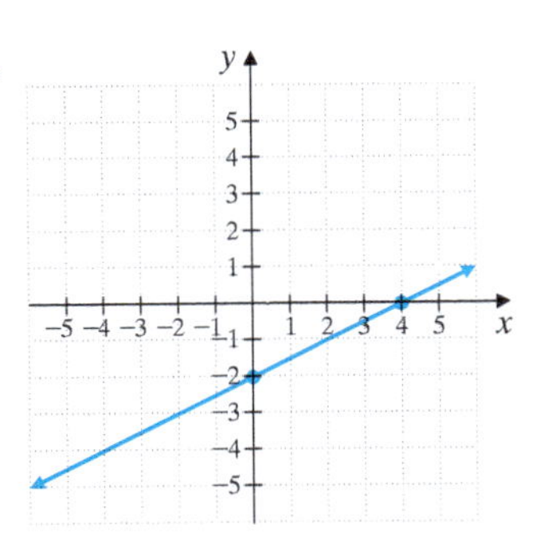

20.

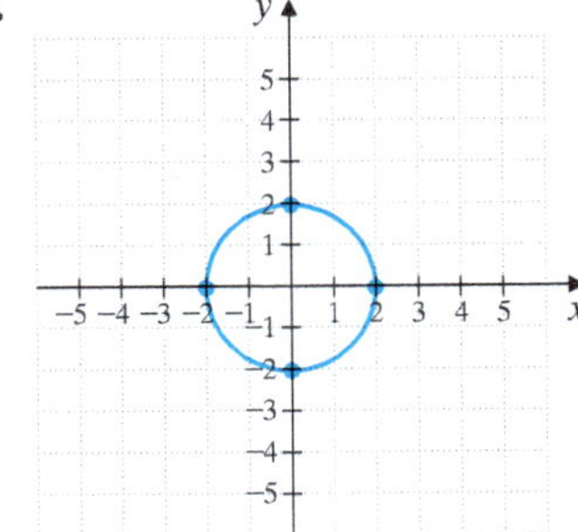

Graph each linear equation.

21. $y = -3$

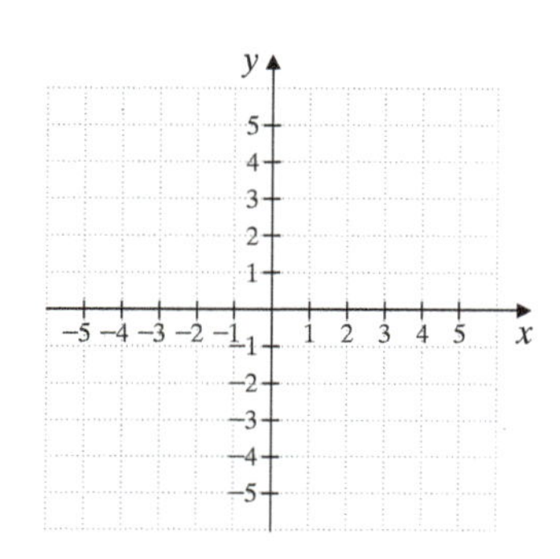

22. $x = 5$

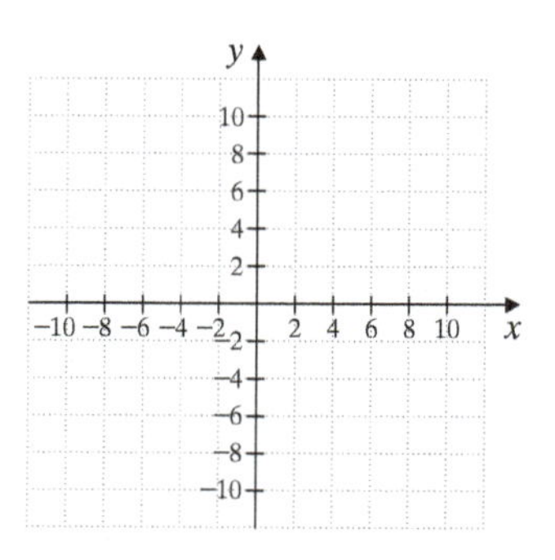

Find the intercepts of each equation.

23. $x - 3y = 12$

24. $-4x + y = 8$

(6.4) *Find the slope of each line.*

25.

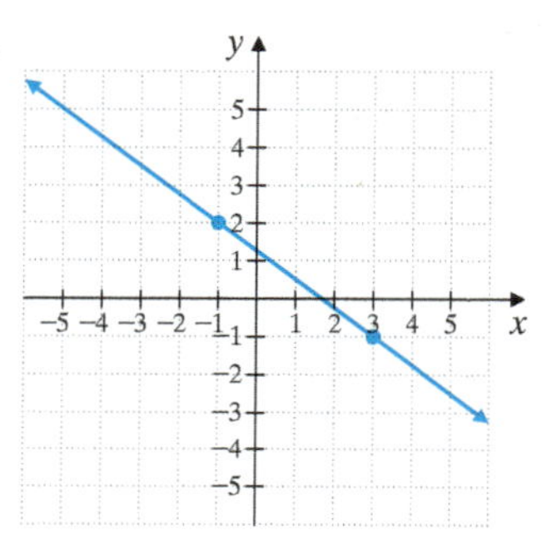

26.

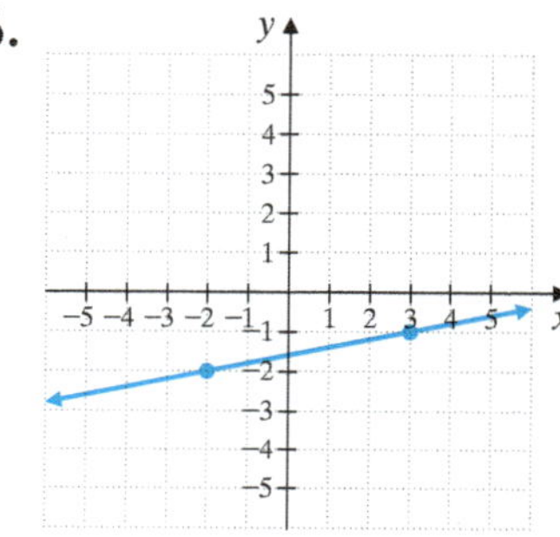

Match each line with its slope.

a.

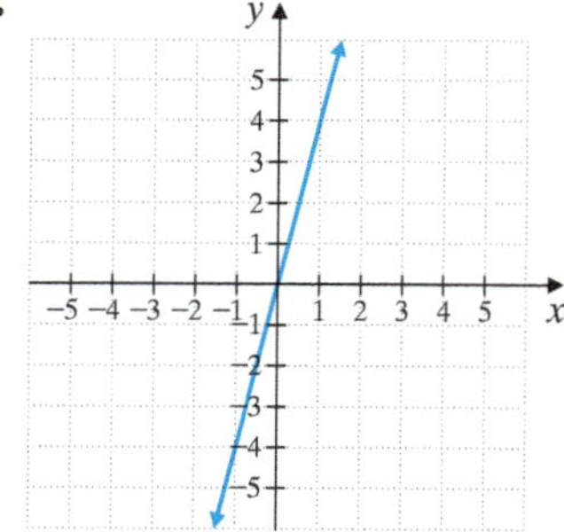

b.

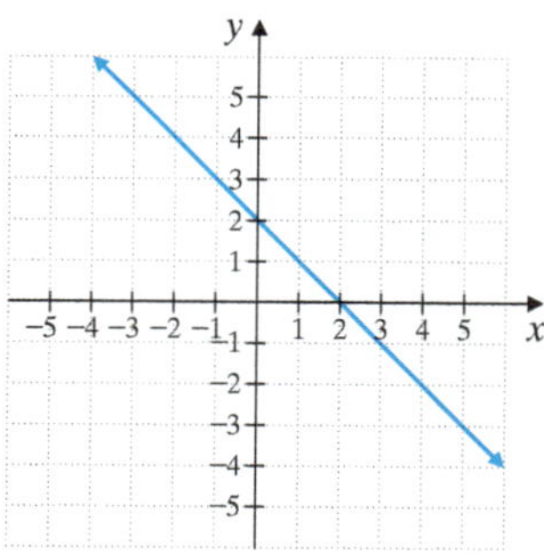

c.

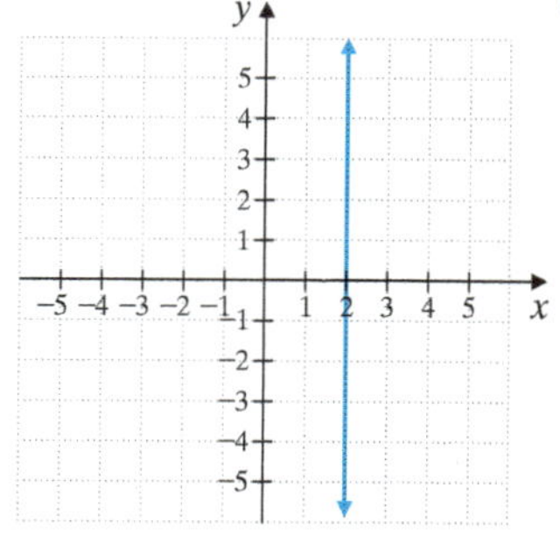

d.

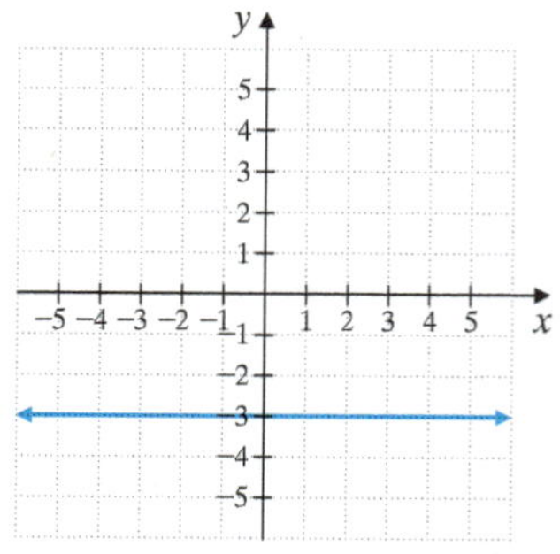

27. $m = 0$

28. $m = -1$

29. undefined slope

30. $m = 4$

Find the slope of the line that passes through each pair of points.

31. $(2, 5)$ and $(6, 8)$

32. $(4, 7)$ and $(1, 2)$

33. $(1, 3)$ and $(-2, -9)$

34. $(-4, 1)$ and $(3, -6)$

Find the slope of each line.

35. $y = 3x + 7$

36. $x - 2y = 4$

37. $y = -2$

38. $x = 0$

Determine whether each pair of lines is parallel, perpendicular, or neither.

39. $x - y = -6$
$x + y = 3$

40. $3x + y = 7$
$-3x - y = 10$

41. $y = 4x + \frac{1}{2}$
$4x + 2y = 1$

42. $y = 6x - \frac{1}{3}$
$x + 6y = 6$

Find the slope of each line and write the slope as a rate of change. Don't forget to attach the proper units.

43. The graph below approximates the total circulation of U. S. daily newspapers (print and digital) for each year x.

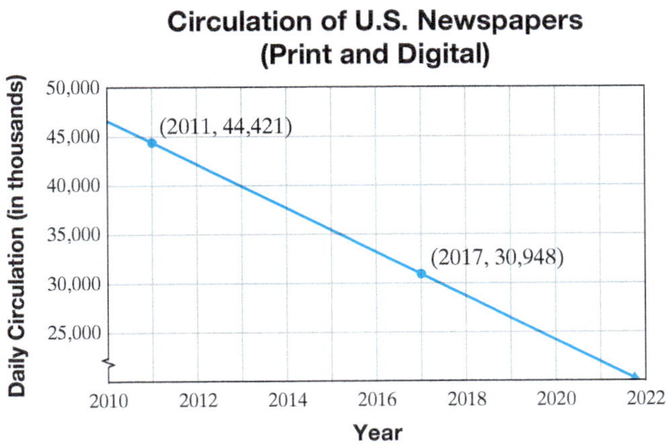

Source: MPA—The Association of Magazine Media

44. The graph below approximates the number of transplants y in the United States for year x. Round to the nearest whole.

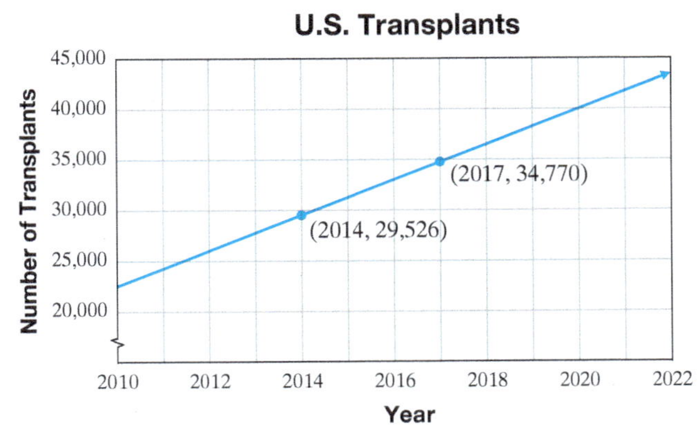

Source: Organ Procurement and Transplantation Network

(6.5) *Determine the slope and the y-intercept of the graph of each equation.*

45. $x - 6y = -1$

46. $3x + y = 7$

Write an equation of each line.

47. slope -5; y-intercept $\left(0, \frac{1}{2}\right)$

48. slope $\frac{2}{3}$; y-intercept $(0, 6)$

Match each equation with its graph.

49. $y = 2x + 1$

50. $y = -4x$

51. $y = 2x$

52. $y = 2x - 1$

a.

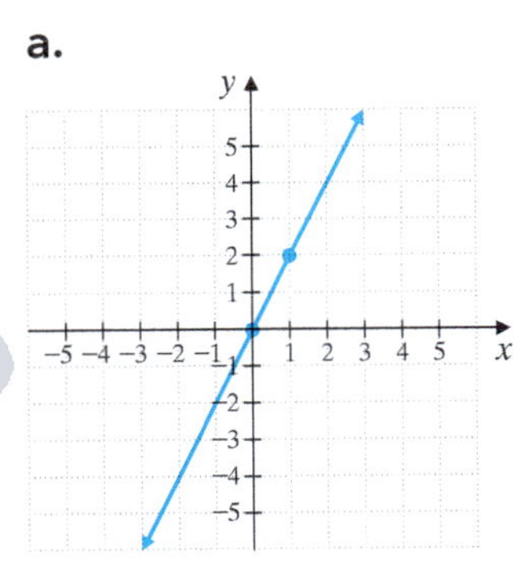

b.

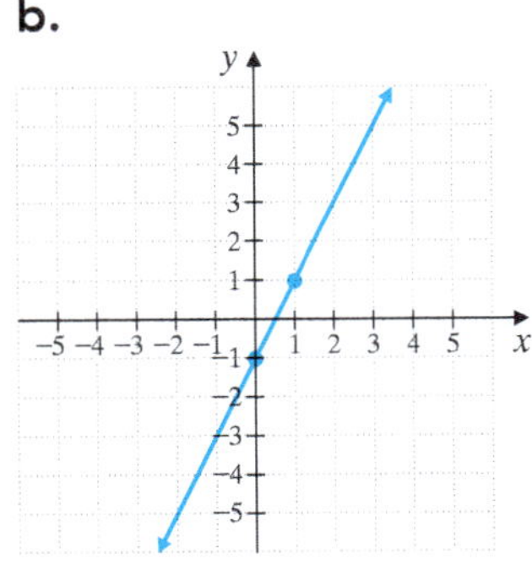

c.

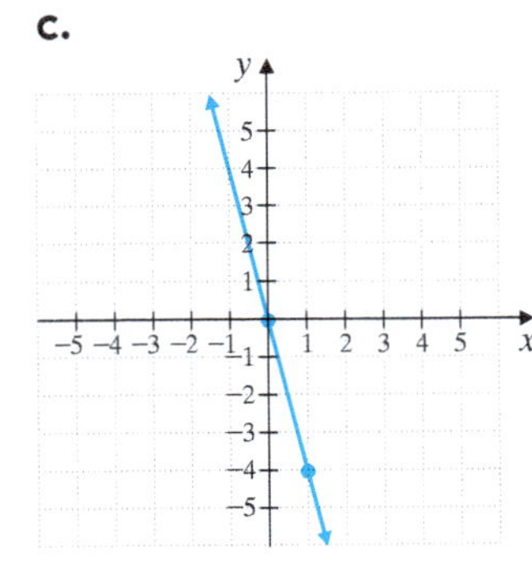

d.

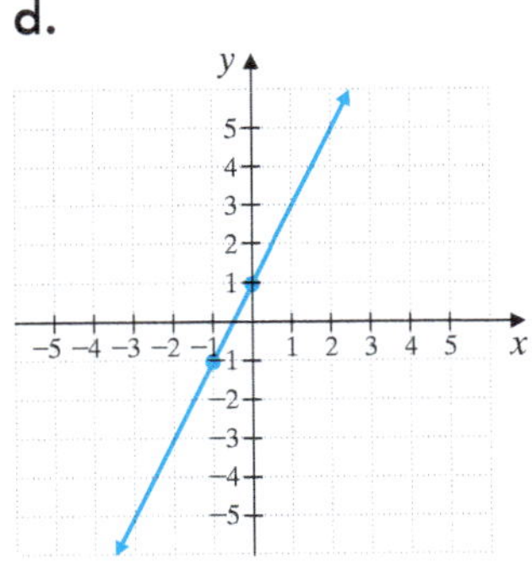

Write an equation of the line with the given slope that passes through the given point. Write the equation in the form $Ax + By = C$.

53. $m = 4$; $(2, 0)$ **54.** $m = -3$; $(0, -5)$ **55.** $m = \frac{3}{5}$; $(1, 4)$ **56.** $m = -\frac{1}{3}$; $(-3, 3)$

Write an equation of the line passing through each pair of points. Write the equation in the form $y = mx + b$.

57. $(1, 7)$ and $(2, -7)$ **58.** $(-2, 5)$ and $(-4, 6)$

(6.6) *Determine whether each relation or graph is a function.*

59. $\{(7, 1), (7, 5), (2, 6)\}$ **60.** $\{(0, -1), (5, -1), (2, 2)\}$

61.

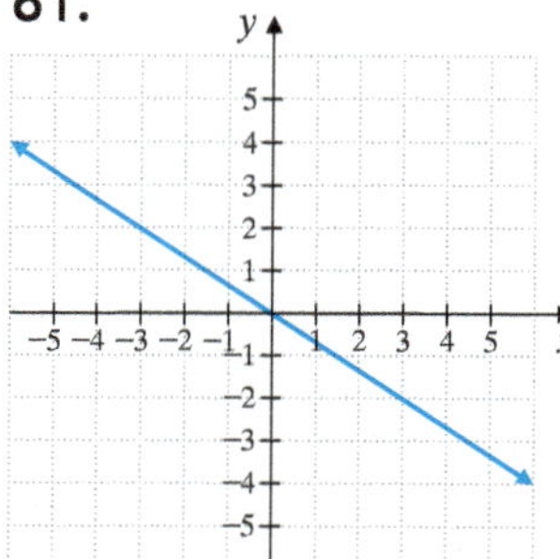

62.

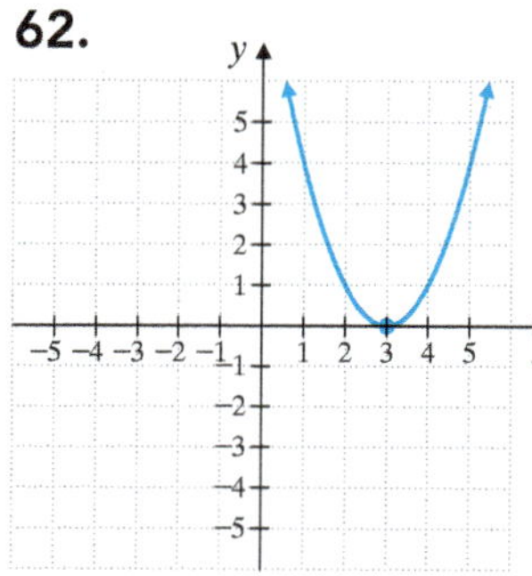

63.

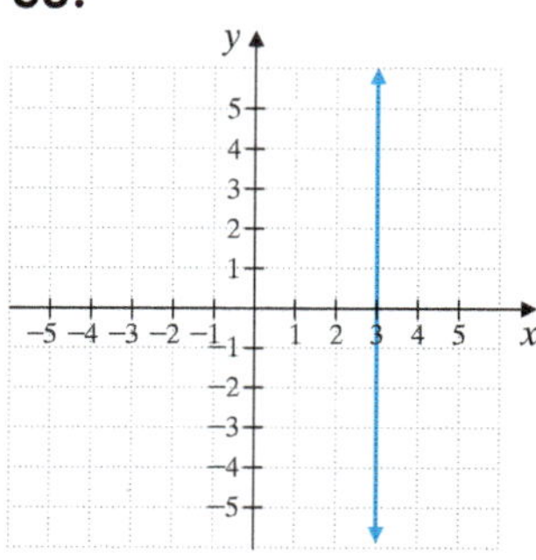

64.

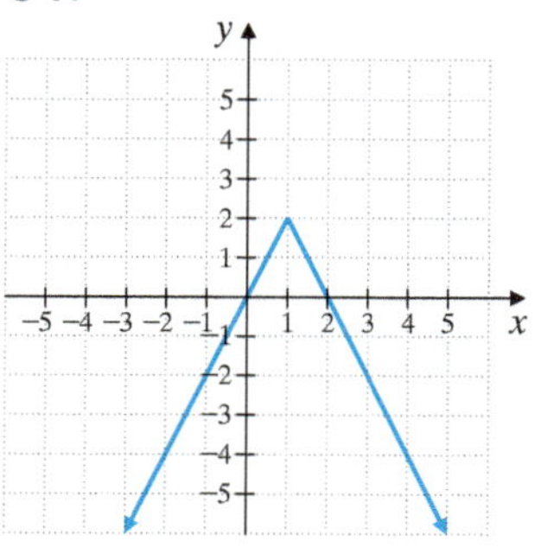

Find each indicated function value for the function $f(x) = -2x + 6$.

65. $f(0)$ **66.** $f(-2)$ **67.** $f\left(\frac{1}{2}\right)$ **68.** $f\left(-\frac{1}{2}\right)$

(6.7) *Graph each inequality.*

69. $x + 6y < 6$

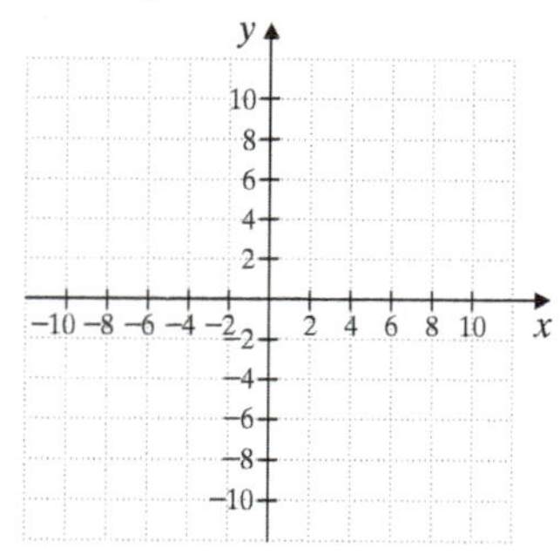

70. $x + y > -2$

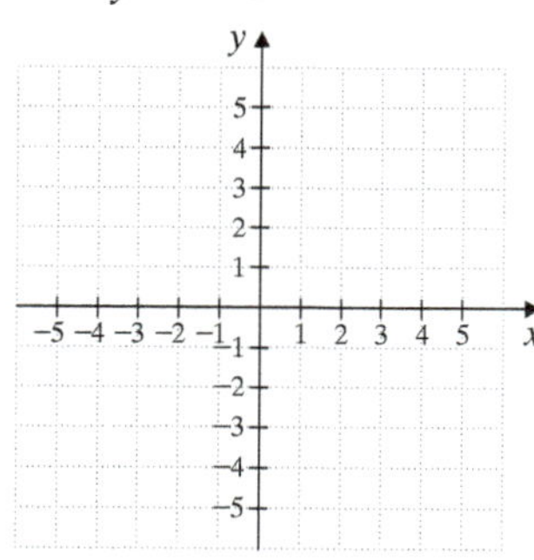

71. $y \geq -7$

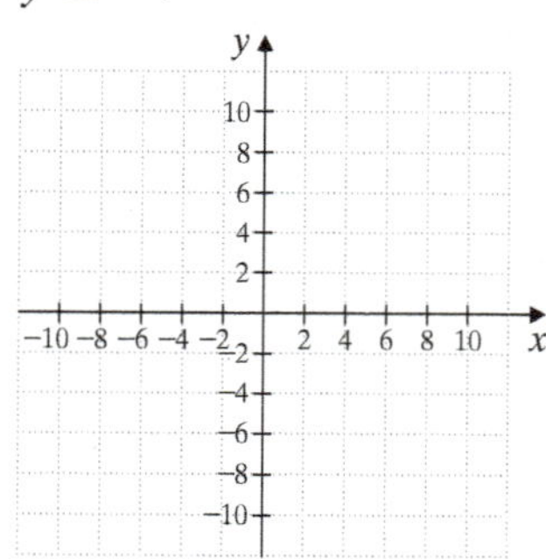

72. $y \leq -4$

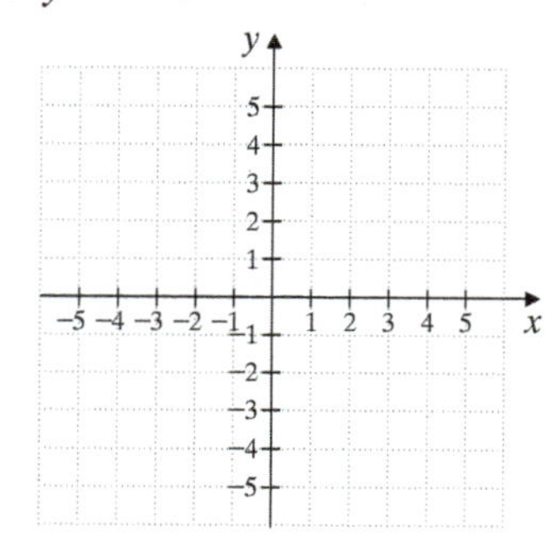

73. $-x \leq y$

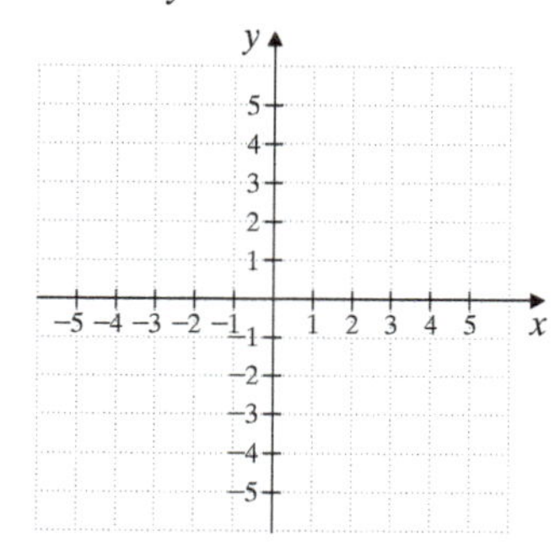

74. $x \geq -y$

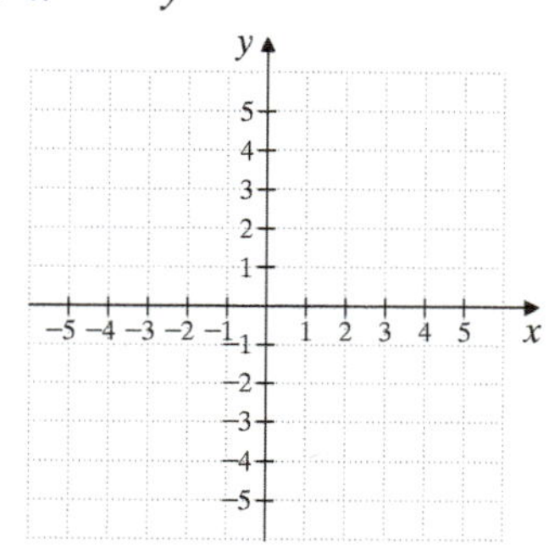

Mixed Review

Complete the table of values for each given equation.

75. $2x - 5y = 9$

x	y
	1
2	
	−3

76. $x = -3y$

x	y
0	
	1
6	

Find the intercepts for each equation.

77. $2x - 3y = 6$

78. $-5x + y = 10$

Graph each linear equation.

79. $x - 5y = 10$

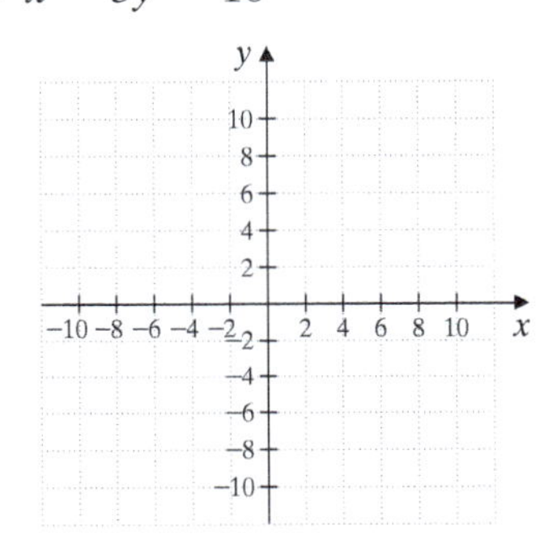

80. $x + y = 4$

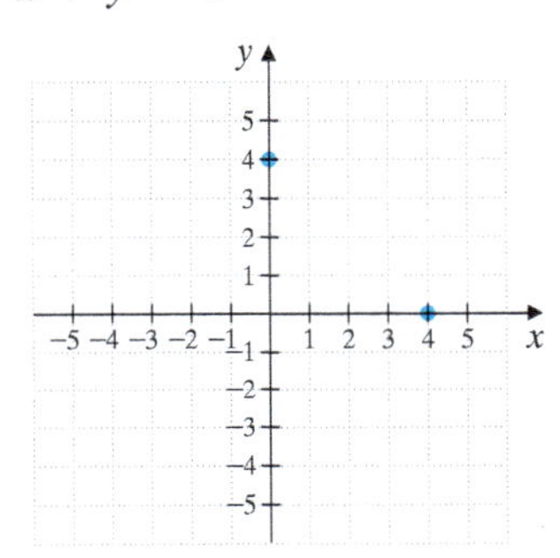

81. $y = -4x$

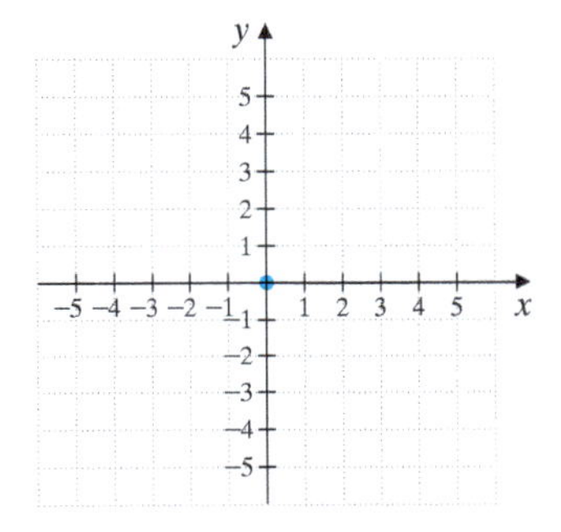

82. $2x + 3y = -6$

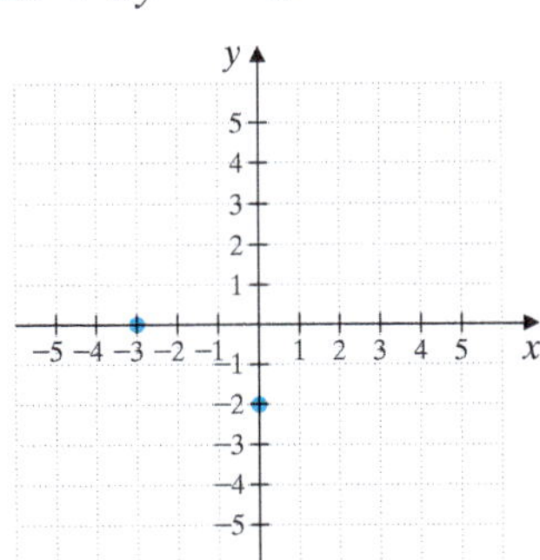

83. $x = 3$

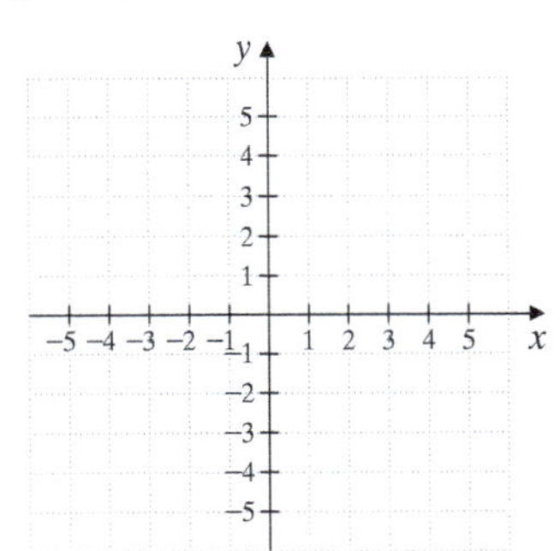

84. $y = -2$

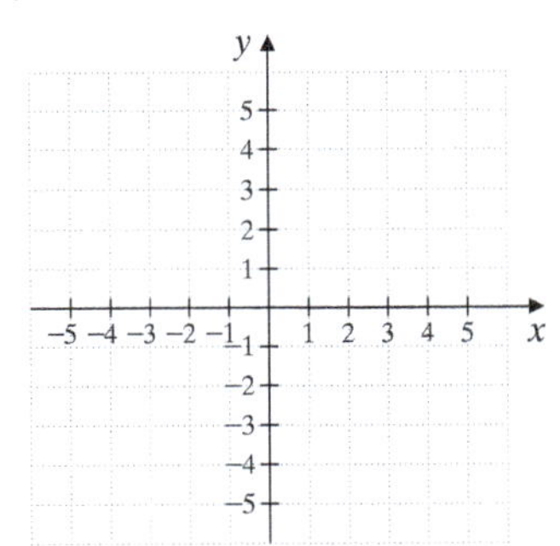

Find the slope of the line that passes through each pair of points.

85. $(3, -5)$ and $(-4, 2)$

86. $(1, 3)$ and $(-6, -8)$

Find the slope of each line.

87.

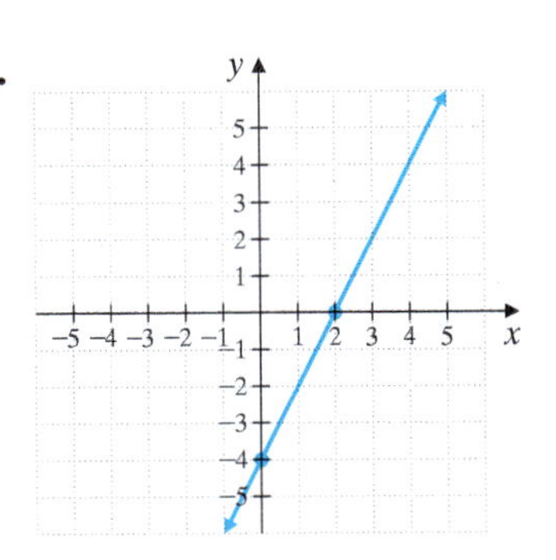

88.

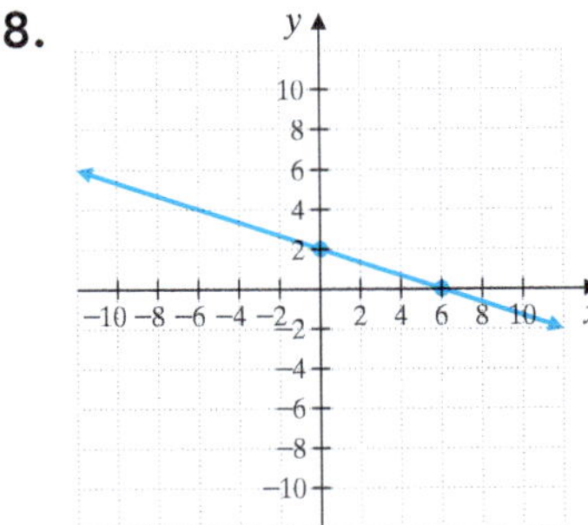

Determine the slope and y-intercept of the graph of each equation.

89. $-2x + 3y = -15$

90. $6x + y - 2 = 0$

Write an equation of the line with the given slope that passes through the given point. Write the equation in the form $Ax + By = C$.

91. $m = -5$; $(3, -7)$

92. $m = 3$; $(0, 6)$

Write an equation of the line passing through each pair of points. Write the equation in the form $Ax + By = C$.

93. $(-3, 9)$ and $(-2, 5)$

94. $(3, 1)$ and $(5, -9)$

Chapter 6 Getting Ready for the Test LC

MULTIPLE CHOICE Exercises 1 through 4 are **Multiple Choice**. Choose the correct letter.

For Exercises 1 and 2, choose the ordered pair that is NOT a solution of the linear equation.

1. $x - y = 5$

A. $(7, 2)$ **B.** $(0, -5)$ **C.** $(-2, 3)$ **D.** $(-2, -7)$

2. $y = 4$

A. $(4, 0)$ **B.** $(0, 4)$ **C.** $(2, 4)$ **D.** $(100, 4)$

3. What is the most and then the fewest number of intercepts a line may have?

A. most: 2; fewest: 1 **B.** most: infinite number; fewest: 1 **C.** most: 2; fewest: 0
D. most: infinite number; fewest; 0

4. Choose the linear equation:

A. $\sqrt{x} - 3y = 7$ **B.** $2x = 6^2$ **C.** $4x^3 + 6y^3 = 5^3$ **D.** $y = |x|$

MATCHING For Exercises 5 through 8, **Match** each numbered line in the rectangular system with its slope to the right. Each slope may be used only once.

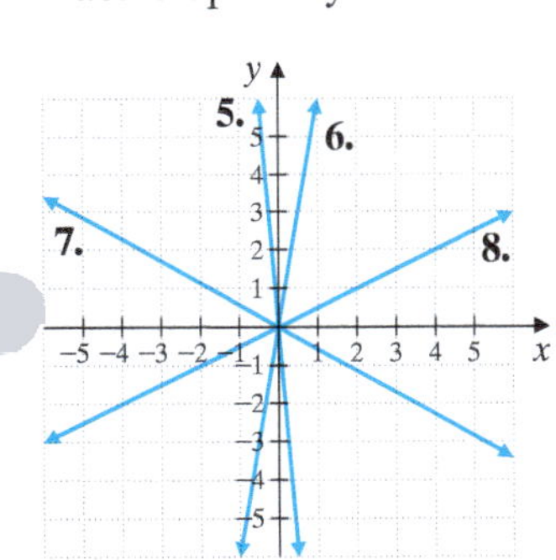

5. **A.** $m = 5$

6. **B.** $m = -10$

7. **C.** $m = \frac{1}{2}$

8. **D.** $m = -\frac{4}{7}$

MULTIPLE CHOICE Exercises 9 through 14 are **Multiple Choice**. Choose the correct letter.

9. An ordered pair solution for the function $f(x)$ is $(0, 5)$. This solution using function notation is:

A. $f(5) = 0$ **B.** $f(5) = f(0)$ **C.** $f(0) = 5$ **D.** $0 = 5$

10. Given: $(2, 3)$ and $(0, 9)$. Final Answer: $y = -3x + 9$. Select the correct directions:

A. Find the slope of the line through the two points.
B. Find an equation of the line through the two points. Write the equation in standard form.
C. Find an equation of the line through the two points. Write the equation in slope-intercept form.

For Exercises 11 through 14, use the graph to fill in each blank using the choices below.

A. -2 **B.** 2 **C.** 4 **D.** 0 **E.** 3

11. $f(0) =$ ________.

12. $f(4) =$ ________.

13. If $f(x) = 0$, then $x =$ ________ or $x =$ ________.

14. $f(1) =$ ________.

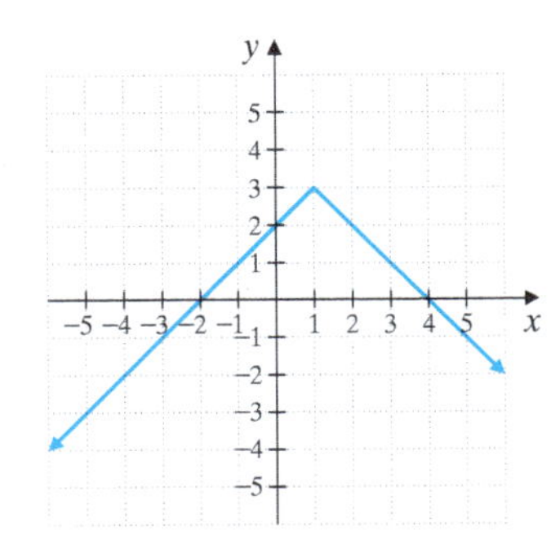

Chapter 6 Test MyLab Math

For additional practice go to your study plan in MyLab Math.

Answers

1. ______
2. ______
3. ______
4. ______
5. ______
6. ______
7. ______
8. ______
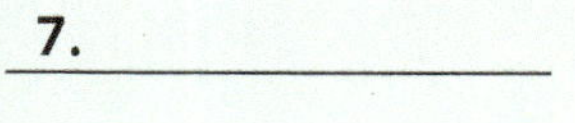
9. ______
10. ______
11. ______
12. ______
13. ______
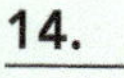
14. ______

Complete each ordered pair so that it is a solution of the given equation.

1. $12y - 7x = 5$; $(1, \quad)$

2. $y = 17$; $(-4, \quad)$

Find the slope of each line.

3.

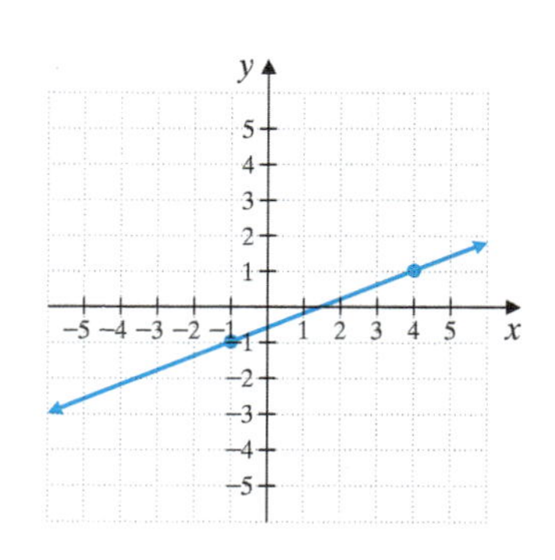

4.

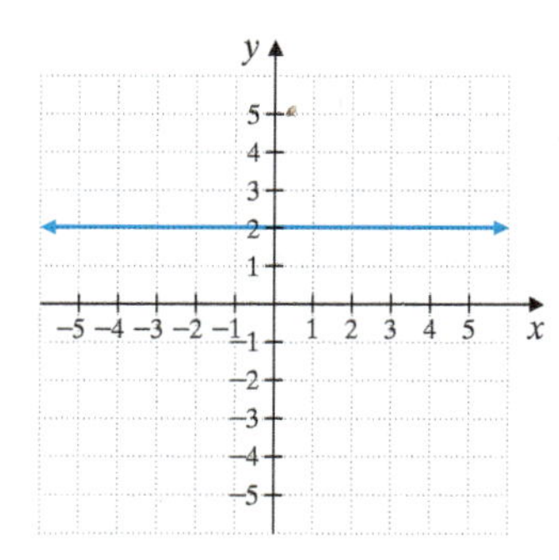

5. Passes through $(6, -5)$ and $(-1, 2)$

6. Passes through $(0, -8)$ and $(-1, -1)$

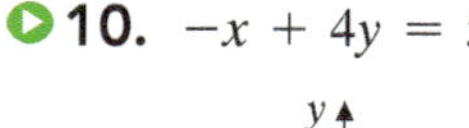

7. $-3x + y = 5$

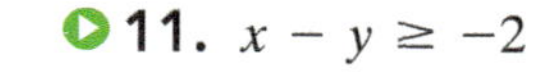

8. $x = 6$

Graph.

9. $2x + y = 8$

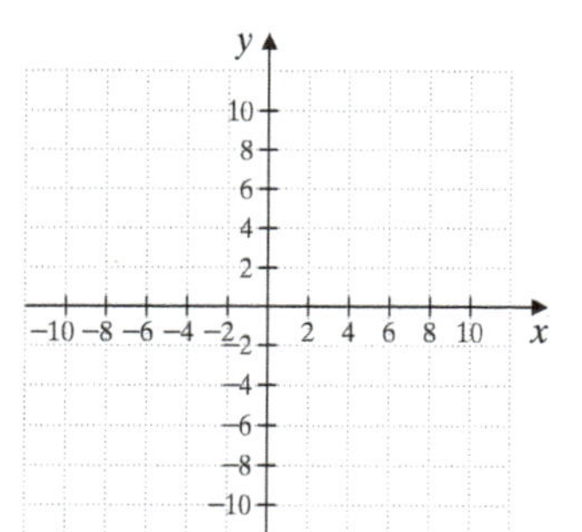

10. $-x + 4y = 5$

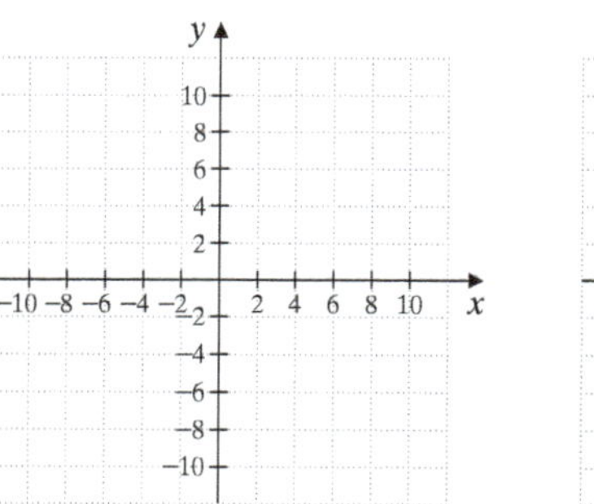

11. $x - y \geq -2$

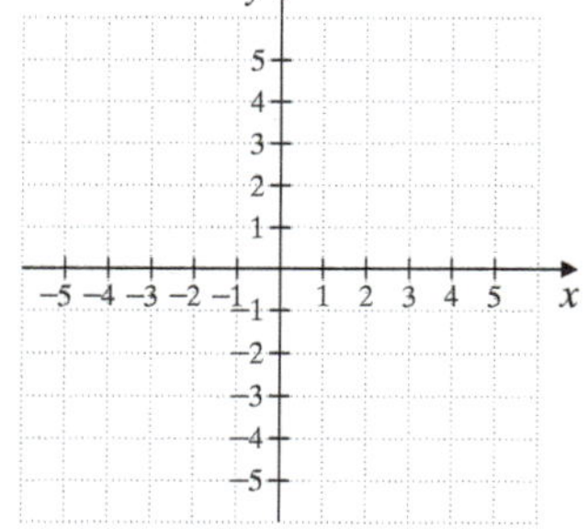

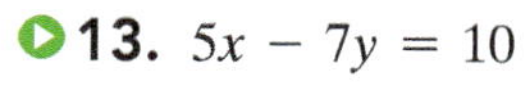

12. $y \geq -4x$

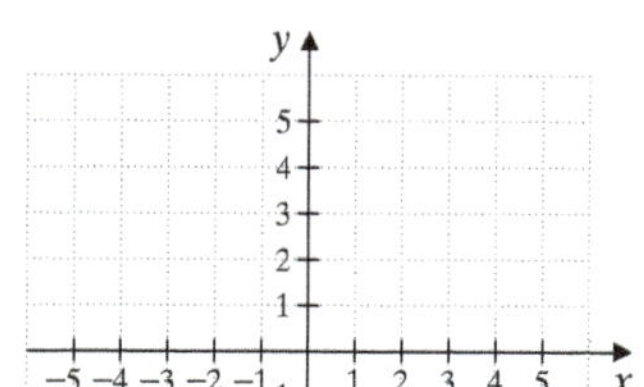

13. $5x - 7y = 10$

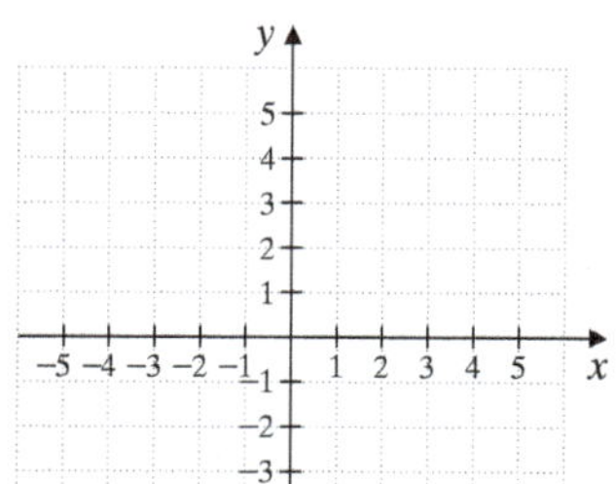

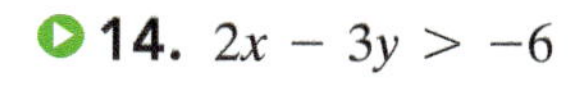

14. $2x - 3y > -6$

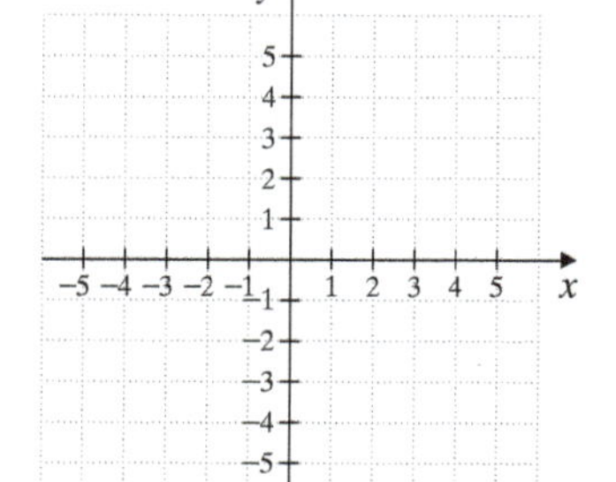

15. $6x + y > -1$

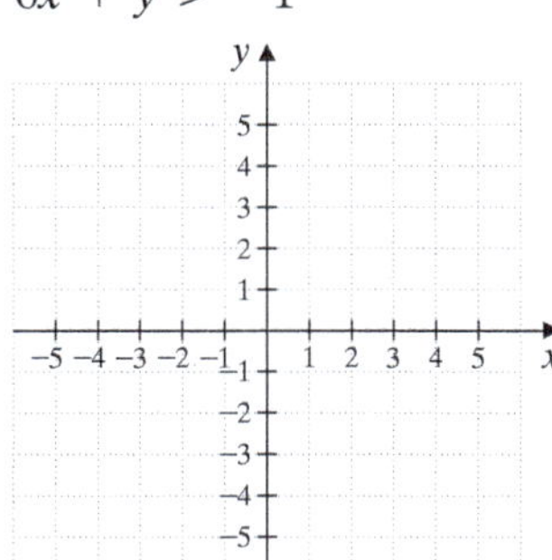

16. $y = -1$

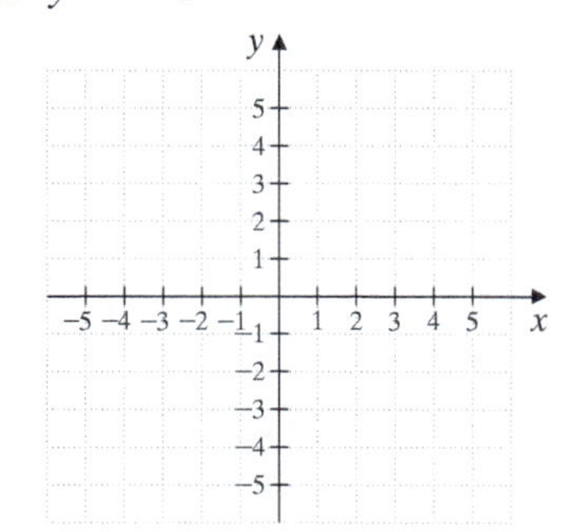

17. Determine whether the graphs of $y = 2x - 6$ and $-4x = 2y$ are parallel lines, perpendicular lines, or neither.

Find an equation of each line. Write the equation in the form $Ax + By = C$.

18. Slope $-\frac{1}{4}$, passes through $(2, 2)$

19. Passes through the origin and $(6, -7)$

20. Passes through $(2, -5)$ and $(1, 3)$

21. Slope $\frac{1}{8}$; y-intercept $(0, 12)$

Determine whether each relation is a function.

22. $\{(-1, 2), (-2, 4), (-3, 6), (-4, 8)\}$

23. $\{(-3, -3), (0, 5), (-3, 2), (0, 0)\}$

24. The graph shown in Exercise **3**

25. The graph shown in Exercise **4**

Find the indicated function values for each function.

26. $f(x) = 2x - 4$
- **a.** $f(-2)$
- **b.** $f(0.2)$
- **c.** $f(0)$

27. $f(x) = x^3 - x$
- **a.** $f(-1)$
- **b.** $f(0)$
- **c.** $f(4)$

28. The perimeter of the parallelogram below is 42 meters. Write a linear equation in two variables for the perimeter. Use this equation to find x when y is 8 meters.

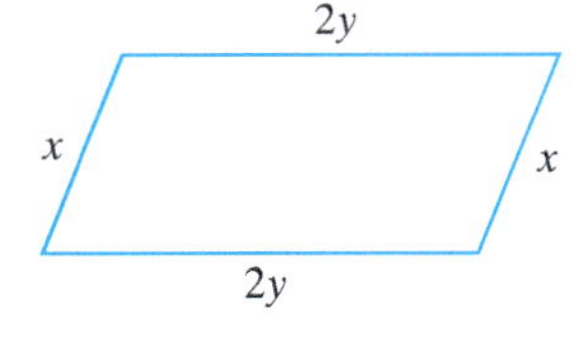

15. ________
16. ________
17. ________
18. ________
19. ________
20. ________
21. ________
22. ________
23. ________
24. ________
25. ________
26. a. ________
b. ________
c. ________
27. a. ________
b. ________
c. ________
28. ________

29. The table gives the percent of total U.S. music revenue derived from streaming music for the years shown. (*Source:* Recording Industry Association of America)

Year	Percent of Music Revenue from Streaming
2011	9
2012	15
2013	21
2014	27
2015	34
2016	51
2017	65

a. Write this data as a set of ordered pairs of the form (year, percent of music revenue from streaming).

b. Create a scatter diagram of the data. Be sure to label the axes properly.

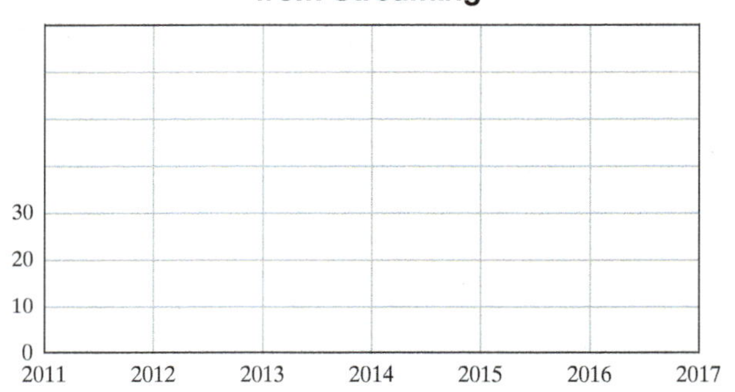

29. a. ______________

b. ______________

30. This graph approximates the gross box office sales y (in billions) for Canada and the U.S. for the year x. Find the slope of the line and write the slope as a rate of change. Don't forget to attach the proper units.

30. ______________

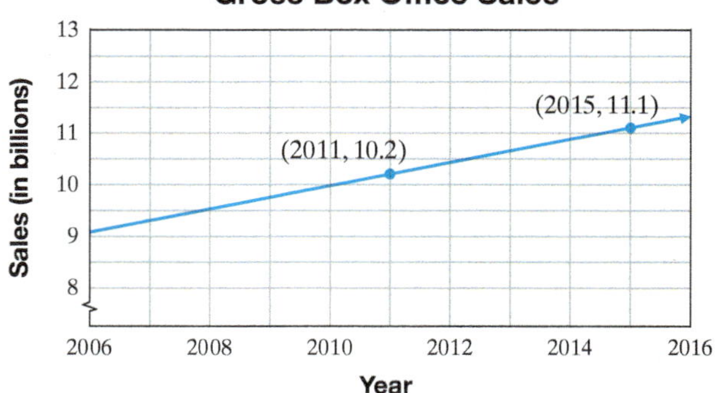

Source: National Association of Theater Owners

Cumulative Review Chapters 1–6

Simplify each expression.

1. $6 \div 3 + 5^2$

2. $\frac{10}{3} + \frac{5}{21}$

3. $1 + 2[5(2 \cdot 3 + 1) - 10]$

4. $16 - 3 \cdot 3 + 2^4$

5. The highest point in the United States is the top of Denali, at a height of 20,320 feet above sea level. The lowest point is Death Valley, California, which is 282 feet below sea level. How much higher is Denali than Death Valley? (*Source:* U.S. Geological Society)

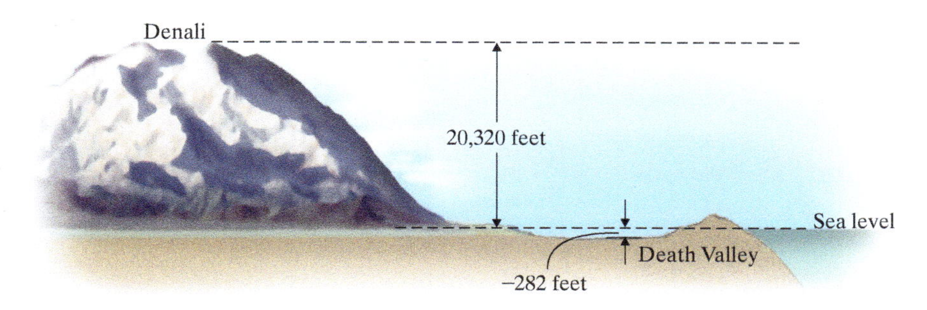

6. Simplify: $1.7x - 11 - 0.9x - 25$

Write each phrase as an algebraic expression and simplify if possible. Let x represent the unknown number.

7. Twice a number, plus 6.

8. The product of −15 and the sum of a number and $\frac{2}{3}$.

9. The difference of a number and 4, divided by 7.

10. The quotient of −9 and twice a number.

11. Five plus the sum of a number and 1.

12. A number subtracted from −86.

13. Solve for x: $\frac{5}{2}x = 15$

14. Solve for x: $\frac{x}{4} - 1 = -7$

15. Solve $2x < -4$. Graph the solutions.

−5 −4 −3 −2 −1 0 1 2 3 4 5

16. Solve: $5(x + 4) \geq 4(2x + 3)$

17. Find the degree of each polynomial and tell whether the polynomial is a monomial, binomial, trinomial, or none of these.

a. $-2t^2 + 3t + 6$

b. $15x - 10$

c. $7x + 3x^3 + 2x^2 - 1$

18. Solve $x + 2y = 6$ for y.

Answers

1. ______
2. ______
3. ______
4. ______
5. ______
6. ______
7. ______
8. ______
9. ______
10. ______
11. ______
12. ______
13. ______
14. ______
15. ______
16. ______
17. a. ______
b. ______
c. ______
18. ______

19.

20.

21.

22.

23.

24.

25.

26.

27.

28.

29.

30.

31.

32.

33.

34.

35.

36.

37.

38.

19. Add: $(-2x^2 + 5x - 1) + (-2x^2 + x + 3)$

20. Subtract: $(-2x^2 + 5x - 1) - (-2x^2 + x + 3)$

21. Multiply: $(3y + 1)^2$

22. Multiply: $(x - 12)^2$

23. Factor: $-9a^5 + 18a^2 - 3a$

24. Factor: $4x^2 - 36$

25. Factor: $x^2 + 4x - 12$

26. Factor: $3x^2 - 20xy - 7y^2$

27. Factor: $8x^2 - 22x + 5$

28. Factor: $18x^2 + 35x - 2$

29. Solve: $x^2 - 9x - 22 = 0$

30. Solve: $x^2 = x$

31. Divide: $\dfrac{2x^2 - 11x + 5}{5x - 25} \div \dfrac{4x - 2}{10}$

32. Simplify: $\dfrac{2x^2 - 50}{4x^4 - 20x^3}$

Write each rational expression as an equivalent rational expression with the given denominator.

33. $\dfrac{4b}{9a} = \dfrac{}{27a^2b}$

34. $\dfrac{1}{2x} = \dfrac{}{14x^3}$

35. Add: $1 + \dfrac{m}{m + 1}$

36. Subtract: $\dfrac{2x + 1}{x - 6} - \dfrac{x - 4}{x - 6}$

37. Solve: $3 - \dfrac{6}{x} = x + 8$

38. Solve: $3x^2 + 5x = 2$

39. Simplify: $\dfrac{\dfrac{x+1}{y}}{\dfrac{x}{y}+2}$

40. Simplify: $\dfrac{\dfrac{x}{2}-\dfrac{y}{6}}{\dfrac{x}{12}-\dfrac{y}{3}}$

41. Complete each ordered pair so that it is a solution of the equation $3x + y = 12$.

a. $(0, \quad)$
b. $(\quad, 6)$
c. $(-1, \quad)$

42. Complete the table for $y = -5x$.

x	y
	0
−1	
	10

43. Graph: $2x + y = 5$

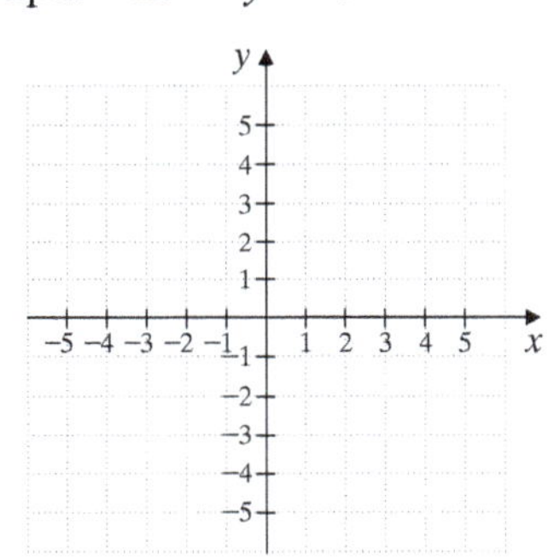

44. Find the slope of the line through $(0, 5)$ and $(-5, 4)$.

45. Find the slope of the line $-2x + 3y = 11$.

46. Find the slope of the line $x = -10$.

47. Find an equation of the line with slope -2 that passes through $(-1, 5)$. Write the equation in slope-intercept form, $y = mx + b$, and in standard form, $Ax + By = C$.

48. Find the slope and y-intercept of the line whose equation is $2x - 5y = 10$.

49. Given $g(x) = x^2 - 3$, find each function value and list the corresponding ordered pair.

a. $g(2)$
b. $g(-2)$
c. $g(0)$

50. Write an equation of the line through $(2, 3)$ and $(0, 0)$. Write the equation in standard form.

39. ______
40. ______
41. a. ______
b. ______
c. ______
42. ______
43. ______
44. ______
45. ______
46. ______
47. ______
48. ______
49. a. ______
b. ______
c. ______
50. ______

7 Systems of Equations

In Chapter 6, we graphed equations containing two variables. As we have seen, equations like these are often needed to represent relationships between two different quantities. There are also many opportunities to compare and contrast two such equations, called a *system of equations*. This chapter presents *linear systems* and ways we solve these systems and apply them to real-life situations.

Sections

7.1 Solving Systems of Linear Equations by Graphing

7.2 Solving Systems of Linear Equations by Substitution

7.3 Solving Systems of Linear Equations by Addition

Integrated Review— Summary on Solving Systems of Equations

7.4 Systems of Linear Equations and Problem Solving

Check Your Progress

Vocabulary Check

Chapter Highlights

Chapter Review

Getting Ready for the Test

Chapter Test

Cumulative Review

U.S. Ad Spending by Media Type—2021 Projected

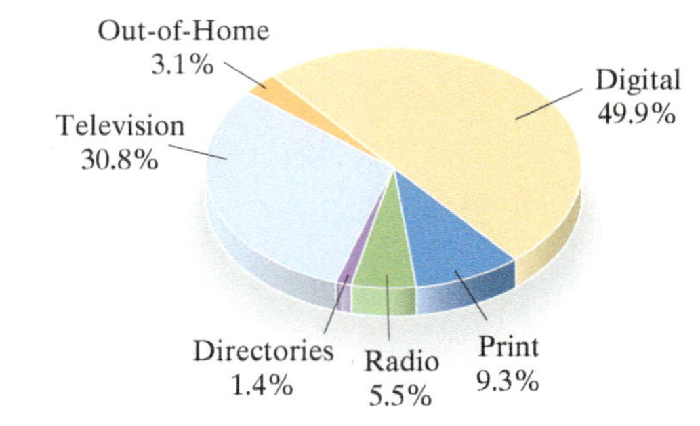

Source: eMarketer

Where Are You Exposed to More Advertising: Internet or Television?

Advertising is big business. Notice the largest sectors above correspond to TV and Internet (or digital). This means that most advertising money spent in the United States is spent on TV advertising and digital (or Internet) advertising. In fact, the 2017 total for these two types of advertising is estimated to be $151 billion. As you may imagine, the fastest-growing U.S. consumer segment is digital users. Since advertisers follow consumers, digital advertising is the fastest-growing sector above.

Budgets for digital advertising are increasing at a faster rate than budgets for TV advertising as shown on the double line graph below. In Section 7.3, Exercise 59, we study these two types of advertising further.

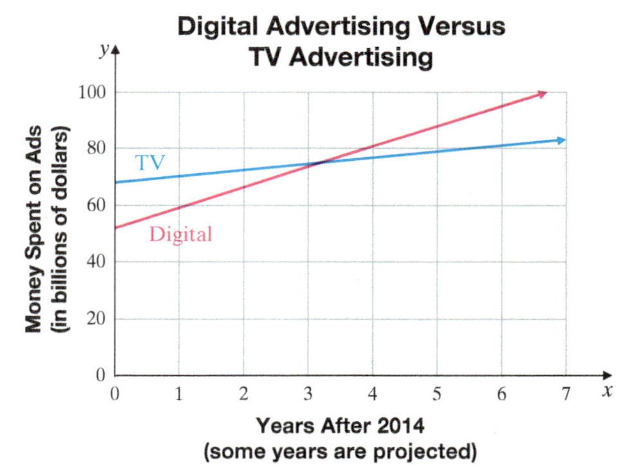

Source: PricewaterhouseCoopers Global

7.1 Solving Systems of Linear Equations by Graphing

A **system of linear equations** consists of two or more linear equations. In this section, we focus on solving systems of linear equations containing two equations in two variables. Examples of such linear systems are

$$\begin{cases} 3x - 3y = 0 \\ x = 2y \end{cases} \quad \begin{cases} x - y = 0 \\ 2x + y = 10 \end{cases} \quad \begin{cases} y = 7x - 1 \\ y = 4 \end{cases}$$

Objectives

A Decide Whether an Ordered Pair Is a Solution of a System of Linear Equations.

B Solve a System of Linear Equations by Graphing.

C Identify Special Systems of Linear Equations.

D Without Graphing, Determine the Number of Solutions of a System.

Objective A Deciding Whether an Ordered Pair Is a Solution

A **solution** of a system of two equations in two variables is an ordered pair of numbers that is a solution of both equations in the system.

Example 1 Determine whether $(12, 6)$ is a solution of the system

$$\begin{cases} 2x - 3y = 6 \\ x = 2y \end{cases}$$

Solution: To determine whether $(12, 6)$ is a solution of the system, we replace x with 12 and y with 6 in both equations.

$2x - 3y = 6$	First equation	$x = 2y$	Second equation
$2(12) - 3(6) \stackrel{?}{=} 6$	Let $x = 12$ and $y = 6$.	$12 \stackrel{?}{=} 2(6)$	Let $x = 12$ and $y = 6$.
$24 - 18 \stackrel{?}{=} 6$	Simplify.	$12 = 12$	True
$6 = 6$	True		

Since $(12, 6)$ is a solution of both equations, it is a solution of the system.

Work Practice 1

Practice 1

Determine whether $(3, 9)$ is a solution of the system

$$\begin{cases} 5x - 2y = -3 \\ y = 3x \end{cases}$$

Example 2 Determine whether $(-1, 2)$ is a solution of the system

$$\begin{cases} x + 2y = 3 \\ 4x - y = 6 \end{cases}$$

Solution: We replace x with -1 and y with 2 in both equations.

$x + 2y = 3$	First equation	$4x - y = 6$	Second equation
$-1 + 2(2) \stackrel{?}{=} 3$	Let $x = -1$ and $y = 2$.	$4(-1) - 2 \stackrel{?}{=} 6$	Let $x = -1$ and $y = 2$.
$-1 + 4 \stackrel{?}{=} 3$	Simplify.	$-4 - 2 \stackrel{?}{=} 6$	Simplify.
$3 = 3$	True	$-6 = 6$	False

$(-1, 2)$ is not a solution of the second equation, $4x - y = 6$, so it is not a solution of the system.

Work Practice 2

Practice 2

Determine whether $(3, -2)$ is a solution of the system

$$\begin{cases} 2x - y = 8 \\ x + 3y = 4 \end{cases}$$

Objective B Solving Systems of Equations by Graphing

Since a solution of a system of two equations in two variables is a solution common to both equations, it is also a point common to the graphs of both equations. Let's practice finding solutions of both equations in a system—that is, solutions of the system—by graphing and identifying points of intersection.

Answers

1. $(3, 9)$ is a solution of the system.
2. $(3, -2)$ is not a solution of the system.

Practice 3

Solve the system of equations by graphing.

$$\begin{cases} -3x + y = -10 \\ x - y = 6 \end{cases}$$

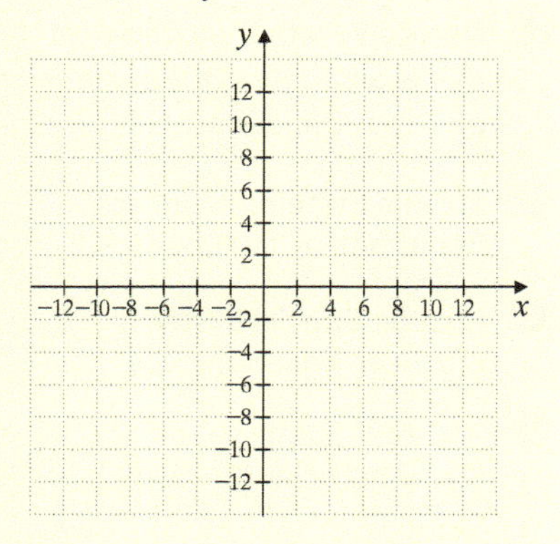

Practice 4

Solve the system of equations by graphing.

$$\begin{cases} x + 3y = -1 \\ y = 1 \end{cases}$$

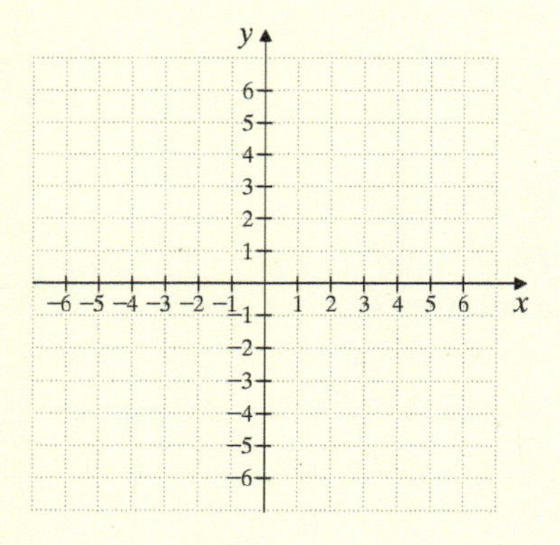

Answers

3. $(2, -4)$;

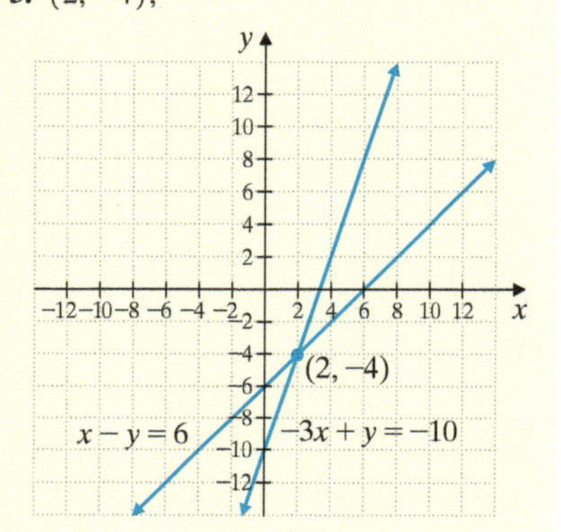

4. See page 532.

Example 3 Solve the system of equations by graphing.

$$\begin{cases} -x + 3y = 10 \\ x + y = 2 \end{cases}$$

Solution: On a single set of axes, graph each linear equation.

$-x + 3y = 10$

x	y
0	$\frac{10}{3}$
−4	2
2	4

$x + y = 2$

x	y
0	2
2	0
1	1

Helpful Hint The point of intersection gives the solution of the system.

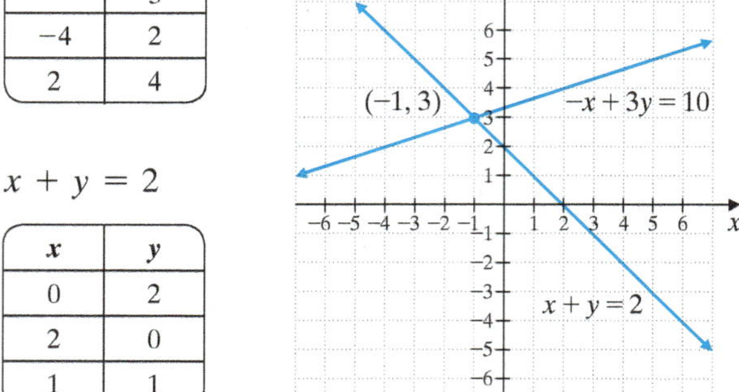

The two lines appear to intersect at the point $(-1, 3)$. To check, we replace x with -1 and y with 3 in both equations.

$-x + 3y = 10$ First equation
$-(-1) + 3(3) \stackrel{?}{=} 10$ Let $x = -1$ and $y = 3$.
$1 + 9 \stackrel{?}{=} 10$ Simplify.
$10 = 10$ True

$x + y = 2$ Second equation
$-1 + 3 \stackrel{?}{=} 2$ Let $x = -1$ and $y = 3$.
$2 = 2$ True

$(-1, 3)$ checks, so it is the solution of the system.

Work Practice 3

Helpful Hint

Neatly drawn graphs can help when "guessing" the solution of a system of linear equations by graphing.

Example 4 Solve the system of equations by graphing.

$$\begin{cases} 2x + 3y = -2 \\ x = 2 \end{cases}$$

Solution: We graph each linear equation on a single set of axes.

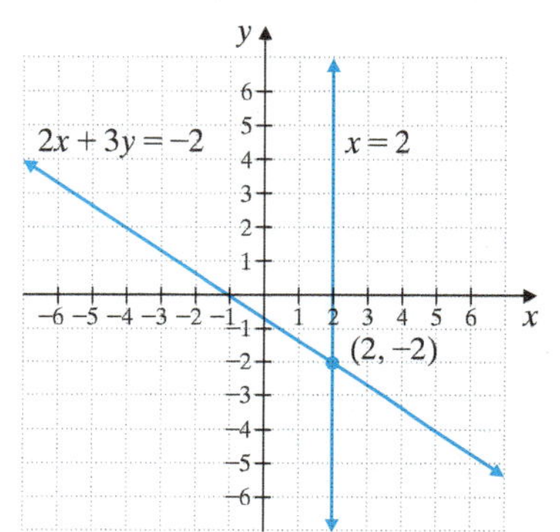

The two lines appear to intersect at the point $(2, -2)$. To determine whether $(2, -2)$ is the solution, we replace x with 2 and y with -2 in both equations.

$2x + 3y = -2$	First equation	$x = 2$	Second equation
$2(2) + 3(-2) \stackrel{?}{=} -2$	Let $x = 2$ and $y = -2$.	$2 \stackrel{?}{=} 2$	Let $x = 2$.
$4 + (-6) \stackrel{?}{=} -2$	Simplify.	$2 = 2$	True
$-2 = -2$	True		

Since a true statement results in both equations, $(2, -2)$ is the solution of the system.

Work Practice 4

Objective C Identifying Special Systems of Linear Equations

Not all systems of linear equations have a single solution. Some systems have no solution and some have an infinite number of solutions. See the next two examples for these special cases.

Example 5 Solve the system of equations by graphing.

$$\begin{cases} 2x + y = 7 \\ 2y = -4x \end{cases}$$

Solution: We graph the two equations in the system. The equations in slope-intercept form are $y = -2x + 7$ and $y = -2x$. Notice from the equations that the lines have the same slope, -2, and different y-intercepts. This means that the lines are parallel.

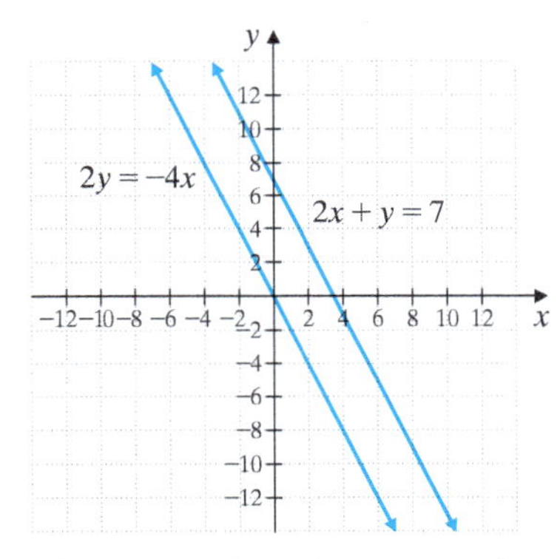

Since the lines are parallel, they do not intersect. This means that the system has no solution.

Work Practice 5

Practice 5

Solve the system of equations by graphing.

$$\begin{cases} 3x - y = 6 \\ 6x = 2y \end{cases}$$

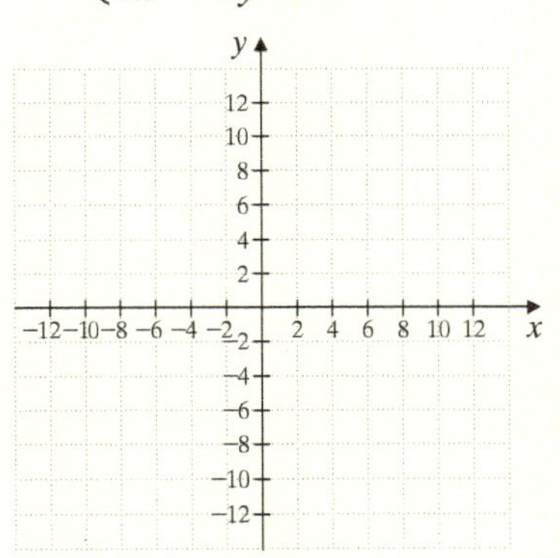

Example 6 Solve the system of equations by graphing.

$$\begin{cases} x - y = 3 \\ -x + y = -3 \end{cases}$$

Solution: We graph each equation. The graphs of the equations are the same line. To see this, notice that if both sides of the first equation in the system are multiplied by -1, the result is the second equation.

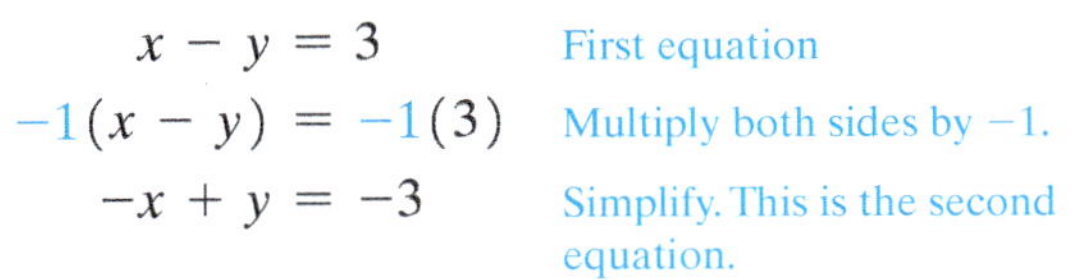

$x - y = 3$	First equation
$-1(x - y) = -1(3)$	Multiply both sides by -1.
$-x + y = -3$	Simplify. This is the second equation.

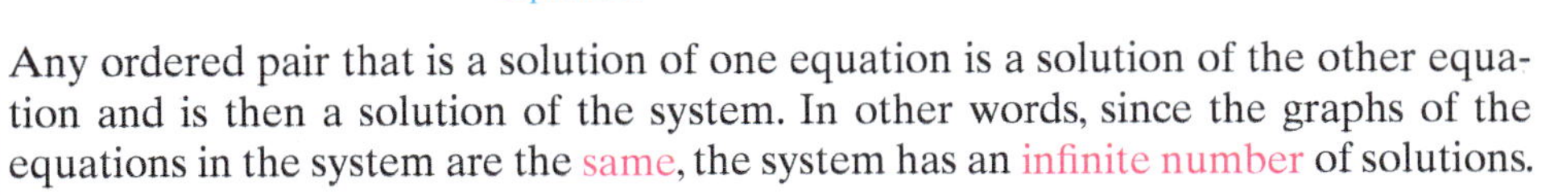

Any ordered pair that is a solution of one equation is a solution of the other equation and is then a solution of the system. In other words, since the graphs of the equations in the system are the same, the system has an infinite number of solutions.

Work Practice 6

Practice 6

Solve the system of equations by graphing.

$$\begin{cases} x + y = -4 \\ -2x - 2y = 8 \end{cases}$$

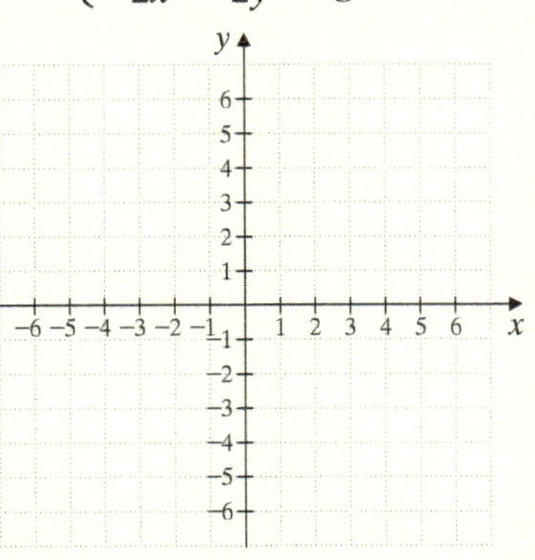

Answers

5. See page 532. 6. See page 532.

Examples 5 and 6 are special cases of systems of linear equations. A system that has no solution is said to be an **inconsistent system.** If the graphs of the two equations of a system are identical, we call the equations **dependent equations.** Thus, the system in Example 5 is an inconsistent system and the equations in the system in Example 6 are dependent equations.

As we have seen, three different situations can occur when graphing the two lines associated with the equations in a linear system. These situations are shown in the figures.

One point of intersection: one solution

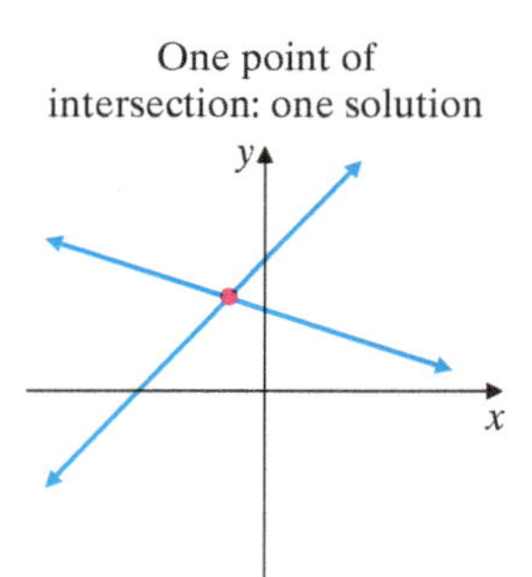

Consistent system
(at least one solution)
Independent equations
(graphs of equations differ)

Parallel lines: no solution

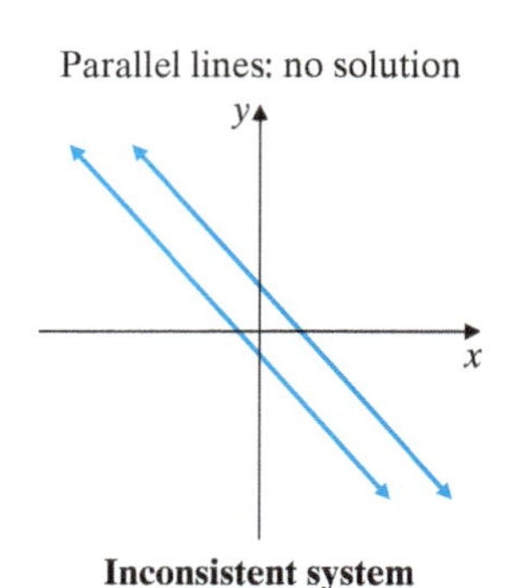

Inconsistent system
(no solution)
Independent equations
(graphs of equations differ)

Same line: infinite number of solutions

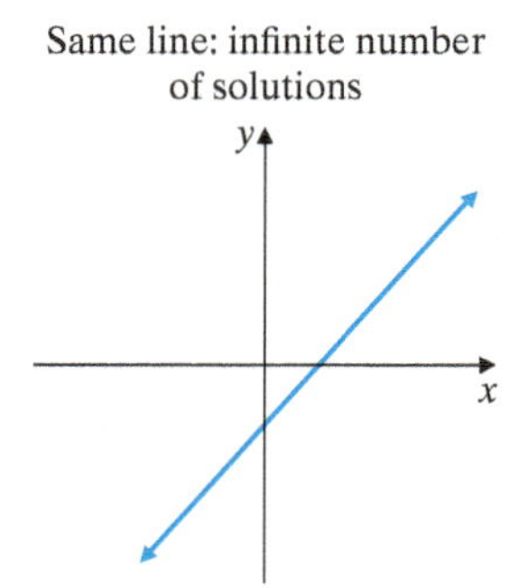

Consistent system
(at least one solution)
Dependent equations
(graphs of equations identical)

Practice 7

Without graphing, determine the number of solutions of the system.

$$\begin{cases} 5x + 4y = 6 \\ x - y = 3 \end{cases}$$

Answers

4. $(-4, 1)$;

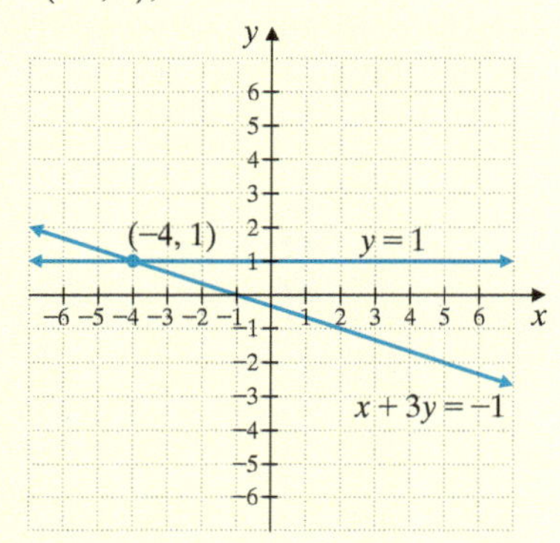

5. no solution;

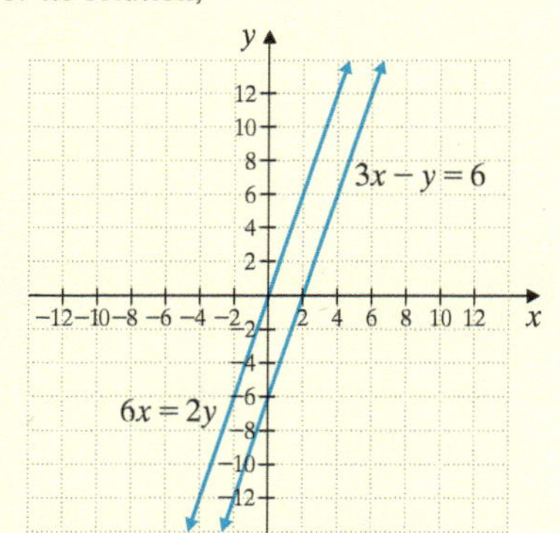

6. infinite number of solutions;

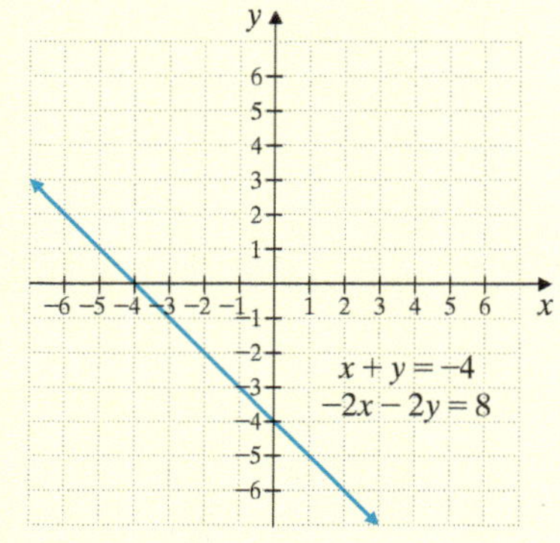

7. one solution

Objective D Finding the Number of Solutions of a System Without Graphing

You may have suspected by now that graphing alone is not an accurate way to solve a system of linear equations. For example, a solution of $\left(\frac{1}{2}, \frac{2}{9}\right)$ is unlikely to be read correctly from a graph. The next two sections present two accurate methods of solving these systems. In the meantime, we can decide how many solutions a system has by writing each equation in slope-intercept form.

Example 7

Without graphing, determine the number of solutions of the system.

$$\begin{cases} \frac{1}{2}x - y = 2 \\ x = 2y + 5 \end{cases}$$

Solution: First write each equation in slope-intercept form.

$\frac{1}{2}x - y = 2$	First equation	$x = 2y + 5$	Second equation
		$x - 5 = 2y$	Subtract 5 from both sides.
$\frac{1}{2}x = y + 2$	Add y to both sides.	$\frac{x}{2} - \frac{5}{2} = \frac{2y}{2}$	Divide both sides by 2.
$\frac{1}{2}x - 2 = y$	Subtract 2 from both sides.	$\frac{1}{2}x - \frac{5}{2} = y$	Simplify.

The slope of each line is $\frac{1}{2}$, but they have different y-intercepts. This tells us that the lines representing these equations are parallel. Since the lines are parallel, the system has no solution and is inconsistent.

Work Practice 7

Example 8 Without graphing, determine the number of solutions of the system.

$$\begin{cases} 3x - y = 4 \\ x + 2y = 8 \end{cases}$$

Solution: Once again, the slope-intercept form helps determine how many solutions this system has.

$3x - y = 4$	First equation	$x + 2y = 8$	Second equation
$3x = y + 4$	Add y to both sides.	$x = -2y + 8$	Subtract $2y$ from both sides.
$3x - 4 = y$	Subtract 4 from both sides.	$x - 8 = -2y$	Subtract 8 from both sides.
		$\frac{x}{-2} - \frac{8}{-2} = \frac{-2y}{-2}$	Divide both sides by -2.
		$-\frac{1}{2}x + 4 = y$	Simplify.

The slope of the second line is $-\frac{1}{2}$, whereas the slope of the first line is 3. Since the slopes are not equal, the two lines are neither parallel nor identical and must intersect. Therefore, this system has one solution and is consistent.

Work Practice 8

Practice 8

Without graphing, determine the number of solutions of the system.

$$\begin{cases} -\frac{2}{3}x + y = 6 \\ 3y = 2x + 5 \end{cases}$$

Answer

8. no solution

Calculator Explorations Graphing

A graphing calculator may be used to approximate solutions of systems of equations. For example, to approximate the solution of the system

$$\begin{cases} y = -3.14x - 1.35 \\ y = 4.88x + 5.25, \end{cases}$$

first graph each equation on the same set of axes. Then use the Intersect feature of your calculator to approximate the point of intersection.

The approximate point of intersection is $(-0.82, 1.23)$.

Solve each system of equations. Approximate the solutions to two decimal places.

1. $\begin{cases} y = -2.68x + 1.21 \\ y = 5.22x - 1.68 \end{cases}$

2. $\begin{cases} y = 4.25x + 3.89 \\ y = -1.88x + 3.21 \end{cases}$

3. $\begin{cases} 4.3x - 2.9y = 5.6 \\ 8.1x + 7.6y = -14.1 \end{cases}$

4. $\begin{cases} -3.6x - 8.6y = 10 \\ -4.5x + 9.6y = -7.7 \end{cases}$

Vocabulary, Readiness & Video Check

Fill in each blank with one of the words or phrases listed below.

system of linear equations	solution	consistent
dependent	inconsistent	independent

1. In a system of linear equations in two variables, if the graphs of the equations are the same, the equations are ____________ equations.
2. Two or more linear equations are called a(n) ____________.
3. A system of equations that has at least one solution is called a(n) ____________ system.
4. A(n) ____________ of a system of two equations in two variables is an ordered pair of numbers that is a solution of both equations in the system.
5. A system of equations that has no solution is called a(n) ____________ system.
6. In a system of linear equations in two variables, if the graphs of the equations are different, the equations are ____________ equations.

Each rectangular coordinate system shows the graph of the equations in a system of equations. Use each graph to determine the number of solutions for each associated system. If the system has only one solution, give its coordinates.

7.

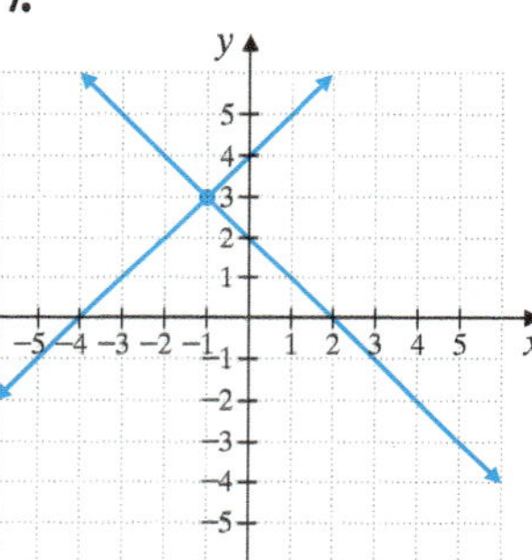

8.

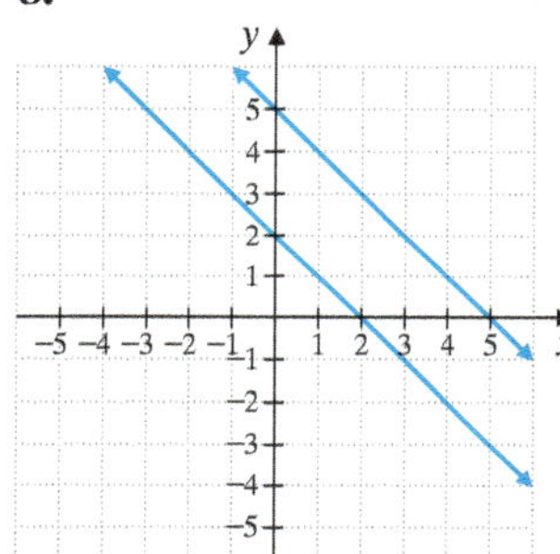

9.

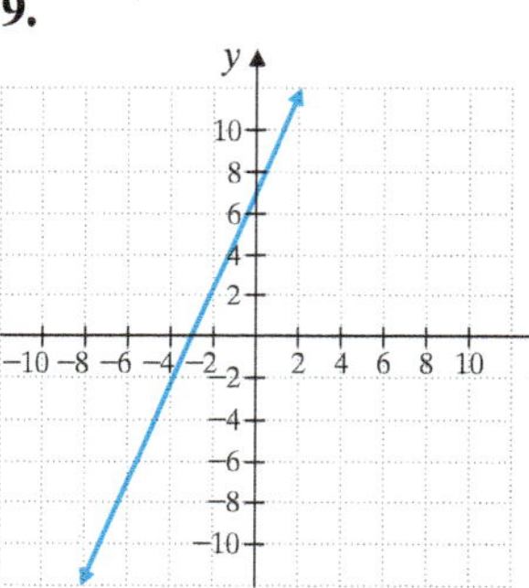

10.

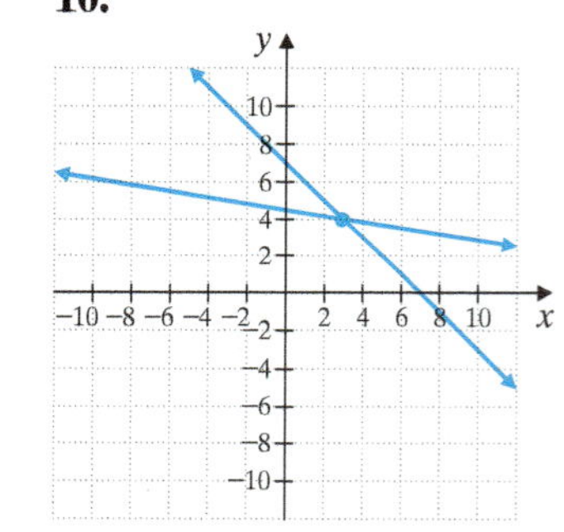

Martin-Gay Interactive Videos

See Video 7.1

Watch the section lecture video and answer the following questions.

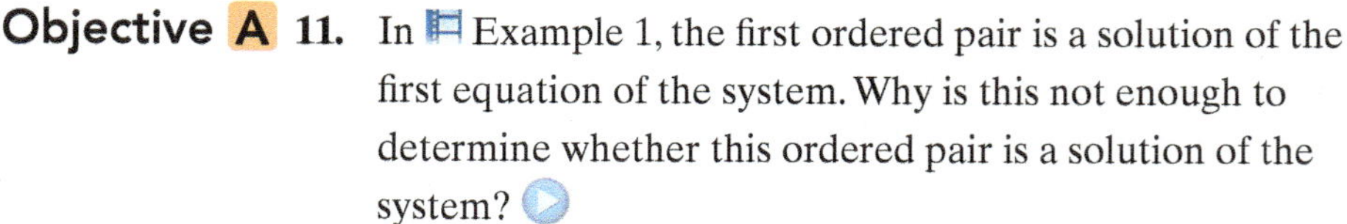

Objective A **11.** In Example 1, the first ordered pair is a solution of the first equation of the system. Why is this not enough to determine whether this ordered pair is a solution of the system?

Objective B C **12.** From Examples 2 and 3, why is finding the solution of a system of equations from a graph considered "guessing" and this proposed solution checked algebraically?

Objective D **13.** From Examples 5–7, explain how the slope-intercept form tells us how many solutions a system of equations has.

7.1 Exercise Set MyLab Math 

Objective A *Determine whether each ordered pair is a solution of the system of linear equations. See Examples 1 and 2.*

1. $\begin{cases} x + y = 8 \\ 3x + 2y = 21 \end{cases}$

a. $(2, 4)$

b. $(5, 3)$

2. $\begin{cases} 2x + y = 5 \\ x + 3y = 5 \end{cases}$

a. $(5, 0)$

b. $(2, 1)$

3. $\begin{cases} 3x - y = 5 \\ x + 2y = 11 \end{cases}$

a. $(3, 4)$

b. $(0, -5)$

4. $\begin{cases} 2x - 3y = 8 \\ x - 2y = 6 \end{cases}$

a. $(-2, -4)$

b. $(7, 2)$

5. $\begin{cases} 2y = 4x + 6 \\ 2x - y = -3 \end{cases}$

a. $(-3, -3)$

b. $(0, 3)$

6. $\begin{cases} x + 5y = -4 \\ -2x = 10y + 8 \end{cases}$

a. $(-4, 0)$

b. $(6, -2)$

7. $\begin{cases} -2 = x - 7y \\ 6x - y = 13 \end{cases}$

a. $(-2, 0)$

b. $\left(\frac{1}{2}, \frac{5}{14}\right)$

8. $\begin{cases} 4x = 1 - y \\ x - 3y = -8 \end{cases}$

a. $(0, 1)$

b. $\left(\frac{1}{6}, \frac{1}{3}\right)$

Objectives B C *Solve each system of linear equations by graphing. See Examples 3 through 6.*

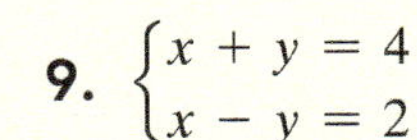

9. $\begin{cases} x + y = 4 \\ x - y = 2 \end{cases}$

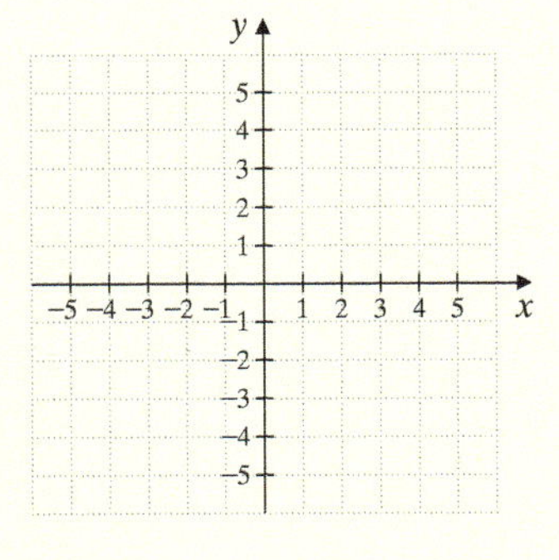

10. $\begin{cases} x + y = 3 \\ x - y = 5 \end{cases}$

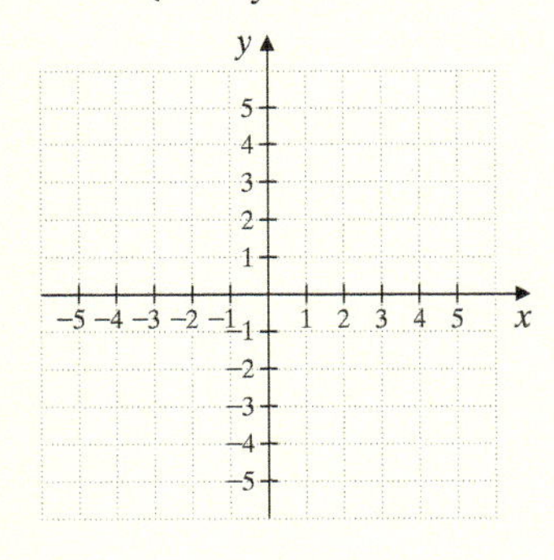

11. $\begin{cases} x + y = 6 \\ -x + y = -6 \end{cases}$

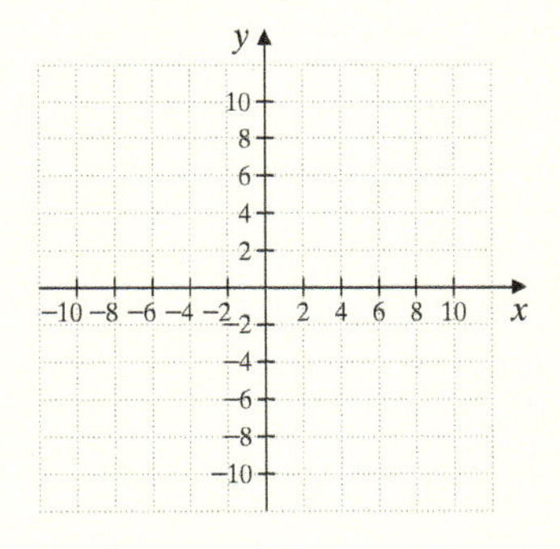

12. $\begin{cases} x + y = 1 \\ -x + y = -3 \end{cases}$

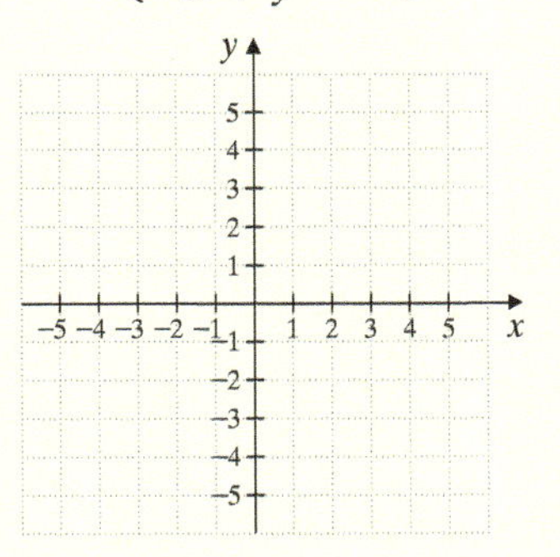

13. $\begin{cases} y = 2x \\ 3x - y = -2 \end{cases}$

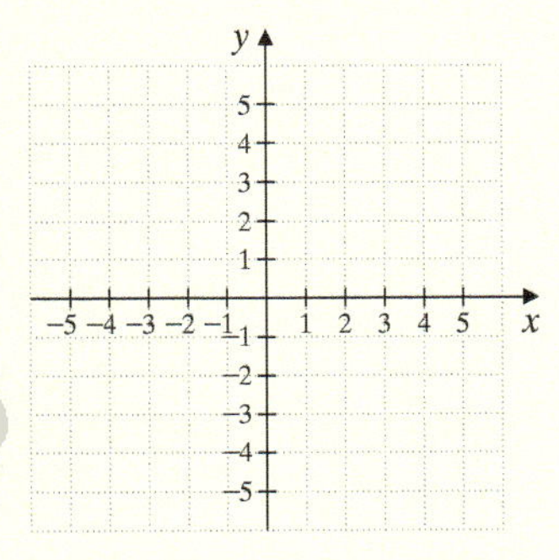

14. $\begin{cases} y = -3x \\ 2x - y = -5 \end{cases}$

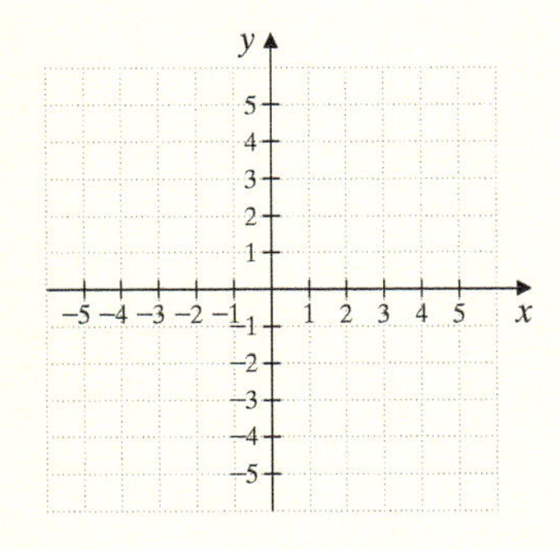

15. $\begin{cases} y = x + 1 \\ y = 2x - 1 \end{cases}$

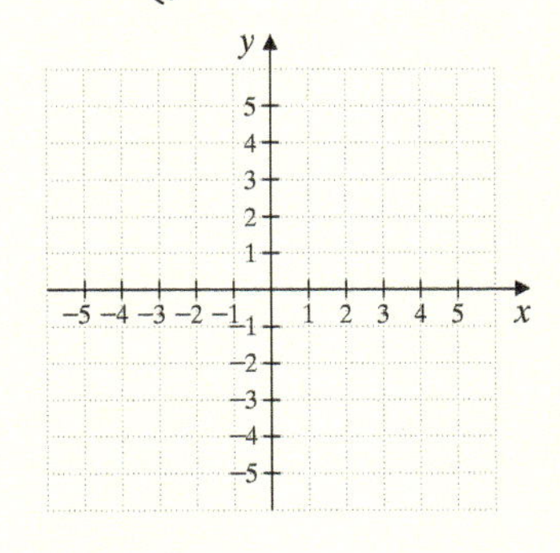

16. $\begin{cases} y = 3x - 4 \\ y = x + 2 \end{cases}$

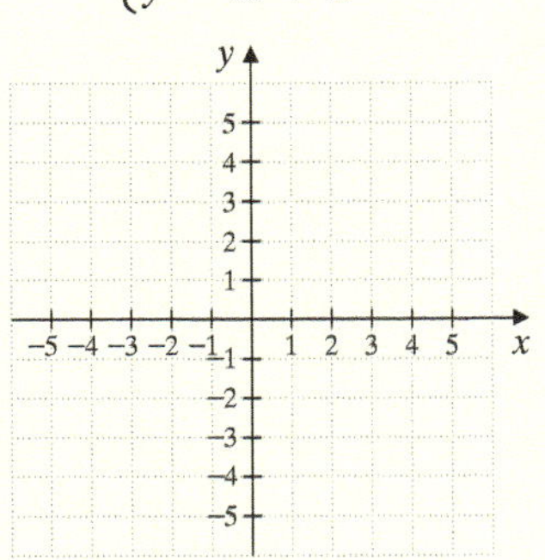

17. $\begin{cases} 2x + y = 0 \\ 3x + y = 1 \end{cases}$

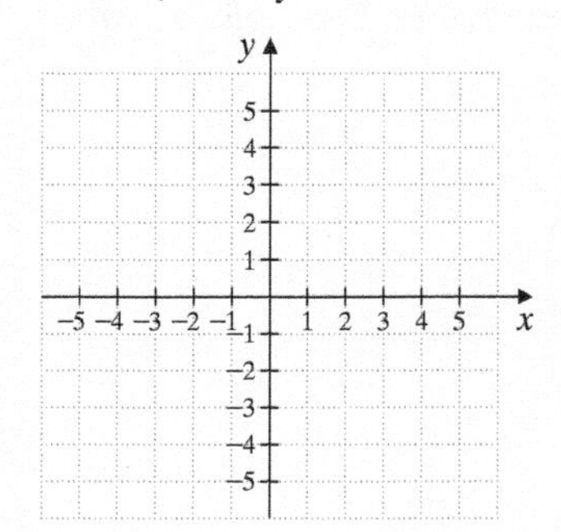

18. $\begin{cases} 2x + y = 1 \\ 3x + y = 0 \end{cases}$

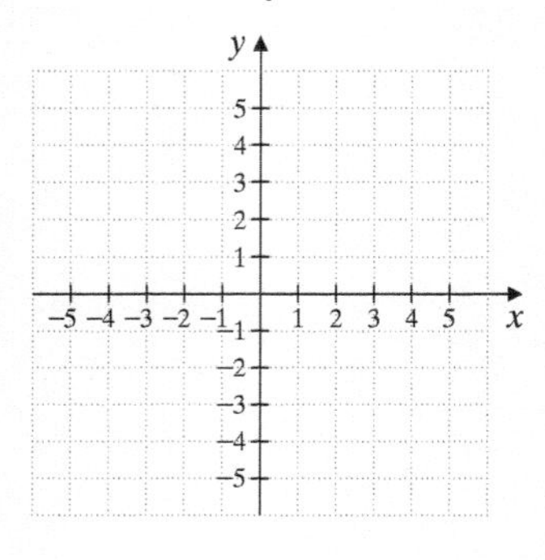

19. $\begin{cases} y = -x - 1 \\ y = 2x + 5 \end{cases}$

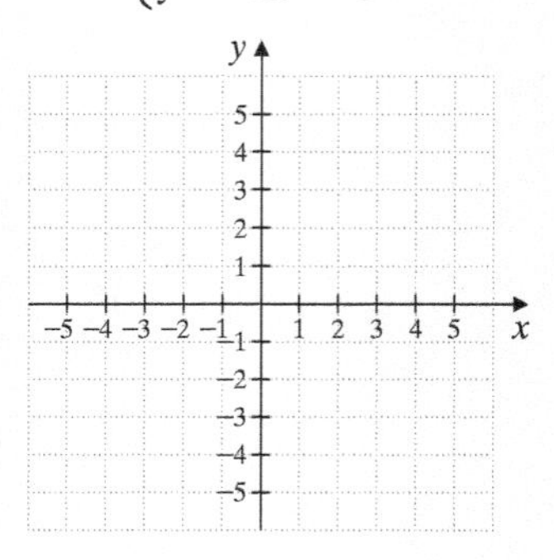

20. $\begin{cases} y = x - 1 \\ y = -3x - 5 \end{cases}$

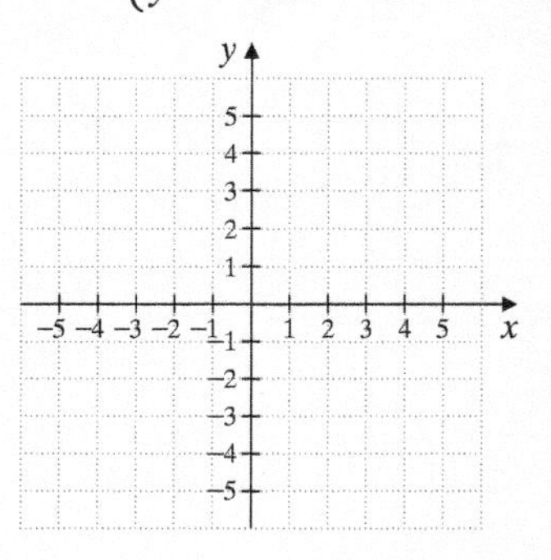

21. $\begin{cases} x + y = 5 \\ x + y = 6 \end{cases}$

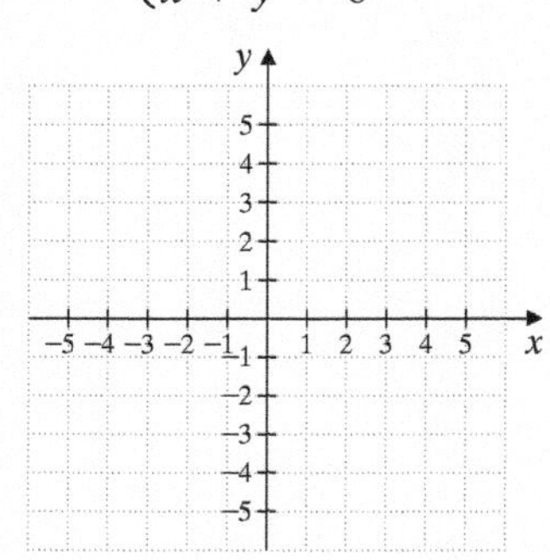

22. $\begin{cases} x - y = 4 \\ x - y = 1 \end{cases}$

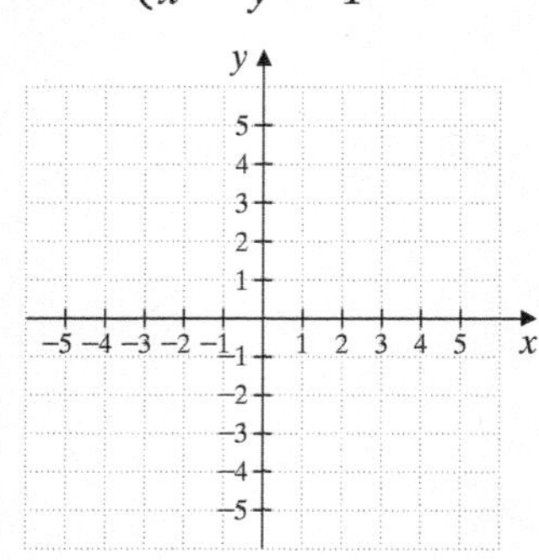

23. $\begin{cases} 2x - y = 6 \\ y = 2 \end{cases}$

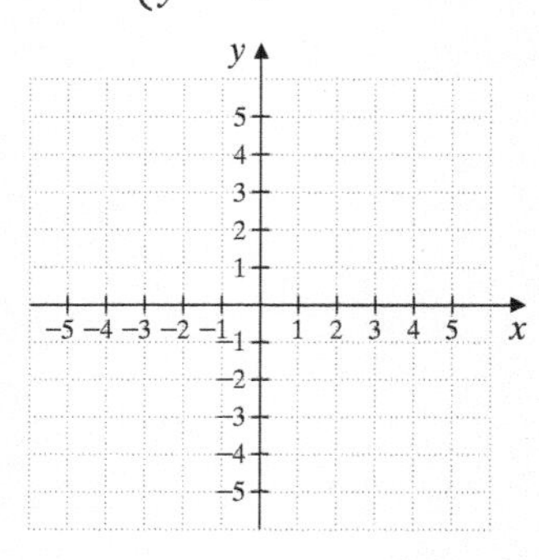

24. $\begin{cases} x + y = 5 \\ x = 4 \end{cases}$

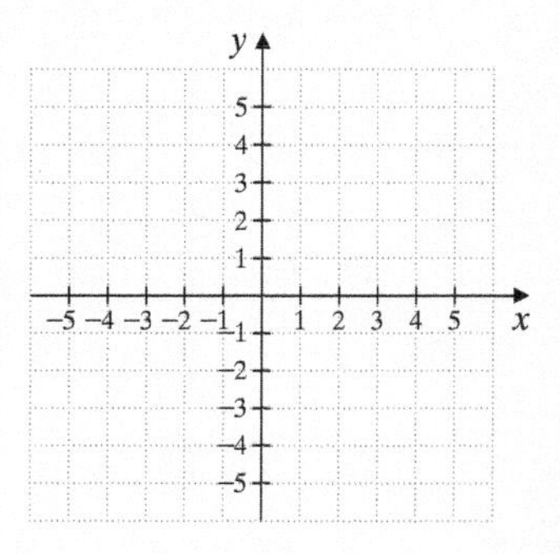

25. $\begin{cases} x - 2y = 2 \\ 3x + 2y = -2 \end{cases}$

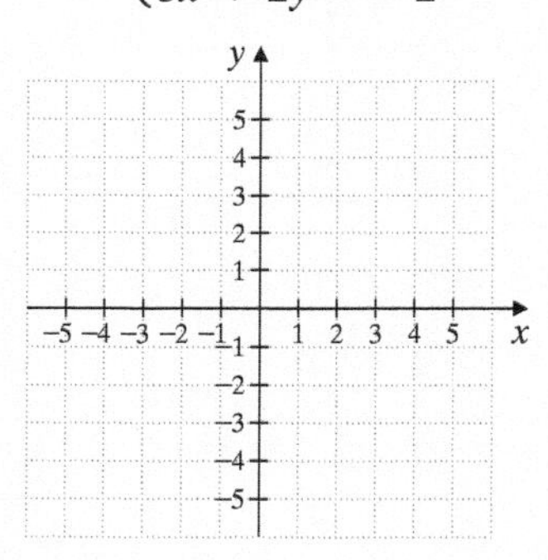

26. $\begin{cases} x + 3y = 7 \\ 2x - 3y = -4 \end{cases}$

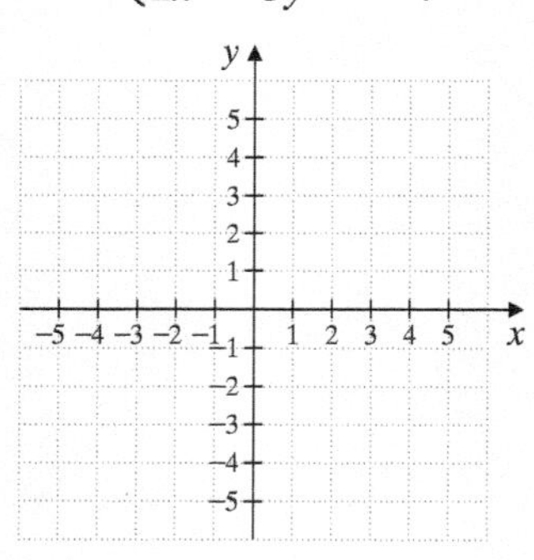

27. $\begin{cases} 2x + y = 4 \\ 6x = -3y + 6 \end{cases}$

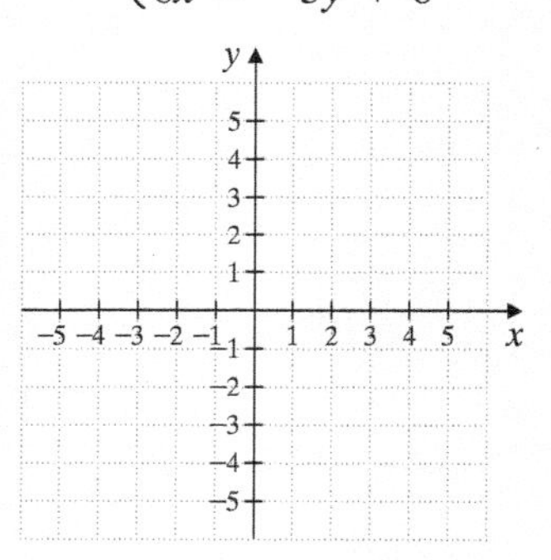

28. $\begin{cases} y + 2x = 3 \\ 4x = 2 - 2y \end{cases}$

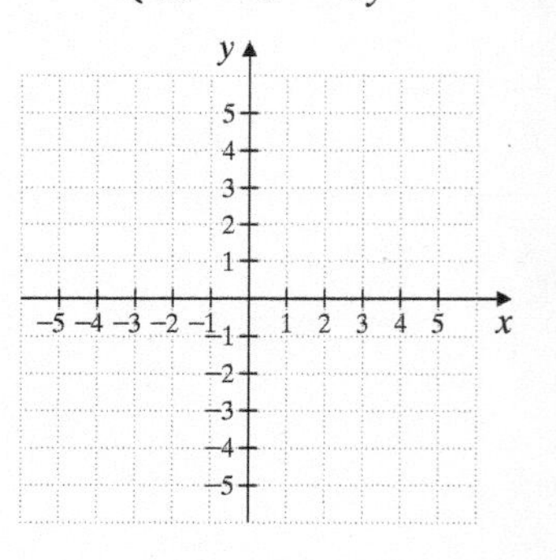

29. $\begin{cases} y - 3x = -2 \\ 6x - 2y = 4 \end{cases}$

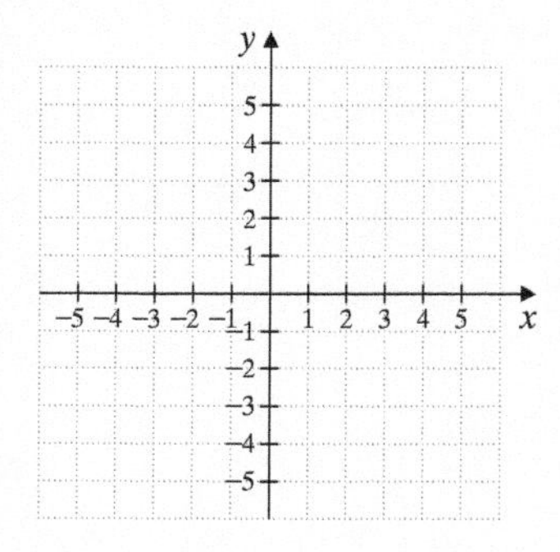

30. $\begin{cases} x - 2y = -6 \\ -2x + 4y = 12 \end{cases}$

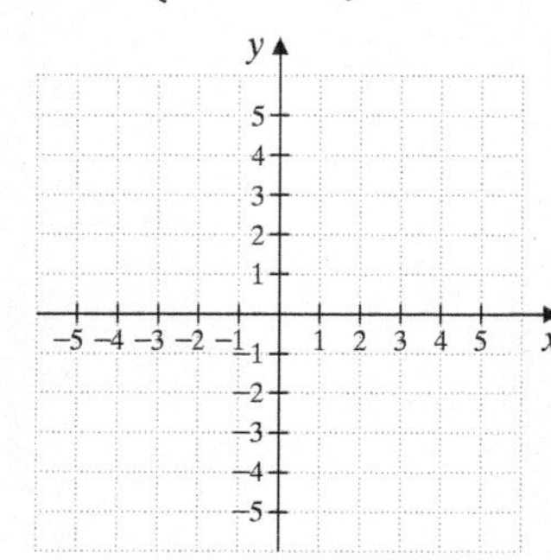

31. $\begin{cases} x = 3 \\ y = -1 \end{cases}$

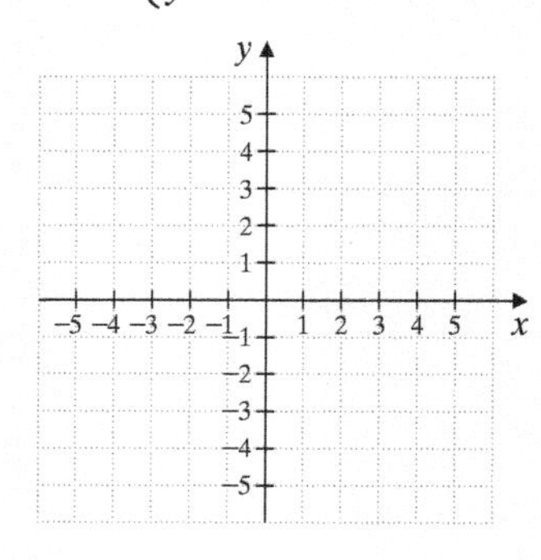

32. $\begin{cases} x = -5 \\ y = 3 \end{cases}$

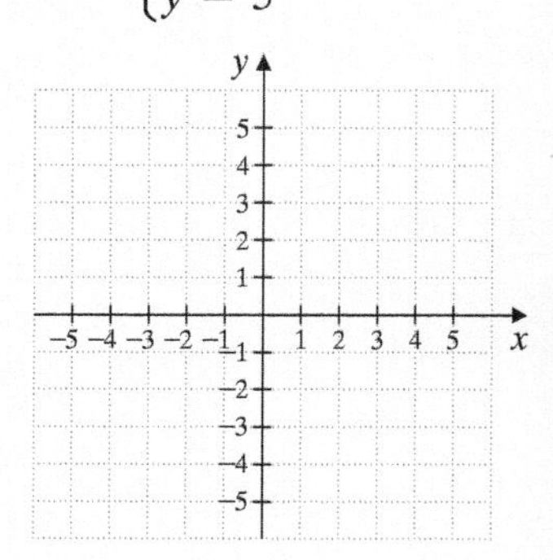

33. $\begin{cases} y = x - 2 \\ y = 2x + 3 \end{cases}$

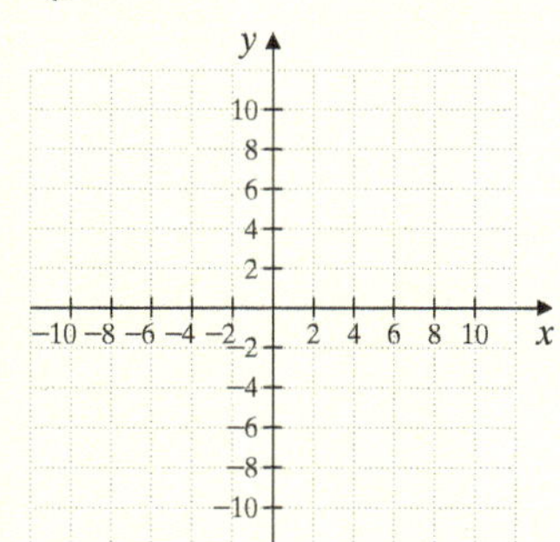

34. $\begin{cases} y = x + 5 \\ y = -2x - 4 \end{cases}$

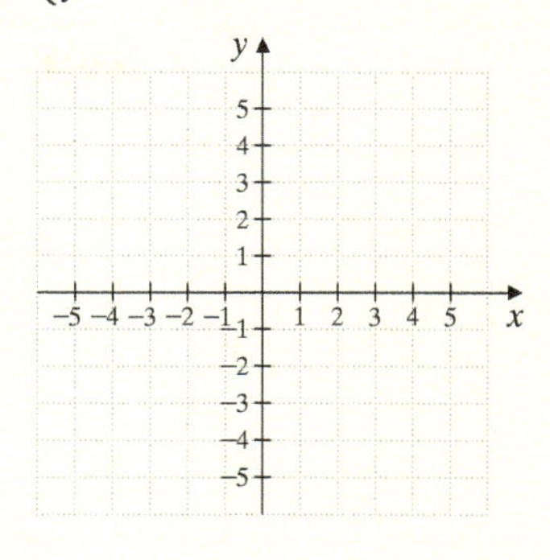

35. $\begin{cases} 2x - 3y = -2 \\ -3x + 5y = 5 \end{cases}$

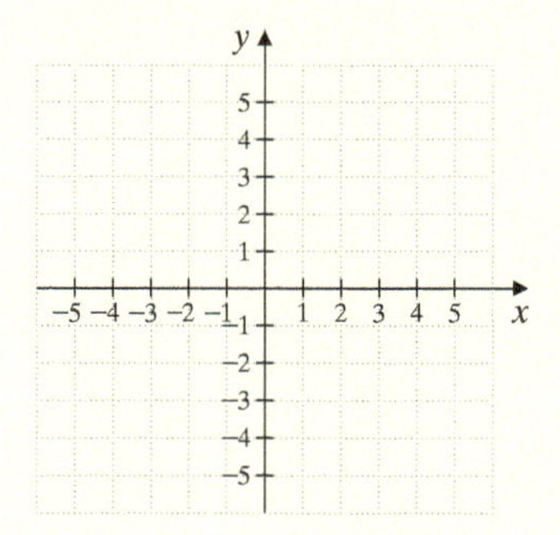

36. $\begin{cases} 4x - y = 7 \\ 2x - 3y = -9 \end{cases}$

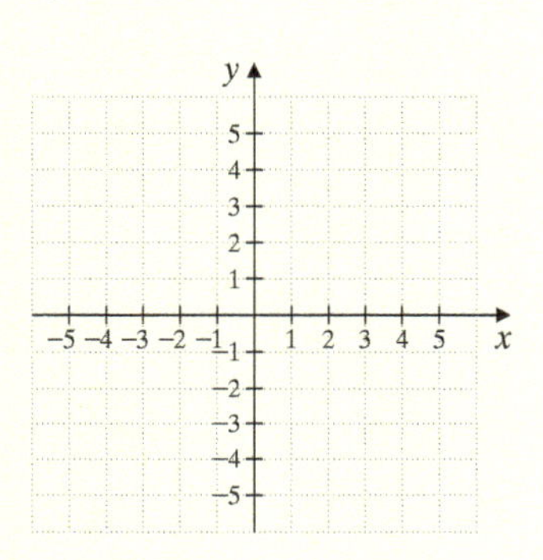

37. $\begin{cases} 6x - y = 4 \\ \frac{1}{2}y = -2 + 3x \end{cases}$

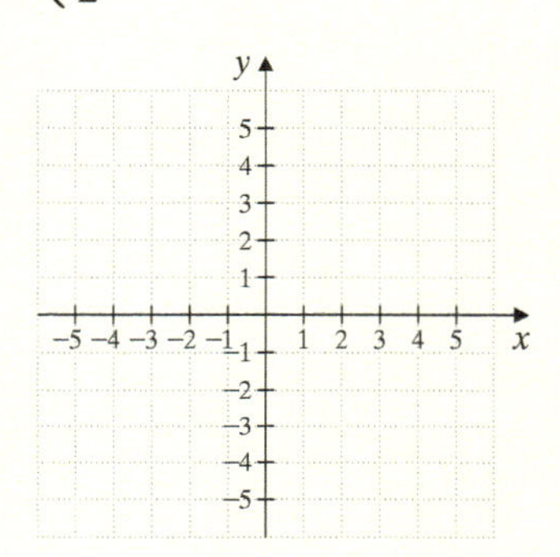

38. $\begin{cases} 3x - y = 6 \\ \frac{1}{3}y = -2 + x \end{cases}$

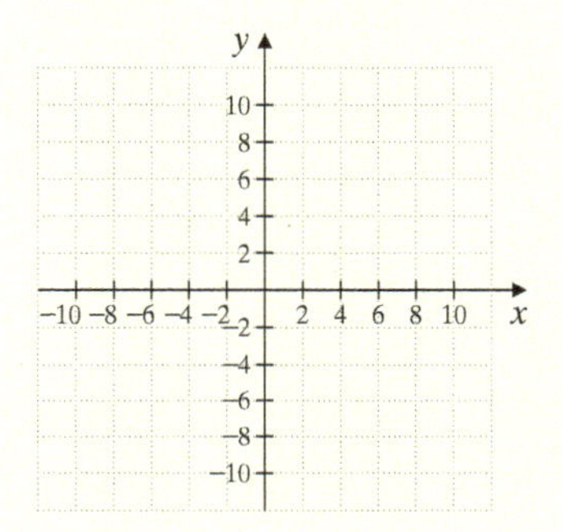

Objective D *Without graphing, decide:*

a. *Are the graphs of the equations identical lines, parallel lines, or lines intersecting at a single point?*

b. *How many solutions does the system have? See Examples 7 and 8.*

39. $\begin{cases} 4x + y = 24 \\ x + 2y = 2 \end{cases}$

40. $\begin{cases} 3x + y = 1 \\ 3x + 2y = 6 \end{cases}$

41. $\begin{cases} 2x + y = 0 \\ 2y = 6 - 4x \end{cases}$

42. $\begin{cases} 3x + y = 0 \\ 2y = -6x \end{cases}$

43. $\begin{cases} 6x - y = 4 \\ \frac{1}{2}y = -2 + 3x \end{cases}$

44. $\begin{cases} 3x - y = 2 \\ \frac{1}{3}y = -2 + 3x \end{cases}$

45. $\begin{cases} x = 5 \\ y = -2 \end{cases}$

46. $\begin{cases} y = 3 \\ x = -4 \end{cases}$

47. $\begin{cases} 3y - 2x = 3 \\ x + 2y = 9 \end{cases}$

48. $\begin{cases} 2y = x + 2 \\ y + 2x = 3 \end{cases}$

49. $\begin{cases} 6y + 4x = 6 \\ 3y - 3 = -2x \end{cases}$

50. $\begin{cases} 8y + 6x = 4 \\ 4y - 2 = 3x \end{cases}$

51. $\begin{cases} x + y = 4 \\ x + y = 3 \end{cases}$

52. $\begin{cases} 2x + y = 0 \\ y = -2x + 1 \end{cases}$

Review

Solve each equation. See Section 2.3.

53. $5(x - 3) + 3x = 1$

54. $-2x + 3(x + 6) = 17$

55. $4\left(\frac{y + 1}{2}\right) + 3y = 0$

56. $-y + 12\left(\frac{y - 1}{4}\right) = 3$

57. $8a - 2(3a - 1) = 6$

58. $3z - (4z - 2) = 9$

Concept Extensions

59. Draw a graph of two linear equations whose associated system has the solution $(-1, 4)$.

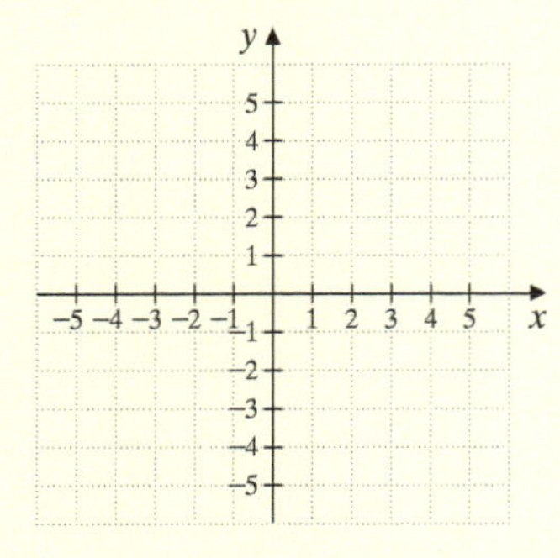

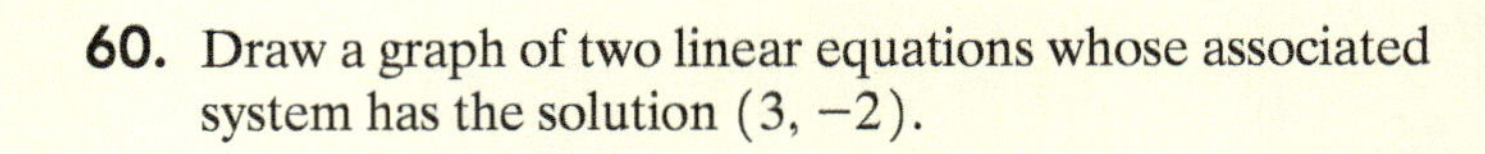

60. Draw a graph of two linear equations whose associated system has the solution $(3, -2)$.

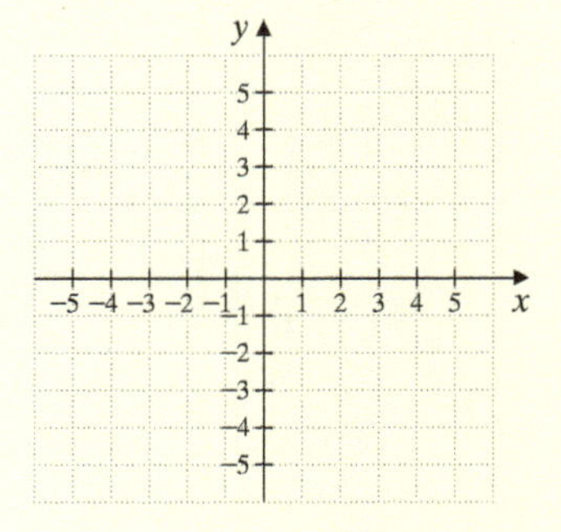

61. Draw a graph of two linear equations whose associated system has no solution.

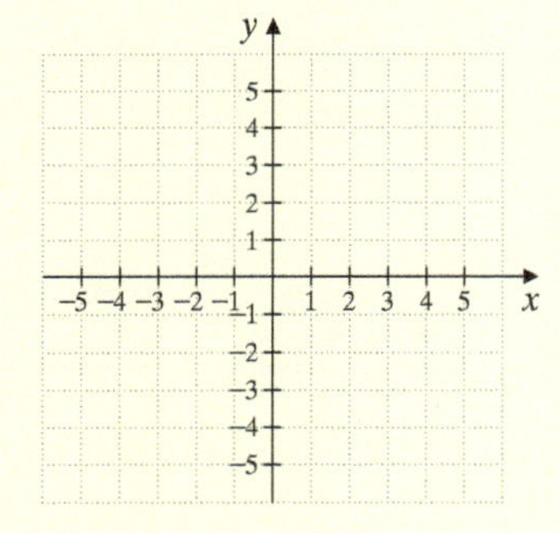

62. Draw a graph of two linear equations whose associated system has an infinite number of solutions.

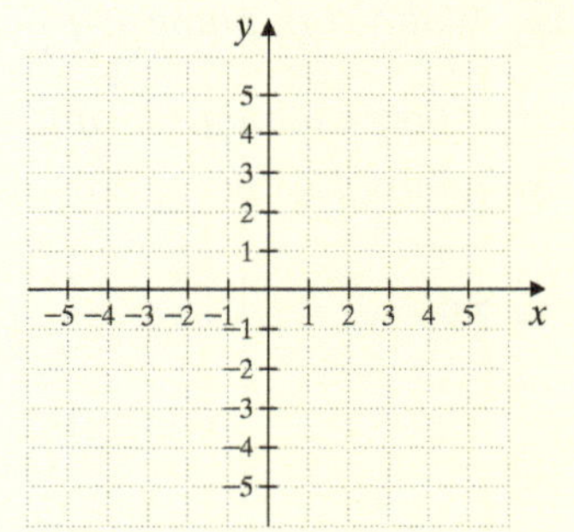

The double line graph below shows the number of digital 3-D and analog movie screens in U.S. cinemas for the years shown. Use this graph to answer Exercises 63 and 64. (Source: Motion Picture Association of America, Inc.)

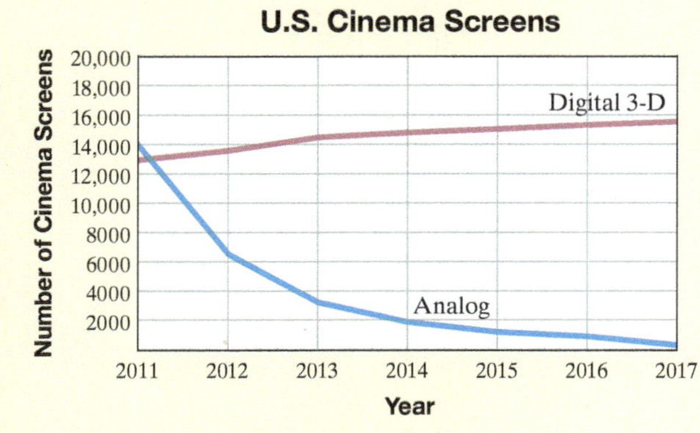

Source: Motion Picture Association of America, Inc.

63. Between what pairs of years did the number of digital 3-D cinema screens equal the number of analog cinema screens?

64. For what year was the number of digital 3-D cinema screens less than the number of analog cinema screens?

The double line graph below shows the number of pounds of fresh Pacific salmon imported to or exported from the United States during the given years. Use this graph to answer Exercises 65 and 66.

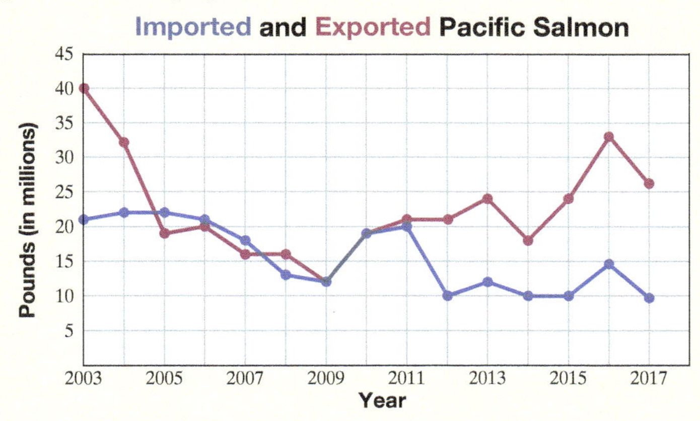

Source: USDA Economic Research Service

65. During which year(s) shown did the number of pounds of imported Pacific salmon equal the number of pounds of exported Pacific salmon?

66. For what year(s) shown was the number of pounds of exported Pacific salmon less than the number of pounds of imported Pacific salmon?

Solve.

67. Construct a system of two linear equations that has $(2, 5)$ as a solution.

68. Construct a system of two linear equations that has $(0, 1)$ as a solution.

69. The ordered pair $(-2, 3)$ is a solution of the three linear equations below:

$$x + y = 1$$
$$2x - y = -7$$
$$x + 3y = 7$$

If each equation has a distinct graph, describe the graph of all three equations on the same axes.

70. Explain how to use a graph to determine the number of solutions of a system.

71. Below are tables of values for two linear equations.

a. Find a solution of the corresponding system.

b. Graph several ordered pairs from each table and sketch the two lines.

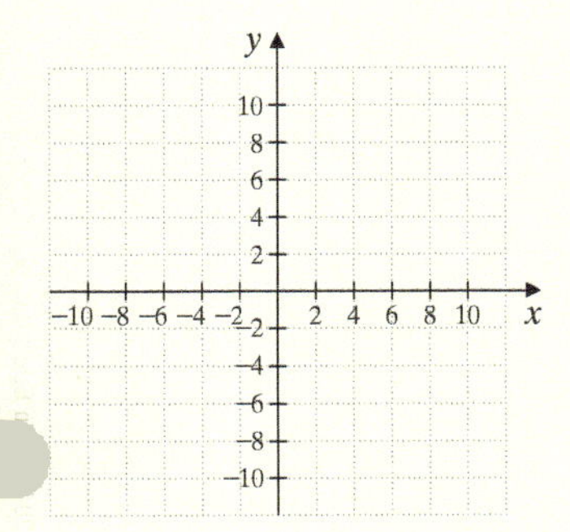

x	y
1	3
2	5
3	7
4	9
5	11

x	y
1	6
2	7
3	8
4	9
5	10

c. Does your graph confirm the solution from part **a**?

72. Below are tables of values for two linear equations.

a. Find a solution of the corresponding system.

b. Graph several ordered pairs from each table and sketch the two lines.

x	y
−3	5
−1	1
0	−1
1	−3
2	−5

x	y
−3	7
−1	1
0	−2
1	−5
2	−8

c. Does your graph confirm the solution from part **a**?

7.2 Solving Systems of Linear Equations by Substitution

Objective

A Use the Substitution Method to Solve a System of Linear Equations.

Objective A Using the Substitution Method

You may have suspected by now that graphing alone is not an accurate way to solve a system of linear equations. For example, a solution of $\left(\frac{1}{2}, \frac{2}{9}\right)$ is unlikely to be read correctly from a graph. In this section, we discuss a second, more accurate method for solving systems of equations. This method is called the **substitution method** and is introduced in the next example.

Practice 1

Use the substitution method to solve the system:

$$\begin{cases} 2x + 3y = 13 \\ x = y + 4 \end{cases}$$

Example 1 Solve the system:

$$\begin{cases} 2x + y = 10 & \text{First equation} \\ x = y + 2 & \text{Second equation} \end{cases}$$

Solution: The second equation in this system is $x = y + 2$. This tells us that x and $y + 2$ have the same value. This means that we may substitute $y + 2$ for x in the first equation.

$$2x + y = 10 \quad \text{First equation}$$

$$2(y + 2) + y = 10 \quad \text{Substitute } y + 2 \text{ for } x \text{ since } x = y + 2.$$

Notice that this equation now has one variable, y. Let's now solve this equation for y.

Helpful Hint
Don't forget the distributive property.

$$\begin{aligned} 2(y + 2) + y &= 10 & \\ 2y + 4 + y &= 10 & \text{Apply the distributive property.} \\ 3y + 4 &= 10 & \text{Combine like terms.} \\ 3y &= 6 & \text{Subtract 4 from both sides.} \\ y &= 2 & \text{Divide both sides by 3.} \end{aligned}$$

Now we know that the y-value of the ordered pair solution of the system is 2. To find the corresponding x-value, we replace y with 2 in the second equation, $x = y + 2$, and solve for x.

$$\begin{aligned} x &= y + 2 & \text{Second equation} \\ x &= 2 + 2 & \text{Let } y = 2. \\ x &= 4 & \end{aligned}$$

The solution of the system is the ordered pair $(4, 2)$. Since an ordered pair solution must satisfy both linear equations in the system, we could have chosen the equation $2x + y = 10$ to find the corresponding x-value. The resulting x-value is the same.

Check: We check to see that $(4, 2)$ satisfies both equations of the original system.

First Equation		Second Equation	
$2x + y = 10$		$x = y + 2$	
$2(4) + 2 \stackrel{?}{=} 10$		$4 \stackrel{?}{=} 2 + 2$	Let $x = 4$ and $y = 2$.
$10 = 10$	True	$4 = 4$	True

Answer

1. $(5, 1)$

The solution of the system is $(4, 2)$.

A graph of the two equations shows the two lines intersecting at the point $(4, 2)$.

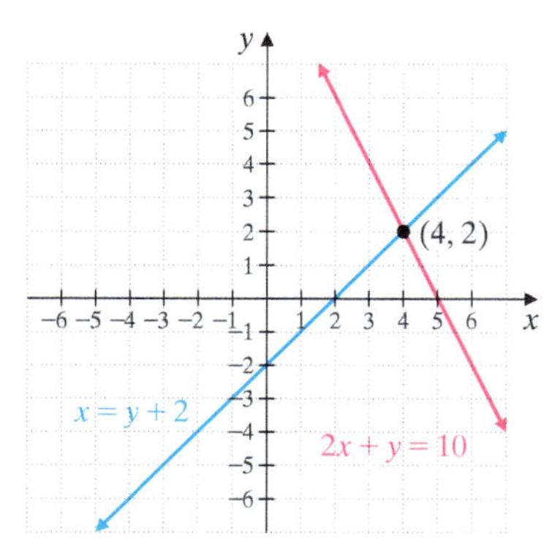

Work Practice 1

Example 2 Solve the system:

$$\begin{cases} 5x - y = -2 \\ y = 3x \end{cases}$$

Solution: The second equation is solved for y in terms of x. We substitute $3x$ for y in the first equation.

$5x - y = -2$ First equation

$5x - (3x) = -2$ Substitute $3x$ for y.

Now we solve for x.

$5x - 3x = -2$

$2x = -2$ Combine like terms.

$x = -1$ Divide both sides by 2.

The x-value of the ordered pair solution is -1. To find the corresponding y-value, we replace x with -1 in the second equation, $y = 3x$.

$y = 3x$ Second equation

$y = 3(-1)$ Let $x = -1$.

$y = -3$

Check to see that the solution of the system is $(-1, -3)$.

Work Practice 2

Practice 2

Use the substitution method to solve the system:

$$\begin{cases} 4x - y = 2 \\ y = 5x \end{cases}$$

To solve a system of equations by substitution, we first need an equation solved for one of its variables, as in Examples 1 and 2. If neither equation in a system is solved for x or y, this will be our first step.

Example 3 Solve the system:

$$\begin{cases} x + 2y = 7 \\ 2x + 2y = 13 \end{cases}$$

Solution: Notice that neither equation is solved for x or y. Thus, we choose one of the equations and solve for x or y. We will solve the first equation for x so that we will not introduce tedious fractions when solving. To solve the first equation for x, we subtract $2y$ from both sides.

$x + 2y = 7$ First equation

$x = 7 - 2y$ Subtract $2y$ from both sides.

(*Continued on next page*)

Practice 3

Solve the system:

$$\begin{cases} 3x + y = 5 \\ 3x - 2y = -7 \end{cases}$$

Answers

2. $(-2, -10)$ **3.** $\left(\frac{1}{3}, 4\right)$

Since $x = 7 - 2y$, we now substitute $7 - 2y$ for x in the second equation and solve for y.

$$\begin{aligned} 2x + 2y &= 13 && \text{Second equation} \\ 2(7 - 2y) + 2y &= 13 && \text{Let } x = 7 - 2y. \\ 14 - 4y + 2y &= 13 && \text{Apply the distributive property.} \\ 14 - 2y &= 13 && \text{Simplify.} \\ -2y &= -1 && \text{Subtract 14 from both sides.} \\ y &= \frac{1}{2} && \text{Divide both sides by } -2. \end{aligned}$$

Helpful Hint Don't forget to insert parentheses when substituting $7 - 2y$ for x.

Helpful Hint To find x, any equation in two variables equivalent to the original equations of the system may be used. We used this equation since it is solved for x.

To find x, we let $y = \frac{1}{2}$ in the equation $x = 7 - 2y$.

$$\begin{aligned} x &= 7 - 2y \\ x &= 7 - 2\left(\frac{1}{2}\right) && \text{Let } y = \frac{1}{2}. \\ x &= 7 - 1 \\ x &= 6 \end{aligned}$$

Check the solution in both equations of the original system. The solution is $\left(6, \frac{1}{2}\right)$.

Work Practice 3

The following steps summarize how to solve a system of equations by the substitution method.

To Solve a System of Two Linear Equations by the Substitution Method

Step 1: Solve one of the equations for one of its variables.

Step 2: Substitute the expression for the variable found in Step 1 into the other equation.

Step 3: Solve the equation from Step 2 to find the value of one variable.

Step 4: Substitute the value found in Step 3 into any equation containing both variables to find the value of the other variable.

Step 5: Check the proposed solution in the original system.

✓**Concept Check** As you solve the system

$$\begin{cases} 2x + y = -5 \\ x - y = 5 \end{cases}$$

you find that $y = -5$. Is this the solution of the system?

Example 4 Solve the system:

$$\begin{cases} 7x - 3y = -14 \\ -3x + y = 6 \end{cases}$$

Solution: Since the coefficient of y is 1 in the second equation, we will solve the second equation for y. This way, we avoid introducing tedious fractions.

$$\begin{aligned} -3x + y &= 6 && \text{Second equation} \\ y &= 3x + 6 \end{aligned}$$

Practice 4

Solve the system:

$$\begin{cases} 5x - 2y = 6 \\ -3x + y = -3 \end{cases}$$

Answer

4. $(0, -3)$

✓**Concept Check Answer**

no, the solution will be an ordered pair

Next, we substitute $3x + 6$ for y in the first equation.

$$\begin{aligned} 7x - 3y &= -14 && \text{First equation} \\ 7x - 3(3x + 6) &= -14 && \text{Let } y = 3x + 6. \\ 7x - 9x - 18 &= -14 && \text{Use the distributive property.} \\ -2x - 18 &= -14 && \text{Simplify.} \\ -2x &= 4 && \text{Add 18 to both sides.} \\ x &= -2 && \text{Divide both sides by } -2. \end{aligned}$$

To find the corresponding y-value, we substitute -2 for x in the equation $y = 3x + 6$. Then $y = 3(-2) + 6$ or $y = 0$. The solution of the system is $(-2, 0)$. Check this solution in both equations of the system.

Work Practice 4

✓**Concept Check** To avoid fractions, which of the equations below would you use to solve for x?

a. $3x - 4y = 15$ **b.** $14 - 3y = 8x$ **c.** $7y + x = 12$

Helpful Hint

When solving a system of equations by the substitution method, begin by solving an equation for one of its variables. If possible, solve for a variable that has a coefficient of 1 or -1 to avoid working with time-consuming fractions.

Example 5 Solve the system: $\begin{cases} \frac{1}{2}x - y = 3 \\ x = 6 + 2y \end{cases}$

Solution: The second equation is already solved for x in terms of y. Thus we substitute $6 + 2y$ for x in the first equation and solve for y.

$$\begin{aligned} \frac{1}{2}x - y &= 3 && \text{First equation} \\ \frac{1}{2}(6 + 2y) - y &= 3 && \text{Let } x = 6 + 2y. \\ 3 + y - y &= 3 && \text{Apply the distributive property.} \\ 3 &= 3 && \text{Simplify.} \end{aligned}$$

The true statement $3 = 3$ indicates that the two linear equations in the original system are equivalent. This means that their graphs are identical, as shown in the figure. There is an infinite number of solutions to the system, and any solution of one equation is also a solution of the other.

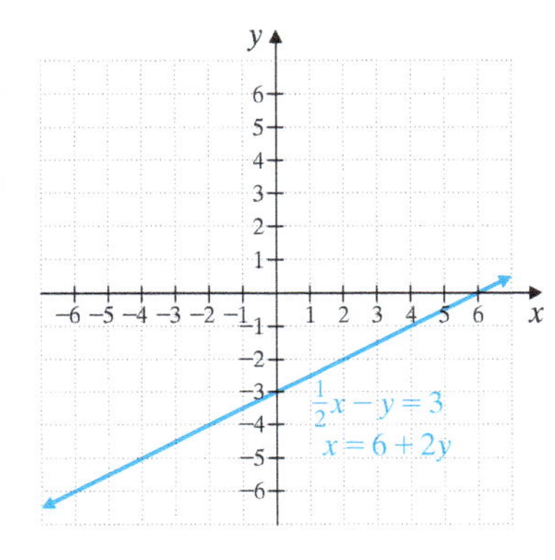

Work Practice 5

Practice 5

Solve the system:

$\begin{cases} -x + 3y = 6 \\ y = \frac{1}{3}x + 2 \end{cases}$

Answer

5. infinite number of solutions

✓**Concept Check Answer**

c

Helpful Hint

An infinite number of solutions does *not* mean that all ordered pairs are solutions of both equations of the system.

An infinite number of solutions for Example 5 means that any of the infinite number of ordered pairs that is a solution of one equation in the system is also a solution of the other and is thus a solution of the system.

For Example 5,

$(2, 0)$ is *not* a solution of the system, but
$(6, 0)$ is a solution of the system.

Practice 6

Solve the system:

$$\begin{cases} 2x - 3y = 6 \\ -4x + 6y = 12 \end{cases}$$

Example 6

Solve the system:

$$\begin{cases} 6x + 12y = 5 \\ -4x - 8y = 0 \end{cases}$$

Solution: We choose the second equation and solve for y. (*Note:* Although you might not see this beforehand, if you solve the second equation for x, the result is $x = -2y$ and no fractions are introduced. Either way will lead to the correct solution.)

$$-4x - 8y = 0 \quad \text{Second equation}$$
$$-8y = 4x \quad \text{Add } 4x \text{ to both sides.}$$
$$\frac{-8y}{-8} = \frac{4x}{-8} \quad \text{Divide both sides by } -8.$$
$$y = -\frac{1}{2}x \quad \text{Simplify.}$$

Now we replace y with $-\frac{1}{2}x$ in the first equation.

$$6x + 12y = 5 \quad \text{First equation}$$
$$6x + 12\left(-\frac{1}{2}x\right) = 5 \quad \text{Let } y = -\frac{1}{2}x.$$
$$6x + (-6x) = 5 \quad \text{Simplify.}$$
$$0 = 5 \quad \text{Combine like terms.}$$

The false statement $0 = 5$ indicates that this system has no solution. The graph of the linear equations in the system is a pair of parallel lines, as shown in the figure.

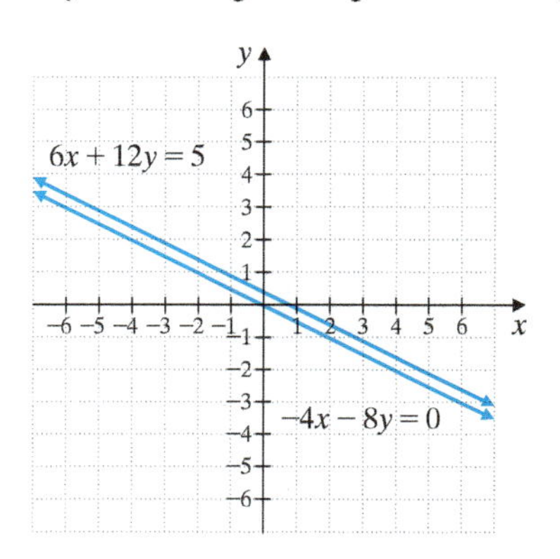

Work Practice 6

✓Concept Check Describe how the graphs of the equations in a system appear if the system has

a. no solution **b.** one solution **c.** an infinite number of solutions

Answer

6. no solution

✓Concept Check Answer

a. parallel lines **b.** intersect at one point **c.** identical graphs

Vocabulary, Readiness & Video Check

Give the solution of each system. If the system has no solution or an infinite number of solutions, say so. If the system has one solution, find it.

1. $\begin{cases} y = 4x \\ -3x + y = 1 \end{cases}$

When solving, you obtain $x = 1$.

2. $\begin{cases} 4x - y = 17 \\ -8x + 2y = 0 \end{cases}$

When solving, you obtain $0 = 34$.

3. $\begin{cases} 4x - y = 17 \\ -8x + 2y = -34 \end{cases}$

When solving, you obtain $0 = 0$.

4. $\begin{cases} 5x + 2y = 25 \\ x = y + 5 \end{cases}$

When solving, you obtain $y = 0$.

5. $\begin{cases} x + y = 0 \\ 7x - 7y = 0 \end{cases}$

When solving, you obtain $x = 0$.

6. $\begin{cases} y = -2x + 5 \\ 4x + 2y = 10 \end{cases}$

When solving, you obtain $0 = 0$.

Martin-Gay Interactive Videos *Watch the section lecture video and answer the following question.*

See Video 7.2

Objective A **7.** The systems in Examples 2–4 all need one of their equations solved for a variable as a first step. What important part of the substitution method is emphasized in each example?

7.2 Exercise Set MyLab Math

Objective A *Solve each system of equations by the substitution method. See Examples 1 and 2.*

1. $\begin{cases} x + y = 3 \\ x = 2y \end{cases}$

2. $\begin{cases} x + y = 20 \\ x = 3y \end{cases}$

3. $\begin{cases} x + y = 6 \\ y = -3x \end{cases}$

4. $\begin{cases} x + y = 6 \\ y = -4x \end{cases}$

5. $\begin{cases} y = 3x + 1 \\ 4y - 8x = 12 \end{cases}$

6. $\begin{cases} y = 2x + 3 \\ 5y - 7x = 18 \end{cases}$

7. $\begin{cases} y = 2x + 9 \\ y = 7x + 10 \end{cases}$

8. $\begin{cases} y = 5x - 3 \\ y = 8x + 4 \end{cases}$

Solve each system of equations by the substitution method. See Examples 1 through 6.

9. $\begin{cases} 3x - 4y = 10 \\ y = x - 3 \end{cases}$

10. $\begin{cases} 4x - 3y = 10 \\ y = x - 5 \end{cases}$

11. $\begin{cases} x + 2y = 6 \\ 2x + 3y = 8 \end{cases}$

12. $\begin{cases} x + 3y = -5 \\ 2x + 2y = 6 \end{cases}$

13. $\begin{cases} 3x + 2y = 16 \\ x = 3y - 2 \end{cases}$

14. $\begin{cases} 2x + 3y = 18 \\ x = 2y - 5 \end{cases}$

15. $\begin{cases} 2x - 5y = 1 \\ 3x + y = -7 \end{cases}$

16. $\begin{cases} 3y - x = 6 \\ 4x + 12y = 0 \end{cases}$

17. $\begin{cases} 4x + 2y = 5 \\ -2x = y + 4 \end{cases}$

18. $\begin{cases} 2y = x + 2 \\ 6x - 12y = 0 \end{cases}$

19. $\begin{cases} 4x + y = 11 \\ 2x + 5y = 1 \end{cases}$

20. $\begin{cases} 3x + y = -14 \\ 4x + 3y = -22 \end{cases}$

21. $\begin{cases} x + 2y + 5 = -4 + 5y - x \\ 2x + x = y + 4 \end{cases}$
(*Hint:* First simplify each equation.)

22. $\begin{cases} 5x + 4y - 2 = -6 + 7y - 3x \\ 3x + 4x = y + 3 \end{cases}$
(*Hint:* See Exercise **21**.)

23. $\begin{cases} 6x - 3y = 5 \\ x + 2y = 0 \end{cases}$

24. $\begin{cases} 10x - 5y = -21 \\ x + 3y = 0 \end{cases}$

25. $\begin{cases} 3x - y = 1 \\ 2x - 3y = 10 \end{cases}$

26. $\begin{cases} 2x - y = -7 \\ 4x - 3y = -11 \end{cases}$

27. $\begin{cases} -x + 2y = 10 \\ -2x + 3y = 18 \end{cases}$

28. $\begin{cases} -x + 3y = 18 \\ -3x + 2y = 19 \end{cases}$

29. $\begin{cases} 5x + 10y = 20 \\ 2x + 6y = 10 \end{cases}$

30. $\begin{cases} 6x + 3y = 12 \\ 9x + 6y = 15 \end{cases}$

31. $\begin{cases} 3x + 6y = 9 \\ 4x + 8y = 16 \end{cases}$

32. $\begin{cases} 2x + 4y = 6 \\ 5x + 10y = 16 \end{cases}$

33. $\begin{cases} \frac{1}{3}x - y = 2 \\ x - 3y = 6 \end{cases}$

34. $\begin{cases} \frac{1}{4}x - 2y = 1 \\ x - 8y = 4 \end{cases}$

35. $\begin{cases} x = \frac{3}{4}y - 1 \\ 8x - 5y = -6 \end{cases}$

36. $\begin{cases} x = \frac{5}{6}y - 2 \\ 12x - 5y = -9 \end{cases}$

Review

Write equivalent equations by multiplying both sides of each given equation by the given nonzero number. See Section 2.2.

37. $3x + 2y = 6$ by -2

38. $-x + y = 10$ by 5

39. $-4x + y = 3$ by 3

40. $5a - 7b = -4$ by -4

Simplify each expression by combining any like terms. See Section 3.4.

41. $3n + 6m + 2n - 6m$

42. $-2x + 5y + 2x + 11y$

43. $-5a - 7b + 5a - 8b$

44. $9q + p - 9q - p$

Concept Extensions

Solve each system by the substitution method. First simplify each equation by combining like terms.

45. $\begin{cases} -5y + 6y = 3x + 2(x - 5) - 3x + 5 \\ 4(x + y) - x + y = -12 \end{cases}$

46. $\begin{cases} 5x + 2y - 4x - 2y = 2(2y + 6) - 7 \\ 3(2x - y) - 4x = 1 + 9 \end{cases}$

47. Explain how to identify a system with no solution when using the substitution method.

48. Occasionally, when using the substitution method, we obtain the equation $0 = 0$. Explain how this result indicates that the graphs of the equations in the system are identical.

Solve. See a Concept Check in this section.

49. As you solve the system $\begin{cases} 3x - y = -6 \\ -3x + 2y = 7 \end{cases}$, you find that $y = 1$. Is this the solution of the system?

50. As you solve the system $\begin{cases} x = 5y \\ y = 2x \end{cases}$, you find that $x = 0$ and $y = 0$. What is the solution of this system?

51. To avoid fractions, which of the equations below would you use if solving for y? Explain why.

a. $\frac{1}{2}x - 4y = \frac{3}{4}$ **b.** $8x - 5y = 13$ **c.** $7x - y = 19$

52. Give the number of solutions for a system if the graphs of the equations in the system are

a. lines intersecting in one point
b. parallel lines
c. same line

Use a graphing calculator to solve each system.

53. $\begin{cases} y = 5.1x + 14.56 \\ y = -2x - 3.9 \end{cases}$

54. $\begin{cases} y = 3.1x - 16.35 \\ y = -9.7x + 28.45 \end{cases}$

55. $\begin{cases} 3x + 2y = 14.04 \\ 5x + y = 18.5 \end{cases}$

56. $\begin{cases} x + y = -15.2 \\ -2x + 5y = -19.3 \end{cases}$

57. U.S. consumer spending y (in billions of dollars) on DVD- or Blu-ray-format home entertainment from 2014 to 2016 is given by $y = -0.6x + 6.8$, where x is the number of years after 2014. U.S. consumer spending y (in billions of dollars) on streaming services home entertainment from 2014 to 2016 is given by $y = 0.9x + 4.6$, where x is the number of years after 2014. (*Source:* Based on data from Variety)

a. Use the substitution method to solve this system of equations.

$$\begin{cases} y = -0.6x + 6.8 \\ y = 0.9x + 4.6 \end{cases}$$

Round x to the nearest tenth and y to the nearest whole.

b. Explain the meaning of your answer to part **a**.

c. Sketch a graph of the system of equations. Write a sentence describing the trends in the popularity of these types of home entertainment formats.

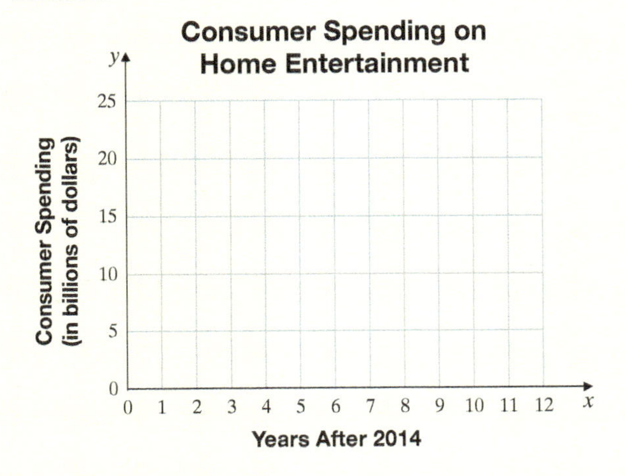

58. Consumption of diet soda in the United States continues to decline as the consumption of bottled water increases. For that reason, more beverage companies are adding bottled water to their stable of products. From 2012 through 2016, the amount y (in gallons per person) of diet soda consumed in the United States is given by the equation $y = -1.4x + 36$, where x is the number of years after 2012. For the same period, the consumption of bottled water (in gallons per person) in the United States can be represented by the equation $y = 2.1x + 30$, where x is the number of years after 2012. (*Source: Beverage Digest* and *Forbes*)

a. Use the substitution method to solve this system of equations.

$$\begin{cases} y = -1.4x + 36 \\ y = 2.1x + 30 \end{cases}$$

Round x and y to the nearest tenth.

b. Explain the meaning of your answer to part **a**.

c. Sketch a graph of the system of equations. Write a sentence describing the trends in diet soda and bottled water consumption in the United States.

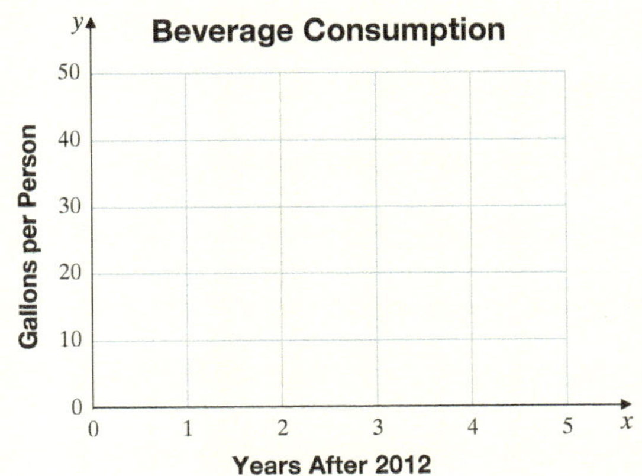

Objective

A Use the Addition Method to Solve a System of Linear Equations.

Practice 1

Use the addition method to solve the system:

$$\begin{cases} x + y = 13 \\ x - y = 5 \end{cases}$$

7.3 Solving Systems of Linear Equations by Addition

Objective A Using the Addition Method

We have seen that substitution is an accurate method for solving a system of linear equations. Another accurate method is the **addition** or **elimination method.** The addition method is based on the addition property of equality: Adding equal quantities to both sides of an equation does not change the solution of the equation. In symbols,

if $A = B$ and $C = D$, then $A + C = B + D$

To see how we use this to solve a system of equations, study Example 1.

Example 1 Solve the system: $\begin{cases} x + y = 7 \\ x - y = 5 \end{cases}$

Solution: Since the left side of each equation is equal to its right side, we are adding equal quantities when we add the left sides of the equations together and add the right sides of the equations together. This adding eliminates the variable y and gives us an equation in one variable, x. We can then solve for x.

$x + y = 7$ First equation

$x - y = 5$ Second equation

$2x = 12$ Add the equations to eliminate y.

$x = 6$ Divide both sides by 2.

Helpful Hint Notice that our goal when solving by the addition method is to eliminate a variable when adding the equations.

The x-value of the solution is 6. To find the corresponding y-value, we let $x = 6$ in either equation of the system. We will use the first equation.

$x + y = 7$ First equation

$6 + y = 7$ Let $x = 6$.

$y = 1$ Solve for y.

The solution is $(6, 1)$.

Check: Check the solution in both equations of the original system.

First Equation		**Second Equation**	
$x + y = 7$		$x - y = 5$	
$6 + 1 \stackrel{?}{=} 7$	Let $x = 6$ and $y = 1$.	$6 - 1 \stackrel{?}{=} 5$	Let $x = 6$ and $y = 1$
$7 = 7$	True	$5 = 5$	True

Thus, the solution of the system is $(6, 1)$.

If we graph the two equations in the system, we have two lines that intersect at the point $(6, 1)$, as shown.

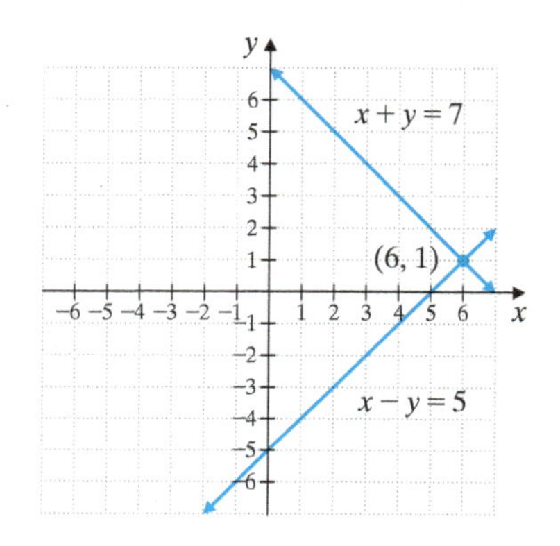

Work Practice 1

Answer
1. $(9, 4)$

Example 2 Solve the system: $\begin{cases} -2x + y = 2 \\ -x + 3y = -4 \end{cases}$

Solution: If we simply add these two equations, the result is still an equation in two variables. However, from Example 1, remember that our goal is to eliminate one of the variables so that we have an equation in the other variable. To do this, notice what happens if we multiply *both sides* of the first equation by -3. We are allowed to do this by the multiplication property of equality. Then the system

$$\begin{cases} -3(-2x + y) = -3(2) \\ -x + 3y = -4 \end{cases} \quad \text{simplifies to} \quad \begin{cases} 6x - 3y = -6 \\ -x + 3y = -4 \end{cases}$$

When we add the resulting equations, the y-variable is eliminated.

$$\begin{aligned} 6x - 3y &= -6 \\ -x + 3y &= -4 \\ \hline 5x \qquad &= -10 && \text{Add.} \\ x &= -2 && \text{Divide both sides by 5.} \end{aligned}$$

To find the corresponding y-value, we let $x = -2$ in either of the original equations. We use the first equation of the original system.

$$\begin{aligned} -2x + y &= 2 && \text{First equation} \\ -2(-2) + y &= 2 && \text{Let } x = -2. \\ 4 + y &= 2 \\ y &= -2 \end{aligned}$$

Check the ordered pair $(-2, -2)$ in both equations of the *original* system. The solution is $(-2, -2)$.

Work Practice 2

Practice 2

Solve the system:

$\begin{cases} 2x - y = -6 \\ -x + 4y = 17 \end{cases}$

Helpful Hint

When finding the second value of an ordered pair solution, any equation equivalent to one of the original equations in the system may be used.

In Example 2, the decision to multiply the first equation by -3 was no accident. **To eliminate a variable** when adding two equations, **the coefficient of the variable in one equation must be the opposite of its coefficient in the other equation.**

Helpful Hint

- Be sure to multiply *both sides* of an equation by the chosen number when solving by the addition method.
- A common mistake is to multiply only the side containing the variables.

Example 3 Solve the system: $\begin{cases} 2x - y = 7 \\ 8x - 4y = 1 \end{cases}$

Solution: When we multiply both sides of the first equation by -4, the resulting coefficient of x is -8. This is the opposite of 8, the coefficient of x in the second equation. Then the system

$$\begin{cases} -4(2x - y) = -4(7) \\ 8x - 4y = 1 \end{cases} \quad \text{simplifies to}$$

Helpful Hint Don't forget to multiply both sides by -4.

$$\begin{cases} -8x + 4y = -28 \\ 8x - 4y = 1 \end{cases}$$
$$0 = -27 \quad \text{Add the equations.}$$

(*Continued on next page*)

Practice 3

Solve the system:

$\begin{cases} x - 3y = -2 \\ -3x + 9y = 5 \end{cases}$

Answers

2. $(-1, 4)$ **3.** no solution

When we add the equations, both variables are eliminated and we have $0 = -27$, a false statement. This means that the system has no solution. The equations, if graphed, would represent parallel lines.

Work Practice 3

Practice 4

Solve the system:

$$\begin{cases} 2x + 5y = 1 \\ -4x - 10y = -2 \end{cases}$$

Example 4 Solve the system: $\begin{cases} 3x - 2y = 2 \\ -9x + 6y = -6 \end{cases}$

Solution: First we multiply both sides of the first equation by 3 and then we add the resulting equations.

$$\begin{cases} 3(3x - 2y) = 3(2) \\ -9x + 6y = -6 \end{cases} \quad \text{simplifies to} \quad \begin{cases} 9x - 6y = 6 \\ -9x + 6y = -6 \end{cases} \quad \text{Add the equations.}$$

$$0 = 0$$

Both variables are eliminated and we have $0 = 0$, a true statement. This means that the system has an infinite number of solutions. The equations, if graphed, would be the same line.

Work Practice 4

✓Concept Check Suppose you are solving the system

$$\begin{cases} 3x + 8y = -5 \\ 2x - 4y = 3 \end{cases}$$

You decide to use the addition method by multiplying both sides of the second equation by 2. In which of the following was the multiplication performed correctly? Explain.

a. $4x - 8y = 3$ **b.** $4x - 8y = 6$

In the next example, we multiply both equations by numbers so that coefficients of a variable are opposites.

Practice 5

Solve the system:

$$\begin{cases} 4x + 5y = 14 \\ 3x - 2y = -1 \end{cases}$$

Example 5 Solve the system: $\begin{cases} 3x + 4y = 13 \\ 5x - 9y = 6 \end{cases}$

Solution: We can eliminate the variable y by multiplying the first equation by 9 and the second equation by 4. Then we add the resulting equations.

$$\begin{cases} 9(3x + 4y) = 9(13) \\ 4(5x - 9y) = 4(6) \end{cases} \quad \text{simplifies to} \quad \begin{cases} 27x + 36y = 117 \\ 20x - 36y = 24 \end{cases}$$

$$47x = 141 \quad \text{Add the equations.}$$

$$x = 3 \quad \text{Solve for } x.$$

To find the corresponding y-value, we let $x = 3$ in one of the original equations of the system. Doing so in either of these equations will give $y = 1$. Check to see that $(3, 1)$ satisfies each equation in the original system. The solution is $(3, 1)$.

Work Practice 5

If we had decided to eliminate x instead of y in Example 5, the first equation could have been multiplied by 5 and the second by -3. Try solving the original system this way to check that the solution is $(3, 1)$.

The following steps summarize how to solve a system of linear equations by the addition method.

Answers

4. infinite number of solutions

5. $(1, 2)$

✓Concept Check Answer

b; answers may vary

To Solve a System of Two Linear Equations by the Addition Method

Step 1: Rewrite each equation in standard form, $Ax + By = C$.

Step 2: If necessary, multiply one or both equations by a nonzero number so that the coefficients of a chosen variable in the system are opposites.

Step 3: Add the equations.

Step 4: Find the value of one variable by solving the resulting equation from Step 3.

Step 5: Find the value of the second variable by substituting the value found in Step 4 into either of the original equations.

Step 6: Check the proposed solution in the original system.

✓**Concept Check** Suppose you are solving the system

$$\begin{cases} -4x + 7y = 6 \\ x + 2y = 5 \end{cases}$$

by the addition method.

a. What step(s) should you take if you wish to eliminate x when adding the equations?

b. What step(s) should you take if you wish to eliminate y when adding the equations?

Example 6 Solve the system: $\begin{cases} -x - \frac{y}{2} = \frac{5}{2} \\ \frac{x}{6} - \frac{y}{2} = 0 \end{cases}$

Solution: We begin by clearing each equation of fractions. To do so, we multiply both sides of the first equation by the LCD, 2, and both sides of the second equation by the LCD, 6. Then the system

$$\begin{cases} 2\left(-x - \frac{y}{2}\right) = 2\left(\frac{5}{2}\right) \\ 6\left(\frac{x}{6} - \frac{y}{2}\right) = 6(0) \end{cases} \quad \text{simplifies to} \quad \begin{cases} -2x - y = 5 \\ x - 3y = 0 \end{cases}$$

We can now eliminate the variable x by multiplying the second equation by 2.

$$\begin{cases} -2x - y = 5 \\ 2(x - 3y) = 2(0) \end{cases} \quad \text{simplifies to} \quad \begin{cases} -2x - y = 5 \\ 2x - 6y = 0 \end{cases}$$

$$-7y = 5 \quad \text{Add the equations.}$$

$$y = -\frac{5}{7} \quad \text{Solve for } y.$$

To find x, we could replace y with $-\frac{5}{7}$ in one of the equations with two variables. Instead, let's go back to the simplified system and multiply by appropriate factors to eliminate the variable y and solve for x. To do this, we multiply the first equation by -3. Then the system

$$\begin{cases} -3(-2x - y) = -3(5) \\ x - 3y = 0 \end{cases} \quad \text{simplifies to} \quad \begin{cases} 6x + 3y = -15 \\ x - 3y = 0 \end{cases}$$

$$7x = -15 \quad \text{Add the equations.}$$

$$x = -\frac{15}{7} \quad \text{Solve for } x.$$

Check the ordered pair $\left(-\frac{15}{7}, -\frac{5}{7}\right)$ in both equations of the original system. The solution is $\left(-\frac{15}{7}, -\frac{5}{7}\right)$.

Work Practice 6

Practice 6

Solve the system:

$$\begin{cases} -\frac{x}{3} + y = \frac{4}{3} \\ \frac{x}{2} - \frac{5}{2}y = -\frac{1}{2} \end{cases}$$

Answer

6. $\left(-\frac{17}{2}, -\frac{3}{2}\right)$

✓**Concept Check Answer**

a. multiply the second equation by 4

b. possible answer: multiply the first equation by -2 and the second equation by 7

Vocabulary, Readiness & Video Check

Given the system $\begin{cases} 3x - 2y = -9 \\ x + 5y = 14 \end{cases}$ *read each row (Step 1, Step 2, and Result). Then answer whether the result is true or false.*

	Step 1	Step 2	Result	True or False?
1.	Multiply 2nd equation through by -3.	Add the resulting equation to the 1st equation.	The y's are eliminated.	
2.	Multiply 2nd equation through by -3.	Add the resulting equation to the 1st equation.	The x's are eliminated.	
3.	Multiply 1st equation by 5 and 2nd equation by 2.	Add the two new equations.	The y's are eliminated.	
4.	Multiply 1st equation by 5 and 2nd equation by -2.	Add the two new equations.	The y's are eliminated.	

Martin-Gay Interactive Videos

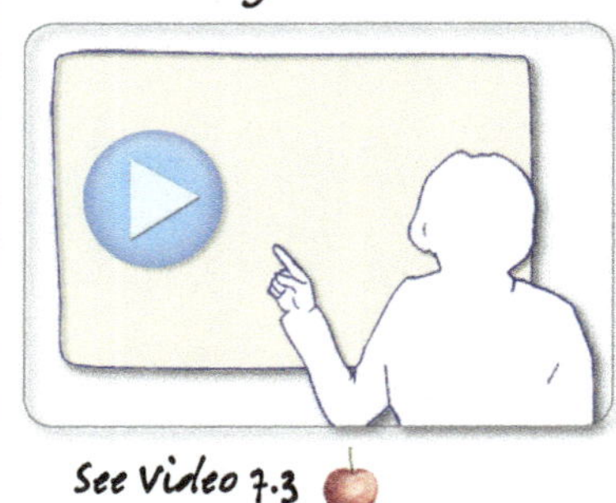

See Video 7.3

Watch the section lecture video and answer the following question.

Objective A **5.** For the addition/elimination method, sometimes we need to multiply an equation through by a nonzero number so that the coefficients of a variable are opposites, as is shown in Example 2. What property allows us to do this? What important reminder is made at this step?

7.3 Exercise Set MyLab Math

Objective A *Solve each system of equations by the addition method. See Example 1.*

1. $\begin{cases} 3x + y = 5 \\ 6x - y = 4 \end{cases}$

2. $\begin{cases} 4x + y = 13 \\ 2x - y = 5 \end{cases}$

3. $\begin{cases} x - 2y = 8 \\ -x + 5y = -17 \end{cases}$

4. $\begin{cases} x - 2y = -11 \\ -x + 5y = 23 \end{cases}$

Solve each system of equations by the addition method. If a system contains fractions or decimals, you may want to first clear each equation of the fractions or decimals. See Examples 1 through 6.

5. $\begin{cases} 3x + y = -11 \\ 6x - 2y = -2 \end{cases}$

6. $\begin{cases} 4x + y = -13 \\ 6x - 3y = -15 \end{cases}$

7. $\begin{cases} 3x + 2y = 11 \\ 5x - 2y = 29 \end{cases}$

8. $\begin{cases} 4x + 2y = 2 \\ 3x - 2y = 12 \end{cases}$

9. $\begin{cases} x + 5y = 18 \\ 3x + 2y = -11 \end{cases}$

10. $\begin{cases} x + 4y = 14 \\ 5x + 3y = 2 \end{cases}$

11. $\begin{cases} x + y = 6 \\ x - y = 6 \end{cases}$

12. $\begin{cases} x - y = 1 \\ -x + 2y = 0 \end{cases}$

13. $\begin{cases} 2x + 3y = 0 \\ 4x + 6y = 3 \end{cases}$

14. $\begin{cases} 3x + y = 4 \\ 9x + 3y = 6 \end{cases}$

15. $\begin{cases} -x + 5y = -1 \\ 3x - 15y = 3 \end{cases}$

16. $\begin{cases} 2x + y = 6 \\ 4x + 2y = 12 \end{cases}$

17. $\begin{cases} 3x - 2y = 7 \\ 5x + 4y = 8 \end{cases}$

18. $\begin{cases} 6x - 5y = 25 \\ 4x + 15y = 13 \end{cases}$

19. $\begin{cases} 8x = -11y - 16 \\ 2x + 3y = -4 \end{cases}$

20. $\begin{cases} 10x + 3y = -12 \\ 5x = -4y - 16 \end{cases}$

21. $\begin{cases} 4x - 3y = 7 \\ 7x + 5y = 2 \end{cases}$

22. $\begin{cases} -2x + 3y = 10 \\ 3x + 4y = 2 \end{cases}$

23. $\begin{cases} 4x - 6y = 8 \\ 6x - 9y = 12 \end{cases}$

24. $\begin{cases} 9x - 3y = 12 \\ 12x - 4y = 18 \end{cases}$

25. $\begin{cases} 2x - 5y = 4 \\ 3x - 2y = 4 \end{cases}$

26. $\begin{cases} 6x - 5y = 7 \\ 4x - 6y = 7 \end{cases}$

27. $\begin{cases} \dfrac{x}{3} + \dfrac{y}{6} = 1 \\ \dfrac{x}{2} - \dfrac{y}{4} = 0 \end{cases}$

28. $\begin{cases} \dfrac{x}{2} + \dfrac{y}{8} = 3 \\ x - \dfrac{y}{4} = 0 \end{cases}$

29. $\begin{cases} \dfrac{10}{3}x + 4y = -4 \\ 5x + 6y = -6 \end{cases}$

30. $\begin{cases} \dfrac{3}{2}x + 4y = 1 \\ 9x + 24y = 5 \end{cases}$

31. $\begin{cases} x - \dfrac{y}{3} = -1 \\ -\dfrac{x}{2} + \dfrac{y}{8} = \dfrac{1}{4} \end{cases}$

32. $\begin{cases} 2x - \dfrac{3y}{4} = -3 \\ x + \dfrac{y}{9} = \dfrac{13}{3} \end{cases}$

33. $\begin{cases} -4(x + 2) = 3y \\ 2x - 2y = 3 \end{cases}$

34. $\begin{cases} -9(x + 3) = 8y \\ 3x - 3y = 8 \end{cases}$

35. $\begin{cases} \dfrac{x}{3} - y = 2 \\ -\dfrac{x}{2} + \dfrac{3y}{2} = -3 \end{cases}$

36. $\begin{cases} \dfrac{x}{2} + \dfrac{y}{4} = 1 \\ -\dfrac{x}{4} - \dfrac{y}{8} = 1 \end{cases}$

37. $\begin{cases} \dfrac{3}{5}x - y = -\dfrac{4}{5} \\ 3x + \dfrac{y}{2} = -\dfrac{9}{5} \end{cases}$

38. $\begin{cases} 3x + \dfrac{7}{2}y = \dfrac{3}{4} \\ -\dfrac{x}{2} + \dfrac{5}{3}y = -\dfrac{5}{4} \end{cases}$

39. $\begin{cases} 3.5x + 2.5y = 17 \\ -1.5x - 7.5y = -33 \end{cases}$

40. $\begin{cases} -2.5x - 6.5y = 47 \\ 0.5x - 4.5y = 37 \end{cases}$

41. $\begin{cases} 0.02x + 0.04y = 0.09 \\ -0.1x + 0.3y = 0.8 \end{cases}$

42. $\begin{cases} 0.04x - 0.05y = 0.105 \\ 0.2x - 0.6y = 1.05 \end{cases}$

Review

Translating *Rewrite each sentence using mathematical symbols. Do not solve the equations. See Section 2.4.*

43. Twice a number, added to 6, is 3 less than the number.

44. The sum of three consecutive integers is 66.

45. Three times a number, subtracted from 20, is 2.

46. Twice the sum of 8 and a number is the difference of the number and 20.

47. The product of 4 and the sum of a number and 6 is twice the number.

48. If the quotient of twice a number and 7 is subtracted from the reciprocal of the number, the result is 2.

Concept Extensions

Solve. See a Concept Check in this section.

49. To solve this system by the addition method and eliminate the variable y,

$$\begin{cases} 4x + 2y = -7 \\ 3x - y = -12 \end{cases}$$

by what value would you multiply the second equation? What do you get when you complete the multiplication?

Given the system of linear equations $\begin{cases} 3x - y = -8 \\ 5x + 3y = 2 \end{cases}$:

50. Use the addition method and

a. Solve the system by eliminating x.

b. Solve the system by eliminating y.

51. Suppose you are solving the system

$$\begin{cases} 3x + 8y = -5 \\ 2x - 4y = 3 \end{cases}$$

You decide to use the addition method by multiplying both sides of the second equation by 2. In which of the following was the multiplication performed correctly? Explain.

a. $4x - 8y = 3$

b. $4x - 8y = 6$

52. Suppose you are solving the system

$$\begin{cases} -2x - y = 0 \\ -2x + 3y = 6 \end{cases}$$

You decide to use the addition method by multiplying both sides of the first equation by 3, then adding the resulting equation to the second equation. Which of the following is the correct sum? Explain.

a. $-8x = 6$

b. $-8x = 9$

53. When solving a system of equations by the addition method, how do we know when the system has no solution?

54. Explain why the addition method might be preferred over the substitution method for solving the system $\begin{cases} 2x - 3y = 5 \\ 5x + 2y = 6. \end{cases}$

55. Use the system of linear equations below to answer the questions.

$$\begin{cases} x + y = 5 \\ 3x + 3y = b \end{cases}$$

a. Find the value of b so that the system has an infinite number of solutions.

b. Find a value of b so that there are no solutions to the system.

56. Use the system of linear equations below to answer the questions.

$$\begin{cases} x + y = 4 \\ 2x + by = 8 \end{cases}$$

a. Find the value of b so that the system has an infinite number of solutions.

b. Find a value of b so that the system has a single solution.

Solve each system by the addition method.

57. $\begin{cases} 2x + 3y = 14 \\ 3x - 4y = -69.1 \end{cases}$

58. $\begin{cases} 5x - 2y = -19.8 \\ -3x + 5y = -3.7 \end{cases}$

59. In recent years, budgets for digital advertising on the Internet have been increasing at a faster rate than budgets for advertising on television. The amount of money y (in billions) budgeted for digital advertising from 2014 through 2017 can be approximated by $-43x + 5y = 251$, where x is the number of years after 2014. The amount of money y (in billions) budgeted for television advertising can be approximated by $21x - 10y = -685$, where x is the number of years after 2014.

U.S. Ad Spending by Media Type—2021 Projected

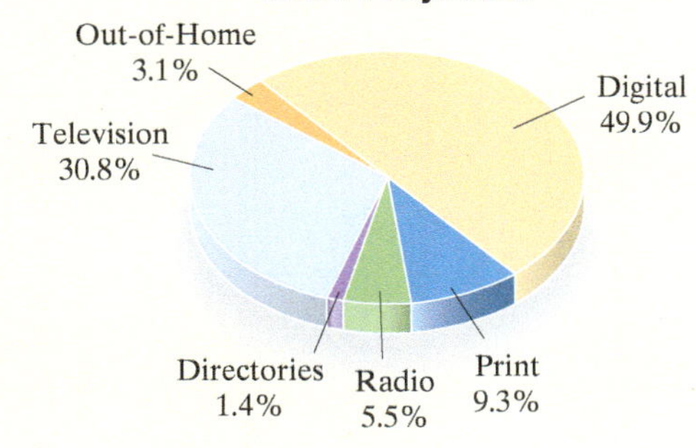

a. Use the addition method to solve this system of equations.

$$\begin{cases} -43x + 5y = 251 \\ 21x - 10y = -685 \end{cases}$$

(Eliminate y first and solve for x. Round x to the nearest whole number. Because of rounding, the y-value of your ordered pair solution may vary.)

b. Interpret your solution from part **a**.

c. Use the year in your answer to part **b** to find how much money was spent on each type of advertising that year.

60. As the economy and job marketplace change, demand for certain types of workers changes. The number of jobs for audiologists that is predicted for 2014 through 2024 can be approximated by $-38x + 10y = 112$. The number of jobs for exercise physiologists that is predicted for the same period can be approximated by $15x - 10y = -155$. For both equations, x is the number of years after 2014, and y is the number of jobs in thousands. (*Source:* Based on data from the U.S. Bureau of Labor Statistics)

a. Use the addition method to solve this system of equations.

$$\begin{cases} -38x + 10y = 112 \\ 15x - 10y = -155 \end{cases}$$

(Eliminate y first and solve for x. Round x to the nearest whole number. Because of rounding, the y-value of your ordered pair solution may vary.)

b. Interpret your solution from part **a**.

c. Using the year in your answer to part **b**, estimate the number of audiologist jobs and exercise physiologist jobs in that year.

Answers

1. ______
2. ______
3. ______
4. ______
5. ______
6. ______
7. ______
8. ______
9. ______
10. ______
11. ______
12. ______
13. ______
14. ______
15. ______
16. ______
17. ______
18. ______
19. ______
20. ______
21. ______
22. ______

Summary on Solving Systems of Equations

Solve each system by either the addition method or the substitution method.

1. $\begin{cases} 2x - 3y = -11 \\ y = 4x - 3 \end{cases}$

2. $\begin{cases} 4x - 5y = 6 \\ y = 3x - 10 \end{cases}$

3. $\begin{cases} x + y = 3 \\ x - y = 7 \end{cases}$

4. $\begin{cases} x - y = 20 \\ x + y = -8 \end{cases}$

5. $\begin{cases} x + 2y = 1 \\ 3x + 4y = -1 \end{cases}$

6. $\begin{cases} x + 3y = 5 \\ 5x + 6y = -2 \end{cases}$

7. $\begin{cases} y = x + 3 \\ 3x = 2y - 6 \end{cases}$

8. $\begin{cases} y = -2x \\ 2x - 3y = -16 \end{cases}$

9. $\begin{cases} y = 2x - 3 \\ y = 5x - 18 \end{cases}$

10. $\begin{cases} y = 6x - 5 \\ y = 4x - 11 \end{cases}$

11. $\begin{cases} x + \frac{1}{6}y = \frac{1}{2} \\ 3x + 2y = 3 \end{cases}$

12. $\begin{cases} x + \frac{1}{3}y = \frac{5}{12} \\ 8x + 3y = 4 \end{cases}$

13. $\begin{cases} x - 5y = 1 \\ -2x + 10y = 3 \end{cases}$

14. $\begin{cases} -x + 2y = 3 \\ 3x - 6y = -9 \end{cases}$

15. $\begin{cases} 0.2x - 0.3y = -0.95 \\ 0.4x + 0.1y = 0.55 \end{cases}$

16. $\begin{cases} 0.08x - 0.04y = -0.11 \\ 0.02x - 0.06y = -0.09 \end{cases}$

17. $\begin{cases} x = 3y - 7 \\ 2x - 6y = -14 \end{cases}$

18. $\begin{cases} y = \frac{x}{2} - 3 \\ 2x - 4y = 0 \end{cases}$

19. $\begin{cases} 2x + 5y = -1 \\ 3x - 4y = 33 \end{cases}$

20. $\begin{cases} 7x - 3y = 2 \\ 6x + 5y = -21 \end{cases}$

21. Which method, substitution or addition, would you prefer to use to solve the system below? Explain your reasoning.

$$\begin{cases} 3x + 2y = -2 \\ y = -2x \end{cases}$$

22. Which method, substitution or addition, would you prefer to use to solve the system below? Explain your reasoning.

$$\begin{cases} 3x - 2y = -3 \\ 6x + 2y = 12 \end{cases}$$

7.4 Systems of Linear Equations and Problem Solving

Objective A Using a System of Equations for Problem Solving

Objective

A Use a System of Equations to Solve Problems.

Many of the word problems solved earlier with one-variable equations can also be solved with two equations in two variables. We use the same problem-solving steps that we have used throughout this text. The only difference is that two variables are assigned to represent the two unknown quantities and that the problem is translated into two equations.

Problem-Solving Steps

1. UNDERSTAND the problem. During this step, become comfortable with the problem. Some ways of doing this are to

   Read and reread the problem.
   Choose two variables to represent the two unknowns.
   Construct a drawing.
   Propose a solution and check. Pay careful attention to how you check your proposed solution. This will help when writing equations to model the problem.

2. TRANSLATE the problem into two equations.
3. SOLVE the system of equations.
4. INTERPRET the results: *Check* the proposed solution in the stated problem and *state* your conclusion.

Example 1 Finding Unknown Numbers

Find two numbers whose sum is 37 and whose difference is 21.

Solution:

1. UNDERSTAND. Read and reread the problem. Suppose that one number is 20. If their sum is 37, the other number is 17 because $20 + 17 = 37$. Is their difference 21? No; $20 - 17 = 3$. Our proposed solution is incorrect, but we now have a better understanding of the problem.

   Since we are looking for two numbers, we let

   $x =$ first number and

   $y =$ second number

2. TRANSLATE. Since we have assigned two variables to this problem, we translate our problem into two equations.

In words:	two number whose sum	is	37
	↓	↓	↓
Translate:	$x + y$	$=$	37
In words:	two number whose difference	is	21
	↓	↓	↓
Translate:	$x - y$	$=$	21

(*Continued on next page*)

Practice 1

Find two numbers whose sum is 50 and whose difference is 22.

Answer

1. 36 and 14

3. SOLVE. Now we solve the system.

$$\begin{cases} x + y = 37 \\ x - y = 21 \end{cases}$$

Notice that the coefficients of the variable y are opposites. Let's then solve by the addition method and begin by adding the equations.

$$\begin{aligned} x + y &= 37 \\ x - y &= 21 \quad \text{Add the equations.} \\ \hline 2x \quad &= 58 \\ x &= 29 \quad \text{Divide both sides by 2.} \end{aligned}$$

Now we let $x = 29$ in the first equation to find y.

$$\begin{aligned} x + y &= 37 \quad \text{First equation} \\ 29 + y &= 37 \\ y &= 8 \quad \text{Subtract 29 from both sides.} \end{aligned}$$

4. INTERPRET. The solution of the system is $(29, 8)$.

Check: Notice that the sum of 29 and 8 is $29 + 8 = 37$, the required sum. Their difference is $29 - 8 = 21$, the required difference.

State: The numbers are 29 and 8.

Work Practice 1

Practice 2

Admission prices at a local weekend fair were $5 for children and $7 for adults. The total money collected was $3379, and 587 people attended the fair. How many children and how many adults attended the fair?

Example 2 Solving a Problem About Prices

The Cirque du Soleil show Ovo is performing locally. Matinee admission for 4 adults and 2 children is $374, while admission for 2 adults and 3 children is $285.

a. What is the price of an adult's ticket?

b. What is the price of a child's ticket?

c. Suppose that a special rate of $1000 is offered for groups of 20 persons. Should a group of 4 adults and 16 children use the group rate? Why or why not?

Solution:

1. UNDERSTAND. Read and reread the problem and guess a solution. Let's suppose that the price of an adult's ticket is $50 and the price of a child's ticket is $40. To check our proposed solution, let's see if admission for 4 adults and 2 children is $374. Admission for 4 adults is 4($50) or $200 and admission for 2 children is 2($40) or $80. This gives a total admission of $200 + $80 = $280, not the required $374. Again, though, we have accomplished the purpose of this process: We have a better understanding of the problem. To continue, we let

A = the price of an adult's ticket and

C = the price of a child's ticket

Answer

2. 365 children and 222 adults

2. TRANSLATE. We translate the problem into two equations using both variables.

In words:	admission for 4 adults	and	admission for 2 children	is	\$374
	↓	↓	↓	↓	↓
Translate:	$4A$	$+$	$2C$	$=$	374
In words:	admission for 2 adults	and	admission for 3 children	is	\$285
	↓	↓	↓	↓	↓
Translate:	$2A$	$+$	$3C$	$=$	285

3. SOLVE. We solve the system.

$$\begin{cases} 4A + 2C = 374 \\ 2A + 3C = 285 \end{cases}$$

Since both equations are written in standard form, we solve by the addition method. First we multiply the second equation by -2 so that when we add the equations, we eliminate the variable A. Then the system

$$\begin{cases} 4A + 2C = 374 \\ -2(2A + 3C) = -2(285) \end{cases} \text{ simplifies to } \begin{cases} 4A + 2C = 374 \\ -4A - 6C = -570 \end{cases}$$

Add the equations.

$$-4C = -196$$

$C = 49$ or \$49, the child's ticket price

To find A, we replace C with 49 in the first equation.

$$4A + 2C = 374 \quad \text{First equation}$$
$$4A + 2(49) = 374 \quad \text{Let } C = 49.$$
$$4A + 98 = 374$$
$$4A = 276$$

$A = 69$ or \$69, the adult's ticket price

4. INTERPRET.

Check: Notice that 4 adults and 2 children will pay $4(\$69) + 2(\$49) = \$276 + \$98 = \$374$, the required amount. Also, the price for 2 adults and 3 children is $2(\$69) + 3(\$49) = \$138 + \$147 = \$285$, the required amount.

State: Answer the three original questions.

a. Since $A = 69$, the price of an adult's ticket is \$69.

b. Since $C = 49$, the price of a child's ticket is \$49.

c. The regular admission price for 4 adults and 16 children is

$$4(\$69) + 16(\$49) = \$276 + \$784$$
$$= \$1060$$

This is \$60 more than the special group rate of \$1000, so they should request the group rate.

Work Practice 2

Practice 3

Two cars are 440 miles apart and traveling toward each other. They meet in 3 hours. If one car's speed is 10 miles per hour faster than the other car's speed, find the speed of each car.

	r	t	d
Faster Car			
Slower Car			

Example 3 Finding Rates

As part of an exercise program, two students, Louisa and Alfredo, start walking each morning. They live 15 miles away from each other. They decide to meet one day by walking toward one another. After 2 hours they meet. If Louisa walks one mile per hour faster than Alfredo, find both walking speeds.

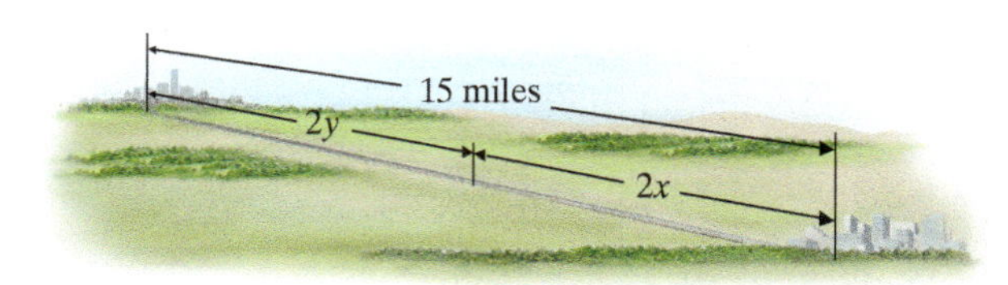

Solution:

1. UNDERSTAND. Read and reread the problem. Let's propose a solution and use the formula $d = r \cdot t$ to check. Suppose that Louisa's rate is 4 miles per hour. Since Louisa's rate is 1 mile per hour faster, Alfredo's rate is 3 miles per hour. To check, see if they can walk a total of 15 miles in 2 hours. Louisa's distance is rate · time $= 4(2) = 8$ miles and Alfredo's distance is rate · time $= 3(2) = 6$ miles. Their total distance is 8 miles + 6 miles = 14 miles, not the required 15 miles. Now that we have a better understanding of the problem, let's model it with a system of equations.
   First, we let

   x = Alfredo's rate in miles per hour and
   y = Louisa's rate in miles per hour

   Now we use the facts stated in the problem and the formula $d = rt$ to fill in the following chart.

	r	t	d
Alfredo	x	2	$2x$
Louisa	y	2	$2y$

2. TRANSLATE. We translate the problem into two equations using both variables.

In words:	Alfredo's distance	+	Louisa's distance	=	15 miles
	↓	↓	↓	↓	↓
Translate:	$2x$	+	$2y$	=	15

In words:	Louisa's rate	is	1 mile per hour faster than Alfredo's
	↓	↓	↓
Translate:	y	=	$x + 1$

3. SOLVE. The system of equations we are solving is

$$\begin{cases} 2x + 2y = 15 \\ y = x + 1 \end{cases}$$

Answer

3. One car's speed is $68\frac{1}{3}$ mph and the other car's speed is $78\frac{1}{3}$ mph.

Let's use substitution to solve the system since the second equation is solved for y.

$$2x + 2y = 15 \quad \text{First equation}$$
$$2x + 2(x + 1) = 15 \quad \text{Replace } y \text{ with } x + 1.$$
$$2x + 2x + 2 = 15$$
$$4x = 13$$
$$x = \frac{13}{4} = 3\frac{1}{4} \text{ or } 3.25$$
$$y = x + 1 = 3\frac{1}{4} + 1 = 4\frac{1}{4} \text{ or } 4.25$$

4. INTERPRET. Alfredo's proposed rate is $3\frac{1}{4}$ miles per hour and Louisa's proposed rate is $4\frac{1}{4}$ miles per hour.

Check: Use the formula $d = rt$ and find that in 2 hours, Alfredo's distance is $(3.25)(2)$ miles or 6.5 miles. In 2 hours, Louisa's distance is $(4.25)(2)$ miles or 8.5 miles. The total distance walked is 6.5 miles + 8.5 miles or 15 miles, the given distance.

State: Alfredo walks at a rate of 3.25 miles per hour and Louisa walks at a rate of 4.25 miles per hour.

Work Practice 3

Example 4 Finding Amounts of Solutions

A chemistry teaching assistant needs 10 liters of a 20% saline solution (salt water) for his 2 p.m. laboratory class. Unfortunately, the only mixtures on hand are a 5% saline solution and a 25% saline solution. How much of each solution should he mix to produce the 20% solution?

Solution:

1. UNDERSTAND. Read and reread the problem. Suppose that we need 4 liters of the 5% solution. Then we need $10 - 4 = 6$ liters of the 25% solution. To see if this gives us 10 liters of a 20% saline solution, let's find the amount of pure salt in each solution.

	concentration rate	×	amount of solution	=	amount of pure salt
	↓		↓		↓
5% solution:	0.05	×	4 liters	=	0.2 liter
25% solution:	0.25	×	6 liters	=	1.5 liters
20% solution:	0.20	×	10 liters	=	2 liters

Since 0.2 liter + 1.5 liters = 1.7 liters, not 2 liters, our proposed solution is incorrect. But we have gained some insight into how to model and check this problem.

We let

x = number of liters of 5% solution and

y = number of liters of 25% solution

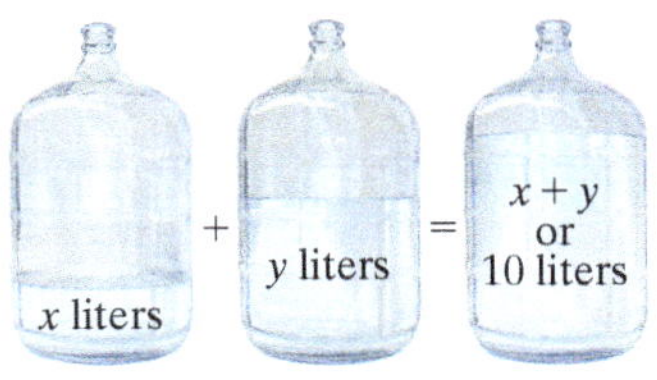

5% saline solution 25% saline solution 20% saline solution

(*Continued on next page*)

Practice 4

A pharmacist needs 50 liters of a 60% alcohol solution. She currently has available a 20% solution and a 70% solution. How many liters of each must she use to make the needed 50 liters of 60% alcohol solution?

Answer

4. 10 liters of the 20% alcohol solution and 40 liters of the 70% alcohol solution

Now we use a table to organize the given data.

	Concentration Rate	Liters of Solution	Liters of Pure Salt
First Solution	5%	x	$0.05x$
Second Solution	25%	y	$0.25y$
Mixture Needed	20%	10	$(0.20)(10)$

2. TRANSLATE. We translate into two equations using both variables.

In words:	liters of 5% solution	+	liters of 25% solution	=	10 liters
	↓		↓		↓
Translate:	x	+	y	=	10
In words:	salt in 5% solution	+	salt in 25% solution	=	salt in mixture
	↓		↓		↓
Translate:	$0.05x$	+	$0.25y$	=	$(0.20)(10)$

3. SOLVE. Here we solve the system

$$\begin{cases} x + y = 10 \\ 0.05x + 0.25y = 2 \end{cases}$$

To solve by the addition method, we first multiply the first equation by -25 and the second equation by 100. Then the system

$$\begin{cases} -25(x + y) = -25(10) \\ 100(0.05x + 0.25y) = 100(2) \end{cases} \quad \text{simplifies to} \quad \begin{cases} -25x - 25y = -250 \\ 5x + 25y = 200 \end{cases}$$

$$-20x = -50 \quad \text{Add.}$$

$$x = 2.5$$

To find y, we let $x = 2.5$ in the first equation of the original system.

$$x + y = 10$$

$$2.5 + y = 10 \quad \text{Let } x = 2.5.$$

$$y = 7.5$$

4. INTERPRET. Thus, we propose that he needs to mix 2.5 liters of 5% saline solution with 7.5 liters of 25% saline solution.

Check: Notice that $2.5 + 7.5 = 10$, the required number of liters. Also, the sum of the liters of salt in the two solutions equals the liters of salt in the required mixture:

$$0.05(2.5) + 0.25(7.5) = 0.20(10)$$

$$0.125 + 1.875 = 2$$

State: He needs 2.5 liters of the 5% saline solution and 7.5 liters of the 25% saline solution.

Work Practice 4

✓Concept Check Suppose you mix an amount of a 30% acid solution with an amount of a 50% acid solution. Which of the following acid strengths would be possible for the resulting acid mixture?

a. 22% **b.** 44% **c.** 63%

✓Concept Check Answer

b

Vocabulary, Readiness & Video Check

Martin-Gay Interactive Videos

See Video 7.4

Watch the section lecture video and answer the following question.

Objective A **1.** In the lecture before Example 1, the problem-solving steps for solving applications involving systems are discussed. How do these steps differ from the general problem-solving strategy steps?

7.4 Exercise Set MyLab Math

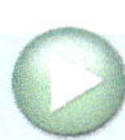

Without actually solving each problem, choose the correct solution by deciding which choice satisfies the given conditions.

1. The length of a rectangle is 3 feet longer than the width. The perimeter is 30 feet. Find the dimensions of the rectangle.

a. length = 8 feet; width = 5 feet
b. length = 8 feet; width = 7 feet
c. length = 9 feet; width = 6 feet

2. An isosceles triangle, a triangle with two sides of equal length, has a perimeter of 20 inches. Each of the equal sides is one inch longer than the third side. Find the lengths of the three sides.

a. 6 inches, 6 inches, and 7 inches
b. 7 inches, 7 inches, and 6 inches
c. 6 inches, 7 inches, and 8 inches

3. Two computer disks and three notebooks cost \$17. However, five computer disks and four notebooks cost \$32. Find the price of each.

a. notebook = \$4; computer disk = \$3
b. notebook = \$3; computer disk = \$4
c. notebook = \$5; computer disk = \$2

4. Two music CDs and four DVDs cost a total of \$40. However, three music CDs and five DVDs cost \$55. Find the price of each.

a. CD = \$12; DVD = \$4
b. CD = \$15; DVD = \$2
c. CD = \$10; DVD = \$5

5. Kesha has a total of 100 coins, all of which are either dimes or quarters. The total value of the coins is \$13.00. Find the number of each type of coin.

a. 80 dimes; 20 quarters
b. 20 dimes; 44 quarters
c. 60 dimes; 40 quarters

6. Samuel has 28 gallons of saline solution available in two large containers at his pharmacy. One container holds three times as much as the other container. Find the capacity of each container.

a. 15 gallons; 5 gallons
b. 20 gallons; 8 gallons
c. 21 gallons; 7 gallons

Objective **A** *Write a system of equations describing each situation. Do not solve the system. See Example 1.*

7. Two numbers add up to 15 and have a difference of 7.

8. The total of two numbers is 16. The first number plus 2 more than 3 times the second equals 18.

9. Keiko has a total of \$6500, which she has invested in two accounts. The larger account is \$800 greater than the smaller account.

10. Dominique has four times as much money in his savings account as in his checking account. The total amount is \$2300.

Solve. See Examples 1 through 4.

11. Two numbers total 83 and have a difference of 17. Find the two numbers.

12. The sum of two numbers is 76 and their difference is 52. Find the two numbers.

13. A first number plus twice a second number is 8. Twice the first number plus the second totals 25. Find the numbers.

14. One number is 4 more than twice a second number. Their total is 25. Find the numbers.

15. Javier Báez of the Chicago Cubs led Major League Baseball in runs batted in for the 2018 regular season. Eugenio Suárez of the Cincinnati Reds, who came in sixth, had 7 fewer runs batted in for the 2018 regular season. Together, these two players brought home 215 runs during the 2018 regular season. How many runs batted in did Báez and Suárez each account for? (*Source:* Major League Baseball)

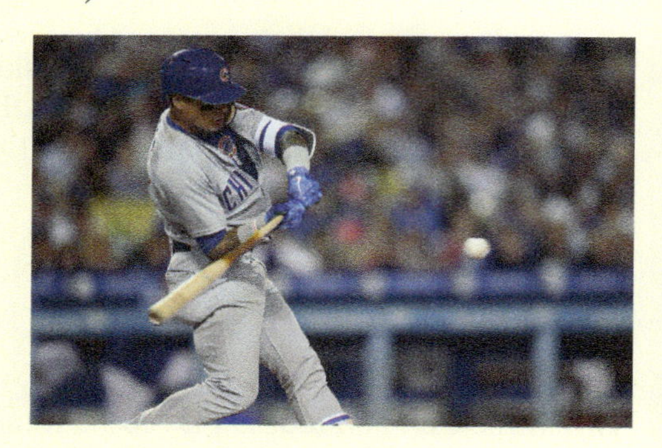

16. The highest scorer during the WNBA 2018 regular season was Breanna Stewart of the Seattle Storm. Over the season, Stewart scored 1 more point than the second-highest scorer, Liz Cambage of the Dallas Wings. Together, Stewart and Cambage scored 1453 points during the regular 2018 season. How many points did each player score over the course of the season? (*Source:* Women's National Basketball Association)

17. Ann Marie Jones has been pricing Amtrak train fares for a group trip to New York. Three adults and four children must pay $159. Two adults and three children must pay $112. Find the price of an adult's ticket, and find the price of a child's ticket.

18. Last month, Jerry Papa purchased five DVDs and two CDs at Wall-to-Wall Sound for $65. This month he bought three DVDs and four CDs for $81. Find the price of each DVD, and find the price of each CD.

19. Lily and Eleanor Saucier have a jar containing 80 coins, all of which are either quarters or nickels. The total value of the coins is $14.60. How many of each type of coin do they have?

20. Sarah and Keith Robinson purchased 40 stamps, a mixture of 50¢ and 35¢ stamps. Find the number of each type of stamp if they spent $19.25.

21. Steve and Katy Scarpulla own 30 shares of McDonald's Corporation stock and 68 shares of Mattel, Inc. stock. As the New York Stock Exchanges opened on October 8, 2018, their stock portfolio consisting of these two stocks was worth $5984.46. The McDonald's stock was worth $152.05 more per share than the Mattel, Inc. stock. What was the price of each stock on that day? (*Source:* YAHOO Finance)

22. Lakeesha Tarewan has investments in Google and Facebook stock. As the NASDAQ exchange opened on October 8, 2018, Google stock was at $1167.83 per share, and Facebook stock was at $157.33 per share. Lakeesha's portfolio made up of these two stocks was worth $45,345.81 at that time. If Lakeesha owns 15 more shares of Google stock than she owns of Facebook stock, how many shares of each type of stock does she own? (*Source:* Scottrade, Inc.)

23. Twice last month, Judy Carter rented a car from Enterprise in Fresno, California, and traveled around the Southwest on business. Enterprise rents this car for a daily fee, plus an additional charge per mile driven. Judy recalls that her first trip lasted 4 days, she drove 450 miles, and the rental cost her \$240.50. On her second business trip she drove the same level of car 200 miles in 3 days, and paid \$146.00 for the rental. Find the daily fee and the mileage charge.

24. Joan Gundersen rented the same car model twice from Hertz, which rents this car model for a daily fee plus an additional charge per mile driven. Joan recalls that the car rented for 5 days and driven for 300 miles cost her \$178, while the same model car rented for 4 days and driven for 500 miles cost \$197. Find the daily fee and the mileage charge.

25. Pratap Puri rowed 18 miles down the Delaware River in 2 hours, but the return trip took him $4\frac{1}{2}$ hours. Find the rate Pratap can row in still water and find the rate of the current.

Let x = rate Pratap can row in still water and
y = rate of the current

	d =	r ·	t
Downstream		$x + y$	
Upstream		$x - y$	

26. The Jonathan Schultz family took a canoe 10 miles down the Allegheny River in $1\frac{1}{4}$ hours. After lunch it took them 4 hours to return. Find the rate of the current.

Let x = rate the family can row in still water and
y = rate of the current

	d =	r ·	t
Downstream		$x + y$	
Upstream		$x - y$	

27. Dave and Sandy Hartranft are frequent flyers with Delta Airlines. They often fly from Philadelphia to Chicago, a distance of 780 miles. On one particular trip they fly into the wind, and the flight takes 2 hours. The return trip, with the wind behind them, takes only $1\frac{1}{2}$ hours. Find the speed of the wind and find the speed of the plane in still air.

28. With a strong wind behind it, a United Airlines jet flies 2400 miles from Los Angeles to Orlando in $4\frac{3}{4}$ hours. The return trip takes 6 hours, as the plane flies into the wind. Find the speed of the plane in still air and find the wind speed to the nearest tenth of a mile per hour.

29. Kevin Briley began a 186-mile bicycle trip to build up stamina for a triathlon competition. Unfortunately, his bicycle chain broke, so he finished the trip walking. The whole trip took 6 hours. If Kevin walks at a rate of 4 miles per hour and rides at 40 miles per hour, find the amount of time he spent on the bicycle.

30. In Canada, eastbound and westbound trains travel along the same track, with sidings to pull onto to avoid accidents. Two trains are now 150 miles apart, with the westbound train traveling twice as fast as the eastbound train. A warning must be issued to pull one train onto a siding, or else the trains will crash in $1\frac{1}{4}$ hours. Find the speed of the eastbound train and the speed of the westbound train.

31. Dorren Schmidt is a chemist with Gemco Pharmaceutical. She needs to prepare 12 ounces of a 9% hydrochloric acid solution. Find the amount of a 4% solution and the amount of a 12% solution she should mix to get this solution.

Concentration Rate	Liters of Solution	Liters of Pure Acid
0.04	x	$0.04x$
0.12	y	?
0.09	12	?

32. A chemistry student is preparing 15 liters of a 25% saline solution. She has two other saline solutions with strengths of 40% and 10%. Find the amount of 40% solution and the amount of 10% solution she should mix to get 15 liters of a 25% solution.

Concentration Rate	Liters of Solution	Liters of Pure Salt
0.40	x	$0.40x$
0.10	y	?
0.25	15	?

33. A barista blends coffee for a local coffee café. He needs to prepare 200 pounds of blended coffee beans selling for $3.95 per pound. He intends to do this by blending together a high-quality bean costing $4.95 per pound and a cheaper bean costing $2.65 per pound. To the nearest pound, find how much of the high-quality coffee beans and how much of the cheaper coffee beans he should blend.

34. Macadamia nuts cost an astounding $16.50 per pound, but research by an independent firm says that mixed nuts sell better if macadamias are included. The standard mix costs $9.25 per pound. Find how many pounds of macadamias and how many pounds of the standard mix should be combined to produce 40 pounds that will cost $10 per pound. Find the amounts to the nearest tenth of a pound.

△ **35.** Recall that two angles are complementary if the sum of their measures is 90°. Find the measures of two complementary angles if one angle is twice the other.

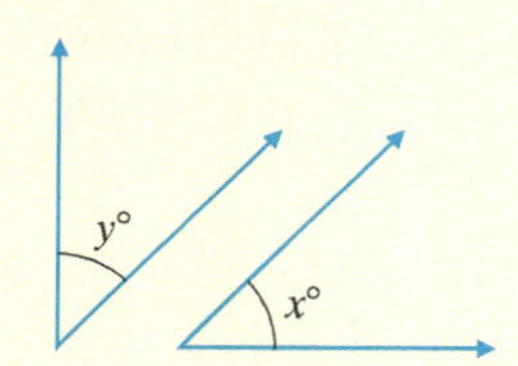

△ **36.** Recall that two angles are supplementary if the sum of their measures is 180°. Find the measures of two supplementary angles if one angle is 20° more than four times the other.

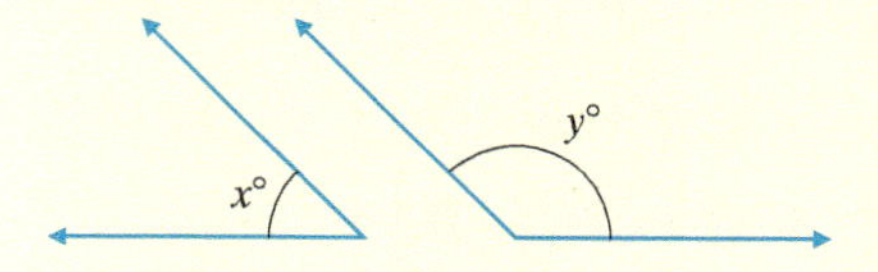

△ **37.** Find the measures of two complementary angles if one angle is 10° more than three times the other.

△ **38.** Find the measures of two supplementary angles if one angle is 18° more than twice the other.

39. Keith and Mandy McQueen had a pottery stand at the annual Skippack Craft Fair. They sold some of their pottery at the original price of $9.50 each, but later decreased the price of each by $2. If they sold all 90 pieces and took in $721, find how many they sold at the original price and how many they sold at the reduced price.

40. A charity fundraiser consisted of a spaghetti supper where a total of 387 people were fed. They charged $6.80 for adults and half price for children. If they took in $2444.60, find how many adults and how many children attended the supper.

41. The Santa Fe National Historic Trail is approximately 1200 miles between Old Franklin, Missouri, and Santa Fe, New Mexico. Suppose that a group of hikers start from each town and walk the trail toward each other. They meet after a total hiking time of 240 hours. If one group travels $\frac{1}{2}$ mile per hour slower than the other group, find the rate of each group. (*Source:* National Park Service)

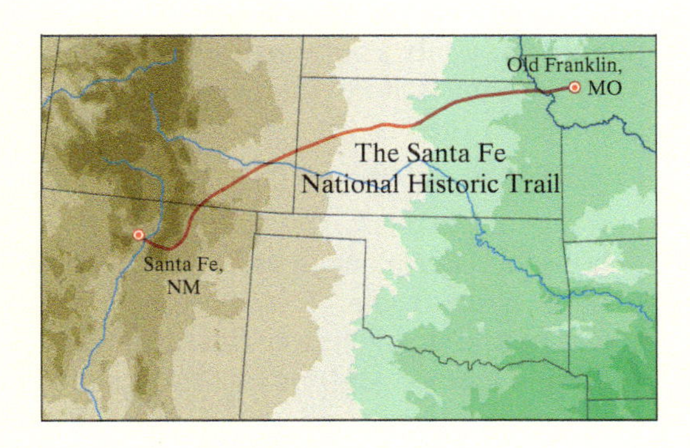

42. California 1 South is a historic highway that stretches 123 miles along the coast from Monterey to Morro Bay. Suppose that two cars start driving this highway, one from each town. They meet after 3 hours. Find the rate of each car if one car travels 1 mile per hour faster than the other car. (*Source: National Geographic*)

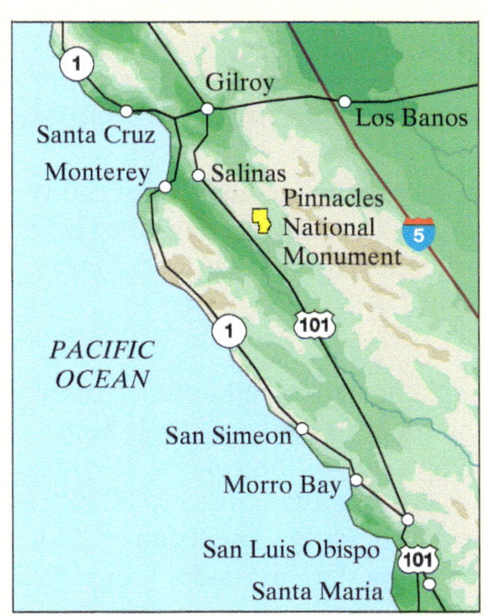

43. A 30% solution of fertilizer is to be mixed with a 60% solution of fertilizer in order to get 150 gallons of a 50% solution. How many gallons of the 30% solution and 60% solution should be mixed?

44. A 10% acid solution is to be mixed with a 50% acid solution in order to get 120 ounces of a 20% acid solution. How many ounces of the 10% solution and 50% solution should be mixed?

45. Traffic signs are regulated by the *Manual on Uniform Traffic Control Devices* (MUTCD). According to this manual, if the sign below is placed on a freeway, its perimeter must be 144 inches. Also, its length must be 12 inches longer than its width. Find the dimensions of this sign.

46. According to the MUTCD (see Exercise **45**), this sign must have a perimeter of 60 inches. Also, its length must be 6 inches longer than its width. Find the perimeter of this sign.

Review

Find the square of each number. For example, the square of 7 is 7^2 or 49. See Section 3.1.

47. 4 **48.** 3 **49.** 6 **50.** 11 **51.** 10 **52.** 8

Concept Extensions

Solve. See the Concept Check in this section.

53. Suppose you mix an amount of candy costing \$0.49 a pound with candy costing \$0.65 a pound. Which of the following costs per pound could result?

a. \$0.58 **b.** \$0.72 **c.** \$0.29

54. Suppose you mix a 50% acid solution with pure acid (100%). Which of the following acid strengths are possible for the resulting acid mixture?

a. 25% **b.** 150% **c.** 62% **d.** 90%

△ **55.** A couple have decided to fence off a garden plot behind their house, using their house as the "fence" along one side of the garden. The length (which runs parallel to the house) is 3 feet less than twice the width. Find the dimensions if 33 feet of fencing is used along the three sides requiring it.

△ **56.** A horse breeder plans to erect 152 feet of fencing to make a rectangular horse pasture. A river bank serves as one side length of the rectangle. If each width is 4 feet longer than half the length, find the dimensions.

Chapter 7 Group Activity

Break-Even Point

Sections 7.1, 7.2, 7.3, 7.4

When a business sells a new product, it generally does not start making a profit right away. There are usually many expenses associated with creating a new product. These expenses might include an advertising blitz to introduce the product to the public. These start-up expenses might also include the cost of market research and product development or any brand-new equipment needed to manufacture the product. Start-up costs like these are generally called *fixed costs* because they don't depend on the number of items manufactured. Expenses that do depend on the number of items manufactured, such as the cost of materials and shipping, are called *variable costs*. The total cost of manufacturing the new product is given by the cost equation Total cost = Fixed costs + Variable costs.

For instance, suppose a greeting card company is launching a new line of greeting cards. The company spent \$7000 doing product research and development for the new line and spent \$15,000 advertising the new line. The company does not need to buy any new equipment to manufacture the cards, but the paper and ink needed to make each card will cost \$0.20 per card. The total cost y in dollars for manufacturing x cards is $y = 22{,}000 + 0.20x$.

Once a business sets a price for a new product, the company can find the product's expected *revenue*. Revenue is the amount of money the company takes in from the sales of its product. The revenue from selling a product is given by the revenue equation Revenue = Price per item × Number of items sold.

For instance, suppose that the card company plans to sell its new cards for \$1.50 each. The revenue y, in dollars, that the company can expect to receive from the sales of x cards is $y = 1.50x$.

If the total cost and revenue equations are graphed on the same coordinate system, the graphs should intersect. The point of intersection is where total cost equals revenue and is called the *break-even point*. The break-even point gives the number of items x that must be manufactured and sold for the company to recover its expenses. If fewer than this number of items are produced and sold, the company loses money. If more than this number of items are produced and sold, the company makes a profit. In the case of the greeting card company, approximately 16,923 cards must be manufactured and sold for the company to break even on this new card line. The total cost and revenue of producing and selling 16,923 cards is the same. It is approximately \$25,385.

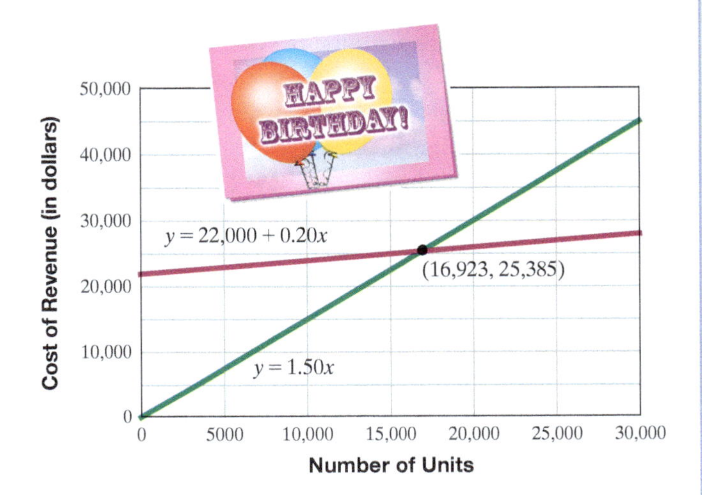

Group Activity

Suppose your group is starting a small business near your campus.

a. Choose a business and decide what campus-related product or service you will provide.

b. Research the fixed costs of starting up such a business.

c. Research the variable costs of producing such a product or providing such a service.

d. Decide how much you will charge per unit of your product or service.

e. Find a system of equations for the total cost and revenue of your product or service.

f. How many units of your product or service must be sold before your business will break even?

Chapter 7 Vocabulary Check

Fill in each blank with one of the words or phrases listed below.

system of linear equations	solution	consistent	independent
dependent	inconsistent	substitution	addition

1. In a system of linear equations in two variables, if the graphs of the equations are the same, the equations are ______________________ equations.
2. Two or more linear equations are called a(n) ______________________.
3. A system of equations that has at least one solution is called a(n) ______________________ system.
4. A(n) ______________________ of a system of two equations in two variables is an ordered pair of numbers that is a solution of both equations in the system.
5. Two algebraic methods for solving systems of equations are ______________________ and ______________________.
6. A system of equations that has no solution is called a(n) ______________________ system.
7. In a system of linear equations in two variables, if the graphs of the equations are different, the equations are ______________________ equations.

Helpful Hint

Are you preparing for your test? To help, don't forget to take these:

- Chapter 7 Getting Ready for the Test on page 576
- Chapter 7 Test on page 577

Then check all of your answers at the back of the text. For further review, the step-by-step video solutions to any of these exercises are located in MyLab Math.

7 Chapter Highlights

Definitions and Concepts	Examples
Section 7.1 Solving Systems of Linear Equations by Graphing	
A **system of linear equations** consists of two or more linear equations.	$\begin{cases} 2x + y = 6 \\ x = -3y \end{cases}$ $\begin{cases} -3x + 5y = 10 \\ x - 4y = -2 \end{cases}$
A **solution** of a system of two equations in two variables is an ordered pair of numbers that is a solution of both equations in the system.	Determine whether $(-1, 3)$ is a solution of the system. $\begin{cases} 2x - y = -5 \\ x = 3y - 10 \end{cases}$ Replace x with -1 and y with 3 in both equations. $2x - y = -5$ $2(-1) - 3 \stackrel{?}{=} -5$ $-5 = -5$ True $x = 3y - 10$ $-1 \stackrel{?}{=} 3(3) - 10$ $-1 = -1$ True $(-1, 3)$ is a solution of the system.
Graphically, a solution of a system is a point common to the graphs of both equations.	Solve by graphing: $\begin{cases} 3x - 2y = -3 \\ x + y = 4 \end{cases}$ 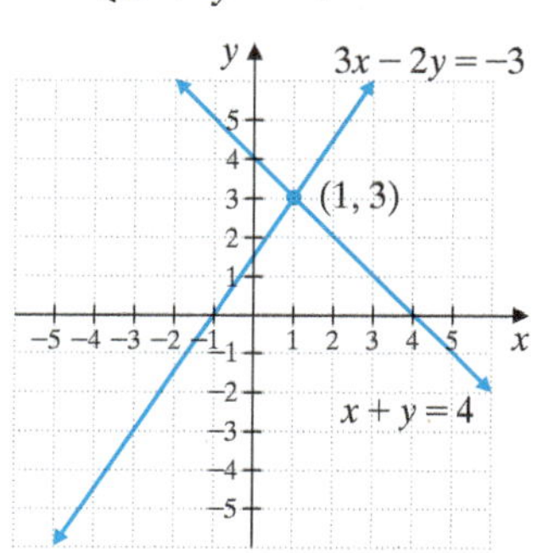

Definitions and Concepts	Examples

Section 7.1 Solving Systems of Linear Equations by Graphing (*continued*)

Three different situations can occur when graphing the two lines associated with the equations in a linear system.

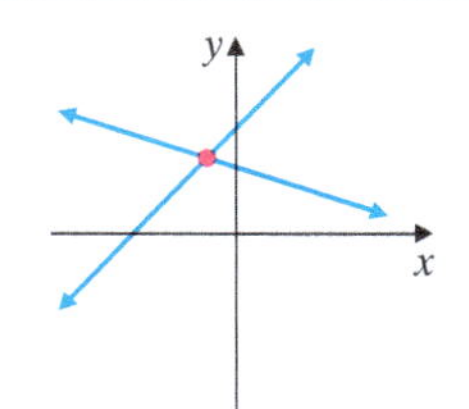

One point of intersection; one solution

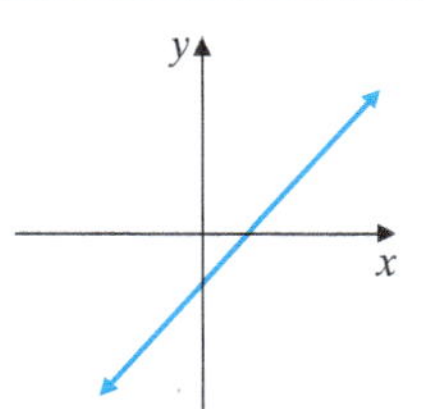

Same line; infinite number of solutions

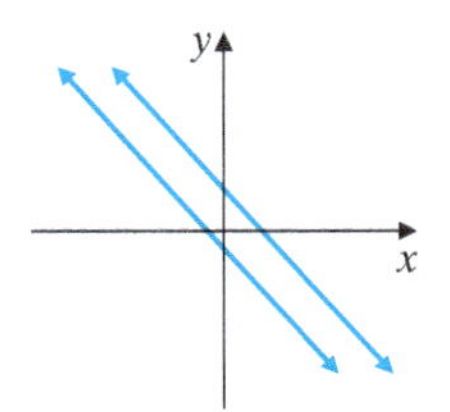

Parallel lines; no solution

Section 7.2 Solving Systems of Linear Equations by Substitution

To Solve a System of Linear Equations by the Substitution Method

Step 1: Solve one equation for one of its variables.

Step 2: Substitute the expression for the variable into the other equation.

Step 3: Solve the equation from Step 2 to find the value of one variable.

Step 4: Substitute the value from Step 3 into any equation containing both variables to find the value of the other variable.

Step 5: Check the proposed solution in the original system.

Solve by substitution.

$$\begin{cases} 3x + 2y = 1 \\ x = y - 3 \end{cases}$$

Substitute $y - 3$ for x in the first equation.

$$\begin{aligned} 3x + 2y &= 1 \\ 3(y - 3) + 2y &= 1 \\ 3y - 9 + 2y &= 1 \\ 5y &= 10 \\ y &= 2 \quad \text{Divide by 5.} \end{aligned}$$

To find x, substitute 2 for y in $x = y - 3$ so that $x = 2 - 3$ or -1. The solution $(-1, 2)$ checks.

Section 7.3 Solving Systems of Linear Equations by Addition

To Solve a System of Linear Equations by the Addition Method

Step 1: Rewrite each equation in standard form, $Ax + By = C$.

Step 2: Multiply one or both equations by a nonzero number so that the coefficients of a chosen variable in the system are opposites.

Step 3: Add the equations.

Step 4: Find the value of one variable by solving the resulting equation from Step 3.

Step 5: Substitute the value from Step 4 into either original equation to find the value of the other variable.

Step 6: Check the proposed solution in the original system.

Solve by addition.

$$\begin{cases} x - 2y = 8 \\ 3x + y = -4 \end{cases}$$

Multiply both sides of the first equation by -3.

$$\begin{cases} -3x + 6y = -24 \\ \underline{3x + y = -4} \end{cases}$$

$$\begin{aligned} 7y &= -28 \quad \text{Add.} \\ y &= -4 \quad \text{Divide by 7.} \end{aligned}$$

To find x, let $y = -4$ in an original equation.

$$\begin{aligned} x - 2(-4) &= 8 \quad \text{First equation} \\ x + 8 &= 8 \\ x &= 0 \end{aligned}$$

The solution $(0, -4)$ checks.

(*continued*)

Definitions and Concepts	Examples
Section 7.3 Solving Systems of Linear Equations by Addition (*continued*)	
If solving a system of linear equations by substitution or addition yields a true statement such as $-2 = -2$, then the graphs of the equations in the system are identical and the system has an infinite number of solutions.	Solve: $\begin{cases} 2x - 6y = -2 \\ x = 3y - 1 \end{cases}$ Substitute $3y - 1$ for x in the first equation. $2(3y - 1) - 6y = -2$ $6y - 2 - 6y = -2$ $-2 = -2$ True The system has an infinite number of solutions.
If solving a system of linear equations yields a false statement such as $0 = 3$, the graphs of the equations in the system are parallel lines and the system has no solution.	Solve: $\begin{cases} 5x - 2y = 6 \\ -5x + 2y = -3 \end{cases}$ $0 = 3$ False The system has no solution.
Section 7.4 Systems of Linear Equations and Problem Solving	
Problem-Solving Steps **1.** UNDERSTAND. Read and reread the problem.	Two angles are supplementary if the sum of their measures is 180°. The larger of two supplementary angles is three times the smaller, decreased by twelve. Find the measure of each angle. Let x = measure of smaller angle and y = measure of larger angle 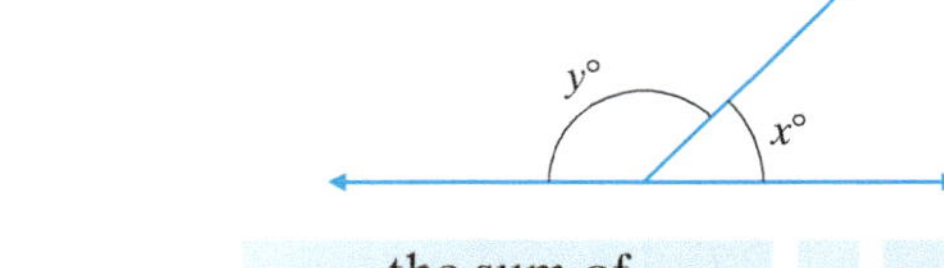
2. TRANSLATE.	In words: the sum of supplementary angles ↓ is ↓ 180° ↓ Translate: $x + y = 180$ In words: larger angle ↓ is ↓ 3 times smaller ↓ decreased by ↓ 12 ↓ Translate: $y = 3x - 12$
3. SOLVE.	Solve the system. $\begin{cases} x + y = 180 \\ y = 3x - 12 \end{cases}$ Use the substitution method and replace y with $3x - 12$ in the first equation. $x + y = 180$ $x + (3x - 12) = 180$ $4x = 192$ $x = 48$
4. INTERPRET.	Since $y = 3x - 12$, then $y = 3 \cdot 48 - 12$ or 132. The solution checks. The smaller angle measures 48° and the larger angle measures 132°.

Chapter 7 Review

(7.1) *Determine whether each ordered pair is a solution of the system of linear equations.*

1. $\begin{cases} 2x - 3y = 12 \\ 3x + 4y = 1 \end{cases}$

a. $(12, 4)$
b. $(3, -2)$

2. $\begin{cases} 2x + 3y = 1 \\ 3y - x = 4 \end{cases}$

a. $(2, 2)$
b. $(-1, 1)$

3. $\begin{cases} 5x - 6y = 18 \\ 2y - x = -4 \end{cases}$

a. $(-6, -8)$
b. $\left(3, \frac{5}{2}\right)$

4. $\begin{cases} 4x + y = 0 \\ -8x - 5y = 9 \end{cases}$

a. $\left(\frac{3}{4}, -3\right)$
b. $(-2, 8)$

Solve each system of equations by graphing.

5. $\begin{cases} x + y = 5 \\ x - y = 1 \end{cases}$

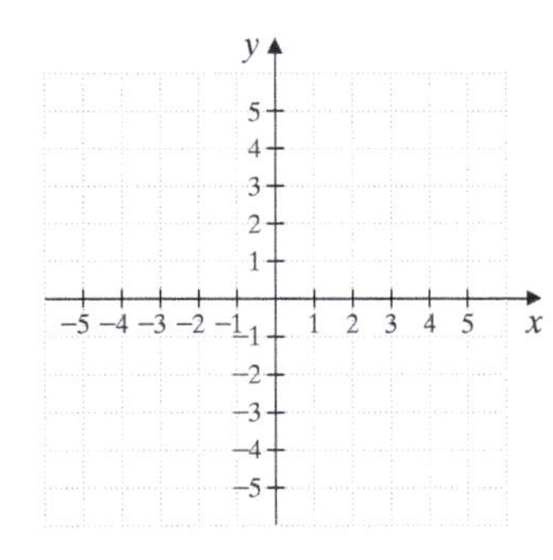

6. $\begin{cases} x + y = 3 \\ x - y = -1 \end{cases}$

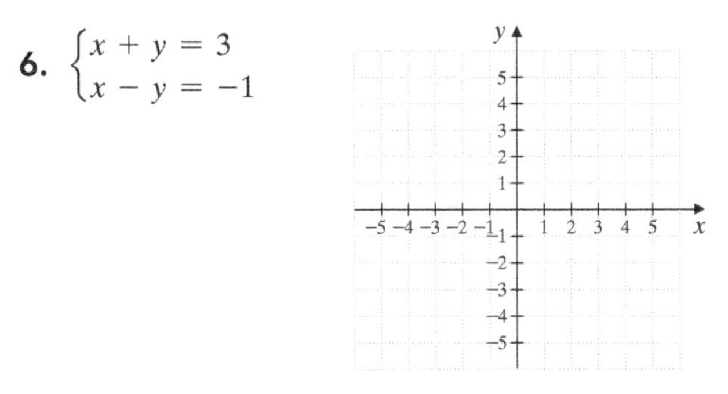

7. $\begin{cases} x = 5 \\ y = -1 \end{cases}$

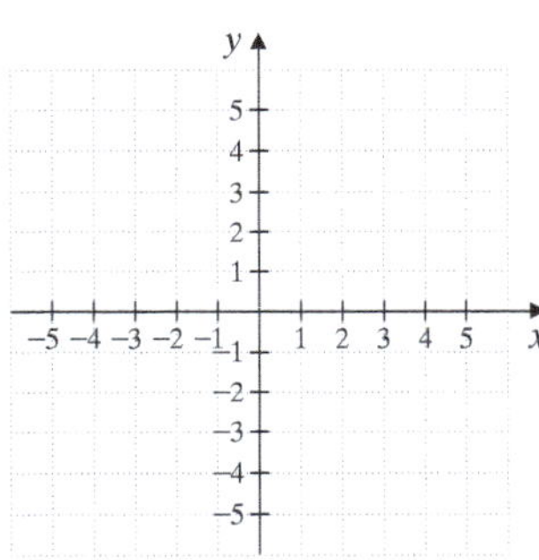

8. $\begin{cases} x = -3 \\ y = 2 \end{cases}$

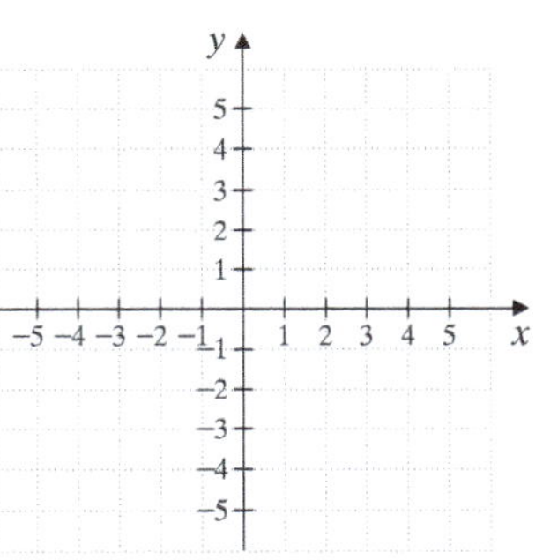

9. $\begin{cases} 2x + y = 5 \\ x = -3y \end{cases}$

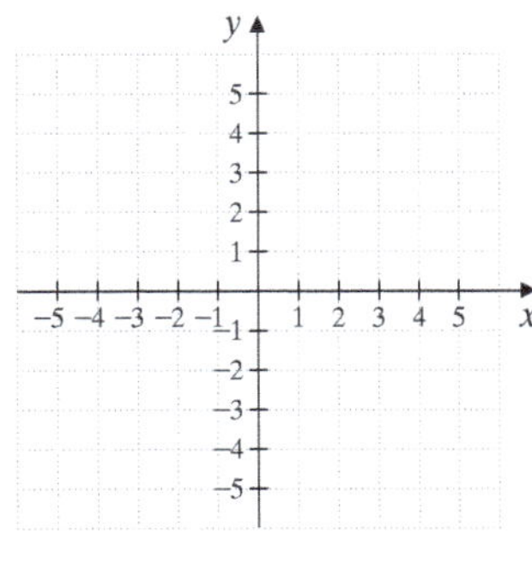

10. $\begin{cases} 3x + y = -2 \\ y = -5x \end{cases}$

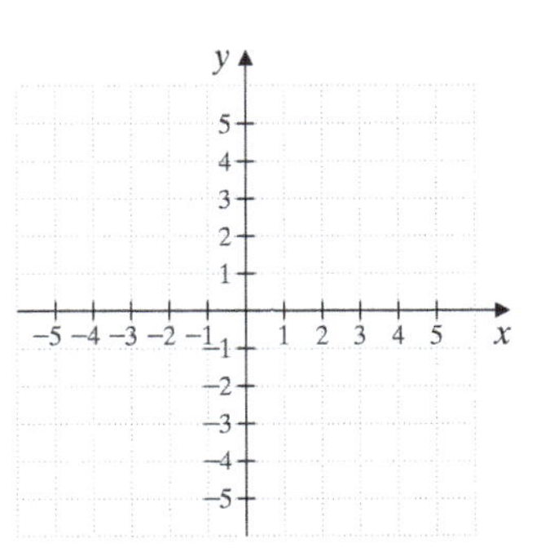

11. $\begin{cases} y = 3x \\ -6x + 2y = 6 \end{cases}$

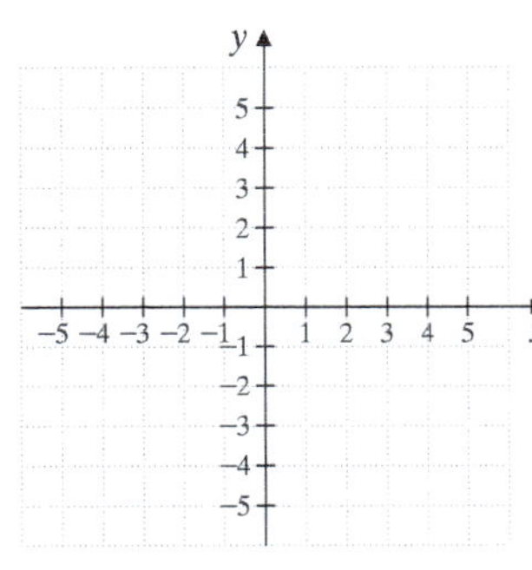

12. $\begin{cases} x - 2y = 2 \\ -2x + 4y = -4 \end{cases}$

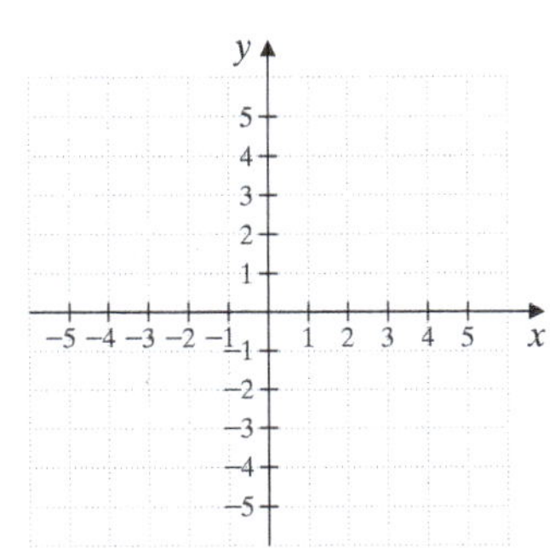

(7.2) *Solve each system of equations by the substitution method.*

13. $\begin{cases} y = 2x + 6 \\ 3x - 2y = -11 \end{cases}$

14. $\begin{cases} y = 3x - 7 \\ 2x - 3y = 7 \end{cases}$

15. $\begin{cases} x + 3y = -3 \\ 2x + y = 4 \end{cases}$

16. $\begin{cases} 3x + y = 11 \\ x + 2y = 12 \end{cases}$

17. $\begin{cases} 4y = 2x + 6 \\ x - 2y = -3 \end{cases}$

18. $\begin{cases} 9x = 6y + 3 \\ 6x - 4y = 2 \end{cases}$

19. $\begin{cases} x + y = 6 \\ y = -x - 4 \end{cases}$

20. $\begin{cases} -3x + y = 6 \\ y = 3x + 2 \end{cases}$

(7.3) *Solve each system of equations by the addition method.*

21. $\begin{cases} 2x + 3y = -6 \\ x - 3y = -12 \end{cases}$

22. $\begin{cases} 4x + y = 15 \\ -4x + 3y = -19 \end{cases}$

23. $\begin{cases} 2x - 3y = -15 \\ x + 4y = 31 \end{cases}$

24. $\begin{cases} x - 5y = -22 \\ 4x + 3y = 4 \end{cases}$

25. $\begin{cases} 2x - 6y = -1 \\ -x + 3y = \dfrac{1}{2} \end{cases}$

26. $\begin{cases} 0.6x - 0.3y = -1.5 \\ 0.04x - 0.02y = -0.1 \end{cases}$

27. $\begin{cases} \dfrac{3}{4}x + \dfrac{2}{3}y = 2 \\ x + \dfrac{y}{3} = 6 \end{cases}$

28. $\begin{cases} 10x + 2y = 0 \\ 3x + 5y = 33 \end{cases}$

(7.4) *Solve each problem by writing and solving a system of linear equations.*

29. The sum of two numbers is 16. Three times the larger number decreased by the smaller number is 72. Find the two numbers.

30. The Forrest Theater can seat a total of 360 people. They take in \$15,150 when every seat is sold. If orchestra section tickets cost \$45 and balcony tickets cost \$35, find the number of seats in the orchestra section and the number of seats in the balcony.

31. A riverboat can go 340 miles upriver in 19 hours, but the return trip takes only 14 hours. Find the current of the river and find the speed of the riverboat in still water to the nearest tenth of a mile.

	d =	r ·	t
Upriver		$x - y$	
Downriver		$x + y$	

32. Find the amount of a 6% acid solution and the amount of a 14% acid solution Pat Mayfield should combine to prepare 50 cc (cubic centimeters) of a 12% solution.

33. A deli charges \$3.80 for a breakfast of three eggs and four strips of bacon. The charge is \$2.75 for two eggs and three strips of bacon. Find the cost of each egg and the cost of each strip of bacon.

34. An exercise enthusiast alternates between jogging and walking. He traveled 15 miles during the past 3 hours. He jogs at a rate of 7.5 miles per hour and walks at a rate of 4 miles per hour. Find how much time, to the nearest hundredth of an hour, he actually spent jogging and how much time he spent walking.

Mixed Review

Solve each system of equations by graphing.

35. $\begin{cases} x - 2y = 1 \\ 2x + 3y = -12 \end{cases}$

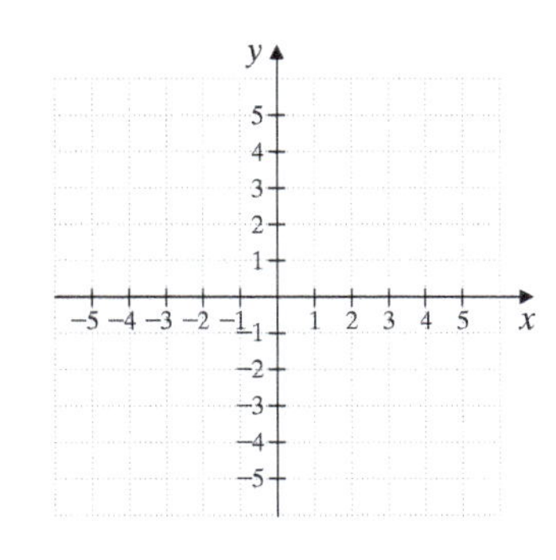

36. $\begin{cases} 3x - y = -4 \\ 6x - 2y = -8 \end{cases}$

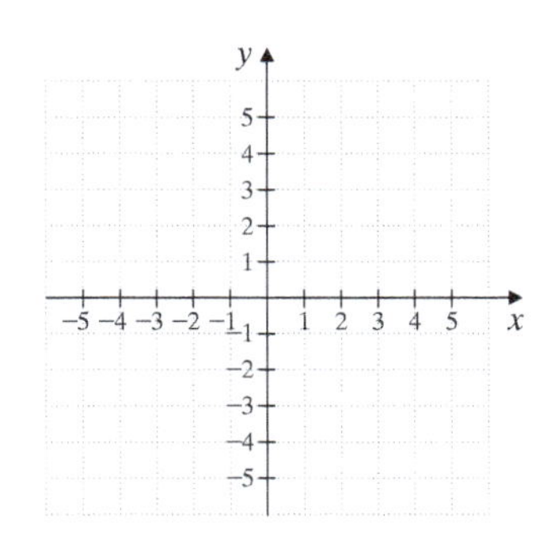

Solve each system of equations.

37. $\begin{cases} x + 4y = 11 \\ 5x - 9y = -3 \end{cases}$

38. $\begin{cases} x + 9y = 16 \\ 3x - 8y = 13 \end{cases}$

39. $\begin{cases} y = -2x \\ 4x + 7y = -15 \end{cases}$

40. $\begin{cases} 3y = 2x + 15 \\ -2x + 3y = 21 \end{cases}$

41. $\begin{cases} 3x - y = 4 \\ 4y = 12x - 16 \end{cases}$

42. $\begin{cases} x + y = 19 \\ x - y = -3 \end{cases}$

43. $\begin{cases} x - 3y = -11 \\ 4x + 5y = -10 \end{cases}$

44. $\begin{cases} -x - 15y = 44 \\ 2x + 3y = 20 \end{cases}$

45. $\begin{cases} 2x + y = 3 \\ 6x + 3y = 9 \end{cases}$

46. $\begin{cases} -3x + y = 5 \\ -3x + y = -2 \end{cases}$

Solve each problem by writing and solving a system of linear equations.

47. The sum of two numbers is 12. Three times the smaller number increased by the larger number is 20. Find the numbers.

48. The difference of two numbers is −18. Twice the smaller decreased by the larger is −23. Find the two numbers.

49. Emma Hodges has a jar containing 65 coins, all of which are either nickels or dimes. The total value of the coins is \$5.30. How many of each type does she have?

50. Sarah and Owen Hebert purchased 26 stamps, a mixture of \$1.15 and 50¢ stamps. Find the number of each type of stamp if they spent \$19.50.

Chapter 7 Getting Ready for the Test LC

1. MULTIPLE CHOICE The ordered pair $(-1, 2)$ is a solution to what system?

A. $\begin{cases} 5x - y = -7 \\ x - y = 3 \end{cases}$ **B.** $\begin{cases} 3x - y = -5 \\ x + y = 1 \end{cases}$

C. $\begin{cases} x = 2 \\ x + y = 1 \end{cases}$ **D.** $\begin{cases} y = -1 \\ x + y = -3 \end{cases}$

MATCHING *For Exercises 2 through 5,* ***Match*** *each system with the graph of its equations.*

2. $\begin{cases} x = 3 \\ y = -3 \end{cases}$

3. $\begin{cases} x = -3 \\ y = 3 \end{cases}$

4. $\begin{cases} x = 3 \\ y = 3 \end{cases}$

5. $\begin{cases} x = -3 \\ y = -3 \end{cases}$

A.

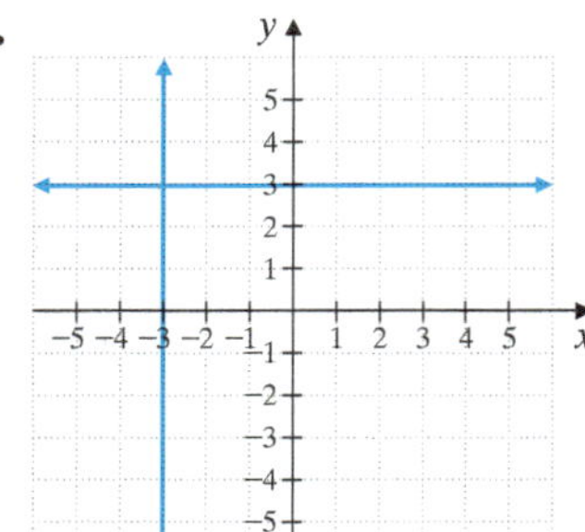

B.

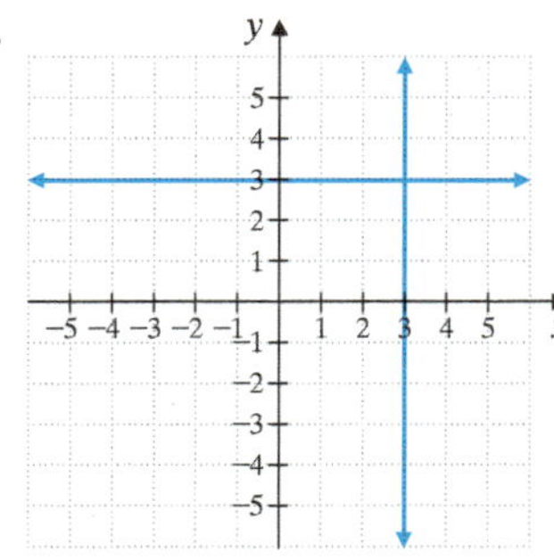

C.

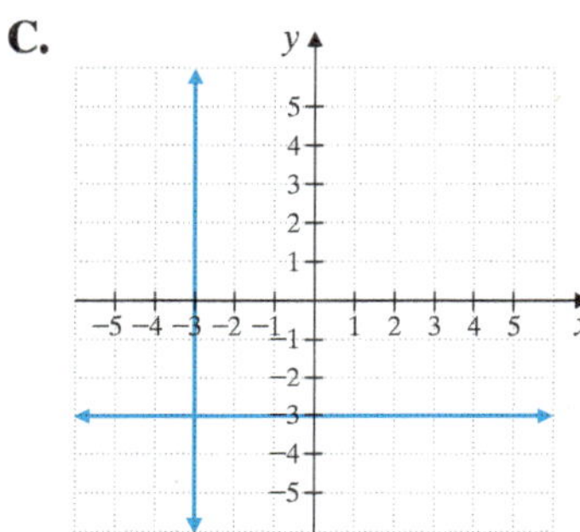

D.

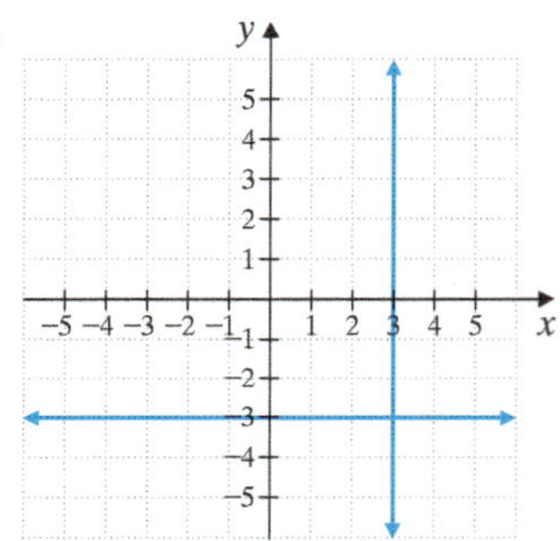

6. MULTIPLE CHOICE When solving a system of two linear equations in two variables, all variables subtract out and the resulting equation is $0 = 5$. What does this mean?

A. the solution is $(0, 5)$ **B.** the system has an infinite number of solutions **C.** the system has no solution

MATCHING *For Exercises 7 through 10,* ***Match*** *each system with its solution. Letter choices may be used more than once or not at all.*

7. $\begin{cases} y = 5x + 2 \\ y = -5x + 2 \end{cases}$

8. $\begin{cases} y = \frac{1}{2}x - 3 \\ y = \frac{1}{2}x + 7 \end{cases}$

9. $\begin{cases} y = 4x + 2 \\ 8x - 2y = -4 \end{cases}$

10. $\begin{cases} y = 6x \\ y = -\frac{1}{6}x \end{cases}$

A. no solution

B. one solution

C. two solutions

D. an infinite number of solutions

MULTIPLE CHOICE *Choose the correct choice for Exercises 11 and 12. The system for these exercises is:*

$$\begin{cases} 5x - y = -8 \\ 2x + 3y = 1 \end{cases}$$

11. When solving, if we decide to multiply the first equation above by 3, the result of the multiplication is:

A. $15x - 3y = -8$ **B.** $6x + 9y = 1$ **C.** $6x + 9y = 3$ **D.** $15x - 3y = -24$

12. When solving, if we decide to multiply the second equation above by -5, the result of the multiplication is:

A. $-10x - 15y = 1$ **B.** $-25x + 5y = 40$ **C.** $-10x - 15y = -5$ **D.** $-25x + 5y = -8$

For additional practice go to your study plan in MyLab Math.

MyLab Math

Chapter 7 Test

Answer each question true or false.

1. A system of two linear equations in two variables can have exactly two solutions.

2. Although $(1, 4)$ is not a solution of $x + 2y = 6$, it can still be a solution of the system $\begin{cases} x + 2y = 6 \\ x + y = 5 \end{cases}$.

3. If the two equations in a system of linear equations are added and the result is $3 = 0$, the system has no solution.

4. If the two equations in a system of linear equations are added and the result is $3x = 0$, the system has no solution.

Is the ordered pair a solution of the given linear system?

5. $\begin{cases} 2x - 3y = 5 \\ 6x + y = 1 \end{cases}$; $(1, -1)$

6. $\begin{cases} 4x - 3y = 24 \\ 4x + 5y = -8 \end{cases}$; $(3, -4)$

Solve each system by graphing.

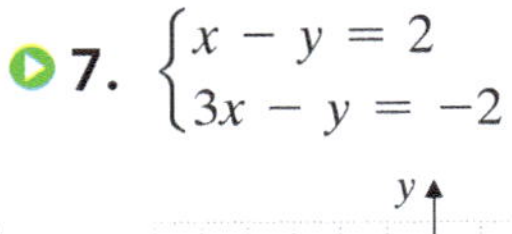

7. $\begin{cases} x - y = 2 \\ 3x - y = -2 \end{cases}$

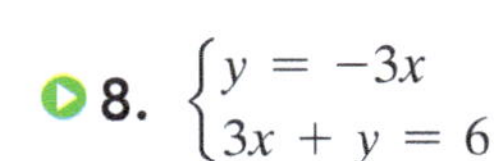

8. $\begin{cases} y = -3x \\ 3x + y = 6 \end{cases}$

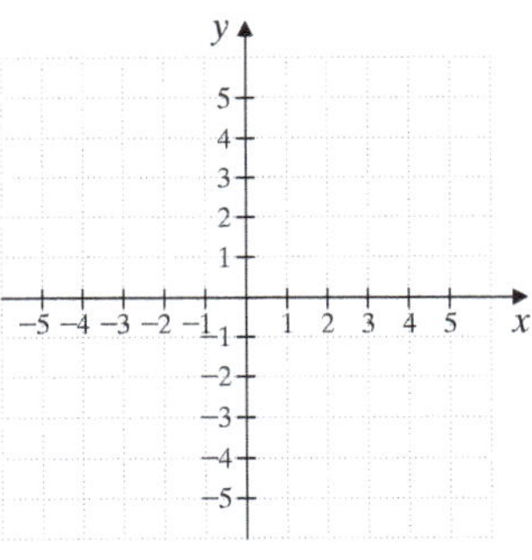

Solve each system by the substitution method.

9. $\begin{cases} 3x - 2y = -14 \\ y = x + 5 \end{cases}$

10. $\begin{cases} \frac{1}{2}x + 2y = -\frac{15}{4} \\ 4x = -y \end{cases}$

Solve each system by the addition method.

11. $\begin{cases} x + y = 28 \\ x - y = 12 \end{cases}$

12. $\begin{cases} 4x - 6y = 7 \\ -2x + 3y = 0 \end{cases}$

Solve each system using the substitution method or the addition method.

13. $\begin{cases} 3x + y = 7 \\ 4x + 3y = 1 \end{cases}$

14. $\begin{cases} 3(2x + y) = 4x + 20 \\ x - 2y = 3 \end{cases}$

Answers

1. ______
2. ______
3. ______
4. ______
5. ______
6. ______
7. ______
8. ______
9. ______
10. ______
11. ______
12. ______
13. ______
14. ______

15. $\begin{cases} \dfrac{x-3}{2} = \dfrac{2-y}{4} \\ \dfrac{7-2x}{3} = \dfrac{y}{2} \end{cases}$

16. $\begin{cases} 8x - 4y = 12 \\ y = 2x - 3 \end{cases}$

17. $\begin{cases} 0.01x - 0.06y = -0.23 \\ 0.2x + 0.4y = 0.2 \end{cases}$

18. $\begin{cases} x - \dfrac{2}{3}y = 3 \\ -2x + 3y = 10 \end{cases}$

Solve each problem by writing and using a system of linear equations.

19. Two numbers have a sum of 124 and a difference of 32. Find the numbers.

20. Find the amount of a 12% saline solution a lab assistant should add to 80 cc (cubic centimeters) of a 22% saline solution in order to have a 16% solution.

21. Texas and Missouri are the states with the most farms. Texas has 140 thousand more farms than Missouri and the total number of farms for these two states is 356 thousand. Find the number of farms for each state.

22. Two hikers start at opposite ends of the St. Tammany Trails and walk toward each other. The trail is 36 miles long and they meet in 4 hours. If one hiker is twice as fast as the other, find both hiking speeds.

The double line graph below shows the average attendance per game for the years shown for the Minnesota Twins and the Texas Rangers baseball teams. Use this for Exercises 23 and 24.

Average Game Attendance

Average Attendance (in thousands)

44, 42, 40, 38, 36, 34, 32, 30, 28, 26, 24, 22

Minnesota Twins

Texas Rangers

2006 2007 2008 2009 2010 2011 2012 2013 2014 2015 2016

Year

Source: Baseball Almanac

23. In what year(s) was the average attendance per game for the Texas Rangers greater than the average attendance per game for the Minnesota Twins?

24. In what year was the average attendance per game for the Texas Rangers closest to the average attendance per game for the Minnesota Twins, 2011 or 2016?

15. ____
16. ____
17. ____
18. ____
19. ____
20. ____
21. ____
22. ____
23. ____
24. ____

Cumulative Review — Chapters 1–7

1. Simplify each expression.
   a. $-14 - 8 + 10 - (-6)$
   b. $1.6 - (-10.3) + (-5.6)$

2. Evaluate:
   a. 5^2
   b. 2^5

Find the reciprocal or opposite of each number.

3. reciprocal of 22

4. opposite of 22

5. reciprocal of $\frac{3}{16}$

6. opposite of $\frac{3}{16}$

7. reciprocal of -10

8. opposite of -10

9. reciprocal of $-\frac{9}{13}$

10. opposite of $-\frac{9}{13}$

11. reciprocal of 1.7

12. opposite of 1.7

13. a. The sum of two numbers is 8. If one number is 3, find the other number.
    b. The sum of two numbers is 8. If one number is x, write an expression representing the other number.

14. Five times the sum of a number and -1 is the same as 6 times the number. Find the number.

15. Solve:
    $-2(x - 5) + 10 = -3(x + 2) + x$

16. Solve: $5(y - 5) = 5y + 10$

17. Solve: $\frac{x}{2} - 1 = \frac{2}{3}x - 3$

18. Solve: $7(x - 2) - 6(x + 1) = 20$

Answers

1. a. ______
   b. ______
2. a. ______
   b. ______
3. ______
4. ______
5. ______
6. ______
7. ______
8. ______
9. ______
10. ______
11. ______
12. ______
13. a. ______
    b. ______
14. ______
15. ______
16. ______
17. ______
18. ______

19. Solve $-5x + 7 < 2(x - 3)$. Graph the solution set.

−5 −4 −3 −2 −1 0 1 2 3 4 5

20. Solve $P = a + b + c$ for b.

Simplify each expression.

21. $\left(\frac{m}{n}\right)^7$

22. $\frac{a^7b^{10}}{ab^{15}}$

23. $\left(\frac{2x^4}{3y^5}\right)^4$

24. $(7a^2b^{-3})^2$

25. Subtract: $(2x^3 + 8x^2 - 6x) - (2x^3 - x^2 + 1)$

26. Add: $\left(5x^2 + 6x + \frac{1}{2}\right) + \left(x^2 - \frac{4}{3}x - \frac{10}{21}\right)$

27. Divide $6x^2 + 10x - 5$ by $3x - 1$ using long division.

28. Find the GCF of $9x^2$, $6x^3$, and $21x^5$.

29. Solve: $x(2x - 7) = 4$

30. Solve: $x(2x - 7) = 0$

△ 31. Find the lengths of the sides of a right triangle if the lengths can be expressed as three consecutive even integers.

△ 32. The height of a parallelogram is 5 feet more than three times its base. If the area of the parallelogram is 182 square feet, find the length of its base and height.

33. Subtract: $\frac{2y}{2y - 7} - \frac{7}{2y - 7}$

34. Add: $\frac{2}{x - 6} + \frac{3}{x + 1}$

35. Simplify: $\dfrac{\frac{x}{y} + \frac{3}{2x}}{\frac{x}{2} + y}$

36. Find the slope of a line parallel to the line passing through $(-1, 3)$ and $(2, -8)$.

37. Find the slope of the line $y = -1$.

19. ______
20. ______
21. ______
22. ______
23. ______
24. ______
25. ______
26. ______
27. ______
28. ______
29. ______
30. ______
31. ______
32. ______
33. ______
34. ______
35. ______
36. ______
37. ______

38. Find the slope of the line $x = 2$.

39. Find an equation of the line through $(2, 5)$ and $(-3, 4)$. Write the equation in the form $Ax + By = C$.

40. Write an equation of the line with slope -5 through $(-2, 3)$.

41. Find the domain and the range of the relation $\{(0, 2), (3, 3), (-1, 0), (3, -2)\}$.

42. If $f(x) = 5x^2 - 6$, find $f(0)$ and $f(-2)$.

43. Determine whether $(12, 6)$ is a solution of the system $\begin{cases} 2x - 3y = 6 \\ x = 2y \end{cases}$

44. Determine whether each ordered pair is a solution of the given system.
$\begin{cases} 2x - y = 6 \\ 3x + 2y = -5 \end{cases}$

a. $(1, -4)$

b. $(0, 6)$

c. $(3, 0)$

45. Solve the system: $\begin{cases} x + 2y = 7 \\ 2x + 2y = 13 \end{cases}$

Solve each system.

46. $\begin{cases} 3x - 4y = 10 \\ y = 2x \end{cases}$

47. $\begin{cases} -x - \dfrac{y}{2} = \dfrac{5}{2} \\ \dfrac{x}{6} - \dfrac{y}{2} = 0 \end{cases}$

48. $\begin{cases} x = 5y - 3 \\ x = 8y + 4 \end{cases}$

49. Find two numbers whose sum is 37 and whose difference is 21.

50. Determine which graph(s) are graphs of functions.

a.

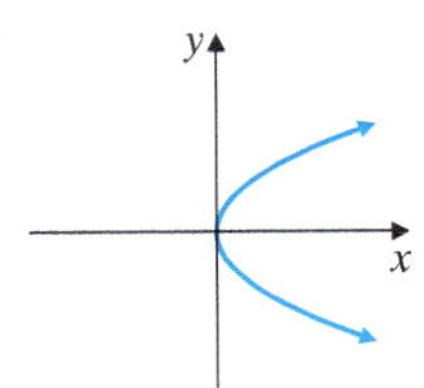

b.

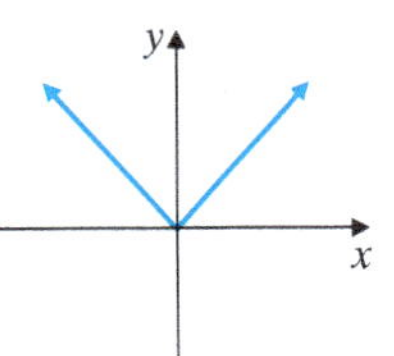

c.

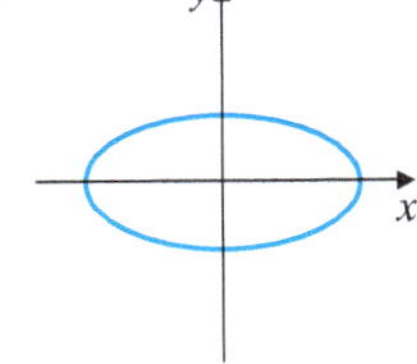

38. ______

39. ______

40. ______

41. ______

42. ______

43. ______

44. a. ______

b. ______

c. ______

45. ______

46. ______

47. ______

48. ______

49. ______

50. a. ______

b. ______

c. ______

8 Roots and Radicals

Having spent the last chapter studying equations, we return now to algebraic expressions. We expand on our skills of operating on expressions—adding, subtracting, multiplying, dividing, and raising to powers—to include finding roots. Just as subtraction is defined by addition and division by multiplication, finding roots is defined by raising to powers. As we master finding roots, we will work with equations that contain roots and solve problems that can be modeled by such equations.

Sections

8.1 Introduction to Radicals

8.2 Simplifying Radicals

8.3 Adding and Subtracting Radicals

8.4 Multiplying and Dividing Radicals

Integrated Review—Simplifying Radicals

8.5 Solving Equations Containing Radicals

8.6 Radical Equations and Problem Solving

8.7 Direct and Inverse Variation Including Radical Applications

Check Your Progress

Vocabulary Check

Chapter Highlights

Chapter Review

Getting Ready for the Test

Chapter Test

Cumulative Review

Did You Know That Pendulums Can Be Used to Demonstrate That the Earth Rotates on Its Axis?

In 1851, French physicist Léon Foucault developed a special pendulum in an experiment to demonstrate that the earth rotates on its axis. He connected his tall pendulum, capable of running for many hours, to the roof of the Paris Observatory. The pendulum's bob was able to swing back and forth in one plane but not to twist in other directions. So, when the pendulum bob appeared to move in a circle over time, he demonstrated that it was not the pendulum but the building that moved. And since the building was firmly attached to the earth, it must be the earth rotating that created the apparent circular motion of the bob. In Section 8.1, Exercise 93, roots are used to explore the time it takes Foucault's pendulum to complete one swing of its bob.

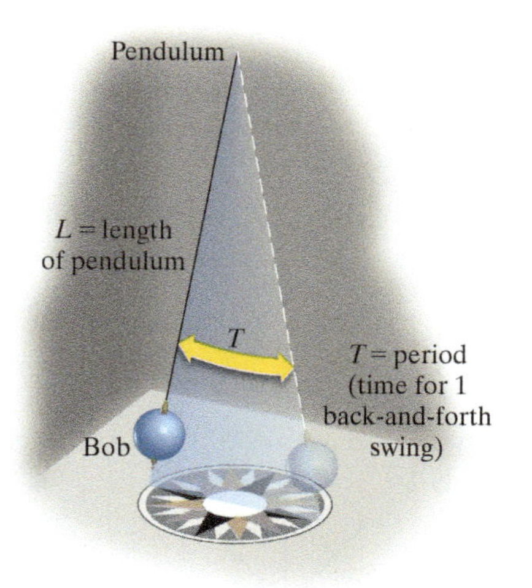

8.1 Introduction to Radicals

Objective A Finding Square Roots

In this section, we define finding the **root** of a number by its reverse operation, raising a number to a power. We begin with squares and square roots.

The *square* of 5 is $5^2 = 25$.
The *square* of -5 is $(-5)^2 = 25$.
The *square* of $\frac{1}{2}$ is $\left(\frac{1}{2}\right)^2 = \frac{1}{4}$.

The reverse operation of squaring a number is finding a **square root** of a number. For example,

A *square root* of 25 is 5, because $5^2 = 25$.
A *square root* of 25 is also -5, because $(-5)^2 = 25$.
A *square root* of $\frac{1}{4}$ is $\frac{1}{2}$, because $\left(\frac{1}{2}\right)^2 = \frac{1}{4}$.

In general, the number b is a square root of a number a if $b^2 = a$.

The symbol $\sqrt{}$ is used to denote the **positive** or **principal square root** of a number. For example,

$\sqrt{25} = 5$ only, since $5^2 = 25$ and 5 is positive.

The symbol $-\sqrt{}$ is used to denote the **negative square root.** For example,

$-\sqrt{25} = -5$

The symbol $\sqrt{}$ is called a **radical** or **radical sign.** The expression within or under a radical sign is called the **radicand.** An expression containing a radical is called a **radical expression.**

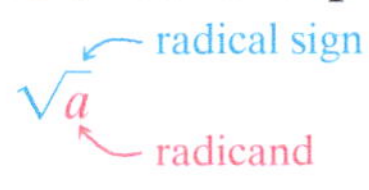

Square Root

If a is a positive number, then

$\sqrt{a}$ is the **positive square root** of a and
$-\sqrt{a}$ is the **negative square root** of a.

Also, $\sqrt{0} = 0$.

Objectives

A Find Square Roots.
B Find Cube Roots.
C Find *n*th Roots.
D Approximate Square Roots.
E Simplify Radicals Containing Variables.

Examples Find each square root.

1. $\sqrt{36} = 6$, because $6^2 = 36$ and 6 is positive.
2. $-\sqrt{16} = -4$. The negative sign in front of the radical indicates the negative square root of 16.
3. $\sqrt{\frac{9}{100}} = \frac{3}{10}$ because $\left(\frac{3}{10}\right)^2 = \frac{9}{100}$ and $\frac{3}{10}$ is positive.
4. $\sqrt{0} = 0$ because $0^2 = 0$.
5. $\sqrt{0.64} = 0.8$ because $(0.8)^2 = 0.64$ and 0.8 is positive.

Work Practice 1–5

Practice 1–5

Find each square root.

1. $\sqrt{100}$
2. $-\sqrt{81}$
3. $\sqrt{\frac{25}{81}}$
4. $\sqrt{1}$
5. $\sqrt{0.81}$

Answers

1. 10 2. -9 3. $\frac{5}{9}$ 4. 1 5. 0.9

Is the square root of a negative number a real number? For example, is $\sqrt{-4}$ a real number? To answer this question, we ask ourselves, is there a real number whose square is -4? Since there is no real number whose square is -4, we say that $\sqrt{-4}$ is not a real number. In general,

A square root of a negative number is not a real number.

Study the following table to make sure you understand the differences discussed earlier.

Number	Square Roots of Number	$\sqrt{\text{number}}$	$-\sqrt{\text{number}}$
25	$-5, 5$	$\sqrt{25} = 5$ only	$-\sqrt{25} = -5$
$\frac{1}{4}$	$-\frac{1}{2}, \frac{1}{2}$	$\sqrt{\frac{1}{4}} = \frac{1}{2}$ only	$-\sqrt{\frac{1}{4}} = -\frac{1}{2}$
-9	No real square roots.	$\sqrt{-9}$ is not a real number.	

Objective B Finding Cube Roots

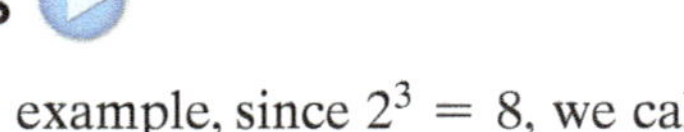

We can find roots other than square roots. For example, since $2^3 = 8$, we call 2 the **cube root** of 8. In symbols, we write

$\sqrt[3]{8} = 2$ The number 3 is called the **index.**

Also,

$\sqrt[3]{-64} = -4$ Since $(-4)^3 = -64$

Notice that unlike the square root of a negative number, the cube root of a negative number is a real number. This is so because while we cannot find a real number whose *square* is negative, we *can* find a real number whose *cube* is negative. In fact, the cube of a negative number is a negative number. Therefore, the cube root of a negative number is a negative number.

Practice 6–8

Find each cube root.

6. $\sqrt[3]{27}$
7. $\sqrt[3]{-8}$
8. $\sqrt[3]{\frac{1}{64}}$

Examples Find each cube root.

6. $\sqrt[3]{1} = 1$ because $1^3 = 1$.
7. $\sqrt[3]{-27} = -3$ because $(-3)^3 = -27$.
8. $\sqrt[3]{\frac{1}{125}} = \frac{1}{5}$ because $\left(\frac{1}{5}\right)^3 = \frac{1}{125}$.

Work Practice 6–8

Objective C Finding *n*th Roots

Just as we can raise a real number to powers other than 2 or 3, we can find roots other than square roots and cube roots. In fact, we can take the nth root of a number where n is any natural number. An ***n*th root** of a number a is a number whose nth power is a.

In symbols, the nth root of a is written as $\sqrt[n]{a}$. Recall that n is called the **index.** The index 2 is usually omitted for square roots.

Helpful Hint

If the index is even, as it is in $\sqrt{}$, $\sqrt[4]{}$, $\sqrt[6]{}$, and so on, the radicand must be nonnegative for the root to be a real number. For example,

$\sqrt[4]{81} = 3$ but $\sqrt[4]{-81}$ is not a real number.
$\sqrt[6]{64} = 2$ but $\sqrt[6]{-64}$ is not a real number.

Answers

6. 3 7. −2 8. $\frac{1}{4}$

✓**Concept Check** Which of the following is a real number?

a. $\sqrt{-64}$ **b.** $\sqrt[4]{-64}$ **c.** $\sqrt[5]{-64}$ **d.** $\sqrt[6]{-64}$

Examples Find each root.

9. $\sqrt[4]{16} = 2$ because $2^4 = 16$ and 2 is positive.

10. $\sqrt[5]{-32} = -2$ because $(-2)^5 = -32$.

11. $-\sqrt[6]{1} = -1$ because $\sqrt[6]{1} = 1$.

12. $\sqrt[4]{-81}$ is not a real number since the index, 4, is even and the radicand, -81, is negative. In other words, there is no real number that when raised to the 4th power gives -81.

Work Practice 9–12

Practice 9–12

Find each root.

9. $\sqrt[4]{-16}$

10. $\sqrt[5]{-1}$

11. $\sqrt[4]{256}$

12. $\sqrt[6]{-1}$

Objective D Approximating Square Roots

Recall that numbers such as 1, 4, 9, 25, and $\frac{4}{25}$ are called **perfect squares,** since $1^2 = 1$, $2^2 = 4$, $3^2 = 9$, $5^2 = 25$, and $\left(\frac{2}{5}\right)^2 = \frac{4}{25}$. Square roots of perfect square radicands simplify to rational numbers.

What happens when we try to simplify a root such as $\sqrt{3}$? Since 3 is not a perfect square, $\sqrt{3}$ is not a rational number. It cannot be written as a quotient of integers. It is called an **irrational number** and we can find a decimal **approximation** of it. To find decimal approximations, use a calculator or Appendix A.3. (For calculator help, see the next example or the box at the end of this section.)

Example 13 Use a calculator or Appendix A.3 to approximate $\sqrt{3}$ to three decimal places.

Solution: We may use Appendix A.3 or a calculator to approximate $\sqrt{3}$. To use a calculator, find the square root key $\boxed{\sqrt{}}$.

$$\sqrt{3} \approx 1.732050808$$

To three decimal places, $\sqrt{3} \approx 1.732$.

Work Practice 13

Practice 13

Use a calculator or Appendix A.3 to approximate $\sqrt{22}$ to three decimal places.

From Example 13, we found that

$$\sqrt{3} \approx 1.732$$

To see if the approximation is reasonable, notice that since

$1 < 3 < 4$, then

$\sqrt{1} < \sqrt{3} < \sqrt{4}$, or

$1 < \sqrt{3} < 2$.

Since $\sqrt{3}$ is a number between 1 and 2, our result of $\sqrt{3} \approx 1.732$ is reasonable.

Objective E Simplifying Radicals Containing Variables

Radicals can also contain variables. To simplify radicals containing variables, special care must be taken. To see how we simplify $\sqrt{x^2}$, let's look at a few examples in this form.

If $x = 3$, we have $\sqrt{3^2} = \sqrt{9} = 3$, or x.

If x is 5, we have $\sqrt{5^2} = \sqrt{25} = 5$, or x.

From these two examples, you may think that $\sqrt{x^2}$ simplifies to x. Let's now look at an example where x is a negative number. If $x = -3$, we have $\sqrt{(-3)^2} = \sqrt{9} = 3$, not -3, our original x. To make sure that $\sqrt{x^2}$ simplifies to a nonnegative number, we have the following.

Answers

9. not a real number **10.** -1 **11.** 4 **12.** not a real number **13.** 4.690

✓**Concept Check Answer**

c

For any real number a,

$$\sqrt{a^2} = |a|$$

Thus,

$$\sqrt{x^2} = |x|,$$
$$\sqrt{(-8)^2} = |-8| = 8$$
$$\sqrt{(7y)^2} = |7y|, \quad \text{and so on.}$$

To avoid this confusion, for the rest of the chapter we assume that **if a variable appears in the radicand of a radical expression, it represents positive numbers only.** Then

$\sqrt{x^2} = |x| = x$ since x is a positive number.

$\sqrt{y^2} = y$ — Because $(y)^2 = y^2$

$\sqrt{x^8} = x^4$ — Because $(x^4)^2 = x^8$

$\sqrt{9x^2} = 3x$ — Because $(3x)^2 = 9x^2$

Practice 14–19

Simplify each expression. Assume that all variables represent positive numbers.

14. $\sqrt{z^8}$
15. $\sqrt{x^{20}}$
16. $\sqrt{4x^6}$
17. $\sqrt[3]{8y^{12}}$
18. $\sqrt{\frac{z^8}{81}}$
19. $\sqrt[3]{-64x^9y^{24}}$

Answers

14. z^4 15. x^{10} 16. $2x^3$ 17. $2y^4$ 18. $\frac{z^4}{9}$ 19. $-4x^3y^8$

Examples

Simplify each expression. Assume that all variables represent positive numbers.

14. $\sqrt{z^2} = z$ because $(z)^2 = z^2$.
15. $\sqrt{x^6} = x^3$ because $(x^3)^2 = x^6$.
16. $\sqrt[3]{27y^6} = 3y^2$ because $(3y^2)^3 = 27y^6$.
17. $\sqrt{16x^{16}} = 4x^8$ because $(4x^8)^2 = 16x^{16}$.
18. $\sqrt{\frac{x^4}{25}} = \frac{x^2}{5}$ because $\left(\frac{x^2}{5}\right)^2 = \frac{x^4}{25}$.
19. $\sqrt[3]{-125a^{12}b^{15}} = -5a^4b^5$ because $(-5a^4b^5)^3 = -125a^{12}b^{15}$.

Work Practice 14–19

Calculator Explorations Simplifying Square Roots

To simplify or approximate square roots using a calculator, locate the key marked $\boxed{\sqrt{\ }}$. To simplify $\sqrt{25}$ using a scientific calculator, press $\boxed{25}$ $\boxed{\sqrt{\ }}$. The display should read $\boxed{5}$. To simplify $\sqrt{25}$ using a graphing calculator, press $\boxed{\sqrt{\ }}$ $\boxed{25}$ $\boxed{\text{ENTER}}$.

To approximate $\sqrt{30}$, press $\boxed{30}$ $\boxed{\sqrt{\ }}$ (or $\boxed{\sqrt{\ }}$ $\boxed{30}$). The display should read $\boxed{5.477225575}$. This is an approximation for $\sqrt{30}$. A three-decimal-place approximation is

$$\sqrt{30} \approx 5.477$$

Is this answer reasonable? Since 30 is between perfect squares 25 and 36, $\sqrt{30}$ is between $\sqrt{25} = 5$ and $\sqrt{36} = 6$. The calculator result is then reasonable since 5.477225575 is between 5 and 6.

Use a calculator to approximate each expression to three decimal places. Decide whether each result is reasonable.

1. $\sqrt{6}$
2. $\sqrt{14}$
3. $\sqrt{11}$
4. $\sqrt{200}$
5. $\sqrt{82}$
6. $\sqrt{46}$

Many scientific calculators have a key, such as $\boxed{\sqrt[x]{y}}$, that can be used to approximate roots other than square roots. To approximate these roots using a graphing calculator, look under the $\boxed{\text{MATH}}$ menu or consult your manual. To use a $\boxed{\sqrt[x]{y}}$ key to find $\sqrt[3]{8}$, press $\boxed{3}$ $\boxed{\sqrt[x]{y}}$ $\boxed{8}$ (press $\boxed{\text{ENTER}}$ if needed). The display should read $\boxed{2}$.

Use a calculator to approximate each expression to three decimal places. Decide whether each result is reasonable.

7. $\sqrt[3]{40}$
8. $\sqrt[3]{71}$
9. $\sqrt[4]{20}$
10. $\sqrt[4]{15}$
11. $\sqrt[5]{18}$
12. $\sqrt[6]{2}$

Vocabulary, Readiness & Video Check

Use the choices below to fill in each blank.

positive	index	radical sign	power
negative	principal	square root	radicand

1. The symbol $\sqrt{\ }$ is used to denote the positive, or ____________, square root.
2. In the expression $\sqrt[4]{16}$, the number 4 is called the ____________, the number 16 is called the ____________, and $\sqrt{\ }$ is called the ____________.
3. The reverse operation of squaring a number is finding a(n) ____________ of a number.
4. For a positive number a,
   $-\sqrt{a}$ is the ____________ square root of a and
   $\sqrt{a}$ is the ____________ square root of a.
5. An nth root of a number a is a number whose nth ____________ is a.

Answer each true or false.

6. $\sqrt{4} = -2$ ____________
7. $\sqrt{-9} = -3$ ____________
8. $\sqrt{1000} = 100$ ____________
9. $\sqrt{1} = 1$ and $\sqrt{0} = 0$ ____________
10. $\sqrt{64} = 8$ and $\sqrt[3]{64} = 4$ ____________

Martin-Gay Interactive Videos

See Video 8.1

Watch the section lecture video and answer the following questions.

Objective A 11. Explain the differences between Examples 1 and 2, including how we know which one simplifies to a positive number and which one simplifies to a negative number.

Objective B 12. From Example 11, what is an important difference between the square root and the cube root of a negative number?

Objective C 13. From Examples 12–15, given a negative radicand, what kind of index must we have to be a real number?

Objective D 14. From Example 16, how do we determine if an approximate answer is reasonable?

Objective E 15. As explained in Example 19, when simplifying radicals containing variables, what is a shortcut we can use when dealing with exponents?

8.1 Exercise Set MyLab Math

Objective A *Find each square root. See Examples 1 through 5.*

1. $\sqrt{16}$
2. $\sqrt{64}$
3. $\sqrt{\frac{1}{25}}$
4. $\sqrt{\frac{1}{64}}$
5. $-\sqrt{100}$
6. $-\sqrt{36}$
7. $\sqrt{-4}$
8. $\sqrt{-25}$
9. $-\sqrt{121}$
10. $-\sqrt{49}$

11. $\sqrt{\frac{9}{25}}$ 12. $\sqrt{\frac{4}{81}}$ 13. $\sqrt{900}$ 14. $\sqrt{400}$ 15. $\sqrt{144}$

16. $\sqrt{169}$ 17. $\sqrt{\frac{1}{100}}$ 18. $\sqrt{\frac{1}{121}}$ 19. $\sqrt{0.25}$ 20. $\sqrt{0.49}$

Objective B *Find each cube root. See Examples 6 through 8.*

21. $\sqrt[3]{125}$ 22. $\sqrt[3]{64}$ 23. $\sqrt[3]{-64}$ 24. $\sqrt[3]{-27}$ 25. $-\sqrt[3]{8}$

26. $-\sqrt[3]{27}$ 27. $\sqrt[3]{\frac{1}{8}}$ 28. $\sqrt[3]{\frac{1}{64}}$ 29. $\sqrt[3]{-125}$ 30. $\sqrt[3]{-1}$

Objectives A B C Mixed Practice *Find each root. See Examples 1 through 12.*

31. $\sqrt[5]{32}$ 32. $\sqrt[4]{81}$ 33. $\sqrt{81}$ 34. $\sqrt{49}$

35. $\sqrt[4]{-16}$ 36. $\sqrt{-9}$ 37. $\sqrt[3]{-\frac{27}{64}}$ 38. $\sqrt[3]{-\frac{8}{27}}$

39. $-\sqrt[4]{625}$ 40. $-\sqrt[5]{32}$ 41. $\sqrt[6]{1}$ 42. $\sqrt[5]{1}$

Objective D *Approximate each square root to three decimal places. See Example 13.*

43. $\sqrt{7}$ 44. $\sqrt{10}$ 45. $\sqrt{37}$ 46. $\sqrt{27}$ 47. $\sqrt{136}$ 48. $\sqrt{8}$

49. A standard baseball diamond is a square with 90-foot sides connecting the bases. The distance from home plate to second base is $90 \cdot \sqrt{2}$ feet. Approximate $\sqrt{2}$ to two decimal places and use your result to approximate the distance $90 \cdot \sqrt{2}$ feet.

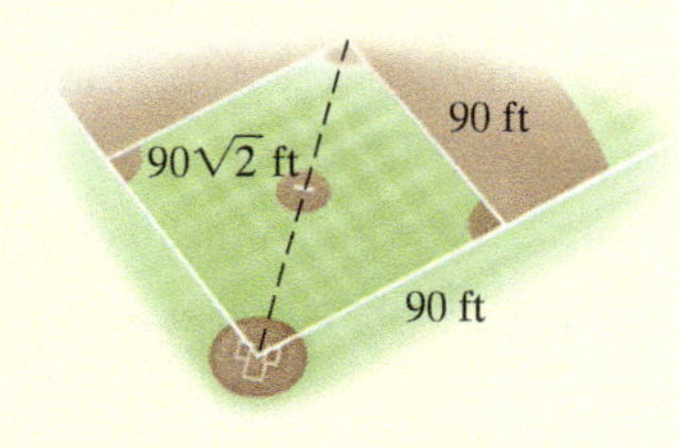

50. The roof of the warehouse shown needs to be shingled. The total area of the roof is exactly $480 \cdot \sqrt{29}$ square feet. Approximate $\sqrt{29}$ to two decimal places and use your result to approximate the area $480 \cdot \sqrt{29}$ square feet. Approximate this area to the nearest whole number.

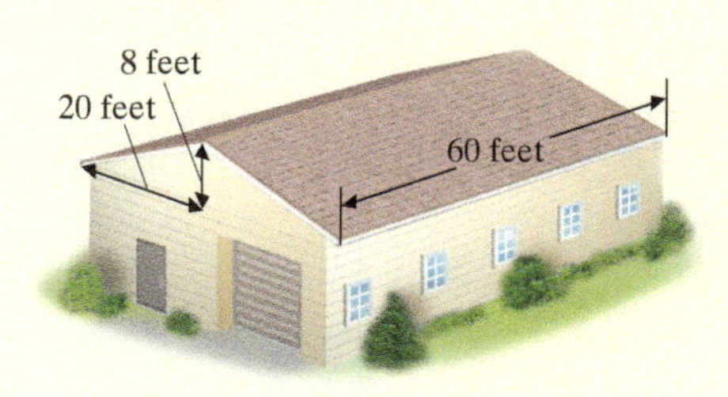

Objective E *Find each root. Assume that all variables represent positive numbers. See Examples 14 through 19.*

51. $\sqrt{m^2}$ 52. $\sqrt{y^{10}}$ 53. $\sqrt{x^4}$ 54. $\sqrt{x^6}$

55. $\sqrt{9x^8}$ 56. $\sqrt{36x^{12}}$ 57. $\sqrt{81x^2}$ 58. $\sqrt{100z^4}$

59. $\sqrt{a^2b^4}$ 60. $\sqrt{x^{12}y^{20}}$ 61. $\sqrt{16a^6b^4}$ 62. $\sqrt{4m^{14}n^2}$

63. $\sqrt[3]{a^6b^{18}}$ **64.** $\sqrt[3]{x^{12}y^{18}}$ **65.** $\sqrt[3]{-8x^3y^{27}}$ **66.** $\sqrt[3]{-27a^6b^{30}}$

67. $\sqrt{\frac{x^6}{36}}$ **68.** $\sqrt{\frac{y^8}{49}}$ **69.** $\sqrt{\frac{25y^2}{9}}$ **70.** $\sqrt{\frac{4x^2}{81}}$

Review

Write each integer as a product of two integers such that one of the factors is a perfect square. For example, we can write $18 = 9 \cdot 2$, where 9 is a perfect square. See Section R.1.

71. 50 **72.** 8 **73.** 32 **74.** 75

75. 28 **76.** 44 **77.** 27 **78.** 90

Concept Extensions

Solve. See the Concept Check in this section.

79. Which of the following is a real number?
- **a.** $\sqrt[7]{-1}$
- **b.** $\sqrt[3]{-125}$
- **c.** $\sqrt[6]{-128}$
- **d.** $\sqrt[8]{-1}$

80. Which of the following is a real number?
- **a.** $\sqrt{-1}$
- **b.** $\sqrt[3]{-1}$
- **c.** $\sqrt[4]{-1}$
- **d.** $\sqrt[5]{-1}$

The length of a side of a square is given by the expression $\sqrt{A}$, where A is the square's area. Use this expression for Exercises 81 through 84. Be sure to attach the appropriate units.

△ **81.** The area of a square is 49 square miles. Find the length of a side of the square.

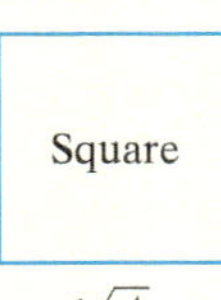

△ **82.** The area of a square is $\frac{1}{81}$ square meters. Find the length of a side of the square.

△ **83.** The world's smallest Rubik's Cube was built by Tony Fisher of the United Kingdom. The area of one square face of this fully functional Rubik's Cube is 31.36 square millimeters. Find the length of a side of the square face. (*Source: Guinness World Records*)

△ **84.** A parking lot is in the shape of a square with area 2500 square yards. Find the length of a side.

85. Simplify $\sqrt{\sqrt{81}}$.

86. Simplify $\sqrt[3]{\sqrt[3]{1}}$.

87. Simplify $\sqrt{\sqrt{10{,}000}}$.

88. Simplify $\sqrt{\sqrt{1{,}600{,}000{,}000}}$.

For each square root below, give two whole numbers that the square root lies between. For example, since 11 is between 9 and 16, then $\sqrt{11}$ is between $\sqrt{9}$ and $\sqrt{16}$ or between 3 and 4.

89. $\sqrt{18}$ **90.** $\sqrt{28}$ **91.** $\sqrt{80}$ **92.** $\sqrt{98}$

93. The formula for calculating the period (one back-and-forth swing) of a pendulum is $T = 2\pi\sqrt{\frac{L}{g}}$, where T is time of the period of the swing, L is the length of the pendulum, and g is the acceleration of gravity. At the California Academy of Sciences, one can see a Foucault's pendulum with length $= 30$ ft and $g = 32\text{ ft/sec}^2$. Using $\pi \approx 3.14$, find the period of this pendulum. (Round to the nearest tenth of a second.)

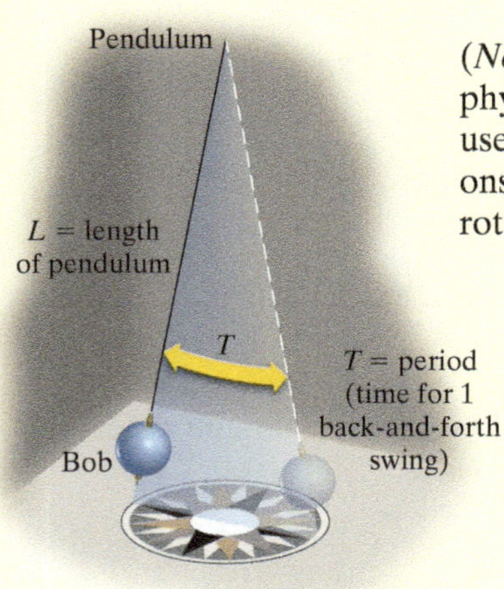

(*Note:* In 1851, French physicist Léon Foucault used a pendulum to demonstrate that the earth rotates on its axis.)

94. If the amount of gold discovered by humankind could be assembled in one place, it would be a cube with a volume of 195,112 cubic feet. Each side of the cube would be $\sqrt[3]{195{,}112}$ feet long. How long would one side of the cube be? (*Source: Reader's Digest*)

95. Explain why the square root of a negative number is not a real number.

96. Explain why the cube root of a negative number is a real number.

97. Graph $y = \sqrt{x}$. (Complete the table below, plot the ordered pair solutions, and draw a smooth curve through the points. Remember that since the radicand cannot be negative, this particular graph begins at the point with coordinates $(0, 0)$.)

x	y	
0	0	
1		
3		(approximate)
4		
9		

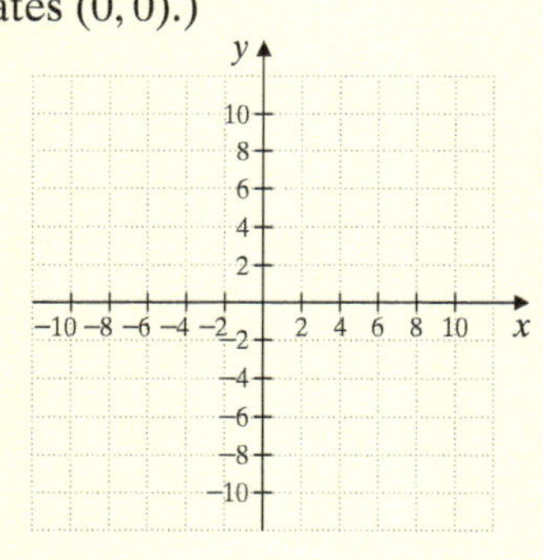

98. Graph $y = \sqrt[3]{x}$. (Complete the table below, plot the ordered pair solutions, and draw a smooth curve through the points.)

x	y	
−8		
−2		(approximate)
−1		
0		
1		
2		(approximate)
8		

Recall from this section that $\sqrt{a^2} = |a|$ for any real number a. Simplify the following given that x represents any real number.

99. $\sqrt{x^2}$

100. $\sqrt{4x^2}$

101. $\sqrt{(x + 2)^2}$

102. $\sqrt{x^2 + 6x + 9}$
(*Hint:* First factor $x^2 + 6x + 9$.)

Use a graphing calculator and graph each function. Observe the graph from left to right and give the ordered pair that corresponds to the "beginning" of the graph. Then tell why the graph starts at that point.

103. $y = \sqrt{x - 2}$

104. $y = \sqrt{x + 3}$

105. $y = \sqrt{x + 4}$

106. $y = \sqrt{x - 5}$

8.2 Simplifying Radicals

Objective A Simplifying Radicals Using the Product Rule

Objectives

A Use the Product Rule to Simplify Radicals.

B Use the Quotient Rule to Simplify Radicals.

C Use Both Rules to Simplify Radicals Containing Variables.

D Simplify Cube Roots.

A square root is simplified when the radicand contains no perfect square factors (other than 1). For example, $\sqrt{20}$ is not simplified because $\sqrt{20} = \sqrt{4 \cdot 5}$ and 4 is a perfect square.

To begin simplifying square roots, we notice the following pattern.

$$\sqrt{9 \cdot 16} = \sqrt{144} = 12$$
$$\sqrt{9} \cdot \sqrt{16} = 3 \cdot 4 = 12$$

Since both expressions simplify to 12, we can write

$$\sqrt{9 \cdot 16} = \sqrt{9} \cdot \sqrt{16}$$

This suggests the following product rule for square roots.

Product Rule for Square Roots

If $\sqrt{a}$ and $\sqrt{b}$ are real numbers, then

$$\sqrt{a \cdot b} = \sqrt{a} \cdot \sqrt{b}$$

In other words, the square root of a product is equal to the product of the square roots.

To simplify $\sqrt{45}$, for example, we factor 45 so that one of its factors is a perfect square factor.

$\sqrt{45} = \sqrt{9 \cdot 5}$ Factor 45.

$= \sqrt{9} \cdot \sqrt{5}$ Use the product rule.

$= 3\sqrt{5}$ Write $\sqrt{9}$ as 3.

The notation $3\sqrt{5}$ means $3 \cdot \sqrt{5}$. Since the radicand 5 has no perfect square factor other than 1, the expression $3\sqrt{5}$ is in simplest form.

Helpful Hint Remember that the notation

$3\sqrt{5}$ means $3 \cdot \sqrt{5}$.

Helpful Hint

A radical expression in simplest form *does not mean* a decimal approximation. The simplest form of a radical expression is an exact form and may still contain a radical.

$\sqrt{45} = 3\sqrt{5}$ (exact) $\qquad \sqrt{45} \approx 6.71$ (decimal approximation)

Examples Simplify.

1. $\sqrt{54} = \sqrt{9 \cdot 6}$ Factor 54 so that one factor is a perfect square. 9 is a perfect square.

$= \sqrt{9} \cdot \sqrt{6}$ Use the product rule.

$= 3\sqrt{6}$ Write $\sqrt{9}$ as 3.

2. $\sqrt{12} = \sqrt{4 \cdot 3}$ Factor 12 so that one factor is a perfect square. 4 is a perfect square.

$= \sqrt{4} \cdot \sqrt{3}$ Use the product rule.

$= 2\sqrt{3}$ Write $\sqrt{4}$ as 2.

(Continued on next page)

Practice 1–4

Simplify.

1. $\sqrt{40}$
2. $\sqrt{18}$
3. $\sqrt{500}$
4. $\sqrt{15}$

Answers

1. $2\sqrt{10}$ 2. $3\sqrt{2}$ 3. $10\sqrt{5}$ 4. $\sqrt{15}$

3. $\sqrt{200} = \sqrt{100 \cdot 2}$ — Factor 200 so that one factor is a perfect square. 100 is a perfect square.
$= \sqrt{100} \cdot \sqrt{2}$ — Use the product rule.
$= 10\sqrt{2}$ — Write $\sqrt{100}$ as 10.

4. $\sqrt{35}$ The radicand 35 contains no perfect square factors other than 1. Thus $\sqrt{35}$ is in simplest form.

Work Practice 1–4

In Example 3, 100 is the largest perfect square factor of 200. What happens if we don't use the largest perfect square factor? Although using the largest perfect square factor saves time, the result is the same no matter what perfect square factor is used. For example, it is also true that $200 = 4 \cdot 50$. Then

$$\begin{aligned}\sqrt{200} &= \sqrt{4} \cdot \sqrt{50}\\ &= 2 \cdot \sqrt{50}\end{aligned}$$

Since $\sqrt{50}$ is not in simplest form, we continue.

$$\begin{aligned}\sqrt{200} &= 2 \cdot \sqrt{50}\\ &= 2 \cdot \sqrt{25 \cdot 2}\\ &= 2 \cdot \sqrt{25} \cdot \sqrt{2}\\ &= 2 \cdot 5 \cdot \sqrt{2}\\ &= 10\sqrt{2}\end{aligned}$$

Practice 5

Simplify $7\sqrt{75}$.

Example 5 Simplify $3\sqrt{8}$.

Solution: Remember that $3\sqrt{8}$ means $3 \cdot \sqrt{8}$.

$3 \cdot \sqrt{8} = 3 \cdot \sqrt{4 \cdot 2}$ — Factor 8 so that one factor is a perfect square.
$= 3 \cdot \sqrt{4} \cdot \sqrt{2}$ — Use the product rule.
$= 3 \cdot 2 \cdot \sqrt{2}$ — Write $\sqrt{4}$ as 2.
$= 6 \cdot \sqrt{2}$ or $6\sqrt{2}$ — Write $3 \cdot 2$ as 6.

Work Practice 5

Objective B Simplifying Radicals Using the Quotient Rule

Next, let's examine the square root of a quotient.

$$\sqrt{\frac{16}{4}} = \sqrt{4} = 2$$

Also,

$$\frac{\sqrt{16}}{\sqrt{4}} = \frac{4}{2} = 2$$

Since both expressions equal 2, we can write

$$\sqrt{\frac{16}{4}} = \frac{\sqrt{16}}{\sqrt{4}}$$

This suggests the following quotient rule.

Answer

5. $35\sqrt{3}$

Quotient Rule for Square Roots

If $\sqrt{a}$ and $\sqrt{b}$ are real numbers and $b \neq 0$, then

$$\sqrt{\frac{a}{b}} = \frac{\sqrt{a}}{\sqrt{b}}$$

In other words, the square root of a quotient is equal to the quotient of the square roots.

Examples Use the quotient rule to simplify.

6. $\sqrt{\frac{25}{36}} = \frac{\sqrt{25}}{\sqrt{36}} = \frac{5}{6}$

7. $\sqrt{\frac{3}{64}} = \frac{\sqrt{3}}{\sqrt{64}} = \frac{\sqrt{3}}{8}$

8. $\sqrt{\frac{40}{81}} = \frac{\sqrt{40}}{\sqrt{81}}$ Use the quotient rule.

$= \frac{\sqrt{4} \cdot \sqrt{10}}{9}$ Use the product rule and write $\sqrt{81}$ as 9.

$= \frac{2\sqrt{10}}{9}$ Write $\sqrt{4}$ as 2.

Work Practice 6–8

Practice 6–8

Use the quotient rule to simplify.

6. $\sqrt{\frac{16}{81}}$

7. $\sqrt{\frac{2}{25}}$

8. $\sqrt{\frac{45}{49}}$

Objective C Simplifying Radicals Containing Variables 

Recall that $\sqrt{x^6} = x^3$ because $(x^3)^2 = x^6$. If a variable radicand in a square root has an odd exponent, we write the exponential expression so that one factor is the greatest even power contained in the expression. Then we use the product rule to simplify.

Examples Simplify each radical. Assume that all variables represent positive numbers.

9. $\sqrt{x^5} = \sqrt{x^4 \cdot x} = \sqrt{x^4} \cdot \sqrt{x} = x^2\sqrt{x}$

10. $\sqrt{8y^2} = \sqrt{4 \cdot 2 \cdot y^2} = \sqrt{4y^2 \cdot 2} = \sqrt{4y^2} \cdot \sqrt{2} = 2y\sqrt{2}$; 4 and y^2 are both perfect square factors so we grouped them under one radical.

11. $\sqrt{\frac{45}{x^6}} = \frac{\sqrt{45}}{\sqrt{x^6}} = \frac{\sqrt{9 \cdot 5}}{x^3} = \frac{\sqrt{9} \cdot \sqrt{5}}{x^3} = \frac{3\sqrt{5}}{x^3}$

12. $\sqrt{\frac{5p^3}{9}} = \frac{\sqrt{5p^3}}{\sqrt{9}} = \frac{\sqrt{p^2 \cdot 5p}}{3} = \frac{\sqrt{p^2} \cdot \sqrt{5p}}{3} = \frac{p\sqrt{5p}}{3}$

Work Practice 9–12

Practice 9–12

Simplify each radical. Assume that all variables represent positive numbers.

9. $\sqrt{x^{11}}$ 10. $\sqrt{18x^4}$

11. $\sqrt{\frac{27}{x^8}}$ 12. $\sqrt{\frac{7y^7}{25}}$

Answers

6. $\frac{4}{9}$ 7. $\frac{\sqrt{2}}{5}$ 8. $\frac{3\sqrt{5}}{7}$ 9. $x^5\sqrt{x}$ 10. $3x^2\sqrt{2}$ 11. $\frac{3\sqrt{3}}{x^4}$ 12. $\frac{y^3\sqrt{7y}}{5}$

Objective D Simplifying Cube Roots

The product and quotient rules also apply to roots other than square roots. For example, to simplify cube roots, we look for perfect cube factors of the radicand. Recall that 8 is a perfect cube since $2^3 = 8$. Therefore, to simplify $\sqrt[3]{80}$, we factor 80 as $8 \cdot 10$.

$$\sqrt[3]{80} = \sqrt[3]{8 \cdot 10} \quad \text{Factor 80.}$$
$$= \sqrt[3]{8} \cdot \sqrt[3]{10} \quad \text{Use the product rule.}$$
$$= 2\sqrt[3]{10} \quad \text{Write } \sqrt[3]{8} \text{ as 2.}$$

$2\sqrt[3]{10}$ is in simplest form since the radicand, 10, contains no perfect cube factors other than 1.

Practice 13-16

Simplify each radical.

13. $\sqrt[3]{88}$ **14.** $\sqrt[3]{50}$

15. $\sqrt[3]{\frac{10}{27}}$ **16.** $\sqrt[3]{\frac{81}{8}}$

Answers

13. $2\sqrt[3]{11}$ **14.** $\sqrt[3]{50}$ **15.** $\frac{\sqrt[3]{10}}{3}$ **16.** $\frac{3\sqrt[3]{3}}{2}$

Examples

Simplify each radical.

13. $\sqrt[3]{54} = \sqrt[3]{27 \cdot 2} = \sqrt[3]{27} \cdot \sqrt[3]{2} = 3\sqrt[3]{2}$

14. $\sqrt[3]{18}$ The number 18 contains no perfect cube factors, so $\sqrt[3]{18}$ cannot be simplified further.

15. $\sqrt[3]{\frac{7}{8}} = \frac{\sqrt[3]{7}}{\sqrt[3]{8}} = \frac{\sqrt[3]{7}}{2}$

16. $\sqrt[3]{\frac{40}{27}} = \frac{\sqrt[3]{40}}{\sqrt[3]{27}} = \frac{\sqrt[3]{8 \cdot 5}}{3} = \frac{\sqrt[3]{8} \cdot \sqrt[3]{5}}{3} = \frac{2\sqrt[3]{5}}{3}$

Work Practice 13–16

Vocabulary, Readiness & Video Check

Use the choices below to fill in each blank. Not all choices will be used.

$a \cdot b$ $\quad \frac{a}{b}$ $\quad \frac{\sqrt{a}}{\sqrt{b}}$ $\quad \sqrt{a} \cdot \sqrt{b}$

1. If $\sqrt{a}$ and $\sqrt{b}$ are real numbers, then $\sqrt{a \cdot b} =$ ________.

2. If $\sqrt{a}$ and $\sqrt{b}$ are real numbers, then $\sqrt{\frac{a}{b}} =$ ________.

For Exercises 3 and 4, fill in the blanks using the example: $\sqrt{4 \cdot 9} = \underline{\sqrt{4}} \cdot \underline{\sqrt{9}} = \underline{2} \cdot \underline{3} = \underline{6}$.

3. $\sqrt{16 \cdot 25} = \sqrt{\_\_} \cdot \sqrt{\_\_} = \_ \cdot \_ = \_\_$.

4. $\sqrt{36 \cdot 3} = \sqrt{\_\_} \cdot \sqrt{\_} = \_ \cdot \sqrt{\_} = \_\_\_$.

Yes or No? Decide whether each radical is completely simplified.

5. $\sqrt{48} = 2\sqrt{12}$. Is $2\sqrt{12}$ completely simplified? ______

6. $\sqrt[3]{48} = 2\sqrt[3]{6}$. Is $2\sqrt[3]{6}$ completely simplified? ______

Martin-Gay Interactive Videos

See Video 8.2

Watch the section lecture video and answer the following questions.

Objective A 7. From Example 3, if we have trouble finding a perfect square factor in the radicand, what is recommended?

Objective B 8. Based on the lecture before Example 5, complete the following statement: In words, the quotient rule for square roots says that the square root of a quotient is equal to the square root of the ____________ over the square root of the ____________.

Objective C 9. From Examples 6–8, we know that even powers of a variable are perfect square factors of the variable. Therefore, what must be true about the power of any variable left in the radicand of a simplified square root? Explain.

Objective D 10. From Example 9, how does factoring the radicand as a product of primes help simplify higher roots also?

8.2 Exercise Set MyLab Math 

Objective A *Use the product rule to simplify each radical. See Examples 1 through 4.*

1. $\sqrt{20}$
2. $\sqrt{44}$
3. $\sqrt{50}$
4. $\sqrt{28}$
5. $\sqrt{33}$
6. $\sqrt{21}$
7. $\sqrt{98}$
8. $\sqrt{125}$
9. $\sqrt{60}$
10. $\sqrt{90}$
11. $\sqrt{180}$
12. $\sqrt{150}$
13. $\sqrt{52}$
14. $\sqrt{75}$

Use the product rule to simplify each radical. See Example 5.

15. $3\sqrt{25}$
16. $9\sqrt{36}$
17. $7\sqrt{63}$
18. $11\sqrt{99}$
19. $-5\sqrt{27}$
20. $-6\sqrt{75}$

Objective B *Use the quotient rule and the product rule to simplify each radical. See Examples 6 through 8.*

21. $\sqrt{\frac{8}{25}}$
22. $\sqrt{\frac{63}{16}}$
23. $\sqrt{\frac{27}{121}}$
24. $\sqrt{\frac{24}{169}}$
25. $\sqrt{\frac{9}{4}}$
26. $\sqrt{\frac{100}{49}}$
27. $\sqrt{\frac{125}{9}}$
28. $\sqrt{\frac{27}{100}}$
29. $\sqrt{\frac{11}{36}}$
30. $\sqrt{\frac{30}{49}}$
31. $-\sqrt{\frac{27}{64}}$
32. $-\sqrt{\frac{84}{121}}$

Objective C *Simplify each radical. Assume that all variables represent positive numbers. See Examples 9 through 12.*

33. $\sqrt{x^7}$ **34.** $\sqrt{y^3}$ **35.** $\sqrt{x^{13}}$ **36.** $\sqrt{y^{17}}$

37. $\sqrt{36a^3}$ **38.** $\sqrt{81b^5}$ **39.** $\sqrt{96x^4}$ **40.** $\sqrt{40y^{10}}$

41. $\sqrt{\frac{12}{m^2}}$ **42.** $\sqrt{\frac{63}{p^2}}$ **43.** $\sqrt{\frac{9x}{y^{10}}}$ **44.** $\sqrt{\frac{6y^2}{z^{16}}}$

45. $\sqrt{\frac{88}{x^{12}}}$ **46.** $\sqrt{\frac{500}{y^{22}}}$

Objectives A B C Mixed Practice *Simplify each radical. See Examples 1 through 12.*

47. $8\sqrt{4}$ **48.** $6\sqrt{49}$ **49.** $\sqrt{\frac{36}{121}}$ **50.** $\sqrt{\frac{25}{144}}$

51. $\sqrt{175}$ **52.** $\sqrt{700}$ **53.** $\sqrt{\frac{20}{9}}$ **54.** $\sqrt{\frac{45}{64}}$

55. $\sqrt{24m^7}$ **56.** $\sqrt{50n^{13}}$ **57.** $\sqrt{\frac{23y^3}{4x^6}}$ **58.** $\sqrt{\frac{41x^5}{9y^8}}$

Objective D *Simplify each radical. See Examples 13 through 16.*

59. $\sqrt[3]{24}$ **60.** $\sqrt[3]{81}$ **61.** $\sqrt[3]{250}$ **62.** $\sqrt[3]{56}$

63. $\sqrt[3]{\frac{5}{64}}$ **64.** $\sqrt[3]{\frac{32}{125}}$ **65.** $\sqrt[3]{\frac{23}{8}}$ **66.** $\sqrt[3]{\frac{37}{27}}$

67. $\sqrt[3]{\frac{15}{64}}$ **68.** $\sqrt[3]{\frac{4}{27}}$ **69.** $\sqrt[3]{270}$ **70.** $\sqrt[3]{108}$

Review

Perform each indicated operation. See Sections 3.4 and 3.5.

71. $6x + 8x$ **72.** $(6x)(8x)$ **73.** $(2x + 3)(x - 5)$

74. $(2x + 3) + (x - 5)$ **75.** $9y^2 - 9y^2$ **76.** $(9y^2)(-8y^2)$

Concept Extensions

Simplify each radical. Assume that all variables represent positive numbers.

77. $\sqrt{x^6y^3}$ **78.** $\sqrt{a^{13}b^{14}}$ **79.** $\sqrt{98x^5y^4}$

80. $\sqrt{27x^8y^{11}}$ **81.** $\sqrt[3]{-8x^6}$ **82.** $\sqrt[3]{27x^{12}}$

83. If a cube is to have a volume of 80 cubic inches, then each side must be $\sqrt[3]{80}$ inches long. Simplify the radical representing the side length.

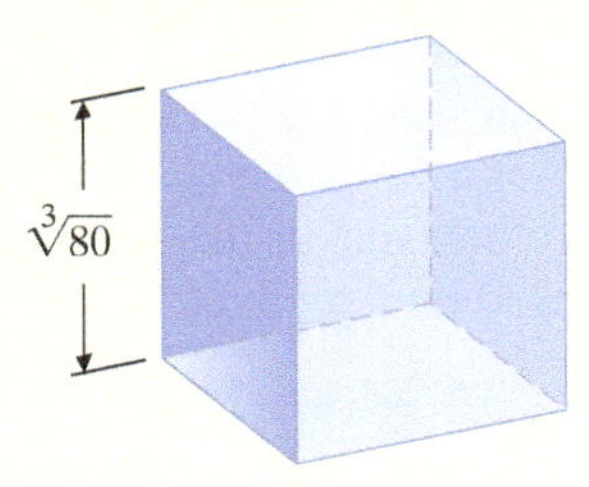

84. Jeannie Boswell is swimming across a 40-foot-wide river, trying to head straight across to the opposite shore. However, the current is strong enough to move her downstream 100 feet by the time she reaches land. (See the figure.) Because of the current, the actual distance she swims is $\sqrt{11{,}600}$ feet. Simplify this radical.

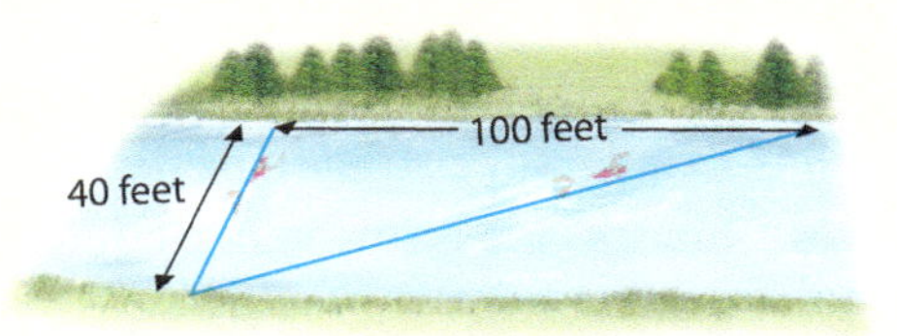

85. By using replacement values for a and b, show that $\sqrt{a^2 + b^2}$ does not equal $a + b$.

86. By using replacement values for a and b, show that $\sqrt{a + b}$ does not equal $\sqrt{a} + \sqrt{b}$.

87. The "Water Cube" was the swimming and diving venue for the 2008 Beijing Summer Olympics. It is not actually a cube, because it is only 31 meters tall, which is not the same as its width and length. However, the roof of it is a square. If the area of the roof of the Water Cube is 31,329 square meters, find the dimensions of the roof of the Water Cube.

△ **88.** The competition diving pool in the Water Cube at the Beijing Summer Olympics is not a cube either. It has a square footprint, but is only 5 meters deep. If the volume of the diving pool is 3125 cubic meters, find the length and width of the competition diving pool.

The length of a side of a cube is given by the expression $\frac{\sqrt{6A}}{6}$, *where A is the cube's surface area. Use this expression for Exercises 89 through 92. Be sure to attach the appropriate units.*

△ **89.** The surface area of a cube is 120 square inches. Find the exact length of a side of the cube.

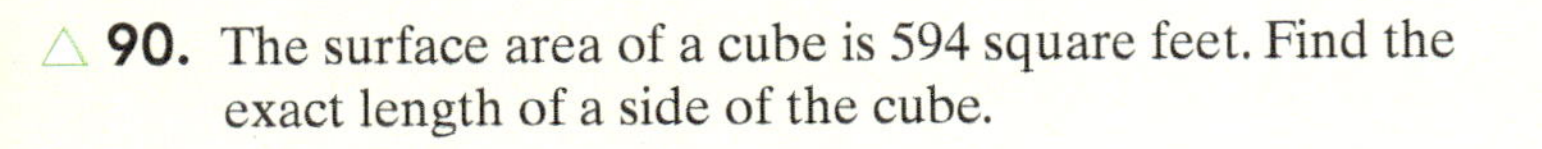

△ **90.** The surface area of a cube is 594 square feet. Find the exact length of a side of the cube.

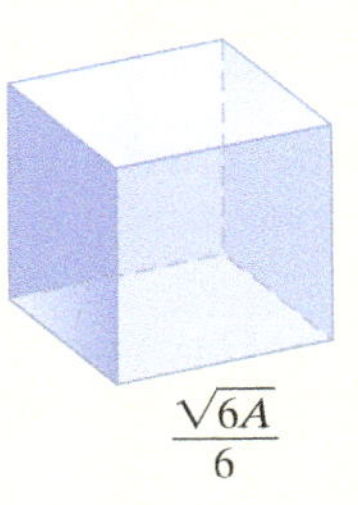

Solve.

91. Rubik's Cube, named after its inventor, Erno Rubik, was first imagined by him in 1974, and by 1980 was a worldwide phenomenon. A standard Rubik's Cube has a surface area of 30.375 square inches. Find the length of one side of a Rubik's Cube. (A few world records are listed below. *Source: Guinness World Records*)

- Fastest time to solve 1 Rubik's Cube: 4.73 sec by Feliks Zemdegs (Australia) in 2016.
- Most Rubik's Cubes solved in 1 hour: 65 by Aashik Madhav G.S. (India) in 2017.

92. Apple renovated its flagship Apple Store on Fifth Avenue in New York City, taking advantage of advances in glass manufacturing to simplify the giant glass cube that serves as the store's entrance. A cube of this size has a surface area of 6144 square feet. Find the length of one side of the Apple Store glass cube. (*Source:* Based on data from AppleInsider.com)

The cost C in dollars per day to operate a small delivery service is given by $C = 100\sqrt[3]{n} + 700$, *where n is the number of deliveries per day.*

93. Find the cost if the number of deliveries is 1000.

94. Approximate the cost if the number of deliveries is 500.

The Mosteller formula for calculating body surface area is $B = \sqrt{\frac{hw}{3600}}$, *where B is an individual's body surface area in square meters, h is the individual's height in centimeters, and w is the individual's weight in kilograms. Use this formula in Exercises 95 and 96. Round answers to the nearest tenth.*

95. Find the body surface area of a person who is 169 cm tall and weighs 64 kilograms.

96. Approximate the body surface area of a person who is 183 cm tall and weighs 85 kilograms.

8.3 Adding and Subtracting Radicals 

Objective A Adding and Subtracting Radicals

Recall that to combine like terms, we use the distributive property.

$$5x + 3x = (5 + 3)x = 8x$$

The distributive property can also be applied to expressions containing the same radicals. For example,

$$5\sqrt{2} + 3\sqrt{2} = (5 + 3)\sqrt{2} = 8\sqrt{2}$$

Also,

$$9\sqrt{5} - 6\sqrt{5} = (9 - 6)\sqrt{5} = 3\sqrt{5}$$

Radical terms such as $5\sqrt{2}$ and $3\sqrt{2}$ are **like radicals,** as are $9\sqrt{5}$ and $6\sqrt{5}$. Like radicals have the same index and the same radicand.

Examples Add or subtract as indicated.

1. $4\sqrt{5} + 3\sqrt{5} = (4 + 3)\sqrt{5} = 7\sqrt{5}$
2. $\sqrt{10} - 6\sqrt{10} = 1\sqrt{10} - 6\sqrt{10} = (1 - 6)\sqrt{10} = -5\sqrt{10}$
3. $2\sqrt{6} + 2\sqrt{5}$ cannot be simplified further since the radicands are not the same.
4. $\sqrt{15} + \sqrt{15} - \sqrt{2} = 1\sqrt{15} + 1\sqrt{15} - \sqrt{2}$
   $= (1 + 1)\sqrt{15} - \sqrt{2}$
   $= 2\sqrt{15} - \sqrt{2}$

   This expression cannot be simplified further since the radicands are not the same.

Work Practice 1–4

✓**Concept Check** Which is true?

a. $2 + 3\sqrt{5} = 5\sqrt{5}$

b. $2\sqrt{3} + 2\sqrt{7} = 2\sqrt{10}$

c. $\sqrt{3} + \sqrt{5} = \sqrt{8}$

d. $\sqrt{3} + \sqrt{3} = 3$

e. None of the above is true. In each case, the left-hand side cannot be simplified further.

Objective B Simplifying Square Root Radicals Before Adding or Subtracting

At first glance, it appears that the expression $\sqrt{50} + \sqrt{8}$ cannot be simplified further because the radicands are different. However, the product rule can be used to simplify each radical, and then further simplification might be possible.

Examples Simplify each radical expression.

5. $\sqrt{50} + \sqrt{8} = \sqrt{25 \cdot 2} + \sqrt{4 \cdot 2}$ Factor radicands.
   $= \sqrt{25} \cdot \sqrt{2} + \sqrt{4} \cdot \sqrt{2}$ Use the product rule.
   $= 5\sqrt{2} + 2\sqrt{2}$ Simplify $\sqrt{25}$ and $\sqrt{4}$.
   $= 7\sqrt{2}$ Add like radicals.

(Continued on next page)

Objectives

A Add or Subtract Like Radicals.

B Simplify Square Root Radical Expressions, and Then Add or Subtract Any Like Radicals.

C Simplify Cube Root Radical Expressions, and Then Add or Subtract Any Like Radicals.

Practice 1–4

Add or subtract as indicated.

1. $6\sqrt{11} + 9\sqrt{11}$
2. $\sqrt{7} - 3\sqrt{7}$
3. $\sqrt{2} + \sqrt{2} - \sqrt{15}$
4. $3\sqrt{3} - 3\sqrt{2}$

Practice 5–8

Simplify each radical expression.

5. $\sqrt{27} + \sqrt{75}$
6. $3\sqrt{20} - 7\sqrt{45}$
7. $\sqrt{36} - \sqrt{48} - 4\sqrt{3} - \sqrt{9}$
8. $\sqrt{9x^4} - \sqrt{36x^3} + \sqrt{x^3}$

Answers

1. $15\sqrt{11}$ **2.** $-2\sqrt{7}$ **3.** $2\sqrt{2} - \sqrt{15}$ **4.** $3\sqrt{3} - 3\sqrt{2}$ **5.** $8\sqrt{3}$ **6.** $-15\sqrt{5}$ **7.** $3 - 8\sqrt{3}$ **8.** $3x^2 - 5x\sqrt{x}$

✓**Concept Check Answer**

e

6. $7\sqrt{12} - 2\sqrt{75} = 7\sqrt{4 \cdot 3} - 2\sqrt{25 \cdot 3}$ Factor radicands.
$= 7\sqrt{4} \cdot \sqrt{3} - 2\sqrt{25} \cdot \sqrt{3}$ Use the product rule.
$= 7 \cdot 2\sqrt{3} - 2 \cdot 5\sqrt{3}$ Simplify $\sqrt{4}$ and $\sqrt{25}$.
$= 14\sqrt{3} - 10\sqrt{3}$ Multiply.
$= 4\sqrt{3}$ Subtract like radicals.

7. $\sqrt{25} - \sqrt{27} - 2\sqrt{18} - \sqrt{16}$
$= 5 - \sqrt{9 \cdot 3} - 2\sqrt{9 \cdot 2} - 4$ Factor radicands and simplify $\sqrt{25}$ and $\sqrt{16}$.
$= 5 - \sqrt{9} \cdot \sqrt{3} - 2\sqrt{9} \cdot \sqrt{2} - 4$ Use the product rule.
$= 5 - 3\sqrt{3} - 2 \cdot 3\sqrt{2} - 4$ Simplify $\sqrt{9}$.
$= 1 - 3\sqrt{3} - 6\sqrt{2}$ Write $5 - 4$ as 1 and $2 \cdot 3$ as 6.

Helpful Hint Don't forget for the remainder of this chapter, that if a variable appears in a radicand, the variable represents positive numbers only.

8. $2\sqrt{x^2} - \sqrt{25x^5} + \sqrt{x^5}$
$= 2x - \sqrt{25x^4 \cdot x} + \sqrt{x^4 \cdot x}$ Factor radicands so that one factor is a perfect square. Simplify $\sqrt{x^2}$.
$= 2x - \sqrt{25x^4} \cdot \sqrt{x} + \sqrt{x^4} \cdot \sqrt{x}$ Use the product rule.
$= 2x - 5x^2\sqrt{x} + x^2\sqrt{x}$ Write $\sqrt{25x^4}$ as $5x^2$ and $\sqrt{x^4}$ as x^2.
$= 2x - 4x^2\sqrt{x}$ Add like radicals.

Work Practice 5–8

Objective C Simplifying Cube Root Radicals Before Adding or Subtracting

Practice 9

Simplify the radical expression.
$10\sqrt[3]{81p^6} - \sqrt[3]{24p^6}$

Example 9 Simplify the radical expression.

$5\sqrt[3]{16x^3} - \sqrt[3]{54x^3}$
$= 5\sqrt[3]{8x^3 \cdot 2} - \sqrt[3]{27x^3 \cdot 2}$ Factor radicands so that one factor is a perfect cube.
$= 5 \cdot \sqrt[3]{8x^3} \cdot \sqrt[3]{2} - \sqrt[3]{27x^3} \cdot \sqrt[3]{2}$ Use the product rule.
$= 5 \cdot 2x\sqrt[3]{2} - 3x\sqrt[3]{2}$ Write $\sqrt[3]{8x^3}$ as $2x$ and $\sqrt[3]{27x^3}$ as $3x$.
$= 10x\sqrt[3]{2} - 3x\sqrt[3]{2}$ Write $5 \cdot 2x$ as $10x$.
$= 7x\sqrt[3]{2}$ Subtract like radicands.

Work Practice 9

Answer
9. $28p^2\sqrt[3]{3}$

Vocabulary, Readiness & Video Check

Fill in each blank.

1. Radicals that have the same index and same radicand are called __________.

2. The expressions $7\sqrt[3]{2x}$ and $-\sqrt[3]{2x}$ are called __________.

3. $11\sqrt{2} + 6\sqrt{2} =$ __________.
a. $66\sqrt{2}$ **b.** $17\sqrt{2}$ **c.** $17\sqrt{4}$

4. $\sqrt{5}$ is the same as __________.
a. $0\sqrt{5}$ **b.** $1\sqrt{5}$ **c.** $5\sqrt{5}$

5. $\sqrt{5} + \sqrt{5} =$ __________
a. $\sqrt{10}$ **b.** 5 **c.** $2\sqrt{5}$

6. $9\sqrt{7} - \sqrt{7} =$ __________
a. $8\sqrt{7}$ **b.** 9 **c.** 0

Watch the section lecture video and answer the following questions.

Objective A 7. From Examples 1–4, how is combining like radicals similar to combining like terms?

Objective B 8. From Example 5, why should we always check to see if all radical terms in our expression are simplified before attempting to add or subtract the radicals?

Objective C 9. In Example 8, what property is used during the simplification of the expression?

8.3 Exercise Set MyLab Math 

Objective A *Add or subtract as indicated. See Examples 1 through 4.*

1. $4\sqrt{3} - 8\sqrt{3}$
2. $\sqrt{5} - 9\sqrt{5}$
3. $3\sqrt{6} + 8\sqrt{6} - 2\sqrt{6} - 5$
4. $12\sqrt{2} - 3\sqrt{2} + 8\sqrt{2} + 10$
5. $6\sqrt{5} - 5\sqrt{5} + \sqrt{2}$
6. $4\sqrt{3} + \sqrt{5} - 3\sqrt{3}$
7. $2\sqrt{3} + 5\sqrt{3} - \sqrt{2}$
8. $8\sqrt{14} + 2\sqrt{14} + \sqrt{5}$
9. $2\sqrt{2} - 7\sqrt{2} - 6$
10. $5\sqrt{7} + 2 - 11\sqrt{7}$ + 2

Objective B *Add or subtract by first simplifying each radical and then combining any like radicals. See Examples 5 through 8.*

11. $\sqrt{12} + \sqrt{27}$
12. $\sqrt{50} + \sqrt{18}$
13. $\sqrt{45} + 3\sqrt{20}$
14. $5\sqrt{32} - \sqrt{72}$
15. $2\sqrt{54} - \sqrt{20} + \sqrt{45} - \sqrt{24}$
16. $2\sqrt{8} - \sqrt{128} + \sqrt{48} + \sqrt{18}$
17. $4x - 3\sqrt{x^2} + \sqrt{x}$
18. $x - 6\sqrt{x^2} + 2\sqrt{x}$
19. $\sqrt{25x} + \sqrt{36x} - 11\sqrt{x}$
20. $\sqrt{9x} - \sqrt{16x} + 2\sqrt{x}$
21. $\sqrt{\frac{5}{9}} + \sqrt{\frac{5}{81}}$
22. $\sqrt{\frac{3}{64}} + \sqrt{\frac{3}{16}}$
23. $\sqrt{\frac{3}{4}} - \sqrt{\frac{3}{64}}$
24. $\sqrt{\frac{2}{25}} + \sqrt{\frac{2}{9}}$

Objectives A B Mixed Practice *See Examples 1 through 8.*

25. $12\sqrt{5} - \sqrt{5} - 4\sqrt{5}$
26. $\sqrt{6} + 3\sqrt{6} + \sqrt{6}$
27. $\sqrt{75} + \sqrt{48}$
28. $2\sqrt{80} - \sqrt{45}$
29. $\sqrt{5} + \sqrt{15}$
30. $\sqrt{5} + \sqrt{5}$
31. $3\sqrt{x^3} - x\sqrt{4x}$
32. $x\sqrt{16x} - \sqrt{x^3}$
33. $\sqrt{8} + \sqrt{9} + \sqrt{18} + \sqrt{81}$
34. $\sqrt{6} + \sqrt{16} + \sqrt{24} + \sqrt{25}$
35. $4 + 8\sqrt{2} - 9$
36. $11 - 5\sqrt{7} - 8$

37. $2\sqrt{45} - 2\sqrt{20}$
38. $5\sqrt{18} + 2\sqrt{32}$
39. $\sqrt{35} - \sqrt{140}$
40. $\sqrt{15} - \sqrt{135}$
41. $6 - 2\sqrt{3} - \sqrt{3}$
42. $8 - \sqrt{2} - 5\sqrt{2}$
43. $3\sqrt{9x} + 2\sqrt{x}$
44. $5\sqrt{2x} + \sqrt{98x}$
45. $\sqrt{9x^2} + \sqrt{81x^2} - 11\sqrt{x}$
46. $\sqrt{100x^2} + 3\sqrt{x} - \sqrt{36x^2}$
47. $\sqrt{3x^3} + 3x\sqrt{x}$
48. $x\sqrt{4x} + \sqrt{9x^3}$
49. $\sqrt{32x^2} + \sqrt{32x^2} + \sqrt{4x^2}$
50. $\sqrt{18x^2} + \sqrt{24x^3} + \sqrt{2x^2}$
51. $\sqrt{40x} + \sqrt{40x^4} - 2\sqrt{10x} - \sqrt{5x^4}$
52. $\sqrt{72x^2} + \sqrt{54x} - x\sqrt{50} - 3\sqrt{2x}$

Objective C *Simplify each radical expression. See Example 9.*

53. $2\sqrt[3]{9} + 5\sqrt[3]{9} - \sqrt[3]{25}$
54. $8\sqrt[3]{4} + 2\sqrt[3]{4} - \sqrt[3]{49}$
55. $2\sqrt[3]{2} - 7\sqrt[3]{2} - 6$
56. $5\sqrt[3]{9} + 2 - 11\sqrt[3]{9}$
57. $\sqrt[3]{81} + \sqrt[3]{24}$
58. $\sqrt[3]{32} + \sqrt[3]{4}$
59. $\sqrt[3]{8} + \sqrt[3]{54} - 5$
60. $\sqrt[3]{64} + \sqrt[3]{14} - 9$
61. $2\sqrt[3]{8x^3} + 2\sqrt[3]{16x^3}$
62. $3\sqrt[3]{27z^3} + 3\sqrt[3]{81z^3}$
63. $12\sqrt[3]{y^7} - y^2\sqrt[3]{8y}$
64. $19\sqrt[3]{z^{11}} - z^3\sqrt[3]{125z^2}$
65. $\sqrt{60x} + x\sqrt[3]{40} - 2\sqrt{15x} - x\sqrt[3]{5}$
66. $\sqrt{98x^2} + \sqrt[3]{54} - x\sqrt{72} - 3\sqrt[3]{2}$

Review

Square each binomial. See Sections 3.6.

67. $(x + 6)^2$
68. $(3x + 2)^2$
69. $(2x - 1)^2$
70. $(x - 5)^2$

Concept Extensions

71. In your own words, describe like radicals.

72. In the expression $\sqrt{5} + 2 - 3\sqrt{5}$, explain why 2 and -3 cannot be combined.

73. Find the perimeter of the rectangular picture frame.

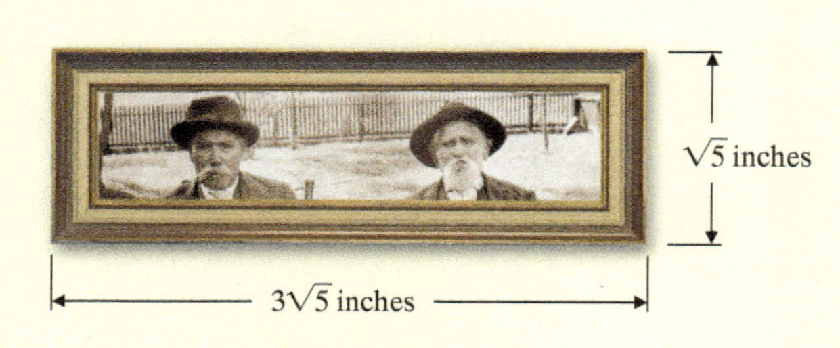

74. Find the perimeter of the plot of land.

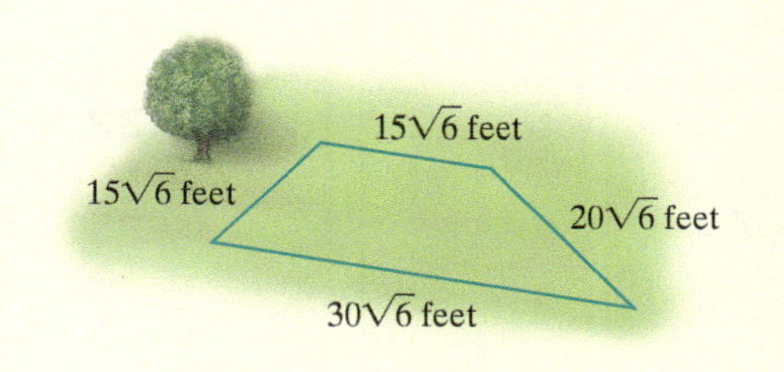

△ **75.** A water trough is to be made of wood. Each of the two triangular end pieces has an area of $\frac{3\sqrt{27}}{4}$ square feet. The two side panels are both rectangular. In simplest radical form, find the total area of the wood needed.

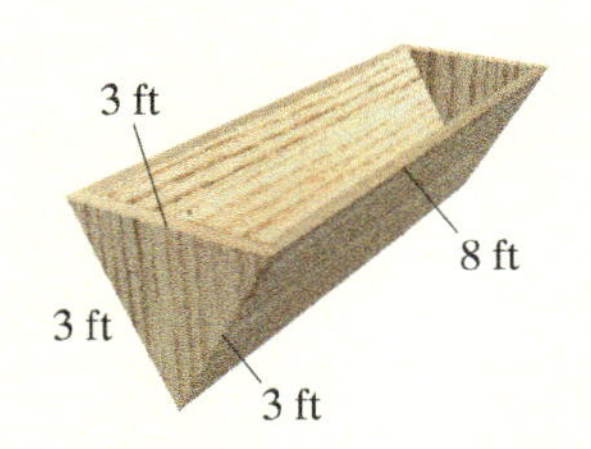

76. Eight wooden braces are to be attached along the diagonals of the vertical sides of a storage bin. Each of four of these diagonals has a length of $\sqrt{52}$ feet, while each of the other four has a length of $\sqrt{80}$ feet. In simplest radical form, find the total length of the wood needed for these braces.

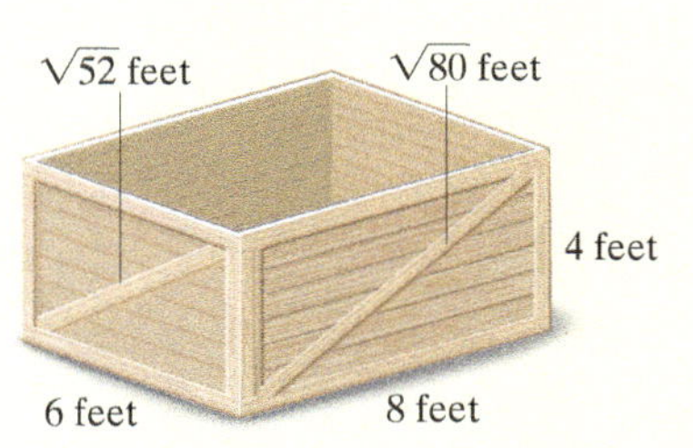

Determine whether each expression can be simplified. If yes, then simplify. See the Concept Check in this section.

77. $4\sqrt{2} + 3\sqrt{2}$

78. $3\sqrt{7} + 3\sqrt{6}$

79. $6 + 7\sqrt{6}$

80. $5x\sqrt{2} + 8x\sqrt{2}$

81. $\sqrt{7} + \sqrt{7} + \sqrt{7}$

82. $6\sqrt{5} - \sqrt{5}$

Simplify.

83. $\sqrt{\frac{x^3}{16}} - x\sqrt{\frac{9x}{25}} + \frac{\sqrt{81x^3}}{2}$

84. $7\sqrt{x^{11}y^7} - x^2y\sqrt{25x^7y^5} + \sqrt{8x^8y^2}$

8.4 Multiplying and Dividing Radicals

Objective A Multiplying Radicals 

Objectives

- **A** Multiply Radicals.
- **B** Divide Radicals.
- **C** Rationalize Denominators.
- **D** Rationalize Denominators Using Conjugates.

In Section 8.2, we used the product and quotient rules for radicals to help us simplify radicals. In this section, we use these rules to simplify products and quotients of radicals.

Product Rule for Radicals

If $\sqrt{a}$ and $\sqrt{b}$ are real numbers, then

$$\sqrt{a} \cdot \sqrt{b} = \sqrt{a \cdot b}$$

In other words, the product of the square roots of two numbers is the square root of the product of the two numbers. For example,

$$\sqrt{3} \cdot \sqrt{2} = \sqrt{3 \cdot 2} = \sqrt{6}$$

Practice 1–4

Multiply. Then simplify each product if possible.

1. $\sqrt{5} \cdot \sqrt{2}$
2. $\sqrt{7} \cdot \sqrt{7}$
3. $\sqrt{6} \cdot \sqrt{3}$
4. $\sqrt{10x} \cdot \sqrt{2x}$

Examples

Multiply. Then simplify each product if possible.

1. $\sqrt{5} \cdot \sqrt{3} = \sqrt{5 \cdot 3}$
$= \sqrt{15}$

2. $\sqrt{3} \cdot \sqrt{3} = \sqrt{3 \cdot 3} = \sqrt{9} = 3$

3. $\sqrt{3} \cdot \sqrt{15} = \sqrt{45}$ Use the product rule.
$= \sqrt{9 \cdot 5}$ Factor the radicand.
$= \sqrt{9} \cdot \sqrt{5}$ Use the product rule.
$= 3\sqrt{5}$ Simplify $\sqrt{9}$.

4. $\sqrt{2x^3} \cdot \sqrt{6x} = \sqrt{2x^3 \cdot 6x}$ Use the product rule.
$= \sqrt{12x^4}$ Multiply.
$= \sqrt{4x^4 \cdot 3}$ Write $12x^4$ so that one factor is a perfect square.
$= \sqrt{4x^4} \cdot \sqrt{3}$ Use the product rule.
$= 2x^2\sqrt{3}$ Simplify.

Work Practice 1–4

From Example 2, we found that

$\sqrt{3} \cdot \sqrt{3} = 3$ or $(\sqrt{3})^2 = 3$

This is true in general.

> If a is a positive number,
> $\sqrt{a} \cdot \sqrt{a} = a$ or $(\sqrt{a})^2 = a$

✓Concept Check

Identify the true statement(s).

a. $\sqrt{7} \cdot \sqrt{7} = 7$
b. $\sqrt{2} \cdot \sqrt{3} = 6$
c. $(\sqrt{131})^2 = 131$
d. $\sqrt{5x} \cdot \sqrt{5x} = 5x$ (Here x is a positive number.)

When multiplying radical expressions containing more than one term, we use the same techniques we use to multiply other algebraic expressions with more than one term.

Practice 5

Multiply.

a. $\sqrt{7}(\sqrt{7} - \sqrt{3})$
b. $\sqrt{5x}(\sqrt{x} - 3\sqrt{5})$
c. $(\sqrt{x} + \sqrt{5})(\sqrt{x} - \sqrt{3})$

Example 5

Multiply.

a. $\sqrt{5}(\sqrt{5} - \sqrt{2})$
b. $\sqrt{3x}(\sqrt{x} - 5\sqrt{3})$
c. $(\sqrt{x} + \sqrt{2})(\sqrt{x} - \sqrt{7})$

Solution:

a. Using the distributive property, we have

$\sqrt{5}(\sqrt{5} - \sqrt{2}) = \sqrt{5} \cdot \sqrt{5} - \sqrt{5} \cdot \sqrt{2}$
$= 5 - \sqrt{10}$ Since $\sqrt{5} \cdot \sqrt{5} = 5$ and $\sqrt{5} \cdot \sqrt{2} = \sqrt{10}$

b. $\sqrt{3x}(\sqrt{x} - 5\sqrt{3}) = \sqrt{3x} \cdot \sqrt{x} - \sqrt{3x} \cdot 5\sqrt{3}$ Use the distributive property.
$= \sqrt{3x \cdot x} - 5\sqrt{3x \cdot 3}$ Use the product rule.
$= \sqrt{3 \cdot x^2} - 5\sqrt{9 \cdot x}$ Factor each radicand so that one factor is a perfect square.
$= \sqrt{3} \cdot \sqrt{x^2} - 5 \cdot \sqrt{9} \cdot \sqrt{x}$ Use the product rule.
$= x\sqrt{3} - 5 \cdot 3 \cdot \sqrt{x}$ Simplify.
$= x\sqrt{3} - 15\sqrt{x}$ Simplify.

Answers

1. $\sqrt{10}$ 2. 7 3. $3\sqrt{2}$ 4. $2x\sqrt{5}$
5. a. $7 - \sqrt{21}$ b. $x\sqrt{5} - 15\sqrt{x}$
c. $x - \sqrt{3x} + \sqrt{5x} - \sqrt{15}$

✓Concept Check Answer

a, c, d

c. Using the FOIL method of multiplication, we have

$$(\sqrt{x}+\sqrt{2})(\sqrt{x}-\sqrt{7})$$

$$\overset{F}{=} \sqrt{x}\cdot\sqrt{x} \overset{O}{-} \sqrt{x}\cdot\sqrt{7} \overset{I}{+} \sqrt{2}\cdot\sqrt{x} \overset{L}{-} \sqrt{2}\cdot\sqrt{7}$$

$$= x - \sqrt{7x} + \sqrt{2x} - \sqrt{14}$$ Use the product rule.

Work Practice 5

The special product formulas also can be used to multiply expressions containing radicals.

Example 6 Multiply.

a. $(\sqrt{5}-7)(\sqrt{5}+7)$ **b.** $(\sqrt{7x}+2)^2$

Solution:

a. $(\sqrt{5}-7)(\sqrt{5}+7) = (\sqrt{5})^2 - 7^2$ Recall that $(a-b)(a+b) = a^2 - b^2$.

$= 5 - 49$

$= -44$

b. $(\sqrt{7x}+2)^2$

$= (\sqrt{7x})^2 + 2(\sqrt{7x})(2) + (2)^2$ Recall that $(a+b)^2 = a^2 + 2ab + b^2$.

$= 7x + 4\sqrt{7x} + 4$

Work Practice 6

Practice 6

Multiply.

a. $(\sqrt{3}+8)(\sqrt{3}-8)$

b. $(\sqrt{5x}+4)^2$

Objective B Dividing Radicals

To simplify quotients of rational expressions, we use the quotient rule.

Quotient Rule for Radicals

If $\sqrt{a}$ and $\sqrt{b}$ are real numbers and $b \neq 0$, then

$$\frac{\sqrt{a}}{\sqrt{b}} = \sqrt{\frac{a}{b}}$$

Examples Divide. Then simplify the quotient if possible.

7. $\frac{\sqrt{14}}{\sqrt{2}} = \sqrt{\frac{14}{2}} = \sqrt{7}$

8. $\frac{\sqrt{100}}{\sqrt{5}} = \sqrt{\frac{100}{5}} = \sqrt{20} = \sqrt{4\cdot 5} = \sqrt{4}\cdot\sqrt{5} = 2\sqrt{5}$

9. $\frac{\sqrt{12x^3}}{\sqrt{3x}} = \sqrt{\frac{12x^3}{3x}} = \sqrt{4x^2} = 2x$

Work Practice 7–9

Practice 7–9

Divide. Then simplify the quotient if possible.

7. $\frac{\sqrt{21}}{\sqrt{3}}$

8. $\frac{\sqrt{90}}{\sqrt{2}}$

9. $\frac{\sqrt{125x^3}}{\sqrt{5x}}$

Objective C Rationalizing Denominators

It is sometimes easier to work with radical expressions if the denominator does not contain a radical. To rewrite an expression so that the denominator does not contain a radical expression, we use the fact that we can multiply the numerator and the denominator of a fraction by the same nonzero number without changing the value

Answers

6. a. -61 **b.** $5x + 8\sqrt{5x} + 16$
7. $\sqrt{7}$ **8.** $3\sqrt{5}$ **9.** $5x$

of the expression. This is the same as multiplying the fraction by 1. For example, to get rid of the radical in the denominator of $\frac{\sqrt{5}}{\sqrt{2}}$, we multiply by 1 in the form of $\frac{\sqrt{2}}{\sqrt{2}}$. Then

$$\frac{\sqrt{5}}{\sqrt{2}} = \frac{\sqrt{5}}{\sqrt{2}} \cdot 1 = \frac{\sqrt{5}}{\sqrt{2}} \cdot \frac{\sqrt{2}}{\sqrt{2}} = \frac{\sqrt{5} \cdot \sqrt{2}}{\sqrt{2} \cdot \sqrt{2}} = \frac{\sqrt{10}}{2}$$

This process is called **rationalizing** the denominator.

Practice 10

Rationalize the denominator of $\frac{5}{\sqrt{3}}$.

Example 10 Rationalize the denominator of $\frac{2}{\sqrt{7}}$.

Solution: To rewrite $\frac{2}{\sqrt{7}}$ so that there is no radical in the denominator, we multiply by 1 in the form of $\frac{\sqrt{7}}{\sqrt{7}}$.

$$\frac{2}{\sqrt{7}} = \frac{2}{\sqrt{7}} \cdot \frac{\sqrt{7}}{\sqrt{7}} = \frac{2 \cdot \sqrt{7}}{\sqrt{7} \cdot \sqrt{7}} = \frac{2\sqrt{7}}{7}$$

Work Practice 10

Practice 11

Rationalize the denominator of $\frac{\sqrt{7}}{\sqrt{20}}$.

Example 11 Rationalize the denominator of $\frac{\sqrt{5}}{\sqrt{12}}$.

Solution: We can multiply by $\frac{\sqrt{12}}{\sqrt{12}}$, but see what happens if we simplify first.

$$\frac{\sqrt{5}}{\sqrt{12}} = \frac{\sqrt{5}}{\sqrt{4 \cdot 3}} = \frac{\sqrt{5}}{2\sqrt{3}}$$

To rationalize the denominator now, we multiply by $\frac{\sqrt{3}}{\sqrt{3}}$.

$$\frac{\sqrt{5}}{2\sqrt{3}} = \frac{\sqrt{5}}{2\sqrt{3}} \cdot \frac{\sqrt{3}}{\sqrt{3}} = \frac{\sqrt{5} \cdot \sqrt{3}}{2\sqrt{3} \cdot \sqrt{3}} = \frac{\sqrt{15}}{2 \cdot 3} = \frac{\sqrt{15}}{6}$$

Work Practice 11

Practice 12

Rationalize the denominator of $\sqrt{\frac{2}{45x}}$.

Example 12 Rationalize the denominator of $\sqrt{\frac{1}{18x}}$.

Solution: First we simplify.

$$\sqrt{\frac{1}{18x}} = \frac{\sqrt{1}}{\sqrt{18x}} = \frac{1}{\sqrt{9} \cdot \sqrt{2x}} = \frac{1}{3\sqrt{2x}}$$

Now to rationalize the denominator, we multiply by $\frac{\sqrt{2x}}{\sqrt{2x}}$.

$$\frac{1}{3\sqrt{2x}} = \frac{1}{3\sqrt{2x}} \cdot \frac{\sqrt{2x}}{\sqrt{2x}} = \frac{1 \cdot \sqrt{2x}}{3\sqrt{2x} \cdot \sqrt{2x}} = \frac{\sqrt{2x}}{3 \cdot 2x} = \frac{\sqrt{2x}}{6x}$$

Work Practice 12

Objective D Rationalizing Denominators Using Conjugates

To rationalize a denominator that is a sum or a difference, such as the denominator in

$$\frac{2}{4 + \sqrt{3}}$$

we multiply the numerator and the denominator by $4 - \sqrt{3}$. The expressions $4 + \sqrt{3}$ and $4 - \sqrt{3}$ are called conjugates of each other. When a radical expression

Answers

10. $\frac{5\sqrt{3}}{3}$ **11.** $\frac{\sqrt{35}}{10}$ **12.** $\frac{\sqrt{10x}}{15x}$

such as $4 + \sqrt{3}$ is multiplied by its conjugate, $4 - \sqrt{3}$, the product simplifies to an expression that contains no radicals.

In general, the expressions $a + b$ and $a - b$ are **conjugates** of each other.

$$(a + b)(a - b) = a^2 - b^2$$
$$(4 + \sqrt{3})(4 - \sqrt{3}) = 4^2 - (\sqrt{3})^2 = 16 - 3 = 13$$

Then

$$\frac{2}{4 + \sqrt{3}} = \frac{2(4 - \sqrt{3})}{(4 + \sqrt{3})(4 - \sqrt{3})} = \frac{2(4 - \sqrt{3})}{13}$$

Example 13 Rationalize the denominator of $\frac{2}{1 + \sqrt{3}}$.

Solution: We multiply the numerator and the denominator of this fraction by the conjugate of $1 + \sqrt{3}$, that is, by $1 - \sqrt{3}$.

$$\frac{2}{1 + \sqrt{3}} = \frac{2(1 - \sqrt{3})}{(1 + \sqrt{3})(1 - \sqrt{3})}$$
$$= \frac{2(1 - \sqrt{3})}{1^2 - (\sqrt{3})^2}$$
$$= \frac{2(1 - \sqrt{3})}{1 - 3}$$
$$= \frac{2(1 - \sqrt{3})}{-2}$$
$$= -\frac{2(1 - \sqrt{3})}{2} \quad \frac{a}{-b} = -\frac{a}{b}$$
$$= -1(1 - \sqrt{3}) \quad \text{Simplify.}$$
$$= -1 + \sqrt{3} \quad \text{Multiply.}$$

Helpful Hint
Don't forget that $(\sqrt{3})^2 = 3$.

Work Practice 13

Practice 13
Rationalize the denominator of $\frac{3}{2 + \sqrt{7}}$.

Example 14 Rationalize the denominator of $\frac{\sqrt{5} + 4}{\sqrt{5} - 1}$.

Solution:

$$\frac{\sqrt{5} + 4}{\sqrt{5} - 1} = \frac{(\sqrt{5} + 4)(\sqrt{5} + 1)}{(\sqrt{5} - 1)(\sqrt{5} + 1)} \quad \text{Multiply the numerator and denominator by } \sqrt{5} + 1 \text{, the conjugate of } \sqrt{5} - 1.$$
$$= \frac{5 + \sqrt{5} + 4\sqrt{5} + 4}{5 - 1} \quad \text{Multiply.}$$
$$= \frac{9 + 5\sqrt{5}}{4} \quad \text{Simplify.}$$

Work Practice 14

Practice 14
Rationalize the denominator of $\frac{\sqrt{2} + 5}{\sqrt{2} - 1}$.

Example 15 Rationalize the denominator of $\frac{3}{1 + \sqrt{x}}$.

Solution:

$$\frac{3}{1 + \sqrt{x}} = \frac{3(1 - \sqrt{x})}{(1 + \sqrt{x})(1 - \sqrt{x})} \quad \text{Multiply the numerator and denominator by } 1 - \sqrt{x} \text{, the conjugate of } 1 + \sqrt{x}.$$
$$= \frac{3(1 - \sqrt{x})}{1 - x}$$

Work Practice 15

Practice 15
Rationalize the denominator of $\frac{7}{2 - \sqrt{x}}$.

Answers
13. $-2 + \sqrt{7}$ **14.** $7 + 6\sqrt{2}$
15. $\frac{7(2 + \sqrt{x})}{4 - x}$

Vocabulary, Readiness & Video Check

Fill in each blank.

1. $\sqrt{7} \cdot \sqrt{3} =$ ______________
2. $\sqrt{10} \cdot \sqrt{10} =$ ______________
3. $\dfrac{\sqrt{15}}{\sqrt{3}} =$ ______________
4. The process of eliminating the radical in the denominator of a radical expression is called ______________.
5. The conjugate of $2 + \sqrt{3}$ is ______________.

Martin-Gay Interactive Videos

See Video 8.4

Watch the section lecture video and answer the following questions.

Objective A 6. In Examples 1 and 3, the product rule for radicals is applied twice, but in different ways. Explain.

7. Starting with Example 2, what important reminder is made repeatedly about the square root of a positive number that is squared?

Objective B 8. From Examples 5 and 6, when we're looking at a quotient of two radicals, what would make us think to apply the quotient rule in order to simplify?

Objective C 9. From the lecture before Example 7, what is the goal of rationalizing a denominator?

Objective D 10. From Example 9, why will multiplying a denominator by its conjugate rationalize the denominator?

8.4 Exercise Set MyLab Math

Objective A *Multiply and simplify. Assume that all variables represent positive numbers. See Examples 1 through 6.*

1. $\sqrt{8} \cdot \sqrt{2}$
2. $\sqrt{3} \cdot \sqrt{12}$
3. $\sqrt{10} \cdot \sqrt{5}$
4. $\sqrt{2} \cdot \sqrt{14}$
5. $(\sqrt{6})^2$
6. $(\sqrt{10})^2$
7. $\sqrt{2x} \cdot \sqrt{2x}$
8. $\sqrt{5y} \cdot \sqrt{5y}$
9. $(2\sqrt{5})^2$
10. $(3\sqrt{10})^2$
11. $(6\sqrt{x})^2$
12. $(8\sqrt{y})^2$
13. $\sqrt{3x^5} \cdot \sqrt{6x}$
14. $\sqrt{21y^7} \cdot \sqrt{3y}$
15. $\sqrt{2xy^2} \cdot \sqrt{8xy}$
16. $\sqrt{18x^2y^2} \cdot \sqrt{2x^2y}$
17. $\sqrt{6}(\sqrt{5} + \sqrt{7})$
18. $\sqrt{10}(\sqrt{3} - \sqrt{7})$
19. $\sqrt{10}(\sqrt{2} + \sqrt{5})$
20. $\sqrt{6}(\sqrt{3} + \sqrt{2})$
21. $\sqrt{7y}(\sqrt{y} - 2\sqrt{7})$
22. $\sqrt{5b}(2\sqrt{b} + \sqrt{5})$
23. $(\sqrt{3} + 6)(\sqrt{3} - 6)$
24. $(\sqrt{5} + 2)(\sqrt{5} - 2)$
25. $(\sqrt{3} + \sqrt{5})(\sqrt{2} - \sqrt{5})$
26. $(\sqrt{7} + \sqrt{5})(\sqrt{2} - \sqrt{5})$
27. $(2\sqrt{11} + 1)(\sqrt{11} - 6)$
28. $(5\sqrt{3} + 2)(\sqrt{3} - 1)$

29. $(\sqrt{x}+6)(\sqrt{x}-6)$
30. $(\sqrt{y}+5)(\sqrt{y}-5)$
31. $(\sqrt{x}-7)^2$
32. $(\sqrt{x}+4)^2$
33. $(\sqrt{6y}+1)^2$
34. $(\sqrt{3y}-2)^2$

Objective B *Divide and simplify. Assume that all variables represent positive numbers. See Examples 7 through 9.*

35. $\frac{\sqrt{32}}{\sqrt{2}}$
36. $\frac{\sqrt{40}}{\sqrt{10}}$
37. $\frac{\sqrt{21}}{\sqrt{3}}$
38. $\frac{\sqrt{55}}{\sqrt{5}}$
39. $\frac{\sqrt{90}}{\sqrt{5}}$
40. $\frac{\sqrt{96}}{\sqrt{8}}$
41. $\frac{\sqrt{75y^5}}{\sqrt{3y}}$
42. $\frac{\sqrt{24x^7}}{\sqrt{6x}}$
43. $\frac{\sqrt{150}}{\sqrt{2}}$
44. $\frac{\sqrt{120}}{\sqrt{3}}$
45. $\frac{\sqrt{72y^5}}{\sqrt{3y^3}}$
46. $\frac{\sqrt{54x^3}}{\sqrt{2x}}$
47. $\frac{\sqrt{24x^3y^4}}{\sqrt{2xy}}$
48. $\frac{\sqrt{96x^5y^3}}{\sqrt{3x^2y}}$

Objective C *Rationalize each denominator and simplify. Assume that all variables represent positive numbers. See Examples 10 through 12.*

49. $\frac{\sqrt{3}}{\sqrt{5}}$
50. $\frac{\sqrt{2}}{\sqrt{3}}$
51. $\frac{7}{\sqrt{2}}$
52. $\frac{8}{\sqrt{11}}$
53. $\frac{1}{\sqrt{6y}}$
54. $\frac{1}{\sqrt{10z}}$
55. $\sqrt{\frac{5}{18}}$
56. $\sqrt{\frac{7}{12}}$
57. $\sqrt{\frac{3}{x}}$
58. $\sqrt{\frac{5}{x}}$
59. $\sqrt{\frac{1}{8}}$
60. $\sqrt{\frac{1}{27}}$
61. $\sqrt{\frac{2}{15}}$
62. $\sqrt{\frac{11}{14}}$
63. $\sqrt{\frac{3}{20}}$
64. $\sqrt{\frac{3}{50}}$
65. $\frac{3x}{\sqrt{2x}}$
66. $\frac{5y}{\sqrt{3y}}$
67. $\frac{8y}{\sqrt{5}}$
68. $\frac{7x}{\sqrt{2}}$
69. $\sqrt{\frac{x}{36y}}$
70. $\sqrt{\frac{z}{49y}}$
71. $\sqrt{\frac{y}{12x}}$
72. $\sqrt{\frac{x}{20y}}$

Objective D *Rationalize each denominator and simplify. Assume that all variables represent positive numbers. See Examples 13 through 15.*

73. $\frac{3}{\sqrt{2}+1}$
74. $\frac{6}{\sqrt{5}+2}$
75. $\frac{4}{2-\sqrt{5}}$
76. $\frac{2}{\sqrt{10}-3}$
77. $\frac{\sqrt{5}+1}{\sqrt{6}-\sqrt{5}}$
78. $\frac{\sqrt{3}+1}{\sqrt{3}-\sqrt{2}}$
79. $\frac{\sqrt{3}+1}{\sqrt{2}-1}$
80. $\frac{\sqrt{2}-2}{2-\sqrt{3}}$

81. $\dfrac{5}{2+\sqrt{x}}$

82. $\dfrac{9}{3+\sqrt{x}}$

83. $\dfrac{3}{\sqrt{x}-4}$

84. $\dfrac{4}{\sqrt{x}-1}$

Review

Solve each equation. See Sections 2.3 and 4.6.

85. $x + 5 = 7^2$

86. $2y - 1 = 3^2$

87. $4z^2 + 6z - 12 = (2z)^2$

88. $16x^2 + x + 9 = (4x)^2$

89. $9x^2 + 5x + 4 = (3x + 1)^2$

90. $x^2 + 3x + 4 = (x + 2)^2$

Concept Extensions

△ **91.** Find the area of a rectangular room whose length is $13\sqrt{2}$ meters and width is $5\sqrt{6}$ meters.

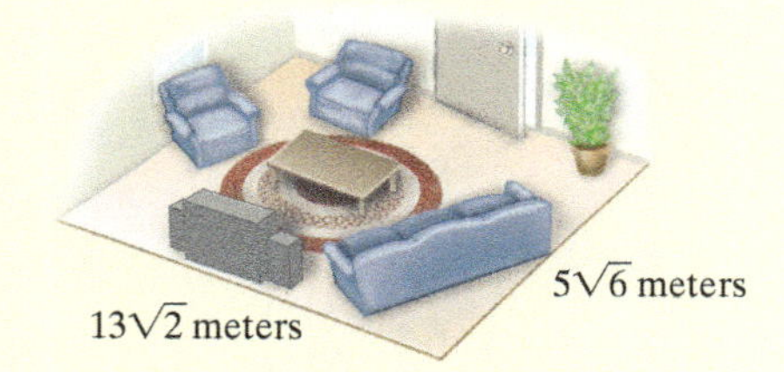

△ **92.** Find the volume of a microwave oven whose length is $\sqrt{3}$ feet, width is $\sqrt{2}$ feet, and height is $\sqrt{2}$ feet.

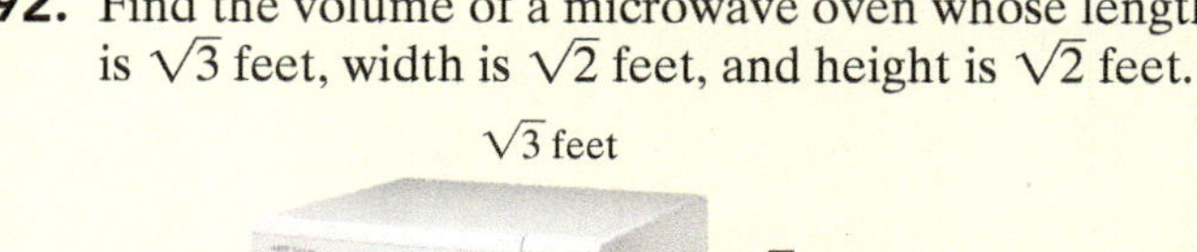

△ **93.** If a circle has area A, then the formula for the radius r of the circle is

$$r = \sqrt{\frac{A}{\pi}}$$

Rationalize the denominator of this expression.

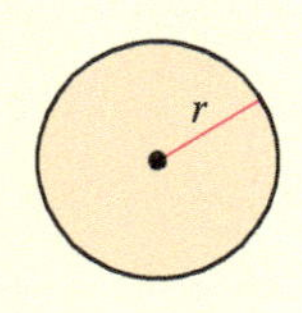

△ **94.** If the surface area of a round ball is S, then the formula for the radius r of the ball is

$$r = \sqrt{\frac{S}{4\pi}}$$

Simplify this expression by rationalizing the denominator.

Identify each statement as true or false. See the Concept Check in this section.

95. $\sqrt{5}\cdot\sqrt{5} = 5$

96. $\sqrt{5}\cdot\sqrt{3} = 15$

97. $\sqrt{3x}\cdot\sqrt{3x} = 2\sqrt{3x}$

98. $\sqrt{3x} + \sqrt{3x} = 2\sqrt{3x}$

99. $\sqrt{11} + \sqrt{2} = \sqrt{13}$

100. $\sqrt{11}\cdot\sqrt{2} = \sqrt{22}$

101. When rationalizing the denominator of $\frac{\sqrt{2}}{\sqrt{3}}$, explain why both the numerator and the denominator must be multiplied by $\sqrt{3}$.

102. In your own words, explain why $\sqrt{6} + \sqrt{2}$ cannot be simplified further, but $\sqrt{6} \cdot \sqrt{2}$ can be.

103. To rationalize the denominator of $\frac{\sqrt[3]{2}}{\sqrt[3]{3}}$, multiply the numerator and the denominator by $\sqrt[3]{9}$. Then simplify. Explain why this works.

104. When rationalizing the denominator of $\frac{5}{1 + \sqrt{2}}$, explain why multiplying by $\frac{\sqrt{2}}{\sqrt{2}}$ will not accomplish this, but multiplying by $\frac{1 - \sqrt{2}}{1 - \sqrt{2}}$ will.

It is often more convenient to work with a radical expression whose numerator is rationalized. Rationalize the numerator of each expression by multiplying the numerator and denominator by the conjugate of the numerator.

105. $\frac{\sqrt{3} + 1}{\sqrt{2} - 1}$

106. $\frac{\sqrt{2} - 2}{2 - \sqrt{3}}$

Answers

1. ______

2. ______

3. ______

4. ______

5. ______

6. ______

7. ______

8. ______

9. ______

10. ______

11. ______

12. ______

13. ______

14. ______

15. ______

16. ______

17. ______

18. ______

19. ______

20. ______

21. ______

22. ______

Simplifying Radicals

Simplify. Assume that all variables represent positive numbers.

1. $\sqrt{36}$ **2.** $\sqrt{48}$ **3.** $\sqrt{x^4}$

4. $\sqrt{y^7}$ **5.** $\sqrt{16x^2}$ **6.** $\sqrt{18x^{11}}$

7. $\sqrt[3]{8}$ **8.** $\sqrt[4]{81}$ **9.** $\sqrt[3]{-27}$

10. $\sqrt{-4}$ **11.** $\sqrt{\frac{11}{9}}$ **12.** $\sqrt[3]{\frac{7}{64}}$

13. $-\sqrt{16}$ **14.** $-\sqrt{25}$ **15.** $\sqrt{\frac{9}{49}}$

16. $\sqrt{\frac{1}{64}}$ **17.** $\sqrt{a^8b^2}$ **18.** $\sqrt{x^{10}y^{20}}$

19. $\sqrt{25m^6}$ **20.** $\sqrt{9n^{16}}$

Add or subtract as indicated.

21. $5\sqrt{7} + \sqrt{7}$ **22.** $\sqrt{50} - \sqrt{8}$

23. $5\sqrt{2} - 5\sqrt{3}$

24. $2\sqrt{x} + \sqrt{25x} - \sqrt{36x} + 3x$

Multiply and simplify if possible.

25. $\sqrt{2} \cdot \sqrt{15}$

26. $\sqrt{3} \cdot \sqrt{3}$

27. $(2\sqrt{7})^2$

28. $(3\sqrt{5})^2$

29. $\sqrt{3}(\sqrt{11} + 1)$

30. $\sqrt{6}(\sqrt{3} - 2)$

31. $\sqrt{8y} \cdot \sqrt{2y}$

32. $\sqrt{15x^2} \cdot \sqrt{3x^2}$

33. $(\sqrt{x} - 5)(\sqrt{x} + 2)$

34. $(3 + \sqrt{2})^2$

Divide and simplify if possible.

35. $\frac{\sqrt{8}}{\sqrt{2}}$

36. $\frac{\sqrt{45}}{\sqrt{15}}$

37. $\frac{\sqrt{24x^5}}{\sqrt{2x}}$

38. $\frac{\sqrt{75a^4b^5}}{\sqrt{5ab}}$

Rationalize each denominator.

39. $\sqrt{\frac{1}{6}}$

40. $\frac{x}{\sqrt{20}}$

41. $\frac{4}{\sqrt{6} + 1}$

42. $\frac{\sqrt{2} + 1}{\sqrt{x} - 5}$

23. ______

24. ______

25. ______

26. ______

27. ______

28. ______

29. ______

30. ______

31. ______

32. ______

33. ______

34. ______

35. ______

36. ______

37. ______

38. ______

39. ______

40. ______

41. ______

42. ______

8.5 Solving Equations Containing Radicals 

Objectives

A Solve Radical Equations by Using the Squaring Property of Equality Once.

B Solve Radical Equations by Using the Squaring Property of Equality Twice.

Objective A Using the Squaring Property of Equality Once

In this section, we solve **radical equations** such as

$$\sqrt{x+3} = 5 \quad \text{and} \quad \sqrt{2x+1} = \sqrt{3x}$$

Radical equations contain variables in the radicand. To solve these equations, we rely on the following squaring property.

The Squaring Property of Equality

If $a = b$, then $a^2 = b^2$.

Unfortunately, this squaring property does not guarantee that all solutions of the new equation are solutions of the original equation. For example, if we square both sides of the equation

$$x = 2$$

we have

$$x^2 = 4$$

This new equation has two solutions, 2 and -2, while the original equation, $x = 2$, has only one solution. For this reason, we must **always check proposed solutions of radical equations in the original equation.**

Practice 1

Solve: $\sqrt{x-2} = 7$

Example 1 Solve: $\sqrt{x+3} = 5$

Solution: To solve this radical equation, we use the squaring property of equality and square both sides of the equation.

$$\sqrt{x+3} = 5$$
$$\left(\sqrt{x+3}\right)^2 = 5^2 \quad \text{Square both sides.}$$
$$x + 3 = 25 \quad \text{Simplify.}$$
$$x = 22 \quad \text{Subtract 3 from both sides.}$$

Check: We replace x with 22 in the original equation.

$$\sqrt{x+3} = 5 \quad \text{Original equation}$$
$$\sqrt{22+3} \stackrel{?}{=} 5 \quad \text{Let } x = 22.$$
$$\sqrt{25} \stackrel{?}{=} 5$$
$$5 = 5 \quad \text{True}$$

Since a true statement results, 22 is the solution.

Helpful Hint Don't forget to check the proposed solutions of a radical equation in the original equation.

Work Practice 1

When solving radical equations, if possible, move radicals so that at least one radical is by itself on one side of the equation.

Practice 2

Solve: $\sqrt{6x-1} = \sqrt{x}$

Example 2 Solve: $\sqrt{x} = \sqrt{5x-2}$

Solution: Each radical is by itself on one side of the equation. Let's begin solving by squaring both sides.

$$\sqrt{x} = \sqrt{5x-2} \quad \text{Original equation}$$
$$\left(\sqrt{x}\right)^2 = \left(\sqrt{5x-2}\right)^2 \quad \text{Square both sides.}$$
$$x = 5x - 2 \quad \text{Simplify.}$$
$$-4x = -2 \quad \text{Subtract } 5x \text{ from both sides.}$$
$$x = \frac{-2}{-4} = \frac{1}{2} \quad \text{Divide both sides by } -4 \text{ and simplify.}$$

Answers

1. $x = 51$ 2. $x = \frac{1}{5}$

Check: We replace x with $\frac{1}{2}$ in the original equation.

$$\sqrt{x} = \sqrt{5x - 2} \quad \text{Original equation}$$
$$\sqrt{\frac{1}{2}} \stackrel{?}{=} \sqrt{5 \cdot \frac{1}{2} - 2} \quad \text{Let } x = \frac{1}{2}.$$
$$\sqrt{\frac{1}{2}} \stackrel{?}{=} \sqrt{\frac{5}{2} - 2} \quad \text{Multiply.}$$
$$\sqrt{\frac{1}{2}} \stackrel{?}{=} \sqrt{\frac{5}{2} - \frac{4}{2}} \quad \text{Write 2 as } \frac{4}{2}.$$
$$\sqrt{\frac{1}{2}} = \sqrt{\frac{1}{2}} \quad \text{True}$$

This statement is true, so the solution is $\frac{1}{2}$.

Work Practice 2

Example 3 Solve: $\sqrt{x} + 6 = 4$

Solution: First we write the equation so that the radical is by itself on one side of the equation.

$$\sqrt{x} + 6 = 4$$
$$\sqrt{x} = -2 \quad \text{Subtract 6 from both sides to get the radical by itself.}$$

Normally we would now square both sides. Recall, however, that $\sqrt{x}$ is the principal or nonnegative square root of x, so $\sqrt{x}$ cannot equal -2 and thus this equation has no solution. We arrive at the same conclusion if we continue by applying the squaring property.

$$\sqrt{x} = -2$$
$$(\sqrt{x})^2 = (-2)^2 \quad \text{Square both sides.}$$
$$x = 4 \quad \text{Simplify.}$$

Check: We replace x with 4 in the original equation.

$$\sqrt{x} + 6 = 4 \quad \text{Original equation}$$
$$\sqrt{4} + 6 \stackrel{?}{=} 4 \quad \text{Let } x = 4.$$
$$2 + 6 = 4 \quad \text{False}$$

Since 4 *does not* satisfy the original equation, this equation has no solution.

Work Practice 3

Practice 3

Solve: $\sqrt{x} + 9 = 2$

Example 3 makes it very clear that we *must* check proposed solutions in the original equation to determine if they are truly solutions. If a proposed solution does not work, we say that the value is an **extraneous solution.**

The following steps can be used to solve radical equations containing square roots.

To Solve a Radical Equation Containing Square Roots

Step 1: Arrange terms so that one radical is by itself on one side of the equation. That is, isolate a radical.

Step 2: Square both sides of the equation.

Step 3: Simplify both sides of the equation.

Step 4: If the equation still contains a radical term, repeat Steps 1 through 3.

Step 5: Solve the equation.

Step 6: Check all solutions in the original equation for extraneous solutions.

Answer

3. no solution

Practice 4

Solve: $\sqrt{9y^2 + 2y - 10} = 3y$

Example 4 Solve: $\sqrt{4y^2 + 5y - 15} = 2y$

Solution: The radical is already isolated, so we start by squaring both sides.

$$\sqrt{4y^2 + 5y - 15} = 2y$$
$$\left(\sqrt{4y^2 + 5y - 15}\right)^2 = (2y)^2 \quad \text{Square both sides.}$$
$$4y^2 + 5y - 15 = 4y^2 \quad \text{Simplify.}$$
$$5y - 15 = 0 \quad \text{Subtract } 4y^2 \text{ from both sides.}$$
$$5y = 15 \quad \text{Add 15 to both sides.}$$
$$y = 3 \quad \text{Divide both sides by 5.}$$

Check: We replace y with 3 in the original equation.

$$\sqrt{4y^2 + 5y - 15} = 2y \quad \text{Original equation}$$
$$\sqrt{4 \cdot 3^2 + 5 \cdot 3 - 15} \stackrel{?}{=} 2 \cdot 3 \quad \text{Let } y = 3.$$
$$\sqrt{4 \cdot 9 + 15 - 15} \stackrel{?}{=} 6 \quad \text{Simplify.}$$
$$\sqrt{36} \stackrel{?}{=} 6$$
$$6 = 6 \quad \text{True}$$

This statement is true, so the solution is 3.

Work Practice 4

Practice 5

Solve: $\sqrt{x + 1} - x = -5$

Example 5 Solve: $\sqrt{x + 3} - x = -3$

Solution: First we isolate the radical by adding x to both sides. Then we square both sides.

$$\sqrt{x + 3} - x = -3$$
$$\sqrt{x + 3} = x - 3 \quad \text{Add } x \text{ to both sides.}$$
$$\left(\sqrt{x + 3}\right)^2 = (x - 3)^2 \quad \text{Square both sides.}$$
$$x + 3 = x^2 - 6x + 9 \quad \text{Simplify.}$$

Helpful Hint Don't forget that $(x-3)^2 = (x-3)(x-3) = x^2 - 6x + 9$.

To solve the resulting quadratic equation, we write the equation in standard form by subtracting x and 3 from both sides.

$$x + 3 = x^2 - 6x + 9$$
$$3 = x^2 - 7x + 9 \quad \text{Subtract } x \text{ from both sides.}$$
$$0 = x^2 - 7x + 6 \quad \text{Subtract 3 from both sides.}$$
$$0 = (x - 6)(x - 1) \quad \text{Factor.}$$
$$0 = x - 6 \quad \text{or} \quad 0 = x - 1 \quad \text{Set each factor equal to zero.}$$
$$6 = x \qquad 1 = x \quad \text{Solve for } x.$$

Check: We replace x with 6 and then x with 1 in the original equation.

Let $x = 6$.

$$\sqrt{x + 3} - x = -3$$
$$\sqrt{6 + 3} - 6 \stackrel{?}{=} -3$$
$$\sqrt{9} - 6 \stackrel{?}{=} -3$$
$$3 - 6 \stackrel{?}{=} -3$$
$$-3 = -3 \quad \text{True}$$

Let $x = 1$.

$$\sqrt{x + 3} - x = -3$$
$$\sqrt{1 + 3} - 1 \stackrel{?}{=} -3$$
$$\sqrt{4} - 1 \stackrel{?}{=} -3$$
$$2 - 1 \stackrel{?}{=} -3$$
$$1 = -3 \quad \text{False}$$

Since replacing x with 1 resulted in a false statement, 1 is an extraneous solution. The only solution is 6.

Work Practice 5

Answers

4. $y = 5$ **5.** $x = 8$

Objective B Using the Squaring Property of Equality Twice

If a radical equation contains two radicals, we may need to use the squaring property twice.

Example 6 Solve: $\sqrt{x-4} = \sqrt{x} - 2$

Solution:

$$\sqrt{x-4} = \sqrt{x} - 2$$

$$(\sqrt{x-4})^2 = (\sqrt{x} - 2)^2 \quad \text{Square both sides.}$$

$$x - 4 = x - 4\sqrt{x} + 4$$

$$-8 = -4\sqrt{x} \quad \text{To get the radical term alone, subtract } x \text{ and 4 from both sides.}$$

$$2 = \sqrt{x} \quad \text{Divide both sides by } -4.$$

$$4 = x \quad \text{Square both sides again.}$$

Check the proposed solution in the original equation. The solution is 4.

Work Practice 6

Helpful Hint

Don't forget:

$$(\sqrt{x} - 2)^2 = (\sqrt{x} - 2)(\sqrt{x} - 2)$$
$$= \sqrt{x} \cdot \sqrt{x} - 2\sqrt{x} - 2\sqrt{x} + 4$$
$$= x - 4\sqrt{x} + 4$$

Practice 6

Solve: $\sqrt{x} + 3 = \sqrt{x + 15}$

Answer

6. $x = 1$

Vocabulary, Readiness & Video Check

Martin-Gay Interactive Videos

See Video 8.5

Watch the section lecture video and answer the following questions.

Objective A **1.** From Examples 1 and 2, why must we be sure to check our proposed solution(s) in the original equation?

Objective B **2.** Solving Example 5 requires using the squaring property twice. Is anything else done differently to solve these equations as compared to equations where the property is used only once?

8.5 Exercise Set MyLab Math 

Objective A *Solve each equation. See Examples 1 through 3.*

1. $\sqrt{x} = 9$
2. $\sqrt{x} = 4$
3. $\sqrt{x + 5} = 2$
4. $\sqrt{x + 12} = 3$
5. $\sqrt{x} - 2 = 5$
6. $4\sqrt{x} - 7 = 5$
7. $3\sqrt{x} + 5 = 2$
8. $3\sqrt{x} + 8 = 5$
9. $\sqrt{x} = \sqrt{3x - 8}$
10. $\sqrt{x} = \sqrt{4x - 3}$
11. $\sqrt{4x - 3} = \sqrt{x + 3}$
12. $\sqrt{5x - 4} = \sqrt{x + 8}$

Solve each equation. See Examples 4 and 5.

13. $\sqrt{9x^2 + 2x - 4} = 3x$
14. $\sqrt{4x^2 + 3x - 9} = 2x$
15. $\sqrt{x} = x - 6$
16. $\sqrt{x} = x - 2$
17. $\sqrt{x + 7} = x + 5$
18. $\sqrt{x + 5} = x - 1$
19. $\sqrt{3x + 7} - x = 3$
20. $x = \sqrt{4x - 7} + 1$
21. $\sqrt{16x^2 + 2x + 2} = 4x$
22. $\sqrt{4x^2 + 3x + 2} = 2x$
23. $\sqrt{2x^2 + 6x + 9} = 3$
24. $\sqrt{3x^2 + 6x + 4} = 2$

Objective B *Solve each equation. See Example 6.*

25. $\sqrt{x - 7} = \sqrt{x} - 1$
26. $\sqrt{x - 8} = \sqrt{x} - 2$
27. $\sqrt{x} + 2 = \sqrt{x + 24}$
28. $\sqrt{x} + 5 = \sqrt{x + 55}$
29. $\sqrt{x + 8} = \sqrt{x} + 2$
30. $\sqrt{x} + 1 = \sqrt{x + 15}$

Objectives A B Mixed Practice *Solve each equation. See Examples 1 through 6.*

31. $\sqrt{2x + 6} = 4$
32. $\sqrt{3x + 7} = 5$
33. $\sqrt{x + 6} + 1 = 3$
34. $\sqrt{x + 5} + 2 = 5$
35. $\sqrt{x + 6} + 5 = 3$
36. $\sqrt{2x - 1} + 7 = 1$
37. $\sqrt{16x^2 - 3x + 6} = 4x$
38. $\sqrt{9x^2 - 2x + 8} = 3x$
39. $-\sqrt{x} = -6$
40. $-\sqrt{y} = -8$
41. $\sqrt{x + 9} = \sqrt{x} - 3$
42. $\sqrt{x} - 6 = \sqrt{x + 36}$
43. $\sqrt{2x + 1} + 3 = 5$
44. $\sqrt{3x - 1} + 1 = 4$
45. $\sqrt{x} + 3 = 7$
46. $\sqrt{x} + 5 = 10$
47. $\sqrt{4x} = \sqrt{2x + 6}$
48. $\sqrt{5x + 6} = \sqrt{8x}$
49. $\sqrt{2x + 1} = x - 7$
50. $\sqrt{2x + 5} = x - 5$
51. $x = \sqrt{2x - 2} + 1$
52. $\sqrt{2x - 4} + 2 = x$
53. $\sqrt{1 - 8x} - x = 4$
54. $\sqrt{2x + 5} - 1 = x$

Review

Translating *Translate each sentence into an equation and then solve. See Section 2.4.*

55. If 8 is subtracted from the product of 3 and x, the result is 19. Find x.

56. If 3 more than x is subtracted from twice x, the result is 11. Find x.

57. The length of a rectangle is twice the width. The perimeter is 24 inches. Find the length.

58. The length of a rectangle is 2 inches longer than the width. The perimeter is 24 inches. Find the length.

Concept Extensions

Solve each equation.

59. $\sqrt{x-3}+3=\sqrt{3x+4}$

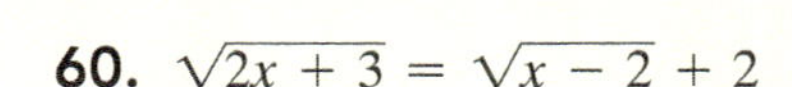

60. $\sqrt{2x+3}=\sqrt{x-2}+2$

61. Explain why proposed solutions of radical equations must be checked in the original equation.

62. Is 8 a solution of the equation $\sqrt{x-4}-5=\sqrt{x+1}$? Explain why or why not.

63. The formula $b=\sqrt{\frac{V}{2}}$ can be used to determine the length b of a side of the base of a square-based pyramid with height 6 units and volume V cubic units.

a. Find the length of the side of the base that produces a pyramid with each volume. (Round to the nearest tenth of a unit.)

V	20	200	2000
b			

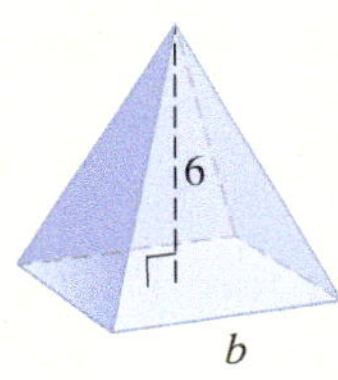

b. Notice in the table that volume V has been increased by a factor of 10 each time. Does the corresponding length b of a side increase by a factor of 10 each time also?

64. The formula $r=\sqrt{\frac{V}{2\pi}}$ can be used to determine the radius r of a cylinder with height 2 units and volume V cubic units.

a. Find the radius needed to manufacture a cylinder with each volume. (Round to the nearest tenth of a unit.)

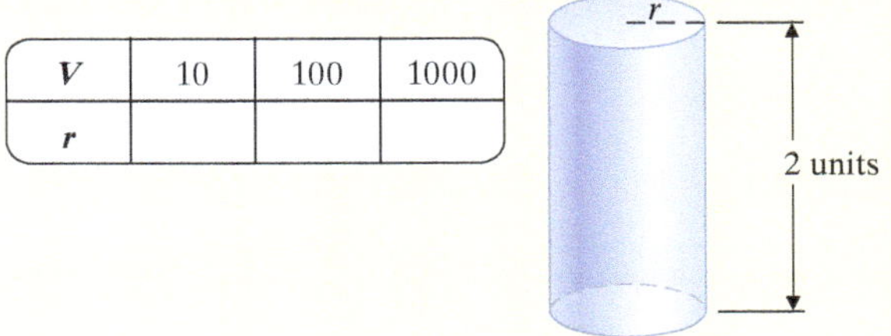

V	10	100	1000
r			

b. Notice in the table that volume V has been increased by a factor of 10 each time. Does the corresponding radius increase by a factor of 10 each time also?

The formula for the radius of a sphere is $r=\sqrt[3]{\frac{3V}{4\pi}}$, where V is the volume in cubic millimeters and r is the radius in millimeters. Use 3.14 as an approximation for π, and round Exercises 65 and 66 to the nearest millimeter.

65. Find the radius of a table tennis ball whose volume is 33,494 cubic millimeters.

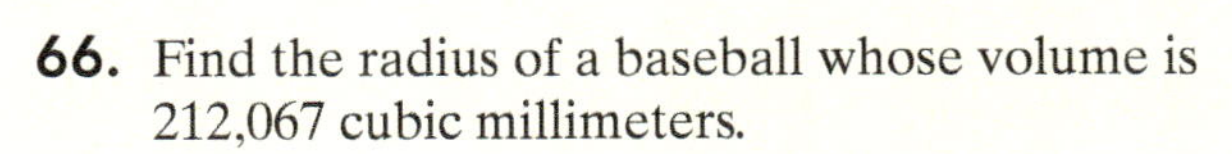

66. Find the radius of a baseball whose volume is 212,067 cubic millimeters.

Graphing calculators can be used to solve equations. To solve $\sqrt{x-2}=x-5$, for example, graph $y_1=\sqrt{x-2}$ and $y_2=x-5$ on the same set of axes. Use the Trace and Zoom features or an Intersect feature to find the point of intersection of the graphs. The x-value of the point is the solution of the equation. Use a graphing calculator to solve the equations below. Approximate solutions to the nearest hundredth.

67. $\sqrt{x-2}=x-5$

68. $\sqrt{x+1}=2x-3$

69. $-\sqrt{x+4}=5x-6$

70. $-\sqrt{x+5}=-7x+1$

8.6 Radical Equations and Problem Solving

Objectives

A Use the Pythagorean Theorem to Solve Problems.

B Solve Problems Using Formulas Containing Radicals.

Objective A Using the Pythagorean Theorem

Applications of radicals can be found in geometry, finance, science, and other areas of technology. Our first application involves the Pythagorean theorem, which gives a formula that relates the lengths of the three sides of a right triangle. We first studied the Pythagorean theorem in Chapter 4, and we review it here.

The Pythagorean Theorem

If a and b are lengths of the legs of a right triangle and c is the length of the hypotenuse, then $a^2 + b^2 = c^2$.

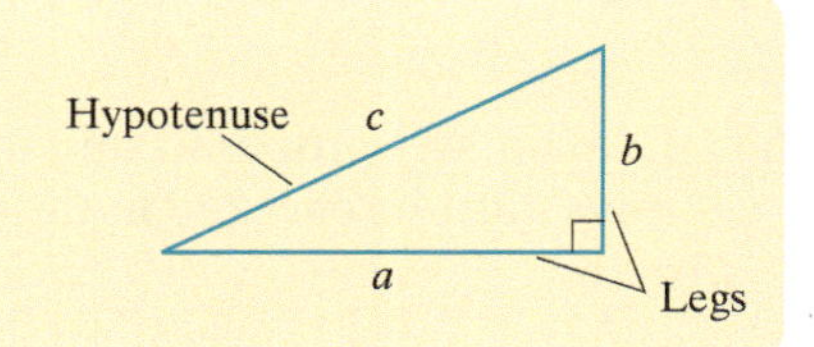

Practice 1

Find the length of the hypotenuse of the right triangle shown.

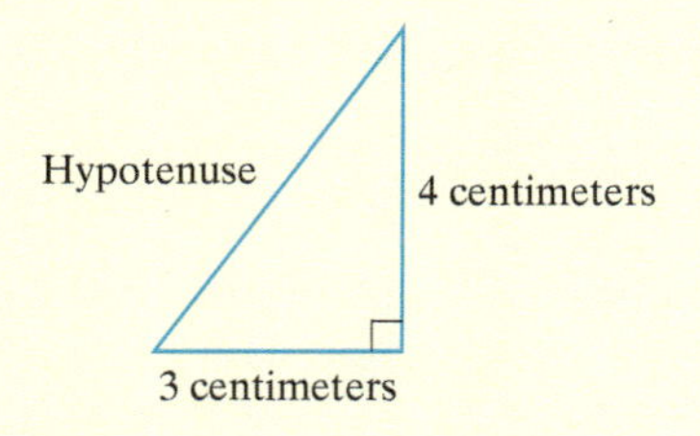

Example 1 Find the length of the hypotenuse of a right triangle whose legs are 6 inches and 8 inches long.

Solution: Because this is a right triangle, we use the Pythagorean theorem. We let $a = 6$ inches and $b = 8$ inches. Length c must be the length of the hypotenuse.

$a^2 + b^2 = c^2$ Use the Pythagorean theorem.

$6^2 + 8^2 = c^2$ Substitute the lengths of the legs.

$36 + 64 = c^2$ Simplify.

$100 = c^2$

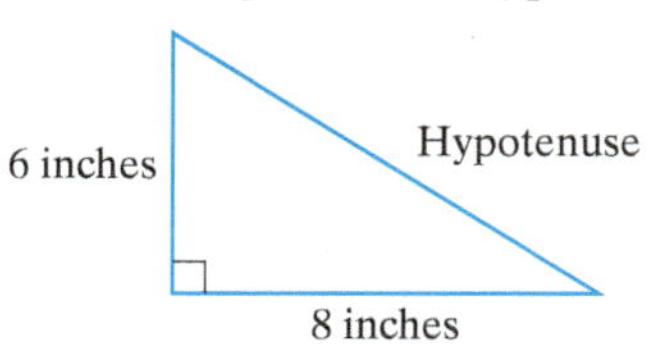

Since $100 = c^2$, then c is a square root of 100. Also, c represents a length, thus we know that c is positive and is the principal square root of 100.

$100 = c^2$

$\sqrt{100} = c$ Use the definition of principal square root.

$10 = c$ Simplify.

The hypotenuse has a length of 10 inches.

Work Practice 1

Helpful Hint

In Section 9.1, we formally introduce the Square Root Property: If $x^2 = a$ (for a nonnegative a), then $x = \pm\sqrt{a}$. Use these examples to prepare yourselves for this property.

Practice 2

Find the length of the leg of the right triangle shown. Give the exact length and a two-decimal-place approximation.

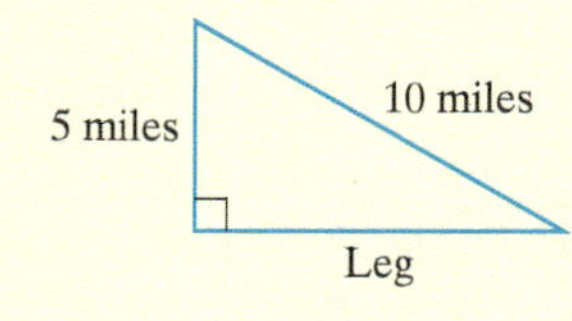

Example 2 Find the length of the leg of the right triangle shown. Give the exact length and a two-decimal-place approximation.

Solution: We let $a = 2$ meters and b be the unknown length of the other leg. The hypotenuse is $c = 5$ meters.

$a^2 + b^2 = c^2$ Use the Pythagorean theorem.

$2^2 + b^2 = 5^2$ Let $a = 2$ and $c = 5$.

$4 + b^2 = 25$

$b^2 = 21$

$b = \sqrt{21} \approx 4.58$ meters

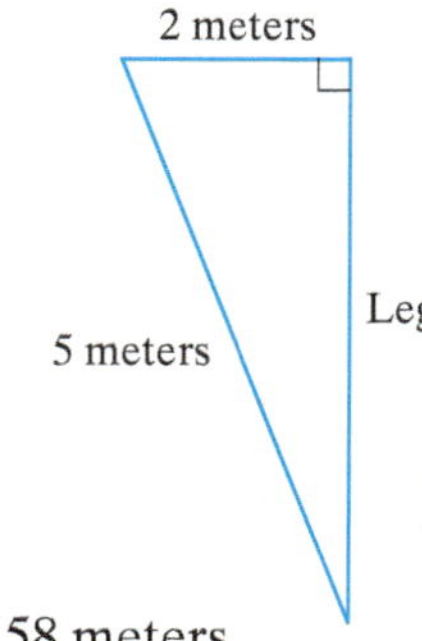

The length of the leg is exactly $\sqrt{21}$ meters or approximately 4.58 meters.

Work Practice 2

Answers
1. 5 cm 2. $5\sqrt{3}$ mi; 8.66 mi

Example 3 Finding a Distance

A surveyor must determine the distance across a lake at points P and Q, as shown in the figure. To do this, she finds a third point, R, such that line QR is perpendicular to line PQ. If the length of $\overline{PR}$ is 320 feet and the length of $\overline{QR}$ is 240 feet, what is the distance across the lake? Approximate this distance to the nearest whole foot.

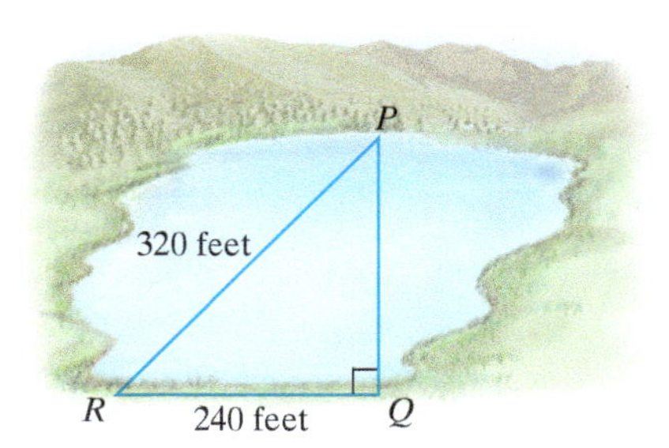

Solution:

1. UNDERSTAND. Read and reread the problem. We will set up the problem using the Pythagorean theorem. By creating a line perpendicular to line PQ, the surveyor deliberately constructed a right triangle. The hypotenuse, $\overline{PR}$, has a length of 320 feet, so we let $c = 320$ in the Pythagorean theorem. The side $\overline{QR}$ is one of the legs, so we let $a = 240$ and $b =$ the unknown length.

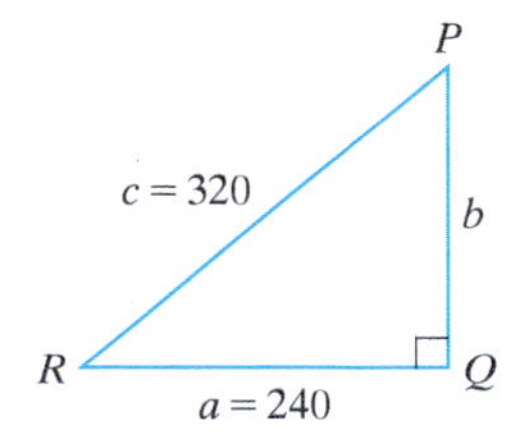

2. TRANSLATE.

$a^2 + b^2 = c^2$ Use the Pythagorean theorem.

$240^2 + b^2 = 320^2$ Let $a = 240$ and $c = 320$.

3. SOLVE.

$57{,}600 + b^2 = 102{,}400$

$b^2 = 44{,}800$ Subtract 57,600 from both sides.

$b = \sqrt{44{,}800}$ Use the definition of principal square root.

$= 80\sqrt{7}$ Simplify.

4. INTERPRET.

Check: See that $240^2 + (\sqrt{44{,}800})^2 = 320^2$.

State: The distance across the lake is *exactly* $\sqrt{44{,}800}$ or $80\sqrt{7}$ feet. The surveyor can now use a calculator to find that $80\sqrt{7}$ feet is *approximately* 211.6601 feet, so the distance across the lake is roughly 212 feet.

Work Practice 3

Practice 3

Evan Saacks wants to determine the distance at certain points across a pond on his property. He is able to measure the distances shown on the following diagram. Find how wide the pond is to the nearest tenth of a foot.

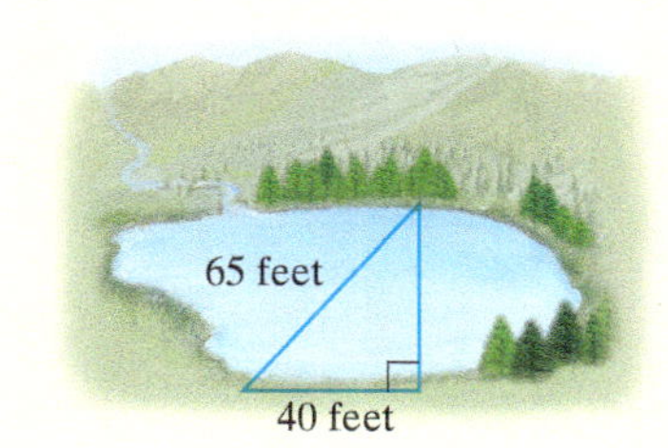

Objective B Using Formulas Containing Radicals

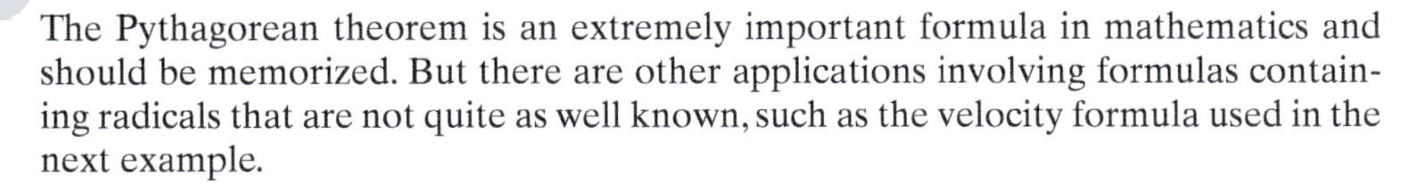

The Pythagorean theorem is an extremely important formula in mathematics and should be memorized. But there are other applications involving formulas containing radicals that are not quite as well known, such as the velocity formula used in the next example.

Answer

3. 51.2 feet

Practice 4

Use the formula from Example 4 to find the velocity of an object after it has fallen 20 feet. Find the exact value and an approximation rounded to the nearest tenth.

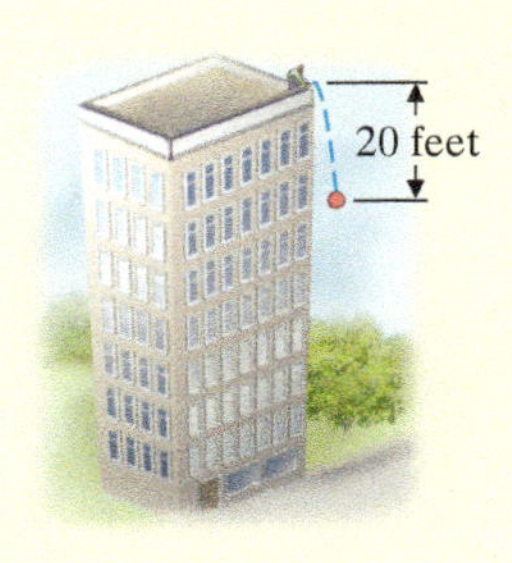

Answer

4. $16\sqrt{5}$ ft per sec $\approx$ 35.8 ft per sec

Example 4 Finding the Velocity of an Object

A formula used to determine the velocity v, in feet per second, of an object after it has fallen a certain height (neglecting air resistance) is $v = \sqrt{2gh}$, where g is the acceleration due to gravity and h is the height the object has fallen. On Earth, the acceleration g due to gravity is approximately 32 feet per second per second. Find the velocity of a person after falling 5 feet. Round to the nearest tenth.

Solution: We are told that $g = 32$ feet per second per second. To find the velocity v when $h = 5$ feet, we use the velocity formula.

$$\begin{aligned} v &= \sqrt{2gh} && \text{Use the velocity formula.} \\ &= \sqrt{2 \cdot 32 \cdot 5} && \text{Substitute known values.} \\ &= \sqrt{320} \\ &= 8\sqrt{5} && \text{Simplify the radicand.} \end{aligned}$$

The velocity of the person after falling 5 feet is *exactly* $8\sqrt{5}$ feet per second, or *approximately* 17.9 feet per second.

Work Practice 4

Vocabulary, Readiness & Video Check

Martin-Gay Interactive Videos

See Video 8.6

Watch the section lecture video and answer the following questions.

Objective A

1. From Examples 1 and 2, when solving exercises using the Pythagorean theorem, what two things must we keep in mind?
2. What very important point is made as the final answer to Example 1 is being found?

Objective B

3. In Example 4, how do we know to give an estimated answer instead of an exact answer? In what form would the exact answer be given?

8.6 Exercise Set MyLab Math 

Objective A *Use the Pythagorean theorem to find the length of the unknown side of each right triangle. Give the exact answer and a two-decimal-place approximation. See Examples 1 and 2.*

1.

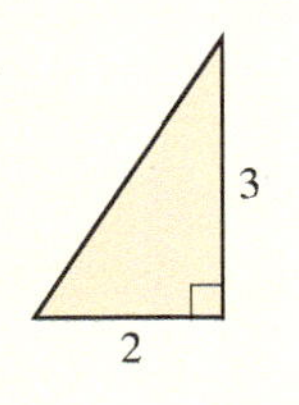

2.

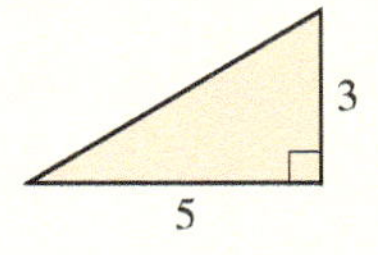

3.

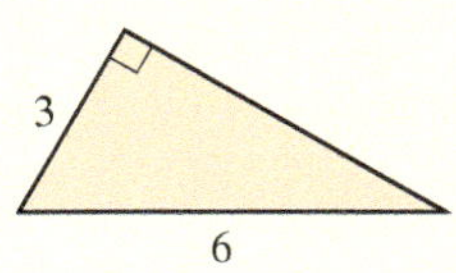

4.

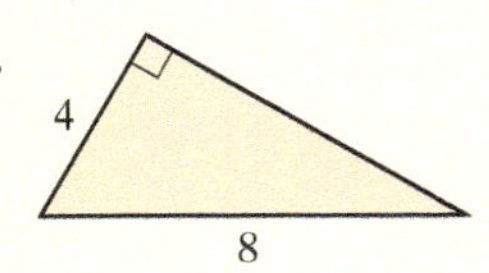

5.

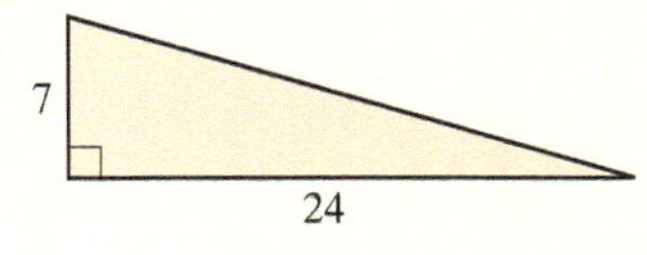

6.

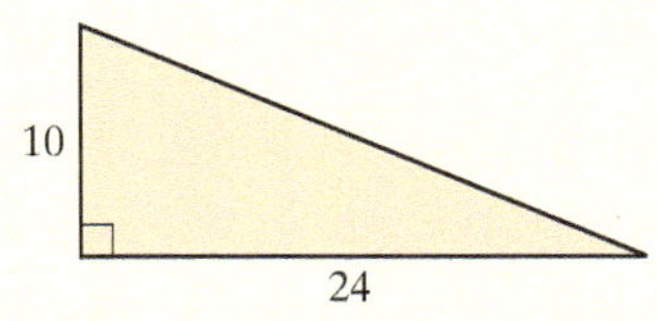

7.

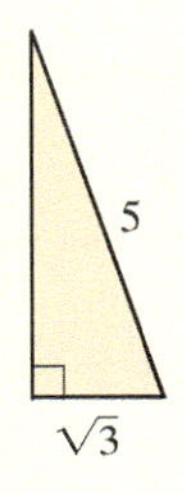

8.

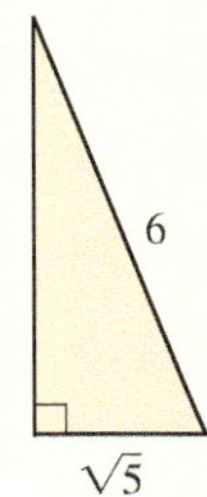

9.

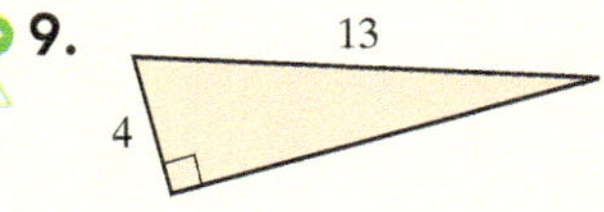

10.

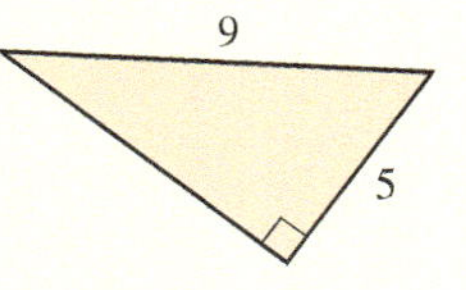

Find the length of the unknown side of each right triangle with sides a, b, and c, where c is the hypotenuse. See Examples 1 and 2. Give the exact answer and a two-decimal-place approximation.

11. $a = 4, b = 5$

12. $a = 2, b = 7$

13. $b = 2, c = 6$

14. $b = 1, c = 5$

15. $a = \sqrt{10}, c = 10$

16. $a = \sqrt{7}, c = \sqrt{35}$

Solve each problem. See Example 3.

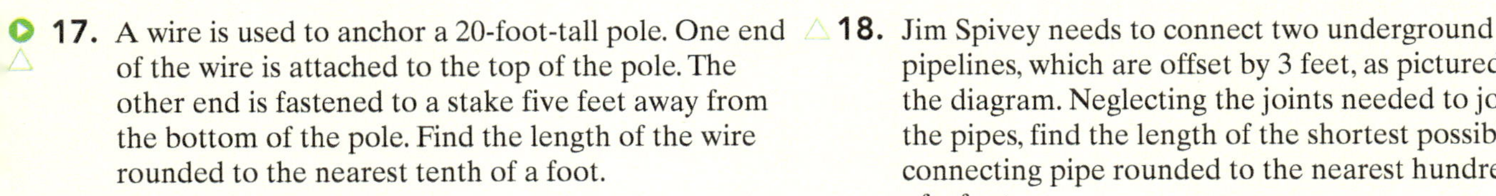

17. A wire is used to anchor a 20-foot-tall pole. One end of the wire is attached to the top of the pole. The other end is fastened to a stake five feet away from the bottom of the pole. Find the length of the wire rounded to the nearest tenth of a foot.

18. Jim Spivey needs to connect two underground pipelines, which are offset by 3 feet, as pictured in the diagram. Neglecting the joints needed to join the pipes, find the length of the shortest possible connecting pipe rounded to the nearest hundredth of a foot.

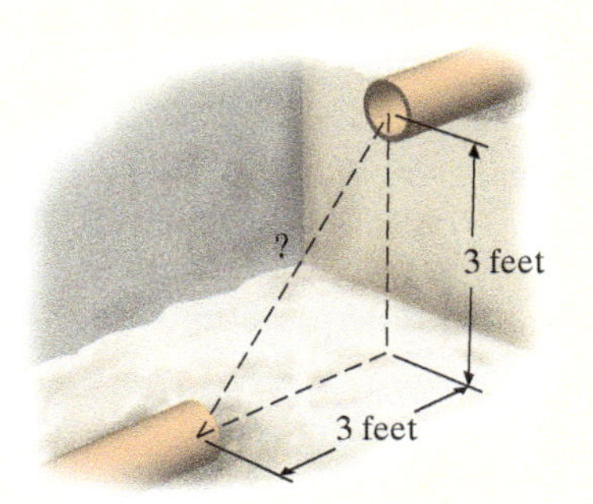

△ **19.** Robert Weisman needs to attach a diagonal brace to a rectangular frame in order to make it structurally sound. If the framework is 6 feet by 10 feet, find how long the brace needs to be to the nearest tenth of a foot.

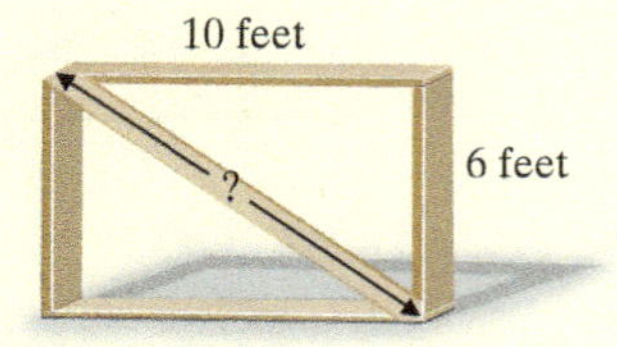

△ **20.** Elizabeth Kaster is flying a kite. She let out 80 feet of string and attached the string to a stake in the ground. The kite is now directly above her brother Mike, who is 32 feet away from the stake. Find the height of the kite to the nearest foot.

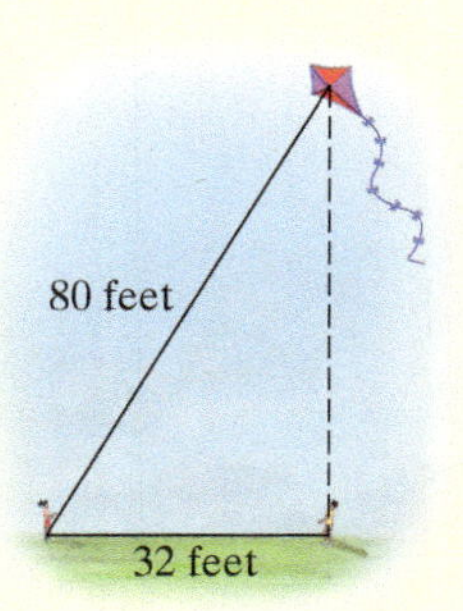

Objective B *Solve each problem. See Example 4.*

△ **21.** For a square-based pyramid, the formula $b = \sqrt{\frac{3V}{h}}$ describes the relationship among the length b of one side of the base, the volume V, and the height h. Find the volume if each side of the base is 6 feet long, and the pyramid is 2 feet high.

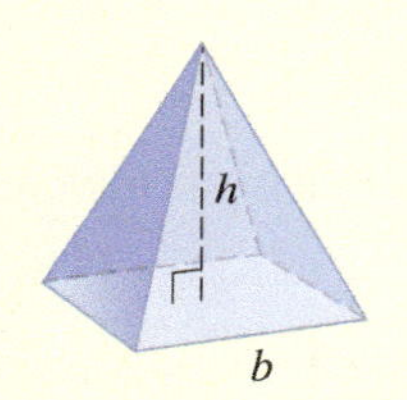

22. The formula $t = \frac{\sqrt{d}}{4}$ relates the distance d, in feet, that an object falls in t seconds, assuming that air resistance does not slow down the object. Find how long, to the nearest hundredth of a second, it takes an object to reach the ground from the top of the Willis Tower in Chicago, a distance of 1730 feet. (*Source:* Council on Tall Buildings and Urban Habitat)

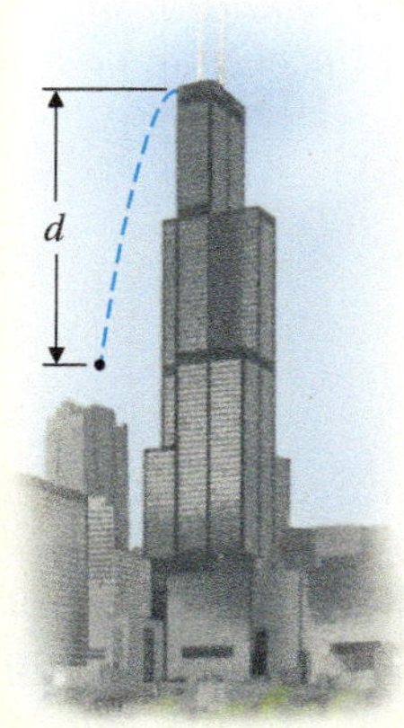

23. Police use the formula $s = \sqrt{30fd}$ to estimate the speed s of a car just before it skidded. In this formula, the speed s is measured in miles per hour, d represents the distance the car skidded in feet, and f represents the coefficient of friction. The value of f depends on the type of road surface, and for wet concrete f is 0.35. Find how fast a car was moving if it skidded 280 feet on wet concrete. Round your result to the nearest mile per hour.

24. The coefficient of friction of a certain dry road is 0.95. Use the formula in Exercise **23** to find how far a car will skid on this dry road if it is traveling at a rate of 60 mph. Round the length to the nearest foot.

25. The formula $v = \sqrt{2.5r}$ can be used to estimate the maximum safe velocity, v, in miles per hour, at which a car can travel if it is driven along a curved road with a **radius of curvature** r in feet. Find the maximum safe speed to the nearest whole number if a cloverleaf exit on an expressway has a radius of curvature of 300 feet.

26. Use the formula from Exercise **25** to find the radius of curvature if the safe velocity is 30 mph.

The maximum distance d in kilometers that you can see from a height of h meters is given by $d = 3.5\sqrt{h}$. *Use this equation for Exercises 27 through 30.*

27. Find how far you can see from the top of the Comcast Building in New York City, a height of 259.1 meters. Round to the nearest tenth of a kilometer. (*Source:* Council on Tall Buildings and Urban Habitat)

28. Find how far you can see from the top of Great American Tower at Queen City Square in Cincinnati, Ohio, a height of 202.7 meters. Round to the nearest tenth of a kilometer. (*Source:* Council on Tall Buildings and Urban Habitat)

29. The newly built One World Trade Center, in New York City, is the tallest building in the Western Hemisphere. Its height, including the spire at the top of the building, is 541.3 meters. Find how far you could see from the top of One World Trade Center's spire. Round to the nearest tenth of a kilometer. (*Source:* Council on Tall Buildings and Urban Habitat)

30. Guests can take in the views from One World Trade Center by visiting the building's One World Observatory, located at a height of 386.1 meters. Find how far a visitor to One World Observatory could see. Round to the nearest tenth of a kilometer. (*Source:* Council on Tall Buildings and Urban Habitat)

Review

Find two numbers whose square is the given number. See Section 8.1.

31. 9

32. 25

33. 100

34. 49

35. 64

36. 121

Concept Extensions

For each triangle, find the length of y, and then x.

△ **37.**

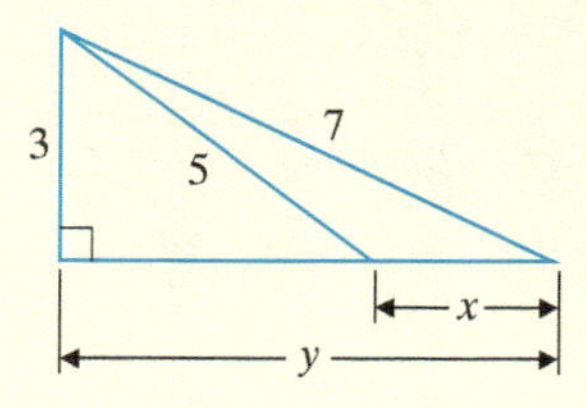

△ **38.**

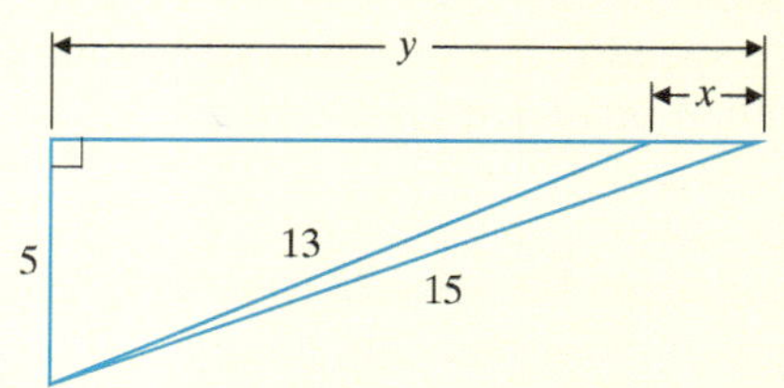

△ **39.** Mike and Sandra Hallahan leave the seashore at the same time. Mike drives northward at a rate of 30 miles per hour, while Sandra drives west at 60 mph. Find how far apart they are after 3 hours to the nearest mile.

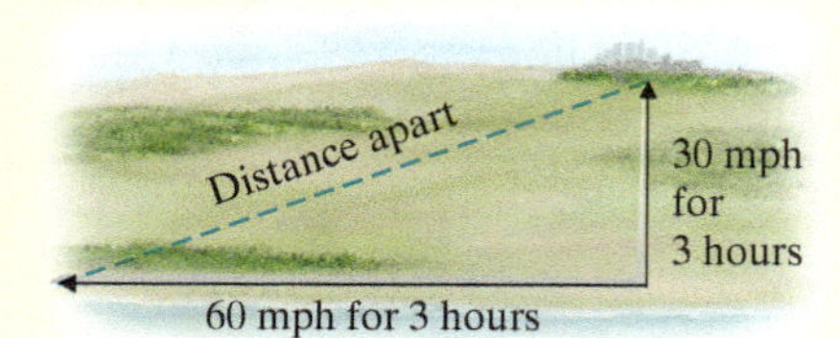

△ **40.** Railroad tracks are invariably made up of relatively short sections of rail connected by expansion joints. To see why this construction is necessary, consider a single rail 100 feet long (or 1200 inches). On an extremely hot day, suppose it expands 1 inch in the hot sun to a new length of 1201 inches. Theoretically, the track would bow upward as pictured.

Let us approximate the bulge in the railroad this way.

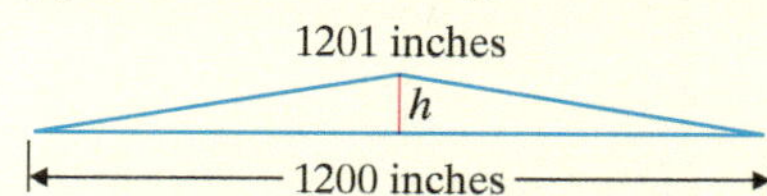

Calculate the height h of the bulge to the nearest tenth of an inch.

41. Based on the results of Exercise **40**, explain why railroads use short sections of rail connected by expansion joints.

42. In your own words, explain the Pythagorean theorem, and include the type(s) of triangle this theorem applies to.

8.7 Direct and Inverse Variation Including Radical Applications

Objectives

A Solve Problems Involving Direct Variation.

B Solve Problems Involving Inverse Variation.

C Solve Problems Involving Other Types of Direct and Inverse Variation.

D Solve Applications of Variation.

Thus far, we have studied linear equations in two variables. Recall that such an equation can be written in the form $Ax + By = C$, where A and B are not both 0. Also recall that the graph of a linear equation in two variables is a line. In this section, we begin by looking at a particular family of linear equations—those that can be written in the form

$$y = kx$$

where k is a constant. This family of equations is called *direct variation.*

Objective A Solving Direct Variation Problems 

Let's suppose that you are earning minimum wage, \$7.25 per hour, at a part-time job. The amount of money you earn depends on the number of hours you work. This is illustrated by the following table:

Hours Worked	0	1	2	3	4
Money Earned (before deductions)	0	7.25	14.50	21.75	29.00

and so on

In general, to calculate your earnings (before deductions), multiply the constant \$7.25 by the number of hours you work. If we let y represent the amount of money earned and x represent the number of hours worked, we get the direct variation equation

$$y = 7.25 \cdot x$$

earnings = \$7.25 · hours worked

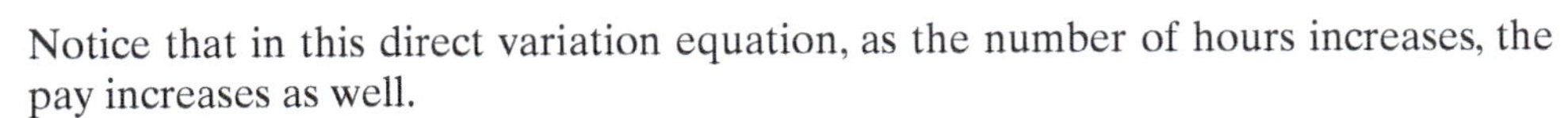

Notice that in this direct variation equation, as the number of hours increases, the pay increases as well.

Direct Variation

y varies directly as x, or y is directly proportional to x, if there is a nonzero constant k such that

$$y = kx$$

The number k is called the **constant of variation** or the **constant of proportionality.**

In our direct variation example, $y = 7.25x$, the constant of variation is 7.25.

Let's use the previous table to graph $y = 7.25x$. We begin our graph at the ordered pair solution (0, 0). Why? We assume that the least amount of hours worked is 0. If 0 hours are worked, then the pay is \$0.

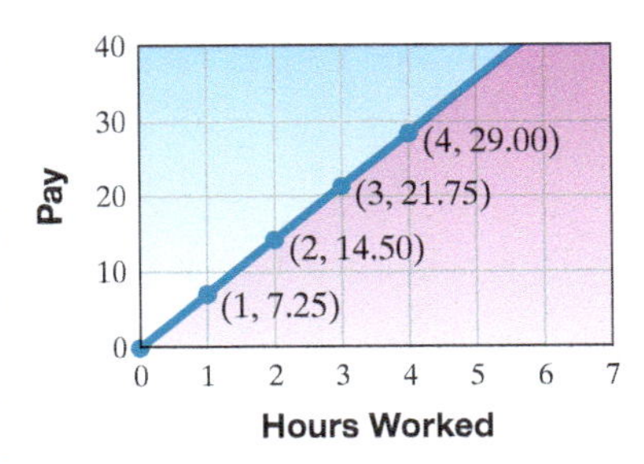

As illustrated in this graph, a direct variation equation $y = kx$ is linear. Also notice that $y = 7.25x$ is a function since its graph passes the vertical line test.

Example 1

Write a direct variation equation of the form $y = kx$ that satisfies the ordered pairs in the table below.

x	2	9	1.5	−1
y	6	27	4.5	−3

Solution: We are given that there is a direct variation relationship between x and y. This means that

$$y = kx$$

By studying the given values, you may be able to mentally calculate k. If not, to find k, we simply substitute one given ordered pair into this equation and solve for k. We'll use the given pair (2, 6).

$$y = kx$$
$$6 = k \cdot 2$$
$$\frac{6}{2} = \frac{k \cdot 2}{2}$$
$$3 = k \quad \text{Solve for } k.$$

Since $k = 3$, we have the equation $y = 3x$.

To check, see that each given y is 3 times the given x.

Work Practice 1

Practice 1

Write a direct variation equation that satisfies:

x	4	$\frac{1}{2}$	1.5	6
y	8	1	3	12

Answer

1. $y = 2x$

Practice 2

Suppose that y varies directly as x. If y is 15 when x is 45, find the constant of variation and the direct variation equation. Then find y when x is 3.

Let's try another type of direct variation example.

Example 2 Suppose that y varies directly as x. If y is 17 when x is 34, find the constant of variation and the direct variation equation. Then find y when x is 12.

Solution: Let's use the same method as in Example 1 to find k. Since we are told that y varies directly as x, we know the relationship is of the form

$y = kx$

Let $y = 17$ and $x = 34$ and solve for k.

$17 = k \cdot 34$

$$\frac{17}{34} = \frac{k \cdot 34}{34}$$

$$\frac{1}{2} = k \quad \text{Solve for } k.$$

Thus, the constant of variation is $\frac{1}{2}$ and the equation is $y = \frac{1}{2}x$.

To find y when $x = 12$, use $y = \frac{1}{2}x$ and replace x with 12.

$$y = \frac{1}{2}x$$

$$y = \frac{1}{2} \cdot 12 \quad \text{Replace } x \text{ with 12.}$$

$y = 6$

Thus, when x is 12, y is 6.

Work Practice 2

Let's review a few facts about linear equations of the form $y = kx$.

Direct Variation: $y = kx$

- There is a direct variation relationship between x and y.
- The graph is a line.
- The line will always go through the origin $(0, 0)$. Why?
  Let $x = 0$. Then $y = k \cdot 0$ or $y = 0$.
- The slope of the graph of $y = kx$ is k, the constant of variation. Why? Remember that the slope of an equation of the form $y = mx + b$ is m, the coefficient of x.
- The equation $y = kx$ describes a function. Each x has a unique y and its graph passes the vertical line test.

Practice 3

Find the constant of variation and the direct variation equation for the line below.

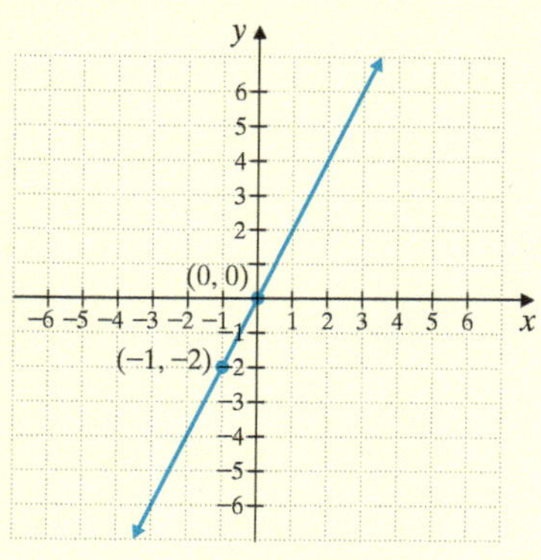

Example 3 The line is the graph of a direct variation equation. Find the constant of variation and the direct variation equation.

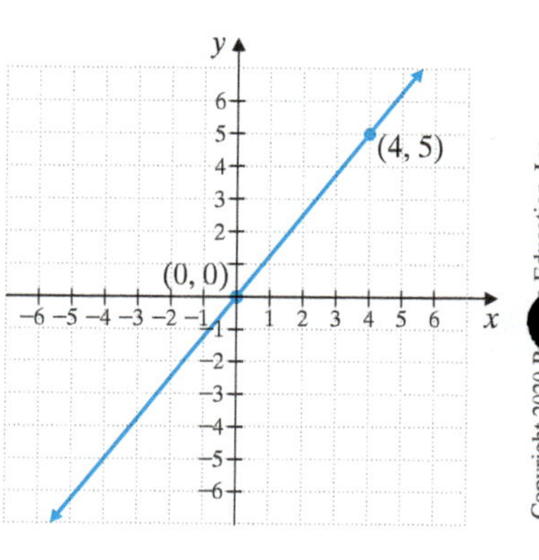

Answers

2. $k = \frac{1}{3}; y = \frac{1}{3}x; y = 1$
3. $k = 2; y = 2x$

Solution: Recall that k, the constant of variation, is the same as the slope of the line. Thus, to find k, we use the slope formula and find slope.

Using the given points (0, 0) and (4, 5), we have

$$\text{slope} = \frac{5-0}{4-0} = \frac{5}{4}$$

Thus, $k = \frac{5}{4}$ and the variation equation is $y = \frac{5}{4}x$.

Work Practice 3

Objective B Solving Inverse Variation Problems 

In this objective, we introduce another type of variation called inverse variation.

Let's suppose you need to drive a distance of 40 miles. You know that the faster you drive the distance, the sooner you arrive at your destination. Recall that there is a mathematical relationship (formula) between distance, rate, and time. It is $d = r \cdot t$. In our example, distance is a constant 40 miles, so we have $40 = r \cdot t$ or $t = \frac{40}{r}$.

For example, if you drive 10 mph, the time to drive the 40 miles is

$$t = \frac{40}{r} = \frac{40}{10} = 4 \text{ hours}$$

If you drive 20 mph, the time is

$$t = \frac{40}{r} = \frac{40}{20} = 2 \text{ hours}$$

Again, notice that as speed increases, time decreases. Below are some ordered pair solutions of $t = \frac{40}{r}$ and its graph.

Rate (mph)	r	5	10	20	40	60	80
Time (hr)	t	8	4	2	1	$\frac{2}{3}$	$\frac{1}{2}$

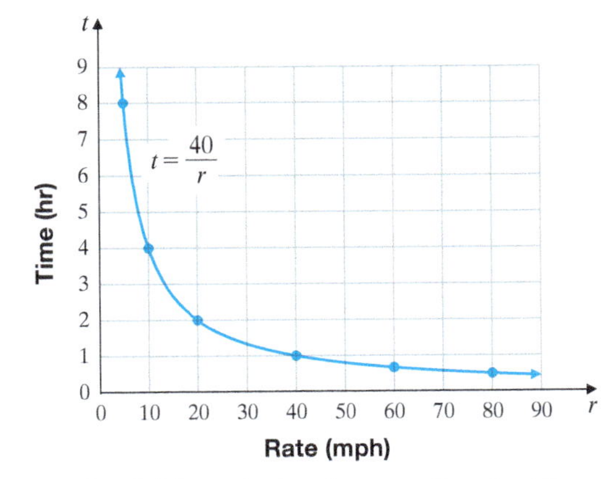

Notice that the graph of this variation is not a line, but it passes the vertical line test so $t = \frac{40}{r}$ does describe a function. This is an example of inverse variation.

> **Inverse Variation**
>
> ***y* varies inversely as *x*,** or ***y* is inversely proportional to *x*,** if there is a nonzero constant k such that
>
> $$y = \frac{k}{x}$$
>
> The number k is called the **constant of variation** or the **constant of proportionality.**

In our inverse variation example, $t = \frac{40}{r}$ or $y = \frac{40}{x}$, the constant of variation is 40.

We can immediately see differences and similarities in direct variation and inverse variation.

Direct Variation	$y = kx$	linear equation	both functions
Inverse Variation	$y = \frac{k}{x}$	rational equation	

In Chapter 5, we called $y = \frac{k}{x}$ a rational equation and not a linear equation. Notice that the graph of our inverse variation $t = \frac{40}{r}$ is thus not a line. Also notice that because x is in the denominator, x can be any value except 0.

We can still derive an inverse variation equation from a table of values.

Practice 4

Write an inverse variation equation of the form $y = \frac{k}{x}$ that satisfies:

x	4	10	40	−2
y	5	2	$\frac{1}{2}$	−10

Example 4 Write an inverse variation equation of the form $y = \frac{k}{x}$ that satisfies the ordered pairs in the table below.

x	2	4	$\frac{1}{2}$
y	6	3	24

Solution: Since there is an inverse variation relationship between x and y, we know that $y = \frac{k}{x}$.

To find k, choose one given ordered pair and substitute the values into the equation. We'll use (2, 6).

$$y = \frac{k}{x}$$

$$6 = \frac{k}{2}$$

$$2 \cdot 6 = 2 \cdot \frac{k}{2} \quad \text{Multiply both sides by 2.}$$

$$12 = k \quad \text{Solve.}$$

Since $k = 12$, we have the equation $y = \frac{12}{x}$.

Work Practice 4

Helpful Hint

Multiply both sides of the inverse variation relationship equation $y = \frac{k}{x}$ by x (as long as x is not 0), and we have $xy = k$. This means that if y varies inversely as x, their product is always the constant of variation k. For an example of this, check the table from Example 4:

x	2	4	$\frac{1}{2}$
y	6	3	24

$2 \cdot 6 = 12$ $4 \cdot 3 = 12$ $\frac{1}{2} \cdot 24 = 12$

Answer

4. $y = \frac{20}{x}$

Example 5 Suppose that y varies inversely as x. If $y = 0.02$ when $x = 75$, find the constant of variation and the inverse variation equation. Then find y when x is 30.

Solution: Since y varies inversely as x, the constant of variation may be found by simply finding the product of the given x and y.

$$k = xy = 75(0.02) = 1.5$$

To check, we will use the inverse variation equation

$$y = \frac{k}{x}$$

Let $y = 0.02$ and $x = 75$ and solve for k.

$$0.02 = \frac{k}{75}$$

$$75(0.02) = 75 \cdot \frac{k}{75} \quad \text{Multiply both sides by 75.}$$

$$1.5 = k \quad \text{Solve for } k.$$

Thus, the constant of variation is 1.5 and the equation is $y = \frac{1.5}{x}$. To find y when $x = 30$, use $y = \frac{1.5}{x}$ and replace x with 30.

$$y = \frac{1.5}{x}$$

$$y = \frac{1.5}{30} \quad \text{Replace } x \text{ with 30.}$$

$$y = 0.05$$

Thus, when x is 30, y is 0.05.

Work Practice 5

Practice 5

Suppose that y varies inversely as x. If y is 4 when x is 0.8, find the constant of variation and the inverse variation equation. Then find y when x is 20.

Objective C Solving Other Types of Direct and Inverse Variation Problems

It is possible for y to vary directly or inversely as powers (or roots) of x.

Direct and Inverse Variation as *n*th Powers (or Roots) of *x*

y* varies directly as a power of *x if there is a nonzero constant k and a natural number n such that

$$y = kx^n$$

y* varies inversely as a power of *x if there is a nonzero constant k and a natural number n such that

$$y = \frac{k}{x^n}$$

Example 6 The surface area of a cube A varies directly as the square of a length of its sides. If A is 54 when s is 3, find A when $s = 4.2$.

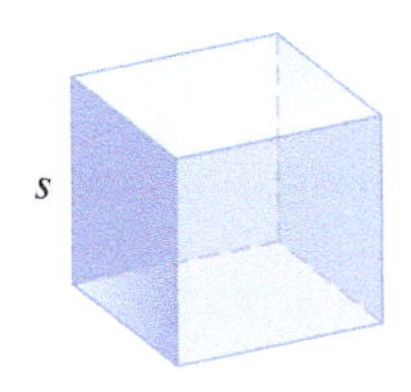

(*Continued on next page*)

Practice 6

The area of a circle varies directly as the square of its radius. A circle with radius 7 inches has an area of 49π square inches. Find the area of a circle whose radius is 4 feet.

Answers

5. $k = 3.2; y = \frac{3.2}{x}; y = 0.16$

6. 16π sq ft

Solution: Since the surface area A varies directly as the square of side s, we have

$A = ks^2$

To find k, let $A = 54$ and $s = 3$.

$A = k \cdot s^2$

$54 = k \cdot 3^2$ Let $A = 54$ and $s = 3$.

$54 = 9k$ $3^2 = 9$.

$6 = k$ Divide by 9.

The formula for surface area of a cube is then

$A = 6s^2$, where s is the length of a side.

To find the surface area when $s = 4.2$, substitute.

$A = 6s^2$

$A = 6 \cdot (4.2)^2$

$A = 105.84$

The surface area of a cube whose side measures 4.2 units is 105.84 sq units.

Work Practice 6

Below are additional examples of direct and inverse variation.

Additional Examples of Direct and Inverse Variation			
Direct		**Inverse**	
F varies directly as the square of r	$F = kr^2$	y varies inversely as the cube of x	$y = \frac{k}{x^3}$
y varies directly as the square root of x	$y = k\sqrt{x}$	m varies inversely as the square root of n	$m = \frac{k}{\sqrt{n}}$

Objective D Solving Applications of Variation 

There are many real-life applications of direct and inverse variation.

Practice 7

The distance d that an object falls is directly proportional to the square of the time of the fall, t. If an object falls 144 feet in 3 seconds, find how far the object falls in 5 seconds.

Example 7 The weight of a body w varies inversely with the square of its distance from the center of Earth, d. If a person weighs 160 pounds on the surface of Earth, what is the person's weight 200 miles above the surface? (Assume that the radius of Earth is 4000 miles.)

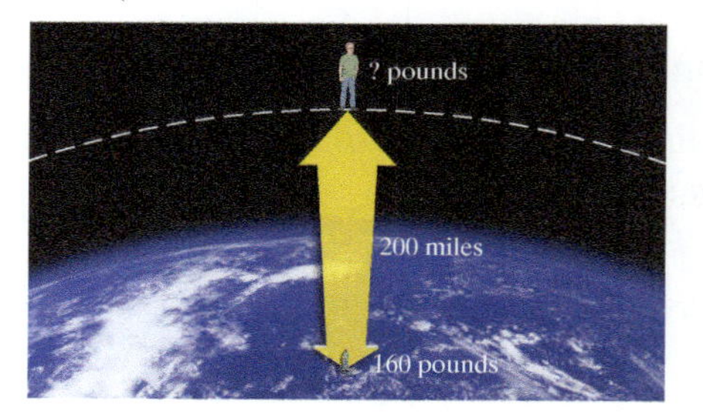

Solution:

1. UNDERSTAND. Make sure you read and reread the problem.
2. TRANSLATE. Since we are told that weight w varies inversely with the square of its distance from the center of Earth, d, we have

$$w = \frac{k}{d^2}$$

3. SOLVE. To solve the problem, we first find k. To do so, we use the fact that the person weighs 160 pounds on Earth's surface, which is a distance of 4000 miles from the Earth's center.

Answer

7. 400 feet

$$w = \frac{k}{d^2}$$

$$160 = \frac{k}{(4000)^2}$$

$$2{,}560{,}000{,}000 = k$$

Thus, we have $w = \frac{2{,}560{,}000{,}000}{d^2}$.

Since we want to know the person's weight 200 miles above the Earth's surface, we let $d = 4200$ and find w.

$$w = \frac{2{,}560{,}000{,}000}{d^2}$$

$$w = \frac{2{,}560{,}000{,}000}{(4200)^2}$$ A person 200 miles above the Earth's surface is 4200 miles from the Earth's center.

$$w \approx 145$$ Simplify.

4. INTERPRET.

Check: Your answer is reasonable since the farther a person is from Earth, the less the person weighs.

State: Thus, 200 miles above the surface of the Earth, a 160-pound person weighs approximately 145 pounds.

Work Practice 7

Vocabulary, Readiness & Video Check

State whether each equation represents direct or inverse variation.

1. $y = \frac{k}{x}$, where k is a constant. ________

2. $y = kx$, where k is a constant. ________

3. $y = 5x$ ________

4. $y = \frac{5}{x}$ ________

5. $y = \frac{7}{x^2}$ ________

6. $y = 6.5x^4$ ________

7. $y = \frac{11}{x}$ ________

8. $y = 18x$ ________

9. $y = 12x^2$ ________

10. $y = \frac{20}{x^3}$ ________

Martin-Gay Interactive Videos

See Video 8.7

Watch the section lecture video and answer the following questions.

Objective A 11. Based on the lecture before Example 1, what kind of equation is a direct variation equation? What does k represent in this equation?

Objective B 12. In Example 4, why is it not necessary to place the given values of x and y into the inverse variation equation in order to find k?

Objective C 13. From Examples 5–7, does a power on x change the basic direct and inverse variation formula relationships?

Objective D 14. In Example 8, why is it reasonable to expect our answer to be a greater distance than the original distance given in the problem?

8.7 Exercise Set MyLab Math

Objective A *Write a direct variation equation, $y = kx$, that satisfies the ordered pairs in each table. See Example 1.*

1.

x	0	6	10
y	0	3	5

2.

x	0	2	−1	3
y	0	14	−7	21

3.

x	−2	2	4	5
y	−12	12	24	30

4.

x	3	9	−2	12
y	1	3	$-\frac{2}{3}$	4

Write a direct variation equation, $y = kx$, that describes each graph. See Example 3.

5.

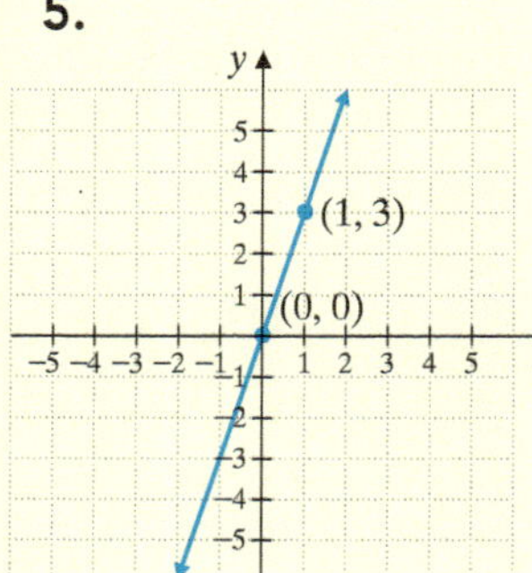

6.

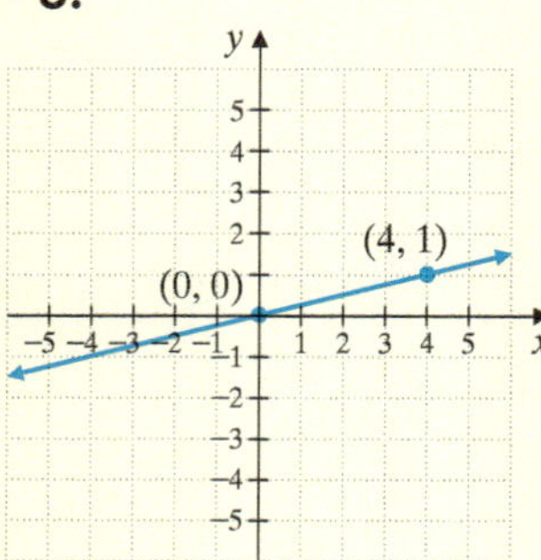

7.

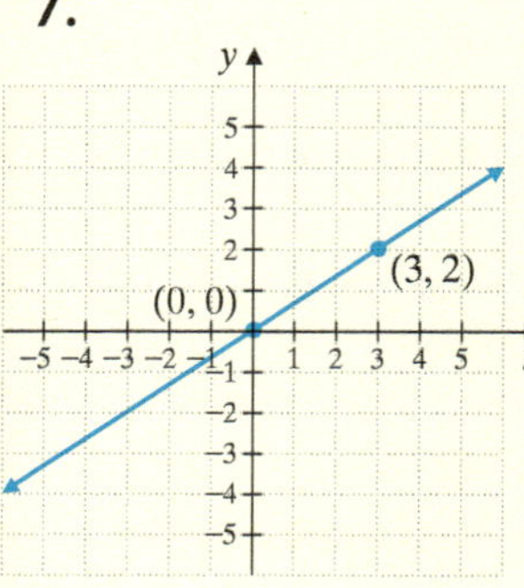

8.

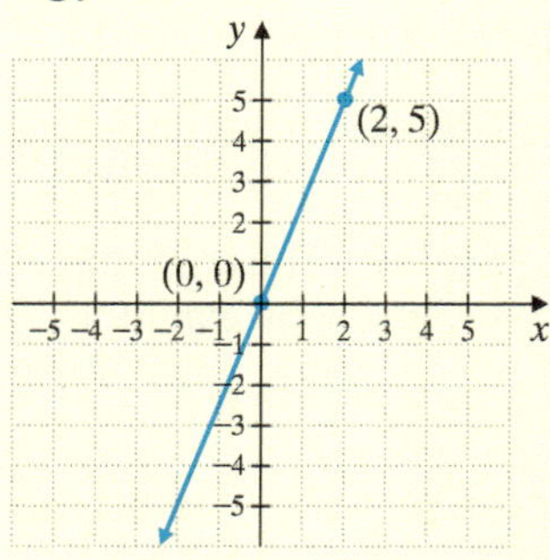

Objective B *Write an inverse variation equation, $y = \frac{k}{x}$, that satisfies the ordered pairs in each table. See Example 4.*

9.

x	1	−7	3.5	−2
y	7	−1	2	−3.5

10.

x	2	−11	4	−4
y	11	−2	5.5	−5.5

11.

x	10	$\frac{1}{2}$	$-\frac{1}{4}$
y	0.05	1	−2

12.

x	4	$\frac{1}{5}$	−8
y	0.1	2	−0.05

Objectives A B C Translating *Write an equation to describe each variation. Use k for the constant of proportionality. See Examples 1 through 6.*

13. y varies directly as x.

14. a varies directly as b.

15. h varies inversely as t.

16. s varies inversely as t.

17. z varies directly as x^2.

18. p varies inversely as x^2.

19. y varies inversely as z^3.

20. x varies directly as y^4.

21. y varies directly as $\sqrt{x}$.

22. x varies inversely as $\sqrt{y}$.

Objectives A B C Mixed Practice *Solve. See Examples 2, 5, and 6.*

23. y varies directly as x. If $y = 20$ when $x = 5$, find y when x is 10.

24. y varies directly as x. If $y = 27$ when $x = 3$, find y when x is 2.

25. y varies inversely as x. If $y = 5$ when $x = 60$, find y when x is 100.

26. y varies inversely as x. If $y = 200$ when $x = 5$, find y when x is 4.

27. z varies directly as $\sqrt{x}$. If $z = 96$ when $x = 4$, find z when $x = 9$.

28. s varies directly as $\sqrt{t}$. If $s = 270$ when $t = 25$, find s when $t = 1$.

29. a varies inversely as b^3. If $a = \frac{3}{2}$ when $b = 2$, find a when b is 3.

30. p varies inversely as q^2. If $p = \frac{5}{16}$ when $q = 8$, find p when $q = \frac{1}{2}$.

Objectives A B C D Mixed Practice *Solve. See Examples 1 through 7.*

31. Your paycheck (before deductions) varies directly as the number of hours you work. If your paycheck is \$166.50 for 18 hours, find your pay for 10 hours.

32. If your paycheck (before deductions) is \$304.50 for 30 hours, find your pay for 34 hours. (See Exercise **31.**)

33. The cost of manufacturing a certain type of headphone varies inversely as the number of headphones increases. If 5000 headphones can be manufactured for \$9.00 each, find the cost per headphone to manufacture 7500 headphones.

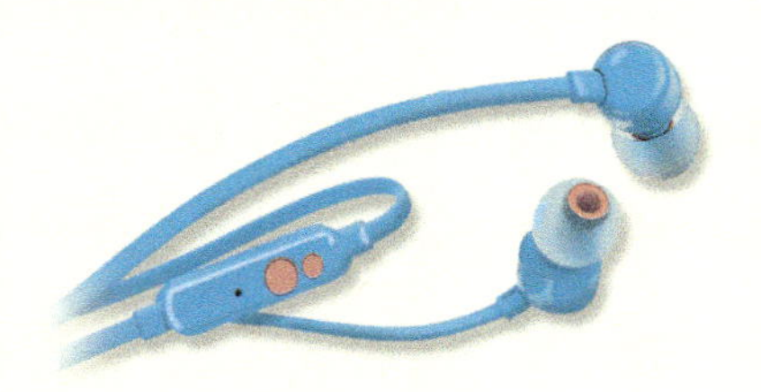

34. The cost of manufacturing a certain composition notebook varies inversely as the number of notebooks increases. If 10,000 notebooks can be manufactured for \$0.50 each, find the cost per notebook to manufacture 18,000 notebooks. Round your answer to the nearest cent.

35. The distance a spring stretches varies directly with the weight attached to the spring. If a 60-pound weight stretches the spring 4 inches, find the distance that an 80-pound weight stretches the spring.

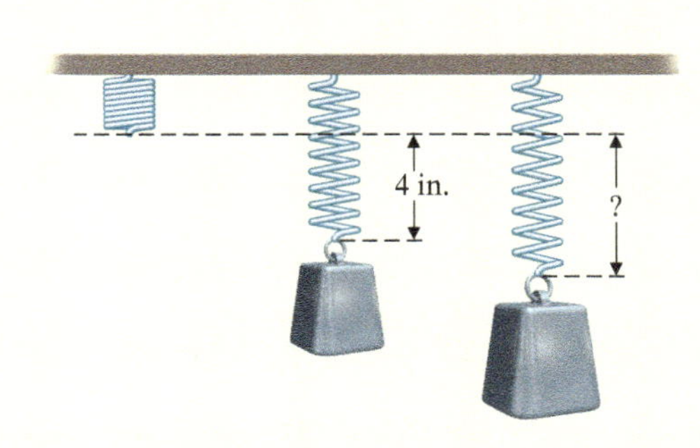

36. If a 30-pound weight stretches a spring 10 inches, find the distance a 20-pound weight stretches the spring. (See Exercise **35.**)

37. The weight of an object varies inversely as the square of its distance from the center of Earth. If a person weighs 180 pounds on Earth's surface, what is his weight 10 miles above the surface of Earth? (Assume that Earth's radius is 4000 miles and round your answer to one decimal place.)

38. For a constant distance, the rate of travel varies inversely as the time traveled. If a family travels 55 mph and arrives at a destination in 4 hours, how long will the return trip take traveling at 60 mph?

39. The distance d that an object falls is directly proportional to the square of the time of the fall, t. A person who is parachuting for the first time is told to wait 10 seconds before opening the parachute. If the person falls 64 feet in 2 seconds, find how far he falls in 10 seconds.

40. The distance needed for a car to stop, d, is directly proportional to the square of its rate of travel, r. Under certain driving conditions, a car traveling 60 mph needs 300 feet to stop. With these same driving conditions, how long does it take a car to stop if the car is traveling 30 mph when the brakes are applied?

Review

Add the equations by adding the left sides of the equations, bringing down an equal sign, and then adding the right sides of the equations. See Section 1.8.

41. $\begin{aligned} -3x + 4y &= 7 \\ 3x - 2y &= 9 \end{aligned}$

42. $\begin{aligned} x - y &= -9 \\ -x - y &= -14 \end{aligned}$

43. $\begin{aligned} 5x - 0.4y &= 0.7 \\ -9x + 0.4y &= -0.2 \end{aligned}$

44. $\begin{aligned} 1.9x - 2y &= 5.7 \\ -1.9x - 0.1y &= 2.3 \end{aligned}$

Concept Extensions

45. Suppose that y varies directly as x. If x is tripled, what is the effect on y?

46. Suppose that y varies directly as x^2. If x is tripled, what is the effect on y?

47. The period of a pendulum p (the time of one complete back-and-forth swing) varies directly with the square root of its length, ℓ. If the length of the pendulum is quadrupled, what is the effect on the period, p?

48. For a constant distance, the rate of travel r varies inversely with the time traveled, t. If a car traveling 100 mph completes a test track in 6 minutes, find the rate needed to complete the same test track in 4 minutes. (*Hint:* Convert minutes to hours.)

Chapter 8 Group Activity

Graphing and the Distance Formula

One application of radicals is finding the distance between two points in the coordinate plane. This can be very useful in graphing.

The distance d between two points with coordinates (x_1, y_1) and (x_2, y_2) is given by the **distance formula** $d = \sqrt{(x_2 - x_1)^2 + (y_2 - y_1)^2}$.

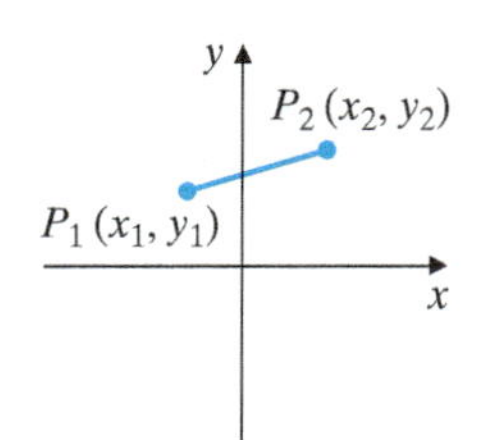

Suppose we want to find the distance between the two points $(-1, 9)$ and $(3, 5)$. We can use the distance formula with $(x_1, y_1) = (-1, 9)$ and $(x_2, y_2) = (3, 5)$. Then we have

$$\begin{aligned} d &= \sqrt{(x_2 - x_1)^2 + (y_2 - y_1)^2} \\ &= \sqrt{[3 - (-1)]^2 + (5 - 9)^2} \\ &= \sqrt{(4)^2 + (-4)^2} \\ &= \sqrt{16 + 16} \\ &= \sqrt{32} = 4\sqrt{2} \end{aligned}$$

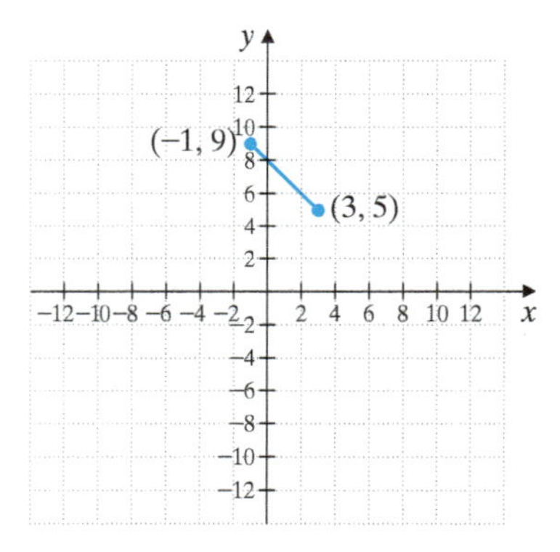

The distance between the two points is exactly $4\sqrt{2}$ units or approximately 5.66 units.

Group Activity

Brainstorm to come up with several disciplines or activities in which the distance formula might be useful. Make up an example that shows how the distance formula would be used in one of the activities on your list. Then present your example to the rest of the class.

Chapter 8 Vocabulary Check

Fill in each blank with one of the words or phrases listed below. Not all choices will be used.

index	radicand	like radicals	hypotenuse
rationalizing the denominator	conjugate	leg	radical
principal square root	direct	inverse	

1. The expressions $5\sqrt{x}$ and $7\sqrt{x}$ are examples of ______________.

2. In the expression $\sqrt[3]{45}$, the number 3 is the ______________, the number 45 is the ______________, and $\sqrt{}$ is called the ______________ sign.

3. The ______________ of $a + b$ is $a - b$.

4. The ______________ of 25 is 5.

5. The process of eliminating the radical in the denominator of a radical expression is called ______________.

6. The Pythagorean theorem states that for a right triangle, $(\text{leg})^2 + (\text{leg})^2 = (\_\_\_\_\_\_\_\_)^2$.

7. The equation $y = kx$ is an example of ______________ variation.

8. The equation $y = \frac{k}{x}$ is an example of ______________ variation.

Are you preparing for your test? To help, don't forget to take these:

- Chapter 8 Getting Ready for the Test on page 644
- Chapter 8 Test on page 645

Then check all of your answers at the back of this text. For further review, the step-by-step video solutions to any of these exercises are located in MyLab Math.

8 Chapter Highlights

Definitions and Concepts	Examples
Section 8.1 Introduction to Radicals	
The **positive** or **principal square root** of a positive number a is written as $\sqrt{a}$. The **negative square root** of a is written as $-\sqrt{a}$. $\sqrt{a} = b$ only if $b^2 = a$ and $b > 0$.	$\sqrt{25} = 5$ $\quad \sqrt{100} = 10$ $-\sqrt{9} = -3$ $\quad \sqrt{\frac{4}{49}} = \frac{2}{7}$
A square root of a negative number is not a real number.	$\sqrt{-4}$ is not a real number.
The **cube root** of a real number a is written as $\sqrt[3]{a}$. $\sqrt[3]{a} = b$ only if $b^3 = a$. The ***n*th root** of a number a is written as $\sqrt[n]{a}$. $\sqrt[n]{a} = b$ only if $b^n = a$. In $\sqrt[n]{a}$, the natural number n is called the **index**, the symbol $\sqrt{}$ is called a **radical**, and the expression within the radical is called the **radicand**. (*Note:* If the index is even, the radicand must be nonnegative for the root to be a real number.)	$\sqrt[3]{64} = 4$ $\quad \sqrt[3]{-8} = -2$ $\sqrt[4]{81} = 3$ $\sqrt[5]{-32} = -2$ index ↓ $\sqrt[n]{a}$ ↑ radicand
Section 8.2 Simplifying Radicals	
Product Rule for Radicals If $\sqrt{a}$ and $\sqrt{b}$ are real numbers, then $\sqrt{a \cdot b} = \sqrt{a} \cdot \sqrt{b}$ A square root is in **simplified form** if the radicand contains no perfect square factors other than 1. To simplify a square root, factor the radicand so that one of its factors is a perfect square factor.	$\sqrt{45} = \sqrt{9 \cdot 5}$ $= \sqrt{9} \cdot \sqrt{5}$ $= 3\sqrt{5}$

(Continued)

Definitions and Concepts	Examples
Section 8.2 Simplifying Radicals (*continued*)	
Quotient Rule for Radicals If $\sqrt{a}$ and $\sqrt{b}$ are real numbers and $b \neq 0$, then $\sqrt{\frac{a}{b}} = \frac{\sqrt{a}}{\sqrt{b}}$	$\sqrt{\frac{18}{x^6}} = \frac{\sqrt{9 \cdot 2}}{\sqrt{x^6}} = \frac{\sqrt{9} \cdot \sqrt{2}}{x^3} = \frac{3\sqrt{2}}{x^3}$
Section 8.3 Adding and Subtracting Radicals	
Like radicals are radical expressions that have the same index and the same radicand. To **combine like radicals,** use the distributive property.	$5\sqrt{2}, -7\sqrt{2}, \sqrt{2}$ $2\sqrt{7} - 13\sqrt{7} = (2 - 13)\sqrt{7} = -11\sqrt{7}$ $\sqrt{8} + \sqrt{50} = 2\sqrt{2} + 5\sqrt{2} = 7\sqrt{2}$
Section 8.4 Multiplying and Dividing Radicals	
The product and quotient rules for radicals may be used to simplify products and quotients of radicals.	Perform each indicated operation and simplify. Multiply. $\sqrt{2} \cdot \sqrt{8} = \sqrt{16} = 4$ $(\sqrt{3x} + 1)(\sqrt{5} - \sqrt{3})$ $= \sqrt{15x} - \sqrt{9x} + \sqrt{5} - \sqrt{3}$ $= \sqrt{15x} - 3\sqrt{x} + \sqrt{5} - \sqrt{3}$ Divide. $\frac{\sqrt{20}}{\sqrt{2}} = \sqrt{\frac{20}{2}} = \sqrt{10}$
The process of eliminating the radical in the denominator of a radical expression is called **rationalizing the denominator.**	Rationalize the denominator. $\frac{5}{\sqrt{11}} = \frac{5 \cdot \sqrt{11}}{\sqrt{11} \cdot \sqrt{11}} = \frac{5\sqrt{11}}{11}$
The **conjugate** of $a + b$ is $a - b$. To rationalize a denominator that is a sum or difference of radicals, multiply the numerator and the denominator by the conjugate of the denominator.	The conjugate of $2 + \sqrt{3}$ is $2 - \sqrt{3}$. Rationalize the denominator. $\frac{5}{6 - \sqrt{5}} = \frac{5(6 + \sqrt{5})}{(6 - \sqrt{5})(6 + \sqrt{5})}$ $= \frac{5(6 + \sqrt{5})}{36 - 5}$ $= \frac{5(6 + \sqrt{5})}{31}$
Section 8.5 Solving Equations Containing Radicals	
To Solve a Radical Equation Containing Square Roots **Step 1:** Get one radical by itself on one side of the equation. **Step 2:** Square both sides of the equation. **Step 3:** Simplify both sides of the equation.	Solve: $\sqrt{2x - 1} - x = -2$ $\sqrt{2x - 1} = x - 2$ $(\sqrt{2x - 1})^2 = (x - 2)^2$ Square both sides. $2x - 1 = x^2 - 4x + 4$ $0 = x^2 - 6x + 5$

Definitions and Concepts	Examples
Section 8.5 Solving Equations Containing Radicals *(continued)*	
Step 4: If the equation still contains a radical term, repeat Steps 1 through 3. **Step 5:** Solve the equation. **Step 6:** Check solutions in the original equation.	$0 = (x - 1)(x - 5)$ Factor. $x - 1 = 0$ or $x - 5 = 0$ $x = 1$ $\quad x = 5$ Solve. Check both proposed solutions in the original equation. Here, 5 checks but 1 does not. The only solution is 5.
Section 8.6 Radical Equations and Problem Solving	
Problem-Solving Steps **1.** UNDERSTAND. Read and reread the problem.	A rain gutter is to be mounted on the eaves of a house 15 feet above the ground. A garden is adjacent to the house, so the closest a ladder can be placed to the house is 6 feet. How long a ladder is needed for installing the gutter? Let $x =$ the length of the ladder. 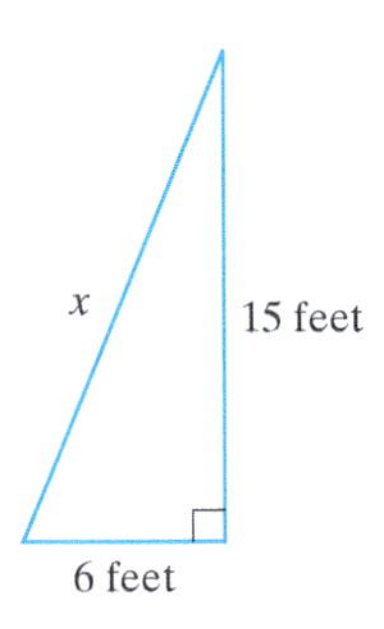 Here, we use the Pythagorean theorem. The unknown length x is the hypotenuse. In words: $(\text{leg})^2 + (\text{leg})^2 = (\text{hypotenuse})^2$
2. TRANSLATE.	Translate:
3. SOLVE.	$6^2 + 15^2 = x^2$ $36 + 225 = x^2$ $261 = x^2$ $\sqrt{261} = x$ or $x = 3\sqrt{29}$
4. INTERPRET.	Check and state. The ladder needs to be $3\sqrt{29}$ feet or approximately 16.2 feet long.
Section 8.7 Direct and Inverse Variation Including Radical Applications	
y **varies directly as** x, or y is **directly proportional to** x, if there is a nonzero constant k such that $y = kx$ y **varies inversely as** x, or y is **inversely proportional to** x, if there is a nonzero constant k such that $y = \frac{k}{x}$	The circumference of a circle C varies directly as its radius r. $C = \underbrace{2\pi}_{k} r$ Pressure P varies inversely with volume V. $P = \frac{k}{V}$

Chapter 8 Review

(8.1) *Find each root.*

1. $\sqrt{81}$ **2.** $-\sqrt{49}$ **3.** $\sqrt[3]{27}$ **4.** $\sqrt[4]{81}$

5. $-\sqrt{\frac{9}{64}}$ **6.** $\sqrt{\frac{36}{81}}$ **7.** $\sqrt[4]{16}$ **8.** $\sqrt[3]{-8}$

9. Which radical(s) is not a real number?
a. $\sqrt{4}$ **b.** $-\sqrt{4}$ **c.** $\sqrt{-4}$ **d.** $\sqrt[3]{-4}$

10. Which radical(s) is not a real number?
a. $\sqrt{-5}$ **b.** $\sqrt[3]{-5}$ **c.** $\sqrt[4]{-5}$ **d.** $\sqrt[5]{-5}$

Find each root. Assume that all variables represent positive numbers.

11. $\sqrt{x^{12}}$ **12.** $\sqrt{x^8}$ **13.** $\sqrt{9y^2}$ **14.** $\sqrt{25x^4}$

(8.2) *Simplify each expression using the product rule. Assume that all variables represent positive numbers.*

15. $\sqrt{40}$ **16.** $\sqrt{24}$ **17.** $\sqrt{54}$ **18.** $\sqrt{88}$ **19.** $\sqrt{x^5}$

20. $\sqrt{y^7}$ **21.** $\sqrt{20x^2}$ **22.** $\sqrt{50y^4}$ **23.** $\sqrt[3]{54}$ **24.** $\sqrt[3]{88}$

Simplify each expression using the quotient rule. Assume that all variables represent positive numbers.

25. $\sqrt{\frac{18}{25}}$ **26.** $\sqrt{\frac{75}{64}}$ **27.** $-\sqrt{\frac{50}{9}}$ **28.** $-\sqrt{\frac{12}{49}}$

29. $\sqrt{\frac{11}{x^2}}$ **30.** $\sqrt{\frac{7}{y^4}}$ **31.** $\sqrt{\frac{y^5}{100}}$ **32.** $\sqrt{\frac{x^3}{81}}$

(8.3) *Add or subtract by combining like radicals.*

33. $5\sqrt{2} - 8\sqrt{2}$ **34.** $\sqrt{3} - 6\sqrt{3}$

35. $6\sqrt{5} + 3\sqrt{6} - 2\sqrt{5} + \sqrt{6}$ **36.** $-\sqrt{7} + 8\sqrt{2} - \sqrt{7} - 6\sqrt{2}$

Add or subtract by simplifying each radical and then combining like terms. Assume that all variables represent positive numbers.

37. $\sqrt{28} + \sqrt{63} + \sqrt{56}$ **38.** $\sqrt{75} + \sqrt{48} - \sqrt{16}$

39. $\sqrt{\frac{5}{9}} - \sqrt{\frac{5}{36}}$ **40.** $\sqrt{\frac{11}{25}} + \sqrt{\frac{11}{16}}$

41. $\sqrt{45x^2} + 3\sqrt{5x^2} - 7x\sqrt{5} + 10$ **42.** $\sqrt{50x} - 9\sqrt{2x} + \sqrt{72x} - \sqrt{3x}$

(8.4) *Multiply and simplify if possible. Assume that all variables represent positive numbers.*

43. $\sqrt{3} \cdot \sqrt{6}$ **44.** $\sqrt{5} \cdot \sqrt{15}$

45. $\sqrt{2}(\sqrt{5} - \sqrt{7})$

46. $\sqrt{5}(\sqrt{11} + \sqrt{3})$

47. $(\sqrt{3} + 2)(\sqrt{6} - 5)$

48. $(\sqrt{5} + 1)(\sqrt{5} - 3)$

49. $(\sqrt{x} - 2)^2$

50. $(\sqrt{y} + 4)^2$

Divide and simplify if possible. Assume that all variables represent positive numbers.

51. $\frac{\sqrt{27}}{\sqrt{3}}$

52. $\frac{\sqrt{20}}{\sqrt{5}}$

53. $\frac{\sqrt{160}}{\sqrt{8}}$

54. $\frac{\sqrt{96}}{\sqrt{3}}$

55. $\frac{\sqrt{30x^6}}{\sqrt{2x^3}}$

56. $\frac{\sqrt{54x^5y^2}}{\sqrt{3xy^2}}$

Rationalize each denominator and simplify.

57. $\frac{\sqrt{2}}{\sqrt{11}}$

58. $\frac{\sqrt{3}}{\sqrt{13}}$

59. $\sqrt{\frac{5}{6}}$

60. $\sqrt{\frac{7}{10}}$

61. $\frac{1}{\sqrt{5x}}$

62. $\frac{5}{\sqrt{3y}}$

63. $\sqrt{\frac{3}{x}}$

64. $\sqrt{\frac{6}{y}}$

65. $\frac{3}{\sqrt{5} - 2}$

66. $\frac{8}{\sqrt{10} - 3}$

67. $\frac{\sqrt{2} + 1}{\sqrt{3} - 1}$

68. $\frac{\sqrt{3} - 2}{\sqrt{5} + 2}$

69. $\frac{10}{\sqrt{x} + 5}$

70. $\frac{8}{\sqrt{x} - 1}$

(8.5) *Solve each radical equation.*

71. $\sqrt{2x} = 6$

72. $\sqrt{x + 3} = 4$

73. $\sqrt{x} + 3 = 8$

74. $\sqrt{x} + 8 = 3$

75. $\sqrt{2x + 1} = x - 7$

76. $\sqrt{3x + 1} = x - 1$

77. $\sqrt{x} + 3 = \sqrt{x + 15}$

78. $\sqrt{x - 5} = \sqrt{x} - 1$

(8.6) *Use the Pythagorean theorem to find the length of each unknown side. Give the exact answer and a two-decimal-place approximation.*

△ **79.**

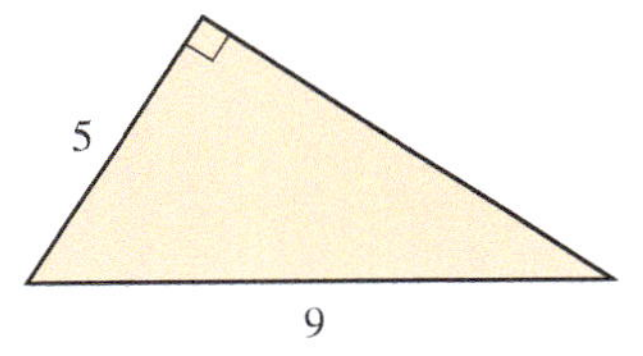

△ **80.**

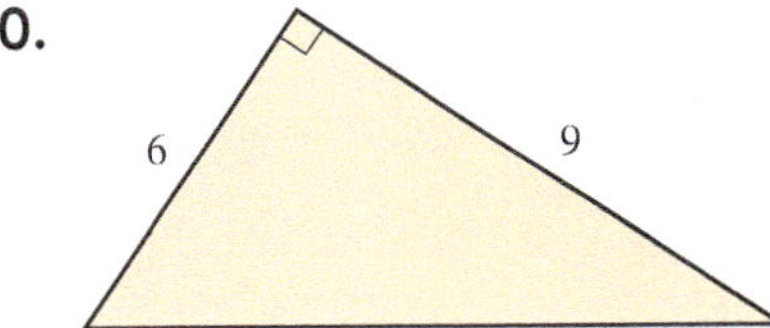

△ **81.** Romeo is standing 20 feet away from the wall below Juliet's balcony during a school play. Juliet is on the balcony, 12 feet above the ground. Find how far apart Romeo and Juliet are.

△ **82.** The diagonal of a rectangle is 10 inches long. If the width of the rectangle is 5 inches, find the length of the rectangle.

For Exercises 83 and 84, use the formula $r = \sqrt{\frac{S}{4\pi}}$, *where* r = *the radius of a sphere and* S = *the surface area of the sphere.*

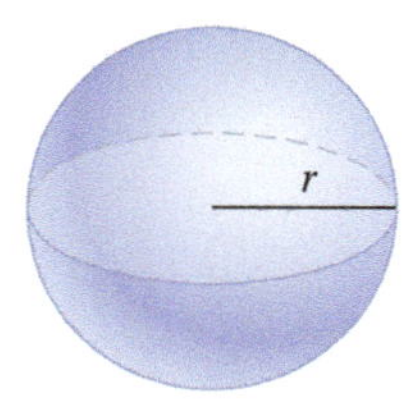

△ **83.** Find the radius of a sphere to the nearest tenth of an inch if the surface area is 72 square inches.

△ **84.** Find the exact surface area of a sphere if its radius is 6 inches. (Do not approximate π.)

(8.7) *Solve.*

85. y varies directly as x. If $y = 40$ when $x = 4$, find y when x is 11.

86. y varies inversely as x. If $y = 4$ when $x = 6$, find y when x is 48.

87. y varies inversely as x^3. If $y = 12.5$ when $x = 2$, find y when x is 3.

88. y varies directly as x^2. If $y = 175$ when $x = 5$, find y when $x = 10$.

89. y varies inversely as $\sqrt{x}$. If y is 2 when x is 25, find y when x is 36.

90. y varies directly as $\sqrt{x}$. If y is 33 when x is 9, find y when x is 4.

91. The cost of manufacturing a certain medicine varies inversely as the amount of medicine manufactured increases. If 3000 milliliters can be manufactured for \$6600, find the cost to manufacture 5000 milliliters.

92. The distance a spring stretches varies directly with the weight attached to the spring. If a 150-pound weight stretches the spring 8 inches, find the distance that a 90-pound weight stretches the spring.

Mixed Review

Find each root. Assume all variables represent positive numbers.

93. $\sqrt{144}$

94. $-\sqrt[3]{64}$

95. $\sqrt{16x^{16}}$

96. $\sqrt{4x^{24}}$

Simplify each expression. Assume all variables represent positive numbers.

97. $\sqrt{18x^7}$

98. $\sqrt{48y^6}$

99. $\sqrt{\frac{y^4}{81}}$

100. $\sqrt{\frac{x^9}{9}}$

Add or subtract by simplifying and then combining like terms. Assume all variables represent positive numbers.

101. $\sqrt{12} + \sqrt{75}$

102. $\sqrt{63} + \sqrt{28} - \sqrt{9}$

103. $\sqrt{\frac{3}{16}} - \sqrt{\frac{3}{4}}$

104. $\sqrt{45x^3} + x\sqrt{20x} - \sqrt{5x^3}$

Multiply and simplify if possible. Assume all variables represent positive numbers.

105. $\sqrt{7} \cdot \sqrt{14}$

106. $\sqrt{3}(\sqrt{9} - \sqrt{2})$

107. $(\sqrt{2} + 4)(\sqrt{5} - 1)$

108. $(\sqrt{x} + 3)^2$

Divide and simplify if possible. Assume all variables represent positive numbers.

109. $\dfrac{\sqrt{120}}{\sqrt{5}}$

110. $\dfrac{\sqrt{60x^9}}{\sqrt{15x^7}}$

Rationalize each denominator and simplify.

111. $\sqrt{\dfrac{2}{7}}$

112. $\dfrac{3}{\sqrt{2x}}$

113. $\dfrac{3}{\sqrt{x} - 6}$

114. $\dfrac{\sqrt{7} - 5}{\sqrt{5} + 3}$

Solve each radical equation.

115. $\sqrt{4x} = 2$

116. $\sqrt{x - 4} = 3$

117. $\sqrt{4x + 8} + 6 = x$

118. $\sqrt{x - 8} = \sqrt{x} - 2$

Solve. Give the exact answer and a two-decimal-place approximation.

119. Use the Pythagorean theorem to find the length of the unknown side.

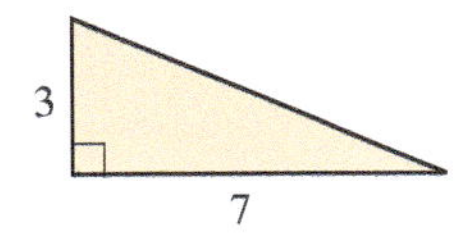

120. The diagonal of a rectangle is 6 inches long. If the width of the rectangle is 2 inches, find the length of the rectangle.

Chapter 8 Getting Ready for the Test LC

MULTIPLE CHOICE Exercises 1 through 11 are **Multiple Choice**. Select the correct choice.

1. Choose the expression that simplifies to -4.

A. $\sqrt{-16}$ **B.** $-\sqrt{16}$ **C.** $\sqrt[3]{8}$ **D.** $\sqrt[3]{-8}$

2. $7\sqrt{3} - \sqrt{3} =$

A. 7 **B.** 6 **C.** $6\sqrt{3}$ **D.** cannot be simplified

3. $7\sqrt{3} \cdot \sqrt{2} =$

A. 35 **B.** $7\sqrt{6}$ **C.** 42 **D.** $8\sqrt{6}$ **E.** cannot be simplified

4. $(4\sqrt{5})^2 =$

A. 40 **B.** 80 **C.** $8\sqrt{5}$ **D.** $16\sqrt{5}$ **E.** cannot be simplified

5. Simplify: $\sqrt{\dfrac{28}{25}}$

A. $\dfrac{14}{5}$ **B.** $\dfrac{14}{\sqrt{25}}$ **C.** $\dfrac{2\sqrt{7}}{5}$ **D.** cannot be simplified

6. Simplify: $\sqrt{18x^{16}}$

A. $9x^8$ **B.** $9x^4$ **C.** $3x^4\sqrt{2}$ **D.** $3x^8\sqrt{2}$

7. Simplify: $\sqrt[3]{64}$

A. 8 **B.** 4 **C.** 192 **D.** cannot be simplified

8. Simplify: $\sqrt[3]{x^{27}}$

A. x^3 **B.** x^9 **C.** $x^{13}\sqrt[3]{x}$ **D.** cannot be simplified

9. To rationalize the denominator of $\dfrac{\sqrt{5}}{\sqrt{2}}$, we multiply by:

A. $\dfrac{\sqrt{5}}{\sqrt{2}}$ **B.** $\dfrac{\sqrt{10}}{\sqrt{10}}$ **C.** $\dfrac{\sqrt{2}}{\sqrt{2}}$ **D.** $\dfrac{\sqrt{5}}{\sqrt{5}}$

10. Square both sides of the equation $3\sqrt{x} = \sqrt{10x - 9}$. The result is:

A. $3x = 10x - 9$ **B.** $3x^2 = 10x - 9$ **C.** $9x = 10x - 9$ **D.** $3x^2 = 100x^2 - 38x + 81$

11. Square both sides of the equation $x + 1 = \sqrt{9x - 9}$. The result is:

A. $x^2 + 2x + 1 = 9x - 9$ **B.** $x^2 + 1 = 9x - 9$ **C.** $x^2 + x + 1 = 9x - 9$ **D.** $x + 1 = 9x - 9$

Simplify each radical. Indicate if the radical is not a real number. Assume that x represents a positive number.

1. $\sqrt{16}$
2. $\sqrt[3]{125}$
3. $\sqrt[4]{81}$
4. $\sqrt{\frac{9}{16}}$
5. $\sqrt[4]{-81}$
6. $\sqrt{x^{10}}$

Simplify each radical. Assume that all variables represent positive numbers.

7. $\sqrt{54}$
8. $\sqrt{92}$
9. $\sqrt{y^7}$
10. $\sqrt{24x^8}$
11. $\sqrt[3]{27}$
12. $\sqrt[3]{16}$
13. $\sqrt{\frac{5}{16}}$
14. $\sqrt{\frac{y^3}{25}}$

Perform each indicated operation. Assume that all variables represent positive numbers.

15. $\sqrt{13} + \sqrt{13} - 4\sqrt{13}$
16. $\sqrt{18} - \sqrt{75} + 7\sqrt{3} - \sqrt{8}$
17. $\sqrt{\frac{3}{4}} + \sqrt{\frac{3}{25}}$
18. $\sqrt{7} \cdot \sqrt{14}$
19. $\sqrt{2}(\sqrt{6} - \sqrt{5})$

Answers

1. ______
2. ______
3. ______
4. ______
5. ______
6. ______
7. ______
8. ______
9. ______
10. ______
11. ______
12. ______
13. ______
14. ______
15. ______
16. ______
17. ______
18. ______
19. ______

20. ______

21. ______

22. ______

23. ______

24. ______

25. ______

26. ______

27. ______

28. ______

29. ______

30. ______

31. ______

32. ______

33. ______

20. $(\sqrt{x} + 2)(\sqrt{x} - 3)$ **21.** $\dfrac{\sqrt{50}}{\sqrt{10}}$ **22.** $\dfrac{\sqrt{40x^4}}{\sqrt{2x}}$

Rationalize each denominator. Assume that all variables represent positive numbers.

23. $\sqrt{\dfrac{2}{3}}$ **24.** $\dfrac{8}{\sqrt{5y}}$ **25.** $\dfrac{8}{\sqrt{6} + 2}$ **26.** $\dfrac{1}{3 - \sqrt{x}}$

Solve each radical equation.

27. $\sqrt{x} + 8 = 11$ **28.** $\sqrt{3x - 6} = \sqrt{x + 4}$ **29.** $\sqrt{2x - 2} = x - 5$

30. Find the length of the unknown leg of the right triangle shown. Give the exact answer.

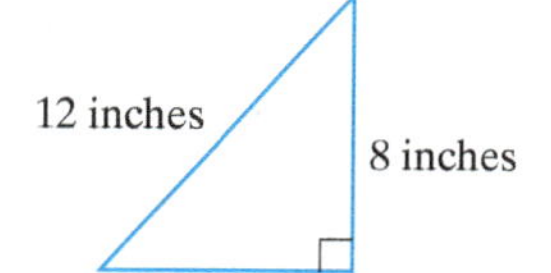

31. The formula $r = \sqrt{\dfrac{A}{\pi}}$ can be used to find the radius r of a circle given its area A. Use this formula to approximate the radius of the given circle. Round to two decimal places.

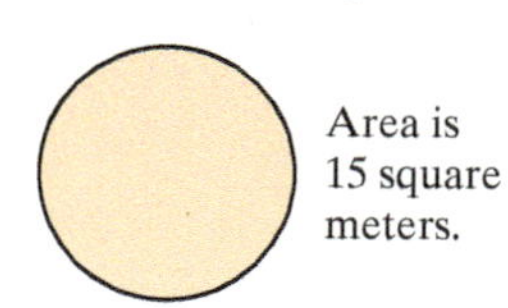

32. y varies directly as x. If $y = 10$ when $x = 15$, find y when x is 42.

33. y varies inversely as x^2. If $y = 8$ when $x = 5$, find y when x is 15.

Cumulative Review

Multiply.

1. $-2(-14)$

2. $9(-5.2)$

3. $-\frac{2}{3} \cdot \frac{4}{7}$

4. $-3\frac{3}{8} \cdot 5\frac{1}{3}$

5. Solve: $4(2x - 3) + 7 = 3x + 5$

6. Solve: $6y - 11 + 4 + 2y = 8 + 15y - 8y$

7. The circle graph below shows the purpose of trips made by American travelers. Use this graph to answer the questions below.

a. What percent of trips made by American travelers is solely for the purpose of business?

b. What percent of trips made by American travelers is for the purpose of business or combined business/pleasure?

c. On an airplane flight of 253 Americans, how many of these people might we expect to be traveling solely for business?

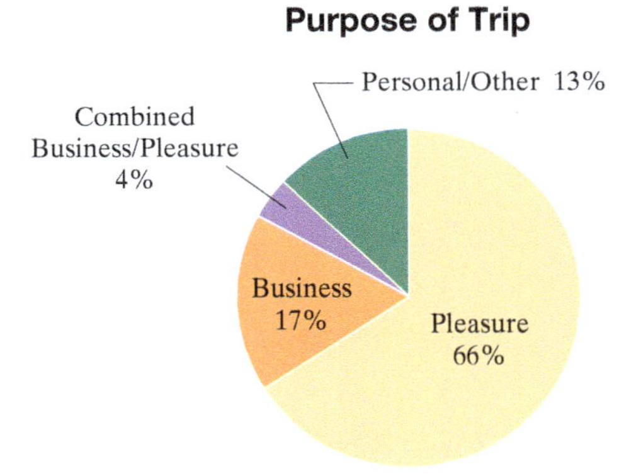

Source: Travel Industry Association of America

8. Simplify each expression.

a. $\dfrac{4(-3) - (-6)}{-8 + 4}$

b. $\dfrac{3 + (-3)(-2)^3}{-1 - (-4)}$

9. Write each number in standard form, without exponents.

a. 1.02×10^5

b. 7.358×10^{-3}

c. 8.4×10^7

d. 3.007×10^{-5}

10. Write the following numbers in scientific notation.

a. 7,200,000

b. 0.000308

11. Multiply: $(3x + 2)(2x - 5)$

Answers

1. ______
2. ______
3. ______
4. ______
5. ______
6. ______
7. a. ______
   b. ______
   c. ______
8. a. ______
   b. ______
9. a. ______
   b. ______
   c. ______
   d. ______
10. a. ______
    b. ______
11. ______

12. ______
13. ______
14. ______
15. ______
16. ______
17. a. ______
b. ______
c. ______
18. ______
19. ______
20. ______
21. a. ______
b. ______
22. ______
23. ______
24. ______
25. ______
26. ______
27. ______
28. ______
29. ______
30. ______

12. Multiply: $(7x + 1)^2$

13. Factor $xy + 2x + 3y + 6$ by grouping.

14. Factor $xy^2 + 5x - y^2 - 5$ by grouping.

15. Factor: $3x^2 + 11x + 6$

16. Factor: $3x^2 + 15x + 18$

17. Are there any values for x for which each expression is undefined?

a. $\dfrac{x}{x - 3}$

b. $\dfrac{x^2 + 2}{x^2 - 3x + 2}$

c. $\dfrac{x^3 - 6x^2 - 10x}{3}$

18. Simplify: $\dfrac{2x^2 + 7x + 3}{x^2 - 9}$

19. Simplify: $\dfrac{x^2 + 4x + 4}{x^2 + 2x}$

20. Divide: $\dfrac{12x^2y^3}{5} \div \dfrac{3y^3}{x}$

21. Perform each indicated operation.

a. $\dfrac{a}{4} - \dfrac{2a}{8}$ **b.** $\dfrac{3}{10x^2} + \dfrac{7}{25x}$

22. Find an equation of the line with y-intercept $(0, 4)$ and slope of -2.

23. Solve: $\dfrac{4x}{x^2 + x - 30} + \dfrac{2}{x - 5} = \dfrac{1}{x + 6}$

24. Combine like terms to simplify. $4a^2 + 3a - 2a^2 + 7a - 5$

25. Graph $y = -3$.

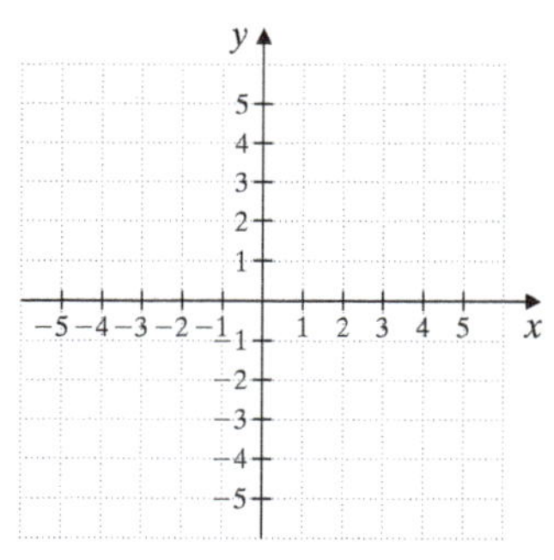

26. Complete the table for the equation $2x + y = 6$.

x	y
0	
	−2
3	

27. Find an equation of the line with y-intercept $(0, -3)$ and slope of $\dfrac{1}{4}$.

28. Find an equation of the line perpendicular to the line $y = 4$ and passing through $(1, 5)$.

29. Solve the system:

$$\begin{cases} 3x + 4y = 13 \\ 5x - 9y = 6 \end{cases}$$

30. Solve the system:

$$\begin{cases} \dfrac{x}{2} + y = \dfrac{5}{6} \\ 2x - y = \dfrac{5}{6} \end{cases}$$

31. As part of an exercise program, two students, Louisa and Alfredo, start walking each morning. They live 15 miles away from each other. They decide to meet one day by walking toward one another. After 2 hours they meet. If Louisa walks one mile per hour faster than Alfredo, find both walking speeds.

32. Two streetcars are 11 miles apart and traveling toward each other on parallel tracks. They meet in 12 minutes. Find the speed of each streetcar if one travels 15 miles per hour faster than the other.

Find each root.

33. $\sqrt[3]{1}$

34. $\sqrt{121}$

35. $\sqrt[3]{-27}$

36. $\sqrt{\frac{1}{4}}$

37. $\sqrt[3]{\frac{1}{125}}$

38. $\sqrt{\frac{25}{144}}$

Simplify.

39. $\sqrt{54}$

40. $\sqrt{63}$

41. $\sqrt{200}$

42. $\sqrt{500}$

Perform the indicated operations. If possible, first simplify each radical.

43. $7\sqrt{12} - 2\sqrt{75}$

44. $(\sqrt{x} + 5)(\sqrt{x} - 5)$

45. $2\sqrt{x^2} - \sqrt{25x^5} + \sqrt{x^5}$

46. $(\sqrt{6} + 2)^2$

47. Rationalize the denominator of $\frac{2}{\sqrt{7}}$.

48. Simplify: $\dfrac{x + 3}{\frac{1}{x} + \frac{1}{3}}$

49. Solve: $\sqrt{x} = \sqrt{5x - 2}$

50. Solve: $\sqrt{x + 4} = \sqrt{3x - 1}$

31. ______

32. ______

33. ______

34. ______

35. ______

36. ______

37. ______

38. ______

39. ______

40. ______

41. ______

42. ______

43. ______

44. ______

45. ______

46. ______

47. ______

48. ______

49. ______

50. ______

9 Quadratic Equations

An important part of the study of algebra is learning to use methods for solving equations. Starting in Chapter 2, we presented techniques for solving linear equations in one variable. In Chapter 4, we solved quadratic equations in one variable by factoring the quadratic expressions. We now present other methods for solving quadratic equations in one variable.

Sections

9.1 Solving Quadratic Equations by the Square Root Property

9.2 Solving Quadratic Equations by Completing the Square

9.3 Solving Quadratic Equations by the Quadratic Formula

Integrated Review—Summary on Solving Quadratic Equations

9.4 Graphing Quadratic Equations in Two Variables

Check Your Progress

Vocabulary Check

Chapter Highlights

Chapter Review

Getting Ready for the Test

Chapter Test

Cumulative Review

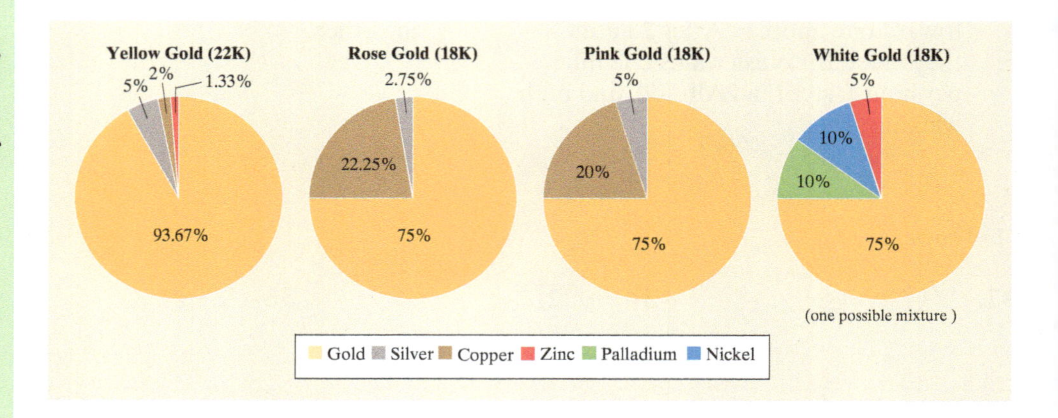

Does Gold Really Come in Colors?

How are rose gold, pink gold, and white gold formed? As we see above, they come from mixing gold with other metals. Most of the gold jewelry that we buy is a mixture (or an alloy) of metals. The purity of gold is given by its karat (K). Twenty-four karat gold is called fine gold and is greater than 99.7% pure gold.

Gold has many interesting qualities. For example, one ounce of gold can be stretched into a golden thread of about 5 miles in length, or it can be made into a thin sheet of about 300 square feet in area.

Below is a bar graph of the price of gold at the beginning of each year shown. In Section 9.2, Exercise 47, we will explore the current and possible future price of gold.

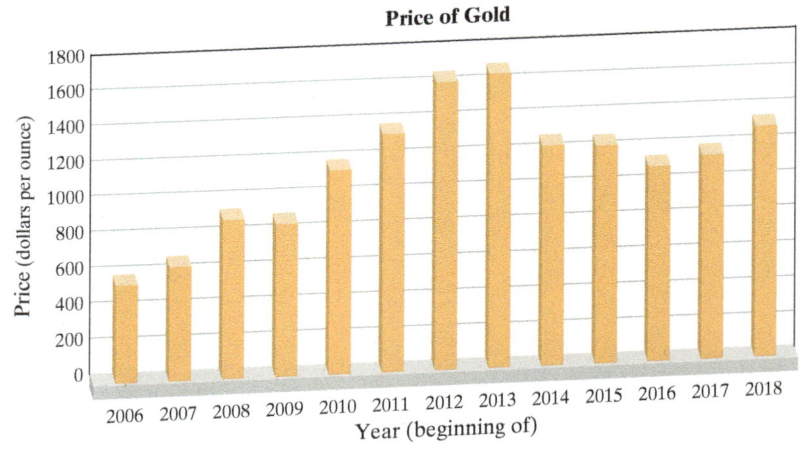

Source: goldprice.org

9.1 Solving Quadratic Equations by the Square Root Property 

Recall that a quadratic equation is an equation that can be written in the form

$$ax^2 + bx + c = 0$$

where a, b, and c are real numbers and $a \neq 0$.

Solving Quadratic Equations by Factoring 

To solve quadratic equations by factoring, we use the **zero-factor property:**

Zero Factor Property

If a and b are real numbers and
if $ab = 0$, then $a = 0$ or $b = 0$.

Examples 1 and 2 review the process of solving quadratic equations by factoring.

Example 1 Solve: $x^2 - 4 = 0$

Solution:

$$x^2 - 4 = 0$$

$$(x + 2)(x - 2) = 0 \quad \text{Factor.}$$

$$x + 2 = 0 \quad \text{or} \quad x - 2 = 0 \quad \text{Use the zero-factor property.}$$

$$x = -2 \qquad\qquad x = 2 \quad \text{Solve each equation.}$$

The solutions are -2 and 2.

Work Practice 1

Example 2 Solve: $3y^2 + 13y = 10$

Solution: Recall that to use the zero-factor property, one side of the equation must be 0 and the other side must be factored.

$$3y^2 + 13y = 10$$

$$3y^2 + 13y - 10 = 0 \quad \text{Subtract 10 from both sides.}$$

$$(3y - 2)(y + 5) = 0 \quad \text{Factor.}$$

$$3y - 2 = 0 \quad \text{or} \quad y + 5 = 0 \quad \text{Use the zero-factor property.}$$

$$3y = 2 \qquad\qquad y = -5 \quad \text{Solve each equation.}$$

$$y = \frac{2}{3}$$

The solutions are $\frac{2}{3}$ and -5.

Work Practice 2

Objective A Using the Square Root Property 

Consider solving Example 1, $x^2 - 4 = 0$, another way. First, add 4 to both sides of the equation.

$$x^2 - 4 = 0$$

$$x^2 = 4 \quad \text{Add 4 to both sides.}$$

Now we see that the value for x must be a number whose square is 4. Therefore $x = \sqrt{4} = 2$ or $x = -\sqrt{4} = -2$. This reasoning is an example of the square root property.

Objectives

A Use the Square Root Property to Solve Quadratic Equations.

B Use the Square Root Property to Solve Applications.

Practice 1

Solve: $x^2 - 25 = 0$

Practice 2

Solve: $2x^2 - 3x = 9$

Answers

1. 5 and -5 **2.** $-\frac{3}{2}$ and 3

Square Root Property

If $x^2 = a$ for $a \geq 0$, then

$x = \sqrt{a}$ or $x = -\sqrt{a}$

Practice 3

Use the square root property to solve $x^2 - 16 = 0$.

Example 3 Use the square root property to solve $x^2 - 9 = 0$.

Solution: First we solve for x^2 by adding 9 to both sides.

$x^2 - 9 = 0$

$x^2 = 9$ Add 9 to both sides.

Next we use the square root property.

$x = \sqrt{9}$ or $x = -\sqrt{9}$

$x = 3$ $\quad x = -3$

Check:

$x^2 - 9 = 0$ Original equation
$3^2 - 9 \stackrel{?}{=} 0$ Let $x = 3$.
$0 = 0$ True

$x^2 - 9 = 0$ Original equation
$(-3)^2 - 9 \stackrel{?}{=} 0$ Let $x = -3$.
$0 = 0$ True

The solutions are 3 and -3.

Work Practice 3

Practice 4

Use the square root property to solve $3x^2 = 11$.

Example 4 Use the square root property to solve $2x^2 = 7$.

Solution: First we solve for x^2 by dividing both sides by 2. Then we use the square root property.

$2x^2 = 7$

$x^2 = \frac{7}{2}$ Divide both sides by 2.

$x = \sqrt{\frac{7}{2}}$ or $x = -\sqrt{\frac{7}{2}}$ Use the square root property.

$x = \frac{\sqrt{7} \cdot \sqrt{2}}{\sqrt{2} \cdot \sqrt{2}}$ $\quad x = -\frac{\sqrt{7} \cdot \sqrt{2}}{\sqrt{2} \cdot \sqrt{2}}$ Rationalize the denominator.

$x = \frac{\sqrt{14}}{2}$ $\quad x = -\frac{\sqrt{14}}{2}$ Simplify.

Remember to check both solutions in the original equation. The solutions are $\frac{\sqrt{14}}{2}$ and $-\frac{\sqrt{14}}{2}$.

Work Practice 4

Practice 5

Use the square root property to solve $(x - 4)^2 = 49$.

Example 5 Use the square root property to solve $(x - 3)^2 = 16$.

Solution: Instead of x^2, here we have $(x - 3)^2$. But the square root property can still be used.

$(x - 3)^2 = 16$

$x - 3 = \sqrt{16}$ or $x - 3 = -\sqrt{16}$ Use the square root property.

$x - 3 = 4$ $\quad x - 3 = -4$ Write $\sqrt{16}$ as 4 and $-\sqrt{16}$ as -4.

$x = 7$ $\quad x = -1$ Solve.

Answers

3. 4 and -4 4. $\frac{\sqrt{33}}{3}$ and $-\frac{\sqrt{33}}{3}$

5. 11 and -3

Check:

$(x-3)^2 = 16$	Original equation	$(x-3)^2 = 16$	Original equation
$(7-3)^2 \stackrel{?}{=} 16$	Let $x = 7$.	$(-1-3)^2 \stackrel{?}{=} 16$	Let $x = -1$.
$4^2 \stackrel{?}{=} 16$	Simplify.	$(-4)^2 \stackrel{?}{=} 16$	Simplify.
$16 = 16$	True	$16 = 16$	True

Both 7 and -1 are solutions.

Work Practice 5

Example 6 Use the square root property to solve $(x+1)^2 = 8$.

Solution: $(x+1)^2 = 8$

$x + 1 = \sqrt{8}$ or $x + 1 = -\sqrt{8}$ Use the square root property.

$x + 1 = 2\sqrt{2}$ $\quad x + 1 = -2\sqrt{2}$ Simplify the radical.

$x = -1 + 2\sqrt{2}$ $\quad x = -1 - 2\sqrt{2}$ Solve for x.

Check both solutions in the original equation. The solutions are $-1 + 2\sqrt{2}$ and $-1 - 2\sqrt{2}$. This can be written compactly as $-1 \pm 2\sqrt{2}$. The notation $\pm$ is read as "plus or minus."

Work Practice 6

Practice 6

Use the square root property to solve $(x-5)^2 = 18$.

read "plus or minus"

The notation $-1 \pm \sqrt{5}$, for example, is just a shorthand notation for both $-1 + \sqrt{5}$ and $-1 - \sqrt{5}$.

Example 7 Use the square root property to solve $(x-1)^2 = -2$.

Solution: This equation has no real solution because the square root of -2 is not a real number.

Work Practice 7

Practice 7

Use the square root property to solve $(x+3)^2 = -5$.

Example 8 Use the square root property to solve $(5x-2)^2 = 10$.

Solution: $(5x-2)^2 = 10$

$5x - 2 = \sqrt{10}$ or $5x - 2 = -\sqrt{10}$ Use the square root property.

$5x = 2 + \sqrt{10}$ $\quad 5x = 2 - \sqrt{10}$ Add 2 to both sides.

$x = \dfrac{2 + \sqrt{10}}{5}$ $\quad x = \dfrac{2 - \sqrt{10}}{5}$ Divide both sides by 5.

Check both solutions in the original equation. The solutions are $\dfrac{2+\sqrt{10}}{5}$ and $\dfrac{2-\sqrt{10}}{5}$, which can be written as $\dfrac{2 \pm \sqrt{10}}{5}$.

Work Practice 8

Practice 8

Use the square root property to solve $(4x+1)^2 = 15$.

Helpful Hint

For some applications and graphing purposes, decimal approximations of exact solutions to quadratic equations may be desired.

Exact Solutions from Example 8		Decimal Approximations
$\dfrac{2+\sqrt{10}}{5}$	$\approx$	1.032
$\dfrac{2-\sqrt{10}}{5}$	$\approx$	-0.232

Answers

6. $5 \pm 3\sqrt{2}$ **7.** no real solution

8. $\dfrac{-1 \pm \sqrt{15}}{4}$

Objective B Using the Square Root Property to Solve Applications

Many real-world applications are modeled by quadratic equations. In the next example, we use the quadratic formula $h = 16t^2$. This formula gives the distance h traveled by a free-falling object in time t. One important note is that this formula does not take into account any air resistance.

Practice 9

Use the formula $h = 16t^2$ (see Example 9) to find how long, to the nearest tenth of a second, it takes a free-falling body to fall 650 feet.

Example 9 Finding the Length of Time of a Dive

The record for the highest dive into a lake remains to be held by Harry Froboess of Switzerland. In 1936 he dove 394 feet from the airship Hindenburg into Lake Constance. To the nearest tenth of a second, how long did his dive take? (*Source*: *Guinness World Records*)

Solution:

1. UNDERSTAND. To approximate the time of the dive, we use the formula* $h = 16t^2$, where t is time in seconds and h is the distance in feet traveled by a free-falling body or object. For example, to find the distance traveled in 1 second, or 3 seconds, we let $t = 1$ and then $t = 3$.

   If $t = 1$, $h = 16(1)^2 = 16 \cdot 1 = 16$ feet.

   If $t = 3$, $h = 16(3)^2 = 16 \cdot 9 = 144$ feet.

   Since a body travels 144 feet in 3 seconds, we now know the dive of 394 feet lasted longer than 3 seconds.

2. TRANSLATE. Use the formula $h = 16t^2$, let the distance $h = 394$, and we have the equation $394 = 16t^2$.

3. SOLVE. To solve $394 = 16t^2$ for t, we will use the square root property.

$$394 = 16t^2$$

$$\frac{394}{16} = t^2 \qquad \text{Divide both sides by 16.}$$

$$24.625 = t^2 \qquad \text{Simplify.}$$

$$\sqrt{24.625} = t \quad \text{or} \quad -\sqrt{24.625} = t \qquad \text{Use the square root property.}$$

$$5.0 \approx t \quad \text{or} \quad -5.0 \approx t \qquad \text{Approximate.}$$

4. INTERPRET.

   Check: We reject the solution -5.0 since the length of the dive is not a negative number.

   State: The dive lasted approximately 5 seconds.

Work Practice 9

Answer

9. 6.4 sec

*The formula $h = 16t^2$ does not take into account air resistance.

Vocabulary, Readiness & Video Check

Martin-Gay Interactive Videos

See Video 9.1

Watch the section lecture video and answer the following questions.

Objective A **1.** As explained in Example 2, why is $a \geq 0$ in the statement of the square root property?

Objective B **2.** In Example 6, how can we tell by looking at the translated equation that the square root property can be used to solve it? Why is the negative square root not considered?

9.1 Exercise Set MyLab Math

Solve each equation by factoring. See Examples 1 and 2.

1. $k^2 - 49 = 0$ **2.** $k^2 - 9 = 0$ **3.** $m^2 + 2m = 15$ **4.** $m^2 + 6m = 7$ **5.** $2x^2 - 32 = 0$

6. $2x^2 - 98 = 0$ **7.** $4a^2 - 36 = 0$ **8.** $7a^2 - 175 = 0$ **9.** $x^2 + 7x = -10$ **10.** $x^2 + 10x = -24$

Objective A *Use the square root property to solve each quadratic equation. See Examples 3 and 4.*

11. $x^2 = 64$ **12.** $x^2 = 121$ **13.** $x^2 = 21$ **14.** $x^2 = 22$ **15.** $x^2 = \frac{1}{25}$

16. $x^2 = \frac{1}{16}$ **17.** $x^2 = -4$ **18.** $x^2 = -25$ **19.** $3x^2 = 13$

20. $5x^2 = 2$ **21.** $7x^2 = 4$ **22.** $2x^2 = 9$ **23.** $2x^2 - 10 = 0$ **24.** $3x^2 - 45 = 0$

Use the square root property to solve each quadratic equation. See Examples 5 through 8.

25. $(x - 5)^2 = 49$ **26.** $(x + 2)^2 = 25$ **27.** $(x + 2)^2 = 7$ **28.** $(x - 7)^2 = 2$

29. $\left(m - \frac{1}{2}\right)^2 = \frac{1}{4}$ **30.** $\left(m + \frac{1}{3}\right)^2 = \frac{1}{9}$ **31.** $(p + 2)^2 = 10$ **32.** $(p - 7)^2 = 13$

33. $(3y + 2)^2 = 100$ **34.** $(4y - 3)^2 = 81$ **35.** $(z - 4)^2 = -9$ **36.** $(z + 7)^2 = -20$

37. $(2x - 11)^2 = 50$ **38.** $(3x - 17)^2 = 28$ **39.** $(3x - 7)^2 = 32$ **40.** $(5x - 11)^2 = 54$

Use the square root property to solve. See Examples 3 through 8.

41. $x^2 - 29 = 0$

42. $x^2 - 35 = 0$

43. $(x + 6)^2 = 24$

44. $(x + 5)^2 = 20$

45. $\frac{1}{2}n^2 = 5$

46. $\frac{1}{5}y^2 = 2$

47. $(4x - 1)^2 = 5$

48. $(7x - 2)^2 = 11$

49. $3z^2 = 36$

50. $3z^2 = 24$

51. $(8 - 3x)^2 - 45 = 0$

52. $(10 - 9x)^2 - 75 = 0$

Objective B *The formula for the area of a square is $A = s^2$, where s is the length of a side. Use this formula for Exercises 53 through 56. For each exercise, give the exact answer and a two-decimal-place approximation.*

△ **53.** If the area of a square is 20 square inches, find the length of a side.

△ **54.** If the area of a square is 32 square meters, find the length of a side.

△ **55.** The "Water Cube" National Swimming Center was constructed in Beijing for the 2008 Summer Olympics. Its square base has an area of 31,329 square meters. Find the length of a side of this building. (*Source:* ARUP East Asia)

Note: The Beijing Water Cube was converted to an indoor water park and recently reopened.

△ **56.** The Washington Monument has a square base whose area is approximately 3039 square feet. Find the length of a side. (*Source: The World Almanac*)

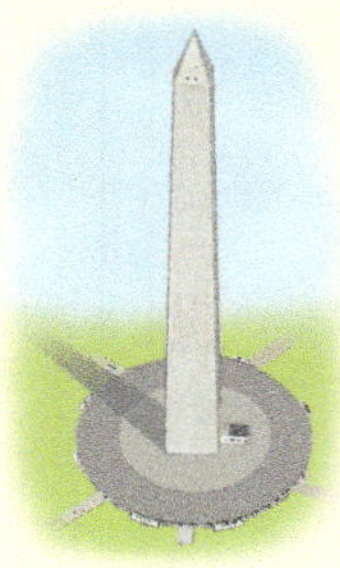

Solve. For Exercises 57 through 60, use the formula $h = 16t^2$. See Example 9. Round each answer to the nearest tenth of a second.

57. The highest regularly performed dives are made by professional divers from La Quebrada. If this cliff in Acapulco has a height of 87.6 feet, determine the time of a dive. (*Source: Guinness World Records*)

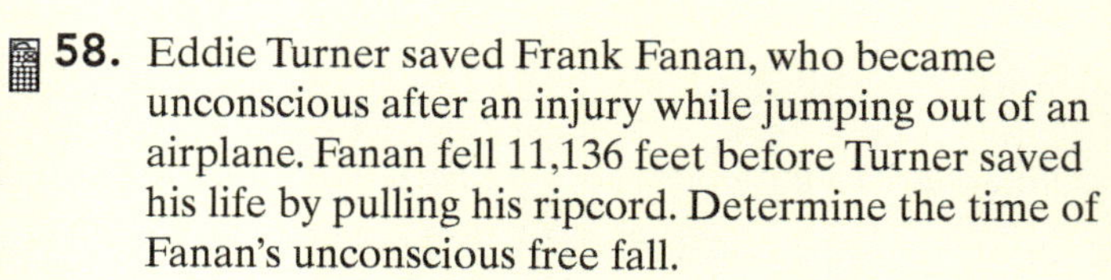

58. Eddie Turner saved Frank Fanan, who became unconscious after an injury while jumping out of an airplane. Fanan fell 11,136 feet before Turner saved his life by pulling his ripcord. Determine the time of Fanan's unconscious free fall.

59. The Hualapai Indian Tribe allowed the Grand Canyon Skywalk to be built over the rim of the Grand Canyon on its tribal land. The skywalk extends 70 feet beyond the canyon's edge and is 4000 feet above the canyon floor. Determine the time, to the nearest tenth of a second, it would take an object, dropped off the skywalk, to land at the bottom of the Grand Canyon. (*Source: Boston Globe*)

60. If a sandblaster drops his goggles from a bridge 400 feet from the water below, find how long it takes for the goggles to hit the water.

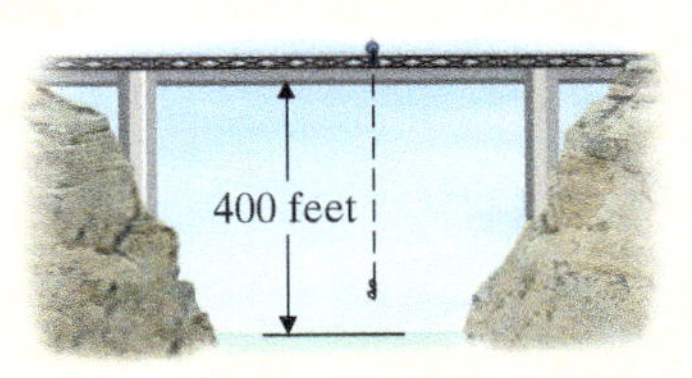

△ **61.** The area of a circle is found by the equation $A = \pi r^2$. If the area A of a certain circle is 36π square inches, find its radius r.

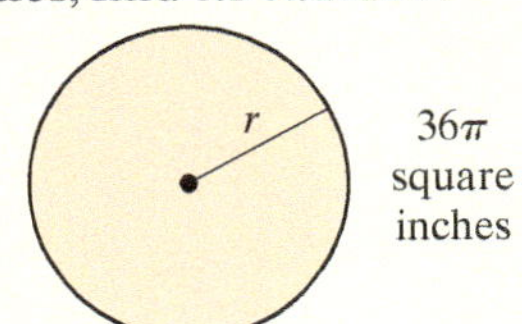

62. If the area of the circle below is 10π square units, find its exact radius. (See Exercise **61**.)

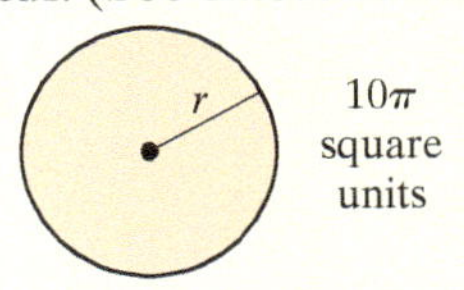

Review

Factor each perfect square trinomial. See Section 4.5.

63. $x^2 + 6x + 9$

64. $y^2 + 10y + 25$

65. $x^2 - 4x + 4$

66. $x^2 - 20x + 100$

Concept Extensions

67. Explain why the equation $x^2 = -9$ has no real solution.

68. Explain why the equation $x^2 = 9$ has two solutions.

Solve each quadratic equation by first factoring the perfect square trinomial on the left side. Then apply the square root property.

69. $x^2 + 4x + 4 = 16$

70. $y^2 - 10y + 25 = 11$

Solve each quadratic equation by using the square root property. Use a calculator and round each solution to the nearest hundredth.

71. $x^2 = 1.78$

72. $(x - 1.37)^2 = 5.71$

73. The number of U.S. highway bridges for the years 2006 to 2016 can be modeled by the equation $y = 110(x - 2)^2 + 596{,}680$, where $x = 0$ represents the year 2006. Assume that this trend continues and find the first year in which there will be 650,000 highway bridges in the United States. (*Hint:* Replace y with 650,000 in the equation and solve for x. Round to the nearest year.) (*Source:* Based on data from the U.S. Department of Transportation, Federal Highway Administration)

74. The annual wireless data usage (in trillions of megabytes) in the United States for the years 2005 through 2016 can be estimated by $y = 0.5(x - 6)^2 + 1$, where $x = 0$ represents the year 2005. Assume that this trend continues, and determine the first year in which the annual wireless data usage will surpass 30 trillion megabytes. (*Hint:* Replace y with 30 in the equation and solve for x. Round to the nearest year.) (*Source:* Based on data from CTIA—The Wireless Association)

9.2 Solving Quadratic Equations by Completing the Square

Objectives

A Solve Quadratic Equations of the Form $x^2 + bx + c = 0$ by Completing the Square.

B Solve Quadratic Equations of the Form $ax^2 + bx + c = 0$ by Completing the Square.

Objective A Completing the Square to Solve $x^2 + bx + c = 0$

In the last section, we used the square root property to solve equations such as

$$(x + 1)^2 = 8 \quad \text{and} \quad (5x - 2)^2 = 3$$

Notice that one side of each equation is a quantity squared and that the other side is a constant. To solve

$$x^2 + 2x = 4$$

notice that if we add 1 to both sides of the equation, the left side is a perfect square trinomial that can be factored.

$$x^2 + 2x + 1 = 4 + 1 \quad \text{Add 1 to both sides.}$$
$$(x + 1)^2 = 5 \quad \text{Factor.}$$

Now we can solve this equation as we did in the previous section, by using the square root property.

$$x + 1 = \sqrt{5} \quad \text{or} \quad x + 1 = -\sqrt{5} \quad \text{Use the square root property.}$$
$$x = -1 + \sqrt{5} \quad\quad x = -1 - \sqrt{5} \quad \text{Solve.}$$

The solutions are $-1 \pm \sqrt{5}$.

Adding a number to $x^2 + 2x$ to form a perfect square trinomial is called **completing the square** on $x^2 + 2x$.

In general, we have the following:

Completing the Square

To complete the square on $x^2 + bx$, add $\left(\frac{b}{2}\right)^2$. To find $\left(\frac{b}{2}\right)^2$, **find half the coefficient of x, and then square the result.**

Practice 1

Solve $x^2 + 8x + 1 = 0$ by completing the square.

Example 1 Solve $x^2 + 6x + 3 = 0$ by completing the square.

Solution: First we get the variable terms alone by subtracting 3 from both sides of the equation.

$$x^2 + 6x + 3 = 0$$
$$x^2 + 6x = -3 \quad \text{Subtract 3 from both sides.}$$

Next we find half the coefficient of the x-term, and then square it. We add this result to *both sides* of the equation. This will make the left side a perfect square trinomial. The coefficient of x is 6, and half of 6 is 3. So we add 3^2 or 9 to both sides.

$$x^2 + 6x + 9 = -3 + 9 \quad \text{Complete the square.}$$
$$(x + 3)^2 = 6 \quad \text{Factor the trinomial } x^2 + 6x + 9.$$
$$x + 3 = \sqrt{6} \quad \text{or} \quad x + 3 = -\sqrt{6} \quad \text{Use the square root property.}$$
$$x = -3 + \sqrt{6} \quad\quad x = -3 - \sqrt{6} \quad \text{Subtract 3 from both sides.}$$

Check by substituting $-3 + \sqrt{6}$ and $-3 - \sqrt{6}$ in the original equation. The solutions are $-3 \pm \sqrt{6}$.

Work Practice 1

Answer

1. $-4 \pm \sqrt{15}$

Helpful Hint

Remember, when completing the square, add the number that completes the square to **both sides of the equation.**

Example 2 Solve $x^2 - 10x = -14$ by completing the square.

Solution: The variable terms are already alone on one side of the equation. The coefficient of x is -10. Half of -10 is -5, and $(-5)^2 = 25$. So we add 25 to both sides.

$$x^2 - 10x = -14$$

$$x^2 - 10x + 25 = -14 + 25$$

Helpful Hint Add 25 to *both* sides of the equation.

$$(x - 5)^2 = 11 \quad \text{Factor the trinomial and simplify } -14 + 25.$$

$$x - 5 = \sqrt{11} \quad \text{or} \quad x - 5 = -\sqrt{11} \quad \text{Use the square root property.}$$

$$x = 5 + \sqrt{11} \qquad x = 5 - \sqrt{11} \quad \text{Add 5 to both sides.}$$

The solutions are $5 \pm \sqrt{11}$.

Work Practice 2

Practice 2

Solve $x^2 - 14x = -32$ by completing the square.

Objective B Completing the Square to Solve $ax^2 + bx + c = 0$

The method of completing the square can be used to solve *any* quadratic equation whether the coefficient of the squared variable is 1 or not. When the coefficient of the squared variable is not 1, we first divide both sides of the equation by the coefficient of the squared variable so that the new coefficient is 1. Then we complete the square.

Example 3 Solve $4x^2 - 8x - 5 = 0$ by completing the square.

Solution: Since the coefficient of x^2 is 4, not 1, we first divide both sides of the equation by 4 so that the coefficient of x^2 is 1.

$$4x^2 - 8x - 5 = 0$$

$$x^2 - 2x - \frac{5}{4} = 0 \quad \text{Divide both sides by 4.}$$

$$x^2 - 2x = \frac{5}{4} \quad \text{Get the variable terms alone on one side of the equation.}$$

The coefficient of x is -2. Half of -2 is -1, and $(-1)^2 = 1$. So we add 1 to both sides.

$$x^2 - 2x + 1 = \frac{5}{4} + 1$$

$$(x - 1)^2 = \frac{9}{4} \quad \text{Factor } x^2 - 2x + 1 \text{ and simplify } \frac{5}{4} + 1.$$

$$x - 1 = \sqrt{\frac{9}{4}} \quad \text{or} \quad x - 1 = -\sqrt{\frac{9}{4}} \quad \text{Use the square root property.}$$

$$x = 1 + \frac{3}{2} \qquad x = 1 - \frac{3}{2} \quad \text{Add 1 to both sides and simplify the radical.}$$

$$x = \frac{5}{2} \qquad x = -\frac{1}{2} \quad \text{Simplify.}$$

Both $\frac{5}{2}$ and $-\frac{1}{2}$ are solutions.

Work Practice 3

Practice 3

Solve $4x^2 - 16x - 9 = 0$ by completing the square.

Answers

2. $7 \pm \sqrt{17}$ **3.** $\frac{9}{2}$ and $-\frac{1}{2}$

The following steps may be used to solve a quadratic equation in x by completing the square.

To Solve a Quadratic Equation in x by Completing the Square

Step 1: If the coefficient of x^2 is 1, go to Step 2. If not, divide both sides of the equation by the coefficient of x^2.

Step 2: Get all terms with variables on one side of the equation and constants on the other side.

Step 3: Find half the coefficient of x and then square the result. Add this number to both sides of the equation.

Step 4: Factor the resulting perfect square trinomial.

Step 5: Use the square root property to solve the equation.

Practice 4

Solve $2x^2 + 10x = -13$ by completing the square.

Example 4 Solve $2x^2 + 6x = -7$ by completing the square.

Solution: The coefficient of x^2 is not 1. We divide both sides by 2, the coefficient of x^2.

$$2x^2 + 6x = -7$$

$$x^2 + 3x = -\frac{7}{2} \quad \text{Divide both sides by 2.}$$

$$x^2 + 3x + \frac{9}{4} = -\frac{7}{2} + \frac{9}{4} \quad \text{Add } \left(\frac{3}{2}\right)^2 \text{ or } \frac{9}{4} \text{ to both sides.}$$

$$\left(x + \frac{3}{2}\right)^2 = -\frac{5}{4} \quad \text{Factor the left side and simplify the right.}$$

There is no real solution to this equation since the square root of a negative number is not a real number.

Work Practice 4

Practice 5

Solve $2x^2 = -6x + 5$ by completing the square.

Example 5 Solve $2x^2 = 10x + 1$ by completing the square.

Solution: First we divide both sides of the equation by 2, the coefficient of x^2.

$$2x^2 = 10x + 1$$

$$x^2 = 5x + \frac{1}{2} \quad \text{Divide both sides by 2.}$$

Next we get the variable terms alone by subtracting $5x$ from both sides.

$$x^2 - 5x = \frac{1}{2}$$

$$x^2 - 5x + \frac{25}{4} = \frac{1}{2} + \frac{25}{4} \quad \text{Add } \left(-\frac{5}{2}\right)^2 \text{ or } \frac{25}{4} \text{ to both sides.}$$

$$\left(x - \frac{5}{2}\right)^2 = \frac{27}{4} \quad \text{Factor the left side and simplify the right side.}$$

$$x - \frac{5}{2} = \sqrt{\frac{27}{4}} \quad \text{or} \quad x - \frac{5}{2} = -\sqrt{\frac{27}{4}} \quad \text{Use the square root property.}$$

$$x - \frac{5}{2} = \frac{3\sqrt{3}}{2} \qquad x - \frac{5}{2} = -\frac{3\sqrt{3}}{2} \quad \text{Simplify.}$$

$$x = \frac{5}{2} + \frac{3\sqrt{3}}{2} \qquad x = \frac{5}{2} - \frac{3\sqrt{3}}{2}$$

The solutions are $\frac{5 \pm 3\sqrt{3}}{2}$.

Work Practice 5

Answers

4. no real solution 5. $\frac{-3 \pm \sqrt{19}}{2}$

Vocabulary, Readiness & Video Check

Use the choices below to fill in each blank. Not all choices will be used, and these exercises come from Sections 9.1 and 9.2.

$\sqrt{a}$	linear equation	zero	$\left(\frac{b}{2}\right)^2$	$\frac{b}{2}$	6
$\pm\sqrt{a}$	quadratic equation	one	completing the square	9	3

1. By the zero-factor property, if the product of two numbers is zero, then at least one of these two numbers must be ______________.
2. If a is a positive number, and if $x^2 = a$, then $x =$ ______________.
3. An equation that can be written in the form $ax^2 + bx + c = 0$ where a, b, and c are real numbers and a is not zero is called a(n) ______________.
4. The process of solving a quadratic equation by writing it in the form $(x + a)^2 = c$ is called ______________.
5. To complete the square on $x^2 + 6x$, add ______________.
6. To complete the square on $x^2 + bx$, add ______________.

Fill in the blank with the number needed to make each expression a perfect square trinomial. See Example 1.

7. $p^2 + 8p +$ ______________
8. $p^2 + 6p +$ ______________
9. $x^2 + 20x +$ ______________
10. $x^2 + 18x +$ ______________
11. $y^2 + 14y +$ ______________
12. $y^2 + 2y +$ ______________

Martin-Gay Interactive Videos

See Video 9.2

Watch the section lecture video and answer the following questions.

Objective A 13. In Examples 3 and 4, explain why the constant that completes the square is added to both sides of the equation.

Objective B 14. In Example 5, why is the equation first divided through by 2?

9.2 Exercise Set MyLab Math

Objective A *Solve each quadratic equation by completing the square. See Examples 1 and 2.*

1. $x^2 + 8x = -12$
2. $x^2 - 10x = -24$
3. $x^2 + 2x - 7 = 0$
4. $z^2 + 6z - 9 = 0$
5. $x^2 - 6x = 0$
6. $y^2 + 4y = 0$
7. $y^2 + 5y + 4 = 0$
8. $y^2 - 5y + 6 = 0$
9. $x^2 - 2x - 1 = 0$
10. $x^2 - 4x + 2 = 0$
11. $z^2 + 5z = 7$
12. $x^2 - 7x = 5$

Objective B *Solve each quadratic equation by completing the square. See Examples 3 through 5.*

13. $3x^2 - 6x = 24$
14. $2x^2 + 18x = -40$
15. $5x^2 + 10x + 6 = 0$
16. $3x^2 - 12x + 14 = 0$

17. $2x^2 = 6x + 5$

18. $4x^2 = -20x + 3$

19. $2y^2 + 8y + 5 = 0$

20. $4z^2 - 8z + 1 = 0$

Objectives A B Mixed Practice *Solve each quadratic equation by completing the square. See Examples 1 through 5.*

21. $x^2 + 6x - 25 = 0$

22. $x^2 - 6x + 7 = 0$

23. $x^2 - 3x - 3 = 0$

24. $x^2 - 9x + 3 = 0$

25. $2y^2 - 3y + 1 = 0$

26. $2y^2 - y - 1 = 0$

27. $x(x + 3) = 18$ (*Hint:* First use the distributive property and multiply.)

28. $x(x - 3) = 18$ (See hint for Exercise **27**.)

29. $3z^2 + 6z + 4 = 0$

30. $2y^2 + 8y + 9 = 0$

31. $4x^2 + 16x = 48$

32. $6x^2 - 30x = -36$

Review

Simplify each expression. See Section 8.2.

33. $\frac{3}{4} - \sqrt{\frac{25}{16}}$

34. $\frac{3}{5} + \sqrt{\frac{16}{25}}$

35. $\frac{1}{2} + \sqrt{\frac{9}{4}}$

36. $\frac{9}{10} - \sqrt{\frac{49}{100}}$

Simplify each expression. See Section 8.4.

37. $\frac{6 + 4\sqrt{5}}{2}$

38. $\frac{10 + 20\sqrt{3}}{2}$

39. $\frac{3 - 9\sqrt{2}}{6}$

40. $\frac{12 - 8\sqrt{7}}{16}$

Concept Extensions

41. In your own words, describe a perfect square trinomial.

42. Describe how to find the number to add to $x^2 - 7x$ to make a perfect square trinomial.

43. Write your own quadratic equation to be solved by completing the square. Write it in the form

perfect square trinomial = a number that is not a perfect square. (An example is shown.)

$$x^2 + 6x + 9 = 11$$

a. Solve the example above: $x^2 + 6x + 9 = 11$.

b. Write your own quadratic equation using the same format.

c. Solve your quadratic equation by completing the square.

44. Follow the directions of Exercise **43**, except

a. Write your quadratic equation in the form

perfect square trinomial = negative number

b. Solve your quadratic equation by completing the square.

45. Find a value of k that will make $x^2 + kx + 16$ a perfect square trinomial.

46. Find a value of k that will make $x^2 + kx + 25$ a perfect square trinomial.

47. The average price y of gold (in dollars per ounce) from 2013 through 2017 is given by the equation $y = 63x^2 - 364x + 1622$, where x is the number of years after 2013. Assume that this trend continued and find the year after 2013 in which the price of gold was \$1706 per ounce. (*Source:* Based on data from goldprice.org)

48. The revenues from product sales y (in millions of dollars) of Abiomed, Inc., maker of the AbioCor artificial heart, during fiscal years 2012 through 2016 can be modeled by the equation $y = 12x^2 + 3x + 132$, where $x = 0$ represents 2012. Assume that this trend continues and predict the year after 2012 in which Abiomed's revenues from product sales will be \$540 million. (Round to the nearest whole number.) (*Source:* Based on data from Abiomed, Inc.)

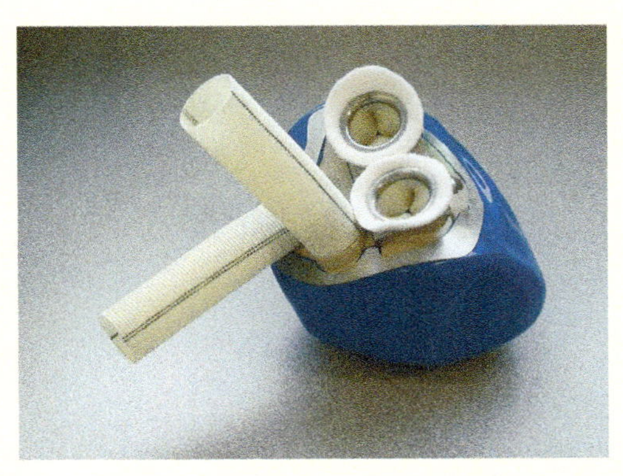

*Recall that a graphing calculator may be used to solve an equation. For example, to solve $x^2 + 8x = -12$ (Exercise **1**), graph*

$$y_1 = x^2 + 8x$$
$$y_2 = -12$$

The x-coordinates of the points of intersection of the graphs is the solution. Use a graphing calculator to solve each equation. Round solutions to the nearest hundredth.

49. Exercise **1**

50. Exercise **2**

51. Exercise **17**

52. Exercise **12**

9.3 Solving Quadratic Equations by the Quadratic Formula

Objective A Using the Quadratic Formula

Objectives

A Use the Quadratic Formula to Solve Quadratic Equations.

B Approximate Solutions to Quadratic Equations.

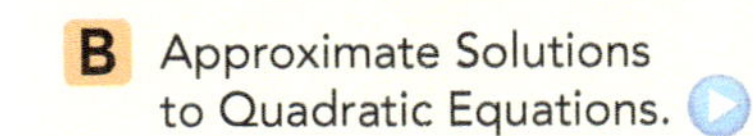

We can use the technique of completing the square to develop a formula to find solutions of any quadratic equation. We develop and use the **quadratic formula** in this section.

Recall that a quadratic equation in **standard form** is

$$ax^2 + bx + c = 0, \quad a \neq 0$$

To develop the quadratic formula, let's complete the square for this quadratic equation in standard form.

First we divide both sides of the equation by the coefficient of x^2 and then get the variable terms alone on one side of the equation.

$$x^2 + \frac{b}{a}x + \frac{c}{a} = 0 \qquad \text{Divide by } a\text{; recall that } a \text{ cannot be 0.}$$

$$x^2 + \frac{b}{a}x = -\frac{c}{a} \qquad \text{Get the variable terms alone on one side of the equation.}$$

The coefficient of x is $\frac{b}{a}$. Half of $\frac{b}{a}$ is $\frac{b}{2a}$ and $\left(\frac{b}{2a}\right)^2 = \frac{b^2}{4a^2}$. So we add $\frac{b^2}{4a^2}$ to both sides of the equation.

$$x^2 + \frac{b}{a}x + \frac{b^2}{4a^2} = -\frac{c}{a} + \frac{b^2}{4a^2}$$ Add $\frac{b^2}{4a^2}$ to both sides.

$$\left(x + \frac{b}{2a}\right)^2 = -\frac{c}{a} + \frac{b^2}{4a^2}$$ Factor the left side.

$$\left(x + \frac{b}{2a}\right)^2 = -\frac{4ac}{4a^2} + \frac{b^2}{4a^2}$$ Multiply $-\frac{c}{a}$ by $\frac{4a}{4a}$ so that the terms on the right side have a common denominator.

$$\left(x + \frac{b}{2a}\right)^2 = \frac{b^2 - 4ac}{4a^2}$$ Simplify the right side.

Now we use the square root property.

$$x + \frac{b}{2a} = \sqrt{\frac{b^2 - 4ac}{4a^2}} \quad \text{or} \quad x + \frac{b}{2a} = -\sqrt{\frac{b^2 - 4ac}{4a^2}}$$ Use the square root property.

$$x + \frac{b}{2a} = \frac{\sqrt{b^2 - 4ac}}{2a} \qquad x + \frac{b}{2a} = -\frac{\sqrt{b^2 - 4ac}}{2a}$$ Simplify the radical.

$$x = -\frac{b}{2a} + \frac{\sqrt{b^2 - 4ac}}{2a} \qquad x = -\frac{b}{2a} - \frac{\sqrt{b^2 - 4ac}}{2a}$$ Subtract $\frac{b}{2a}$ from both sides.

$$x = \frac{-b + \sqrt{b^2 - 4ac}}{2a} \qquad x = \frac{-b - \sqrt{b^2 - 4ac}}{2a}$$ Simplify.

The solutions are $\frac{-b \pm \sqrt{b^2 - 4ac}}{2a}$. This final equation is called the **quadratic formula** and gives the solutions of any quadratic equation.

Quadratic Formula

If a, b, and c are real numbers and $a \neq 0$, a quadratic equation written in the standard form $ax^2 + bx + c = 0$ has solutions

$$x = \frac{-b \pm \sqrt{b^2 - 4ac}}{2a}$$

Helpful Hint

Don't forget that to correctly identify a, b, and c in the quadratic formula, you should write the equation in standard form.

Quadratic Equations in Standard Form

$5x^2 - 6x + 2 = 0$	$a = 5, b = -6, c = 2$
$4y^2 - 9 = 0$	$a = 4, b = 0, c = -9$
$x^2 + x = 0$	$a = 1, b = 1, c = 0$
$\sqrt{2}x^2 + \sqrt{5}x + \sqrt{3} = 0$	$a = \sqrt{2}, b = \sqrt{5}, c = \sqrt{3}$

Example 1 Solve $3x^2 + x - 3 = 0$ using the quadratic formula.

Solution: This equation is in standard form with $a = 3$, $b = 1$, and $c = -3$. By the quadratic formula, we have

$$x = \frac{-b \pm \sqrt{b^2 - 4ac}}{2a}$$

$$x = \frac{-1 \pm \sqrt{1^2 - 4 \cdot 3 \cdot (-3)}}{2 \cdot 3} \quad \text{Let } a = 3, b = 1, \text{ and } c = -3.$$

$$= \frac{-1 \pm \sqrt{1 + 36}}{6} \quad \text{Simplify.}$$

$$= \frac{-1 \pm \sqrt{37}}{6}$$

Check both solutions in the original equation. The solutions are $\frac{-1 + \sqrt{37}}{6}$ and $\frac{-1 - \sqrt{37}}{6}$.

Work Practice 1

Practice 1

Solve $2x^2 - x - 5 = 0$ using the quadratic formula.

Example 2 Solve $2x^2 - 9x = 5$ using the quadratic formula.

Solution: First we write the equation in standard form by subtracting 5 from both sides.

$$2x^2 - 9x = 5$$
$$2x^2 - 9x - 5 = 0$$

Next we note that $a = 2$, $b = -9$, and $c = -5$. We substitute these values into the quadratic formula.

$$x = \frac{-b \pm \sqrt{b^2 - 4ac}}{2a}$$

$$x = \frac{-(-9) \pm \sqrt{(-9)^2 - 4 \cdot 2 \cdot (-5)}}{2 \cdot 2} \quad \text{Substitute into the formula.}$$

$$= \frac{9 \pm \sqrt{81 + 40}}{4} \quad \text{Simplify.}$$

$$= \frac{9 \pm \sqrt{121}}{4} = \frac{9 \pm 11}{4}$$

Then,

$$x = \frac{9 - 11}{4} = -\frac{1}{2} \quad \text{or} \quad x = \frac{9 + 11}{4} = 5$$

Check $-\frac{1}{2}$ and 5 in the original equation. Both $-\frac{1}{2}$ and 5 are solutions.

Work Practice 2

Practice 2

Solve $3x^2 + 8x = 3$ using the quadratic formula.

Helpful Hint Notice that the fraction bar is under the entire numerator $-b \pm \sqrt{b^2 - 4ac}$.

The following steps may be useful when solving a quadratic equation by the quadratic formula.

Answers

1. $\frac{1 + \sqrt{41}}{4}$ and $\frac{1 - \sqrt{41}}{4}$

2. $\frac{1}{3}$ and -3

To Solve a Quadratic Equation by the Quadratic Formula

Step 1: Write the quadratic equation in standard form: $ax^2 + bx + c = 0$.

Step 2: If necessary, clear the equation of fractions to simplify calculations.

Step 3: Identify a, b, and c.

Step 4: Replace a, b, and c in the quadratic formula with the identified values, and simplify.

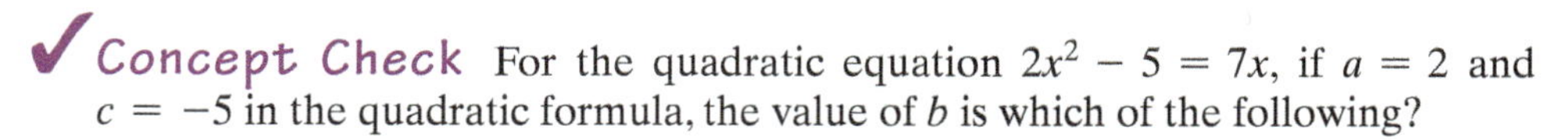

✓Concept Check For the quadratic equation $2x^2 - 5 = 7x$, if $a = 2$ and $c = -5$ in the quadratic formula, the value of b is which of the following?

a. $\frac{7}{2}$ **b.** 7 **c.** -5 **d.** -7

Practice 3

Solve $5x^2 = 2$ using the quadratic formula.

Example 3 Solve $7x^2 = 1$ using the quadratic formula.

Solution: First we write the equation in standard form by subtracting 1 from both sides.

$$7x^2 = 1$$
$$7x^2 - 1 = 0$$

Helpful Hint

$7x^2 - 1 = 0$ can be written as $7x^2 + 0x - 1 = 0$. This form helps you see that $b = 0$.

Next we replace a, b, and c with the identified values: $a = 7, b = 0, c = -1$.

$$x = \frac{0 \pm \sqrt{0^2 - 4 \cdot 7 \cdot (-1)}}{2 \cdot 7} \quad \text{Substitute into the formula.}$$

$$= \frac{\pm\sqrt{28}}{14} \quad \text{Simplify.}$$

$$= \frac{\pm 2\sqrt{7}}{14}$$

$$= \pm\frac{2\sqrt{7}}{2 \cdot 7}$$

$$= \pm\frac{\sqrt{7}}{7}$$

The solutions are $\frac{\sqrt{7}}{7}$ and $-\frac{\sqrt{7}}{7}$.

Work Practice 3

Notice that we could have solved the equation $7x^2 = 1$ in Example 3 by dividing both sides by 7 and then using the square root property. We solved the equation by the quadratic formula to show that this formula can be used to solve any quadratic equation.

Practice 4

Solve $x^2 = -2x - 3$ using the quadratic formula.

Example 4 Solve $x^2 = -x - 1$ using the quadratic formula.

Solution: First we write the equation in standard form.

$$x^2 + x + 1 = 0$$

Next we replace a, b, and c in the quadratic formula with $a = 1, b = 1$, and $c = 1$.

Answers

3. $\frac{\sqrt{10}}{5}$ and $-\frac{\sqrt{10}}{5}$

4. no real solution

✓Concept Check Answer

d

$$x = \frac{-1 \pm \sqrt{1^2 - 4 \cdot 1 \cdot 1}}{2 \cdot 1} \quad \text{Substitute into the formula.}$$

$$= \frac{-1 \pm \sqrt{-3}}{2} \quad \text{Simplify.}$$

There is no real number solution because $\sqrt{-3}$ is not a real number.

Work Practice 4

Example 5 Solve $\frac{1}{2}x^2 - x = 2$ using the quadratic formula.

Practice 5

Solve $\frac{1}{3}x^2 - x = 1$ using the quadratic formula.

Solution: We write the equation in standard form and then clear the equation of fractions by multiplying both sides by the LCD, 2.

$$\frac{1}{2}x^2 - x = 2$$

$$\frac{1}{2}x^2 - x - 2 = 0 \quad \text{Write in standard form.}$$

$$x^2 - 2x - 4 = 0 \quad \text{Multiply both sides by 2.}$$

Here, $a = 1$, $b = -2$, and $c = -4$, so we substitute these values into the quadratic formula.

$$x = \frac{-(-2) \pm \sqrt{(-2)^2 - 4 \cdot 1 \cdot (-4)}}{2 \cdot 1}$$

$$= \frac{2 \pm \sqrt{20}}{2} = \frac{2 \pm 2\sqrt{5}}{2} \quad \text{Simplify.}$$

$$= \frac{2(1 \pm \sqrt{5})}{2} = 1 \pm \sqrt{5} \quad \text{Factor and simplify.}$$

The solutions are $1 - \sqrt{5}$ and $1 + \sqrt{5}$.

Work Practice 5

Notice that in Example 5, although we cleared the equation of fractions, using the coefficients $a = \frac{1}{2}$, $b = -1$, and $c = -2$ will give the same results.

Helpful Hint

When simplifying an expression such as

$$\frac{3 \pm 6\sqrt{2}}{6}$$

first factor out a common factor from the terms of the numerator and then simplify.

$$\frac{3 \pm 6\sqrt{2}}{6} = \frac{3(1 \pm 2\sqrt{2})}{2 \cdot 3} = \frac{1 \pm 2\sqrt{2}}{2}$$

Answer

5. $\frac{3 + \sqrt{21}}{2}$ and $\frac{3 - \sqrt{21}}{2}$

Objective B Approximating Solutions to Quadratic Equations

Sometimes approximate solutions for quadratic equations are appropriate.

Practice 6

Approximate the exact solutions of the quadratic equation in Practice 1. Round the approximations to the nearest tenth.

Example 6 Approximate the exact solutions of the quadratic equation in Example 1. Round the approximations to the nearest tenth.

Solution: From Example 1, we have exact solutions $\frac{-1 \pm \sqrt{37}}{6}$. Thus,

$$\frac{-1 + \sqrt{37}}{6} \approx 0.847127088 \approx 0.8 \text{ to the nearest tenth.}$$

$$\frac{-1 - \sqrt{37}}{6} \approx -1.180460422 \approx -1.2 \text{ to the nearest tenth.}$$

Thus approximate solutions to the quadratic equation in Example 1 are 0.8 and -1.2.

Work Practice 6

Answer

6. $\frac{1 + \sqrt{41}}{4} \approx 1.9, \frac{1 - \sqrt{41}}{4} \approx -1.4$

Vocabulary, Readiness & Video Check

Fill in each blank.

1. The quadratic formula is ________________.

Identify the values of a, b, and c in each quadratic equation.

2. $5x^2 - 7x + 1 = 0$; $a =$ ________, $b =$ ________, $c =$ ________

3. $x^2 + 3x - 7 = 0$; $a =$ ________, $b =$ ________, $c =$ ________

4. $x^2 - 6 = 0$; $a =$ ________, $b =$ ________, $c =$ ________

5. $x^2 + x - 1 = 0$; $a =$ ________, $b =$ ________, $c =$ ________

6. $9x^2 - 4 = 0$; $a =$ ________, $b =$ ________, $c =$ ________

Simplify the following.

7. $\frac{-1 \pm \sqrt{1^2 - 4(1)(-2)}}{2(1)}$

8. $\frac{-(-5) \pm \sqrt{(-5)^2 - 4(2)(3)}}{2(2)}$

9. $\frac{-5 \pm \sqrt{5^2 - 4(1)(2)}}{2(1)}$

10. $\frac{-7 \pm \sqrt{7^2 - 4(2)(1)}}{2(2)}$

Martin-Gay Interactive Videos

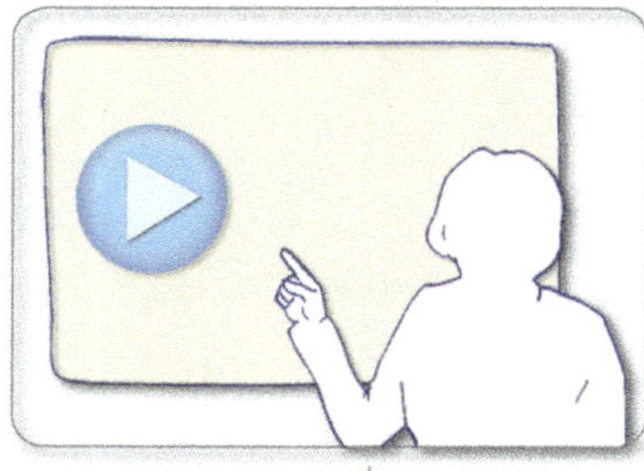

See Video 9.3

Watch the section lecture video and answer the following questions.

Objective A 11. Based on the lectures and Examples 1–3, answer the following.

a. Must a quadratic equation be written in standard form in order to use the quadratic formula? Why or why not?

b. Must fractions be cleared from the equation before using the quadratic formula? Why or why not?

Objective B 12. From Example 4, how are approximate solutions found?

9.3 Exercise Set MyLab Math

Objective A *Use the quadratic formula to solve each quadratic equation. See Examples 1 through 4.*

1. $x^2 - 3x + 2 = 0$ **2.** $x^2 - 5x - 6 = 0$ **3.** $3k^2 + 7k + 1 = 0$ **4.** $7k^2 + 3k - 1 = 0$

5. $4x^2 - 3 = 0$ **6.** $25x^2 - 15 = 0$ **7.** $5z^2 - 4z + 3 = 0$ **8.** $3x^2 + 2x + 1 = 0$

9. $y^2 = 7y + 30$ **10.** $y^2 = 5y + 36$ **11.** $2x^2 = 10$ **12.** $5x^2 = 15$

13. $m^2 - 12 = m$ **14.** $m^2 - 14 = 5m$ **15.** $3 - x^2 = 4x$ **16.** $10 - x^2 = 2x$

17. $6x^2 + 9x = 2$ **18.** $3x^2 - 9x = 8$ **19.** $7p^2 + 2 = 8p$ **20.** $11p^2 + 2 = 10p$

21. $x^2 - 6x + 2 = 0$ **22.** $x^2 - 10x + 19 = 0$ **23.** $2x^2 - 6x + 3 = 0$ **24.** $5x^2 - 8x + 2 = 0$

25. $3x^2 = 1 - 2x$ **26.** $5y^2 = 4 - y$ **27.** $4y^2 = 6y + 1$ **28.** $6z^2 = 2 - 3z$

29. $x^2 + x + 2 = 0$ **30.** $k^2 + 2k + 5 = 0$ **31.** $20y^2 = 3 - 11y$ **32.** $2z^2 = z + 3$

33. $x^2 - 5x - 2 = 0$ **34.** $x^2 - 2x - 5 = 0$ **35.** $3x^2 - x - 14 = 0$ **36.** $5x^2 - 13x - 6 = 0$

Use the quadratic formula to solve each quadratic equation. See Example 5.

37. $\frac{m^2}{2} = m + \frac{1}{2}$ **38.** $\frac{m^2}{2} = 3m - 1$ **39.** $3p^2 - \frac{2}{3}p + 1 = 0$ **40.** $\frac{5}{2}p^2 - p + \frac{1}{2} = 0$

41. $4p^2 + \frac{3}{2} = -5p$ **42.** $4p^2 + \frac{3}{2} = 5p$ **43.** $5x^2 = \frac{7}{2}x + 1$ **44.** $2x^2 = \frac{5}{2}x + \frac{7}{2}$

45. $x^2 - \frac{11}{2}x - \frac{1}{2} = 0$ **46.** $\frac{2}{3}x^2 - 2x - \frac{2}{3} = 0$ **47.** $5z^2 - 2z = \frac{1}{5}$ **48.** $3z^2 + 4z = -\frac{1}{3}$

Objectives A B Mixed Practice *Use the quadratic formula to solve each quadratic equation. Find the exact solutions; then approximate these solutions to the nearest tenth. See Examples 1 through 6.*

49. $3x^2 = 21$ **50.** $2x^2 = 26$ **51.** $x^2 + 6x + 1 = 0$ **52.** $x^2 + 4x + 2 = 0$

53. $x^2 = 9x + 4$ **54.** $x^2 = 7x + 5$ **55.** $3x^2 - 2x - 2 = 0$ **56.** $5x^2 - 3x - 1 = 0$

Review

Graph the following linear equations in two variables. See Sections 6.2 and 6.3.

57. $y = -3$

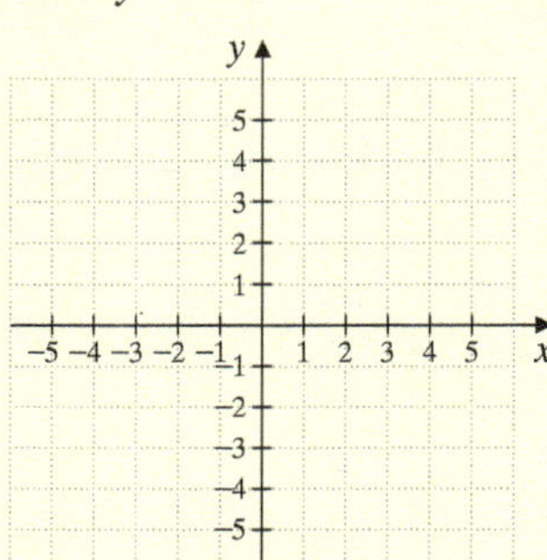

58. $x = 4$

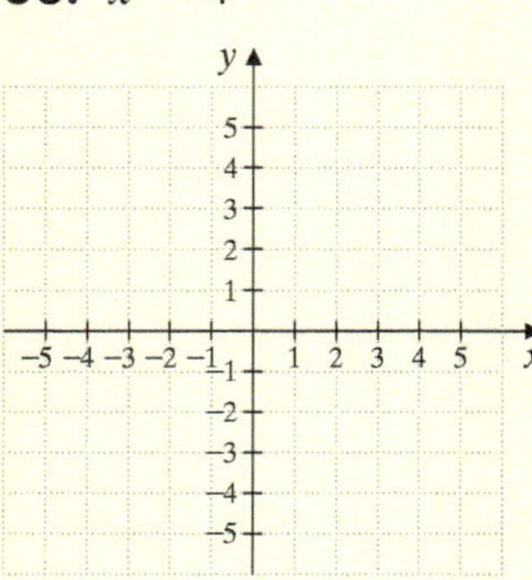

59. $y = 3x - 2$

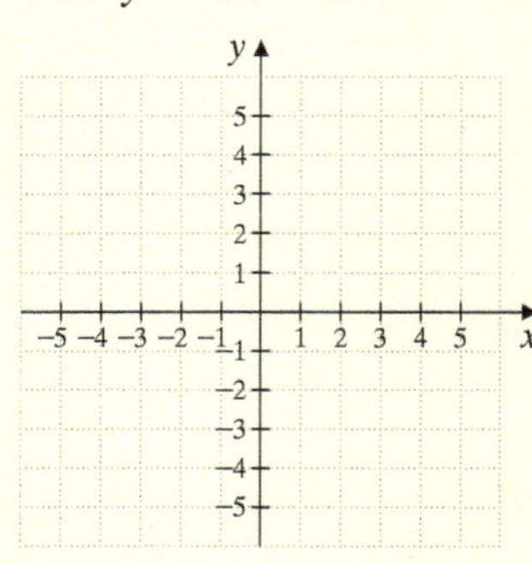

60. $y = 2x + 3$

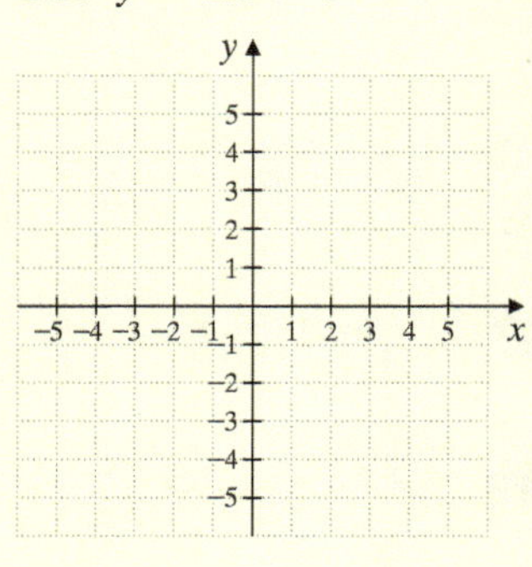

Concept Extensions

Solve. See the Concept Check in this section. For the quadratic equation $5x^2 + 2 = x$, *if* $a = 5$,

61. What is the value of b?

a. $\frac{1}{5}$ **b.** 0 **c.** -1 **d.** 1

62. What is the value of c?

a. 5 **b.** x **c.** -2 **d.** 2

For the quadratic equation $7y^2 = 3y$, *if* $b = 3$,

63. What is the value of a?

a. 7 **b.** -7 **c.** 0 **d.** 1

64. What is the value of c?

a. 7 **b.** 3 **c.** 0 **d.** 1

△ **65.** In a recent year, Nestle created a chocolate bar that the company claimed weighed more than 2 tons. The rectangular bar had a base area of approximately 34.65 square feet, and its length was 0.6 foot shorter than three times its width. Find the length and width of the bar. (*Source:* Nestle)

△ **66.** The area of a rectangular conference room table is 95 square feet. If its length is six feet longer than its width, find the dimensions of the table. Round each dimension to the nearest tenth.

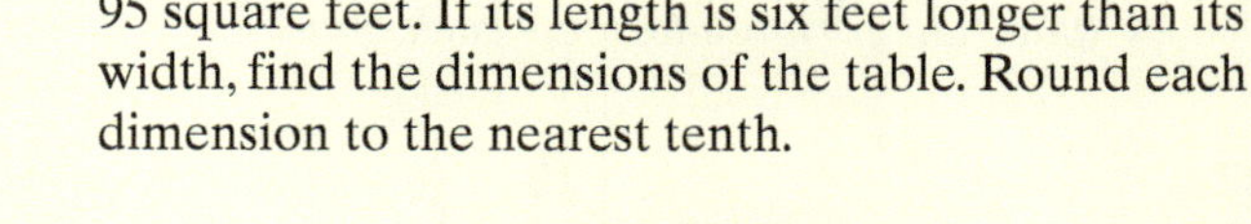

Solve each equation using the quadratic formula.

67. $x^2 + 3\sqrt{2}x - 5 = 0$

68. $y^2 - 2\sqrt{5}y - 1 = 0$

69. Explain how to identify a, b, and c correctly when solving a quadratic equation by the quadratic formula.

70. Explain how the quadratic formula is developed and why it is useful.

Use the quadratic formula and a calculator to solve each equation. Round solutions to the nearest tenth.

71. $7.3z^2 + 5.4z - 1.1 = 0$

72. $1.2x^2 - 5.2x - 3.9 = 0$

A rocket is launched from the top of an 80-foot cliff with an initial velocity of 120 feet per second. The height, h, of the rocket after t seconds is given by the equation $h = -16t^2 + 120t + 80$.

73. How long after the rocket is launched will it be 30 feet from the ground? Round to the nearest tenth of a second.

74. How long after the rocket is launched will it strike the ground? Round to the nearest tenth of a second. (*Hint:* The rocket will strike the ground when its height $h = 0$.)

75. Restaurant industry food and drink sales y (in billions of dollars) in the United States from 2008 through 2017 can be approximated by the equation $y = 2x^2 + 9.6x + 565$, where x is the number of years since 2008. Assume this trend continues and predict the year in which the restaurant industry food and drink sales will be \$1160 billion. (Round to the nearest whole number.) (*Source:* National Restaurant Association)

76. Retail sales y (in billions of dollars) for Target Corporation for the years 2012 through 2016 is approximated by the equation $y = -0.06x^2 + x + 70$, where $x = 0$ represents the year 2012. Assume that this trend continues and predict the year after 2012 in which Target's retail sales will be approximately \$68 billion. (Round to the nearest whole number.) (*Source:* Based on data from Target Corporation)

Integrated Review Sections 9.1–9.3

Summary on Solving Quadratic Equations

An important skill in mathematics is learning when to use one technique in favor of another. We now practice this by deciding which method to use when solving quadratic equations. Although both the quadratic formula and completing the square can be used to solve any quadratic equation, the quadratic formula is usually less tedious and thus preferred. The following steps may be used to solve a quadratic equation.

To Solve a Quadratic Equation

Step 1: If the equation is in the form $ax^2 = c$ or $(ax + b)^2 = c$, use the square root property and solve. If not, go to Step 2.

Step 2: Write the equation in standard form: $ax^2 + bx + c = 0$.

Step 3: Try to solve the equation by factoring. If not possible, go to Step 4.

Step 4: Solve the equation by the quadratic formula.

Study the examples below to help you review these steps.

Practice 1

Solve $y^2 - 4y - 6 = 0$.

Example 1 Solve $m^2 - 2m - 7 = 0$.

Solution: The equation is in standard form, but the quadratic expression $m^2 - 2m - 7$ is not factorable, so use the quadratic formula with $a = 1, b = -2$, and $c = -7$.

$$m^2 - 2m - 7 = 0$$

$$m = \frac{-(-2) \pm \sqrt{(-2)^2 - 4 \cdot 1 \cdot (-7)}}{2 \cdot 1} = \frac{2 \pm \sqrt{32}}{2}$$

$$m = \frac{2 \pm 4\sqrt{2}}{2} = \frac{2(1 \pm 2\sqrt{2})}{2} = 1 \pm 2\sqrt{2}$$

The solutions are $1 - 2\sqrt{2}$ and $1 + 2\sqrt{2}$.

Work Practice 1

Practice 2

Solve $(2x + 5)^2 = 45$.

Example 2 Solve $(3x + 1)^2 = 20$.

Solution: This equation is in a form that makes the square root property easy to apply.

$$(3x + 1)^2 = 20$$

$$3x + 1 = \pm\sqrt{20} \quad \text{Apply the square root property.}$$

$$3x + 1 = \pm 2\sqrt{5} \quad \text{Simplify } \sqrt{20}.$$

$$3x = -1 \pm 2\sqrt{5}$$

$$x = \frac{-1 \pm 2\sqrt{5}}{3}$$

The solutions are $\frac{-1 - 2\sqrt{5}}{3}$ and $\frac{-1 + 2\sqrt{5}}{3}$.

Work Practice 2

Answers

1. $2 \pm \sqrt{10}$
2. $\frac{-5 \pm 3\sqrt{5}}{2}$

Example 3 Solve $x^2 - \frac{11}{2}x = -\frac{5}{2}$.

Solution: The fractions make factoring more difficult and complicate the calculations for using the quadratic formula. Clear the equation of fractions by multiplying both sides of the equation by the LCD, 2.

$$x^2 - \frac{11}{2}x = -\frac{5}{2}$$

$$x^2 - \frac{11}{2}x + \frac{5}{2} = 0 \quad \text{Write in standard form.}$$

$$2x^2 - 11x + 5 = 0 \quad \text{Multiply both sides by 2.}$$

$$(2x - 1)(x - 5) = 0 \quad \text{Factor.}$$

$$2x - 1 = 0 \quad \text{or} \quad x - 5 = 0 \quad \text{Apply the zero factor property.}$$

$$2x = 1 \qquad x = 5$$

$$x = \frac{1}{2} \qquad x = 5$$

The solutions are $\frac{1}{2}$ and 5.

Work Practice 3

Practice 3

Solve $x^2 - \frac{5}{2}x = -\frac{3}{2}$.

Answer

3. $\frac{3}{2}, 1$

Choose and use a method to solve each equation.

1. $5x^2 - 11x + 2 = 0$

2. $5x^2 + 13x - 6 = 0$

3. $x^2 - 1 = 2x$

4. $x^2 + 7 = 6x$

5. $a^2 = 20$

6. $a^2 = 72$

7. $x^2 - x + 4 = 0$

8. $x^2 - 2x + 7 = 0$

9. $3x^2 - 12x + 12 = 0$

10. $5x^2 - 30x + 45 = 0$

11. $9 - 6p + p^2 = 0$

12. $49 - 28p + 4p^2 = 0$

13. $4y^2 - 16 = 0$

14. $3y^2 - 27 = 0$

15. $x^2 - 3x + 2 = 0$

16. $x^2 + 7x + 12 = 0$

Answers

1. ______
2. ______
3. ______
4. ______
5. ______
6. ______
7. ______
8. ______
9. ______
10. ______
11. ______
12. ______
13. ______
14. ______
15. ______
16. ______

17. ______
18. ______
19. ______
20. ______
21. ______
22. ______
23. ______
24. ______
25. ______
26. ______
27. ______
28. ______
29. ______
30. ______
31. ______
32. ______
33. ______
34. ______
35. ______
36. ______
37. ______
38. ______
39. ______
40. ______
41. ______

17. $(2z + 5)^2 = 25$

18. $(3z - 4)^2 = 16$

19. $30x = 25x^2 + 2$

20. $12x = 4x^2 + 4$

21. $\frac{2}{3}m^2 - \frac{1}{3}m - 1 = 0$

22. $\frac{5}{8}m^2 + m - \frac{1}{2} = 0$

23. $x^2 - \frac{1}{2}x - \frac{1}{5} = 0$

24. $x^2 + \frac{1}{2}x - \frac{1}{8} = 0$

25. $4x^2 - 27x + 35 = 0$

26. $9x^2 - 16x + 7 = 0$

27. $(7 - 5x)^2 = 18$

28. $(5 - 4x)^2 = 75$

29. $3z^2 - 7z = 12$

30. $6z^2 + 7z = 6$

31. $x = x^2 - 110$

32. $x = 56 - x^2$

33. $\frac{3}{4}x^2 - \frac{5}{2}x - 2 = 0$

34. $x^2 - \frac{6}{5}x - \frac{8}{5} = 0$

35. $x^2 - 0.6x + 0.05 = 0$

36. $x^2 - 0.1x - 0.06 = 0$

37. $10x^2 - 11x + 2 = 0$

38. $20x^2 - 11x + 1 = 0$

39. $\frac{1}{2}z^2 - 2z + \frac{3}{4} = 0$

40. $\frac{1}{5}z^2 - \frac{1}{2}z - 2 = 0$

41. Explain how you will decide what method to use when solving quadratic equations.

9.4 Graphing Quadratic Equations in Two Variables

Recall from Section 6.2 that the graph of a linear equation in two variables, $Ax + By = C$, is a straight line. In this section, we will find that the graph of a quadratic equation in the form $y = ax^2 + bx + c$ is a parabola.

Objectives

A Graph Quadratic Equations of the Form $y = ax^2$.

B Graph Quadratic Equations of the Form $y = ax^2 + bx + c$.

C Use the Vertex Formula to Determine the Vertex of a Parabola.

Objective A Graphing $y = ax^2$

We begin our work by graphing $y = x^2$. To do so, we will find and plot ordered pair solutions of this equation. Let's select a few values for x, find the corresponding y-values, and record them in a table of values to keep track. Then we can plot the points corresponding to these solutions on a coordinate plane.

If $x = -3$, then $y = (-3)^2$, or 9.
If $x = -2$, then $y = (-2)^2$, or 4.
If $x = -1$, then $y = (-1)^2$, or 1.
If $x = 0$, then $y = 0^2$, or 0.
If $x = 1$, then $y = 1^2$, or 1.
If $x = 2$, then $y = 2^2$, or 4.
If $x = 3$, then $y = 3^2$, or 9.

$y = x^2$

x	y
−3	9
−2	4
−1	1
0	0
1	1
2	4
3	9

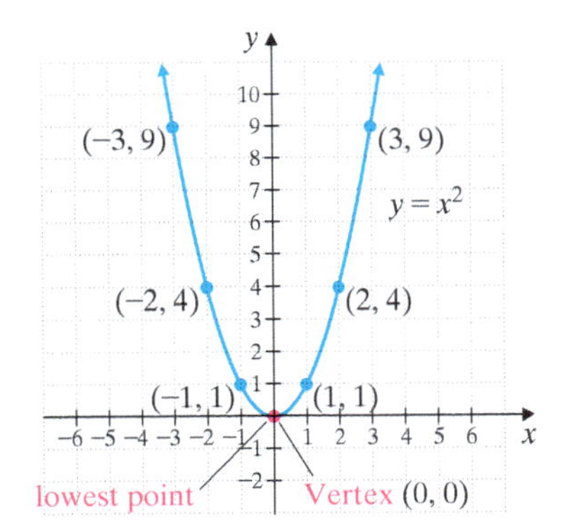

Practice 1

Graph: $y = -3x^2$

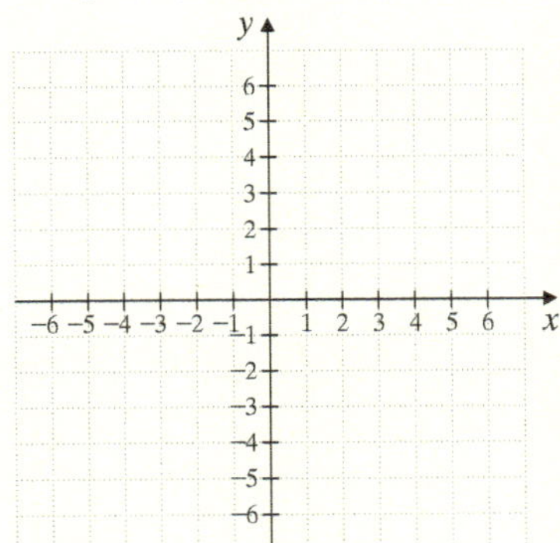

The graph of $y = x^2$ is a smooth curve through the plotted points. This curve is called a **parabola.** The lowest point on a parabola opening upward is called the **vertex.** The vertex is (0, 0) for the parabola $y = x^2$. If we fold the graph along the y-axis, the two pieces of the parabola match perfectly. For this reason, we say the graph is **symmetric about the y-axis,** and we call the y-axis the **axis of symmetry.**

Notice that the parabola that corresponds to the equation $y = x^2$ opens upward. This happens when the coefficient of x^2 is positive. In the equation $y = x^2$, the coefficient of x^2 is 1. Example 1 shows the graph of a quadratic equation where the coefficient of x^2 is negative.

Answer

1.

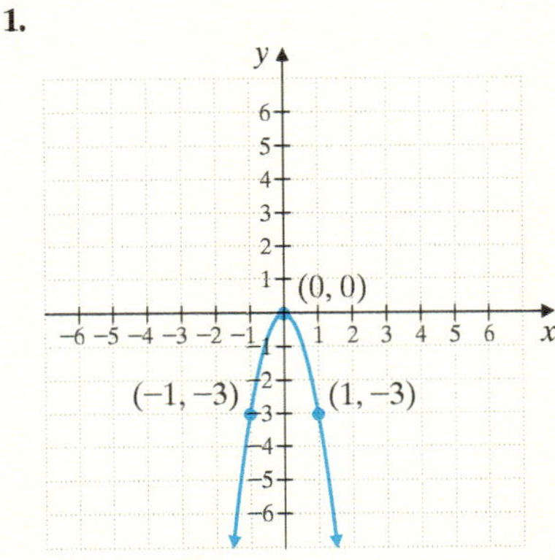

Example 1

Graph: $y = -2x^2$

Solution: We begin by selecting x-values and calculating the corresponding y-values. Then we plot the ordered pairs found and draw a smooth curve through those points. Notice that when the coefficient of x^2 is negative, the corresponding

(Continued on next page)

parabola opens downward. When a parabola opens downward, the vertex is the highest point of the parabola. The vertex of this parabola is $(0, 0)$, and the axis of symmetry is again the y-axis.

$y = -2x^2$

x	y
0	0
1	−2
2	−8
3	−18
−1	−2
−2	−8
−3	−18

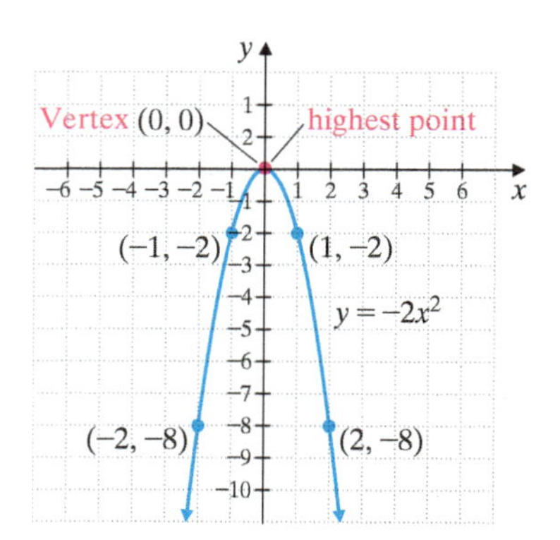

Work Practice 1

Objective B Graphing $y = ax^2 + bx + c$

Just as for linear equations, we can use x- and y-intercepts to help graph quadratic equations. Recall from Chapter 6 that an x-intercept is the point where the graph crosses the x-axis. A y-intercept is the point where the graph crosses the y-axis. We find intercepts just as we did in Chapter 6.

Helpful Hint

Recall that:

To find x-intercepts, let $y = 0$ and solve for x.
To find y-intercepts, let $x = 0$ and solve for y.

Practice 2

Graph: $y = x^2 - 9$

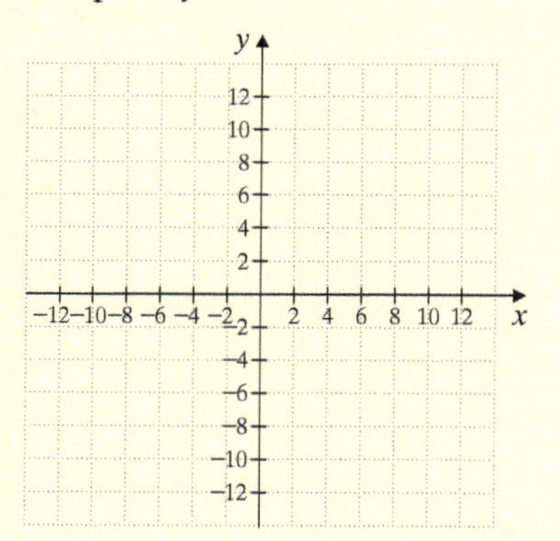

Answer

2.

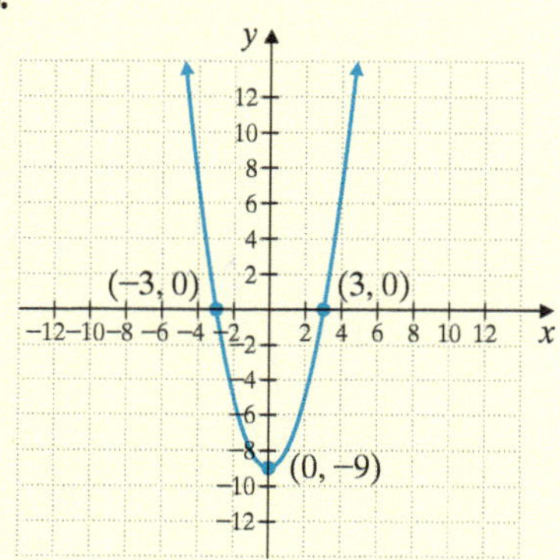

Example 2 Graph: $y = x^2 - 4$

Solution: If we write this equation as $y = x^2 + 0x + (-4)$, we can see that it is in the form $y = ax^2 + bx + c$. To graph it, we first find the intercepts. To find the y-intercept, we let $x = 0$. Then

$$y = 0^2 - 4 = -4$$

To find x-intercepts, we let $y = 0$.

$$0 = x^2 - 4$$
$$0 = (x - 2)(x + 2)$$
$$x - 2 = 0 \quad \text{or} \quad x + 2 = 0$$
$$x = 2 \qquad\qquad x = -2$$

Thus far, we have the y-intercept $(0, -4)$ and the x-intercepts $(2, 0)$ and $(-2, 0)$. Now we can select additional x-values, find the corresponding y-values, plot the points, and draw a smooth curve through the points.

$y = x^2 - 4$

x	y
0	−4
1	−3
2	0
3	5
−1	−3
−2	0
−3	5

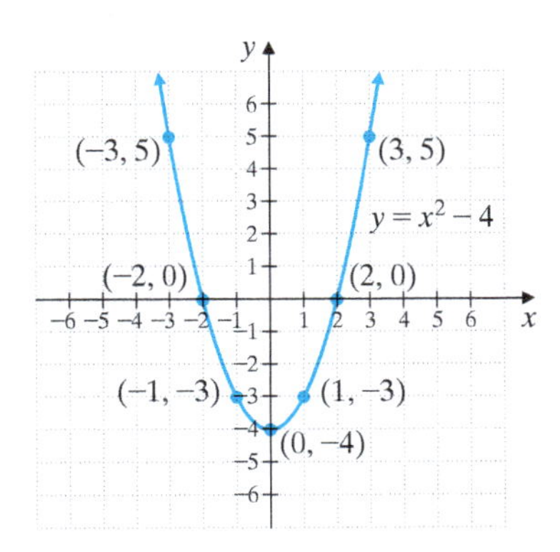

Notice that the vertex of this parabola is $(0, -4)$.

Work Practice 2

✓**Concept Check** Tell whether the graph of each equation opens upward or downward.

a. $y = 2x^2$ **b.** $y = 3x^2 + 4x - 5$ **c.** $y = -5x^2 + 2$

Helpful Hint

For the graph of $y = ax^2 + bx + c$,

If a is positive, the parabola opens upward.
If a is negative, the parabola opens downward.

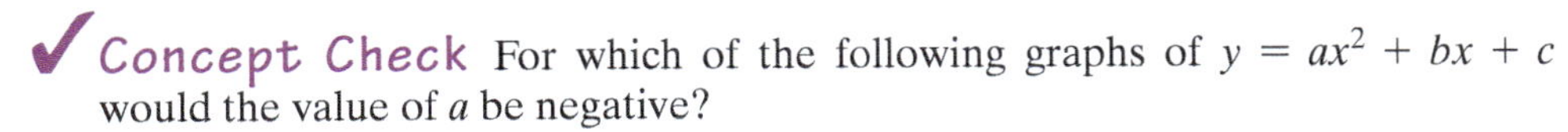

✓**Concept Check** For which of the following graphs of $y = ax^2 + bx + c$ would the value of a be negative?

a.

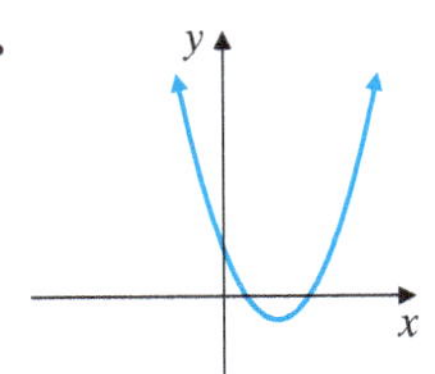

b.

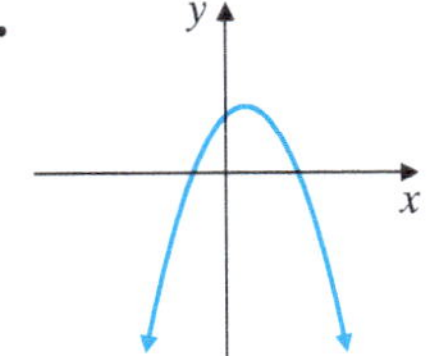

Objective C Using the Vertex Formula 

Thus far, we have accidentally stumbled upon the vertex of each parabola that we have graphed. However, our choice of values for x may not yield an ordered pair for the vertex of the parabola. It would be helpful if we could first find the vertex of a parabola. Next we would determine whether the parabola opens upward or downward. Finally we would calculate additional points such as x- and y-intercepts as needed. In fact, there is a formula that may be used to find the vertex of a parabola.

Vertex Formula

The vertex of the parabola $y = ax^2 + bx + c$ has x-coordinate

$$\frac{-b}{2a}$$

The corresponding y-coordinate of the vertex is obtained by substituting the x-coordinate into the equation and finding y.

✓**First Concept Check Answer**
a. upward **b.** upward **c.** downward

✓**Second Concept Check Answer**
b

One way to develop this formula is to notice that the x-value of the vertex of the parabolas that we are considering lies halfway between its x-intercepts. Another way to develop this formula is to complete the square on the general form of a quadratic equation: $y = ax^2 + bx + c$. We will not show the development of this formula here.

Practice 3

Graph: $y = x^2 - 2x - 3$

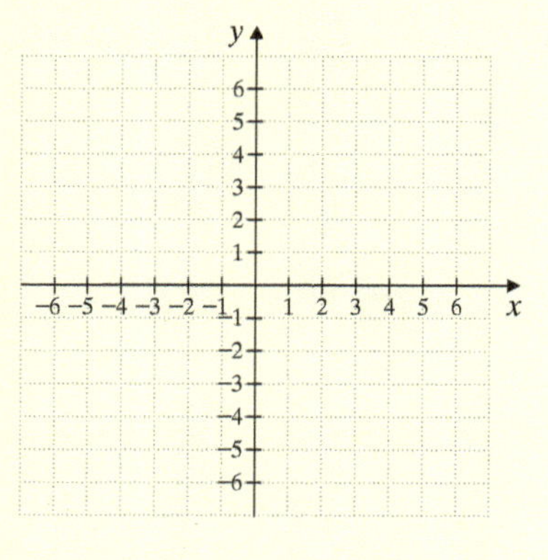

Example 3 Graph: $y = x^2 - 6x + 8$

Solution: In the equation $y = x^2 - 6x + 8$, $a = 1$ and $b = -6$.

Vertex: The x-coordinate of the vertex is

$$\frac{-b}{2a} = \frac{-(-6)}{2 \cdot 1} = 3 \quad \text{Use the vertex formula, } \frac{-b}{2a}.$$

To find the corresponding y-coordinate, we let $x = 3$ in the original equation.

$$y = x^2 - 6x + 8 = 3^2 - 6 \cdot 3 + 8 = -1$$

The vertex is $(3, -1)$ and the parabola opens upward since a is positive. We now find and plot the intercepts.

Intercepts: To find the x-intercepts, we let $y = 0$.

$$0 = x^2 - 6x + 8$$

We factor the expression $x^2 - 6x + 8$ to find $(x - 4)(x - 2) = 0$. The x-intercepts are $(4, 0)$ and $(2, 0)$.

If we let $x = 0$ in the original equation, then $y = 8$ gives us the y-intercept $(0, 8)$. Now we plot the vertex $(3, -1)$ and the intercepts $(4, 0)$, $(2, 0)$, and $(0, 8)$. Then we can sketch the parabola.

These and two additional points are shown in the table.

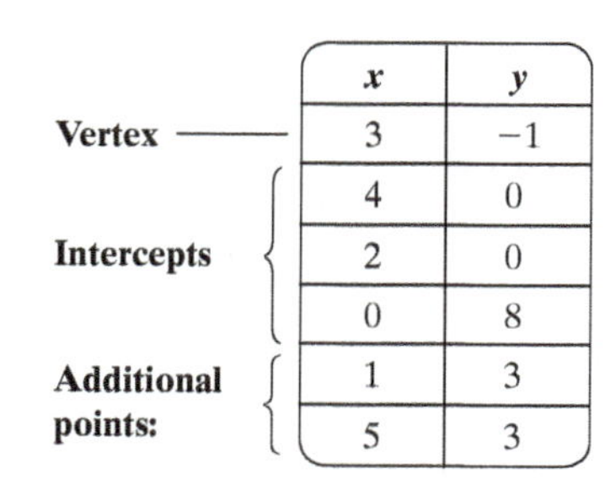

	x	y
Vertex	3	−1
Intercepts	4	0
	2	0
	0	8
Additional points:	1	3
	5	3

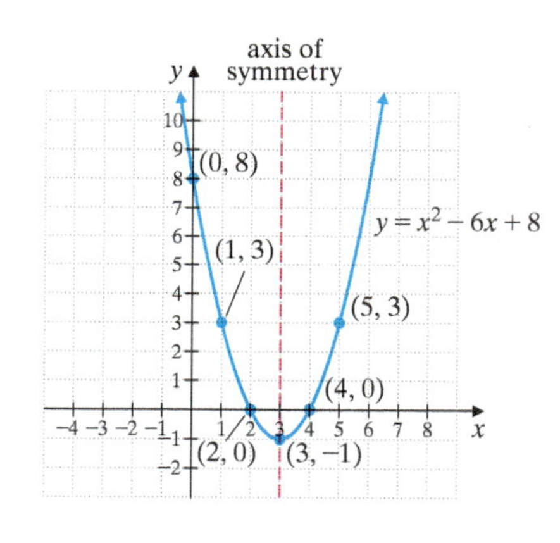

Work Practice 3

Study Example 3 and let's use it to write down a general procedure for graphing quadratic equations.

Answer

3.

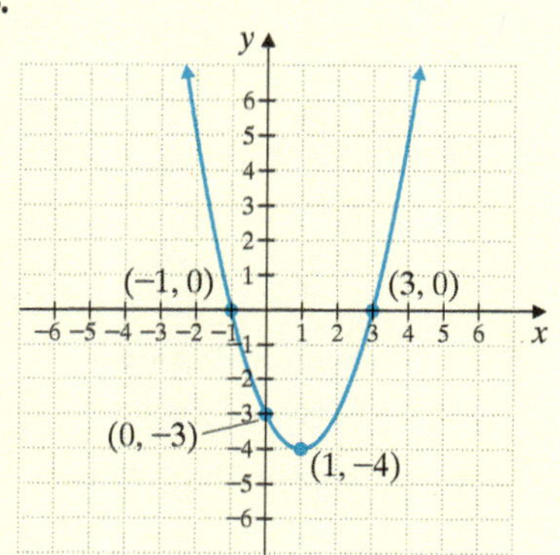

Graphing Parabolas Defined by $y = ax^2 + bx + c$

1. **Find the vertex by using the formula $x = \frac{-b}{2a}$.** Don't forget to find the y-value of the vertex.
2. **Find the intercepts.**
   - Let $x = 0$ and solve for y to find the y-intercept. There will be only one.
   - Let $y = 0$ and solve for x to find any x-intercepts. There may be 0, 1, or 2.
3. **Plot the vertex and the intercepts.**
4. **Find and plot additional points on the graph.** Then draw a smooth curve through the plotted points. Keep in mind that if $a > 0$, the parabola opens upward and that if $a < 0$, the parabola opens downward.

Example 4 Graph: $y = x^2 + 2x - 5$

Solution: In the equation $y = x^2 + 2x - 5$, $a = 1$ and $b = 2$. Using the vertex formula, we find that the x-coordinate of the vertex is

$$x = \frac{-b}{2a} = \frac{-2}{2 \cdot 1} = -1$$

The y-coordinate is

$$y = (-1)^2 + 2(-1) - 5 = -6$$

Thus the vertex is $(-1, -6)$.

To find the x-intercepts, we let $y = 0$.

$$0 = x^2 + 2x - 5$$

This cannot be solved by factoring, so we use the quadratic formula.

$$x = \frac{-2 \pm \sqrt{2^2 - 4(1)(-5)}}{2 \cdot 1} \quad \text{Let } a = 1, b = 2, \text{ and } c = -5.$$

$$= \frac{-2 \pm \sqrt{24}}{2}$$

$$= \frac{-2 \pm 2\sqrt{6}}{2} \quad \text{Simplify the radical.}$$

$$= \frac{2(-1 \pm \sqrt{6})}{2} = -1 \pm \sqrt{6}$$

The x-intercepts are $(-1 + \sqrt{6}, 0)$ and $(-1 - \sqrt{6}, 0)$. We use a calculator to approximate these so that we can easily graph these intercepts.

$$-1 + \sqrt{6} \approx 1.4 \quad \text{and} \quad -1 - \sqrt{6} \approx -3.4$$

To find the y-intercept, we let $x = 0$ in the original equation and find that $y = -5$. Thus the y-intercept is $(0, -5)$. You will find, because of symmetry, that $(-2, -5)$ is also an ordered pair solution.

x	y
-1	-6
$-1 + \sqrt{6} \approx 1.4$	0
$-1 - \sqrt{6} \approx -3.4$	0
0	-5
-2	-5

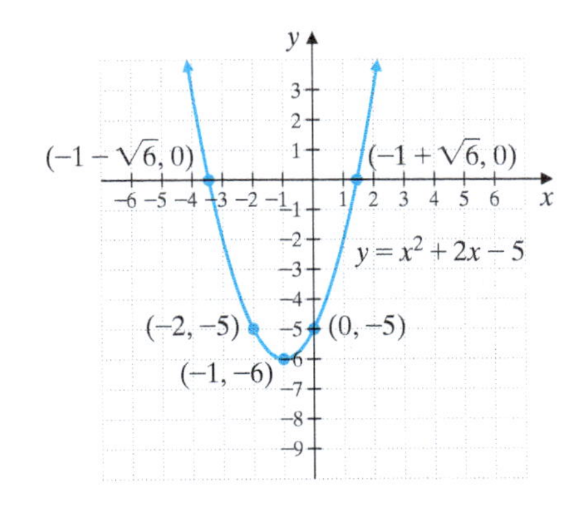

Work Practice 4

Practice 4

Graph: $y = x^2 - 4x + 1$

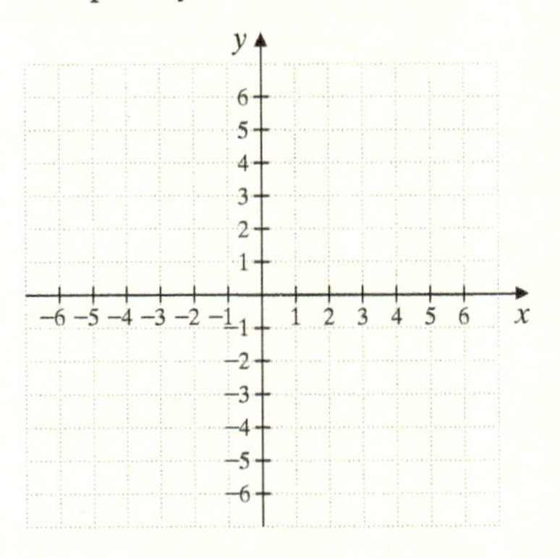

Answer

4.

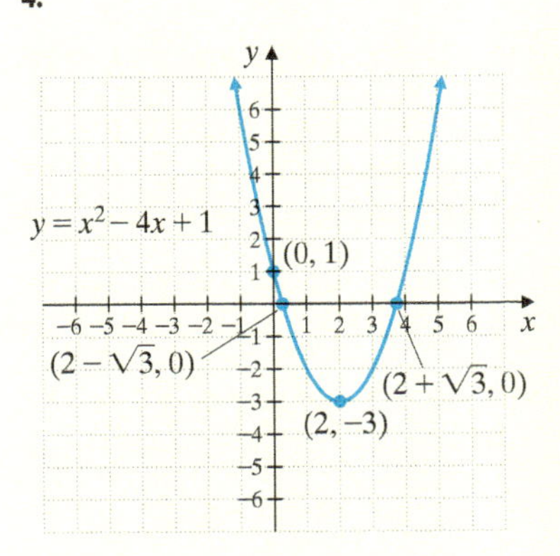

Helpful Hint

Notice that the number of x-intercepts of the graph of the parabola $y = ax^2 + bx + c$ is the same as the number of real solutions of $0 = ax^2 + bx + c$.

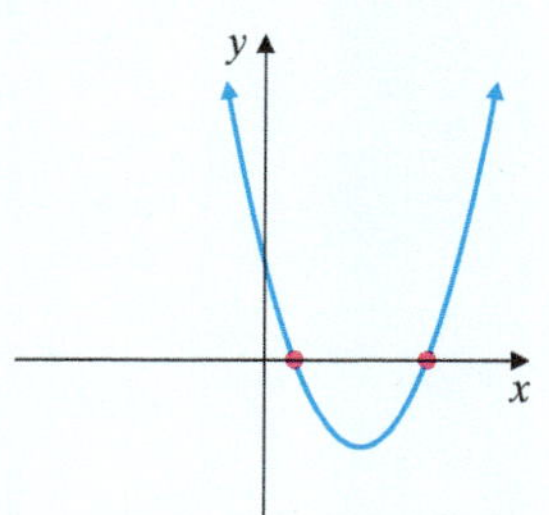

$y = ax^2 + bx + c$
$a > 0$

Two x-intercepts
Two real solutions of
$0 = ax^2 + bx + c$

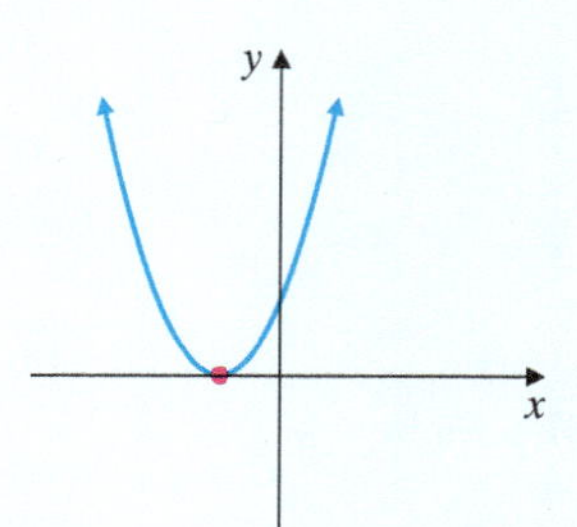

$y = ax^2 + bx + c$
$a > 0$

One x-intercept
One real solution of
$0 = ax^2 + bx + c$

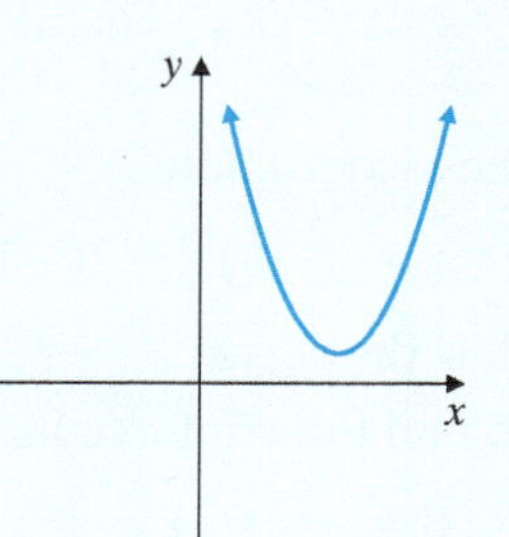

$y = ax^2 + bx + c$
$a > 0$

No x-intercepts
No real solutions of
$0 = ax^2 + bx + c$

Calculator Explorations Graphing

Recall that a graphing calculator may be used to solve quadratic equations. The x-intercepts of the graph of $y = ax^2 + bx + c$ are solutions of $0 = ax^2 + bx + c$. To solve $x^2 - 7x - 3 = 0$, for example, graph $y_1 = x^2 - 7x - 3$. The x-intercepts of the graph are the solutions of the equation.

Use a graphing calculator to solve each quadratic equation. Round solutions to two decimal places.

1. $x^2 - 7x - 3 = 0$

2. $2x^2 - 11x - 1 = 0$

3. $-1.7x^2 + 5.6x - 3.7 = 0$

4. $-5.8x^2 + 2.3x - 3.9 = 0$

5. $5.8x^2 - 2.6x - 1.9 = 0$

6. $7.5x^2 - 3.7x - 1.1 = 0$

Vocabulary, Readiness & Video Check

Martin-Gay Interactive Videos

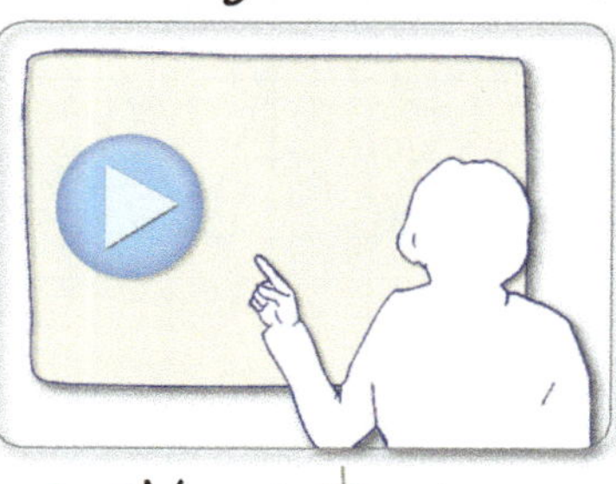

See Video 9.4

Watch the section lecture video and answer the following questions.

Objective A **1.** In Example 1, how are the vertex and line of symmetry of a parabola explained?

Objective B **2.** In Example 2, what important point was accidentally found? Why would it be useful to have an algebraic way to find this point?

Objective C **3.** From Example 3, how can finding the vertex and noting whether the parabola opens up or down possibly help save us time and work? Explain using an example.

9.4 Exercise Set MyLab Math 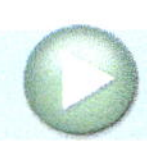

Objective A *Graph each quadratic equation by finding and plotting ordered pair solutions. See Example 1.*

1. $y = 2x^2$

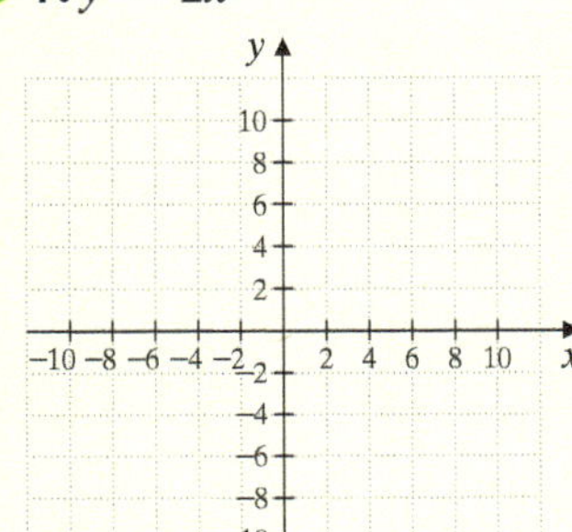

2. $y = 3x^2$

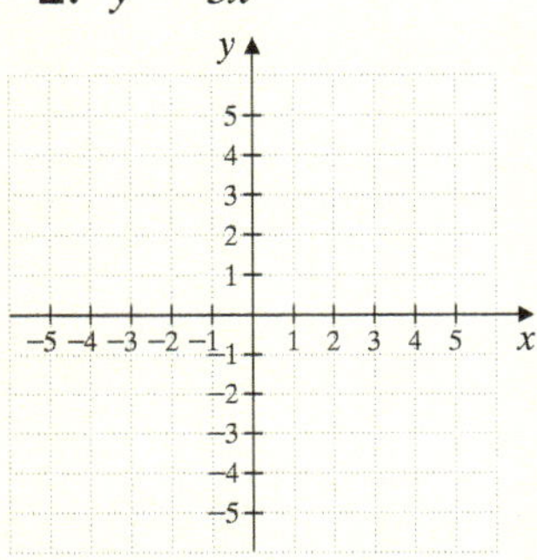

3. $y = -x^2$

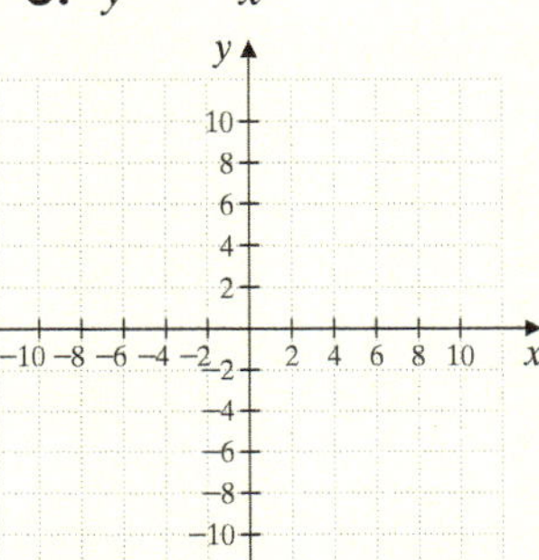

4. $y = -4x^2$

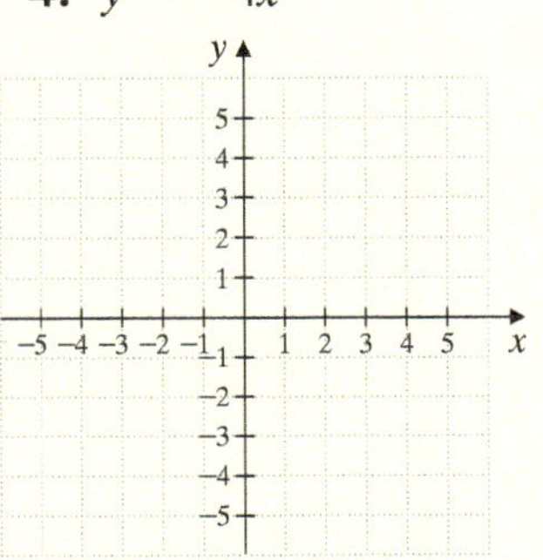

Objective B *Sketch the graph of each equation. Label the vertex and the intercepts. See Example 2.*

5. $y = x^2 - 1$

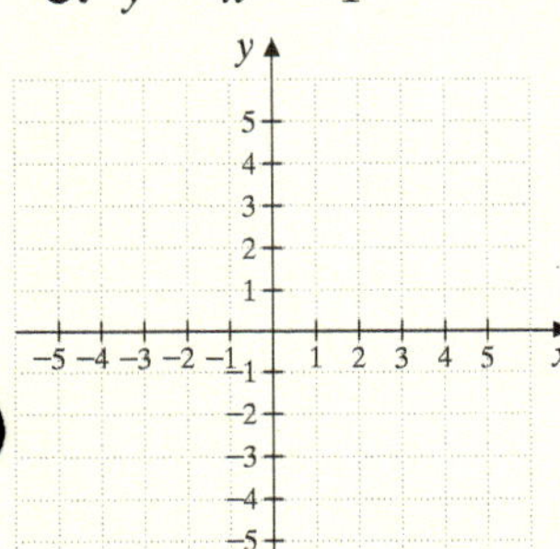

6. $y = x^2 - 16$

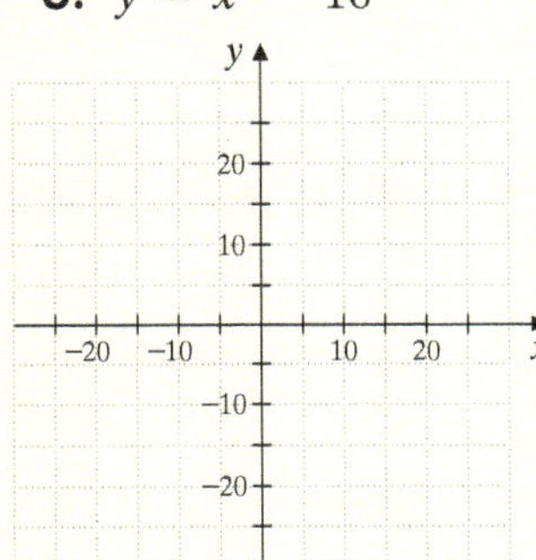

7. $y = x^2 + 4$

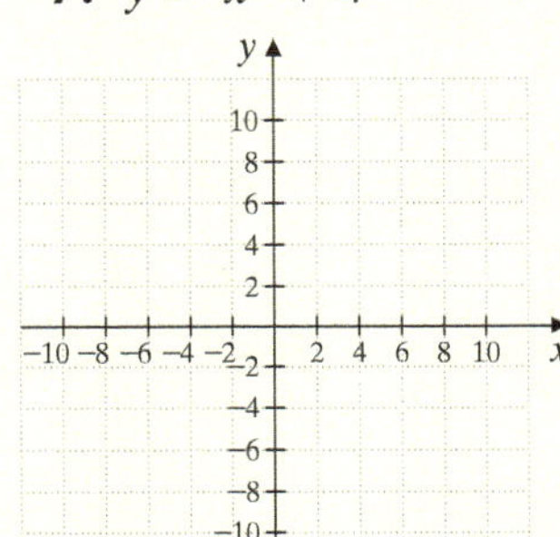

8. $y = x^2 + 9$

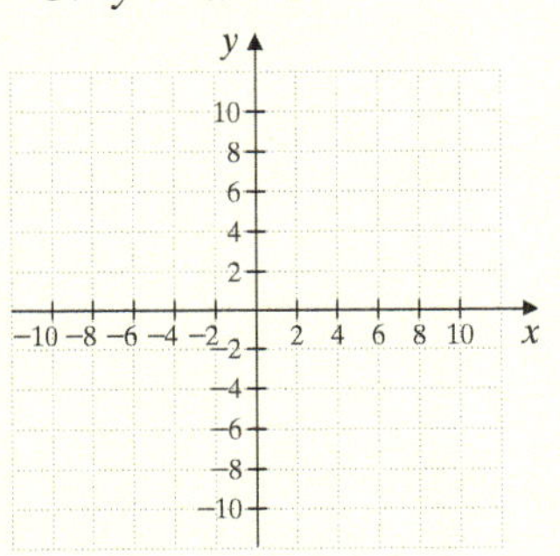

Objectives A B C *Sketch the graph of each equation. Label the vertex and the intercepts. See Examples 1 through 4.*

9. $y = -x^2 + 4x - 4$

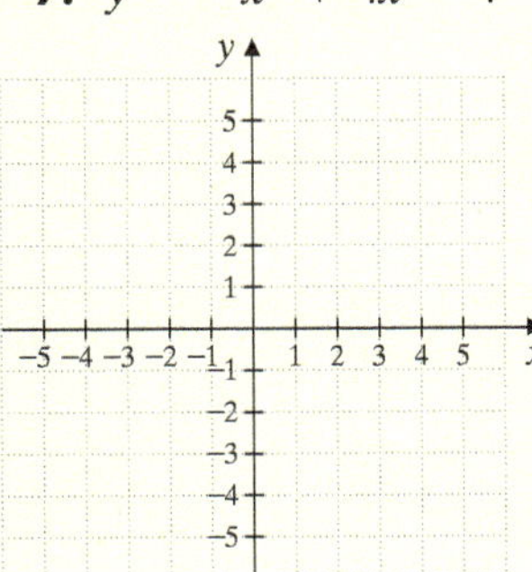

10. $y = -x^2 - 2x - 1$

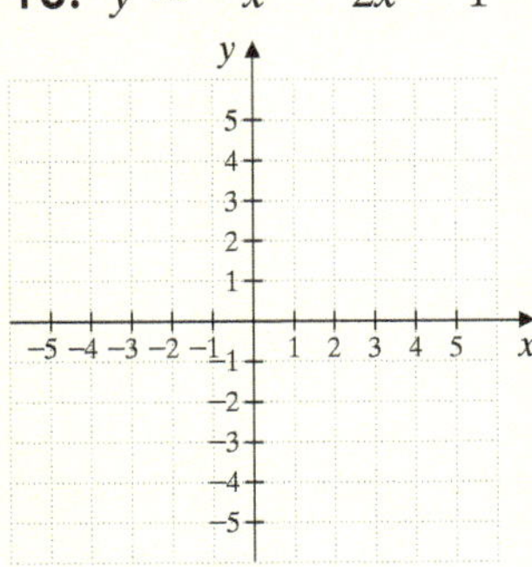

11. $y = x^2 + 5x + 4$

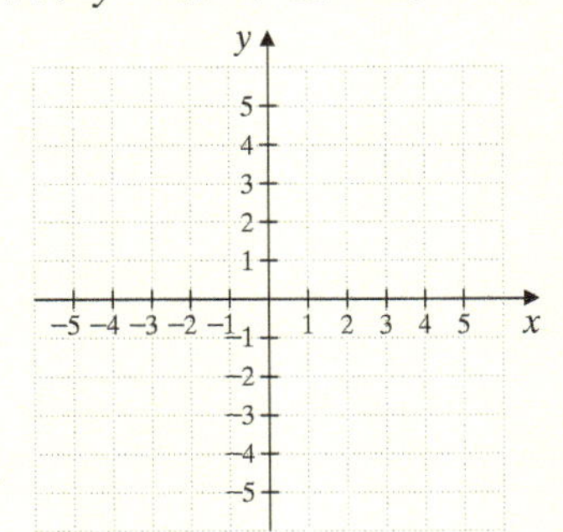

12. $y = x^2 + 7x + 10$

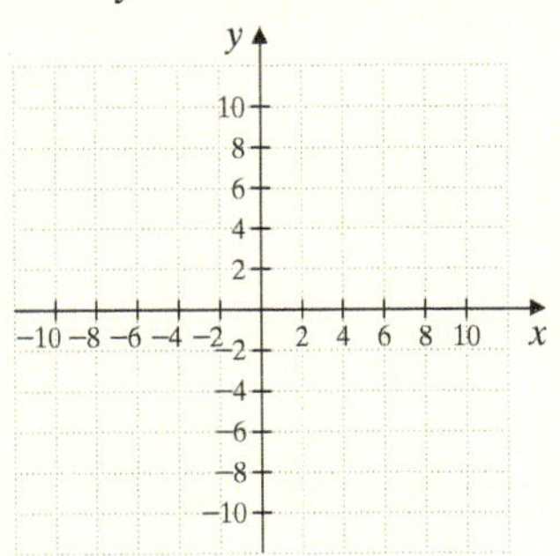

13. $y = x^2 - 4x + 5$

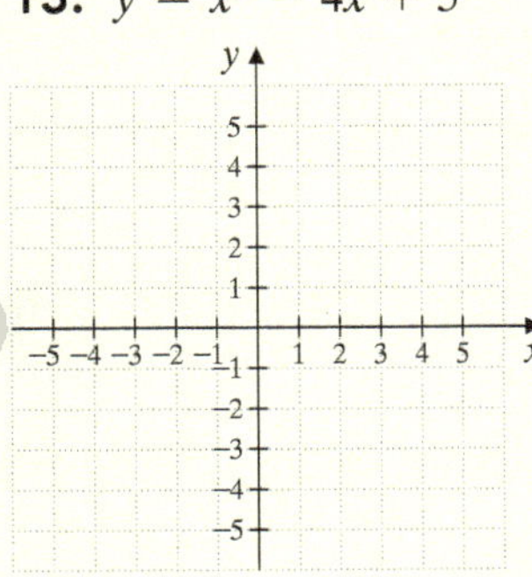

14. $y = x^2 - 6x + 10$

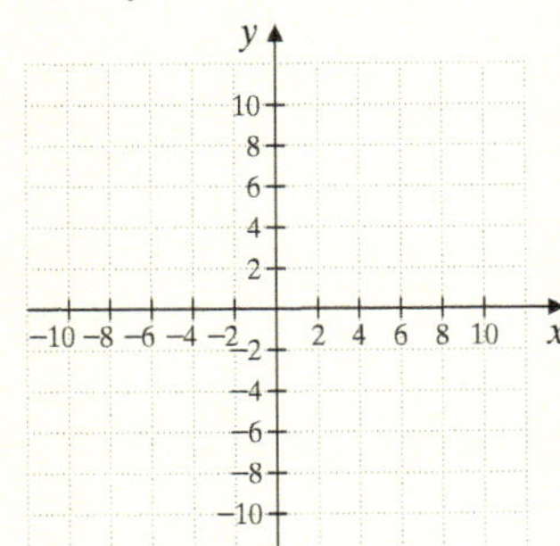

15. $y = 2 - x^2$

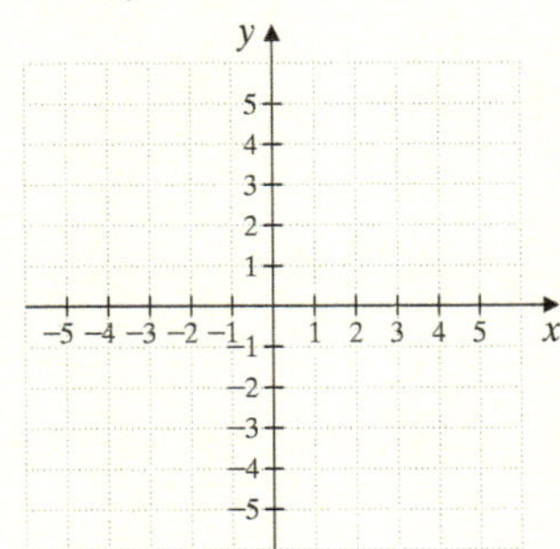

16. $y = 3 - x^2$

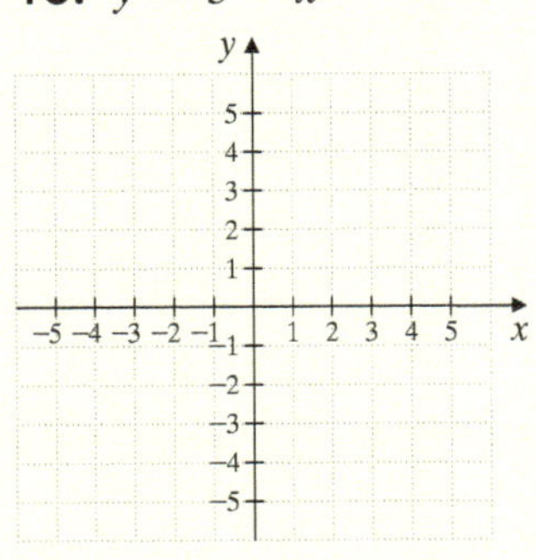

17. $y = \frac{1}{3}x^2$

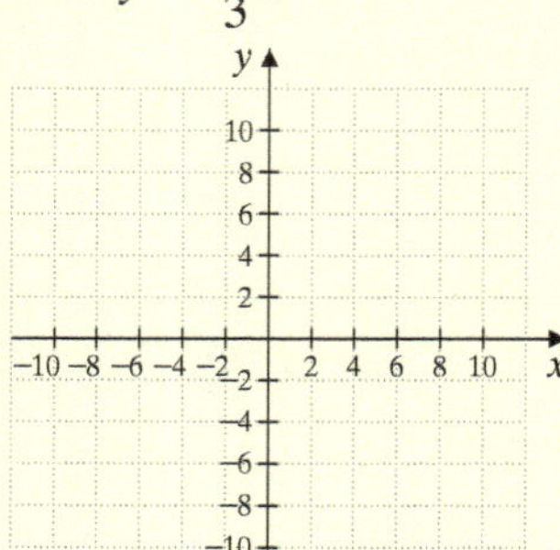

18. $y = \frac{1}{2}x^2$

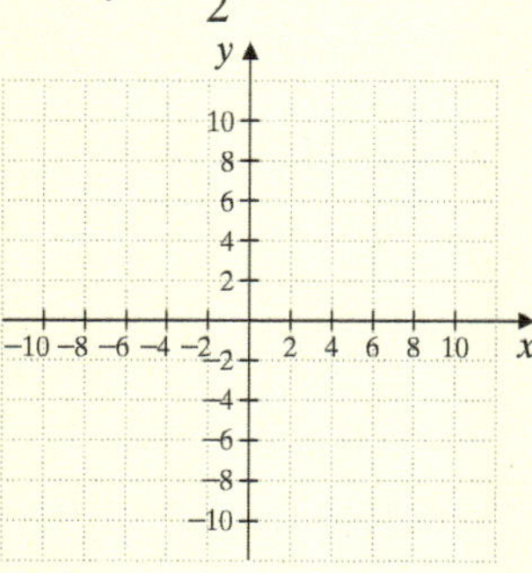

19. $y = x^2 + 6x$

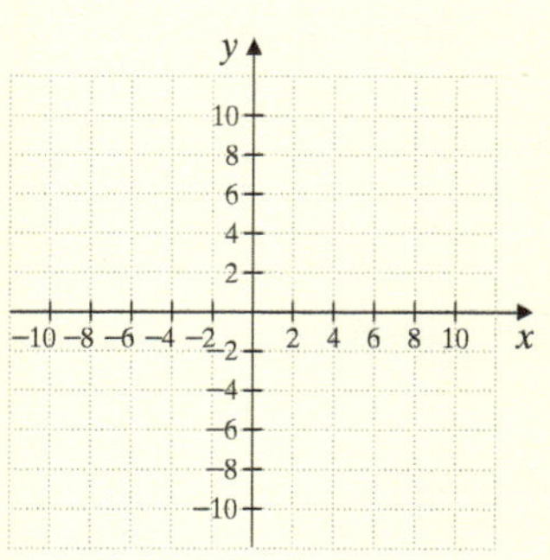

20. $y = x^2 - 4x$

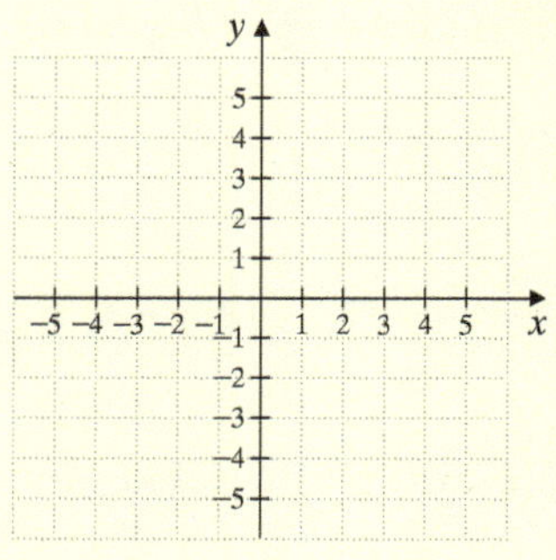

21. $y = x^2 + 2x - 8$

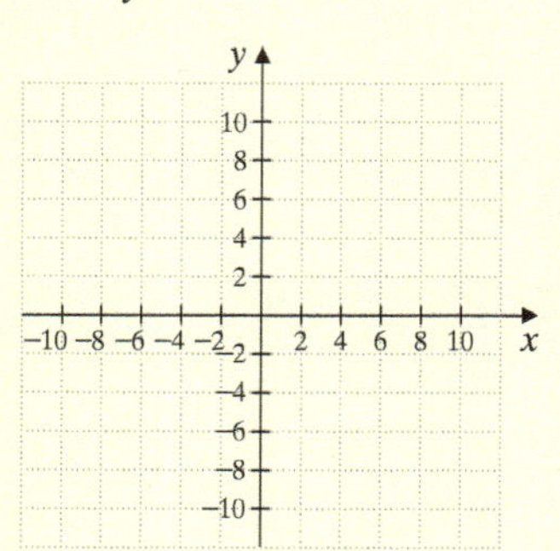

22. $y = x^2 - 2x - 3$

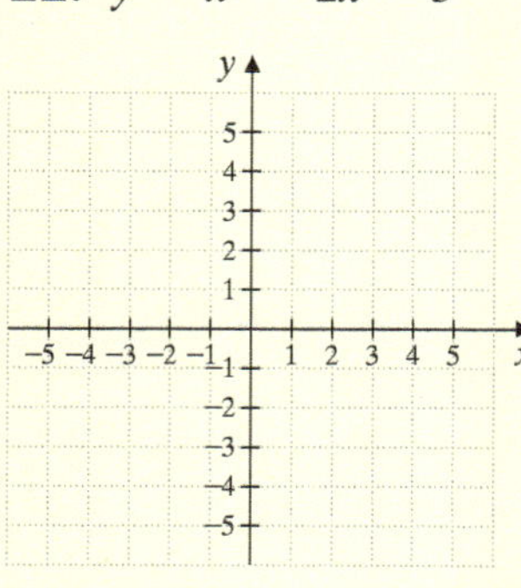

23. $y = -\frac{1}{2}x^2$

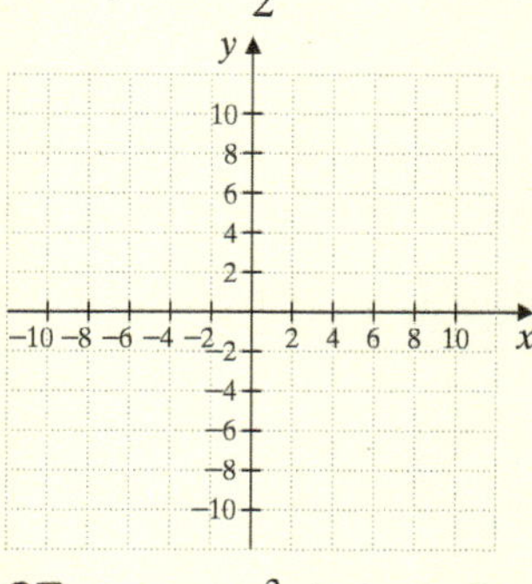

24. $y = -\frac{1}{3}x^2$

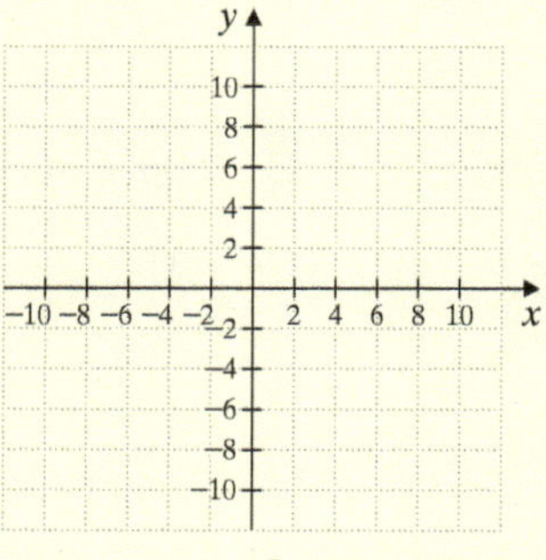

25. $y = 2x^2 - 11x + 5$

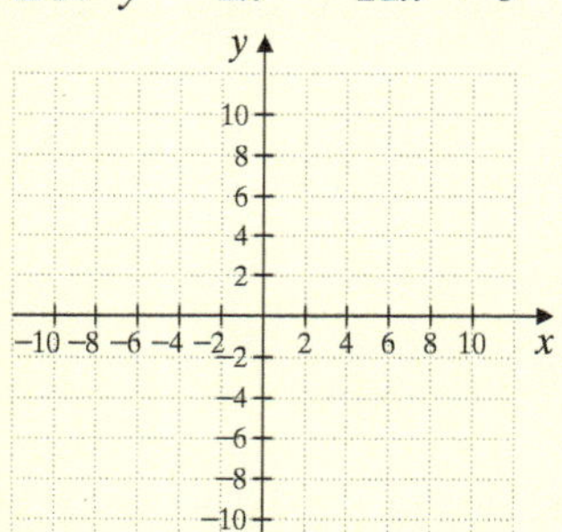

26. $y = 2x^2 + x - 3$

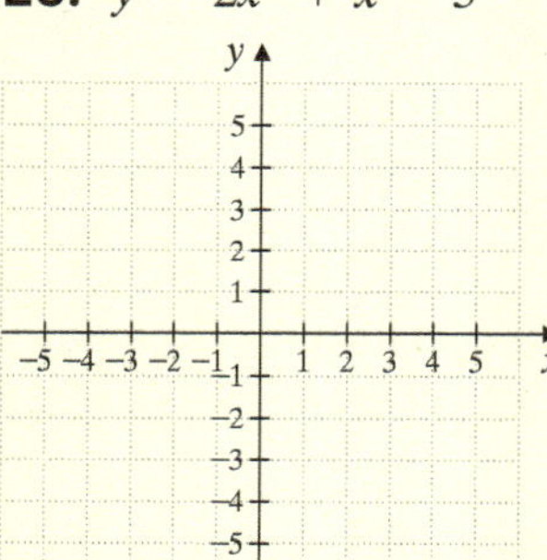

27. $y = -x^2 + 4x - 3$

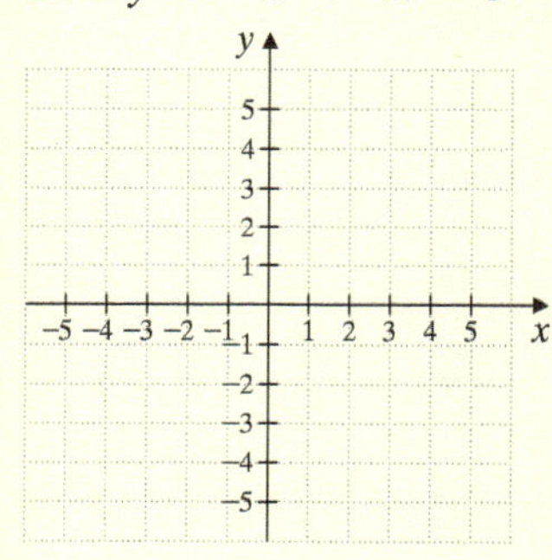

28. $y = -x^2 + 6x - 8$

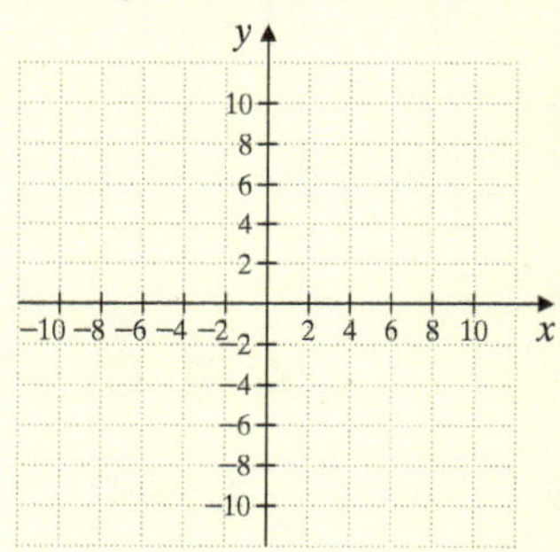

Review

Simplify each complex fraction. See Section 5.7.

29. $\dfrac{\frac{1}{7}}{\frac{2}{5}}$ **30.** $\dfrac{\frac{3}{8}}{\frac{1}{7}}$ **31.** $\dfrac{\frac{1}{x}}{\frac{2}{x^2}}$ **32.** $\dfrac{\frac{x}{5}}{\frac{2}{x}}$ **33.** $\dfrac{2x}{1 - \frac{1}{x}}$ **34.** $\dfrac{x}{x - \frac{1}{x}}$ **35.** $\dfrac{\frac{a - b}{2b}}{\frac{b - a}{8b^2}}$ **36.** $\dfrac{\frac{2a^2}{a - 3}}{\frac{a}{3 - a}}$

Concept Extensions

For Exercises 37 through 40, sketch the graph of each equation. Label the vertex and intercepts. Use the quadratic formula to locate the exact x-intercepts.

37. $y = x^2 + 2x - 2$

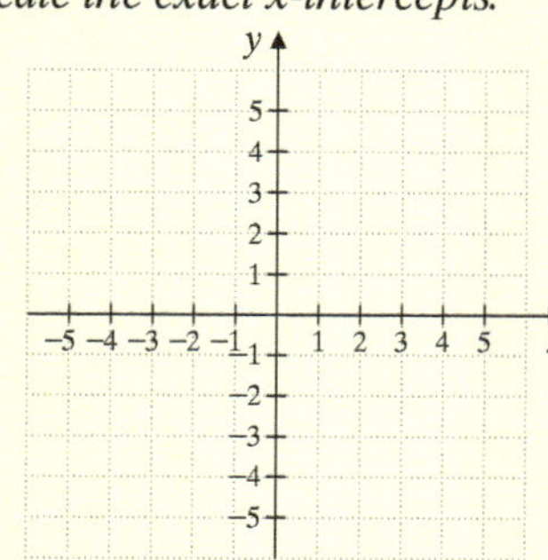

38. $y = x^2 - 4x - 3$

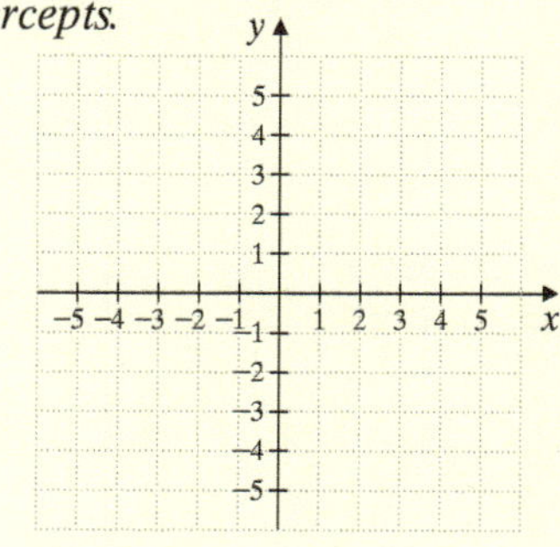

39. $y = x^2 - 3x + 1$

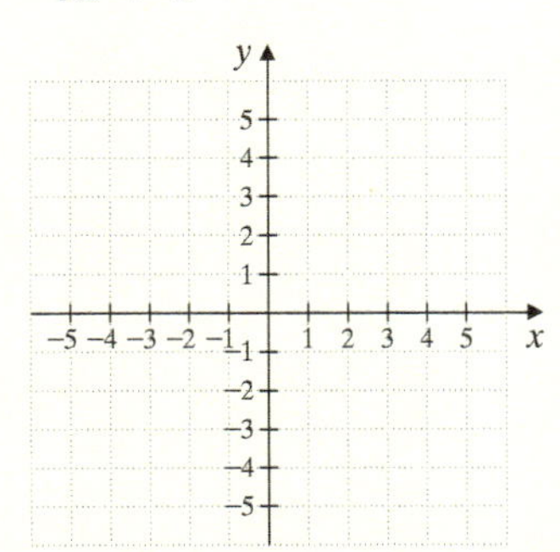

40. $y = x^2 - 2x - 5$

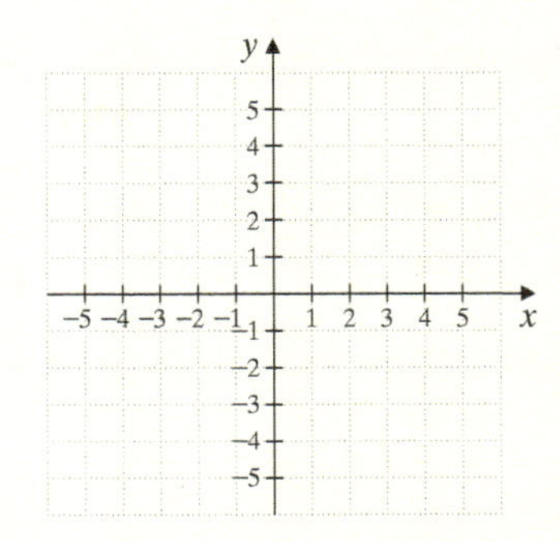

41. The height h of a fireball launched from a Roman candle with an initial velocity of 128 feet per second is given by the equation $h = -16t^2 + 128t$, where t is time in seconds after launch.

Use the graph of this equation to answer each question.

a. Estimate the maximum height of the fire ball.

b. Estimate the time when the fireball is at its maximum height.

c. Estimate the time when the fireball would return to the ground.

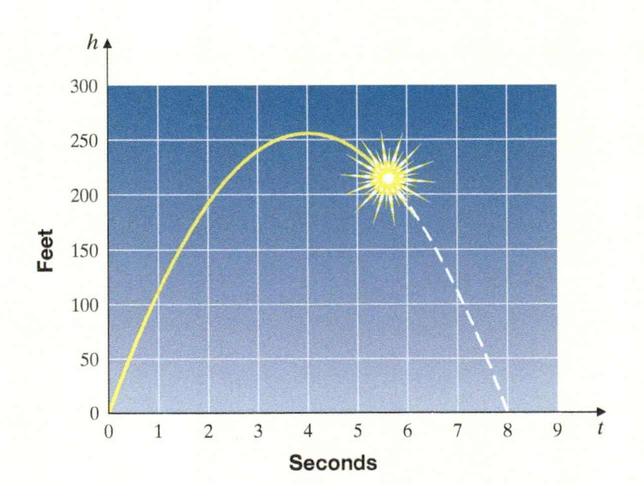

42. Determine the maximum number and the minimum number of x-intercepts for a parabola. Explain your answers.

Match the graph of each equation of the form $y = ax^2 + bx + c$ with the given description.

43. $a > 0$, two x-intercepts

44. $a < 0$, one x-intercept

45. $a < 0$, no x-intercept

46. $a > 0$, no x-intercept

47. $a > 0$, one x-intercept

48. $a < 0$, two x-intercepts

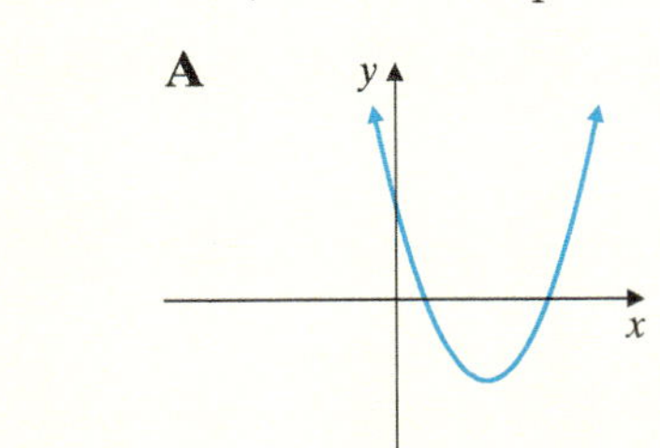

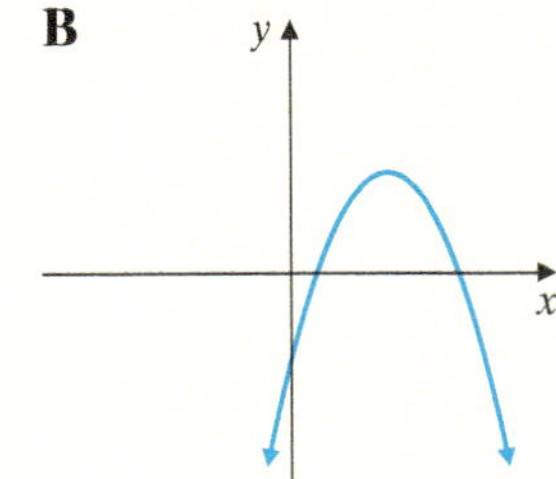

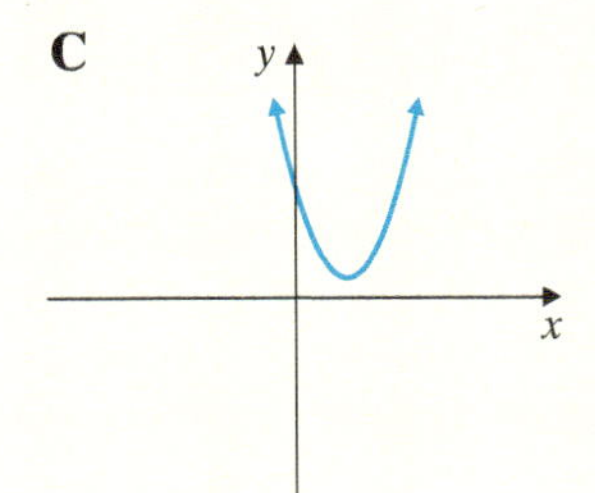

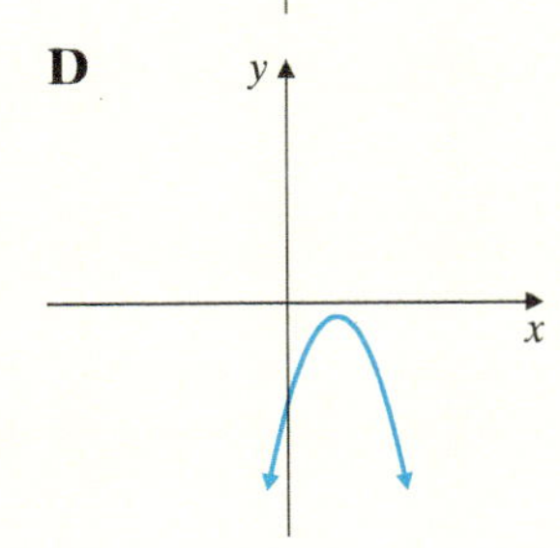

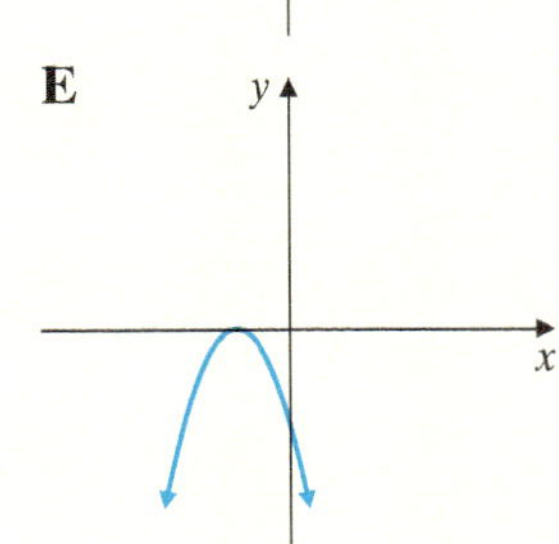

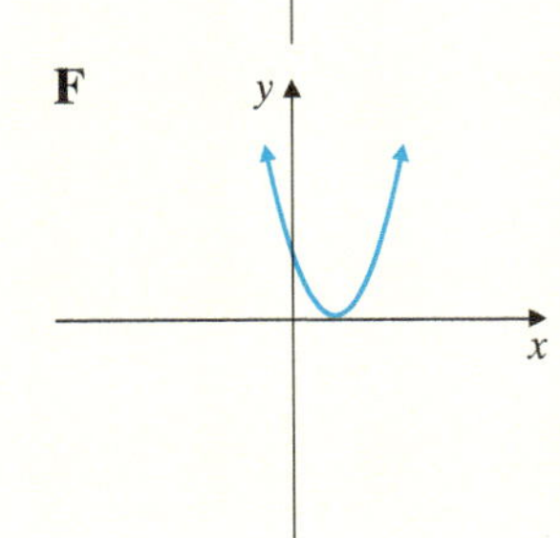

Chapter 9 Group Activity

Uses of Parabolas

In this chapter, we learned that the graph of a quadratic equation in two variables of the form $y = ax^2 + bx + c$ is a shape called a **parabola.** The figure to the right shows the general shape of a parabola.

The shape of a parabola shows up in many situations, both natural and human-made, in the world around us.

Natural Situations

- **Hurricanes** The paths of many hurricanes are roughly shaped like parabolas. In the Northern Hemisphere, hurricanes generally begin moving to the northwest. Then, as they move farther from the equator, they swing around to head in a northeasterly direction.

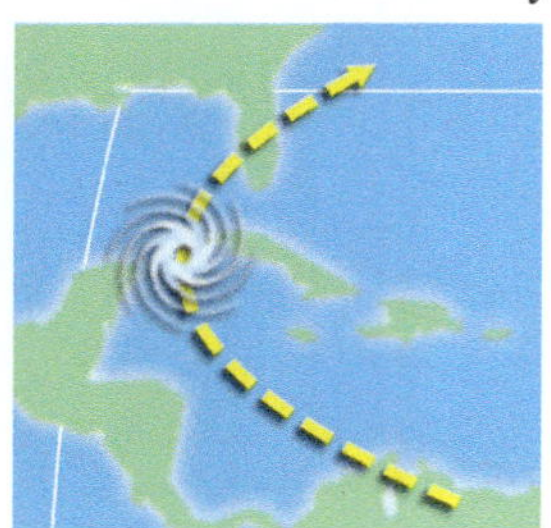

- **Projectiles** The force of the Earth's gravity acts on a projectile launched into the air. The resulting path of the projectile, anything from a bullet to a football, is generally shaped like a parabola.

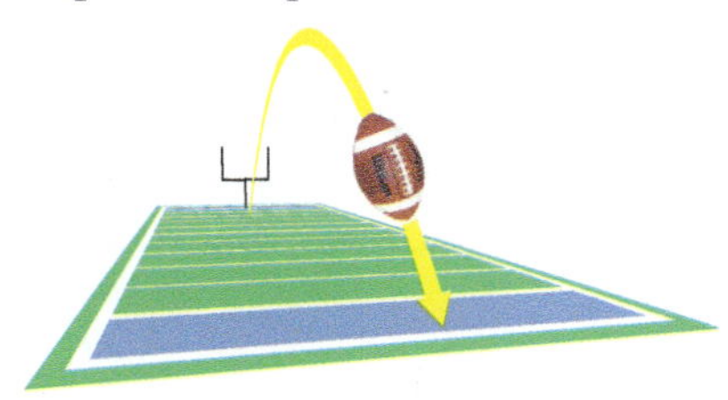

- **Orbits** There are several different possible shapes for orbits of satellites, planets, moons, and comets in outer space. One of the possible types of orbits is in the shape of a parabola. A parabolic orbit is most often seen with comets.

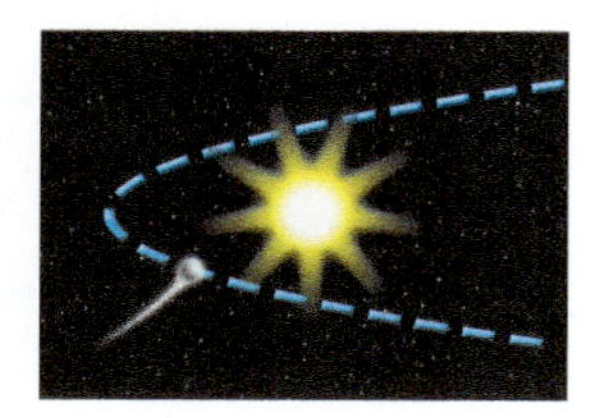

Human-Made Situations

- **Telescopes** Because a parabola has nice reflecting properties, its shape is used in many kinds of telescopes. The largest nonsteerable radio telescope is the Arecibo Observatory in Puerto Rico. This telescope consists of a huge parabolic dish built into a valley. The dish is about 1000 feet across.

- **Training Astronauts** Astronauts must be able to work in zero-gravity conditions on missions in space. However, it's nearly impossible to escape the force of gravity on Earth. To help astronauts train to work in weightlessness, a specially modified jet can be flown in a parabolic path. At the top of the parabola, weightlessness can be simulated for up to 30 seconds at a time.

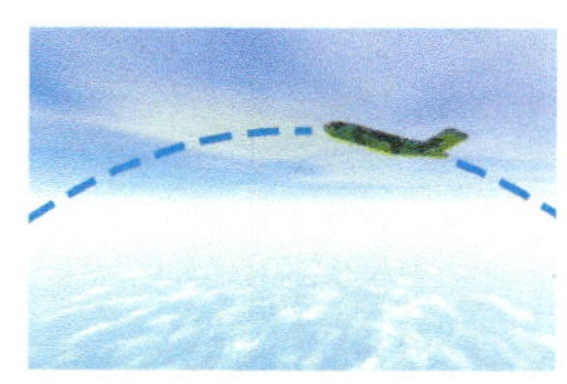

- **Architecture** The reinforced concrete arches used in many modern buildings are based on the shape of a parabola.

- **Music** The design of the modern flute incorporates a parabolic head joint.

Group Activity

There are many other physical applications of parabolas. For example, satellite dishes often have parabolic shapes. Choose a physical example of a parabola given here or use one of your own and write a report (with diagrams).

Chapter 9 Vocabulary Check

Fill in each blank with one of the words or phrases listed below. Some choices may be used more than once and some may not be used at all.

square root	vertex	one	parabola
completing the square	quadratic	zero	

1. If $x^2 = a$, then $x = \sqrt{a}$ or $x = -\sqrt{a}$. This property is called the ______________ property.
2. The graph of $y = x^2$ is called a(n) ______________.
3. The formula $x = \frac{-b}{2a}$, where $y = ax^2 + bx + c$, is called the ______________ formula.
4. The process of solving a quadratic equation by writing it in the form $(x + a)^2 = c$ is called ______________
5. The formula $x = \frac{-b \pm \sqrt{b^2 - 4ac}}{2a}$ is called the ______________ formula.
6. The lowest point on a parabola that opens upward is called the ______________.
7. The zero-factor property states that if the product of two numbers is zero, then at least one of the two numbers is ______________.

Are you preparing for your test? To help, don't forget to take these:

- Chapter 9 Getting Ready for the Test on page 691
- Chapter 9 Test on page 692

Then check all of your answers at the back of this text. For further review, the step-by-step video solutions to any of these exercises are located in MyLab Math.

9 Chapter Highlights

Definitions and Concepts	Examples
Section 9.1 Solving Quadratic Equations by the Square Root Property	
Square Root Property If $x^2 = a$ for $a \ge 0$, then $x = \sqrt{a}$ or $x = -\sqrt{a}$.	Solve the equation. $(x - 1)^2 = 15$ $x - 1 = \sqrt{15}$ or $x - 1 = -\sqrt{15}$ $x = 1 + \sqrt{15}$ $\quad$ $x = 1 - \sqrt{15}$
Section 9.2 Solving Quadratic Equations by Completing the Square	
To Solve a Quadratic Equation by Completing the Square **Step 1:** If the coefficient of x^2 is not 1, divide both sides of the equation by the coefficient. **Step 2:** Get all terms with variables alone on one side. **Step 3:** Complete the square by adding the square of half of the coefficient of x to both sides. **Step 4:** Factor the perfect square trinomial. **Step 5:** Use the square root property to solve.	Solve $2x^2 + 12x - 10 = 0$ by completing the square. $\frac{2x^2}{2} + \frac{12x}{2} - \frac{10}{2} = \frac{0}{2}$ Divide by 2. $x^2 + 6x - 5 = 0$ Simplify. $x^2 + 6x = 5$ Add 5. The coefficient of x is 6. Half of 6 is 3 and $3^2 = 9$. Add 9 to both sides. $x^2 + 6x + 9 = 5 + 9$ $(x + 3)^2 = 14$ Factor. $x + 3 = \sqrt{14}$ or $x + 3 = -\sqrt{14}$ $x = -3 + \sqrt{14}$ $\quad$ $x = -3 - \sqrt{14}$

Definitions and Concepts	Examples

Section 9.3 Solving Quadratic Equations by the Quadratic Formula

Quadratic Formula

If a, b, and c are real numbers and $a \neq 0$, the quadratic equation $ax^2 + bx + c = 0$ has solutions

$$x = \frac{-b \pm \sqrt{b^2 - 4ac}}{2a}$$

Identify a, b, and c in the quadratic equation

$$4x^2 - 6x = 5$$

First, subtract 5 from both sides.

$$4x^2 - 6x - 5 = 0$$

$a = 4$, $b = -6$, and $c = -5$

To Solve a Quadratic Equation by the Quadratic Formula

Step 1: Write the equation in standard form: $ax^2 + bx + c = 0$.

Step 2: If necessary, clear the equation of fractions.

Step 3: Identify a, b, and c.

Step 4: Replace a, b, and c in the quadratic formula with the identified values, and simplify.

Solve $3x^2 - 2x - 2 = 0$.

In this equation, $a = 3$, $b = -2$, and $c = -2$.

$$x = \frac{-(-2) \pm \sqrt{(-2)^2 - 4(3)(-2)}}{2 \cdot 3}$$

$$= \frac{2 \pm \sqrt{4 - (-24)}}{6}$$

$$= \frac{2 \pm \sqrt{28}}{6} = \frac{2 \pm \sqrt{4 \cdot 7}}{6} = \frac{2 \pm 2\sqrt{7}}{6}$$

$$= \frac{2(1 \pm \sqrt{7})}{2 \cdot 3} = \frac{1 \pm \sqrt{7}}{3}$$

Section 9.4 Graphing Quadratic Equations in Two Variables

The graph of a quadratic equation $y = ax^2 + bx + c$, $a \neq 0$, is called a **parabola.** The lowest point on a parabola opening upward or the highest point on a parabola opening downward is called the **vertex.** The vertical line through the vertex is the **axis of symmetry.**

Vertex Formula

The vertex of the parabola $y = ax^2 + bx + c$ has x-coordinate $\frac{-b}{2a}$. To find the corresponding y-coordinate, substitute the x-coordinate into the original equation and solve for y.

Graph: $y = 2x^2 - 6x + 4$

The x-coordinate of the vertex is

$$x = \frac{-b}{2a} = \frac{-(-6)}{2(2)} = \frac{6}{4} = \frac{3}{2}$$

The y-coordinate is

$$y = 2\left(\frac{3}{2}\right)^2 - 6\left(\frac{3}{2}\right) + 4 = 2\left(\frac{9}{4}\right) - 9 + 4 = -\frac{1}{2}$$

The vertex is $\left(\frac{3}{2}, -\frac{1}{2}\right)$.

The y-intercept is

$$y = 2 \cdot 0^2 - 6 \cdot 0 + 4 = 4$$

The x-intercepts are

$$0 = 2x^2 - 6x + 4$$
$$0 = 2(x - 2)(x - 1)$$
$$x = 2 \quad \text{or} \quad x = 1$$

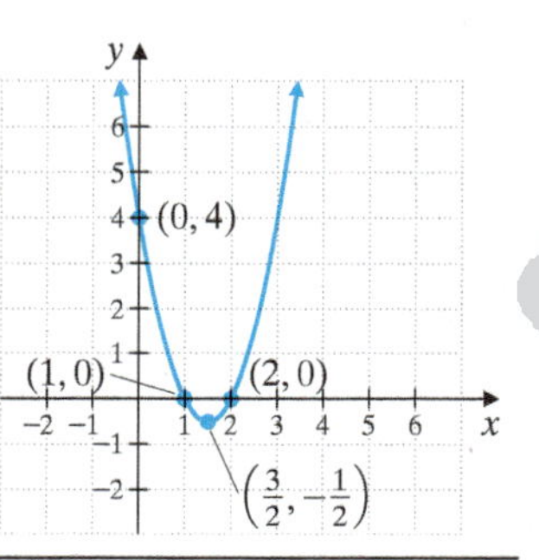

Chapter 9 Review

(9.1) *Solve each quadradic equation by factoring.*

1. $x^2 - 121 = 0$

2. $y^2 - 100 = 0$

3. $3m^2 - 5m = 2$

4. $7m^2 + 2m = 5$

Use the square root property to solve each quadratic equation.

5. $x^2 = 36$

6. $x^2 = 81$

7. $k^2 = 50$

8. $k^2 = 45$

9. $(x - 11)^2 = 49$

10. $(x + 3)^2 = 100$

11. $(4p + 5)^2 = 41$

12. $(3p + 7)^2 = 37$

Solve. For Exercises 13 and 14, use the formula $h = 16t^2$, where h is the height in feet at time t seconds.

13. If Kara Washington dives from a height of 100 feet, how long before she hits the water?

14. How long does a 5-mile free fall take? Round your result to the nearest tenth of a second. (*Hint:* 1 mi = 5280 ft)

(9.2) *Solve each quadratic equation by completing the square.*

15. $x^2 - 9x = -8$

16. $x^2 + 8x = 20$

17. $x^2 + 4x = 1$

18. $x^2 - 8x = 3$

19. $x^2 - 6x + 7 = 0$

20. $x^2 + 6x + 7 = 0$

21. $2y^2 + y - 1 = 0$

22. $4y^2 + 3y - 1 = 0$

(9.3) *Use the quadratic formula to solve each quadratic equation.*

23. $9x^2 + 30x + 25 = 0$

24. $16x^2 - 72x + 81 = 0$

25. $7x^2 = 35$

26. $11x^2 = 33$

27. $x^2 - 10x + 7 = 0$

28. $x^2 + 4x - 7 = 0$

29. $3x^2 + x - 1 = 0$

30. $x^2 + 3x - 1 = 0$

31. $2x^2 + x + 5 = 0$

32. $7x^2 - 3x + 1 = 0$

For the exercise numbers given, approximate the exact solutions to the nearest tenth.

33. Exercise **29**

34. Exercise **30**

35. The annual number of visitors y (in thousands) to Yosemite National Park in California is modeled by the equation $y = 78x^2 - 267x + 3975$. In this equation, x is the number of years since 2011. Assume that this trend continued and find the year after 2011 in which 3882 thousand people visited Yosemite National Park. (*Source:* Based on data from the National Park Service)

36. The amount y of electricity generated by solar power (in thousand megawatt hours per day) in the United States from 2012 through 2016 is modeled by the equation $y = 2x^2 + 13x + 12$, where x represents the number of years after 2012. Assume that this trend continues and find the year after 2012 in which the amount of electricity generated by solar power is 127 thousand megawatt hours per day. (*Source:* Based on information from the Energy Information Administration)

(9.4) *Graph each quadratic equation and find and plot any intercepts.*

37. $y = 5x^2$

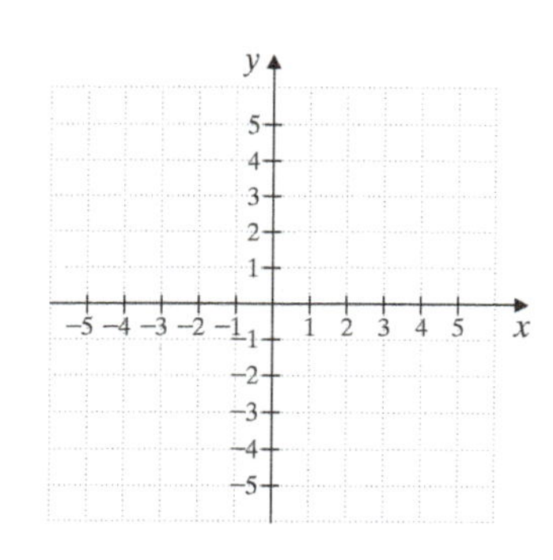

38. $y = -\frac{1}{2}x^2$

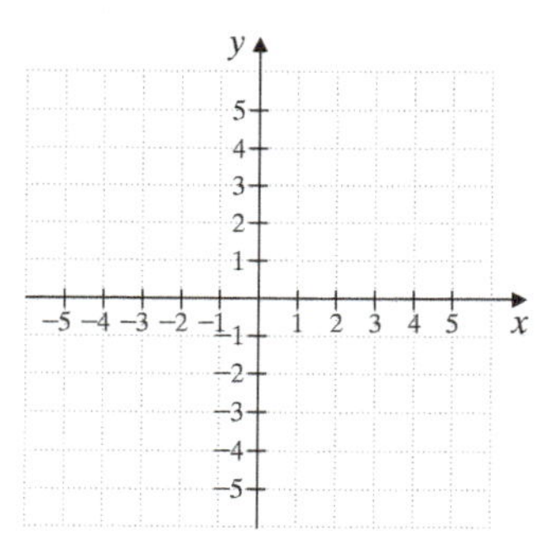

Graph each quadratic equation. Label the vertex and the intercepts with their coordinates.

39. $y = x^2 - 25$

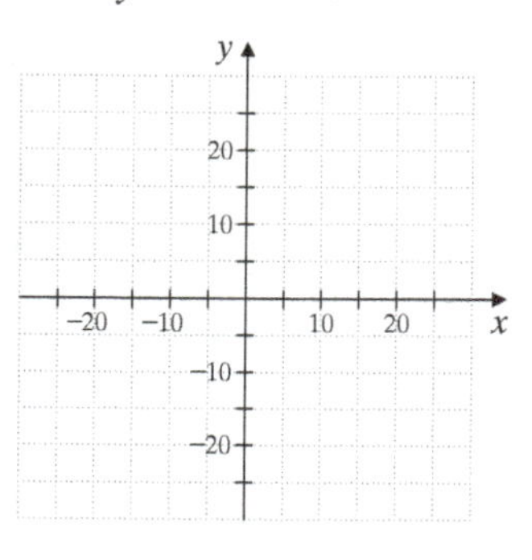

40. $y = x^2 - 36$

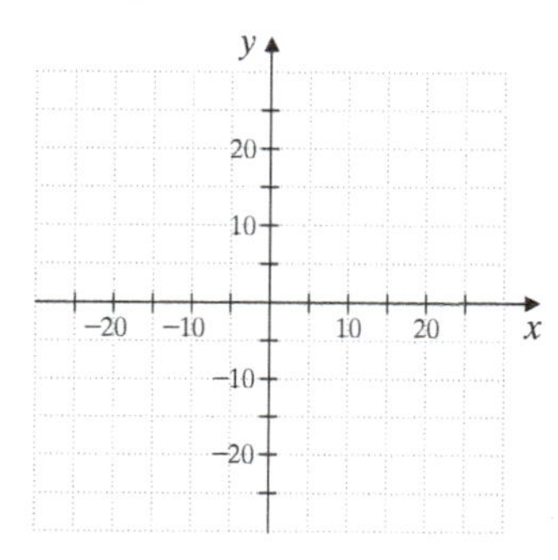

41. $y = x^2 + 3$

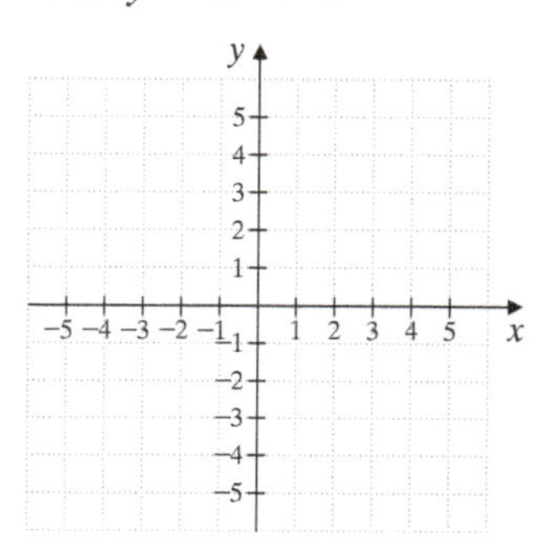

42. $y = x^2 + 8$

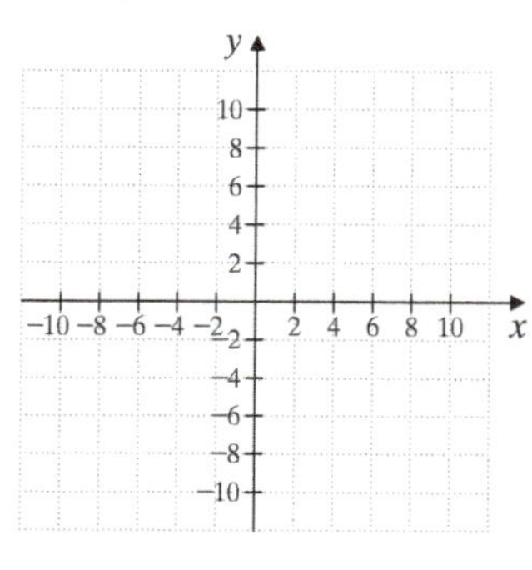

43. $y = -4x^2 + 8$

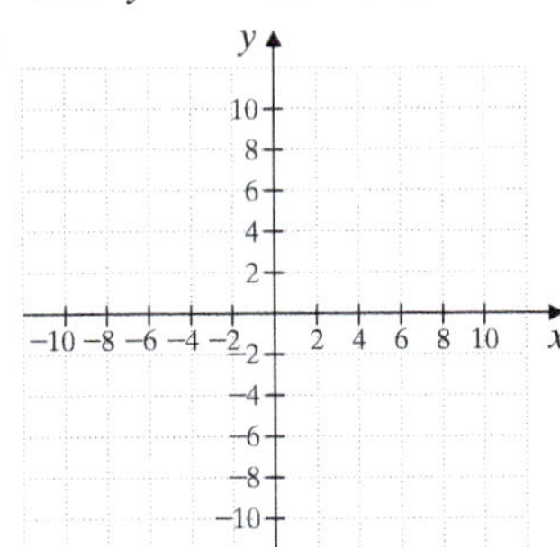

44. $y = -3x^2 + 9$

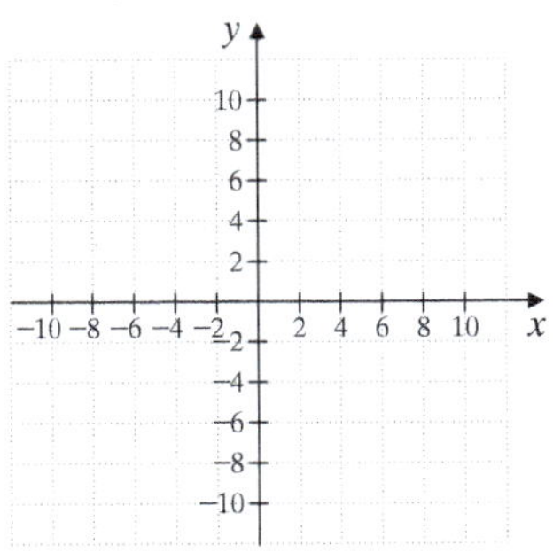

45. $y = x^2 + 3x - 10$

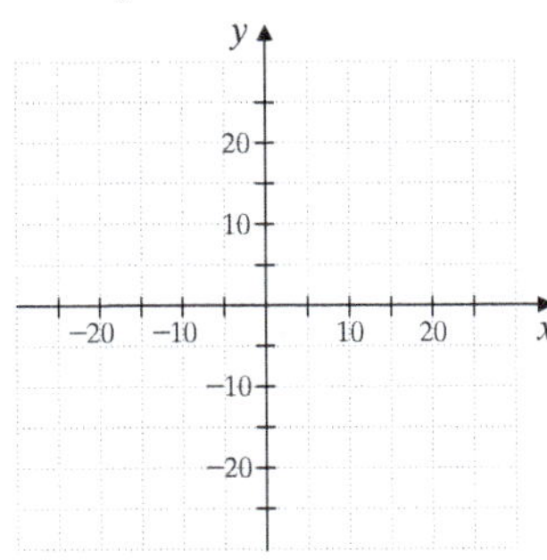

46. $y = x^2 + 3x - 4$

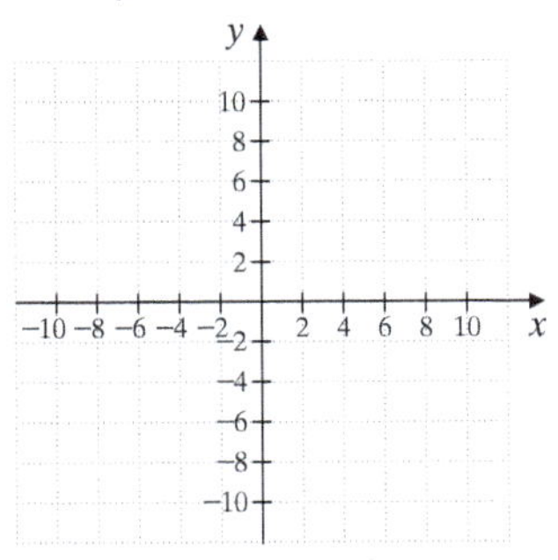

47. $y = -x^2 - 5x - 6$

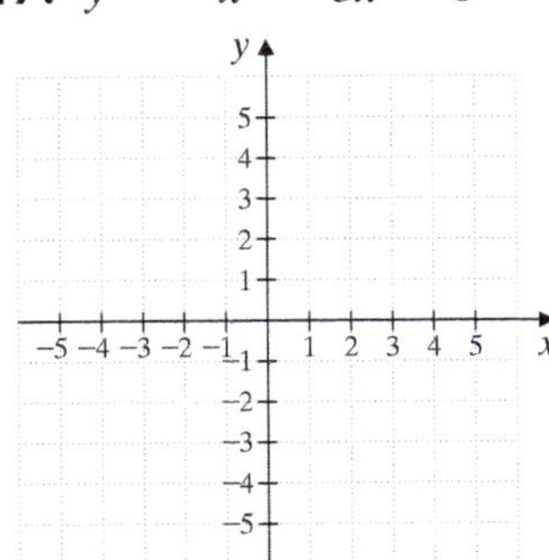

48. $y = 3x^2 - x - 2$

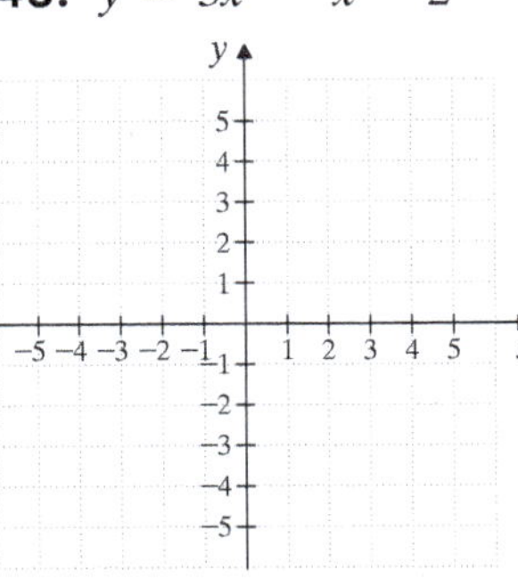

49. $y = 2x^2 - 11x - 6$

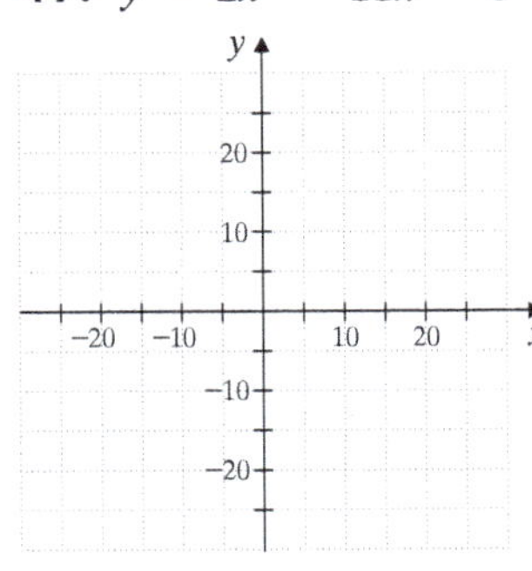

50. $y = -x^2 + 4x + 8$

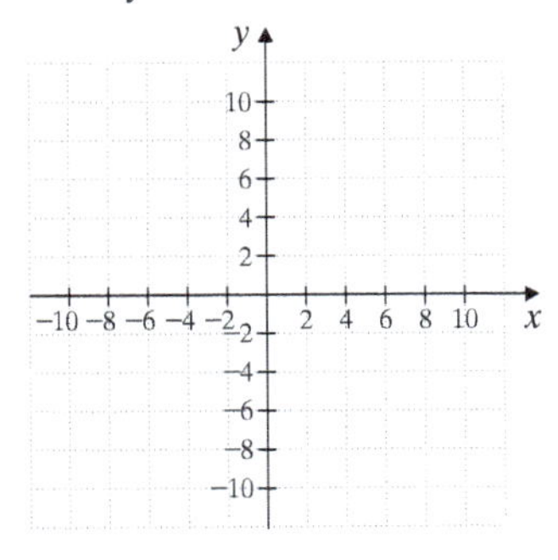

Match each quadratic equation with its graph.

51. $y = 2x^2$

52. $y = -x^2$

53. $y = x^2 + 4x + 4$

54. $y = x^2 + 5x + 4$

A

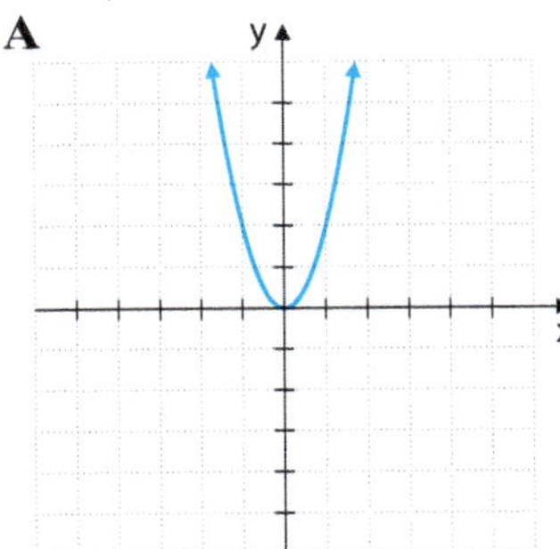

B

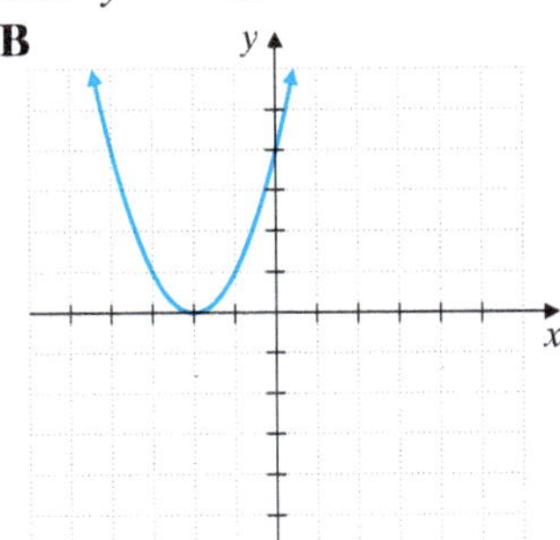

C

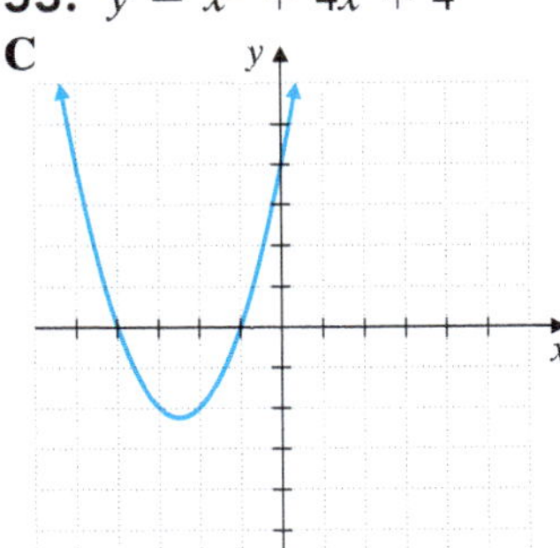

D

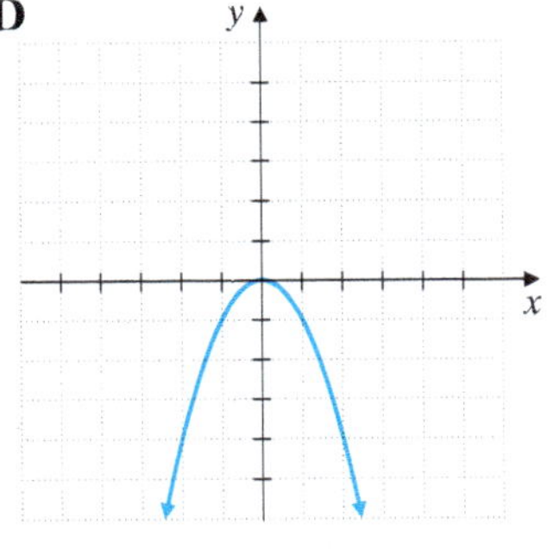

Quadratic equations in the form $y = ax^2 + bx + c$ are graphed below. Determine the number of real solutions for the related equation $0 = ax^2 + bx + c$ from each graph.

55.

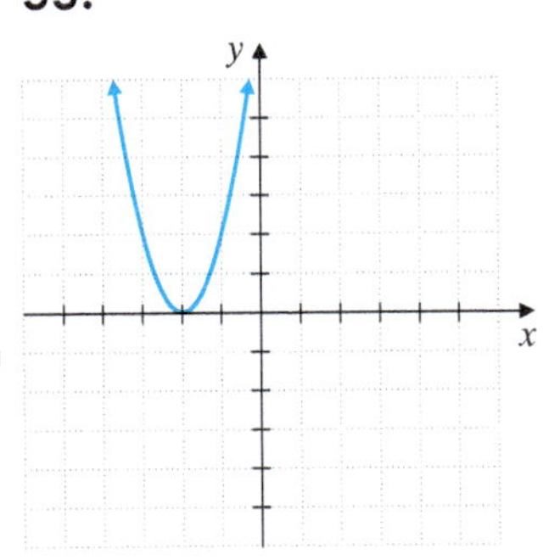

56.

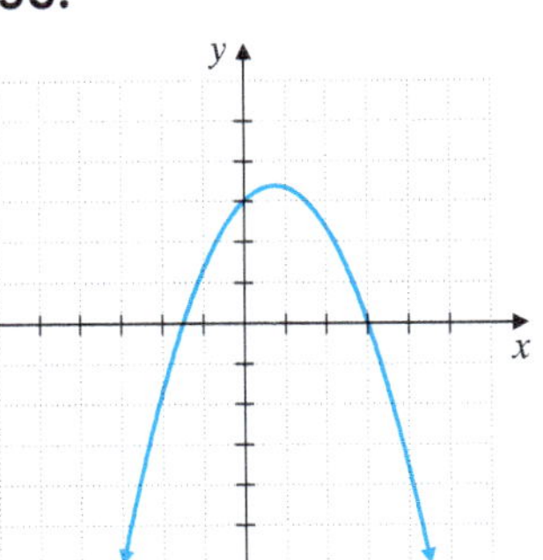

57.

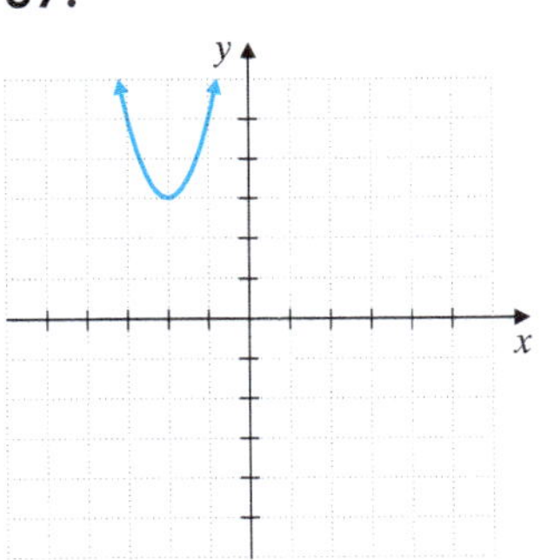

58.

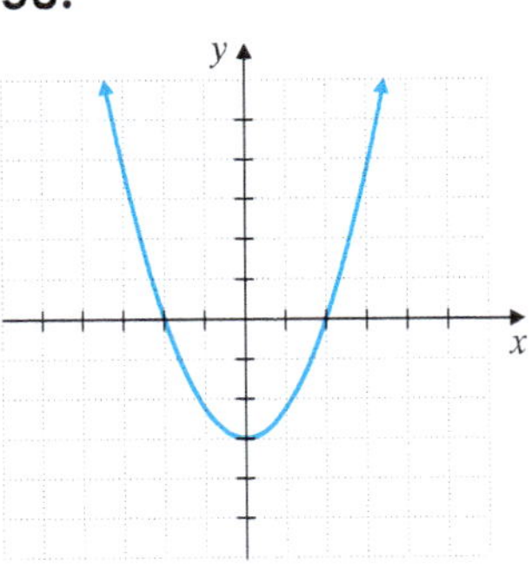

Mixed Review

Use the square root property to solve each quadratic equation.

59. $x^2 = 49$

60. $y^2 = 75$

61. $(x - 7)^2 = 64$

Solve each quadratic equation by completing the square.

62. $x^2 + 4x = 6$

63. $3x^2 + x = 2$

64. $4x^2 - x - 2 = 0$

Use the quadratic formula to solve each quadratic equation.

65. $4x^2 - 3x - 2 = 0$

66. $5x^2 + x - 2 = 0$

67. $4x^2 + 12x + 9 = 0$

68. $2x^2 + x + 4 = 0$

Graph each quadratic equation. Label the vertex and the intercepts with their coordinates.

69. $y = 4 - x^2$

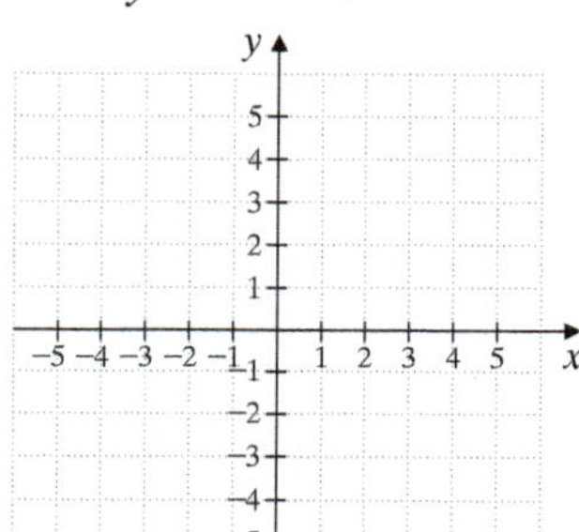

70. $y = x^2 + 4$

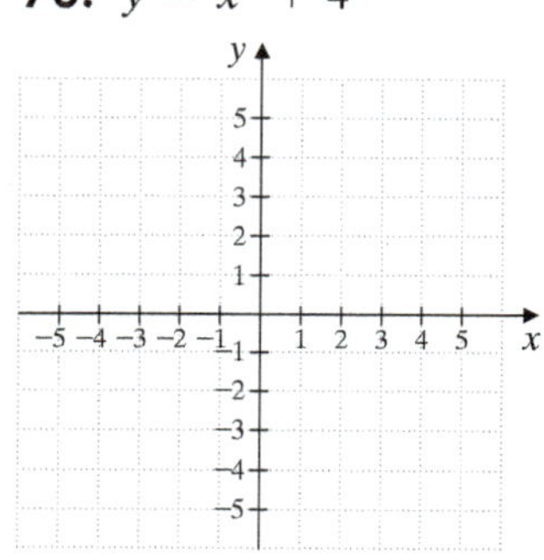

71. $y = x^2 + 6x + 8$

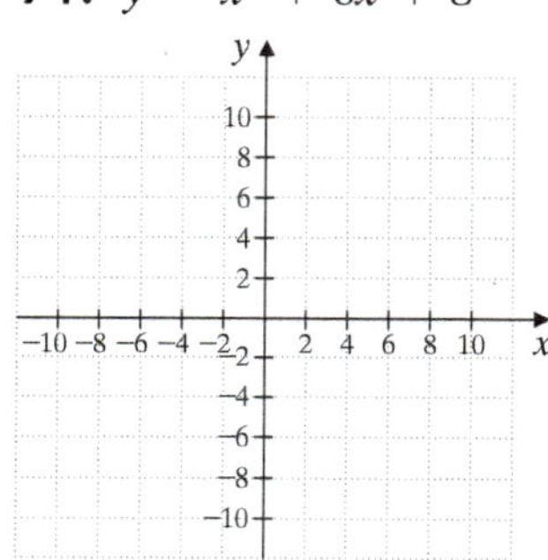

72. $y = x^2 - 2x - 4$

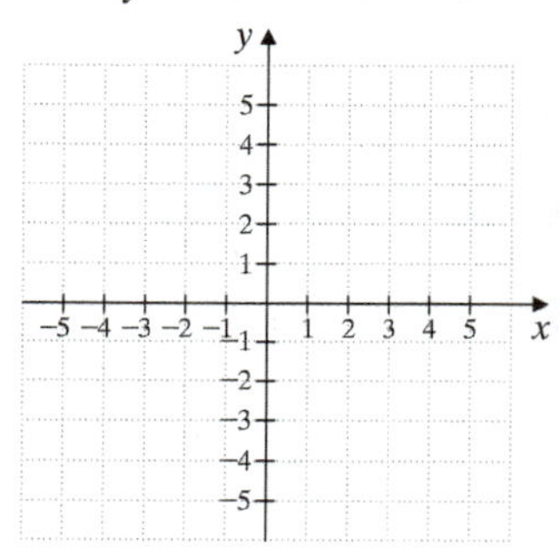

Chapter 9 Getting Ready for the Test LC

MULTIPLE CHOICE All the exercises below are **Multiple Choice**. Choose the correct letter.

1. To solve $x^2 - 6x = -1$ by completing the square, choose the next correct step.

A. $x^2 - 6x + 9 = -1$ **B.** $x^2 - 6x + 9 = -1 + 9$ **C.** $x^2 - 6x - 9 = -1 - 9$ **D.** $(x - 6)^2 = (-1)^2$

2. To solve $x^2 + 5x = 4$ by completing the square, choose the next correct step.

A. $x^2 + 5x + 25 = 4$ **B.** $x^2 + 5x + 25 = 4 + 25$ **C.** $x^2 + 5x + \frac{25}{4} = 4$ **D.** $x^2 + 5x + \frac{25}{4} = 4 + \frac{25}{4}$

3. The expression $\frac{12 \pm 3\sqrt{7}}{9}$ simplifies to:

A. $\frac{\pm 15\sqrt{7}}{9}$ **B.** $\frac{4 \pm 3\sqrt{7}}{3}$ **C.** $\frac{4 \pm \sqrt{7}}{3}$ **D.** $4 \pm \sqrt{7}$

4. The expression $\frac{5 \pm 10\sqrt{2}}{5}$ simplifies to:

A. $1 \pm 2\sqrt{2}$ **B.** $\pm 10\sqrt{2}$ **C.** $\pm 3\sqrt{2}$ **D.** $1 \pm 10\sqrt{2}$

For Exercises 5 and 6, the quadratic equation is $7x^2 = 3 - x$, with $a = 7$ in the quadratic formula.

5. Choose the value of c.

A. 3 **B.** −3 **C.** 1 **D.** −1 **E.** x

6. Choose the value of b.

A. 3 **B.** −3 **C.** 1 **D.** −1 **E.** x

For Exercises 7 through 10, choose the correct letter.

7. Select the vertex of the graph of $y = x^2 + 3$.

A. $(0, 0)$ **B.** $(0, 3)$ **C.** $(3, 0)$ **D.** $\left(\frac{9}{2}, 3\right)$

8. Select the vertex of the graph of $y = x^2 - 2x$.

A. $(1, -2)$ **B.** $(-1, 3)$ **C.** $(2, 0)$ **D.** $(1, -1)$

9. Select the vertex of the graph of $y = -x^2 + 2x - 4$.

A. $(-1, -7)$ **B.** $(2, -4)$ **C.** $(1, -1)$ **D.** $(1, -3)$

10. Select the intercept(s) of the graph of $y = x^2 + x + 1$.

A. $(0, 1), (-1, 0)$ **B.** $(0, 1)$ **C.** $(1, 0), (-1, 0)$ **D.** $(0, 1), (1, 0), (-1, 0)$

Chapter 9 Test

MyLab Math

For additional practice go to your study plan in MyLab Math.

Answers

1. ______

2. ______

3. ______

4. ______

5. ______

6. ______

7. ______

8. ______

9. ______

10. ______

11. ______

12. ______

13. ______

14. ______

15. ______

16. ______

17. ______

Solve by factoring.

1. $x^2 - 400 = 0$

2. $2x^2 - 11x = 21$

Solve using the square root property.

3. $5k^2 = 80$

4. $(3m - 5)^2 = 8$

Solve by completing the square.

5. $x^2 - 26x + 160 = 0$

6. $3x^2 + 12x - 4 = 0$

Solve using the quadratic formula.

7. $x^2 - 3x - 10 = 0$

8. $p^2 - \frac{5}{3}p - \frac{1}{3} = 0$

Solve by the most appropriate method.

9. $(3x - 5)(x + 2) = -6$

10. $(3x - 1)^2 = 16$

11. $3x^2 - 7x - 2 = 0$

12. $x^2 - 4x - 5 = 0$

13. $3x^2 - 7x + 2 = 0$

14. $2x^2 - 6x + 1 = 0$

15. The height of a triangle is 4 times the length of the base. The area of the triangle is 18 square feet. Find the height and base of the triangle.

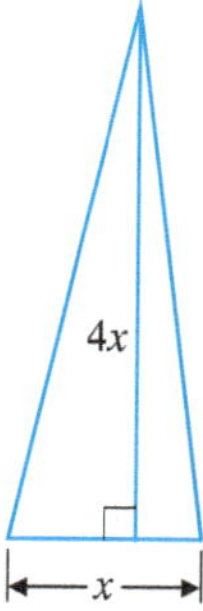

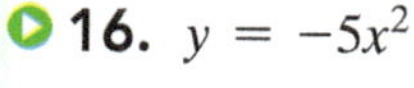

Graph each quadratic equation. Label the vertex and the intercepts with their coordinates.

16. $y = -5x^2$

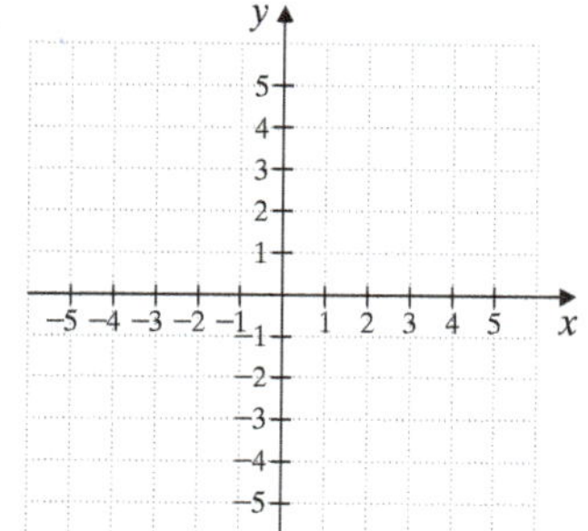

17. $y = x^2 - 4$

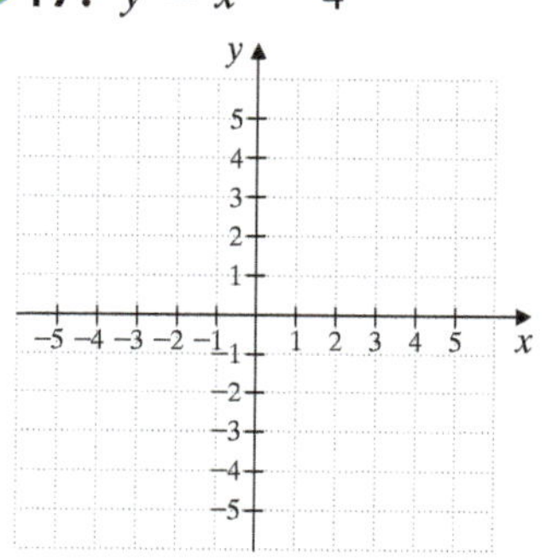

18. $y = x^2 - 7x + 10$

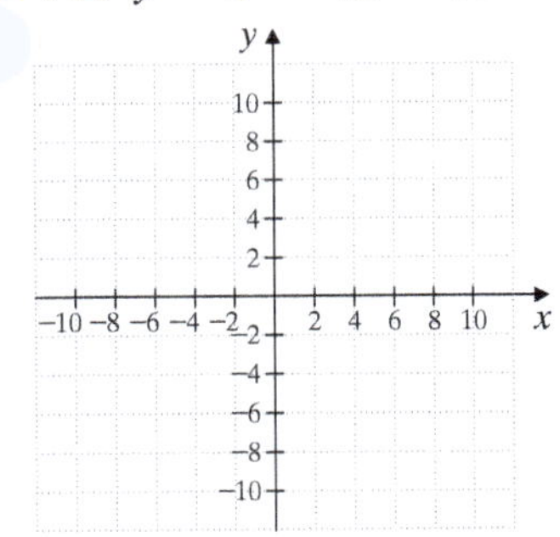

19. $y = 2x^2 + 4x - 1$

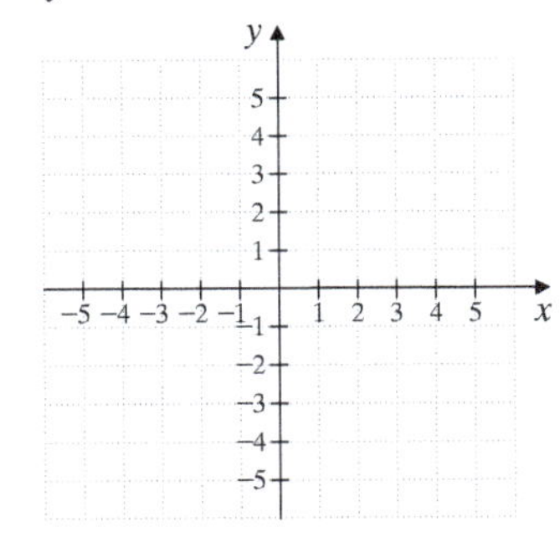

20. The number of diagonals d that a polygon with n sides has is given by the formula

$$d = \frac{n^2 - 3n}{2}$$

Find the number of sides of a polygon if it has 9 diagonals.

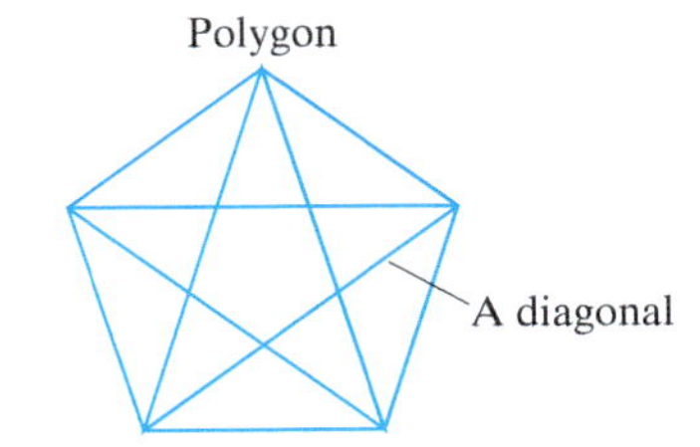

18. ______

19. ______

20. ______

21. ______

Solve.

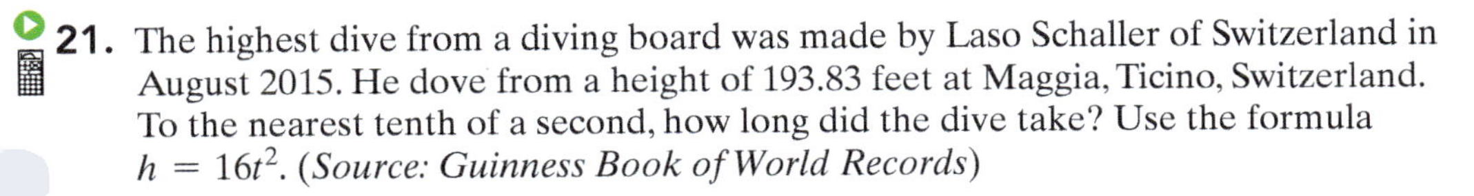

21. The highest dive from a diving board was made by Laso Schaller of Switzerland in August 2015. He dove from a height of 193.83 feet at Maggia, Ticino, Switzerland. To the nearest tenth of a second, how long did the dive take? Use the formula $h = 16t^2$. (*Source: Guinness Book of World Records*)

Chapters 1–9 Cumulative Review

Answers

1. ______
2. ______
3. ______
4. ______
5. ______
6. ______
7. ______
8. ______
9. ______
10. ______
11. ______
12. ______
13. ______
14. ______
15. ______
16. ______
17. ______
18. a. ______
 b. ______
19. ______
20. ______
21. ______

Solve each equation.

1. $y + 0.6 = -1.0$

2. $8x - 14 = 6x - 20$

3. $8(2 - t) = -5t$

4. $2(x + 7) = 5(2x - 3)$

5. Find two numbers whose sum is 37 and whose difference is 21.

6. The sum of three consecutive integers is 438. Find the integers.

Simplify the following expressions.

7. 3^0

8. $\left(\frac{-6x}{y^3}\right)^3$

9. $(5x^3y^2)^0$

10. $\frac{a^2b^7}{(2b^2)^5}$

11. -4^0

12. $\frac{(3y)^2}{y^2}$

13. Multiply: $(3y + 2)^2$

14. Multiply: $(x^2 + 5)(y - 1)$

15. Divide $x^2 + 7x + 12$ by $x + 3$ using long division.

16. Simplify by combining like terms: $2 + 8.1a + a - 6$

17. Factor: $r^2 - r - 42$

18. Find the value of each expression when $x = -4$ and $y = 7$.

a. $\frac{x - y}{7 - x}$ **b.** $x^2 + 2y$

19. Factor: $10x^2 - 13xy - 3y^2$

20. Add: $\frac{1}{x + 2} + \frac{7}{x - 1}$

21. Factor $8x^2 - 14x + 5$ by grouping.

22. Multiply: $\dfrac{x^2 + 7x}{5x} \cdot \dfrac{10x + 25}{x^2 - 49}$

23. Factor each binomial.

a. $4x^3 - 49x$ **b.** $162x^4 - 2$

24. Solve: $\dfrac{2x + 7}{3} = \dfrac{x - 6}{2}$

25. Solve: $(5x - 1)(2x^2 + 15x + 18) = 0$

26. Simplify each expression by combining like terms.

a. $4x - 3 + 7 - 5x$ **b.** $-6y + 3y - 8 + 8y$

c. $7 + 10.1a - a - 11$ **d.** $2x^2 - 2x$

27. Simplify: $\dfrac{x^2 + 8x + 7}{x^2 - 4x - 5}$

28. Solve $2x^2 + 5x = 7$.

29. The quotient of a number and 6, minus $\frac{5}{3}$, is the quotient of the number and 2. Find the number.

30. Find the distance between $(-7, 4)$ and $(2, 5)$. (See the Chapter 8 Group Activity.)

31. Complete the table for the equation $y = 3x$.

	x	y
a.	−1	
b.		0
c.		−9

32. Identify the x- and y-intercepts.

a.

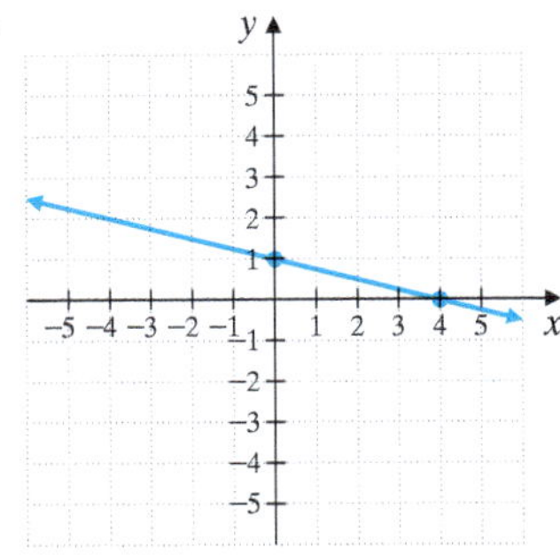

b.

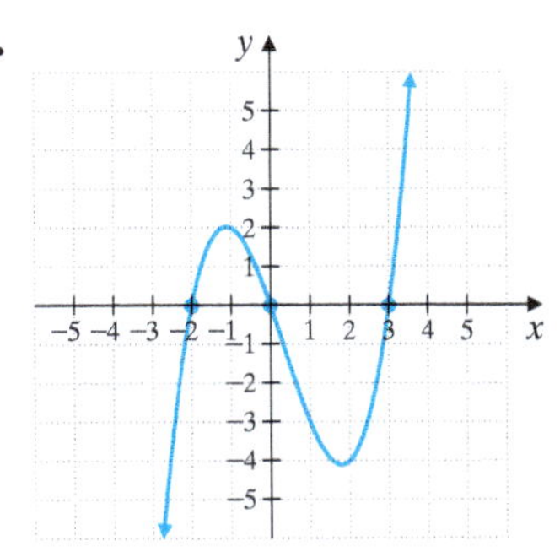

33. Determine whether each pair of lines is parallel, perpendicular, or neither.

a. $y = -\frac{1}{5}x + 1$
$2x + 10y = 3$

b. $x + y = 3$
$-x + y = 4$

c. $3x + y = 5$
$2x + 3y = 6$

34. Determine whether the graphs of $y = 3x + 7$ and $x + 3y = -15$ are parallel lines, perpendicular lines, or neither.

35. Which of the following relations is also a function?

a. $\{(-1, 1), (2, 3), (7, 3), (8, 6)\}$

b. $\{(0, -2), (1, 5), (0, 3), (7, 7)\}$

22. ______
23. a. ______
b. ______
24. ______
25. ______
26. a. ______
b. ______
c. ______
d. ______
27. ______
28. ______
29. ______
30. ______
31. ______
32. a. ______
b. ______
33. a. ______
b. ______
c. ______
34. ______
35. a. ______
b. ______

36. a. ______

b. ______

c. ______

37. ______

38. ______

39. ______

40. ______

41. ______

42. ______

43. ______

44. ______

45. ______

46. ______

47. ______

48. ______

49. ______

50. ______

36. Add or subtract by first simplifying each radical.

a. $\sqrt{80} + \sqrt{20}$

b. $2\sqrt{98} - 2\sqrt{18}$

c. $\sqrt{32} + \sqrt{121} - \sqrt{12}$

37. Solve the system: $\begin{cases} 2x + y = 10 \\ x = y + 2 \end{cases}$

38. Solve the system. $\begin{cases} 5x + y = 3 \\ y = -5x \end{cases}$

39. Solve the system: $\begin{cases} 2x - y = 7 \\ 8x - 4y = 1 \end{cases}$

40. Solve the system. $\begin{cases} -2x + y = 7 \\ 6x - 3y = -21 \end{cases}$

Find each square root.

41. $\sqrt{36}$

42. $\sqrt{\frac{4}{25}}$

43. $\sqrt{\frac{9}{100}}$

44. $\sqrt{\frac{16}{121}}$

45. Rationalize the denominator of $\frac{2}{1 + \sqrt{3}}$.

46. Rationalize the denominator of $\frac{5}{\sqrt{8}}$.

47. Use the square root property to solve $(x - 3)^2 = 16$.

48. Use the square root property to solve $3(x - 4)^2 = 9$.

49. Solve $\frac{1}{2}x^2 - x = 2$ using the quadratic formula.

50. Solve $x^2 + 4x = 8$ using the quadratic formula.

Appendix

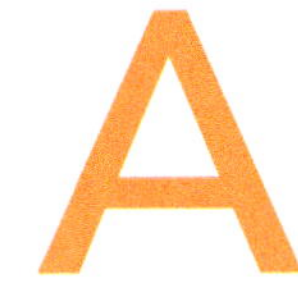

Tables

A.1 Table of Percents, Decimals, and Fraction Equivalents

Percent	Decimal	Fraction
1%	0.01	$\frac{1}{100}$
5%	0.05	$\frac{1}{20}$
10%	0.1	$\frac{1}{10}$
12.5% or $12\frac{1}{2}\%$	0.125	$\frac{1}{8}$
$16.\overline{6}\%$ or $16\frac{2}{3}\%$	$0.1\overline{6}$	$\frac{1}{6}$
20%	0.2	$\frac{1}{5}$
25%	0.25	$\frac{1}{4}$
30%	0.3	$\frac{3}{10}$
$33.\overline{3}\%$ or $33\frac{1}{3}\%$	$0.\overline{3}$	$\frac{1}{3}$
37.5% or $37\frac{1}{2}\%$	0.375	$\frac{3}{8}$
40%	0.4	$\frac{2}{5}$
50%	0.5	$\frac{1}{2}$
60%	0.6	$\frac{3}{5}$
62.5% or $62\frac{1}{2}\%$	0.625	$\frac{5}{8}$
$66.\overline{6}\%$ or $66\frac{2}{3}\%$	$0.\overline{6}$	$\frac{2}{3}$
70%	0.7	$\frac{7}{10}$
75%	0.75	$\frac{3}{4}$
80%	0.8	$\frac{4}{5}$
$83.\overline{3}\%$ or $83\frac{1}{3}\%$	$0.8\overline{3}$	$\frac{5}{6}$
87.5% or $87\frac{1}{2}\%$	0.875	$\frac{7}{8}$
90%	0.9	$\frac{9}{10}$
100%	1.0	1
110%	1.1	$1\frac{1}{10}$
125%	1.25	$1\frac{1}{4}$
$133.\overline{3}\%$ or $133\frac{1}{3}\%$	$1.\overline{3}$	$1\frac{1}{3}$
150%	1.5	$1\frac{1}{2}$
$166.\overline{6}\%$ or $166\frac{2}{3}\%$	$1.\overline{6}$	$1\frac{2}{3}$
175%	1.75	$1\frac{3}{4}$
200%	2.0	2

A.2 Table on Finding Common Percents of a Number

Common Percent Equivalences*	Shortcut Method for Finding Percent	Examples
$1\% = 0.01 \left(\text{or } \frac{1}{100}\right)$	To find 1% of a number, multiply by 0.01. To do so, move the decimal point 2 places to the left.	1% of 210 is 2.10 or 2.1. 1% of 1500 is 15. 1% of 8.6 is 0.086.
$10\% = 0.1 \left(\text{or } \frac{1}{10}\right)$	To find 10% of a number, multiply by 0.1, or move the decimal point of the number 1 place to the left.	10% of 140 is 14. 10% of 30 is 3. 10% of 17.6 is 1.76.
$25\% = \frac{1}{4}$	To find 25% of a number, find $\frac{1}{4}$ of the number, or divide the number by 4.	25% of 20 is $\frac{20}{4}$ or 5. 25% of 8 is 2. 25% of 10 is $\frac{10}{4}$ or $2\frac{1}{2}$.
$50\% = \frac{1}{2}$	To find 50% of a number, find $\frac{1}{2}$ of the number, or divide the number by 2.	50% of 64 is $\frac{64}{2}$ or 32. 50% of 1000 is 500. 50% of 9 is $\frac{9}{2}$ or $4\frac{1}{2}$.
$100\% = 1$	To find 100% of a number, multiply the number by 1. In other words, 100% of a number is the number.	100% of 98 is 98. 100% of 1407 is 1407. 100% of 18.4 is 18.4.
$200\% = 2$	To find 200% of a number, multiply the number by 2.	200% of 31 is $31 \cdot 2$ or 62. 200% of 750 is 1500. 200% of 6.5 is 13.

*See Appendix A.1.

A.3 Table of Squares and Square Roots 

n	n^2	$\sqrt{n}$	n	n^2	$\sqrt{n}$
1	1	1.000	51	2601	7.141
2	4	1.414	52	2704	7.211
3	9	1.732	53	2809	7.280
4	16	2.000	54	2916	7.348
5	25	2.236	55	3025	7.416
6	36	2.449	56	3136	7.483
7	49	2.646	57	3249	7.550
8	64	2.828	58	3364	7.616
9	81	3.000	59	3481	7.681
10	100	3.162	60	3600	7.746
11	121	3.317	61	3721	7.810
12	144	3.464	62	3844	7.874
13	169	3.606	63	3969	7.937
14	196	3.742	64	4096	8.000
15	225	3.873	65	4225	8.062
16	256	4.000	66	4356	8.124
17	289	4.123	67	4489	8.185
18	324	4.243	68	4624	8.246
19	361	4.359	69	4761	8.307
20	400	4.472	70	4900	8.367
21	441	4.583	71	5041	8.426
22	484	4.690	72	5184	8.485
23	529	4.796	73	5329	8.544
24	576	4.899	74	5476	8.602
25	625	5.000	75	5625	8.660
26	676	5.099	76	5776	8.718
27	729	5.196	77	5929	8.775
28	784	5.292	78	6084	8.832
29	841	5.385	79	6241	8.888
30	900	5.477	80	6400	8.944
31	961	5.568	81	6561	9.000
32	1024	5.657	82	6724	9.055
33	1089	5.745	83	6889	9.110
34	1156	5.831	84	7056	9.165
35	1225	5.916	85	7225	9.220
36	1296	6.000	86	7396	9.274
37	1369	6.083	87	7569	9.327
38	1444	6.164	88	7744	9.381
39	1521	6.245	89	7921	9.434
40	1600	6.325	90	8100	9.487
41	1681	6.403	91	8281	9.539
42	1764	6.481	92	8464	9.592
43	1849	6.557	93	8649	9.644
44	1936	6.633	94	8836	9.695
45	2025	6.708	95	9025	9.747
46	2116	6.782	96	9216	9.798
47	2209	6.856	97	9409	9.849
48	2304	6.928	98	9604	9.899
49	2401	7.000	99	9801	9.950
50	2500	7.071	100	10,000	10.000

A.4 Geometric Formulas

Rectangle

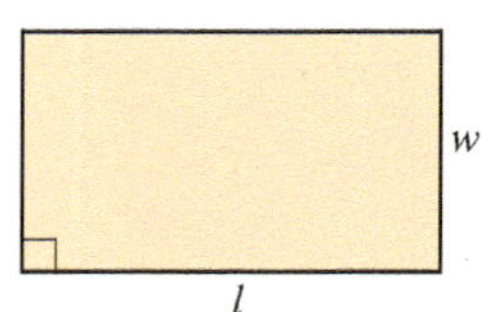

Perimeter: $P = 2l + 2w$
Area: $A = lw$

Square

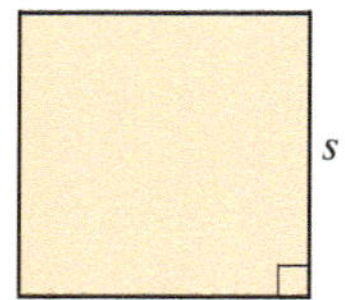

Perimeter: $P = 4s$
Area: $A = s^2$

Triangle

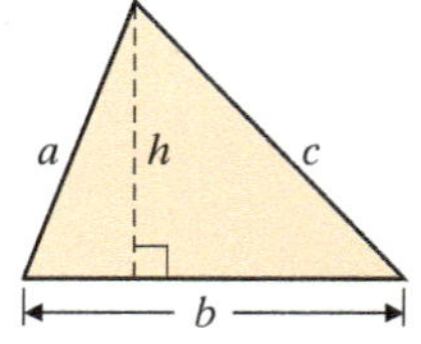

Perimeter: $P = a + b + c$
Area: $A = \frac{1}{2}bh$

Sum of Angles of Triangle

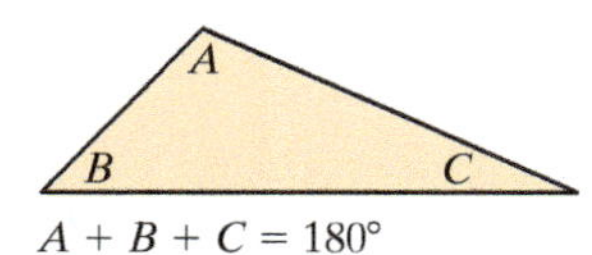

$A + B + C = 180°$
The sum of the measures of the three angles is 180°.

Pythagorean Theorem (for right triangles)

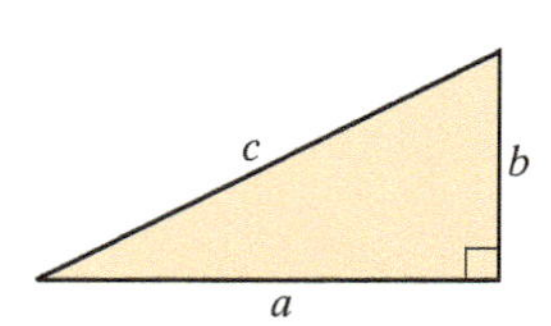

Perimeter: $P = a + b + c$
Area: $A = \frac{1}{2}ab$
One 90° (right) angle

Isosceles Triangle

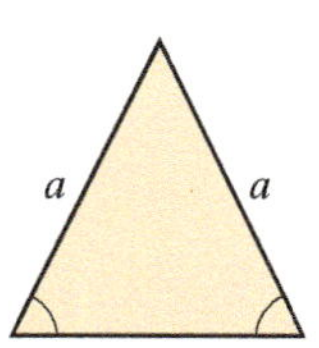

Triangle has:
two equal sides and
two equal angles.

Equilateral Triangle

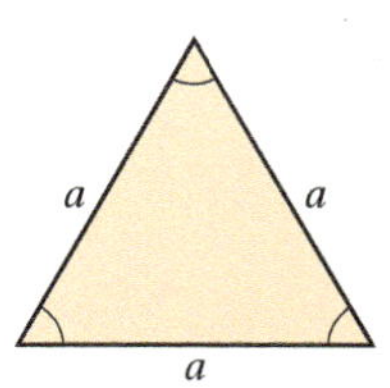

Triangle has:
three equal sides and
three equal angles.
Measure of each angle is 60°.

Trapezoid

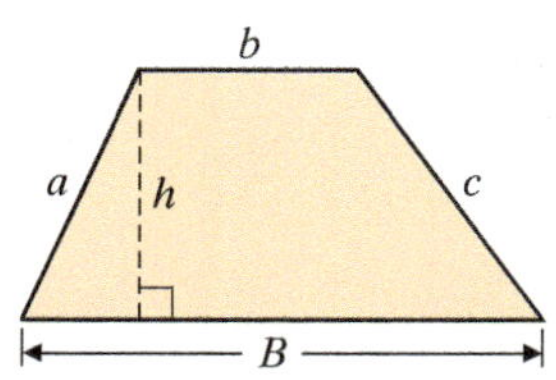

Perimeter: $P = a + b + c + B$
Area: $A = \frac{1}{2}h(B + b)$

Parallelogram

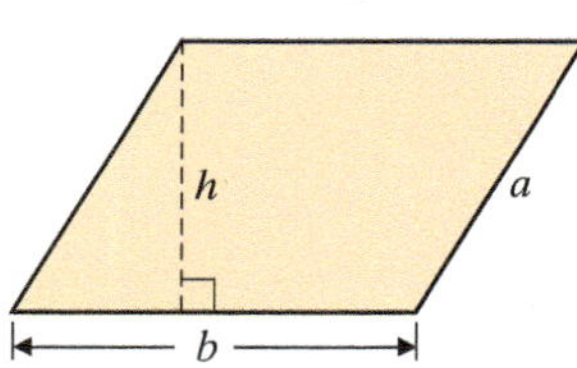

Perimeter: $P = 2a + 2b$
Area: $A = bh$

Circle

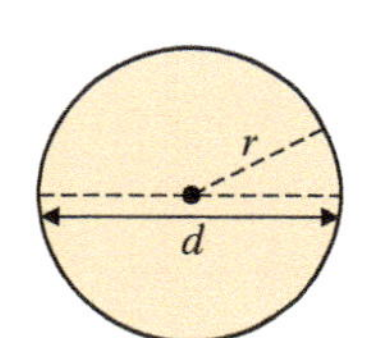

Circumference: $C = \pi d$
$C = 2\pi r$
Area: $A = \pi r^2$

Rectangular Solid

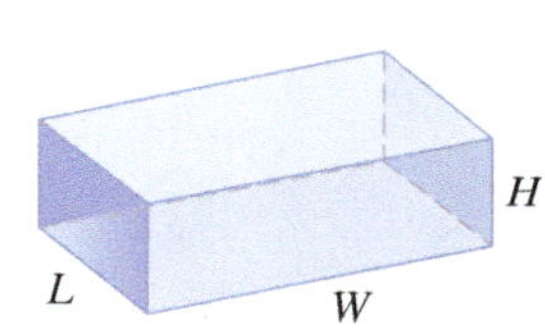

Volume: $V = LWH$
Surface Area:
$S = 2LW + 2HL + 2HW$

Cube

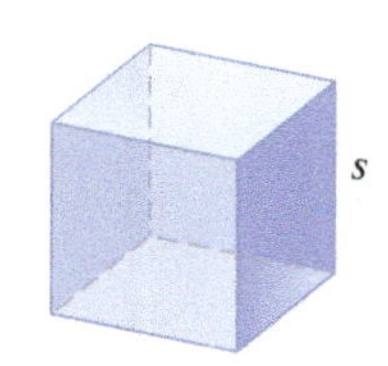

Volume: $V = s^3$
Surface Area: $S = 6s^2$

Cone

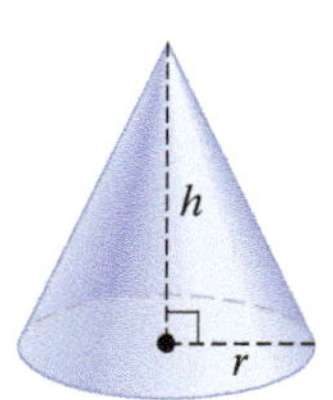

Volume: $V = \frac{1}{3}\pi r^2 h$
Surface Area:
$S = \pi r\sqrt{r^2 + h^2} + \pi r^2$

Right Circular Cylinder

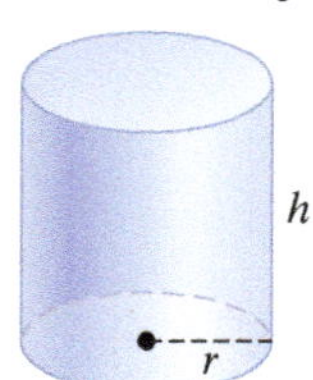

Volume: $V = \pi r^2 h$
Surface Area: $S = 2\pi r^2 + 2\pi rh$

Sphere

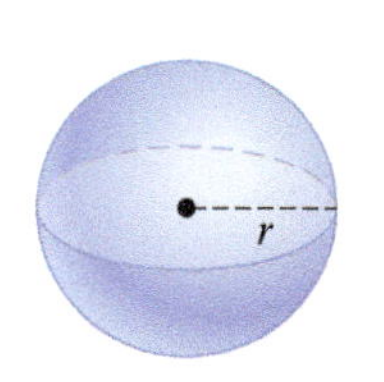

Volume: $V = \frac{4}{3}\pi r^3$
Surface Area: $S = 4\pi r^2$

Square-Based Pyramid

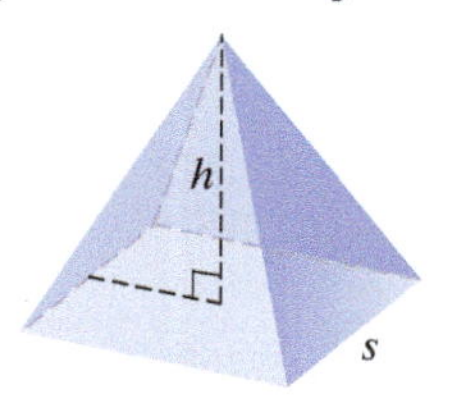

Volume: $V = \frac{1}{3}s^2 h$

Appendix

B

Factoring Sums and Differences of Cubes

Objective A Factoring the Sum or Difference of Cubes

Objective

A Factor the Sum or Difference of Cubes.

Although the sum of two squares usually does not factor, the sum or difference of two cubes can be factored and reveal factoring patterns. The pattern for the sum of cubes can be checked by multiplying the binomial $x + y$ and the trinomial $x^2 - xy + y^2$. The pattern for the difference of two cubes can be checked by multiplying the binomial $x - y$ by the trinomial $x^2 + xy + y^2$.

Sum or Difference of Two Cubes

$$a^3 + b^3 = (a + b)(a^2 - ab + b^2)$$
$$a^3 - b^3 = (a - b)(a^2 + ab + b^2)$$

Example 1 Factor $x^3 + 8$.

Solution:

First, write the binomial in the form $a^3 + b^3$.

$x^3 + 8 = x^3 + 2^3$ Write in the form $a^3 + b^3$.

If we replace a with x and b with 2 in the formula above, we have

$$x^3 + 2^3 = (x + 2)[x^2 - (x)(2) + 2^2]$$
$$= (x + 2)(x^2 - 2x + 4)$$

Work Practice 1

Practice 1

Factor $x^3 + 27$.

Helpful Hint

When factoring sums or differences of cubes, notice the sign patterns.

$$x^3 + y^3 = (x + y)(x^2 - xy + y^2)$$

same sign; opposite signs; always positive

$$x^3 - y^3 = (x - y)(x^2 + xy + y^2)$$

same sign; opposite signs; always positive

Example 2 Factor $y^3 - 27$.

Solution:

$$y^3 - 27 = y^3 - 3^3 \quad \text{Write in the form } a^3 - b^3.$$
$$= (y - 3)[y^2 + (y)(3) + 3^2]$$
$$= (y - 3)(y^2 + 3y + 9)$$

Work Practice 2

Practice 2

Factor $z^3 - 8$.

Answers

1. $(x + 3)(x^2 - 3x + 9)$
2. $(z - 2)(z^2 + 2z + 4)$

Practice 3

Factor $125a^3 + 1$.

Practice 4

Factor $54x^3 - 128y^3$

Answers

3. $(5a + 1)(25a^2 - 5a + 1)$
4. $2(3x - 4y)(9x^2 + 12xy + 16y^2)$

Example 3 Factor $64x^3 + 1$.

Solution:

$$\begin{aligned} 64x^3 + 1 &= (4x)^3 + 1^3 \\ &= (4x + 1)[(4x)^2 - (4x)(1) + 1^2] \\ &= (4x + 1)(16x^2 - 4x + 1) \end{aligned}$$

Work Practice 3

Example 4 Factor $54a^3 - 16b^3$.

Solution: Remember to factor out common factors first before using other factoring methods.

$$\begin{aligned} 54a^3 - 16b^3 &= 2(27a^3 - 8b^3) && \text{Factor out the GCF, 2.} \\ &= 2[(3a)^3 - (2b)^3] && \text{Difference of two cubes} \\ &= 2(3a - 2b)[(3a)^2 + (3a)(2b) + (2b)^2] \\ &= 2(3a - 2b)(9a^2 + 6ab + 4b^2) \end{aligned}$$

Work Practice 4

B Exercise Set MyLab Math

Objective A *Factor the binomials completely. See Examples 1 through 4.*

1. $a^3 + 27$
2. $b^3 - 8$
3. $8a^3 + 1$
4. $64x^3 - 1$
5. $5k^3 + 40$
6. $6r^3 - 162$
7. $x^3y^3 - 64$
8. $8x^3 - y^3$
9. $x^3 + 125$
10. $a^3 - 216$
11. $24x^4 - 81xy^3$
12. $375y^6 - 24y^3$
13. $27 - t^3$
14. $125 + r^3$
15. $8r^3 - 64$
16. $54r^3 + 2$
17. $t^3 - 343$
18. $s^3 + 216$
19. $s^3 - 64t^3$
20. $8t^3 + s^3$

Appendix C

Mean, Median, Mode, Range, and Introduction to Statistics

We are certainly familiar with the word "average." The next examples show real-life averages. For example,

- The average cost of a pie (or whole pizza) is $16.73.
- Adults employed in the United States report working an **average** of 47 hours per week, according to a Gallup poll.
- The U.S. miles per gallon **average** for light vehicles is 25.5, according to autonews.com (*Automotive News*).

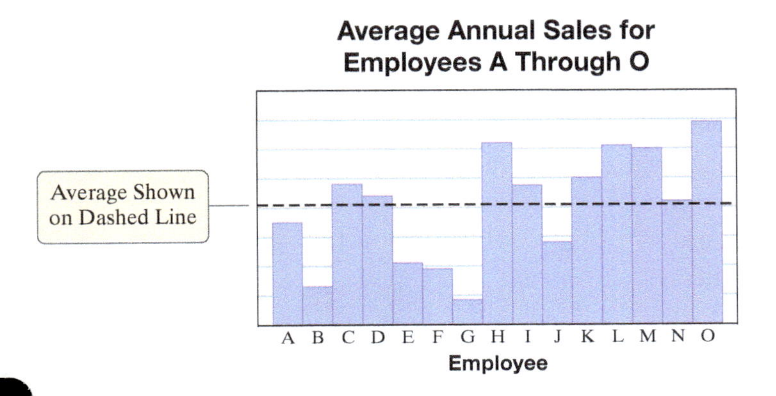

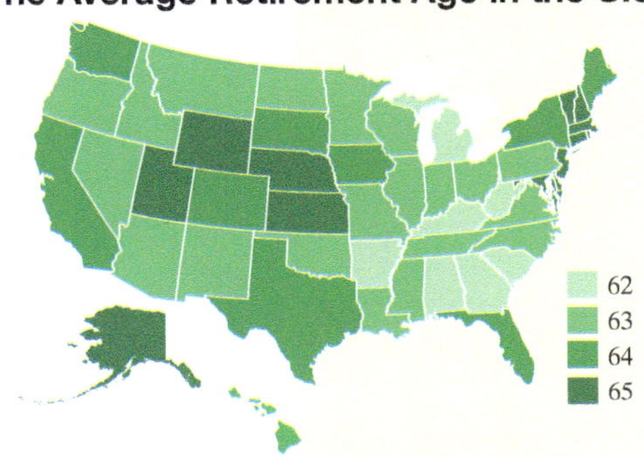

Based on U.S. Census Bureau labor force participation data.

As our accumulation of data increases, our ability to gather, store, and present these tremendous amounts of data increases. Sometimes it is desirable to be able to describe a set of data by a single "middle" number or a measure of central tendency. Three of the most common **measures of central tendency** are the **mean** (or average), the **median**, and the **mode**.

Objectives

A Find the Mean (or Average) of a List of Numbers.

B Find the Median of a List of Numbers.

C Find the Mode of a List of Numbers.

D Calculate Range, Mean, Median, and Mode from a Frequency Distribution Table or Graph.

Objective A Finding the Mean 

The most common measure of central tendency is the mean (sometimes called the "arithmetic mean" or the "average").

The **mean (average)** of a set of number items is the sum of the items divided by the number of items.

$$\text{mean} = \frac{\text{sum of items}}{\text{number of items}}$$

For example: To find the mean of four test scores—86, 82, 93, and 75—we find the sum of the scores and then divide by the number of scores, 4.

$$\text{mean} = \frac{86 + 82 + 93 + 75}{4} = \frac{336}{4} = 84$$

Notice that by looking at a bar graph of the scores with a dashed line at the mean, it is reasonable that 84 is one *measure of central tendency*.

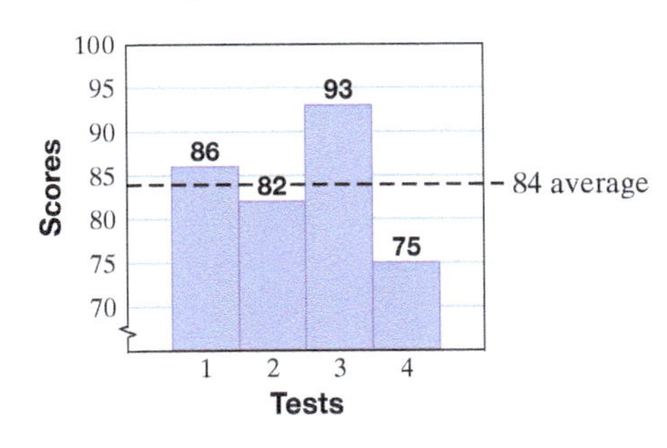

Practice 1

Find the mean of the following test scores: 87, 75, 96, 91, and 78.

Example 1 Finding the Mean Time in an Experiment

Seven students in a psychology class conducted an experiment on mazes. Each student was given a pencil and asked to successfully complete the same maze. The timed results are below:

Student	Ann	Thanh	Carlos	Jesse	Melinda	Ramzi	Dayni
Time (Seconds)	13.2	11.8	10.7	16.2	15.9	13.8	18.5

a. Who completed the maze in the shortest time? Who completed the maze in the longest time?

b. Find the mean time.

c. How many students took longer than the mean time? How many students took shorter than the mean time?

Solution:

a. Carlos completed the maze in 10.7 seconds, the shortest time. Dayni completed the maze in 18.5 seconds, the longest time.

b. To find the mean (or average), we find the sum of the items and divide by 7, the number of items.

$$\text{mean} = \frac{\text{sum of items}}{\text{number of items}} = \frac{13.2 + 11.8 + 10.7 + 16.2 + 15.9 + 13.8 + 18.5}{7}$$

$$= \frac{100.1}{7} = 14.3$$

c. Three students, Jesse, Melinda, and Dayni, had times longer than the mean time. Four students, Ann, Thanh, Carlos, and Ramzi, had times shorter than the mean time.

Work Practice 1

✓**Concept Check** Estimate the mean of the following set of data:

5, 10, 10, 10, 10, 15

The mean has one main disadvantage. This measure of central tendency can be greatly affected by *outliers*. (Outliers are values that are especially large or small when compared with the rest of the data set.) Let's see an example of this next.

Practice 2

Use the table in Example 2 and find the mean salary of all staff members except G. Round thousands of dollars to 2 decimal places.

Example 2

The table lists the rounded salary of 10 staff numbers.

Staff	A	B	C	D	E	F	G	H	I	J
Salary (in thousands)	$32	$34	$46	$38	$42	$95	$102	$50	$42	$41

a. Find the mean of all 10 staff members.

b. Find the mean of all staff members except F and G.

Solution

a. $$\text{mean} = \frac{32 + 34 + 46 + 38 + 42 + 95 + 102 + 50 + 42 + 41}{10} = \frac{522}{10}$$

$$= 52.2$$

The mean salary is $52.2 thousand or $52,200.

Answers

1. 85.4 **2.** $46.67 thousand or $46,670

✓**Concept Check Answer**

10

b. $\text{mean} = \dfrac{32 + 34 + 46 + 38 + 42 + 50 + 42 + 41}{8} = \dfrac{325}{8}$

$= 40.625$

Now, the mean salary is \$40.625 thousand or \$40,625.

Work Practice 2

The mean in part **a** does not appear to be a measure of central tendency because this mean, \$52.2 thousand, is greater than all salaries except 2 of the 10. Also, notice the difference in the means for parts **a** and **b.** By removing the 2 outliers, the mean was greatly reduced.

Although the mean was calculated correctly each time, parts **a** and **b** of Example 2 show one disadvantage of the mean. That is, a few numerical outliers can greatly affect the mean.

Later in this section, we will discuss the range of a data set as well as calculate measures of central tendency from frequency distribution tables and graphs.

Helpful Hint

Remember an important disadvantage of the mean:
If our data set has a few outliers, the mean may not be the best measure of central tendency.

Often in college, the calculation of a **grade point average** (GPA) is a **weighted mean** and is calculated as shown in Example 3.

Example 3 Calculating Grade Point Average (GPA)

The following grades were earned by a student during one semester. Find the student's grade point average.

Course	Grade	Credit Hours
College mathematics	A	3
Biology	B	3
English	A	3
PE	C	1
Social studies	D	2

Solution: To calculate the grade point average, we need to know the point values for the different possible grades. The point values of grades commonly used in colleges and universities are given below:

A: 4, B: 3, C: 2, D: 1, F: 0

Now, to find the grade point average, we multiply the number of credit hours for each course by the point value of each grade. The grade point average is the sum of these products divided by the sum of the credit hours.

Course	Grade	Point Value of Grade	Credit Hours	Point Value of Credit Hours
College mathematics	A	4	3	12
Biology	B	3	3	9
English	A	4	3	12
PE	C	2	1	2
Social studies	D	1	2	2
		Totals:	12	37

(Continued on next page)

Practice 3

Find the grade point average if the following grades were earned in one semester. Round to 2 decimal places.

Grade	Credit Hours
A	2
B	4
C	5
D	2
A	2

Answer

3. 2.67

$$\text{grade point average} = \frac{37}{12} \approx 3.08 \text{ rounded to two decimal places}$$

The student earned a grade point average of 3.08.

Work Practice 3

Objective B Finding the Median

You may have noticed that a very low number or a very high number can affect the mean of a list of numbers. Because of this, you may sometimes want to use another measure of central tendency. A second measure of central tendency is called the **median**. The median of a list of numbers is not affected by a low or high number in the list.

The **median** of a set of numbers in numerical order is the middle number. If the number of items is odd, the median is the middle number. If the number of items is even, the median is the mean of the two middle numbers.

Practice 4

Find the median of the list of numbers: 5, 11, 14, 23, 24, 35, 38, 41, 43

Example 4 Find the median of the following list of numbers:

25, 54, 56, 57, 60, 71, 98

Solution: Because this list is in numerical order, the median is the middle number, 57.

Work Practice 4

Practice 5

Find the median of the list of scores:
36, 91, 78, 65, 95, 95, 88, 71

Example 5 Find the median of the following list of scores: 67, 91, 75, 86, 55, 91

Solution: First we list the scores in numerical order and then we find the middle number.

55, 67, 75, 86, 91, 91

Since there is an even number of scores, there are two middle numbers, 75 and 86. The median is the mean of the two middle numbers.

$$\text{median} = \frac{75 + 86}{2} = 80.5$$

The median is 80.5.

Helpful Hint Don't forget to write the numbers in order from smallest to largest before finding the median.

Work Practice 5

Objective C Finding the Mode

The last common measure of central tendency is called the **mode**.

The **mode** of a set of numbers is the number that occurs most often. (It is possible for a set of numbers to have more than one mode or to have no mode.)

Practice 6

Find the mode of the list of numbers:
14, 10, 10, 13, 15, 15, 15, 17, 18, 18, 20

Example 6 Find the mode of the list of numbers:

11, 14, 14, 16, 31, 56, 65, 77, 77, 78, 79

Solution: There are two numbers that occur the most often. They are 14 and 77. This list of numbers has two modes, 14 and 77.

Work Practice 6

Answers
4. 24 **5.** 83 **6.** 15

Example 7 Find the median and the mode of the following set of numbers. These numbers were high temperatures for 14 consecutive days in a city in Montana.

76, 80, 85, 86, 89, 87, 82, 77, 76, 79, 82, 89, 89, 92

Solution: First we write the numbers in numerical order.

76, 76, 77, 79, 80, 82, 82, 85, 86, 87, 89, 89, 89, 92

Since there is an even number of items, the median is the mean of the two middle numbers, 82 and 85.

$$\text{median} = \frac{82 + 85}{2} = 83.5$$

The mode is 89, since 89 occurs most often.

Work Practice 7

Practice 7

Find the median and the mode of the list of numbers:
26, 31, 15, 15, 26, 30, 16, 18, 15, 35

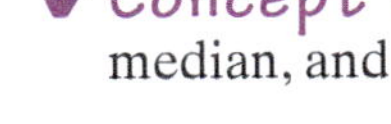

Concept Check True or false? Every set of numbers *must* have a mean, median, and mode. Explain your answer.

Helpful Hint

Don't forget that it is possible for a list of numbers to have no mode. For example, the list

2, 4, 5, 6, 8, 9

has no mode. There is no number or numbers that occur more often than the others.

Objective D Finding the Range of a Data Set and Reviewing Mean, Median, and Mode

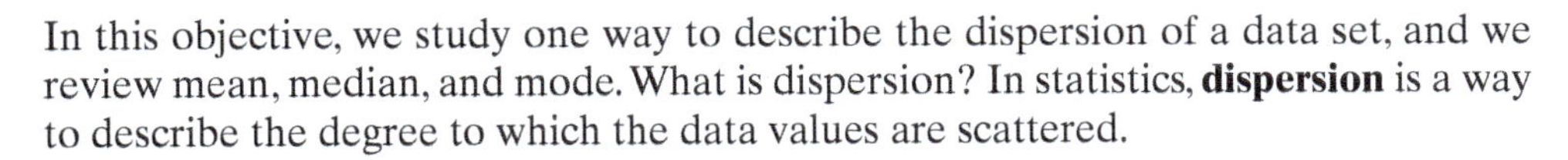

In this objective, we study one way to describe the dispersion of a data set, and we review mean, median, and mode. What is dispersion? In statistics, **dispersion** is a way to describe the degree to which the data values are scattered.

Range

The range of a data set is the difference between the largest data value and the smallest data value.

range = largest data value − smallest data value

Example 8 The following pulse rates (for 1 minute) were recorded for a group of 15 students. Find the range.

78, 80, 66, 68, 71, 64, 82, 71, 70, 65, 70, 75, 77, 86, 72.

Solution: range = largest data value − smallest data value
= 86 − 64
= 22

The range of this data set is 22.

Work Practice 8

Practice 8

The table lists the rounded salary of 9 staff members. Find the range.

Staff	*Salary (in thousands)*
A	$32
B	$34
C	$46
D	$38
E	$42
F	$95
G	$102
H	$50
J	$42

Answers

7. median: 22; mode: 15 **8.** $70 thousand

Concept Check Answer

false; a set of numbers may have no mode

Let's recall a few facts about the median, and then we will introduce a formula for finding the *position* of the median.

- The **median** of a set of numbers in numerical order is the middle number.
- If the number of items is odd, the median is the middle number.
- If the number of items is even, the median is the *mean* (average) of the two middle numbers.

For a long list of data items, this formula gives us the **position** of the median.

Position of the Median

For n data items in order from smallest to largest, the median is the item in the

$$\frac{n+1}{2} \text{ position.}$$

Note:
If n is an even number, then the position formula, $\frac{n+1}{2}$, will not be a whole number.
In this case, simply find the average of the two data items whose positions are closest to, but before and after $\frac{n+1}{2}$.

Helpful Hint

The formula above, $\frac{n+1}{2}$ does not give the *value* of the median, just the **position of the median.**

Practice 9

One state with a young retirement age is Michigan. The table below is from a poll of retirement ages from that state.

Age	*Frequency*
50	1
59	3
60	3
62	5
63	2
67	1

Find the (a) range, (b) mean, (c) median, and (d) mode of these data. If needed, round answers to 1 decimal place.

Example 9 Find the (a) range, (b) mean, (c) median, and (d) mode from this frequency distribution table of retirement ages. If needed, round answers to 1 decimal place.

Age	*Frequency*
60	3
61	1
62	1
63	2
64	2
65	2

Solution: Study the table for a moment. From the frequency column, we see that there are 11 data items $(3 + 1 + 1 + 2 + 2 + 2)$.

a. range = largest data value − smallest data value

$$= 65 - 60$$
$$= 5$$

The range of this data set is 5.

Answers

9. a. range: 17 **b.** mean: 60.7, **c.** median: 62, **d.** mode: 62

b. To find the mean, we use our mean formula:

$$\text{mean} = \frac{\text{sum of items}}{\text{number of items}} = \frac{3 \cdot 60 + 61 + 62 + 2 \cdot 63 + 2 \cdot 64 + 2 \cdot 65}{11}$$

$$= \frac{687}{11} \approx 62.5$$

Helpful Hint Since there are three 60's, for example, we can either use:

60 + 60 + 60 or 3 · 60.

The mean of the data set is approximately 62.5.

c. Since there are 11 data items and the items are arranged in numerical order in the table, we find $\frac{n+1}{2}$ to locate the middle item. This is $\frac{11+1}{2} = \frac{12}{2} = 6$, or the sixth item.

The median is the sixth number, or 63.

d. The mode has the greatest frequency, so the mode is 60.

Work Practice 9

C Exercise Set MyLab Math

Objectives A B C Mixed Practice *For each set of numbers, find the mean, median, and mode. If necessary, round the mean to one decimal place. See Examples 1, 2, and 4 through 6.*

1. 15, 23, 24, 18, 25

2. 45, 36, 28, 46, 52

3. 7.6, 8.2, 8.2, 9.6, 5.7, 9.1

4. 4.9, 7.1, 6.8, 6.8, 5.3, 4.9

5. 0.5, 0.2, 0.2, 0.6, 0.3, 1.3, 0.8, 0.1, 0.5

6. 0.6, 0.6, 0.8, 0.4, 0.5, 0.3, 0.7, 0.8, 0.1

7. 231, 543, 601, 293, 588, 109, 334, 268

8. 451, 356, 478, 776, 892, 500, 467, 780

The 10 tallest buildings in the world, completed as of 2017, are listed in the following table. Use this table to answer Exercises 9 through 14. If necessary, round results to one decimal place. See Examples 1, 2, and 4 through 6.

9. Find the mean height of the five tallest buildings.

10. Find the median height of the five tallest buildings.

11. Find the median height of the six tallest buildings.

12. Find the mean height of the six tallest buildings.

Building	Height in Feet
Burj Khalifa, Dubai	2717
Shanghai Tower, Shanghai	2073
Makkah Royal Clock Tower, Mecca	1972
Ping An Finance Center	1965
Lotte World Tower	1819
One World Trade Center, New York City	1776
Guangzhou CTF Finance Center, Guangzhou	1739
Taipei 101, Taipei	1667
Shanghai World Financial Center, Shanghai	1614
International Commerce Center, Hong Kong	1588

Source: Council on Tall Buildings and Urban Habitat

13. Given the building heights, explain how you know, without calculating, that the answer to Exercise **10** is greater than the answer to Exercise **11**.

14. Given the building heights, explain how you know, without calculating, that the answer to Exercise **12** is less than the answer to Exercise **9**.

For Exercises 15 through 18, the grades are given for a student for a particular semester. Find the grade point average. If necessary, round the grade point average to the nearest hundredth. See Example 3.

15.

Grade	Credit Hours
B	3
C	3
A	4
C	4

16.

Grade	Credit Hours
D	1
F	1
C	4
B	5

17.

Grade	Credit Hours
A	3
A	3
A	4
B	3
C	1

18.

Grade	Credit Hours
B	2
B	2
C	3
A	3
B	3

During an experiment, the following times (in seconds) were recorded:

7.8, 6.9, 7.5, 4.7, 6.9, 7.0.

19. Find the mean.

20. Find the median.

21. Find the mode.

In a mathematics class, the following test scores were recorded for a student:

93, 85, 89, 79, 88, 91.

22. Find the mean.

23. Find the median.

24. Find the mode.

The following pulse rates were recorded for a group of 15 students:

78, 80, 66, 68, 71, 64, 82, 71, 70, 65, 70, 75, 77, 86, 72.

25. Find the mean.

26. Find the median.

27. Find the mode.

28. How many pulse rates were higher than the mean?

29. How many pulse rates were lower than the mean?

30. Explain how to find the position of the median.

Below are lengths for the six longest rivers in the world.

Name	*Length (miles)*
Nile	4160
Amazon	4000
Yangtze	3915
Mississippi-Missouri	3709
Ob-Irtysh	3459
Huang Ho	3395

Find the mean and the median for each of the following.

31. the six longest rivers

32. the three longest rivers

Objective D *Find the range for each data set. See Example 8.*

33. 14, 16, 8, 10, 20

34. 25, 15, 11, 40, 37

35. 206, 206, 555, 556

36. 129, 188, 188, 276

37. 9, 9, 9, 9, 11

38. 7, 7, 7, 7, 10

39.

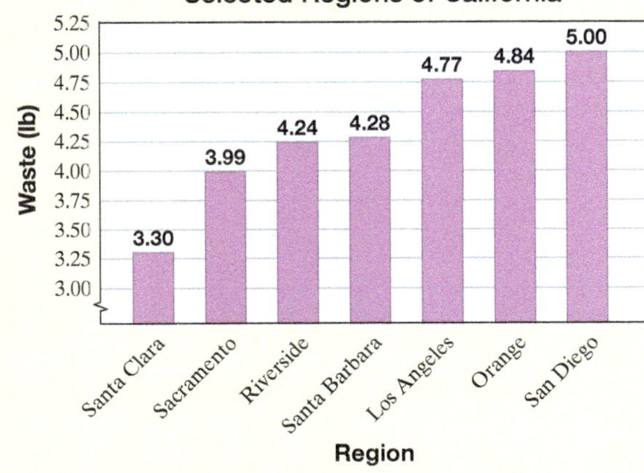

Sources: Equinox Center, 2013; Calrecycle, 2013

40.

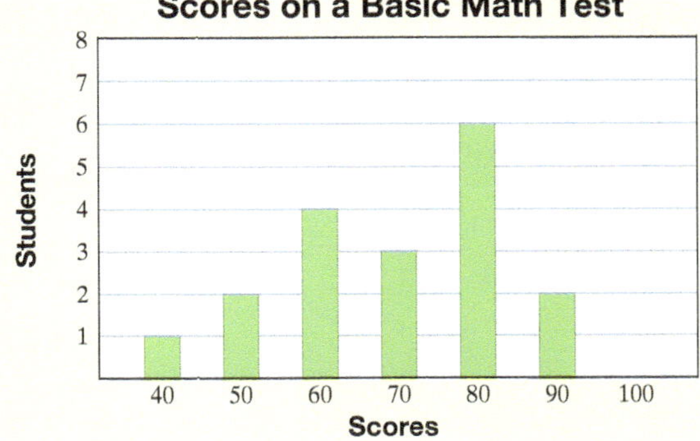

Use each frequency distribution table to find the **a.** *mean,* **b.** *median, and* **c.** *mode. If needed, round the mean to 1 decimal place. See Example 9.*

41.

Data Item	*Frequency*
5	1
6	1
7	2
8	5
9	6
10	2

42.

Data Item	*Frequency*
3	2
4	1
5	4
6	7
7	2
8	1

43.

Data Item	*Frequency*
2	5
3	7
4	4
5	7
6	8
7	8
8	8
9	6
10	5

44.

Data Item	*Frequency*
4	3
5	8
6	5
7	8
8	2

*Use each graph of data items to find the **a.** mean, **b.** median, and **c.** mode. If needed round the mean to one decimal place.*

45.

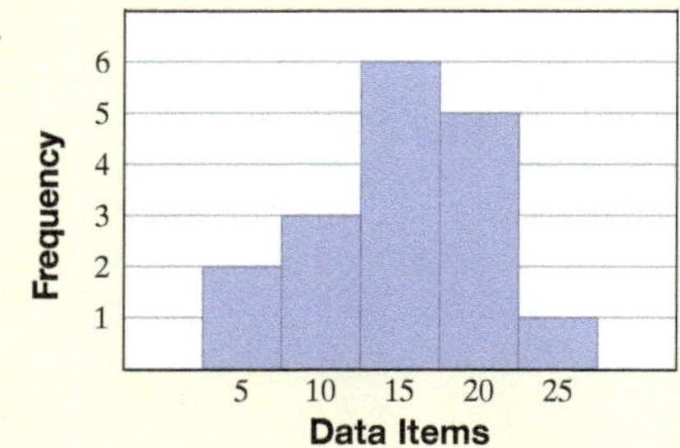

46.

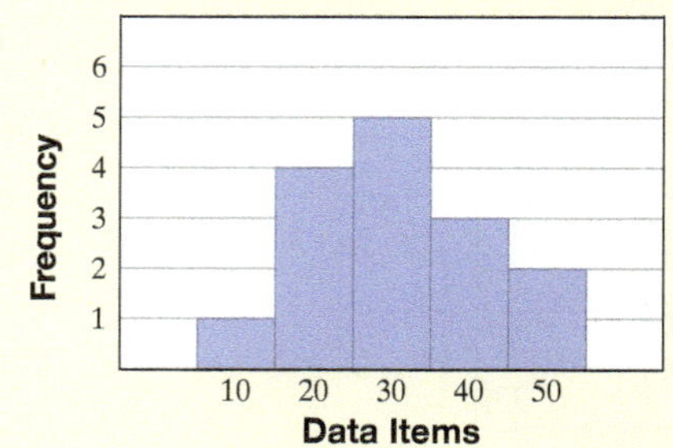

47.

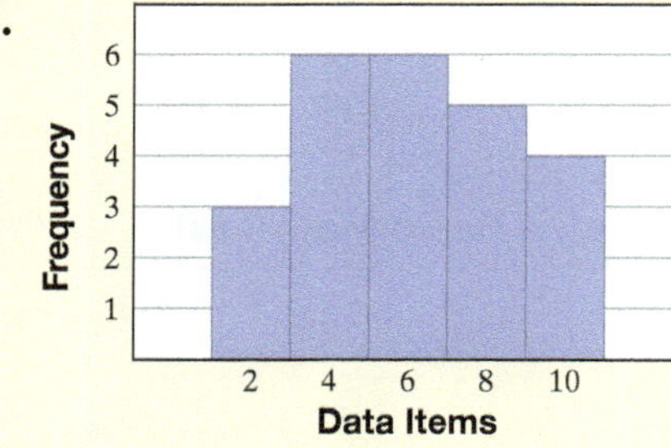

48.

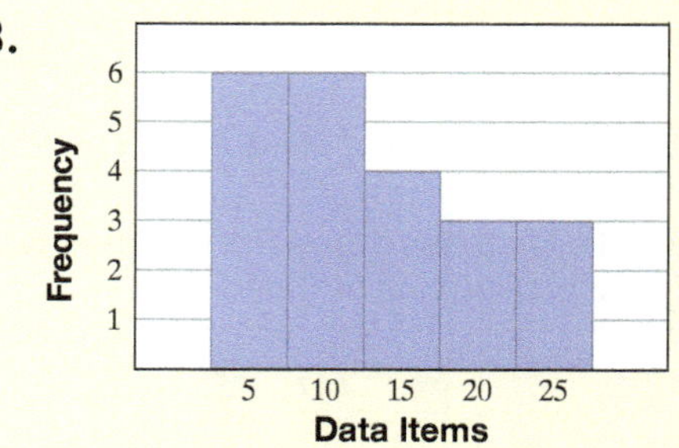

Appendix

Sets

Objectives

A Determine Whether an Object Is an Element of a Set.

B Determine Whether a Set Is a Subset of Another Set.

C Find Unions and Intersections of Sets.

Objective A Determining Whether an Object Is an Element of a Set

A **set** is a collection of objects, called **elements** or **members**. The elements of a set are listed or described between a pair of braces, { }. Two common ways of representing a set are by **roster form** or by **set builder** notation. Examples of each are as follows:

$A = \{1, 2, 3, 4, 5\}$ Roster form—elements are listed.

$B = \{x \mid x \leq 3\}$ Set builder notation

Set B is read as "the set of all x such that x is less than or equal to 3." The symbol $\in$ means "is an element of." For set A above, we can write

$3 \in A$,

which means the number 3 is an element of set A.

Also,

$6 \notin A$

means the number 6 is not an element of set A.

A set that has no elements is called the **empty set** or the **null set**. The empty set (or null set) is symbolized by { } or $\varnothing$. For example, if set C is the set of all positive numbers less than 0, then

$C = \{\ \}$ or $C = \varnothing$.

Helpful Hint

The set $\{\varnothing\}$ is *not* the empty set. It is a set with one element, $\varnothing$.

Example 1 Given the following sets, determine whether each statement is true or false.

$N = \{0, 1, 2, 3, 4, 5\}$ $E = \{0, 2, 4, 6, 8, 10\}$ $O = \{x \mid x \text{ is an odd number}\}$

a. $2 \in N$
b. $5 \in E$
c. $5 \in O$
d. $10 \in O$

Solution:

a. True, since 2 is a listed element of N
b. False, since 5 is not listed as an element of E
c. True, since 5 is an odd number
d. False, since 10 is not an odd number

Work Practice 1

Practice 1

Use the sets from Example 1 and determine whether each statement is true or false.

a. $2 \in O$
b. $2 \in E$
c. $6 \in N$
d. $6 \in E$

Answers

1. a. false **b.** true **c.** false **d.** true

Objective B Determining Whether a Set Is a Subset of Another Set

Set A is a subset of set B if every element of A is also an element of B. In symbols, we write $A \subseteq B$, and this is illustrated to the right. The empty set is a subset of every set. In symbols, $\varnothing \subseteq A$ is true for every set A.

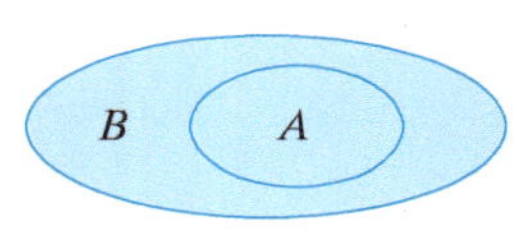

$A \subseteq B$ means set A is a subset of set B. The symbol $\not\subseteq$ means "is not a subset."

Thus,

$A \not\subseteq B$ means set A is not a subset of set B.

Helpful Hint

If $A \not\subseteq B$ is true, then there must be at least one element of set A that is not in set B.

Practice 2

Use the same sets in Example 2, parts **a**, **b**, **c**, and **d**, but this time, determine whether

$B \subseteq A$

Example 2 Determine whether $A \subseteq B$.

a. $A = \{0, 7\}, B = \{0, 1, 2, 3, 4, 5, 6, 7\}$
b. $A = \{2, 4, 6\}, B = \{2, 4, 6\}$
c. $A = \{x|x < 5\}, B = \{x|x < 1\}$
d. $A = \{\bigcirc, \triangle, \square\}, B = \{\bigcirc, \triangle\}$

Solution:

a. $A \subseteq B$ since every element of A is also an element of B.
b. $A \subseteq B$ since every element of A is also an element of B.
c. $A \not\subseteq B$ because there are elements of A that are not in B. For example, 3 is an element of A but is not an element of B.
d. $A \not\subseteq B$ because $\square \in A$, but $\square \notin B$.

Work Practice 2

Helpful Hint

From Example 2b, we see that every set is a subset of itself. For example, $A \subseteq A$. Why? Because every element of A is always an element of A.

Objective C Finding Unions and Intersections of Sets

The **union** of two sets A and B, written as $A \cup B$, is the set of all elements that are in set A or in set B (or in both sets).

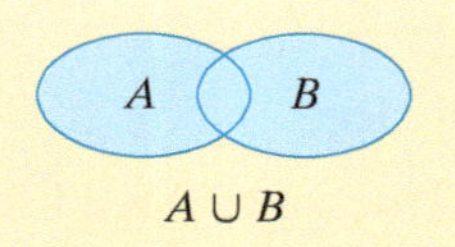

The **intersection** of two sets A and B, written as $A \cap B$, is the set of all elements that are common to both set A and set B.

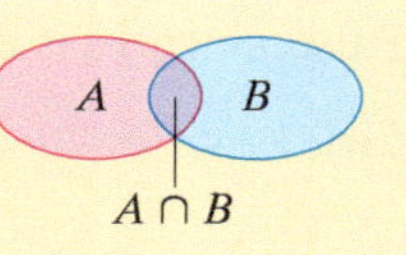

Answers
2. a. $B \not\subseteq A$ **b.** $B \subseteq A$ **c.** $B \subseteq A$ **d.** $B \subseteq A$

Example 3 Let $R = \{0, 2, 5\}$ and $S = \{0, 1, 2, 3, 4\}$. Find

a. $R \cup S$
b. $R \cap S$

Solution:

a. $R \cup S = \{0, 1, 2, 3, 4, 5\}$, all the numbers in set R or set S (or both)
b. $R \cap S = \{0, 2\}$, the only numbers in both sets

Work Practice 3

Practice 3

Let $P = \{2, 4, 6, 9, 12\}$ and $Q = \{9, 12, 15\}$. Find

a. $P \cup Q$
b. $P \cap Q$

Example 4 Let $M = \{10, 20\}$ and $N = \{5, 15, 25\}$. Find

a. $M \cup N$
b. $M \cap N$

Solution:

a. $M \cup N = \{5, 10, 15, 20, 25\}$ Any order of the elements is fine.
b. $M \cap N = \{\ \}$ or $\varnothing$ There are no elements common to both sets. The intersection is the empty set.

Work Practice 4

Practice 4

Let $T = \{3, 30\}$ and $W = \{40, 50, 60\}$. Find

a. $T \cup W$
b. $T \cap W$

Helpful Hint

- You may list the elements of a set in any order.
- There is no need to list an element of a set more than once.

Answers

3. a. $P \cup Q = \{2, 4, 6, 9, 12, 15\}$ **b.** $P \cap Q = \{9, 12\}$
4. a. $T \cup W = \{3, 30, 40, 50, 60\}$ **b.** $T \cap W = \{\ \}$ or $\varnothing$

D Exercise Set MyLab Math

For each set described, write the set in roster form.

1. The set of negative integers from -10 to -5.

2. The set of integers between -1 and 1.

3. The set of the days of the week starting with the letter T.

4. The set of the first five letters of the alphabet

5. The set of whole numbers

6. The set of natural numbers

7. $\{x \mid x \text{ is an integer between 1 and 2}\}$

8. $\{x \mid x \text{ is a number that is both even and odd}\}$

Objectives A B Mixed Practice *Determine whether each statement is true or false. See Examples 1 and 2.*

9. $3 \in \{1, 3, 5, 7, 9\}$

10. $6 \in \{1, 3, 5, 7, 9\}$

11. $\{3\} \subseteq \{1, 3, 5, 7, 9\}$

12. $\{6\} \subseteq \{1, 3, 5, 7, 9\}$

13. $\{a, e, i, o, u\} \subseteq \{a, e, i, o, u\}$

14. $\{\triangle, \square, \bigcirc\} \subseteq \{\triangle, \square, \bigcirc\}$

15. $\{\text{May}\} \subseteq$ the set of days of the week

16. $\{\text{Sunday}\} \subseteq$ the set of months of the year

17. $9 \notin \{x \mid x \text{ is an even number}\}$

18. $10 \notin \{x \mid x \text{ is an odd number}\}$

19. $\{a\} \not\subseteq$ the set of vowels

20. $\{\triangle\} \not\subseteq$ the set of polygons

Objective C *Given the sets, find each union or intersection. See Examples 3 and 4.*

$A = \{1, 2, 3, 4, 5, 6\}$ $B = \{2, 4, 6\}$ $C = \{1, 3, 5\}$ $D = \{7\}$

21. $A \cup B$

22. $A \cup C$

23. $A \cap B$

24. $A \cap C$

25. $C \cup D$

26. $C \cap D$

27. $B \cap D$

28. $B \cup D$

Objectives B C Mixed Practice *Use sets A, B, C, and D above to determine whether each statement is true or false. See Examples 2 through 4.*

29. $B \subseteq A$

30. $A \subseteq B$

31. $D \subseteq C$

32. $C \subseteq D$

33. $A \subseteq C$

34. $C \subseteq A$

35. $B \not\subseteq C$

36. $C \not\subseteq B$

37. $\varnothing \subseteq A$

38. $\varnothing \subseteq D$

39. $A \cup D$ is $\{1, 2, 3, 4, 5, 6, 7\}$

40. $A \cap D$ is $\varnothing$

41. $\{a, b, c\} \cup \{\ \}$ is $\{a, b, c\}$

42. $\{a, b, c\} \cap \{\ \}$ is $\{\ \}$

Appendix

Review of Angles, Lines, and Special Triangles

The word **geometry** is formed from the Greek words, **geo**, meaning earth, and **metron**, meaning measure. Geometry literally means to measure the earth.

This appendix contains a review of some basic geometric ideas. It will be assumed that fundamental ideas of geometry such as point, line, ray, and angle are known. In this appendix, the notation $\angle 1$ is read "angle 1" and the notation $m\angle 1$ is read "the measure of angle 1."

Objectives

A Identify Angles.

B Identify Lines.

C Identify Special Triangles.

D Use the Pythagorean Theorem.

Objective A Identifying Angles 

We first review types of angles.

Angles

An angle whose measure is greater than 0° but less than 90° is called an **acute angle**.

A **right angle** is an angle whose measure is 90°. A right angle can be indicated by a square drawn at the vertex of the angle, as shown below.

An angle whose measure is greater than 90° but less than 180° is called an **obtuse angle**.

An angle whose measure is 180° is called a **straight angle**.

Two angles are said to be **complementary** if the sum of their measures is 90°. Each angle is called the **complement** of the other.

Two angles are said to be **supplementary** if the sum of their measures is 180°. Each angle is called the **supplement** of the other.

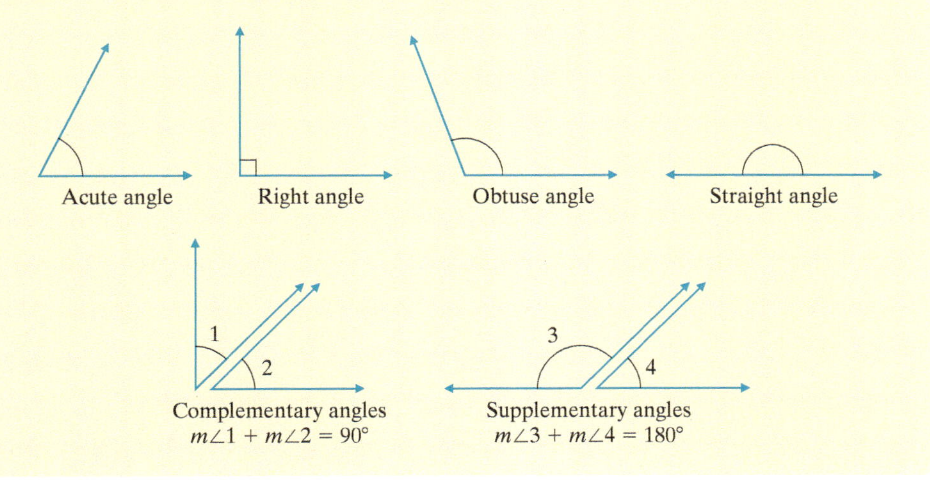

Example 1 If an angle measures 28°, find its complement.

Solution: Two angles are complementary if the sum of their measures is 90°. The complement of a 28° angle is an angle whose measure is 90° − 28° = 62°. To check, notice that 28° + 62° = 90°.

Work Practice 1

Practice 1

Find the supplement of a 67° angle.

Objective B Identifying Lines 

Plane is an undefined term that we will describe. A plane can be thought of as a flat surface with infinite length and width but no thickness. A plane is two dimensional. The arrows in the following diagram indicate that a plane extends indefinitely and has no boundaries.

Answer

1. 113°

Figures that lie on a plane are called **plane figures**. Lines that lie in the same plane are called **coplanar**.

Lines

Two lines are **parallel** if they lie in the same plane but never meet. **Intersecting lines** meet or cross in one point.

Two lines that form right angles when they intersect are **perpendicular**.

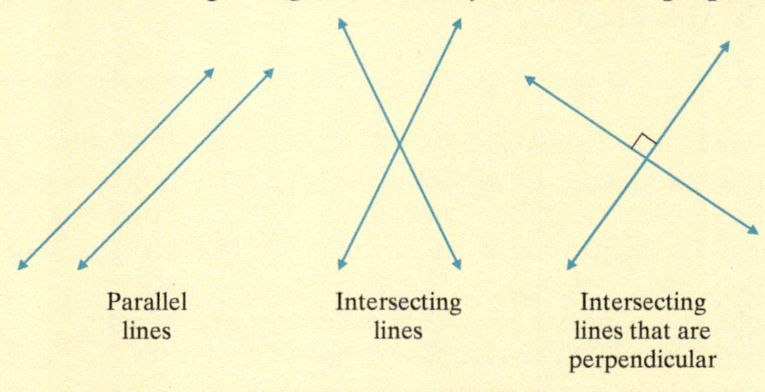

Two intersecting lines form **vertical angles**. Angles 1 and 3 are vertical angles. Also angles 2 and 4 are vertical angles. It can be shown that **vertical angles have equal measures**.

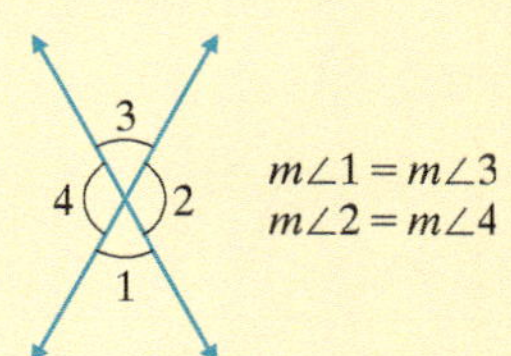

Adjacent angles have the same vertex and share a side. Angles 1 and 2 are adjacent angles. Other pairs of adjacent angles are angles 2 and 3, angles 3 and 4, and angles 4 and 1.

A **transversal** is a line that intersects two or more lines in the same plane. Line l is a transversal that intersects lines m and n. The eight angles formed are numbered and certain pairs of these angles are given special names.

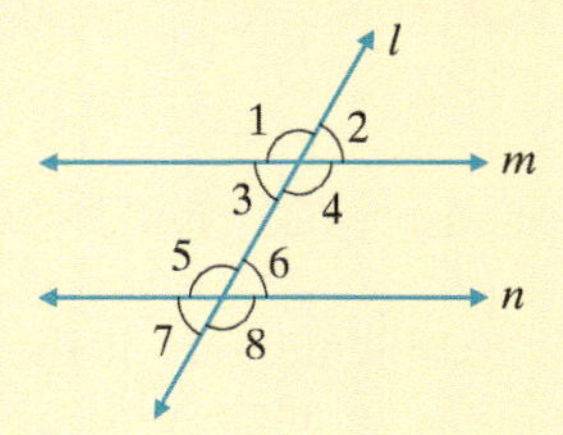

Corresponding angles: $\angle 1$ and $\angle 5$, $\angle 3$ and $\angle 7$, $\angle 2$ and $\angle 6$, and $\angle 4$ and $\angle 8$.

Exterior angles: $\angle 1$, $\angle 2$, $\angle 7$, and $\angle 8$.

Interior angles: $\angle 3$, $\angle 4$, $\angle 5$, and $\angle 6$.

Alternate interior angles: $\angle 3$ and $\angle 6$, $\angle 4$ and $\angle 5$.

These angles and parallel lines are related in the following manner.

Parallel Lines Cut by a Transversal

1. If two parallel lines are cut by a transversal, then
   a. **corresponding angles are equal** and
   b. **alternate interior angles are equal.**
2. If corresponding angles formed by two lines and a transversal are equal, then the lines are parallel.
3. If alternate interior angles formed by two lines and a transversal are equal, then the lines are parallel.

Example 2 Given that lines m and n are parallel and that the measure of angle 1 is 100°, find the measures of angles 2, 3, and 4.

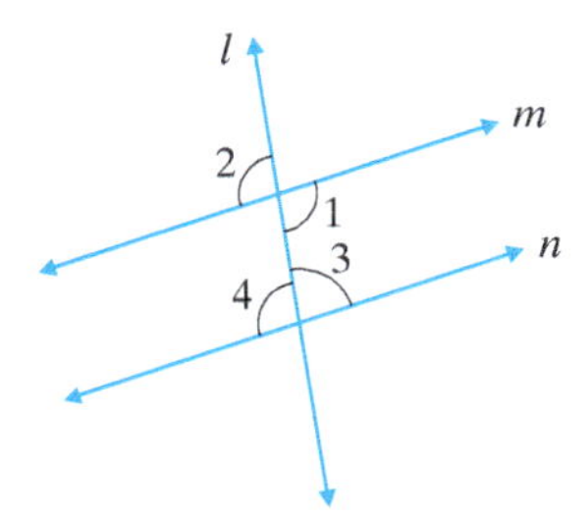

Solution:

$m\angle 2 = 100°$ Since angles 1 and 2 are vertical angles

$m\angle 4 = 100°$ Since angles 1 and 4 are alternate interior angles

$m\angle 3 = 180° - 100° = 80°$ Since angles 4 and 3 are supplementary angles

Work Practice 2

Practice 2

Given that $m \| n$ and that the measure of $\angle w = 45°$, find the measures of all the angles shown.

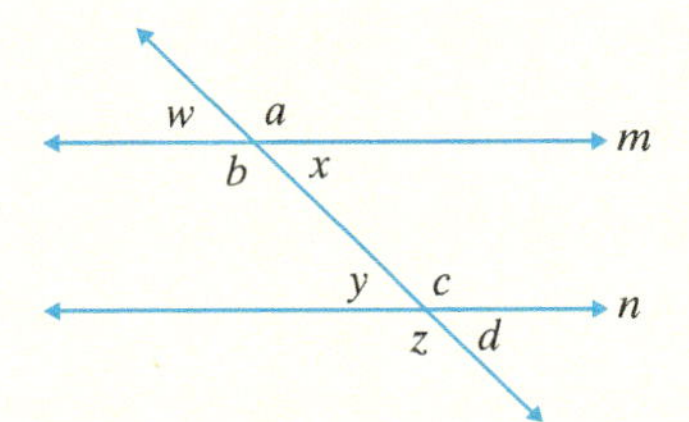

Objective C Identifying Special Triangles 

A **polygon** is the union of three or more coplanar line segments that intersect each other only at each endpoint, with each endpoint shared by exactly two segments.

A **triangle** is a polygon with three sides. The sum of the measures of the three angles of a triangle is 180°. In the following figure, $m\angle 1 + m\angle 2 + m\angle 3 = 180°$.

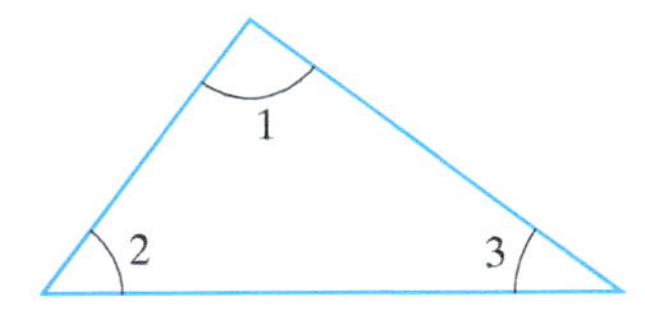

Example 3 Find the measure of the third angle of the triangle shown.

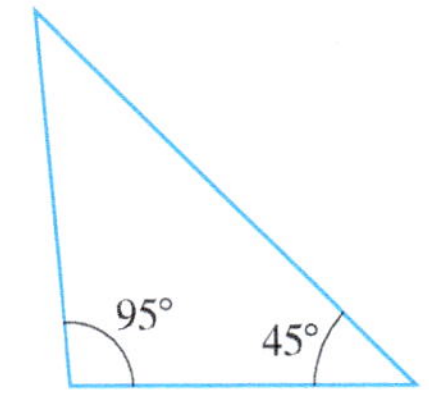

Solution: The sum of the measures of the angles of a triangle is 180°. Since one angle measures 45° and the other angle measures 95°, the third angle measures $180° - 45° - 95° = 40°$.

Work Practice 3

Practice 3

Find the measure of the third angle of the triangle shown.

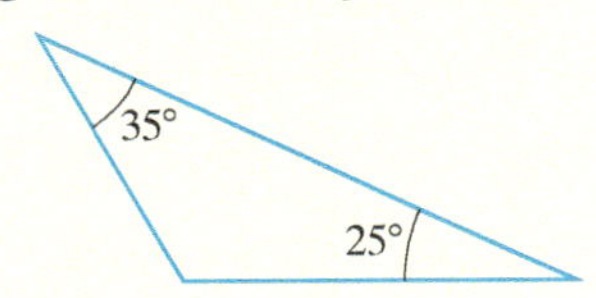

Answers

2. $m\angle x = 45°$; $m\angle y = 45°$; $m\angle z = 135°$; $m\angle a = 135°$; $m\angle b = 135°$; $m\angle c = 135°$; $m\angle d = 45°$
3. 120°

Two triangles are **congruent** if they have the same size and the same shape. In congruent triangles, the measures of corresponding angles are equal and the lengths of corresponding sides are equal. The following triangles are congruent.

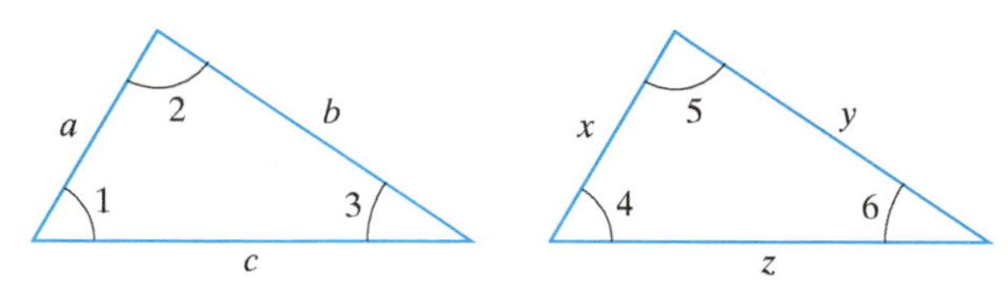

Corresponding angles are equal: $m\angle 1 = m\angle 4$, $m\angle 2 = m\angle 5$, and $m\angle 3 = m\angle 6$. Also, lengths of corresponding sides are equal: $a = x, b = y$, and $c = z$.

Any one of the following may be used to determine whether two triangles are congruent.

Congruent Triangles

1. If the measures of two angles of a triangle equal the measures of two angles of another triangle and the lengths of the sides between each pair of angles are equal, the triangles are congruent.

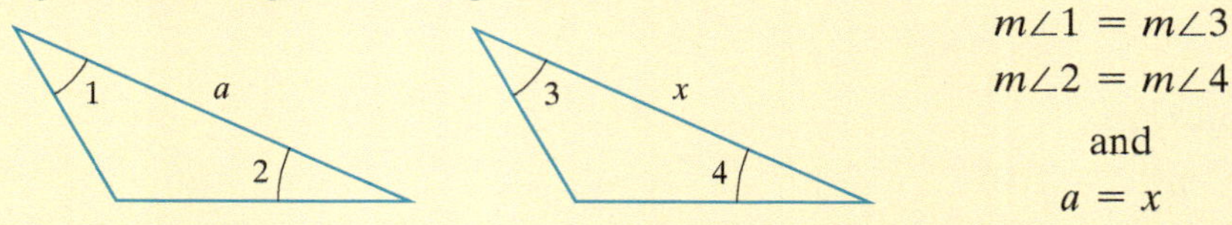

$m\angle 1 = m\angle 3$
$m\angle 2 = m\angle 4$
and
$a = x$

2. If the lengths of the three sides of a triangle equal the lengths of corresponding sides of another triangle, the triangles are congruent.

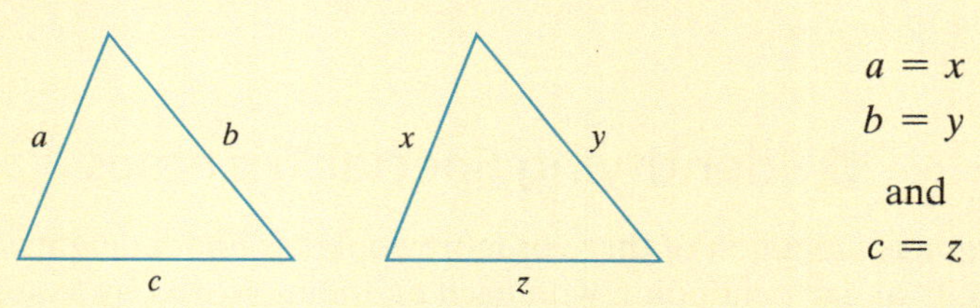

$a = x$
$b = y$
and
$c = z$

3. If the lengths of two sides of a triangle equal the lengths of corresponding sides of another triangle, and the measures of the angles between each pair of sides are equal, the triangles are congruent.

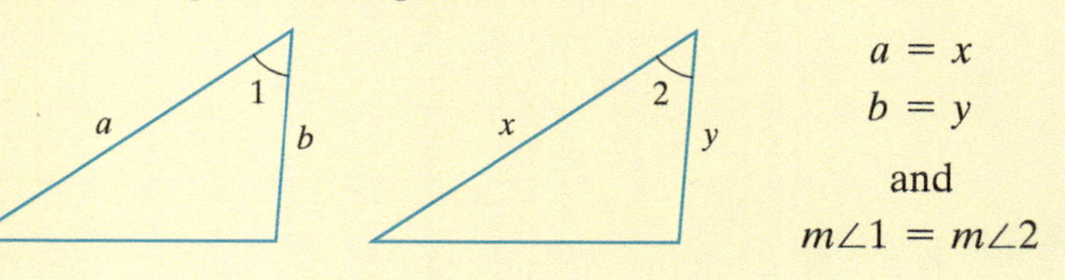

$a = x$
$b = y$
and
$m\angle 1 = m\angle 2$

Two triangles are **similar** if they have the same shape but not necessarily the same size. In similar triangles, the measures of corresponding angles are equal and corresponding sides are in proportion. The following triangles are similar. (All pairs of similar triangles drawn in this appendix will be oriented the same.)

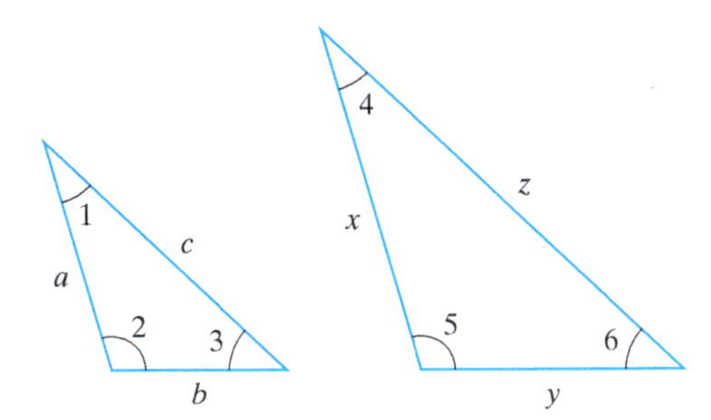

Corresponding angles are equal: $m\angle 1 = m\angle 4$, $m\angle 2 = m\angle 5$, and $m\angle 3 = m\angle 6$. Also, corresponding sides are proportional: $\frac{a}{x} = \frac{b}{y} = \frac{c}{z}$.

Any one of the following may be used to determine whether two triangles are similar.

Similar Triangles

1. If the measures of two angles of a triangle equal the measures of two angles of another triangle, the triangles are similar.

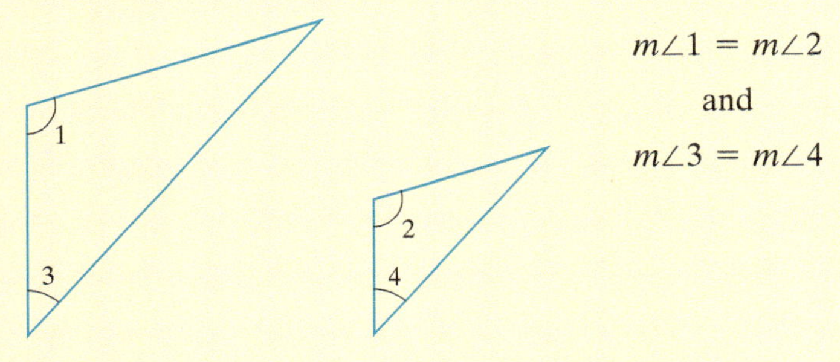

$m\angle 1 = m\angle 2$

and

$m\angle 3 = m\angle 4$

2. If three sides of one triangle are proportional to three sides of another triangle, the triangles are similar.

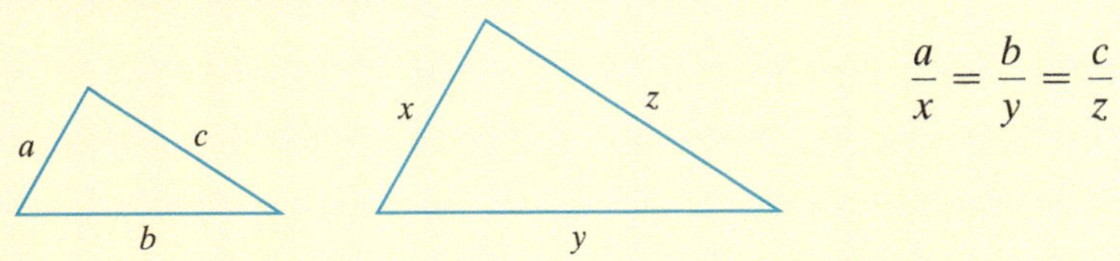

$$\frac{a}{x} = \frac{b}{y} = \frac{c}{z}$$

3. If two sides of a triangle are proportional to two sides of another triangle and the measures of the included angles are equal, the triangles are similar.

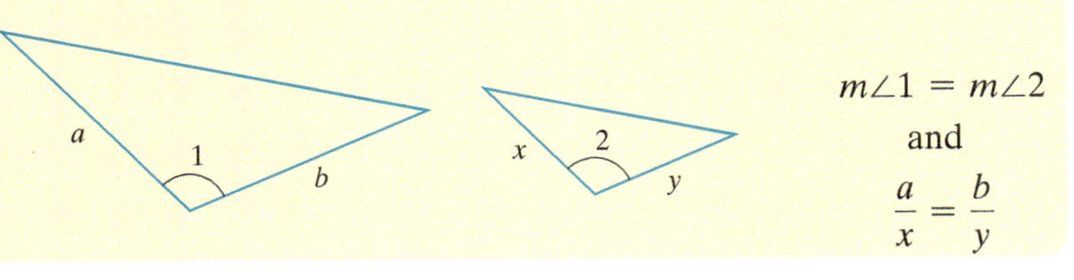

$m\angle 1 = m\angle 2$

and

$$\frac{a}{x} = \frac{b}{y}$$

Example 4 Given that the following triangles are similar, find the missing length x.

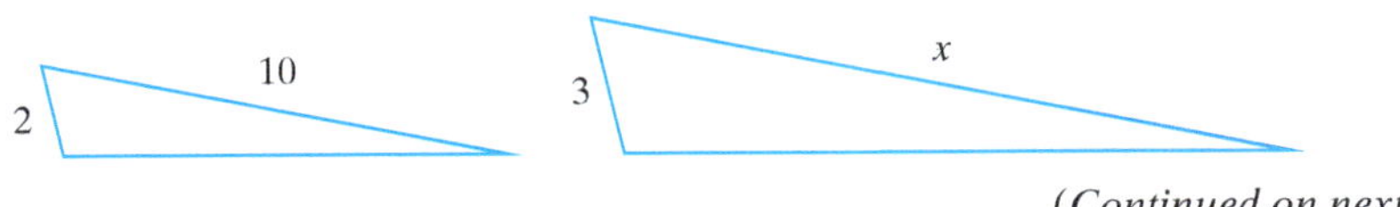

(*Continued on next page*)

Practice 4

Given that the triangles are similar, find the missing length n.

a.

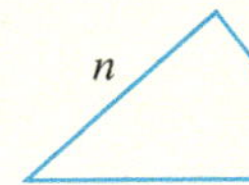

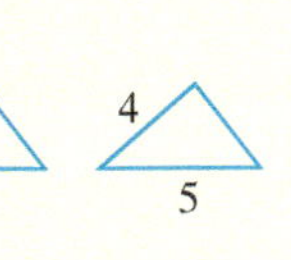

b.

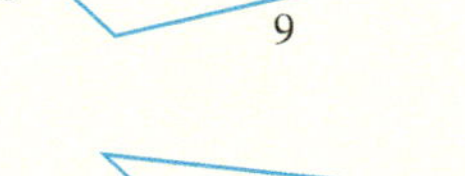

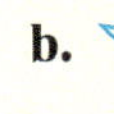

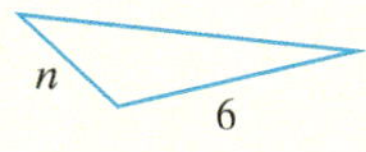

Answers

4. a. $n = 8$ **b.** $n = \frac{10}{3}$ or $3\frac{1}{3}$

Solution: Since the triangles are similar, corresponding sides are in proportion. Thus, $\frac{2}{3} = \frac{10}{x}$. To solve this equation for x, we cross multiply.

$$\frac{2}{3} = \frac{10}{x}$$
$$2x = 30$$
$$x = 15$$

The missing length is 15 units.

Work Practice 4

Objective D Using the Pythagorean Theorem 

A **right triangle** contains a right angle. The side opposite the right angle is called the **hypotenuse**, and the other two sides are called the **legs**. The **Pythagorean theorem** gives a formula that relates the lengths of the three sides of a right triangle.

The Pythagorean Theorem

If a and b are the lengths of the legs of a right triangle, and c is the length of the hypotenuse, then $a^2 + b^2 = c^2$.

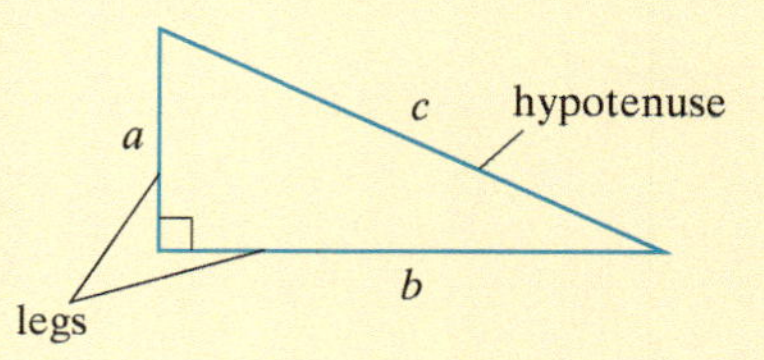

Practice 5

Find the length of the hypotenuse of the given right triangle.

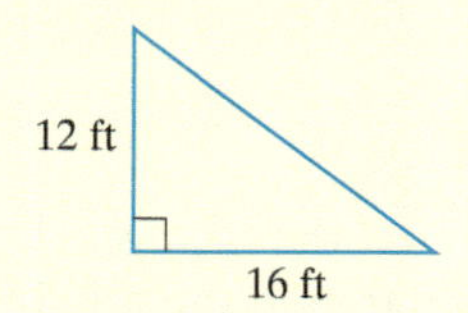

Answer

5. 20 ft

Example 5 Find the length of the hypotenuse of a right triangle whose legs have lengths of 3 centimeters and 4 centimeters.

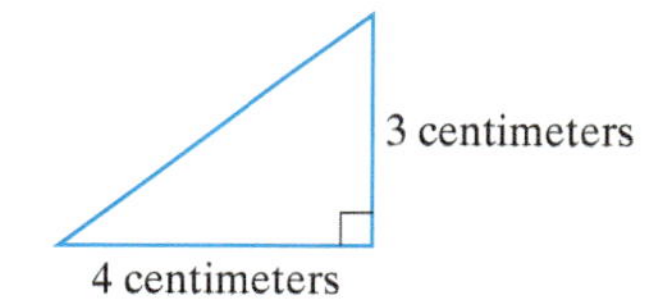

Solution: Because we have a right triangle, we use the Pythagorean theorem. The legs are 3 centimeters and 4 centimeters, so let $a = 3$ and $b = 4$ in the formula.

$$a^2 + b^2 = c^2$$
$$3^2 + 4^2 = c^2$$
$$9 + 16 = c^2$$
$$25 = c^2$$

Since c represents a length, we assume that c is positive. Thus, if c^2 is 25, c must be 5. The hypotenuse has a length of 5 centimeters.

Work Practice 5

E Exercise Set MyLab Math 

Objective A *Find the complement of each angle. See Example 1.*

1. 19°

2. 65°

3. 70.8°

4. $45\frac{2}{3}°$

5. $11\frac{1}{4}°$

6. 19.6°

Find the supplement of each angle.

7. 150°

8. 90°

9. 30.2°

10. 81.9°

11. $79\frac{1}{2}°$

12. $165\frac{8}{9}°$

Objective B *Find the measures of the numbered angles. See Example 2.*

13. If lines m and n are parallel, find the measures of angles 1 through 7.

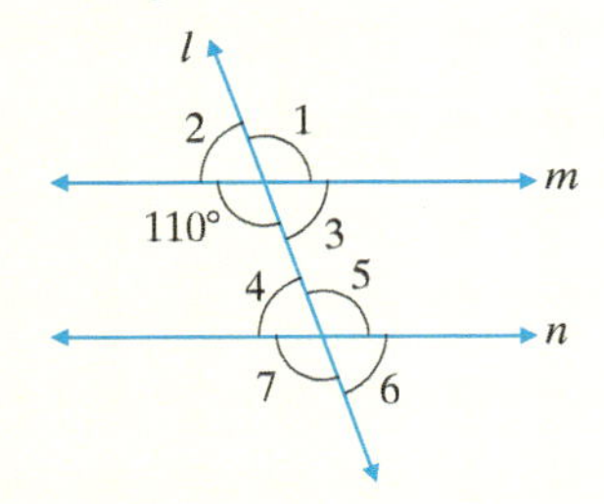

14. If lines m and n are parallel, find the measures of angles 1 through 5.

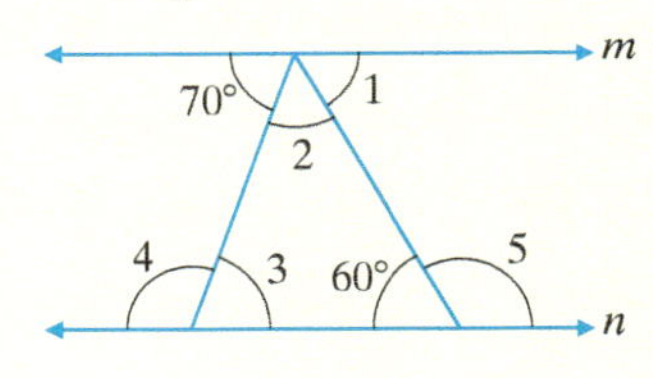

Objective C *In each of the following, the measures of two angles of a triangle are given. Find the measure of the third angle. See Example 3.*

15. 11°, 79°

16. 8°, 102°

17. 25°, 65°

18. 44°, 19°

19. 30°, 60°

20. 67°, 23°

In each of the following, the measure of one angle of a right triangle is given. Find the measures of the other two angles.

21. 45°

22. 60°

23. 17°

24. 30°

25. $39\frac{3}{4}°$

26. 72.6°

Given that each of the following pairs of triangles is similar, find the missing length x. See Example 4.

27.

4 x

28.

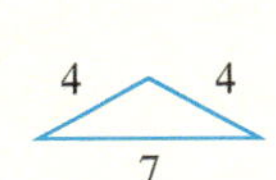

x x 14

29.

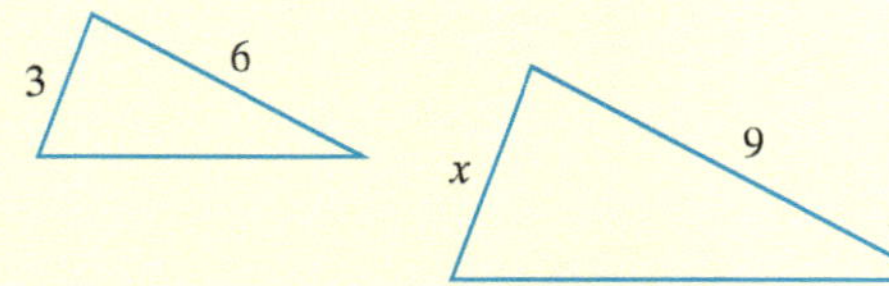

30.

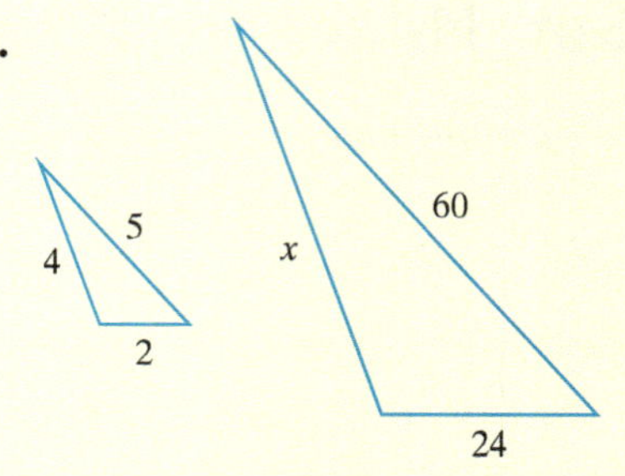

Objective D *Use the Pythagorean theorem to find the missing lengths in the right triangles. See Example 5.*

31.

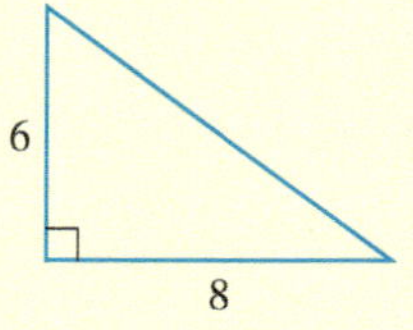

32.

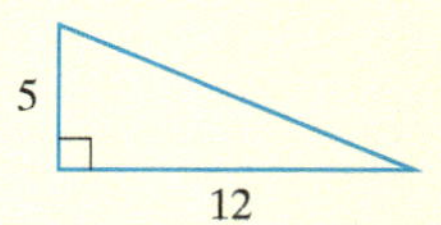

33.

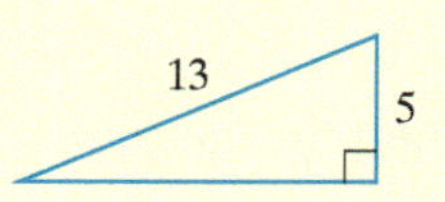

34.

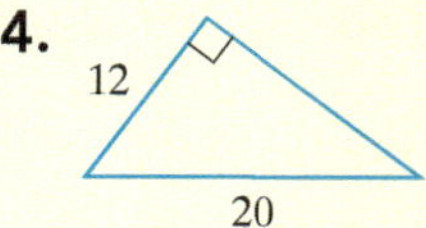

Appendix F

Interval Notation and Finding Domains and Ranges from Graphs

Objectives

A  

Use Interval Notation.

B Find the Domain and Range from a Graph.

Objective A Using Interval Notation

Recall that a **solution** of an inequality is a value of the variable that makes the inequality a true statement. The **solution set** of an inequality is the set of all solutions. Notice that the solution set of the inequality $x > 2$, for example, contains all numbers greater than 2. Its graph is an interval on the number line since an infinite number of values satisfy the variable. If we use open/closed-circle notation, the graph of $\{x|x > 2\}$ looks like:

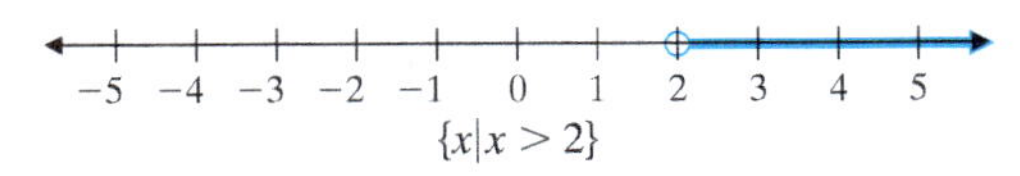

In this section, a different graphing notation will be used to help us understand **interval notation**. Instead of an open circle, we use a parenthesis; instead of a closed circle, we use a bracket. With this new notation, the graph of $\{x|x > 2\}$ now looks like:

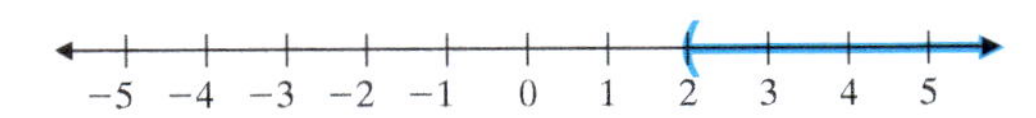

and can be represented in interval notation as $(2, \infty)$. The symbol ∞ is read "infinity" and indicates that the interval includes *all* numbers greater than 2. The left parenthesis indicates that 2 *is not* included in the interval. Using a left bracket, [, would indicate that 2 *is* included in the interval. The following table shows three equivalent ways to describe an interval: in set notation, as a graph, and in interval notation.

Set Notation	Graph	Interval Notation
$\{x\|x < a\}$	a	$(-\infty, a)$
$\{x\|x > a\}$	a	(a, ∞)
$\{x\|x \le a\}$	a	$(-\infty, a]$
$\{x\|x \ge a\}$	a	$[a, \infty)$
$\{x\|a < x < b\}$	a b	(a, b)
$\{x\|a \le x \le b\}$	a b	$[a, b]$
$\{x\|a < x \le b\}$	a b	$(a, b]$
$\{x\|a \le x < b\}$	a b	$[a, b)$
$\{x\|x$ is a real number$\}$		$(-\infty, \infty)$

Helpful Hint

Notice that a parenthesis is always used to enclose ∞ and $-\infty$.

Practice 1–3

Graph each set on a number line and then write it in interval notation.

1. $\{x|x > -3\}$

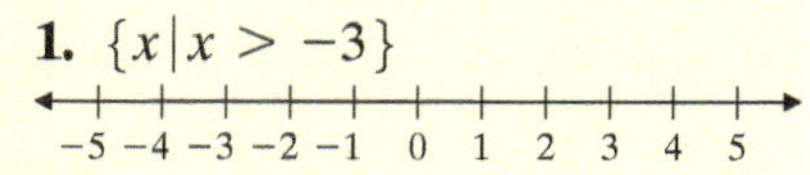

2. $\{x|x \le 0\}$

−5 −4 −3 −2 −1 0 1 2 3 4 5

3. $\{x|-0.5 \le x < 2\}$

−5 −4 −3 −2 −1 0 1 2 3 4 5

Practice 4–7

Find the domain and range of each relation.

4.

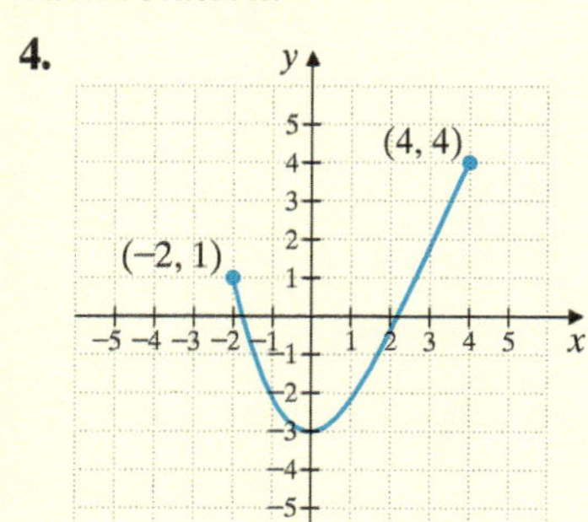

5.

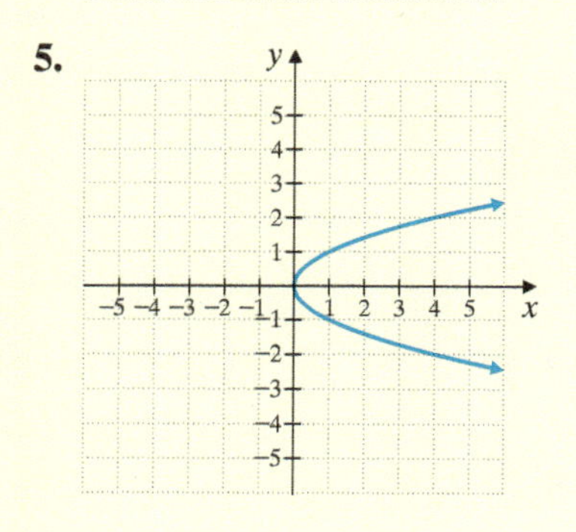

Answers

1. $(-3, \infty)$

−5 −4 −3 −2 −1 0 1 2 3 4 5

2. $(-\infty, 0]$

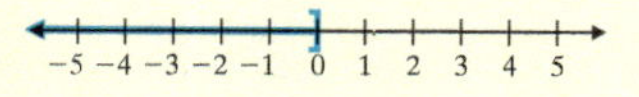

3. $[-0.5, 2)$

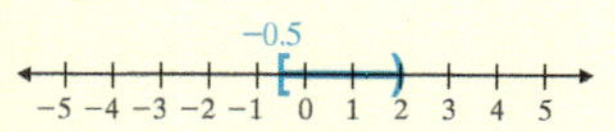

4. domain: $[-2, 4]$; range: $[-3, 4]$

5. domain: $[0, \infty)$; range: $(-\infty, \infty)$

✓Concept Check Answer

should be $(5, \infty)$ since a parenthesis is always used to enclose ∞

Examples Graph each set on a number line and then write it in interval notation.

1. $\{x|x \ge 2\}$

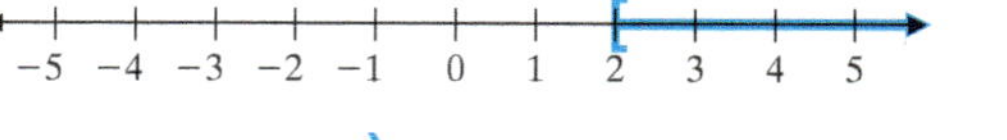

$[2, \infty)$

2. $\{x|x < -1\}$

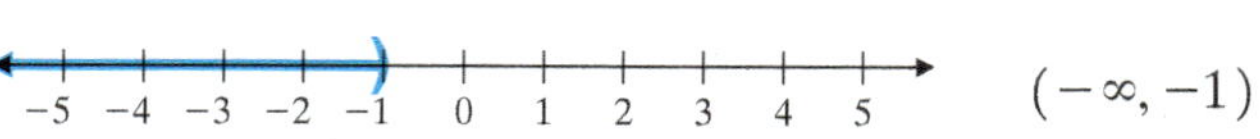

$(-\infty, -1)$

3. $\{x|0.5 < x \le 3\}$

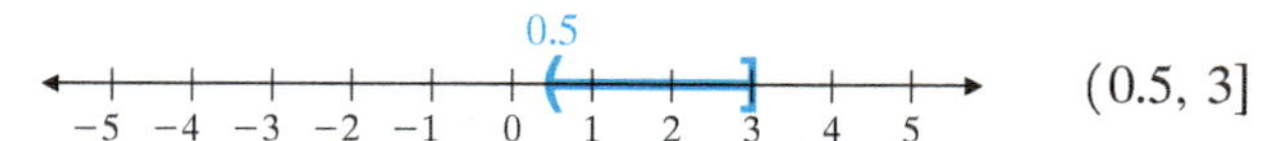

$(0.5, 3]$

Work Practice 1–3

✓Concept Check Explain what is wrong with writing the interval $(5, \infty]$.

Objective B Finding the Domain and Range from a Graph

Recall from Section 6.6 that the

> **domain** of a relation is the set of all first components of the ordered pairs of the relation and the
>
> **range** of a relation is the set of all second components of the ordered pairs of the relation.

In this section we use the graph of a relation to find its domain and range. Let's use interval notation to write these domains and ranges. Remember, we use a parenthesis to indicate that a number is not part of the domain and we use a bracket to indicate that a number is part of the domain. Of course, as usual, parentheses are placed about infinity symbols indicating that we approach but never reach infinity.

To find the domain of a function (or relation) from its graph, recall that on the rectangular coordinate system, "domain" is the set of first components of the ordered pairs, so this means the x-values that are graphed. Similarly, "range" is the set of second components of the ordered pairs, so this means the y-values that are graphed.

Examples Find the domain and range of each relation.

4.

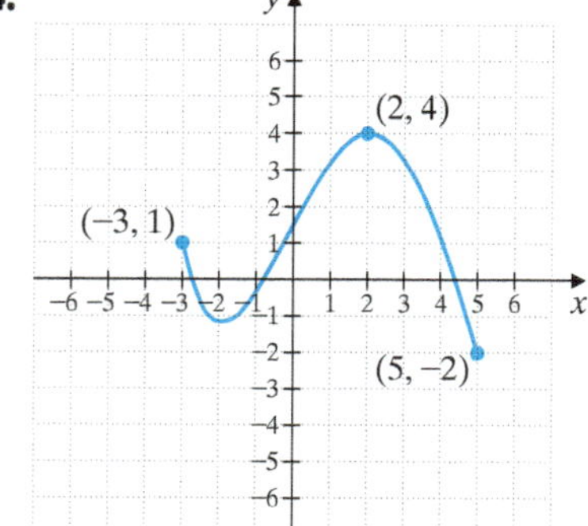

5.

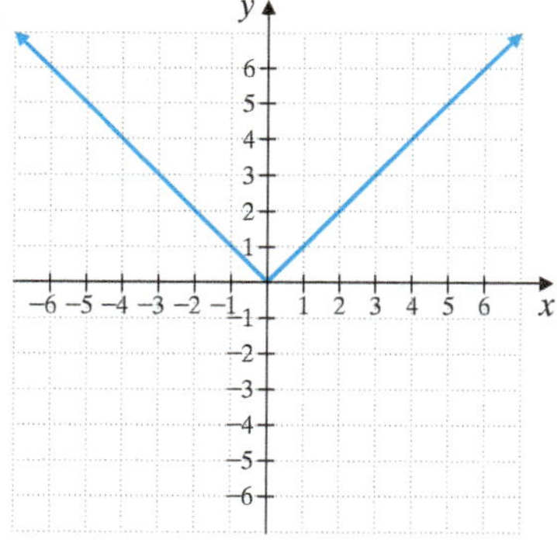

6.

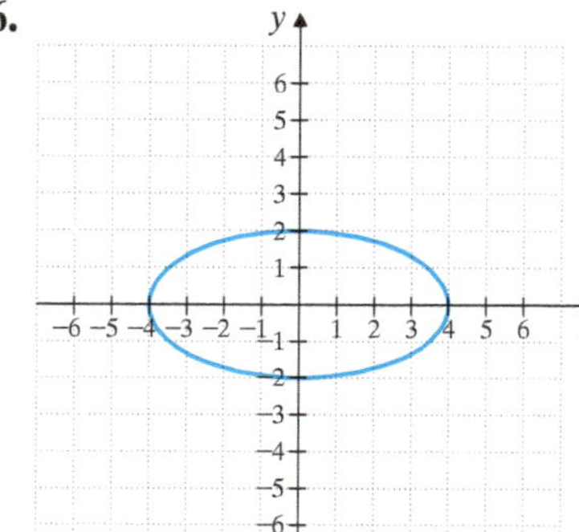

7.

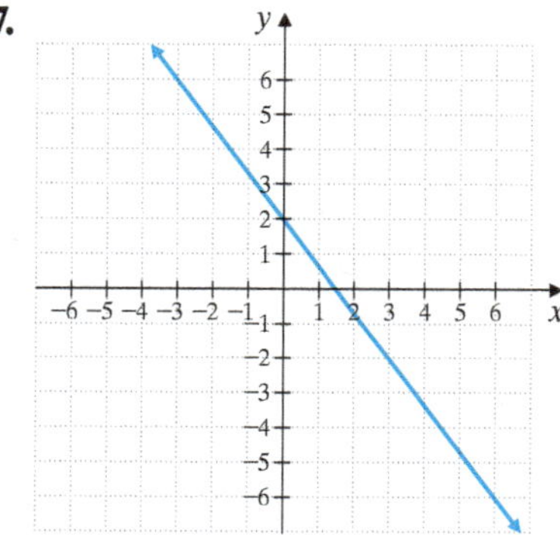

Solution: Notice that the graphs for Examples 4, 5, and 7 are graphs of functions because each passes the vertical line test.

4.

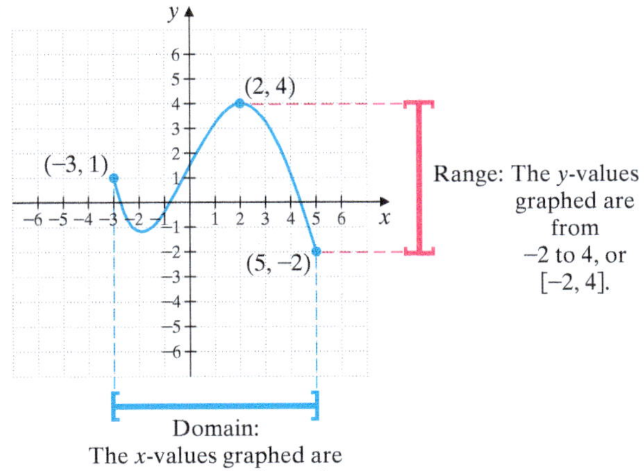

5.

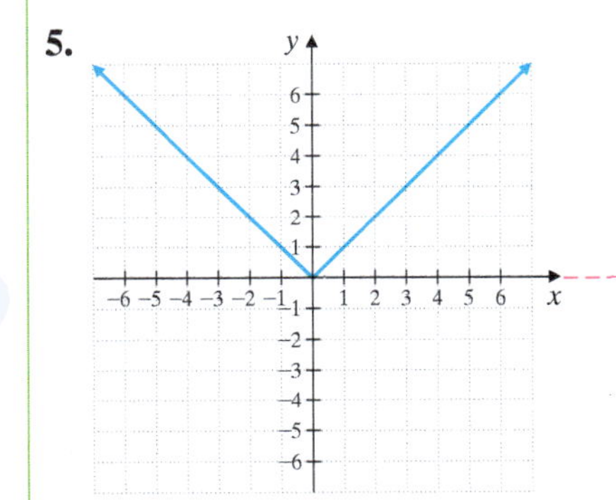

Range: $[0, \infty)$

Domain: $(-\infty, \infty)$

6.

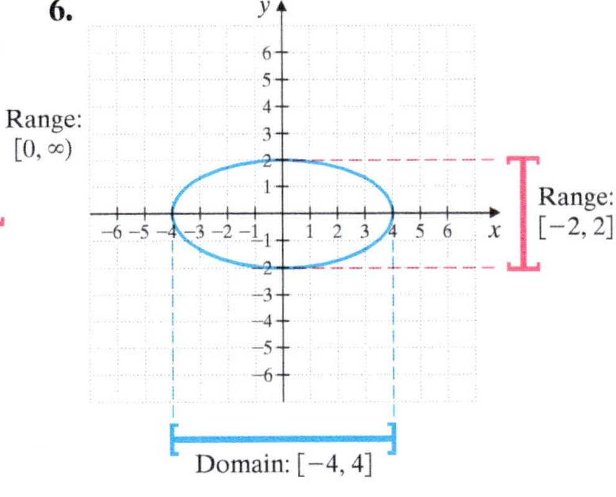

7.

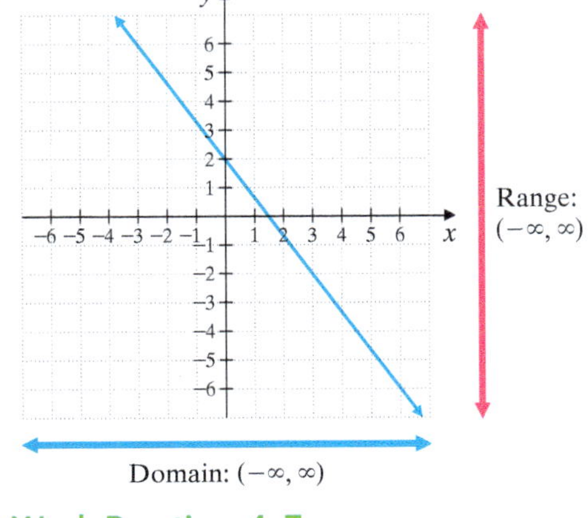

Work Practice 4–7

6.

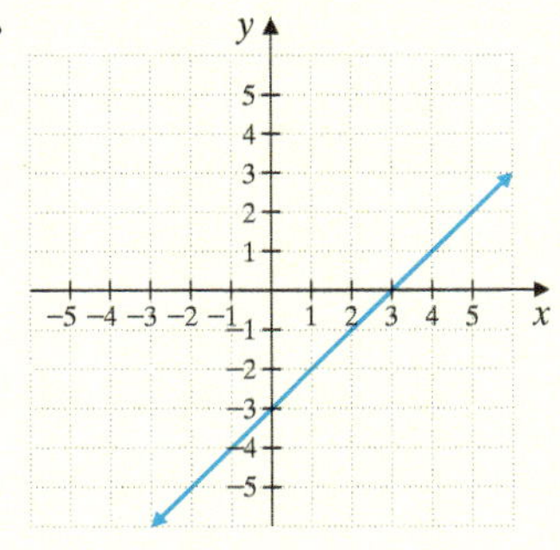

7.

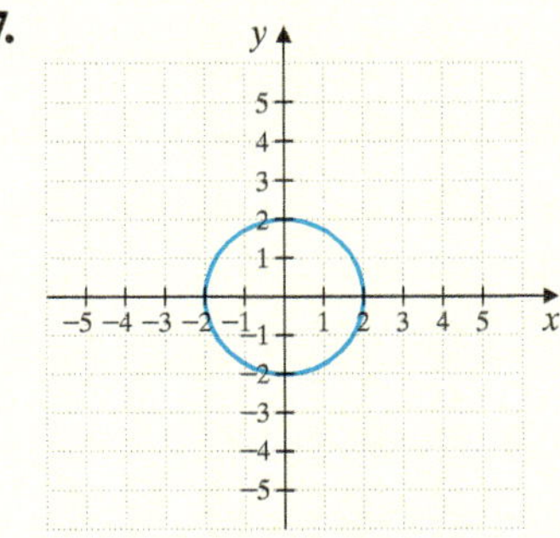

Answers

6. domain: $(-\infty, \infty)$; range: $(-\infty, \infty)$
7. domain: $[-2, 2]$; range: $[-2, 2]$

F Exercise Set MyLab Math

Objective A *Graph the solution set of each inequality on a number line and then write it in interval notation. See Examples 1 through 3.*

1. $\{x|x < -3\}$

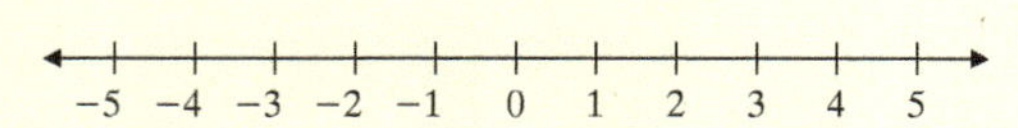

2. $\{x|x > 5\}$

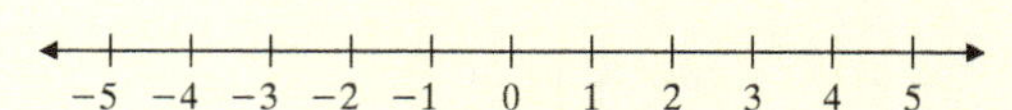

3. $\{x|x \geq 0.3\}$

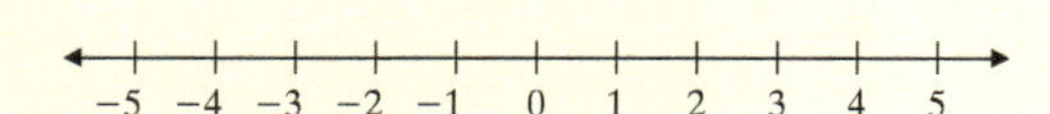

4. $\{x|x < -0.2\}$

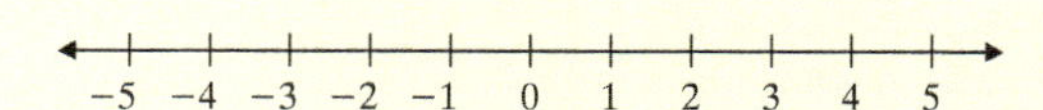

5. $\{x|-7 \leq x\}$

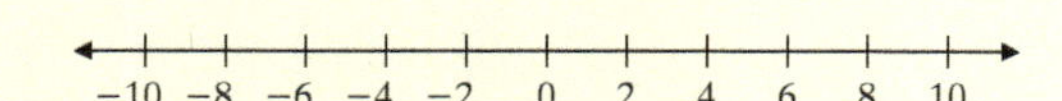

6. $\{x|-7 \geq x\}$

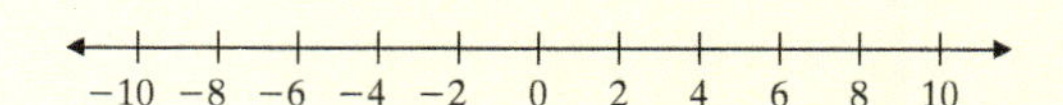

7. $\{x|-2 < x < 5\}$

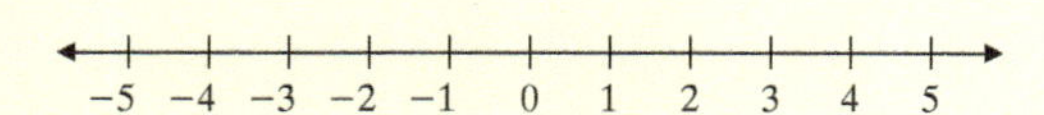

8. $\{x|-5 \leq x \leq -1\}$

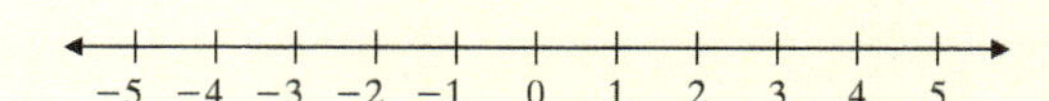

9. $\{x|5 \geq x > -1\}$

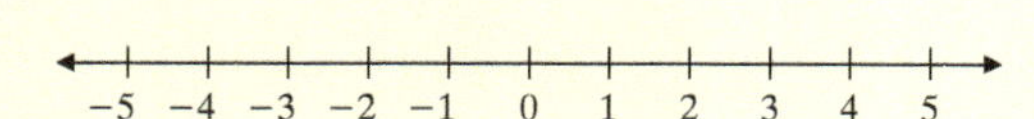

10. $\{x|-3 > x \geq -7\}$

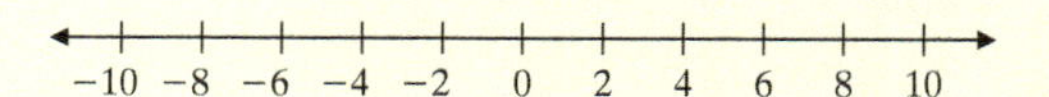

Objective B *Find the domain and the range of each relation. See Examples 4 through 7.*

11.

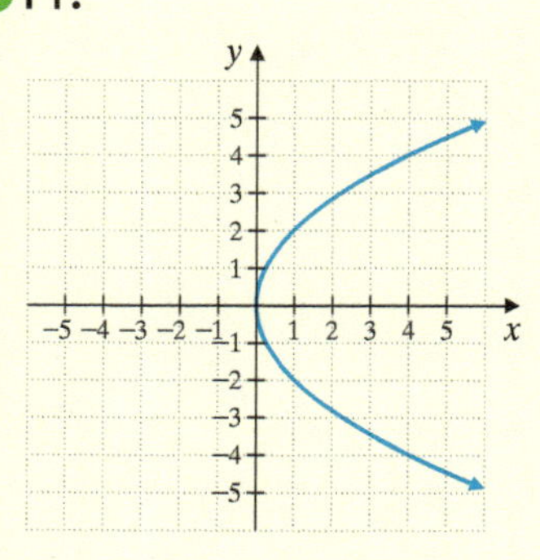

12.

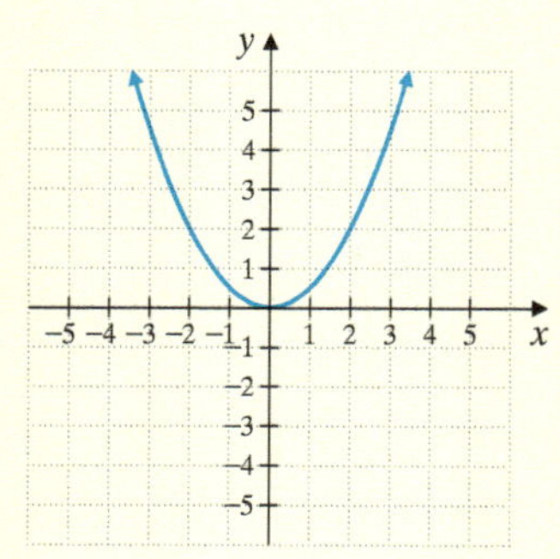

13.

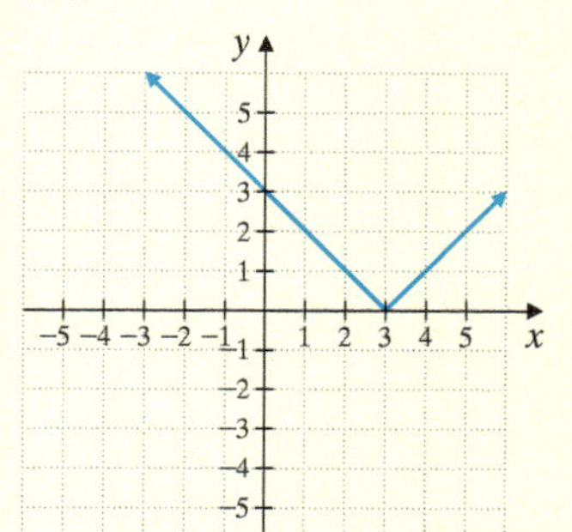

14.

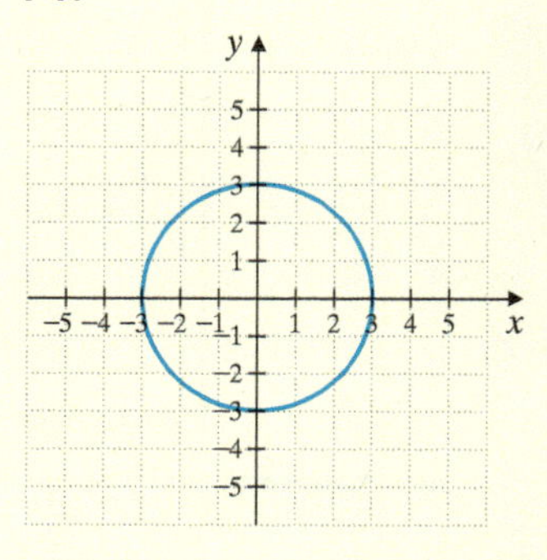

15.

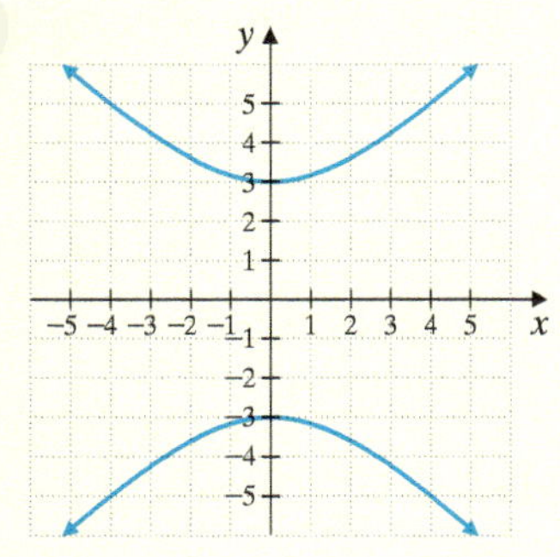

16.

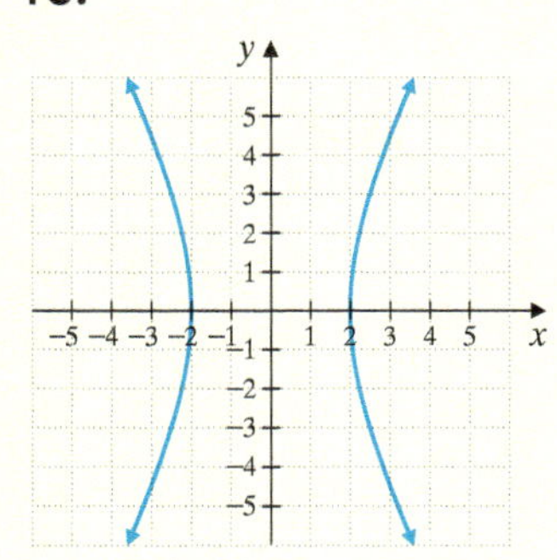

17.

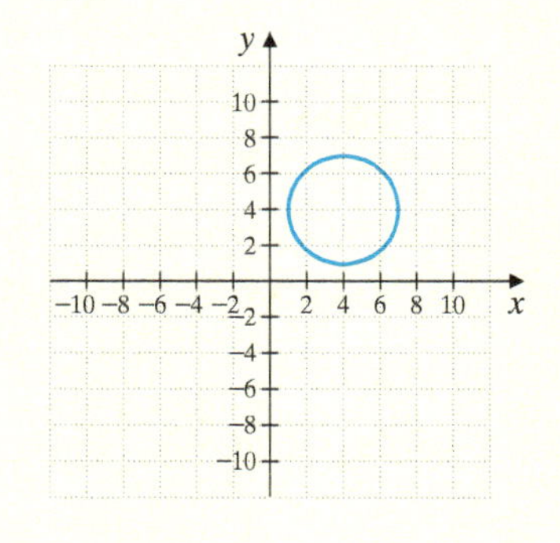

18.

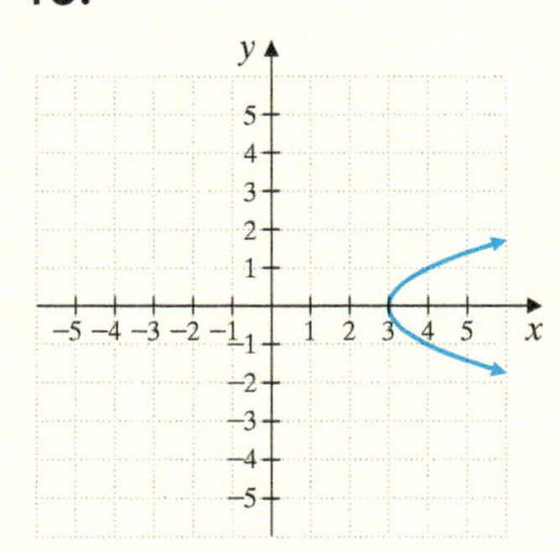

19.

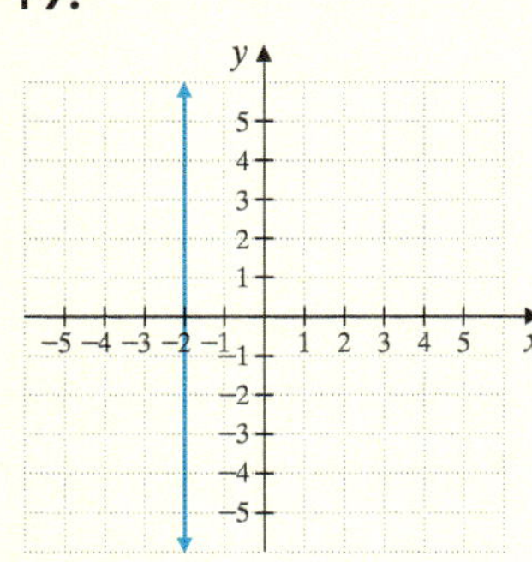

20.

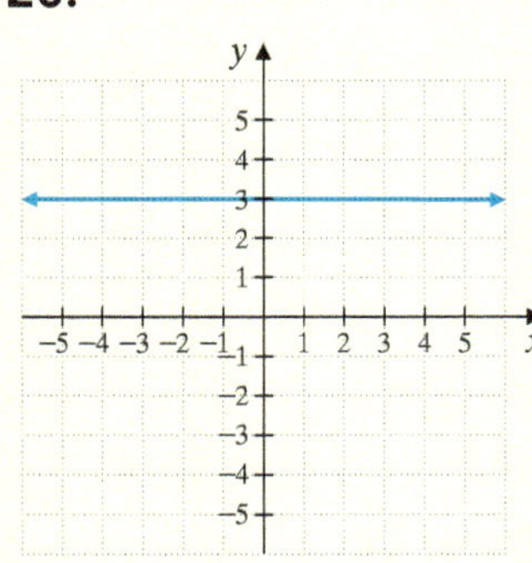

21.

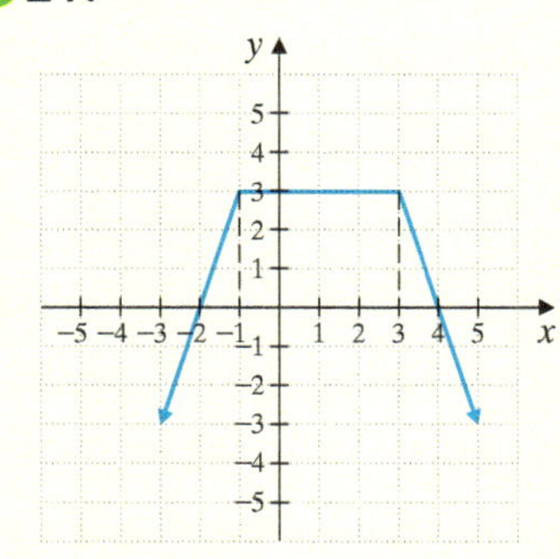

22.

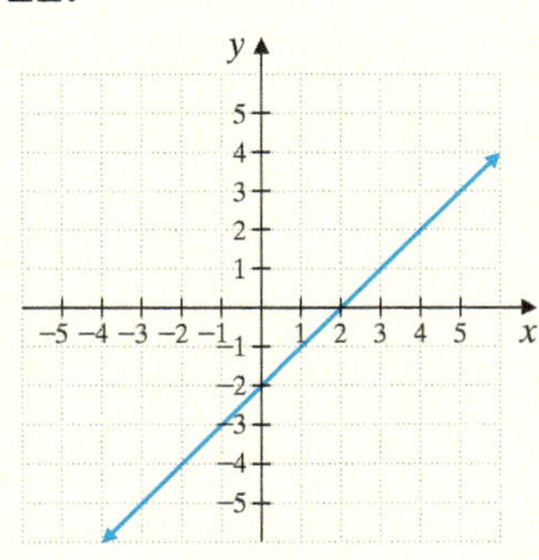

23.

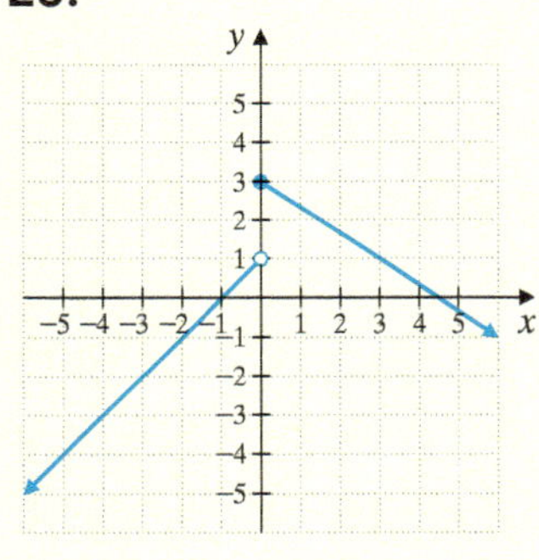

24.

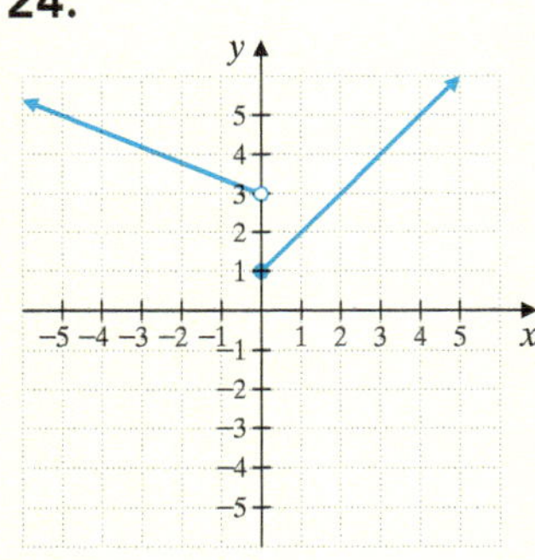

25.

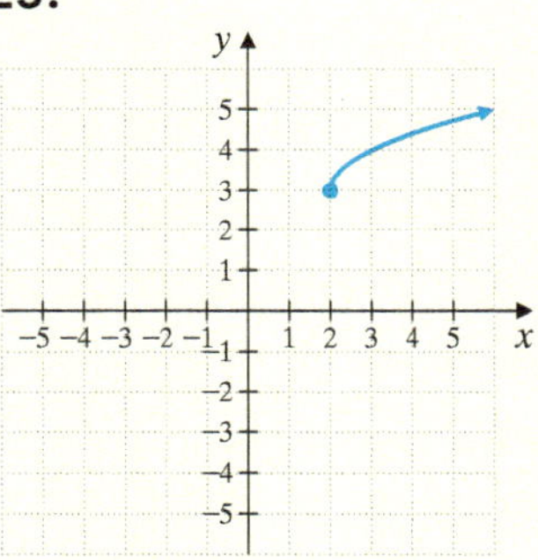

26.

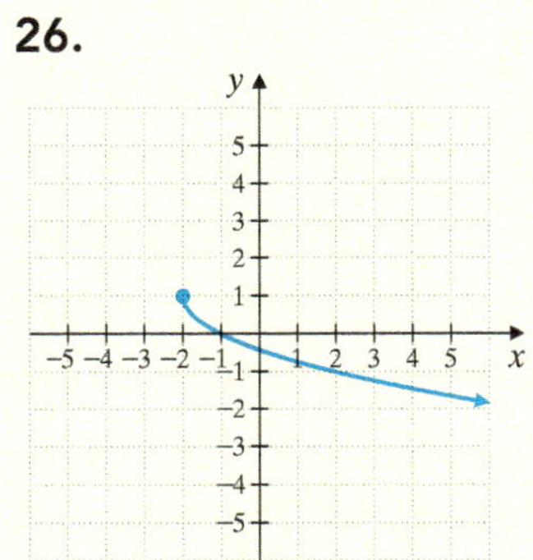

Contents of Student Resources

Study Skills Builders

Attitude and Study Tips:

1. Have You Decided to Complete This Course Successfully?
2. Tips for Studying for an Exam
3. What to Do the Day of an Exam
4. Are You Satisfied with Your Performance on a Particular Quiz or Exam?
5. How Are You Doing?
6. Are You Preparing for Your Final Exam?

Organizing Your Work:

7. Learning New Terms
8. Are You Organized?
9. Organizing a Notebook
10. How Are Your Homework Assignments Going?

MyLab Math and MathXL:

11. Tips for Turning in Your Homework on Time
12. Tips for Doing Your Homework Online
13. Organizing Your Work
14. Getting Help with Your Homework Assignments
15. Tips for Preparing for an Exam
16. How Well Do You Know the Resources Available to You in MyLab Math?

Additional Help Inside and Outside Your Textbook:

17. How Well Do You Know Your Textbook?
18. Are You Familiar with Your Textbook Supplements?
19. Are You Getting All the Mathematics Help That You Need?

Bigger Picture—Study Guide Outline

Practice Final Exam

Answers to Selected Exercises

Student Resources

Study Skills Builders

Attitude and Study Tips

Study Skills Builder 1

Have You Decided to Complete This Course Successfully?

Ask yourself if one of your current goals is to complete this course successfully.

If it is not a goal of yours, ask yourself why. One common reason is fear of failure. Amazingly enough, fear of failure alone can be strong enough to keep many of us from doing our best in any endeavor.

Another common reason is that you simply haven't taken the time to think about or write down your goals for this course. To help accomplish this, answer the questions below.

Exercises

1. Write down your goal(s) for this course.
2. Now list steps you will take to make sure your goal(s) in Exercise **1** are accomplished.
3. Rate your commitment to this course with a number between 1 and 5. Use the diagram below to help.

High Commitment		Average Commitment		Not Committed at All
5	4	3	2	1

4. If you have rated your personal commitment level (from the exercise above) as a 1, 2, or 3, list the reasons why this is so. Then determine whether it is possible to increase your commitment level to a 4 or 5.

Good luck, and don't forget that a positive attitude will make a big difference.

Study Skills Builder 2

Tips for Studying for an Exam

To prepare for an exam, try the following study techniques:

- Start the study process days before your exam.
- Make sure that you are up to date on your assignments.
- If there is a topic that you are unsure of, use one of the many resources that are available to you. For example,

    See your instructor.
    View a lecture video on the topic.
    Visit a learning resource center on campus.
    Read the textbook material and examples on the topic.

- Reread your notes and carefully review the Chapter Highlights at the end of any chapter.
- Work the review exercises at the end of the chapter.
- Find a quiet place to take the Chapter Test found at the end of the chapter. Do not use any resources when taking this sample test. This way, you will have a clear indication of how prepared you are for your exam. Check your answers and use the Chapter Test Prep Videos to make sure that you correct any missed exercises.

Good luck, and keep a positive attitude.

Exercises

Let's see how you did on your last exam.

1. How many days before your last exam did you start studying for that exam?

2. Were you up to date on your assignments at that time or did you need to catch up on assignments?

3. List the most helpful text supplement (if you used one).

4. List the most helpful campus supplement (if you used one).

5. List your process for preparing for a mathematics test.

6. Was this process helpful? In other words, were you satisfied with your performance on your exam?

7. If not, what changes can you make in your process that will make it more helpful to you?

Study Skills Builder 3

What to Do the Day of an Exam

Your first exam may be soon. On the day of an exam, don't forget to try the following:

- Allow yourself plenty of time to arrive.
- Read the directions on the test carefully.
- Read each problem carefully as you take your test. Make sure that you answer the question asked.
- Watch your time and pace yourself so that you may attempt each problem on your test.
- Check your work and answers.
- ***Do not turn your test in early.*** If you have extra time, spend it double-checking your work.

Good luck!

Exercises

Answer the following questions based on your most recent mathematics exam, whenever that was.

1. How soon before class did you arrive?

2. Did you read the directions on the test carefully?

3. Did you make sure you answered the question asked for each problem on the exam?

4. Were you able to attempt each problem on your exam?

5. If your answer to Exercise **4** is no, list reasons why.

6. Did you have extra time on your exam?

7. If your answer to Exercise **6** is yes, describe how you spent that extra time.

Study Skills Builder 4

Are You Satisfied with Your Performance on a Particular Quiz or Exam?

If not, don't forget to analyze your quiz or exam and look for common errors. Were most of your errors a result of:

- *Carelessness?* Did you turn in your quiz or exam before the allotted time expired? If so, resolve to use any extra time to check your work.
- *Running out of time?* Answer the questions you are sure of first. Then attempt the questions you are unsure of, and delay checking your work until all questions have been answered.
- *Not understanding a concept?* If so, review that concept and correct your work so that you make sure you understand it before the next quiz or the final exam.
- *Test conditions?* When studying for a quiz or exam, make sure you place yourself in conditions similar to test conditions. For example, before your next quiz or exam, take a sample test without the aid of your notes or text.

(For a sample test, see your instructor or use the Chapter Test at the end of each chapter.)

Exercises

1. Have you corrected all your previous quizzes and exams?
2. List any errors you have found common to two or more of your graded papers.
3. Is one of your common errors not understanding a concept? If so, are you making sure you understand all the concepts for the next quiz or exam?
4. Is one of your common errors making careless mistakes? If so, are you now taking all the time allotted to check over your work so that you can minimize the number of careless mistakes?
5. Are you satisfied with your grades thus far on quizzes and tests?
6. If your answer to Exercise **5** is no, are there any more suggestions you can make to your instructor or yourself to help? If so, list them here and share these with your instructor.

Study Skills Builder 5

How Are You Doing?

If you haven't done so yet, take a few moments and think about how you are doing in this course. Are you working toward your goal of successfully completing this course? Is your performance on homework, quizzes, and tests satisfactory? If not, you might want to see your instructor to see if he/she has any suggestions on how you can improve your performance. Reread Section 1.1 for ideas on places to get help with your mathematics course.

Exercises

Answer the following.

1. List any textbook supplements you are using to help you through this course.
2. List any campus resources you are using to help you through this course.
3. Write a short paragraph describing how you are doing in your mathematics course.
4. If improvement is needed, list ways that you can work toward improving your situation as described in Exercise **3**.

Study Skills Builder 6

Are You Preparing for Your Final Exam?

To prepare for your final exam, try the following study techniques:

- Review the material that you will be responsible for on your exam. This includes material from your textbook, your notebook, and any handouts from your instructor.
- Review any formulas that you may need to memorize.
- Check to see if your instructor or mathematics department will be conducting a final exam review.
- Check with your instructor to see whether final exams from previous semesters/quarters are available to students for review.
- Use your previously taken exams as a practice final exam. To do so, rewrite the test questions in mixed order on blank sheets of paper. This will help you prepare for exam conditions.
- If you are unsure of a few concepts, see your instructor or visit a learning lab for assistance. Also, view the video segment of any troublesome sections.
- If you need further exercises to work, try the Cumulative Reviews at the end of the chapters.

Once again, good luck! I hope you are enjoying this textbook and your mathematics course.

Organizing Your Work

Study Skills Builder 7

Learning New Terms

Many of the terms used in this text may be new to you. It will be helpful to make a list of new mathematical terms and symbols as you encounter them and to review them frequently. Placing these new terms (including page references) on 3×5 index cards might help you later when you're preparing for a quiz.

Exercises

1. Name one way you might place a word and its definition on a 3×5 card.
2. How do new terms stand out in this text so that they can be found?

Study Skills Builder 8

Are You Organized?

Have you ever had trouble finding a completed assignment? When it's time to study for a test, are your notes neat and organized? Have you ever had trouble reading your own mathematics handwriting? (Be honest—I have.)

When any of these things happens, it's time to get organized. Here are a few suggestions:

- Write your notes and complete your homework assignments in a notebook with pockets (spiral or ring binder).
- Take class notes in this notebook, and then follow the notes with your completed homework assignment.
- When you receive graded papers or handouts, place them in the notebook pocket so that you will not lose them.
- Mark (possibly with an exclamation point) any note(s) that seem extra important to you.
- Mark (possibly with a question mark) any notes or homework that you are having trouble with.
- See your instructor or a math tutor for help with the concepts or exercises that you are having trouble understanding.
- If you are having trouble reading your own handwriting, *slow down* and write your mathematics work clearly!

Exercises

1. Have you been completing your assignments on time?
2. Have you been correcting any exercises you may be having difficulty with?
3. If you are having trouble understanding a mathematical concept or correcting any homework exercises, have you visited your instructor, a tutor, or your campus math lab?
4. Are you taking lecture notes in your mathematics course? (By the way, these notes should include worked-out examples solved by your instructor.)
5. Is your mathematics course material (handouts, graded papers, lecture notes) organized?
6. If your answer to Exercise **5** is no, take a moment and review your course material. List at least two ways that you might better organize it.

Study Skills Builder 9

Organizing a Notebook

It's never too late to get organized. If you need ideas about organizing a notebook for your mathematics course, try some of these:

- Use a spiral or ring binder notebook with pockets and use it for mathematics only.
- Start each page by writing the book's section number you are working on at the top.
- When your instructor is lecturing, take notes. *Always* include any examples your instructor works for you.
- Place your worked-out homework exercises in your notebook immediately after the lecture notes from that section. This way, a section's worth of material is together.
- Homework exercises: Attempt and check all assigned homework.
- Place graded quizzes in the pockets of your notebook or a special section of your binder.

Exercises

Check your notebook organization by answering the following questions.

1. Do you have a spiral or ring binder notebook for your mathematics course only?
2. Have you ever had to flip through several sheets of notes and work in your mathematics notebook to determine what section's work you are in?
3. Are you now writing the textbook's section number at the top of each notebook page?
4. Have you ever lost or had trouble finding a graded quiz or test?
5. Are you now placing all your graded work in a dedicated place in your notebook?
6. Are you attempting all of your homework and placing all of your work in your notebook?
7. Are you checking and correcting your homework in your notebook? If not, why not?
8. Are you writing in your notebook the examples your instructor works for you in class?

Study Skills Builder 10

How Are Your Homework Assignments Going?

It is very important in mathematics to keep up with homework. Why? Many concepts build on each other. Often your understanding of a day's concepts depends on an understanding of the previous day's material.

Remember that completing your homework assignment involves a lot more than attempting a few of the problems assigned.

To complete a homework assignment, remember these four things:

- Attempt all of it.
- Check it.
- Correct it.
- If needed, ask questions about it.

Exercises

Take a moment and review your completed homework assignments. Answer the questions below based on this review.

1. Approximate the fraction of your homework you have attempted.
2. Approximate the fraction of your homework you have checked (if possible).
3. If you are able to check your homework, have you corrected it when errors have been found?
4. When working homework, if you do not understand a concept, what do you do?

MyLab Math and MathXL

Study Skills Builder 11

Tips for Turning in Your Homework on Time

It is very important to keep up with your mathematics homework assignments. Why? Many concepts in mathematics build upon each other.

Remember these four tips to help ensure your work is completed on time:

- Know the assignments and due dates set by your instructor.
- Do not wait until the last minute to submit your homework.
- Set a goal to submit your homework 6–8 hours before the scheduled due date in case you have unexpected technology trouble.
- Schedule enough time to complete each assignment.

Following the tips above will also help you avoid potentially losing points for late or missed assignments.

Exercises

Take a moment to consider your work on your homework assignments to date and answer the following questions:

1. What percentage of your assignments have you turned in on time?
2. Why might it be a good idea to submit your homework 6–8 hours before the scheduled deadline?
3. If you have missed submitting any homework by the due date, list some of the reasons why this occurred.
4. What steps do you plan to take in the future to ensure your homework is submitted on time?

Study Skills Builder 12

Tips for Doing Your Homework Online

Practice is one of the main keys to success in any mathematics course. Did you know that MyLab Math/MathXL provides you with **immediate feedback** for each exercise? If you are incorrect, you are given hints to work the exercise correctly. You have **unlimited practice opportunities** and can rework any exercises you have trouble with until you master them, and submit homework assignments unlimited times before the deadline.

Remember these success tips when doing your homework online:

- Attempt all assigned exercises.
- Write down (neatly) your step-by-step work for each exercise before entering your answer.
- Use the immediate feedback provided by the program to help you check and correct your work for each exercise.
- Rework any exercises you have trouble with until you master them.
- Work through your homework assignment as many times as necessary until you are satisfied.

Exercises

Take a moment to think about your homework assignments to date and answer the following:

1. Have you attempted all assigned exercises?
2. Of the exercises attempted, have you also written out your work before entering your answer—so that you can check it?
3. Are you familiar with how to enter answers using the MathXL player so that you avoid answer-entry-type errors?
4. List some ways the immediate feedback and practice supports have helped you with your homework. If you have not used these supports, how do you plan to use them with the success tips above on your next assignment?

Study Skills Builder 13

Organizing Your Work

Have you ever used any readily available paper (such as the back of a flyer, another course assignment, Post-its, etc.) to work out homework exercises before entering the answer in MathXL? To save time, have you ever entered answers directly into MathXL without working the exercises on paper? When it's time to study, have you ever been unable to find your completed work or read and follow your own mathematics handwriting?

When any of these things happen, it's time to get organized. Here are some suggestions:

- Write your step-by-step work for each homework exercise, (neatly) on lined, loose-leaf paper and keep this in a 3-ring binder.
- Refer to your step-by-step work when you receive feedback that your answer is incorrect in MathXL. Double-check using the steps and hints provided by the program and correct your work accordingly.
- Keep your written homework with your class notes for that section.
- Identify any exercises you are having trouble with and ask questions about them.
- Keep all graded quizzes and tests in this binder as well to study later.

If you follow the suggestions above, you and your instructor or tutor will be able to follow your steps and correct any mistakes. You will have a written copy of your work to refer to later to ask questions and study for tests.

Exercises

1. Why is it important that you write out your step-by-step work for homework exercises and keep a hard copy of all work submitted online?
2. If you have gotten an incorrect answer, are you able to follow your steps and find your error?
3. If you were asked today to review your previous homework assignments and first test, could you find them? If not, list some ways you might better organize your work.

Study Skills Builder 14

Getting Help with Your Homework Assignments

There are many resources available to you through MathXL to help you work through any homework exercises you may have trouble with. It is important that you know what these resources are and know when and how to use them.

Let's review the features found in the homework exercises:

- **Help Me Solve This**—provides step-by-step help for the exercise you are working. You must work an additional exercise of the same type (without this help) before you can get credit for having worked it correctly.
- **View an Example**—allows you to view a correctly worked exercise similar to the one you are having trouble with. You can go back to your original exercise and work it on your own.
- **E-Book**—allows you to read examples from your text and find similar exercises.
- **Video**—your text author, Elayn Martin-Gay, works an exercise similar to the one you need help with. **Not all exercises have an accompanying video clip.
- **Ask My Instructor**—allows you to e-mail your instructor for help with an exercise.

Exercises

1. How does the "Help Me Solve This" feature work?
2. If the "View an Example" feature is used, is it necessary to work an additional problem before continuing the assignment?
3. When might be a good time to use the "Video" feature? Do all exercises have an accompanying video clip?
4. Which of the features above have you used? List those you found the most helpful to you.
5. If you haven't used the features discussed, list those you plan to try on your next homework assignment.

Study Skills Builder 15

Tips for Preparing for an Exam

Did you know that you can rework your previous homework assignments in MyLab Math and MathXL? This is a great way to prepare for tests. To do this, open a previous homework assignment and click "similar exercise." This will generate new exercises similar to the homework you have submitted. You can then rework the exercises and assignments until you feel confident that you understand them.

To prepare for an exam, follow these tips:

- Review your written work for your previous homework assignments along with your class notes.
- Identify any exercises or topics that you have questions on or have difficulty understanding.
- Rework your previous assignments in MyLab Math and MathXL until you fully understand them and can do them without help.
- Get help for any topics you feel unsure of or for which you have questions.

Exercises

1. Are your current homework assignments up to date and is your written work for them organized in a binder or notebook? If the answer is no, it's time to get organized. For tips on this, see Study Skills Builder 13—Organizing Your Work.
2. How many days in advance of an exam do you usually start studying?
3. List some ways you think that working previous homework assignments can help you prepare for your test.
4. List two or three resources you can use to get help for any topics you are unsure of or have questions on.

Good luck!

Study Skills Builder 16

How Well Do You Know the Resources Available to You in MyLab Math?

There are many helpful resources available to you in MyLab Math. Let's take a moment to locate and explore a few of them now. Go into your MyLab Math course, and visit the Multimedia Library, Tools for Success, and E-Book.

Let's see what you found.

Exercises

1. List the resources available to you in the Multimedia Library.
2. List the resources available to you in the Tools for Success folder.
3. Where did you find the English/Spanish Audio Glossary?
4. Can you view videos from the E-Book?
5. Did you find any resources you did not know about? If so, which ones?
6. Which resources have you used most often or find most helpful?

Additional Help Inside and Outside Your Textbook

Study Skills Builder 17

How Well Do You Know Your Textbook?

The questions below will help determine whether you are familiar with your textbook. For additional information, see Section 1.1 in this text.

1. What does the ▶ icon mean?
2. What does the ✎ icon mean?
3. What does the △ icon mean?
4. Where can you find a review for each chapter? What answers to this review can be found in the back of your text?
5. Each chapter contains an overview of the chapter along with examples. What is this feature called?
6. Each chapter contains a review of vocabulary. What is this feature called?
7. There are practice exercises that are contained in this text. What are they and how can they be used?
8. This text contains a student section in the back entitled Student Resources. List the contents of this section and how they might be helpful.
9. What exercise answers are available in this text? Where are they located?

Study Skills Builder 18

Are You Familiar with Your Textbook Supplements?

Below is a review of some of the student supplements available for additional study. Check to see if you are using the ones most helpful to you.

- Chapter Test Prep Videos. These videos provide video clip solutions to the Chapter Test exercises in this text. You will find them extremely useful when studying for tests or exams.
- Interactive DVD Lecture Series. These are keyed to each section of the text. The material is presented by me, Elayn Martin-Gay, and I have placed a ▶ by the exercises in the text that I have worked on the video.
- The *Student Solutions Manual*. This contains worked-out solutions to odd-numbered exercises as well as every exercise in the Integrated Reviews, Chapter Reviews, Getting Ready for the Tests, Chapter Tests, and Cumulative Reviews.
- Pearson Tutor Center. Mathematics questions may be phoned, faxed, or e-mailed to this center.
- MyLab Math is a text-specific online course. MathXL is an online homework, tutorial, and assessment system. Take a moment and determine whether these are available to you.

As usual, your instructor is your best source of information.

Exercises

Let's see how you are doing with textbook supplements.

1. Name one way the Lecture Videos can be helpful to you.
2. Name one way the Chapter Test Prep Videos can help you prepare for a chapter test.
3. List any textbook supplements that you have found useful.
4. Have you located and visited a learning resource lab located on your campus?
5. List the textbook supplements that are currently housed in your campus' learning resource lab.

Study Skills Builder 19

Are You Getting All the Mathematics Help That You Need?

Remember that, in addition to your instructor, there are many places to get help with your mathematics course. For example:

- This text has an accompanying video lesson for every section, and the CD in this text contains worked-out solutions to every Chapter Test exercise.
- The back of the book contains answers to odd-numbered exercises.
- A *Student Solutions Manual* is available that contains worked-out solutions to odd-numbered exercises as well as solutions to every exercise in the Integrated Reviews, Chapter Reviews, Getting Ready for the Tests, Chapter Tests, and Cumulative Reviews.
- Don't forget to check with your instructor for other local resources available to you, such as a tutor center.

Exercises

1. List items you find helpful in the text and all student supplements to this text.
2. List all the campus help that is available to you for this course.
3. List any help (besides the textbook) from Exercises **1** and **2** above that you are using.
4. List any help (besides the textbook) that you feel you should try.
5. Write a goal for yourself that includes trying everything you listed in Exercise **4** during the next week.

Study Guide

Bigger Picture—Study Guide Outline

Simplifying Expressions and Solving Equations

I. Simplifying Expressions

A. Real Numbers

1. Add: (Sec. 1.4)

$-1.7 + (-0.21) = -1.91$ Adding like signs. Add absolute values. Attach common sign.

$-7 + 3 = -4$ Adding different signs. Subtract absolute values. Attach the sign of the number with the larger absolute value.

2. Subtract: Add the first number to the opposite of the second number. (Sec. 1.5)

$17 - 25 = 17 + (-25) = -8$

3. Multiply or divide: Multiply or divide the two numbers as usual. If the signs are the same, the answer is positive. If the signs are different, the answer is negative. (Sec. 1.6)

$-10 \cdot 3 = -30, \quad -81 \div (-3) = 27$

B. Exponents (Sec. 3.1, 3.2)

$$x^7 \cdot x^5 = x^{12};\ (x^7)^5 = x^{35};\ \frac{x^7}{x^5} = x^2;\ x^0 = 1;\ 8^{-2} = \frac{1}{8^2} = \frac{1}{64}$$

C. Polynomials

1. Add: Combine like terms. (Sec. 3.4)

$$\begin{aligned}(3y^2 + 6y + 7) + (9y^2 - 11y - 15) &= 3y^2 + 6y + 7 + 9y^2 - 11y - 15\\ &= 12y^2 - 5y - 8\end{aligned}$$

2. Subtract: Change the sign of the terms of the polynomial being subtracted, then add. (Sec. 3.4)

$$\begin{aligned}(3y^2 + 6y + 7) - (9y^2 - 11y - 15) &= 3y^2 + 6y + 7 - 9y^2 + 11y + 15\\ &= -6y^2 + 17y + 22\end{aligned}$$

3. Multiply: Multiply each term of one polynomial by each term of the other polynomial. (Sec. 3.5)

$$\begin{aligned}(x + 5)(2x^2 - 3x + 4) &= x(2x^2 - 3x + 4) + 5(2x^2 - 3x + 4)\\ &= 2x^3 - 3x^2 + 4x + 10x^2 - 15x + 20\\ &= 2x^3 + 7x^2 - 11x + 20\end{aligned}$$

4. Divide: (Sec. 3.7)

a. To divide by a monomial, divide each term of the polynomial by the monomial.

$$\frac{8x^2 + 2x - 6}{2x} = \frac{8x^2}{2x} + \frac{2x}{2x} - \frac{6}{2x} = 4x + 1 - \frac{3}{x}$$

b. To divide by a polynomial other than a monomial, use long division.

$$\begin{array}{r} x - 6 + \dfrac{40}{2x+5} \\ 2x + 5\overline{)2x^2 - 7x + 10} \\ \underline{-2x^2 \mp 5x} \\ -12x + 10 \\ \underline{\pm 12x \pm 30} \\ 40 \end{array}$$

D. Factoring Polynomials

See the Chapter 4 Integrated Review for steps.

$3x^4 - 78x^2 + 75 = 3(x^4 - 26x^2 + 25)$ — Factor out GCF—always first step.

$= 3(x^2 - 25)(x^2 - 1)$ — Factor trinomial.

$= 3(x + 5)(x - 5)(x + 1)(x - 1)$ — Factor further—each difference of squares.

E. Rational Expressions

1. **Simplify:** Factor the numerator and denominator. Then remove factors of 1 by dividing out common factors in the numerator and denominator. (Sec. 5.1)

$$\frac{x^2 - 9}{7x^2 - 21x} = \frac{(x + 3)(x - 3)}{7x(x - 3)} = \frac{x + 3}{7x}$$

2. **Multiply:** Multiply numerators, then multiply denominators. (Sec. 5.2)

$$\frac{5z}{2z^2 - 9z - 18} \cdot \frac{22z + 33}{10z} = \frac{5 \cdot z}{(2z + 3)(z - 6)} \cdot \frac{11(2z + 3)}{2 \cdot 5 \cdot z} = \frac{11}{2(z - 6)}$$

3. **Divide:** First fraction times the reciprocal of the second fraction. (Sec. 5.2)

$$\frac{14}{x + 5} \div \frac{x + 1}{2} = \frac{14}{x + 5} \cdot \frac{2}{x + 1} = \frac{28}{(x + 5)(x + 1)}$$

4. **Add or subtract:** Must have same denominator. If not, find the LCD and write each fraction as an equivalent fraction with the LCD as denominator. (Sec. 5.4)

$$\frac{9}{10} - \frac{x + 1}{x + 5} = \frac{9(x + 5)}{10(x + 5)} - \frac{10(x + 1)}{10(x + 5)}$$

$$= \frac{9x + 45 - 10x - 10}{10(x + 5)} = \frac{-x + 35}{10(x + 5)}$$

F. Radicals

1. **Simplify square roots:** If possible, factor the radicand so that one factor is a perfect square. Then use the product rule and simplify. (Sec. 8.2)

$$\sqrt{75} = \sqrt{25 \cdot 3} = \sqrt{25} \cdot \sqrt{3} = 5\sqrt{3}$$

2. **Add or subtract:** Only like radicals (same index and radicand) can be added or subtracted. (Sec. 8.3)

$$8\sqrt{10} - \sqrt{40} + \sqrt{5} = 8\sqrt{10} - 2\sqrt{10} + \sqrt{5} = 6\sqrt{10} + \sqrt{5}$$

3. Multiply or divide: $\sqrt{a} \cdot \sqrt{b} = \sqrt{ab}$; $\dfrac{\sqrt{a}}{\sqrt{b}} = \sqrt{\dfrac{a}{b}}$. (Sec. 8.4)

$$\sqrt{11} \cdot \sqrt{3} = \sqrt{33};\ \frac{\sqrt{140}}{\sqrt{7}} = \sqrt{\frac{140}{7}} = \sqrt{20} = \sqrt{4 \cdot 5} = 2\sqrt{5}$$

4. Rationalizing the denominator: (Sec. 8.4)

a. If denominator is one term,

$$\frac{5}{\sqrt{11}} = \frac{5 \cdot \sqrt{11}}{\sqrt{11} \cdot \sqrt{11}} = \frac{5\sqrt{11}}{11}$$

b. If denominator is two terms, multiply by 1 in the form of $\dfrac{\text{conjugate of denominator}}{\text{conjugate of denominator}}$.

$$\frac{13}{3 + \sqrt{2}} = \frac{13}{3 + \sqrt{2}} \cdot \frac{3 - \sqrt{2}}{3 - \sqrt{2}} = \frac{13(3 - \sqrt{2})}{9 - 2} = \frac{39 - 13\sqrt{2}}{7}$$

II. Solving Equations and Inequalities

A. Linear Equations: Power on variable is 1 and there are no variables in denominator. (Sec. 2.3)

$7(x - 3) = 4x + 6$	Linear equation. (If fractions, multiply by LCD.)
$7x - 21 = 4x + 6$	Use the distributive property.
$7x = 4x + 27$	Add 21 to both sides.
$3x = 27$	Subtract $4x$ from both sides.
$x = 9$	Divide both sides by 3.

B. Linear Inequalities: Same as linear equation except if you multiply or divide by a negative number, then reverse direction of inequality. (Sec. 2.7)

$-4x + 11 \le -1$	Linear inequality.
$-4x \le -12$	Subtract 11 from both sides.
$\dfrac{-4x}{-4} \ge \dfrac{-12}{-4}$	Divide both sides by -4 and reverse the direction of the inequality symbol.
$x \ge 3$	Simplify.

C. Quadratic and Higher Degree Equations: Solve: first write the equation in standard form (one side is 0).

1. If the polynomial on one side factors, solve by factoring. (Sec. 4.6)

2. If the polynomial does not factor, solve by the quadratic formula. (Sec. 9.3)

By factoring:

$$x^2 + x = 6$$
$$x^2 + x - 6 = 0$$
$$(x - 2)(x + 3) = 0$$
$$x - 2 = 0 \text{ or } x + 3 = 0$$
$$x = 2 \quad \text{or} \quad x = -3$$

By quadratic formula:

$$x^2 + x = 5$$
$$x^2 + x - 5 = 0$$
$$a = 1, b = 1, c = -5$$
$$x = \frac{-1 \pm \sqrt{1^2 - 4(1)(-5)}}{2 \cdot 1}$$
$$= \frac{-1 \pm \sqrt{21}}{2}$$

D. Equations with Rational Expressions: Make sure the proposed solution does not make any denominator 0. (Sec. 5.5)

$$\frac{3}{x} - \frac{1}{x-1} = \frac{4}{x-1}$$ Equation with rational expressions

$$x(x-1)\cdot\frac{3}{x} - x(x-1)\cdot\frac{1}{x-1} = x(x-1)\cdot\frac{4}{x-1}$$ Multiply through by $x(x-1)$.

$$3(x-1) - x\cdot 1 = x\cdot 4$$ Simplify.

$$3x - 3 - x = 4x$$ Use the distributive property.

$$-3 = 2x$$ Simplify and move variable terms to right side.

$$-\frac{3}{2} = x$$ Divide both sides by 2.

E. Proportions: An equation with two ratios equal. Set cross products equal, then solve. Make sure the proposed solution does not make any denominator 0. (Sec. 5.6)

$$\frac{5}{x} = \frac{9}{2x-3}$$

$$5(2x-3) = 9\cdot x$$ Set cross products equal.

$$10x - 15 = 9x$$ Multiply.

$$x = 15$$ Write equation with variable terms on one side and constants on the other.

F. Equations with Radicals: To solve, isolate a radical, then square both sides. You may have to repeat this. Check possible solution(s) in the original equation. (Sec. 8.5)

$$\sqrt{x+49} + 7 = x$$

$$\sqrt{x+49} = x - 7$$ Subtract 7 from both sides.

$$x + 49 = x^2 - 14x + 49$$ Square both sides.

$$0 = x^2 - 15x$$ Set terms equal to 0.

$$0 = x(x-15)$$ Factor.

$$\cancel{x = 0} \text{ or } x = 15$$ Set each factor equal to 0 and solve.

Check the proposed solutions for extraneous solutions.

This proposed solution ($x = 0$) does not check in the original equation. It is an extraneous solution and not a solution to the equation.

Study Guide

Practice Final Exam

1. ______

2. ______

3. ______

4. ______

5. ______

6. ______

7. ______

8. ______

9. ______

10. ______

11. ______

12. ______

13. ______

14. ______

15. ______

16. ______

17. ______

Preparing for your Final Exam? Take this Practice Final and watch the full video solutions to any of the exercises you want to review. You will find the Practice Final video in the Video Lecture Series. The video also provides you with an overview to help you approach different problem types just as you will need to do on a Final Exam. To build your own study guide use the previous Bigger Picture—Study Guide Outline feature as an example.

Evaluate.

1. -3^4

2. 4^{-3}

3. $6[5 + 2(3 - 8) - 3]$

Perform the indicated operations and simplify if possible.

4. $(5x^3 + x^2 + 5x - 2) - (8x^3 - 4x^2 + x - 7)$

5. $(4x - 2)^2$

6. $(3x + 7)(x^2 + 5x + 2)$

Factor.

7. $6t^2 - t - 5$

8. $3x^3 - 21x^2 + 30x$

9. $180 - 5x^2$

10. $3a^2 + 3ab - 7a - 7b$

11. $x - x^5$

Simplify. Write answers with positive exponents only.

12. $\left(\dfrac{4x^2y^3}{x^3y^{-4}}\right)^2$

13. $\dfrac{5 - \dfrac{1}{y^2}}{\dfrac{1}{y} + \dfrac{2}{y^2}}$

Perform the indicated operations and simplify if possible.

14. $\dfrac{x^2 - 13x + 42}{x^2 + 10x + 21} \div \dfrac{x^2 - 4}{x^2 + x - 6}$

15. $\dfrac{5a}{a^2 - a - 6} - \dfrac{2}{a - 3}$

Solve each equation or inequality.

16. $4(n - 5) = -(4 - 2n)$

17. $x(x + 6) = 7$

18. $3x - 5 \geq 7x + 3$

19. $2x^2 - 6x + 1 = 0$

20. $\frac{4}{y} - \frac{5}{3} = -\frac{1}{5}$

21. $\frac{5}{y+1} = \frac{4}{y+2}$

22. $\frac{a}{a-3} = \frac{3}{a-3} - \frac{3}{2}$

23. $\sqrt{2x - 2} = x - 5$

Graph the following.

24. $5x - 7y = 10$

25. $y = -1$

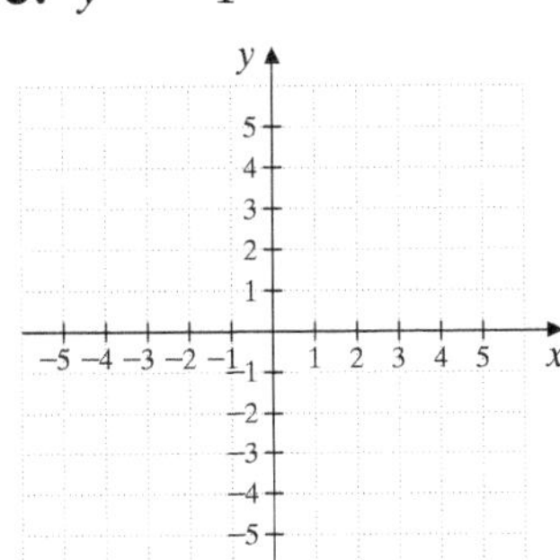

26. $y \geq -4x$

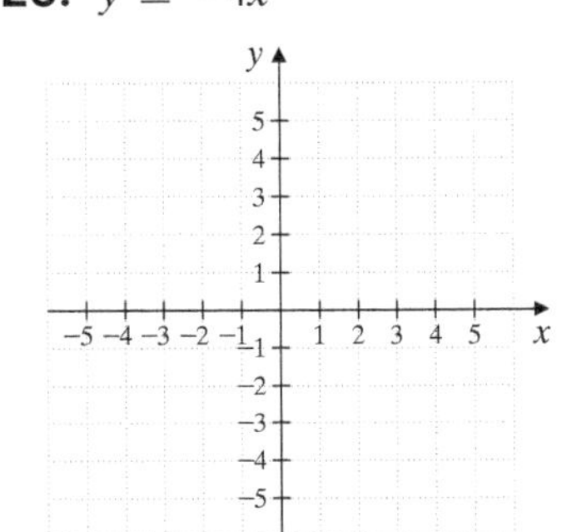

Find the slope of each line.

27. through $(6, -5)$ and $(-1, 2)$

28. $-3x + y = 5$

Write equations of the following lines. Write each equation in standard form.

29. through $(2, -5)$ and $(1, 3)$

30. slope $\frac{1}{8}$; y-intercept $(0, 12)$

Solve each system of equations.

31. $\begin{cases} 3x - 2y = -14 \\ y = x + 5 \end{cases}$

32. $\begin{cases} 4x - 6y = 7 \\ -2x + 3y = 0 \end{cases}$

Answer the questions about functions.

33. If $f(x) = x^3 - x$, find

a. $f(-1)$ **b.** $f(0)$ **c.** $f(4)$

18. ______
19. ______
20. ______
21. ______
22. ______
23. ______
24. ______
25. ______
26. ______
27. ______
28. ______
29. ______
30. ______
31. ______
32. ______
33. a. ______
b. ______
c. ______

34. ______
35. ______
36. ______
37. ______
38. ______
39. ______
40. ______
41. ______
42. ______
43. ______
44. ______
45. ______
46. ______
47. ______
48. ______

34. Determine whether the relation is also a function. If a function, find its domain and range.

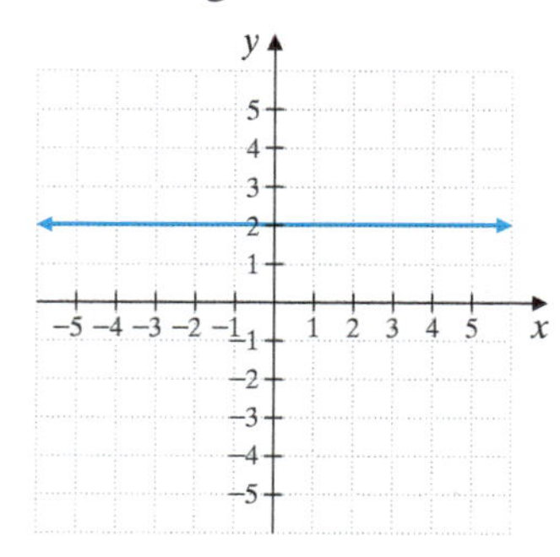

Evaluate.

35. $\sqrt{16}$

36. $\sqrt[3]{125}$

37. $\sqrt{\frac{9}{16}}$

Simplify.

38. $\sqrt{54}$

39. $\sqrt{24x^8}$

Perform the indicated operations and simplify if possible.

40. $\sqrt{18} - \sqrt{75} + 7\sqrt{3} - \sqrt{8}$

41. $\frac{\sqrt{40x^4}}{\sqrt{2x}}$

42. $\sqrt{2}(\sqrt{6} - \sqrt{5})$

Rationalize each denominator.

43. $\frac{8}{\sqrt{5y}}$

44. $\frac{8}{\sqrt{6} + 2}$

Solve each application.

45. One number plus five times its reciprocal is equal to six. Find the number.

46. Some states have a single area code for the entire state. Two such states have area codes where one is double the other. If the sum of these integers is 1203, find the two area codes.

47. Two hikers start at opposite ends of the St. Tammany Trails and walk toward each other. The trail is 36 miles long and they meet in 4 hours. If one hiker is twice as fast as the other, find both hiking speeds.

48. Find the amount of a 12% saline solution a lab assistant should add to 80 cc (cubic centimeters) of a 22% saline solution in order to have a 16% solution.

Answers to Selected Exercises

Chapter R Prealgebra Review

Section R.1

Vocabulary, Readiness & Video Check **1.** prime factorization **3.** prime **5.** factor **7.** No, the natural number 1 is neither prime nor composite. **9.** smallest

Exercise Set R.1 **1.** 1, 3, 9 **3.** 1, 2, 3, 4, 6, 8, 12, 24 **5.** 1, 2, 3, 6, 7, 14, 21, 42 **7.** 1, 2, 4, 5, 8, 10, 16, 20, 40, 80 **9.** 1, 19 **11.** prime **13.** composite **15.** prime **17.** composite **19.** composite **21.** $2 \cdot 3 \cdot 3$ **23.** $2 \cdot 2 \cdot 5$ **25.** $2 \cdot 2 \cdot 2 \cdot 7$ **27.** $3 \cdot 3 \cdot 3 \cdot 3$ **29.** $2 \cdot 2 \cdot 3 \cdot 5 \cdot 5$ **31.** $2 \cdot 2 \cdot 3 \cdot 7 \cdot 7$ **33.** d **35.** 12 **37.** 42 **39.** 60 **41.** 35 **43.** 36 **45.** 80 **47.** 360 **49.** 72 **51.** 120 **53.** 42 **55.** 48 **57.** 360 **59.** **a.** $2 \cdot 2 \cdot 2 \cdot 5$ **b.** $2 \cdot 2 \cdot 2 \cdot 5$ **c.** answers may vary **61.** every 35 days **63.** 2520

Section R.2

Vocabulary, Readiness & Video Check **1.** fraction; denominator; numerator **3.** simplest form **5.** $\frac{a \cdot c}{b \cdot d}$ **7.** $\frac{a \cdot d}{b \cdot c}$ **9.** least common denominator (LCD) **11.** The fraction is equal to 1. **13.** wrote both the numerator and denominator as products of prime numbers **15.** When adding or subtracting fractions, we must have a common denominator. When multiplying or dividing fractions, we do not.

Exercise Set R.2 **1.** 1 **3.** 10 **5.** 13 **7.** 0 **9.** undefined **11.** $\frac{21}{30}$ **13.** $\frac{4}{18}$ **15.** $\frac{16}{20}$ **17.** $\frac{1}{2}$ **19.** $\frac{2}{3}$ **21.** $\frac{3}{7}$ **23.** $\frac{3}{5}$ **25.** $\frac{4}{5}$ **27.** $\frac{11}{8}$ **29.** $\frac{30}{61}$ **31.** $\frac{8}{11}$ **33.** $\frac{3}{8}$ **35.** $\frac{1}{2}$ **37.** $\frac{6}{7}$ **39.** 15 **41.** $18\frac{20}{27}$ **43.** $2\frac{28}{29}$ **45.** 1 **47.** $\frac{11}{60}$ **49.** $\frac{23}{21}$ **51.** $\frac{65}{21}$ **53.** $1\frac{3}{4}$ **55.** $5\frac{1}{6}$ **57.** $\frac{9}{35}$ **59.** $\frac{1}{3}$ **61.** $\frac{1}{6}$ **63.** $\frac{3}{80}$ **65.** $\frac{5}{66}$ **67.** $48\frac{1}{15}$ **69.** 37 **71.** $10\frac{5}{11}$ **73.** $\frac{7}{5}$ **75.** $7\frac{1}{12}$ **77.** $\frac{17}{18}$ **79.** incorrect; $\frac{12}{24} = \frac{2 \cdot 2 \cdot 3}{2 \cdot 2 \cdot 2 \cdot 3} = \frac{1}{2}$ **81.** incorrect; $\frac{2}{7} + \frac{9}{7} = \frac{11}{7}$ **83.** answers may vary **85.** $\frac{1}{12}$ **87.** $\frac{6}{11}$ **89.** $7\frac{13}{20}$ in. **91.** $\frac{3}{50}$ **93.** answers may vary **95.** $\frac{2}{25}$ **97.** answers may vary **99.** $\frac{6}{55}$ sq m

Section R.3

Vocabulary, Readiness & Video Check **1.** decimal **3.** vertically **5.** Percent **7.** percent **9.** right **11.** do; do not **13.** numerator; denominator

Exercise Set R.3 **1.** $\frac{6}{10}$ **3.** $\frac{186}{100}$ **5.** $\frac{114}{1000}$ **7.** $\frac{1231}{10}$ **9.** 6.83 **11.** 27.0578 **13.** 6.5 **15.** 15.22 **17.** 0.12 **19.** 0.2646 **21.** 1.68 **23.** 5.8 **25.** 56.431 **27.** 67.5 **29.** 70 **31.** 598.23 **33.** 43.274 **35.** 840 **37.** 84.97593 **39.** 0.6 **41.** 0.23 **43.** 0.595 **45.** 98,207.2 **47.** 12.35 **49.** 0.75 **51.** $0.\overline{3} \approx 0.33$ **53.** 0.4375 **55.** $0.\overline{54} \approx 0.55$ **57.** $4.8\overline{3} \approx 4.83$ **59.** 0.28 **61.** 0.031 **63.** 1.35 **65.** 2 **67.** 0.9655 **69.** 0.001 **71.** 0.158 **73.** 68% **75.** 87.6% **77.** 100% **79.** 50% **81.** 192% **83.** 0.4% **85.** 78.1% **87.** 0.005; $\frac{1}{200}$ **89.** 0.142; $\frac{71}{500}$ **91.** **a.** tenths **b.** thousandths **c.** ones **93.** answers may vary **95.** 109.9 lb **97.** **a.** 52.9% **b.** 52.86% **99.** b, d **101.** 4% **103.** occupational therapy assistant **105.** 0.30 or 0.3 **107.** answers may vary

Section R.4

Vocabulary, Readiness & Video Check **1.** bar **3.** line **5.** circle **7.** 100 **9.** Count the number of symbols and multiply this number by how much each symbol stands for (from the key). **11.** bar graph **13.** 100%

Exercise Set R.4 **1.** Kansas **3.** 4.5 million or 4,500,000 acres **5.** South Dakota, Colorado, and Washington **7.** North Dakota **9.** 48,000 **11.** 2016 **13.** 18,000 **15.** 60,000 wildfires/year **17.** September **19.** 79 (exact); or approximately 80 **21.** $\frac{1}{18}$ **23.** Tokyo, Japan; about 38 million or 38,000,000 **25.** New York; 21.4 million or 21,400,000 **27.** approximately 3 million

29.

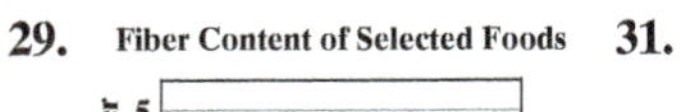

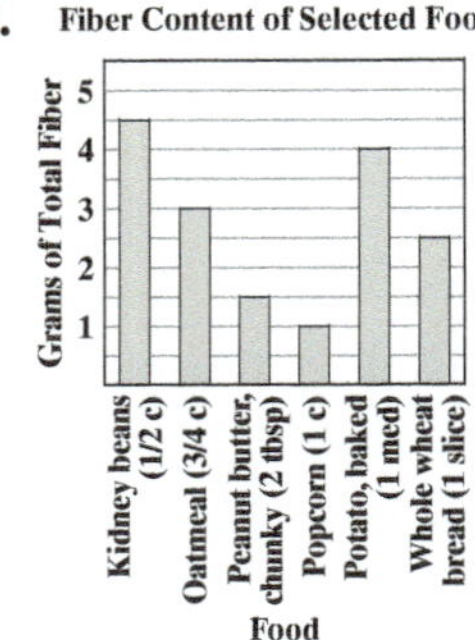

31.

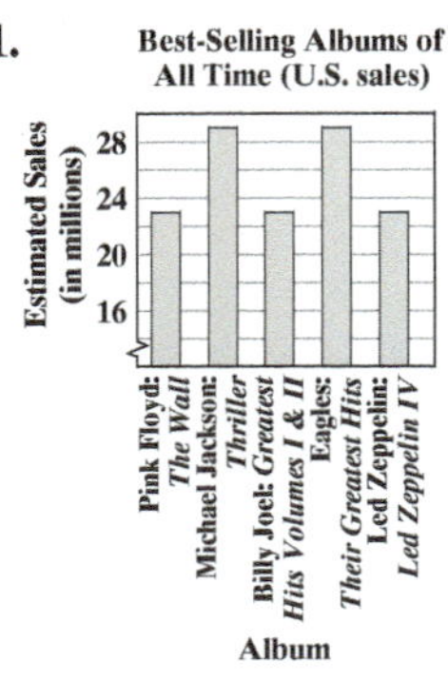

33. 15 adults **35.** 61 adults **37.** 24 adults **39.** 12 adults **41.** $\frac{9}{100}$ **43.** 20 to 44 **45.** 109 million **47.** 23 million **49.** answers may vary

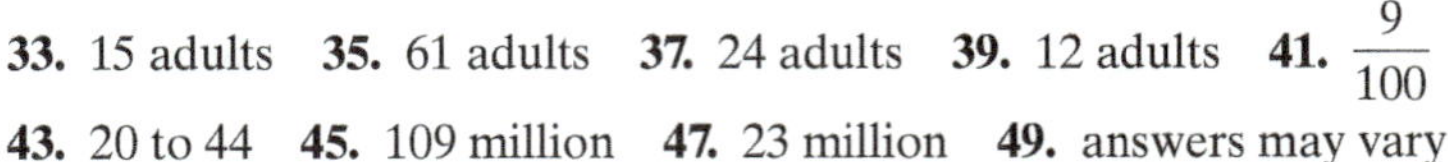

51. |; 1 **53.** ~~||||~~ |||; 8 **55.** ~~||||~~ |; 6 **57.** ~~||||~~ |; 6 **59.** ||; 2

61.

Number of Games: 8, 6, 4, 2; Golf Scores: 70–79, 80–89, 90–99, 100–109

63. 8.3 goals/game **65.** 2009 **67.** increase **69.** 2007, 2013 **71.** parent or guardian's home **73.** $\frac{9}{35}$ **75.** $\frac{9}{16}$ **77.** Asia **79.** 37% **81.** 17,100,000 sq mi **83.** 2,850,000 sq mi **85.** 55% **87.** nonfiction **89.** 31,400 books **91.** 27,632 books **93.** 25,120 books **95.** 83°F **97.** Sunday; 68°F **99.** Tuesday; 13°F **101.** answers may vary **103.** Pacific; answers may vary **105.** 129,600,002 sq km **107.** 55,542,858 sq km **109.** 672 respondents **111.** 2408 respondents **113.** $\frac{12}{31}$ **115.** no; answers may vary **117.** answers may vary

Chapter R Vocabulary Check **1.** factor **2.** multiple **3.** composite number **4.** percent **5.** equivalent **6.** improper fraction **7.** prime number **8.** simplified **9.** proper fraction **10.** mixed number **11.** bar **12.** pictograph **13.** line **14.** circle **15.** histogram; class interval; class frequency

Chapter R Review **1.** $2 \cdot 3 \cdot 7$ **2.** $2 \cdot 2 \cdot 2 \cdot 2 \cdot 2 \cdot 5 \cdot 5$ **3.** 60 **4.** 42 **5.** 60 **6.** 70 **7.** $\frac{15}{24}$ **8.** $\frac{40}{60}$ **9.** $\frac{2}{5}$ **10.** $\frac{3}{20}$ **11.** 2 **12.** 1 **13.** $\frac{8}{77}$ **14.** $\frac{11}{20}$ **15.** $\frac{1}{20}$ **16.** $\frac{11}{18}$ **17.** $14\frac{11}{32}$ **18.** $\frac{1}{2}$ **19.** $20\frac{17}{30}$ **20.** $2\frac{6}{7}$ **21.** $\frac{11}{20}$ sq mi **22.** $\frac{5}{16}$ sq m **23.** $\frac{181}{100}$ **24.** $\frac{35}{1000}$ **25.** 95.118 **26.** 36.785 **27.** 13.38 **28.** 691.573 **29.** 91.2 **30.** 46.816 **31.** 28.6 **32.** 230 **33.** 0.77 **34.** 25.6 **35.** 0.5 **36.** 0.375 **37.** $0.\overline{36} \approx 0.364$ **38.** $0.8\overline{3} \approx 0.833$ **39.** 0.29 **40.** 0.014 **41.** 39% **42.** 120% **43.** 0.643 **44.** b **45.** 175,000 **46.** 600,000 **47.** South **48.** Northeast **49.** South, West **50.** Northeast, Midwest **51.** 30% **52.** 2017 **53.** 1990, 2000, 2010, 2017 **54.** answers may vary **55.** 962 (exact number) **56.** 927 (exact number) **57.** 958 (exact number) **58.** 842 (exact number) **59.** 31 **60.** answers may vary **61.** 1 employee **62.** 4 employees **63.** 18 employees **64.** 9 employees **65.** ~~||||~~; 5 **66.** |||; 3

67. ||||; 4 **68.**

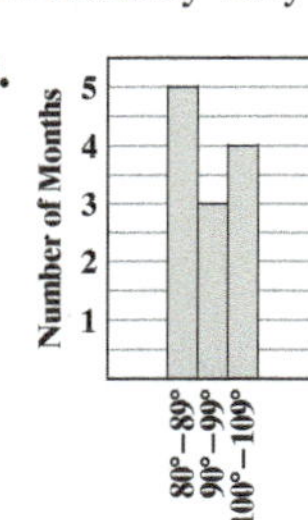

69. mortgage payment **70.** utilities **71.** \$1225 **72.** \$700 **73.** $\frac{39}{160}$ **74.** $\frac{7}{40}$ **75.** 75 **76.** 28 **77.** 2 **78.** 9

Chapter R Getting Ready for the Test **1.** A **2.** B **3.** D **4.** C **5.** C **6.** B **7.** C **8.** D **9.** B **10.** A **11.** F **12.** H **13.** B **14.** D **15.** C **16.** B **17.** E **18.** D **19.** D **20.** A **21.** B **22.** C **23.** C **24.** D **25.** A **26.** C **27.** C **28.** D **29.** C **30.** C

Chapter R Test **1.** $2 \cdot 2 \cdot 2 \cdot 3 \cdot 3$ **2.** 180 **3.** $\frac{25}{60}$ **4.** $\frac{3}{4}$ **5.** $\frac{12}{25}$ **6.** $\frac{13}{10}$ **7.** $\frac{53}{40}$ **8.** $\frac{18}{49}$ **9.** $\frac{1}{20}$ **10.** $\frac{29}{36}$ **11.** $4\frac{8}{9}$ **12.** $2\frac{5}{22}$ **13.** 55 **14.** $13\frac{13}{20}$ **15.** 45.11 **16.** 65.88 **17.** 12.688 **18.** 320 **19.** 23.73 **20.** 0.875 **21.** $0.1\overline{6} \approx 0.167$ **22.** 0.632 **23.** 9% **24.** 75% **25.** $\frac{3}{4}$ **26.** $\frac{1}{200}$ **27.** $\frac{49}{200}$ **28.** $\frac{199}{200}$ **29.** $\frac{1}{8}$ sq ft **30.** $\frac{63}{64}$ sq cm **31.** \$225 **32.** 3rd week; \$350 **33.** \$1100 **34.** June, August, September **35.** February; 3 cm **36.** March and November **37.** 1.6% **38.** 2008, 2011 **39.** 2008–2009, 2011–2012, 2012–2013, 2014–2015 **40.** $\frac{17}{40}$ **41.** $\frac{31}{22}$ **42.** 74 million **43.** 90 million **44.** 9 students **45.** 11 students

Chapter 1 Real Numbers and Introduction to Algebra

Section 1.2

Vocabulary, Readiness & Video Check **1.** whole **3.** inequality **5.** real **7.** 0 **9.** absolute value **11.** To form a true statement: $0 < 7$. **13.** 0 belongs to the whole numbers, the integers, the rational numbers, and the real numbers; since 0 is a rational number, it cannot also be an irrational number.

Exercise Set 1.2 **1.** $<$ **3.** $>$ **5.** $=$ **7.** $<$ **9.** $32 < 212$ **11.** $30 \leq 45$ **13.** true **15.** false **17.** true **19.** false **21.** $20 \leq 25$ **23.** $6 > 0$ **25.** $-12 < -10$ **27.** $7 < 11$ **29.** $5 \geq 4$ **31.** $15 \neq -2$ **33.** 14,494; −282 **35.** −17,813 **37.** 475; −195

39. −4 −3 −2 −1 0 1 2 3 4 5 **41.** $-\frac{1}{4}$ $\frac{1}{3}$ −4 −3 −2 −1 0 1 2 3 4 **43.** −4.5 $-\frac{3}{2}$ $\frac{7}{4}$ 3.25 −5 −4 −3 −2 −1 0 1 2 3 4

45. whole, integers, rational, real **47.** integers, rational, real **49.** natural, whole, integers, rational, real **51.** rational, real **53.** false **55.** true **57.** false **59.** false **61.** 8.9 **63.** 20 **65.** $\frac{9}{2}$ **67.** $\frac{12}{13}$ **69.** $>$ **71.** $=$ **73.** $<$ **75.** $<$ **77.** 2250 million $<$ 2750 million, or 2,250,000,000 $<$ 2,750,000,000 **79.** 550 million bushels less, or −550,000,000 **81.** $-0.04 > -26.7$ **83.** sun **85.** sun **87.** answers may vary

Section 1.3

Calculator Explorations **1.** 125 **3.** 59,049 **5.** 30 **7.** 9857 **9.** 2376

Vocabulary, Readiness & Video Check **1.** base; exponent **3.** multiplication **5.** subtraction **7.** expression **9.** expression; variables **11.** equation **13.** The order in which we perform operations does matter! We came up with an order of operations to avoid getting more than one answer when evaluating an expression. **15.** No; the variable was replaced with 0 in the equation to see if a true statement occurred, and it did not.

Exercise Set 1.3 **1.** 243 **3.** 27 **5.** 1 **7.** 5 **9.** 49 **11.** $\frac{16}{81}$ **13.** $\frac{1}{125}$ **15.** 1.44 **17.** 0.343 **19.** 5^2 sq m **21.** 17 **23.** 20 **25.** 12 **27.** 21 **29.** 45 **31.** 0 **33.** $\frac{2}{7}$ **35.** 30 **37.** 2 **39.** $\frac{7}{18}$ **41.** $\frac{14}{5}$ **43.** $\frac{7}{5}$ **45.** 32 **47.** $\frac{23}{27}$ **49.** 9 **51.** 1 **53.** 1 **55.** 11 **57.** 8 **59.** 45 **61.** 27 **63.** 132 **65.** $\frac{37}{18}$ **67.** solution **69.** not a solution **71.** not a solution **73.** solution **75.** not a solution **77.** solution **79.** $x + 15$ **81.** $x - 5$ **83.** $\frac{x}{4}$ **85.** $3x + 22$ **87.** $1 + 2 = 9 \div 3$ **89.** $3 \neq 4 \div 2$ **91.** $5 + x = 20$ **93.** $7.6x = 17$ **95.** $13 - 3x = 13$ **97.** no; answers may vary **99.** **a.** 64 **b.** 43 **c.** 19 **d.** 22 **101.** 14 in.; 12 sq in. **103.** 14 in.; 9.01 sq in. **105.** Rectangles with the same perimeter can have different areas. **107.** $(20 - 4) \cdot 4 \div 2$ **109.** **a.** expression **b.** equation **c.** equation **d.** expression **e.** expression **111.** answers may vary **113.** answers may vary, for example, $-2(5) - 1$.

Section 1.4

Vocabulary, Readiness & Video Check **1.** 0 **3.** *a* **5.** absolute values **7.** Example 12 is an example of the opposite of the *absolute value* of $-a$, not the opposite of $-a$. The absolute value of $-a$ is positive, so its opposite is negative. Therefore the answers to Examples 11 and 12 have different signs. **9.** Depths below the surface; the diver's position is 231 feet below the surface.

Exercise Set 1.4 **1.** 3 **3.** −14 **5.** 1 **7.** −12 **9.** −5 **11.** −12 **13.** −4 **15.** 7 **17.** −2 **19.** 0 **21.** −19 **23.** 31 **25.** −47 **27.** −2.1 **29.** 38 **31.** −8 **33.** $\frac{1}{4}$ **35.** $-\frac{3}{16}$ **37.** $-\frac{13}{10}$ **39.** $-13\frac{1}{10}$ **41.** −8 **43.** −59 **45.** −9 **47.** 5 **49.** 11 **51.** −18 **53.** 19 **55.** −7 **57.** −26 **59.** −6 **61.** 2 **63.** 0 **65.** −6 **67.** −2 **69.** 7 **71.** 7.9 **73.** $5z$ **75.** $\frac{2}{3}$ **77.** −70 **79.** 3 **81.** 19 **83.** −10 **85.** $0 + (-215) + (-16) = -231$; 231 ft below the surface **87.** 107°F **89.** −95 m **91.** −15 **93.** −\$1053.8 million **95.** July **97.** October **99.** 4.7°F **101.** answers may vary **103.** −3 **105.** −22 **107.** true **109.** false **111.** answers may vary

Section 1.5

Vocabulary, Readiness & Video Check **1.** $a + (-b)$; **b** **3.** $-10 - (-14)$; **d** **5.** addition, opposite **7.** There's a minus sign in the numerator and the replacement value is negative (notice parentheses are used around the replacement value), and it's always good to be careful when working with negative signs. **9.** This means that the overall vertical altitude change of the jet is actually a decrease in altitude from when the Example started.

Exercise Set 1.5 **1.** −10 **3.** −5 **5.** 19 **7.** 11 **9.** −8 **11.** −11 **13.** 37 **15.** 5 **17.** −71 **19.** 0 **21.** $\frac{2}{11}$ **23.** −6.4 **25.** 4.1 **27.** $-\frac{1}{6}$ **29.** $-\frac{11}{12}$ **31.** 8.92 **33.** −8.92 **35.** 13 **37.** −5 **39.** −1 **41.** −23 **43.** −26 **45.** −24 **47.** 3 **49.** −45 **51.** −4 **53.** 13 **55.** 6 **57.** 9 **59.** −9 **61.** $\frac{7}{5}$ **63.** −7 **65.** 21 **67.** $\frac{1}{4}$ **69.** not a solution **71.** not a solution **73.** solution **75.** 263°F **77.** −308 ft **79.** 30° **81.** 35,653 ft **83.** 19,852 ft **85.** 130° **87.** $-5 + x$ **89.** $-20 - x$ **91.** −4.4°, 2.6°, 12°, 23.5°, 15.3° **93.** May **95.** answers may vary **97.** 16 **99.** −20 **101.** true; answers may vary **103.** false; answers may vary **105.** negative, −30,387

Integrated Review **1.** negative **2.** negative **3.** positive **4.** 0 **5.** positive **6.** 0 **7.** positive **8.** positive **9.** $-\frac{1}{7}; \frac{1}{7}$ **10.** $\frac{12}{5}; \frac{12}{5}$ **11.** 3; 3 **12.** $-\frac{9}{11}; \frac{9}{11}$ **13.** −42 **14.** 10 **15.** 2 **16.** −18 **17.** −7 **18.** −39 **19.** −2 **20.** −9 **21.** −3.4 **22.** −9.8

23. $-\frac{25}{28}$ **24.** $-\frac{5}{24}$ **25.** -4 **26.** -24 **27.** 6 **28.** 20 **29.** 6 **30.** 61 **31.** -6 **32.** -16 **33.** -19 **34.** -13 **35.** -4 **36.** -1 **37.** $\frac{13}{20}$ **38.** $-\frac{29}{40}$ **39.** 4 **40.** 9 **41.** -1 **42.** -3 **43.** 8 **44.** 10 **45.** 47 **46.** $\frac{2}{3}$

Section 1.6

Calculator Explorations **1.** 38 **3.** -441 **5.** 490 **7.** 54,499 **9.** 15,625

Vocabulary, Readiness & Video Check **1.** negative **3.** positive **5.** 0 **7.** 0 **9.** The parentheses, or lack of them, determine the base of the expression. In Example 6, $(-2)^4$, the base is -2 and all of -2 is raised to the 4th power. In Example 7, -2^4, the base is 2 and only 2 is raised to the 4th power. **11.** Yes; because division of real numbers is defined in terms of multiplication. **13.** Yes; a true statement results when x is replaced with 5.

Exercise Set 1.6 **1.** -24 **3.** -2 **5.** 50 **7.** -45 **9.** $\frac{3}{10}$ **11.** -7 **13.** -15 **15.** 0 **17.** 16 **19.** -16 **21.** $\frac{9}{16}$ **23.** -0.49 **25.** $\frac{3}{2}$ **27.** $-\frac{1}{14}$ **29.** $-\frac{11}{3}$ **31.** $\frac{1}{0.2}$ **33.** -9 **35.** -4 **37.** 0 **39.** undefined **41.** $-\frac{18}{7}$ **43.** 160 **45.** 64 **47.** $-\frac{8}{27}$ **49.** 3 **51.** -15 **53.** -125 **55.** -0.008 **57.** $\frac{2}{3}$ **59.** $\frac{20}{27}$ **61.** 0.84 **63.** -40 **65.** 81 **67.** -1 **69.** -121 **71.** -1 **73.** -19 **75.** 90 **77.** -84 **79.** -5 **81.** $-\frac{9}{2}$ **83.** 18 **85.** 17 **87.** -20 **89.** 16 **91.** 2 **93.** $-\frac{34}{7}$ **95.** 0 **97.** $\frac{6}{5}$ **99.** $\frac{3}{2}$ **101.** $-\frac{5}{38}$ **103.** 3 **105.** -1 **107.** undefined **109.** $-\frac{22}{9}$ **111.** solution **113.** not a solution **115.** solution **117.** $-71 \cdot x$ or $-71x$ **119.** $-16 - x$ **121.** $-29 + x$ **123.** $\frac{x}{-33}$ or $x \div (-33)$ **125.** $3 \cdot (-4) = -12$; a loss of 12 yd **127.** $5(-20) = -100$; a depth of 100 ft **129.** true **131.** false **133.** $-162°$F **135.** $-\$17$ million per month **137.** answers may vary **139.** 1, -1; answers may vary **141.** $\frac{0}{5} - 7 = -7$ **143.** $-8(-5) + (-1) = 39$

Section 1.7

Vocabulary, Readiness & Video Check **1.** commutative property of addition **3.** distributive property **5.** associative property of addition **7.** opposites or additive inverses **9.** 2 is outside the parentheses, so the point is made that we should only distribute the -9 to the terms within the parentheses and not also to the 2.

Exercise Set 1.7 **1.** $16 + x$ **3.** $y \cdot (-4)$ **5.** yx **7.** $13 + 2x$ **9.** $x \cdot (yz)$ **11.** $(2 + a) + b$ **13.** $(4a) \cdot b$ **15.** $a + (b + c)$ **17.** $17 + b$ **19.** $24y$ **21.** y **23.** $26 + a$ **25.** $-72x$ **27.** s **29.** $-\frac{5}{2}x$ **31.** $4x + 4y$ **33.** $9x - 54$ **35.** $6x + 10$ **37.** $28x - 21$ **39.** $18 + 3x$ **41.** $-2y + 2z$ **43.** $-y - \frac{5}{3}$ **45.** $5x + 20m + 10$ **47.** $8m - 4n$ **49.** $-5x - 2$ **51.** $-r + 3 + 7p$ **53.** $3x + 4$ **55.** $-x + 3y$ **57.** $6r + 8$ **59.** $-36x - 70$ **61.** $-1.6x - 2.5$ **63.** $4(1 + y)$ **65.** $11(x + y)$ **67.** $-1(5 + x)$ **69.** $30(a + b)$ **71.** commutative property of multiplication **73.** associative property of addition **75.** commutative property of addition **77.** associative property of multiplication **79.** identity element for addition **81.** distributive property **83.** multiplicative inverse property **85.** identity element for multiplication **87.** $-8; \frac{1}{8}$ **89.** $-x; \frac{1}{x}$ **91.** $2x; -2x$ **93.** false **95.** no **97.** yes **99.** yes **101.** yes **103. a.** commutative property of addition **b.** commutative property of addition **c.** associative property of addition **105.** answers may vary **107.** answers may vary

Section 1.8

Vocabulary, Readiness & Video Check **1.** expression; term **3.** combine like terms **5.** like; unlike **7.** Although these terms have exactly the same variables, the exponents on each are not exactly the same—the exponents on x differ in each term. **9.** -1

Exercise Set 1.8 **1.** -7 **3.** 1 **5.** 17 **7.** like **9.** unlike **11.** like **13.** $15y$ **15.** $13w$ **17.** $-7b - 9$ **19.** $-m - 6$ **21.** -8 **23.** $7.2x - 5.2$ **25.** $k - 6$ **27.** $-15x + 18$ **29.** $4x - 3$ **31.** $5x^2$ **33.** -11 **35.** $1.3x + 3.5$ **37.** $5y + 20$ **39.** $-2x - 4$ **41.** $-10x + 15y - 30$ **43.** $-3x + 2y - 1$ **45.** $7d - 11$ **47.** 16 **49.** $x + 5$ **51.** $x + 2$ **53.** $2k + 10$ **55.** $-3x + 5$ **57.** $2x + 14$ **59.** $3y + \frac{5}{6}$ **61.** $-22 + 24x$ **63.** $0.9m + 1$ **65.** $10 - 6x - 9y$ **67.** $-x - 38$ **69.** $5x - 7$ **71.** $10x - 3$ **73.** $-4x - 9$ **75.** $-4m - 3$ **77.** $2x - 4$ **79.** $\frac{3}{4}x + 12$ **81.** $12x - 2$ **83.** $8x + 48$ **85.** $x - 10$ **87.** balanced **89.** balanced **91.** answers may vary **93.** $(18x - 2)$ ft **95.** $(15x + 23)$ in. **97.** answers may vary

Chapter 1 Vocabulary Check **1.** inequality symbols **2.** equation **3.** absolute value **4.** variable **5.** opposites **6.** numerator **7.** solution **8.** reciprocals **9.** base; exponent **10.** numerical coefficient **11.** denominator **12.** grouping symbols **13.** term **14.** like terms **15.** unlike terms

Chapter 1 Review **1.** $<$ **2.** $>$ **3.** $>$ **4.** $>$ **5.** $<$ **6.** $>$ **7.** $=$ **8.** $=$ **9.** $>$ **10.** $<$ **11.** $4 \geq -3$ **12.** $6 \neq 5$ **13.** $0.03 < 0.3$ **14.** $1579 < 3004$ **15. a.** 1, 3 **b.** 0, 1, 3 **c.** $-6, 0, 1, 3$ **d.** $-6, 0, 1, 1\frac{1}{2}, 3, 9.62$ **e.** π **f.** all numbers in set **16. a.** 2, 5 **b.** 2, 5 **c.** $-3, 2, 5$ **d.** $-3, -1.6, 2, 5, \frac{11}{2}, 15.1$ **e.** $\sqrt{5}, 2\pi$ **f.** all numbers in set **17.** Friday **18.** Wednesday **19.** c **20.** b **21.** 37 **22.** 41 **23.** $\frac{18}{7}$ **24.** 80 **25.** $20 - 12 = 2 \cdot 4$ **26.** $\frac{9}{2} > -5$ **27.** 18 **28.** 108 **29.** 5 **30.** 24 **31.** $63°$ **32.** $105°$ **33.** solution **34.** not a solution **35.** 9 **36.** $-\frac{2}{3}$ **37.** -2 **38.** 7 **39.** -11 **40.** -17 **41.** $-\frac{3}{16}$ **42.** -5 **43.** -13.9 **44.** 3.9 **45.** -14 **46.** -11.5 **47.** 5 **48.** -11 **49.** -19 **50.** 4 **51.** a **52.** a **53.** \$51 **54.** \$54 **55.** $-\frac{1}{6}$ **56.** $\frac{5}{3}$ **57.** -48 **58.** 28 **59.** 3 **60.** -14 **61.** -36 **62.** 0 **63.** undefined **64.** $-\frac{1}{2}$ **65.** commutative property of addition **66.** identity element for multiplication **67.** distributive property **68.** additive inverse property **69.** associative property of addition **70.** commutative property of multiplication **71.** distributive property **72.** associative property of multiplication **73.** multiplicative inverse property **74.** identity element for addition **75.** commutative property of addition **76.** distributive property **77.** $6x$ **78.** $-11.8z$ **79.** $4x - 2$ **80.** $2y + 3$ **81.** $3n - 18$ **82.** $4w - 6$ **83.** $-6x + 7$ **84.** $-0.4y + 2.3$ **85.** $3x - 7$ **86.** $5x + 5.6$ **87.** $<$ **88.** $>$ **89.** -15.3 **90.** -6 **91.** -80 **92.** -5 **93.** $-\frac{1}{4}$ **94.** 0.15 **95.** 16 **96.** 16 **97.** -5 **98.** 9 **99.** $-\frac{5}{6}$ **100.** undefined **101.** $16x - 41$ **102.** $18x - 12$

Chapter 1 Getting Ready for the Test **1.** A **2.** C **3.** B **4.** A **5.** B **6.** A **7.** A **8.** B **9.** C **10.** D **11.** B **12.** C **13.** B **14.** C **15.** A **16.** B **17.** C **18.** A **19.** B **20.** C

Chapter 1 Test **1.** $|-7| > 5$ **2.** $9 + 5 \geq 4$ **3.** -5 **4.** -11 **5.** -14 **6.** -39 **7.** 12 **8.** -2 **9.** undefined **10.** -8 **11.** $-\frac{1}{3}$ **12.** $4\frac{5}{8}$ **13.** $\frac{51}{40}$ **14.** -32 **15.** -48 **16.** 3 **17.** 0 **18.** $>$ **19.** $>$ **20.** $>$ **21.** $=$ **22. a.** 1, 7 **b.** 0, 1, 7 **c.** $-5, -1, 0, 1, 7$ **d.** $-5, -1, \frac{1}{4}, 0, 1, 7, 11.6$ **e.** $\sqrt{7}, 3\pi$ **f.** $-5, -1, \frac{1}{4}, 0, 1, 7, 11.6, \sqrt{7}, 3\pi$ **23.** 40 **24.** 12 **25.** 22 **26.** -1 **27.** associative property of addition **28.** commutative property of multiplication **29.** distributive property **30.** multiplicative inverse **31.** 9 **32.** -3 **33.** second down **34.** yes **35.** $17°$ F **36.** \$420 **37.** $y - 10$ **38.** $5.9x + 1.2$ **39.** $-2x + 10$ **40.** $-15y + 1$

Chapter 2 Equations, Inequalities, and Problem Solving

Section 2.1

Vocabulary, Readiness & Video Check **1.** expression **3.** equation **5.** expression; equation **7.** Equivalent **9.** 2 **11.** 12 **13.** 17 **15.** both sides **17.** $\frac{1}{7}x$

Exercise Set 2.1 **1.** 3 **3.** -2 **5.** -14 **7.** 0.5 **9.** $\frac{1}{4}$ **11.** $\frac{5}{12}$ **13.** -3 **15.** -9 **17.** -10 **19.** 2 **21.** -7 **23.** -1 **25.** -9 **27.** -12 **29.** $-\frac{1}{2}$ **31.** 11 **33.** 21 **35.** 25 **37.** -3 **39.** -0.7 **41.** 11 **43.** 13 **45.** -30 **47.** -0.4 **49.** -7 **51.** $-\frac{1}{3}$ **53.** -17.9 **55.** $20 - p$ **57.** $(10 - x)$ ft **59.** $(180 - x)°$ **61.** $(m + 139)$ mi **63.** $7x$ sq mi **65.** $\frac{8}{5}$ **67.** $\frac{1}{2}$ **69.** -9 **71.** x **73.** y **75.** x **77.** answers may vary **79.** 4 **81.** answers may vary **83.** $(173 - 3x)°$ **85.** answers may vary **87.** -145.478

Section 2.2

Vocabulary, Readiness & Video Check **1.** multiplication **3.** equation; expression **5.** Equivalent **7.** 9 **9.** 2 **11.** -5 **13.** same **15.** $(x + 1) + (x + 3) = 2x + 4$

Exercise Set 2.2 **1.** 4 **3.** 0 **5.** 12 **7.** -12 **9.** 3 **11.** 2 **13.** 0 **15.** 6.3 **17.** 10 **19.** -20 **21.** 0 **23.** -9 **25.** 1 **27.** -30 **29.** 3 **31.** $\frac{10}{9}$ **33.** -1 **35.** -4 **37.** $-\frac{1}{2}$ **39.** 0 **41.** 4 **43.** $-\frac{1}{14}$ **45.** 0.21 **47.** 5 **49.** 6 **51.** -5.5 **53.** -5 **55.** 0 **57.** -3 **59.** $-\frac{9}{28}$ **61.** $\frac{14}{3}$ **63.** -9 **65.** -2 **67.** $\frac{11}{2}$ **69.** $-\frac{1}{4}$ **71.** $\frac{9}{10}$ **73.** $-\frac{17}{20}$ **75.** -16 **77.** $2x + 2$ **79.** $2x + 2$ **81.** $5x + 20$ **83.** $7x - 12$ **85.** $12z + 44$ **87.** 1 **89.** -48 **91.** answers may vary **93.** answers may vary **95.** 2

Section 2.3

Calculator Explorations **1.** solution **3.** not a solution **5.** solution

Vocabulary, Readiness & Video Check **1.** equation **3.** expression **5.** expression **7.** equation **9.** 3; distributive property, addition property of equality, multiplication property of equality **11.** The number of decimal places in each number helps us determine the smallest power of 10 we can multiply through by so we are no longer dealing with decimals.

Exercise Set 2.3 **1.** -6 **3.** 3 **5.** 1 **7.** $\frac{3}{2}$ **9.** 0 **11.** -1 **13.** 4 **15.** -4 **17.** -3 **19.** 2 **21.** 50 **23.** 1 **25.** $\frac{7}{3}$ **27.** 0.2 **29.** all real numbers **31.** no solution **33.** no solution **35.** all real numbers **37.** 18 **39.** $\frac{19}{9}$ **41.** $\frac{14}{3}$ **43.** 13 **45.** 4 **47.** all real numbers **49.** $-\frac{3}{5}$ **51.** -5 **53.** 10 **55.** no solution **57.** 3 **59.** -17 **61.** $\frac{7}{5}$ **63.** $-\frac{1}{50}$ **65.** $(6x - 8)$ m **67.** $-8 - x$ **69.** $-3 + 2x$ **71.** $9(x + 20)$ **73.** **a.** all real numbers **b.** answers may vary **c.** answers may vary **75.** **a** **77.** **b** **79.** **c** **81.** answers may vary **83.** **a.** $x + x + x + 2x + 2x = 28$ **b.** $x = 4$ **c.** x cm $= 4$ cm; $2x$ cm $= 8$ cm **85.** answers may vary **87.** 15.3 **89.** -0.2

Integrated Review **1.** 6 **2.** -17 **3.** 12 **4.** -26 **5.** -3 **6.** -1 **7.** $\frac{27}{2}$ **8.** $\frac{25}{2}$ **9.** 8 **10.** -64 **11.** 2 **12.** -3 **13.** 5 **14.** -1 **15.** 2 **16.** 2 **17.** -2 **18.** -2 **19.** $-\frac{5}{6}$ **20.** $\frac{1}{6}$ **21.** 1 **22.** 6 **23.** 4 **24.** 1 **25.** $\frac{9}{5}$ **26.** $-\frac{6}{5}$ **27.** all real numbers **28.** all real numbers **29.** 0 **30.** -1.6 **31.** $\frac{4}{19}$ **32.** $-\frac{5}{19}$ **33.** $\frac{7}{2}$ **34.** $-\frac{1}{4}$ **35.** no solution **36.** no solution **37.** $\frac{7}{6}$ **38.** $\frac{1}{15}$

Section 2.4

Vocabulary, Readiness & Video Check **1.** $2x$; $2x - 31$ **3.** $x + 5$; $2(x + 5)$ **5.** $20 - y$; $\frac{20 - y}{3}$ or $(20 - y) \div 3$ **7.** in the statement of the application **9.** That the three angle measures are consecutive even integers and that they sum to 180°.

Exercise Set 2.4 **1.** $2x + 7 = x + 6$; -1 **3.** $3x - 6 = 2x + 8$; 14 **5.** -25 **7.** $-\frac{3}{4}$ **9.** 3 in.; 6 in.; 16 in. **11.** 1st piece: 5 in.; 2nd piece: 10 in.; 3rd piece: 25 in. **13.** Pennsylvania: 528 million pounds; New York: 1300 million pounds **15.** 172 mi **17.** 25 mi **19.** 1st angle: 37.5°; 2nd angle: 37.5°; 3rd angle: 105° **21.** A: 60°; B: 120°; C: 120°; D: 60° **23.** $3x + 3$ **25.** $x + 2$; $x + 4$; $2x + 4$ **27.** $x + 1$; $x + 2$; $x + 3$; $4x + 6$ **29.** $x + 2$; $x + 4$; $2x + 6$ **31.** 234, 235 **33.** Belgium: 32; France: 33; Spain: 34 **35.** 5 ft, 12 ft **37.** CRH380A: 302 mph; Transrapid TR-09: 279 mph **39.** 43°, 137° **41.** 58°, 60°, 62° **43.** 1 **45.** 280 mi **47.** Michigan: 39; Ohio: 62 **49.** Montana: 56 counties; California: 58 counties **51.** Neptune: 14 satellites; Uranus: 27 satellites; Saturn: 62 satellites **53.** -16 **55.** Sahara: 3,500,000 sq mi; Gobi: 500,000 sq mi **57.** France: 15; Austria: 14; Japan: 13 **59.** females: 239 thousand; males: 191 thousand **61.** 34.5°; 34.5°; 111° **63.** California **65.** Florida: \$82.7 million; Hawaii: \$93.2 million **67.** answers may vary **69.** 34 **71.** 225π **73.** 15 ft by 24 ft **75.** 5400 chirps per hour; 129,600 chirps per day; 47,304,000 chirps per year **77.** answers may vary **79.** answers may vary **81.** c

Section 2.5

Vocabulary, Readiness & Video Check **1.** relationships **3.** That the process of solving this equation for x—dividing both sides by 5, the coefficient of x—is the same process used to solve a formula for a specific variable. Treat whatever is multiplied by that specific variable as the coefficient—the coefficient is all the factors except that specific variable.

Exercise Set 2.5 **1.** $h = 3$ **3.** $h = 3$ **5.** $h = 20$ **7.** $c = 12$ **9.** $r = 2.5$ **11.** $h = \frac{f}{5g}$ **13.** $w = \frac{V}{lh}$ **15.** $y = 7 - 3x$ **17.** $R = \frac{A - P}{PT}$ **19.** $A = \frac{3V}{h}$ **21.** $a = P - b - c$ **23.** $h = \frac{S - 2\pi r^2}{2\pi r}$ **25.** 120 ft **27.** **a.** area: 480 sq in.; perimeter: 120 in. **b.** frame: perimeter; glass: area **29.** **a.** area: 103.5 sq ft; perimeter: 41 ft **b.** baseboard: perimeter; carpet: area **31.** -10°C **33.** 6.25 hr **35.** length: 78 ft; width: 52 ft **37.** 18 ft, 36 ft, 48 ft **39.** 306 mi **41.** 61.5°F **43.** 60 chirps per minute **45.** increases **47.** 96 piranhas **49.** 2 bags **51.** one 16-in. pizza **53.** x m $= 6$ m; $2.5x$ m $= 15$ m **55.** 22 hr **57.** 13 in. **59.** 2.25 hr **61.** 12,090 ft **63.** 50°C **65.** 686,664 cu in. **67.** 449 cu in. **69.** 333°F **71.** 0.32 **73.** 2.00 or 2 **75.** 17% **77.** 720% **79.** $V = G(N - R)$ **81.** multiplies the volume by 8; answers may vary **83.** $53\frac{1}{3}$ **85.** ○ = (△ − □)/▯ **87.** 44.3 sec **89.** $P = 3{,}200{,}000$ **91.** $V = 113.1$

Section 2.6

Vocabulary, Readiness & Video Check **1.** no **3.** yes **5.** **a.** equals; = **b.** multiplication; · **c.** Drop the percent symbol and move the decimal point two places to the left. **7.** We must first find the actual amount of increase in price by subtracting the original price from the new price.

Exercise Set 2.6 **1.** 11.2 **3.** 55% **5.** 180 **7.** 17% **9.** 304,080 **11.** discount: \$1480; new price: \$17,020 **13.** \$46.58 **15.** 21.7% **17.** 30% **19.** \$104 **21.** \$42,500 **23.** 2 gal **25.** 7 lb **27.** 4.6 **29.** 50 **31.** 30% **33.** 20% **35.** 93.2 million **37.** 62%, 5%, 24%, 2% **39.** 115% **41.** \$3900 **43.** 300% **45.** mark-up: \$0.11; new price: \$2.31 **47.** 400 oz **49.** 374.6% **51.** 120 employees

53. decrease: \$64; sale price: \$192 **55.** 854 thousand Scoville units **57.** 67.3 million households **59.** 400 oz **61.** $>$ **63.** $=$ **65.** $>$ **67.** no; answers may vary **69.** 9.6% **71.** 26.9%; yes **73.** 17.1%

Section 2.7

Vocabulary, Readiness & Video Check **1.** expression **3.** inequality **5.** equation **7.** -5 **9.** 4.1 **11.** An open circle indicates $>$ or $<$; a closed circle indicates $\geq$ or $\leq$. **13.** $\{x \mid x \geq -2\}$ **15.** is greater than; $>$

Exercise Set 2.7 **1.**  -1 **3.** 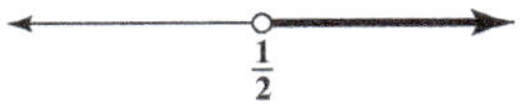 $\frac{1}{2}$ **5.**  4

7.  -2 **9.**  -1, 3 **11.**  0, 2 **13.** $\{x \mid x \geq -5\}$ -5

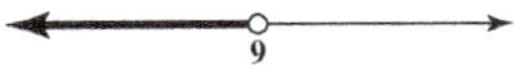

15. $\{y \mid y < 9\}$ 9 **17.** $\{x \mid x > -3\}$ -3 **19.** $\{x \mid x \leq 1\}$ 1

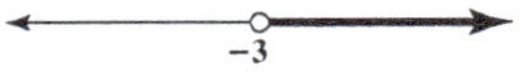

21. $\{x \mid x < -3\}$ -3 **23.** $\{x \mid x \geq -2\}$ -2 **25.** $\{x \mid x < 0\}$ 0

27. $\left\{y \middle| y \geq -\frac{8}{3}\right\}$ $-\frac{8}{3}$ **29.** $\{y \mid y > 3\}$ 3 **31.** $\{x \mid x > -15\}$ **33.** $\{x \mid x \geq -11\}$

35. $\left\{x \middle| x > \frac{1}{4}\right\}$ **37.** $\{y \mid y \geq -12\}$ **39.** $\{z \mid z < 0\}$ **41.** $\{x \mid x > -3\}$ **43.** $\left\{x \middle| x \geq -\frac{2}{3}\right\}$ **45.** $\{x \mid x \leq -2\}$

47. $\{x \mid x > -13\}$ **49.** $\{x \mid x \leq -8\}$ **51.** $\{x \mid x > 4\}$ **53.** $\left\{x \middle| x \leq \frac{5}{4}\right\}$ **55.** $\left\{x \middle| x > \frac{8}{3}\right\}$ **57.** $\{x \mid x \geq 0\}$ **59.** all numbers greater than -10 **61.** 35 cm **63.** at least 193 **65.** 86 people **67.** at least 52 min **69.** 81 **71.** 1 **73.** $\frac{49}{64}$ **75.** about 19,750 **77.** 2016 and 2017 **79.** 2012 **81.** $>$ **83.** $\geq$ **85.** when multiplying or dividing by a negative number **87.** final exam score $\geq$ 78.5 **89.** answers may vary

Chapter 2 Vocabulary Check **1.** linear equation in one variable **2.** equivalent equations **3.** formula **4.** inequality **5.** all real numbers **6.** no solution **7.** the same **8.** reversed

Chapter 2 Review **1.** 4 **2.** -3 **3.** 6 **4.** -6 **5.** 0 **6.** -9 **7.** -23 **8.** 28 **9.** b **10.** a **11.** b **12.** c **13.** -12 **14.** 4 **15.** 0 **16.** -7 **17.** 0.75 **18.** -3 **19.** -6 **20.** -1 **21.** -1 **22.** $\frac{3}{2}$ **23.** $-\frac{1}{5}$ **24.** 7 **25.** $3x + 3$ **26.** $2x + 6$ **27.** -4 **28.** -4 **29.** 2 **30.** -3 **31.** no solution **32.** no solution **33.** $\frac{3}{4}$ **34.** $-\frac{8}{9}$ **35.** 20 **36.** $-\frac{6}{23}$ **37.** $\frac{23}{7}$ **38.** $-\frac{2}{5}$ **39.** 102 **40.** 0.25 **41.** 6665.5 in. **42.** short piece: 4 ft; long piece: 8 ft **43.** national battlefields: 11; national memorials: 30 **44.** $-39, -38, -37$ **45.** 3 **46.** -4 **47.** $w = 9$ **48.** $h = 4$ **49.** $m = \frac{y - b}{x}$ **50.** $s = \frac{r + 5}{vt}$ **51.** $x = \frac{2y - 7}{5}$ **52.** $y = \frac{2 + 3x}{6}$ **53.** $\pi = \frac{C}{d}$ **54.** $\pi = \frac{C}{2r}$ **55.** 15 m **56.** 18 ft by 12 ft **57.** 1 hr 20 min **58.** 56.7°C **59.** 20% **60.** 70% **61.** 110 **62.** 1280 **63.** mark-up: \$209; new price: \$2109 **64.** 91,800 businesses **65.** 40% solution: 10 gal; 10% solution: 20 gal **66.** 13.7% increase **67.** 18% **68.** swerving into another lane **69.** 966 customers **70.** no; answers may vary **71.**  -2

72.  0, 5 **73.** $\{x \mid x \leq 1\}$ **74.** $\{x \mid x > -5\}$ **75.** $\{x \mid x \leq 10\}$ **76.** $\{x \mid x < -4\}$ **77.** $\{x \mid x < -4\}$

78. $\{x \mid x \leq 4\}$ **79.** $\{y \mid y > 9\}$ **80.** $\{y \mid y \geq -15\}$ **81.** $\left\{x \middle| x < \frac{7}{4}\right\}$ **82.** $\left\{x \middle| x \leq \frac{19}{3}\right\}$ **83.** \$2500 **84.** score must be less than 83

85. 4 **86.** -14 **87.** $-\frac{3}{2}$ **88.** 21 **89.** all real numbers **90.** no solution **91.** -13 **92.** shorter piece: 4 in.; longer piece: 19 in.

93. $h = \frac{3V}{A}$ **94.** 22.1 **95.** 160 **96.** 20% **97.** $\{x \mid x > 9\}$ 9 **98.** $\{x \mid x > -4\}$ -4

99. $\{x \mid x \leq 0\}$ 0

Chapter 2 Getting Ready for the Test **1.** C **2.** A **3.** D **4.** B **5.** B **6.** C **7.** B **8.** A **9.** C **10.** C **11.** B **12.** D

Chapter 2 Test **1.** -5 **2.** 8 **3.** $\frac{7}{10}$ **4.** 0 **5.** 27 **6.** $-\frac{19}{6}$ **7.** 3 **8.** $\frac{3}{11}$ **9.** 0.25 **10.** $\frac{25}{7}$ **11.** no solution **12.** 21 **13.** 7 gal **14.** $x = 6$ **15.** $h = \frac{V}{\pi r^2}$ **16.** $y = \frac{3x - 10}{4}$ **17.** $\{x \mid x \leq -2\}$ -2 **18.** $\{x \mid x < 4\}$ 4

19. $\{x \mid x \leq -8\}$ **20.** $\{x \mid x \geq 11\}$ **21.** $\left\{x \middle| x > \frac{2}{5}\right\}$ **22.** 552 **23.** 40% **24.** 401,802 **25.** California: 1107; Ohio: 720

Cumulative Review **1.** true; Sec. 1.2, Ex. 3 **2.** false; Sec. 1.2 **3.** true; Sec. 1.2, Ex. 4 **4.** true; Sec. 1.2 **5.** false; Sec. 1.2, Ex. 5 **6.** true; Sec. 1.2 **7.** true; Sec. 1.2, Ex. 6 **8.** true; Sec. 1.2 **9.** **a.** $<$ **b.** $=$ **c.** $>$ **d.** $<$ **e.** $>$; Sec. 1.2, Ex. 13 **10.** **a.** 5 **b.** 8 **c.** $\frac{2}{3}$; Sec. 1.2 **11.** $\frac{8}{3}$; Sec. 1.3, Ex. 6 **12.** 33; Sec. 1.3 **13.** -19; Sec. 1.4, Ex. 6 **14.** -10; Sec. 1.4 **15.** 8; Sec. 1.4, Ex. 7 **16.** 10; Sec. 1.4

17. -0.3; Sec. 1.4, Ex. 8 **18.** 0; Sec. 1.4 **19. a.** -12 **b.** -3; Sec. 1.5, Ex. 7 **20. a.** 5 **b.** $\frac{2}{3}$ **c.** a **d.** -3; Sec. 1.2 **21. a.** 0 **b.** -24 **c.** 90; Sec. 1.6, Ex. 7 **22. a.** -11.1 **b.** $-\frac{1}{5}$ **c.** $\frac{3}{4}$; Sec. 1.5 **23. a.** -6 **b.** 7 **c.** -5; Sec. 1.6, Ex. 10 **24. a.** -0.36 **b.** $\frac{6}{17}$; Sec. 1.6 **25.** $15 - 10z$; Sec. 1.7, Ex. 8 **26.** $2y - 6x + 8$; Sec. 1.7 **27.** $3x + 17$; Sec. 1.7, Ex. 12 **28.** $2x + 8$; Sec. 1.7 **29. a.** unlike **b.** like **c.** like **d.** like **e.** like; Sec. 1.8, Ex. 2 **30. a.** -4 **b.** 9 **c.** $\frac{10}{63}$; Sec. 1.6 **31.** $-2x - 1$; Sec. 1.8, Ex. 15 **32.** $-15x - 2$; Sec. 1.8 **33.** 17; Sec. 2.1, Ex. 1 **34.** $-\frac{1}{6}$; Sec. 2.1 **35.** -10; Sec. 2.2, Ex. 7 **36.** 3; Sec. 2.3 **37.** 0; Sec. 2.3, Ex. 4 **38.** 72; Sec. 2.2 **39.** Republicans: 241; Democrats: 194; Sec. 2.4, Ex. 4 **40.** 5; Sec. 2.3 **41.** 79.2 yr; Sec. 2.5, Ex. 1 **42.** 6; Sec. 2.4 **43.** 87.5%; Sec. 2.6, Ex. 1 **44.** $\frac{C}{2\pi} = r$; Sec. 2.5 **45.** $-\frac{9}{10}$; Sec. 2.2, Ex. 10 **46.** $\{x \mid x > 5\}$; Sec. 2.7 **47.** -1 Sec. 2.7, Ex. 2 **48.** $\{x \mid x \le -10\}$; Sec. 2.7 **49.** $\{x \mid x \ge 1\}$; Sec. 2.7, Ex. 9 **50.** $\{x \mid x \le -3\}$; Sec. 2.7

Chapter 3 Exponents and Polynomials

Section 3.1

Vocabulary, Readiness & Video Check **1.** exponent **3.** add **5.** 1 **7.** base: 3; exponent: 2 **9.** base: 4; exponent: 2 **11.** base: x; exponent: 2 **13.** Example 4 can be written as $-4^2 = -1 \cdot 4^2$, which is similar to Example 7, $4 \cdot 3^2$, and shows why the negative sign should not be considered part of the base when there are no parentheses. **15.** Be careful not to confuse the power rule with the product rule. The power rule involves a power raised to a power (exponents are multiplied), and the product rule involves a product (exponents are added). **17.** the quotient rule

Exercise Set 3.1 **1.** 49 **3.** -5 **5.** -16 **7.** 16 **9.** $\frac{1}{27}$ **11.** 112 **13.** 4 **15.** 135 **17.** 150 **19.** $\frac{32}{5}$ **21.** x^7 **23.** $(-3)^{12}$ **25.** $15y^5$ **27.** $x^{19}y^6$ **29.** $-72m^3n^8$ **31.** $-24z^{20}$ **33.** $20x^5$ sq ft **35.** x^{36} **37.** p^8q^8 **39.** $8a^{15}$ **41.** $x^{10}y^{15}$ **43.** $49a^4b^{10}c^2$ **45.** $\frac{r^9}{s^9}$ **47.** $\frac{m^9p^9}{n^9}$ **49.** $\frac{4x^2z^2}{y^{10}}$ **51.** $64z^{10}$ sq dm **53.** $27y^{12}$ cu ft **55.** x^2 **57.** -64 **59.** p^6q^5 **61.** $\frac{xy^3}{2}$ **63.** 1 **65.** 1 **67.** -7 **69.** 2 **71.** -81 **73.** $\frac{1}{64}$ **75.** b^6 **77.** a^9 **79.** $-16x^7$ **81.** $a^{11}b^{20}$ **83.** $26m^9n^7$ **85.** z^{40} **87.** $64a^3b^3$ **89.** $36x^2y^2z^6$ **91.** z^8 **93.** $3x^4$ **95.** 1 **97.** $81x^2y^2$ **99.** 40 **101.** $\frac{y^{15}}{8x^{12}}$ **103.** $2x^2y$ **105.** -2 **107.** 5 **109.** -7 **111.** c **113.** e **115.** answers may vary **117.** answers may vary **119.** 343 cu m **121.** volume **123.** answers may vary **125.** answers may vary **127.** x^{9a} **129.** a^{5b} **131.** x^{5a}

Section 3.2

Calculator Explorations **1.** 5.31 EE 3 **3.** 6.6 EE -9 **5.** 1.5×10^{13} **7.** 8.15×10^{19}

Vocabulary, Readiness & Video Check **1.** $\frac{1}{x^3}$ **3.** scientific notation **5.** $\frac{5}{x^2}$ **7.** y^6 **9.** $4y^3$ **11.** A negative exponent has nothing to do with the sign of the simplified result. **13.** When the decimal point is moved to the left, the sign of the exponent will be positive; when the decimal point is moved to the right, the sign of the exponent will be negative. **15.** the quotient rule

Exercise Set 3.2 **1.** $\frac{1}{64}$ **3.** $\frac{7}{x^3}$ **5.** -64 **7.** $\frac{5}{6}$ **9.** p^3 **11.** $\frac{q^4}{p^5}$ **13.** $\frac{1}{x^3}$ **15.** z^3 **17.** $\frac{4}{9}$ **19.** $\frac{1}{9}$ **21.** $-p^4$ **23.** -2 **25.** x^4 **27.** p^4 **29.** m^{11} **31.** r^6 **33.** $\frac{1}{x^{15}y^9}$ **35.** $\frac{1}{x^4}$ **37.** $\frac{1}{a^2}$ **39.** $4k^3$ **41.** $3m$ **43.** $-\frac{4a^5}{b}$ **45.** $-\frac{6}{7y^2}$ **47.** $\frac{27a^6}{b^{12}}$ **49.** $\frac{a^{30}}{b^{12}}$ **51.** $x^{10}y^6$ **53.** $\frac{z^2}{4}$ **55.** $\frac{x^{11}}{81}$ **57.** $\frac{49a^4}{b^6}$ **59.** $-\frac{3m^7}{n^4}$ **61.** $a^{24}b^8$ **63.** 200 **65.** x^9y^{19} **67.** $-\frac{y^8}{8x^2}$ **69.** $\frac{25b^{33}}{a^{16}}$ **71.** $\frac{27}{z^3x^6}$ cu in. **73.** 7.8×10^4 **75.** 1.67×10^{-6} **77.** 6.35×10^{-3} **79.** 1.16×10^6 **81.** 4.2×10^3 **83.** 0.0000000008673 **85.** 0.033 **87.** 20,320 **89.** 700,000,000 **91.** 2.415×10^{12} **93.** 5,500,000,000,000 **95.** 9,000,000,000,000; 9×10^{12} **97.** 0.000036 **99.** 0.00000000000000000028 **101.** 0.0000005 **103.** 200,000 **105.** 2.7×10^9 gal **107.** $-2x + 7$ **109.** $2y - 10$ **111.** $-x - 4$ **113.** 435,300,000; 4.353×10^8 **115.** 14,056,000; 1.4056×10^7 **117.** 2.5×10^{-9} m **119.** 0.00000031 m; 3.1×10^{-7} m **121.** $9a^{13}$ **123.** -5 **125.** answers may vary **127. a.** 1.3×10^1 **b.** 4.4×10^7 **c.** 6.1×10^{-2} **129.** answers may vary **131.** $\frac{1}{x^{9s}}$ **133.** a^{4m+5}

Section 3.3

Vocabulary, Readiness & Video Check **1.** binomial **3.** trinomial **5.** constant **7.** 3; $x^2, -3x, 5$ **9.** the replacement value for the variable **11.** 2; $9ab$

Exercise Set 3.3 **1.** 1; $-3x$; 5 **3.** -5; 3.2; 1; -5 **5.** 1; binomial **7.** 3; none of these **9.** 6; trinomial **11.** 4; binomial **13. a.** -6 **b.** -11 **15. a.** -2 **b.** 4 **17. a.** -15 **b.** -10 **19.** 184 ft **21.** 595.84 ft **23.** 1394 thousand **25.** 33.76 million **27.** $-11x$ **29.** $23x^3$

31. $16x^2 - 7$ **33.** $12x^2 - 13$ **35.** $7s$ **37.** $-1.1y^2 + 4.8$ **39.** $\frac{5}{6}x^4 - 7x^3 - 19$ **41.** $\frac{3}{20}x^3 + 6x^2 - \frac{13}{20}x - \frac{1}{10}$ **43.** $4x^2 + 7x + x^2 + 5x; 5x^2 + 12x$ **45.** $5x + 3 + 4x + 3 + 2x + 6 + 3x + 7x; 21x + 12$ **47.** 2, 1, 1, 0; 2 **49.** 4, 0, 4, 3; 4 **51.** $9ab - 11a$ **53.** $4x^2 - 7xy + 3y^2$ **55.** $-3xy^2 + 4$ **57.** $14y^3 - 19 - 16a^2b^2$ **59.** $7x^2 + 0x + 3$ **61.** $x^3 + 0x^2 + 0x - 64$ **63.** $5y^3 + 0y^2 + 2y - 10$ **65.** $2y^4 + 0y^3 + 0y^2 + 8y + 0y^0$ or $2y^4 + 0y^3 + 0y^2 + 8y + 0$ **67.** $6x^5 + 0x^4 + x^3 + 0x^2 - 3x + 15$ **69.** $10x + 19$ **71.** $-x + 5$ **73.** answers may vary **75.** answers may vary **77.** x^{13} **79.** a^3b^{10} **81.** $2y^{20}$ **83.** answers may vary **85.** answers may vary **87.** $11.1x^2 - 7.97x + 10.76$

Section 3.4

Vocabulary, Readiness & Video Check **1.** $-14y$ **3.** $7x$ **5.** $5m^2 + 2m$ **7.** $-3y^2$ and $2y^2$; $-4y$ and y **9.** We're translating a subtraction problem. Order matters when subtracting, so we need to be careful that the order of the expressions is correct.

Exercise Set 3.4 **1.** $12x + 12$ **3.** $-3x^2 + 10$ **5.** $-3x^2 + 4$ **7.** $-y^2 - 3y - 1$ **9.** $7.9x^3 + 4.4x^2 - 3.4x - 3$ **11.** $\frac{1}{2}m^2 - \frac{7}{10}m + \frac{13}{16}$ **13.** $8t^2 - 4$ **15.** $15a^3 + a^2 - 3a + 16$ **17.** $-x + 14$ **19.** $7x^2 - 2$ **21.** $-2x + 9$ **23.** $2x^2 + 7x - 16$ **25.** $2x^2 + 11x$ **27.** $-0.2x^2 + 0.2x - 2.2$ **29.** $\frac{2}{5}z^2 - \frac{3}{10}z + \frac{7}{20}$ **31.** $-2z^2 - 16z + 6$ **33.** $2u^5 - 10u^2 + 11u - 9$ **35.** $5x - 9$ **37.** $4x - 3$ **39.** $11y + 7$ **41.** $-2x^2 + 8x - 1$ **43.** $14x + 18$ **45.** $3a^2 - 6a + 11$ **47.** $3x - 3$ **49.** $7x^2 - 4x + 2$ **51.** $7x^2 - 2x + 2$ **53.** $4y^2 + 12y + 19$ **55.** $-15x + 7$ **57.** $-2a - b + 1$ **59.** $3x^2 + 5$ **61.** $6x^2 - 2xy + 19y^2$ **63.** $8r^2s + 16rs - 8 + 7r^2s^2$ **65.** $(x^2 + 7x + 4)$ ft **67.** $\left(\frac{19}{2}x + 3\right)$ units **69.** $(3y^2 + 4y + 11)$ m **71.** $-6.6x^2 - 1.8x - 1.8$ **73.** $6x^2$ **75.** $-12x^8$ **77.** $200x^3y^2$ **79.** 2; 2 **81.** 4; 3; 3; 4 **83.** b **85.** e **87. a.** $4z$ **b.** $3z^2$ **c.** $-4z$ **d.** $3z^2$; answers may vary **89. a.** m^3 **b.** $3m$ **c.** $-m^3$ **d.** $-3m$; answers may vary **91.** $-4052x^2 + 34{,}684x + 144{,}536$

Section 3.5

Vocabulary, Readiness & Video Check **1.** distributive **3.** $(5y - 1)(5y - 1)$ **5.** x^8 **7.** cannot simplify **9.** x^{14} **11.** $2x^7$ **13.** No. The monomials are unlike terms. **15.** Three times: First $(a - 2)$ is distributed to a and 7, and then a is distributed to $(a - 2)$ and 7 is distributed to $(a - 2)$.

Exercise Set 3.5 **1.** $24x^3$ **3.** x^4 **5.** $-28n^{10}$ **7.** $-12.4x^{12}$ **9.** $-\frac{2}{15}y^3$ **11.** $-24x^8$ **13.** $6x^2 + 15x$ **15.** $7x^3 + 14x^2 - 7x$ **17.** $-2a^2 - 8a$ **19.** $6x^3 - 9x^2 + 12x$ **21.** $12a^5 + 45a^2$ **23.** $-6a^4 + 4a^3 - 6a^2$ **25.** $6x^5y - 3x^4y^3 + 24x^2y^4$ **27.** $-4x^3y + 7x^2y^2 - xy^3 - 3y^4$ **29.** $4x^4 - 3x^3 + \frac{1}{2}x^2$ **31.** $x^2 + 7x + 12$ **33.** $a^2 + 5a - 14$ **35.** $x^2 + \frac{1}{3}x - \frac{2}{9}$ **37.** $12x^4 + 25x^2 + 7$ **39.** $12x^2 - 29x + 15$ **41.** $1 - 7a + 12a^2$ **43.** $4y^2 - 16y + 16$ **45.** $x^3 - 5x^2 + 13x - 14$ **47.** $x^4 + 5x^3 - 3x^2 - 11x + 20$ **49.** $10a^3 - 27a^2 + 26a - 12$ **51.** $49x^2y^2 - 14xy^2 + y^2$ **53.** $12x^2 - 64x - 11$ **55.** $2x^3 + 10x^2 + 11x - 3$ **57.** $2x^4 + 3x^3 - 58x^2 + 4x + 63$ **59.** $8.4y^7$ **61.** $-3x^3 - 6x^2 + 24x$ **63.** $2x^2 + 39x + 19$ **65.** $x^2 - \frac{2}{7}x - \frac{3}{49}$ **67.** $9y^2 + 30y + 25$ **69.** $a^3 - 2a^2 - 18a + 24$ **71.** $(4x^2 - 25)$ sq yd **73.** $(6x^2 - 4x)$ sq in. **75.** $5a + 15a = 20a; 5a - 15a = -10a; 5a \cdot 15a = 75a^2; \frac{5a}{15a} = \frac{1}{3}$ **77.** $-3y^5 + 9y^4$, cannot be simplified; $-3y^5 - 9y^4$, cannot be simplified; $-3y^5 \cdot 9y^4 = -27y^9; \frac{-3y^5}{9y^4} = -\frac{y}{3}$ **79. a.** $6x + 12$ **b.** $9x^2 + 36x + 35$; answers may vary **81.** $13x - 7$ **83.** $30x^2 - 28x + 6$ **85.** $-7x + 5$ **87.** $x^2 + 3x$ **89.** $x(1 + 2x); x + 2x^2$ **91.** $11a$ **93.** $25x^2 + 4y^2$ **95. a.** $a^2 - b^2$ **b.** $4x^2 - 9y^2$ **c.** $16x^2 - 49$ **d.** answers may vary

Section 3.6

Vocabulary, Readiness & Video Check **1.** false **3.** false **5.** a binomial times a binomial **7.** Multiplying gives you four terms, and the two like terms will always subtract out.

Exercise Set 3.6 **1.** $x^2 + 7x + 12$ **3.** $x^2 + 5x - 50$ **5.** $5x^2 + 4x - 12$ **7.** $4y^2 - 25y + 6$ **9.** $6x^2 + 13x - 5$ **11.** $6y^3 + 4y^2 + 42y + 28$ **13.** $x^2 + \frac{1}{3}x - \frac{2}{9}$ **15.** $0.08 - 2.6a + 15a^2$ **17.** $2x^2 + 9xy - 5y^2$ **19.** $x^2 + 4x + 4$ **21.** $4a^2 - 12a + 9$ **23.** $9a^2 - 30a + 25$ **25.** $x^4 + x^2 + 0.25$ **27.** $y^2 - \frac{4}{7}y + \frac{4}{49}$ **29.** $4x^2 - 4x + 1$ **31.** $25x^2 + 90x + 81$ **33.** $9x^2 - 42xy + 49y^2$ **35.** $16m^2 + 40mn + 25n^2$ **37.** $25x^8 - 30x^4 + 9$ **39.** $a^2 - 49$ **41.** $x^2 - 36$ **43.** $9x^2 - 1$ **45.** $x^4 - 25$

47. $4y^4 - 1$ **49.** $16 - 49x^2$ **51.** $9x^2 - \frac{1}{4}$ **53.** $81x^2 - y^2$ **55.** $4m^2 - 25n^2$ **57.** $a^2 + 9a + 20$ **59.** $a^2 - 14a + 49$ **61.** $12a^2 - a - 1$ **63.** $x^2 - 4$ **65.** $9a^2 + 6a + 1$ **67.** $4x^2 + 3xy - y^2$ **69.** $\frac{1}{9}a^4 - 49$ **71.** $6b^2 - b - 35$ **73.** $x^4 - 100$ **75.** $16x^2 - 25$ **77.** $25x^2 - 60xy + 36y^2$ **79.** $4r^2 - 9s^2$ **81.** $(4x^2 + 4x + 1)$ sq ft **83.** $\frac{5b^5}{7}$ **85.** $-\frac{2a^{10}}{b^5}$ **87.** $\frac{2y^8}{3}$ **89.** c **91.** d **93.** 2; 2 **95.** $(x^4 - 3x^2 + 1)$ sq m **97.** $(24x^2 - 32x + 8)$ sq m **99.** answers may vary **101.** answers may vary

Integrated Review **1.** $35x^5$ **2.** $-32y^9$ **3.** -16 **4.** 16 **5.** $2x^2 - 9x - 5$ **6.** $3x^2 + 13x - 10$ **7.** $3x - 4$ **8.** $4x + 3$ **9.** $7x^6y^2$ **10.** $\frac{10b^6}{7}$ **11.** $144m^{14}n^{12}$ **12.** $64y^{27}z^{30}$ **13.** $16y^2 - 9$ **14.** $49x^2 - 1$ **15.** $\frac{y^{45}}{x^{63}}$ **16.** $\frac{1}{64}$ **17.** $\frac{x^{27}}{27}$ **18.** $\frac{r^{58}}{16s^{14}}$ **19.** $2x^2 - 2x - 6$ **20.** $6x^2 + 13x - 11$ **21.** $2.5y^2 - 6y - 0.2$ **22.** $8.4x^2 - 6.8x - 4.2$ **23.** $2y^2 - 6y - 1$ **24.** $6z^2 + 2z + \frac{11}{2}$ **25.** $x^2 + 8x + 16$ **26.** $y^2 - 18y + 81$ **27.** $2x + 8$ **28.** $2y - 18$ **29.** $7x^2 - 10xy + 4y^2$ **30.** $-a^2 - 3ab + 6b^2$ **31.** $x^3 + 2x^2 - 16x + 3$ **32.** $x^3 - 2x^2 - 5x - 2$ **33.** $6x^2 - x - 70$ **34.** $20x^2 + 21x - 5$ **35.** $2x^3 - 19x^2 + 44x - 7$ **36.** $5x^3 + 9x^2 - 17x + 3$ **37.** $4x^2 - \frac{25}{81}$ **38.** $144y^2 - \frac{9}{49}$

Section 3.7

Vocabulary, Readiness & Video Check **1.** dividend; quotient; divisor **3.** a^2 **5.** y **7.** the common denominator

Exercise Set 3.7 **1.** $12x^3 + 3x$ **3.** $4x^3 - 6x^2 + x + 1$ **5.** $5p^2 + 6p$ **7.** $-\frac{3}{2x} + 3$ **9.** $-3x^2 + x - \frac{4}{x^3}$ **11.** $-1 + \frac{3}{2x} - \frac{7}{4x^4}$ **13.** $x + 1$ **15.** $2x + 3$ **17.** $2x + 1 + \frac{7}{x - 4}$ **19.** $3a^2 - 3a + 1 + \frac{2}{3a + 2}$ **21.** $4x + 3 - \frac{2}{2x + 1}$ **23.** $2x^2 + 6x - 5 - \frac{2}{x - 2}$ **25.** $x + 6$ **27.** $x^2 + 3x + 9$ **29.** $-3x + 6 - \frac{11}{x + 2}$ **31.** $2b - 1 - \frac{6}{2b - 1}$ **33.** $ab - b^2$ **35.** $4x + 9$ **37.** $x + 4xy - \frac{y}{2}$ **39.** $2b^2 + b + 2 - \frac{12}{b + 4}$ **41.** $y^2 + 5y + 10 + \frac{24}{y - 2}$ **43.** $-6x - 12 - \frac{19}{x - 2}$ **45.** $x^3 - x^2 + x$ **47.** 3 **49.** -4 **51.** $3x$ **53.** $9x$ **55.** $(3x^3 + x - 4)$ ft **57.** $(2x + 5)$ m **59.** answers may vary **61.** c

Chapter 3 Vocabulary Check **1.** term **2.** FOIL **3.** trinomial **4.** degree of a polynomial **5.** binomial **6.** coefficient **7.** degree of a term **8.** monomial **9.** polynomials **10.** distributive

Chapter 3 Review **1.** base: 3; exponent: 4 **2.** base: -5; exponent: 4 **3.** base: 5; exponent: 4 **4.** base: x; exponent: 4 **5.** 512 **6.** 36 **7.** -36 **8.** -65 **9.** 1 **10.** 1 **11.** y^9 **12.** x^{14} **13.** $-6x^{11}$ **14.** $-20y^7$ **15.** x^8 **16.** y^{15} **17.** $81y^{24}$ **18.** $8x^9$ **19.** x^5 **20.** z^7 **21.** a^4b^3 **22.** x^3y^5 **23.** $\frac{x^3y^4}{4}$ **24.** $\frac{x^6y^6}{4}$ **25.** $40a^{19}$ **26.** $36x^3$ **27.** 3 **28.** 9 **29.** b **30.** c **31.** $\frac{1}{49}$ **32.** $-\frac{1}{49}$ **33.** $\frac{2}{x^4}$ **34.** $\frac{1}{16x^4}$ **35.** 125 **36.** $\frac{9}{4}$ **37.** $\frac{17}{16}$ **38.** $\frac{1}{42}$ **39.** x^8 **40.** z^8 **41.** r **42.** y^3 **43.** c^4 **44.** $\frac{x^3}{y^3}$ **45.** $\frac{1}{x^6y^{13}}$ **46.** $\frac{a^{10}}{b^{10}}$ **47.** 2.7×10^{-4} **48.** 8.868×10^{-1} **49.** 8.08×10^7 **50.** 8.68×10^5 **51.** 1.37×10^8 **52.** 1.5×10^5 **53.** 867,000 **54.** 0.00386 **55.** 0.00086 **56.** 893,600 **57.** 1,431,280,000,000,000 **58.** 0.0000000001 **59.** 0.016 **60.** 400,000,000,000 **61.** 5 **62.** 2 **63.** 5 **64.** 6 **65.** 4000 ft; 3984 ft; 3856 ft; 3600 ft **66.** 22; 78; 154.02; 400 **67.** $2a^2$ **68.** $-4y$ **69.** $15a^2 + 4a$ **70.** $22x^2 + 3x + 6$ **71.** $-6a^2b - 3b^2 - q^2$ **72.** cannot be combined **73.** $8x^2 + 3x + 6$ **74.** $2x^5 + 3x^4 + 4x^3 + 9x^2 + 7x + 6$ **75.** $-7y^2 - 1$ **76.** $-6m^7 - 3x^4 + 7m^6 - 4m^2$ **77.** $-x^2 - 6xy - 2y^2$ **78.** $x^6 + 4xy + 2y^2$ **79.** $-5x^2 + 5x + 1$ **80.** $-2x^2 - x + 20$ **81.** $6x + 30$ **82.** $9x - 63$ **83.** $8a + 28$ **84.** $54a - 27$ **85.** $-7x^3 - 35x$ **86.** $-32y^3 + 48y$ **87.** $-2x^3 + 18x^2 - 2x$ **88.** $-3a^3b - 3a^2b - 3ab^2$ **89.** $-6a^4 + 8a^2 - 2a$ **90.** $42b^4 - 28b^2 + 14b$ **91.** $2x^2 - 12x - 14$ **92.** $6x^2 - 11x - 10$ **93.** $4a^2 + 27a - 7$ **94.** $42a^2 + 11a - 3$ **95.** $x^4 + 7x^3 + 4x^2 + 23x - 35$ **96.** $x^6 + 2x^5 + x^2 + 3x + 2$ **97.** $x^4 + 4x^3 + 4x^2 - 16$ **98.** $x^6 + 8x^4 + 16x^2 - 16$ **99.** $x^3 + 21x^2 + 147x + 343$ **100.** $8x^3 - 60x^2 + 150x - 125$ **101.** $x^2 + 14x + 49$ **102.** $x^2 - 10x + 25$ **103.** $9x^2 - 42x + 49$ **104.** $16x^2 + 16x + 4$ **105.** $25x^2 - 90x + 81$ **106.** $25x^2 - 1$ **107.** $49x^2 - 16$ **108.** $a^2 - 4b^2$ **109.** $4x^2 - 36$ **110.** $16a^4 - 4b^2$ **111.** $(9x^2 - 6x + 1)$ sq m **112.** $(5x^2 - 3x - 2)$ sq mi **113.** $\frac{1}{7} + \frac{3}{x} + \frac{7}{x^2}$ **114.** $-a^2 + 3b - 4$ **115.** $a + 1 + \frac{6}{a - 2}$ **116.** $4x + \frac{7}{x + 5}$ **117.** $a^2 + 3a + 8 + \frac{22}{a - 2}$ **118.** $3b^2 - 4b - \frac{1}{3b - 2}$ **119.** $2x^3 - x^2 + 2 - \frac{1}{2x - 1}$ **120.** $-x^2 - 16x - 117 - \frac{684}{x - 6}$ **121.** $\left(5x - 1 + \frac{20}{x^2}\right)$ ft **122.** $(7a^3b^6 + a - 1)$ units **123.** 27 **124.** $-\frac{1}{8}$ **125.** $4x^4y^7$ **126.** $\frac{2x^6}{3}$ **127.** $\frac{27a^{12}}{b^6}$ **128.** $\frac{x^{16}}{16y^{12}}$ **129.** $9a^2b^8$ **130.** $2y^2 - 10$ **131.** $11x - 5$ **132.** $5x^2 + 3x - 2$ **133.** $5y^2 - 3y - 1$ **134.** $6x^2 + 11x - 10$ **135.** $28x^3 + 12x$ **136.** $28x^2 - 71x + 18$ **137.** $x^3 + x^2 - 18x + 18$ **138.** $25x^2 + 40x + 16$ **139.** $36x^2 - 9$ **140.** $4a - 1 + \frac{2}{a^2} - \frac{5}{2a^3}$ **141.** $x - 3 + \frac{25}{x + 5}$ **142.** $2x^2 + 7x + 5 + \frac{19}{2x - 3}$

Chapter 3 Getting Ready for the Test **1.** C **2.** A **3.** E **4.** D **5.** F **6.** C **7.** E **8.** I **9.** C **10.** D **11.** C **12.** B **13.** F **14.** D

Chapter 3 Test **1.** 32 **2.** 81 **3.** −81 **4.** $\frac{1}{64}$ **5.** $-15x^{11}$ **6.** y^5 **7.** $\frac{1}{r^5}$ **8.** $\frac{16y^{14}}{x^2}$ **9.** $\frac{1}{6xy^8}$ **10.** 5.63×10^5 **11.** 8.63×10^{-5} **12.** 0.0015 **13.** 62,300 **14.** 0.036 **15. a.** 4, 3; 7, 3; 1, 4; −2, 0 **b.** 4 **16.** $-2x^2 + 12x + 11$ **17.** $16x^3 + 7x^2 - 3x - 13$ **18.** $-3x^3 + 5x^2 + 4x + 5$ **19.** $x^3 + 8x^2 + 3x - 5$ **20.** $3x^3 + 22x^2 + 41x + 14$ **21.** $6x^4 - 9x^3 + 21x^2$ **22.** $3x^2 + 16x - 35$ **23.** $9x^2 - \frac{1}{25}$ **24.** $16x^2 - 16x + 4$ **25.** $64x^2 + 48x + 9$ **26.** $x^4 - 81b^2$ **27.** 1001 ft; 985 ft; 857 ft; 601 ft **28.** $(4x^2 - 9)$ sq in. **29.** $\frac{x}{2y} + \frac{1}{4} - \frac{7}{8y}$ **30.** $x + 2$ **31.** $9x^2 - 6x + 4 - \frac{16}{3x + 2}$

Cumulative Review **1. a.** 11, 112 **b.** 0, 11, 112 **c.** −3, −2, 0, 11, 112 **d.** $-3, -2, 0, \frac{1}{4}, 11, 112$ **e.** $\sqrt{2}$ **f.** All numbers in the given set.; Sec. 1.2, Ex. 11 **2. a.** 7.2 **b.** 0 **c.** $\frac{1}{2}$; Sec. 1.2 **3. a.** 9 **b.** 125 **c.** 16 **d.** 7 **e.** $\frac{9}{49}$ **f.** 0.36; Sec. 1.3, Ex. 1 **4. a.** $\frac{1}{4}$ **b.** $2\frac{5}{12}$; Sec. R.2 **5.** $\frac{1}{4}$; Sec. 1.3, Ex. 4 **6.** $\frac{3}{25}$; Sec. 1.3 **7. a.** $x + 3$ **b.** $3x$ **c.** $7.3 \div x$ or $\frac{7.3}{x}$ **d.** $10 - x$ **e.** $5x + 7$; Sec. 1.3, Ex. 9 **8.** 41; Sec. 1.3 **9.** 6.7; Sec. 1.4, Ex. 10 **10.** no; Sec. 1.5 **11. a.** $\frac{1}{2}$ **b.** 9; Sec. 1.5, Ex. 8 **12. a.** −33 **b.** 5; Sec. 1.5 **13.** 3; Sec. 1.6, Ex. 11a **14.** −8; Sec. 1.6 **15.** −70; Sec. 1.6, Ex. 11d **16.** 150; Sec. 1.6 **17.** $15x + 10$; Sec. 1.8, Ex. 8 **18.** $-6x + 9$; Sec. 1.8 **19.** $-2y - 0.6z + 2$; Sec. 1.8, Ex. 9 **20.** $-4x^2 + 24x - 4$; Sec. 1.8 **21.** $-9x - y + 2z - 6$; Sec. 1.8, Ex. 10 **22.** $4xy - 6y + 2$; Sec. 1.8 **23.** $a = 19$; Sec. 2.1, Ex. 6 **24.** $x = -\frac{1}{2}$; Sec. 2.1 **25.** $y = 140$; Sec. 2.2, Ex. 4 **26.** $j = \frac{12}{5}$; Sec. 2.2 **27.** $x = 4$; Sec. 2.3, Ex. 5 **28.** $x = 1$; Sec. 2.3 **29.** 10; Sec. 2.4, Ex. 2 **30.** $(x + 7) - 2x$; $-x + 7$; Sec. 2.1 **31.** 40 ft; Sec. 2.5, Ex. 2 **32.** undefined; Sec. 1.6 **33.** 800; Sec. 2.6, Ex. 2 **34.**  ; Sec. 2.7 **35.**  $\{x \mid x \le 4\}$; Sec. 2.7, Ex. 7 **36. a.** 25 **b.** −25 **c.** 50; Sec. 3.1 **37. a.** x^{11} **b.** $\frac{t^4}{16}$ **c.** $81y^{10}$; Sec. 3.1, Ex. 33 **38.** z^4; Sec. 3.1 **39.** $\frac{b^3}{27a^6}$; Sec. 3.2, Ex. 10 **40.** $-15x^{16}$; Sec. 3.1 **41.** $\frac{1}{25y^6}$; Sec. 3.2, Ex. 14 **42.** $\frac{1}{9}$; Sec. 3.2 **43.** $10x^3$; Sec. 3.3, Ex. 8 **44.** $4y^2 - 8$; Sec. 3.4 **45.** $5x^2 - 3x - 3$; Sec. 3.3, Ex. 9 **46.** $100x^4 - 9$; Sec. 3.6 **47.** $7x^3 + 14x^2 + 35x$; Sec. 3.5, Ex. 4 **48.** $100x^4 + 60x^2 + 9$; Sec. 3.6 **49.** $3x^3 - 4 + \frac{1}{x}$; Sec. 3.7, Ex. 2

Chapter 4 Factoring Polynomials

Section 4.1

Vocabulary, Readiness & Video Check **1.** factors **3.** least **5.** false **7.** $2 \cdot 7$ **9.** 3 **11.** 5 **13.** The GCF of a list of numbers is the largest number that is a factor of all numbers in the list. **15.** When factoring out a GCF, the number of terms in the other factor should have the same number of terms as our original polynomial.

Exercise Set 4.1 **1.** 4 **3.** 6 **5.** 1 **7.** y^2 **9.** z^7 **11.** xy^2 **13.** 7 **15.** $4y^3$ **17.** $5x^2$ **19.** $3x^3$ **21.** $9x^2y$ **23.** $10a^6b$ **25.** $3(a + 2)$ **27.** $15(2x - 1)$ **29.** $x^2(x + 5)$ **31.** $2y^3(3y + 1)$ **33.** $2x(16y - 9x)$ **35.** $4(x - 2y + 1)$ **37.** $3x(2x^2 - 3x + 4)$ **39.** $a^2b^2(a^5b^4 - a + b^3 - 1)$ **41.** $5xy(x^2 - 3x + 2)$ **43.** $4(2x^5 + 4x^4 - 5x^3 + 3)$ **45.** $\frac{1}{3}x(x^3 + 2x^2 - 4x^4 + 1)$ **47.** $(x^2 + 2)(y + 3)$ **49.** $(y + 4)(z + 3)$ **51.** $(z^2 - 6)(r + 1)$ **53.** $-2(x + 7)$ **55.** $-x^5(2 - x^2)$ **57.** $-3a^2(2a^2 - 3a + 1)$ **59.** $(x + 2)(x^2 + 5)$ **61.** $(x + 3)(5 + y)$ **63.** $(3x - 2)(2x^2 + 5)$ **65.** $(5m^2 + 6n)(m + 1)$ **67.** $(y - 4)(2 + x)$ **69.** $(2x + 1)(x^2 + 4)$ **71.** not factorable by grouping **73.** $(x - 2y)(4x - 3)$ **75.** $(5q - 4p)(q - 1)$ **77.** $2(2y - 7)(3x^2 - 1)$ **79.** $3(2a + 3b^2)(a + b)$ **81.** $x^2 + 7x + 10$ **83.** $b^2 - 3b - 4$ **85.** 2, 6 **87.** −1, −8 **89.** −2, 5 **91.** −8, 3 **93.** d **95.** factored **97.** not factored **99. a.** 3594.2 thousand **b.** 3845.6 thousand **c.** $\frac{1}{10}(16x^2 + 67x + 34{,}382)$ **d.** 2515.94 thousand **101. a.** 5916 thousand tons **b.** 5568 thousand tons **c.** $87(x^2 - 13x + 104)$ **103.** $4x^2 - \pi x^2$; $x^2(4 - \pi)$ **105.** $(x^3 - 1)$ units **107.** answers may vary **109.** answers may vary

Section 4.2

Vocabulary, Readiness & Video Check **1.** true **3.** false **5.** +5 **7.** −3 **9.** +2 **11.** 15 is positive, so its factors would have to be either both positive or both negative. Since the factors need to sum to −8, both factors must be negative.

Exercise Set 4.2 **1.** $(x + 6)(x + 1)$ **3.** $(y - 9)(y - 1)$ **5.** $(x - 3)(x - 3)$ or $(x - 3)^2$ **7.** $(x - 6)(x + 3)$ **9.** $(x + 10)(x - 7)$ **11.** prime **13.** $(x + 5y)(x + 3y)$ **15.** $(a^2 - 5)(a^2 + 3)$ **17.** $(m + 13)(m + 1)$ **19.** $(t - 2)(t + 12)$ **21.** $(a - 2b)(a - 8b)$ **23.** $2(z + 8)(z + 2)$ **25.** $2x(x - 5)(x - 4)$ **27.** $(x - 4y)(x + y)$ **29.** $(x + 12)(x + 3)$ **31.** $(x^2 - 2)(x^2 + 1)$ **33.** $(r - 12)(r - 4)$ **35.** $(x + 2y)(x - y)$ **37.** $3(x + 5)(x - 2)$

39. $3(x^2 - 18)(x^2 - 2)$ **41.** $(x - 24)(x + 6)$ **43.** prime **45.** $(x - 5)(x - 3)$ **47.** $6x(x + 4)(x + 5)$ **49.** $4y(x^2 + x - 3)$ **51.** $(x - 7)(x + 3)$ **53.** $(x + 5y)(x + 2y)$ **55.** $2(t + 8)(t + 4)$ **57.** $x(x - 6)(x + 4)$ **59.** $2t^3(t - 4)(t - 3)$ **61.** $5xy(x - 8y)(x + 3y)$ **63.** $3(m - 9)(m - 6)$ **65.** $-1(x - 11)(x - 1)$ **67.** $\frac{1}{2}(y - 11)(y + 2)$ **69.** $x(xy - 4)(xy + 5)$ **71.** $2x^2 + 11x + 5$ **73.** $15y^2 - 17y + 4$ **75.** $9a^2 + 23ab - 12b^2$ **77.** $x^2 + 5x - 24$ **79.** answers may vary **81.** $2x^2 + 28x + 66$; $2(x + 3)(x + 11)$ **83.** $-16(t - 5)(t + 1)$ **85.** $\left(x + \frac{1}{4}\right)\left(x + \frac{1}{4}\right)$ or $\left(x + \frac{1}{4}\right)^2$ **87.** $(x + 1)(z - 10)(z + 7)$ **89.** 15; 28; 39; 48; 55; 60; 63; 64 **91.** 9; 12; 21 **93.** $(x^n + 10)(x^n - 2)$

Section 4.3

Vocabulary, Readiness & Video Check **1.** d **3.** c **5.** Consider the factors of the first and last terms and the signs of the trinomial. Continue to check possible factors by multiplying until we get the middle term of the trinomial.

Exercise Set 4.3 **1.** $(x + 4)$ **3.** $(10x - 1)$ **5.** $(4x - 3)$ **7.** $(2x + 3)(x + 5)$ **9.** $(y - 1)(8y - 9)$ **11.** $(2x + 1)(x - 5)$ **13.** $(4r - 1)(5r + 8)$ **15.** $(10x + 1)(x + 3)$ **17.** $(3x - 2)(x + 1)$ **19.** $(3x - 5y)(2x - y)$ **21.** $(3m - 5)(5m + 3)$ **23.** $(x - 4)(x - 5)$ **25.** $(2x + 11)(x - 9)$ **27.** $(7t + 1)(t - 4)$ **29.** $(3a + b)(a + 3b)$ **31.** $(7p + 1)(7p - 2)$ **33.** $(6x - 7)(3x + 2)$ **35.** prime **37.** $(8x + 3)(3x + 4)$ **39.** $x(3x + 2)(4x + 1)$ **41.** $3(7b + 5)(b - 3)$ **43.** $(3z + 4)(4z - 3)$ **45.** $2y^2(3x - 10)(x + 3)$ **47.** $(2x - 7)(2x + 3)$ **49.** $3(x^2 - 14x + 21)$ **51.** $(4x + 9y)(2x - 3y)$ **53.** $-1(x - 6)(x + 4)$ **55.** $x(4x + 3)(x - 3)$ **57.** $(4x - 9)(6x - 1)$ **59.** $b(8a - 3)(5a + 3)$ **61.** $2x(3x + 2)(5x + 3)$ **63.** $2y(3y + 5)(y - 3)$ **65.** $5x^2(2x - y)(x + 3y)$ **67.** $-1(2x - 5)(7x - 2)$ **69.** $p^2(4p - 5)(4p - 5)$ or $p^2(4p - 5)^2$ **71.** $-1(2x + 1)(x - 5)$ **73.** $-4(12x - 1)(x - 1)$ **75.** $(2t^2 + 9)(t^2 - 3)$ **77.** prime **79.** $a(6a^2 + b^2)(a^2 + 6b^2)$ **81.** $x^2 - 16$ **83.** $x^2 + 4x + 4$ **85.** $4x^2 - 4x + 1$ **87.** 18–24 **89.** answers may vary **91.** no **93.** $4x^2 + 21x + 5$; $(4x + 1)(x + 5)$ **95.** $\left(2x + \frac{1}{2}\right)\left(2x + \frac{1}{2}\right)$ or $\left(2x + \frac{1}{2}\right)^2$ **97.** $(y - 1)^2(4x + 5)(x + 5)$ **99.** 2; 14 **101.** 2 **103.** answers may vary

Section 4.4

Vocabulary, Readiness & Video Check **1.** a **3.** b **5.** This gives us a four-term polynomial, which may be factored by grouping.

Exercise Set 4.4 **1.** $(x + 3)(x + 2)$ **3.** $(y + 8)(y - 2)$ **5.** $(8x - 5)(x - 3)$ **7.** $(5x^2 - 3)(x^2 + 5)$ **9.** **a.** 9, 2 **b.** $9x + 2x$ **c.** $(2x + 3)(3x + 1)$ **11.** **a.** $-20, -3$ **b.** $-20x - 3x$ **c.** $(3x - 4)(5x - 1)$ **13.** $(3y + 2)(7y + 1)$ **15.** $(7x - 11)(x + 1)$ **17.** $(5x - 2)(2x - 1)$ **19.** $(2x - 5)(x - 1)$ **21.** $(2x + 3)(2x + 3)$ or $(2x + 3)^2$ **23.** $(2x + 3)(2x - 7)$ **25.** $(5x - 4)(2x - 3)$ **27.** $x(2x + 3)(x + 5)$ **29.** $2(8y - 9)(y - 1)$ **31.** $(2x - 3)(3x - 2)$ **33.** $3(3a + 2)(6a - 5)$ **35.** $a(4a + 1)(5a + 8)$ **37.** $3x(4x + 3)(x - 3)$ **39.** $y(3x + y)(x + y)$ **41.** prime **43.** $6(a + b)(4a - 5b)$ **45.** $p^2(15p + q)(p + 2q)$ **47.** $(7 + x)(5 + x)$ or $(x + 7)(x + 5)$ **49.** $(6 - 5x)(1 - x)$ or $(5x - 6)(x - 1)$ **51.** $x^2 - 4$ **53.** $y^2 + 8y + 16$ **55.** $81z^2 - 25$ **57.** $16x^2 - 24x + 9$ **59.** $10x^2 + 45x + 45$; $5(2x + 3)(x + 3)$ **61.** $(x^n + 2)(x^n + 3)$ **63.** $(3x^n - 5)(x^n + 7)$ **65.** answers may vary

Section 4.5

Calculator Explorations

	$x^2 - 2x + 1$	$x^2 - 2x - 1$	$(x - 1)^2$
$x = 5$	16	14	16
$x = -3$	16	14	16
$x = 2.7$	2.89	0.89	2.89
$x = -12.1$	171.61	169.61	171.61
$x = 0$	1	-1	1

Vocabulary, Readiness & Video Check **1.** perfect square trinomial **3.** perfect square trinomial **5.** $(x + 5y)^2$ **7.** false **9.** 8^2 **11.** $(11a)^2$ **13.** $(6p^2)^2$ **15.** No, it just means it won't factor into a binomial squared. It may or may not be factorable. **17.** In order to recognize the binomial as a difference of squares and also to identify the terms to use in the special factoring formula.

Exercise Set 4.5 **1.** yes **3.** no **5.** yes **7.** no **9.** no **11.** yes **13.** $(x + 11)^2$ **15.** $(x - 8)^2$ **17.** $(4a - 3)^2$ **19.** $(x^2 + 2)^2$ **21.** $2(n - 7)^2$ **23.** $(4y + 5)^2$ **25.** $(xy - 5)^2$ **27.** $m(m + 9)^2$ **29.** prime **31.** $(3x - 4y)^2$ **33.** $(x + 2)(x - 2)$ **35.** $(9 + p)(9 - p)$ or $-1(p + 9)(p - 9)$ **37.** $-1(2r + 1)(2r - 1)$ **39.** $(3x + 4)(3x - 4)$ **41.** prime **43.** $-1(6 + x)(6 - x)$ or $(x + 6)(x - 6)$ **45.** $(m^2 + 1)(m + 1)(m - 1)$ **47.** $(x + 13y)(x - 13y)$ **49.** $2(3r + 2)(3r - 2)$ **51.** $x(3y + 2)(3y - 2)$ **53.** $16x^2(x + 2)(x - 2)$ **55.** $xy(y - 3z)(y + 3z)$ **57.** $4(3x - 4y)(3x + 4y)$ **59.** $9(4 - 3x)(4 + 3x)$ **61.** $(5y - 3)(5y + 3)$ **63.** $(11m + 10n)(11m - 10n)$ **65.** $(xy - 1)(xy + 1)$ **67.** $\left(x - \frac{1}{2}\right)\left(x + \frac{1}{2}\right)$ **69.** $\left(7 - \frac{3}{5}m\right)\left(7 + \frac{3}{5}m\right)$ **71.** $(9a + 5b)(9a - 5b)$ **73.** $(x + 7y)^2$ **75.** $2(4n^2 - 7)^2$ **77.** $x^2(x^2 + 9)(x + 3)(x - 3)$ **79.** $pq(8p + 9q)(8p - 9q)$ **81.** 6

Answers to Selected Exercises

83. -2 **85.** $\frac{1}{5}$ **87.** $\left(x - \frac{1}{3}\right)^2$ **89.** $(x + 2 - y)(x + 2 + y)$ **91.** $(b - 4)(a + 4)(a - 4)$ **93.** $(x + 3 - 2y)(x + 3 + 2y)$ **95.** $(x^n + 10)(x^n - 10)$ **97.** 8 **99.** answers may vary **101.** $(x + 6)$ **103.** $a^2 + 2ab + b^2$ **105. a.** 2560 ft **b.** 1920 ft **c.** 13 sec **d.** $16(13 - t)(13 + t)$ **107. a.** 2160 feet **b.** 1520 feet **c.** 12 seconds **d.** $16(12 + t)(12 - t)$

Integrated Review **1.** $(x - 3)(x + 4)$ **2.** $(x - 8)(x - 2)$ **3.** $(x + 1)^2$ **4.** $(x - 3)^2$ **5.** $(x + 2)(x - 3)$ **6.** $(x + 2)(x - 1)$ **7.** $(x + 3)(x - 2)$ **8.** $(x + 3)(x + 4)$ **9.** $(x - 5)(x - 2)$ **10.** $(x - 6)(x + 5)$ **11.** $2(x - 7)(x + 7)$ **12.** $3(x - 5)(x + 5)$ **13.** $(x + 3)(x + 5)$ **14.** $(y - 7)(3 + x)$ **15.** $(x + 8)(x - 2)$ **16.** $(x - 7)(x + 4)$ **17.** $4x(x + 7)(x - 2)$ **18.** $6x(x - 5)(x + 4)$ **19.** $2(3x + 4)(2x + 3)$ **20.** $3(2a - b)(4a + 5b)$ **21.** $(2a + b)(2a - b)$ **22.** $(x + 5y)(x - 5y)$ **23.** $(4 - 3x)(7 + 2x)$ **24.** $(5 - 2x)(4 + x)$ **25.** prime **26.** prime **27.** $(3y + 5)(2y - 3)$ **28.** $(4x - 5)(x + 1)$ **29.** $9x(2x^2 - 7x + 1)$ **30.** $4a(3a^2 - 6a + 1)$ **31.** $(4a - 7)^2$ **32.** $(5p - 7)^2$ **33.** $(7 - x)(2 + x)$ **34.** $(3 + x)(1 - x)$ **35.** $3x^2y(x + 6)(x - 4)$ **36.** $2xy(x + 5y)(x - y)$ **37.** $3xy(4x^2 + 81)$ **38.** $2xy^2(3x^2 + 4)$ **39.** $2xy(1 + 6x)(1 - 6x)$ **40.** $2x(x - 3)(x + 3)$ **41.** $(x + 6)(x + 2)(x - 2)$ **42.** $(x - 2)(x + 6)(x - 6)$ **43.** $2a^2(3a + 5)$ **44.** $2n(2n - 3)$ **45.** $(3x - 1)(x^2 + 4)$ **46.** $(x - 2)(x^2 + 3)$ **47.** $6(x + 2y)(x + y)$ **48.** $2(x + 4y)(6x - y)$ **49.** $(x + y)(5 + x)$ **50.** $(x - y)(7 + y)$ **51.** $(7t - 1)(2t - 1)$ **52.** prime **53.** $-1(3x + 5)(x - 1)$ **54.** $-1(7x - 2)(x + 3)$ **55.** $(1 - 10a)(1 + 2a)$ **56.** $(1 + 5a)(1 - 12a)$ **57.** $(x + 3)(x - 3)(x - 1)(x + 1)$ **58.** $(x + 3)(x - 3)(x + 2)(x - 2)$ **59.** $(x - 15)(x - 8)$ **60.** $(y + 16)(y + 6)$ **61.** $(5p - 7q)^2$ **62.** $(4a - 7b)^2$ **63.** prime **64.** $(7x + 3y)(x + 3y)$ **65.** $-1(x - 5)(x + 6)$ **66.** $-1(x - 2)(x - 4)$ **67.** $(3r - 1)(s + 4)$ **68.** $(x - 2)(x^2 + 1)$ **69.** $(x - 2y)(4x - 3)$ **70.** $(2x - y)(2x + 7z)$ **71.** $(x + 12y)(x - 3y)$ **72.** $(3x - 2y)(x + 4y)$ **73.** $(x^2 + 2)(x + 4)(x - 4)$ **74.** $(x^2 + 3)(x + 5)(x - 5)$ **75.** answers may vary **76.** yes; $9(x^2 + 9y^2)$

Section 4.6

Vocabulary, Readiness & Video Check **1.** quadratic **3.** 3, −5 **5.** One side of the equation must be a factored polynomial and the other side must be zero.

Exercise Set 4.6 **1.** 2, −1 **3.** 6, 7 **5.** −9, −17 **7.** 0, −6 **9.** 0, 8 **11.** $-\frac{3}{2}, \frac{5}{4}$ **13.** $\frac{7}{2}, -\frac{2}{7}$ **15.** $\frac{1}{2}, -\frac{1}{3}$ **17.** −0.2, −1.5 **19.** 9, 4 **21.** −4, 2 **23.** 0, 7 **25.** 0, −20 **27.** 4, −4 **29.** 8, −4 **31.** −3, 12 **33.** $\frac{7}{3}, -2$ **35.** $\frac{8}{3}, -9$ **37.** $0, -\frac{1}{2}, \frac{1}{2}$ **39.** $\frac{17}{2}$ **41.** $\frac{3}{4}$ **43.** $-\frac{1}{2}, \frac{1}{2}$ **45.** $-\frac{3}{2}, -\frac{1}{2}, 3$ **47.** −5, 3 **49.** $-\frac{5}{6}, \frac{6}{5}$ **51.** $2, -\frac{4}{5}$ **53.** $-\frac{4}{3}, 5$ **55.** −4, 3 **57.** 0, 8, 4 **59.** −7 **61.** $0, \frac{3}{2}$ **63.** 0, 1, −1 **65.** $-6, \frac{4}{3}$ **67.** $\frac{6}{7}, 1$ **69.** $\frac{47}{45}$ **71.** $\frac{17}{60}$ **73.** $\frac{7}{10}$ **75.** didn't write equation in standard form; should be $x = 4$ or $x = -2$ **77.** answers may vary, for example, $(x - 6)(x + 1) = 0$ **79.** answers may vary, for example, $x^2 - 12x + 35 = 0$ **81. a.** 300; 304; 276; 216; 124; 0; −156 **b.** 5 sec **c.** 304 ft **83.** $0, \frac{1}{2}$ **85.** 0, −15

Section 4.7

Vocabulary, Readiness & Video Check **1.** In applications, the context of the stated application needs to be considered. Each translated equation resulted in both a positive and a negative solution, and a negative solution is not appropriate for any of the stated applications.

Exercise Set 4.7 **1.** width: x; length: $x + 4$ **3.** x and $x + 2$ if x is an odd integer **5.** base: x; height: $4x + 1$ **7.** 11 units **9.** 15 cm, 13 cm, 22 cm, 70 cm **11.** base: 16 mi; height: 6 mi **13.** 5 sec **15.** width: 5 cm; length: 6 cm **17.** 54 diagonals **19.** 10 sides **21.** −12 or 11 **23.** 14, 15 **25.** 13 feet **27.** 5 in. **29.** 12 mm, 16 mm, 20 mm **31.** 10 km **33.** 36 ft **35.** 9.5 sec **37.** 15, 16; Golden Gate visitors: 15 million; Blue Ridge Parkway visitors: 16 million **39.** length: 15 mi; width: 8 mi **41.** 105 units **43.** 2.8 million or 2,800,000 **45.** 3.5 million or 3,500,000 **47.** 2016 **49.** answers may vary **51.** $\frac{4}{7}$ **53.** $\frac{3}{2}$ **55.** $\frac{1}{3}$ **57.** 8 m **59.** 10 and 15 **61.** width of pool: 29 m; length of pool: 35 m **63.** answers may vary

Chapter 4 Vocabulary Check **1.** quadratic equation **2.** Factoring **3.** greatest common factor **4.** perfect square trinomial **5.** hypotenuse **6.** leg **7.** hypotenuse

Chapter 4 Review **1.** $5(m + 6)$ **2.** $3x(2x - 5)$ **3.** $2x(2x^4 + 1 - 5x^3)$ **4.** $4x(5x^2 + 3x + 6)$ **5.** $(2x + 3)(3x - 5)$ **6.** $(x + 1)(5x - 1)$ **7.** $(x - 1)(3x + 2)$ **8.** $(a + 3b)(3a + b)$ **9.** $(2a + b)(5a + 7b)$ **10.** $(3x + 5)(2x - 1)$ **11.** $(x + 4)(x + 2)$ **12.** $(x - 8)(x - 3)$ **13.** prime **14.** $(x - 6)(x + 1)$ **15.** $(x + 4)(x - 2)$ **16.** $(x + 6y)(x - 2y)$ **17.** $(x + 5y)(x + 3y)$ **18.** $-2(x - 3)(x + 12)$ **19.** $-4(x^2 - 3x - 8)$ **20.** $5y(y - 6)(y - 4)$ **21.** −48; 2 **22.** factor out the GCF, 3 **23.** $(2x + 1)(x + 6)$ **24.** $(2x + 3)(2x - 1)$ **25.** $(3x + 4y)(2x - y)$ **26.** prime **27.** $(2x + 3)(x - 13)$ **28.** $(6x + 5y)(3x - 4y)$ **29.** $5y(2y - 3)(y + 4)$ **30.** $3y(4y - 1)(5y - 2)$ **31.** $5x^2 - 9x - 2$; $(5x + 1)(x - 2)$ **32.** $16x^2 - 28x + 6$; $2(4x - 1)(2x - 3)$ **33.** yes **34.** no **35.** no **36.** yes **37.** yes **38.** no **39.** yes **40.** no **41.** $(x + 9)(x - 9)$ **42.** $(x + 6)^2$ **43.** $(2x + 3)(2x - 3)$ **44.** $(3t + 5s)(3t - 5s)$ **45.** prime **46.** $(n - 9)^2$ **47.** $3(r + 6)^2$ **48.** $(3y - 7)^2$ **49.** $5m^6(m + 1)(m - 1)$ **50.** $(2x - 7y)^2$ **51.** $3y(x + y)^2$ **52.** $(4x^2 + 1)(2x + 1)(2x - 1)$ **53.** −6, 2

54. $-11, 7$ **55.** $0, -1, \frac{2}{7}$ **56.** $-\frac{1}{5}, -3$ **57.** $-7, -1$ **58.** $-4, 6$ **59.** -5 **60.** $2, 8$ **61.** $\frac{1}{3}$ **62.** $-\frac{2}{7}, \frac{3}{8}$ **63.** $0, 6$ **64.** $5, -5$ **65.** $x^2 - 9x + 20 = 0$ **66.** $x^2 + 2x + 1 = 0$ **67.** c **68.** d **69.** 9 units **70.** 8 units, 13 units, 16 units, 10 units **71.** width: 20 in.; length: 25 in. **72.** 36 yd **73.** 19 and 20 **74.** 20 and 22 **75. a.** 17.5 sec and 10 sec; answers may vary **b.** 27.5 sec **76.** 32 cm **77.** $6(x + 4)$ **78.** $7(x - 9)$ **79.** $(4x - 3)(11x - 6)$ **80.** $(x - 5)(2x - 1)$ **81.** $(3x - 4)(x^2 + 2)$ **82.** $(y + 2)(x - 1)$ **83.** $2(x + 4)(x - 3)$ **84.** $3x(x - 9)(x - 1)$ **85.** $(2x + 9)(2x - 9)$ **86.** $2(x + 3)(x - 3)$ **87.** $(4x - 3)^2$ **88.** $5(x + 2)^2$ **89.** $-\frac{7}{2}, 4$ **90.** $-3, 5$ **91.** $0, -7, -4$ **92.** $3, 2$ **93.** $0, 16$ **94.** 19 in.; 8 in.; 21 in. **95.** length: 6 in.; width: 2 in.

Chapter 4 Getting Ready for the Test **1.** B **2.** D **3.** A **4.** A **5.** B **6.** B **7.** B **8.** A **9.** C

Chapter 4 Test **1.** $3x(3x - 1)$ **2.** $(x + 7)(x + 4)$ **3.** $(7 + m)(7 - m)$ **4.** $(y + 11)^2$ **5.** $(x^2 + 4)(x + 2)(x - 2)$ **6.** $(a + 3)(4 - y)$ **7.** prime **8.** $(y - 12)(y + 4)$ **9.** $(a + b)(3a - 7)$ **10.** $(3x - 2)(x - 1)$ **11.** $5(6 + x)(6 - x)$ **12.** $3x(x - 5)(x - 2)$ **13.** $(6t + 5)(t - 1)$ **14.** $(x - 7)(y - 2)(y + 2)$ **15.** $x(1 + x^2)(1 + x)(1 - x)$ **16.** $(x + 12y)(x + 2y)$ **17.** $3, -9$ **18.** $-7, 2$ **19.** $-7, 1$ **20.** $0, \frac{3}{2}, -\frac{4}{3}$ **21.** $0, 3, -3$ **22.** $-3, 5$ **23.** $0, \frac{5}{2}$ **24.** 17 ft **25.** width: 6 units; length: 9 units **26.** 7 sec **27.** hypotenuse: 25 cm; legs: 15 cm, 20 cm **28.** 8.25 sec

Cumulative Review **1. a.** $9 \le 11$ **b.** $8 > 1$ **c.** $3 \ne 4$; Sec. 1.2, Ex. 7 **2. a.** $>$ **b.** $<$; Sec. 1.2 **3.** solution; Sec. 1.3, Ex. 8 **4.** 102; Sec. 1.6 **5.** -12; Sec. 1.5, Ex. 5a **6.** -102; Sec. 1.6 **7. a.** $\frac{3}{4}$ **b.** -24 **c.** 1; Sec. 1.6, Ex. 16 **8.** -98; Sec. 1.6 **9.** $5x + 7$; Sec. 1.8, Ex. 4 **10.** $19 - 6x$; Sec. 1.8 **11.** $-4a - 1$; Sec. 1.8, Ex. 5 **12.** $-13x - 21$; Sec. 1.8 **13.** $7.3x - 6$; Sec. 1.8, Ex. 7 **14.** 2; Sec. 2.1 **15.** -11; Sec. 2.2, Ex. 3 **16.** 28; Sec. 2.2 **17.** all real numbers are solutions; Sec. 2.3, Ex. 7 **18.** 33; Sec. 2.2 **19.** $l = \frac{V}{wh}$; Sec. 2.5, Ex. 5 **20.** $y = \frac{-3x - 7}{2}$ or $y = -\frac{3}{2}x - \frac{7}{2}$; Sec. 2.5 **21.** 5^{18}; Sec. 3.1, Ex. 16 **22.** 30; Sec. 3.1 **23.** y^{16}; Sec. 3.1, Ex. 17 **24.** y^{10}; Sec. 3.1 **25.** x^6; Sec. 3.2, Ex. 9 **26.** $\frac{1}{9}$; Sec. 3.2 **27.** $\frac{y^{18}}{z^{36}}$; Sec. 3.2, Ex. 11 **28.** x^4; Sec. 3.2 **29.** $\frac{1}{x^{19}}$; Sec. 3.2, Ex. 13 **30.** $25a^9$; Sec. 3.2 **31.** $4x$; Sec. 3.3, Ex. 6 **32.** $\frac{5}{6}x - 77$; Sec. 3.3 **33.** $13x^2 - 2$; Sec. 3.3, Ex. 7 **34.** $-0.5x + 1.2$; Sec. 3.3 **35.** $4x^2 - 4xy + y^2$; Sec. 3.5, Ex. 8 **36.** $9x^2 - 42xy + 49y^2$; Sec. 3.5 **37.** $t^2 + 4t + 4$; Sec. 3.6, Ex. 5 **38.** $x^2 - 26x + 169$; Sec. 3.6 **39.** $x^4 - 14x^2y + 49y^2$; Sec. 3.6, Ex. 8 **40.** $49x^2 + 14xy + y^2$; Sec. 3.6 **41.** $2xy - 4 + \frac{1}{2y}$; Sec. 3.7, Ex. 3 **42.** $(z^2 + 7)(z + 1)$; Sec. 4.1 **43.** $(x + 3)(5 + y)$; Sec. 4.1, Ex. 9 **44.** $2x(x + 7)(x - 6)$; Sec. 4.2 **45.** $(x^2 + 2)(x^2 + 3)$; Sec. 4.2, Ex. 7 **46.** $-1(4x - 1)(x + 6)$; Sec. 4.3 **47.** $2(x - 2)(3x + 5)$; Sec. 4.4, Ex. 2 **48.** $x(3y + 4)(3y - 4)$; Sec. 4.5 **49.** 3 sec; Sec. 4.7, Ex. 1 **50.** 9, 4; Sec. 4.6

Chapter 5 Rational Expressions

Section 5.1

Vocabulary, Readiness & Video Check **1.** rational expression **3.** -1 **5.** 2 **7.** $\frac{-a}{b}, \frac{a}{-b}$ **9.** yes **11.** no **13.** Rational expressions are fractions and are therefore undefined if the denominator is zero; if a denominator contains variables, set it equal to zero and solve. **15.** We would need to write parentheses around the numerator or denominator if it had more than one term because the negative sign needs to apply to the entire numerator or denominator.

Exercise Set 5.1 **1.** $\frac{7}{4}$ **3.** 3 **5.** $-\frac{8}{3}$ **7.** $-\frac{11}{2}$ **9.** $x = 0$ **11.** $x = -2$ **13.** $x = \frac{5}{2}$ **15.** $x = 0, x = -2$ **17.** none **19.** $x = 6, x = -1$ **21.** $x = -2, x = -\frac{7}{3}$ **23.** 1 **25.** -1 **27.** $\frac{1}{4(x + 2)}$ **29.** $\frac{1}{x + 2}$ **31.** can't simplify **33.** -5 **35.** $\frac{7}{x}$ **37.** $\frac{1}{x - 9}$ **39.** $5x + 1$ **41.** $\frac{x^2}{x - 2}$ **43.** $7x$ **45.** $\frac{x + 5}{x - 5}$ **47.** $\frac{x + 2}{x + 4}$ **49.** $\frac{x + 2}{2}$ **51.** $-(x + 2)$ **53.** $\frac{x + 1}{x - 1}$ **55.** $x + y$ **57.** $\frac{5 - y}{2}$ **59.** $\frac{2y + 5}{3y + 4}$ **61.** $\frac{-(x - 10)}{x + 8}; \frac{-x + 10}{x + 8}; \frac{x - 10}{-(x + 8)}; \frac{x - 10}{-x - 8}$ **63.** $\frac{-(5y - 3)}{y - 12}; \frac{-5y + 3}{y - 12}; \frac{5y - 3}{-(y - 12)}; \frac{5y - 3}{-y + 12}$ **65.** correct **67.** correct **69.** $\frac{3}{11}$ **71.** $\frac{4}{3}$ **73.** $\frac{117}{40}$ **75.** 0.567 **77.** 0.640 **79.** 0.629 **81.** Mookie Betts **83.** correct **85.** incorrect; $\frac{1 + 2}{1 + 3} = \frac{3}{4}$ **87.** answers may vary **89.** answers may vary **91. a.** \$403 **b.** \$7 **c.** decrease; answers may vary **93.** 400 mg **95.** $C = 78.125$; medium

Section 5.2

Vocabulary, Readiness & Video Check **1.** reciprocals **3.** $\frac{a \cdot d}{b \cdot c}$ **5.** $\frac{6}{7}$ **7.** fractions; reciprocal **9.** We're converting to cubic feet so we want cubic feet in the numerator. We want cubic yards to divide out, so cubic yards is in the denominator.

Exercise Set 5.2 **1.** $\frac{21}{4y}$ **3.** x^4 **5.** $-\frac{b^2}{6}$ **7.** $\frac{x^2}{10}$ **9.** $\frac{1}{3}$ **11.** $\frac{m+n}{m-n}$ **13.** $\frac{x+5}{x}$ **15.** $\frac{(x+2)(x-3)}{(x-4)(x+4)}$ **17.** $\frac{2x^4}{3}$ **19.** $\frac{12}{y^6}$ **21.** $x(x+4)$ **23.** $\frac{3(x+1)}{x^3(x-1)}$ **25.** $(m-n)(m+n)$ or m^2-n^2 **27.** $-\frac{x+2}{x-3}$ **29.** $\frac{x+2}{x-3}$ **31.** $\frac{5}{6}$ **33.** $\frac{3x}{8}$ **35.** $\frac{3}{2}$ **37.** $\frac{3x+4y}{2(x+2y)}$ **39.** $\frac{2(x+2)}{x-2}$ **41.** $-\frac{y(x+2)}{4}$ **43.** $\frac{(a+5)(a+3)}{(a+2)(a+1)}$ **45.** $\frac{5}{x}$ **47.** $\frac{2(n-8)}{3n-1}$ **49.** 1440 **51.** 5 **53.** 81 **55.** 73 **57.** 56.7 **59.** 1,201,500 sq ft **61.** 62.1 miles/hour **63.** 1 **65.** $-\frac{10}{9}$ **67.** $-\frac{1}{5}$ **69.** true **71.** false; $\frac{x^2+3x}{20}$ **73.** $\frac{2}{9(x-5)}$ sq ft **75.** $\frac{x}{2}$ **77.** $\frac{5a(2a+b)(3a-2b)}{b^2(a-b)(a+2b)}$ **79.** answers may vary **81.** 1616.81 euros

Section 5.3

Vocabulary, Readiness & Video Check **1.** $\frac{9}{11}$ **3.** $\frac{a+c}{b}$ **5.** $\frac{5-(6+x)}{x}$ **7.** We completely factor denominators—including coefficients—so we can determine the greatest number of times each unique factor occurs in any one denominator for the LCD.

Exercise Set 5.3 **1.** $\frac{a+9}{13}$ **3.** $\frac{3m}{n}$ **5.** 4 **7.** $\frac{y+10}{3+y}$ **9.** $5x+3$ **11.** $\frac{4}{a+5}$ **13.** $\frac{1}{x-6}$ **15.** $\frac{5x+7}{x-3}$ **17.** $x+5$ **19.** 3 **21.** $4x^3$ **23.** $8x(x+2)$ **25.** $(x+3)(x-2)$ **27.** $3(x+6)$ **29.** $5(x-6)^2$ **31.** $6(x+1)^2$ **33.** $x-8$ or $8-x$ **35.** $(x-1)(x+4)(x+3)$ **37.** $(3x+1)(x+1)(x-1)(2x+1)$ **39.** $2x^2(x+4)(x-4)$ **41.** $\frac{6x}{4x^2}$ **43.** $\frac{24b^2}{12ab^2}$ **45.** $\frac{9y}{2y(x+3)}$ **47.** $\frac{9ab+2b}{5b(a+2)}$ **49.** $\frac{x^2+x}{x(x+4)(x+2)(x+1)}$ **51.** $\frac{18y-2}{30x^2-60}$ **53.** $2x$ **55.** $\frac{x+3}{2x-1}$ **57.** $x+1$ **59.** $\frac{3}{x}$ **61.** $\frac{3x+1}{5x+1}$ **63.** $\frac{29}{21}$ **65.** $-\frac{5}{12}$ **67.** $\frac{7}{30}$ **69.** d **71.** answers may vary **73.** c **75.** b **77.** $-\frac{5}{x-2}$ **79.** $\frac{7+x}{x-2}$ **81.** $\frac{20}{x-2}$ m **83.** answers may vary **85.** 95,304 Earth days **87.** answers may vary **89.** answers may vary

Section 5.4

Vocabulary, Readiness & Video Check **1.** b **3.** The exercise is adding two rational expressions with denominators that are opposites of each other. Recognizing this special case can save us time and effort. If we recognize that one denominator is -1 times the other denominator, we may save many steps.

Exercise Set 5.4 **1.** $\frac{5}{x}$ **3.** $\frac{75a+6b^2}{5b}$ **5.** $\frac{6x+5}{2x^2}$ **7.** $\frac{11}{x+1}$ **9.** $\frac{x-6}{(x-2)(x+2)}$ **11.** $\frac{35x-6}{4x(x-2)}$ **13.** $-\frac{2}{x-3}$ **15.** 0 **17.** $-\frac{1}{x^2-1}$ **19.** $\frac{5+2x}{x}$ **21.** $\frac{6x-7}{x-2}$ **23.** $-\frac{y+4}{y+3}$ **25.** $\frac{-5x+14}{4x}$ or $-\frac{5x-14}{4x}$ **27.** 2 **29.** $\frac{9x^4-4x^2}{21}$ **31.** $\frac{x+2}{(x+3)^2}$ **33.** $\frac{9b-4}{5b(b-1)}$ **35.** $\frac{2+m}{m}$ **37.** $\frac{x(x+3)}{(x-7)(x-2)}$ **39.** $\frac{10}{1-2x}$ **41.** $\frac{15x-1}{(x+1)^2(x-1)}$ **43.** $\frac{x^2-3x-2}{(x-1)^2(x+1)}$ **45.** $\frac{a+2}{2(a+3)}$ **47.** $\frac{y(2y+1)}{(2y+3)^2}$ **49.** $\frac{x-10}{2(x-2)}$ **51.** $\frac{2x+21}{(x+3)^2}$ **53.** $\frac{-5x+23}{(x-2)(x-3)}$ **55.** $\frac{7}{2(m-10)}$ **57.** $\frac{2(x^2-x-23)}{(x+1)(x-6)(x-5)}$ **59.** $\frac{n+4}{4n(n-1)(n-2)}$ **61.** 10 **63.** 2 **65.** $\frac{25a}{9(a-2)}$ **67.** $\frac{x+4}{(x-2)(x-1)}$ **69.** $\frac{2}{3}$ **71.** $-\frac{1}{2}, 1$ **73.** $-\frac{15}{2}$ **75.** $\frac{6x^2-5x-3}{x(x+1)(x-1)}$ **77.** $\frac{4x^2-15x+6}{(x-2)^2(x+2)(x-3)}$ **79.** $\frac{-2x^2+14x+55}{(x+2)(x+7)(x+3)}$ **81.** $\frac{2(x-8)}{(x+4)(x-4)}$ in. **83.** $\frac{P-G}{P}$ **85.** answers may vary **87.** $\left(\frac{90x-40}{x}\right)^\circ$ **89.** answers may vary

Section 5.5

Vocabulary, Readiness & Video Check **1.** c **3.** b **5.** a **7.** These equations are solved in very different ways, so we need to determine the next correct step to make. For a linear equation, we first "move" variable terms to one side and numbers to the other; for a quadratic equation, we first set the equation equal to 0. **9.** the steps for solving an equation containing rational expressions; as if it's the only variable in the equation

Exercise Set 5.5 **1.** 30 **3.** 0 **5.** −2 **7.** −5, 2 **9.** 5 **11.** 3 **13.** 1 **15.** 5 **17.** no solution **19.** 4 **21.** −8 **23.** 6, −4 **25.** 1 **27.** 3, −4 **29.** −3 **31.** 0 **33.** −2 **35.** 8, −2 **37.** no solution **39.** 3 **41.** −11, 1 **43.** $I = \frac{E}{R}$ **45.** $B = \frac{2U - TE}{T}$ **47.** $w = \frac{Bh^2}{705}$ **49.** $G = \frac{V}{N - R}$ **51.** $r = \frac{C}{2\pi}$ **53.** $x = \frac{3y}{3 + y}$ **55.** $\frac{1}{x}$ **57.** $\frac{1}{x} + \frac{1}{2}$ **59.** $\frac{1}{3}$ **61.** answers may vary **63.** $\frac{9 + 5x}{9x}$ **65.** no solution **67.** 100°, 80° **69.** 22.5°, 67.5° **71.** 5

Integrated Review **1.** expression; $\frac{3 + 2x}{3x}$ **2.** expression; $\frac{18 + 5a}{6a}$ **3.** equation; 3 **4.** equation; 18 **5.** expression; $\frac{x - 1}{x(x + 1)}$ **6.** expression; $\frac{3(x + 1)}{x(x - 3)}$ **7.** equation; no solution **8.** equation; 1 **9.** expression; 10 **10.** expression; $\frac{z}{3(9z - 5)}$ **11.** expression; $\frac{5x + 7}{x - 3}$ **12.** expression; $\frac{7p + 5}{2p + 7}$ **13.** equation; 23 **14.** equation; 3 **15.** expression; $\frac{25a}{9(a - 2)}$ **16.** expression; $\frac{9}{4(x - 1)}$ **17.** expression; $\frac{3x^2 + 5x + 3}{(3x - 1)^2}$ **18.** expression; $\frac{2x^2 - 3x - 1}{(2x - 5)^2}$ **19.** expression; $\frac{4x - 37}{5x}$ **20.** expression; $\frac{29x - 23}{3x}$ **21.** equation; $\frac{8}{5}$ **22.** equation; $-\frac{7}{3}$ **23.** answers may vary **24.** answers may vary

Section 5.6

Vocabulary, Readiness & Video Check **1.** c **3.** $\frac{1}{x}; \frac{1}{x} - 3$ **5.** $z + 5; \frac{1}{z + 5}$ **7.** $2y; \frac{11}{2y}$ **9.** No. Proportions are actually equations containing rational expressions, so they can also be solved by using the steps to solve those equations. **11.** divided by, quotient **13.**

	d	$= r$	$\cdot\ t$
car	325	$x + 7$	$\frac{325}{x + 7}$
motorcycle	290	x	$\frac{290}{x}$

$\frac{325}{x + 7} = \frac{290}{x}$

Exercise Set 5.6 **1.** 4 **3.** $\frac{50}{9}$ **5.** −3 **7.** $\frac{14}{9}$ **9.** 123 lb **11.** 165 cal **13.** $y = 21.25$ **15.** $y = 5\frac{5}{7}$ ft **17.** −3 **19.** 2 or 7 **21.** $2\frac{2}{9}$ hr **23.** $1\frac{1}{2}$ min **25.** trip to park rate: r; to park time: $\frac{12}{r}$; return trip rate: r; return time: $\frac{18}{r} = \frac{12}{r} + 1$; $r = 6$ mph **27.** 1st portion: 10 mph; cooldown: 8 mph **29.** 360 sq ft **31.** 2 **33.** $108.00 **35.** 20 mph **37.** $y = 37\frac{1}{2}$ ft **39.** 41 mph; 51 mph **41.** 5 **43.** 217 mph **45.** 9 gal **47.** 8 mph **49.** 2.2 mph; 3.3 mph **51.** 3 hr **53.** $26\frac{2}{3}$ ft **55.** 216 nuts **57.** $666\frac{2}{3}$ mi **59.** 20 hr **61.** car: 70 mph; motorcycle: 60 mph **63.** $5\frac{1}{4}$ hr **65.** 8 **67.** 35 mph; 75 mph **69.** 510 mph **71.** $x = 5$ **73.** $x = 13.5$ **75.** $\frac{1}{2}$ **77.** $\frac{3}{7}$ **79.** faster pump: 28 min; slower pump: 84 min **81.** answers may vary **83.** $R = \frac{D}{T}$ **85.** 3.75 min

Section 5.7

Vocabulary, Readiness & Video Check **1.** $\frac{y}{5x}$ **3.** $\frac{3x}{5}$ **5.** c **7.** a **9.** a single fraction in the numerator and in the denominator

Exercise Set 5.7 **1.** $\frac{2}{3}$ **3.** $\frac{2}{3}$ **5.** $\frac{1}{2}$ **7.** $-\frac{21}{5}$ **9.** $\frac{27}{16}$ **11.** $\frac{4}{3}$ **13.** $\frac{1}{21}$ **15.** $-\frac{4x}{15}$ **17.** $\frac{m - n}{m + n}$ **19.** $\frac{2x(x - 5)}{7x^2 + 10}$ **21.** $\frac{1}{y - 1}$ **23.** $\frac{1}{6}$ **25.** $\frac{x + y}{x - y}$ **27.** $\frac{3}{7}$ **29.** $\frac{a}{x + b}$ **31.** $\frac{7(y - 3)}{8 + y}$ **33.** $\frac{3x}{x - 4}$ **35.** $-\frac{x + 8}{x - 2}$ **37.** $\frac{s^2 + r^2}{s^2 - r^2}$ **39.** $\frac{(x - 6)(x + 4)}{x - 2}$ **41.** Serena Williams **43.** about $13 million **45.** answers may vary **47.** $\frac{13}{24}$ **49.** $4\frac{1}{4}$ ft or 4.25 ft **51.** $\frac{R_1R_2}{R_2 + R_1}$ **53.** $\frac{2x}{2 - x}$ **55.** $\frac{1}{y^2 - 1}$ **57.** 12 hr

Chapter 5 Vocabulary Check **1.** rational expression **2.** complex fraction **3.** $\frac{-a}{b}; \frac{a}{-b}$ **4.** denominator **5.** simplifying **6.** reciprocals **7.** proportion **8.** least common denominator **9.** unit **10.** rate

Chapter 5 Review **1.** $x = 2, x = -2$ **2.** $x = \frac{5}{2}, x = -\frac{3}{2}$ **3.** $\frac{4}{3}$ **4.** $\frac{11}{12}$ **5.** $\frac{2}{x}$ **6.** $\frac{3}{x}$ **7.** $\frac{1}{x - 5}$ **8.** $\frac{1}{x + 1}$ **9.** $\frac{x(x - 2)}{x + 1}$ **10.** $\frac{5(x - 5)}{x - 3}$ **11.** $\frac{x - 3}{x - 5}$ **12.** $\frac{x}{x + 4}$ **13.** $\frac{x + a}{x - c}$ **14.** $\frac{x + 5}{x - 3}$ **15.** $\frac{3x^2}{y}$ **16.** $-\frac{9x^2}{8}$ **17.** $\frac{x - 3}{x + 2}$ **18.** $\frac{-2x(2x + 5)}{(x - 6)^2}$ **19.** $\frac{x + 3}{x - 4}$

20. $\frac{4x}{3y}$ **21.** $(x-6)(x-3)$ **22.** $\frac{2}{3}$ **23.** $\frac{1}{2}$ **24.** $\frac{3(x+2)}{3x+y}$ **25.** $\frac{1}{x+2}$ **26.** $\frac{1}{x-3}$ **27.** $\frac{2(x-5)}{3x^2}$ **28.** $\frac{2x+1}{2x^2}$ **29.** $14x$ **30.** $(x-8)(x+8)(x+3)$ **31.** $\frac{10x^2y}{14x^3y}$ **32.** $\frac{36y^2x}{16y^3x}$ **33.** $\frac{x^2-3x-10}{(x+2)(x-5)(x+9)}$ **34.** $\frac{3x^2+4x-15}{(x+2)^2(x+3)}$ **35.** $\frac{4y+30x^2}{5x^2y}$ **36.** $\frac{-2x+10}{(x-3)(x-1)}$ **37.** $\frac{-2x-2}{x+3}$ **38.** $\frac{5(x+1)}{(x+4)(x-2)(x-1)}$ **39.** $\frac{x-4}{3x}$ **40.** $-\frac{x}{x-1}$ **41.** 30 **42.** 3, −4 **43.** no solution **44.** 5 **45.** $\frac{9}{7}$ **46.** −6, 1 **47.** 9 **48.** no solution **49.** 675 parts **50.** \$33.75 **51.** 3 **52.** 2 **53.** faster car speed: 30 mph; slower car speed: 20 mph **54.** 20 mph **55.** $17\frac{1}{2}$ hr **56.** $8\frac{4}{7}$ days **57.** $x = 15$ **58.** $x = 6$ **59.** $-\frac{7}{18y}$ **60.** $\frac{6}{7}$ **61.** $\frac{3y-1}{2y-1}$ **62.** $-\frac{7+2x}{2x}$ **63.** $\frac{1}{2x}$ **64.** $\frac{x(x-3)}{x+7}$ **65.** $\frac{x-4}{x+4}$ **66.** $\frac{(x-9)(x+8)}{(x+5)(x+9)}$ **67.** $\frac{1}{x-6}$ **68.** $\frac{2x+1}{4x}$ **69.** $\frac{2}{(x+3)(x-2)}$ **70.** $-\frac{3x}{(x+2)(x-3)}$ **71.** $\frac{1}{2}$ **72.** no solution **73.** 1 **74.** $1\frac{5}{7}$ days **75.** $x = 6$ **76.** $x = 12$ **77.** $\frac{3}{10}$ **78.** $\frac{2}{3}$ **79.** 16.2 **80.** 5

Chapter 5 Getting Ready for the Test **1.** B **2.** C **3.** D **4.** D **5.** A **6.** D **7.** A **8.** B **9.** B **10.** A **11.** C **12.** A

Chapter 5 Test **1.** $x = -1, x = -3$ **2. a.** \$115 **b.** \$103 **3.** $\frac{3}{5}$ **4.** $\frac{1}{x+6}$ **5.** −1 **6.** $-\frac{1}{x+y}$ **7.** $\frac{2m(m+2)}{m-2}$ **8.** $\frac{a+2}{a+5}$ **9.** $\frac{(x-6)(x-7)}{(x+7)(x+2)}$ **10.** 15 **11.** $\frac{y-2}{4}$ **12.** $-\frac{1}{2x+5}$ **13.** $\frac{3a-4}{(a-3)(a+2)}$ **14.** $\frac{3}{x-1}$ **15.** $\frac{2(x+3)(x+5)}{x(x^2+4x+1)}$ **16.** $\frac{x^2+2x+35}{(x+9)(x+2)(x-5)}$ **17.** $\frac{4y^2+13y-15}{(y+5)(y+1)(y+4)}$ **18.** $\frac{30}{11}$ **19.** −6 **20.** no solution **21.** no solution **22.** −2, 5 **23.** $\frac{xz}{2y}$ **24.** $b - a$ **25.** $\frac{5y^2-1}{y+2}$ **26.** 1 or 5 **27.** 30 mph **28.** $6\frac{2}{3}$ hr **29.** $x = 12$ **30.** 18 bulbs

Cumulative Review **1. a.** $\frac{15}{x} = 4$ **b.** $12 - 3 = x$ **c.** $4x + 17 = 21$; Sec. 1.3, Ex. 10 **2. a.** $12 - x = -45$ **b.** $12x = -45$ **c.** $x - 10 = 2x$; Sec. 1.3 **3. a.** −12 **b.** −9; Sec. 1.4, Ex. 12 **4. a.** −8 **b.** −17; Sec. 1.5 **5.** distributive property; Sec. 1.7, Ex. 15 **6.** commutative property of addition; Sec. 1.7 **7.** associative property of addition; Sec. 1.7, Ex. 16 **8.** associative property of multiplication; Sec. 1.7 **9.** $x = -4$; Sec. 2.1, Ex. 7 **10.** $x = 0$; Sec. 2.1 **11.** shorter piece, 2 ft; longer piece, 8 ft; Sec. 2.4, Ex. 3 **12.** 190, 192; Sec. 2.4 **13.** $\frac{y-b}{m} = x$; Sec. 2.5, Ex. 6 **14.** $x = \frac{2y+6}{3}$; Sec. 2.5 **15.** $x \le -10$; 

; Sec. 2.7, Ex. 4 **16.** $\{x \mid x < -1\}$; Sec. 2.7 **17.** x^3; Sec. 3.1, Ex. 24 **18.** 1; Sec. 3.1 **19.** 256; Sec. 3.1, Ex. 25 **20.** $x^{15}y^6$; Sec. 3.1 **21.** −27; Sec. 3.1, Ex. 26 **22.** $x^{18}y^4$; Sec. 3.1 **23.** $2x^4y$; Sec. 3.1, Ex. 27 **24.** $-15a^5b^2$; Sec. 3.1 **25.** $\frac{2}{x^3}$; Sec. 3.2, Ex. 2 **26.** $\frac{1}{49}$; Sec. 3.2 **27.** $\frac{1}{16}$; Sec. 3.2, Ex. 4 **28.** $\frac{5}{z^7}$; Sec. 3.2 **29.** $10x^4 + 30x$; Sec. 3.5, Ex. 5 **30.** $x^2 + 18x + 81$; Sec. 3.5 **31.** $-15x^4 - 18x^3 + 3x^2$; Sec. 3.5, Ex. 6 **32.** $4x^2 - 1$; Sec. 3.5 **33.** $4x^2 - 4x + 6 - \frac{11}{2x+3}$; Sec. 3.7, Ex. **34.** $4x^2 + 16x + 55 + \frac{222}{x-4}$; Sec. 3.7 **35.** $(x+3)(x+4)$; Sec. 4.2, Ex. 1 **36.** $-2(a+1)(a-6)$; Sec. 4.2 **37.** $(5x+2y)^2$; Sec. 4.5, Ex. 5 **38.** $(x+2)(x-2)$; Sec. 4.5 **39.** −2, 11; Sec. 4.6, Ex. 4 **40.** $-2, \frac{1}{3}$; Sec. 4.6 **41.** $\frac{2}{5}$; Sec. 5.2, Ex. 2 **42.** $\frac{x+5}{2x^3}$; Sec. 5.1 **43.** $3x - 5$; Sec. 5.3, Ex. 3 **44.** $7x^4(x^2 - x + 1)$; Sec. 4.1 **45.** $\frac{3}{x-2}$; Sec. 5.4, Ex. 2 **46.** $(2x+3)^2$; Sec. 4.5 **47.** $t = 5$; Sec. 5.5, Ex. 2 **48.** $\frac{30}{x+3}$; Sec. 5.2 **49.** $2\frac{1}{10}$ hr; Sec. 5.6, Ex. 6 **50.** $\frac{4m+2n}{m+n}$ or $\frac{2(2m+n)}{m+n}$; Sec. 5.7

Chapter 6 Graphing Equations and Inequalities

Section 6.1

Vocabulary, Readiness & Video Check **1.** x-axis **3.** origin **5.** x-coordinate; y-coordinate **7.** solution **9.** origin; left or right; up or down **11.** Replace both values of the ordered pair in the linear equation and see if a true statement results.

Exercise Set 6.1 **1.**

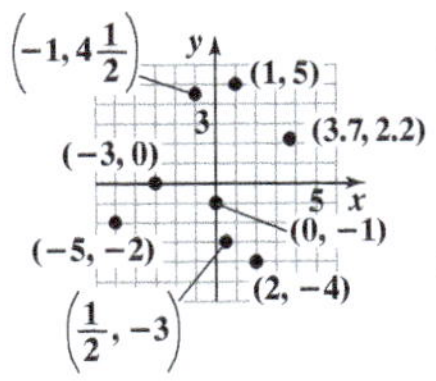

(1, 5) and (3.7, 2.2) are in quadrant I, $\left(-1, 4\frac{1}{2}\right)$ is in quadrant II, (−5, −2) is in quadrant III, (2, −4) and $\left(\frac{1}{2}, -3\right)$ are in quadrant IV, (−3, 0) lies on the x-axis, (0, −1) lies on the y-axis **3.** (0, 0) **5.** (3, 2) **7.** (−2, −2) **9.** (2, −1) **11.** (0, −3) **13.** (1, 3) **15.** (−3, −1)

17. a. (2010, 10.6), (2011, 10.2), (2012, 10.8), (2013, 10.9), (2014, 10.4), (2015, 11.1), (2016, 11.4), (2017, 11.1) **b.** In the year 2017, the domestic box office was \$11.1 billion. **c.**

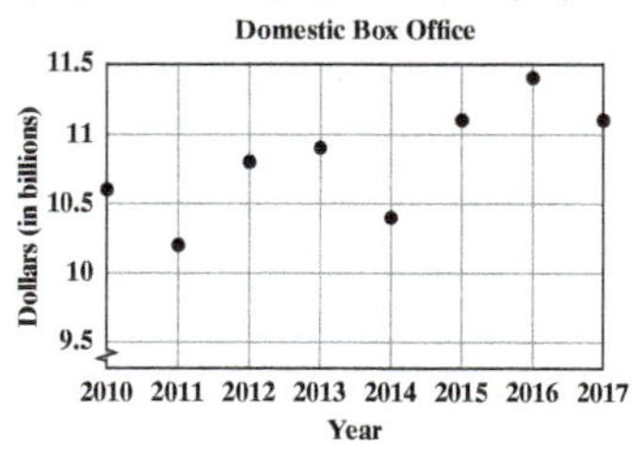

d. answers may vary **19. a.** (0.50, 10), (0.75, 12), (1.00, 15), (1.25, 16), (1.50, 18), (1.50, 19), (1.75, 19), (2.00, 20) **b.** When Minh studied 1.25 hours, her quiz score was 16. **c.**

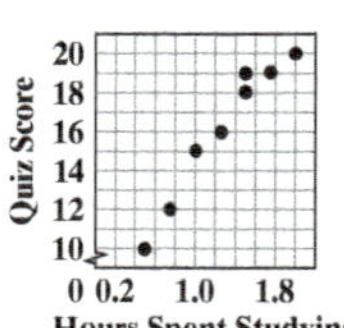

d. answers may vary

21. $(-4, -2), (4, 0)$ **23.** $(-8, -5), (16, 1)$ **25.** $0; 7; -\frac{2}{7}$ **27.** 2; 2; 5 **29.** 0; −3; 2 **31.** 2; 6; 3 **33.** −12; 5; −6 **35.** $\frac{5}{7}; \frac{5}{2}; -1$

37. 0; −5; −2

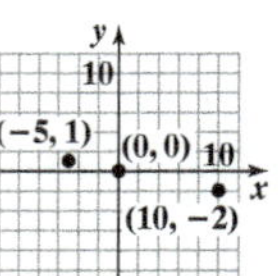

39. 2; 1; −6

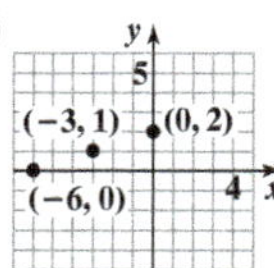

41. a. 13,000; 21,000; 29,000 **b.** 45 desks **43. a.** 7.74; 7.89; 8.79 **b.** 2012 **c.** 2025

45. China **47.** Spain, France, United States, and China **49.** 53 million **51.** $y = 5 - x$ **53.** $y = \frac{5 - 2x}{4}$ **55.** $y = -2x$ **57.** false **59.** true **61.** negative; negative **63.** positive; negative **65.** 0; 0 **67.** y **69.** no; answers may vary **71.** answers may vary **73.** answers may vary **75.** $(4, -7)$ **77.** 26 units **79.** 29 million; 32 million; 34 million; 37 million

Section 6.2

Calculator Explorations **1.**

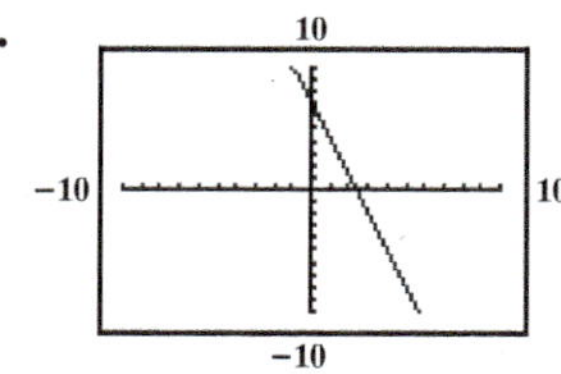

3.

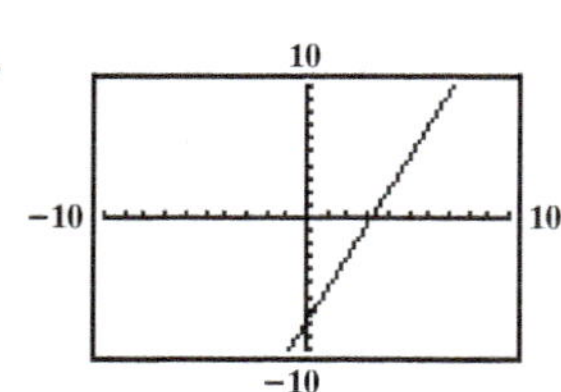

5.

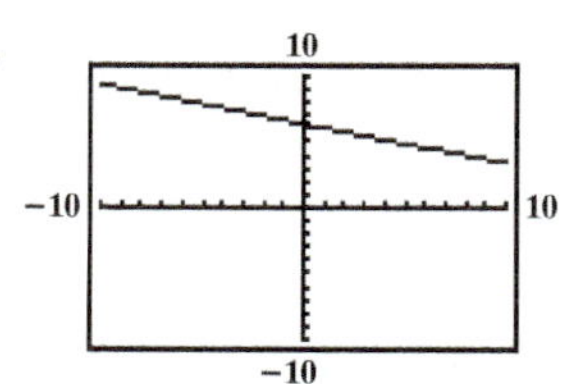

Vocabulary, Readiness & Video Check **1.** It is always good practice to use a third point as a check to see that our points lie along a straight line.

Exercise Set 6.2 **1.** 6; −2; 5

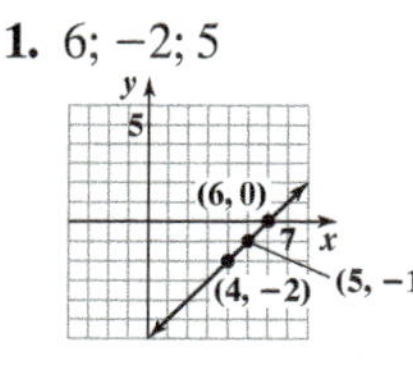

3. −4; 0; 4

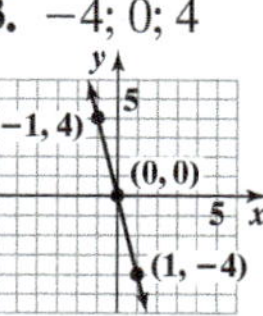

5. 0; 2; −1

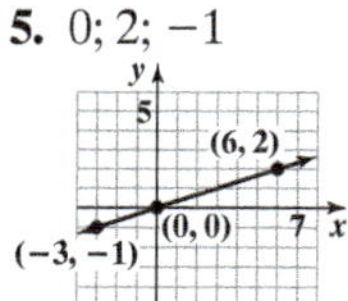

7. 3; −1; −5

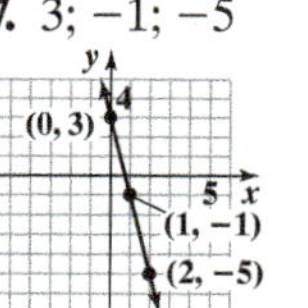

9.

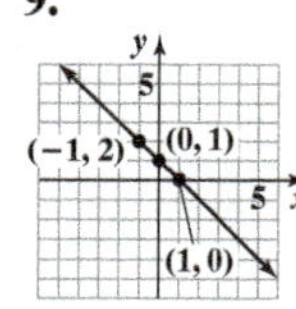

11.

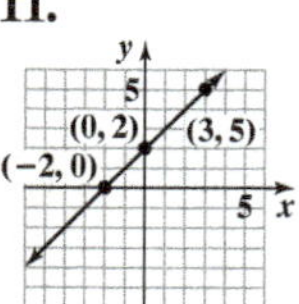

13.

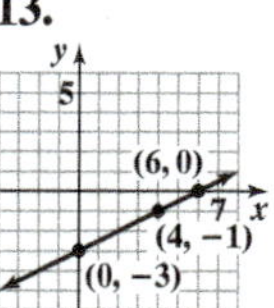

15.

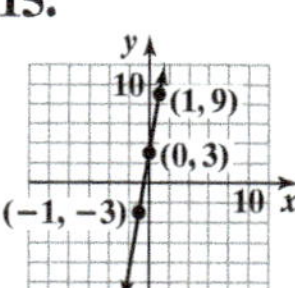

17.

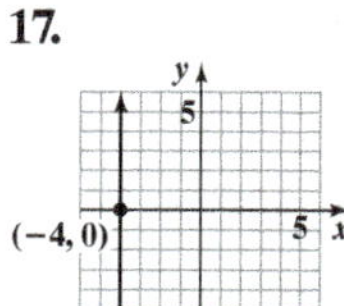

19.

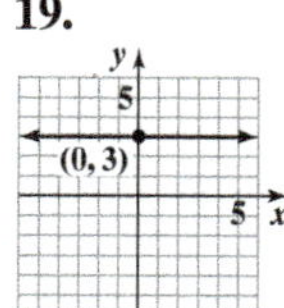

21.

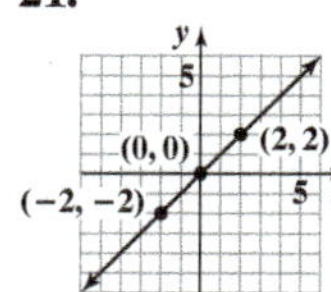

23.

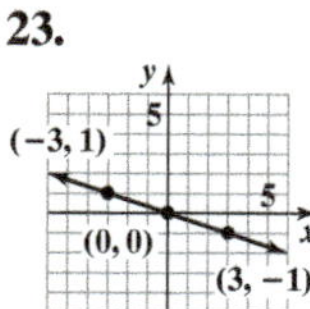

25.

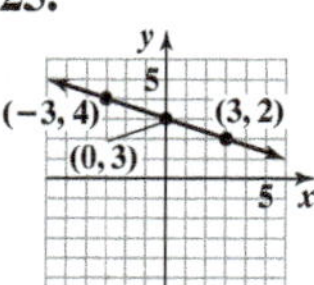

27.

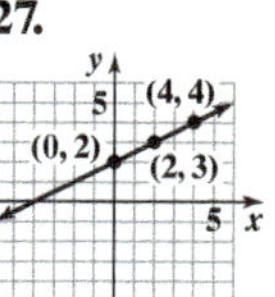

29.

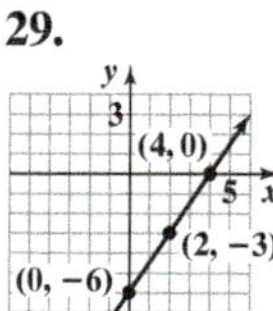

31.

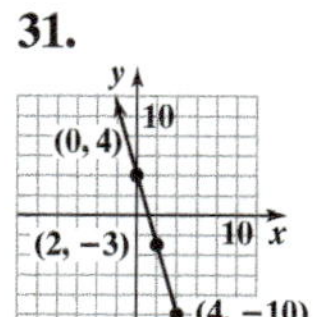

33. a. 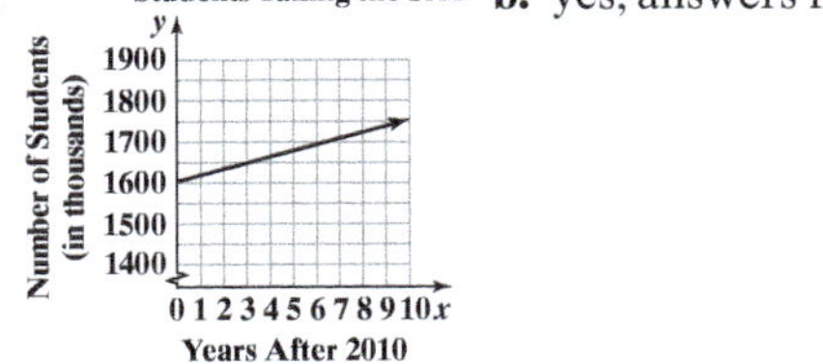

b. yes; answers may vary

35. a. 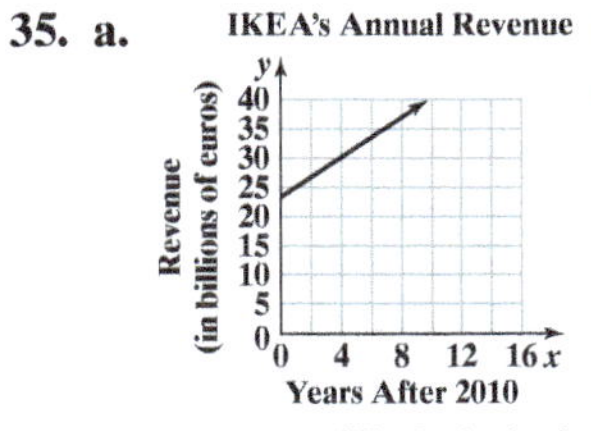

b. (6, 33.7) **c.** In 2016, IKEA's total annual revenue was 33.7 billion euros.

37. $(4, -1)$ **39.** $3; -3$ **41.** $0; 0$ **43.** 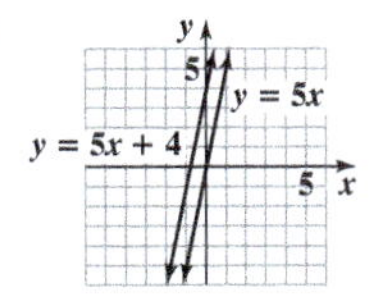

45. 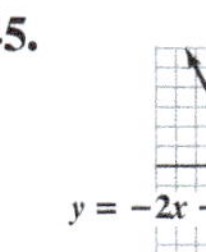

47. $0; 1; 1; 4; 4$ 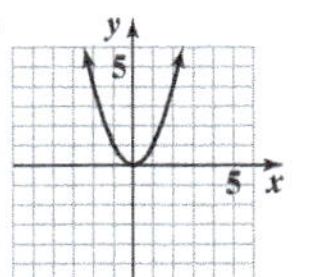

49. $x + y = 12$; 9 cm **51.** yes; answers may vary

Section 6.3

Calculator Explorations

1. 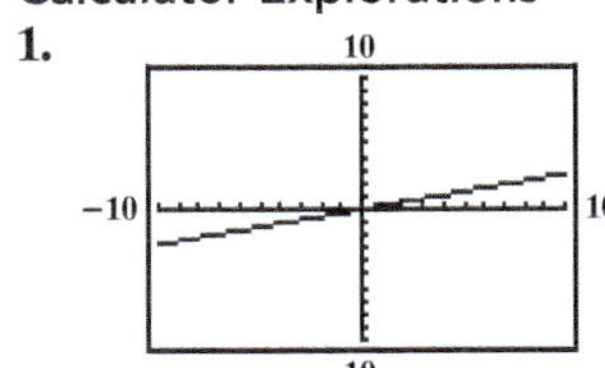

3. 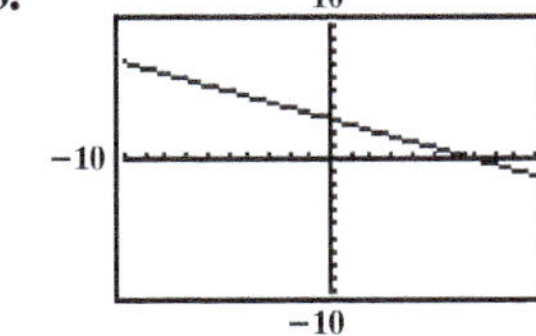

5. 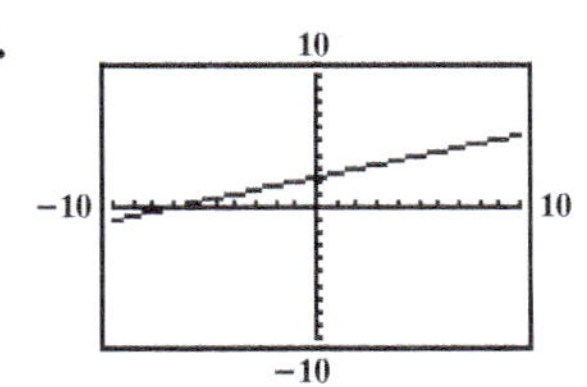

Vocabulary, Readiness & Video Check **1.** linear **3.** horizontal **5.** y-intercept **7.** $y; x$ **9.** false **11.** true **13.** because x-intercepts lie on the x-axis; because y-intercepts lie on the y-axis. **15.** For a horizontal line, the coefficient of x will be 0; for a vertical line, the coefficient of y will be 0.

Exercise Set 6.3 **1.** $(-1, 0); (0, 1)$ **3.** $(-2, 0); (2, 0); (0, -2)$ **5.** $(-2, 0); (1, 0); (3, 0); (0, 3)$ **7.** $(-1, 0); (1, 0); (0, 1); (0, -2)$

9.

11.  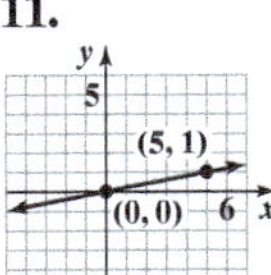

13. 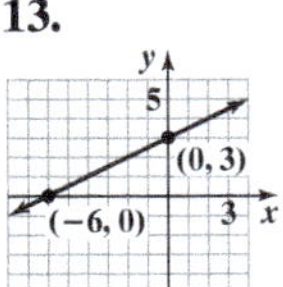

15. 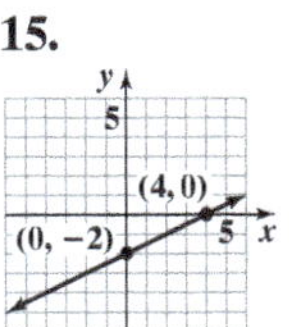

17. 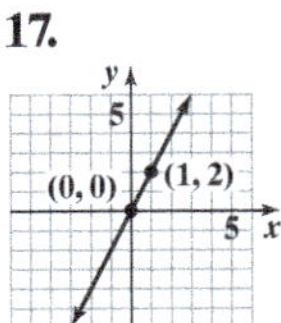

19. 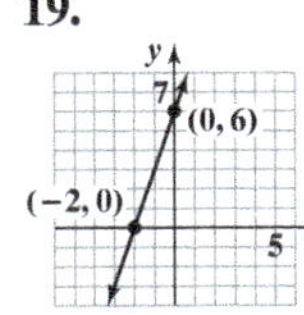

21.

23.  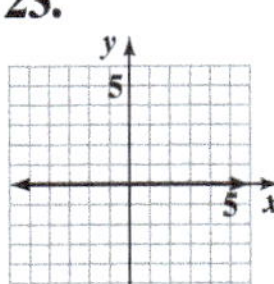

25. 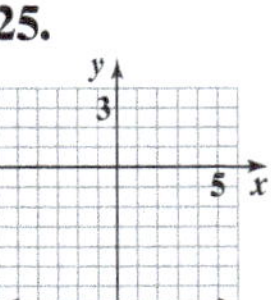

27. 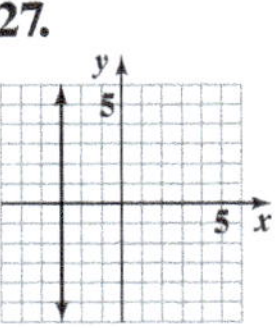

29. 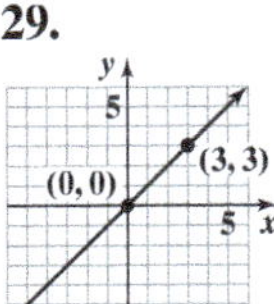

31. 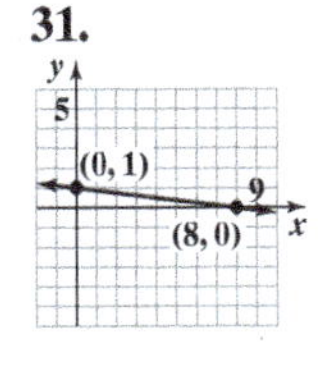

33.

35. 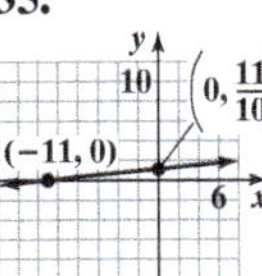

37. 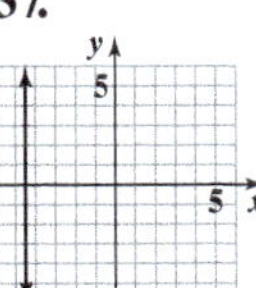

39. 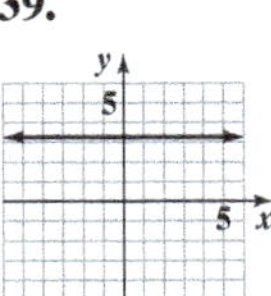

41. 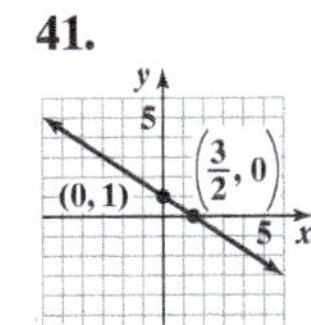

43. 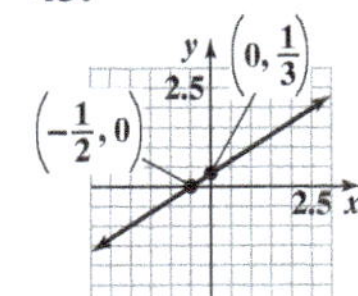

45. $\frac{3}{2}$ **47.** 6 **49.** $-\frac{6}{5}$ **51.** c **53.** a **55.** infinite **57.** 0 **59.** answers may vary **61.** $(0, 200)$; no chairs and 200 desks are manufactured. **63.** 300 chairs **65.** $y = -4$ **67. a.** $(24.1, 0)$ **b.** 24.1 years after 2010, there may be no print newspaper employees.

Section 6.4

Calculator Explorations

1. 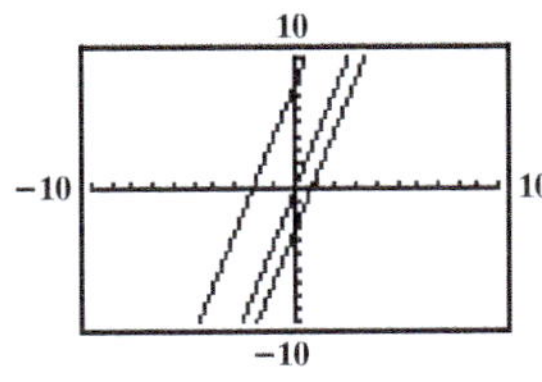

3. 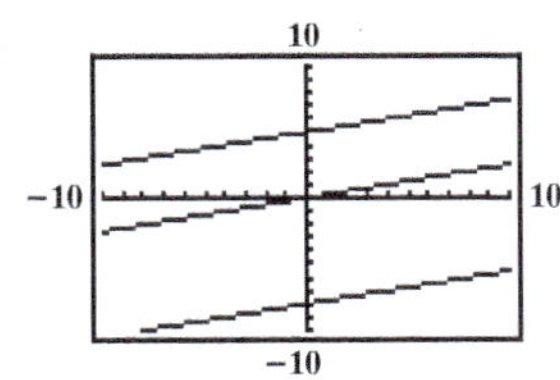

Vocabulary, Readiness & Video Check **1.** slope **3.** 0 **5.** positive **7.** $y; x$ **9.** upward **11.** horizontal **13.** Solve the equation for y; the slope is the coefficient of x. **15.** Slope-intercept form; this form makes the slope easy to see, and we need to compare slopes to determine if two lines are parallel or perpendicular.

Exercise Set 6.4 **1.** $m = -1$ **3.** $m = -\frac{1}{4}$ **5.** $m = 0$ **7.** undefined slope **9.** $m = -\frac{4}{3}$ **11.** $m = \frac{5}{2}$ **13.** negative **15.** undefined **17.** upward **19.** horizontal **21.** line 1 **23.** line 2 **25.** $m = 5$ **27.** $m = -0.3$ **29.** $m = -2$ **31.** undefined slope **33.** $m = \frac{2}{3}$ **35.** undefined slope **37.** $m = \frac{1}{2}$ **39.** $m = 0$ **41.** $m = -\frac{3}{4}$ **43.** $m = 4$ **45. a.** 1 **b.** -1 **47. a.** $\frac{9}{11}$ **b.** $-\frac{11}{9}$ **49.** neither **51.** neither **53.** parallel **55.** perpendicular **57.** $\frac{3}{5}$ **59.** 12.5% **61.** 40% **63.** 37%; 35% **65.** $m = 0.53$; Every year, the number of U.S. households with televisions increases by 0.53 million households. **67.** $m = 0.14$; Every year, the median age of automobiles in the United States increases by 0.14 year. **69.** $y = 2x - 14$ **71.** $y = -6x - 11$ **73.** D **75.** B **77.** E **79.** $m = \frac{1}{2}$ **81.** answers may vary **83.** 31.5 mi per gal **85.** 2007; 31.2 mi per gallon **87.** from 2011 to 2012 **89.** $x = 20$ **91. a.** (2012, 2378), (2017, 3244) **b.** 173.2 **c.** For the years 2012 through 2017, the number of heart transplants increased at a rate of 173.2 per year. **93.** Opposite sides are parallel since their slopes are equal, so the figure is a parallelogram. **95.** 2.0625 **97.** -1.6 **99.** The line becomes steeper.

Section 6.5

Calculator Explorations

1.

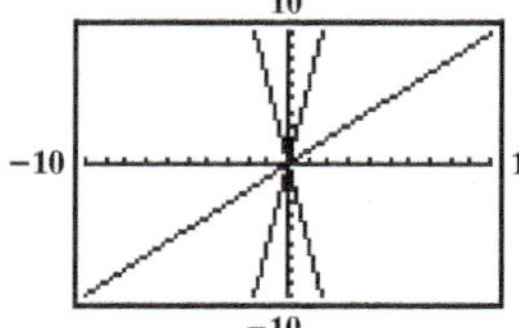

3.

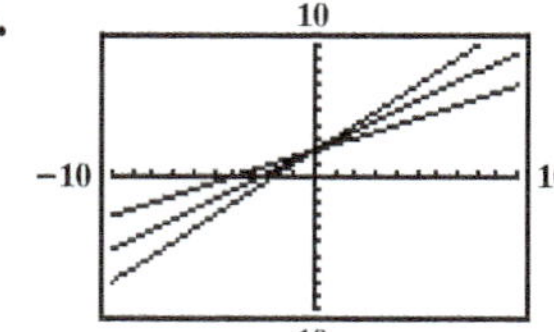

Vocabulary, Readiness & Video Check **1.** slope-intercept; m; b **3.** point-slope **5.** slope-intercept **7.** horizontal **9.** y-intercept; fraction **11.** Write the equation with x- and y-terms on one side of the equal sign and a constant on the other side. **13.** Example 6: $y = -3$; Example 7: $x = -2$

Exercise Set 6.5 **1.**

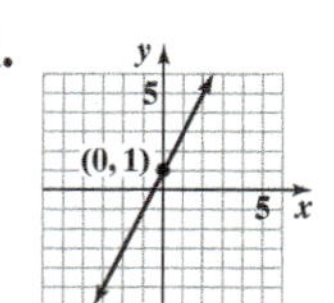

3.

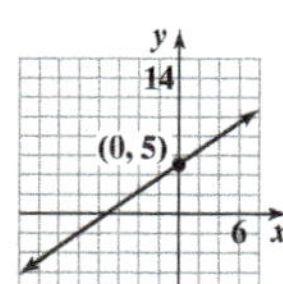

5.

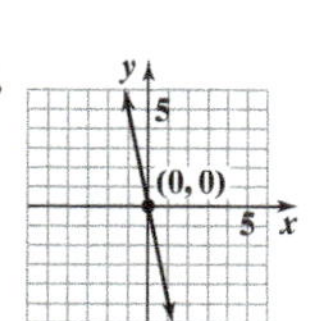

7. (0, 6)

9.

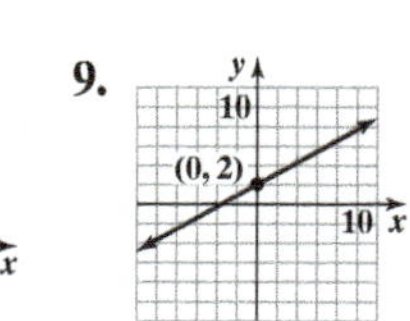

11.

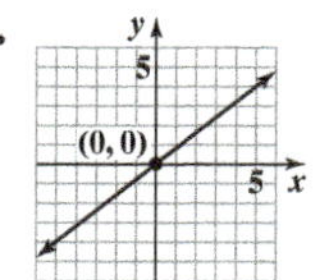

13. $y = 5x + 3$ **15.** $y = -4x - \frac{1}{6}$ **17.** $y = \frac{2}{3}x$ **19.** $y = -8$ **21.** $y = -\frac{1}{5}x + \frac{1}{9}$ **23.** $-6x + y = -10$ **25.** $8x + y = -13$ **27.** $3x - 2y = 27$ **29.** $x + 2y = -3$ **31.** $2x - y = 4$ **33.** $8x - y = -11$ **35.** $4x - 3y = -1$ **37.** $8x + 13y = 0$ **39.** $x = 0$ **41.** $y = 3$ **43.** $x = -\frac{7}{3}$ **45.** $y = 2$ **47.** $y = 5$ **49.** $x = 6$ **51.** $y = -\frac{1}{2}x + \frac{5}{3}$ **53.** $y = -x + 17$ **55.** $x = -\frac{3}{4}$ **57.** $y = x + 16$ **59.** $y = -5x + 7$ **61.** $x = -8$ **63.** $y = \frac{3}{2}x$ **65.** $y = -3$ **67.** $y = -\frac{4}{7}x - \frac{18}{7}$ **69. a.** (0, 7162), (6, 7216) **b.** $y = 9x + 7162$ **c.** 7207 magazines **71. a.** $s = 32t$ **b.** 128 ft/sec **73. a.** $y = -13{,}200x + 434{,}000$ **b.** 381,200 vehicles **75. a.** $y = 27x + 5320$ **b.** 5617 indoor cinema sites **77. a.** $S = -1000p + 13{,}000$ **b.** 9500 Fun Noodles **79.** -1 **81.** 5 **83.** B **85.** D **87.** $2x - y = -8$ **89. a.** $3x - y = -5$ **b.** $x + 3y = 5$

Integrated Review **1.** $m = 2$ **2.** $m = 0$ **3.** $m = -\frac{2}{3}$ **4.** slope is undefined

5.

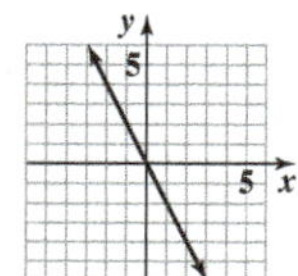

6.

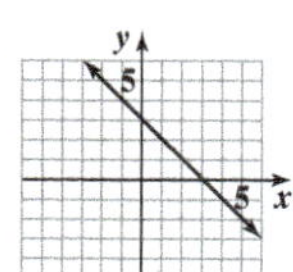

7.

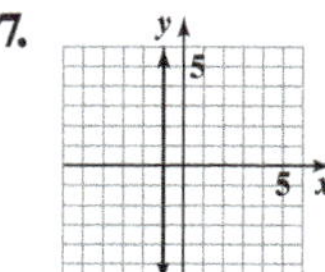

8.

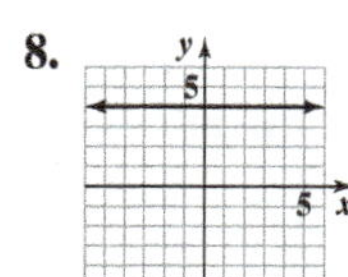

9. 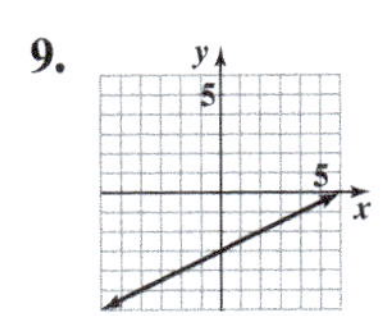

10. 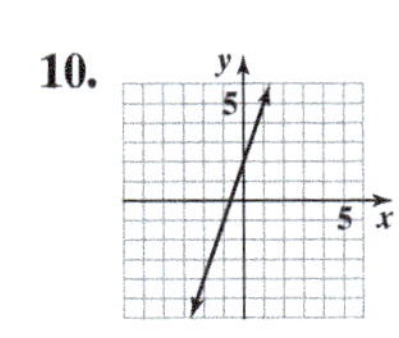

11. 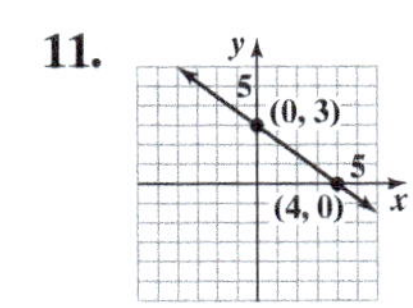

12. 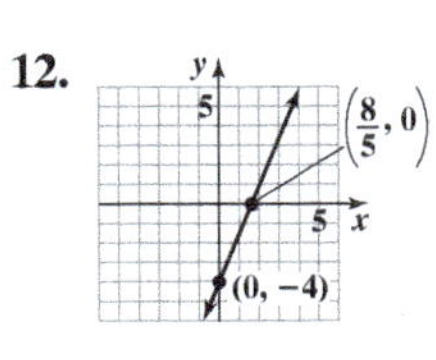

13. $m = 3$ **14.** $m = -6$ **15.** $m = -\frac{7}{2}$ **16.** $m = 2$ **17.** undefined slope **18.** $m = 0$ **19.** $y = 2x - \frac{1}{3}$ **20.** $y = -4x - 1$ **21.** $-x + y = -2$ **22.** neither **23.** perpendicular **24. a.** (2012, 4418), (2017, 4478) **b.** 12 **c.** For the years 2012 through 2017, the amount of yogurt produced increased at a rate of 12 million pounds per year.

Section 6.6

Vocabulary, Readiness & Video Check **1.** relation **3.** range **5.** vertical **7.** (3, 7) **9.** $y; x$ **11.** Yes, this is a function. The definition restricts x-values to be assigned to exactly one y-value, but it makes no such restriction on the y-values. **13.** $f(-2) = 6$ corresponds to $(-2, 6)$ and $f(3) = 11$ corresponds to (3, 11).

Exercise Set 6.6 **1.** domain: $\{-7, 0, 2, 10\}$; range: $\{-7, 0, 4, 10\}$ **3.** domain: $\{0, 1, 5\}$; range: $\{-2\}$ **5.** yes **7.** no **9.** no **11.** yes **13.** yes **15.** no **17.** a **19.** yes **21.** yes **23.** no **25.** no **27.** 9:30 p.m. **29.** January 1 and December 1 **31.** yes; it passes the vertical line test **33.** $4.25 per hour **35.** 2009 **37.** yes; answers may vary **39.** $1.84 **41.** more than 1 ounce and less than or equal to 2 ounces **43.** yes; answers may vary **45.** $-9, -5, 1$ **47.** 6, 2, 11 **49.** $-6, 0, 9$ **51.** 2, 0, 3 **53.** $5, 0, -20$ **55.** 5, 3, 35 **57.** (3, 6) **59.** $\left(0, -\frac{1}{2}\right)$ **61.** $(-2, 9)$ **63.** all real numbers **65.** all real number except -5 **67.** domain: all real numbers; range: $y \geq -4$ **69.** domain: all real numbers; range; all real numbers **71.** domain: all real numbers; range: $\{2\}$ **73.** -1 **75.** -1 **77.** $-1, 5$ **79.** $x < 1$ **81.** $x \geq -3$ **83.** $\frac{19}{2x}$ m **85.** $f(-5) = 12$ **87.** $(3, -4)$ **89.** $f(5) = 0$ **91.** answers may vary **93.** $f(x) = x + 7$ **95. a.** 190.4 mg **b.** 380.8 mg

Section 6.7

Vocabulary, Readiness & Video Check **1.** linear inequality in two variables **3.** false **5.** true **7.** An ordered pair is a solution of an inequality if replacing the variables with the coordinates of the ordered pair results in a true statement.

Exercise Set 6.7 **1.** no; no **3.** yes; no **5.** no; yes

7. 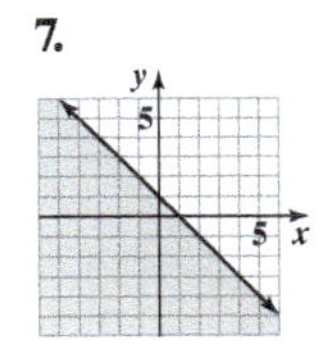

9. 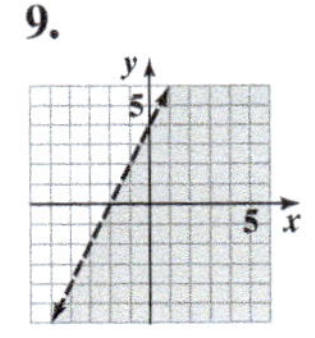

11. 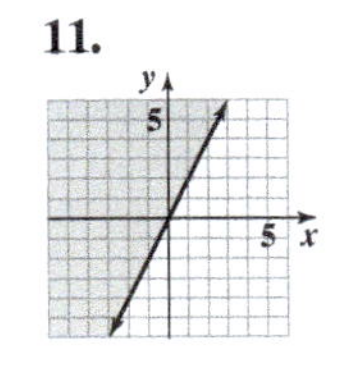

13. 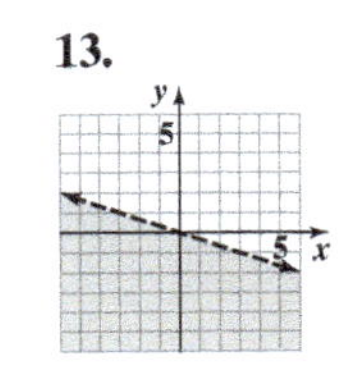

15. 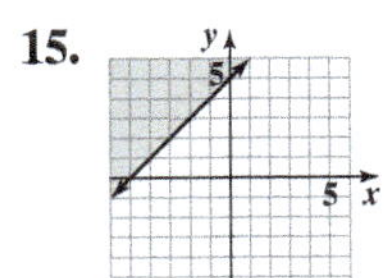

17. 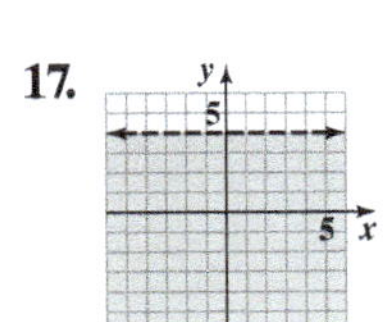

19. 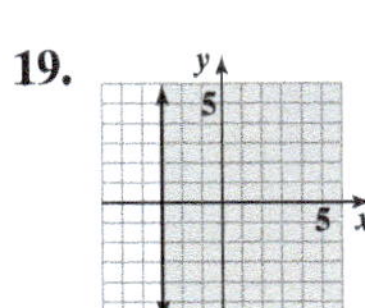

21. 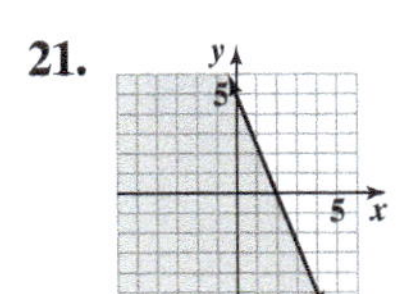

23. 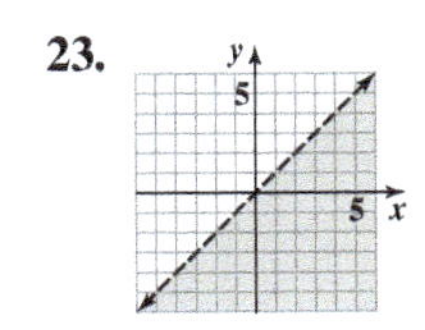

25. 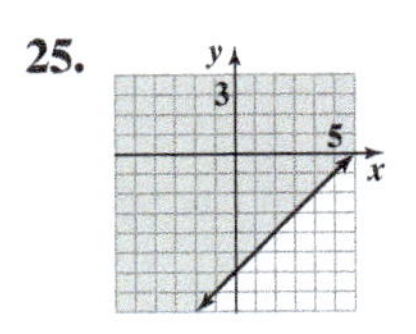

27. 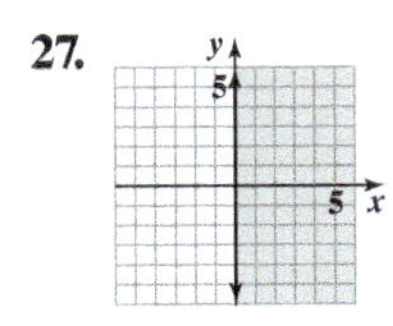

29. 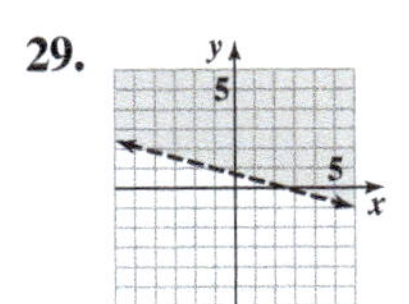

31. 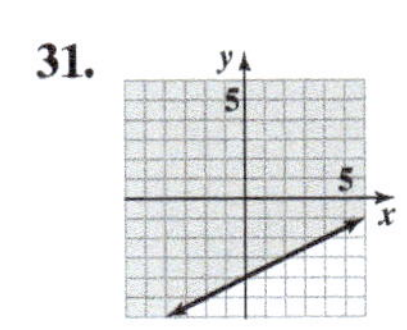

33. 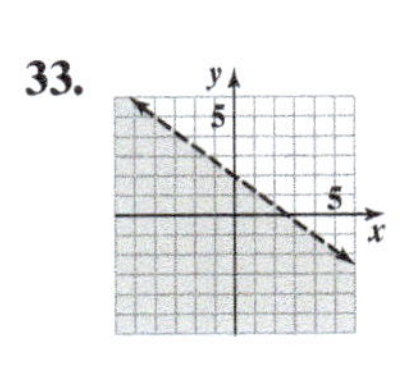

35. $(-2, 1)$ **37.** $(-3, -1)$ **39.** A **41.** B **43.** answers may vary **45.** yes **47.** yes **49. a.** $30x + 0.15y \leq 500$

b. 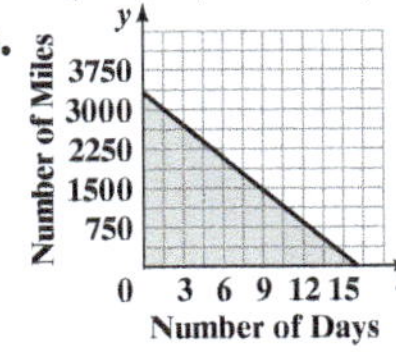

c. answers may vary

Chapter 6 Vocabulary Check **1.** solution **2.** y-axis **3.** linear **4.** x-intercept **5.** standard **6.** y-intercept **7.** function **8.** slope-intercept **9.** domain **10.** range **11.** relation **12.** point-slope **13.** y **14.** x-axis **15.** x **16.** slope

Chapter 6 Review **1-6.**

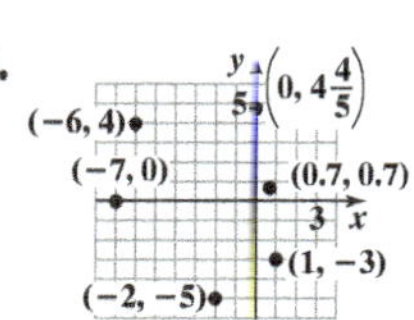

7. $(7, 44)$ **8.** $\left(-\frac{13}{3}, -8\right)$ **9.** $-3; 1; 9$ **10.** $5; 5; 5$ **11.** $0; 10; -10$ **12. a.** 2005; 2500; 7000 **b.** 886 compact disc holders

13.
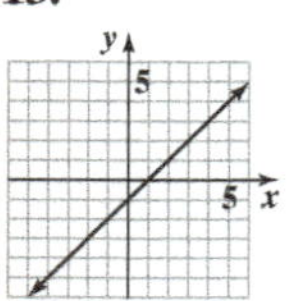

14.
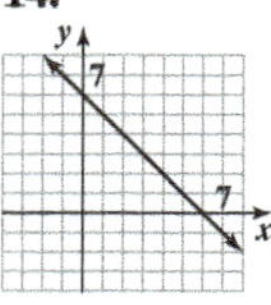

15.
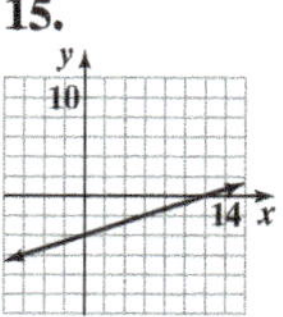

16.
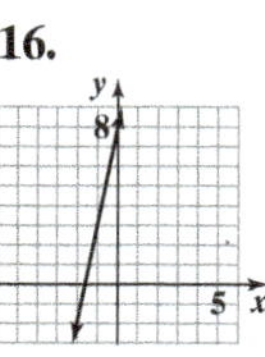

17.
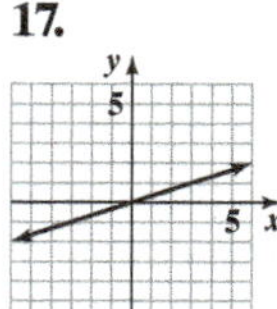

18.
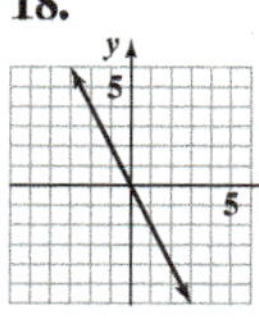

19. $(4, 0); (0, -2)$ **20.** $(-2, 0); (2, 0); (0, 2); (0, -2)$ **21.**
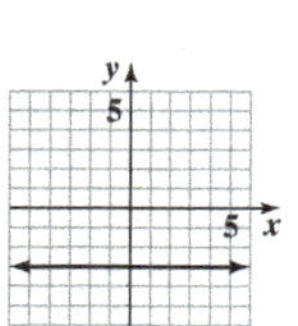

22.
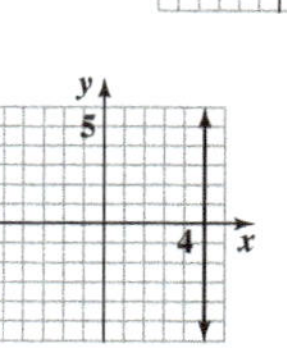

23. $(12, 0), (0, -4)$

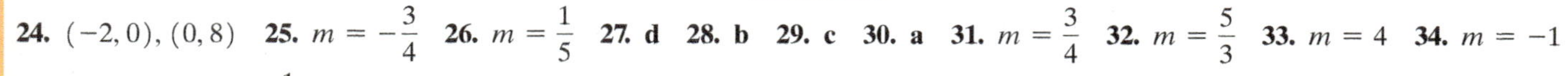

24. $(-2, 0), (0, 8)$ **25.** $m = -\frac{3}{4}$ **26.** $m = \frac{1}{5}$ **27.** d **28.** b **29.** c **30.** a **31.** $m = \frac{3}{4}$ **32.** $m = \frac{5}{3}$ **33.** $m = 4$ **34.** $m = -1$

35. $m = 3$ **36.** $m = \frac{1}{2}$ **37.** $m = 0$ **38.** undefined slope **39.** perpendicular **40.** parallel **41.** neither **42.** perpendicular **43.** $m = -2245.5$; The total number of U.S. daily newspaper subscriptions decreases by 2245.5 thousand per year. **44.** $m = 1748$; The number of U.S. transplants increases by 1748 transplants per year. **45.** $m = \frac{1}{6}; \left(0, \frac{1}{6}\right)$ **46.** $m = -3; (0, 7)$ **47.** $y = -5x + \frac{1}{2}$ **48.** $y = \frac{2}{3}x + 6$ **49.** d **50.** c **51.** a **52.** b **53.** $-4x + y = -8$ **54.** $3x + y = -5$ **55.** $-3x + 5y = 17$ **56.** $x + 3y = 6$ **57.** $y = -14x + 21$ **58.** $y = -\frac{1}{2}x + 4$ **59.** no **60.** yes **61.** yes **62.** yes **63.** no **64.** yes **65.** 6 **66.** 10 **67.** 5 **68.** 7

69.
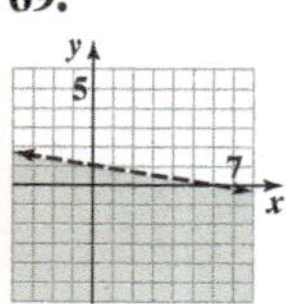

70.
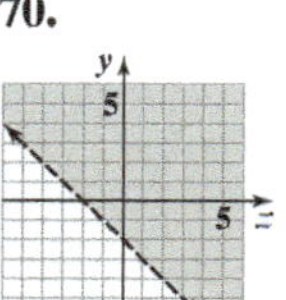

71.
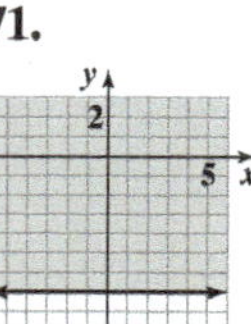

72.
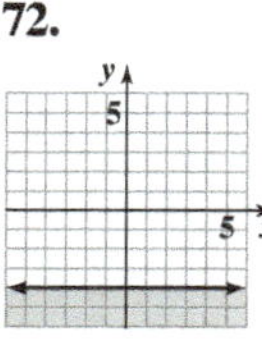

73.
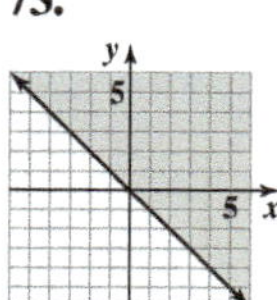

74.
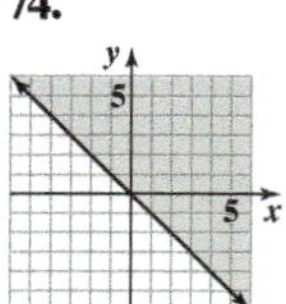

75. $7; -1; -3$ **76.** $0; -3; -2$ **77.** $(3, 0); (0, -2)$ **78.** $(-2, 0); (0, 10)$ **79.**
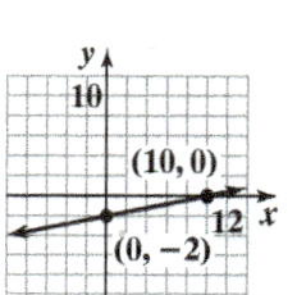

80.
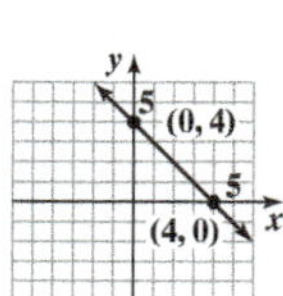

81.
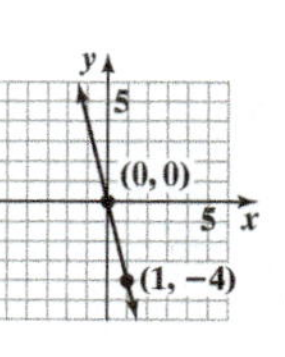

82.
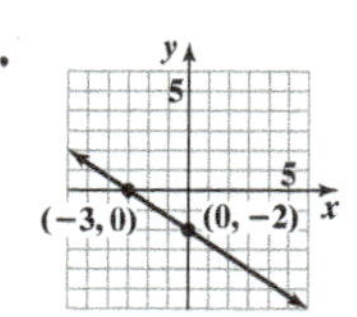

83.
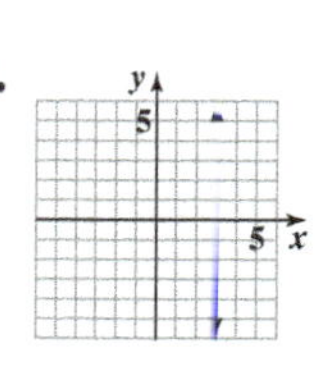

84.
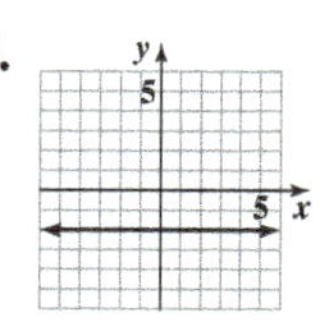

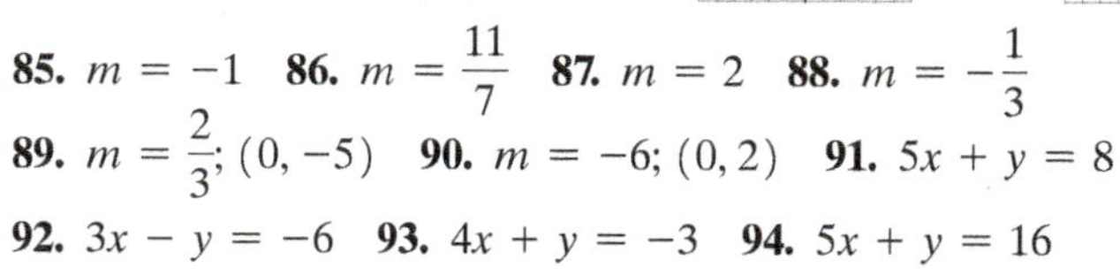

85. $m = -1$ **86.** $m = \frac{11}{7}$ **87.** $m = 2$ **88.** $m = -\frac{1}{3}$ **89.** $m = \frac{2}{3}; (0, -5)$ **90.** $m = -6; (0, 2)$ **91.** $5x + y = 8$ **92.** $3x - y = -6$ **93.** $4x + y = -3$ **94.** $5x + y = 16$

Chapter 6 Getting Ready for the Test **1.** C **2.** A **3.** B **4.** B **5.** B **6.** A **7.** D **8.** C **9.** C **10.** C **11.** B **12.** D **13.** A or C; A or C **14.** E

Chapter 6 Test **1.** $(1, 1)$ **2.** $(-4, 17)$ **3.** $m = \frac{2}{5}$ **4.** $m = 0$ **5.** $m = -1$ **6.** $m = -7$ **7.** $m = 3$ **8.** undefined slope

9.

10.
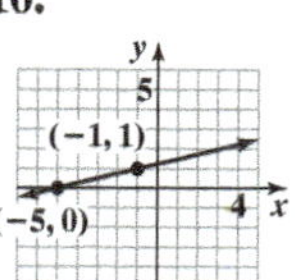

11.
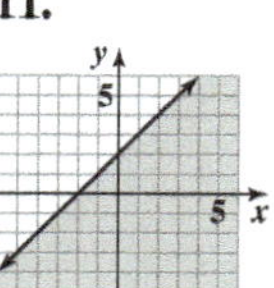

12.
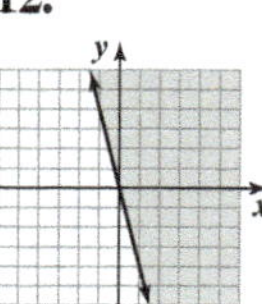

13.
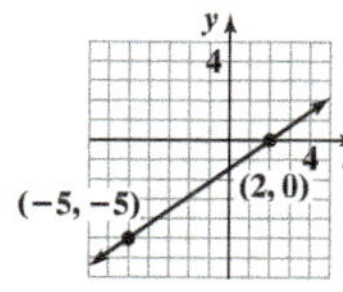

14.

15. 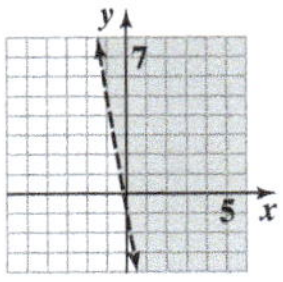

16. 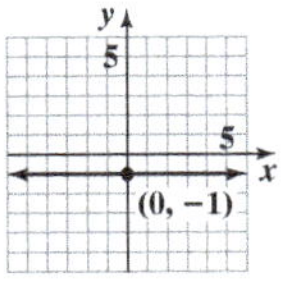

17. neither **18.** $x + 4y = 10$ **19.** $7x + 6y = 0$ **20.** $8x + y = 11$ **21.** $x - 8y = -96$ **22.** yes **23.** no **24.** yes **25.** yes **26. a.** −8 **b.** −3.6 **c.** −4 **27. a.** 0 **b.** 0 **c.** 60 **28.** $x + 2y = 21$; $x = 5$ m **29. a.** (2011, 9), (2012, 15), (2013, 21), (2014, 27), (2015, 34), (2016, 51), (2017, 65) **b.**

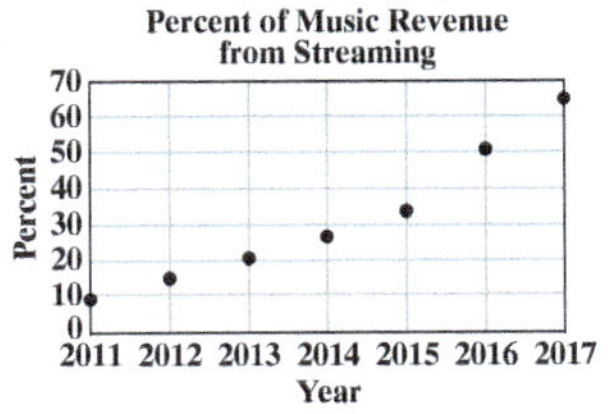

30. $m = 0.225$; For every 1 year, the box office gross sales increase by \$0.225 billion.

Cumulative Review **1.** 27; Sec. 1.3, Ex. 2 **2.** $\frac{25}{7}$; Sec. R.2 **3.** 51; Sec. 1.3, Ex. 5 **4.** 23; Sec. 1.3 **5.** 20,602 ft; Sec. 1.5, Ex. 10 **6.** $0.8x - 36$; Sec. 1.8 **7.** $2x + 6$; Sec. 1.8, Ex. 16 **8.** $-15\left(x + \frac{2}{3}\right) = -15x - 10$; Sec. 1.8 **9.** $(x - 4) \div 7$ or $\frac{x-4}{7}$; Sec. 1.8, Ex. 17 **10.** $\frac{-9}{2x}$; Sec. 1.8 **11.** $5 + (x + 1) = 6 + x$; Sec. 1.8, Ex. 18 **12.** $-86 - x$; Sec. 1.8 **13.** 6; Sec. 2.2, Ex. 1 **14.** −24; Sec. 2.3 **15.** 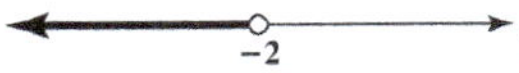 $\{x \mid x < -2\}$; Sec. 2.7, Ex. 6 **16.** $\left\{x \middle| x \le \frac{8}{3}\right\}$; Sec. 2.7 **17. a.** 2; trinomial **b.** 1; binomial **c.** 3; none of these; Sec. 3.3, Ex. 3 **18.** $y = \frac{6-x}{2}$; Sec. 2.5 **19.** $-4x^2 + 6x + 2$; Sec. 3.4, Ex. 2 **20.** $4x - 4$; Sec. 3.4 **21.** $9y^2 + 6y + 1$; Sec. 3.6, Ex. 4 **22.** $x^2 - 24x + 144$; Sec. 3.6 **23.** $3a(-3a^4 + 6a - 1)$; Sec. 4.1, Ex. 5 **24.** $4(x + 3)(x - 3)$; Sec. 4.5 **25.** $(x - 2)(x + 6)$; Sec. 4.2, Ex. 3 **26.** $(3x + y)(x - 7y)$; Sec. 4.3 **27.** $(4x - 1)(2x - 5)$; Sec. 4.3, Ex. 2 **28.** $(18x - 1)(x + 2)$; Sec. 4.3 **29.** $x = 11, x = -2$; Sec. 4.6, Ex. 4 **30.** $x = 0, x = 1$; Sec. 4.6 **31.** 1; Sec. 5.2, Ex. 7 **32.** $\frac{x+5}{2x^3}$; Sec. 5.1 **33.** $\frac{12ab^2}{27a^2b}$; Sec. 5.3, Ex. 9a **34.** $\frac{7x^2}{14x^3}$; Sec. 5.3 **35.** $\frac{2m+1}{m+1}$; Sec. 5.4, Ex. 5 **36.** $\frac{x+5}{x-6}$; Sec. 5.3 **37.** $x = -3, x = -2$; Sec. 5.5, Ex. 3 **38.** $x = -2, x = \frac{1}{3}$; Sec. 4.6 **39.** $\frac{x+1}{x+2y}$; Sec. 5.7, Ex. 5 **40.** $\frac{6x-2y}{x-4y}$ or $\frac{2(3x-y)}{x-4y}$; Sec. 5.7 **41. a.** (0, 12) **b.** (2, 6) **c.** (−1, 15); Sec. 6.1, Ex. 3 **42.** 0; 5; −2; Sec. 6.1 **43.**

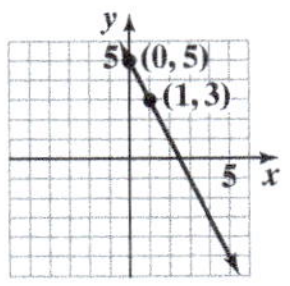

Sec. 6.2, Ex. 1 **44.** $m = \frac{1}{5}$; Sec. 6.4 **45.** $m = \frac{2}{3}$; Sec. 6.4, Ex. 3 **46.** undefined slope; Sec. 6.4 **47.** $y = -2x + 3$; $2x + y = 3$; Sec. 6.5, Ex. 4 **48.** $m = \frac{2}{5}$, y-intercept: $(0, -2)$; Sec. 6.4 **49. a.** 1; (2, 1) **b.** 1; (−2, 1) **c.** −3; (0, −3); Sec. 6.6, Ex. 7 **50.** $3x - 2y = 0$; Sec. 6.5

Chapter 7 Systems of Equations

Section 7.1

Calculator Explorations **1.** (0.37, 0.23) **3.** (0.03, −1.89)

Vocabulary, Readiness & Video Check **1.** dependent **3.** consistent **5.** inconsistent **7.** 1 solution, (−1, 3) **9.** infinite number of solutions **11.** The ordered pair must satisfy all equations of the system in order to be a solution of the system, so we must check that the ordered pair is a solution of both equations. **13.** Writing the equations of a system in slope-intercept form lets us see and compare their slopes and y-intercepts. Different slopes mean one solution; same slopes with different y-intercepts mean no solution; same slopes with same y-intercepts mean an infinite number of solutions.

Exercise Set 7.1 **1. a.** no **b.** yes **3. a.** yes **b.** no **5. a.** yes **b.** yes **7. a.** no **b.** no

9. 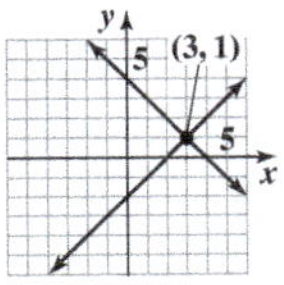

11. 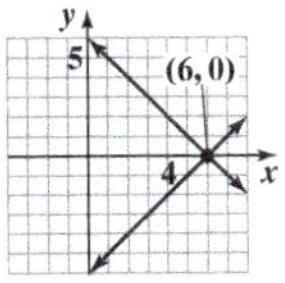

13. 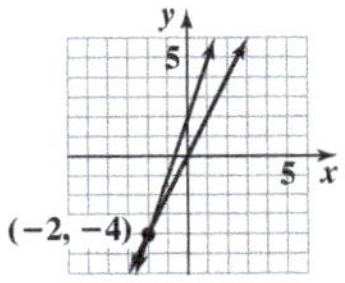

15. 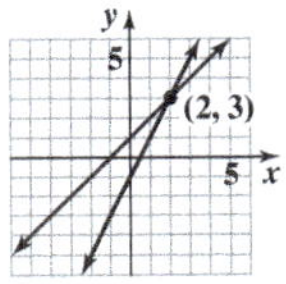

17. 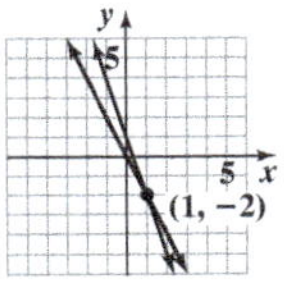

19. 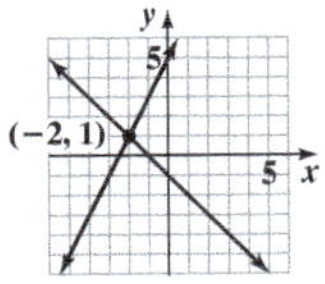

21. no solution 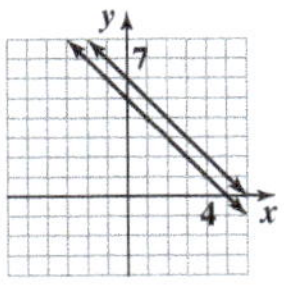

23. 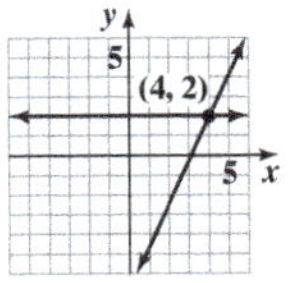

25. 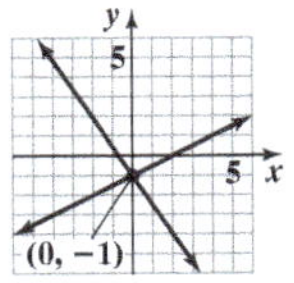

27. no solution 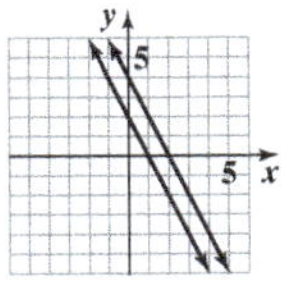

29. infinite number of solutions

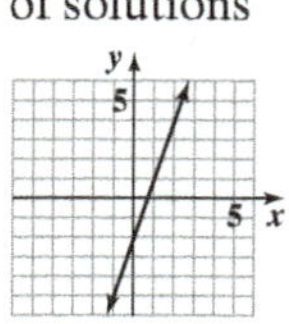

31.

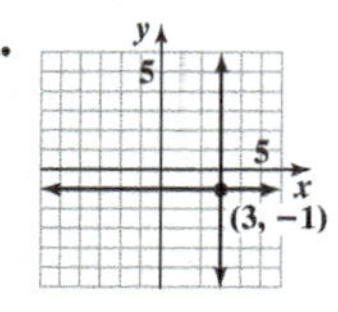

33.

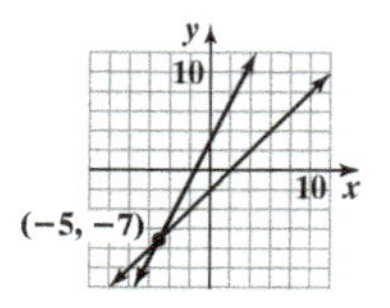

35.

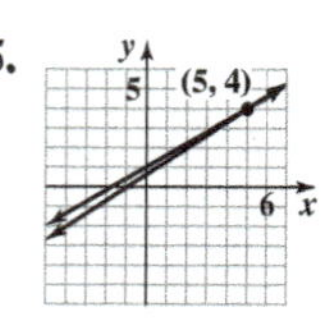

37. infinite number of solutions

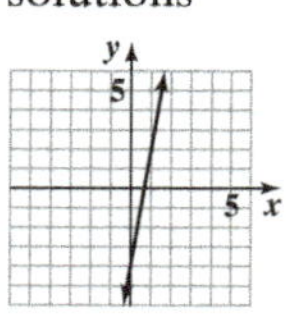

For Exercises **39–51**, the first answer given is the answer for part **a**, and the second answer given is the answer for part **b**. **39.** intersecting; one solution **41.** parallel; no solution **43.** identical lines; infinite number of solutions **45.** intersecting; one solution **47.** intersecting; one solution **49.** identical lines; infinite number of solutions **51.** parallel; no solution **53.** 2 **55.** $-\frac{2}{5}$ **57.** 2 **59.** answers may vary **61.** answers may vary **63.** 2011–2012 **65.** 2009, 2010 **67.** answers may vary **69.** answers may vary **71.** **a.** $(4, 9)$ **b.**

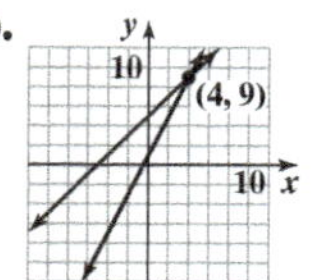

c. yes

Section 7.2

Vocabulary, Readiness & Video Check **1.** $(1, 4)$ **3.** infinite number of solutions **5.** $(0, 0)$ **7.** We solved one equation for a variable. Next, be sure to substitute this expression for the variable into the *other* equation.

Exercise Set 7.2 **1.** $(2, 1)$ **3.** $(-3, 9)$ **5.** $(2, 7)$ **7.** $\left(-\frac{1}{5}, \frac{43}{5}\right)$ **9.** $(2, -1)$ **11.** $(-2, 4)$ **13.** $(4, 2)$ **15.** $(-2, -1)$ **17.** no solution **19.** $(3, -1)$ **21.** $(3, 5)$ **23.** $\left(\frac{2}{3}, -\frac{1}{3}\right)$ **25.** $(-1, -4)$ **27.** $(-6, 2)$ **29.** $(2, 1)$ **31.** no solution **33.** infinite number of solutions **35.** $\left(\frac{1}{2}, 2\right)$ **37.** $-6x - 4y = -12$ **39.** $-12x + 3y = 9$ **41.** $5n$ **43.** $-15b$ **45.** $(1, -3)$ **47.** answers may vary **49.** no **51.** **c**; answers may vary **53.** $(-2.6, 1.3)$ **55.** $(3.28, 2.1)$ **57.** **a.** $(1.5, 6)$ **b.** In about 1.5 years after 2014, U.S. consumer spending on DVD- or Blu-ray-format home entertainment was the same as spending on streaming services home entertainment at approximately \$6 billion for each. **c.** answers may vary;

Section 7.3

Vocabulary, Readiness & Video Check **1.** false **3.** true **5.** The multiplication property of equality; be sure to multiply *both* sides of the equation by the nonzero number chosen.

Exercise Set 7.3 **1.** $(1, 2)$ **3.** $(2, -3)$ **5.** $(-2, -5)$ **7.** $(5, -2)$ **9.** $(-7, 5)$ **11.** $(6, 0)$ **13.** no solution **15.** infinite number of solutions **17.** $\left(2, -\frac{1}{2}\right)$ **19.** $(-2, 0)$ **21.** $(1, -1)$ **23.** infinite number of solutions **25.** $\left(\frac{12}{11}, -\frac{4}{11}\right)$ **27.** $\left(\frac{3}{2}, 3\right)$ **29.** infinite number of solutions **31.** $(1, 6)$ **33.** $\left(-\frac{1}{2}, -2\right)$ **35.** infinite number of solutions **37.** $\left(-\frac{2}{3}, \frac{2}{5}\right)$ **39.** $(2, 4)$ **41.** $(-0.5, 2.5)$ **43.** $2x + 6 = x - 3$ **45.** $20 - 3x = 2$ **47.** $4(x + 6) = 2x$ **49.** 2; $6x - 2y = -24$ **51.** **b**; answers may vary **53.** answers may vary **55.** **a.** $b = 15$ **b.** any real number except 15 **57.** $(-8.9, 10.6)$ **59.** **a.** $(3, 76)$ or $(3, 74.8)$ **b.** In 2017 $(2014 + 3)$, the amount of money spent on digital advertising was approximately equal to the amount of money spent on television advertising. **c.** \$76 billion or \$74.8 billion

Integrated Review **1.** $(2, 5)$ **2.** $(4, 2)$ **3.** $(5, -2)$ **4.** $(6, -14)$ **5.** $(-3, 2)$ **6.** $(-4, 3)$ **7.** $(0, 3)$ **8.** $(-2, 4)$ **9.** $(5, 7)$ **10.** $(-3, -23)$ **11.** $\left(\frac{1}{3}, 1\right)$ **12.** $\left(-\frac{1}{4}, 2\right)$ **13.** no solution **14.** infinite number of solutions **15.** $(0.5, 3.5)$ **16.** $(-0.75, 1.25)$ **17.** infinite number of solutions **18.** no solution **19.** $(7, -3)$ **20.** $(-1, -3)$ **21.** answers may vary **22.** answers may vary

Section 7.4

Vocabulary, Readiness & Video Check **1.** Up to now we've been working with one variable/unknown and one equation. Because systems involve two equations with two unknowns, for these applications we need to choose two variables to represent two unknowns and translate the problem into two equations.

Exercise Set 7.4 **1.** c **3.** b **5.** a **7.** $\begin{cases} x + y = 15 \\ x - y = 7 \end{cases}$ **9.** $\begin{cases} x + y = 6500 \\ x = y + 800 \end{cases}$ **11.** 33 and 50 **13.** 14 and −3 **15.** Báez: 111; Suárez: 104 **17.** child's ticket: \$18; adult's ticket: \$29 **19.** quarters: 53; nickels: 27 **21.** McDonald's: \$166.57; Mattel: \$14.52 **23.** daily fee: \$32; mileage charge: \$0.25 per mi **25.** distance downstream = distance upstream = 18 mi; time downstream: 2 hr; time upstream: $4\frac{1}{2}$ hr; still water: 6.5 mph; current: 2.5 mph **27.** still air: 455 mph; wind: 65 mph **29.** $4\frac{1}{2}$ hr **31.** 12% solution: $7\frac{1}{2}$ oz; 4% solution: $4\frac{1}{2}$ oz **33.** \$4.95 beans: 113 lb; \$2.65 beans: 87 lb **35.** 60°, 30° **37.** 20°, 70° **39.** number sold at \$9.50: 23; number sold at \$7.50: 67 **41.** $2\frac{1}{4}$ mph and $2\frac{3}{4}$ mph **43.** 30%: 50 gal; 60%: 100 gal **45.** length: 42 in.; width: 30 in. **47.** 16 **49.** 36 **51.** 100 **53.** a **55.** width: 9 ft; length: 15 ft

Chapter 7 Vocabulary Check **1.** dependent **2.** system of linear equations **3.** consistent **4.** solution **5.** addition; substitution **6.** inconsistent **7.** independent

Chapter 7 Review **1. a.** no **b.** yes **2. a.** no **b.** yes **3. a.** no **b.** no **4. a.** yes **b.** no

5. 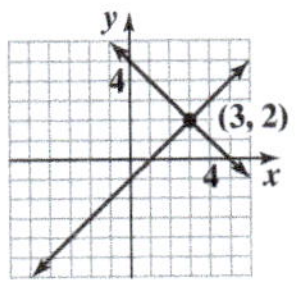

6. 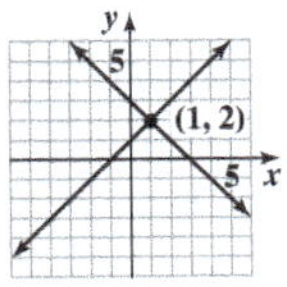

7. 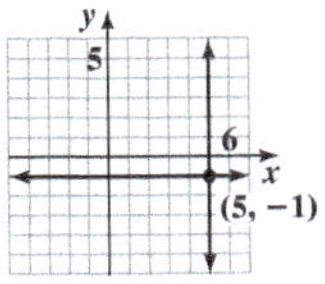

8. 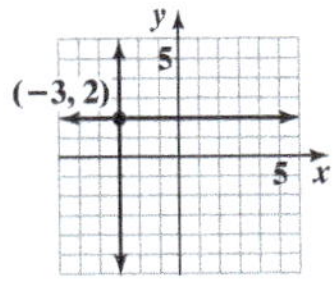

9. 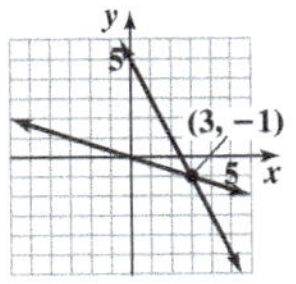

10. 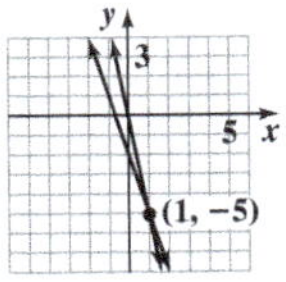

11. no solution 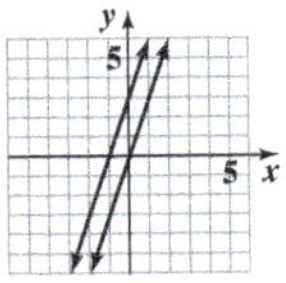
12. infinite number of solutions 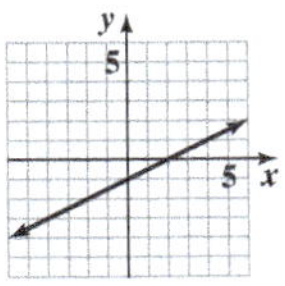
13. $(-1, 4)$ **14.** $(2, -1)$ **15.** $(3, -2)$ **16.** $(2, 5)$ **17.** infinite number of solutions **18.** infinite number of solutions **19.** no solution **20.** no solution **21.** $(-6, 2)$ **22.** $(4, -1)$ **23.** $(3, 7)$ **24.** $(-2, 4)$ **25.** infinite number of solutions **26.** infinite number of solutions

27. $(8, -6)$ **28.** $\left(-\frac{3}{2}, \frac{15}{2}\right)$ **29.** −6 and 22 **30.** orchestra: 255 seats; balcony: 105 seats **31.** current of river: 3.2 mph; speed in still water: 21.1 mph **32.** 6% solution: $12\frac{1}{2}$ cc; 14% solution: $37\frac{1}{2}$ cc **33.** egg: \$0.40; strip of bacon: \$0.65 **34.** jogging: 0.86 hr; walking: 2.14 hr

35. 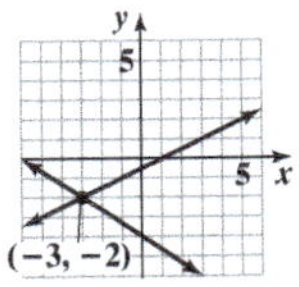

36. infinite number of solutions 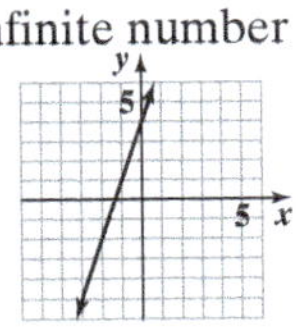
37. $(3, 2)$ **38.** $(7, 1)$ **39.** $\left(\frac{3}{2}, -3\right)$ **40.** no solution **41.** infinite number of solutions **42.** $(8, 11)$ **43.** $(-5, 2)$ **44.** $(16, -4)$ **45.** infinite number of solutions **46.** no solution **47.** 4 and 8 **48.** −5 and 13 **49.** 24 nickels and 41 dimes **50.** \$1.15 stamps: 10; 47¢ stamps: 16

Chapter 7 Getting Ready for the Test **1.** B **2.** D **3.** A **4.** B **5.** C **6.** C **7.** B **8.** A **9.** D **10.** B **11.** D **12.** C

Chapter 7 Test **1.** false **2.** false **3.** true **4.** false **5.** no **6.** yes **7.** 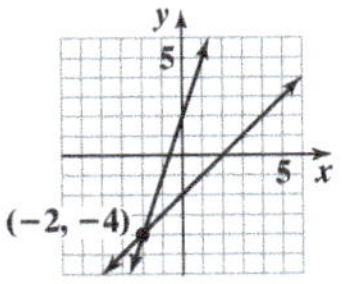

8. no solution **9.** $(-4, 1)$ 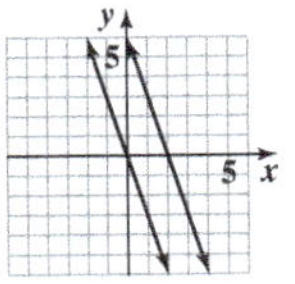
10. $\left(\frac{1}{2}, -2\right)$ **11.** $(20, 8)$ **12.** no solution **13.** $(4, -5)$ **14.** $(7, 2)$ **15.** $(5, -2)$ **16.** infinite number of solutions **17.** $(-5, 3)$ **18.** $\left(\frac{47}{5}, \frac{48}{5}\right)$ **19.** 78, 46 **20.** 120 cc **21.** Texas: 248 thousand; Missouri: 108 thousand **22.** 3 mph; 6 mph **23.** 2006, 2007, 2012, 2013, 2014, 2015, 2016 **24.** 2011

Cumulative Review **1. a.** −6 **b.** 6.3; Sec. 1.5, Ex. 6 **2. a.** 25 **b.** 32; Sec. 1.3 **3.** $\frac{1}{22}$; Sec. 1.6, Ex. 9a **4.** −22; Sec. 1.4 **5.** $\frac{16}{3}$; Sec. 1.6, Ex. 9b **6.** $-\frac{3}{16}$; Sec. 1.4 **7.** $-\frac{1}{10}$; Sec. 1.6, Ex. 9c **8.** 10; Sec. 1.4 **9.** $-\frac{13}{9}$; Sec. 1.6, Ex. 9d **10.** $\frac{9}{13}$; Sec. 1.4 **11.** $\frac{1}{1.7}$; Sec. 1.6, Ex. 9e **12.** −1.7; Sec. 1.4 **13. a.** 5 **b.** $8 - x$; Sec. 2.1, Ex. 8 **14.** −5; Sec. 2.4 **15.** no solution; Sec. 2.3, Ex. 6 **16.** no solution; Sec. 2.3 **17.** 12; Sec. 2.3, Ex. 3 **18.** 40; Sec. 2.1 **19.** $\left\{x \middle| x > \frac{13}{7}\right\}$ 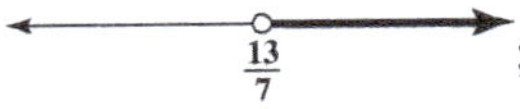

; Sec. 2.7, Ex. 8 **20.** $b = P - a - c$; Sec. 2.5 **21.** $\frac{m^7}{n^7}$, $n \neq 0$; Sec. 3.1, Ex. 22 **22.** $\frac{a^6}{b^5}$, $b \neq 0$; Sec. 3.2 **23.** $\frac{16x^{16}}{81y^{20}}$, $y \neq 0$; Sec. 3.1, Ex. 23 **24.** $\frac{49a^4}{b^6}$, $b \neq 0$; Sec. 3.2 **25.** $9x^2 - 6x - 1$; Sec. 3.4, Ex. 5 **26.** $6x^2 + \frac{14}{3}x + \frac{1}{42}$; Sec. 3.4 **27.** $2x + 4 - \frac{1}{3x - 1}$; Sec. 3.7, Ex. 5 **28.** $3x^2$; Sec. 4.1 **29.** $-\frac{1}{2}$, 4; Sec. 4.6, Ex. 6 **30.** 0, $\frac{7}{2}$; Sec. 4.6 **31.** 6 units, 8 units, 10 units; Sec. 4.7, Ex. 5 **32.** base: 7 ft; height: 26 ft; Sec. 4.7 **33.** 1; Sec. 5.3, Ex. 2

Answers to Selected Exercises

34. $\frac{5x-16}{(x-6)(x+1)}$; Sec. 5.4 **35.** $\frac{2x^2+3y}{x^2y+2xy^2}$ or $\frac{2x^2+3y}{xy(x+2y)}$; Sec. 5.7, Ex. 6 **36.** $-\frac{11}{3}$; Sec. 6.4 **37.** $m = 0$; Sec. 6.4, Ex. 5 **38.** undefined slope; Sec. 6.4 **39.** $-x + 5y = 23$; Sec. 6.5, Ex. 5 **40.** $y = -5x - 7$; Sec. 6.5 **41.** domain: $\{-1, 0, 3\}$ range: $\{-2, 0, 2, 3\}$; Sec. 6.6, Ex. 1 **42.** -6; 14; Sec. 6.6 **43.** It is a solution; Sec. 7.1, Ex. 1 **44. a.** yes **b.** no **c.** no; Sec. 7.1 **45.** $\left(6, \frac{1}{2}\right)$; Sec. 7.2, Ex. 3 **46.** $(-2, -4)$; Sec. 7.2 **47.** $\left(-\frac{15}{7}, -\frac{5}{7}\right)$; Sec. 7.3, Ex. 6 **48.** $\left(-\frac{44}{3}, -\frac{7}{3}\right)$; Sec. 7.3 **49.** 29 and 8; Sec. 7.4, Ex. 1 **50. a.** not a function **b.** function **c.** not a function; Sec. 6.6

Chapter 8 Roots and Radicals

Section 8.1

Calculator Explorations **1.** 2.449 **3.** 3.317 **5.** 9.055 **7.** 3.420 **9.** 2.115 **11.** 1.783

Vocabulary, Readiness & Video Check **1.** principal **3.** square root **5.** power **7.** false **9.** true **11.** The radical sign, $\sqrt{\ }$, indicates a positive square root only. A negative sign before the radical sign, $-\sqrt{\ }$, indicates a negative square root. **13.** an odd-numbered index **15.** Divide the index into each exponent in the radicand—but still check by raising our answer to a power equal to the index.

Exercise Set 8.1 **1.** 4 **3.** $\frac{1}{5}$ **5.** -10 **7.** not a real number **9.** -11 **11.** $\frac{3}{5}$ **13.** 30 **15.** 12 **17.** $\frac{1}{10}$ **19.** 0.5 **21.** 5 **23.** -4 **25.** -2 **27.** $\frac{1}{2}$ **29.** -5 **31.** 2 **33.** 9 **35.** not a real number **37.** $-\frac{3}{4}$ **39.** -5 **41.** 1 **43.** 2.646 **45.** 6.083 **47.** 11.662 **49.** $\sqrt{2} \approx 1.41$; 126.90 ft **51.** m **53.** x^2 **55.** $3x^4$ **57.** $9x$ **59.** ab^2 **61.** $4a^3b^2$ **63.** a^2b^6 **65.** $-2xy^9$ **67.** $\frac{x^3}{6}$ **69.** $\frac{5y}{3}$ **71.** $25 \cdot 2$ **73.** $16 \cdot 2$ or $4 \cdot 8$ **75.** $4 \cdot 7$ **77.** $9 \cdot 3$ **79.** a, b **81.** 7 mi **83.** 5.6 mm **85.** 3 **87.** 10 **89.** 4, 5 **91.** 8, 9 **93.** $T \approx 6.1$ sec **95.** answers may vary **97.** 1; 1.7; 2; 3 **99.** $|x|$ **101.** $|x+2|$ **103.** $(2, 0)$ **105.** $(-4, 0)$

Section 8.2

Vocabulary, Readiness & Video Check **1.** $\sqrt{a} \cdot \sqrt{b}$ **3.** 16; 25; 4; 5; 20 **5.** no **7.** Factor until we have a product of primes. A repeated prime factor means a perfect square—if more than one factor is repeated, we can multiply all the repeated factors together to get one larger perfect square factor. **9.** The power must be 1. Any even power is a perfect square and can be simplified; any higher odd power is the product of an even power times the variable with a power of 1.

Exercise Set 8.2 **1.** $2\sqrt{5}$ **3.** $5\sqrt{2}$ **5.** $\sqrt{33}$ **7.** $7\sqrt{2}$ **9.** $2\sqrt{15}$ **11.** $6\sqrt{5}$ **13.** $2\sqrt{13}$ **15.** 15 **17.** $21\sqrt{7}$ **19.** $-15\sqrt{3}$ **21.** $\frac{2\sqrt{2}}{5}$ **23.** $\frac{3\sqrt{3}}{11}$ **25.** $\frac{3}{2}$ **27.** $\frac{5\sqrt{5}}{3}$ **29.** $\frac{\sqrt{11}}{6}$ **31.** $-\frac{3\sqrt{3}}{8}$ **33.** $x^3\sqrt{x}$ **35.** $x^6\sqrt{x}$ **37.** $6a\sqrt{a}$ **39.** $4x^2\sqrt{6}$ **41.** $\frac{2\sqrt{3}}{m}$ **43.** $\frac{3\sqrt{x}}{y^5}$ **45.** $\frac{2\sqrt{22}}{x^6}$ **47.** 16 **49.** $\frac{6}{11}$ **51.** $5\sqrt{7}$ **53.** $\frac{2\sqrt{5}}{3}$ **55.** $2m^3\sqrt{6m}$ **57.** $\frac{y\sqrt{23y}}{2x^3}$ **59.** $2\sqrt[3]{3}$ **61.** $5\sqrt[3]{2}$ **63.** $\frac{\sqrt[3]{5}}{4}$ **65.** $\frac{\sqrt[3]{23}}{2}$ **67.** $\frac{\sqrt[3]{15}}{4}$ **69.** $3\sqrt[3]{10}$ **71.** $14x$ **73.** $2x^2 - 7x - 15$ **75.** 0 **77.** $x^3y\sqrt{y}$ **79.** $7x^2y^2\sqrt{2x}$ **81.** $-2x^2$ **83.** $2\sqrt[3]{10}$ in. **85.** answers may vary **87.** 177 m by 177 m **89.** $2\sqrt{5}$ in. **91.** 2.25 in. **93.** \$1700 **95.** 1.7 sq m

Section 8.3

Vocabulary, Readiness & Video Check **1.** like radicals **3. b.** $17\sqrt{2}$ **5. c.** $2\sqrt{5}$ **7.** Both like terms and like radicals are combined using the distributive property; also, only like (vs. unlike) terms can be combined, as with like radicals (same index and same radicand). **9.** the product rule for radicals

Exercise Set 8.3 **1.** $-4\sqrt{3}$ **3.** $9\sqrt{6} - 5$ **5.** $\sqrt{5} + \sqrt{2}$ **7.** $7\sqrt{3} - \sqrt{2}$ **9.** $-5\sqrt{2} - 6$ **11.** $5\sqrt{3}$ **13.** $9\sqrt{5}$ **15.** $4\sqrt{6} + \sqrt{5}$ **17.** $x + \sqrt{x}$ **19.** 0 **21.** $\frac{4\sqrt{5}}{9}$ **23.** $\frac{3\sqrt{3}}{8}$ **25.** $7\sqrt{5}$ **27.** $9\sqrt{3}$ **29.** $\sqrt{5} + \sqrt{15}$ **31.** $x\sqrt{x}$ **33.** $5\sqrt{2} + 12$ **35.** $8\sqrt{2} - 5$ **37.** $2\sqrt{5}$ **39.** $-\sqrt{35}$ **41.** $6 - 3\sqrt{3}$ **43.** $11\sqrt{x}$ **45.** $12x - 11\sqrt{x}$ **47.** $x\sqrt{3x} + 3x\sqrt{x}$ **49.** $8x\sqrt{2} + 2x$ **51.** $2x^2\sqrt{10} - x^2\sqrt{5}$ **53.** $7\sqrt[3]{9} - \sqrt[3]{25}$ **55.** $-5\sqrt[3]{2} - 6$ **57.** $5\sqrt[3]{3}$ **59.** $-3 + 3\sqrt[3]{2}$ **61.** $4x + 4x\sqrt[3]{2}$ **63.** $10y^2\sqrt[3]{y}$ **65.** $x\sqrt[3]{5}$

67. $x^2 + 12x + 36$ **69.** $4x^2 - 4x + 1$ **71.** answers may vary **73.** $8\sqrt{5}$ in. **75.** $\left(48 + \frac{9\sqrt{3}}{2}\right)$ sq ft **77.** yes; $7\sqrt{2}$ **79.** no **81.** yes; $3\sqrt{7}$ **83.** $\frac{83x\sqrt{x}}{20}$

Section 8.4

Vocabulary, Readiness & Video Check **1.** $\sqrt{21}$ **3.** $\sqrt{\frac{15}{3}}$ or $\sqrt{5}$ **5.** $2 - \sqrt{3}$ **7.** The square root of a positive number times the square root of the same positive number (or the square root of a positive number squared) is that positive number. **9.** To write an equivalent expression without a radical in the denominator.

Exercise Set 8.4 **1.** 4 **3.** $5\sqrt{2}$ **5.** 6 **7.** $2x$ **9.** 20 **11.** $36x$ **13.** $3x^3\sqrt{2}$ **15.** $4xy\sqrt{y}$ **17.** $\sqrt{30} + \sqrt{42}$ **19.** $2\sqrt{5} + 5\sqrt{2}$ **21.** $y\sqrt{7} - 14\sqrt{y}$ **23.** -33 **25.** $\sqrt{6} - \sqrt{15} + \sqrt{10} - 5$ **27.** $16 - 11\sqrt{11}$ **29.** $x - 36$ **31.** $x - 14\sqrt{x} + 49$ **33.** $6y + 2\sqrt{6y} + 1$ **35.** 4 **37.** $\sqrt{7}$ **39.** $3\sqrt{2}$ **41.** $5y^2$ **43.** $5\sqrt{3}$ **45.** $2y\sqrt{6}$ **47.** $2xy\sqrt{3y}$ **49.** $\frac{\sqrt{15}}{5}$ **51.** $\frac{7\sqrt{2}}{2}$ **53.** $\frac{\sqrt{6y}}{6y}$ **55.** $\frac{\sqrt{10}}{6}$ **57.** $\frac{\sqrt{3x}}{x}$ **59.** $\frac{\sqrt{2}}{4}$ **61.** $\frac{\sqrt{30}}{15}$ **63.** $\frac{\sqrt{15}}{10}$ **65.** $\frac{3\sqrt{2x}}{2}$ **67.** $\frac{8y\sqrt{5}}{5}$ **69.** $\frac{\sqrt{xy}}{6y}$ **71.** $\frac{\sqrt{3xy}}{6x}$ **73.** $3\sqrt{2} - 3$ **75.** $-8 - 4\sqrt{5}$ **77.** $\sqrt{30} + 5 + \sqrt{6} + \sqrt{5}$ **79.** $\sqrt{6} + \sqrt{3} + \sqrt{2} + 1$ **81.** $\frac{10 - 5\sqrt{x}}{4 - x}$ **83.** $\frac{3\sqrt{x} + 12}{x - 16}$ **85.** 44 **87.** 2 **89.** 3 **91.** $130\sqrt{3}$ sq m **93.** $\frac{\sqrt{A\pi}}{\pi}$ **95.** true **97.** false **99.** false **101.** answers may vary **103.** answers may vary **105.** $\frac{2}{\sqrt{6}-\sqrt{2}-\sqrt{3}+1}$

Integrated Review **1.** 6 **2.** $4\sqrt{3}$ **3.** x^2 **4.** $y^3\sqrt{y}$ **5.** $4x$ **6.** $3x^5\sqrt{2x}$ **7.** 2 **8.** 3 **9.** -3 **10.** not a real number **11.** $\frac{\sqrt{11}}{3}$ **12.** $\frac{\sqrt[3]{7}}{4}$ **13.** -4 **14.** -5 **15.** $\frac{3}{7}$ **16.** $\frac{1}{8}$ **17.** a^4b **18.** x^5y^{10} **19.** $5m^3$ **20.** $3n^8$ **21.** $6\sqrt{7}$ **22.** $3\sqrt{2}$ **23.** cannot be simplified **24.** $\sqrt{x} + 3x$ **25.** $\sqrt{30}$ **26.** 3 **27.** 28 **28.** 45 **29.** $\sqrt{33} + \sqrt{3}$ **30.** $3\sqrt{2} - 2\sqrt{6}$ **31.** $4y$ **32.** $3x^2\sqrt{5}$ **33.** $x - 3\sqrt{x} - 10$ **34.** $11 + 6\sqrt{2}$ **35.** 2 **36.** $\sqrt{3}$ **37.** $2x^2\sqrt{3}$ **38.** $ab^2\sqrt{15a}$ **39.** $\frac{\sqrt{6}}{6}$ **40.** $\frac{x\sqrt{5}}{10}$ **41.** $\frac{4\sqrt{6} - 4}{5}$ **42.** $\frac{\sqrt{2x} + 5\sqrt{2} + \sqrt{x} + 5}{x - 25}$

Section 8.5

Vocabulary, Readiness & Video Check **1.** The squaring property can result in extraneous solutions, so we need to check our solutions in the original equation—before the squaring property was applied—to make sure they are actual solutions.

Exercise Set 8.5 **1.** 81 **3.** -1 **5.** 49 **7.** no solution **9.** 4 **11.** 2 **13.** 2 **15.** 9 **17.** -3 **19.** $-1, -2$ **21.** no solution **23.** $0, -3$ **25.** 16 **27.** 25 **29.** 1 **31.** 5 **33.** -2 **35.** no solution **37.** 2 **39.** 36 **41.** no solution **43.** $\frac{3}{2}$ **45.** 16 **47.** 3 **49.** 12 **51.** 3, 1 **53.** -1 **55.** $3x - 8 = 19; x = 9$ **57.** $2(2x) + 2x = 24$; length: 8 in. **59.** 4, 7 **61.** answers may vary **63. a.** 3.2; 10; 31.6 **b.** no **65.** 20 mm **67.** 7.30 **69.** 0.76

Section 8.6

Vocabulary, Readiness & Video Check **1.** The Pythagorean theorem applies to right triangles only, and in the formula $a^2 + b^2 = c^2$, c is the length of the hypotenuse. **3.** This example asks for an answer rounded to a given place, meaning an estimated answer is expected rather than an exact answer. An exact answer would be given in radical form.

Exercise Set 8.6 **1.** $\sqrt{13}$; 3.61 **3.** $3\sqrt{3}$; 5.20 **5.** 25 **7.** $\sqrt{22}$; 4.69 **9.** $3\sqrt{17}$; 12.37 **11.** $\sqrt{41}$; 6.40 **13.** $4\sqrt{2}$; 5.66 **15.** $3\sqrt{10}$; 9.49 **17.** 20.6 ft **19.** 11.7 ft **21.** 24 cu ft **23.** 54 mph **25.** 27 mph **27.** 56.3 km **29.** 81.4 km **31.** 3, -3 **33.** 10, -10 **35.** 8, -8 **37.** $y = 2\sqrt{10}; x = 2\sqrt{10} - 4$ **39.** 201 mi **41.** answers may vary

Section 8.7

Vocabulary, Readiness & Video Check **1.** inverse **3.** direct **5.** inverse **7.** inverse **9.** direct **11.** linear; slope **13.** No. The direct relationship is the power of x times a constant, and the inverse relationship is the reciprocal of the power of x times a constant.

Exercise Set 8.7 **1.** $y = \frac{1}{2}x$ **3.** $y = 6x$ **5.** $y = 3x$ **7.** $y = \frac{2}{3}x$ **9.** $y = \frac{7}{x}$ **11.** $y = \frac{0.5}{x}$ **13.** $y = kx$ **15.** $h = \frac{k}{t}$ **17.** $z = kx^2$ **19.** $y = \frac{k}{z^3}$ **21.** $y = k\sqrt{x}$ **23.** $y = 40$ **25.** $y = 3$ **27.** $z = 144$ **29.** $a = \frac{4}{9}$ **31.** \$92.50 **33.** \$6 **35.** $5\frac{1}{3}$ in. **37.** 179.1 lb **39.** 1600 ft **41.** $2y = 16$ **43.** $-4x = 0.5$ **45.** multiplied by 3 **47.** It is doubled.

Chapter 8 Vocabulary Check **1.** like radicals **2.** index; radicand; radical **3.** conjugate **4.** principal square root **5.** rationalizing the denominator **6.** hypotenuse **7.** direct **8.** inverse

Chapter 8 Review **1.** 9 **2.** -7 **3.** 3 **4.** 3 **5.** $-\frac{3}{8}$ **6.** $\frac{2}{3}$ **7.** 2 **8.** -2 **9.** c **10.** a, c **11.** x^6 **12.** x^4 **13.** $3y$ **14.** $5x^2$ **15.** $2\sqrt{10}$ **16.** $2\sqrt{6}$ **17.** $3\sqrt{6}$ **18.** $2\sqrt{22}$ **19.** $x^2\sqrt{x}$ **20.** $y^3\sqrt{y}$ **21.** $2x\sqrt{5}$ **22.** $5y^2\sqrt{2}$ **23.** $3\sqrt[3]{2}$ **24.** $2\sqrt[3]{11}$ **25.** $\frac{3\sqrt{2}}{5}$ **26.** $\frac{5\sqrt{3}}{8}$ **27.** $-\frac{5\sqrt{2}}{3}$ **28.** $-\frac{2\sqrt{3}}{7}$ **29.** $\frac{\sqrt{11}}{x}$ **30.** $\frac{\sqrt{7}}{y^2}$ **31.** $\frac{y^2\sqrt{y}}{10}$ **32.** $\frac{x\sqrt{x}}{9}$ **33.** $-3\sqrt{2}$ **34.** $-5\sqrt{3}$ **35.** $4\sqrt{5} + 4\sqrt{6}$ **36.** $-2\sqrt{7} + 2\sqrt{2}$ **37.** $5\sqrt{7} + 2\sqrt{14}$ **38.** $9\sqrt{3} - 4$ **39.** $\frac{\sqrt{5}}{6}$ **40.** $\frac{9\sqrt{11}}{20}$ **41.** $-x\sqrt{5} + 5$ **42.** $2\sqrt{2x} - \sqrt{3x}$ **43.** $3\sqrt{2}$ **44.** $5\sqrt{3}$ **45.** $\sqrt{10} - \sqrt{14}$ **46.** $\sqrt{55} + \sqrt{15}$ **47.** $3\sqrt{2} - 5\sqrt{3} + 2\sqrt{6} - 10$ **48.** $2 - 2\sqrt{5}$ **49.** $x - 4\sqrt{x} + 4$ **50.** $y + 8\sqrt{y} + 16$ **51.** 3 **52.** 2 **53.** $2\sqrt{5}$ **54.** $4\sqrt{2}$ **55.** $x\sqrt{15x}$ **56.** $3x^2\sqrt{2}$ **57.** $\frac{\sqrt{22}}{11}$ **58.** $\frac{\sqrt{39}}{13}$ **59.** $\frac{\sqrt{30}}{6}$ **60.** $\frac{\sqrt{70}}{10}$ **61.** $\frac{\sqrt{5x}}{5x}$ **62.** $\frac{5\sqrt{3y}}{3y}$ **63.** $\frac{\sqrt{3x}}{x}$ **64.** $\frac{\sqrt{6y}}{y}$ **65.** $3\sqrt{5} + 6$ **66.** $8\sqrt{10} + 24$ **67.** $\frac{\sqrt{6} + \sqrt{2} + \sqrt{3} + 1}{2}$ **68.** $\sqrt{15} - 2\sqrt{3} - 2\sqrt{5} + 4$ **69.** $\frac{10\sqrt{x} - 50}{x - 25}$ **70.** $\frac{8\sqrt{x} + 8}{x - 1}$ **71.** 18 **72.** 13 **73.** 25 **74.** no solution **75.** 12 **76.** 5 **77.** 1 **78.** 9 **79.** $2\sqrt{14}$; 7.48 **80.** $3\sqrt{13}$; 10.82 **81.** $4\sqrt{34}$ ft; 23.32 ft **82.** $5\sqrt{3}$ in.; 8.66 in. **83.** 2.4 in. **84.** 144π sq in. **85.** $y = 110$ **86.** $y = \frac{1}{2}$ **87.** $y = \frac{100}{27}$ **88.** $y = 700$ **89.** $y = \frac{5}{3}$ **90.** $y = 22$ **91.** \$3960 **92.** $4\frac{4}{5}$ in. **93.** 12 **94.** -4 **95.** $4x^8$ **96.** $2x^{12}$ **97.** $3x^3\sqrt{2x}$ **98.** $4y^3\sqrt{3}$ **99.** $\frac{y^2}{9}$ **100.** $\frac{x^4\sqrt{x}}{3}$ **101.** $7\sqrt{3}$ **102.** $5\sqrt{7} - 3$ **103.** $-\frac{\sqrt{3}}{4}$ **104.** $4x\sqrt{5x}$ **105.** $7\sqrt{2}$ **106.** $3\sqrt{3} - \sqrt{6}$ **107.** $\sqrt{10} - \sqrt{2} + 4\sqrt{5} - 4$ **108.** $x + 6\sqrt{x} + 9$ **109.** $2\sqrt{6}$ **110.** $2x$ **111.** $\frac{\sqrt{14}}{7}$ **112.** $\frac{3\sqrt{2x}}{2x}$ **113.** $\frac{3\sqrt{x} + 18}{x - 36}$ **114.** $-\frac{\sqrt{35} - 3\sqrt{7} - 5\sqrt{5} + 15}{4}$ **115.** 1 **116.** 13 **117.** 14 **118.** 9 **119.** $\sqrt{58}$; 7.62 **120.** $4\sqrt{2}$ in.; 5.66 in.

Chapter 8 Getting Ready for the Test **1.** B **2.** C **3.** B **4.** B **5.** C **6.** D **7.** B **8.** B **9.** C **10.** C **11.** A

Chapter 8 Test **1.** 4 **2.** 5 **3.** 3 **4.** $\frac{3}{4}$ **5.** not a real number **6.** x^5 **7.** $3\sqrt{6}$ **8.** $2\sqrt{23}$ **9.** $y^3\sqrt{y}$ **10.** $2x^4\sqrt{6}$ **11.** 3 **12.** $2\sqrt[3]{2}$ **13.** $\frac{\sqrt{5}}{4}$ **14.** $\frac{y\sqrt{y}}{5}$ **15.** $-2\sqrt{13}$ **16.** $\sqrt{2} + 2\sqrt{3}$ **17.** $\frac{7\sqrt{3}}{10}$ **18.** $7\sqrt{2}$ **19.** $2\sqrt{3} - \sqrt{10}$ **20.** $x - \sqrt{x} - 6$ **21.** $\sqrt{5}$ **22.** $2x\sqrt{5x}$ **23.** $\frac{\sqrt{6}}{3}$ **24.** $\frac{8\sqrt{5y}}{5y}$ **25.** $4\sqrt{6} - 8$ **26.** $\frac{3 + \sqrt{x}}{9 - x}$ **27.** 9 **28.** 5 **29.** 9 **30.** $4\sqrt{5}$ in. **31.** 2.19 m **32.** $y = 28$ **33.** $y = \frac{8}{9}$

Cumulative Review **1.** 28; Sec. 1.6, Ex. 3 **2.** -46.8; Sec. 1.6 **3.** $-\frac{8}{21}$; Sec. 1.6, Ex. 4 **4.** -18; Sec. 1.6 **5.** 2; Sec. 2.3, Ex. 1 **6.** 15; Sec. 2.3 **7. a.** 17% **b.** 21% **c.** 43 American travelers; Sec. 2.6, Ex. 3 **8. a.** $\frac{3}{2}$ **b.** 9; Sec. 1.3 **9. a.** 102,000 **b.** 0.007358 **c.** 84,000,000 **d.** 0.00003007; Sec. 3.2, Ex. 18 **10. a.** 7.2×10^6 **b.** 3.08×10^{-4}; Sec. 3.2 **11.** $6x^2 - 11x - 10$; Sec. 3.5, Ex. 7b **12.** $49x^2 + 14x + 1$; Sec. 3.5 **13.** $(y + 2)(x + 3)$; Sec. 4.1; Ex. 11 **14.** $(y^2 + 5)(x - 1)$; Sec. 4.1 **15.** $(3x + 2)(x + 3)$; Sec. 4.3, Ex. 1 **16.** $3(x + 2)(x + 3)$; Sec. 4.2 **17. a.** $x = 3$ **b.** $x = 2, x = 1$ **c.** none; Sec. 5.1, Ex. 2 **18.** $\frac{2x + 1}{x - 3}$; Sec. 5.1 **19.** $\frac{x + 2}{x}$; Sec. 5.1, Ex. 5 **20.** $\frac{4x^3}{5}$; Sec. 5.2 **21. a.** 0 **b.** $\frac{15 + 14x}{50x^2}$; Sec. 5.4, Ex. 1 **22.** $y = -2x + 4$; Sec. 6.5 **23.** $-\frac{17}{5}$; Sec. 5.5, Ex. 4 **24.** $2a^2 + 10a - 5$; Sec. 3.3 **25.** ; Sec. 6.3, Ex. 10 **26.** 6; 4; 0; Sec. 6.1 **27.** $y = \frac{1}{4}x - 3$; Sec. 6.5, Ex. 1 **28.** $x = 1$; Sec. 6.5 **29.** (3, 1); Sec. 7.3, Ex. 5 **30.** $\left(\frac{2}{3}, \frac{1}{2}\right)$; Sec. 7.3 **31.** Alfredo: 3.25 mph; Louisa: 4.25 mph; Sec. 7.4, Ex. 3 **32.** 20 mph, 35 mph; Sec. 7.4 **33.** 1; Sec. 8.1, Ex. 6 **34.** 11; Sec. 8.1 **35.** -3; Sec. 8.1, Ex. 7 **36.** $\frac{1}{2}$; Sec. 8.1 **37.** $\frac{1}{5}$; Sec. 8.1, Ex. 8 **38.** $\frac{5}{12}$; Sec. 8.1 **39.** $3\sqrt{6}$; Sec. 8.2, Ex. 1 **40.** $3\sqrt{7}$; Sec. 8.2 **41.** $10\sqrt{2}$; Sec. 8.2, Ex. 3 **42.** $10\sqrt{5}$; Sec. 8.2 **43.** $4\sqrt{3}$; Sec. 8.3, Ex. 6 **44.** $x - 25$; Sec. 8.4 **45.** $2x - 4x^2\sqrt{x}$; Sec. 8.3, Ex. 8 **46.** $10 + 4\sqrt{6}$; Sec. 8.4 **47.** $\frac{2\sqrt{7}}{7}$; Sec. 8.4, Ex. 10 **48.** $3x$; Sec. 5.7 **49.** $\frac{1}{2}$; Sec. 8.5, Ex. 2 **50.** $\frac{5}{2}$; Sec. 8.5

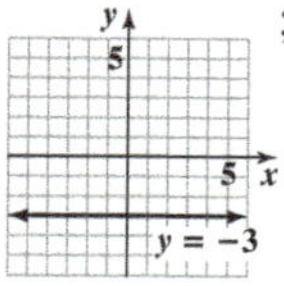

Chapter 9 Quadratic Equations

Section 9.1

Vocabulary, Readiness & Video Check **1.** To solve, a becomes the radicand and the square root of a negative number is not a real number.

Exercise Set 9.1 **1.** ± 7 **3.** $-5, 3$ **5.** ± 4 **7.** ± 3 **9.** $-5, -2$ **11.** ± 8 **13.** $\pm\sqrt{21}$ **15.** $\pm\frac{1}{5}$ **17.** no real solution **19.** $\pm\frac{\sqrt{39}}{3}$ **21.** $\pm\frac{2\sqrt{7}}{7}$ **23.** $\pm\sqrt{5}$ **25.** $12, -2$ **27.** $-2 \pm \sqrt{7}$ **29.** $1, 0$ **31.** $-2 \pm \sqrt{10}$ **33.** $\frac{8}{3}, -4$ **35.** no real solution **37.** $\frac{11 \pm 5\sqrt{2}}{2}$ **39.** $\frac{7 \pm 4\sqrt{2}}{3}$ **41.** $\pm\sqrt{29}$ **43.** $-6 \pm 2\sqrt{6}$ **45.** $\pm\sqrt{10}$ **47.** $\frac{1 \pm \sqrt{5}}{4}$ **49.** $\pm 2\sqrt{3}$ **51.** $\frac{-8 \pm 3\sqrt{5}}{-3}$ or $\frac{8 \pm 3\sqrt{5}}{3}$ **53.** $2\sqrt{5}$ in. ≈ 4.47 in. **55.** 177 m **57.** 2.3 sec **59.** 15.8 sec **61.** 6 in. **63.** $(x + 3)^2$ **65.** $(x - 2)^2$ **67.** answers may vary **69.** $2, -6$ **71.** ± 1.33 **73.** $x = 24$, which is 2030

Section 9.2

Vocabulary, Readiness & Video Check **1.** zero **3.** quadratic equation **5.** 9 **7.** 16 **9.** 100 **11.** 49 **13.** When working with equations, whatever is added to one side must also be added to the other side to keep equality.

Exercise Set 9.2 **1.** $-6, -2$ **3.** $-1 \pm 2\sqrt{2}$ **5.** $0, 6$ **7.** $-1, -4$ **9.** $1 \pm \sqrt{2}$ **11.** $\frac{-5 \pm \sqrt{53}}{2}$ **13.** $-2, 4$ **15.** no real solution **17.** $\frac{3 \pm \sqrt{19}}{2}$ **19.** $-2 \pm \frac{\sqrt{6}}{2}$ **21.** $-3 \pm \sqrt{34}$ **23.** $\frac{3 \pm \sqrt{21}}{2}$ **25.** $\frac{1}{2}, 1$ **27.** $-6, 3$ **29.** no real solution **31.** $2, -6$ **33.** $-\frac{1}{2}$ **35.** 2 **37.** $3 + 2\sqrt{5}$ **39.** $\frac{1 - 3\sqrt{2}}{2}$ **41.** answers may vary **43. a.** $-3 \pm \sqrt{11}$ **b., c.** answers may vary **45.** $k = 8$ or $k = -8$ **47.** $x = 6$, or 2019 **49.** $-6, -2$ **51.** $\approx -0.68, 3.68$

Section 9.3

Vocabulary, Readiness & Video Check **1.** $x = \frac{-b \pm \sqrt{b^2 - 4ac}}{2a}$ **3.** $1; 3; -7$ **5.** $1; 1; -1$ **7.** $-2, 1$ **9.** $\frac{-5 \pm \sqrt{17}}{2}$ **11. a.** Yes, in order to make sure we have correct values for a, b, and c. **b.** No; it simplifies calculations, but we would still get a correct answer using fractional values in the formula.

Exercise Set 9.3 **1.** $2, 1$ **3.** $\frac{-7 \pm \sqrt{37}}{6}$ **5.** $\pm\frac{\sqrt{3}}{2}$ **7.** no real solution **9.** $10, -3$ **11.** $\pm\sqrt{5}$ **13.** $-3, 4$ **15.** $-2 \pm \sqrt{7}$ **17.** $\frac{-9 \pm \sqrt{129}}{12}$ **19.** $\frac{4 \pm \sqrt{2}}{7}$ **21.** $3 \pm \sqrt{7}$ **23.** $\frac{3 \pm \sqrt{3}}{2}$ **25.** $\frac{1}{3}, -1$ **27.** $\frac{3 \pm \sqrt{13}}{4}$ **29.** no real solution **31.** $\frac{1}{5}, -\frac{3}{4}$ **33.** $\frac{5 \pm \sqrt{33}}{2}$ **35.** $-2, \frac{7}{3}$ **37.** $1 \pm \sqrt{2}$ **39.** no real solution **41.** $-\frac{1}{2}, -\frac{3}{4}$ **43.** $\frac{7 \pm \sqrt{129}}{20}$ **45.** $\frac{11 \pm \sqrt{129}}{4}$ **47.** $\frac{1 \pm \sqrt{2}}{5}$ **49.** $\pm\sqrt{7}$; $-2.6, 2.6$ **51.** $-3 \pm 2\sqrt{2}$; $-5.8, -0.2$ **53.** $\frac{9 \pm \sqrt{97}}{2}$; $9.4, -0.4$ **55.** $\frac{1 \pm \sqrt{7}}{3}$; $1.2, -0.5$

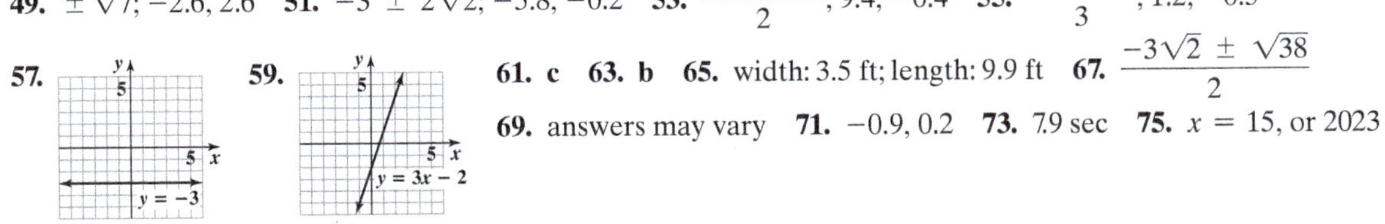

61. c **63.** b **65.** width: 3.5 ft; length: 9.9 ft **67.** $\frac{-3\sqrt{2} \pm \sqrt{38}}{2}$ **69.** answers may vary **71.** $-0.9, 0.2$ **73.** 7.9 sec **75.** $x = 15$, or 2023

Integrated Review **1.** $2, \frac{1}{5}$ **2.** $\frac{2}{5}, -3$ **3.** $1 \pm \sqrt{2}$ **4.** $3 \pm \sqrt{2}$ **5.** $\pm 2\sqrt{5}$ **6.** $\pm 6\sqrt{2}$ **7.** no real solution **8.** no real solution **9.** 2 **10.** 3 **11.** 3 **12.** $\frac{7}{2}$ **13.** ± 2 **14.** ± 3 **15.** $1, 2$ **16.** $-3, -4$ **17.** $0, -5$ **18.** $\frac{8}{3}, 0$ **19.** $\frac{3 \pm \sqrt{7}}{5}$ **20.** $\frac{3 \pm \sqrt{5}}{2}$ **21.** $\frac{3}{2}, -1$ **22.** $\frac{2}{5}, -2$ **23.** $\frac{5 \pm \sqrt{105}}{20}$ **24.** $\frac{-1 \pm \sqrt{3}}{4}$ **25.** $5, \frac{7}{4}$ **26.** $1, \frac{7}{9}$ **27.** $\frac{-7 \pm 3\sqrt{2}}{-5}$ or $\frac{7 \pm 3\sqrt{2}}{5}$ **28.** $\frac{-5 \pm 5\sqrt{3}}{-4}$ or $\frac{5 \pm 5\sqrt{3}}{4}$ **29.** $\frac{7 \pm \sqrt{193}}{6}$ **30.** $\frac{-7 \pm \sqrt{193}}{12}$ **31.** $11, -10$ **32.** $7, -8$ **33.** $4, -\frac{2}{3}$ **34.** $2, -\frac{4}{5}$ **35.** $0.5, 0.1$ **36.** $0.3, -0.2$ **37.** $\frac{11 \pm \sqrt{41}}{20}$ **38.** $\frac{11 \pm \sqrt{41}}{40}$ **39.** $\frac{4 \pm \sqrt{10}}{2}$ **40.** $\frac{5 \pm \sqrt{185}}{4}$ **41.** answers may vary

Section 9.4

Calculator Explorations **1.** $-0.41, 7.41$ **3.** $0.91, 2.38$ **5.** $-0.39, 0.84$

Vocabulary, Readiness & Video Check **1.** If a parabola opens upward, the lowest point is called the vertex; if a parabola opens downward, the highest point is called the vertex. If a graph can be folded along a line such that the two sides coincide or form mirror images of each other, we say the graph is symmetric about that line and that line is the line of symmetry. **3.** For example, if the vertex is in quadrant III or IV and the parabola opens downward, then there won't be any x-intercepts and there's no need to let $y = 0$ and solve the equation for x.

Exercise Set 9.4

1. 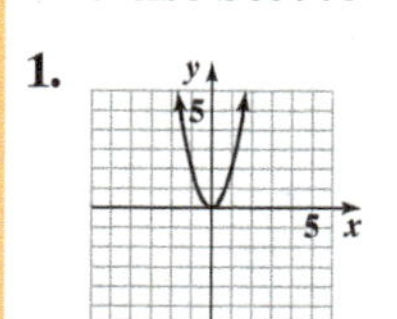

3. 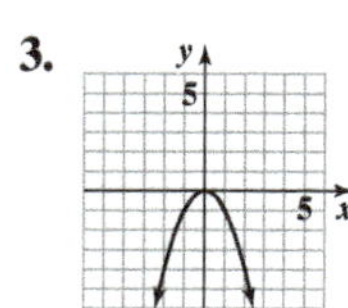

5. 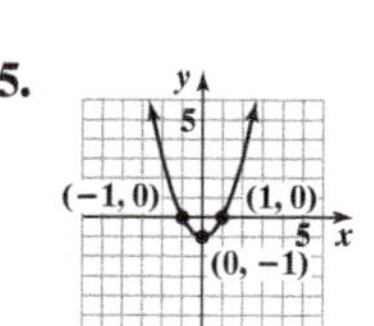

7. 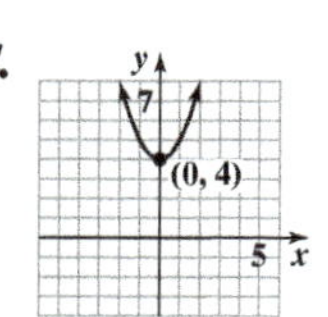

9. 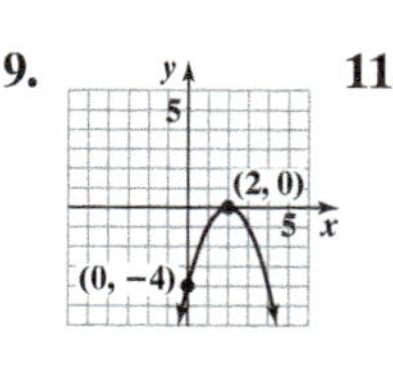

11. 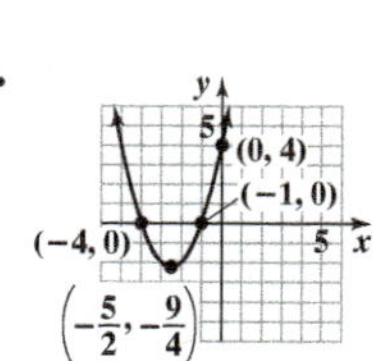

13. 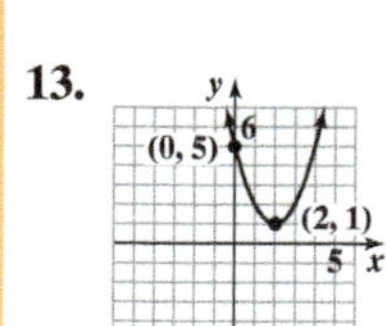

15. 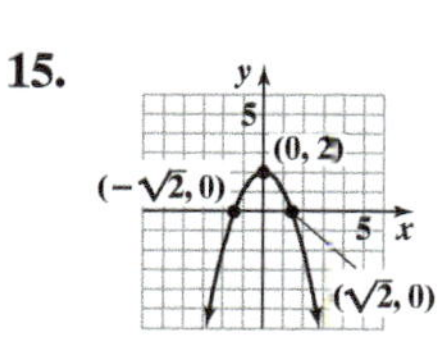

17. 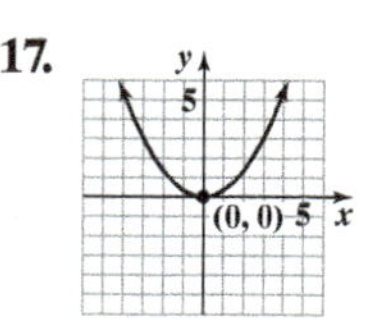

19. 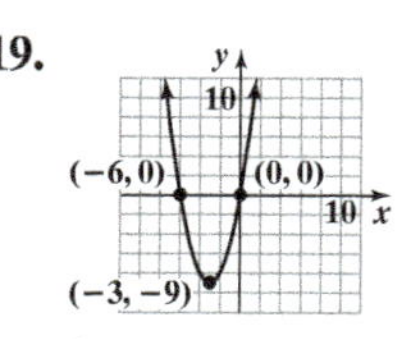

21. 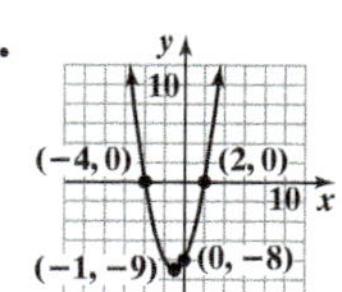

23. 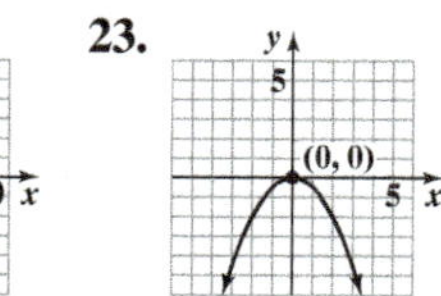

25. 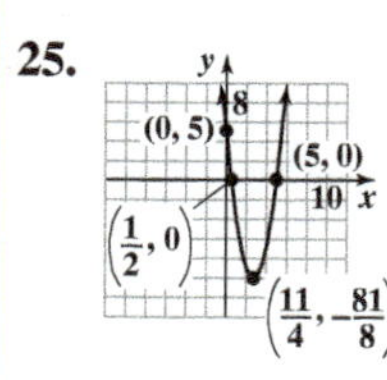

27. 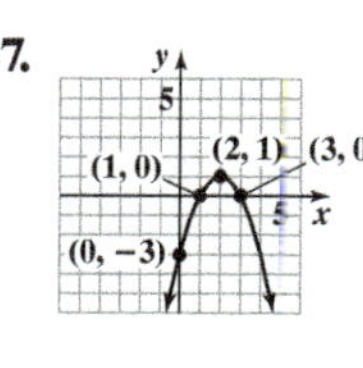

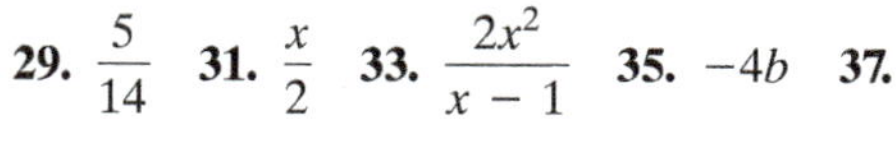

29. $\frac{5}{14}$ **31.** $\frac{x}{2}$ **33.** $\frac{2x^2}{x-1}$ **35.** $-4b$ **37.**

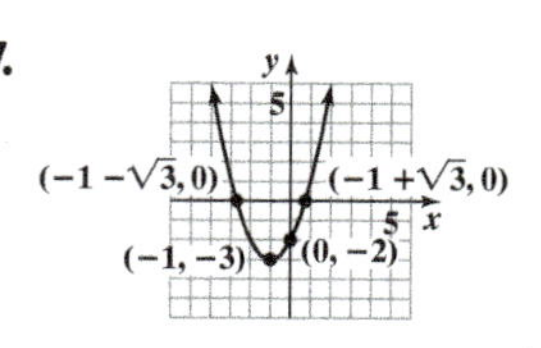

39. 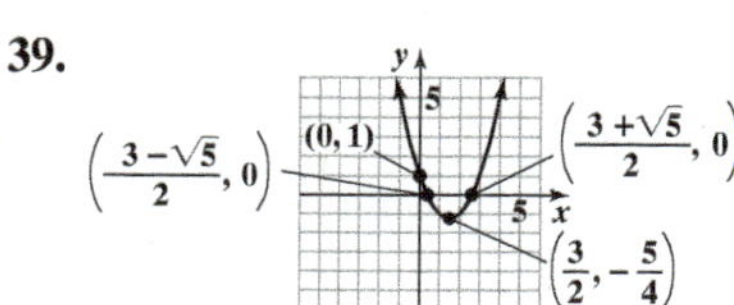

41. a. 256 ft (exact height) **b.** $t = 4$ sec **c.** $t = 8$ sec **43.** A **45.** D **47.** F

Chapter 9 Vocabulary Check **1.** square root **2.** parabola **3.** vertex **4.** completing the square **5.** quadratic **6.** vertex **7.** zero

Chapter 9 Review **1.** ± 11 **2.** ± 10 **3.** $-\frac{1}{3}, 2$ **4.** $\frac{5}{7}, -1$ **5.** ± 6 **6.** ± 9 **7.** $\pm 5\sqrt{2}$ **8.** $\pm 3\sqrt{5}$ **9.** $4, 18$ **10.** $7, -13$ **11.** $\frac{-5 \pm \sqrt{41}}{4}$ **12.** $\frac{-7 \pm \sqrt{37}}{3}$ **13.** 2.5 sec **14.** 40.6 sec **15.** $1, 8$ **16.** $-10, 2$ **17.** $-2 \pm \sqrt{5}$ **18.** $4 \pm \sqrt{19}$ **19.** $3 \pm \sqrt{2}$ **20.** $-3 \pm \sqrt{2}$ **21.** $\frac{1}{2}, -1$ **22.** $\frac{1}{4}, -1$ **23.** $-\frac{5}{3}$ **24.** $\frac{9}{4}$ **25.** $\pm\sqrt{5}$ **26.** $\pm\sqrt{3}$ **27.** $5 \pm 3\sqrt{2}$ **28.** $-2 \pm \sqrt{11}$ **29.** $\frac{-1 \pm \sqrt{13}}{6}$ **30.** $\frac{-3 \pm \sqrt{13}}{2}$ **31.** no real solution **32.** no real solution **33.** $0.4, -0.8$ **34.** $0.3, -3.3$ **35.** $x = 3$, or 2014 **36.** $x = 5$, or 2017

37. 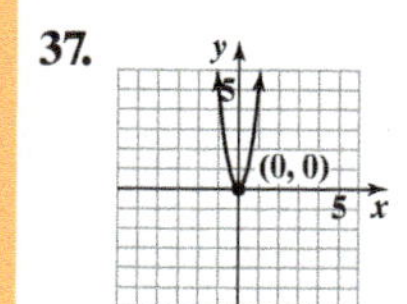

38. 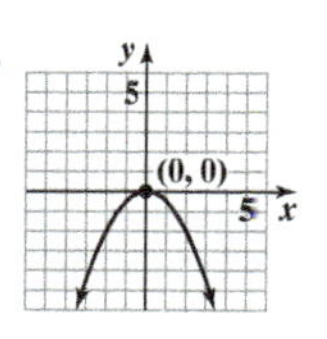

39. 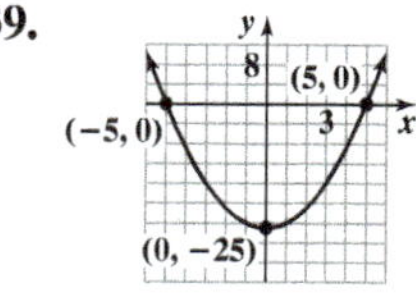

40. 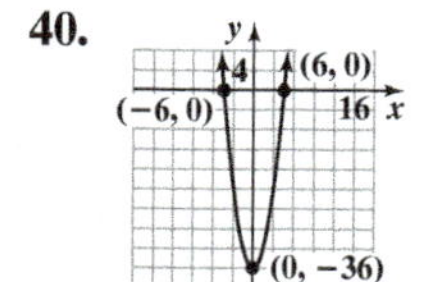

41. 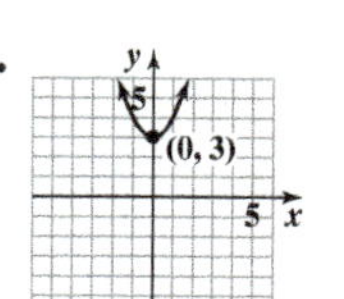

42. 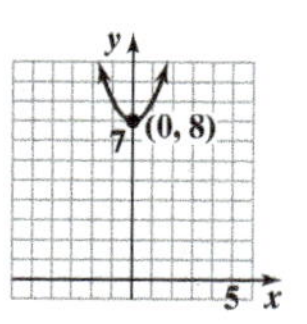

43. 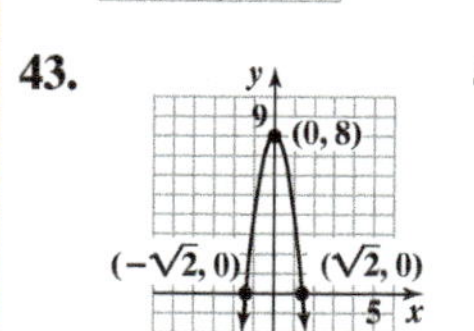

44. 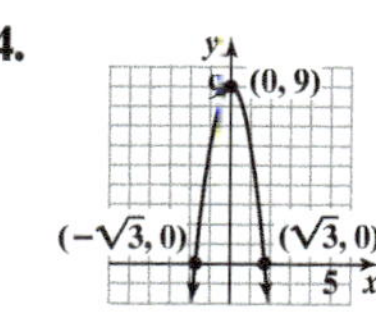

45. 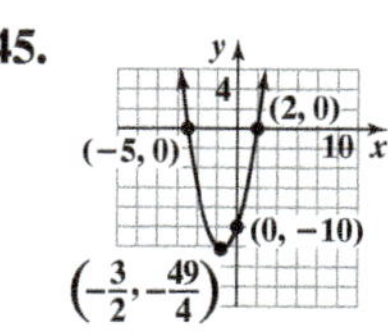

46. 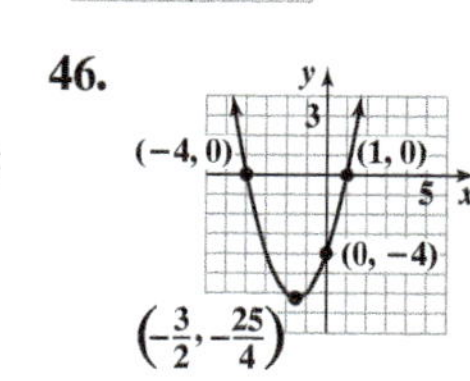

47. 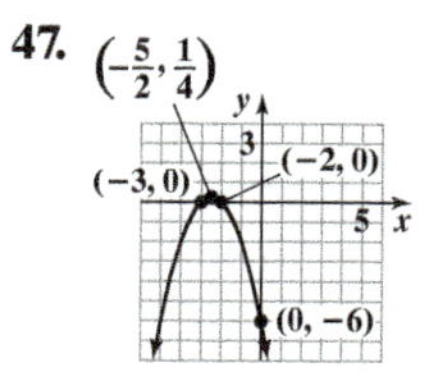

48. 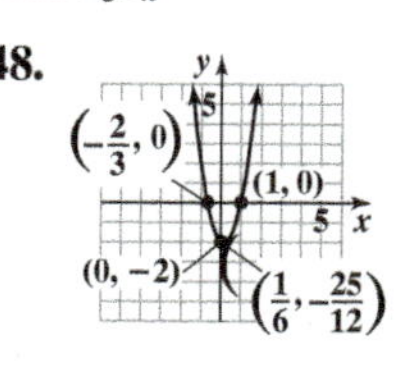

49. 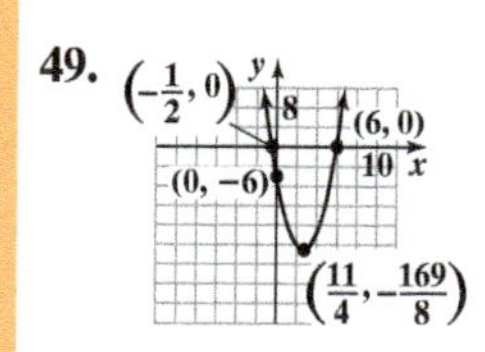

50. 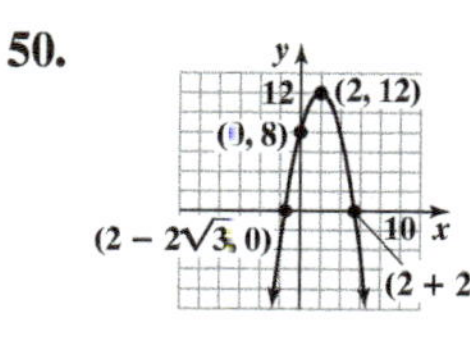

51. A **52.** D **53.** B **54.** C **55.** one real solution **56.** two real solutions **57.** no real solution **58.** two real solutions **59.** ± 7 **60.** $\pm 5\sqrt{3}$

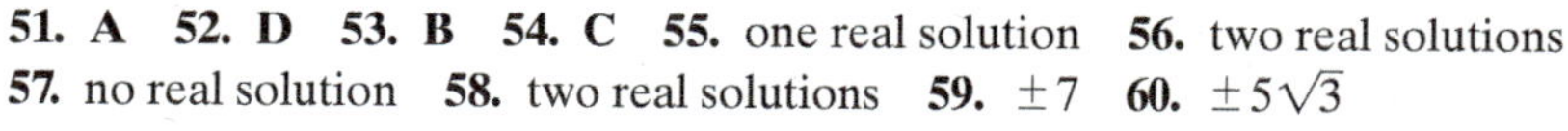

61. $15, -1$ **62.** $-2 \pm \sqrt{10}$ **63.** $\frac{2}{3}, -1$ **64.** $\frac{1 \pm \sqrt{33}}{8}$ **65.** $\frac{3 \pm \sqrt{41}}{8}$

66. $\frac{-1 \pm \sqrt{41}}{10}$ **67.** $-\frac{3}{2}$ **68.** no real solution **69.**

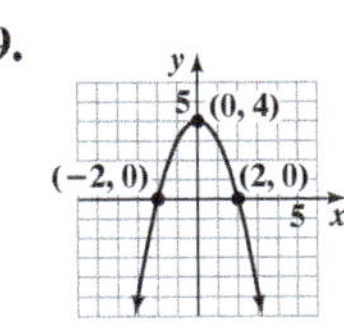

70. **71.**

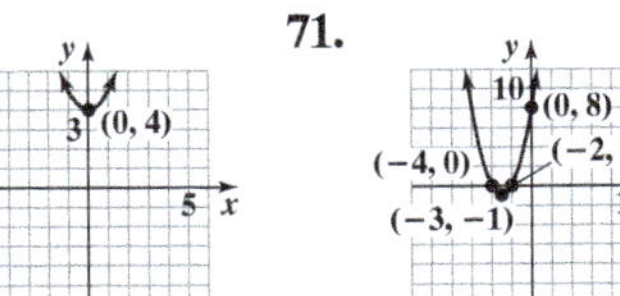

72.

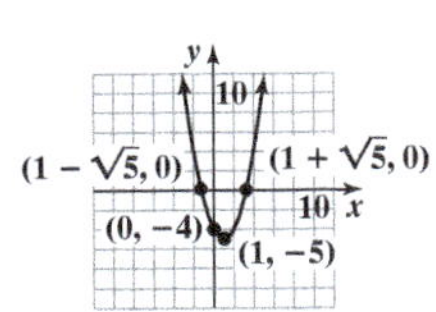

Chapter 9 Getting Ready for the Test

1. B **2.** D **3.** C **4.** A **5.** B **6.** C **7.** B **8.** D **9.** D **10.** B

Chapter 9 Test **1.** ± 20 **2.** $-\frac{3}{2}, 7$ **3.** ± 4 **4.** $\frac{5 \pm 2\sqrt{2}}{3}$ **5.** 10, 16 **6.** $-2 \pm \frac{4\sqrt{3}}{3}$ **7.** $-2, 5$ **8.** $\frac{5 \pm \sqrt{37}}{6}$ **9.** $1, -\frac{4}{3}$

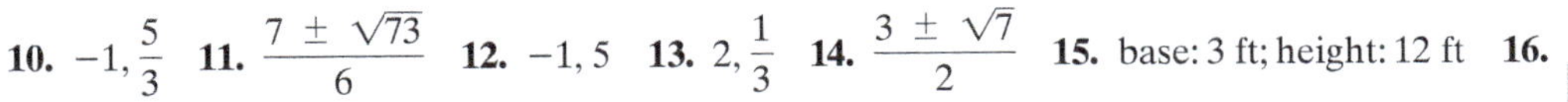

10. $-1, \frac{5}{3}$ **11.** $\frac{7 \pm \sqrt{73}}{6}$ **12.** $-1, 5$ **13.** $2, \frac{1}{3}$ **14.** $\frac{3 \pm \sqrt{7}}{2}$ **15.** base: 3 ft; height: 12 ft **16.**

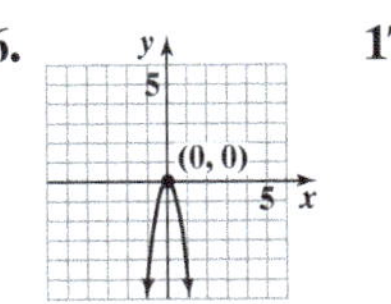

17. (graph: vertex $(0, -4)$; x-intercepts $(-2, 0)$, $(2, 0)$)

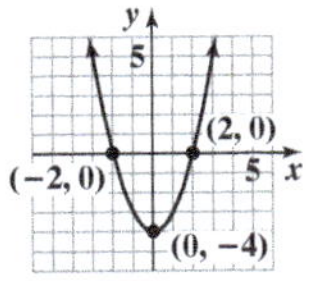

18.

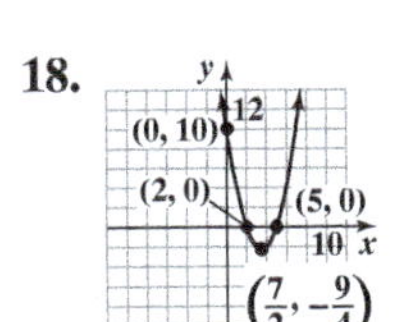

19.

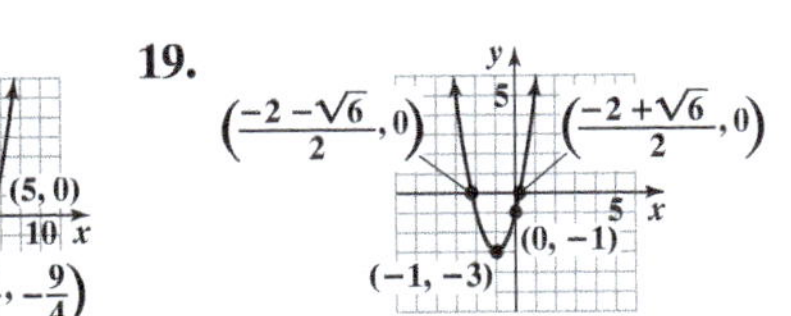

20. 6 sides **21.** 3.5 sec

Cumulative Review **1.** -1.6; Sec. 2.1, Ex. 2 **2.** -3; Sec. 2.3 **3.** $\frac{16}{3}$; Sec. 2.3, Ex. 2 **4.** $\frac{29}{8}$; Sec. 2.3 **5.** 29 and 8; Sec. 7.4, Ex. 1 **6.** 145, 146, 147; Sec. 2.4 **7.** 1; Sec. 3.1, Ex. 28 **8.** $-\frac{216x^3}{y^9}$; Sec. 3.1 **9.** 1; Sec. 3.1, Ex. 29 **10.** $\frac{a^2}{32b^3}$; Sec. 3.2 **11.** -1; Sec. 3.1, Ex. 31 **12.** 9; Sec. 3.1 **13.** $9y^2 + 12y + 4$; Sec. 3.6, Ex. 15 **14.** $x^2y - x^2 + 5y - 5$; Sec. 3.5 **15.** $x + 4$; Sec. 3.7, Ex. 4 **16.** $9.1a - 4$; Sec. 3.3 **17.** $(r + 6)(r - 7)$; Sec. 4.2, Ex. 4 **18. a.** -1 **b.** 30; Sec. 1.6 **19.** $(2x - 3y)(5x + y)$; Sec. 4.3, Ex. 4 **20.** $\frac{8x + 13}{(x + 2)(x - 1)}$; Sec. 5.4 **21.** $(2x - 1)(4x - 5)$; Sec. 4.4, Ex. 1 **22.** $\frac{2x + 5}{x - 7}$; Sec. 5.2 **23. a.** $x(2x + 7)(2x - 7)$; Sec. 4.5, Ex. 16 **b.** $2(9x^2 + 1)(3x + 1)(3x - 1)$; Sec. 4.5, Ex. 17 **24.** -32; Sec. 5.5 **25.** $\frac{1}{5}, -\frac{3}{2}, -6$; Sec. 4.6, Ex. 8 **26. a.** $-x + 4$ **b.** $5y - 8$ **c.** $9.1a - 4$ **d.** $2x^2 - 2x$; Sec. 3.3 **27.** $\frac{x + 7}{x - 5}$; Sec. 5.1, Ex. 4 **28.** $-\frac{7}{2}, 1$; Sec. 4.6 **29.** -5; Sec. 5.6, Ex. 5 **30.** $\sqrt{82}$ units; Ch. 8 Group Activity **31. a.** -3 **b.** 0 **c.** -3; Sec. 6.1, Ex. 4 **32. a.** x-int: $(4, 0)$; y-int: $(0, 1)$ **b.** x-int: $(-2, 0), (0, 0), (3, 0)$; y-int: $(0, 0)$; Sec. 6.3 **33. a.** parallel **b.** perpendicular **c.** neither; Sec. 6.4, Ex. 7 **34.** perpendicular; Sec. 6.4 **35. a.** function **b.** not a function; Sec. 6.6, Ex. 2 **36. a.** $6\sqrt{5}$; **b.** $8\sqrt{2}$ **c.** $4\sqrt{2} + 11 - 2\sqrt{3}$; Sec. 8.3 **37.** $(4, 2)$; Sec. 7.2, Ex. 1 **38.** no solution; Sec. 7.2 **39.** no solution; Sec. 7.3, Ex. 3 **40.** infinite number of solutions; Sec. 7.3 **41.** 6; Sec. 8.1, Ex. 1 **42.** $\frac{2}{5}$; Sec. 8.1 **43.** $\frac{3}{10}$; Sec. 8.1, Ex. 3 **44.** $\frac{4}{11}$; Sec. 8.1 **45.** $-1 + \sqrt{3}$; Sec. 8.4, Ex. 13 **46.** $\frac{5\sqrt{2}}{4}$; Sec. 8.4 **47.** $7, -1$; Sec. 9.1, Ex. 5 **48.** $4 \pm \sqrt{3}$; Sec. 9.1 **49.** $1 \pm \sqrt{5}$; Sec. 9.3, Ex. 5 **50.** $-2 \pm 2\sqrt{3}$; Sec. 9.3

Appendices

Exercise Set Appendix B **1.** $(a + 3)(a^2 - 3a + 9)$ **3.** $(2a + 1)(4a^2 - 2a + 1)$ **5.** $5(k + 2)(k^2 - 2k + 4)$ **7.** $(xy - 4)(x^2y^2 + 4xy + 16)$ **9.** $(x + 5)(x^2 - 5x + 25)$ **11.** $3x(2x - 3y)(4x^2 + 6xy + 9y^2)$ **13.** $(3 - t)(9 + 3t + t^2)$ **15.** $8(r - 2)(r^2 + 2r + 4)$ **17.** $(t - 7)(t^2 + 7t + 49)$ **19.** $(s - 4t)(s^2 + 4st + 16t^2)$

Exercise Set Appendix C **1.** mean: 21; median: 23; no mode **3.** mean: 8.1; median: 8.2; mode: 8.2 **5.** mean: 0.5; median: 0.5; mode: 0.2 and 0.5 **7.** mean: 370.9; median: 313.5; no mode **9.** 2109.2 ft **11.** 1968.5 ft **13.** answers may vary **15.** 2.79 **17.** 3.64 **19.** 6.8 **21.** 6.9 **23.** 88.5 **25.** 73 **27.** 70 and 71 **29.** 9 rates **31.** mean: 3773 mi; median: 3812 mi **33.** 12 **35.** 350 **37.** 2 **39.** 1.7 **41. a.** 8.2 **b.** 8 **c.** 9 **43. a.** 6.1 **b.** 6 **c.** 6, 7, 8 **45. a.** 15 **b.** 15 **c.** 15 **47. a.** 6.1 **b.** 6 **c.** 4, 6

Exercise Set Appendix D **1.** $\{-9, -8, -7, -6\}$ **3.** {Tuesday, Thursday} **5.** $\{0, 1, 2, 3, 4, \ldots\}$ **7.** { } or $\varnothing$ **9.** true **11.** true **13.** true **15.** false **17.** true **19.** false **21.** $\{1, 2, 3, 4, 5, 6\}$ **23.** $\{2, 4, 6\}$ **25.** $\{1, 3, 5, 7\}$ **27.** $\varnothing$ **29.** true **31.** false **33.** false **35.** true **37.** true **39.** true **41.** true

Exercise Set Appendix E **1.** 71° **3.** 19.2° **5.** $78\frac{3}{4}°$ **7.** 30° **9.** 149.8° **11.** $100\frac{1}{2}°$ **13.** $m\angle 1 = m\angle 5 = m\angle 7 = 110°$, $m\angle 2 = m\angle 3 = m\angle 4 = m\angle 6 = 70°$ **15.** 90° **17.** 90° **19.** 90° **21.** 45°, 90° **23.** 73°, 90° **25.** $50\frac{1}{4}°$, 90° **27.** $x = 6$ **29.** $x = 4.5$ **31.** 10 **33.** 12

Exercise Set Appendix F **1.** 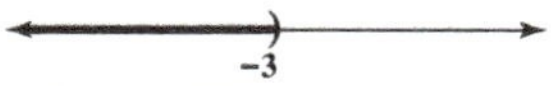

$(-\infty, -3)$ **3.** 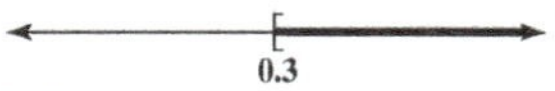

$[0.3, \infty)$
5. 

$[-7, \infty)$ **7.** 

$(-2, 5)$ **9.** 

$(-1, 5]$
11. domain: $[0, \infty)$; range: $(-\infty, \infty)$ **13.** domain: $(-\infty, \infty)$; range: $[0, \infty)$ **15.** domain: $(-\infty, \infty)$; range: $[-\infty, -3) \cup [3, \infty)$ **17.** domain: $[1, 7]$; range: $[1, 7]$ **19.** domain: $\{-2\}$; range: $(-\infty, \infty)$ **21.** domain: $(-\infty, \infty)$; range: $(-\infty, 3]$ **23.** domain: $(-\infty, \infty)$; range: $(-\infty, 3]$ **25.** domain: $[2, \infty)$; range: $[3, \infty)$

Practice Final Exam **1.** -81 **2.** $\frac{1}{64}$ **3.** -48 **4.** $-3x^3 + 5x^2 + 4x + 5$ **5.** $16x^2 - 16x + 4$ **6.** $3x^3 + 22x^2 + 41x + 14$ **7.** $(6t + 5)(t - 1)$ **8.** $3x(x - 5)(x - 2)$ **9.** $5(6 + x)(6 - x)$ **10.** $(a + b)(3a - 7)$ **11.** $x(1 - x)(1 + x)(1 + x^2)$ **12.** $\frac{16y^{14}}{x^2}$ **13.** $\frac{5y^2 - 1}{y + 2}$ **14.** $\frac{(x - 6)(x - 7)}{(x + 7)(x + 2)}$ **15.** $\frac{3a - 4}{(a - 3)(a + 2)}$ **16.** 8 **17.** $-7, 1$ **18.** $\{x \mid x \le -2\}$ **19.** $\frac{3 \pm \sqrt{7}}{2}$ **20.** $\frac{30}{11}$ **21.** -6 **22.** no solution **23.** 9 **24.** 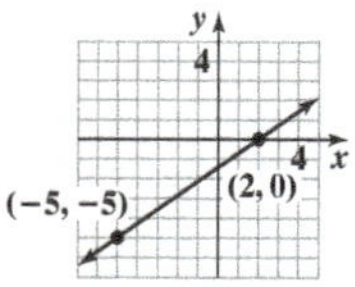

25. 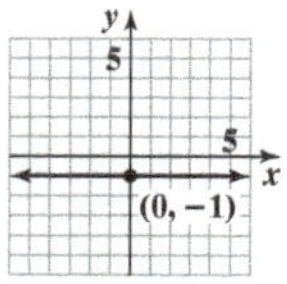

26. 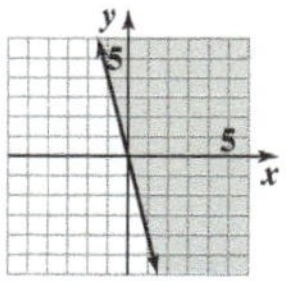

27. $m = -1$ **28.** $m = 3$

29. $8x + y = 11$ **30.** $x - 8y = -96$ **31.** $(-4, 1)$ **32.** no solution **33. a.** 0 **b.** 0 **c.** 60 **34.** function; domain: all real numbers; range: {2} **35.** 4 **36.** 5 **37.** $\frac{3}{4}$ **38.** $3\sqrt{6}$ **39.** $2x^4\sqrt{6}$ **40.** $\sqrt{2} + 2\sqrt{3}$ **41.** $2x\sqrt{5x}$ **42.** $2\sqrt{3} - \sqrt{10}$ **43.** $\frac{8\sqrt{5y}}{5y}$ **44.** $4\sqrt{6} - 8$ **45.** 5 or 1 **46.** 401, 802 **47.** 3 mph; 6 mph **48.** 120 cc

Subject Index

A

Absolute value
  explanation of, 82
  method to find, 14–15
Absolute value bars, 20
Acute angles. *See* Angles acute
Addition
  applications of, 34
  associative property of, 65–66, 84
  commutative property of, 64, 84
  of decimals, R-20–R-21, R-50
  distributive property of multiplication over, 66–67
  of fractions, R-13, R-49
  identities for, 67
  of polynomials, 221–223, 254
  of radicals, 599–600, 638
  of rational expressions, 360, 368–371, 410–411
  of real numbers, 30–34, 83
  of signed numbers, 30, 31
  symbol for, 24
  words/phrases for, 24
Addition method
  explanation of, 548
  to solve systems of linear equations, 548–551, 571–572
Addition property of equality
  explanation of, 93, 174
  use of, 93–96, 105–106
Addition property of inequality, 163
Additive inverse. *See also* Opposites
  explanation of, 32, 67, 68, 84
  method to find, 32–34
Adjacent angles, 718
Algebraic expressions. *See also* Expressions
  evaluation of, 22, 34, 41–42, 55–57, 82, 301
  explanation of, 22, 82
  method to write, 22, 76, 97–98, 106–107
Alternate interior angles, 718
Angle notation, 717
Angles acute, 717
  adjacent, 718
  alternate interior, 718
  complementary, 43, 717
  corresponding, 718
  explanation of, 717
  exterior, 718
  finding measures of, 126–127
  interior, 718
  obtuse, 717
  right, 717
  straight, 717
  supplementary, 43, 717
  vertical, 718
Applications. *See also* Applications index; Words/phrases
  of addition, 34
  of consecutive integers, 127–128
  of discounting, 151
  of division, 57–58
  of formulas, 136–142, 175
  of inequalities, 166–167
  introduction to, 122–128
  of mark-up, 151–152
  of mixtures, 153–154, 175–176, 561–562
  of multiplication, 57–58
  of percent equations, 149–151
  of percent of increase and decrease, 152–153
  of point-slope form, 477–479
  of proportions, 387–389, 411–412
  of Pythagorean theorem, 319–320, 620–621, 722
  of quadratic equations, 316–320, 328, 654
  of radicals, 621–622
  of rational equations, 387–393, 411–412
  steps in, 122, 149, 175, 328
  of subtraction, 42–43
  of systems of linear equations, 557–562, 572
Approximately equal to sign, R-24
Approximation
  decimal, 585, R-24
  of solutions to quadratic equations, 668
  of square roots, 585
Associative property
  of addition, 65–66, 84
  of multiplication, 65–66, 84
Average, 173
Axis of symmetry, 675–676, 686

B

Bar graphs, R-30–R-33, R-51
Base, of exponential expressions, 19, 189
Binomials. *See also* Polynomials
  explanation of, 212, 254
  FOIL method for, 235–236
  squaring, 236–237, 255
Boundary lines, 501–505
Braces, 8, 20
Brackets, 20
Break-even point, 569
Broken line graphs, 424

C

Calculators. *See also* Graphing calculators
  to approximate solutions of systems of equations, 533
  checking solutions to equations on, 115
  discovering patterns on, 479
  equations on, 442, 452
  evaluating expressions on, 301
  exponents on, 25
  features of, 442
  quadratic equations on, 680
  real number operations, 58
  scientific notation on, 58, 206
  to sketch graph of more than one equation on same set of axes, 466
  square roots on, 586
Circle graphs, R-36–R-38, R-51
Class frequency, R-33, R-51
Class interval, R-33, R-51
Coefficients, 72, 84, 211, 254
Common denominators
  adding and subtracting rational expressions with, 360
  explanation of, R-13
Common factors, 268, 343
Commutative property
  of addition, 64, 84
  of multiplication, 64, 84
Complementary angles, 43, 717
Completing the square, 658–659, 685
Complex fractions
  explanation of, 400
  methods to simplify, 400–404, 412
Composite numbers, R-2
Congruent triangles, 720
Conjugates
  explanation of, 606–607
  rationalizing denominators using, 606–607, 638
Consecutive integers
  even and odd, 107
  solving problems with, 127–128, 318–319
  writing expression for, 106–107
Consistent systems, 532
Constant
  explanation of, 211
  of proportionality, 627
  of variation, 627
Coordinate plane, 424. *See also* Rectangular coordinate system
Coplanar, 718
Corresponding angles, 718
Counting numbers. *See* Natural numbers
Cross products, 386–387, 411
Cube roots
  explanation of, 584, 637
  method to find, 584
  method to simplify, 594, 600
Cubes, sum or difference of two, 701–702

D

Decimal approximations, 585, R-24
Decimal point, R-20
Decimals
  addition of, R-20–R-21, R-49
  division of, R-22, R-50
  equations containing, 112–114
  method to round, R-22–R-23
  multiplication of, R-21–R-22, R-50

Decimals (*continued*)
repeating, R-24
solving equations containing, 112–114
subtraction of, R-21, R-50
table of, 697
writing fractions as, R-23–R-24, R-50
writing percents as, R-24–R-25, R-50
written as fractions, R-19–R-20, R-50
written as percents, R-25, R-50
Degree, of polynomials, 212–213, 215, 254
Denominators
common, 360, R-13
explanation of, R-8
least common, 112, 361–363, R-13
of rational expressions, 341, 361–363, 368–371
rationalizing, 605–606
Dependent equations, 532
Dependent variables, 492
Descending powers, 212, 216
Difference of squares
explanation of, 237–238
factoring, 297–300, 327
Direct variation equations, 626–629, 631–633, 639
Discount problems, 151
Dispersion, 707
Distance
applications involving, 136–137, 392–393, 621, 629
formula for, 136, 636
Distributive property of multiplication over addition
explanation of, 66–67, 84
to remove parentheses, 74–75, 84, 106, 113, 540
Dividend, 246, R-22
Divisibility tests, R-4
Division
applications of, 57–58
of decimals, R-22, R-49
of fractions, R-12, R-49
of polynomials, 245–248, 255
of radicals, 605, 638
of rational expressions, 352–354, 410
of real numbers, 54–55, 83
of signed numbers, 54, 55
symbol for, 24
words/phrases for, 24
by zero, 55, 340
Divisor, R-22
Domain
of function, 493–494
of relation, 488, 513, 726

E

Element, of set, 8, 81, 713
Elimination method. *See* Addition method
Ellipsis, 9
Empty set, 713
Equal sign, 24
Equality
addition property of, 93–96, 105–106, 174
multiplication property of, 102–104, 174, 375
squaring property of, 614–617
symbols for, 9, 24, 82
words/phrases for, 24
Equations. *See also* Linear equations in one variable; Linear equations in two variables; Quadratic equations; Systems of linear equations
on calculators, 115
containing decimals, 112–114
containing fractions, 112–114, 140
containing rational expressions, 375–379, 387
dependent, 532
direct variation, 626–629, 631–633
equivalent, 93, 174
explanation of, 23, 82
on graphing calculators, 442, 452
independent, 532
inverse variation, 629–633
of lines, 460–461
methods to simplify, 96–97
with no solution, 114–115
ordered pairs to record solutions of, 427–429
radical, 614–617
rational, 387–393
solutions of, 23, 42, 57, 83
Equivalent equations, 93, 174
Equivalent fractions
explanation of, R-9–R-10, R-49
method to write, R-10
Estimation. *See* Rounding
Exams. *See* Mathematics class
Exponential expressions
base of, 19, 189
evaluation of, 82, 189–190
explanation of, 19, 82, 189
methods to simplify, 191, 202–203
Exponential notation, 19
Exponents
on calculators, 25
explanation of, 19–21, 82, 253
negative, 201–202, 253
power rule for, 192–194
product rule for, 190–192
quotient rule for, 194–196
summary of rules for, 202, 253
zero, 195
Expressions. *See also* Algebraic expressions; Exponential expressions; Rational expressions
methods to simplify, 72–76
undefined, 55, 340
Exterior angles, 718
Extraneous solutions, 615

F

Factored form, 268
Factoring. *See also* Polynomials
difference of two squares, 299–300, 327
explanation of, 268, 326, R-2
by grouping, 272–273, 292–294, 326, 327
perfect square trinomials, 297–300, 327
quadratic equations, 308–311, 328
to solve equations with degree greater than two, 311–312
to solve quadratic equations, 308–311, 328, 651
steps for, 306
trinomials of form $ax^2 + bx + c$, 285–288, 327
trinomials of form $x^2 + bx + c$, 278–281, 327
Factoring out
common factor, 268
explanation of, 268
greatest common factor, 268–272, 281, 288–289, 326
Factors. *See also* Polynomials
common, 268
explanation of, 268, R-2, R-48
greatest common, 268–270
First-degree equations in two variables. *See* Linear equations in two variables
FOIL method
explanation of, 235–236
use of, 235–236, 237, 278
Formulas
containing radicals, 621–622
distance, 136, 137
explanation of, 136, 175
geometric, 700
perimeter, 138, 139–140
temperature conversion, 138–139, 142
used to solve problems, 136–142, 175
for variable, 140–142
Fraction bars, 20, R-8
Fractions. *See also* Rational expressions; Ratios
addition of, R-13, R-49
complex, 400–404, 412
decimals written as, R-19–R-20, R-49
division of, R-12, R-49
equations containing, 112–114, 140
equivalent, R-9–R-10, R-49
explanation of, 12, R-8–R-10
improper, R-11, R-14–R-15
method to simplify, 341, R-10–R-11, R-49
mixed numbers written as improper, R-11, R-14–R-15
multiplication of, R-11–R-12, R-49
proper, R-11
solving equations containing, 112–114
subtraction of, R-13–R-14, R-49
table of, 697
unit, 354–355
written as decimals, R-23–R-24, R-50
Frequency distribution graphs, R-33
Function notation, 492–494, 513
Functions
domain and range of, 493–494
explanation of, 488–489, 513

graphs of, 490–492
as set of ordered pairs, 488–489
vertical line test and, 489–492

G
Geometric formulas, 700
Graphing calculators. *See also* Calculators
to approximate solutions of systems of equations, 533
discovering patterns on, 479
equations on, 442, 452
evaluating expressions on, 301
features of, 442
negative numbers on, 58
quadratic equations on, 680
to sketch graph of more than one equation on same set of axes, 466
Graphs
broken line, 424
of distance formula, 636
of functions, 490–492
of linear equations in two variables, 437–441, 451, 511
of linear inequalities in two variables, 500–505, 513–514
on number line, 12
ordered pairs and, 424–426
of paired data, 426
of quadratic equations, 675–680, 686
slope-intercept form, 474–475
to solve systems of linear equations, 529–531, 570–571
Greatest common factor (GCF)
explanation of, 268
factoring out, 268–272, 281, 288–289
of list of terms, 268–270, 326
method to find, 268–270
Grouping
explanation of, 65
factoring by, 272–273, 292–294, 326, 327
Grouping symbols, 20, 21

H
Half-planes, 501
Histograms, R-33–R-35, R-51
Horizontal lines
explanation of, 452
slope of, 461, 462, 479
Hypotenuse, 319, 722

I
Identities
for addition and multiplication, 67
explanation of, 114
with no solution, 114–115
Identity properties, 67–68, 84
Improper fractions
explanation of, R-11
as mixed numbers, R-11, R-14–R-15
Inconsistent systems of linear equations, 532
Independent variables, 492
Index, 584
Inequalities
addition property of, 163
explanation of, 162
graphs of linear, 500–505
multiplication property of, 163–165
on number line, 162–163
solving problems modeled by, 166–167
steps to solve, 165
symbols for, 9, 24, 82, 162
Integers. *See also* Signed numbers
consecutive, 106–107, 318–319
explanation of, 11, 81
negative, 11
on number line, 11–12
positive, 11
Intercepts. *See also* x-intercepts; y-intercepts
explanation of, 447, 511–512
identification of, 449–450
method to find and plot, 449–450, 511–512
Interior angles, 718
Intersecting lines, 529, 531, 718
Intersection, of sets, 714–715
Interval notation, 725
Inverse variation equations, 629–633, 639
Inverses
additive, 32–34, 67, 68, 84
multiplicative, 53, 67–68, 84
Irrational numbers, 12–13, 82, 585

L
Least common denominator (LCD)
explanation of, R-13, R-48
method to find, 361–363, 410
multiplying by, 112, 375–379
Least common multiple (LCM), R-4–R-6, R-48–R-49
Like radicals, 599
Like terms
explanation of, 73
method to combine, 73–74, 84
simplifying polynomials by combining, 214–215
Line graphs, R-35–R-36, R-51
Linear equations in one variable. *See also* Equations
containing fractions or decimals, 112–114
explanation of, 93, 111, 174
with no solution, 114–115
steps to solve, 111–112, 174
Linear equations in two variables. *See also* Equations; Systems of linear equations
explanation of, 437
graphs of, 437–441, 451, 511
point-slope form of, 479
slope of line and, 460
slope-intercept form of, 479
solutions to, 427–429, 438
standard form of, 437–438, 479, 511
Linear inequalities
explanation of, 513
solving problems modeled by, 166–167
steps to solve, 165, 177
in two variables, 500–505, 513–514
Linear models, 509
Lines
boundary, 501–505
coplanar, 718
equations of, 460–461
horizontal, 452
intersecting, 529, 531, 718
parallel, 462–463, 466, 718
perpendicular, 463
slope of, 457–464, 479
transversal, 718, 719
vertical, 451
Long division, of polynomials, 246–247
Lowest terms, of fractions, R-10–R-11

M
Magic squares, 80
Mark-up problems, 151–152
Mathematical statements, 9, 10
Mathematics class
attitude, 2, 731
exam preparation for, 5–6, 732–733, 738
getting help in, 5, 737, 739
homework assignments for, 735, 736
learning new terms for, 734
notebooks for, 5, 735
online homework, 736
organizational skills for, 734, 737
practice final exam for, 744–746
resources available for, 738
study guide outline for, 740–743
study skills builders for, 730–739
taking exams for, 6
textbook supplements for, 739
textbook use for, 3–4, 738
time management for, 6
tips for success in, 2–3
video resources for, 4–5
Mean, 703–706, 709
Measurement, converting between units of, 354–355
Measures of central tendency, 703. *See also* Median; Mode
Median, 706, 708
Members. *See* Element, of set
Minus sign, 11
Mixed numbers
improper fractions as, R-11
writing improper fractions as, R-14–R-15
Mixture problems, 153–154, 175–176, 561–562
Mode, 706–707, 709
Modeling
explanation of, 441
with inequalities, 166–167
linear, 509
with polynomials, 252
Monomials. *See also* Polynomials
division of, 245–246

Monomials. *See also* Polynomials (*continued*)
explanation of, 212, 254
multiplication of, 228
Multiple, R-4
Multiplication. *See also* Products
applications of, 57–58
associative property of, 65–66, 84
commutative property of, 64, 84
of decimals, R-21–R-22, R-50
of fractions, R-11–R-12, R-49
identities for, 67
order in, R-3
of polynomials, 228–230, 235–236, 254
of radicals, 603–605, 638
of rational expressions, 350–351, 353–354, 409–410
of real numbers, 51–53, 83
of signed numbers, 52, 53
symbol for, 24
words/phrases for, 24
Multiplication property of equality
explanation of, 102, 174
use of, 102–104, 375
Multiplication property of inequality
explanation of, 164
use of, 163–165
Multiplicative inverse, 53, 67–68, 84. *See also* Reciprocals

N

Natural numbers, 8, 81, R-2
Negative exponents, 201–202, 253
Negative integers, 11
Negative numbers, 51
Negative square root, 583, 637
Notation/symbols. *See* Symbols/notation
*n*th root, 584–585, 637
Null set, 713
Number lines
explanation of, 9, 82
inequalities on, 162–163
integers on, 11–12
real numbers on, 30–32
Numbers. *See also* Integers; Real numbers; Signed numbers; Whole numbers
composite, R-2
factoring, R-2
finding unknown, 122–123, 557–558
irrational, 12–13, 82, 585
mixed, R-11, R-14–R-15
multiple of, R-4
natural, 8, 81, R-2
positive and negative, 51
prime, R-2
rational, 12, 81
set of, 8, 13
Numerators, R-8
Numerical coefficients. *See* Coefficients

O

Obtuse angles, 717
Opposites. *See also* Additive inverse
explanation of, 32, 67
method to find, 32–34
Order of operations
to evaluate expressions, 55–56
examples using, 20–21
explanation of, 20, 82
Order property for real numbers, 10, 82
Ordered pairs
data represented as, 488
explanation of, 424
functions as set of, 488–489
method to plot, 424–426, 510
paired data and, 426
to record solutions of equations, 427–429
relations as set of, 488
as solution to linear equations in two variables, 438–440, 510–511
as solution to systems of linear equations in two variables, 529
Origin, 424, 510
Original inequality, 503

P

Paired data, 426
Paired values table, 428
Parabolas
explanation of, 675, 686
graphs of, 675–680, 686
uses of, 684
vertex of, 675–676, 686
Parallel lines
cut by transversal, 718, 719
explanation of, 462–463, 718
slope of, 462–463, 466
Parentheses
distributive property to remove, 74–75, 84, 106, 113, 540
explanation of, 20
Patterns, 479
Percent equations, method to solve, 149–151
Percent problems
with discount and sale price, 151
increase and decrease, 152–153
mixtures, 153–154
strategies to solve, 149, 175–176
Percents
explanation of, R-24–R-25
table of, 697–698
written as decimals, R-24–R-25, R-50
Perfect square trinomials
explanation of, 297–298
factoring, 297–300, 327, 658
Perfect squares, 585
Perimeter, of rectangles, 138, 139–140
Perpendicular lines
explanation of, 463, 718
slope of, 463
π, 13
Pictographs, R-30, R-51
Place value, for decimals, R-19, R-23
Plane, 717–718
Plane figures, 718
Plus sign, 11
Points
on plane, 425
writing equation given slope and, 460–461
writing equation given two, 476–477
Point-slope form
applications of, 477–479
explanation of, 475–476, 479, 513
Polygons, 719
Polynomials. *See also* Binomials; Factoring; Factors; Monomials; Trinomials
addition of, 221–223, 254
degree of, 212–213, 215, 254
division of, 245–248, 255
evaluation of, 213–214
explanation of, 212, 254
inserting "missing" terms in, 216
methods to simplify, 214–215
modeling with, 252
multiplication of, 228–230, 235–236, 254
prime, 280
quotient of, 339
subtraction of, 221–223, 254
types of, 212, 254
Positive integers, 11
Positive numbers, 51
Positive square root, 583, 637
Power, 189
Power rule for exponents
explanation of, 192
for products and quotients, 192–194
Predictions, 441
Prime factorization
explanation of, R-3, R-48
to find least common multiple, R-5–R-6
for large numbers, R-4
Prime numbers, R-2
Prime polynomials, 280
Principal square root. *See* Positive square root
Problem solving. *See* Applications
Product rule
for exponents, 190–192
for radicals, 591–592, 605, 637
when to use, 196
Products. *See also* Multiplication
cross, 386–387, 411
involving zero, 51–52, 83
of signed numbers, 52, 53
use of power rules for, 192–194
Proper fractions, R-11
Proportions. *See also* Ratios
applications of, 387–389, 411–412
explanation of, 386, 411
Pythagorean theorem
applications of, 319–320, 620–621, 722
explanation of, 319, 620, 700

Q

Quadrants, 424
Quadratic equations
applications of, 316–320, 328, 654
approximating solutions to, 668
completing the square to solve, 658–659, 685
with degree greater than two, 311–312

explanation of, 308, 328, 651
factoring to solve, 308–311, 328, 651
on graphing calculator, 680
graphs of, 675–680, 686
method to solve, 666
quadratic formula to solve, 663–667, 686
square root property to solve, 651–654
standard form of, 308, 328, 663, 664
Quadratic formula
explanation of, 664
to solve quadratic equations, 663–667, 686
use of, 663–667
Quotient rule
explanation of, 194
for exponents, 194–196
for radicals, 592–593, 605, 638
when to use, 196
Quotients
explanation of, R-22
involving zero, 53, 83
of polynomials, 339
of two real numbers, 54, 83
use of power rules for, 192–194

R

Radical equations
explanation of, 614
methods to solve, 614–617
with two radicals, 617
Radical expressions, 583
Radical sign, 583
Radicals. *See also* Roots; Square roots
addition and subtraction of, 599–600, 638
applications with, 621–622
containing variables, 585–586, 593
division of, 605, 638
explanation of, 583, 637
formulas containing, 621–622
like, 599
method to simplify, 585–586, 637–638
multiplication of, 603–605, 638
product rule for, 591–592, 605
quotient rule for, 592–593, 605
solving equations containing, 614–617
Radicand, 583, 586, 637
Range
of data set, 707–709
of function, 493–494
of relation, 488, 513, 726
Rate of change, slope as, 457–466, 512
Rates
applications of, 558–559, 560–561
explanation of, 386
Rational equations
applications of, 387–393
explanation of, 387
Rational expressions. *See also* Fractions
addition and subtraction of, 360–361, 368–371, 410–411
complex, 400–404
division of, 352–354, 410
equations containing, 375–379, 387
evaluation of, 339
explanation of, 339, 409
least common denominator of, 361–363
method to simplify, 341–344, 409
multiplication of, 350–351, 353–354, 409–410
solving equations containing, 375–379, 411
undefined, 340
with variables in denominator, 379
writing equivalent, 344, 363–364
Rational numbers, 12, 81. *See also* Fractions
Rationalizing denominators
explanation of, 605–606, 638
using conjugates, 606–607, 638
Ratios, 386, 411. *See also* Fractions; Proportions
Real numbers
addition of, 30–34, 83
division of, 54–55, 83
explanation of, 82
multiplication of, 51–53, 83
order property for, 10, 82
properties of, 64–68, 84
subtraction of, 39–43, 83
Reciprocals
explanation of, 67–68, 84, R-12
method to find, 53, 352
Rectangles, 138, 139–140, 700
Rectangular coordinate system, 424–429, 510–511
Relations, 488, 513
Repeating decimals, R-24
Replacement values, 41–42, 57
Right angles, 717
Right triangles
explanation of, 319
Pythagorean theorem and, 319–320, 620–621, 700
Rise, 457, 458
Roots. *See also* Radicals; Square roots
cube, 584, 594
explanation of, 583
method to find *n*th, 584–585
Roster form, 713
Rounding, R-22–R-23
Run, 457, 458

S

Scatter diagrams, 426
Scientific notation
on calculators, 58, 206
converted to standard form, 204–205
explanation of, 203–204, 253
operations with, 205
writing numbers in, 204
Sentences, 10. *See also* Words/phrases
Set builder notation, 713
Sets
determining if set is subset of another, 714
element of, 8, 81, 713
explanation of, 8
method to find unions and intersections of, 714–715
Signed numbers. *See also* Integers
addition of, 30, 31
division of, 54–55
multiplication of, 52, 53
Similar triangles
explanation of, 389, 720–721
finding unknown lengths of sides in, 721–722
Slope
explanation of, 457, 462, 512
of horizontal and vertical lines, 461–462
method to find, 457–460
of parallel and perpendicular lines, 462–464, 466
as rate of change, 457–466, 512
summary of, 462
writing equation given, 460–461
Slope-intercept form
explanation of, 474, 479, 512–513
graphing equations using, 474–475, 531, 533
writing equations using, 474
Solution set, 725
Solutions
of equations, 23, 42, 57, 83, 510–511
extraneous, 615
of inequalities, 500
infinite number of, 543–544
to linear equations in two variables, 427–429, 438
of system of two equations in two variables, 529, 570
Special products
explanation of, 236, 255
FOIL method and, 235–236, 255
multiplying sum and difference as two terms as, 237–238
use of, 238–239
Square root property
applications of, 654
explanation of, 652
to solve quadratic equations, 651–654, 685
Square roots. *See also* Radicals
approximation of, 585
on calculators, 586
explanation of, 583
method to find, 583–584
method to simplify, 599–600
negative, 583
positive, 583
product rule for, 591–592
quotient rule for, 592–593
table of, 699
Squares
of binomials, 236–237, 255
completing, 658–659, 685
difference of, 237–238, 299–300, 327
factoring difference of two, 299–300, 327
formulas for, 700

Squares (*continued*)
 perfect, 585
 table of, 699
Squaring property of equality
 used once, 614–616
 used twice, 617
Standard form
 of linear equations, 479
 of linear equations in two variables, 437–438, 479, 511
 of quadratic equations, 308, 328, 663, 664
Straight angles, 717
Study guide outline, 740–743. *See also* Mathematics class
Study skills, 2–6, 730–739. *See also* Mathematics class
Subsets, 714
Substitution method
 explanation of, 540
 to solve systems of linear equations, 540–544, 571
Subtraction
 applications of, 42–43
 of decimals, R-21, R-50
 of fractions, R-13–R-14, R-49
 of polynomials, 221–223, 254
 of radicals, 599–600, 638
 of rational expressions, 361, 410–411
 of real numbers, 39–43, 83
 symbol for, 24
 words/phrases for, 24
Sum and difference of two terms, 237–238, 255
Supplementary angles, 43, 717
Symbols/notation. *See also* Words/phrases
 absolute value, 14, 20
 addition, 24
 angle, 717
 braces, 8, 20
 brackets, 20
 decimal, R-20
 division, 24
 ellipsis, 9
 equality/inequality, 9, 24, 82, 162
 exponential, 19
 fraction bar, 20, R-8
 function, 492–494, 513
 grouping, 20, 21
 interval notation, 725
 is approximately equal to, R-24
 minus sign, 11
 multiplication, 24
 parentheses, 20
 π, 13
 plus sign, 11
 radical, 583
 scientific, 203–204
 set builder, 713
 words translated to, 23–25
Systems of linear equations
 addition method to solve, 548–551, 571–572
 applications of, 557–562, 572
 on calculators, 533
 consistent, 532
 explanation of, 529, 570
 graphing to solve, 529–531, 570–571
 identification of special, 531–532
 inconsistent, 532
 infinite solutions, 531–532
 no solution, 531
 nongraph solutions to, 532–533
 solution of, 529, 570
 substitution method to solve, 540–544, 571

T

Temperature conversion, 138–139, 142
Terms
 constant, 211
 explanation of, 72, 84
 like, 73–74, 84
 numerical coefficient of, 72, 84
 of polynomial, 215, 254
 sum and difference of two, 237–238, 255
 unlike, 73
Translation problems, 122–123
Transversal, 718, 719
Triangles
 congruent, 720
 explanation of, 719
 finding dimensions of, 319–320, 389, 719–720
 formulas for, 700
 right, 319–320, 700
 similar, 389, 720–721
Trinomials. *See also* Polynomials
 explanation of, 212, 254
 of form $ax^2 + bx + c$, 285–288, 289–291, 327
 of form $x^2 + bx + c$, 278–281, 327
 grouping method to factor, 292–294
 perfect square, 297–300, 327
True statement, 427

U

Undefined expressions, 55, 340
Union, of sets, 714–715
Unit fractions, 354–355
Unlike terms, 73

V

Variables dependent, 492
 explanation of, 9, 22, 82
 independent, 492
 simplifying polynomials containing several, 215
 simplifying radicals containing, 585–586, 593
 solving a formula for, 140–142
 solving equations for specified, 379
Velocity formula, 622
Vertex, of parabolas, 675–676, 686
Vertex formula, 677–680, 686
Vertical angles, 718
Vertical change, 457
Vertical line test, examples using, 490–492
Vertical lines
 graphs of, 451
 slope of, 462, 479

W

Whole numbers, 81, R-8
Words/phrases. *See also* Applications; Symbols/notation
 for addition, 24
 algebraic expressions in, 76
 for division, 24
 for equality/inequality, 166
 for multiplication, 24
 for subtraction, 24
 translated to symbols, 23–25
Work problems, 391–392

X

x-axis, 424, 510
x-coordinates, 425, 488
x-intercepts
 explanation of, 447, 448, 511
 of graphs of parabolas, 675–676, 678
 method to find, 449–450, 511–512

Y

y-axis, 424, 510, 675
y-coordinates, 425, 488
y-intercepts
 explanation of, 447, 448, 511
 method to find, 449–450, 511–512
 quadratic equations and, 676

Z

Zero
 division by, 55, 340
 products involving, 51–52, 83
 quotients involving, 53, 83
Zero exponents, 195
Zero-factor property, 308–311, 328, 651

Photo Credits

Chapter R **p. 1** Pearson Education, Inc.; **p. 41** Pearson Education, Inc.; **p. 54** Pearson Education, Inc.; **p. 55** Pearson Education, Inc.; **p. 62** Pearson Education, Inc.

Chapter 1 **p. 6** B&M Noskowski/E+/Getty Images; **p. 11** (left): Ron Smith/Shutterstock; **p. 11** (right): Wind Coast/Shutterstock; **p. 22** Kadmy/123RF; **p. 34** Elen_studio/Shutterstock; **p. 37** (left): Nickolay Vinokurov/Shutterstock; **p. 37** (right): John Carnemolla/Shutterstock; **p. 38** (left): Christopher Y.C. Wong/Shutterstock; **p. 38** (right): Pendakisolo/Shutterstock; **p. 47** (left): Oversnap/E+/Getty Images; **p. 47** (right): Paylessimages/Fotolia; **p. 58** Milkos/123RF

Chapter 2 **p. 97** (left): iofoto/Shutterstock; **p. 97** (right): EpicStockMedia/Shutterstock; **p. 100** (bottom left): Denis Burdin/Shutterstock; **p. 100** (bottom right): Grobler du Preez/123RF; **p. 100** (top): U.S. Department of Transportation 1/M Federal Highway Administration; **p. 124** Matthew Cavanaugh/EPA/Shutterstock; **p. 125** Islemount Images/Alamy Stock Photo; **p. 126** (top): Peter Gudella/Shutterstock; **p. 126** (bottom): Zack Frank/Shutterstock; **p. 128** Rido/Shutterstock; **p. 132** Bernd Mellmann/Alamy Stock Photo; **p. 133** (left): AS Food studio/Shutterstock; **p. 133** (right): Valentin Valkov/Shutterstock; **p. 135** Jenifoto/Fotolia; **p. 137** Raymona Pooler/Shutterstock; **p. 145** (bottom): Steve Geer/Getty Images; **p. 145** (top): Nick David/Alamy Stock Photo; **p. 146** (left): Bennymarty/123RF; **p. 146** (right): Paul Hawthorne/Getty Images; **p. 147** (top): Pearson Education, Inc; **p. 147** (bottom): Elayn Martin-Gay; **p. 152** Longimanus/Shutterstock; **p. 153** Steve Debenport/E+/Getty Images; **p. 159** (left): Dmitriy Shironosov/123RF; **p. 159** (right): Phillip Minnis/Shutterstock; **p. 179** Jiawangkun/Shutterstock

Chapter 3 **p. 188** (top): Carballo/Shutterstock; **p. 188** (bottom): Pearson Education, Inc.; **p. 208** Pearson Education, Inc.; **p. 214** Pearson Education, Inc.; **p. 227** (left): Wajan/Fotolia; **p. 227** (right): Inga Spence/Alamy Stock Photo; **p. 257** (top left): Hill Street Studios/Blend Images/Alamy Stock Photo; **p. 257** (top right): Pearson Education, Inc.; **p. 257** (bottom): National Space Science Data Center; **p. 258** Forcdan/Fotolia

Chapter 4 **p. 267** WavebreakMediaMicro/Fotolia; **p. 277** (left): 135pixels/Fotolia; **p. 277** (right): Kurdistan/Shutterstock; **p. 305** (top): Saraporn/Shutterstock; **p. 305** (bottom left): S-F/Shutterstock; **p. 305** (bottom right): Vvoe/Fotolia; **p. 316** (left): Christina Beauchamp/Photo Resource Hawaii/Alamy Stock Photo; **p. 316** (right): NadyaRa/Shutterstock

Chapter 5 **p. 349** (top left): Pearson Education, Inc.; **p. 349** (top right): Pearson Education, Inc.; **p. 349** (bottom): Pearson Education, Inc.; **p. 355** (left): First Class Photography/Shutterstock; **p. 355** (right): Matthew Grant/Alamy Stock Photo; **p. 358** (left): Xinhua/Alamy Stock Photo; **p. 358** (right): Brian E Kushner/Shutterstock; **p. 391**: xPACIFICA/Alamy Stock Photo; **p. 400**: Max Earey/Shutterstock

Chapter 6 **p. 426** Katsiuba Volha/Shutterstock; **p. 429** Shutterstock; **p. 431** Anton Prado PHOTO/Shutterstock; **p. 432** WilleeCole Photography/Shutterstock; **p. 441** Michaeljung/Shutterstock; **p. 445** NomadFra/Shutterstock; **p. 446** Iain Masterton/Alamy Stock Photo; **p. 465** (top): John Hoffman/Shutterstock; **p. 465** (bottom): Guillermo Avello/123RF; **p. 473** (left): Gennadiy Poznyakov/Fotolia; **p. 473** (right): Elena Volkova/Fotolia; **p. 478** Paul Sutherland/ Photodisc/Getty Images; **p. 484** (left): Michael Doolittle/Alamy Stock Photo; **p. 484** (right): Luckybusiness/Fotolia; **p. 485** Dmitry Naumov/Fotolia

Chapter 7 **p. 547** (left): Piotr Adamowicz/Fotolia; **p. 547** (right): Kanchana thipmontian/123RF; **p. 555** Wavebreakmedia/Shutterstock; **p. 558** Nils Jorgensen/Shutterstock; **p. 564** (left): Peter Joneleit/CSM/Shutterstock; **p. 564** (right): Keeton Gale/Shutterstock; **p. 565** Zhao jiankang/Fotolia; **p. 578** Dean Fikar/Shutterstock

Chapter 8 **p. 582** Nice to meet you/Shutterstock; **p. 597** Henry Westheim Photography/Top Photo/Asia Photo Connection/Alamy Stock Photo; **p. 598** (top left): Pearson Education, Inc; **p. 598** (top right): Judie Long/Age fotostock/Alamy Stock Photo; **p. 598** (bottom): Kurhan/123RF; **p. 602** Pearson Education, Inc; **p. 610** Pearson Education, Inc; **p. 622** Vetal1983/Fotolia; **p. 625** (top): Pearson Education, Inc; **p. 625** (bottom left): Pearson Education, Inc; **p. 625** (bottom right): Pearson Education, Inc; **p. 626** Pearson Education, Inc; **p. 627** Adamgregor/123RF; **p. 629** Cathy Yeulet/123RF

Chapter 9 **p. 654** 504 collection/Alamy Stock Photo; **p. 656** (top left): Jejim/123RF; **p. 656** (top right): Eastphoto/AGE Fotostock; **p. 656** (bottom): Jeremy Woodhouse/Photodisc/Getty Images; **p. 657** Forcdan/Fotolia; **p. 663** (left): Ktsdesign/Shutterstock; **p. 663** (right): Carmat SAS/ABACAPRESS.COM/MCT/Newscom; **p. 671** (left): Shutterstock; **p. 671** (right): Ken Wolter/Shutterstock; **p. 684** (top): KIKE CALVO/Alamy Stock Photo; **p. 684** (bottom): Rudi1976/Fotolia; **p. 688** (left): Chee-Onn Leong/Shutterstock; **p. 688** (right): Jens Ickler/123RF